List of the Elements

Name	Symbol	Atomic Number	Atomic Weight
Actinium	Ac	89	227.028
Aluminum	Al	13	26.982
Americium	Am	95	243
Antimony	Sb	51	121.76
Argon	Ar	18	39.948
Arsenic	As	33	74.922
Astatine	At	85	210
Barium	Ba	56	137.327
Berkelium	Bk	97	247
Beryllium	Be	4	9.012
Bismuth	Bi	83	208.980
Bohrium	Bh	107	262
Boron	B	5	10.811
Bromine	Br	35	79.904
Cadmium	Cd	48	112.411
Calcium	Ca	20	40.078
Californium	Cf	98	251
Carbon	C	6	12.011
Cerium	Ce	58	140.115
Cesium	Cs	55	132.905
Chlorine	Cl	17	35.453
Chromium	Cr	24	51.996
Cobalt	Co	27	58.933
Copper	Cu	29	63.546
Curium	Cm	96	247
Dubnium	Db	105	262
Dysprosium	Dy	66	162.5
Einsteinium	Es	99	252
Erbium	Er	68	167.26
Europium	Eu	63	151.964
Fermium	Fm	100	257
Fluorine	F	9	18.998
Francium	Fr	87	223
Gadolinium	Gd	64	157.25
Gallium	Ga	31	69.723
Germanium	Ge	32	72.61
Gold	Au	79	196.967
Hafnium	Hf	72	178.49
Hassium	Hs	108	265
Helium	He	2	4.003
Holmium	Ho	67	164.93
Hydrogen	H	1	1.0079
Indium	In	49	114.82
Iodine	I	53	126.905
Iridium	Ir	77	192.22
Iron	Fe	26	55.845
Krypton	Kr	36	83.8
Lanthanum	La	57	138.906
Lawrencium	Lr	103	262
Lead	Pb	82	207.2
Lithium	Li	3	6.941
Lutetium	Lu	71	174.967
Magnesium	Mg	12	24.305
Manganese	Mn	25	54.938
Meitnerium	Mt	109	266
Mendelevium	Md	101	258
Mercury	Hg	80	200.59
Molybdenum	Mo	42	95.94
Neodymium	Nd	60	144.24
Neon	Ne	10	20.180
Neptunium	Np	93	237.048
Nickel	Ni	28	58.69
Niobium	Nb	41	92.906
Nitrogen	N	7	14.007
Nobelium	No	102	259
Osmium	Os	76	190.23
Oxygen	O	8	15.999
Palladium	Pd	46	106.42
Phosphorus	P	15	30.974
Platinum	Pt	78	195.08
Plutonium	Pu	94	244
Polonium	Po	84	209
Potassium	K	19	39.098
Praseodymium	Pr	59	140.908
Promethium	Pm	61	145
Protactinium	Pa	91	231.036
Radium	Ra	88	226.025
Radon	Rn	86	222
Rhenium	Re	75	186.207
Rhodium	Rh	45	102.906
Rubidium	Rb	37	85.468
Ruthenium	Ru	44	101.07
Rutherfordium	Rf	104	261
Samarium	Sm	62	150.36
Scandium	Sc	21	44.956
Seaborgium	Sg	106	263
Selenium	Se	34	78.96
Silicon	Si	14	28.086
Silver	Ag	47	107.868
Sodium	Na	11	22.990
Strontium	Sr	38	87.62
Sulfur	S	16	32.066
Tantalum	Ta	73	180.948
Technetium	Tc	43	98
Tellurium	Te	52	127.60
Terbium	Tb	65	158.925
Thallium	Tl	81	204.383
Thorium	Th	90	232.038
Thulium	Tm	69	168.934
Tin	Sn	50	118.71
Titanium	Ti	22	47.88
Tungsten	W	74	183.84
Uranium	U	92	238.029
Vanadium	V	23	50.942
Xenon	Xe	54	131.29
Ytterbium	Yb	70	173.04
Yttrium	Y	39	88.906
Zinc	Zn	30	65.39
Zirconium	Zr	40	91.224
—	Uun	110	269
—	Uuu	111	272
—	Uub	112	277

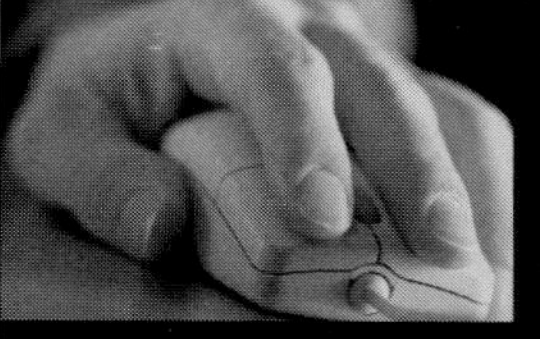

Free Student Aid.

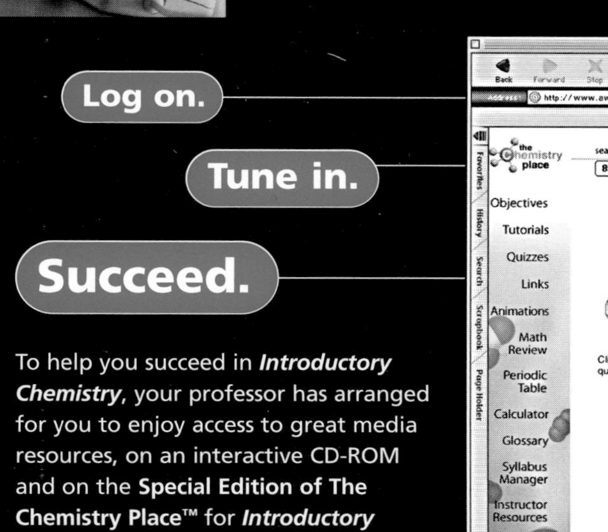

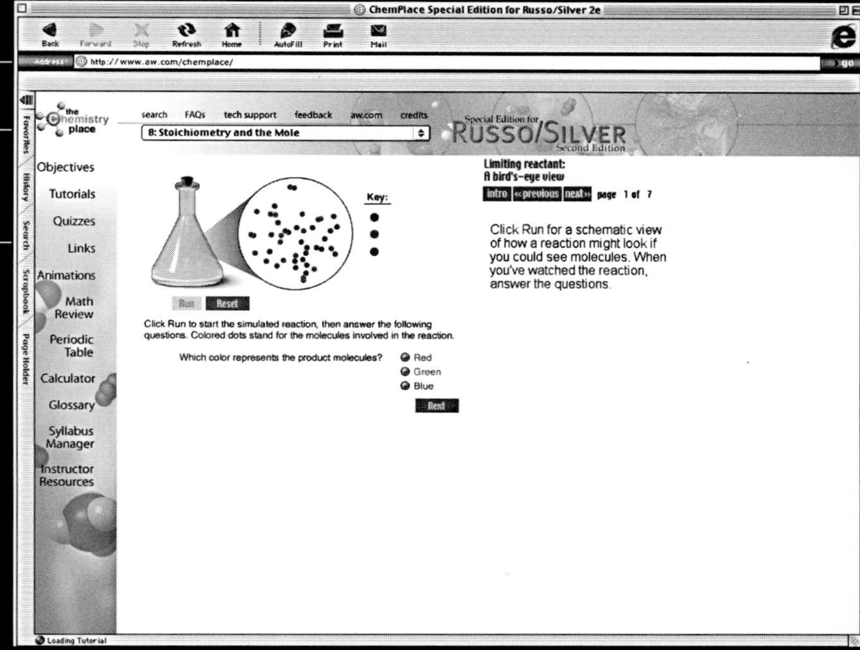

To help you succeed in ***Introductory Chemistry***, your professor has arranged for you to enjoy access to great media resources, on an interactive CD-ROM and on the **Special Edition of The Chemistry Place™** for ***Introductory Chemistry*, Second Edition.** You'll find that these resources that accompany your textbook will enhance your course materials.

What your system needs to use these media resources:

WINDOWS™

- 250 MHz Windows 98/NT/2000/XP
- 32 MB RAM installed, 64 MB preferred
- 800 x 600 screen resolution
- Thousands of colors
- 56K or better internet connection
- Browsers: Internet Explorer 5.0 and higher; Netscape 4.7, 7.0
- Plug Ins: Shockwave 8 and Flash player 5, Chime

NOTE: Use of Netscape 6.0 and 6.1 are not recommended due to a known compatibility issue between Netscape 6.0 and 6.1 and the Flash and Shockwave plug-ins.

MACINTOSH™

- 233 MHz G3
- OS 9.2 or higher
- 32 MB RAM minimum
- 800 x 600 screen resolution
- Thousands of colors
- 56K or better internet connection
- Browsers: Internet Explorer 5.0, Netscape 4.7
- Plug Ins: Shockwave 8 and Flash player 5, Chime

NOTE: Use of Netscape 6.0 and 6.1 are not recommended due to a known compatibility issue between Netscape 6.0 and 6.1 and the Flash and Shockwave plug-ins.

Got technical questions?

For technical support, please visit www.aw.com/techsupport and complete the appropriate online form. You can also call our tech support hotline at 800-6-Pro-Desk (800-677-6337) Monday-Friday, 8 a.m. to 5 p.m. CST.

Here's your personal ticket to success:

How to log on to www.aw.com/chemplace

1. Go to **www.aw.com/chemplace**
2. Click ***Introductory Chemistry*, Second Edition**.
3. Click "Register."
4. Scratch off the silver foil coating below to reveal your pre-assigned access code.
5. Enter your pre-assigned access code exactly as it appears below.
6. Complete the online registration form to create your own personal Login Name and Password.
7. Once your personal Login Name and Password are confirmed by email, go back to www.aw.com/chemplace, type in your new Login Name and Password, and click "Log In."

Your Access Code is:

Record your new Log In and Password on the back of this card.

Cut out this card and keep it handy. It's your ticket to valuable information. Students who purchase a used copy of this book may not have a valid access code. Students may purchase a subscription to this website at www.aw.com/chemplace

Important: Please read the License Agreement, located on the launch screen before using the **Special Edition of The Chemistry Place™** for ***Introductory Chemistry*, Second Edition** or **The Chemistry Place™ CD-ROM**. By using the web site or CD-ROM, you indicate that you have read, understood and accepted the terms of this agreement.

0-321-04634-X

Introductory Chemistry

About the Authors

Mike Silver (left) is a Professor of Chemistry at Hope College. He received his B.S. in chemistry from Farleigh Dickinson University and his Ph.D. in inorganic chemistry from Cornell University. He is also a member of the American Chemical Society, past president of the West Michigan Section, and a member of the Council of Undergraduate Research. He has received the Camille and Henry Dreyfus Teacher-Scholar Award for Excellence in Teaching and Research and the Provost's Award for Teaching Excellence. Currently he is involved in research dealing with novel silicone surfactants, collaborating with various chemical companies.

Steve Russo (right) is a Senior Lecturer in the Department of Chemistry at Cornell University and the Director of Organic Laboratories. Prior to that, he was an Assistant Professor at Indiana University. While there, he designed and implemented a state-of-the-art computer resource center for the undergraduate chemistry curriculum. He received his B.S. in chemistry from St. Francis College, and his Ph.D. in physical organic chemistry from Cornell University. He is a member of the American Chemical Society and has been a recipient of the Dupont Teaching Award, Clark Teaching Award, and the Amoco Distinguished Teaching Award.

Saundra Yancy McGuire is Adjunct Professor of Chemistry and Director of the Center for Academic Success at Louisiana State University. She received her B.S. in chemistry from Southern University in Baton Rouge, LA, and her Ph.D. in Chemical Education from the University of Tennessee, Knoxville. Prior to joining LSU she spent eleven years at Cornell University, where she was a recipient of the Clark Distinguished Teaching Award. Her current interests include improving learning strategies used by university students and improving pedagogical techniques of university faculty. She is a member of the American Chemical Society and the National Organization of Black Chemists and Chemical Engineers (NOBCChE). She has been nationally recognized for her teaching, workshop presentations, and mentoring of students.

INTRODUCTORY CHEMISTRY

Second Edition

STEVE RUSSO
Cornell University

MIKE SILVER
Hope College

San Francisco • Boston • New York
Cape Town • Hong Kong • London • Madrid • Mexico City
Montreal • Munich • Paris • Singapore • Syndey • Tokyo • Toronto

Editor: Ben Roberts
Senior Developmental Editor: Margot Otway
Developmental Editors: Irene Nunes, Becky Strehlow, Susan Weisberg, John Murdzek
Marketing Manager: Christy Lawrence
Director of Marketing: Stacy Treco
Media Producer: Claire Masson
Production Coordination: Joan Marsh, Tony Asaro
Production Management: Dovetail Publishing Services
Composition: Dovetail Publishing Services
Art Director: Blakeley Kim
Cover Image & Chapter Openers: Emi Koike and Quade Paul, fiVth.com
Text Designer: Mark Ong, Side-by-Side Studios
Artists: Emi Koike, Bert Dodson, Blakeley Kim, Ken Probst
Photographer & Photo Research: Cary Groner, Myrna Engler
Manufacturing: Vivian McDougal
Prepress House: H&S Graphics
Cover Printer: Phoenix Color
Printer and Binder: Von Hoffmann Press

About the Artists: Bert Dodson has illustrated over sixty books. He is the author of *Keys to Drawing*, published in eight languages. His distinctive, whimsical cartoons can also be found in *The Way Life Works*, of which he is co-author. Bert lives in Bradford, Vermont. **Emi Koike and Quade Paul** are scientific and medical illustrators. Quade combined digital illustration and 3-D modeling techniques to create the stunning cover images and chapter openers in this book. All of the molecular imagery in the artwork is based on actual data. Quade and Emi's work also includes editorial illustration, scientific animation, and multimedia projects. They can be contacted at fiVth.com.

Photos courtesy of: 1 (right/top) Courtesy of the National Library of Medicine; **1** (right/bottom) © Tony Stone Imaging; **2** (left to right) © 2001 PhotoDisc, Inc.; **2** © Antonio M. Rosario/Image Bank; **2** © Jessica Ehler/Bruce Coleman; **2** © 2001 PhotoDisc, Inc.; **3** NASA; **6** (gold & diamond) © 2001 PhotoDisc, Inc.; **6** (others) Prentice Hall; **7** (top) Prentice Hall; **7** (gold & marble) © 2001 PhotoDisc, Inc.; **25** NASA; **84** Courtesy of the Houston Astrodome; **92–93** (top) Prentice Hall; **99** © Corel; **107** © 2001 PhotoDisc, Inc.; **108** (top) © David Parker/SPL/Photo Researchers, Inc.; **108** (bottom-both) © Rich Treptow/Photo Researchers, Inc.; **119** © Rich Treptow/Photo Researchers, Inc.; **120** (top) NASA; © David Parker/SPL/Photo Researchers, Inc.; **182** (bottom) © SCIMAT/Photo Researchers, Inc.; **195** © 2001 PhotoDisc, Inc.; **208** (coffee cup) Courtesy of Direct Imagination; **241** © AT&T Bell Labs/SPL/Photo Researchers, Inc.; **242** © Marcello Bertinetti/ Photo Researchers, Inc.; **350** © Corel; **352** © 2001 PhotoDisc, Inc.; **381** (left) © 2001 PhotoDisc, Inc.; **381** (right) Prentice Hall; **382** (bottom/left) © 2001 PhotoDisc, Inc.; **383** (center/right) Prentice Hall; **383** (right) © 2001 PhotoDisc, Inc.; **490** Lysozyme based on file 1/mq.pdb; **493** Mary Evan Picture Library/ Photo Researchers, Inc.; **658** © 2001 PhotoDisc, Inc.; **673** (left) © 2001 PhotoDisc, Inc.; **673** (center) © Andrew Syred/SPL/Photo Researchers, Inc.; **727** (top) © Dr. Tony Brian/SPL/Photo Researchers, Inc.; **727** (bottom) © Jackie Lewin/SPL/Photo Researchers, Inc.

Library of Congress Cataloging-in-Publication Data

Russo, Steve
Introductory chemistry / Steve Russo, Mike Silver.—2nd ed.
p. cm.
Includes index.
ISBN 0-321-04634-X (alk. paper)
1. Chemistry. I. Silver, Mike. II. Title.
QD33.2. .R87 2002
540—dc21 2001047684

4 5 6 7 8 9 10—VHP—04 03
www.aw.com/bc

Contents

Preface

Many instructors have told us that the first edition of our text was the most readable introductory chemistry textbook their students had ever used. They also told us that their students need more help learning to solve problems—more worked examples, more tools for mastering problem-solving methods, and, simply, more problems. In this edition, we have incorporated new and stronger tools to help students learn the skills they need, while working to make the text even clearer and more memorable than in the first edition.

Our goal with this text has always been to help students make sense of chemistry. As chemists, we know that chemistry is intrinsically interesting, that its principles make sense, and that its problem-solving skills can be mastered by anyone. But, too often, students see the subject as incomprehensible and the course as a frightening labyrinth. All too frequently, they fall back on rote learning—memorizing algorithms, plugging numbers into formulas, and forgetting the course as quickly as they can. As chemistry instructors, we hate to see that. Therefore, we designed this book to promote comprehension and problem solving as complementary skills. A student who understands the principles of chemistry doesn't have to memorize as much and is more likely to enjoy and retain the material. Equally, practice at solving problems will help students to master the principles. We hope that our book will help students to come out of the course with a body of knowledge and skills that will serve them and that they will want to retain.

Promoting Active Learning

How can we, as textbook authors, promote active learning? First, quite simply, we provide a book that makes sense to students. A student who understands the material is less likely to fall back on passive memorization. As one instructor told us, "This book allows me to spend class time doing hands-on learning versus spending time explaining the book."

Second, we incorporate devices to encourage active reading. A flip through the book will show many sets of Practice Problems. These Practice Problems are located so that students can immediately apply the skill the text has presented. They come in sets of three or four; the first problem is solved in place in the text, and the answers to the others are given at the back of the book. Each chapter contains an average of 25 Practice Problems.

However, we are not naive. We know that many students routinely skip in-chapter practice problems. Therefore, we also include conceptual practice problems called WorkPatches. A Workpatch is a "stealth" problem. It follows smoothly from the preceding text and is not boxed off. What is more, a student who tries to read through a WorkPatch without solving it will find that the subsequent text refers to, and often depends on, the answer—but does not say what it is. (The solutions to all WorkPatches are given at the end of the chapter.) In some cases, a WorkPatch serves as the springboard into the next topic. WorkPatches are denoted by a red stop-sign icon in the margin.

Helping Students Master Problem-Solving Skills

An instructor flipping through our text might be inclined to ask, "Where are all the worked examples?" In fact, this edition has an abundance of worked examples and other problem-solving aids, but we have handled them in a way that preserves the text's coherence. As in the first edition, many worked examples are presented in the text itself. We feel strongly that problem-solving techniques should be explained with the same care and continuity we use for concepts.

When students are working problems, however, they also need access to compact summaries and examples. We have augmented these resources in the following ways:

- Important problem-solving methods are summarized in charts in the text.
- The same charts, accompanied by worked examples and additional material, appear in a special Skills to Know section at the end of the chapter immediately preceding the end-of-chapter problems. This section is intended as a "help center" to which students can refer while working problems. (A few of the more purely conceptual chapters do not have a Skills to Know section.)
- An abundance of additional step-by-step methods, worked examples, and practice problems are provided in the *Problem Solving Guide and Workbook* that accompanies the text, authored by our colleague Saundra Yancy McGuire. For a weak or struggling student, nothing is more important than abundant, guided, confidence-building practice. This workbook enables us to offer a truly realistic amount of help while maintaining a clean, readable textbook. The *Workbook* also contains a generous mathematics review. We came to know Dr. McGuire when she directed the Center for Learning and Teaching at Cornell; she now directs the Center for Academic Success at Louisiana State University. We are extremely glad that she chose to join her expertise with ours.
- As noted earlier, each chapter contains internal Practice Problems and Work-Patches in addition to end-of-chapter problems.

Room to Practice: Revised and Expanded Problem Sets

With the generous aid of several colleagues, we have revised and substantially expanded the end-of-chapter problem sets.

- Each problem set now includes a section of Additional Problems that are not categorized by chapter section.
- We more than doubled the number of end-of-chapter problems.
- We have worked to ensure that each problem set contains a sufficient abundance of each of the types of problems an instructor might require.
- As in the first edition, answers to selected problems are provided at the end of the book. (The full solutions for these selected problems, as well as for all the Practice Problems, are available in the *Study Guide and Selected Solutions* manual.)

Making Chemistry Memorable

Instructors have told us that a surprising number of their students actually read our book, rather than using it mainly as a resource while solving problems. In this second edition, we have worked to improve the features that made the first edition so readable. We explain chemistry using everyday, conversational language, and we tie the concepts and calculations to stories and examples that help bring them to life. We use humor in places. We also ensure that the students know which points are fundamental and which represent additional detail. You will notice that most of the illustrations lack legends. That is because the text and the illustrations work hand-in-hand, and each illustration is placed exactly where it belongs.

WE WANT TO HEAR FROM YOU

One of the pleasures of revising a book is hearing from instructors who use it—learning what works and doesn't work; gathering ideas. If you have any comments or suggestions, please feel free to contact us at the following email addresses: sr19@cornell.edu (Steve Russo); silver@hope.edu (Mike Silver).

CONTENT CHANGES IN THIS EDITION

- **Chapter 2, "The Numerical Side of Chemistry,"** now includes coverage of specific heat and simple calorimetry.
- In **Chapter 4, "The Modern Model of the Atom,"** we expanded the treatment of orbitals.
- In **Chapter 5, "Chemical Bonding and Nomenclature,"** we now give full basic coverage of inorganic nomenclature.
- **Chapter 7, "Chemical Reactions,"** covers equation balancing more thoroughly than in the first edition. It also now covers reaction types. There is a section on solubility and precipitation reactions (including net ionic equations and the solubility rules), and a simple introduction to acid/base reactions.
- The new **Chapter 8, "Stoichiometry and the Mole,"** represents an expanded treatment of material formerly placed in Chapter 7. Stoichiometric calculations and mole/mass conversions are covered at greater length (and more visually) than in the first edition.
- Limiting-reactant problems are now handled in their own expanded section, followed by a section on percent composition and molecular formulas. We placed these sections at the end of the chapter so that instructors who do not teach these topics can easily skip them.
- **Chapter 11, "What If There Were No Intermolecular Forces? The Ideal Gas,"** represents an expanded treatment of a topic formerly covered in the chapter on intermolecular forces. The ideal gas concept and ideal gas law are presented at more length. In addition, the chapter now shows how the ideal gas law can be used to determine the density and molar mass of a gas.
- In **Chapter 12, "Solutions,"** we expanded the treatment of molarity as the connection between molar amount and volume. The discussion of dilution now also includes the $M_1V_1 = M_2V_2$ relationship. We also added sections on solution stoichiometry (including simple titrations) and colligative properties (including vapor pressure). We placed these sections at the end of the chapter, as some instructors may prefer to skip them.

ACKNOWLEDGMENTS

We wish to extend our warmest thanks to the reviewers whose patient and thoughtful comments helped us to shape both the first and the second editions of this book.

Edward Alexander, *San Diego Mesa College*
Melissa Armstrong, *Gaston College*
Joe Asire, *Cuesta College*
Barbara Balko, *Lewis and Clark College*
David Ball, *Cleveland State University*
Dan Bedgood, *Arizona State University*
Jack Benefield, *Valencia Community College*
Kenneth Bennett, *Kalamazoo Valley Community College*
Bill Bornhorst, *Grossmont College*
Simon Bott, *University of Houston*
Joe Burnett, *University of Iowa*
Jeff Cavalieri, *Dutchess Community College*
Jing-Yi Chin, *Suffolk County Community College*
Connie Churchill, *Oakton Community College*
Ana Ciereszko, *Miami-Dade Community College*
John Cullen, *Ricks College*
William Daniel, *Modesto Junior College*
Patrick Daubenmire, *Loyola Blakefield*
Denisha Dawson, *Diablo Valley College*
Walter Dean, *Lawrence Technological University*
Ron Distefano, *Northampton Community College*
David Dollimore, *University of Toledo*
Jerry Driscoll, *University of Utah*
Mary Ann Durick, *Bismarck State College*
Gary Fisher, *De Anza College*
Perry Forman, *University of Alabama at Birmingham*
Roger Frampton, *Tidewater Community College*
Marc Franco, *South Seattle Community College*
Elizabeth Gaillard, *Northern Illinois University*
John Gelder, *Oklahoma State University*
Galen George, *Santa Rosa Junior College*
Angela Glisan-King, *Wake Forest University*
Wendy Gloffke, *Cedar Crest College*
John Goodwin, *Coastal Carolina University*
George Goth, *Skyline College*
Stanley Grenda, *University of Nevada, Las Vegas*
Elizabeth Griffith, *University of South Carolina*
James Hardcastle, *Texas Woman's University*
Leslie Heasley, *University of Alaska, Anchorage*
Claudia Hein, *Diablo Valley College*
Eileen Hinks, *Virginia Military Institute*
R.A. Hoots, *Davis, CA*
Jeffrey Hurlbut, *Metropolitan State College*
James Jacob, *University of Rhode Island*
Fred Johnson, *Brevard Community College*
Ray Johnson, *Hillsdale College*
Stanley Johnson, *Orange Coast College*
Curtis Keedy, *Lewis and Clark College*
Keith Kennedy, *St. Cloud State University*
Roy Kennedy, *Massachusetts Bay Community College*
Christine Kerr, *Montgomery College*
Leslie Kinsland, *University of Southwestern Louisiana*
John Konitzer, *McHenry County College*
Dave Kort, *Mississippi State University*
Glenn Kuehn, *New Mexico State University*
Alfred Lee, *City College of San Francisco*
Nancy Levinger, *Colorado State University*
Jimmy Li, *College of San Mateo*
Ann Loeb, *College of Lake County*
David Macaulay, *William Rainey Harper College*
Christopher Makaroff, *Miami University*
Joe March, *University of Alabama at Birmingham*
Stanley Marcus, *Cornell University*
Don Marshall, *Sonoma State University*
Carol Martinez, *Albuquerque Technical/ Vocational Institute*
Dan McGuire, *Oklahoma State University*
Saundra McGuire, *Louisiana State University*
Valerie Meehan, *City College of San Francisco*
Sara Melford, *DePaul University*
Richard Mitchell, *Arkansas State University*
Wendell Morgan, *Hutchinson Community College*
Karl Mueller, *Pennsylvania State University*
Milica Nedelson, *Oakton Community College*
Tom Neils, *Grand Rapids Community College*
Patricia Perez, *Mt. San Antonio College*
Cortlandt Pierpont, *University of Colorado at Boulder*
Barbara Rainard, *Community College of Allegheny County*
Don Roach, *Miami-Dade Community College*
Mike Rodgers, *Southeast Missouri State University*
Patricia Rogers, *University of California, Irvine*
Rolland Rue, *South Dakota State University*
Ron Rusay, *Diablo Valley College*
Victor Ryzhov, *Northern Illinois University*
Martha Sanner, *Middlesex Community College*
Wesley Smith, *Ricks College*
Dennis Stevens, *University of Nevada, Las Vegas*
Louise Stracener, *Edgewood College*
Laura Stultz, *Birmingham-Southern College*
Christine Sullivan, *Skyline College*
Dave Tanis, *Grand Valley State University*

Michael Tessmer, *Southwestern College*
Vernon Theilmann, *Southwest Missouri State University*
Eric Trump, *Emporia State University*
Kenward Vaughn, *Bakersfield College*
Trudie Jo Slapar Wagner, *Vincennes University*
John Weyh, *Western Washington University*
Thomas Willard, *Florida Southern College*
Don Williams, *Hope College*
Joseph Wilson, *University of Kentucky*
Linda Wilson, *Middle Tennessee State University*

Writing a textbook is a humbling experience. The first edition taught us that no author can produce a good book without the extensive help and support of a large number of dedicated and talented people (senior editors, copy editors, developmental editors, publishing assistants, artists, reviewers, colleagues, . . .). Writing the second edition only increased our humility and our respect for this army of talent.

First, we wish to thank Saundra Yancy McGuire for her invaluable contributions, her incisive advice, her patience, and her rash willingness to get more deeply involved with each edition. In addition to reviewing all the chapters and writing the *Problem Solving Guide and Workbook* and the *Study Guide and Selected Solutions* manual, she stepped into the breach and helped us with revising and expanding the end-of-chapter problem sets. We also extend our grateful thanks to the other people who contributed to, reviewed, critiqued, and tested the problem sets—our colleagues Laura Stultz, Denisha Dawson, Jing-Yi Chin, and Victor Ryzhov, and our skilled editor John Murdzek.

To Margot Otway, our senior developmental editor for both editions, what can we say? Without your exquisite attention to every detail, your passionate drive toward excellence, your belief in our philosophy, and your extraordinary ability to do the work of ten people—this book would be a mere shadow of its present form. You may not retire until we do. To Maureen Kennedy, our editor, whose energy and focus were instrumental in shaping this second edition; to Emiko Koike and Blakeley Kim, the talented artists who are so responsible for giving this book its unique look by always insisting that they could do even better and spending many hours fretting over details in the art that neither author could even discern; to Irene Nunes, our fabulously picky, extremely dedicated, and knowledgeable editor who kept us on the right path and in line; to Joan Keyes and Jonathan Peck at Dovetail Publishing Services, for once again taking on the challenge of laying out an unconventional book; to Lisa Leung, publishing assistant and doer of a thousand tasks, for working directly with Maureen and Saundra and keeping them on task; to Tony Asaro, production coordinator (and music composer) for helping Lisa help everybody else—it goes without saying that this book could not exist without your dedication and efforts. And of course, to Joan Marsh, production editor par excellence, the nicest pest you could ever hope to work with, and Ben Roberts, our senior editor, for keeping the faith with this experiment (a readable chemistry text) and its crazed set of authors, mere words could never express our gratitude (the chocolate is in the mail). Margot Otway, Maureen Kennedy, Joan Marsh, and Ben Roberts—the four chambers of the heart of this book. We will always be in your debt.

Supplements for the Student

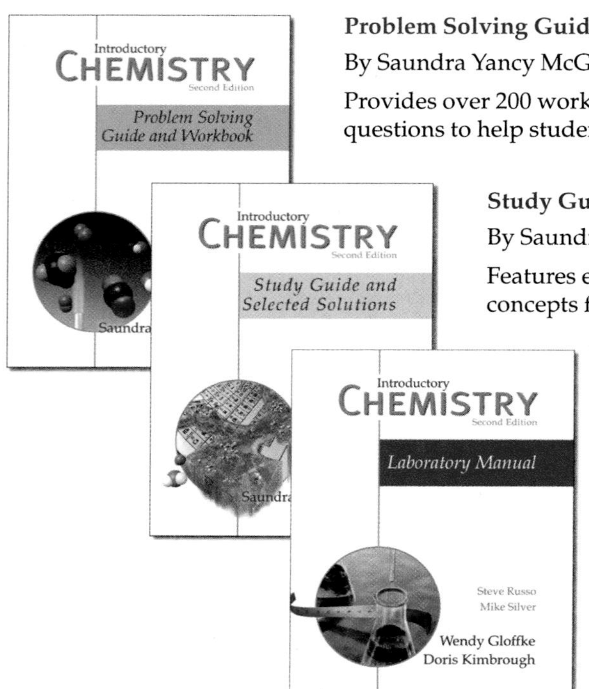

Problem Solving Guide and Workbook (0-321-06866-1)

By Saundra Yancy McGuire of Louisiana State University

Provides over 200 worked examples and more than 550 practice problems and quiz questions to help students develop and practice their problem-solving skills.

Study Guide and Selected Solutions (0-321-05327-3)

By Saundra Yancy McGuire of Louisiana State University

Features examples from each chapter, learning objectives, review of key concepts from the text, and additional problems for student practice. Also provides comprehensive answers and explanations to selected end-of-chapter problems from the text.

Introductory Chemistry Laboratory Manual (0-321-04639-0)

By Wendy Gloffke of Cedar Crest College and Doris Kimbrough of the University of Colorado at Denver

Helps students develop data acquisition, organization, and analysis skills while teaching basic techniques. Written to accompany the Russo/Silver text, this manual offers 25 experiments.

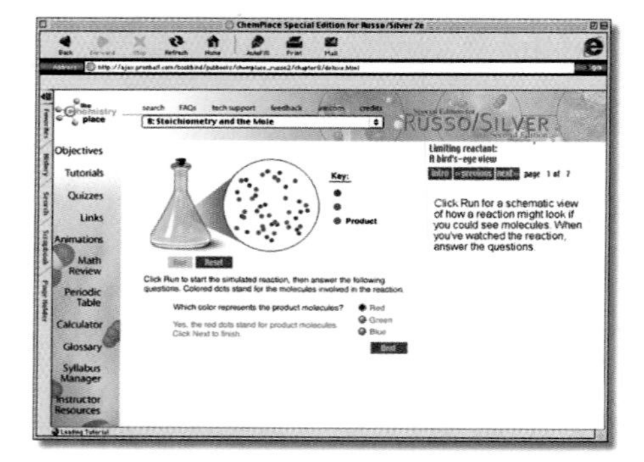

Special Edition of The Chemistry Place
www.chemplace.com/college

This special edition of **The Chemistry Place** engages students in interactive exploration of chemistry concepts and provides a wealth of tutorial support. Tailored to Russo/Silver Second Edition, the site includes detailed objectives for each chapter of the text, interactive tutorials featuring simulations, animations, and 3-D visualization tools, multiple-choice and short-answer quizzes, an extensive set of Web links, and a mathematics review. For instructors, a **Syllabus Manager** makes it easy to create an online syllabus complete with weekly assignments, projects, and test dates that students may access on the ChemPlace site.

The Chemistry of Life for Introductory Chemistry CD-ROM (0-8053-3109-3)

By Robert M. Thornton of the University of California, Davis

This lively tutorial teaches chemistry concepts through animations and interactive activities. The CD-ROM helps students master crucial concepts such as Atoms and Molecules, Reactions and Equilibrium, Water, Acids and Bases, Organic Molecules, Carbohydrates, Lipids, Proteins, Nucleic Acids, and Enzymes and Pathways. Includes diagnostic quizzes, an illustrated glossary, and topics correlated to the Russo/Silver text.

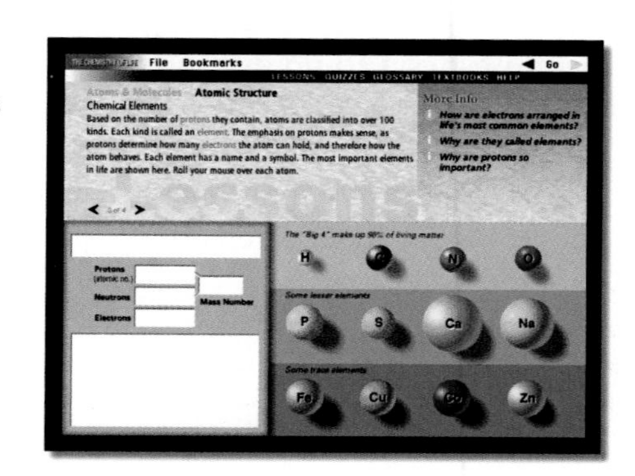

The Chemistry Tutor Center
www.aw.com/tutorcenter

Provides one-to-one tutoring four ways—phone, fax, email, and the Internet—during evening hours and on weekends. Qualified college instructors tutor students by answering questions and providing instruction regarding examples, exercises, and other content found in the text.

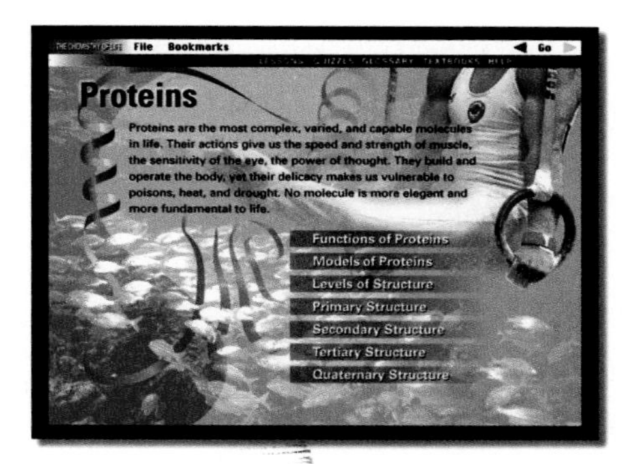

Supplements for the Instructor

Instructor's Teaching Guide (0-321-05332-X)

By Saundra Yancy McGuire of Louisiana State University

Includes chapter summaries, complete descriptions of appropriate chemical demonstrations for lecture, suggestions for addressing common student misconceptions, and examples of everyday applications of selected topics for lecture use.

Printed Test Bank (0-321-05326-5)

By Paris Svoronos and Soraya Svoronos of Queensborough Community College of the City University of New York

This printed test bank includes over 1700 questions that correspond to the major topics in the text.

Computerized Test Bank (0-321-05328-1)

By Paris Svoronos and Soraya Svoronos of Queensborough Community College of the City University of New York

This dual-platform CD-ROM includes over 1700 questions that correspond to the major topics in the text.

Complete Solutions Manual (0-321-05331-1)

By Saundra Yancy McGuire of Louisiana State University

Instructor's Manual for the Lab Manual, Second Edition (0-321-05330-3)

By Wendy Gloffke of Cedar Crest College and Doris Kimbrough of the University of Colorado at Denver

Benjamin Cummings Digital Library for *Introductory Chemistry* CD-ROM (0-321-04637-4)

This cross-platform CD-ROM features all of the visuals from the Russo/Silver text. The CD-ROM provides instructors with a complete set of illustrations for incorporation into lecture presentations, study materials, and tests.

Transparency Acetates (0-321-05329-X)

Includes 125 full-color acetate transparencies.

CourseCompass™

CourseCompass™ combines the strength of Benjamin Cummings content with state-of-the-art eLearning tools! **CourseCompass™** is a nationally hosted, dynamic, interactive online course management system powered by BlackBoard, leaders in the development of Internet-based learning tools. This easy-to-use and customizable program enables professors to tailor content and functionality to meet individual course needs! Every CourseCompass™ course includes a range of preloaded content such as testing and assessment question pools, chapter-level objectives, chapter summaries, photos, illustrations, videos, animations, and Web activities—all designed to help students master core course objectives. Visit www.coursecompass.com for more information.

A Suite of Tools for Learning Problem-Solving Skills

Different students need different tools. Using the example of initial-condition, final-condition gas law problems, these pages illustrate the integrated tools offered by *Introductory Chemistry* to help students master problem-solving skills.

The story begins on page 409 with WorkPatch 11.3, which asks the student to solve *PV/nT* to obtain a value for *R*. Besides emphasizing the importance of using correct units, this WorkPatch sets the stage for initial-condition, final-condition problems.

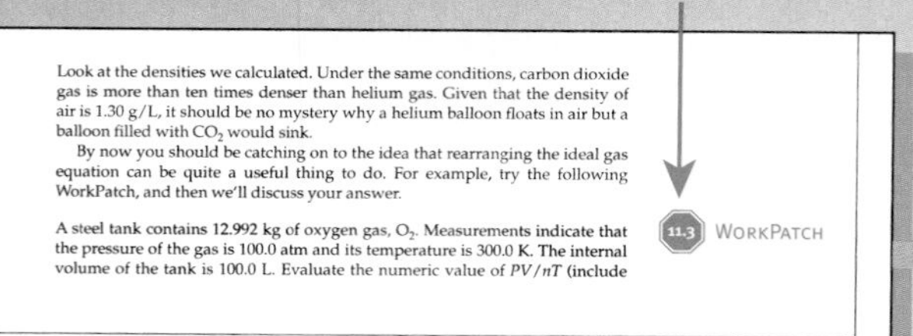

Look at the densities we calculated. Under the same conditions, carbon dioxide gas is more than ten times denser than helium gas. Given that the density of air is 1.30 g/L, it should be no mystery why a helium balloon floats in air but a balloon filled with CO_2 would sink.

By now you should be catching on to the idea that rearranging the ideal gas equation can be quite a useful thing to do. For example, try the following WorkPatch, and then we'll discuss your answer.

A steel tank contains 12.992 kg of oxygen gas, O_2. Measurements indicate that the pressure of the gas is 100.0 atm and its temperature is 300.0 K. The internal volume of the tank is 100.0 L. Evaluate the numeric value of PV/nT (include

11.3 WorkPatch

The text after the WorkPatch discusses the importance of the answer, but does not reveal it. (All WorkPatches are solved at the end of the chapter.)

Using the context of an example, the text explains initial-condition, final-condition problems. The focus here is on ensuring that the student **understands** this type of problem, how it reflects the underlying chemistry, and how to solve it.

At the end of the discussion, the method for solving this type of problem is **summarized in a chart.**

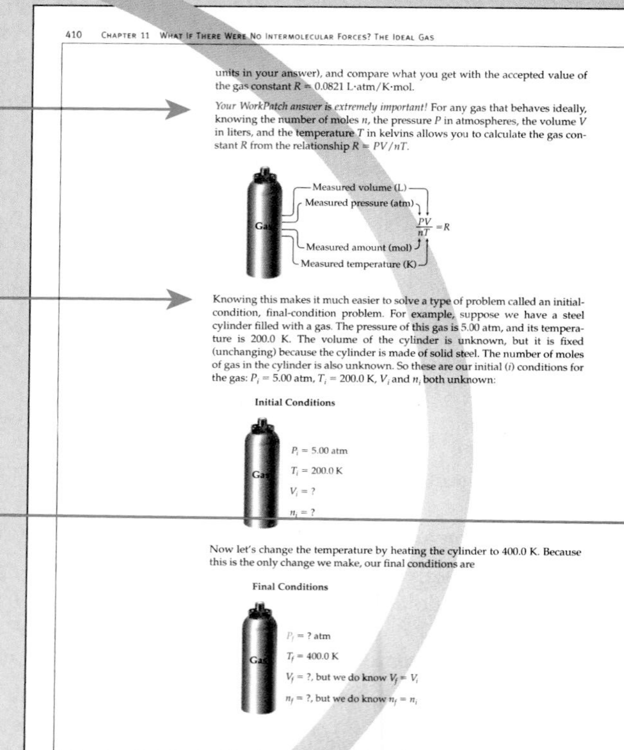

410 Chapter 11 What If There Were No Intermolecular Forces? The Ideal Gas

units in your answer), and compare what you get with the accepted value of the gas constant $R = 0.0821$ L·atm/K·mol.

Your WorkPatch answer is extremely important! For any gas that behaves ideally, knowing the number of moles n, the pressure P in atmospheres, the volume V in liters, and the temperature T in kelvins allows you to calculate the gas constant R from the relationship $R = PV/nT$.

Knowing this makes it much easier to solve a type of problem called an initial-condition, final-condition problem. For example, suppose we have a steel cylinder filled with a gas. The pressure of this gas is 5.00 atm, and its temperature is 200.0 K. The volume of the cylinder is unknown, but it is fixed (unchanging) because the cylinder is made of solid steel. The number of moles of gas in the cylinder is also unknown. So these are our initial (i) conditions for the gas: $P_i = 5.00$ atm, $T_i = 200.0$ K, V_i and n_i both unknown:

Initial Conditions

Now let's change the temperature by heating the cylinder to 400.0 K. Because this is the only change we make, our final conditions are

Final Conditions

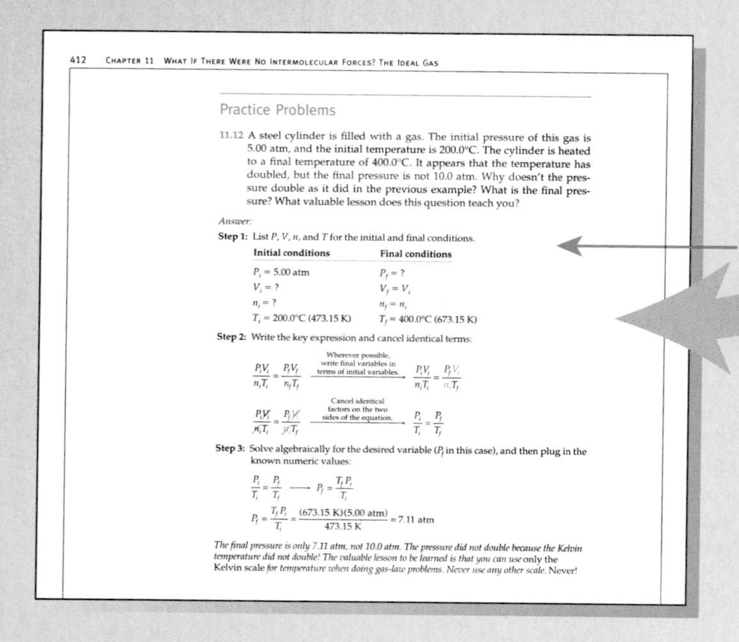

412 Chapter 11 What If There Were No Intermolecular Forces? The Ideal Gas

Practice Problems

11.12 A steel cylinder is filled with a gas. The initial pressure of this gas is 5.00 atm, and the initial temperature is 200.0°C. The cylinder is heated to a final temperature of 400.0°C. It appears that the temperature has doubled, but the final pressure is not 10.0 atm. Why doesn't the pressure double as it did in the previous example? What is the final pressure? What valuable lesson does this question teach you?

Answer:

Step 1: List P, V, n, and T for the initial and final conditions.

Initial conditions	Final conditions
$P_i = 5.00$ atm	$P_f = ?$
$V_i = ?$	$V_f = V_i$
$n_i = ?$	$n_f = n_i$
$T_i = 200.0°C$ (473.15 K)	$T_f = 400.0°C$ (673.15 K)

Step 2: Write the key expression and cancel identical terms:

$$\frac{P_iV_i}{n_iT_i} = \frac{P_fV_f}{n_fT_f} \xrightarrow{\text{Wherever possible, write final variables in terms of initial variables.}} \frac{P_iV_i}{n_iT_i} = \frac{P_fV_i}{n_iT_f}$$

$$\frac{P_i\cancel{V_i}}{\cancel{n_i}T_i} = \frac{P_f\cancel{V_i}}{\cancel{n_i}T_f} \xrightarrow{\text{Cancel identical factors on the two sides of the equation.}} \frac{P_i}{T_i} = \frac{P_f}{T_f}$$

Step 3: Solve algebraically for the desired variable (P_f in this case), and then plug in the known numeric values:

$$\frac{P_i}{T_i} = \frac{P_f}{T_f} \longrightarrow P_f = \frac{T_fP_i}{T_i}$$

$$P_f = \frac{T_fP_i}{T_i} = \frac{(673.15 \text{ K})(5.00 \text{ atm})}{473.15 \text{ K}} = 7.11 \text{ atm}$$

The final pressure is only 7.11 atm, not 10.0 atm. The pressure did not double because the Kelvin temperature did not double! The valuable lesson to be learned is that you can use only the Kelvin scale *for temperature when doing gas-law problems. Never use any other scale. Never!*

Starting on the next page, a set of three **Practice Problems** gives the students a chance to try their hand.

The first problem is solved in place using the three-step method summarized in the chart. The answers to unsolved problems are given in the back of the book. (The solutions are available in the *Study Guide and Selected Solutions* manual.)

The **Skills to Know** section at the end of the chapter gathers together the reference information students need while learning to work problems.

In addition, each chart from the chapter is repeated and accompanied by a new worked example.

value to figure out what the gas must be. Make sure you use units compatible with the units of R. $MM = \frac{}{PV}$

Solving initial-condition, final-condition gas problems

Example: An air bubble rises through the ocean from a depth of 1000 feet to the surface. As it rises, its pressure decreases from 31.3 atm to 1.00 atm and its volume increases. Its initial volume is 5.00 mL. The bubble's temperature remains constant, and it gains and loses no gas. What is its final volume?

1. List P, V, n, and T for the initial and final conditions.
   - Write in numerical values for any variables you know.
   - Where possible, express the final variable in terms of the initial one. For instance, if volume does not change, you can say $V_f = V_i$.

Initial conditions	Final conditions
$P_i = 31.3$ atm	$P_f = 1.00$ atm
$V_i = 5.00$mL	$V_f = ?$
$n_i = ?$	$n_f = n_i$
$T_i = ?$	$T_f = T_i$

2. Write the expression

$$\frac{P_iV_i}{n_iT_i} = \frac{P_fV_f}{n_fT_f}$$

   - Whenever possible, rewrite this expression to show final variables in terms of initial ones (from step 1).

$\frac{P_iV_i}{n_iT_i} = \frac{P_fV_f}{n_iT_i}$ Because $n_f = n_i$ and $T_f = T_i$, we can replace variables on the right side of the equation with those from the left side.

   - Cancel quantities that are identical on the two sides of the expression.

$\frac{P_iV_i}{n_iT_i} = \frac{P_fV_f}{n_iT_i} \longrightarrow P_iV_i = P_fV_f$

3. Solve the equation algebraically for the desired variable.

$P_iV_i = P_fV_f \longrightarrow V_f = \frac{P_iV_i}{P_f}$

Then plug in values for the quantities you know and solve numerically for an answer.

$V_f = \frac{P_iV_i}{P_f} = \frac{(31.3\text{ atm})(5.00\text{ mL})}{1.00\text{ atm}} = 157\text{ mL}$

Notice that although we still do not know the volume of the gas or the number of moles of gas present, we *can* say that $V_f = V_i$ and $n_f = n_i$ because we did not change these quantities on going from the initial conditions to the final conditions. The pressure, however, has changed (it increased as the temperature increased because the volume could not change). What is the new pressure P_f inside the cylinder? This is an important question. If a storage cylinder of gas gets too hot, the increase in pressure can cause it to explode.

To determine P_f, just remember that PV/nT is always equal to R. This means that P_iV_i/n_iT_i is equal to R, and P_fV_f/n_fT_f is also equal to R. And of course because these expressions both equal R, they must be equal to each other:

$$\frac{P_iV_i}{n_iT_i} = \frac{P_fV_f}{n_fT_f}$$

This is an extremely useful result. It is the starting point for solving initial-condition, final-condition problems. The second step is to apply the rule of algebra that allows us to cancel *equal* quantities that appear in the same position on both sides of the equation. In our case, V_i on the left side and V_f on the right side are equal, and both appear in the numerator, which means they cancel each other. Likewise, n_i on the left and n_f on the right are equal and both in the denominator, which means they also cancel. These cancellations simplify our equation to

$$\frac{P_iV_i}{n_iT_i} = \frac{P_fV_f}{n_fT_f} \xrightarrow{\text{simplifies to}} \frac{P_i}{T_i} = \frac{P_f}{T_f}$$

All the variables with unknown values have disappeared! The final step is to rearrange the equation to solve for P_f and then plug in the numeric values for P_i, T_f, and T_i:

$$P_f = \frac{P_iT_f}{T_i} = \frac{(5.00\text{ atm})(400.0\text{ K})}{200.0\text{ K}} = 10.0\text{ atm}$$

This answer tells us that the pressure doubled when we doubled the temperature while holding the volume constant. Make sure you can do this algebraic rearrangement. (Review Chapter 2 if you can't!)

A summary of the method we just described appears at right.

Try the following initial-condition, final-condition practice problems. The first teaches you a valuable lesson about temperature. The second shows you how a seemingly difficult problem is in fact quite solvable.

Solving initial-condition, final-condition gas problems

Step 1: List P, V, n, and T for the initial and final conditions.
- Write in numeric values for any variables you know.
- Where possible, express the final variable in terms of the initial one. For instance, if volume does not change, you can say $V_f = V_i$.

Step 2: Write the expression

$$\frac{P_iV_i}{n_iT_i} = \frac{P_fV_f}{n_fT_f}$$

- Whenever possible, rewrite this expression to show final variables in terms of initial ones (from step 1).
- Cancel factors that are identical on the two sides of the expression.

Step 3: Solve the equation algebraically for the desired variable. Then plug in numeric values for the quantities you know and do the calculation to get your answer.

Students having difficulty will find a wealth of carefully structured, confidence-building **guided practice** in the free *Problem Solving Guide and Workbook* that accompanies the text.

The *Workbook* provides explicit methods for solving a great variety of problems, each accompanied by worked examples and practice problems. The *Workbook* also provides chapter quizzes, cumulative quizzes, and answers to all problems.

Example 2

What is the molar mass of an unknown gas if 12.04 g of the gas occupies 7.40 L at 27.0°C and 980 mm Hg?

Solution

Perform step 1:

$$MM = \frac{m_{\text{sample}}RT}{PV}$$

Perform step 2:

T must be converted from °C to K.

$$K = 27.0°C + 273.15 = 300.2\text{ K}$$

P must be converted from mm Hg to atm.

$$P = 980\text{ mm Hg} \times \frac{1\text{ atm}}{760\text{ mm Hg}} = 1.29\text{ atm}$$

Perform step 3:

$$MM = \frac{(12.04\text{ g})(0.0821\text{ L}\cdot\text{atm / mol}\cdot\text{K})(300.2\text{ K})}{(1.29\text{ atm})(7.40\text{ L})} = 31.1\text{ g / mol}$$

Practice Problems

11.4 Calculate the molar mass of a gas if 1.00 L of the gas has a mass of 5.38 g at 15°C and 736 mm Hg.

11.5 What is the molar mass of a gas if 0.985 g of the gas occupies 3.00 L at a pressure of 178 mm Hg and a temperature of 22.5°?

Chart 11.3 Solving initial-condition, final-condition problems

Step 1
- List P, V, n, and T for the initial and final conditions.
- Write in numeric values for any known variables.
- Where possible, express the final variable in terms of the initial one (e.g., $V_f = V_i$).

Step 2
- Write the expression

$$\frac{P_iV_i}{n_iT_i} = \frac{P_fV_f}{n_fT_f}$$

- Whenever possible, rewrite this expression to show final variables in terms of initial ones (from step 1).
- Cancel factors that are identical on the two sides of the expression.

Step 3
- Solve the equation algebraically for the desired variable and substitute the numerical values for the known quantities.
- Perform the calculation to get the answer.

Example 1

7.62 moles of gas in a container with a moveable piston are heated from 100.0°C to 175.0°C. If the pressure remains constant, what volume will the gas occupy after the heating if it occupied 2.50 L initially?

Solution

Perform step 1:

Initial conditions	Final conditions
$P_i = P_f$ (unchanged)	$P_f = P_i$ (unchanged)
$V_i = 2.50$ L	$V_f = ?$
$n_i = n_f = 7.62$ mol (unchanged)	$n_f = n_i = 7.62$ mol (unchanged)
$T_i = 373.2$ K	$T_f = 448.2$ K

Perform step 2:

$$\frac{P_iV_i}{n_iT_i} = \frac{P_fV_f}{n_fT_f} \longrightarrow \frac{V_i}{T_i} = \frac{V_f}{T_f}$$

Periodic Table of the Elements
Transition metals

CHAPTER 1

What Is Chemistry?

It is the latter part of the nineteenth century, 1896 to be exact. I am overcome with grief. My youngest child is burning with fever. The sickness has spread from her ear to her entire body. Her skin has a scarlet look, and she is in great pain. The doctor has applied some tincture of iodine, but he does not know the cause of what ails her. He has told us to make arrangements. My beloved child will not see her fourth birthday.

It is the latter part of the twentieth century, 1996 to be exact. My daughter was ill yesterday with an earache. Our pediatrician diagnosed a streptococcus infection and administered the antibiotic amoxicillin. My daughter thought it tasted good, and she is back in preschool today, completely free of fever and pain.

It is the early part of the twenty-first century, 2026 to be exact. We have chosen to have a daughter. Unlike most of today's parents, we will not pre-select her IQ. However, we do agree with our genetic counselor that her system should be genetically engineered so that she will be immune to all known bacterial and viral infections.

1.1 Science and Technology

The span from 1896 to 1996 was 100 years. The year 2026 is less than 25 years away. The pace at which things are changing is accelerating at an unbelievable rate. One hundred years from now will be as different from today as today is from 500 years ago. Within the past ten years, a significant portion of the genetic code for the human genome has been unraveled. One hundred years ago we had never even heard of DNA; today we are beginning to modify it. One hundred years ago we burned coal; in less than half that time from now we'll generate energy in fusion reactors powered by hydrogen taken from seawater, duplicating the process that occurs within the cores of stars.

Change this rapid is new for humanity. Until recently, each new generation could expect to live pretty much like the one before it. Not any more. We scarcely have time to get used to one change before ten more are upon us. We've barely had time to think about the ethics of birth control, a development of the 1970s, and now fetal cell research, life prolongation, and cloning are knocking at our door. Your personal computer and its software are almost outdated the day you buy them. What is feeding all this change? The answer is *science*.

Science: The exploration of the structure of the atom

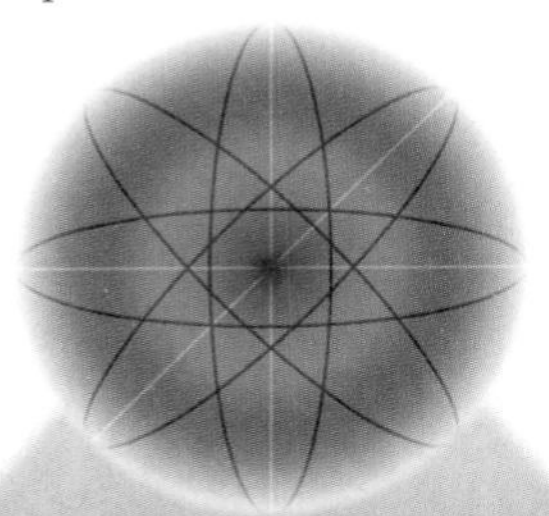

Technology: Applications of our knowledge about atoms

Nuclear medicine — Isotopic dating of Earth's history — Nuclear power — Nuclear weapons

The dictionary defines **science** as "the experimental investigation and explanation of natural phenomena" or "knowledge from experience." Science begins with a simple question, like "how?" or "why?" How are atoms put together? Why do bats fly at night? Science is the pursuit of knowledge for its own sake, because we are curious. But science itself can't cause change unless something is done with the knowledge it uncovers. For that we need **technology**, the *application* of scientific knowledge. So science is also the pipeline for technology—feeding it and supplying it with ideas. For example, scientists asked "How are atoms put together?" and their experiments led them to an answer. Today, technologists can use that knowledge to change our lives by developing nuclear medical technologies to treat cancer and by building nuclear bombs. As is so often the case with scientific knowledge, it can be used to achieve very different ends.

Because science feeds technology, it's not a bad idea to ask the question "Is science always right?" Just consider these headlines:

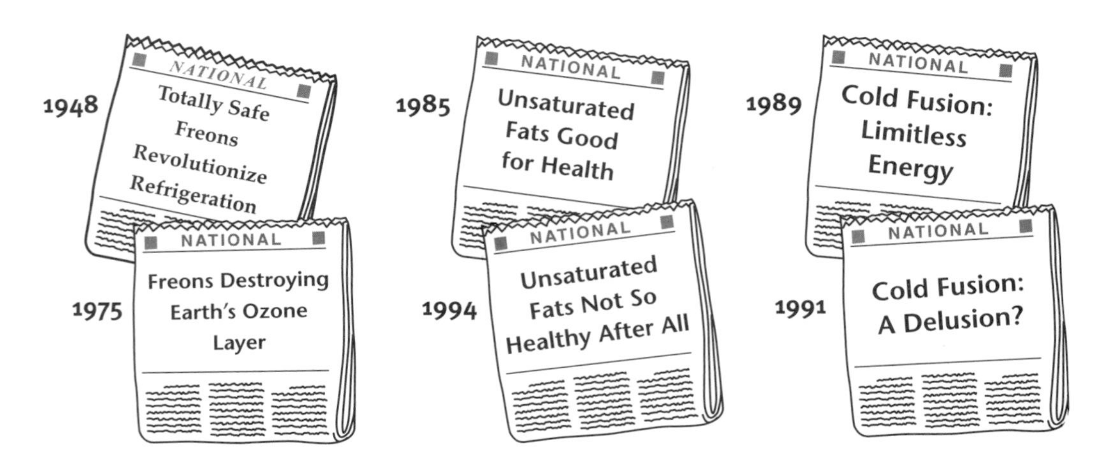
1948
NATIONAL
Totally Safe Freons Revolutionize Refrigeration
1975
NATIONAL
Freons Destroying Earth's Ozone Layer
1985
NATIONAL
Unsaturated Fats Good for Health
1994
NATIONAL
Unsaturated Fats Not So Healthy After All
1989
NATIONAL
Cold Fusion: Limitless Energy
1991
NATIONAL
Cold Fusion: A Delusion?

What is going on? Isn't science only about absolute, fundamental, provable truths? Unfortunately, no. Science, like literature, art, and music, is a human endeavor. And because it is a human endeavor, carried out by human scientists, you would not be wise to bet on science's infallibility. Ego, mistakes, and stubbornness can all get in the way of finding the truth. And what about technology? Is the technological application of scientific knowledge always good? Consider some of the forces that drive technology: the desire to benefit humankind, the desire to make a profit, the desire to be stronger than our enemies. Which motive do you think drives the cigarette industry in its application of knowledge concerning the effects of nicotine?

This brings us to some very important questions. Who should decide which areas of science are explored? Who should decide if a scientific result is real or a hoax? Who should decide what is done with scientific knowledge? In a free society, the answer is supposed to be the collective "you." After all, what scientists discover and technologists put into practice will dramatically affect how you live. In addition, much of the scientific research in this country is paid for by your tax dollars. Does this mean you have to earn a Ph.D. in chemistry, biology, and physics so that you can make informed decisions? For most people that is not a practical or even a desirable solution. But total ignorance of science is not the answer either because science and technology affect us directly in our everyday living.

This book is about the branch of science known as *chemistry*, often considered the "central science" since it forms a bridge between the principles of physics and the practice of biology. **Chemistry** is the study of matter and the transformations it undergoes. To understand this definition, we need to focus on two key concepts: *matter* and *transformation*.

Hydrogen and oxygen *transform* into water vapor and heat in the space shuttle's liquid-fueled engine, helping to propel the shuttle into space.

1.2 Matter

Matter can be simply defined as "stuff"—anything that has mass and occupies space. Some matter you can feel and see, like this book. Other matter, like the air that surrounds you, is difficult to detect, but it still exists. It occupies space (the volume of the room), and it has mass. Defining matter as "stuff" is adequate for most everyday situations. Since science often asks very specific questions, however, it almost always requires much more precise and detailed definitions. Chemistry begins defining matter by dividing it into two broad types, *pure substances* and *mixtures*. In **pure substances**, only a single type of matter is present. **Mixtures** occur when two or more pure substances are intermingled with each other. For example, table salt (chemical name, sodium chloride) is a pure substance. So is water. And so is table sugar (chemical name, sucrose). If you put salt and sugar in a jar together and shake, however, you have a mixture. Dissolve sugar in water and you have another mixture. Some things that you might not think of as mixtures actually do fit the definition—a rock, for example. In most rocks, you'll see a mixture of different minerals, each a different pure substance.

Chemists further subdivide mixtures into two types, *homogeneous* and *heterogeneous*. **Homogeneous mixtures** (*homo-*, meaning "the same") are ones in which the composition of the mixture is identical throughout. A cup of tea with some sugar dissolved in it is a homogeneous mixture. Once well stirred, such a mixture is exactly the same no matter where you sample it.

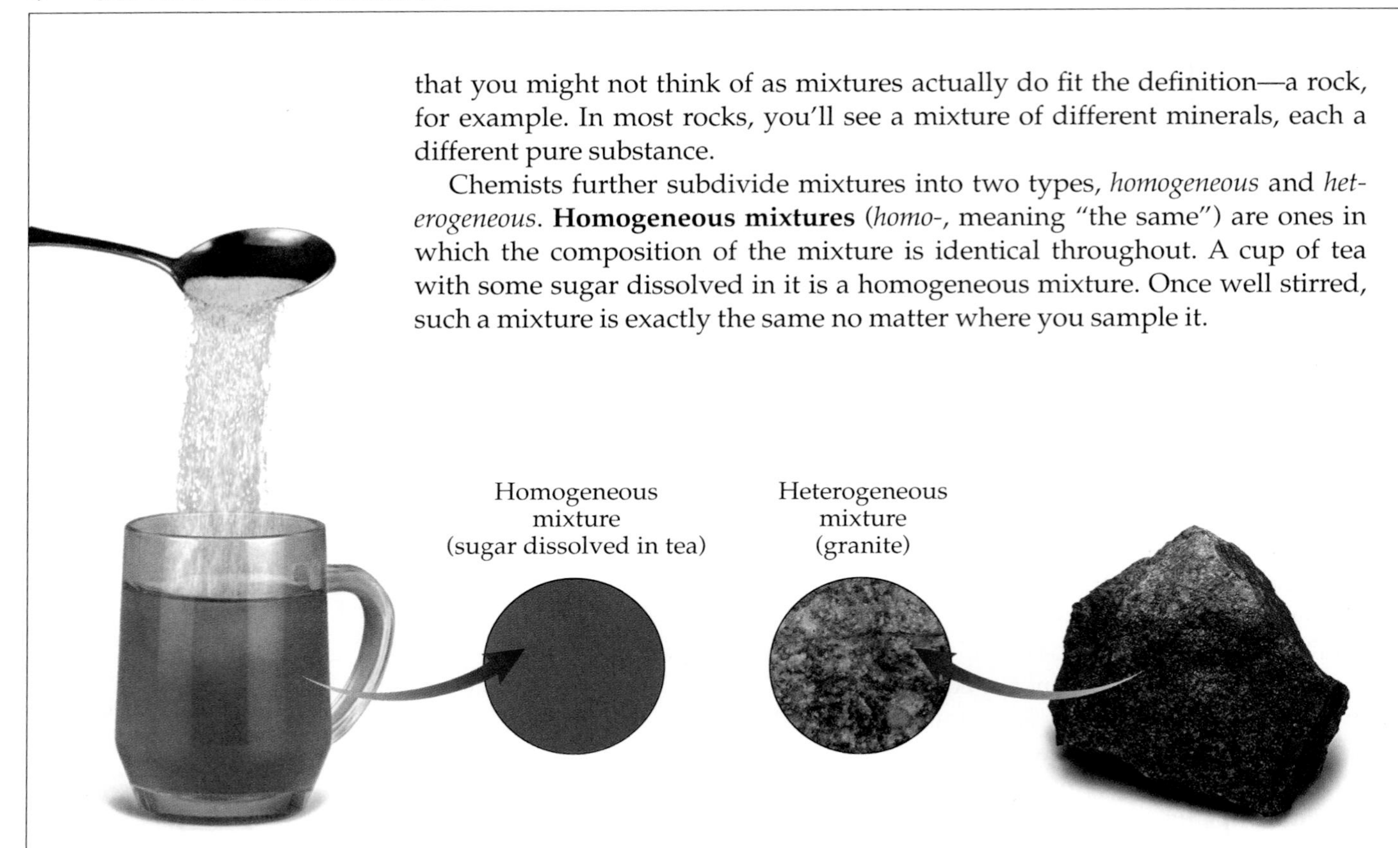

Another name for a homogeneous mixture is a **solution**. Our well-stirred cup of tea with sugar is a solution. The air you are breathing, consisting of a mixture made mostly of nitrogen gas and a smaller amount of oxygen gas, is also a homogeneous mixture, or solution. Under normal conditions, no matter where you sample the air in the room, its composition (percent nitrogen and percent oxygen) is the same.

The rock pictured above is an example of a **heterogeneous mixture** (*hetero-*, meaning "different") because its composition is not the same throughout. Sometimes, heterogeneous mixtures can appear to be homogeneous even when they aren't. A mixture of table salt and table sugar is a heterogeneous mixture even when the two substances are ground into a fine powder. This is true because a tiny sample taken at one place in the mixture might contain a different ratio of salt to sugar than a sample taken at some other place. If the amount taken is small enough, it would even be possible to get a sample from this mixture that was pure sugar or pure salt, no matter how well you mechanically ground the two together. Only when such a tiny sample has the same composition wherever you sample it (as in the tea with sugar) can you call the mixture homogeneous.

A precise dividing line between a heterogeneous mixture and a homogeneous one is difficult to specify. Just how uniform does a mixture have to be in order to be considered homogeneous? Obviously, if the sample analyzed is of atomic size, then no mixture can be considered homogeneous. In a larger sample, containing billions and billions of atoms, the limiting factor is the analysis procedure used and how small a difference in composition is detectable by the experiment. In this text, we shall consider only obvious examples for discussion and classification and leave the more ambiguous determinations to the philosophers in the audience.

1.1 WORKPATCH

Classify the following examples of matter as pure substances, heterogeneous mixtures, or homogeneous mixtures (solutions).

(a) piece of wood
(b) iron nail
(c) rusty iron nail
(d) well-stirred mixture of food dye in water
(e) bee's wax and candle wax mixed together by hand
(f) bee's wax and candle wax melted together, stirred well, then allowed to solidify

Check your answers to WorkPatch 1.1 against the answers given at the end of the chapter. Did you get the right answers for the last two items? These items should make you think. Only one of the mixtures of bee's wax and candle wax represents a solution because only one of the mixing processes (melting) can produce a homogeneous mixture.

Practice Problems

1.1 Which of the following are solutions?
(a) our atmosphere
(b) gold powder and silver powder ground together
(c) piece of 18-karat jewelry made from melting gold and silver metals together
(d) detergent dissolved in water
(e) oil droplets suspended in water

Answer: (a), (c), and (d)

1.2 Examination of "homogenized" milk under a microscope reveals suspended globules of fat. Is milk a heterogeneous mixture or a homogeneous mixture? Explain.

1.3 Fog is a suspension of tiny droplets of water in air. Is fog a heterogeneous mixture or a homogeneous mixture? Explain.

Before leaving the definition of matter, we want to say something about what matter is made of. All the matter that exists on our planet—from the air, to the dust in the air, to the ground we walk on, to the water that covers most of the planet, to the lifeforms that live on it—is made from **elements,** which are the basic building blocks of matter. At the time of writing, there are 113 known elements, 90 that occur naturally and 23 that can be synthetically prepared. All the known elements have been organized in a tabular form known as the *periodic table.*

The periodic table is so important to chemistry that we shall devote most of Chapter 3 to it. (A more complete periodic table and a list of the full names of all the elements, appear inside the front cover of this book.)

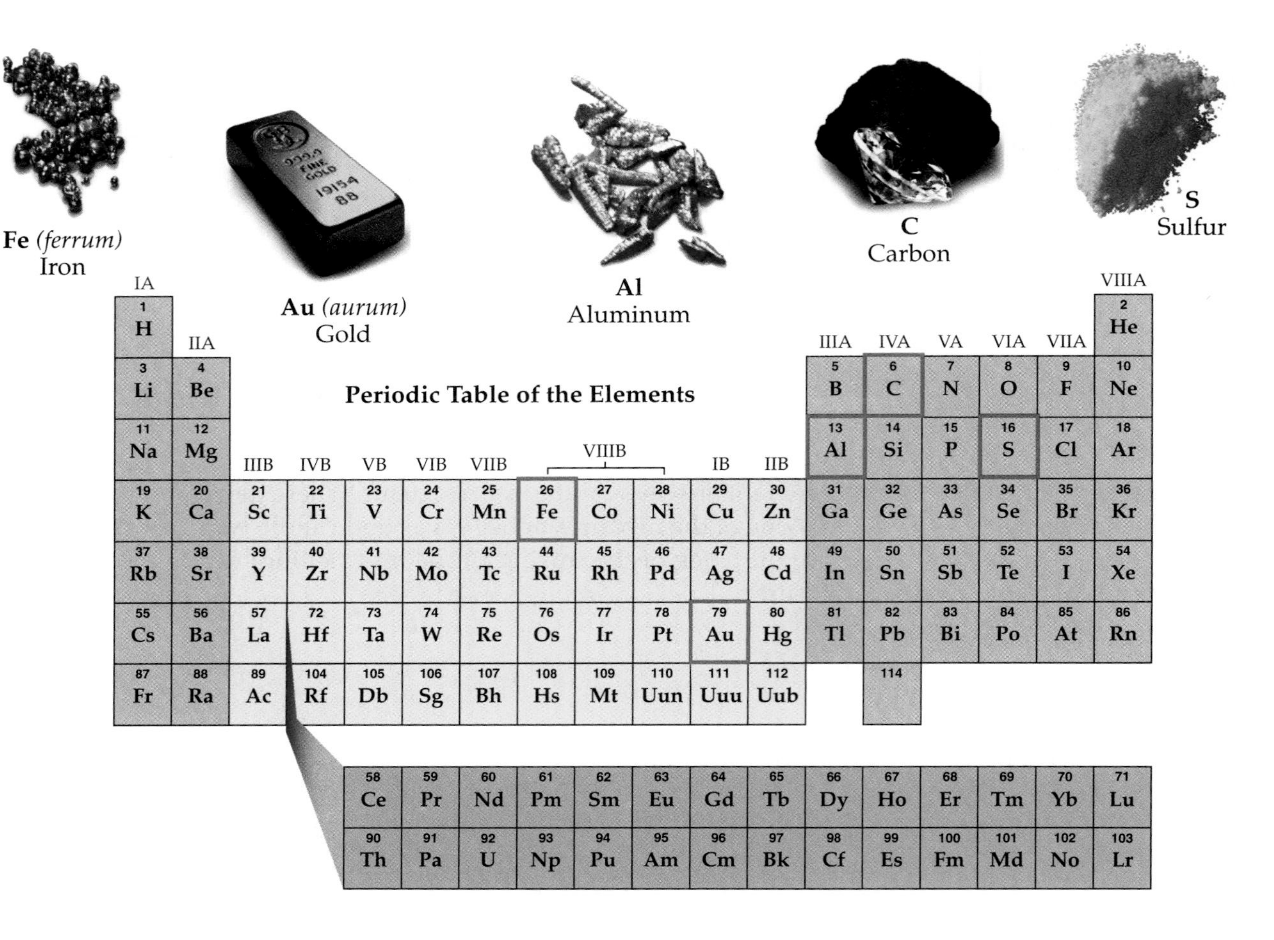

Many elements have names that you are probably quite familiar with—carbon, silver, gold, iron, aluminum, uranium, hydrogen, helium, oxygen, and nitrogen are a few. Chemists represent the elements with one-, two-, or three-letter symbols (as shown in the periodic table), some taken from the English names (like C for carbon, H for hydrogen, Al for aluminum) and others taken from the Latin names (like Fe for iron from the Latin word *ferrum*; Au for gold from the Latin word *aurum*).

When we say that elements are the basic building blocks of matter, how basic is basic? For example, suppose you had a piece of pure elemental gold. What would happen if you cut the piece of gold in half, and then in half again, and then again, and then again? How many times could you divide the piece and still have elemental gold? People have been trying to answer this question for centuries. Around 400 B.C. the Greek philosopher Democritus suggested an atomic theory of the universe. Simply put, this theory said that all things are made up from minute, indivisible, indestructible particles called *atoms*. That this might be true was by no means self-evident. A piece of gold does not appear to be made of individual particles. Nevertheless, Democritus would have said that you can keep dividing your piece of gold only until you get down to a single atom of gold. In fact, Democritus was right, but his theory was not an easy sell. Aristotle, another Greek philosopher living around the same time, placed more trust in his senses and said that matter was continuous and not made up of discrete individual atoms. According to Aristotle, you could keep dividing your gold into smaller and smaller pieces forever; at no point would you reach an indivisible particle. Because Aristotle's proposition seemed more obviously correct, his theory carried the day for more than 2000 years.

As you will see in Chapter 3, scientists did eventually return to the atomic theory. We now know that all the elements exist as atoms, and an **atom** is the smallest possible piece of an element. Atoms are so tiny that, until recently, scientists thought we would never be able to see them. They spoke too soon. Though no one has seen an atom through an ordinary microscope, in the early 1980s a device called a scanning tunneling microscope produced the first images of individual atoms—like the silicon atoms that appear as bumps on the surface of the silicon crystal shown in the photograph at right.

Scanning tunneling microscope image of individual silicon atoms in a crystal of pure silicon. Each division on the scale is approximately 0.5 nanometer; a nanometer is 10^{-9} meter, which is 0.000 000 001 meter.

With the knowledge that matter is made up of atoms of the elements, we can now divide pure substances into two types, *elemental substances* and *compounds*. An **elemental substance** is one that is made from atoms of just one element. For example, our piece of pure gold is made from just gold atoms and nothing else. The same is true of a piece of pure iron (or any other pure metal), the oxygen you breathe, and the helium gas in a balloon.

Compounds, on the other hand, are pure substances made from atoms of two or more different elements. Water, for example, is a compound made from atoms of hydrogen and oxygen. A **chemical formula** for a compound indicates the number of atoms of each element that make up the smallest possible piece of that compound. The chemical formula for water, H_2O, tells us that the smallest possible piece of water is made from two hydrogen atoms and one oxygen atom. Because water is made from two different elements, it is classified as a compound. So is table sugar, whose formula is $C_{12}H_{22}O_{11}$. The oxygen in the air you breathe is a different case. The formula for oxygen, O_2, tells us that the smallest piece of life-sustaining oxygen gas has two oxygen atoms in it. However, though oxygen is made from two atoms, both are atoms of the same element, and so oxygen is not considered to be a compound; oxygen is an elemental substance. To summarize:

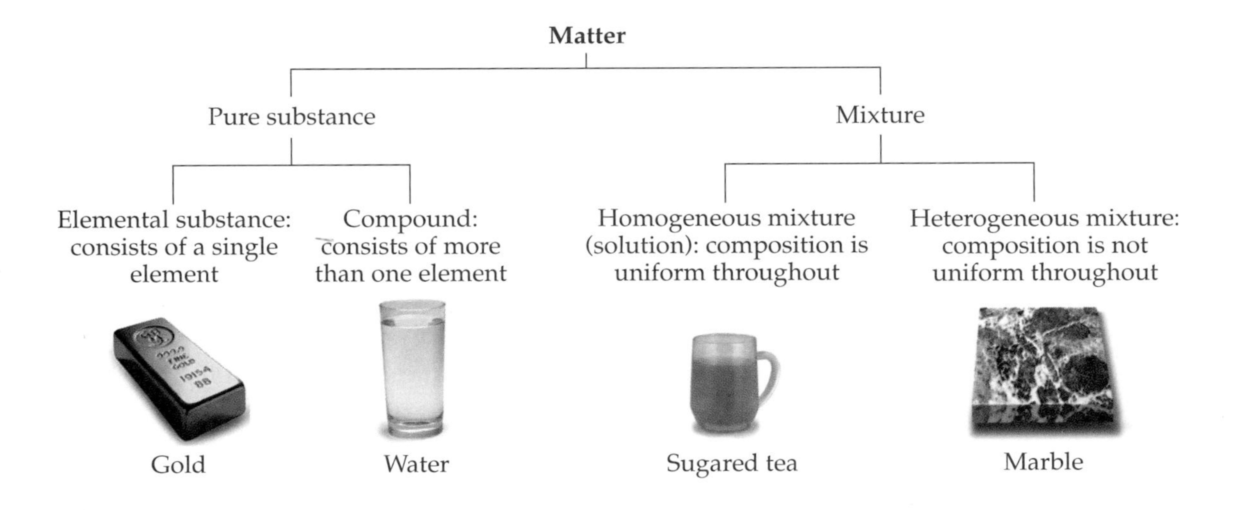
Matter
Pure substance
Mixture
Elemental substance: consists of a single element
Compound: consists of more than one element
Homogeneous mixture (solution): composition is uniform throughout
Heterogeneous mixture: composition is not uniform throughout
Gold
Water
Sugared tea
Marble

Classifying matter according to this scheme sometimes requires careful examination. Consider, for example, a glass of water, a glass of lemonade made from a powdered mix, and a glass of lemonade made with fresh-squeezed lemons.

Pure water — Lemonade from powdered mix — Lemonade from fresh-squeezed lemons

The water is a pure substance and a compound. The lemonade made from the mix is a homogeneous mixture (a solution). The lemonade made from lemons is a heterogeneous mixture because of the lemon pulp bits floating in it.

Practice Problems

1.4 Which of the following are compounds?
(a) iron oxide (Fe_2O_3)
(b) ozone (O_3)
(c) iron (Fe)
(d) carbon monoxide (CO)
(e) propane (C_3H_8)

Answer: (a), (d), and (e) because each is made from more than one kind of element; (b) and (c) are elemental substances, not compounds, because each is made of only one element.

1.5 Which of the following are compounds?
(a) sulfur (S_8)
(b) mixture of iron powder and aluminum powder
(c) mixture of O_2 gas and N_2 gas
(d) sulfur dioxide (SO_2)
(e) ammonia (NH_3)

1.6 True or false? A compound is a pure substance, but a pure substance need not be a compound. Give examples to prove your answer.

1.3 Matter and Its Physical Transformations

Let's get back to the definition of chemistry, which is why we started looking at matter in the first place. We said that chemistry is the study of matter and the transformations it undergoes. Since *transformation* means change, chemistry is the study of changes in matter. And just as the word "matter" had to be subdivided into pure substances and mixtures, we must be very careful with the word "change." In the science of chemistry, there are two kinds of changes matter can undergo: *physical* and *chemical*.

A **physical transformation** of a pure substance is one that leaves it as the same substance but in a different physical state. This leads us to another question: what do we mean by a *state of matter*? You are already familiar with the three most common **states of matter**: solid, liquid, and gas (or vapor). We shall examine these states in a detailed way later in this book, but for now your common knowledge of them will do just fine. For now, let's use water to demonstrate physical changes in matter. If you put a glass of liquid water in a freezer, the water will change to the solid state in the process we call **freezing**. This is an example of a physical change because only the state of the substance has changed. It's still water after the change. The reverse of freezing, called **melting**, is another example of a physical change. Naturally, the temperature below which liquid water freezes is also the temperature above which solid water melts, 32°F or 0°C at normal atmospheric pressure. (The Fahrenheit and Celsius temperature scales are described in Chapter 2.) Thus, this temperature is called either the freezing point or the melting point of water, depending on which way you are going. When heated either to or above its boiling point (212°F or 100°C at normal atmospheric pressure), water boils and undergoes **vaporization**, the change from the liquid to the vapor state. In the reverse of vaporization, **condensation**, the vapor returns to the liquid

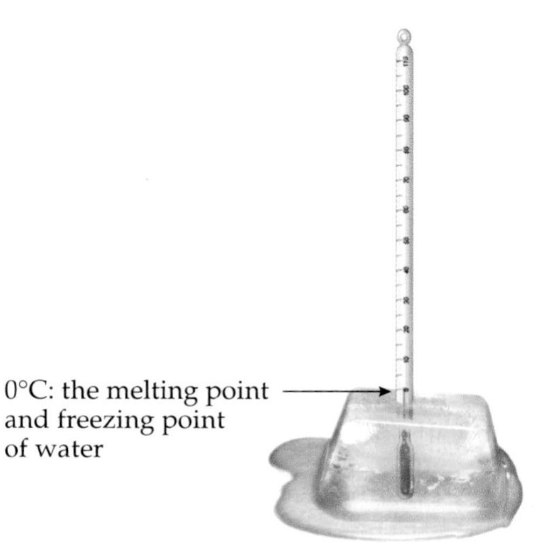

Water melts when its temperature goes above 0°C; it freezes when its temperature goes below 0°C.

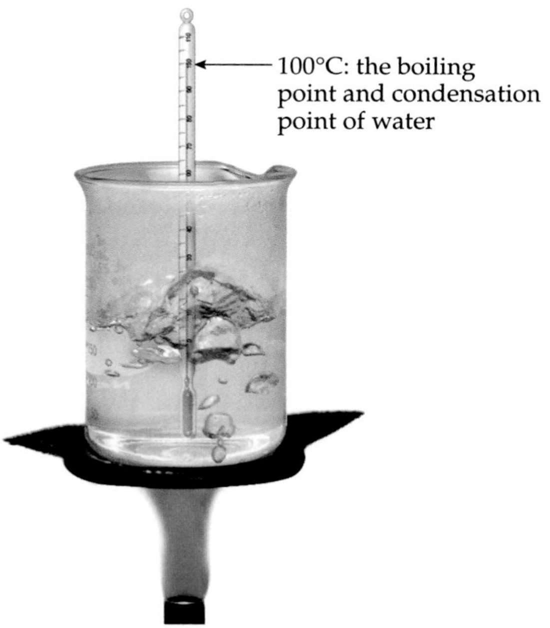

Water boils when its temperature goes above 100°C; it condenses when its temperature goes below 100°C.

state. After either change, we still have water, and so vaporization and condensation are two more examples of physical changes.

The melting point and boiling point of water are among its physical properties. The **physical properties** of a pure substance characterize its physical state and physical behavior. Other physical properties of water include its color (pure water is colorless), its odor (pure water is odorless), and its taste (pure water is tasteless). The following table summarizes some of the physical properties of water:

Physical Properties of Pure Water (at Normal Atmospheric Pressure)

Melting point	0°C
Boiling point	100°C
Color	None
Odor	None
Taste	None

Each pure substance has its own unique set of physical properties. For example, pure ethanol, the component of beer and wine that gets you drunk, is a colorless liquid that looks just like water. However, it melts at −117°C and boils at 78°C. Thus, if you are presented with a colorless liquid and asked to determine whether it is water or ethanol, you can place a thermometer in it and heat it to boiling. If the thermometer reads 78°C, then the liquid is ethanol.

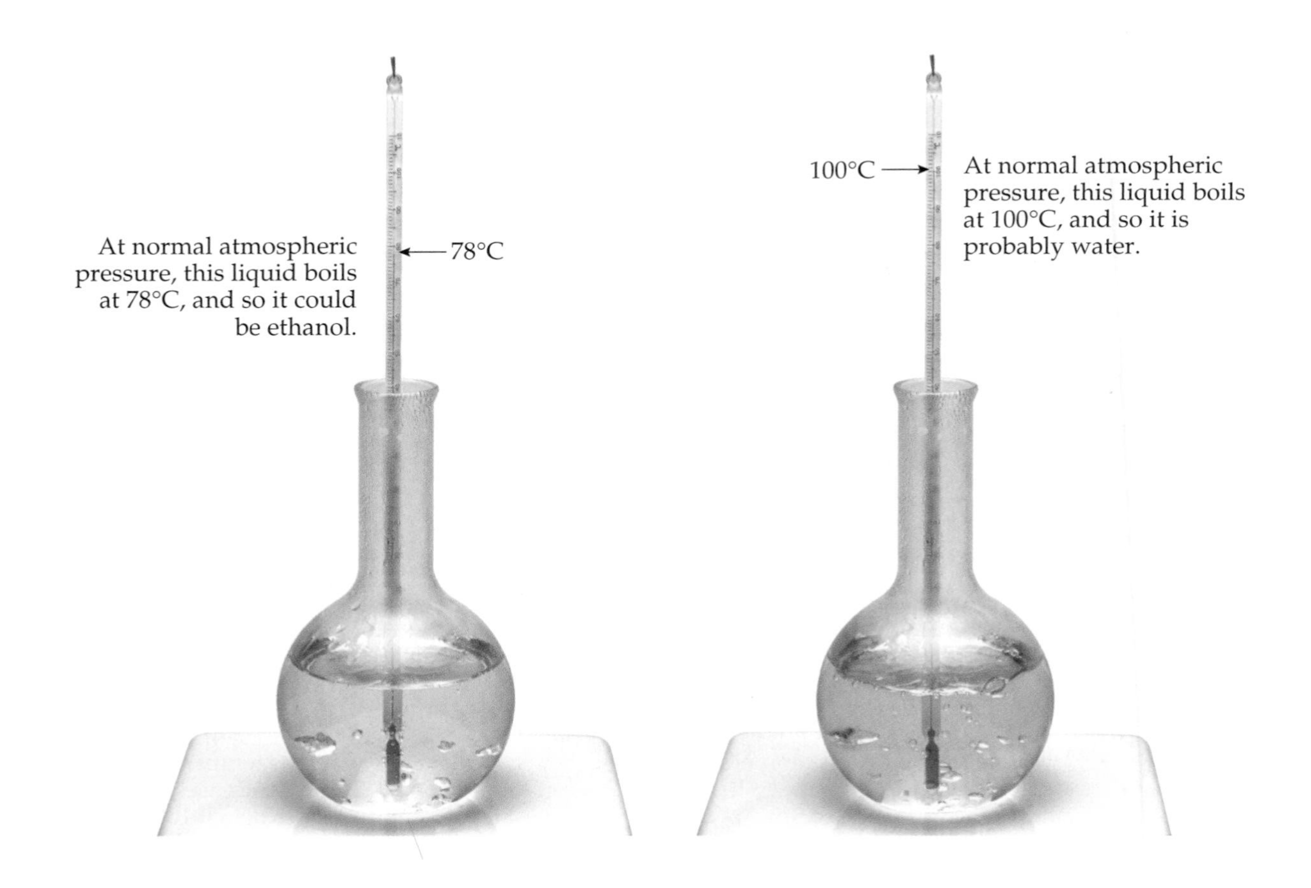

At normal atmospheric pressure, this liquid boils at 78°C, and so it could be ethanol.
78°C
100°C
At normal atmospheric pressure, this liquid boils at 100°C, and so it is probably water.

Another important physical change is **sublimation**, which occurs when a pure substance goes directly from the solid state to the gas state without passing through the liquid state. For example, when you heat an ordinary ice cube, it first melts to liquid water before boiling to enter the vapor state. Carbon dioxide, however, another pure substance, behaves quite differently. Carbon dioxide exists as a solid below −78°C, but if you warm it above this temperature, it slowly vanishes into thin air without ever forming a puddle or displaying any obvious wetness. (For this reason, solid carbon dioxide is called "dry ice.") It goes directly from the solid state to the gas state—it sublimes. Unlike water, carbon dioxide cannot exist in the liquid state unless a great deal of pressure is applied to it.

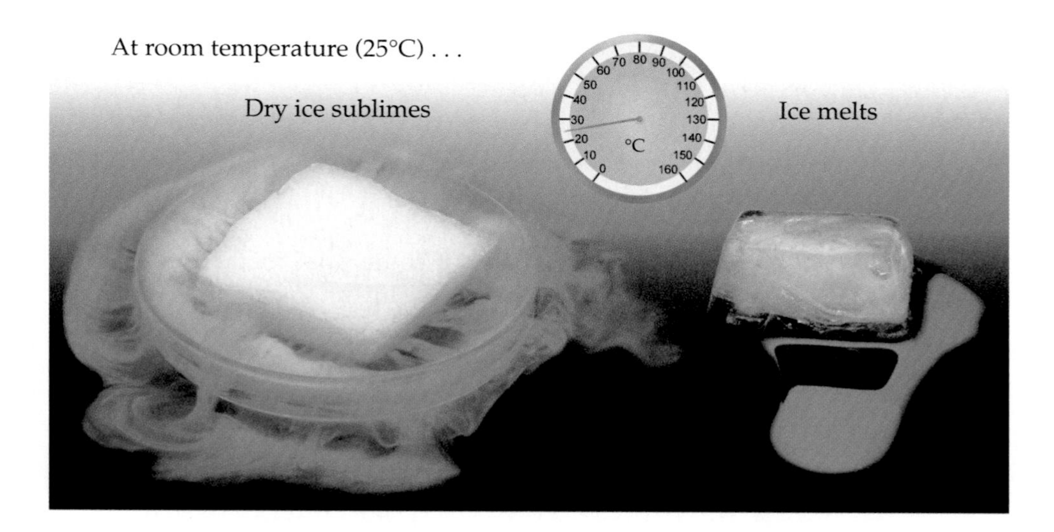

It is even possible for regular ice (solid water) to sublime if conditions are right. Have you ever noticed how snow disappears even when the temperature remains below freezing? The same thing happens inside the freezer compartment of a frost-free refrigerator.

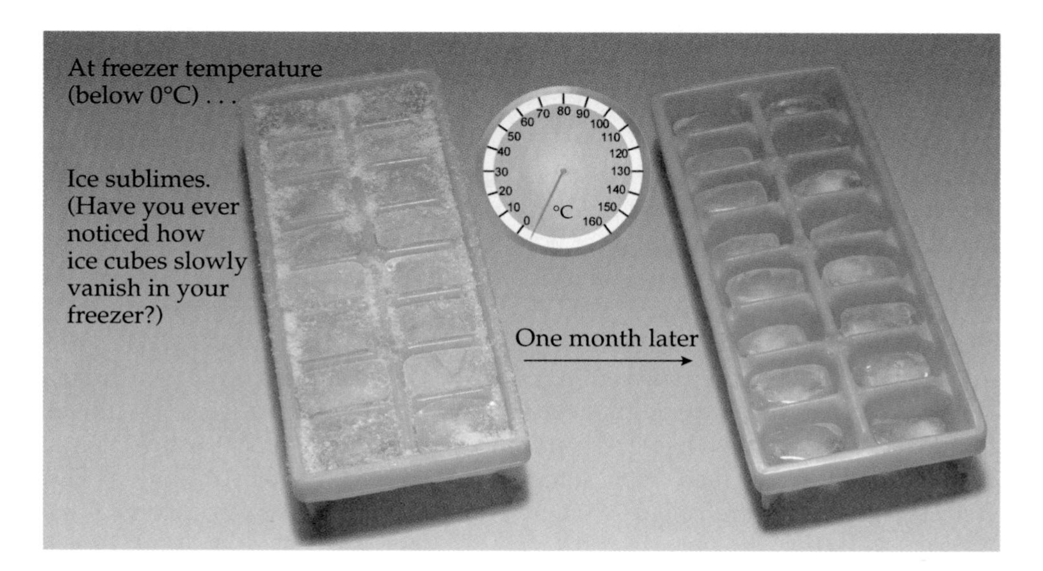

At freezer temperature (below 0°C) . . .
Ice sublimes. (Have you ever noticed how ice cubes slowly vanish in your freezer?)
°C
One month later

Practice Problems

1.7 Ice cubes slowly vanishing while stored in a freezer is an example of
(a) vaporization
(b) condensation
(c) melting
(d) sublimation

Answer: (d); the ice cubes "vanish" by going directly from the solid state to the gas state.

1.8 Molten iron cooling to solid iron is an example of
(a) sublimation
(b) condensation
(c) freezing
(d) melting

1.9 True or false? After being heated to above its boiling point, ethanol is no longer ethanol but something else. Explain.

1.10 You are presented with a block made of some pure metal and told the metal is gold, but you have your doubts. Using a thermometer, how can you determine whether the metal is gold?

1.4 Matter and Its Chemical Transformations

So far we have been concentrating on the physical properties of pure substances. It's time now to switch to *chemical properties*. The **chemical properties** of a pure substance are the ways it behaves when combined with other pure substances. For example, sodium (a soft, shiny metal having the elemental symbol Na, from the Latin *natrium*) catches fire when it is combined with water. This is a chemical property of sodium. A chemical property of chlorine (a greenish gas having the formula Cl_2) is that it "burns" most substances it comes into contact with. If you mix sodium and chlorine together, a rather incredible thing happens. A tremendous amount of heat and light is given off. When things calm down enough for us to take a look, we see that the sodium metal and chlorine gas are both gone, replaced by an entirely new pure substance, the compound sodium chloride (formula NaCl).

The new compound has chemical properties entirely different from those of either of the pure substances from which it was made. It no longer catches fire in water, and it no longer burns things it comes into contact with. It's ordinary table salt. What you have witnessed is a *chemical transformation*. A **chemical transformation** occurs when new substances, having new physical and chemical properties are formed from some starting substance or substances. The starting substances are called **reactants**, and the new substances formed are

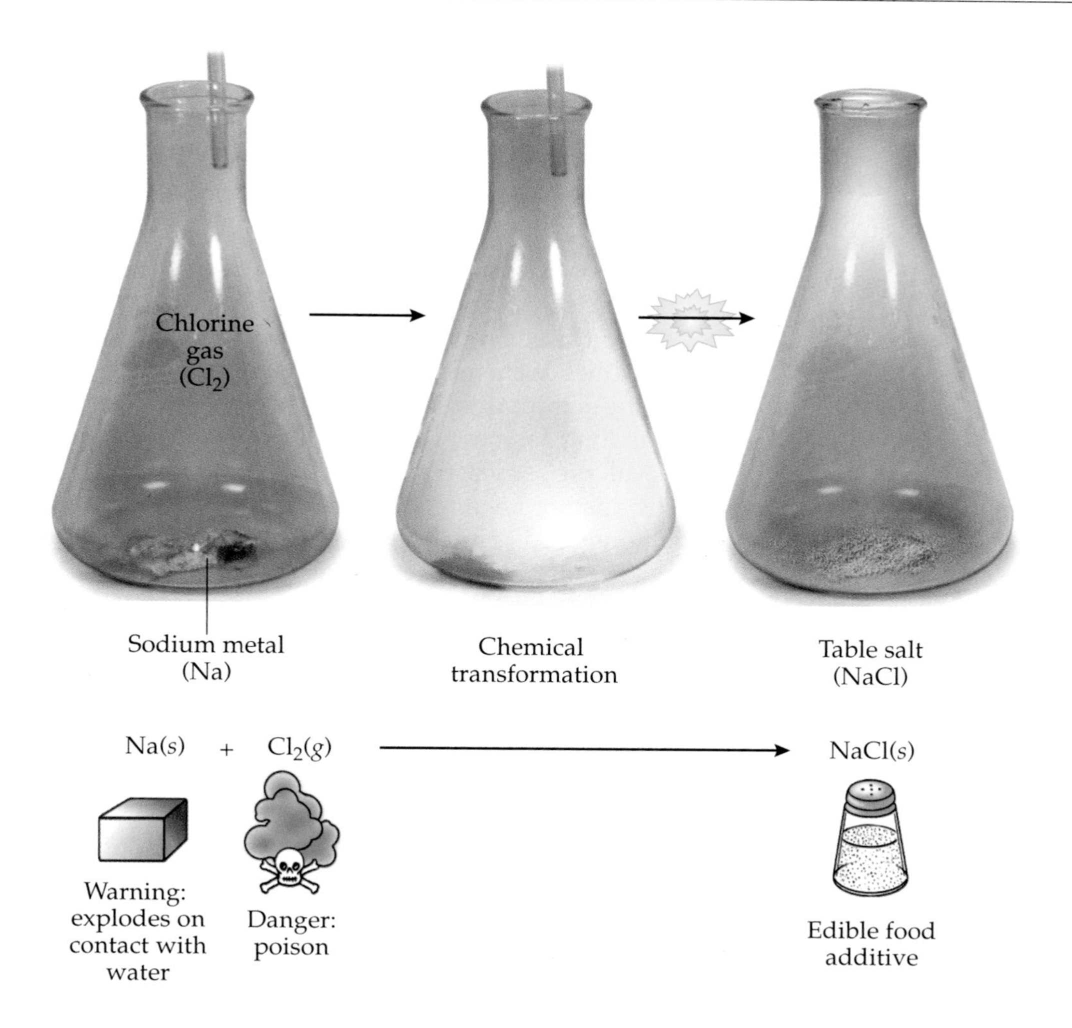

called **products**. The process that goes on during the conversion of reactants to products is called a **chemical reaction**.

Chemists write chemical reactions using a symbolic representation, with the reactant substances to the left of a horizontal arrow and the product substances to the right. The arrow itself means "react to give." For instance, a chemist would write the chemical reaction between sodium and chlorine as

$$\underset{\text{Reactants}}{2\,\text{Na} + \text{Cl}_2} \longrightarrow \underset{\text{Products}}{2\,\text{NaCl}}$$

In English, this says "sodium and chlorine react to give sodium chloride." The matter has undergone a chemical transformation. Often, chemists include the notation (*g*) for gas, (*l*) for liquid, or (*s*) for solid to represent the physical state of each substance in the reaction. Thus, we could have written

$$2\,\text{Na}(s) + \text{Cl}_2(g) \longrightarrow 2\,\text{NaCl}(s)$$

Most chemical transformations are carried out by nature, not by chemists. The rusting of iron, be it in a nail or on your car, is a chemical transformation in

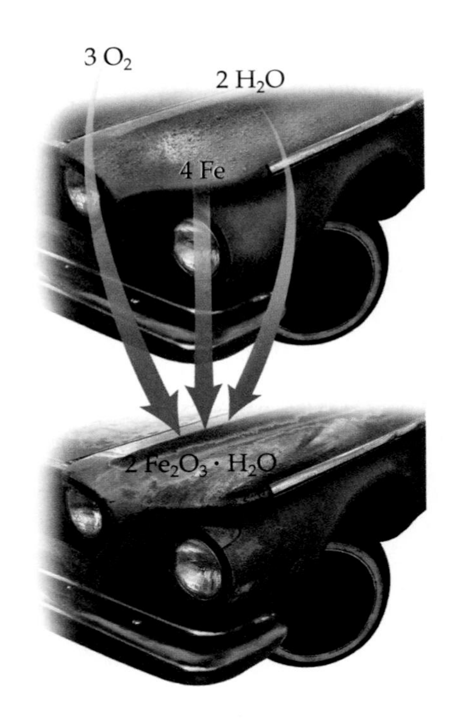

$$4\,Fe + 3\,O_2 + 2\,H_2O \longrightarrow 2\,Fe_2O_3 \cdot H_2O$$

which iron reacts with oxygen in the presence of water to give the new compound iron oxide (formula $Fe_2O_3 \cdot H_2O$), commonly known as rust.

Even more important to humans are the chemical transformations that occur during the digestion of food. Some of the food you eat gets converted to carbon dioxide and water, which you exhale. Your very thought processes involve many chemical reactions.

In spite of the stereotype of the crazed scientist madly mixing test tubes of chemicals for the joy of it, that's not the reason chemists mix substances together. Chemists are constantly searching for new compounds with new chemical and physical properties that can extend our knowledge about how matter behaves and can benefit humankind. New compounds to combat disease, to protect metal from corrosion, to whiten and brighten socks, . . . , the list goes on and on. Chemical transformation is what makes chemistry so interesting and worthy of study.

Chemistry got its start hundreds of years ago in the medieval period. Back then, alchemists, the forerunners of modern-day chemists, were busy trying to turn base metals such as lead into gold. They didn't succeed, but what they discovered while trying laid the foundation for the science of chemistry. Today we know that the alchemists could never have succeeded, for while turning lead into gold is an example of transforming matter from one substance into another, it is not an example of a *chemical* transformation. In a chemical transformation, none of the elements in the reactants change. If your reactants have carbon, hydrogen, and oxygen in them, then your products must have only carbon, hydrogen, and oxygen in them and no other elements. A chemical transformation can't convert one element to another element, and so turning lead into gold by chemical means is impossible. (Modern-day scientists can turn lead into gold via a *nuclear transformation*, which converts one element to another; you'll learn more about this in Chapter 16.)

Practice Problems

Refer to the periodic table inside the front cover for names of elements not yet familiar to you.

1.11 Which of the following represent a chemical transformation?

(a) $4\,P(s) + 5\,O_2(g) \longrightarrow 2\,P_2O_5(s)$

(b) $H_2O(g) \longrightarrow H_2O(l)$

(c) $3\,O_2 \longrightarrow 2\,O_3$

Answer: (a) and (c) because the products are different from the reactants; (b) is a physical change because we have water on both sides of the arrow.

1.12 Water (H_2O) and carbon dioxide (CO_2) are produced in a chemical reaction when methane (CH_4) and oxygen (O_2) are combined and heated. Which are the reactants and which are the products?

1.13 What, if anything, is wrong with the chemical reaction

$$Cl_2(g) + Br_2(l) \longrightarrow 2\,HCl(g) + 2\,HBr(g)$$

(a) It is not a chemical reaction. It is only a physical transformation.
(b) The product includes elements not in the reactants.
(c) There is nothing wrong with the chemical reaction.

1.5 How Science Is Done: The Scientific Method

Concepts such as what atoms are made of and how they react are essential to our understanding of chemistry. Many of the basic ideas that underlie chemistry were developed in the last 150 years, but the process that scientists used to develop these concepts has been around much longer. In fact, any curious person seeking answers to a question might, and probably has, used the *scientific method*. For example, consider Albert. It is a commonly held belief that scientific types have little or no romantic life. This is, in fact, not true. In his younger days, Albert found himself newly arrived at a university, seeking dates.

Being of a curious nature, he decided to study the topic of dating. He went on many dates, making observations, collecting data, and recording it all in his diary (his "laboratory notebook"). These dates were his **experiments**, which are procedures scientists carry out to study some phenomenon. For Albert, some dates were successful and some were not. On each date, he would vary one aspect of his appearance (a racy new tie, a bold set of socks, an eye patch), observe its effect on the outcome, and diligently record the results.

After several months of collecting data from many dates, he came to the conclusion that changes in his appearance above his neckline had great influence on the outcome of his dates, but modifications below his neckline had little or no effect. This led him to postulate the following Law of Dating: "For a successful date, look as good as you can from the neck up." A **law** is a generalization that concisely summarizes the outcome of a series of experiments. At first, Albert thought this law a bit odd, but after much thought, he developed a *theory* that explained why it was true. A **theory** is an attempt to explain *why* a law exists, and a **model** is some kind of physical picture or mathematical expression of a theory. He theorized that in any stressful situation between two people, they watch each other's faces intensely for visual cues about how things are going. This Stress-Induced Intense Facial Watching Theory explained the Law of Dating.

The process Albert used to satisfy his curiosity about dating—collecting data from experiments, using it to fashion a law, then postulating a theory to account for the law—is called the **scientific method**.

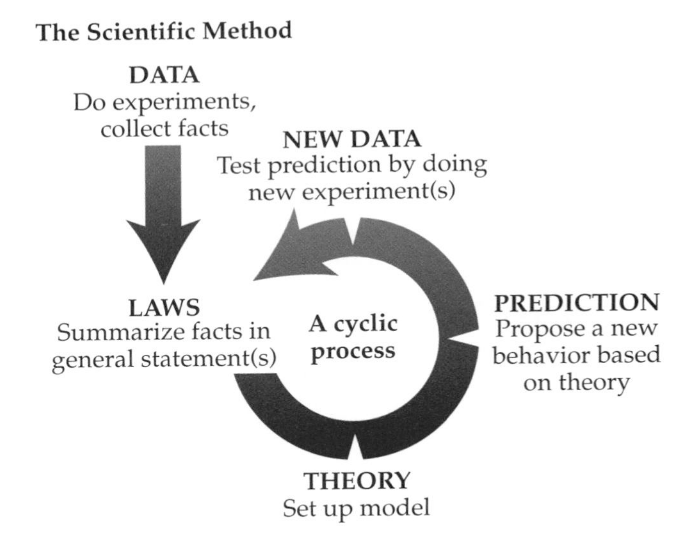
The Scientific Method
DATA
Do experiments, collect facts
NEW DATA
Test prediction by doing new experiment(s)
LAWS
Summarize facts in general statement(s)
A cyclic process
PREDICTION
Propose a new behavior based on theory
THEORY
Set up model

Although the dating scenario we used to illustrate the scientific method is tongue-in-cheek, it does possess all the steps of the scientific method. However, even though Albert was pleased with his procedure, he probably should have discussed it with his advisor because he made some classic beginner's mistakes. Perhaps you noticed them. The most obvious is that he carried out each experiment (date) with a different woman. As a result, his observations of what dates respond to most strongly were based on the likes and dislikes of a different person each time. The only true way to make this study general would be for Albert to go on thousands of dates with thousands of different people in an attempt to average out their responses. This dilemma is one of sampling and is very much like the problem pollsters face when they take a pre-election poll. They must be careful to interview enough people so that they include Democrats, Republicans, independents, and all other affiliations in roughly the percentages that these groups exist in the general population. Only then will the pollsters have a representative sample of the general population and a conclusion that has a chance of being correct.

Working with a large enough sample is rarely a problem in chemical studies because most involve samples that contain millions of billions of atoms or molecules, and we are therefore always measuring the average behavior of very large groups. Quite often, however, care must be taken to be sure the sample being analyzed is representative of the object being studied.

The scientific method is a cyclical process in which scientists continuously uncover new information (data), postulate new laws, and either modify or discard old ones. Based on these new or revised laws, theories are further tested and, if necessary, refined or completely revamped. In repeatedly going through this cycle, sometimes theories and laws are proved to be wrong, and sometimes they hold up. This is how science is supposed to work. Even the great Sherlock Holmes knew this (from *The Yellow Face* by Arthur Conan Doyle):

Practice Problems

1.14 What is the relationship between a theory and a law?

Answer: A theory attempts to explain why *a law is correct.*

1.15 What is the relationship between a law and experimental data?

1.16 What is good evidence that a theory is correct?

If only it were so simple. Unfortunately (or sometimes fortunately), this logical process is carried out by sometimes illogical, emotional (human) scientists. Sometimes investigators believe so strongly in a pet theory that they enter the cycle from the wrong direction and attempt to interpret all they see in terms of their personal *bias*. **Bias** is a strong preference or inclination that inhibits impartial judgment. For example, today scientists believe that atoms are among the fundamental building blocks of matter. As we mentioned earlier, this is not a new idea. Democritus was battling Aristotle over the possibility of the existence of atoms more than 2000 years ago. Now we know that Aristotle, who rejected atoms, was wrong, but we can't be too hard on him—he had no ability to examine matter on a submicroscopic level. The beliefs of the ancient Greeks were founded not on verifiable experimental evidence but on their opinions and convictions about the natural world. These personal biases can really get in the way of scientific development. Indeed, Aristotle went to his grave convinced that an adult man has more teeth in his mouth than an adult woman. As the renowned philosopher Bertrand Russell once commented, "All Aristotle had to do was ask Mrs. Aristotle to open her mouth," and he would have discovered the truth.

The fact that Aristotle never did look into his wife's mouth to check his theory—that he would not even conceive of such a check being necessary—is an ideal illustration of why scientists must always be on guard not to let personal bias, ambition, politics, ideology, or theology divert them from the scientific method. Scientists must be neutral to the point that they are willing to throw out all they were taught and replace it with something else if the scientific method so demands. That is not always easy. It has been observed quite often that "a new theory in science is really accepted only when the last of its opponents dies off."

In addition to human bias, scientific findings can also be compromised by a flaw in the design of an experiment. For example, how would you design an experiment to determine the effects of a vitamin C deficiency in humans? For ethical reasons, you cannot deprive a large group of humans of vitamin C. Instead, you might use animals, but can you be sure the results you see in your animals represent what happens in humans? If you used mice or rats, for instance, you'd see no ill effects because, unlike humans, mice and rats can synthesize their own vitamin C. If you used guinea pigs, on the other hand, the results would be similar to those for humans. (We know what happens to humans deprived of vitamin C because the symptoms were once common during long sea voyages, before sailors learned to carry foods containg vitamin C.)

You might think chemical experiments would be less prone to flaws, but that's not always so. One source of trouble is that we usually can't observe atoms directly. To figure out what is happening on the atomic level, we measure changes we can observe, such as color change or the absorbtion or emission of heat, and then *infer* what must have happened to the atoms. Our inference is based on how we think the observed changes relate to changes occurring on the atomic level. If our understanding of this relationship is correct, our conclusions are likely to be correct. If our understanding is incomplete, however, our conclusions may be wrong.

Without the ability to gather data and do experiments, you can't employ the scientific method. For this reason, Democritus was unable to support his original atomic theory. In contrast, some of the most beautiful examples of the application of the scientific method and the replacement of old theories with new ones come from the development of modern atomic theory, beginning with John Dalton in the early 1800s. When you read about this development in Chapters 3 and 4, keep the scientific method in mind, and you will understand why theories came and went.

1.6 Learning Chemistry with This Book

A common misconception is that chemistry is all math, calculations, and numerical problem-solving. In all honesty, chemistry does have that side to it. But when a chemist is presented with a question about matter, the first thing that comes to mind is *not* a complicated mathematical formula. Instead, chemists use a basic set of fundamental concepts, often best represented with images instead of mathematical equations. For example, a chemist and a nonchemist picture the concept of melting differently:

Nonchemist

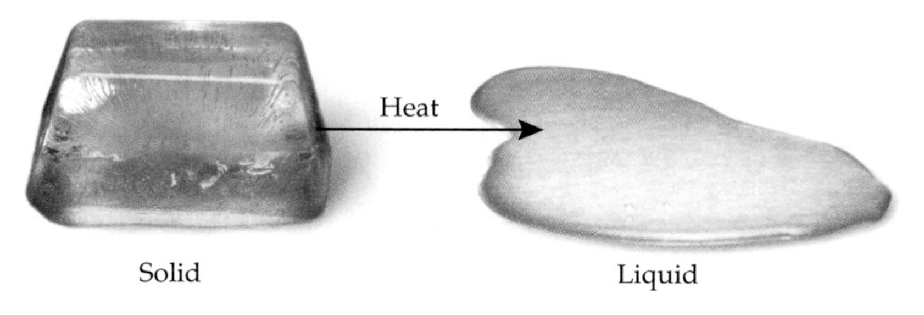

Chemist

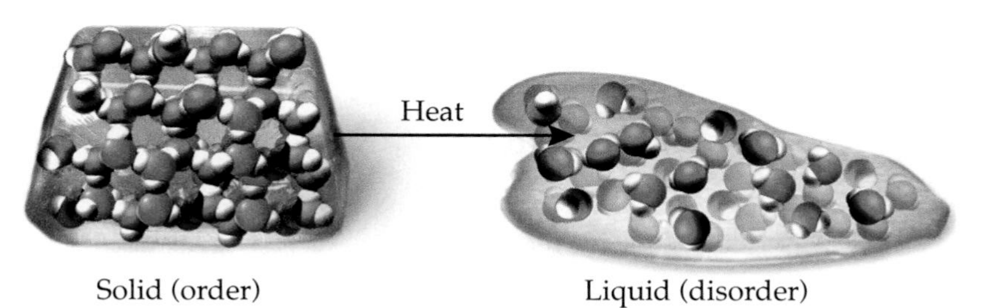

These fundamental concepts and images are the tools chemists use in answering questions about matter and the transformations it undergoes. Though this book tackles the numerical side of chemistry, it focuses on the fundamental concepts, illustrating them with pictures and everyday experiences. Even chemists forget mathematical expressions and memorized equations, but the fundamental concepts, stored as images in their minds, are with them all their lives.

Finally, we want to give you some advice about how to use this textbook. This advice comes from our own experience (after all, we were beginning chemistry students once). Read each chapter slowly, with paper and pencil in hand so you can take notes, draw pictures, and write down questions on points you are not sure about. We have done three things to encourage you to do this. First, each chapter includes a number of WorkPatches marked with a stop sign. When you reach one of these, *stop* reading and try to do the problem or sketch the concept. *Don't go on until you can answer the question.* Check your answer against the answer given at the end of the chapter.

Second, clusters of practice problems appear throughout each chapter. In each cluster, the worked-out solution is given for the first problem. The other problems in the cluster are similar to the first. In some cases we provide simple answers to these problems, but it is up to you to actually work out the solution. Blue numbers indicate practice problems for which answers are given at the back of the book; complete solutions for these problems are provided in the *Student Study Guide and Solutions Manual* that accompanies this text (see Preface). Now, you could skip over the WorkPatches and practice problems, but you would be doing yourself a disservice. The best time to find out if you understand what you read is just after you read it. In addition, working the practice problems and WorkPatches as you encounter them will help you retain the material. Keep that paper and pencil handy, and do the WorkPatches and practice problems *as you encounter them*.

Finally, each chapter ends with a section called "Have You Learned This?" and provides a list of the key concepts followed by questions that represent the essence of the chapter. Go over the list of key concepts, and if something doesn't ring a bell, go back to the indicated page and read that section again. Only when you can answer "yes" to the question "Have You Learned This?" will you be ready to go on.

Have You Learned This?

Science (p. 2)
Technology (p. 2)
Chemistry (p. 3)
Matter (p. 3)
Pure substance (p. 3)
Mixture (p. 3)
Homogeneous mixture (p. 4)
Solution (p. 4)
Heterogeneous mixture (p. 4)
Element (p. 5)
Atom (p. 7)
Elemental substance (p. 7)
Compound (p. 7)
Chemical formula (p. 7)
Physical transformation (p. 9)
State of matter (p. 9)
Freezing (p. 9)
Melting (p. 9)
Vaporization (p. 9)
Condensation (p. 9)
Physical property (p. 10)
Sublimation (p. 11)
Chemical property (p. 12)
Chemical transformation (p. 12)
Reactant (p. 12)
Product (p. 13)
Chemical reaction (p. 13)
Experiment (p. 15)
Law (p. 15)
Theory (p. 15)
Model (p. 15)
Scientific method (p. 16)
Bias (p. 17)

Intermolecular Forces
www.chemistryplace.com

SCIENCE AND TECHNOLOGY

Answers to problems numbered in red ink appear at the back of this book. Complete solutions to these selected problems appear in the Study Guide and Selected Solutions Manual *that accompanies this book.*

1.17 What is the difference between science and technology? Give an example to demonstrate what you mean.

1.18 Define chemistry.

1.19 Give an example (not from this book) of a piece of scientific information and cite both a positive and a negative result of its technological application.

1.20 In the United States, most basic chemical research (science) is funded by the government but most of the money spent on applying scientific knowledge (technology) comes from the chemical industry. Why do you suppose the chemical industry is reluctant to fund basic research?

1.21 Suppose study of the human brain (science) leads to the development of a device (technology) that allows a user to both download information from another person's brain ("mind-reading") and upload information into her or his own brain (instant learning). One extreme opinion is to make the device freely accessible to everyone. The other extreme opinion is to outlaw the device and destroy the plans for creating it. Where would you stand on this issue? Give your reasons, and then elaborate on who should decide.

MATTER AND ITS TRANSFORMATIONS

1.22 Define matter.

1.23 Is it possible to have a mixture of compounds? If yes, give an example.

1.24 How does a solution differ from a heterogeneous mixture?

1.25 Flour is mixed with powdered sugar, and then the combination is ground to a fine dust. Is the result a solution or a heterogeneous mixture? Explain.

1.26 Brass is an *alloy* (a solid solution) of copper, Cu, and zinc, Zn. Given this fact, answer the following:
(a) Copper powder and zinc powder are ground together. Is the result a solution or a heterogeneous mixture? Explain.
(b) The Cu and Zn powders are ground together, and then the mixture is heated until it is molten metal. Is the molten metal a solution or a heterogeneous mixture? Explain.
(c) The molten metal from part (b) is allowed to cool and solidify. Is the resulting solid a solution or a heterogeneous mixture? Explain.

1.27 The air you breathe is properly called a solution. Explain why.

1.28 What is an element? How many elements are known today?

1.29 If H is the symbol for the element *h*ydrogen and O is the symbol for the element *o*xygen, why is the symbol for sodium Na rather than S and the symbol for iron Fe rather than I?

1.30 What is the smallest possible piece of an element called?

For Problems 1.31–34, refer to the list of elements that appears on the inside front cover.

1.31 What are the symbols for the elements lead, molybdenum, tungsten, chromium, and mercury?

1.32 What are the symbols for the elements sulfur, chlorine, phosphorus, magnesium, and manganese?

1.33 Give the element name that goes with each symbol: Ti, Zn, Sn, He, Xe, Li.

1.34 Give the element name that goes with each symbol: U, Pu, Cs, Ba, F, Si.

1.35 How does an elemental substance differ from a compound?

1.36 What does its chemical formula tell you about a pure substance?

1.37 Compounds have properties that are different from those of their constituent elements. Discuss a real example to support this statement.

1.38 Which of the following are elemental substances and which are compounds: F_2, $BrCl_3$, P_4, C_2H_2, HCl, Ar, Al_2O_3, Al?

1.39 Which of the following are elemental substances and which are compounds: chlorine, Cl_2; octane, C_8H_{18}; sulfur, S_8; Neon, Ne?

1.40 The smallest possible piece of the compound hydrogen peroxide contains two hydrogen atoms and two oxygen atoms. Write the chemical formula for hydrogen peroxide, with hydrogen first.

1.41 The smallest possible piece of the compound nonane, a component of gasoline, contains 9 carbon atoms and 20 hydrogen atoms. Write the chemical formula for nonane, with carbon first.

1.42 The smallest possible piece of the sugar glucose has twice as many hydrogen atoms as oxygen atoms or carbon atoms. Given that glucose has six carbon atoms, write its chemical formula with carbon first, then hydrogen, then oxygen.

1.43 The air you breathe is made up mostly of nitrogen gas, N_2.
(a) What can you say about the smallest possible piece of nitrogen gas?
(b) Is nitrogen gas a compound? Explain.

1.44 What are the three most common states of matter?

1.45 What is the name of the process by which matter changes directly from the solid state to the gas state?

1.46 What is the name of the process that is the opposite of evaporation?

1.47 Propane has a boiling point below room temperature, and hexane has a boiling point above room temperature. In what physical state would you expect to find each substance at room temperature (which is usually about 20°C)?

1.48 At normal atmospheric pressure, the freezing point of ethanol is −117.3°C and its boiling point is 78.5°C. What is the melting point of ethanol?

1.49 Solid mothballs work by filling a closet with a chemical vapor that is toxic to moths, and the balls shrink and disappear slowly with time. What is going on?

1.50 What are the melting point and boiling point of water in degrees Celsius and degrees Fahrenheit (at normal atmospheric pressure)?

1.51 Even though both are composed entirely of oxygen, the elemental substance O_2 is a life-sustaining gas but the elemental substance O_3 is a toxic gas. What does this say about the chemical properties of substances and their compositions?

1.52 Water, H_2O, is a harmless liquid, but hydrogen peroxide, H_2O_2, is a poison. How is this possible, given that both are compounds of hydrogen and oxygen?

1.53 Sugar dissolves when added to water. Has the sugar undergone a physical change or a chemical change? Explain.

1.54 When ethanol burns, it is converted to carbon dioxide and water. Has the ethanol undergone a physical change or a chemical change? Explain.

1.55 What are chemists representing when they write a chemical reaction?

1.56 When exposed to oxygen in the air, white phosphorus, P_4, catches fire and produces the compound P_4O_{10}. Is this a description of a physical property or a chemical property of phosphorus? Explain.

1.57 Is the tarnishing of silver, Ag, a physical change or a chemical change? (The chemical formula for the tarnish on silver is Ag_2S.)

1.58 Nitrogen, N_2, combines with hydrogen, H_2, to give ammonia, NH_3. Nitrogen and hydrogen are odorless gases that are not very soluble in water. Ammonia is also a gas but it has an extremely powerful odor and is very soluble in water. How can you explain this great difference in properties?

HOW SCIENCE IS DONE—THE SCIENTIFIC METHOD

1.59 How does a law differ from a theory?

1.60 If a theory is false, how will the scientific method reveal that fact?

1.61 Consider some event or activity for which you have collected data or observations over the years (washing clothes, cooking, dating, studying, and so forth).
(a) State the activity and some question about it that your data might answer.
(b) Postulate a law regarding your observations.
(c) Propose a theory.
(d) How could you test your theory?

1.62 The scientific method is often considered foolproof and incapable of giving wrong answers, and yet history shows that scientists are sometimes wrong. What are some factors that cause the scientific method to give invalid results?

ADDITIONAL PROBLEMS

1.63 Which of the following are elemental substances and which are compounds: baking soda, $NaHCO_3$; graphite, C; dry ice, solid CO_2; metalic sodium, Na; metallic mercury, Hg?

1.64 Which of the following are elemental substances and which are compounds: diamond, a solid form of carbon, C(*s*); liquid water, $H_2O(l)$; nitric acid, $HNO_3(l)$; gaseous nitrogen, $N_2(g)$; liquid nitrogen, $N_2(l)$?

1.65 Classify each of the following as science or technology:
(a) Trying to understand where the Sun gets its energy
(b) The search for anticancer drugs
(c) The invention of the transistor
(d) The discovery of the electron
(e) Landing a person on the moon

1.66 Which one of the following is a compound: ozone, O_3; stainless steel, a combination of iron and other metals; sulfuric acid, H_2SO_4; brewed coffee.

1.67 Residential ozonators work by passing an electric current through the air, converting the oxygen in the air to ozone. Is this process a chemical reaction or a physical transformation of oxygen? Explain.

1.68 Classify each of the following as elemental substance, compound, or mixture:
(a) Milk
(b) Chocolate cake
(c) Clean snow; $H_2O(s)$
(d) The simple sugar fructose, $C_6H_{12}O_6$
(e) The mercury, Hg, in a thermometer

1.69 What is the name of the process by which matter changes from the liquid state to the gas state?

1.70 Match each symbol with the correct element name:
(a) Na (b) Fe (c) Co (d) Sn (e) Mn

Possible names: neon, fermium, copper, mercury, cobalt, scandium, magnesium, phosphorus, manganese, iron, sodium, potassium, argon, tin, silver, gold, calcium.

1.71 Match each symbol with the correct element from the list of names in Problem 1.70:
(a) Au (b) Hg (c) K (d) P (e) Ag

1.72 The development of rust spots on an automobile body is an example of a __________ change. (physical or chemical)

1.73 The concise generalization of the results of a number of experiments is known as scientific __________. (experiment, law, or theory)

1.74 Which of the following can form a homogeneous mixture when mixed thoroughly with water: table salt, sand, table sugar, gold?

1.75 What is the difference between a spoonful of sugar dissolved in a glass of water and a spoonful of sugar melted over a candle flame?

1.76 Give three examples of a homogeneous mixture of two liquids.

1.77 Give an example of a homogeneous mixture of a gas in a liquid.

1.78 Brass is an example of a homogeneous mixture that involves two materials that are solids at room temperature, copper and zinc. How would you go about making such a mixture?

1.79 Give an example of a gas dissolved in another gas. Is this mixture homogeneous?

1.80 When dry ice (solid carbon dioxide, CO_2) goes directly from the solid state to the gas state, the process is called __________.

1.81 When nitrogen dioxide gas is cooled, it changes from brown to colorless. Is this more likely a chemical change or a physical change?

1.82 The explosion of a stick of dynamite is most likely a __________ change. (chemical or physical)

1.83 Mixtures can always be separated by physical means. Describe how you would separate the components in a mixture of sand and iron filings.

1.84 Mixtures can always be separated by physical means. Describe how you would separate the components in a mixture of table salt and sand.

1.85 Can you think of a physical means of separating hydrogen from oxygen in a sample of pure water, H_2O?

1.86 Which one of the following is an elemental substance: table salt, NaCl; vinegar, a solution of $C_2H_4O_2$ in water; table sugar, $C_{12}H_{22}O_{11}$; liquid nitrogen, N_2?

1.87 All the metals in the first column of the periodic table react with water because they all have a single electron in their outermost occupied electron shell. This statement is an example of a scientific __________. (law or theory)

1.88 In the chemical reaction for the "burning" of sugar, $C_6H_{12}O_6 + 6\ O_2 \longrightarrow CO_2 + 6\ H_2O$, which substances are the reactants and which are the products? Does this equation represent a chemical change or a physical change?

1.89 Which one of the following is a homogeneous mixture: milk shake, 14-karat gold (made by melting gold and other metals together), fog, blood?

1.90 Which one of the following is a heterogeneous mixture: fog, gasoline, aluminum–magnesium alloy (made by melting these metals together), mixture of $N_2(g)$ and $Ar(g)$?

1.91 True or false? The melting point and freezing point of water are the same temperature at normal atmospheric pressure.

1.92 Which one of the following is an elemental substance: water, H_2O; bronze, made by melting copper and tin together; table salt, NaCl; ozone, O_3; sulfur dioxide, SO_2?

1.93 Which one of the following is a homogeneous mixture: orange juice with pulp, wood, fog, bronze (made by melting copper and tin together)?

1.94 True or false? A theory is an attempt to explain why a law is true.

1.95 Which one of the following is a compound: ozone, 18-karat gold (made by melting gold and other metals together), NaCl, liquid nitrogen, iced tea?

1 WORKPATCH SOLUTIONS

1.1 (a) Heterogeneous mixture; different regions can easily be seen in the wood.
(b) Pure substance; made of a single element, iron.
(c) Heterogeneous mixture; the rust is a layer of iron oxide over the elemental iron of the nail.
(d) Homogeneous mixture (solution); the dye is homogeneously distributed throughout the water.
(e) Heterogeneous mixture; no amount of hand-mixing can result in a homogeneous mixture.
(f) Homogeneous mixture (solution); melting allows the many compounds making up the two waxes to mingle and become a homogeneous mixture.

Density=M÷V
101010100011
8:49

CHAPTER 2

The Numerical Side of Chemistry

2.1 Numbers in Chemistry—Precision and Accuracy

Although we promised in Chapter 1 to dwell on fundamental concepts in this book, chemistry does have its numerical side. Numbers can be very helpful in expressing, understanding, and applying chemical concepts. For example, a question currently under debate is whether modern society's love for burning fossil fuels is causing global warming. Burning such fuels as gasoline in cars and natural gas in homes produces carbon dioxide. This carbon dioxide helps trap the Sun's warmth in our atmosphere. It is possible that we are producing so much carbon dioxide that we are affecting the planet's temperature, causing it to rise to the point where weather patterns are changing in undesirable ways. The question is, are we really doing this or is our output of carbon dioxide negligible

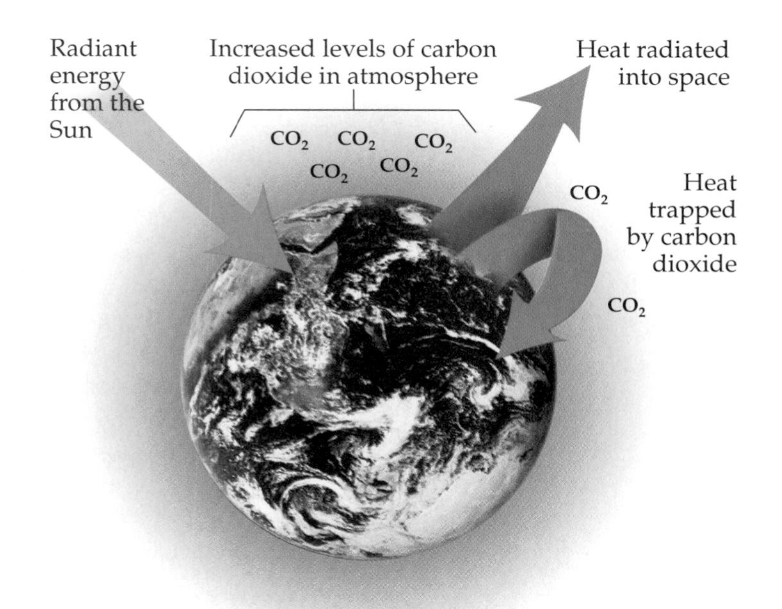

compared with the amounts generated by natural producers of global-warming gases (decomposing vegetation, volcanic activity, and so on)? Only by a careful numerical analysis, at both the level of the chemical reactions involved and the global level, can we hope to find an answer to this question.

Let's tackle this numerical side of chemistry right now so we can use it when we need to. You may already have a feel for some of the fundamental concepts that govern chemistry's numerical side. If you think about it a bit, any numerical quantity is presented to you in one of two ways, either as an *exact number* or as the result of some *measurement*. For example, if someone tells you that there are seven days in a week, the numerical quantity 7 is an **exact number** because the number of days in a week is *universally defined* to be 7, and this number has no uncertainty. Also, the number of coins in your pocket is an exact number because you can't have a partial coin. That is, if you can *count* the number of items, the result is an exact number.

On the other hand, some numerical quantities are the result of a **measurement**. For example, we could use a ruler to measure the diameter of a coin. Let's agree that 2.5 cm is the true diameter of the coin. (The abbreviation "cm" stands for *centimeter*, a length that is a little shorter than 0.5 inch. We'll study this and other units in Section 2.6.) Next, we divide six students into two teams and ask each student to measure the diameter once. Here are the results:

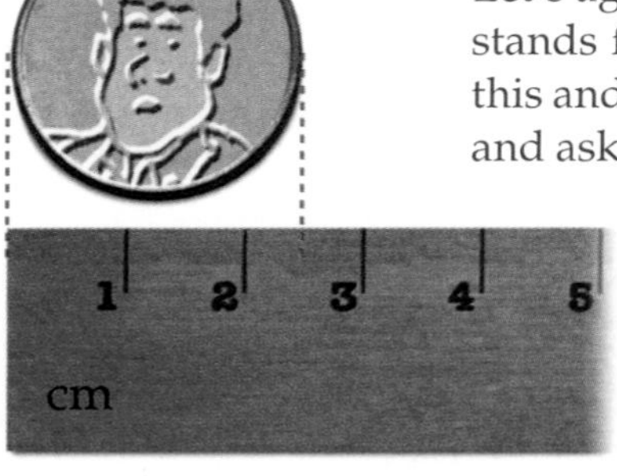

Team A	Team B
2.0 cm	2.2 cm
3.0 cm	2.2 cm
2.5 cm	2.2 cm

It looks like only the third student in team A measured the diameter correctly. How can we explain the other students' results? Well, maybe one of the students in team A was tired, and another had difficulty locating the diameter. In team B, perhaps two of the team members were swayed by the third, who was absolutely convinced that the diameter was 2.2 cm. Bad measurements or not, these are the data presented by the teams, and so let's work with them.

One thing you can do with repetitive measurements is calculate an average by summing the measurements and then dividing by the total number of measurements. For example, (2.0 cm + 3.0 cm + 2.5 cm)/3 = 2.5 cm is the average diameter for team A. The average diameters for the two teams are:

	Team A	Team B
	2.0 cm	2.2 cm
	3.0 cm	2.2 cm
	2.5 cm	2.2 cm
Average diameter:	2.5 cm	2.2 cm

We can use these data to discuss two important aspects of measured numerical quantities, *precision* and *accuracy*. **Precision** refers to the closeness to one another of a series of measurements made on the same object. In our case, team B was precise because its measurements were all the same and thus as close to one another as possible. Team A was not precise because its measurements were all over the place, ranging from 2.0 cm to 3.0 cm. The term "precision" has meaning only when two or more measurements are made. We could not speak of precision if we measured the diameter just once.

If we look at the averages of the measurements, however, team A was more *accurate*. **Accuracy** refers to how close the measured result is to the true value, and team A's average diameter was closer to the true diameter (2.5 cm) than team B's. Unlike "precision," the term "accuracy" can be applied if just a single measurement is made. For example, the third member of team A made the most accurate single measurement. We can summarize team B's results as being precise but not accurate, whereas team A's results are accurate but not precise. Thus, precision and accuracy do not have to go together. Of course, the goal in taking any measurement is to achieve both accuracy and precision.

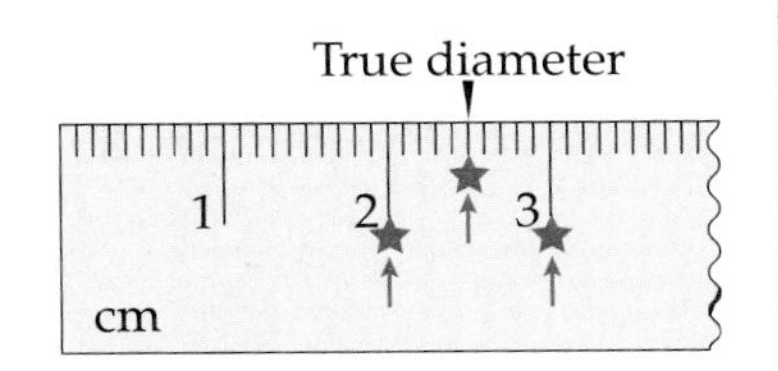

Team A Two students are quite inaccurate, although the *average* of the team's results is accurate.

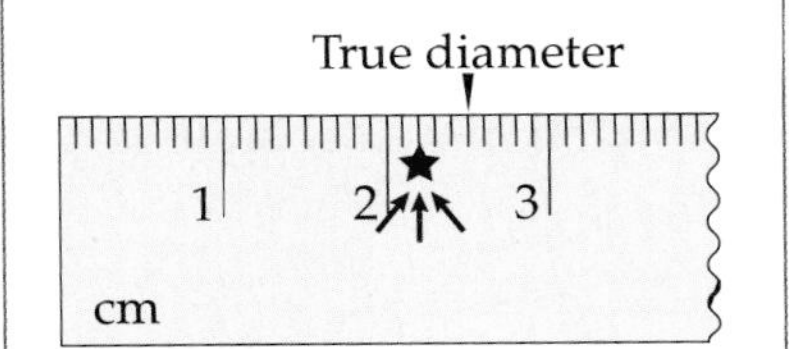

Team B Very precise; quite inaccurate.

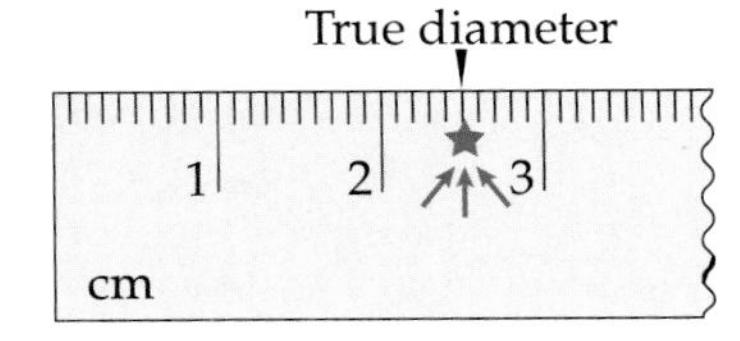

The kind of measurements a scientist likes: accurate *and* precise.

Practice Problems

2.1 Four people listening to a 2800-word speech counted the number of words in the speech as they were listening. They came up with these results:

Fred: 2736 words	Wilma: 2810 words
Barney: 2792 words	Betty: 2734 words

(a) Which person was most accurate?

(b) Which person was most precise?

(c) Would you judge the group's average word count as being accurate or inaccurate?

(d) Four other people heard the same speech and came up with the word counts 2722, 2724, 2719, and 2723. Is this group more or less accurate than the first? Is this group more or less precise than the first?

Answer:

(a) Barney was most accurate, because 2792 is closest to 2800, the actual number of words in the speech.

(b) Because each person made only one count, the term "precision" doesn't apply.

(c) The average word count for the group is (2736 + 2810 + 2792 + 2734)/4 = 2768. This is 32 words fewer than the actual word count of 2800. Because 32 is quite small relative to 2800, the group's determination is pretty accurate.

(d) *The average word count for the second group is (2722 + 2724 + 2719 + 2723)/4 = 2722. This is 78 words fewer than the actual word count, and so the second group is less accurate than the first. As for precision, the first group ranges from a high of 2810 to a low of 2734 (a spread of 76), whereas the second group ranges from a high of 2724 to a low of 2719 (a spread of 5). The smaller spread makes the second group more precise.*

2.2 Two students count the grains of uncooked rice in a small cup. Both students repeat this measurement four times, with the following results:

Mike: 256, 263, 262, 266 Ike: 250, 242, 270, 278

The actual number of grains is 260. Which student is more accurate? Which is more precise? Explain your answers.

2.3 Two students attempt to measure out a quart of water into a bucket. Jack has a half-quart container, and Jill has a 10-gallon container. Which student will probably be more accurate at putting a quart of water into the bucket? Explain.

2.2 Numbers in Chemistry—Uncertainty and Significant Figures

Now, let's take a closer look at numerical quantities that are the result of some measurement. For example, consider the following exchange between two people discussing the age of a dinosaur bone.

Even before you continue reading, we think you will sense something wrong with the logic used in this analysis. The problem is that dating techniques used to determine the age of fossils are accurate to within a few hundred thousand years at best. Therefore the original age determination of 65 million years was approximate. When compared with this uncertainty, the additional 23 years that passed since the dating was carried out mean nothing. The 23 years are simply not significant relative to the error in the original measurement. In other words, while the age stated (65,000,023) has eight digits in it, not all of these digits have equal importance. Some, in fact, have absolutely no importance at all.

Like the fossil age of 65 million years, most numbers in chemistry come from some kind of measurement. You should always be aware that no measuring device can measure without some uncertainty. If you spend a lot of money on the best measuring device, you may lower the uncertainty, but there will always be some uncertainty associated with any measurement. For example, earlier we measured the diameter of a coin with a ruler marked in centimeters:

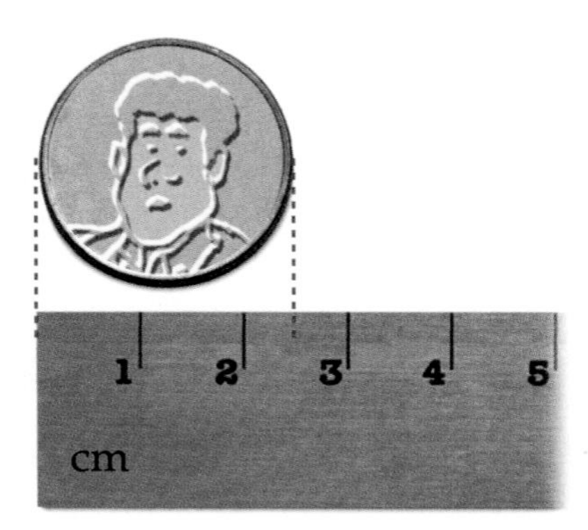

You can see that the diameter is somewhere between 2 cm and 3 cm. The dashed line on the right *seems* to fall halfway between 2 cm and 3 cm, and so you might record the diameter as 2.5 cm. The .5 is only an estimate, however, so someone else might estimate the diameter to be 2.4 cm. Yet another person might estimate 2.6 cm. In other words, there is uncertainty in the last digit.

What if you had not seen the ruler used and were presented with a measured diameter of 2.5 cm? How could you know where the uncertainty in this measured number is? Perhaps a ruler with divisions of millimeters was used, as shown below. (A millimeter is one-tenth of a centimeter). Using this ruler would make the .5 a certainty instead of an estimate.

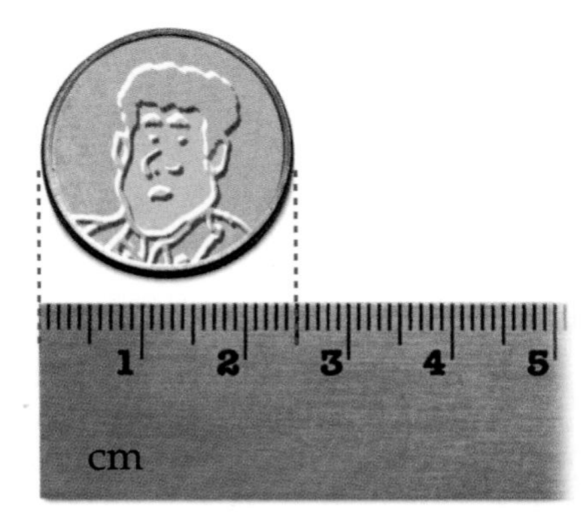

Without access to information about the measuring device (the ruler, in this case), you have no way of knowing where the uncertainty in a reported measurement lies unless some agreed-upon convention is used. Well, not surprisingly, there is just such a convention. The standard convention is to assume that *the uncertainty lies in the last digit written in the number*. Thus, a reported measurement of 2.5 cm is assumed to be uncertain in the tenths place and can also be written 2.5 ±0.05. That is, the actual number could be anywhere from 2.45 cm to 2.55 cm. The ±0.05 is arrived at by placing a 1 in the position of the uncertainty (the tenths place in our example) and dividing by 2 (0.1 ÷ 2 = 0.05). Using the ruler calibrated in millimeters, you would report the coin diameter as 2.50 cm and not as 2.5 cm, because the uncertainty now lies one place farther to the right, in the hundredths position. By this same convention, the measured number 102 cm is considered uncertain to ±0.5 cm because the last digit is in the ones position.

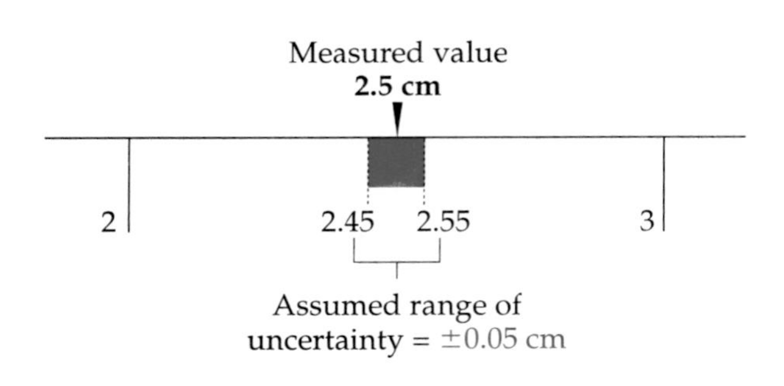

What is the uncertainty associated with the second ruler shown on page 29, the calibrated in millimeters? Explain your answer.

(a) ±0.5 cm (b) ±0.05 cm (c) ±0.005 cm

Practice Problems

2.4 A desk is reported to weigh 185 pounds. What is the assumed uncertainty in this weight?

Answer: The uncertainty is ±0.5 pound because the last digit in the reported number is in the ones column and 1 ÷ 2 = 0.5.

2.5 A volume of liquid is measured as being 16.0 gallons. What is the assumed uncertainty in this measured volume?

2.6 The voltmeter below is used to measure the voltage of a battery. What is the uncertainty in the measured voltage if the digits shown represent volts?

(a) ±0.5 volt (b) ±0.05 volt (c) ±0.005 volt

Now let's get back to the estimated coin diameter of 2.5 cm found by using the ruler calibrated in centimeters. Suppose you are asked to calculate the radius of the coin. To do this, you need to divide the diameter by 2:

Diameter measured with this ruler

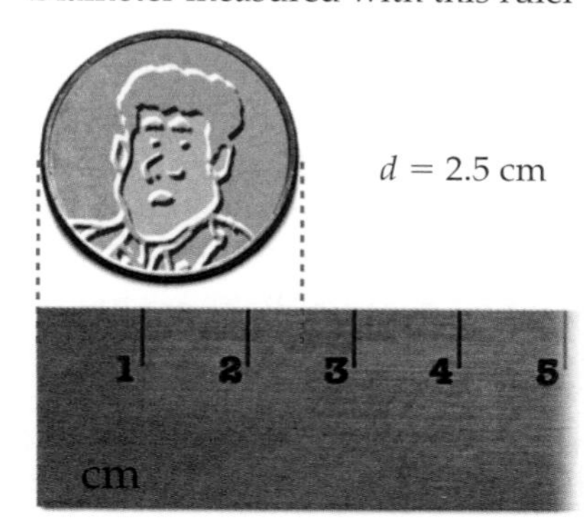

Radius = Diameter divided by 2

$$r = \frac{d}{2} = \frac{2.5\text{ cm}}{2}$$

Your calculator gives 1.25 cm as the result. But wait! The last digit, 5, is in the hundredths column, implying that the radius is known to ±0.005 cm. How can this be? The ruler used to measure the diameter was accurate only to ±0.05 cm. Dividing the diameter by 2 didn't make the ruler more accurate. Even though your calculator gives a radius of 1.25 cm, the 5 is still beyond the ruler's ability to estimate. Instead, the correct value for the radius, rounding up, is 1.3 cm. (The rule for how to round numbers is covered shortly.) This value is once again assumed to be uncertain to ±0.05 cm.

The moral is that you can't always keep all the digits a calculator produces. You can keep only the *significant* ones—the ones that are not beyond the accuracy of the measuring device. These digits are called **significant figures** (often shortened to **sig figs**) or **significant digits**.

2.3 Zeros and Significant Figures

The most important concept from the preceding section is that when you are presented with a measurement and don't know what measuring device was used, you should assume that the uncertainty lies in the last written digit. For example, if you are told that a board is 103.6 inches long, you should assume that the uncertainty is ±0.05 inch (because the 6 is in the tenths place and 0.1 ÷ 2 = 0.05) and all the digits are significant (in other words, there are four significant figures). But what about the number 0.06 inch? Are all three digits, the two 0's and the 6, significant? The answer is no. There is only one significant figure (the 6). The zeros are just placeholders, called *leading zeros*.

0.06 inch
↑↑
Leading (placeholder) zeros are not significant

Leading zeros are all the zeros that precede the first nonzero digit. They are generally found in numbers whose absolute value is less than 1. Our rule for where the uncertainty in the number is has not changed. The last written digit, 6, is in the hundredths column, and so the uncertainty in the measurement 0.01 inch is ±0.005 inch. Notice that while the zeros in 0.06 inch are not significant, the zero in our first example, 103.6 inches, *is* significant. That's because the zero in 103.6 inches is not a leading zero. Leading zeros come before the *first nonzero* digit, never after it. Consider the two numbers:

0.001 02 and 0.001 020

Both have three leading zeros (they are underlined). The first number therefore has only three significant figures, while the second number has four (the digits that are not underlined).

Rule 1: Leading zeros (all zeros before the first nonzero digit) are not significant.

That last zero in the number 0.001 020 is given a special name, a *trailing zero*. **Trailing zeros** are those that appear to the right of the last nonzero digit in a number:

0.001 02 No trailing zeros
0.001 020 One trailing zero

Trailing zeros that follow the decimal point are always significant, and as such have meaning. The trailing zero in 0.001 020 is in the millionths column, and so the number 0.001 020 is known to an uncertainty of ±5 ten-millionths (±0.000 000 5). The trailing zero gives the number 0.001 020 an uncertainty that is smaller than the uncertainty in the number 0.001 02.

Rule 2: Trailing zeros that follow the decimal point are significant.

Try now to determine the number of trailing zeros and significant figures in the following numbers. Don't go on until you get the correct answers.

	Number of trailing zeros	Number of significant figures
20.201	?	?
20.210	?	?
20.000 2	?	?
20.0	?	?
120.	?	?

The last number in this WorkPatch is an interesting case. It has one trailing zero, and its uncertainty is ±0.5. Now, you might think it a bit ridiculous to write the number 120 with a decimal point at the end. Why not just write it without the decimal point? If you wanted to buy one hundred twenty eggs, you would write 120 eggs on your shopping list, not 120. eggs. In fact, writing 120 eggs on your list is OK because it is an exact (counted) number. However, if the number is a measurement, then leaving out the decimal point is the wrong thing to do. Without the decimal point, there are *two* ways to interpret the number: as 120 ±5 (meaning the zero is just a placeholder and not significant) or as 120 ±0.5 (meaning the zero is a significant, trailing zero). In other words, you can't be sure if zeros immediately to the left of the decimal point are trailing zeros unless the decimal point is actually written. With the decimal point included, there is no ambiguity. The decimal point indicates that any trailing zeros to the left of the decimal point *are significant*. Because chemistry is a quantitative science, chemists must make sure that the numbers they write communicate a single clear meaning.

Practice Problems

2.7 How many interpretations are there for the number 600?

Answer: With no decimal point at the end, 600 can be interpreted as having an uncertainty of ±50, ±5, or ±0.5. There is no way to know.

2.8 How would you express 600 inches ±0.5 inch?

2.9 Fill out the following table.

	Number of significant figures	**Uncertainty**
10.0	?	?
0.004 60	?	?
123	?	?

This brings us to an interesting question. Suppose you measured the length of a building as 120 feet using a ruler marked every 100 feet. This puts the uncertainty in the tens column, which means the measurement is uncertain to ±5 feet.

From the picture, it looks as if the building is about 120 feet long. Of course, this is an estimate. Because the ruler is marked in only hundreds of feet, it is safe to say that you could be off by about 5 feet. Another person might read the ruler as 115 feet, and yet another might read it as 125 feet. That is, the length of 120 feet is uncertain to ±5 feet. Now, how are you supposed to write the number 120 and at the same time communicate to the world that the uncertainty in this measured number is ±5? You cannot write 120. because that implies an uncertainty of ±0.5 foot. You cannot write 120 because that is ambiguous. You can put a line under the middle digit (1$\underline{2}$0) and say "this is the last significant figure," but that is not how it is usually done. To express the uncertainty unambiguously, we write the number in a different way, using *scientific notation*.

2.4 Scientific Notation

The accepted way to indicate uncertainty in a measured number unambiguously is to use **scientific notation**—writing the numerical quantity as a number (A) multiplied by 10 raised to an exponent (x).

The general form of a numerical quantity written in scientific notation is

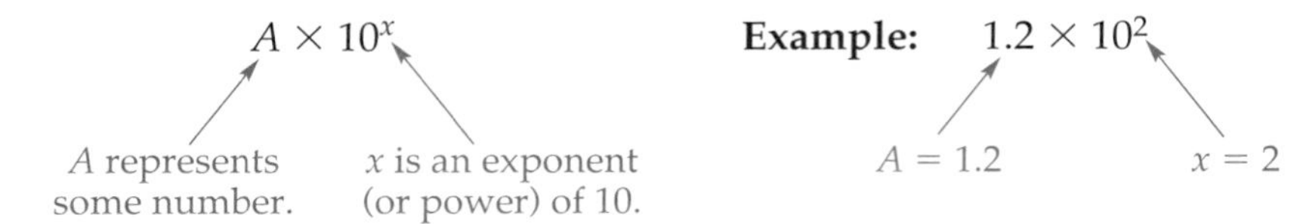

$A \times 10^x$
A represents some number.
x is an exponent (or power) of 10.
Example: 1.2×10^2
$A = 1.2$
$x = 2$

The thing to do now is *not panic*. Using scientific notation is easier than it might first appear. First, you should always include a decimal point in the (A) part of the number. The 10^x part is simply an instruction that tells you what to do with the decimal point in order to determine the real value of the number A. This principle works so long as the exponent (x) is a whole number, which it will always be in this book. That is, x will be 1, 2, 3, . . . , or 0, or −1, −2, −3, . . . , but not something like 1.28. When the exponent x is a positive number, you move the decimal point in A to the right x places. When the exponent is negative, you move the decimal point x places to the left. If the exponent is zero, then leave the decimal point alone.

Scientific notation	Exponent means move decimal point	Normal notation
1.23×10^2	Two places to right	123.
1.23×10^{-1}	One place to left	0.123
1.23×10^0	Don't move it	1.23

How can you remember this? Look at what happened to the number 1.23. When the exponent was *positive* (10^2), the number 1.23 got *bigger* (it became 123.) because $10^2 = 100$. When the exponent was *negative* (10^{-1}), the number 1.23 got *smaller* (it became 0.123) because $10^{-1} = \frac{1}{10}$. When the exponent was zero (10^0), then 1.23 remained unchanged because $10^0 = 1$.

Practice Problems

2.10 Convert 4.68×10^{-1} to normal notation.

Answer: Because the exponent is negative, the number must get smaller. Move the decimal point one place to the left: 0.468

2.11 Convert 47.3×10^{-2} to normal notation.

2.12 Convert 47.325×10^3 to normal notation.

Sometimes in moving the decimal point to convert a number from scientific notation to normal notation, you will find that the decimal point moves beyond the beginning or end of the number. For example, consider the number 4.6×10^4. The exponent is positive, and so we are supposed to move the decimal point *four places to the right* to make the number larger. If the exponent were negative (4.6×10^{-4}), the decimal point would move *four places to the left*. But the starting number has only two digits. What are we to do in these cases? Simple—we let the decimal point move off the number. Each time it moves farther off the number, it creates an empty column into which we place a zero. Look closely at how we handle this for our two examples.

What to do when the decimal point moves off the number on the right side

4.6×10^4 means move the decimal point four places to the right

Three empty columns

4.6 . . . .

Moving the decimal point four places to the right creates three empty columns, and you should fill these empty columns with zeros:

46,000

What to do when the decimal point moves off the number on the left side

4.6×10^{-4} means move the decimal point four places to the left

Three empty columns

. . . . 4.6

Moving the decimal point four places to the left creates three empty columns that must be filled with zeros:

.000 46

Practice Problems

Write the following numbers in normal notation.

2.13 0.400×10^{-6}

Answer: 0.000 000 400

2.14 2.35×10^{-3}

2.15 6.0×10^3

Whichever way the decimal point moves, the procedure is the same. Simply fill the newly created empty columns with zeros.

We have been converting numbers from scientific to normal notation. You should also be able to go the other way and convert normal notation to scientific notation. To do this, you first have to find the decimal point in the number. This is not a problem for a number like 62.8, where the decimal point is clearly written. When no decimal point is written, as in 125, assume there is a decimal point at the far right end.

Once you have located the decimal point, move it to the right of the first nonzero digit. In the number 125., the first nonzero digit is 1, and so move the decimal point two places to the left, making the number 1.25. Next, you must multiply 1.25 by the value of 10^x that would undo what you just did (in order to maintain the same value). Because you moved the decimal point two places to the left, you must set $x = 2$ because multiplying by 10^2 will move the decimal point back two places to the right, giving the original number. So, in scientific notation, 125. is written 1.25×10^2. Likewise, the number 0.000 50 is written 5.0×10^{-4}. (We moved the decimal four places to the right to place it after the first nonzero digit and therefore must multiply by 10^{-4} to undo this.)

2.3 WORKPATCH

Convert the following numbers from normal to scientific notation.
(a) 123 (b) 0.000 06 (c) 0.000 060 (d) 1002.0

Did you notice that (b) and (c) in the WorkPatch look very similar? There's an important difference, however: the number in (c) has a trailing zero. Because this zero is significant, it must show up when the number is written in scientific notation. We never throw away significant figures, even if they are zero.

Now, if you recall, we started this discussion of scientific notation seeking a way to write the number 120 feet ±5 feet. We can do this using scientific notation, but first we need to mention one more critical thing. When converting from scientific to normal notation, *any zeros that are the result of filling in empty columns created as the decimal point was moved off the number are not significant*. For example, consider the measured quantity 1.2×10^2 feet. Let's convert it to normal notation. The 10^2 tells us to move the decimal point two places to the right, creating one empty column that we must fill with a zero to arrive at 120 feet:

Empty column, fill with a zero

$$1.2 \times 10^2 \text{ feet} = 1.2\_\ \text{feet} = 120 \text{ feet}$$

That added zero is not significant, however, as explained earlier. The last significant digit, which is the digit that always carries the uncertainty, is the 2. Thus, by writing the number 120 as 1.2×10^2, we are indicating that we mean 120 ±5.

Last significant digit

120 ±5 is written in scientific notation as 1.2×10^2

What if, when measuring our building length, we had used a ruler calibrated in feet rather than in hundreds of feet? In that case, we would write 1.20×10^2 feet. The 10^2 still moves the decimal point two places to the right, and we still get 120. feet, but now the last significant digit is the zero. Why? Because we included the zero in 1.20×10^2 (a trailing zero) to begin with to indicate the reduced level of uncertainty.

Last significant digit

$120 \pm .5$ is written in scientific notation as 1.20×10^2

So while 1.2×10^2 and 1.20×10^2 both say "one hundred and twenty," they indicate different degrees of uncertainty.

Scientific notation gives us a way to indicate unambiguously where the uncertainty in a measured number is. This is one reason we use this notation.

You try the next problem, in which the diameter of a coin is measured with a ruler marked in both centimeters and millimeters.

What is the uncertainty associated with this ruler?

(a) ± 0.5 mm
(b) ± 0.05 mm
(c) ± 0.005 mm

Which choice for the diameter of the coin is correct?

(a) 5×10^1 mm
(b) 5.0×10^1 mm
(c) 5.00×10^1 mm
(d) 5.000×10^1 mm

Check your answers to the WorkPatch and make sure you understand before going on.

Another advantage of scientific notation is that it provides a convenient way to write numbers that are huge or tiny. For example, suppose you measure the length of some tiny particle under a powerful microscope and find the length to be 0.000 000 120 meter. All the zeros to the left of the 1 are placeholders and not significant (the last zero is a trailing zero and therefore is significant). To avoid having to write all those zeros, a scientist would write this number 1.20×10^{-7}.

Stop, use pencil and paper, and prove to yourself that 1.20×10^{-7} is the same as 0.000 000 120.

Writing 1.20×10^{-7} does away with all the placeholder zeros in this very small number. This strategy also works for very large numbers. Suppose a particular asteroid is measured to be 600,000,000,000 (six hundred billion) miles from us and the uncertainty in the measurement is ± 5 million miles. The uncertainty is in the 10-millions column:

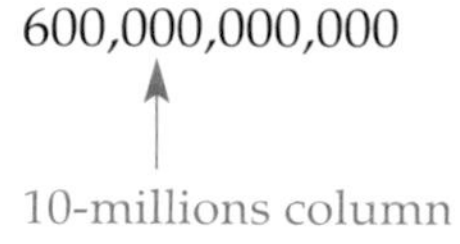

Thus, we want to indicate that all the zeros that come after this column are not significant and are simply placeholders. To do this, we would write this number in scientific notation as 6.0000×10^{11}.

With pencil and paper, prove to yourself that if you were given the number 6.0000×10^{11}, you could (a) come up with the normal notation 600,000,000,000 and (b) explain why the uncertainty is ± 5 million.

If you have trouble with this WorkPatch problem, examine the solution at the end of the chapter.

Practice Problems

Write the following numbers in scientific notation (to prove to yourself that it sometimes saves a lot of effort writing zeros).

2.16 0.000 000 000 020

Answer: 2.0×10^{-11}

2.17 47,100,000,000,000 assuming the uncertainty to be ± 0.5 million

2.18 47,100,000,000,000 assuming the uncertainty to be ± 5 million

---

2.5 How to Handle Significant Figures and Scientific Notation When Doing Math

When you do calculations with measurements, you must follow two rules to ensure that you don't end up with nonsignificant digits in your result. One rule applies to addition and subtraction. The other rule applies to multiplication and division. We'll cover the multiplication/ division rule first.

Recall the earlier example of calculating the radius of a coin by dividing the measured diameter of 2.5 cm by 2. We couldn't keep the digit 5 in the calculated radius of 1.25 cm because it was not significant (it implied an uncertainty in the hundredths place, which was beyond the uncertainty of the ruler used to make the measurement). The rule determining the number of significant figures allowed in the result of multiplying or dividing two measured numbers is simple.

Rule 1: The result of multiplying or dividing two measurements cannot have more significant figures than there are in the measurement that has the smaller number of significant figures.

For example, consider these two multiplications:

2.0 cm	← Two sig figs
× 2 cm	← One sig fig
$4\ cm^2$	← Answer can have only one sig fig

BUT

2.0 cm	← Two sig figs
× 2.0 cm	← Two sig figs
$4.0\ cm^2$	← Answer can have two sig figs

Sometimes the only way to keep your answer from having too many significant figures is to round off. For example:

$$\begin{array}{rl} 2.0 \text{ cm} & \longleftarrow \text{Two sig figs} \\ \underline{\times\, 8 \text{ cm}} & \longleftarrow \text{One sig fig} \\ 16 \text{ cm}^2 & \longleftarrow \text{Answer can have only one sig fig} \end{array}$$

Because you must round to one significant figure, the correct answer is

20 cm^2 or, indicating one significant figure, $2 \times 10^1 \text{ cm}^2$

You must always round off your answer if that is the only way to get the right number of significant figures. You round up when the digit following the last digit to be kept is 5 or greater and down when the digit following the last digit to be kept is 4 or less. Thus to three significant figures, 3625 becomes 3.63×10^3 and 3622 becomes 3.62×10^3. To two significant figures, 36.7 becomes 37 and 36.4 becomes 36.

Before you try the practice problems, we want to make something very clear. The first sentence in this section said, "When you do calculations with *measurements, . . .*" All of the preceding multiplications involved numbers that came from measurements of length. There are times, however, when you will multiply or divide a measurement by an *exact number*. We did this when we took the measured coin diameter of 2.5 cm and divided it by 2 to find the coin's radius. The 2 did not come from some measurement. It is an exact number. The way to deal with an exact number is to treat it as if it has an infinite number of significant figures (that is, 2 is really 2.000 000 00 . . .). Therefore, when we divided 2.5 cm by 2 to get an answer of 1.25 cm on our calculator, we rounded it to 1.3 cm and not 1 cm. Exact numbers have no effect on the number of significant figures allowed in the result of a calculation. Now try your hand at the following practice problems.

Practice Problems

2.19 27.5 inches/2.0 hours = ?

Answer: 27.5 inches/2.0 hours = 13.75 inches/hour, but this is not correct. The answer can have only two significant figures, and so is rounded to 14 inches/hour or 1.4×10^1 inches/hour.

2.20 22.0 miles $\times$ 2.0 miles = ?

2.21 220. hours $\times$ 3 = ? (the 3 is an exact number)

Now try this one.

What is the result of 222 miles/2.0 hours written with the correct number of significant figures?

Obviously, 222 miles/2.0 hours is 111 miles/hour, but your result can have only two significant figures. To make this perfectly clear, you need to write the answer in scientific notation. The answer is 1.1×10^2 miles/hour. It is rounded off (1.1×10^2 miles/hour = 110 miles/hour), and it uses only two significant figures. The moral here is, use scientific notation to get out of significant figure jams.

Practice Problems

Do the following calculations and give your answers to the correct number of significant figures.

2.22 1222 pounds/2 inches = ?

Answer: 6×10^2 pounds/inch (Your calculator gives 611, but your answer can have only one significant figure.)

2.23 Suppose you measure the length of a house to be 60.50 feet. How many yards is that? (1 yard = 3 feet)

Answer: There are exactly 3 feet in 1 yard because that is how a yard is defined. Therefore, the numbers 3 and 1 are exact and have no effect on the number of significant figures in the calculated result. Because the only measurement has four significant figures, your answer should also have four significant figures:

$$60.50 \text{ feet} \times \underbrace{\frac{1 \text{ yard}}{3 \text{ feet}}}_{\text{Exact numbers}} = 20.17 \text{ yards}$$

Measurement (60.50 feet)

2.24 (a) 1222 pounds/2.0 inches = ? (b) 1222 pounds/2.00 inches = ?
(c) What do you get when you quadruple 21.72 cm?

The rule for how many significant figures are allowed in the result of an addition or subtraction is different from the multiplication/division rule.

Rule 2: When adding or subtracting a series of measurements, the result can be no more certain than the least certain measurement in the series.

For example, consider adding the following measurements:

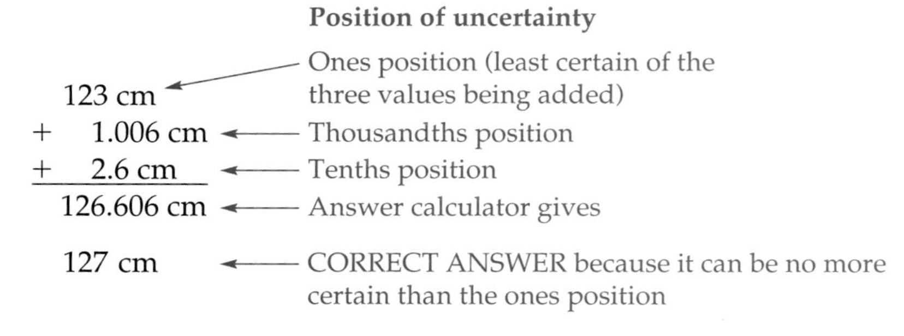

The least certain measurement is 123 cm, which is uncertain in the ones position. The correct answer of 127 cm comes from rounding up the calculated result, 126.606, to the column where the uncertainty is, the ones position.

Try the following practice problems now.

Practice Problems

Do the following problems and report your answers to the correct number of significant figures.

2.25 1555 inches + 0.001 inch + 0.2 inch = ?

Answer: 1555 inches

2.26 1555 cm + 0.001 cm + 0.8 cm = ?

2.27 142 cm − 0.48 cm = ?

If you follow these two rules, you will know how to round off the results your calculator gives you. Ignoring the rules of significant figures would put more certainty into your results than justified.

You will use a scientific calculator to do most of your calculations with numbers expressed in scientific notation. You can simply enter numbers directly in scientific notation and then add, subtract, multiply, or divide by pressing the appropriate keys. The calculator will handle the exponents for you. On most scientific calculators, to enter a number in scientific notation you first punch in the digits, then the [EE] or [EXP] key, then the value of the exponent (x), and then the [+/−] key if you want to make the exponent negative. For example,

Number	Key	How it looks on calculator
1.23×10^{3}	[1] [.] [2] [3] [EE] [3]	1.23 E 03
1.23×10^{-3}	[1] [.] [2] [3] [EE] [3] [+/−]	1.23 E −03

If you have trouble working with scientific notation on your calculator and the instruction manual doesn't help, see your instructor.

Now let's pull out all the stops and do a calculation that involves both the multiplication/division rule and the addition/subtraction rule. Suppose a bowling ball and a bag of dog food are placed on a rectangular shelf. The shelf is measured to be 1.00 foot wide and 2.45 feet long. The weight of the bowling ball is stamped on it, and the weight of the dog food is stamped on the bag. What is the weight pressing down per square foot of shelf surface area? To figure this out, all we have to do is sum up the weights of the ball and bag and divide by the surface area of the shelf (the area of a rectangle is its width times its length).

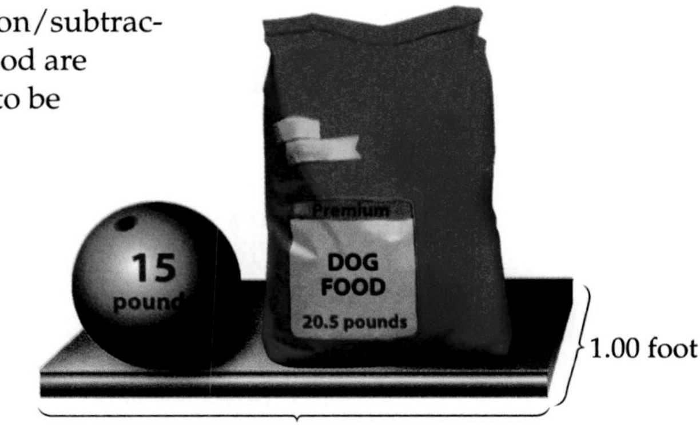

$$\text{weight per square foot} = \frac{20.5 \text{ pounds} + 15 \text{ pounds}}{2.45 \text{ feet} \times 1.00 \text{ foot}}$$

You can do either the addition or the multiplication first—it does not matter. Either way you do it, however, the final answer is allowed to have only two significant figures. Do you see why? When we sum 20.5 pounds and 15 pounds on a calculator, we get 35.5 pounds, but by the addition/subtraction rule, we must round to 36 pounds.

After doing both the addition and the multiplication, we have

$$\text{weight per square foot} = \frac{36 \text{ pounds}}{2.45 \text{ feet}^2}$$

Now we have a division, and the multiplication/division rule clearly limits our result to two significant figures. A calculator gives 14.69387755 pounds/foot^2, but the answer after rounding is 15 pounds/foot^2. *We are not done with this problem yet, however!* Look back and you'll see we rounded numbers twice, once in the middle of our calculations and again at the very end. First we rounded the summed weight of 35.5 pounds to 36 pounds because the addition/subtraction rule limited this intermediate result to two significant figures. But suppose we had not done this intermediate rounding. Rather, suppose we used the result our calculator gave us and waited until all our calculations were done before applying the two-significant-figures restriction. This would change our final step to

$$\text{weight per square foot} = \frac{35.5 \text{ pounds}}{2.45 \text{ feet}^2}$$

Now a calculator gives 14.48979592 pounds/foot^2, which to two significant figures is 14 pounds/foot^2. Which result is more correct—15 pounds/foot^2 or 14 pounds/foot^2? The latter one is more correct because we did less rounding off in obtaining it. In any multistep calculation, it is always best not to do any rounding until the very end. Otherwise you start to accumulate rounding error, with each rounding of an intermediate value putting more and more error into your final result.

2.6 Numbers with a Name—Units of Measure

A measurement makes no sense unless you know its units. For instance, it makes no sense to be told you can get to the airport in 3. Three minutes? Three hours? Three gallons of gas? You need the units.

Deciding which unit of measurement to use can be complicated because there are many possibilities for each type of measurement. Length can be measured in the system of inches, feet, yards, and miles, or it can be measured in the metric system of centimeters, meters, and kilometers. Volume can be measured in the system of quarts and gallons or in the metric system of liters. Mass can be measured in a variety of units. So there's potential for confusion in measurement, sometimes to a disastrous degree.

Consider, for example, the Mars Polar Lander launched by NASA. Instead of landing as planned on September 23, 2000, the $125 million probe burned up in the red planet's atmosphere. The company that built the retrorockets used U.S. Customary units (inches and feet for length, pounds for force) to measure the rockets' thrust. They then communicated these data to NASA. Unfortunately, NASA assumed the data were in metric units (meters for length, newtons for force). Hence the headline "Metric Mars Orbiter Takes a Pounding." The joke is that Martians are currently taking bets on how deep a hole the next Lander will drill into the surface of their planet.

To minimize this kind of confusion, scientists the world over and engineers in every nation except the United States (don't ask!) use a standardized system of units called **SI units**. (SI stands for *Système International*.) Of course, that they were dealing with two systems of units should have been made clear to the NASA scientists, but as one of NASA's directors commented afterwards, "People sometimes make errors."

Table 2.1 presents the base SI units used in chemistry.

Table 2.1 Base SI Units Used in Chemistry

Physical quantity	Name of unit	Abbreviation
Length	meter	m
Mass	kilogram	kg
Time	second	s
Temperature	kelvin	K
Amount of substance	mole	mol

The base SI unit of length is the meter (m). A meter is about $3\frac{1}{3}$ inches longer than a yard (1 yard equals 3 feet).

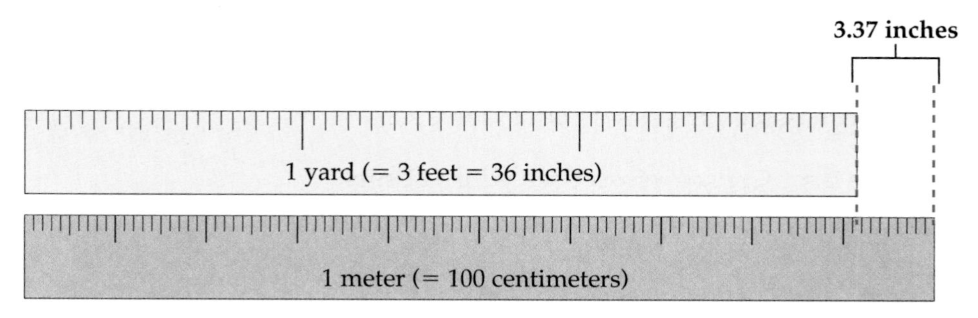

A kilogram (kg) equals a bit more than 2 lb (2.204 lb, to be exact). Notice from Table 2.1 that we call the kilogram a unit of *mass* and not a unit of *weight*. **Mass** is a measure of the quantity of matter in an object and depends only on the quantity present. The **weight** of an object depends on the strength of the gravitational force exerted on the object. If you went to the Moon, your mass would be the same as it is on Earth, but you would weigh only one-sixth of what you weigh on Earth because the Moon has only one-sixth of our gravity.

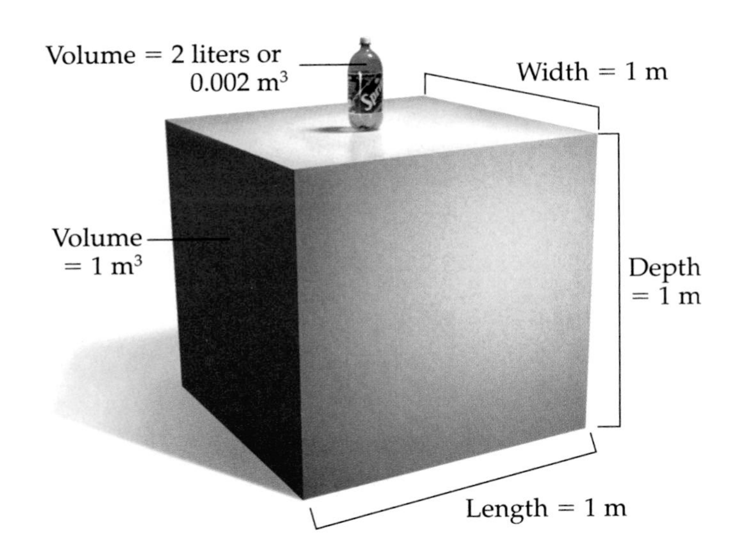

(Nevertheless, many people use the words *mass* and *weight* interchangeably.) The non-SI metric unit of mass, the gram, is used often in chemistry. One gram (g) is equal to 0.001 kg; that is, there are 1000 g in 1 kg. There are 453.6 g in 1 lb.

Derived SI units come from combinations of the base SI units. For instance, the SI unit for **volume** (the amount of space occupied by an object) is the cubic meter (m^3). A cubic meter of volume is the space enclosed by a box whose length, width, and depth are all 1 m. One cubic meter equals approximately 264 gallons. As we'll see in a moment, the liter is another volume unit.

Several derived SI units are listed in Table 2.2.

Table 2.2 Derived SI Units

Physical quantity	Name of unit	Abbreviation
Volume	cubic meter	m^3
Pressure	pascal	Pa
Energy	joule	J
Electrical charge	coulomb	C

We often find the SI units in Tables 2.1 and 2.2 inconvenient because they are either too big or too small for the job at hand. For example, suppose you wanted to measure out a teaspoon of liquid and express this volume in cubic meters. Such a small volume of liquid is only a tiny fraction of a cubic meter (1 teaspoon = 0.000 005 m^3). Would you want to have to ask for 0.000 005 m^3 of cough syrup every time you needed a teaspoon of it? And suppose you wanted to express the thickness of a human hair in the SI unit of meters. A meter is so much larger than the thickness of a human hair that, again, it would be silly to report its thickness in meters. What we do, therefore, is modify the SI units using the Greek prefixes shown in Table 2.3.

Table 2.3 Greek Prefixes Used with SI Units

Greek prefix	Meaning
pico- (p)	one-trillionth (10^{-12})
nano- (n)	one-billionth (10^{-9})
micro- (μ)	one-millionth (10^{-6})
milli- (m)	one-thousandth (.001 or 10^{-3})
centi- (c)	one-hundredth (.01 or 10^{-2})
deci- (d)	one-tenth (.1 or 10^{-1})
kilo- (k)	one thousand (1000 or 10^{3})
mega- (M)	one million (10^{6})
giga- (G)	one billion (10^{9})

Thus, instead of saying that the measured width of a human hair is 0.0002 m, we can say that it is 0.2 millimeter (0.2 mm). Because a millimeter is the same as 10^{-3} m, 0.2 mm is the same as 0.2×10^{-3} m, which is 0.0002 m. Instead of saying that the measured diameter of a blood cell is 0.000 010 0 m, we can say that it is 10.0 micrometers (10.0 μm). Because a micrometer is the same as 10^{-6} m, then 10.0 μm is the same as 10.0×10^{-6} m, which is 0.000 010 0 m. You try it now.

Practice Problems

2.28 Express the distance 24,000,000,000 m $\pm$50 million m using the appropriate Greek prefix.

Answer: The number is 24 billion, so let's use gigameters (Gm). Because 1 Gm is 1 billion meters, 24 billion meters is 24 Gm. However, to indicate an uncertainty of $\pm$50 million, we must write 24.0 Gm (the zero is in the 100 millions place).

2.29 Express the distance 4736 m in kilometers (1 km = 1000 m).

2.30 Express 0.025 m in millimeters (1 mm = 0.001 m).

The Greek prefixes combined with the SI units provide a considerable amount of flexibility for describing a wide range of measurements, from tiny to huge, but scientists needed still more flexibility. For example, the Greek prefixes can be applied to the liter (L), a non-SI metric unit that is used frequently.

There are 1000 liters in a cubic meter, the SI unit of volume. This means that a liter is the volume enclosed in a box that has a length, width, and height of 0.1 m. A liter is just a little bit larger than a quart; 1 L = 1.057 qt.

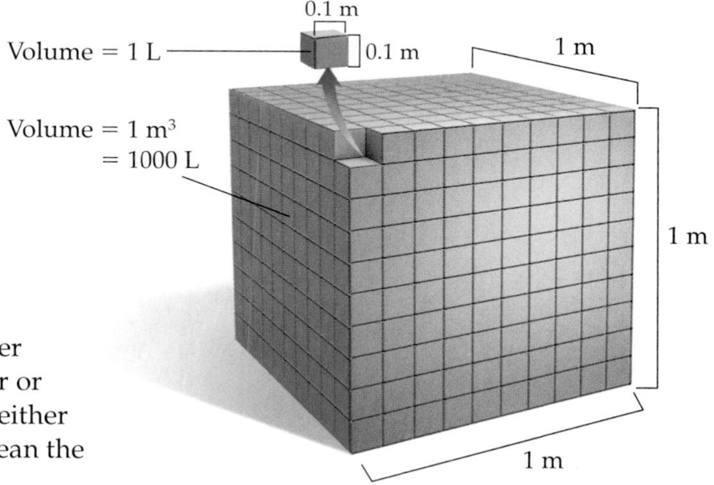

An extremely common unit for describing teaspoon-sized volumes is the milliliter (mL), which is equal to one-thousandth of a liter (1 mL = 0.001 L). Just as 1000 L fit inside 1 m^3, 1000 mL fit inside 1 L. A milliliter is the volume enclosed by a box whose length, width, and height are all 1 cm (0.01 m); thus, a milliliter is also a centimeter cubed (cm^3, commonly called a cubic centimeter or cc, pronounced "cee-cee"). You can always say either milliliter or cubic centimeter because the two mean the same thing.

Non-SI units are also used for pressure and temperature. Rather than using SI units of pascals (Pa), chemists prefer to measure pressure either in millimeters of mercury or in atmospheres, as you will see in Chapter 11. And rather than using the SI unit of temperature, the kelvin (represented as K, *not* °K), chemists typically describe temperatures using metric units of degrees

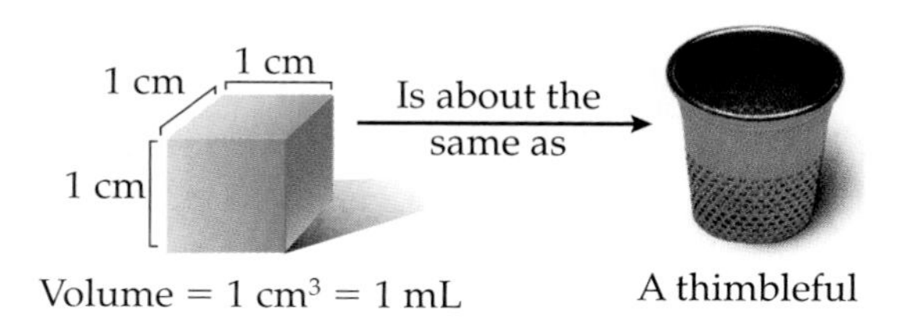

Celsius (°C), which can be converted to kelvins when need be. The Kelvin temperature scale is also called the *absolute temperature scale* because nothing can be colder than 0 K (there is no such thing as a negative Kelvin temperature). You are probably more familiar with the Fahrenheit temperature scale. This unit has now been abandoned by most of the world except the United States. A comparison of the three temperature scales is shown at the top of the next page.

The Celsius scale sets the freezing point of water at 0°C and the boiling point at 100°C, making 100 divisions between these points. The Fahrenheit scale has 180 divisions between water's freezing point of 32°F and its boiling point of 212°F. This makes the Celsius divisions almost twice as large as the Fahrenheit divisions (180/100 = 9/5 as large, to be exact). The Fahrenheit-to-Celsius conversion formulas are

$$°\text{F} = 32 + \frac{9}{5}°\text{C} \qquad \text{or} \qquad °\text{C} = \frac{5}{9}(°\text{F} - 32)$$

Note in the comparison chart that the Kelvin scale is divided the same way the Celsius scale is. Therefore to convert from Celsius to Kelvin, all you have to do is add 273.15:

$$\text{K} = °\text{C} + 273.15$$

In the United States, people are a bit stubborn about using the Fahrenheit scale. Schools teach that water freezes at 32°F and boils at 212°F. Our normal body temperature is 98.6°F, and a pleasant day is about 75°F. Tell an average person in the United States that it is 35 degrees outside, and the reaction would probably be, "Thirty-five degrees! I'd better wear a warm coat." Unfortunately, if that were 35°C, the person would be in for a shock, because 35°C is the same as 95°F!

You should get used to the Celsius scale. Remember, a Celsius degree is about twice as large as a Fahrenheit degree, and so if you go from a pleasant 24°C (75°F) to 30°C, you are increasing the temperature by 6 C°, or almost 12 F°. This takes you from 75°F to almost 87°F, a tad on the warm side. The exact conversion is

$$\begin{aligned} °\text{F} &= 32 + \frac{9}{5} \times 30°\text{C} \\ &= 32 + \left(\frac{9 \times 30}{5}\right) \\ &= 32 + 54 \\ &= 86°\text{C} \end{aligned}$$

If you're used to units such as pounds, miles, quarts, gallons, and degrees Fahrenheit because you've worked with them all your life, it will take some practice to get used to the new units of measure. As you get practice using SI

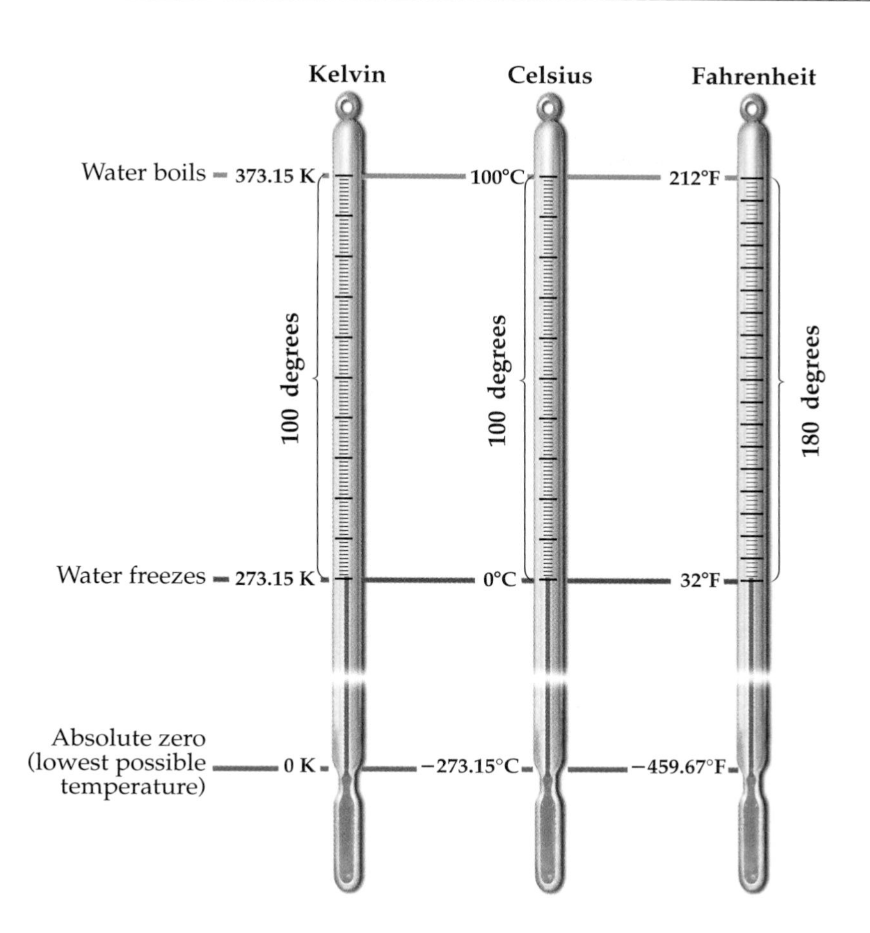

and metric units, you will get more of a feel for them. Try the following WorkPatch to see if you are getting a feel for the metric volume units.

Suppose you want to use a 100-mL beaker to fill a 2.5-L bucket with water. How many times will you have to fill your beaker?

Could you do the WorkPatch in your head? The trick is to realize that 100 mL is one-tenth of a liter. Now try the following practice problems.

Practice Problems

2.31 How many milliliters are there in 1.000 L?

Answer: Because a milliliter is 1/1000 of a liter, there are 1000 mL in a liter.

2.32 How many milliliters are there in 2.500 L?

2.33 How many milliliters are there in 246.7 cm^3?

2.34 The temperature outside is 263.5 K. What is the temperature in degrees Celsius and in degrees Fahrenheit?

2.7 Density: A Useful Physical Property of Matter

Density is a property that tells you how much matter there is in a given volume. The "how much" is expressed as mass, usually in grams. Mathematically, density is defined as the mass per unit volume of the material in question:

$$\text{Density} = \frac{\text{Mass}}{\text{Volume}} = \frac{m}{V}$$

Is there more matter per given volume in clouds or in lead?
Unit volume
Unit volume

There is much less mass per unit volume of cloud than per unit volume of lead. Therefore a cloud has a much lower density than a piece of lead.

Chemists usually express densities in units of g/mL or g/cm^3. Table 2.4 lists the densities of some common substances. Notice how the last three entries in the table, all gases, have much lower densities than the other entries, all liquids and solids.

Table 2.4 Densities of Some Common Substances*

Substance	Density (g/mL)	Substance	Density (g/mL)
Gold (solid)	19.3	Butter (solid)	0.9
Mercury (liquid)	13.6	Gasoline (liquid)	0.7
Lead (solid)	11.4	Oxygen (gas)	0.001 43
Water (liquid)	0.997	Air (gas)	0.001 30
Water (solid ice at 0°C)	0.917	Helium (gas)	0.000 18

*Densities of all liquids and solids except ice measured at 25°C; densities of gases measured at 0°C and 1 atm pressure.

Try the following WorkPatch to see if you have a good feel for density.

Some metals can exist in more than one form, differing in how their atoms are arranged in the solid. Below are shown two forms of the same metal. One form is denser than the other. Which is denser and why?

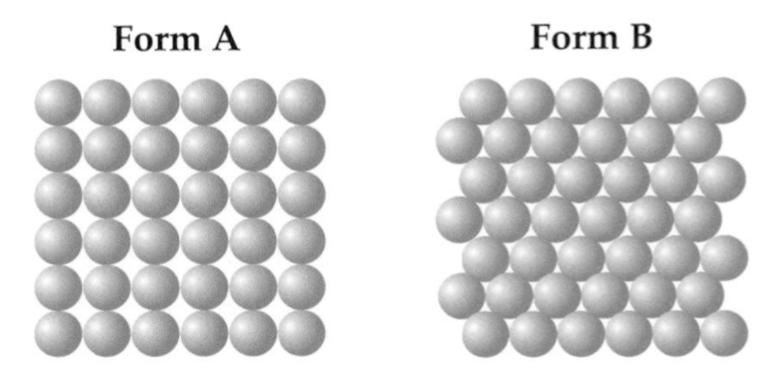

The density of liquid water is 0.997 g/mL, or essentially 1.00 g/mL. This number simply means that 1.00 mL of water has a mass of 1.00 g. Mercury, a liquid metal used in thermometers, has a density of 13.6 g/mL. This means that 1.00 mL of mercury has a mass of 13.6 g. Thus, 1.00 mL of mercury is 13.6 times more massive than the same volume of water.

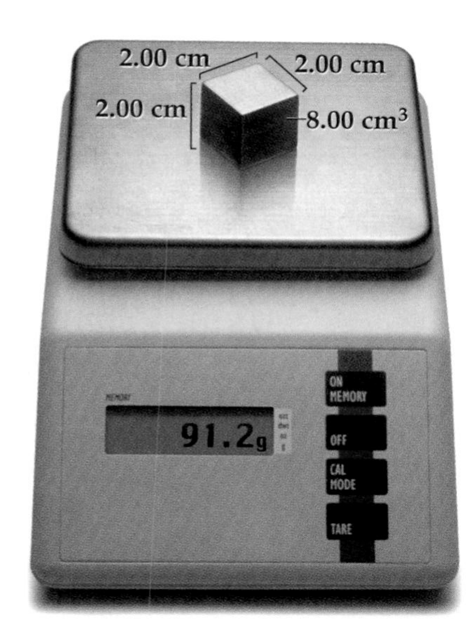

Density can be very useful in identifying an unknown substance. For example, suppose you were given a small cube of golden yellow metal that measured 2.00 cm × 2.00 cm × 2.00 cm and you had to determine whether it was gold or painted lead. One way to do this would be to determine the mass of the cube and then divide this mass by the volume of the cube to get its density. You could then compare this density with the entries in Table 2.4.

$$\text{Volume of cube} = 2.00\ \text{cm} \times 2.00\ \text{cm} \times 2.00\ \text{cm} = 8.00\ \text{cm}^3 = 8.00\ \text{mL}$$

$$\text{Mass of cube} = 91.2\ \text{g}$$

$$\text{Density} = \frac{91.2\ \text{g}}{8.00\ \text{mL}} = \frac{11.4\ \text{g}}{\text{mL}}$$

Because lead has a density of 11.4 g/mL and gold has a density of 19.3 g/mL, the yellow metal block must be lead painted to look like gold. Now you have a very accurate way to determine if the gold someone is selling you is really gold, and it is totally independent of the size of the sample. This is true because density is an **intensive property**, a property that does not depend on the amount of material. Had you been given a larger or smaller cube of metal than the one we just analyzed, the density would have been the same. The opposite of an intensive property is an **extensive property,** one that does depend on the amount present. Examples of extensive properties of matter are mass and volume. Because density equals mass/volume, we see that an intensive property—density—can be derived from two extensive properties.

Measuring the density of an object requires determining its mass and its volume. Determining mass is usually the easy part. Determining volume, however, can be difficult if the object is irregularly shaped. For example, how could you easily measure your own volume? You could not simply multiply your length times your width times your depth as we do for a box, because your width and depth change as you go from feet to head. An easy way to measure the volume of an irregularly shaped object is to immerse it completely in a container filled to the very top with water. Doing so will cause a volume of water equal to the volume of the object to be displaced and spill from the container. Collecting this overflow water and measuring its volume will give you the volume of the object. This is called *determining the volume of an object by displacement*.

Measuring the volume of an irregularly shaped object by displacement

Practice Problems

2.35 Which is denser, a 200-lb block of lead or a 0.1-g piece of gold?

Answer: The piece of gold is more dense (see Table 2.4). The size of the piece does not matter because density is an intensive property.

2.36 A cube that is 10.0 mm × 10.0 mm × 10.0 mm has a mass of 4.70 g. What is its density in grams per milliliter?

2.37 A small statue that has a mass of 500.0 g displaces 150.5 mL of water. What is its density in grams per milliliter?

2.8 Doing Calculations in Chemistry—Unit Analysis

Now we want to review a general strategy for solving many of the numerical problems that come up in chemistry. Often the biggest problem for students is knowing where to start. It can often appear that a blizzard of numbers has blown your way. For example, consider the following problem: If you have 2.00 g of the liquid metal mercury (Hg), what volume will it occupy in milliliters given that the density of mercury is 13.6 g/mL?

We have a measured mass of mercury (2.00 g) and the density of mercury (13.6 g/mL), and we need to find a volume in milliliters. Where should we begin? The answer comes from looking very closely at the units associated with the numerical quantities in the problem. The density of mercury has two units, grams and milliliters. The mass of mercury, 2.00 g, has just one unit. The general strategy for solving a numerical problem makes use of this information in a three-step approach called **unit analysis**.

The first step of unit analysis is to write down the given measurement.

Step 1: Write down the given measurement and its units.

2.00 g mercury

The second step is to use the density of mercury, but we first want you to look at density in a new way—as a *conversion factor*. A **conversion factor** can be recognized because *it always has two different units, one over the other*. For example, you know that there are 60 seconds in a minute,

$$60 \text{ s} = 1 \text{ min}$$

This is a conversion factor. It is often written in ratio form, as

$$\frac{60 \text{ s}}{1 \text{ min}} \quad \text{or} \quad \frac{1 \text{ min}}{60 \text{ s}}$$

60 s per 1 min   1 min per 60 s

You can write it either way. The left ratio says there are 60 s per 1 min (60 s/min), and the right ratio says there is 1 min per 60 s (min/60 s). Both are true. Both ratios have two units, seconds and minutes, one over the other. *This is the hallmark of a conversion factor.*

Because the density of mercury is 13.6 g/mL, we can write

$$1 \text{ mL mercury} = 13.6 \text{ g}$$

Note: The "1" in "1 mL mercury" is treated as an exact number in all calculations.

and we can write this in ratio form as

$$\frac{13.6 \text{ g mercury}}{1 \text{ mL mercury}} \quad \text{or} \quad \frac{1 \text{ mL mercury}}{13.6 \text{ g mercury}}$$

Either form is correct, although traditionally density is written as mass over volume.

So, step 1 was to write down the measurement and units (2.00 g mercury). Step 2 is to multiply the measurement by the appropriate conversion factor in such a way that the units of the measurement cancel out. For our example, this means we do the following:

Step 2: Multiply the measurement by a conversion factor so that the original units of the measurement cancel out.

$$2.00 \text{ g mercury} \times \frac{1 \text{ mL mercury}}{13.6 \text{ g mercury}}$$

The conversion factor is written as 1 mL/13.6 g because this way, the units "grams" cancel out. Identical units cancel out when they appear in both the top (numerator) and the bottom (denominator) of a ratio because anything over itself equals 1.

Identical units on top and bottom cancel:

$$2.00 \, \cancel{\text{g mercury}} \times \frac{1 \text{ mL mercury}}{13.6 \, \cancel{\text{g mercury}}}$$

← The unit that survives (1 mL mercury)

Had we written the density conversion factor the other way (with grams on top and milliliters on the bottom), the units "grams" would appear twice on the top and would not cancel. Step 2 says to multiply by the conversion factor *so as to make the original units of the measurement cancel out!* The unit that survives is "milliliter," a unit of volume. This is what the question asked us for (the volume occupied by the mercury). So we're ready for the final step, calculating our answer:

Step 3: Perform the calculation.

$$2.00 \, \cancel{\text{g mercury}} \times \frac{1 \text{ mL mercury}}{13.6 \, \cancel{\text{g mercury}}} = 0.147 \text{ mL mercury}$$

Practice Problems

2.38 Which of the following can be thought of as conversion factors?
(a) 49.3 kg
(b) 4.184 J/C°
(c) 350 miles per hour
(d) 12 eggs per dozen
(e) 1 dozen grams

Answer: (b), (c), and (d).

2.39 Write two conversion factors (two different ratios) to express the fact that there are 24 h in a day.

2.40 Given a speed of 600.0 miles per hour and a measured distance of 50.0 miles, multiply the conversion factor by the measured distance to make identical units cancel. Calculate the result and include its units.

2.41 Given a speed of 600.0 miles per hour and a measured time of 50.0 h, multiply the conversion factor by the measured time to make identical units cancel. Calculate the result and include its units.

You might be wondering what gives us the right to multiply by conversion factors the way we do. To see why this multiplication is allowed, let's again use the density of mercury as our example. That the density is 13.6 g/mL tells us that 1 mL of mercury = 13.6 g of mercury. This means that the conversion factors

$$\frac{13.6 \text{ g mercury}}{1 \text{ mL mercury}} \quad \text{or} \quad \frac{1 \text{ mL mercury}}{13.6 \text{ g mercury}}$$

both equal 1. In essence, therefore, when we multiply a number by an appropriate conversion factor, we are not changing the value of the number (because we are multiplying it by 1 and multiplying anything by 1 does not change its value). All we are doing is changing its units.

You must always make sure that when you use conversion factors, the units you want to "disappear" are always arranged so that they cancel. Not doing so is one of the most common mistakes students make when doing chemistry calculations.

One of the benefits of unit analysis is that you can use step 2 over and over again, multiplying by conversion factor after conversion factor, to put the final answer into any units you desire. For example, imagine you are told that the density of a liquid is 2.50 g/mL and asked to convert this to pounds per gallon (lb/gal). You have to make *both* g and mL "disappear" and replace them with lb and gal. Here is how it's done:

If you are not at the units you want, keep multiplying by conversion factors:

$$\frac{2.50 \not{\text{g}}}{\not{\text{mL}}} \times \underbrace{\frac{1 \text{ lb}}{453.6 \not{\text{g}}} \times \frac{1000 \not{\text{mL}}}{1 \not{\text{L}}} \times \frac{3.7854 \not{\text{L}}}{1 \text{ gal}}}_{\text{Conversion factors}} = \frac{20.9 \text{ lb}}{\text{gal}}$$

The conversion factors come from the fact that there are 453.6 g per pound, 1000 mL per liter, and 3.7854 L per gallon. Note that each conversion factor is written in such a way as to cancel out some unit so that you are left with lb/gal.

You can string out as many conversion factors as necessary to come up with the unit the problem is asking for. To do the math when many conversion factors are strung together, simply multiply everything on top, then divide by everything on the bottom. For the problem we just did, you would key the following into your calculator: $2.50 \times 1000 \times 3.7854 \div 453.6 =$.

The next example shows you the power of stringing out conversion factors. Suppose it took you a total of 3.5 weeks to read this book. How many seconds did it take you to read it? You could use the following conversion factors:

$$\begin{aligned} 60\text{ s} &= 1\text{ min} \\ 60\text{ min} &= 1\text{ h} \\ 24\text{ h} &= 1\text{ day} \\ 7\text{ days} &= 1\text{ week} \end{aligned}$$

Once again we use the three-step approach of unit analysis: Step 1, write down the measured time it took you to read the book; step 2, multiply it by conversion factors to cancel out its units and arrive at the units you want; step 3, do the calculation. With respect to significant figures, you can treat all of the above conversion factors as exact because they derive not from some measurement but from a definition (for example, a minute is defined as having 60 s in it). Remember, when you are doing math, exact quantities have no effect on the number of significant figures in the answer.

How many seconds did it take to read this book if it took you 3.5 weeks to read it?

$$\underbrace{3.5\ \cancel{\text{weeks}}}_{\substack{\text{The}\\\text{measurement}\\\text{and its units}}} \times \underbrace{\frac{7\ \cancel{\text{days}}}{1\ \cancel{\text{week}}} \times \frac{24\ \cancel{\text{h}}}{1\ \cancel{\text{day}}} \times \frac{60\ \cancel{\text{min}}}{1\ \cancel{\text{h}}} \times \frac{60\ \text{s}}{1\ \cancel{\text{min}}}}_{\text{Conversion factors}} = \underbrace{2{,}116{,}800\ \text{s}}_{\substack{\text{Calculated}\\\text{answer}}}$$

Surviving unit

Because the starting number has only two significant figures, the answer can have only two significant figures and so is 2.1×10^6 s.

The method of unit analysis, starting with a measurement and multiplying by conversion factors to make units cancel, is very simple and yet very powerful. Now let's use it in one more problem. Suppose you want to build a driveway 10.0 ft wide and 60.0 ft long. The Acme Driveway Construction Company quotes you a price of \$2.25 per square meter (m^2) of driveway. (The president of the company is into metric units.) How much will it cost to build your driveway? Unit analysis will give you your answer. Area is just length times width, and so it's no problem to figure out the area of your proposed driveway:

$$10.0\text{ ft} \times 60.0\text{ ft} = 6.00 \times 10^2\text{ ft}^2$$

We have a problem, though. The company quoted you a cost per square meter, not per square foot. Suppose you know the following:

$$\begin{aligned} 12\text{ in.} &= 1\text{ ft} \\ 2.54\text{ cm} &= 1\text{ in.} \\ 100\text{ cm} &= 1\text{ m} \end{aligned}$$

With this information and unit analysis, you are off to the races. Follow it through:

$$6.00 \times 10^2\,\text{ft}^2 \times \frac{12\text{ in.}}{1\text{ ft}} \times \frac{2.54\text{ cm}}{1\text{ in.}} \times \frac{1\text{ m}}{100\text{ cm}} \times \frac{\$2.25}{\text{m}^2}$$

The conversion factors are arranged in such a way that all but one of the units cancel and we are left with units of dollars, but there is a problem! We started with our measurement in ft^2, but the first conversion factor has ft in the denominator rather than ft^2. The unit ft^2 is not the same as the unit ft, and so the two don't cancel! To fix this, all we need to do is square the first conversion factor:

Square it

$$6.00 \times 10^2\,\text{ft}^2 \times \left(\frac{12\text{ in.}}{1\text{ ft}}\right)^2 \times \frac{2.54\text{ cm}}{1\text{ in.}} \times \frac{1\text{ m}}{100\text{ cm}} \times \frac{\$2.25}{\text{m}^2}$$

Using unit analysis to convert a measurement to another set of units

Step 1
- Write the measurement with its units.
- Identify the units to which you need to convert.

↓

Step 2
- Multiply the measurement by a conversion factor that lets you cancel the units you don't want.
- Cancel units. You can cancel identical units on the top and bottom of the expression.

↓

If one conversion factor won't yield the units you need, multiply by additional conversion factors.

↓

Step 3
- Perform the calculation.

This squaring changes the first conversion factor and allows us to cancel ft^2:

$$6.00 \times 10^2\,\cancel{\text{ft}^2} \times \frac{144\text{ in.}^2}{1\,\cancel{\text{ft}^2}} \times \frac{2.54\text{ cm}}{1\text{ in.}} \times \frac{1\text{ m}}{100\text{ cm}} \times \frac{\$2.25}{\text{m}^2}$$

You can always square or cube a conversion factor if necessary, which is a good thing because we are going to have to do it again. When we changed ft to ft^2, we also changed in. to in.^2, and so we must somehow get to in.^2, cm^2, and m^2. Here is what we actually need to do:

$$6.00 \times 10^2\,\cancel{\text{ft}^2} \times \left(\frac{12\,\cancel{\text{in.}}}{1\,\cancel{\text{ft}}}\right)^2 \times \left(\frac{2.54\,\cancel{\text{cm}}}{1\,\cancel{\text{in.}}}\right)^2 \times \left(\frac{1\,\cancel{\text{m}}}{100\,\cancel{\text{cm}}}\right)^2 \times \frac{\$2.25}{\cancel{\text{m}^2}}$$

On your calculator you simply key in

$$600 \times 12 \times 12 \times 2.54 \times 2.54 \times 2.25 \div 100 \div 100 = \$125$$

That's the answer. Whether or not you can afford it, only you know.

The chart at left sums up the method of unit analysis.

Many conversion factors useful for doing numerical calculations in chemistry are given inside the back cover of this book. Try your hand now at the following practice problems.

Practice Problems

2.42 The density of gold is 19.3 g/cm^3. What volume in milliliters will 20.0 g of gold occupy? [*Hint:* Don't be fooled. Remember that 1 cm^3 = 1 mL.]

Answer:

Step 1: The measurement and its units are 20.0 g. The unit conversion needed is grams to milliliters.

Step 2: The conversion factors we have to work with are

$$\frac{19.3\text{ g}}{1\text{ cm}^3} \quad \text{and} \quad \frac{1\text{ cm}^3}{19.3\text{ g}}$$

from the density information and

$$\frac{1\text{ cm}^3}{1\text{ mL}} \quad \text{and} \quad \frac{1\text{ mL}}{1\text{ cm}^3}$$

from what we learned in Section 2.6. Choosing those conversion factors that let us cancel units correctly, we get

$$20.0\text{ g} \times \frac{1\text{ cm}^3}{19.3\text{ g}} \times \frac{1\text{ mL}}{1\text{ cm}^3}$$

Step 3: Cancel and do the math:

$$20.0\cancel{\text{ g}} \times \frac{1\cancel{\text{ cm}^3}}{19.3\cancel{\text{ g}}} \times \frac{1\text{ mL}}{1\cancel{\text{ cm}^3}} = 1.04\text{ mL}$$

2.43 The density of air is 0.001 30 g/mL. What is the mass in grams of 500.0 L of air? What is this mass in kilograms?

2.44 The measured density of lead is 11.4 g/mL. What volume in milliliters will 1.50 lb of lead occupy? [1 lb = 453.6 g]

2.45 It takes six cups of flour to bake one cake; exactly one cup of flour has a mass of 120.0 g. If you have 6955 g of flour, how many cakes can you bake? [*Hint:* Two conversion factors are given in this problem. Find them and write them down first in ratio form. Then use them in the correct form with the measured quantity, which is 6955 g of flour.]

2.46 A floor installer can cover 250.0 m^2 of floor area per hour with floor tiles. How many square feet per minute can he cover?

2.9 Rearranging Equations—Algebraic Manipulations with Density

When you work with mathematical equations, it is often desirable to rearrange them (rewrite them) to make it easier to arrive at the answer. This rearranging of an equation is called **algebraic manipulation** (or **algebra**, for short). You can often avoid having to do this by using unit analysis; however, sometimes it will be to your advantage to do some algebra. The fundamental rule when rearranging equations algebraically is simple: *When rearranging an equation, you are not allowed to do anything that changes the meaning of the equation; you can change only its appearance.*

A good example for demonstrating algebraic manipulation is density. We have seen that density is defined as mass over volume ($d = m/V$). Like any other equation, $d = m/V$ can be algebraically rearranged at will. Let's return to an earlier problem that we already solved by unit analysis. If you have 2.00 g of liquid mercury, what volume will it occupy if its density is 13.6 g/mL? In this problem we are being asked to find a volume. This means we want to algebraically rearrange the density equation $d = m/V$ so that it ends up looking like $V =$ something. This is called *solving the equation for volume*. How can we do this? By following the fundamental rule: do anything you want to this equation so long as what you do doesn't change its meaning. This means you can add anything to the equation, subtract anything from it, multiply it by anything, or divide it by anything, so long as you don't change its meaning. The way to ensure that the meaning isn't changed is *to do the same thing to both sides of the equation*. If we add something on the left side of the equal sign, we must add the same thing on the right side of the equal sign. If we multiply by something on the left side, we must multiply by the same thing on the right side. Let's try this now for our density equation. Suppose we multiply both sides of the equation by volume (V), the quantity we are being asked to solve for:

$$d \times V = \frac{m}{V} \times V$$

In a mathematical equation, anything over itself is equal to 1. On the right side of the equation, we now have V on the top and V on the bottom. This is V/V, which equals 1. Follow the diagram below to see how this changes the appearance (but not the meaning) of our equation.

$$d \times V = \frac{m}{V} \times V$$

V on top

V on bottom

$$d \times V = m \times \frac{V}{V}$$

$V/V = 1$, and $m \times 1 = m$

$$d \times V = m$$

We now have $d \times V = m$. We are almost at our target of $V =$ something. We just have to remove d from the left side. To do this, we divide both sides of the equation by d. Notice how this changes the appearance of our equation:

$$\frac{d \times V}{d} = \frac{m}{d}$$

Dividing both sides by d gives us d/d on the left side, which equals 1, and $V \times 1 = V$

$$V = \frac{m}{d}$$

We have done it! By algebraic manipulation, we have taken the equation for density $d = m/V$ and rewritten it as $V = m/d$ without changing its meaning. We have solved it for V. Now, to calculate the volume asked for, we simply divide the mass by the density:

$$V = \frac{m}{d} = \frac{2.00 \text{ g}}{13.6 \text{ g/mL}} = 0.147 \text{ mL}$$

This is the same answer we arrived at by the method of unit analysis.

Another way to look at algebraic manipulations is to consider yourself an eliminator of variables. The variables are just the quantities in the equation. In our last example, for instance, mass (m), volume (V), and density (d) were the variables. Your job is to eliminate the ones you don't want from one side of an equation, leaving behind only the one you want. For example, given the equation for density $d = m/V$, how could you quickly solve it for m? The answer is to eliminate the V from the right side and get m all by itself. The question then becomes, what do you have to do to eliminate V from the right side? The answer is to multiply both sides by V. Doing so gives $Vd = Vm/V$, and $Vm/V = m$ (the V's cancel). You end up with $Vd = m$, and you have solved for m.

Sometimes you have to add or subtract instead of multiply or divide. For example, if you are given the equation $p = q + r$ and asked to solve for q, you need to get rid of the r from the right side. Because r stands alone (that is, it is not multiplied or divided by anything else), simply subtract r from both sides (*always do the same thing to both sides*). This gives $p - r = q + r - r$. On the right side, $r - r = 0$ (anything minus itself is zero), and so the r's are eliminated from the right side. You end up with $p - r = q$. You have solved the equation for q.

Try your hand at algebraic manipulations in the following problems now. Do whatever it takes to eliminate unwanted variables from one side of an equation. Remember that whatever you do to one side of the equation you must also do to the other side. The first two problems involve multiplying or dividing. The last problem involves adding or subtracting.

Practice Problems

Carry out the following algebraic manipulations. Show how you solved each problem, as is done in the answer to Problem 2.47.

2.47 Given $p/q = r$, solve for p.

Answer: To solve $\frac{p}{q} = r$

for p, you must eliminate q from the denominator of the left side. To do this, multiply both sides by q:

$$\frac{\cancel{q}p}{\cancel{q}} = rq$$

The q cancels from the left side because it is in both the top and the bottom. The result is $p = rq$.

2.48 Given $p/q = r$, solve for q.

Answer: $q = \frac{p}{r}$

2.49 Given $P + Q = z$, solve for P.

Answer: $P = z - Q$

2.10 Quantifying Energy

Another thing chemists quantify is *energy*. We do that because every change that matter undergoes (whether physical or chemical) involves energy. For instance, consider the melting of ice (a physical change). A piece of ice can't melt unless you add energy in the form of heat. Likewise, many chemical reactions, such as the reaction of gasoline and oxygen to form carbon dioxide and water, release lots of energy (heat and light—fire). So, to understand the changes matter undergoes, we must quantify energy.

What exactly is energy? You no doubt have an idea. A bolt of lightning, the light from our Sun, the explosion of an atomic bomb, these are all things you intuitively know have energy. In science, **energy** is defined as the capacity for doing work. The more energy something has, the more work it can do. Energy can also be thought of as having many forms. Two forms we shall dwell on in this book are heat and light. We shall have more to say about these forms later. What we want to do now is to *quantify* energy (which simply means to give it units), just as we did with mass and volume.

A well-known unit of energy is the **calorie** (cal). This is not an SI unit, but it is so well known (who hasn't tried to "cut calories") that we can't ignore it. Actually, there are two types of calories—the "little c" calorie and the "big C" Calorie (Cal), where 1 Cal = 1000 cal. The Calorie is the one used by dietitians and the food industry. When you go to the gym and lift weights, that 250 Cal (= 250,000 cal) you "burned" is the amount of energy you expended. Some of that energy went into the work you performed when you lifted the weights, and some of it went into heating you up (which is why you broke into a sweat).

To give you a feel for how much energy one calorie is, consider 1 g of water at 25°C. A mass of 1 g of water is about 20 drops from an eyedropper, an amount you could easily hold in the palm of your cupped hand. It takes 1 cal of heat energy to warm 1 g of water from 25°C to 26°C. (That is how a **calorie** is defined, in fact.) An increase in temperature of 1 C° would barely be noticeable by your hand, and so on an everyday scale, 1 cal is not much energy. On the other hand, 1 Cal would increase the temperature of that same 1 g of water by 1000 C° (assuming the water stayed a liquid with properties identical to those of 25°C water). You would have little trouble feeling *that* amount of energy!

1 cal 1 Cal

The SI unit of energy is the **joule** (J). It is named in honor of James Prescott Joule (1818–1889), a British physicist who participated in discovering the **law**

of conservation of energy. (This law states that energy is neither created nor destroyed; it just goes from one form to another, such as from light to heat.) There are 4.184 J per calorie (and therefore 4.184 kJ per Calorie). This means that 1 J of heat energy is roughly one-fourth the amount of heat energy in 1 cal. From this information we can write the following conversion factors:

$$\frac{4.184\ \text{J}}{\text{cal}} \quad \text{or} \quad \frac{\text{cal}}{4.184\ \text{J}} \quad \textbf{and} \quad \frac{4.184\ \text{kJ}}{\text{Cal}} \quad \text{or} \quad \frac{\text{Cal}}{4.184\ \text{kJ}}$$

To convert from one energy unit to another, all you do is perform unit analysis using the appropriate conversion factor. For example, suppose that Twinkie you just couldn't resist contained 200.0 Calories. We could convert this to joules as follows:

$$200.0\ \cancel{\text{Cal}} \times \frac{4.184\ \cancel{\text{kJ}}}{\cancel{\text{Cal}}} \times \frac{1000\ \text{J}}{\cancel{\text{kJ}}} = 8.368 \times 10^5\ \text{J}$$

That's almost 837,000 J! Maybe you want to think twice about snacking on Twinkies.

This brings us to a final question about energy—how do you measure it? For example, how did someone determine that a delicious Twinkie has 200.0 Cal? Believe it or not, one way is to use water. Recall that 1 cal is the amount of heat energy needed to warm 1 g of water at 25°C by 1 C° (to 26°C). With this knowledge, we can define the *specific heat* of water. The **specific heat** of a substance is the amount of heat energy needed to warm 1 g of that substance by 1 C°. Thus the specific heat of water is 1 cal/g · C°. *Notice the units.* You read them "1 calorie *per* gram *per* Celsius degree." That is, it takes 1 cal to warm 1 g of water by 1 C°. Table 2.5 lists the specific heats of some common substances.

Table 2.5 Specific Heats of Some Common Substances

	Specific heat	
Substance	**cal/g · C°**	**J/g · C°**
Water	1.000	4.184
Iron	0.107	0.449
Aluminum	0.215	0.901
Ethanol	0.581	2.43

Notice two things: (1) that the values differ quite a lot from one another and (2) that the specific heats expressed in joules are just 4.184 times larger than those expressed in calories. Now it's time to see if you understand what these specific heats are telling you.

Suppose you have 1 g of water and 1 g of aluminum, both at 25°C. You want to warm them to 26°C. Would you need to add the same amount of heat to each? How much heat in joules would you need in each case?

Just from looking at the specific heats in Table 2.5, you can see that it takes almost five times as many calories to warm water by 1 C° as to warm the same mass of aluminum by 1 C°. Make sure you understand this before going on, or what follows won't make much sense.

Once we know a substance's specific heat, we can use that substance to measure energy. For example, let's measure the calorie count for a Twinkie using water. First, a little background. When you eat a Twinkie, your body essentially "burns" it. It does this *slowly* and in many steps, but the result is the same as if the Twinkie had been set on fire in the air. When you burn a Twinkie, you convert its stored chemical energy to heat energy, which is released to the surroundings. Therefore, if we can measure the heat released when we burn a Twinkie, we will have measured its calorie content.

So to measure the calorie content of a Twinkie, we simply burn it in such a way that we can capture and measure the heat released. The picture below shows how. We put the Twinkie in a small metal box that contains enough oxygen to let the Twinkie burn completely. This metal box is called a **bomb**. Then we immerse the bomb in a larger, insulated vessel (called a **calorimeter**) that is filled with a known amount of water. For our Twinkie measurement, we shall use a calorimeter that contains 1.00×10^4 g of water (10,000 g; about 2 gallons). When we burn the Twinkie, the heat flows out into the water, warming it. We carefully measure this change in temperature.

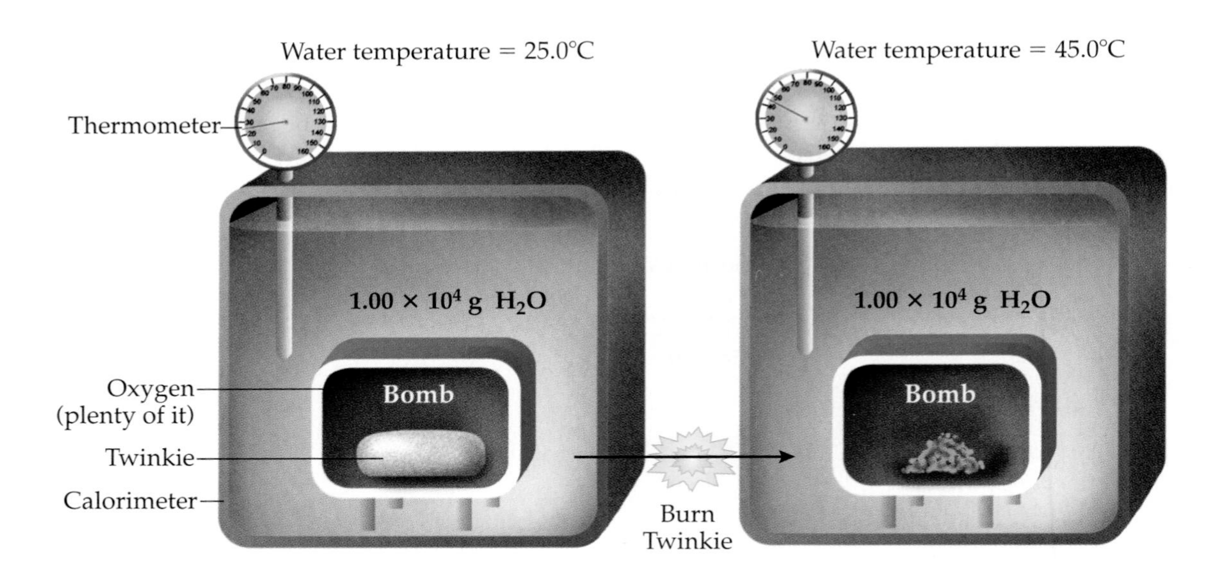

Because we know the specific heat of water and how much water we have, we can use the temperature change to determine the amount of heat the Twinkie released as it burned. (A real calorimeter is also calibrated to allow for the slight amount of heat absorbed by the bomb and the insulated calorimeter walls, but we will ignore that.)

Our calorimeter contains 1.00×10^4 g of water, and the temperature of this water increased by 20.0 C° (from 25.0°C to 45.0°C). We know that the specific heat of water is 1 cal/g · C°. To calculate the heat energy released by burning the Twinkie, all we do is multiply these three things together:

$$\text{Heat} = \begin{matrix}\text{Specific heat}\\ \text{of water}\end{matrix} \times \begin{matrix}\text{Mass of water}\\ \text{in calorimeter}\end{matrix} \times \begin{matrix}\text{Change}\\ \text{in water}\\ \text{temperature}\end{matrix}$$

$$= \frac{1.000\ \text{cal}}{\text{g} \cdot \text{C}^\circ} \times 1.00 \times 10^4\ \text{g} \times 20.0\ \text{C}^\circ = 2.00 \times 10^5\ \text{cal}$$

Notice that the unit that remains after we cancel units is calories. Thus, a Twinkie contains 2.00×10^5 cal of energy. Let's convert this to Calories:

$$2.00 \times 10^5\ \text{cal} \times \frac{\text{Cal}}{1000\ \text{cal}} = 2.00 \times 10^2\ \text{Cal} = 200\ \text{Cal}$$

And there it is, the Calorie content of a Twinkie. Of course, if you were discussing this with a scientist instead of a dieter, you would convert the result to joules or kilojoules so as to be speaking in SI units.

Do you need to memorize the equation for calculating the amount of heat released in a calorimeter? The answer is yes, but if you can visualize the calorimeter (the bomb, the water, the thermometer), then you can reason it out. We want to give you a word of caution, though. Because specific heat has grams as one of its units, the mass of water you enter into the equation must also be in grams (not kilograms, not pounds, not anything else but grams). If you violate this rule, the mass units will not cancel, and your calculated amount of heat will be wrong. Finally, the equation for calculating the heat released in a calorimeter is also useful for calculating heat in other situations. Follow through the solutions for the next two practice problems, and then try the last three on your own.

Practice Problems

2.50 Suppose you have a 2.000-lb block of iron at 50.0°C. How much heat in joules would it take to warm this block to 75.0°C?

Answer: To calculate the amount of heat it takes to warm up any substance, use the equation below—first multiply the specific heat by the mass of your sample and then multiply that product by the temperature change:

$$\textit{Heat} = \begin{matrix}\textit{Specific heat}\\ \textit{of iron}\end{matrix} \times \textit{Mass of iron} \times \begin{matrix}\textit{Change}\\ \textit{in iron}\\ \textit{temperature}\end{matrix}$$

$$= \frac{0.449\ \text{J}}{\text{g} \cdot \text{C}^\circ} \times 907.2\ \text{g} \times 25.0\ \text{C}^\circ = 1.02 \times 10^4\ \text{J}$$

Notice that we converted the mass from pounds to grams before plugging it into the heat equation:

$$2.000\ \text{lb} \times \frac{453.6\ \text{g}}{\text{lb}} = 907.2\ \text{g}$$

2.51 A 20.0-g block of iron initially at 25.0°C has 100.0 J of heat energy added to it. What is its temperature after the heat energy has been added?

Answer: This problem involves doing algebra calisthenics with the equation used in Problem 2.50. All we need to do is rearrange the equation to solve it for "change in temperature" and then plug in the values for everything we know.

Heat = Specific heat of substance × Mass of substance × Change in substance temperature

If we divide both sides by Specific heat of substance × Mass of substance, we get

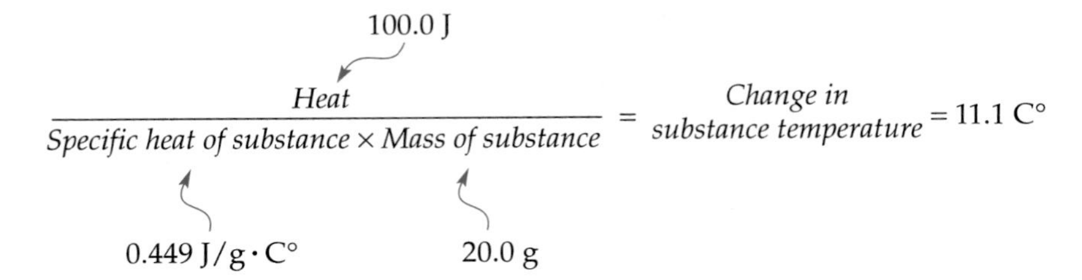

Because the initial temperature is 25.0°C, the final temperature is 25.0°C + 11.1°C = 36.1°C.

2.52 Suppose you have a 50.0-mL sample of ethanol at 22.0°C and want to warm it to 60.0°C. How many kilojoules of heat will this take? The density of ethanol is 0.785 g/mL. [*Hint:* You need to use density to convert the given volume to grams.]

2.53 Convert the answer you calculated in problem 2.52 to joules, Calories, and calories.

2.54 A 0.100-g sample of your favorite candy is burned in a calorimeter that contains 1.00 kg of water initially at 22.0°C. After the candy is burned, the water temperature is 35.5°C. How many Calories are there per gram of your candy?

As we said earlier, you can't avoid the numerical side of chemistry if you hope to study it in any detail. This chapter has prepared you for dealing with this numerical side, and it is important. However, the fundamental concepts of chemistry and the pictures, models, and stories that are discussed in the remainder of this book are just as important, if not more so. After all, you can't use the mathematical tools you just learned about to solve a problem until you understand what the problem is about.

Have You Learned This?

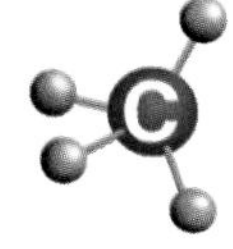

The Numerical Side of Chemistry
www.chemistryplace.com

SKILLS TO KNOW

Uncertainty

1. All numbers derived from measurements have uncertainty. By convention, this uncertainty is assumed to be in the number's final significant digit.	2.46 Certain: 2.4 — Uncertain (estimated): 6
2. To calculate the exact amount of uncertainty, identify the column of the uncertain digit, put a 1 in that column, and divide that 1 by 2. For instance, if the uncertain digit is in the hundredths column, the uncertainty is 0.01/2 = ±0.005.	2.46 ← Uncertain digit in hundredths column Uncertainty = ±0.01/2 = ±0.005

Scientific notation

A number written in scientific notation has the form $A \times 10^x$, where A is a number between 1 and 9.9999 . . . , and x is a power of 10. To convert the number to standard notation, move the decimal point in A by x places. If x is positive, move the decimal point to the right; if x is negative, move the decimal point to the left.

Scientific notation		Standard notation
2.46×10^4	=	24600
2.46×10^0	=	2.46
2.46×10^{-4}	=	0.000246

Rules of significant figures

1. **Leading zeros** are just placeholders; they are not significant. **Trailing zeros** that follow the decimal point are always significant. **Trailing zeros** that precede the decimal point may or may not be significant; use a decimal point to avoid ambiguity.	Not significant — 0.0600 cm — Significant This number has 3 significant figures. 1200. — Decimal point tells you these trailing zeros are significant. 1200 — No way to know whether these two trailing zeros are significant.
2. When you **multiply** or **divide** numbers that have uncertainty, your answer can have no more significant figures than the starting number with the fewest significant figures.	2.22 ← 3 sig figs × 2.0 ← 2 sig figs 4.4 ← Answer can have only two significant figures.
3. When you **add** or **subtract** numbers that have uncertainty, your answer can be no more certain than the least certain number.	2.22 ← Uncertainty in 100ths position + 2.0 ← Uncertainty in 10ths position 4.4 ← Answer as certain as least certain number (uncertainty in 10ths position).
4. An exact number has no effect on the number of significant figures in the result of a calculation.	4 sig figs: 6.525 feet; Exact number: 12 inches/foot; 4 sig figs: 78.30 inches $6.525 \text{ feet} \times \frac{12 \text{ inches}}{\text{foot}} = 78.30 \text{ inches}$

Using unit analysis to numerical problems in chemistry

Example: What is the volume in milliliters of 4.00 g of gold, Au? The density of gold is 19.3 g/mL.

Step 1 • Write the measurement with its units. • Identify the units to which you need to convert.	Measurement: 4.00 g Au Convert from grams to milliliters.
Step 2 • Multiply the measurement by a conversion factor that lets you cancel the units you don't want. • Cancel identical units on the top and bottom of the expression.	$4.00 \text{ g Au} \times \frac{1 \text{ mL Au}}{19.3 \text{ g Au}}$ $4.00 \cancel{\text{g Au}} \times \frac{1 \text{ mL Au}}{19.3 \cancel{\text{g Au}}}$

If one conversion factor is not enough to give the units you want, multiply by additional conversion factors.

Step 3 Perform the calculation.	$4.00 \text{ g Au} = \frac{1 \text{ mL Au}}{19.3 \text{ g Au}} = 0.207 \text{ mL Au}$

Heat

1. The common units of heat energy are calories (cal), Calories (Cal), and joules (J). The joule is the SI unit of energy (including heat energy).	1 Cal = 1000 cal 1 cal = 4.184 J 1 Cal = 4184 J = 4.184 kJ
2. The **specific heat** of a substance is the amount of heat energy you must add to warm 1 g of the substance by 1 C°. For instance, it takes more heat energy to warm an aluminum pan than an equal-mass iron one because aluminum has a larger specific heat than iron.	Units of specific heat: Any of these heat energy energy units may be used . . . $\frac{\text{J, kJ, cal, Cal}}{\text{g} \cdot \text{C}°}$ . . . but mass is always in grams and temperature change is always in C°.
3. To calculate how much heat energy you must add to a sample of a substance to raise its temperature by a certain amount, you must know the specific heat of that substance and the mass of the sample:	Heat energy added = Specific heat × Mass in grams × Change in temperature in C°
Example: How much heat energy does a 140.-g aluminum pan absorb as it heats from 25°C to 175°C?	Specific heat of aluminum = 0.901 J/g • C° Temperature change = 175°C − 25.0°C = 150. C° Heat energy added $= \frac{0.901 \text{ J}}{\text{g} \cdot \text{C}°} \times 140. \text{ g} \times 150. \text{ C}°$ $= 1.89 \times 10^4 \text{ J}$

NUMBERS IN CHEMISTRY—PRECISION AND ACCURACY

2.55 There are 3 feet in a yard. A certain piece of wood is 3 feet long. What is the fundamental difference between the value of *3 feet* in these two statements?

2.56 What is the difference between precision and accuracy? Do they ever mean the same thing?

2.57 Suppose in making repetitive measurements you could have only a precise result or an accurate result, but not both. Which should you choose, and why?

2.58 Three people measure the distance from Main Street to Market Avenue using their car odometers. Their data are 2.3 miles, 2.6 miles, 3.1 miles. Survey charts show the actual distance to be 1.6 miles. Characterize the collected data in terms of accuracy and precision.

2.59 Two people attempt to measure the length in feet of a parking lot they know to be about 100 feet long. One person uses a 6-inch ruler; the other uses a 120-foot tape measure. If both measuring devices are graduated in 1/16 inch, which person is likely to make the more accurate measurement? Explain.

NUMBERS IN CHEMISTRY—UNCERTAINTY AND SIGNIFICANT FIGURES

2.60 A ruler is marked in intervals of 1/8 inch. To what fraction of an inch can you estimate a measurement?

2.61 How is the uncertainty in a measurement determined?

2.62 Why do all measurements have some uncertainty?

2.63 What is the uncertainty in each measured number:
(a) 12.60 cm (b) 12.6 cm
(c) 0.000 000 03 inch (d) 125 feet

2.64 Indicate the number of significant figures in each number in Problem 2.63.

2.65 Using radioactive elements, scientists determined the age of a dinosaur skeleton to be 78.5 million years. In years, what is the uncertainty in this measurement?

2.66 Underline any trailing zeros in these measurements:
(a) 12.202 km (b) 0.01 mL
(c) 205°C (d) 0.010 g

2.67 For the measurements in Problem 2.66, underline all the zeros that are significant.

2.68 What is the uncertainty associated with each measurement in Problem 2.66?

2.69 The measurement 30 feet is ambiguous, but the measurement 30. feet is not. Explain what the ambiguity is and how adding the decimal point eliminates the ambiguity.

2.70 Give all interpretations possible for the measurement 2200 feet.

SCIENTIFIC NOTATION

2.71 Convert the following measured values from scientific notation to normal notation. For each one, indicate the number of significant figures.
(a) 5.60×10^1 kg (b) 2.5×10^{-4} m
(c) 5.600×10^6 miles (d) 0.02×10^2 feet

2.72 What is the uncertainty for each measured value in Problem 2.71?

2.73 Using scientific notation, write the measurement 30 feet as having an uncertainty of
(a) ±5 feet (b) ±0.5 feet (c) ±0.05 feet

2.74 Using scientific notation, write the measurement 2200 feet as having an uncertainty of ±50 feet.

2.75 Convert the following numbers from normal notation to scientific notation:
(a) 226 (b) 226.0 (c) 0.000 000 000 50
(d) 0.3 (e) 0.30
(f) 900,000,574 with an uncertainty of ±0.5 million
(g) 900,000,574 with an uncertainty of ±50

2.76 How many significant figures are there in the following measured values, and what is the uncertainty in each measurement?
(a) 0.001 kg
(b) 0.000 10 m
(c) 102 L
(d) 2.600×10^{-3} m
(e) 1.1×10^6 km

HOW TO HANDLE SIGNIFICANT FIGURES AND SCIENTIFIC NOTATION WHEN DOING MATH

2.77 To the correct number of significant figures, what is the result of adding the measured values 100. inches + 2 inches + 0.001 inch? What is the uncertainty in the result?

2.78 To the correct number of significant figures, what is the product of each multiplication? Use scientific notation when necessary. No units shown means a number is exact.)
(a) 2.3 cm × 2
(b) 2.3 m × 2.0 m
(c) 1000 J × 10
(d) 0.1 mm × 124 mm

2.79 A student walks 20,450.2 feet to school every day. A mile is defined as 5280 feet. How many miles does the student walk to school each day?

2.80 Do these calculations using a scientific calculator and report your answers in scientific notation:
(a) $(3.33 \times 10^4 \text{ km}) + (2.22 \times 10^5 \text{ km})$
(b) $(2.444 \times 10^9 \text{ J}) \div (2.444 \times 10^{-9} \text{ J})$
(c) $(2.34 \times 10^2 \text{ m}) - (2.34 \times 10^1 \text{ m})$
(d) $(4.00 \times 10^4 \text{ L}) + (6.00 \times 10^{-1} \text{ L})$

NUMBERS WITH A NAME—UNITS OF MEASURE

2.81 What is the base SI unit of length? What is the SI unit of volume?

2.82 What are two metric but non-SI units of volume, and why are they more often used than the SI unit of volume?

2.83 When is it correct to use cm^3 instead of mL? Explain.

2.84 Why was the SI unit system developed by scientists?

2.85 Convert each length to meters. Report your answers in scientific notation and watch your significant figures.
(a) 2.31 gigameters (Gm)
(b) 5.00 micrometers (μm)
(c) 1004 millimeters (mm)
(d) 5.00 picometers (pm)
(e) 0.25 kilometer (km)

2.86 Which is larger, a Celsius degree or a Fahrenheit degree? Explain.

2.87 Of the three temperature scales, which can have negative temperatures? For the one(s) that can't, explain why not.

2.88 Convert:
(a) 22.5°C to Fahrenheit and Kelvin
(b) −3.00°F to Celsius and Kelvin
(c) 0.0 K to Celsius and Fahrenheit
(d) 65.1°C to Fahrenheit and Kelvin

2.89 How cold does it have to be for water to freeze in °F? In °C? In kelvins?

2.90 Where does the 9/5 ratio come from in the formula for converting °C to °F? [*Hint:* Consider the number of degrees between the boiling point and freezing point of water in both scales.]

2.91 You fill a bottle with water by pouring into it the contents of these two graduated cylinders:

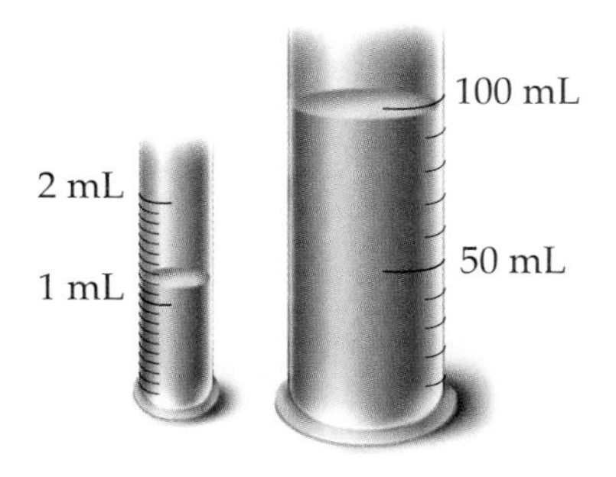

(a) What is the uncertainty in each graduated cylinder?
(b) To the correct number of significant figures, what is the final volume of water in the bottle after you pour the contents of both cylinders into it? What is the uncertainty in this volume? Explain.

2.92 A student measures the length of one side of a perfect cube with the ruler shown here, which is marked in centimeters. She then calculates the volume V by cubing the length l ($V = l^3 = 1 \times 1 \times 1$). What volume should she report? Give the answer to the proper number of significant figures and with the proper units.

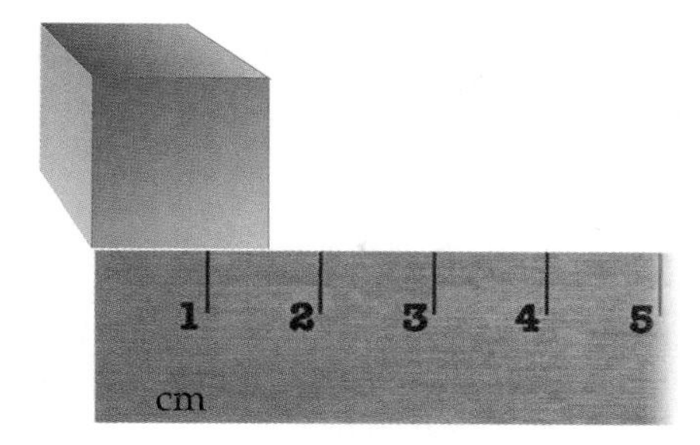

2.93 Using a ruler marked in centimeters and millimeters, a student measures the diameter of a ball to be 1.5 cm. His partner measures the same ball with the same ruler and comes up with 1.50 cm. Which student used the ruler incorrectly? How did that student use the ruler incorrectly?

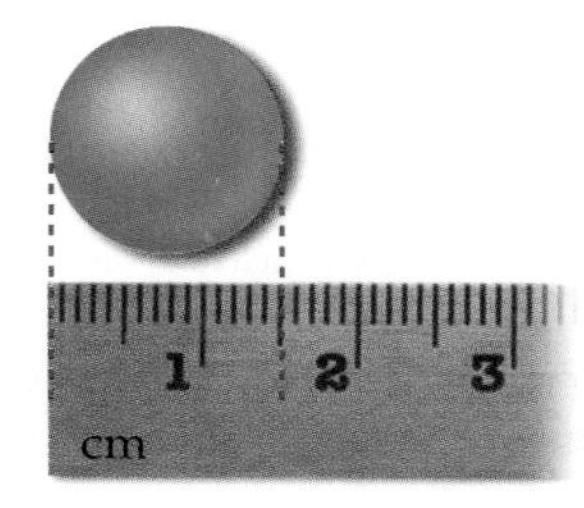

2.94 The students measure another ball with the ruler of Problem 2.93 and determine that its diameter is 2.55 cm. What is the radius of the ball to the correct number of significant figures?

DENSITY

2.95 Define density, and explain why the unit for density is called a derived SI unit.

2.96 At 25°C, what is the mass in grams of 1000.0 mL of water? [*Hint:* Use data from Table 2.4.]

2.97 At 25°C, what is the mass in grams of 2.0 L of mercury? [*Hint:* Use data from Table 2.4.]

2.98 At 25°C, the dimensions of a stick of butter are 10.0 cm × 10.0 cm × 10.0 cm. What is the mass of this butter in grams? [*Hint:* Use data from Table 2.4.]

2.99 Explain how you would measure the density of a pumpkin.

2.100 Two students measure the density of gold. One works with a 100-g bar of pure gold. The other works with a 200-g bar of pure gold. Which student measures the larger density? Explain your answer.

2.101 Any object sinks in a liquid if the object is denser than the liquid but floats on the liquid if less dense than the liquid. Knowing this, consult Table 2.4 and describe a flotation test to distinguish lead from gold.

UNIT ANALYSIS

2.102 Suppose it takes you 1.25 days to drive to your aunt's house. How many seconds does it take you to get there? Use unit analysis to calculate your answer, and show your work.

2.103 A train traveling at 45.0 miles per hour has to make a trip of 100.0 miles. How many minutes will the trip take? Use unit analysis to calculate your answer, and show your work.

2.104 You have a great job in which you earn $25.50 per hour. How many dollars do you earn per second? Use unit analysis to calculate your answer, and show your work.

2.105 Gold has a density of 19.3 g/mL. Suppose you have 100.0 glonkins of gold. What volume in liters will the gold occupy? Here are some conversion factors to help you: 0.911 ounce per glonkin and 28.35 grams per ounce. Use unit analysis to calculate your answer, and show your work. Treat both conversion factors as exact.

2.106 One liter is equal to 0.264 gallon. Suppose you have 1.000×10^3 cm^3 of water. How many gallons do you have? Use unit analysis to calculate your answer, and show your work. Treat all conversion factors as exact.

2.107 You measure the dimensions of a rectangular block to be 10.2 cm $\times$ 43.7 cm $\times$ 95.6 cm. What is its volume in liters?

2.108 The block of Problem 2.107 has a mass of 2.43×10^2 kg. What is the density of the block in grams per milliliter?

2.109 You measure one edge of a cube using a meterstick marked in centimeters. Unfortunately, the edge is longer than 1 m. You mark the 1-m point on the cube edge with a pen and then, using a 15-cm ruler marked in millimeters, measure the remaining distance to be 1.40 cm.
(a) What is the length of the edge in centimeters?
(b) What is the volume of the cube in cubic centimeters? (Remember, the lengths of all edges of a cube are equal.) Watch your significant figures. Use scientific notation if you have to.
(c) The cube has a mass of 111 kg. What is its density in grams per milliliter? Watch your significant figures.

2.110 A rectangular box measures 6.00 inches in length, 7.00 inches in width, and 8.00 inches in height. What is the volume of the box in liters? [2.54 cm = 1 inch]

2.111 An object travels 80.0 m/s. How fast is it traveling in miles per hour? [1 m = 3.28 feet, 1 mile = 5280 feet]

ALGEBRAIC MANIPULATIONS

2.112 Why can't you multiply just one side of an equation by something when algebraically rearranging the equation?

2.113 Solve the equation $y = z/x$ for x.

2.114 Solve the equation $y = z - x$ for x.

2.115 Solve the equation $y = (z/x) + 2$ for x.

2.116 The density of a certain liquid is 1.15 g/mL. What mass in grams of the liquid is needed to fill a 50.00-mL container? Do this problem by the method of algebraic manipulation, beginning with the equation density = mass/volume and showing all steps.

2.117 Repeat Problem 2.116 using unit analysis. Show all steps. Do you get the same answer you got in Problem 2.116?

QUANTIFYING ENERGY

2.118 Define *energy*.

2.119 How much heat energy is 1 cal? Give your answer in terms of changing the temperature of water.

2.120 Convert:
(a) 4.50 Cal to calories
(b) 600.0 Cal to kilojoules
(c) 1.000 J to calories
(d) 50.0 Cal to joules

2.121 Define *specific heat*.

2.122 A block of iron and a block of aluminum of equal mass are both initially at the same low temperature. Both are then warmed to the same high temperature. Does one block require more heat energy than the other to reach the high temperature? If so, how much more?

2.123 How many joules does it take to raise the temperature of 2.00 L of water from 22.0°C to 40.0°C? How many kilojoules? Take the density of water to be 1.00 g/mL.

2.124 Why is it necessary for a calorimeter to be insulated?

2.125 A 2.50-g piece of wood is burned in a calorimeter that contains 0.200 kg of water. The burning causes the water temperature to increase from 22.1°C to

28.7°C. How much heat energy is released in joules? What is the energy content of the wood in joules per gram of wood?

2.126 How many joules of heat energy would it take to raise the temperature of 2.00 lb of iron from 30.0°C to 90.0°C? [453.6 g = 1 lb]

2.127 How many grams of wood from Problem 2.125 would you have to burn to accomplish the task of Problem 2.126?

2.128 A 2.000-ton block of aluminum at room temperature (22.0°C) is struck by lightning. The lightning transfers 8.000×10^6 kJ of energy to the block. How hot does the block get? [1 ton = 2000 lb, 1 lb = 453.6 g]

ADDITIONAL PROBLEMS

2.129 Which one of the following expresses the measured value 1230.0 m with the correct number of significant figures and appropriate Greek prefix?
(a) 1.23 km (b) 1.230 cm
(c) 1.2300 km (d) 1.2300 mm
(e) 12.3 km

2.130 Convert:
(a) 7.98×10^{23} μL to liters
(b) 3.00×10^{-3} mg to grams
(c) 4.21×10^{8} mL to gallons [1 m^3 = 264 gallons]

2.131 A metal sphere has a radius r of 4.00 cm. What is the volume V of this sphere in cubic centimeters? The formula for the volume of a sphere is $V = (4/3)\pi r^3$, where $\pi = 3.141\,59$.

2.132 What is the density, in g/cm^3, of the sphere of Problem 2.131 if its mass is 2.5 kg?

2.133 Which one of the following expresses the measured value 0.000 003 L with the correct number of significant figures?
(a) 3 mL (b) 3 μL
(c) 3.00×10^{-6} L (d) 3.00×10^{-3} mL

2.134 The SI unit of speed is meters per second. In the United States, speed is often expressed in miles per hour. If a car is traveling 60.0 miles/h, what is its speed in meters per second? [1 mile = 1.61 km]

2.135 In the United States, car fuel efficiency is expressed in miles per gallon of gasoline. However, fuel efficiency can also be expressed in kilometers per liter of gasoline. If the fuel efficiency of a car is 11.0 km/L, what is its fuel efficiency in miles per gallon? [1 mile = 1.61 km, 1 gallon = 3.79 L]

2.136 Convert:
(a) 72°F to °C and K
(b) −12°C to °F and K
(c) 178 K to °F and °C

2.137 Use Greek prefixes to express the relationship between the calorie and the Calorie.

2.138 When a 16.74-g rubber stopper is placed in a graduated cylinder containing 25.46 mL of water, the water level rises to 37.42 mL. What is the density of the stopper in grams per milliliter?

2.139 Solve each equation for the underlined quantity:
(a) $°F = 9/5\ \underline{°C} + 32$
(b) $PV = nR\underline{T}$
(c) $E = hc/\underline{\lambda}$

2.140 Convert:
(a) 2.37×10^{2} L to milliliter
(b) 800 kg to grams
(c) 0.592 mm to meters
(d) 8.31 g to kilograms
(e) 9.62×10^{-6} L to microliters
(f) 8000 m to kilometers
(g) 19.3 mg to grams
(h) 0.003 45 mL to liters

2.141 For solids, the amount of material per unit volume is often expressed in grams per milliliter, whereas for gases the amount of material per unit volume is usually expressed in grams per liter. If the amount of matter in air is 1.34 g/L, what is this value in:
(a) g/mL (b) kg/L (c) kg/mL?

2.142 The specific heat of methane gas is 2.20 J/g • C°. If the temperature of a sample of methane gas rises by 15 C° when 8.8 kJ of heat energy is added to the sample, what is the mass of the sample?

2.143 Explain how determining the number of significant figures allowed in an answer when measured values are multiplied or divided is different from determining the number of significant figures allowed in an answer when measured values are added or subtracted.

2.144 Convert:
(a) 23.0°C to K (b) 98.6°F to °C
(c) 296 K to °F (d) 32°F to K
(e) 523 K to °C (f) 38°C to °F

2.145 A student takes three measurements of the mass of an object. If the actual mass is 8.54 g, indicate whether each set of measurements is precise but not accurate, accurate but not precise, both accurate and precise, or neither accurate nor precise:
(a) 6.38 g, 9.23 g, 4.36 g
(b) 8.53 g, 8.59 g, 8.55 g
(c) 9.53 g, 8.54 g, 7.54 g
(d) 6.25 g, 6.27 g, 6.26 g

2.146 A slice of bread is burned in a calorimeter containing 1.000 kg of water. The water temperature rises from 25.0°C to 33.0°C. What is the calorie content of the bread? What is the Calorie content?

2.147 A block of metal measuring 3.0 cm × 4.0 cm × 5.0 cm has a mass of 470.0 g. What is the density of the metal in grams per cubic centimeter?

2.148 A student measures the mass of an object three times and reports the numeric average of her measurements. If her three measurements are 212 g, 260 g, and 233 g and the actual mass is 235 g, which of the following statements is true:
(a) The student is accurate but not precise.
(b) The student is precise but not accurate.
(c) The student is both accurate and precise.
(d) It is impossible to tell whether the student is accurate and/or precise without knowing how she determined the mass.

2.149 The density of an irregularly shaped object is determined by immersing the object in water. If the mass of the object is 8.34 g and the water level rises from 25.00 mL to 28.10 mL, what is the density of the object in grams per milliliter?

2.150 The specific heat of copper is 0.385 J/g • C°. How much heat energy in kilojoules is required to raise the temperature of 454 g of copper from 40.0°C to 75.0°C?

2.151 A friend tells you a quick way to convert °C to °F: double the °C, subtract one-tenth of the doubled value, and add 30 to the result. Use the Cesius–Fahrenheit conversion equation to explain why this is true.

2.152 Express 23,000,000 in scientific notation having
(a) Two significant figures
(b) Three significant figures
(c) Five significant figures
(d) Six significant figures
(e) Eight significant figures

2.153 Dieters are often told that drinking ice-cold water burns more energy than drinking room-temperature water. Why is this true?

2.154 If a graduated cylinder has markings that indicate 0.01 mL, what is the uncertainty in any volumes determined using this graduated cylinder?

2.155 The density of gold is 19.3 g/mL, that of lead is 11.4 g/mL, that of iron 7.8 g/mL, and that of aluminum 2.7 g/mL. A student is given separate samples of three substances A, B, and C, along with a graduated cylinder containing 50.0 mL of water. Each sample has a mass of 200.0 g, and the student finds that the volumes of the samples are A 25.64 mL, B 10.36 mL, and C 17.54 mL. What is the identity of each substance?

2.156 When 10 kJ of heat energy is added to a beaker containing 250 g of water initially at 23.0°C, what is the final temperature of the water?

2.157 Indicate the uncertainty in
(a) 74.8 m (b) 0.0026 g (c) 1.250×10^3 L
(d) 18 cm (e) 18 pennies

2.158 Round each number to three significant digits and express the answer in scientific notation:
(a) 0.592 861 (b) 438 932 (c) 0.000 073 978
(d) 0.235 469 (e) 82.550 (f) 529.8

2.159 Explain the relationship between a calorie and a Calorie.

2.160 Explain what is wrong with the statement "I took one measurement of the mass, and it was a very precise measurement."

2.161 Do the following calculations and express each answer in scientific notation:
(a) $(5.03 \times 10^2) + (8.1 \times 10^1)$
(b) $(8.32 \times 10^{-5}) \times (0.53 \times 10^4)$
(c) $\frac{6.02 \times 10^{23}}{3}$ where the 3 is an exact number
(d) $(3.960 \times 10^3) - (4.62 \times 10^2)$

2.162 A block measures 6.0 cm on each side. What is the volume of the block in cubic meters?

2.163 The mass of an average neon atom is 20.2 atomic mass units, where 1 atomic mass unit = 1.66×10^{-24} g.
(a) What is the mass in atomic mass units of 20 neon atoms?
(b) What is the mass in grams of 20 neon atoms?
(c) What is the mass in grams of 6.022×10^{23} neon atoms?

2.164 If the same amount of heat energy is added to a beaker containing 100 mL of ethanol and a beaker containing 100 mL of water, which liquid experiences the greater rise in temperature?

2.165 What is the volume in milliliters of 15.0 g of (a) liquid water at 25°C, (b) ice at 0°C, (c) gasoline at 25°C, (d) lead at 25°C, (e) mercury at 25°C, (f) helium gas at 0°C and 1 atm pressure? [Use data from Table 2.4.]

2.166 Indicate whether the trailing zero in each value is significant, not significant, or possibly significant:
(a) 540 ± 0.5 (b) 540 ± 5
(c) 0.540 (d) 0.000 540
(e) 540

2.167 Indicate the number of significant figures in
(a) 0.503 200 mL (b) 2000 ± 5
(c) 2000 ± 50 (d) 2000 ± 0.5
(e) 200.1

2.168 As you will learn later in this course, some liquids do not mix with each other, oil and water being an example. When such nonmixing liquids are combined, they form separate layers, with the densest substance at the bottom. Suppose the nonmixing liquids gasoline, liquid mercury, and liquid water are added to an empty graduated cylinder. Using the data in Table 2.4, draw the graduated cylinder and show the positions of the three liquids.

2.169 True or false? If either statement is false, rewrite to make it true.
(a) When multiplying or dividing a series of measured values, the number of significant figures in the answer is determined by the measured value having the fewest significant figures.
(b) When adding or subtracting a series of measured values, the number of significant figures in the answer is determined by the measured value having the fewest significant figures.

2.170 Convert to cal/g • C°:
(a) 1.04 J/g • C° (the specific heat of nitrogen gas)
(b) 0.84 J/g • C° (the specific heat of carbon dioxide gas)

2.171 A one-semester chemistry course meets for 1-h sessions three times a week for 15 weeks. How many milliseconds will a student with perfect attendance spend in class during the semester?

2.172 Describe how the uncertainty in a measured value is determined.

2.173 Write each number in scientific notation, taking all trailing zeros to be nonsignificant:
(a) 502,000 (b) 0.000 038 402
(c) 436,000,000 (d) 8470
(e) 0.005 91 (f) 0.658

2.174 A company wants 800 square feet of carpet, but the carpet store sells only by the square meter. How many square meters does the company need to buy? [1 m = 39.37 inches]

2.175 If 1 U.S. dollar is worth 1.54 Canadian dollars, how many U.S. dollars are needed to purchase an item that costs 350 Canadian dollars?

2.176 Which contains more matter, a 50-g block of lead or a 50-g block of gold? Explain.

2.177 Indicate the number of significant zeros in each value:
(a) 2.300 (b) 2.3003
(c) 0.0023 (d) 2300
(e) 23.000

2.178 Fill in the blanks:
(a) $1.89 \times 10^3 =$ ________ $\times 10^4$
(b) $7.932 \times 10^{-5} =$ ________ $\times 10^{-7}$
(c) $4.68 \times 10^{-12} = 468 \times$ ________
(d) $3.46 \times 10^{-1} =$ ________ $\times 10^0$

2.179 Is reporting a measured mass as 580. g the same as reporting it as 580 g? Explain.

2.180 Which is larger, a kelvin or a Celsius degree? Explain.

2.181 A student reports a series of five length measurements that are accurate but not precise. Is it more likely that his laboratory technique is very good but the measuring instrument is bad, or that his laboratory technique is bad? Explain.

2.182 Write each number in normal notation:
(a) 1.79×10^{-2} (b) 8.76×10^{-9}
(c) 4.88×10^{10} (d) 7.52×10^1
(e) 8.37×10^0 (f) 4.184×10^4

2.183 (a) If 25.0 cm^3 of an unknown substance has a mass of 195 g, what is the density of the substance in grams per cubic centimeter?
(b) How many cubic centimeters does 500.0 g of the substance occupy?
(c) Does this substance sink or float in mercury, which has a density of 13.6 g/mL?

2.184 Express each mass in grams, both in normal notation and in scientific notation:
(a) 536 mg (b) 8.26 dg
(c) 0.0057 μg (d) 139 kg
(e) 836 ng (f) 0.073 Mg

2.185 At 25°C, air has a density of 1.3×10^{-3} g/mL. What is this density in (a) kilograms per liter and (b) pounds per gallon?

2.186 On a hot summer day, you want to cool two glasses of warm lemonade but have no ice. Not wanting to wait until you can make some, you place two small metal blocks in the freezer. One block is pure iron, the other pure aluminum, and each has a mass of exactly 50 g. After both have cooled to −10°C, you put them into separate glasses and add 200 mL of warm lemonade to each. After a few minutes, both blocks have warmed up to +10°C. At this point, is the lemonade in one glass cooler than the lemonade in the other glass? If so, which is cooler and why? [Despite all the numerical information, you should be able to use specific heat values from Table 2.5 to answer without doing any calculations.]

2.187 Use a scientific calculator to do the following calculations. Express each answer in scientific notation and to the correct number of significant figures.
(a) $9.865 \times 10^3 + 8.61 \times 10^2$
(b) $\dfrac{(6.626 \times 10^{23}) \times (3.00 \times 10^8)}{4.5 \times 10^{-7}}$
(c) $\dfrac{5.6200 \times 10^{-9}}{3.821 \times 10^9}$
(d) $\dfrac{4.5600 \times 10^3 - 2.91 \times 10^1}{5}$ where the 5 is an exact number.

2.188 What mass of each substance occupies a volume of 50.0 mL? (Densities are shown in parentheses.)
(a) Lead (11.4 g/mL)
(b) Ethanol (0.785 g/mL)
(c) Oxygen gas (1.4×10^{-3} g/mL)
(d) Hydrogen gas (8.4×10^{-5} g/mL)
(e) Mercury (13.6 g/mL)
(f) Gold (19.3 g/mL)

2.189 Write two conversion factors that express the relationship between
(a) Grams and kilograms, using 1 and 1000
(b) Kilograms and grams, using 1 and 0.001
(c) Yards and feet
(d) Meters and centimeters, using 1 and 100
(e) Meters and centimeters, using 1 and 0.01

2.190 The recommended tire pressure in a bicycle is 125 lb/in.^{2}. What is this tire pressure in atmospheres? [1 atm = 14.70 lb/in.^{2}]

2.191 Do the following calculations and express each answer to the correct number of significant figures. [All values are measurements.]
(a) $\dfrac{5.03 + 7.2}{0.003}$
(b) $\dfrac{8.93 \times 0.054}{1.32}$
(c) $(6.23 \times 0.042) + 9.86$

2.192 You overhear a classmate telling another student that 1 feet equals 12 ± 0.5 inches. Is this statement correct or incorrect? Why?

2.193 Indicate the correct number of significant figures for each answer, given that all values are measurements:
(a) $(6.350 \times 10^{-8}) \times (0.0080)$
(b) $(5.30 \times 10^{2}) + (22.1 \times 10^{2})$
(c) $(5.830 \times 10^{2}) + (22.100 \times 10^{2})$
(d) $\dfrac{100.0 \times 0.1500}{58.443}$
(e) $\dfrac{100.0 \times 0.15}{58.4}$

2.194 If one U.S. dollar is worth 0.690 English pounds, how many U.S. dollars are needed to purchase an item that costs 350 pounds?

2.195 A graduated cylinder was marked incorrectly at the factory and indicates a volume that is 5 mL more than the actual volume of liquid in the cylinder. If an experienced laboratory technician who does not know about the error is using the cylinder, are the volumes she measures more likely to be accurate, precise, or neither? Explain.

2.196 The density of water at 4.00°C is 1.00 g/mL. The density of ice at 0°C is 0.917 g/mL. Water is different from most other substances in that the solid phase (ice) is less dense than the liquid phase. Explain why this characteristic makes ice-fishing possible.

2.197 How many significant figures are there in each number:
(a) 5.300×10^{-2} (b) 3.2×10^{5}
(c) $0.008\,90 \times 10^{-4}$ (d) $7.960\,000\,0 \times 10^{10}$
(e) 8.030×10^{21}

2.198 The speed limit on some interstate highways is 70 mi/h. What is this speed limit in (a) kilometers per hour, (b) kilometers per second, (c) meters per hour, (d) meters per second? [1 mi = 1.61 km]

2.199 Explain what effect compressing a gas has on the density of the gas.

2.200 Calculate the amount of heat energy in joules required to heat 50.0 g of each substance from 25.0°C to 37.0°C. (Specific heats shown in parentheses):
(a) Iron (0.449 J/g • C°)
(b) Aluminum (0.901 J/g • C°)
(c) Mercury (0.14 J/g • C°)
(d) Water (4.18 J/g • C°)

2.201 Which substance in problem 2.200 undergoes the most gradual change in temperature when heated? Which undergoes the fastest change in temperature when heated? Explain.

2 WorkPatch Solutions

2.1 (c) ±0.005 cm. Because of the millimeter divisions (remember, 1 mm = 0.1 cm), you can read a value exactly to 0.1 cm. Then you estimate between two adjacent lines to get a digit for your hundredths place. Because the estimating (uncertainty) is in the hundredths place, you say 0.01 ÷ 2 = 0.005.

2.2

	Number of trailing zeros	Number of significant figures
20.201	None	5
20.210	1	5
20.000 2	None	6
20.0	2	3
120.	1	3

2.3 (a) 1.23×10^{2} (b) 6×10^{-5}
(c) 6.0×10^{-5} (d) $1.002\,0 \times 10^{3}$

2.4 Each small division represents 1 mm. You can estimate between any two adjacent lines, which means you are estimating to 0.1 mm. This gives an uncertainty of (b) ±0.05 mm. The diameter of the coin is (c) 5.00×10^1 mm.

2.5 In 1.20×10^{-7}, the exponent -7 instructs you to move the decimal point seven places to the left, filling in with zeros the empty columns you generate:

Fill with zeros

.1.20 = .000 000 120

(The zero shown to the left of the decimal point in the WorkPatch—0.000 000 120—is there merely to make it easier to spot the decimal point. This zero is usually shown: 0.5, 0.003, and so forth.)

2.6 (a) In 6.0000×10^{11}, the exponent instructs you to move the decimal point 11 places to the right, filling with zeros any empty columns generated:

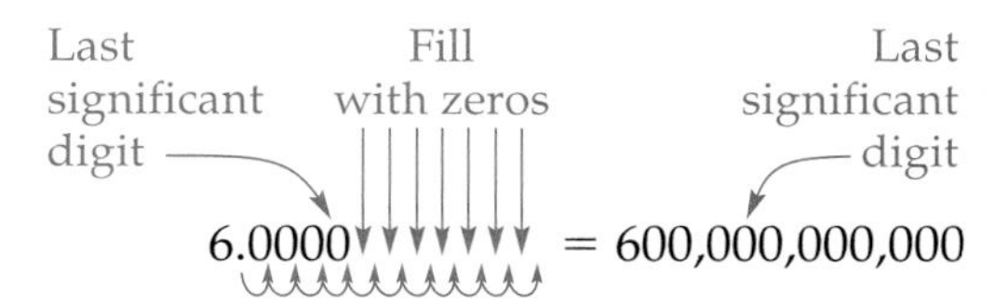

(b) The last significant digit is the fourth zero from the 6. This zero is in the 10-millions column, and so the uncertainty is ±5 million.

2.7 1.1×10^2 miles/hour

2.8 Because it takes 1000 mL to get 1 L, you have to fill your 100-mL beaker ten times to get 1 L. To get 2.5 L, you have to fill your beaker 25 times.

2.9 Form B is denser because the atoms are packed more closely together (notice that the gaps between atoms in B are smaller than the gaps between atoms in A).

2.10 The specific heat of water is much larger than the specific heat of aluminum. Therefore it takes more heat to warm up a given mass of water than it does to warm up the same mass of aluminum. For the water, you need to add

$$1\,\cancel{g} \times \frac{4.184\ J}{\cancel{g} \cdot \cancel{C^\circ}} \times 1\,\cancel{C^\circ} = 4.184\ J$$

of heat energy to raise the temperature from 25°C to 26°C. To increase the temperature of the aluminum from 25°C to 26°C, you must add

$$1\,\cancel{g} \times \frac{0.901\ J}{\cancel{g} \cdot \cancel{C^\circ}} \times 1\,\cancel{C^\circ} = 0.901\ J$$

IB	IIB	IIIA	IVA	VA	VIA	VIIA	VIIIA
							2 He 4.003
		5 B 10.811	6 C 12.011	N 14.007	O 15.999	9 F 18.998	10 Ne 20.180
		13 Al 26.982	14 Si 28.086	15 P 30.974	16 S 32.066	17 Cl 35.453	18 Ar 39.948
29 Cu 63.546	30 Zn 65.39	69.723	32 Ge 72.61	33	34 Se 78.96	35 Br 79.904	36 Kr 83.8
47 Ag 107.868	48 Cd 112.411	49 114.82	50 Sn 118.71	Sb 121.75	52 Te 127.60	53 I 126.906	54 Xe 131.29
79 Au	80 Hg 22.990	81 Tl 204.383	82 Pb 207.2	83 Bi 208.980	84 Po 209	85 At 210	86 Rn 220
111 Uuu 272	112 Uub 277		114				

CHAPTER 3

The Evolution of Atomic Theory

3.1 Dalton's Atomic Theory

Recall from Chapter 1 the battle between the philosophies of Democritus and Aristotle that took place around 400 B.C. Aristotle proposed that matter was continuous and could be infinitely subdivided into smaller and smaller particles. Democritus postulated that matter was ultimately made up of tiny, indivisible particles that he called *atoms*. Aristotle's view won the day, and it took more than 2000 years before the atomic theory of matter made a comeback. Historians generally credit the beginning of modern atomic theory to the work of a mild-mannered English schoolteacher named John Dalton in the early 1800s, and it is here that we shall begin.

John Dalton's atomic model was just that, a model in the true spirit of the scientific method. It was formulated as an explanation for a number of "laws" about how matter behaves in a chemical reaction. These laws were the result of careful experimental work done in the latter part of the 1700s and were based on observations of a great number of chemical substances and their reactions. These laws are the foundation on which modern atomic theory is based.

The first of them is the **law of conservation of matter**:

Law of conservation of matter: When a chemical reaction takes place, matter is neither created nor destroyed.

Today we accept this law as fact (with minor modifications). However, that matter is conserved in any chemical reaction was not at all obvious at first because many reactions *seem* to either create or destroy matter. Two examples are shown in the illustration below.

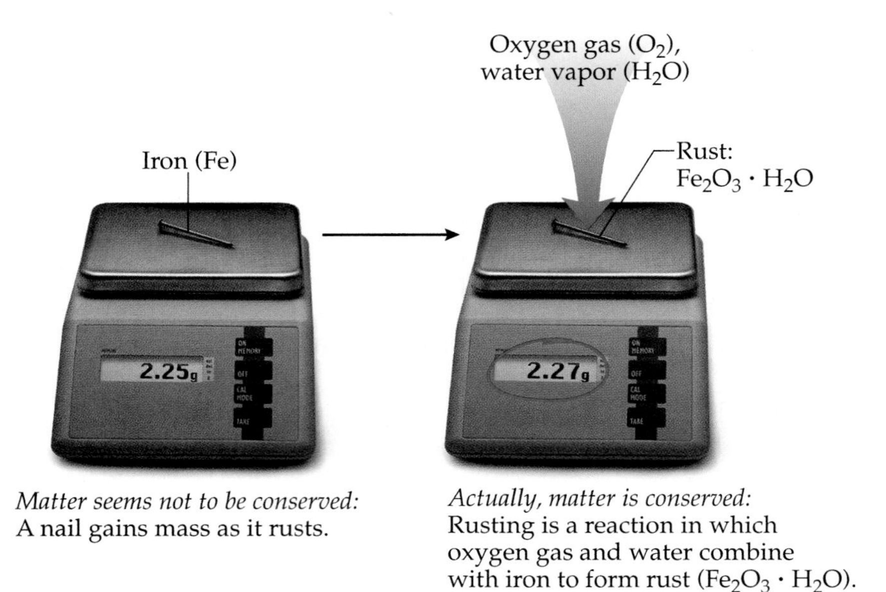

Matter seems not to be conserved: A nail gains mass as it rusts.

Actually, matter is conserved: Rusting is a reaction in which oxygen gas and water combine with iron to form rust ($Fe_2O_3 \cdot H_2O$).

Matter seems not to be conserved: A glass of water with an Alka-Seltzer tablet loses mass as the tablet dissolves.

Actually, matter is conserved: The Alka-Seltzer reacts with water to produce carbon dioxide gas, which escapes from the beaker.

The issue was set straight in 1775 by the French chemist Antoine Lavoisier, along with his wife, Marie, who worked with him in the laboratory. They carried out reactions in sealed vessels that allowed no matter—including gases—to enter or leave during the reactions. Under these conditions, they found that matter is always conserved.

The second law that led Dalton to postulate his atomic theory is known as the **law of constant composition** (also sometimes called the **law of definite proportions**).

Law of constant composition: Multiple samples of any pure chemical compound always contain the same percent by mass of each element making up the compound.

For example, if a 50.0-g sample of pure water (H_2O) is decomposed into its component elements, you will obtain 5.6 g of hydrogen gas and 44.4 g of oxygen gas. The **percent by mass** of these elements is therefore

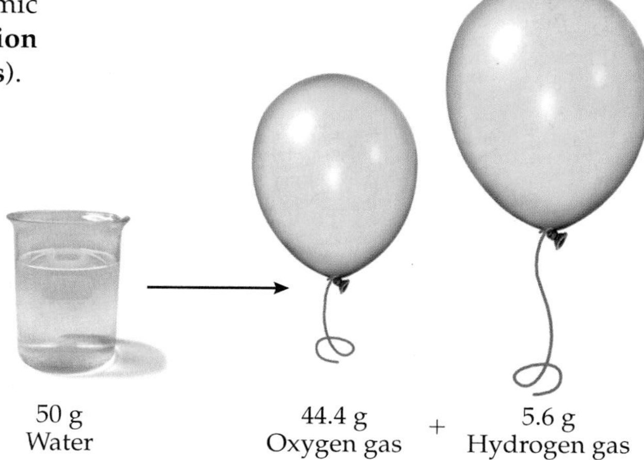

$$\frac{\text{Mass of hydrogen in compound}}{\text{Total mass of compound}} = \frac{5.6\text{ g}}{50.0\text{ g}} \times 100 = 11.2\%$$

$$\frac{\text{Mass of oxygen in compound}}{\text{Total mass of compound}} = \frac{44.4\text{ g}}{50.0\text{ g}} \times 100 = 88.8\%$$

These same percentages are found in any sample of pure water, no matter where it comes from or what the size of the sample is.

The law of constant composition also holds when a chemical compound is synthesized from its elements. For example, no matter how water is made, no matter how much hydrogen and oxygen are combined with each other, the water produced always has the same percentage of oxygen and hydrogen. In the following WorkPatch, water is made first by combining just the right amounts of hydrogen and oxygen (so that they are both used up) and then again by cutting the amount of hydrogen in half.

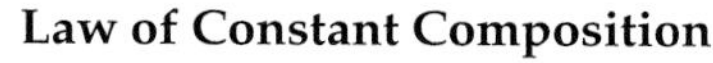

All samples of pure water have the same composition by mass: 11.2% hydrogen; 88.8% oxygen

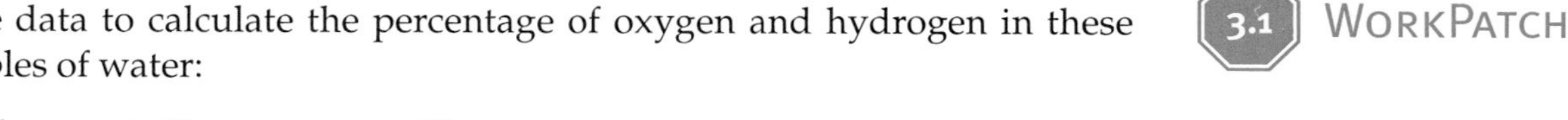

Use these data to calculate the percentage of oxygen and hydrogen in these two samples of water:

Hydrogen	+ Oxygen	$\longrightarrow$ Water
100.0 g	793.6 g	893.6 g

(a) Is the law of conservation of matter obeyed?
(b) What is the percent hydrogen in water?
(c) What is the percent oxygen in water?

Hydrogen	+ Oxygen	$\longrightarrow$ Water
50.0 g	793.6 g	446.8 g + 396.8 g leftover oxygen

(d) Is the law of conservation of matter obeyed?
(e) What is the percent hydrogen in water?
(f) What is the percent oxygen in water?

Did you get the same percentages for both samples? If you did (and you should have), you have demonstrated that the law of constant composition is obeyed.

To Dalton, constant composition was very good evidence that the elements did indeed exist in the form of atoms. When you formed a chemical compound, he reasoned, you were combining atoms of the constituent elements. An atom of one element would always combine with a fixed number of atoms of another element, that fixed number depending on the identity of the elements and on the identity of the compound being made.

Practice Problems

3.1 A student makes a compound of sulfur and oxygen. She uses 5.00 g of sulfur and 4.99 g of oxygen, and all of the elemental substances are completely used up. What is the percent sulfur in the compound?

Answer: (5.00 g S ÷ 9.99 g compound 1) × 100 = 50.1% S

3.2 A student makes two different compounds of nitrogen and oxygen using the masses shown below. In both cases, all of the elemental substances are completely used up.

Compound 1   10.0 g nitrogen + 11.42 g oxygen
Compound 2   10.0 g nitrogen + 22.84 g oxygen

What is the percent oxygen in each compound?

Answer: Compound 1, 53.3% oxygen; compound 2, 69.5% oxygen

3.3 A 101.96-g sample of a compound of aluminum and oxygen is 47.1% by mass oxygen.
(a) What is the percent by mass aluminum in this compound?
(b) Of the 101.96 g of this compound, how many grams are aluminum?

Answer: (a) 52.9% aluminum   (b) 53.9 g aluminum

Based on the law of conservation of matter and the law of constant composition, Dalton formulated his atomic theory, which can be summarized in five short statements:

Dalton's atomic theory

1. All matter is made up of atoms (small, indivisible, indestructible, fundamental particles).
2. Atoms can neither be created nor destroyed (they persist unchanged for all eternity).
3. Atoms of a particular element are alike (in size, mass, and properties).
4. Atoms of different elements are different from one another (different sizes, masses, and properties).
5. A chemical reaction involves either the union or the separation of individual atoms.

Molecules of the chemical compound we call water are always formed from one atom of oxygen and two atoms of hydrogen, H_2O. Why do atoms of one element combine with only certain numbers of atoms of other elements in forming chemical compounds? Dalton wanted a physical model—a picture of the atom—that fit his theory and helped explain it. Chemists of the time ultimately envisioned a ball-and-hook model that allowed atoms to combine with one another in particular ways. This model described atoms of different elements as balls of different sizes (and masses) with a characteristic number of hooks embedded in them.

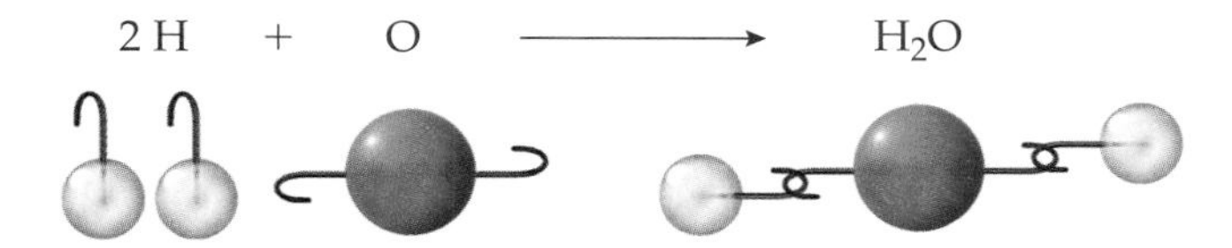

The number of hooks in each type of atom would account for its chemical reactivity and explain why elements of certain atoms characteristically combined with only certain numbers of other atoms. The atoms "reacted" with one another until their hooks were filled.

Today we know that atoms don't have hooks, but for the early 1800s, this was a good model. It made visual sense of the behaviors summarized by the two conservation laws Dalton knew about. Dalton claimed that if you accepted the five statements of his theory as a viable model of reality, the laws of conservation of matter and constant composition could be easily explained and accounted for.

Many scientists at the time did not accept Dalton's theory or his model and continued to argue that atoms did not exist. Needless to say, however, no other proposed theory was able to account for the observed laws.

Law of Conservation of Matter

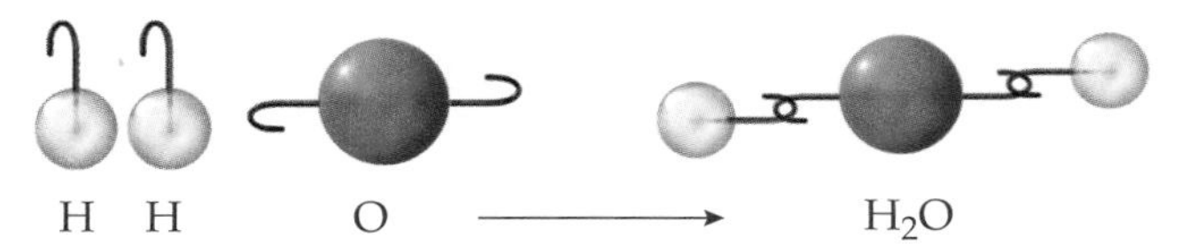

Matter is conserved in a chemical reaction because the reaction consists simply of atoms either hooking together or unhooking.

Nearly 200 years after Dalton proposed his atomic model, we now know that none of the five statements of his theory is *entirely* true. Atoms are not the most fundamental of particles—they are composed of the even smaller particles we call electrons, protons, and neutrons. Atoms can be created and destroyed, but a nuclear process is required to do so. We shall encounter these and other exceptions to Dalton's five statements as we proceed through this book. None of these later findings should be taken as diminishing Dalton's accomplishments, however. For the time, his was a superb model. It adequately explained experimental observations, and it laid the foundation for further developments in atomic theory. Of course, even the inability of Dalton's detractors to come up with any plausible alternatives did not make the acceptance of the atom an overnight phenomenon. It took nearly 100 years for the last of the holdouts to become convinced (or to die off!), but by the beginning of the twentieth century the concept of the atom was firmly established.

Practice Problems

3.4 You visit France, where a local salesperson tries to sell you special water from a mountain spring. This water is supposed to be special because it has the formula H_3O. What law does this claim violate?

Answer: The law of constant composition, which tells us that a compound has the same composition no matter where it comes from. Water is always H_2O.

3.5 When coal (essentially pure carbon) is burned, it is combining with atmospheric oxygen to produce carbon dioxide gas. Nevertheless, when the coal burns, it seems to disappear. Why is this not a violation of the law of conservation of matter?

3.6 Using Dalton's ball-and-hook atomic model, sketch an explanation of how it is possible for hydrogen to combine with oxygen to form two different compounds, water (H_2O) and hydrogen peroxide (H_2O_2).

3.2 Development of a Model for Atomic Structure

Once the existence of atoms was accepted, the obvious next question was, "What do atoms look like?" A series of extremely clever experiments done by physicists J. J. Thomson, James Chadwick, and others decisively showed that atoms are not the fundamental particles of matter but are themselves composed of even smaller particles.

In 1897, Thomson discovered the first *subatomic* particle to be identified, the **electron**. He found that all atoms contained electrons and that the electrons from all atoms were identical. These electrons were very small, lightweight particles indeed, having only 1/1836 the mass of a hydrogen atom, the smallest and lightest of all atoms. In addition, the electron had a negative electrical charge, which for convenience was assigned a numerical value of -1.

Just 10 years later, in 1907, Thomson and E. Goldstein found another subatomic particle that was present in all atoms, the **proton**. This particle was much heavier than the electron, with a mass almost equal to that of a hydrogen atom. The proton was found to have a positive electrical charge equal in magnitude but opposite in sign to that of the electron. This charge was assigned a value of $+1$.

Twenty-five years later, Chadwick demonstrated the existence of a third subatomic particle, the **neutron**. It had about the same mass as a proton but lacked any charge (it was electrically neutral, hence the name neutron). The lack of a charge made the neutron much more difficult to study and helps to explain why its discovery came so much later.

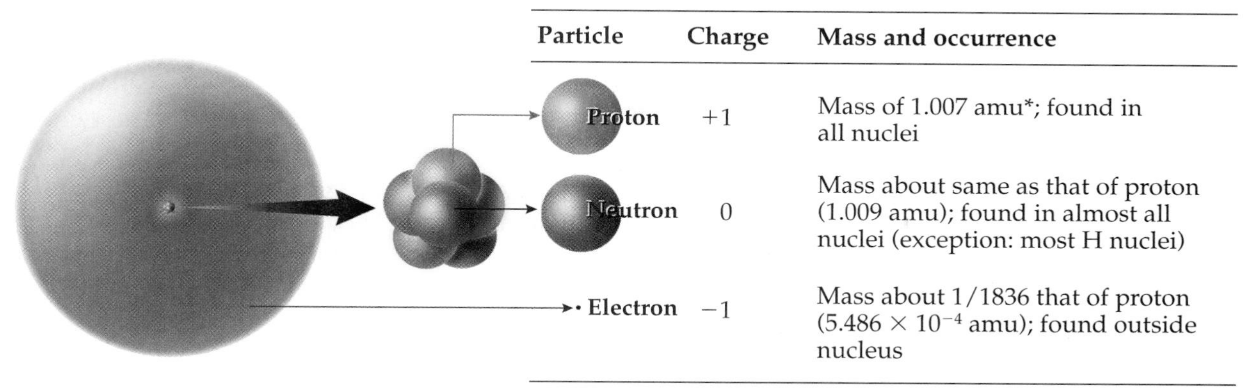

Particle	Charge	Mass and occurrence
Proton	+1	Mass of 1.007 amu*; found in all nuclei
Neutron	0	Mass about same as that of proton (1.009 amu); found in almost all nuclei (exception: most H nuclei)
Electron	−1	Mass about 1/1836 that of proton (5.486×10^{-4} amu); found outside nucleus

*This unit will be explained in Section 3.4.

Once it was established that all atoms are composed of electrons and protons, the path of investigation turned to the exploration of the internal structure of the atom. How could only three subatomic particles be put together to make all the different atoms in the periodic table?

The first model was proposed by Thomson shortly after his discovery of the electron. He knew two basic facts about the atom: (*1*) atoms contain small, negatively charged particles called electrons and (*2*) the atoms of an element behave as if they have no electrical charge at all—they are electrically neutral. Thomson reasoned that there must be something in the atom that carries a positive charge to neutralize the electrons (protons had yet to be discovered). Thomson's model for the atom consisted of a "cloud of positive electricity" in which the negatively charged electrons were embedded. This became known as the plum-pudding model because it reminded many scientists of a classic English pudding.

Thomson Plum-Pudding Model of the Atom

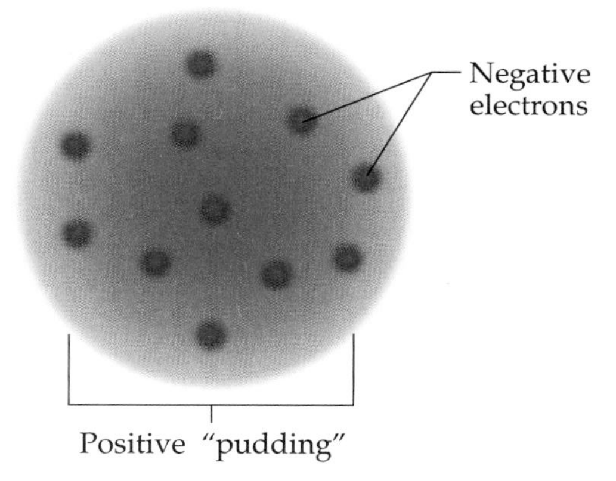

It may seem a silly model today, but it fulfilled all the requirements of a model given the data available at that time. It was consistent with all the known facts about atoms and successfully accounted for their neutral behavior. What ultimately banished this model to the trash heap was an experiment performed just a few years later, in 1909, by the British physicist Ernest Rutherford.

3.3 The Nucleus

Rutherford's experiment, like many of the other great experiments that have shaped modern chemistry and physics, was beautifully simple in its concept. Rutherford had been studying **alpha (α) particles**, small chunks of positively charged matter that are spontaneously and randomly given off by many naturally occurring radioactive elements, such as radium, polonium, and radon. We now know that any alpha particle consists of two protons and two neutrons, giving it a +2 charge. Rutherford did not know this precisely, but he did

know that alpha particles were about 7000 times more massive than electrons, had a positive charge, and were ejected from certain types of matter at very high speeds—about 9300 miles/s (1.5×10^4 km/s), or approximately one-twentieth the speed of light. Rutherford thought of alpha particles as positively charged "bullets" that could be fired at various targets. The targets he chose were thin foils of pure metals. One was made of pure gold, a metal so soft and malleable that it could be hammered into a thin film only a couple of thousand atoms thick. He decided to "fire" the alpha particles at the gold foil and "see" what happened.

Of course, he could not actually see either the alpha particles or the individual atoms of gold in the foil target. However, a screen made of glass coated with zinc sulfide was known to glow (fluoresce) with a tiny point of green light when struck by an alpha particle. Rutherford simply positioned a zinc sulfide screen behind his gold target so that any alpha particle that passed through the gold target would strike the screen and cause it to glow, thus revealing the particle's path. Because the plum-pudding model of the atom was in vogue at the time, Rutherford expected the particles to cruise right through the gold atoms pretty much undisturbed. After all, the electrons in the gold atoms were much too small and lightweight to cause any resistance, and the positive charge in each gold atom was spread out in a thin cloud throughout the entire volume of the atom. If this model of the atom were correct, the fast-moving alpha particles should go straight through the gold foil and hit the zinc sulfide screen in a straight-line trajectory from the particle source. To make sure that he

Rutherford's Alpha-Particle Experiment

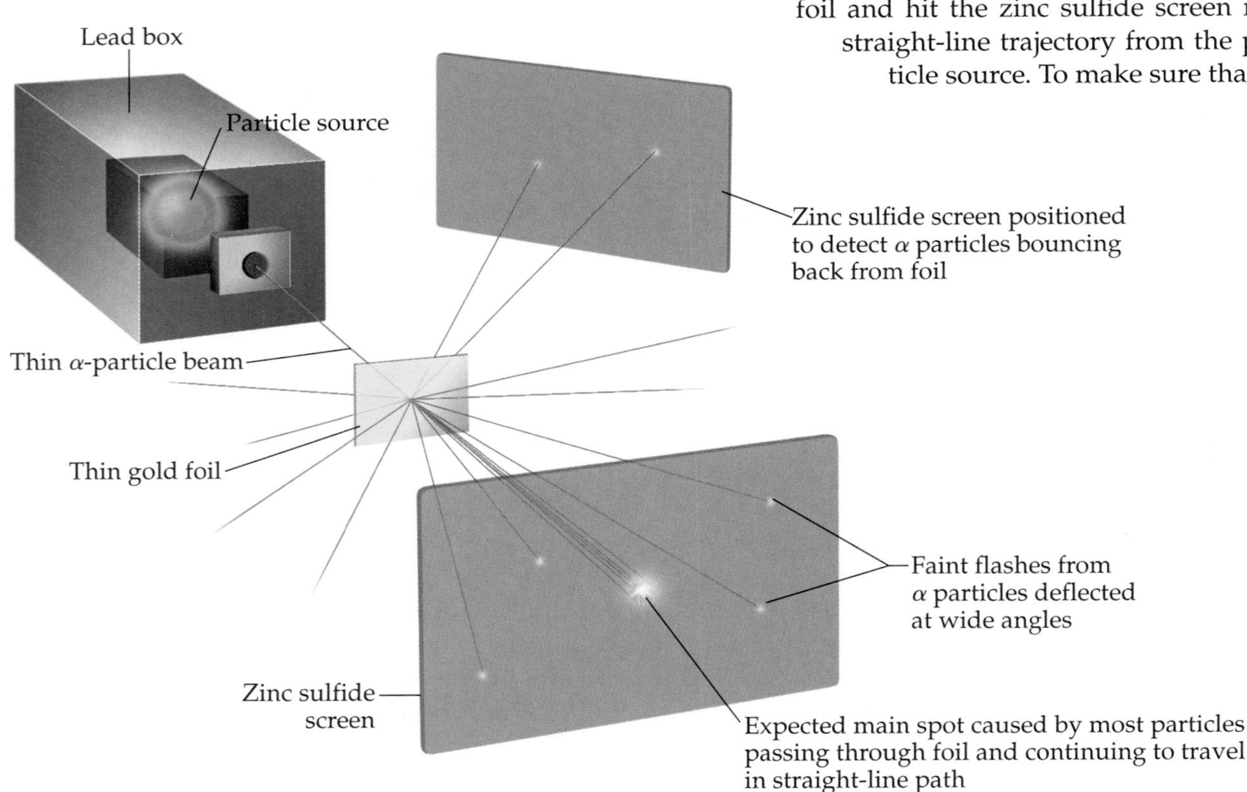

and his students missed nothing, they sat in a pitch-black room for an hour before each experiment so that their eyes would be ultrasensitive to any green flashes on the fluorescent screen. What they saw was *almost* what they expected.

Most of the alpha particles did indeed go straight through the foil as predicted. But a few, barely noticeable as green flashes on the zinc sulfide screen, were deflected at wide angles. Some alpha particles even bounced right back toward the source, causing a glow on a screen positioned alongside the source! This was incredible. In Rutherford's words, "It was . . . as if you fired a 15-inch shell at a piece of tissue paper and it came back and hit you." If the results were to be believed, the plum-pudding model of how the atom was put together could not possibly be correct. A new model for the internal structure of the atom would have to be created to account for the results of Rutherford's experiment. In particular, two observations needed to be explained: (*1*) Most of the alpha particles went straight through the foil, and (2) a few (about 1 in 20,000) were deflected from a straight-line path, some by very extreme angles.

The fact that *any* of the fast-moving alpha particles were deflected by the gold atoms in the foil led Rutherford to conclude that there must be something small and massive inside each gold atom. This conclusion was inescapable. Except for this tiny, massive something inside each atom, the atom must be mostly empty space. (It was through this empty space that all the undeflected alpha particles had passed.) It took Rutherford more than a year to come up with this model of the atom, and close analysis of his results told him that this tiny, massive something must be positively charged.

What led Rutherford to the conclusion that this massive something inside the atom was very small?

Thus in attempting to verify the plum-pudding model of the atom, Rutherford ended up having to discard it and replace it with a model that showed the atom as being mostly empty space but with something tiny, massive, and positively charged inside it. This something he called the atom's **nucleus**, which we now know holds an atom's protons and neutrons. The "empty space" that makes up most of the volume of an atom is occupied by the much less massive electrons that are also present in the atom.

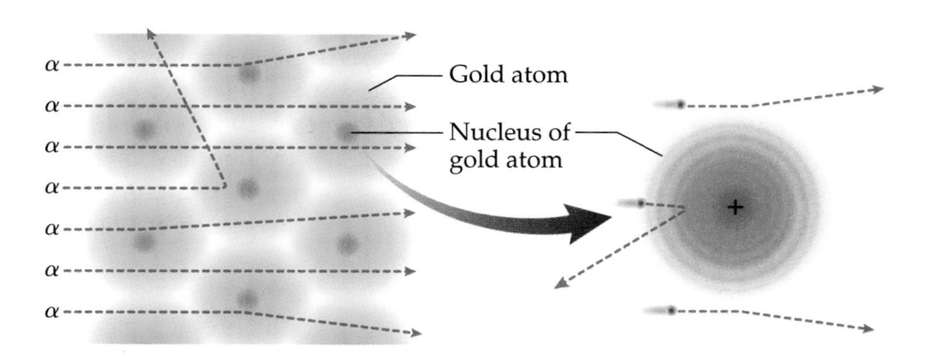

Based on the number of alpha particles deflected and their angles of deflection, Rutherford was able to calculate the relative sizes of the "electron space" and the nucleus. He calculated that the nucleus was about 10^{-13} cm in diameter, while the diameter of the entire atom was about 10^{-8} cm. Thus, the

The Houston Astrodome

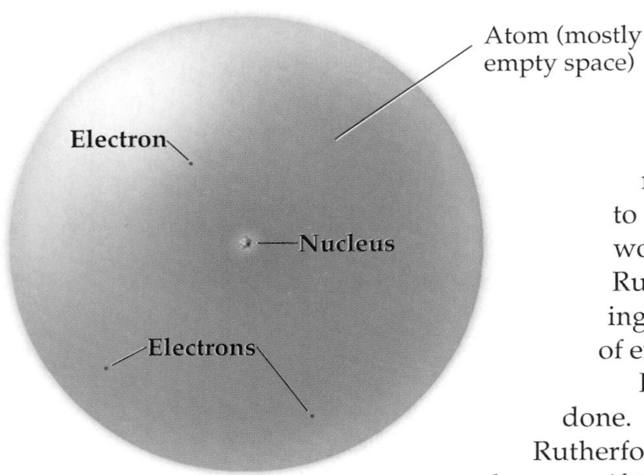

nucleus and the atom differ in size by about 10^5—meaning that an atom is 100,000 times larger than its nucleus.

Rutherford's atom was mostly empty space. At the center of the spherical atom was a tiny, dense nucleus; the rest of the volume was occupied by the electrons, which moved around the nucleus in some fashion. A modern-day comparison to Rutherford's model might be a Ping-Pong ball suspended in the middle of the Houston Astrodome. The Astrodome represents the total volume of the atom, and the Ping-Pong ball represents its nucleus. The electrons would be equivalent to a few mosquitoes flying around inside the stadium.

It certainly was not easy for Rutherford to discard an accepted model. Another, less thorough scientist might have been tempted to ignore the few faint flashes of green light that showed up at large deflection angles or might even have neglected to design the experiment to look for them, because "obviously" all the action would be in a straight-line path. In this case, however, Rutherford and the scientific method triumphed, replacing an old model with a new one and inviting new sets of experiments to challenge and further refine it.

Rutherford realized that more work needed to be done. For instance, the electrons presented a problem. Rutherford assumed they had to be continuously moving because, if they stopped, they would be sucked into the nucleus by its strong positive charge (opposite charges attract each other, even on the atomic level). But *how* were they moving? According to the physics known in Rutherford's day, an electron moving around a nucleus must radiate energy. Thus, the electrons moving about inside an atom should quickly lose energy, slow down, and spiral into the nucleus. In other words, the principles of physics demanded that a Rutherford type of atom should collapse in on itself within a fraction of a second. Because it was pretty clear that all matter in the universe was in no immediate danger of collapsing, Rutherford knew that either something was wrong with his model or something was wrong with physics as it was then understood. He had difficulty accepting either proposition, and so, as you might imagine, he put forth his model of the atom with a great deal of trepidation.

Rutherford knew that his model was in need of refinement and modification. What he did not know was that the solution to the problem was going to be so revolutionary, so inconceivable, that even the scientists who developed it found it almost impossible to accept. Albert Einstein went to his grave not fully believing in it and working feverishly to find the flaw in the analysis. The laws of classical physics, used for centuries to calculate the force of gravity, the motions of the planets, and the trajectory of missiles, just did not work when applied to particles the size of an atom. To solve the problem of the atom, classical physics itself—the foundation of modern science—would have to be tossed away and replaced with something else. Over the next 20 years (1910–1930), a revolution in quantum physics (the physics of atomic and subatomic particles) was born.

3.4 The Structure of the Atom

Let's take stock of where we are with regard to the subatomic pieces that go into making an atom. There are the relatively massive, positively charged (+1) protons (p) that reside at the center of the atom in a tiny nucleus. There are the much less massive, negatively charged (−1) electrons (e^-) that move about outside the positive nucleus. A fundamental law of physics (quantum as well as classical) is that oppositely charged particles attract each other. The significance of this is that the electrons can't leave the atom, at least not without the input of a good deal of energy. They are attracted to the nucleus and so are kept from wandering away.

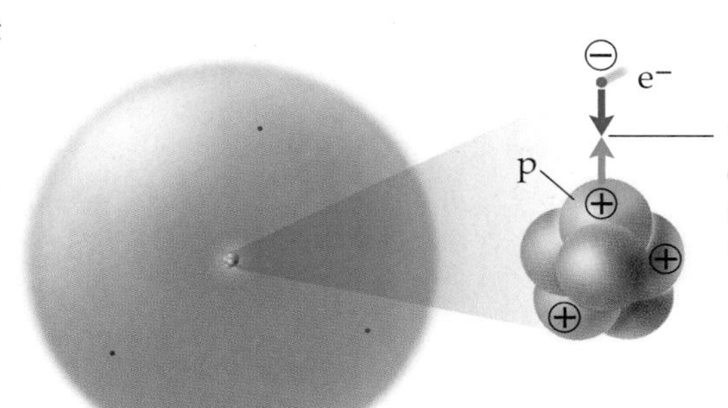

The attraction between ⊖ and ⊕ charges holds the electrons in the atom.

The third subatomic particle is the neutron (n). This particle has no charge; it is electrically neutral. Neutrons have approximately the same mass as protons.

Atomic Number

The number of protons in the nucleus of an atom is called the **atomic number** (represented by the letter *Z*) of that atom. The atomic number is extremely important because it alone determines the identity of the atom. An atom with one proton (atomic number 1) is always hydrogen, an atom with six protons (atomic number 6) is always carbon, and so on.

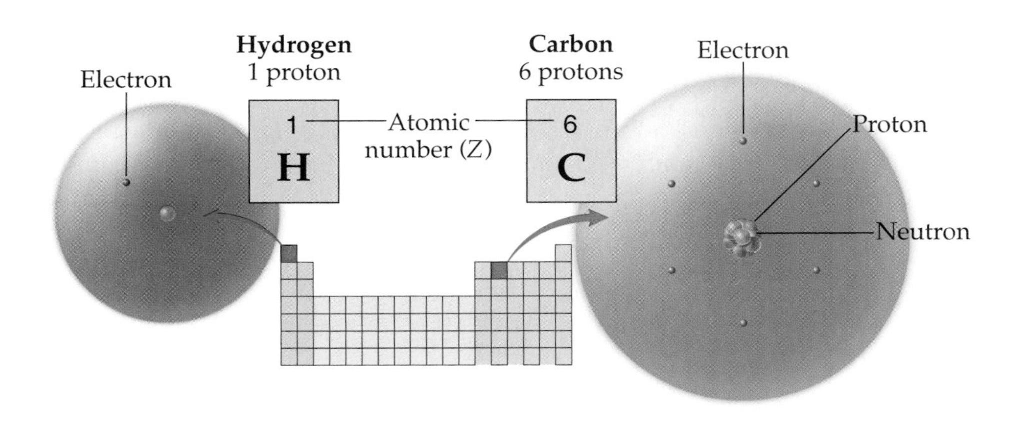

The atomic number is so important that the periodic table of the elements is arranged according to increasing atomic number. Examine the atoms in the figure at the bottom of page 85 and then answer the following WorkPatch.

3-3 WORKPATCH From the figure, what do you think is the rule for determining the number of electrons a neutral atom contains?

Your answer to the WorkPatch should help you understand how the overall neutrality of an atom depends on the number of protons and electrons it contains. (If you are not sure about this, check the answer at the end of the chapter before going on.)

Now, what about neutrons? These are fairly massive subatomic particles that reside in the nucleus. Most hydrogen atoms have no neutrons; all other elements have at least one. Carbon atoms, with six protons, usually have six neutrons, but there is no simple relationship between the number of neutrons and protons in the nucleus of an atom. This is especially true for the heavier atoms in the periodic table. For instance, a typical uranium nucleus, with 92 protons (atomic number 92), contains 146 neutrons.

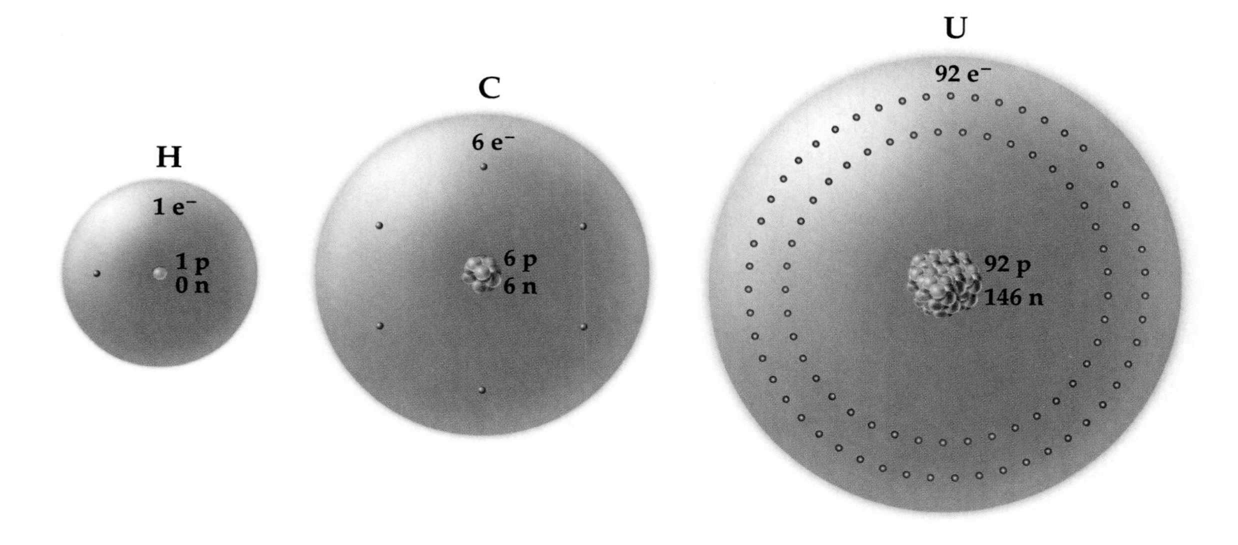

MASS NUMBER

Along with the atomic number (Z), chemists have defined another atomic quantity, called *mass number*. An atom's **mass number** is simply the number of protons plus the number of neutrons in its nucleus. Thus the mass numbers for H, C, and U are

H	1 p	**C**	6 p	**U**	92 p
Mass	+ 0 n	Mass	+ 6 n	Mass	+ 146 n
number ⟶	1	number ⟶	12	number ⟶	238

Let's see if you're catching on to this. Complete the following WorkPatch, and then we'll discuss it.

3·4 WORKPATCH

Sketch the carbon atom in the following three ways, showing the correct numbers of protons, neutrons, and electrons. All three of your carbon atoms should be electrically neutral.

(a) Make the mass number equal to 12.
(b) Make the mass number equal to 13.
(c) Make the mass number equal to 14.

If you did things correctly, all three sketches should have six protons in the nucleus (because these atoms are all carbon) and six electrons outside the nucleus (to keep each atom electrically neutral). The only difference, therefore, must be in the number of neutrons in the nucleus: sketch (a) should have six neutrons, sketch (b) should have seven neutrons, and sketch (c) should have eight neutrons. The question is, what should we call these three slightly different versions of the same element? The answer—in all three cases—is carbon. They must all be carbon because they all have an atomic number of 6. It is the atomic number and *only the atomic number* that determines the elemental identity of an atom. And yet these three carbon atoms are not identical because atom (b) has an extra neutron and atom (c) has two extra neutrons. These three atoms are examples of **isotopes**. Isotopes are different versions of the same element (same atomic number) that contain different numbers of neutrons in their nuclei (different mass numbers). Because they are the same element, they have identical chemical properties (there are some subtle differences that are beyond the scope of this book).

Coal contains all isotopes of carbon

Chemists represent the different isotopes of an element by showing both the mass number and the atomic number along with the symbol for the element. The mass number is written as a superscript to the left of the symbol, and the atomic number is written as a subscript:

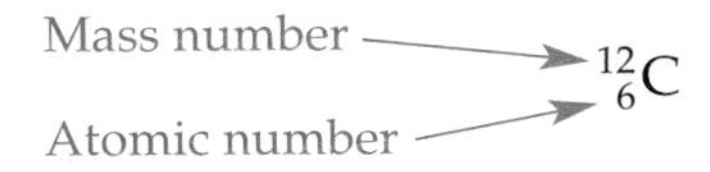

Very often, the atomic number is omitted from this notation. Thus, the three isotopes of carbon can be written ^{12}C, ^{13}C, and ^{14}C. Because these are all carbon atoms, the "missing" atomic number is, by definition, known to be 6. These isotopes can also be written carbon-12, carbon-13, and carbon-14. When you are talking about an isotope, say either the letter or the name of the element plus the mass number. The isotope ^{12}C, for example, is pronounced either "c-twelve" or "carbon-twelve."

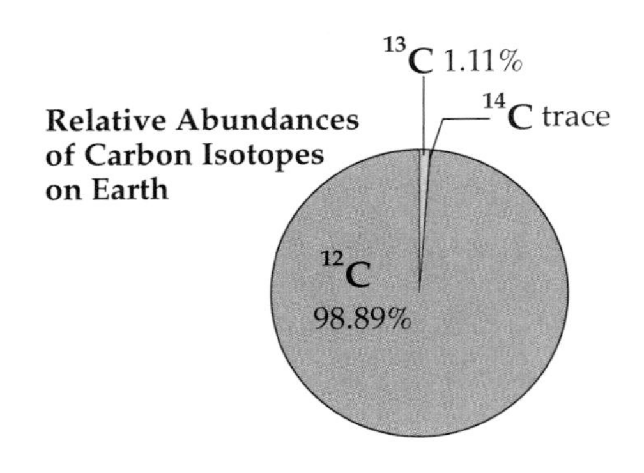

All three of these carbon isotopes exist in nature, although not in equal abundance. On this planet, 98.89% of all the carbon is ^{12}C, 1.11% is ^{13}C, and only a trace is ^{14}C. The relative abundances of these isotopes are most likely different on other planets. Analysis of the percentage of each isotope in a rock is one way chemists have of determining whether the rock is terrestrial or from some extraterrestrial meteorite.

Nearly all the elements exist as two or more naturally occurring isotopes. Even the simplest of the elements, hydrogen, comes in three isotopic forms. Although they are really all hydrogen atoms (because they all have one proton in their nucleus), they are given different names, as shown below. (This multiple naming is not done for isotopes of any other element.)

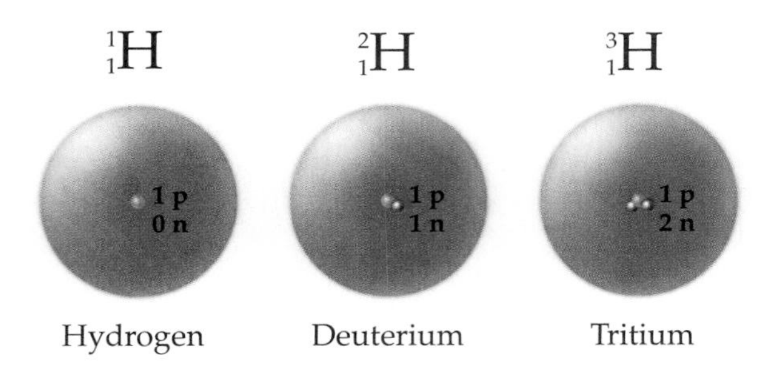

While isotopes of an element have essentially identical chemical properties, there is a single important difference between them. Recall that neutrons and protons carry most of an atom's mass. Because isotopes differ in the numbers of neutrons they contain, isotopes have different masses. Deuterium (^{2}H) is about twice as heavy as ^{1}H, for instance, and tritium (^{3}H) is heavier than deuterium. For this reason, ^{2}H is sometimes referred to as *heavy hydrogen,* and D_2O (the symbol D is used for a deuterium atom) is called *heavy water.*

Together, the mass number and atomic number tell you quite a bit about an atom, but what if you are given only one of these numbers? Try the following WorkPatch.

Suppose you are told that an atom has a mass number of 16. Is this enough information to identify the element ? Why?

If you are not sure of the answer, think about what you need to know in order to determine an atom's identity. Now suppose you are given additional information, as in the next WorkPatch.

Suppose you are told that an atom with a mass number of 16 contains nine neutrons. Can the atom be identified now? Which element is this?

This time you should be able to determine the atom's elemental identity (check the answer at the end of the chapter). The moral of this story is that to identify an atom, you need to know its atomic number, and you can determine this number in a variety of ways provided sufficient information is given.

Practice Problems

3.7 Give the full atomic symbol for an atom that has 16 neutrons and an atomic number of 15.

Answer: The atomic number 15 identifies the element as phosphorus (P). The mass number is 15 + 16 = 31. The full atomic symbol is $^{31}_{15}P$.

3.8 Bromine (Br) has two abundant isotopes, one with 44 neutrons and the other with 46 neutrons. Give the full atomic symbols for both isotopes.

3.9 How many electrons does a neutral atom of each isotope in Practice Problem 3.8 have?

3.10 Fill in this table:

	$^{14}_{7}N$	$^{24}_{12}Mg$	$^{23}_{11}Na$	$^{59}_{26}Fe$
Mass number	?	?	?	?
Atomic number	?	?	?	?
Number of protons	?	?	?	?
Number of neutrons	?	?	?	?
Number of electrons	?	?	?	?

Atomic Mass

We have looked closely at atomic number and mass number, but neither tells us the actual mass of an atom. For example, although the mass number of tritium ($^{3}_{1}H$) is 3, the actual mass of this atom is not 3 (not 3 g, not 3 lb, not 3 anything). So what is the actual mass of a tritium atom? The actual mass of any atom is formally called its **atomic mass**.

To begin our discussion, let's look at the atomic masses for some isotopes of hydrogen, carbon, and magnesium:

Isotope	Atomic mass (amu)	Isotope	Atomic mass (amu)
^{1}H	1.007 83	^{12}C	12
^{2}H	2.014 10	^{13}C	13.003 35
^{3}H	3.016 05	^{24}Mg	23.985 04

Do you notice something about the atomic masses? Look closely. First, for all isotopes except ^{12}C, the mass number does not equal the atomic mass, although they are close. Second, the atomic masses are given in units called *atomic mass units* (abbreviation amu). An **atomic mass unit** (also known as a *dalton*, Da) is defined as exactly one-twelfth the mass of a $^{12}_{6}C$ atom and is equal to $1.660\,54 \times 10^{-24}$ g:

$$1 \text{ amu} = \tfrac{1}{12} \text{ the mass of } ^{12}_{6}\text{C atom} = 1.660\,54 \times 10^{-24} \text{ g}$$

Because the atomic mass unit is defined this way, atomic masses are often called *relative atomic masses* (relative as in "compared to"). In other words, you can think of these masses as telling you how massive an atom is compared to a ^{12}C atom. For example, one ^{1}H atom has an atomic mass of 1.007 83 amu. This means that an ^{1}H atom is 1.007 83 ÷ 12, or roughly one-twelfth, as massive as a ^{12}C atom. One ^{24}Mg atom has an atomic mass of 23.985 04 amu, which means that a ^{24}Mg atom is 23.985 04 ÷ 12, or roughly twice, as massive as a ^{12}C atom.

Of course, because you know how many grams 1 amu equals, you can calculate the mass of an atom in grams. To see this, let's calculate the mass in grams of one ^{12}C atom, *the only atom* whose mass number and atomic mass are equal:

$$12\ \cancel{\text{amu}} \times \frac{1.660\ 54 \times 10^{-24}\ \text{g}}{\cancel{\text{amu}}} = 1.992\ 65 \times 10^{-23}\ \text{g}$$

3.7 WORKPATCH An atom is determined to be 4.015 times more massive than ^{12}C. What is the atomic mass of this atom?

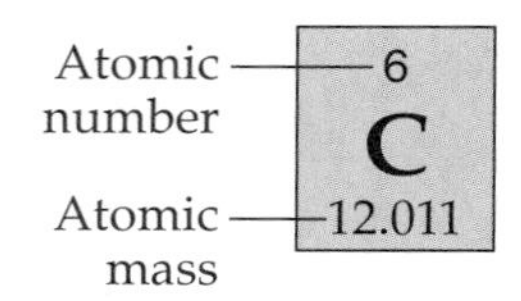

We have one more topic to cover before we end our discussion of atomic mass. We said that an atom of the ^{12}C isotope has an atomic mass of exactly 12 amu by universal agreement. But if you take a look at the periodic table, you will see that carbon's atomic mass is listed as 12.011, not 12. This atomic mass takes into account the fact that not all carbon atoms in nature are ^{12}C atoms. As we have seen, carbon consists predominantly of two isotopes, ^{12}C (98.89% natural abundance) and ^{13}C (1.11% natural abundance).

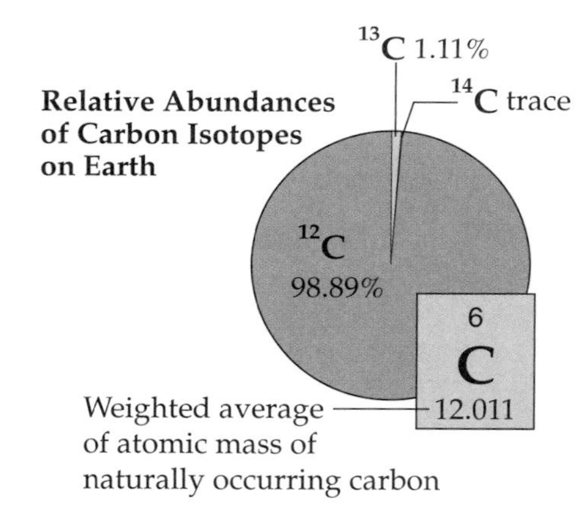

Any naturally occurring sample of carbon atoms will contain these two isotopes in these abundances. The atomic mass of carbon reported in the periodic table is called a *weighted average* of the atomic masses of these two isotopes. Each isotope's atomic mass is multiplied by its percent abundance (in decimal form, meaning the percent divided by 100), and then the results are summed:

$$\begin{aligned}\text{Weighted average atomic mass of naturally occurring carbon} &= \overbrace{(12\ \text{amu})(0.9889)}^{^{12}_{6}\text{C portion}} + \overbrace{(13.003\ 35\ \text{amu})(0.0111)}^{^{13}_{6}\text{C portion}} \\ &= 12.011\ \text{amu}\end{aligned}$$

(The ^{14}C isotope is so scarce that it can be ignored in this calculation.)

All the atomic masses listed in the periodic table of the elements are calculated in this way and are thus average masses for the naturally occurring mixture of isotopes. If you want the atomic mass of an individual isotope (other than ^{12}C), you need to consult a reference book such as the *Handbook of Chemistry and Physics* (CRC Press).

Practice Problems

3.11 Iron (Fe) and silicon (Si) are common elements in Earth's crust. How much more massive is iron than silicon?

Answer: Just take a ratio of the atomic masses from the periodic table:

$$\frac{\text{Fe}}{\text{Si}} \quad \frac{55.845}{28.086} = 1.9884$$

So iron is roughly twice as massive as silicon.

3.12 What is the mass in grams of a billion billion (10^{18}) atoms of ^{24}Mg? The atomic mass of ^{24}Mg is 23.9850 amu.

3.13 Chlorine exists as two isotopes in nature, $^{35}_{17}Cl$ (atomic mass 34.969 amu, abundance 75.77%) and $^{37}_{17}Cl$ (atomic mass 36.966 amu).
(a) What is the percent abundance of the $^{37}_{17}Cl$ isotope?
(b) Calculate the weighted average of the atomic mass of naturally occurring chlorine.
(c) How many times more massive is $^{37}_{17}Cl$ than $^{35}_{17}Cl$?

3.5 The Law of Mendeleev—Chemical Periodicity

Chemists kept very busy during the nineteenth century. A major effort was under way to isolate and characterize all of the elements. Chemists throughout the world set out to decompose various chemical compounds into their component elements in order to determine the properties of these fundamental chemical building blocks. By 1860, as a result of these efforts, nearly 70 of the 113 elements we know about today had been isolated and studied. In the thousands of different chemical compounds and mixtures that chemists decomposed, each with its own unique physical and chemical properties, only these 70 elements were found. This represented a great simplification for the science of chemistry. Any object in the universe could now, at least in principle, be understood in terms of the relatively few elements that made it up.

As elements were discovered and their properties examined, it became necessary to organize the data in some useful way that would help to make sense of it all. One of the greatest advances toward this goal was made in the mid-1800s by a Russian chemist named Dmitri Mendeleev. He listed the elements and their properties on individual cards and then experimented with different arrangements of these cards to look for any patterns. The breakthrough came when he arranged the elements in order of increasing atomic mass (using the atomic mass values known at the time):

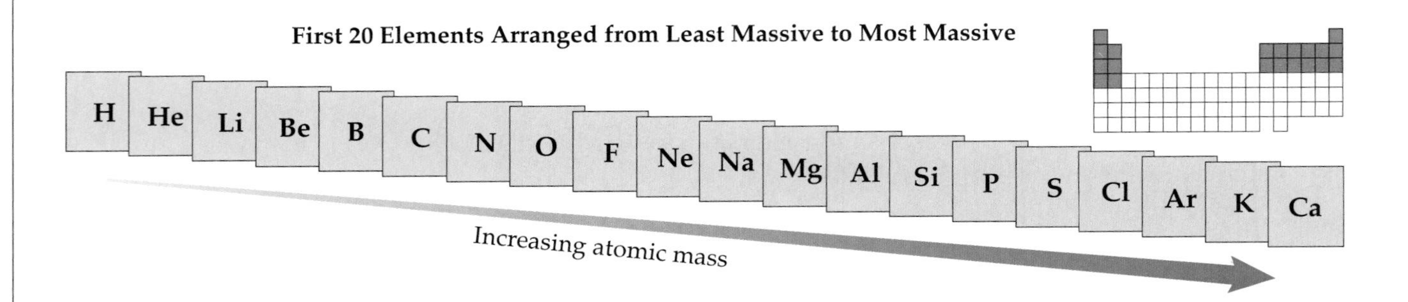

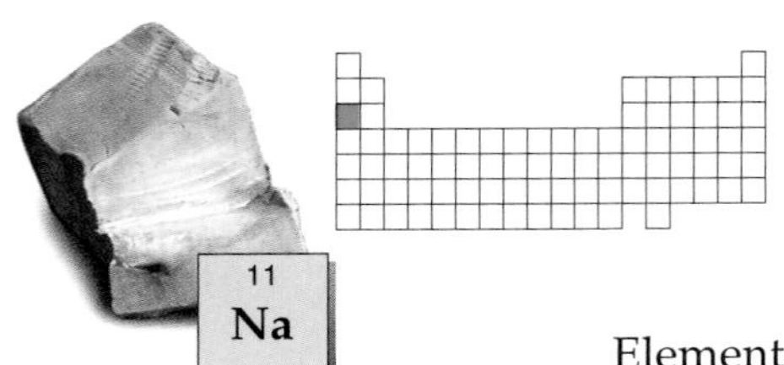

[We have inserted the elements helium (He), neon (Ne), and argon (Ar), which were completely unknown in Mendeleev's time. Mendeleev had other elements in these positions.] With the elements arranged this way, Mendeleev noted that their chemical properties repeated in a regular way. For example, consider the properties of the element sodium (Na) shown at left.

Elemental sodium is too reactive to be found in nature. However, chemists managed to isolate pure sodium from its compounds. It is a soft, silver-colored metal with a low density and a low melting point (for a metal). Like most other metals, it is a good conductor of electricity. It is also highly reactive, which can be readily demonstrated by dropping a piece of sodium metal in water. It reacts violently with the water, producing flammable hydrogen gas, which often ignites. In addition, if a plant dye called litmus is added to the water after the reaction has taken place, the water turns dark blue. An analysis of the reaction products shows that the sodium metal reacted with water to form NaOH, a compound known to make litmus turn blue.

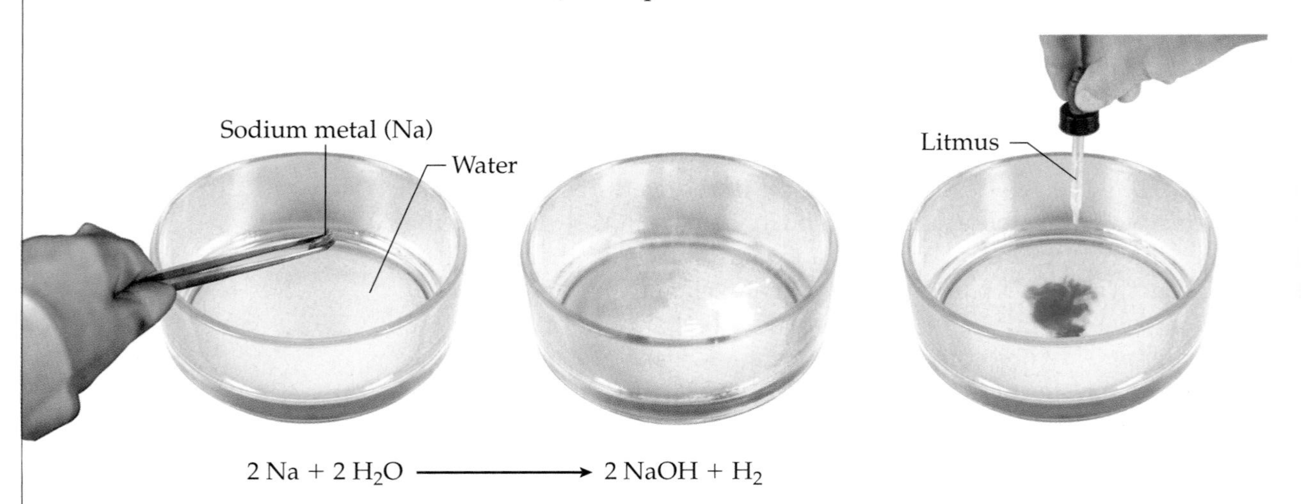

$$2\,Na + 2\,H_2O \longrightarrow 2\,NaOH + H_2$$

Like most other chemists of the time, Mendeleev knew all this about sodium, and by itself this knowledge was nothing to get excited about. But when he examined his arrangement of cards, looking for elements with chemical properties similar to those of sodium, he noticed something interesting.

Eight elements to the left and right of sodium in his list were elements with chemical and physical properties almost identical to those of sodium. Eight elements to the right was potassium (K), and eight elements to the left was lithium (Li):

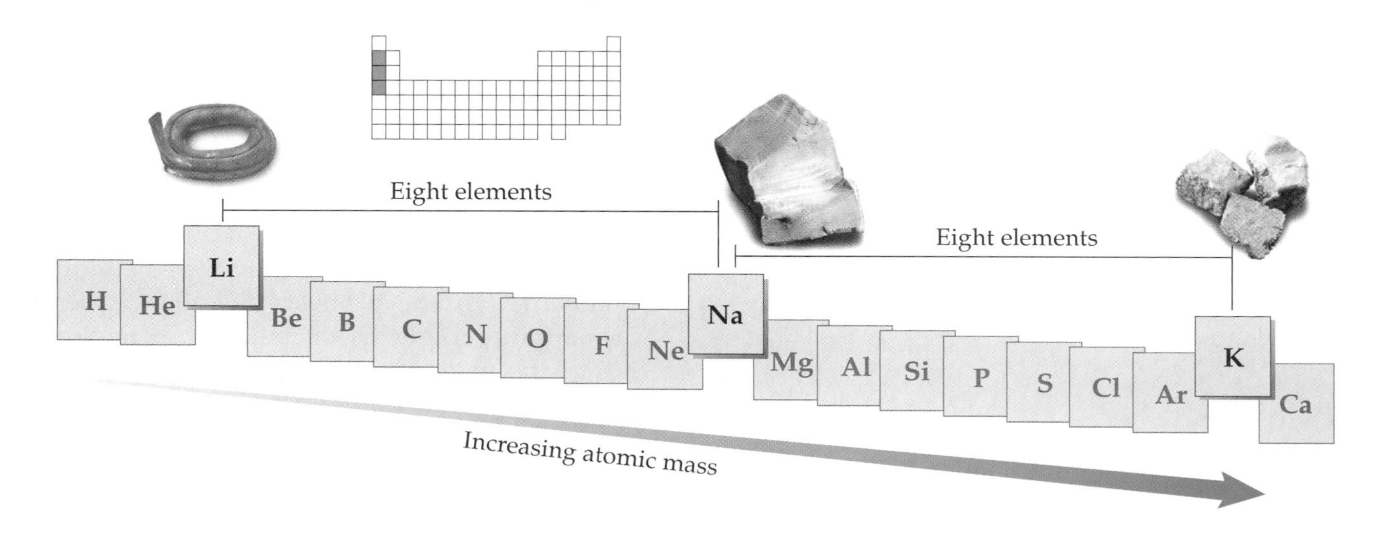

Both of these elements react with oxygen to form oxides (Li_2O and K_2O) that have chemical formulas similar to that of sodium oxide (Na_2O). Both react with water to form hydroxides (LiOH and KOH), just as sodium does (NaOH). All three are silver-colored metals that are too reactive to exist in their elemental form in nature. All are good electrical conductors; all three react vigorously with water, releasing flammable hydrogen gas; and all three produce a blue solution with litmus.

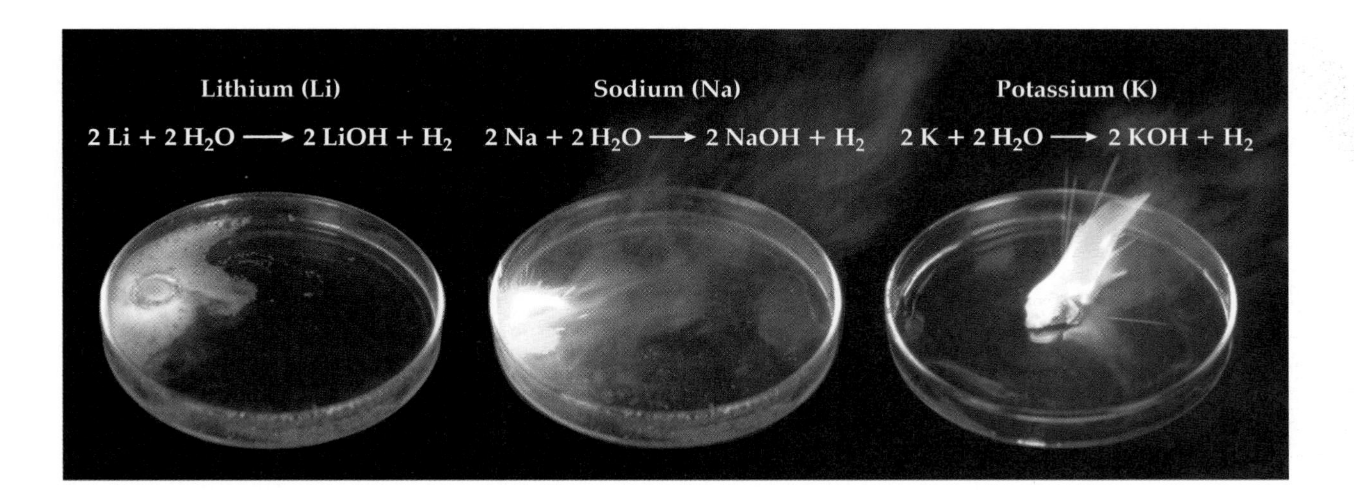

Was this just coincidence? Hardly! Mendeleev observed this same pattern of 8 for other elements in his uniquely arranged list. Magnesium (Mg) reacts with oxygen to give a 1:1 oxide having the chemical formula MgO. Eight elements away in either direction was another element that gave an oxide having a similar formula (beryllium oxide, BeO, and calcium oxide, CaO).

Also, as we now know, neon (Ne) refuses to react with oxygen at all. Sure enough, eight elements away in either direction was an element— helium (He) and argon (Ar)—that also refuses to react with oxygen. There seems to be something almost magical about the number 8. This pattern was sometimes referred to as the *law of octaves*.

I	II	III	IV	V	VI	VII	VIII
1 H H_2O							2 He No oxide
3 Li Li_2O	4 Be BeO	5 B B_2O_3	6 C CO_2	7 N N_2O_5	8 O	9 F F_2O	10 Ne No oxide
11 Na Na_2O	12 Mg MgO	13 Al Al_2O_3	14 Si SiO_2	15 P P_2O_5	16 S SO_2	17 Cl Cl_2O	18 Ar No oxide
19 K K_2O	20 Ca CaO						

When Mendeleev rearranged his cards to stack all elements having similar properties in one column, he got eight columns. No doubt about it. There had to be something special about this number 8. (While the elements of column 8 were unknown at the time, Mendeleev's table did have eight columns.)

This repeating behavior in the chemical properties of the elements is called either **chemical periodicity** or **periodic behavior** (thus the name *periodic table*). In recognition of his efforts, this discovery is often referred to as the **law of Mendeleev**. This same relationship was discovered independently, and at about the same time, by the German chemist Lothar Meyer, but Mendeleev received most of the credit for the periodic table.

Law of Mendeleev: Properties of the elements recur in regular cycles (periodically) when the elements are arranged in order of increasing atomic mass.

Mendeleev was fortunate that his periodic table worked as well as it did. The modern periodic table is arranged according to increasing atomic number, not atomic mass. Fortunately, there are only a few places where Mendeleev's table puts elements in an order different from the order found in the modern table.

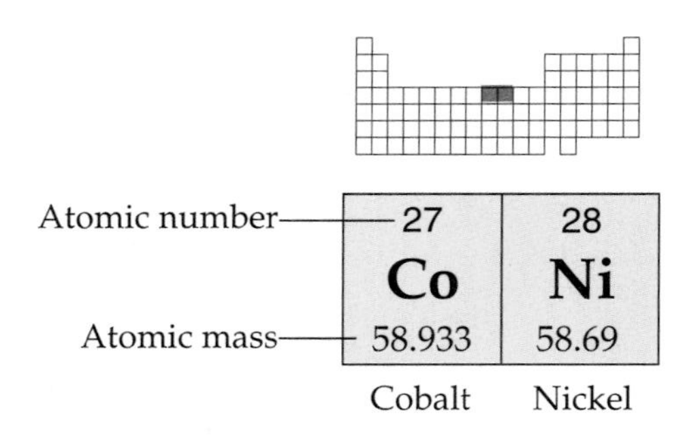

Cobalt comes before nickel in the modern periodic table. Mendeleev would have incorrectly put Co after Ni because the atomic mass of Co is greater than that of Ni.

It was later discovered that 8 was not the only magic number with respect to periodic behavior. In portions of the periodic table, properties repeat every 18 or 32 elements, as we shall soon see.

When Mendeleev began his table, only 70 elements were known. Indeed, when he arranged the elements into columns having similar chemical properties, there were some holes in his table. For example, the column containing the element carbon (C) also contained the elements silicon (Si) and tin (Sn), but there was a hole where an element with a mass somewhere between that of Si and Sn should be.

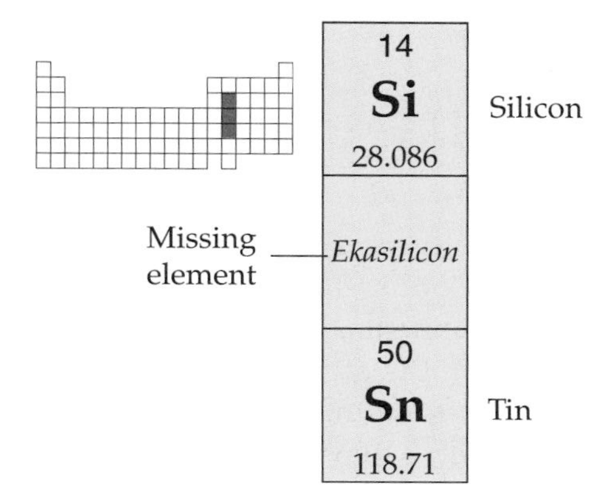

With his newly constructed table, Mendeleev was able to do what no chemist before him could do—predict the properties of elements that had yet to be discovered! This was quite a trick. Mendeleev reasoned that the missing element, which he called *ekasilicon* (*eka* meaning "next in order"), should have properties close to the average of the properties of the elements above and below it. This told chemists what properties to look for and led to the rapid discovery of the missing element. Today we call ekasilicon (represented below by X) germanium, Ge. Look how close Mendeleev's predictions about germanium were:

	Predicted properties (average of Si and Sn)	**Observed properties (measured after discovery)**
Atomic mass	72 amu	72.61 amu
Density	5.5 g/cm^3	5.32 g/cm^3
Melting point	825°C	938°C
Oxide formula	XO_2	GeO_2
Density of oxide	4.7 g/cm^3	4.70 g/cm^3
Chloride formula	XCl_4	$GeCl_4$
Boiling point of chloride	100°C	86°C

Life on our planet is based on the element carbon, but science fiction writers often like to conjure up silicon-based lifeforms. Why have these writers picked on the element Si to create alien life?

Mendeleev's insights greatly advanced our understanding of the chemistry of the elements, but they also presented a new challenge. Any chemically useful model of the atom would have to account for the existence of periodic chemical properties in the elements.

Rutherford's model did not even attempt to address the chemical behavior of the elements, and it did nothing to explain that nasty classical physics problem of why electrons don't spiral into the nucleus. Modifying Rutherford's model was not going to be an easy task. The solution was to come with the new quantum physics and the resulting modern (quantum) model of the atom, which we'll see in the next chapter. In the rest of this chapter, we'll take a closer look at the modern periodic table and then examine some of the clues that pointed the way to the quantum model of the atom.

3.6 The Modern Periodic Table

The **periodic table** is elegant in its simplicity, and yet it says so much. Its neat rows and columns underlie a detailed understanding of the subatomic structure and chemical properties of the elements. Each column is called a **group**, and each row is called a **period**. Both the groups and the periods are numbered. There is no controversy on how to number the periods; they are numbered 1 through 7. There are, however, a few different ways both to number and to name the groups, as shown here. They can be numbered either with roman numerals or with arabic numbers, and the groups colored violet and pale green have alternate names.

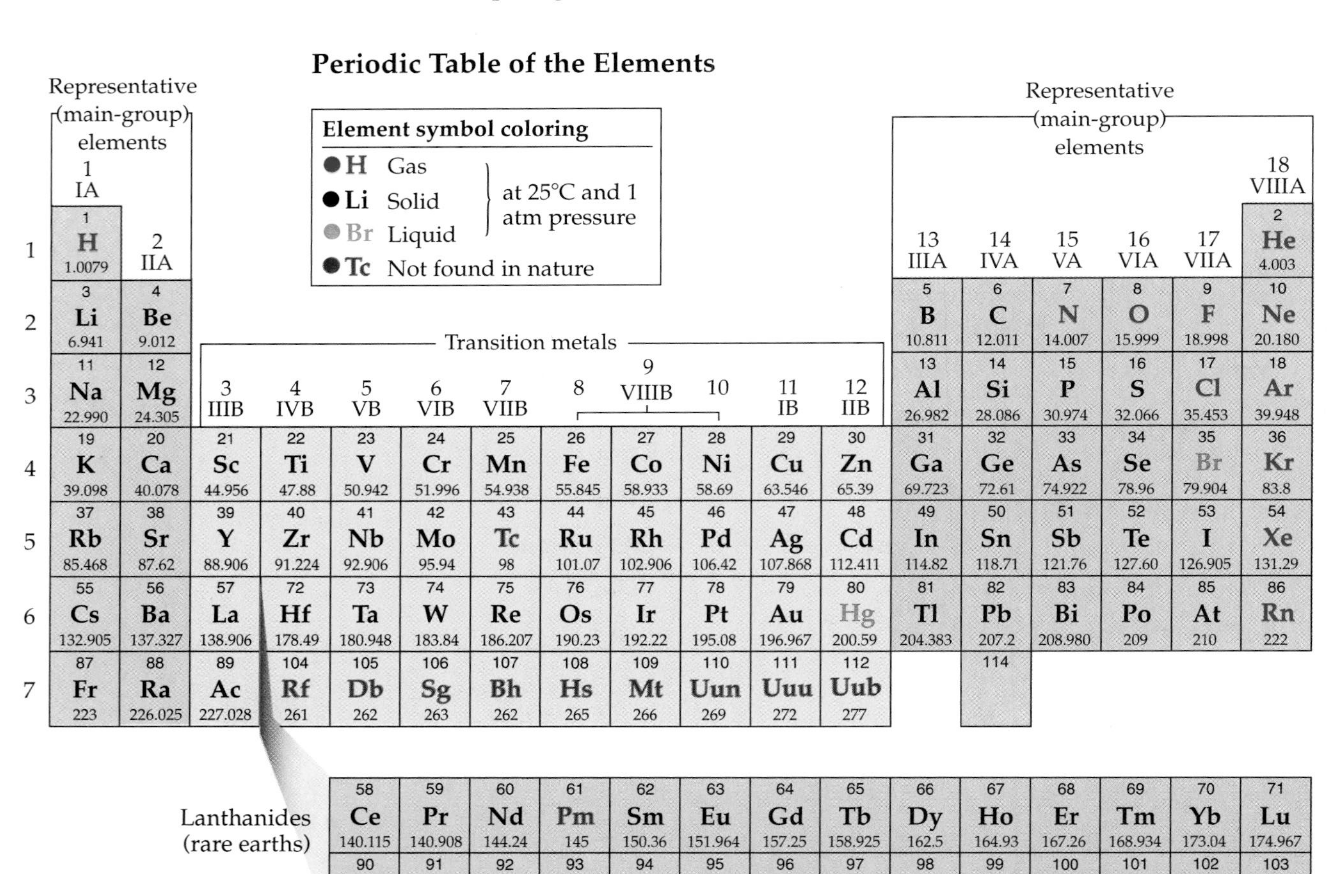

When the groups are numbered with roman numerals, each numeral is followed by either A or B. The large blocks of color in the table on page 96 indicate that the periodic table separates the elements into three broad classes: Violet groups IA–VIIIA are called either the **representative elements** or the **main-group elements**. Tan groups IB–VIIIB are known as **transition metals**. Finally, the elements in the upper pale green row at the bottom are called the **lanthanides** (or **rare earths**), and the elements in the lower pale green row are called the **actinides**. The symbol of each element is accompanied by its atomic number above and its average atomic mass below.

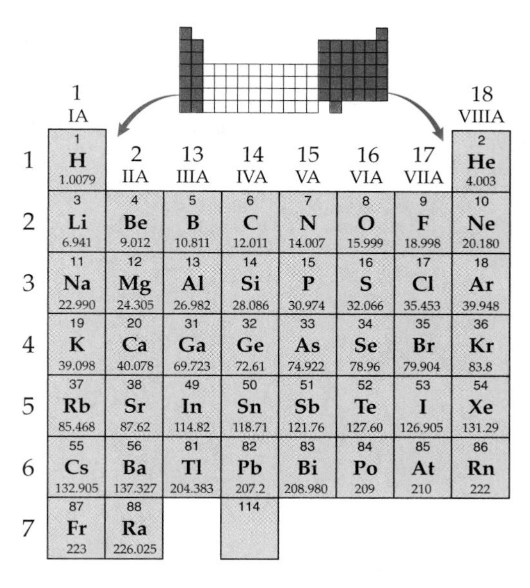

A periodic table showing only the representative elements is eight columns wide. Chemical periodicity occurs every eight elements.

The representative, or main-group, elements are the ones on which much of early chemistry was based. They also show the strongest periodic relationships in the table. When applied only to the representative elements, the law of octaves works magnificently. Picking any element and then moving away eight elements in either direction brings you to another element whose properties are similar to those of the element you started with.

Look again at the entire periodic table, and pay close attention to how the groups are numbered. At first glance, the roman-numeral numbering system looks a bit bizarre. There is a big block of B groups jammed in between the A groups. Even worse, the B groups start with number III, go to VIII (three columns are assigned this number), and then end with I and II. There are good historical reasons for this strange numbering, but its awkwardness is what led IUPAC (the International Union of Pure and Applied Chemistry) to adopt the more straightforward 1-through-18 sequence shown in the table on page 96.

Our periodic table includes useful information about the phase (solid, liquid, or gas) in which the pure element exists under standard conditions (25°C and 1 atm pressure).

Now we shall take a brief tour of the periodic table, starting with the representative elements (groups IA–VIIIA). These groups contain solid, liquid, and gas elements. In fact, all the elements that exist as gases at standard temperature (25°C) and standard pressure (1 atm) are representative elements, including the two that make up most of the air we breathe, nitrogen and oxygen.

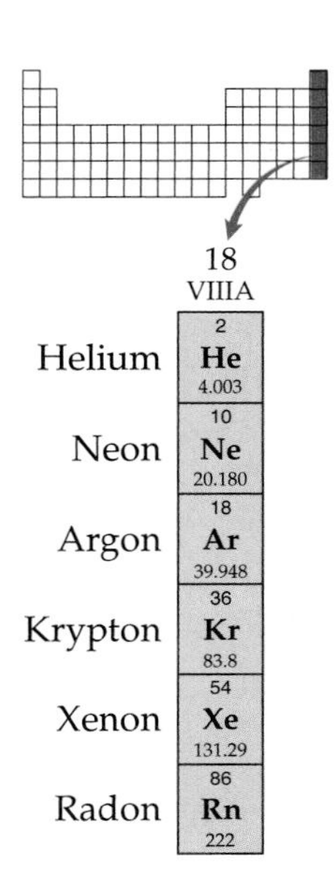

The first representative elements we shall examine are those of group VIIIA (18). This is a very special column of the periodic table. It is the only group whose members are all gases at standard temperature and pressure. In addition, all the elements in this group are extremely unreactive. For this reason the group VIIIA (18) elements are sometimes called the *inert gases*, and for some time chemists believed that these elements were completely chemically unreactive. In 1962, however, the British chemist Neil Bartlett challenged this assumption and prepared the first compound of the gas xenon (Xe). Today we refer to this group as either the **noble gases** or the **rare gases**. With the exception of helium, they are all found in trace amounts in the atmosphere. Earth's gravitational pull is not strong enough to keep any helium that gets into the atmosphere from escaping into space, and as a result all naturally occurring helium comes from deposits trapped underground.

Radon (Rn), the only noble gas that is radioactive, has received some bad press lately. It has been discovered seeping into basements from naturally occurring underground sources. Sustained exposure is thought to be responsible for an increase in the incidence of lung cancer, and many people have purchased radon detectors for their basements.

As we continue to examine the representative elements, we find yet another way to divide them: They can be classified as *metals*, *nonmetals*, or *metalloids*, which have properties intermediate between those of metals and those of nonmetals. (*Semimetals* is an alternate name for metalloids.) This division is indicated in the following illustration. The red stair-step line that starts at boron, B, separates the metals from the nonmetals. The elements bordering this line (except for aluminum, Al, and polonium, Po) are the metalloids. Of the 44 representative elements, 7 are metalloids, 17 are nonmetals (including the noble gases), and the remaining 20 are metals.

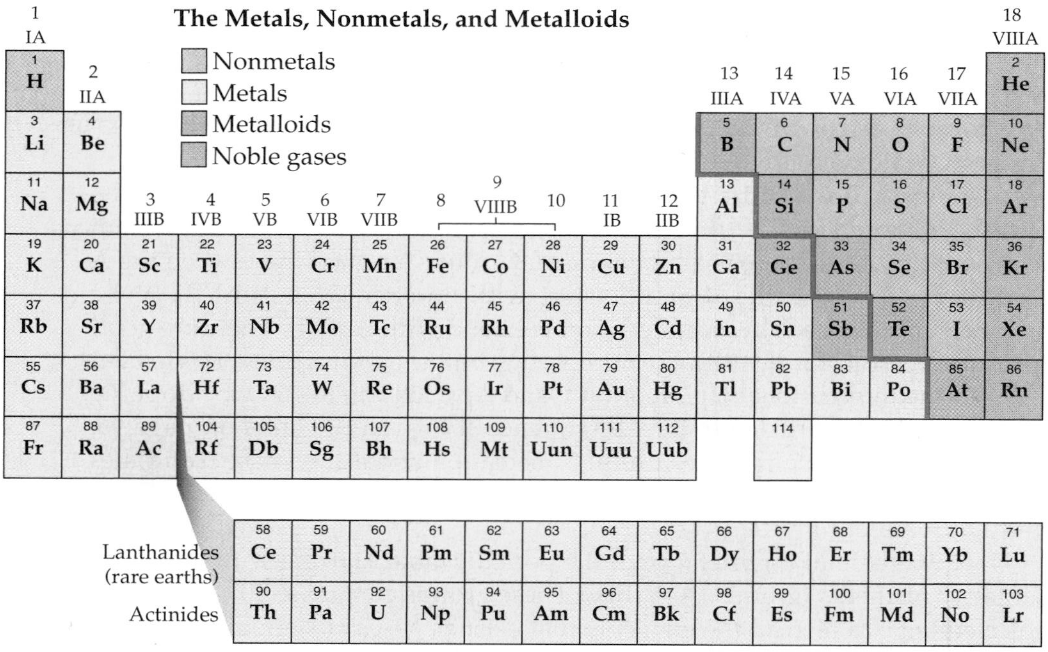

The Metals, Nonmetals, and Metalloids

Nonmetals
Metals
Metalloids
Noble gases

1 IA	2 IIA	3 IIIB	4 IVB	5 VB	6 VIB	7 VIIB	8	9 VIIIB	10	11 IB	12 IIB	13 IIIA	14 IVA	15 VA	16 VIA	17 VIIA	18 VIIIA
1 H																	2 He
3 Li	4 Be											5 B	6 C	7 N	8 O	9 F	10 Ne
11 Na	12 Mg											13 Al	14 Si	15 P	16 S	17 Cl	18 Ar
19 K	20 Ca	21 Sc	22 Ti	23 V	24 Cr	25 Mn	26 Fe	27 Co	28 Ni	29 Cu	30 Zn	31 Ga	32 Ge	33 As	34 Se	35 Br	36 Kr
37 Rb	38 Sr	39 Y	40 Zr	41 Nb	42 Mo	43 Tc	44 Ru	45 Rh	46 Pd	47 Ag	48 Cd	49 In	50 Sn	51 Sb	52 Te	53 I	54 Xe
55 Cs	56 Ba	57 La	72 Hf	73 Ta	74 W	75 Re	76 Os	77 Ir	78 Pt	79 Au	80 Hg	81 Tl	82 Pb	83 Bi	84 Po	85 At	86 Rn
87 Fr	88 Ra	89 Ac	104 Rf	105 Db	106 Sg	107 Bh	108 Hs	109 Mt	110 Uun	111 Uuu	112 Uub		114				

Lanthanides (rare earths)	58 Ce	59 Pr	60 Nd	61 Pm	62 Sm	63 Eu	64 Gd	65 Tb	66 Dy	67 Ho	68 Er	69 Tm	70 Yb	71 Lu
Actinides	90 Th	91 Pa	92 U	93 Np	94 Pu	95 Am	96 Cm	97 Bk	98 Cf	99 Es	100 Fm	101 Md	102 No	103 Lr

While we have not yet formally defined the terms "metal" and "nonmetal," you probably have a pretty good feel for them based on everyday experience. **Metals** tend to be shiny solids that are bendable and malleable (which means they can be pounded into very thin sheets) and conduct heat and electricity well. **Nonmetals** tend to be brittle and do not conduct heat or electricity well. (They are *insulators*, the opposite of conductors.) Lead (Pb), a representative element in group IVA (14) once used for plumbing, is an obvious metal. Diamond, a pure form of carbon (C), is a nonmetal. For now, these everyday "definitions" of metal and nonmetal will suffice. Later we'll give more chemically accurate definitions that relate to the tendency of these elements to gain or lose electrons when forming chemical compounds.

Representative metals such as aluminum (Al) and tin (Sn) are familiar to most people. You've certainly come across aluminum foil and tin cans (which are actually made from steel coated with tin). The metals of groups IA (1) and

IIA (2) are less familiar because they are much too reactive to be found in nature in their elemental forms. However, as parts of compounds, some of them are also quite familiar. The compound sodium chloride (table salt) has the group IA (1) metal sodium (Na) in it. Many common antacids, such as Milk of Magnesia and Tums, contain compounds of the group IIA (2) metals magnesium (Mg) and calcium (Ca).

The **metalloids**, or **semimetals**, can act, depending on circumstances, like either a metal or a nonmetal. The elements silicon (Si), germanium (Ge), and arsenic (As) are good examples of elements that demonstrate classic metalloid character. They neither conduct electricity as well as metals nor insulate electrically as well as nonmetals. They are somewhere in between and are thus called *semiconductors*. These elements are used to make the sophisticated electronic chips that are the "brains" of computers. Indeed, the entire electronics industry owes its existence to the metalloid properties of these elements.

That completes our tour of the representative elements, and so let's move on to the transition metals. These elements form a bridge between the two parts of the representative elements. Perhaps the most familiar transition metal is iron (Fe), the major component of steel, used in everything from bicycles to skyscrapers. Others, such as nickel (Ni), silver (Ag), gold (Au), copper (Cu), and zinc (Zn), are also familiar to most people. Mercury (Hg), commonly used in thermometers, is the only transition metal that is a liquid at room temperature. Chromium (Cr) is used for plating other metals, as is done with car bumpers. Titanium (Ti) is used to strengthen bicycle frames. Platinum (Pt), used in pollution-control catalytic converters on cars, is even more expensive than gold. While the remaining transition metals may be less familiar to you, each has a wide variety of uses in our technological society.

Finally there are the lanthanides and the actinides. These elements are less commonly encountered than those we have already described. Two of them, however, uranium and plutonium, are well known because of their role as fuels for nuclear energy and as materials for producing nuclear weapons.

Of the 113 known elements, approximately 75% are metals and only 16% are nonmetals! Nevertheless, all life that we know of is based on a nonmetallic element, carbon. Had nature built you out of one of the majority elements, a metal, you might have to worry more about rusting than about getting wrinkles in your old age.

Practice Problems

3.14 How many groups constitute the representative (main-group) elements?

Answer: Eight

3.15 Does the stair-step boundary line that separates metals from nonmetals in the periodic table cross into the transition-metal portion of the table?

3.16 What do the elements in a group of the periodic table have in common with one another?

Mendeleev's arrangement of elements by mass resulted in the elements in a particular group (column) having similar chemical properties. Exactly why it worked out this way was not understood at the time. Nevertheless, these similarities invited giving each group a special name. Of the eight representative groups, five have commonly used names, the first two and the last three. When metals from either group IA (1) or group IIA (2) are placed in water, they react with it to make the water alkaline (basic), hence the names **alkali metals** for group IA (1) and **alkaline earth metals** for IIA (2). The group VIA (16) elements, which include oxygen and sulfur, are called the **chalcogens**. The name comes from the Greek words *chalkos* for "copper" and *genes* for "born." Most copper-containing minerals contain oxygen or sulfur. The group VIIA (17) elements, which include chlorine, bromine, and iodine, are called the **halogens** (*halos* is Greek for "salt"). Many salts have a halogen in them, such as sodium chloride. Group VIIIA (18) elements are called either noble gases or rare gases, as mentioned earlier. You should commit these names to memory.

Alkali metals 1 IA	Alkaline earth metals 2 IIA	3 IIIB	4 IVB	5 VB	6 VIB	7 VIIB	8	9 VIIIB	10	11 IB	12 IIB	13 IIIA	14 IVA	15 VA	Chalcogens 16 VIA	Halogens 17 VIIA	Noble gases 18 VIIIA
1 **H** 1.0079																	2 **He** 4.003
3 **Li** 6.941	4 **Be** 9.012											5 **B** 10.811	6 **C** 12.011	7 **N** 14.007	8 **O** 15.999	9 **F** 18.998	10 **Ne** 20.180
11 **Na** 22.990	12 **Mg** 24.305											13 **Al** 26.982	14 **Si** 28.086	15 **P** 30.974	16 **S** 32.066	17 **Cl** 35.453	18 **Ar** 39.948
19 **K** 39.098	20 **Ca** 40.078	21 **Sc** 44.956	22 **Ti** 47.88	23 **V** 50.942	24 **Cr** 51.996	25 **Mn** 54.938	26 **Fe** 55.847	27 **Co** 58.933	28 **Ni** 58.69	29 **Cu** 63.546	30 **Zn** 65.39	31 **Ga** 69.723	32 **Ge** 72.61	33 **As** 74.922	34 **Se** 78.96	35 **Br** 79.904	36 **Kr** 83.8
37 **Rb** 85.468	38 **Sr** 87.62	39 **Y** 88.906	40 **Zr** 91.224	41 **Nb** 92.906	42 **Mo** 95.94	43 **Tc** 98	44 **Ru** 101.07	45 **Rh** 102.906	46 **Pd** 106.42	47 **Ag** 107.868	48 **Cd** 112.411	49 **In** 114.82	50 **Sn** 118.71	51 **Sb** 121.76	52 **Te** 127.60	53 **I** 126.905	54 **Xe** 131.29
55 **Cs** 132.905	56 **Ba** 137.327	57 **La** 138.906	72 **Hf** 178.49	73 **Ta** 180.948	74 **W** 183.85	75 **Re** 186.207	76 **Os** 190.2	77 **Ir** 192.22	78 **Pt** 195.08	79 **Au** 196.967	80 **Hg** 200.59	81 **Tl** 204.383	82 **Pb** 207.2	83 **Bi** 208.980	84 **Po** 209	85 **At** 210	86 **Rn** 222
87 **Fr** 223	88 **Ra** 226.025	89 **Ac** 227.028	104 **Rf** 261	105 **Db** 262	106 **Sg** 263	107 **Bh** 262	108 **Hs** 265	109 **Mt** 266	110 **Uun** 269	111 **Uuu** 272	112 **Uub** 277		114				

Given all we have said so far, you might wonder why hydrogen is included in group IA (1), the alkali metals. After all, hydrogen is not a metal. In fact, Mendeleev would not have put it in this group. We shall wait until the next chapter to explain why it is there. Suffice it to say that while hydrogen is shown in group IA (1), nobody thinks of it as an alkali metal.

As we have seen, the structure of the periodic table reflects what we have learned about the chemistry of the various elements. The law of octaves applies satisfactorily when you are dealing only with the representative elements. The chemical properties of these elements repeat every eight elements, and this part of the table contains eight columns to allow for this. But the law needs some modification if we are to include the rest of the elements in the table. If we continue to place elements into columns based primarily on their similar chemical behaviors, we end up with the form of the periodic table we know today.

The introduction of the transition metals into the table begins with row (period) 4 and causes the number of columns to swell to 18 (because there are 8 representative elements + 10 transition metals per row). In this part of the table, therefore, chemical properties repeat every 18 elements and not every 8.

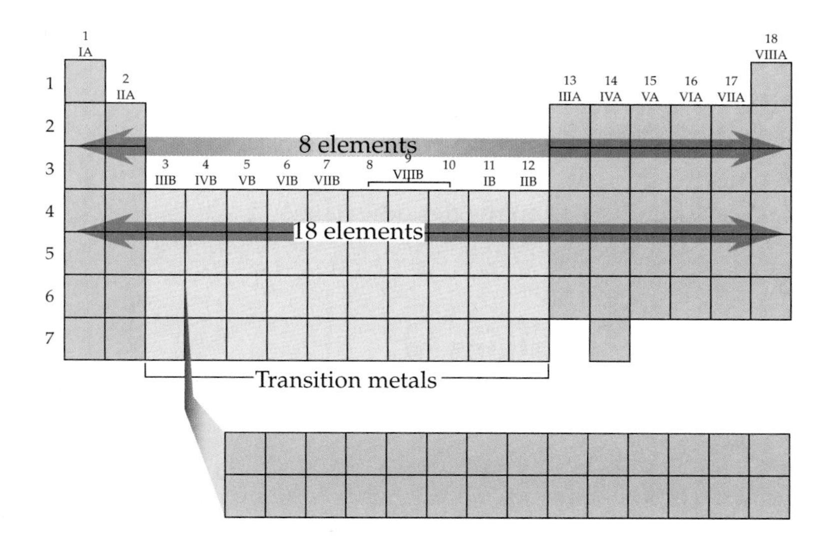

As we continue down the rows of the table, we come to periods 6 and 7, where we must now accommodate the lanthanide and actinide elements. So far, we have shown these elements as a separate block below the main part of the table. That's just a convenience to allow the table to fit better on the printed page. Really, the lanthanide and actinide elements should fit right into periods 6 and 7, like this:

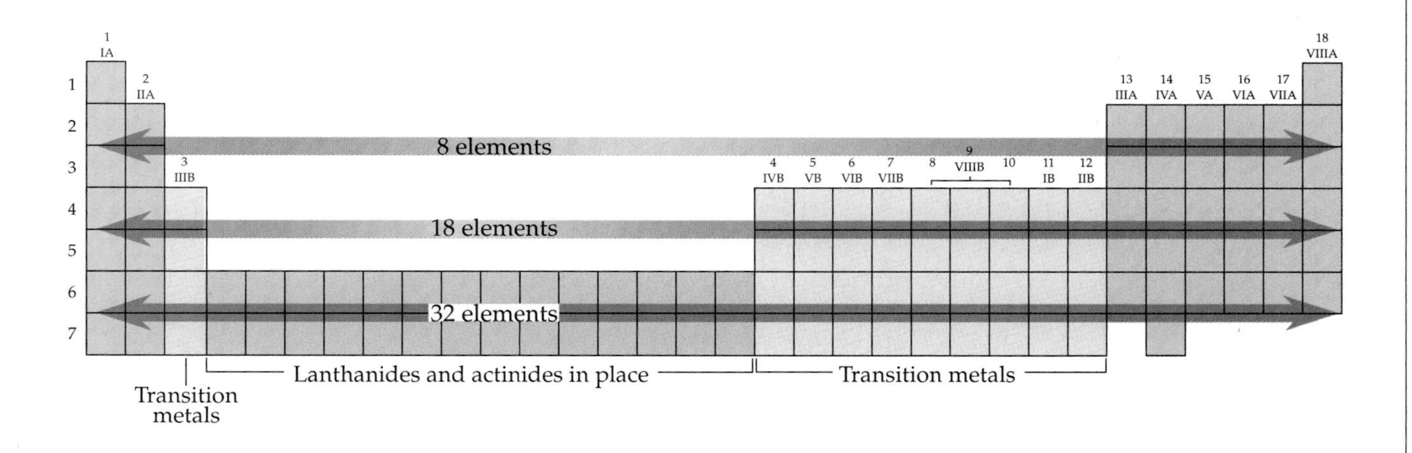

Thus, in periods 6 and 7, the table becomes 32 columns wide, and the chemical properties of the elements in these periods repeat with a cycle of 32, rather than 8 or 18. It's much easier to see how the periodicity changes from 18 to 32 at period 6 when you are looking at the periodic table with the lanthanide and actinide elements inserted.

Practice Problems

3.17 What are the period number and group number of the element that has atomic number 15?

Answer: The element that has atomic number 15 is phosphorus (P). It is in period 3 and group VA (15).

3.18 Which elements with an atomic number greater than 40 can be expected to have chemical properties similar to those of the element bromine (Br)? What is the periodicity of these elements?

3.19 What would the periodicity be if the transition metals, lanthanides, and actinides were removed from the periodic table?

3.7 Other Regular Variations in the Properties of Elements

Chemical behavior is not the only property of the elements that varies in a systematic way. Two others are *atomic size* and *ionization energy*. We'll look at the periodic trend in atomic size first.

ATOMIC SIZE

If we think of atoms as spheres, we can characterize each by its radius, called the **atomic radius**. Atomic radii can be measured by experimental techniques such as X-ray diffraction. They are often reported in angstroms (Å, 1 Å $=10^{-10}$ m), nanometers (10^{-9} m), or picometers (10^{-12} m). The chart at the top of the next page illustrates the relative sizes of the representative elements. Notice that the atoms vary in size in a regular way as you go down a group or across a period. In general,

Atomic radius increases as you go down a group.

Atomic radius decreases as you go from left to right in a period.

There are local exceptions to these trends, but they hold overall.

IONIZATION ENERGY

Another property of atoms that varies regularly in the periodic table is *ionization energy*. Atoms are electrically neutral because the number of negatively charged electrons outside the nucleus equals the number of positively charged protons inside the nucleus. An atom in which this electron–proton balance is not maintained has a net charge and is called an **ion**. The only practical way to create an ion is to add or remove electrons from a neutral atom. If

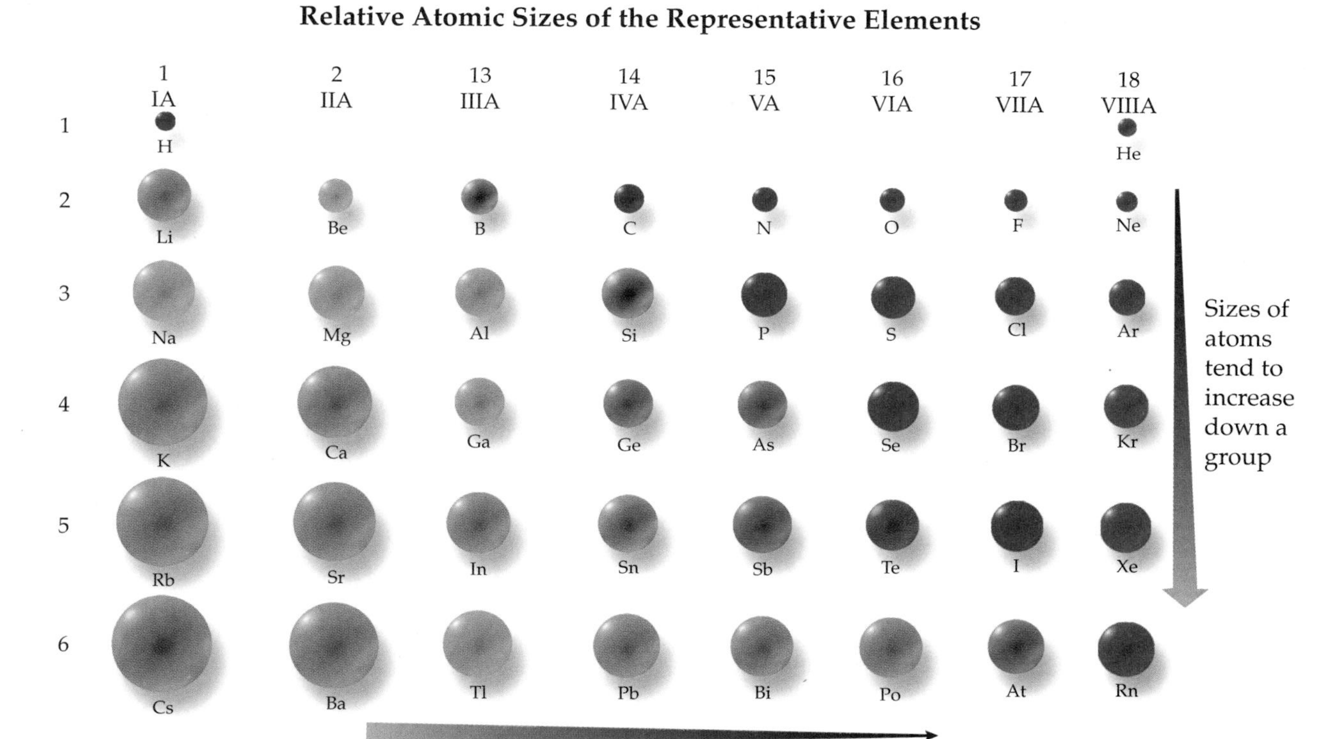

you take a neutral silicon (Si) atom, for instance, and add one electron to it, the resulting ion has one more electron than it has protons and therefore has a net charge of −1. Such negatively charged ions are called **anions**. Likewise, if you remove one electron from a neutral silicon atom, the resulting ion will have one fewer electron than it has protons and a net positive charge of +1. Ions with a positive charge are called **cations**.

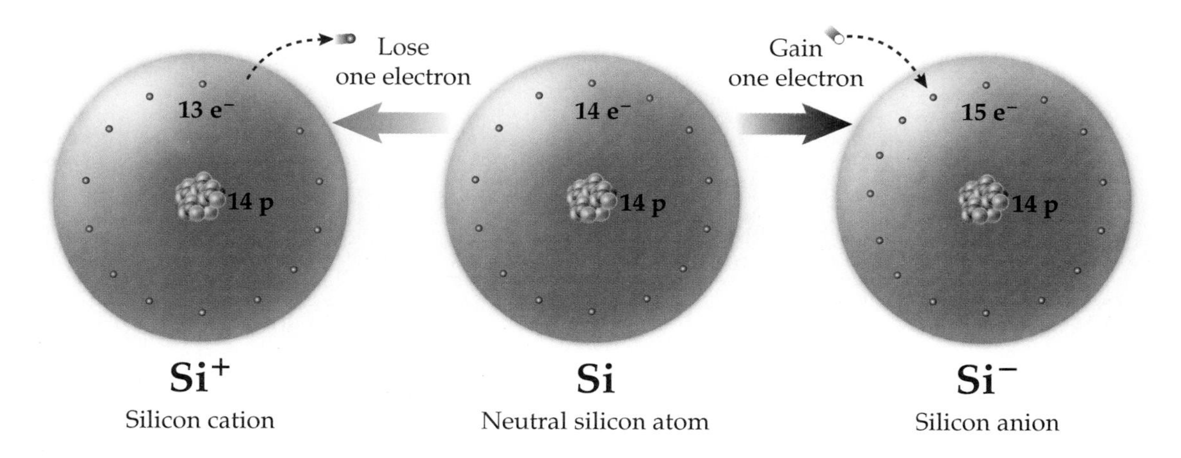

Now, what if we take another atom, say magnesium (Mg), and remove two electrons from the neutral atom? There will now be an excess of two protons, and we end up with a cation with a +2 charge. This ion is written Mg^{2+}. If we instead added two electrons to a magnesium atom (which for magnesium is actually very difficult to do), we would get an ion with a −2 charge, which is written Mg^{2-}. In both cases, the usual convention in writing ions is to give the numerical value of the charge before the + or − sign. In the case of ions with a single charge, the 1 is usually omitted and the ion is written simply, for example, Li^+ or Cl^-. An element symbol written without any charge (or with a zero, for example, Fe^0 or Cu^0) represents a neutral atom of that element.

S^{2-} Cl^- — Anions; C — Neutral atom (no charge indicated); Na^+ Mg^{2+} — Cations

3.9 WORKPATCH

Fill in the table:

	$^{14}_{7}N^{3-}$	$^{24}_{12}Mg^{2+}$	$^{23}_{11}Na^+$	$^{56}_{26}Fe^{3+}$
Mass number	?	?	?	?
Atomic number	?	?	?	?
Number of protons	?	?	?	?
Number of neutrons	?	?	?	?
Number of electrons	?	?	?	?

This WorkPatch is meant to drive home an important point. The only way you can turn a (neutral) atom into an ion is by adding or removing electrons. You do not adjust the number of protons. If the number of protons were changed, the atom would not be the same element anymore.

Now let's look more closely at the removal of electrons from neutral atoms to form cations. Electrons are held in an atom by their attraction to the positive nucleus. To remove an electron from the influence of the nucleus, you have to either pull it out or "kick" it with enough energy to free it. In more formal terms, you must expend energy to remove an electron from an atom. The minimum amount of energy required to remove an electron from an atom is called the **first ionization energy (IE)** of the atom.

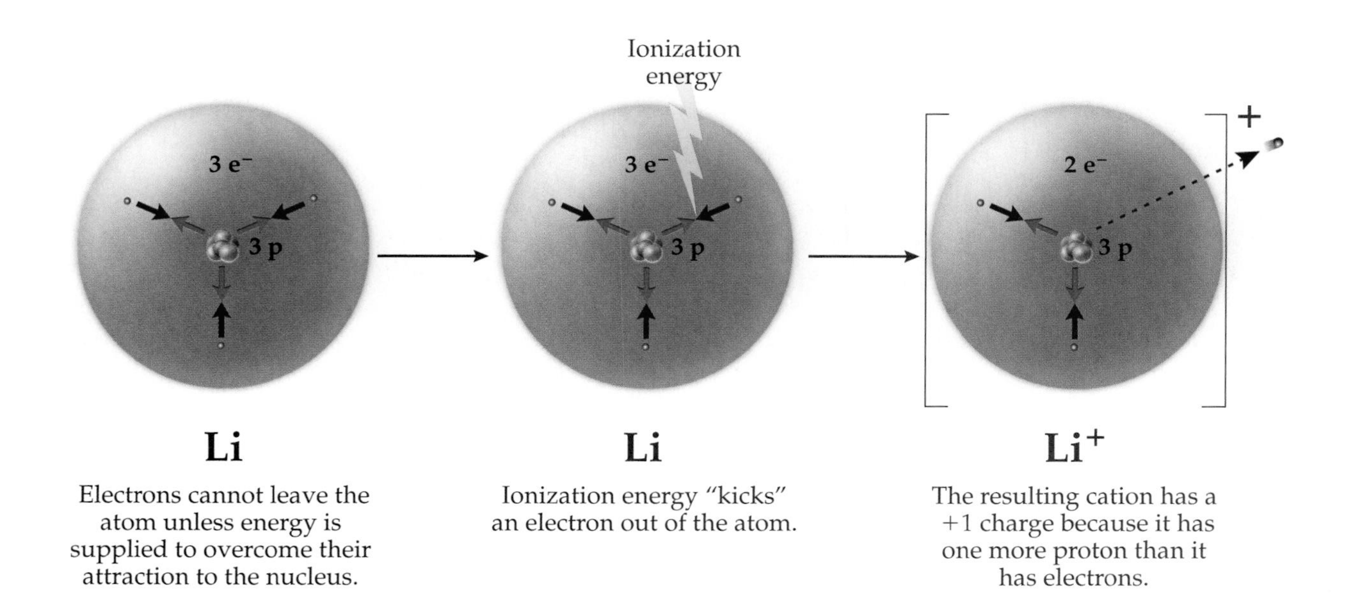

Electrons cannot leave the atom unless energy is supplied to overcome their attraction to the nucleus.

Ionization energy "kicks" an electron out of the atom.

The resulting cation has a +1 charge because it has one more proton than it has electrons.

First Ionization Energies

Representative (main-group) elements | Transition metals | Representative (main-group) elements

Key: atomic number, symbol, first ionization energy (eV)

Period	1 IA	2 IIA	3 IIIB	4 IVB	5 VB	6 VIB	7 VIIB	8	9 VIIIB	10	11 IB	12 IIB	13 IIIA	14 IVA	15 VA	16 VIA	17 VIIA	18 VIIIA
1	1 H 13.6																	2 He 24.6
2	3 Li 5.4	4 Be 9.3											5 B 8.3	6 C 11.3	7 N 14.5	8 O 13.6	9 F 17.4	10 Ne 21.6
3	11 Na 5.1	12 Mg 7.6											13 Al 6.0	14 Si 8.2	15 P 10.5	16 S 10.4	17 Cl 13.0	18 Ar 15.8
4	19 K 4.3	20 Ca 6.1	21 Sc 6.5	22 Ti 6.8	23 V 6.7	24 Cr 6.8	25 Mn 7.4	26 Fe 7.9	27 Co 7.9	28 Ni 7.6	29 Cu 7.7	30 Zn 9.4	31 Ga 6.0	32 Ge 7.9	33 As 9.8	34 Se 9.8	35 Br 11.8	36 Kr 14.0
5	37 Rb 4.2	38 Sr 5.7	39 Y 6.4	40 Zr 6.8	41 Nb 6.9	42 Mo 7.1	43 Tc 7.3	44 Ru 7.4	45 Rh 7.5	46 Pd 8.3	47 Ag 7.6	48 Cd 9.0	49 In 5.8	50 Sn 7.3	51 Sb 8.6	52 Te 9.0	53 I 10.5	54 Xe 12.1
6	55 Cs 3.9	56 Ba 5.2	57 La 5.6	72 Hf 6.7	73 Ta 7.9	74 W 8.0	75 Re 7.9	76 Os 8.7	77 Ir 9.1	78 Pt 9.0	79 Au 9.2	80 Hg 10.4	81 Tl 6.1	82 Pb 7.4	83 Bi 7.3	84 Po 8.4	85 At –	86 Rn 10.7
7	87 Fr –	88 Ra 5.3	89 Ac 5.2	104 Rf –	105 Db –	106 Sg –	107 Bh –	108 Hs –	109 Mt –	110 Uun –	111 Uuu –	112 Uub –		114				

Lanthanides (rare earths)	58 Ce 5.5	59 Pr 5.4	60 Nd 5.5	61 Pm 5.6	62 Sm 5.6	63 Eu 5.7	64 Gd 6.2	65 Tb 5.9	66 Dy 5.9	67 Ho 6.0	68 Er 6.1	69 Tm 6.2	70 Yb 6.3	71 Lu 5.4
Actinides	90 Th 6.1	91 Pa 5.9	92 U 6.1	93 Np 6.2	94 Pu 6.1	95 Am 6.0	96 Cm 6.0	97 Bk 6.2	98 Cf 6.3	99 Es 6.4	100 Fm 6.5	101 Md 6.6	102 No 6.7	103 Lr –

The periodic table above lists the ionization energies of the elements in units of electronvolts (eV), a unit of energy commonly used by physicists. Look at the table closely. Can you see any trends in the ionization energies down a group or across a period? (*Hint*: Take a look at the figure at the top of the next page.)

Just as there are periodic trends in atomic size (radius), there are also periodic trends in ionization energies. In general,

Ionization energy decreases as you go down a group.

Ionization energy increases as you go from left to right in a period.

This means that it becomes easier (takes less energy) to remove an electron from an atom as you go down a group, and it takes more energy to remove an electron as you go from left to right across a period. If we plot the ionization energies as shown at the top of the next page, these trends become even more apparent. Each color represents a different period.

If we focus on the noble-gas elements—He, Ne, Ar, Kr, Xe, and Rn—you can see how the ionization energy decreases as you go down a group. If we then focus on the period 2 elements—Li, Be, B, C, N, O, F, and Ne—you can see how the ionization energy generally increases as you go across a period.

We can understand these period and group trends in ionization energy by looking again at the trends in atomic size. As you go down a group, atoms get bigger, and, at the same time, it becomes easier (requires less energy) to

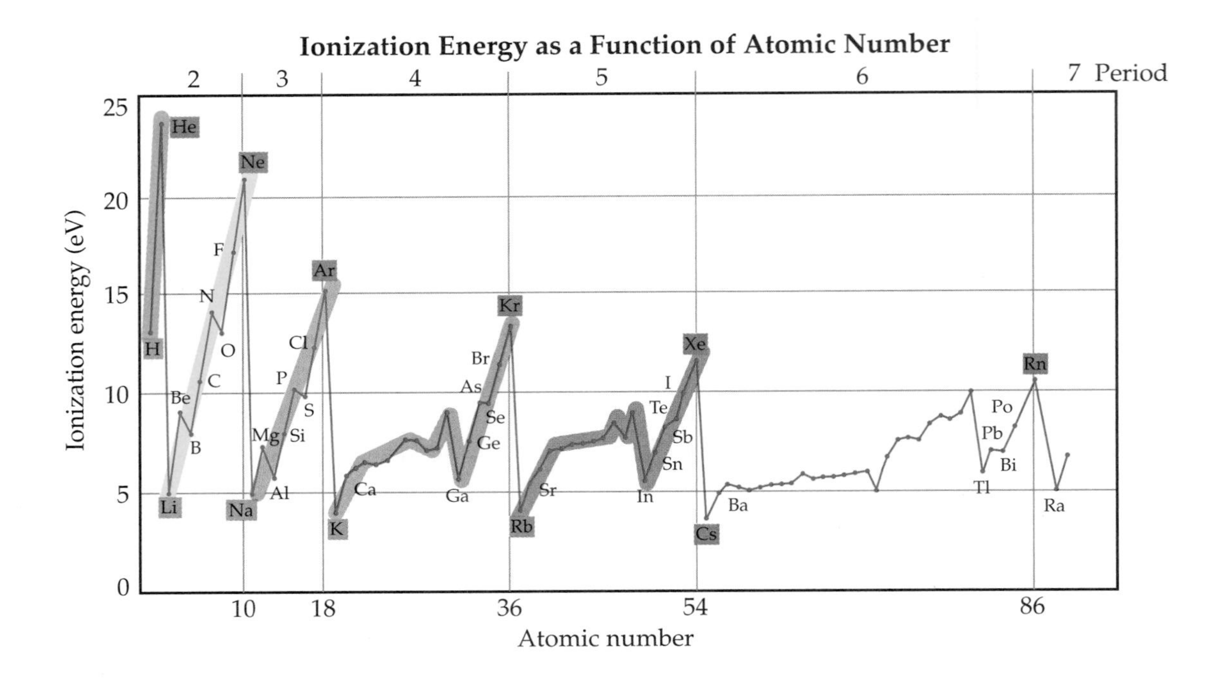

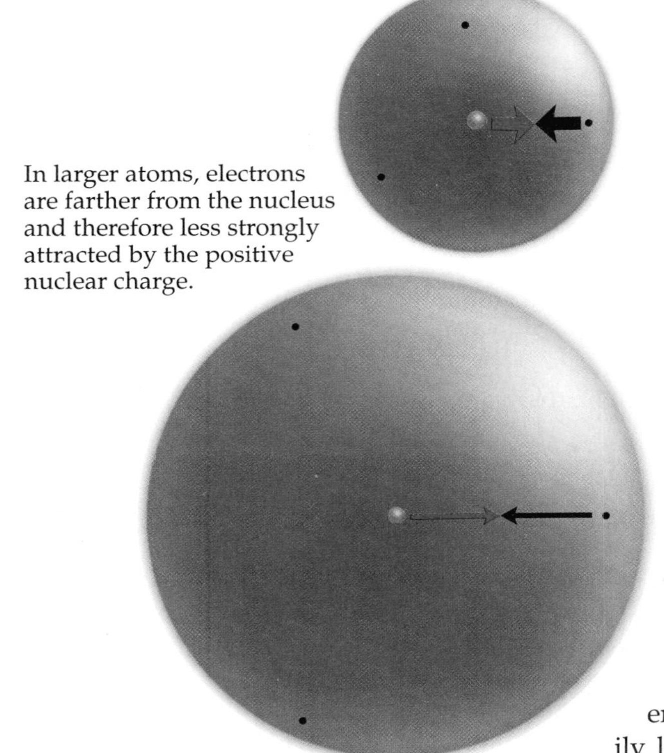

In larger atoms, electrons are farther from the nucleus and therefore less strongly attracted by the positive nuclear charge.

remove an electron. In a larger atom, the outermost electrons are, in general, farther from the nucleus. Also, the inner electrons partly cancel or shield some of the positive charge of the nucleus, with the result that the outer electrons "feel" a reduced attraction to the nucleus. These two effects mean that in a larger atom the outer electrons are less strongly attracted to the nucleus than in a smaller atom. Therefore, the outer electrons are easier to remove from the larger atom—their ionization energy is lower.

Of course, the opposite is also true. As atoms get smaller (either going up a group or left to right across a period), it becomes more difficult to remove an electron, and the ionization energy increases.

These regular variations (periodic trends) in atomic size and ionization energy can be a great help to chemists in explaining the chemical behavior of the elements. For example, the alkali metals in group IA (1) have relatively low ionization energies. This means that an element like sodium easily loses an electron. Now, if you start at sodium and travel across period 3, you come to a much smaller atom that has a relatively large ionization energy, chlorine (Cl). It takes a large amount of energy for chlorine to lose an electron, and this element generally does not give up electrons when it reacts. In fact, just the opposite tends to occur: Cl tends to *gain* an electron when it reacts. Bring sodium and chlorine together, and this is exactly what happens. Each sodium atom loses an electron, becoming a Na^+ cation, and each chlorine atom picks up an electron and becomes a Cl^- anion. The result is the spontaneous and rather spectacular chemical transformation of sodium and chlorine into the compound sodium chloride (NaCl), table salt.

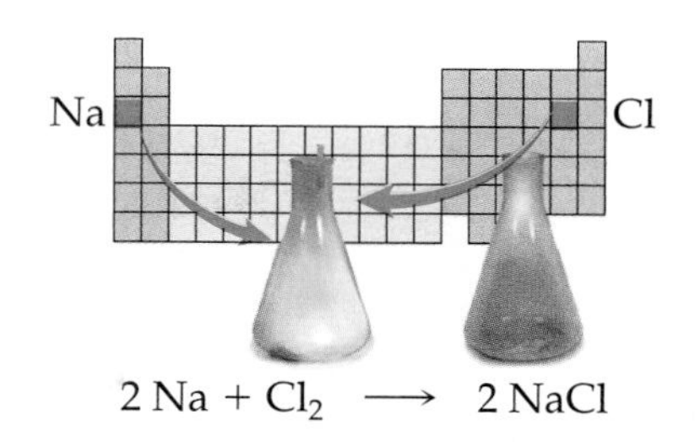

Practice Problems

3.20 The alkali metals all react with water by giving up an electron to it. Based on their relative positions in the periodic table, which do you think would be more reactive with water, lithium or sodium?

Answer: Sodium. It is farther down the group and therefore has a lower ionization energy, meaning it is easier for sodium to give up an electron. Because giving up an electron is how it reacts with water, this should make sodium the more reactive metal.

3.21 Arrange the atoms Mg, Sr, S, O, and Rb from smallest to largest.

3.22 Give the full symbol for the atom or ion that has 26 protons and 30 neutrons in its nucleus and 23 electrons outside its nucleus. Also give the number of the group this element is in.

3.23 Give the full symbol for the atom or ion that has eight protons and eight neutrons in its nucleus and ten electrons outside its nucleus. Also give the name of the group this element is in.

Let's take stock of where chemistry was after Mendeleev's and Rutherford's contributions. Thanks to Mendeleev, chemists had a periodic table that summarized the periodic behavior of the elements. Thanks to Rutherford, chemists knew the basics about atomic structure (massive, tiny nucleus containing protons; electrons somewhere outside the nucleus doing something). But Rutherford's model of the atom could not explain periodicity! An improved model that could explain periodicity was needed. Where was the hint for such a model to be found? Believe it or not, the hint came from examining light. Not just any light, but light given off when atoms of elements in the gas state were provided with large amounts of energy. You are looking at such light every time you read a neon sign. Neon lights are glass tubes filled with neon gas atoms that give off a characteristic orange-red light when supplied with electrical energy. Bright yellow light is given off by sodium-vapor street lamps. Even a tube filled with hydrogen gas glows bluish-pink

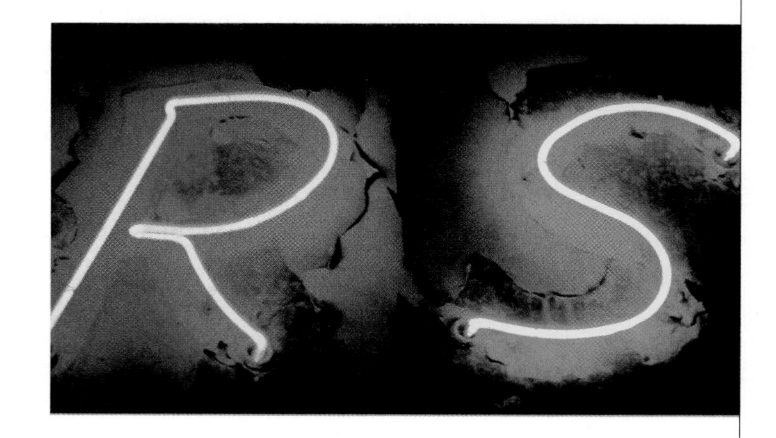

when electrified. Being of a curious nature, physicists passed this light through a prism. Passing light through a prism was nothing new. Sunlight (so-called *white light*) passed through a prism is separated, or *dispersed,* into its component colors. Drops of water in the atmosphere acting like prisms have given all of us a chance to see this dispersion in the form of a rainbow. A rainbow created from white light is called a *continuous spectrum* because the colors smoothly blend into one another without any breaks.

But look at what happened when physicists passed the light given off by a hydrogen or neon lamp through a prism. They saw separate lines of color (first detected for hydrogen by scientists in the mid-1800s). We must tell you that this was astounding! Never before had such a thing been observed. These are called either *discontinuous spectra* or *line spectra* because the colored lines do not blend smoothly into one another. Instead, there are regions of blackness between lines.

Line Spectra of Hydrogen and Neon

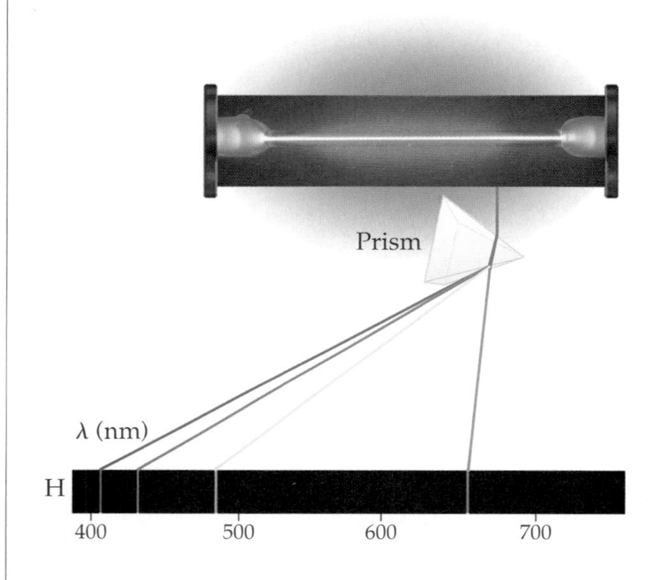

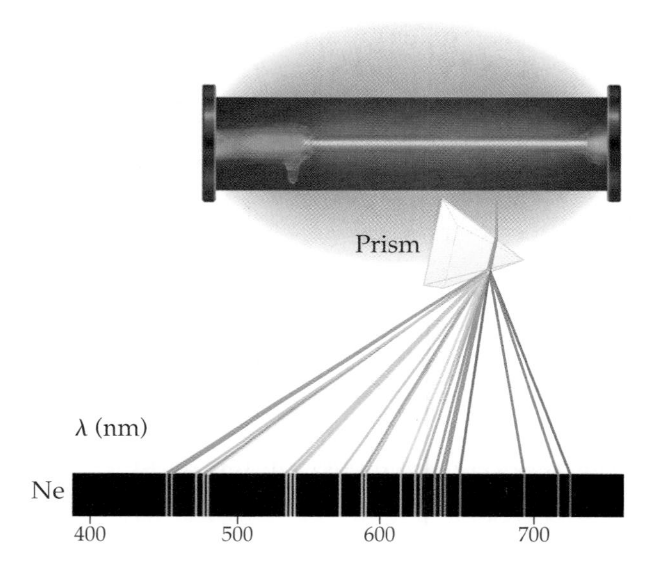

Line spectra were the hint for an improved model of the atom, one that could link periodicity to atomic structure. To understand this hint, we are going to have to take a much closer look at what light is, and so we'll begin the next chapter with light. To keep you interested, however, we shall give a little of the story away now. As it turns out, line spectra would lead to a new model of the atom so unusual, so bizarre, that even those who invented it would have trouble accepting it. Arguments would erupt, scientists would shake their heads in disbelief, and a lot of physics learned over the previous two centuries would be tossed out the window. The areas of chemistry and atomic physics were about to get very exciting.

Have You Learned This?

- Law of conservation of matter (p. 75)
- Law of constant composition (p. 77)
- Percent by mass (p. 77)
- Dalton's atomic theory (p. 78)
- Electron (p. 80)
- Proton (p. 80)
- Neutron (p. 80)
- Alpha particle (p. 81)
- Nucleus (p. 83)
- Atomic number Z (p. 85)
- Mass number (p. 87)
- Isotope (p. 87)
- Atomic mass (p. 89)
- Atomic mass unit (p. 89)
- Chemical periodicity (p. 94)
- Periodic behavior (p. 94)
- Law of Mendeleev (p. 94)
- Periodic table (p. 96)
- Group (p. 96)
- Period (p. 96)
- Representative (main-group) element (p. 97)
- Transition metal (p. 97)
- Lanthanide (rare earth element) (p. 97)
- Actinide (p. 97)
- Noble gas (rare gas) (p. 97)
- Metal (p. 98)
- Nonmetal (p. 98)
- Metalloid (semimetal) (p. 99)
- Alkali metal (p. 100)
- Alkaline earth metal (p. 100)
- Chalcogen (p. 100)
- Halogen (p. 100)
- Atomic radius (p. 102)
- Ion (p. 102)
- Anion (p. 103)
- Cation (p. 103)
- First ionization energy (p. 104)

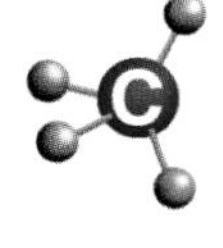

The Evolution of Atomic Theory
www.chemistryplace.com

SKILLS TO KNOW

Calculating a percentage

Often you must calculate what percentage one quantity is of another. What percent of your day is spent commuting? What percent yield can you expect from an investment? These questions all use the basic formula

What percent of y is x?

$$\text{Percent} = \frac{x}{y} \times 100$$

Example: A sample of water consists of 4.435 g of hydrogen and 35.20 g of oxygen. What percent by mass of the sample is hydrogen?

$$\% \text{ Hydrogen} = \frac{\text{Mass of hydrogen}}{\text{Mass of sample}} \times 100$$
$$= \frac{4.435 \text{ g H}}{4.435 \text{ g H} + 35.20 \text{ g O}} \times 100$$
$$= \frac{4.435 \text{ g H}}{34.64 \text{ g H}_2\text{O}} \times 100 = 11.19\%$$

Atomic number, mass number

1. The **atomic number** Z is the number of protons in an atom's nucleus. All atoms of a given element have the same atomic number.

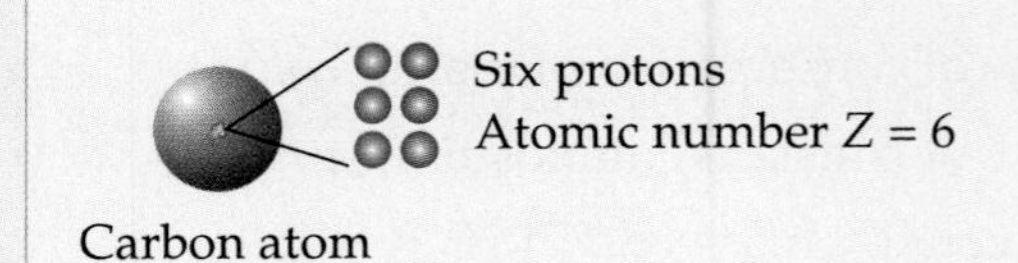

2. The **mass number** of an atom is the total number of protons plus neutrons in the nucleus. Because different isotopes of an element have different numbers of neutrons, each isotope has its own mass number.

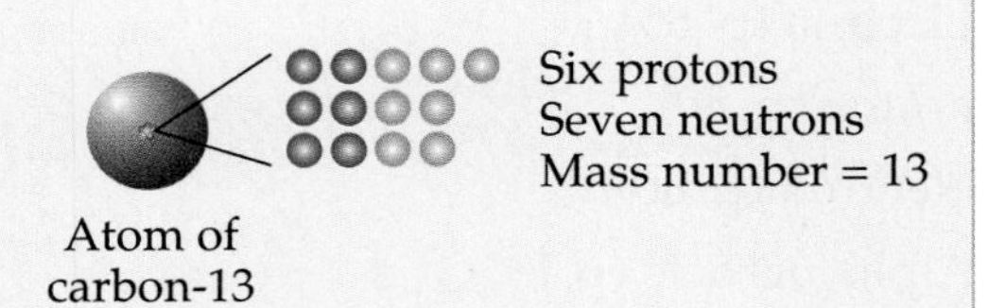

3. An atom's mass number *is not* its atomic mass (except in the case of ^{12}C). However, the mass number is numerically close to the atomic mass because protons and neutrons account for most of an atom's mass and have a mass of about 1 amu each.

Mass number $\neq$ Mass of atom!!
(except in the case of ^{12}C)

Atomic mass

1. Atomic mass is the mass of an atom. The atomic mass *of a specific isotope* is the exact mass of an atom of that isotope.

2. The atomic mass *of an element* reported in the periodic table takes into account the fact that each element consists of isotopes having slightly different masses. The atomic mass reported for an element is a weighted average of the masses of the element's naturally occurring isotopes. It is calculated by multiplying the mass of one atom of each isotope of the element by the relative natural abundance of that isotope.

Example: What is the atomic mass of carbon?

^{13}C
(^{14}C trace)
^{12}C

6
C
12.011

Carbon consists almost entirely of ^{12}C and ^{13}C (the isotope ^{14}C can be neglected). Here is how you calculate the atomic mass of carbon:

$$\leftarrow \text{Atomic mass of C} = \left(\begin{array}{c}\text{Mass of} \\ ^{12}\text{C atom} \\ \text{in amu}\end{array} \times \frac{\%\ \text{abundance of}\ ^{12}\text{C}}{100}\right) + \left(\begin{array}{c}\text{Mass of} \\ ^{13}\text{C atom} \\ \text{in amu}\end{array} \times \frac{\%\ \text{abundance of}\ ^{13}\text{C}}{100}\right)$$

3. Atomic masses are reported in atomic mass units (amu), where 1 amu is exactly 1/12 the mass of an atom of ^{12}C, by definition. Thus, ^{12}C has an atomic mass of 12 amu. (Because atomic mass is defined relative to the mass of a ^{12}C atom, atomic mass is sometimes given without units, as *relative atomic mass*.)

DALTON'S ATOMIC THEORY

3.24 When wood burns, the mass of the ash is less than the mass of the original wood. Yet, the law of conservation of matter says that mass cannot be created or destroyed in a chemical reaction. How do you reconcile the result of burning with this law?

3.25 Volcanoes spew off hydrogen sulfide, a poisonous, bad-smelling gas. A 100.00-g sample of hydrogen sulfide gas was obtained from some strange volcanic planet. Analysis showed that 94.08 g of it is the element sulfur. The rest of the sample consists of the element hydrogen.
(a) What are the percentages by mass of sulfur and hydrogen in this sample?
(b) Hydrogen sulfide from Earth has exactly the same percent compositions for S and H as hydrogen sulfide from the strange volcanic planet. Which law accounts for this?

3.26 If a scientist during Dalton's time found that hydrogen sulfide always had the formula H_2S, how would this have been explained?

3.27 Two compounds of iodine (I) and chlorine (Cl) are analyzed. Compound A consists of 126.9 g of I and 35.45 g of Cl. Compound B consists of 126.9 g of I and 106.4 g of Cl.
(a) What is the percent composition of each element in compound A?
(b) What is the percent composition of each element in compound B?

3.28 Suppose 12.0 g of carbon (C) reacts with 70.0 g of sulfur (S) to give 76.0 g of the compound carbon disulfide (CS_2). In the process, all the carbon gets used up, but some elemental sulfur is left over.
(a) For the law of conservation of matter to be obeyed, how much sulfur is unused?
(b) What is the percent C in CS_2?
(c) What is the percent S in CS_2?
(d) What is the sum of the %C and %S in CS_2?

3.29 Nitrogen and hydrogen form the compound NH_3. How many hooks would Dalton put on the nitrogen atom to explain this compound?

3.30 Two different compounds, both consisting of sodium (Na) and oxygen (O), were analyzed. The data are given below.

Compound	Mass of sample analyzed	Mass of O present	Mass of Na present
A	19.50 g	8.00 g	?
B	61.98 g	16.00 g	?

(a) Fill in the last column of the table.
(b) Calculate the %Na and %O for both compounds.

ATOMIC STRUCTURE

3.31 How did Rutherford interpret the observation that only a very few of the α particles were scattered by the gold foil?

3.32 How did Rutherford know that most of the α particles went through the gold foil without striking anything?

3.33 What was wrong with Rutherford's model of the atom as far as the physics of the day was concerned?

3.34 Suppose all the α particles in Rutherford's experiment went straight through the gold foil with absolutely no deflections. What would this imply about the structure of the atom?

3.35 Why isn't an atom's mass number sufficient to determine the atom's elemental identity?

3.36 Does knowing how many electrons a neutral atom has tell you its elemental identity? Explain.

3.37 Physicists are fond of saying that an atom is mostly empty space. What justifies this statement?

3.38 Fill in the following table for four neutral atoms:

	$^{15}_{8}O$	?	?	?
Mass number	?	16	37	?
Atomic number	?	8	?	?
Number of protons	?	?	?	?
Number of neutrons	?	?	?	12
Number of electrons	?	?	17	11

3.39 Uranium exists mainly as two isotopes in nature, possessing mass number 235 and 238. Write the full atomic symbols for both isotopes.

3.40 What is wrong with this symbol?

$^{12}_{7}C$

3.41 Give the full atomic symbol for this atom:

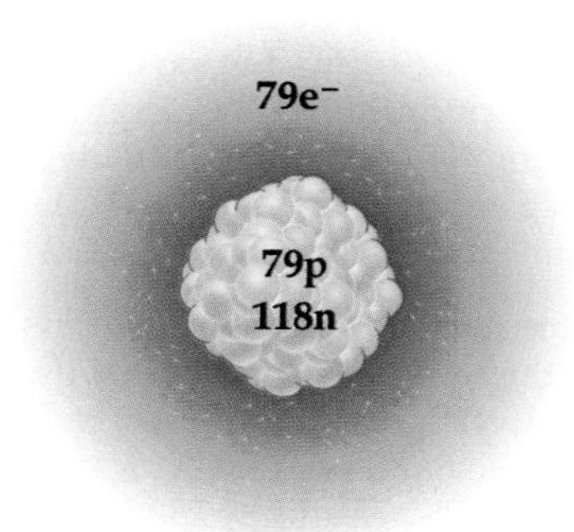

3.42 Write the full atomic symbol for the only atom that has its mass number equal to its atomic number.

3.43 Suppose you wanted to turn atoms of lead into atoms of gold. What would you have to do to the nucleus of the lead atoms?

3.44 What is the difference between an atom's atomic mass and its mass number?

3.45 Why is the atomic mass scale also called the relative atomic mass scale?

3.46 What is so special about the mass number and atomic mass of $^{12}_{6}C$?

3.47 How much more massive is an "average" oxygen atom than a $^{12}_{6}C$ atom? [Use the periodic table for the atomic mass of oxygen.]

3.48 How much more massive is an "average" titanium (Ti) atom than a $^{12}_{6}C$ atom? [Use the periodic table for the atomic mass of titanium.]

3.49 How much more massive is an "average" oxygen atom than a He atom? [Use the atomic masses from the periodic table.]

3.50 Why is the atomic mass of carbon not listed as 12 on the periodic table if the universal standard is that $^{12}_{6}C$ has an atomic mass of exactly 12 amu?

3.51 Suppose we wanted to use one of the more massive atoms, like $^{235}_{92}U$, as the standard reference for the atomic mass scale.
(a) What would be the universally agreed upon atomic mass of the $^{235}_{92}U$ isotope?
(b) Why don't we use $^{235}_{92}U$ instead of $^{12}_{6}C$?

3.52 Bromine exists as only two isotopes in nature, $^{79}_{35}Br$ (atomic mass 78.918 336 amu, % natural abundance = 50.69%), and $^{81}_{35}Br$ (atomic mass 80.916 289 amu).
(a) What is the % natural abundance for $^{81}_{35}Br$?
(b) Calculate the atomic mass of the naturally occurring mixture of isotopes.

3.53 Naturally occurring hydrogen on Earth has an atomic mass of 1.0079 amu. Suppose you were on another planet and found the atomic mass of hydrogen to be 1.2000 amu. How would you explain this? (The atomic mass of $^{1}_{1}H$ is 1.007 825 2 amu; the atomic mass of $^{2}_{1}H$ is 2.104 102 2 amu.)

3.54 While an atom's mass number and its atomic mass are not the same, they are often quite close to one another. Consider the following: Uranium-235 has an atomic mass of 235.043 93 amu and a % abundance of 0.73%. Uranium-238 has an atomic mass of 238.0508 amu and a % abundance of 99.27%. Without doing any calculations, which of the following do you think represents the atomic mass of naturally occurring uranium?
(a) 234.04 amu (b) 236.03 amu
(c) 237.03 amu (d) 238.03 amu
(e) 238.07 amu

Explain how you made your choice.

THE MODERN PERIODIC TABLE

3.55 (a) In what unique way did Mendeleev order the elements to make his discovery?
(b) What was his discovery?
(c) How does his ordering differ from the modern ordering?

3.56 How did Mendeleev's discovery aid in discovering additional elements?

3.57 Define what is meant by chemical periodicity with respect to chemical properties.

3.58 If one considers just the representative elements, how many groups would the periodic table have?

3.59 Considering just the representative elements, what percentage of them are metals, nonmetals, and metalloids? How do your results compare to the same percentages derived using the entire periodic table?

3.60 Give the names associated with groups IA (1), IIA (2), VIA (16), VIIA (17), and VIIIA (18).

3.61 What element seems poorly placed in group IA (1), the alkali metals? Why?

3.62 How does a group differ from a period in the periodic table?

3.63 As you take one step to the right in any period, how many electrons and protons are being added to the atom?

3.64 Magnesium (Mg) reacts with chlorine (Cl_2) to form the compound $MgCl_2$.
(a) Predict the formulas for the compounds formed when all the other alkaline earth metals react with chlorine.
(b) Predict the formulas for the compounds formed when all the alkaline earth metals react with bromine.
(c) Explain what principle you used to make your predictions.

3.65 State two things that are unique about the noble gases.

3.66 Name and give symbols for five of the transition metals.

3.67 How many elements wide are the transition metal and lanthanide (rare earth)/actinide portions of the periodic table?

3.68 Why does chemical periodicity change from 8 to 18 and then to 32?

3.69 Metals are excellent conductors of electricity, whereas nonmetals are poor conductors of electricity (insulators). Silicon (Si) is somewhere in between and is referred to as a semiconductor. Why do you think this is so for Si?

OTHER REGULAR VARIATIONS IN THE PROPERTIES OF ELEMENTS

3.70 Explain what is responsible for the size trend as you go across a period from left to right.

3.71 Order the following atoms from smallest to largest, judging from their relative positions in the periodic table: Cs, Fe, Ti, Hf.

3.72 Order the following atoms from smallest to largest, judging from their relative positions in the periodic table: Ca, Se, F, S.

3.73 Two students are studying using a plastic model of the atom to which they can easily add or remove electrons, protons, and neutrons. They build a model for the neutral $^{14}_{6}C$ atom. Their next job is to make the +2 ion of this carbon isotope. Student X removes 2 electrons from the atom. Student Y adds 2 protons to the atom's nucleus. Did both students successfully construct a +2 carbon-14 cation? Explain.

3.74 Fill in the following table:

	$^{15}_{8}O^{+}$	?	?	?
Mass number	?	27	?	58
Atomic number	?	?	15	?
Number of protons	?	13	?	?
Number of neutrons	?	?	16	30
Number of electrons	?	?	?	27
Charge on ion	?	+3	−3	+1

3.75 What is meant by the term *first ionization energy*?

3.76 True or false? It always takes energy to remove an electron from a neutral atom. Explain your answer.

3.77 What is the trend in first ionization energy as you go down a given group in the periodic table? As you go across a period from left to right?

3.78 Explain the trends in ionization energy in terms of the trends in atomic size.

3.79 Of the atoms Na, Mg, and Al, which should be the most difficult to ionize? Which should have the smallest first ionization energy? Explain fully.

3.80 Of the atoms Na, Mg, and K, which should be the most difficult to ionize? Which should have the smallest first ionization energy? Explain fully.

3.81 Refer to the chart shown at the top of page 106. Do the halogens follow the expected trend in ionization energies? Explain fully.

3.82 Sodium, Na, metal rapidly reacts with chlorine, Cl_2, a nonmetal.
(a) What reason is given for this high reactivity in this chapter?
(b) In what ways is the reaction between lithium, Li, and bromine, Br_2, similar to the reaction between sodium and chlorine?

3.83 In the reaction of lithium, Li, with nitrogen, N_2, three lithium atoms react with one nitrogen atom to give the compound Li_3N. Atoms of one of these elements lose 1 electron; atoms of the other element gain 3 electrons.
(a) Which element gains the 3 electrons? Explain your choice.
(b) Is the element that gained 3 electrons a cation or an anion?
(c) Give the full atomic symbol for the ion of part (b), assuming it has 7 neutrons in its nucleus.

3.84 White light from an incandescent light bulb gives a continuous visible spectrum when the light is passed through a prism. What is meant by the word *continuous* in this context?

3.85 A heated gas made up of individual atoms gives off a line spectrum when the emitted light is passed through a prism. How does a line spectrum differ from a continuous spectrum?

ADDITIONAL PROBLEMS

3.86 A 34.01-g sample of pure hydrogen peroxide produced at a chemical company contains 32.00 g of oxygen. The rest of the mass is due to hydrogen.
(a) What is the percent by mass of oxygen and hydrogen in this sample?
(b) Another company also produces pure hydrogen peroxide. What is the percent by mass hydrogen and oxygen in its hydrogen peroxide? What chemical law allowed you to answer this question?
(c) A 91.83-g sample of pure hydrogen peroxide is obtained. How many grams of it are oxygen?

3.87 Tin, Sn, and oxygen can combine to form two different compounds called oxides. Oxide A contains 78.77% by mass tin. Oxide B contains 88.12% by mass tin. What is the percent by mass of oxygen in each oxide?

3.88 When wood burns, it combines with oxygen from the atmosphere to produce carbon dioxide, water vapor, and ash. What would you have to do to prove that the law of conservation of matter is not disobeyed when wood burns?

3.89 What evidence exists to support the theory that an atom contains a massive nucleus that is very small relative to the size of the whole atom?

3.90 When 5 g of compound A reacts with an unlimited amount of compound B, 7 g of compound C is formed. How many grams of compound B must have reacted, and what law allows you to answer the question?

3.91 A student claims that isotopes of the same element have the same number of electrons and protons. Is she correct? Justify your answer.

3.92 A student claims that isotopes of the same element have the same number of neutrons but different numbers of protons. Is he correct? Justify your answer.

3.93 How do you convert a neutral sulfur atom to an anion carrying a charge of 2−?

3.94 A student says that the easiest way to convert a neutral sodium atom to a Na^+ cation is to add a proton to its nucleus. She reasons that this is so because protons have a 1+ charge. What is wrong with her argument? What does a neutral Na atom become when a proton is added to its nucleus?

3.95 The mass number of a neutral atom is 19, and its atomic number is 9. The atom contains how many protons, electrons, and neutrons?

3.96 The mass number of a 2+ cation is 56, and its atomic number is 26. The ion contains how many protons, electrons, and neutrons?

3.97 How many protons, electrons, and neutrons are there in
(a) ^{79}Br (b) $^{81}Br^-$ (c) $^{23}Na^+$ (d) $^{3}H^+$

3.98 The element nitrogen has only two naturally occurring isotopes: ^{14}N with a mass of 14.003 08 amu and an abundance of 99.635 percent and ^{15}N with a mass of 15.000 11 amu and an abundance of 0.3650 percent. Calculate the atomic mass of nitrogen.

3.99 Silver has only two naturally occurring isotopes: ^{107}Ag with a mass of 106.905 09 amu and an abundance of 51.84 percent and ^{109}Ag with a mass of 108.9047 amu. Calculate the atomic mass of silver.

3.100 Mendeleev arranged the elements by atomic mass and then separated them into various groups. What was his criterion for which elements were grouped together?

3.101 According to the periodic table on the inside front cover, how many main-group elements are known? How many of these are representative elements?

3.102 Arrange Na, Cs, S, Cl in order of (a) increasing atomic size and (b) increasing first ionization energy.

3.103 Arrange calcium, strontium, arsenic, bromine, chlorine in order of (a) increasing atomic size and (b) increasing first ionization energy.

3.104 How many electrons are there in Zr^{4+}? What 2+ cation has the same number of electrons as Zr^{4+}?

3.105 How many electrons are there in Br^-? What 1+ cation has the same number of electrons as Br^-?

3.106 What is the identity of element X if its 2+ cation contains ten electrons?

3.107 Of the atoms Na, Cl, K, Br, which has the largest atomic radius? Which has the largest first ionization energy?

3.108 Use the periodic tables on pages 96 and 98 to complete this table.

Name of group or classification	Period	Group	Elemental symbol	Atomic number	Atomic mass (amu)	Metal, metalloid, or nonmetal
				26		
Noble gas	1					
	5	VIIA (17)				
				19		
Halogen	3					
			Ne			
				8		
——	3					Metalloid
		——		92		

3.109 A 110.99-g sample of a compound containing only calcium and chlorine is found to contain 40.98 g of calcium.
(a) How many grams of chlorine are in the compound?
(b) What is the percent by mass of calcium in the compound?
(c) What is the percent by mass of chlorine in the compound?

3.110 Explain the major difference between Thomson's model of the atom and Rutherford's model. Draw a picture of each model.

3.111 Arrange Ca, Se, F, S, Rb in order of increasing first ionization energy.

3.112 How does the arrangement of elements in the modern periodic table differ from that of Mendeleev's periodic table?

3.113 Write the elemental symbol for nitrogen, silicon, phosphorus, potassium, and gold.

3.114 Indicate atomic number, mass number, and number of protons, neutrons, and electrons in a neutral atom of each isotope:
$^{27}_{13}Al$ $^{60}_{27}Co$ $^{200}_{79}Au$ $^{238}_{92}U$ $^{127}_{53}I$

3.115 List three properties of metals and two properties of nonmetals. Describe the behavior of metalloids relative to the behaviors of metals and nonmetals.

3.116 Explain why Rutherford expected all alpha particles to go through the gold foil undisturbed.

3.117 (a) The elements in group IA (1) are called ________.
(b) The elements in group IIA (2) are called ________.
(c) The elements in group VIIA (17) are called ________.
(d) The elements in group VIIIA (18) are called ________.

3.118 Use Dalton's hook atomic models to sketch a molecule of H_2S.

3.119 Write the full atomic symbol for each isotope:
(a) 8 protons, 9 neutrons
(b) Atomic number 50, mass number 119
(c) 12 neutrons, atomic number 11
(d) 28 protons, mass number 58
(e) 81 neutrons, symbol Ba

3.120 Complete the following statements:
(a) The nucleus contains the ________ charged particles in the atom.
(b) The particles in the nucleus are the ________ and ________.
(c) Almost all of the mass of an atom is contained in the ________.
(d) The two particles whose charges cancel to make an atom neutral are ________ and ________.

3.121 Explain the relationship between the size of an atom and its first ionization energy.

3.122 Calcium reacts with fluorine to form the compound CaF_2. In the reaction, each atom of one of the elements loses two electrons, and each atom of the other element gains one electron.
(a) Atoms of which element gain electrons?
(b) Atoms of which element lose electrons?
(c) What is the charge on the atoms that gain one electron?
(d) What is the charge on the atoms that lose two electrons?
(e) Write the full atomic symbol, including charges, for each element. Give Ca 20 neutrons and F 10 neutrons.

3.123 Use the elements lithium, potassium, and sodium to explain what is meant by chemical periodicity.

3.124 There are three isotopes of hydrogen—protium, deuterium, and tritium. Protium contains one proton and no neutrons. Deuterium contains one more neutron than protium, and tritium contains one more neutron than deuterium. For each isotope, give the atomic number, mass number, number of protons/neutrons/electrons, and full atomic symbol.

3.125 Given that one measure of metallic character is the tendency to lose electrons, arrange Cs, Be, Li, Ne, Na in order of increasing metallic character.

3.126 For each pair, indicate which atom loses an electron more easily:
(a) Na or K (b) Na or Al (c) Rb or Ca
(d) S or Cl (e) Br or Kr

3.127 Explain the difference between mass number and atomic mass.

3.128 Naturally occurring copper, Cu, is composed of two isotopes. The isotope copper-63 is 69.17% of naturally occurring copper and has an atomic mass of 62.94 amu. The isotope copper-65 has an atomic mass of 64.93 amu. (a) What percent of naturally occurring copper is copper-65? (b) Calculate the atomic mass of naturally occurring copper.

3.129 Explain why an alternative name for the vertical columns in the periodic table is *families*.

3.130 Explain why the first and second statements of Dalton's atomic theory are not exactly true.

3.131 Write the name of each element:
(a) Be (b) Mg (c) Fe (d) S (e) Ar (f) Cu

3 WORKPATCH SOLUTIONS

3.1 (a) The law of conservation of matter is obeyed because 100.0 g + 793.6 g = 893.6 g.
(b) The percent hydrogen is (100.0 g/893.6 g) × 100 = 11.19% H.
(c) The percent oxygen is (793.6 g/893.6 g) × 100 = 88.81% O.
(d) The law of conservation of matter is obeyed because 50.0 g + 793.6 g = 446.8 g + 396.8 g.
(e) The percent hydrogen is (50.0 g/446.8 g) × 100 = 11.19% H.
(f) The percent oxygen is (396.8 g/446.8 g) × 100 = 88.81% O. [Notice that the amount of oxygen (396.8 g) in the calculation is the amount actually used, which is the starting amount minus the leftover amount.]

3.2 Because most of the alpha particles passed through the gold atoms undeflected.

3.3 From the figure, it is obvious that an atom has as many electrons outside the nucleus as it has protons inside the nucleus. Because each electron has a −1 charge and each proton has a +1 charge, the overall atom must be neutral because equal numbers of +1 and −1 charges cancel each other.

3.4

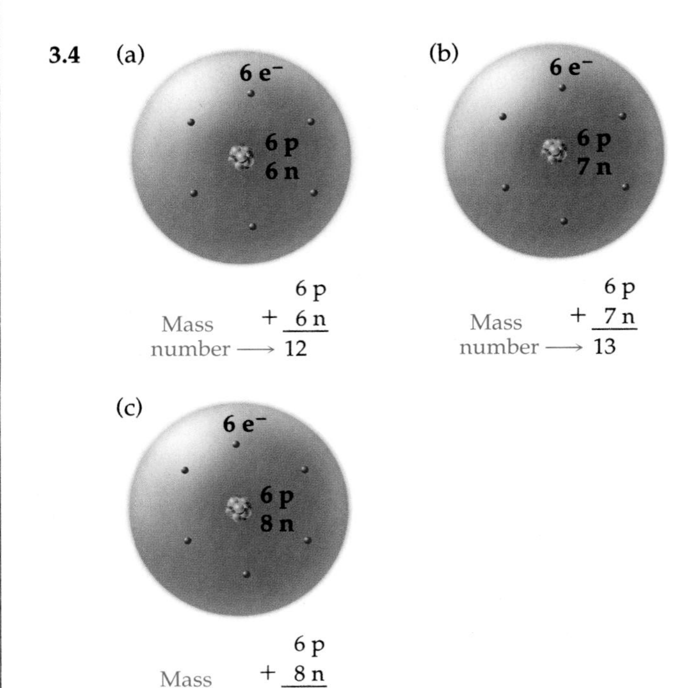

3.5 No. The mass number tells you the number of protons + neutrons. To identify the element, you need to know the number of protons (atomic number) in the nucleus.

3.6 Now you can figure out the atomic number. If the atom has a mass number of 16 and contains nine neutrons, the number of protons is $16 - 9 = 7$. The element with atomic number 7 is nitrogen (N).

3.7 If the mass of an atom is 4.015 times the mass of a $^{12}_{6}C$ atom, the atom's relative atomic mass is $4.015 \times 12 = 48.18$. (Remember that 12 is an exact number, and so we don't need to round off.)

3.8 Because silicon (Si) is in the same group as carbon, and thus its chemical properties should be similar to those of carbon (C).

3.9

	$^{14}_{7}N^{3-}$	$^{24}_{12}Mg^{2+}$	$^{23}_{11}Na^{+}$	$^{56}_{26}Fe^{3+}$
Mass number	14	24	23	56
Atomic number	7	12	11	26
Number of protons	7	12	11	26
Number of neutrons	7	12	12	30
Number of electrons	10	10	10	23

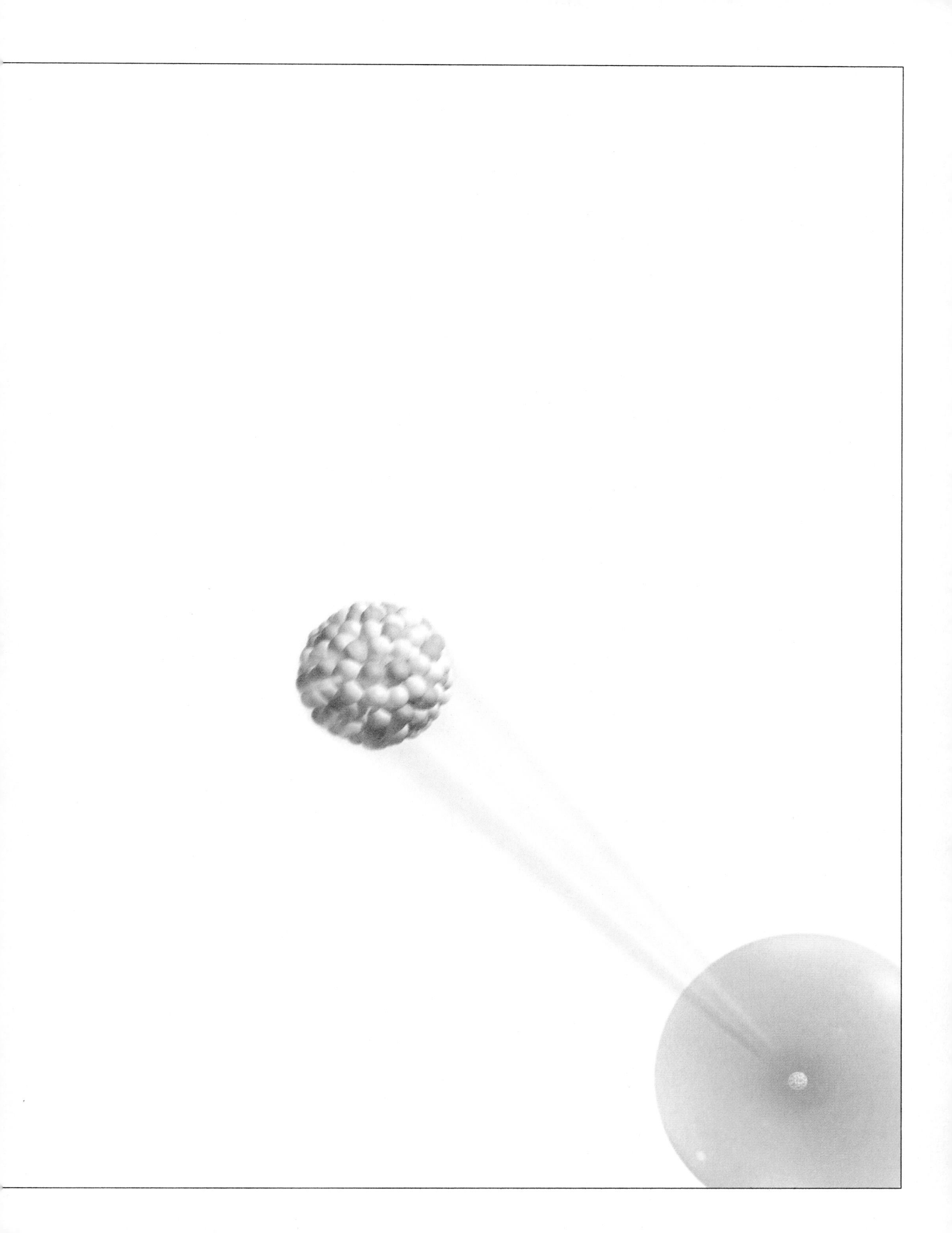

CHAPTER 4

The Modern Model of the Atom

4.1 Seeing the Light—A New Model of the Atom

We ended Chapter 3 with a first look at the line spectrum given off by an electrified tube of hydrogen gas. To explain such a spectrum, we said, science would postulate a model of the atom so strange that even those who proposed it had trouble believing it. Erwin Schrödinger, the physicist most credited with developing the current model of the atom, would be quoted as saying, "I don't like it, and I'm sorry I ever had anything to do with it." The great physicist Albert Einstein refused to believe it. This is pretty exciting stuff for science, but to understand how science explained the hydrogen spectrum, you must first know more about light.

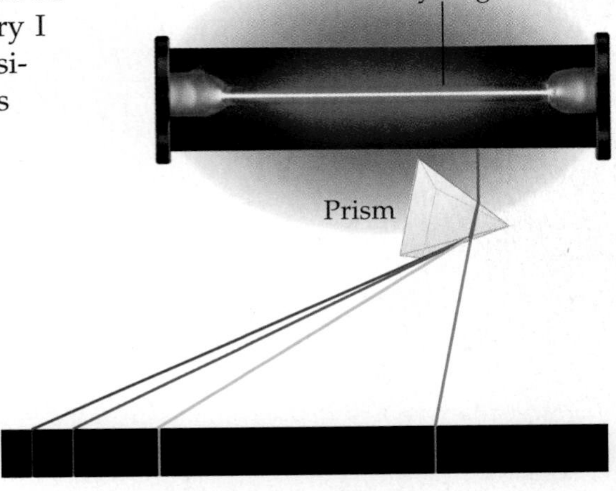

Line spectrum of hydrogen

We are all familiar with light. Flick a switch in a darkened room, and an electric current begins to flow through a tungsten filament until the filament glows white hot inside a glass bulb. The resulting white light illuminates the room.

Light moves through space at the incredible speed of 3.00×10^8 m/s. Called c, or the speed of light, this speed equals 186,000 miles/s or 671 million miles/h! If you were traveling at the speed of light, you could travel around the world at its equator seven and a half times in a single second. But what exactly *is* light? Modern science tells us that light is two different things at once. It consists of particles, or packets, of energy called *photons*, and yet at the same time it acts like a *wave of energy*. Here we want to focus on the view of light as a wave.

Being a wave, light can be characterized by its *wavelength*, symbolized by the Greek letter λ (lambda). **Wavelength** is simply the distance between identical adjacent points on a wave—that is, the distance between one crest (or trough) and the next.

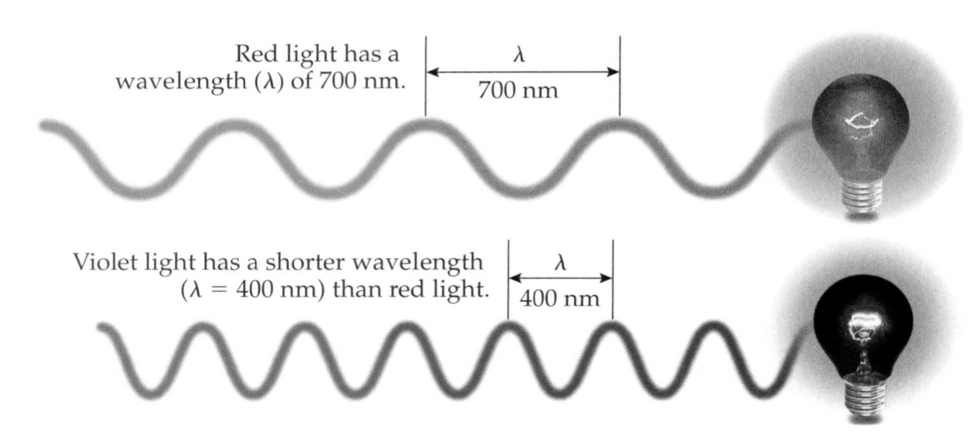

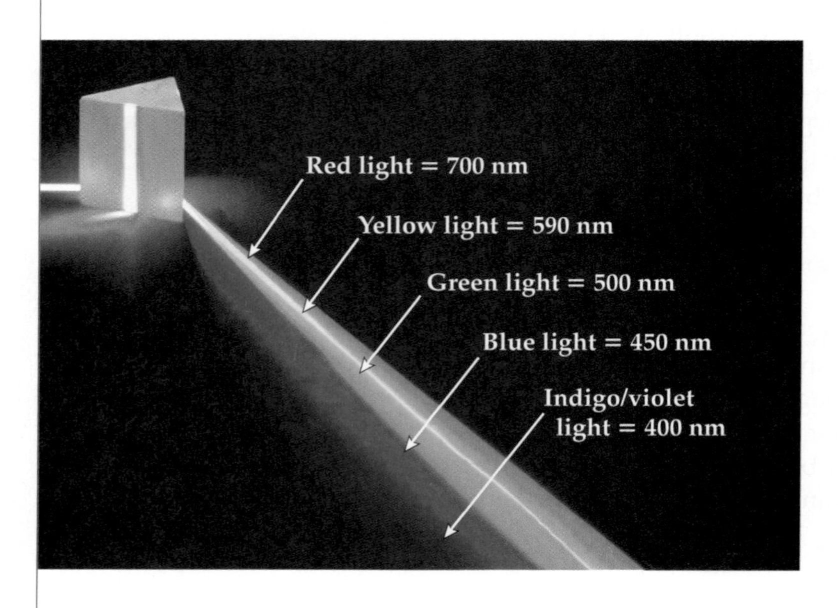

Our brains interpret light of different wavelengths as different colors. Light with a wavelength between 650 and 750 nm (nanometer, 10^{-9} m) is seen as red light. Violet light has shorter wavelengths, around 380–430 nm. The wavelengths between these extremes are interpreted by our brains as the different colors of the rainbow. The wavelengths of light between 380 and 750 nm are collectively called *visible light* because these are the wavelengths our eyes are designed to detect. When all these wavelengths impinge on our eyes at once, we see white light. Either a prism or a diffraction grating will separate, or *disperse*, this white light into its different wavelengths, giving rise to a continuous spectrum, the familiar rainbow of colors. As we said at the end of Chapter 3, it is called continuous because the colors smoothly blend into one another without any breaks.

Though our world seems very bright and colorful, the fact is that we do not see most of the light that hits our eyes. Outside the visible range of 380 to 750 nm, there are many other wavelengths that are invisible to our eyes. All the visible and invisible wavelengths are collectively called **electromagnetic radiation** because they consist of both an electrical component and a magnetic component. The entire spectrum of electromagnetic radiation is known as the **electromagnetic spectrum**.

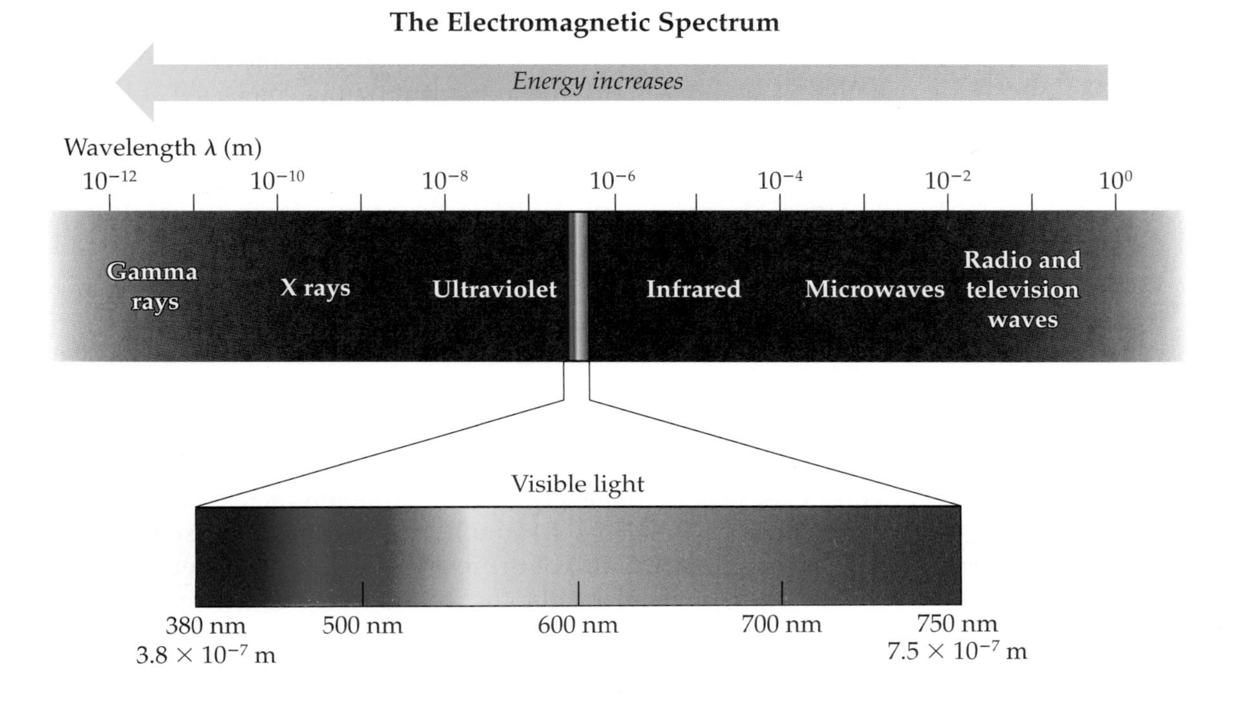

Even though we cannot see any electromagnetic radiation outside the narrow visible region, it can affect us. This is because all electromagnetic radiation has energy, and the amount of energy depends on the wavelength of the radiation. The equation relating the wavelength of a particular beam of electromagnetic radiation to its energy content is

Equation to calculate the energy of electromagnetic radiation

$E = hc/\lambda$ where E = energy of electromagnetic radiation
$c = 3.00 \times 10^8$ m/s (speed of light)
$h = 6.626 \times 10^{-34}$ J · s (Planck's constant)
λ = wavelength of electromagnetic radiation

To use this equation, you simply plug in the values for h and c (both constants) and the wavelength λ of the light. Watch your units, though! Because the speed of light c has units of meters per second, your wavelength must be expressed in meters. Because Planck's constant h has units of joules $\times$ seconds, your calculated energy will be in joules.

Practice Problems

4.1 What is the energy of blue light that has a wavelength of 450.0 nm?

Answer: $E = \frac{hc}{\lambda} = \frac{(6.626 \times 10^{-34}\ \text{J} \cdot \cancel{\text{s}})(3.00 \times 10^{8}\ \cancel{\text{m}}/\cancel{\text{s}})}{(450.0\ \cancel{\text{nm}})(1 \times 10^{-9} \cancel{\text{m}}/\cancel{\text{nm}})} = 4.42 \times 10^{-19}\ \text{J}$

4.2 What is the energy of red light with a wavelength of 660.5 nm?

Answer: 3.01×10^{-19} *J*

4.3 What are the wavelength and color of light that has an energy of 3.50×10^{-19} J? [*Hint:* Use algebraic manipulation to solve the energy equation for λ. You will get an answer in meters. Convert it to nanometers (10^{-9} nm = 1 m) and consult the electromagnetic spectrum shown on page 121.]

Answer: 568 nm, yellow

4.4 Which type of electromagnetic radiation has a wavelength roughly on the order of the height of a person? What can you say about the energy of this radiation?

The equation $E = hc/\lambda$ tells us that the energy of electromagnetic radiation is inversely proportional to its wavelength. As the wavelength of electromagnetic radiation gets smaller, the energy of the radiation increases. Compare the answers to Practice Problems 4.1 and 4.2 to convince yourself of this. Because the wavelength (660.5 nm) of the red light is larger (longer) than the wavelength (450.0 nm) of the blue light, the red light has less energy.

Invisible to us are all forms of electromagnetic radiation having wavelengths shorter than that of violet light (ultraviolet, X rays, and gamma rays). Because they contain so much energy, ultraviolet rays can cause sunburn, X rays can cause cancer, and gamma rays can kill. Also invisible to us are the forms of electromagnetic radiation having wavelengths longer than that of red light (infrared, microwaves, radio and television waves), although we can feel infrared radiation as heat.

So, what should you get out of this discussion? Simply that each color (wavelength λ) of light has a particular amount of energy. Thus the fact that

Four Lines of the Hydrogen Spectrum in the Visible Region

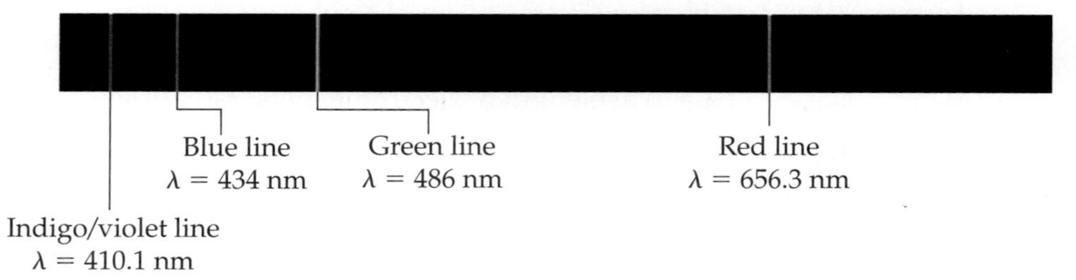

only four distinct lines are seen in the visible region of the spectrum of electrified hydrogen atoms means that these atoms emit only the four particular amounts of energy associated with these four wavelengths and not any intermediate amounts of energy. This is why the spectrum is black between the lines, but why should the atoms behave this way?

4.2 A New Kind of Physics—Energy Is Quantized

In 1913 the Danish physicist Niels Bohr proposed an answer that was revolutionary. He said that the energies of the electrons inside atoms are *quantized*. By **quantized** he meant that the electrons could have only particular discrete amounts of energy. Such an assertion went against all the teachings of classical physics. Both classical physics and everyday observation tell us that an object can have any amount of energy. Consider a tennis ball, for example. It can have any amount of kinetic energy, depending on how hard you hit it. It can have low energy (a gentle lob over the net), high energy (a fierce slam shot), or any energy between these two extremes.

But imagine a tennis ball that could travel at only certain speeds and therefore would have only certain energies. It could never travel at any speed other than the allowed ones, no matter how hard or soft it was hit. That would be quite a tennis ball! Actually, tennis balls *do* have quantized energies; we just can't detect this fact (read on). However, we *can* detect that electrons in an atom have quantized energies.

Before we go on, we want to remind you of the scientific method discussed in Chapter 1. The scientific method can result in a well-respected theory being set aside and replaced with another one. Most often, a major change of this kind is not the result of one person's work. For example, when Bohr proposed that the energy of electrons in atoms is quantized, he was drawing on the work of the German physicist Max Planck. Planck had earlier proposed energy quantization in his efforts to explain the energy characteristics of the light emitted by heated objects. Planck saw this only as a mathematical trick, something that allowed him to arrive at the correct answer in his calculations. He did not believe that energy was actually quantized in any real physical system. Bohr was aware of Planck's work, and he took the concept one step further by applying it to the electrons in an atom. That was a bold thing to do, particularly since Bohr had no explanation for why electrons should exhibit quantization. He just knew that quantization explained certain observations. The explanation of why quantization occurs would come about ten years later from yet another scientist, Louis de Broglie. This is the scientific method in action.

As noted at the end of Section 3.3, scientists today recognize two kinds of physics, classical and quantum. **Classical physics** says that objects can have any energy. It works extremely well for large objects, such as tennis balls, cannon balls, rockets, and planets. **Quantum physics** says that objects can have only certain particular energies. It is the physics that correctly explains the behavior of very small particles the size of atoms and electrons. In fact, quantum physics is the more general of the two. Earlier we said that the motion of a tennis ball is quantized but we just can't detect the quantization. That is because the differences between the allowed energies for a large object like a tennis ball are so small that the energies seemingly blend into one another to yield a smooth continuum.

For example, consider heating a large amount of water on a stove. As you add heat energy to the water, the water's temperature (a measure of its energy content) appears to increase in a smooth, continuous fashion. Even with the world's most sensitive and accurate thermometer, you would not observe any temperature jumps. That is not because there are no temperature jumps but rather because their size is too small to be measured. The jump size is on the order of Planck's constant, which means about 10^{-34} degree! No thermometer could ever detect such small changes. However, the differences between the allowed energies for a tiny electron inside an atom are much larger and therefore detectable. We can therefore get away with using classical physics for large objects, but we must use quantum physics if we want to understand how an electron behaves in an atom.

Imagine that cars were the size of atoms. What would traffic look like on the highway?

4.3 The Bohr Theory of Atomic Structure

Bohr proposed that the electrons in an atom travel around the nucleus in circular orbits much as the planets orbit the Sun. The revolutionary part of his model was that only certain orbits, at certain fixed distances from the nucleus, were allowed. By imposing this restriction, he was saying that electrons in atoms could have only certain energies.

To understand the relationship between the size of an electron's orbit and the energy of the electron, you need only remember that the nucleus is positively charged and an electron is negatively charged, and so they are attracted to each other. Think of this attraction as a spring that connects the electron to the nucleus. When the electron is closest to the nucleus, the spring is relaxed and the electron is at its lowest possible energy. To pull the electron away from

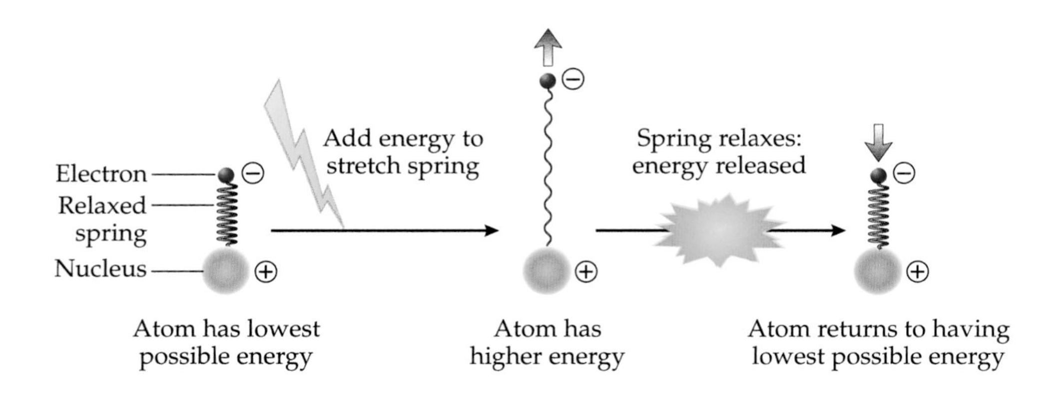

the nucleus into a larger orbit requires putting energy into the electron to overcome the electron–nucleus attraction. In our spring model, that is equivalent to stretching the spring. The energy we put in is stored in the stretched spring. The stretched spring (and thus the atom itself) now contains more energy than before and therefore is in a higher-energy, less stable situation. Unless there is something preventing it, the electron will spontaneously snap back into the lower orbit and release the energy that was stored in the stretched spring.

Therefore, the closer an electron is to the nucleus, the lower the electron's energy and the more stable the atom:

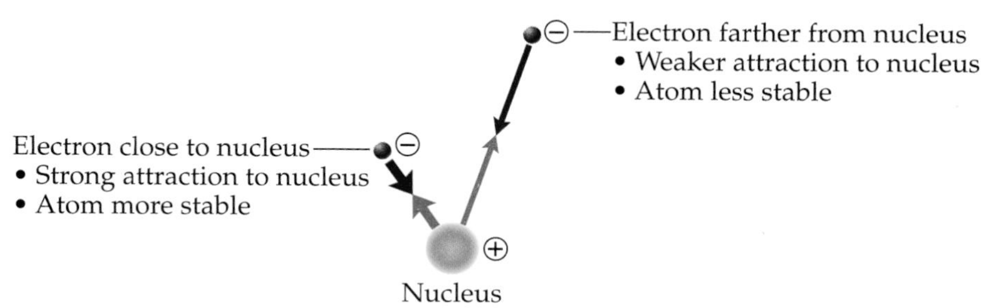

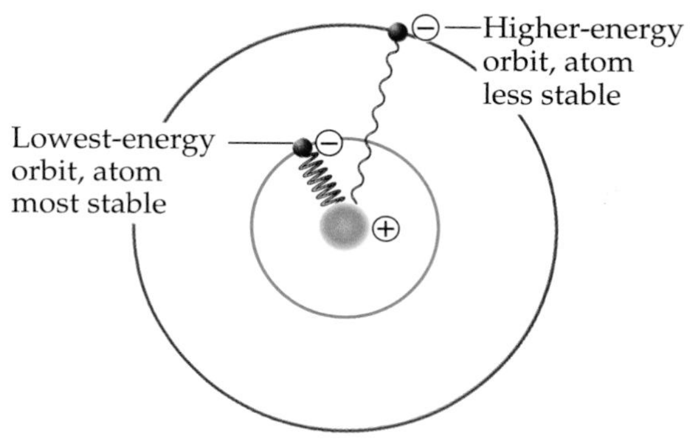

So by saying that an electron could be only at certain distances from the nucleus, Bohr was really saying that the electrons in an atom could have only certain (quantized) energies—as if the electrons were connected to specific orbits.

Before we go on, it is important to tell you that the scientific world quickly moved beyond Bohr's circular orbits. No one today believes that electrons orbit an atom's nucleus on fixed paths. The modern model of the atom has a very different view of how electrons behave in an atom. (We'll introduce you to the modern model at the end of the chapter.) Our new view does not mean that Bohr's model is not worth learning about. Far from it. Bohr's energy quantization and the relationship between the energy of the electron and its distance from the nucleus are all correct and part of the modern model. In addition, Bohr's model is much easier to use than the modern model, and it works well in explaining certain aspects of chemistry, such as the periodic behavior of the elements or why certain compounds have the chemical formulas they do. Also, an acquaintance with Bohr's model is probably the best way to prepare for understanding the modern one.

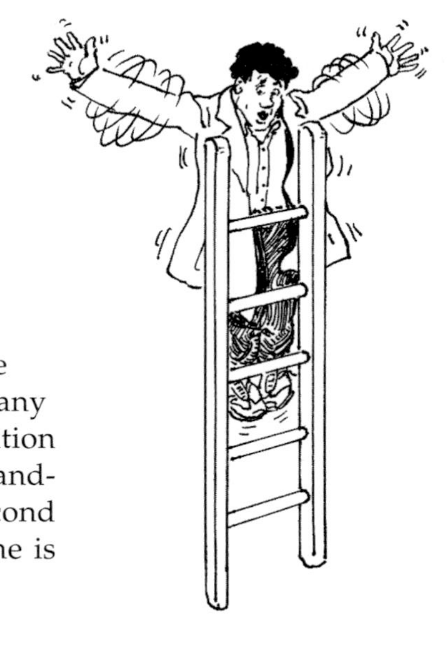

In Bohr's model of the atom, the electrons can jump from one allowed energy level to another (from one orbit to another; these movements are called *quantum jumps*), but they can never have any energy between the allowed ones. In Bohr's view, electrons position themselves around the nucleus in an atom much like a person standing on a ladder. The person can stand on the first rung, the second rung, and so on, but can't stand *between* any two rungs. The same is true for electrons in atoms in Bohr's model.

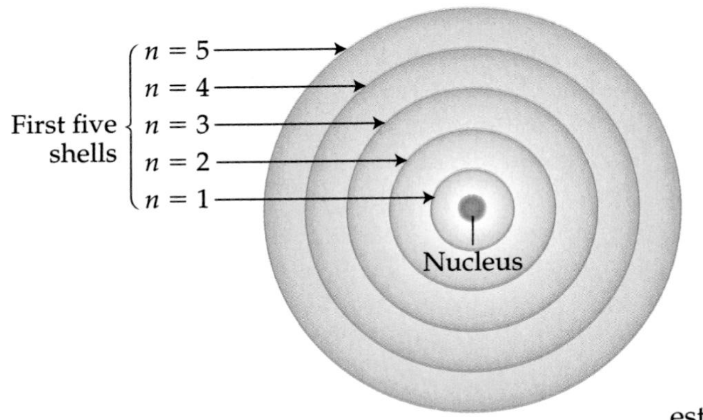

Bohr assigned each of his quantized electron orbits, which were called **shells**, a **principal quantum number** (n), starting with $n = 1$ for the shell closest to the nucleus, $n = 2$ for the next shell out from the nucleus, and so on. The shells continue with $n = 6, 7, 8$, and so on up to infinity, getting larger and larger, taking the electron farther from the nucleus and thus giving it more and more energy.

Ideally, all the electrons in an atom would like to occupy the $n = 1$ shell, because it is the shell of lowest energy (maximum electron–nucleus attraction). But electrons also repel one another, and crowding them into a single shell would quickly raise the energy of the atom dramatically. In order to deal with this problem, and also to explain line spectra and periodic chemical behavior, Bohr maintained that each shell can hold a certain maximum number of electrons. The larger the shell, the more electrons it can hold, because a larger shell provides more room in which the electrons can spread out. Bohr maintained that each shell can hold a maximum of $2n^2$ electrons, where n is the principal quantum number of the shell.

Here, then, are the main features of the Bohr model:

1. Orbits (shells) get larger (their radius increases) as the principal quantum number n increases.
2. Electrons in the $n = 1$ orbit have the lowest energy. Electrons in the $n = 2$ orbit have more energy. The $n = 3$ electrons have even more energy, and so on.
3. Each shell can hold a maximum of $2n^2$ electrons. Therefore the electron capacities of the first four shells are

Maximum Electron Capacities of the First Four Shells

$n = 4$	$2n^2 = 2 \times 4^2$	$= 32$ electrons
$n = 3$	$2n^2 = 2 \times 3^2$	$= 18$ electrons
$n = 2$	$2n^2 = 2 \times 2^2$	$= 8$ electrons
$n = 1$	$2n^2 = 2 \times 1^2$	$= 2$ electrons

4. When an atom is being "built," the innermost shell (the one lowest in energy) is filled first. Electrons are added to each shell until it is filled to capacity ($2n^2$ electrons). For example, the lithium (Li) atom, with atomic number 3, has three protons in its nucleus and three electrons in its electron shells. Two of the electrons go into the most stable $n = 1$ shell, and the third electron must go into the less stable $n = 2$ shell.

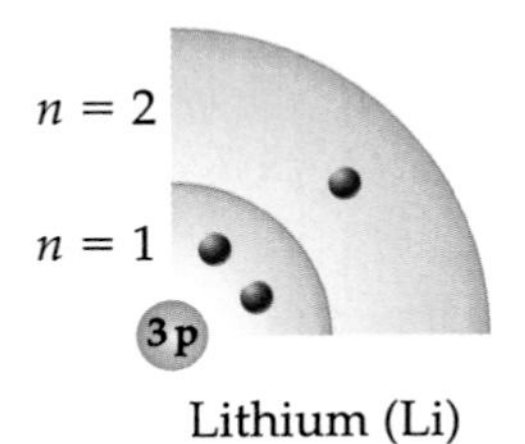

Lithium (Li)

This is the essence of the Bohr atomic model. Now we'll use it to show you why atoms give line spectra and why the elements behave in a periodic fashion.

Practice Problems

4.5 What is wrong with this Bohr model of the beryllium (Be) atom (atomic number 4)?

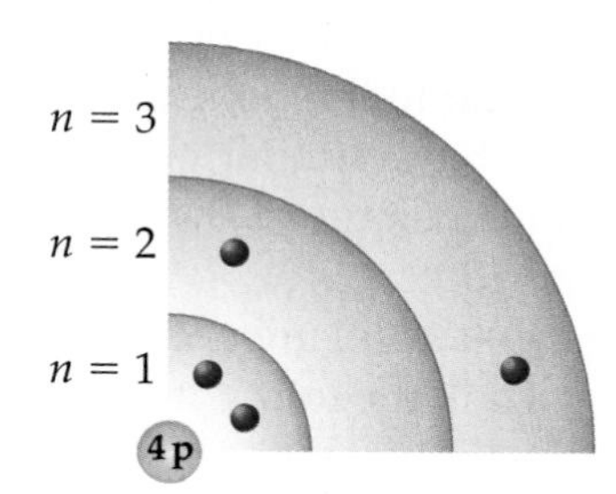

Answer: Don't begin filling the $n = 3$ shell until the $n = 2$ shell is full. Don't skip shells!

4.6 Draw a Bohr model for an atom of sulfur (S). How many additional electrons can fit into the $n = 3$ shell?

4.7 Why is an electron in a shell that has a low value of n in a more stable arrangement than one in a shell that has a higher value of n?

4.8 How many electrons can the $n = 5$ shell in an atom hold?

4.4 Periodicity and Line Spectra Explained

Explaining the periodic nature of the chemical properties of the elements was one of the major triumphs of Bohr's model. Consider the elements lithium (Li) and sodium (Na), both in group IA (1) and possessing similar chemical properties. According to Bohr, lithium, with 3 electrons, and sodium, with 11 electrons, have the structures shown at right. If you look closely at these diagrams, you'll see a similarity between lithium and sodium. They both have completely filled inner shells and only a single electron in the outermost occupied shell. This outermost occupied shell of an atom is called the **valence shell**. Perhaps these elements have similar chemical properties because they have identical valence-shell configurations. Does this pattern hold true throughout the periodic table?

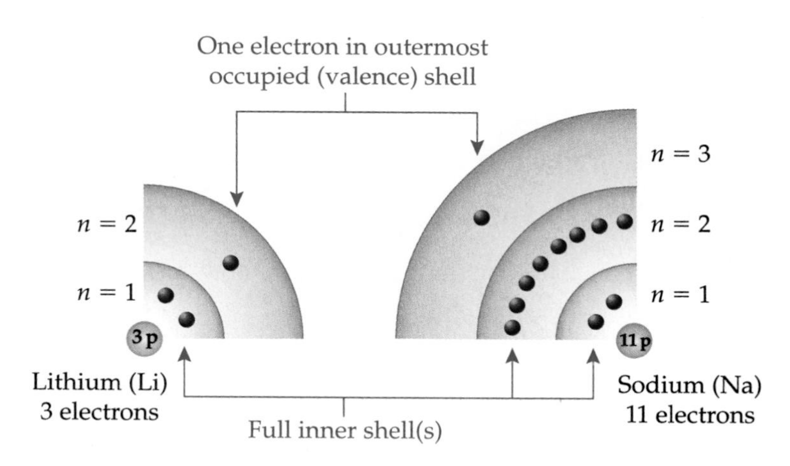

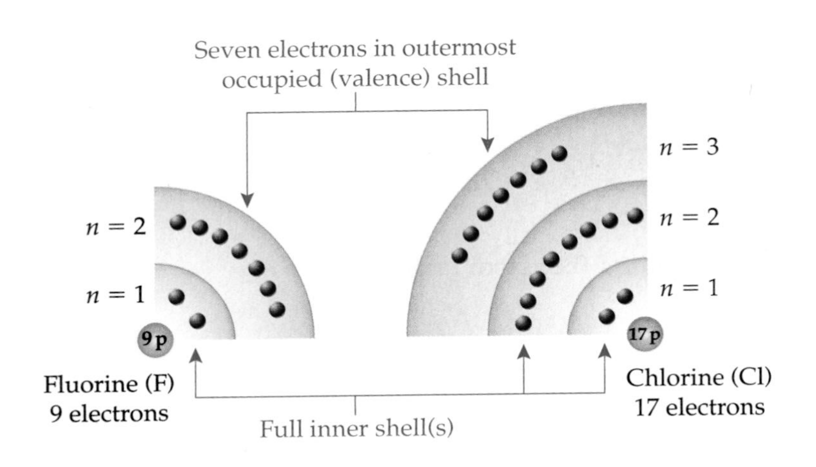

Let's look at another pair of elements from the other side of the table, fluorine (F) and chlorine (Cl). They are both in group VIIA (17), so we know they have similar chemical properties. What about their valence shells? When the electrons are put into the shells following Bohr's prescription, both elements end up with seven electrons in their valence shell. This, according to the Bohr model, is why F and Cl have similar chemical properties and belong in the same group of the periodic table.

The electron-shell configurations for the first 18 elements in the periodic table, filled according to Bohr's rules, are shown in the following chart. Look at the outermost occupied shell in each group—the valence shell—to convince yourself that elements with similar chemical properties have identical valence-shell configurations. Only helium (He) seems not to follow this rule. In a way, though, it does. Both helium and the atom below it, neon (Ne), have a completely filled valence shell. The difference is that the valence shell in helium is the n = 1 shell and can accommodate only two electrons. All atoms after neon in this group contain eight electrons in their valence shell.

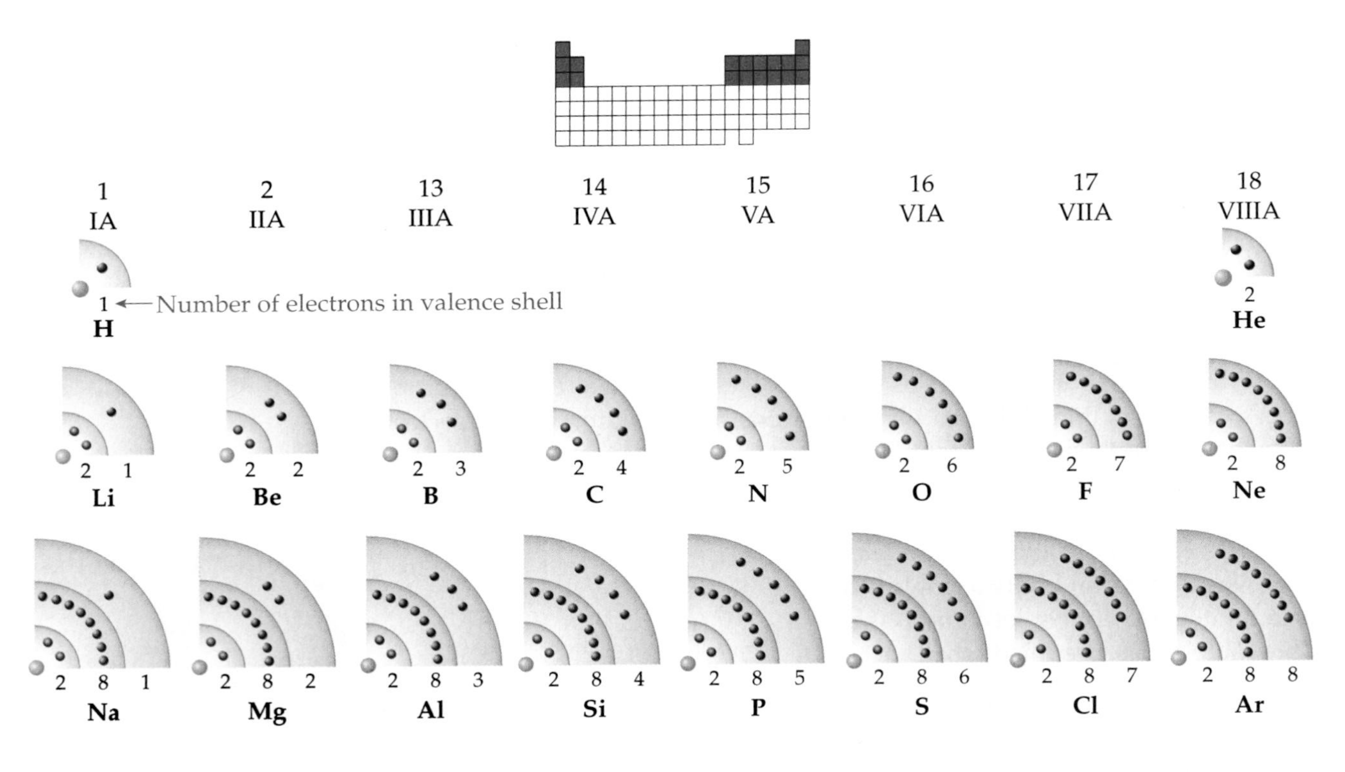

Careful examination of this chart can reveal a shortcut for determining the number of electrons in the valence shell of an atom. Hydrogen, lithium, and sodium each have one valence-shell electron in the Bohr model, and they are also in group IA of the periodic table. Fluorine and chlorine both have seven valence-shell electrons, and are in group VIIA. For the representative elements, the roman-numeral group number gives the number of valence-shell electrons.

The number of electrons in the valence shell of an atom is equal to the roman-numeral group number for the representative (A group) elements.

So, according to Bohr's model, the chemical properties of the elements are periodic because, as successive shells of electrons get filled, the valence-shell configuration repeats.

The other success of the Bohr model was its ability to explain the line spectrum of hydrogen. Not only did the model explain why line spectra exist, but the simple mathematical calculations that Bohr carried out based on the model's assumptions reproduced exactly the experimental line spectrum of hydrogen. From the physicist's perspective, this was the real success of the Bohr model. Bohr's equations could be used to calculate the energy of an electron in any allowed shell of the hydrogen atom. A scaled version of the energy values he calculated for the first six shells of the hydrogen atom is shown at the right in an **energy-level diagram**. As we shall see, the absolute energies of the shells are not important. It is the differences between them that are important. Thus, we have scaled the energies so that the energy of the $n = 1$ shell is 1.0 eV.

Notice how the energies get closer together as n increases. This fact will become important later on.

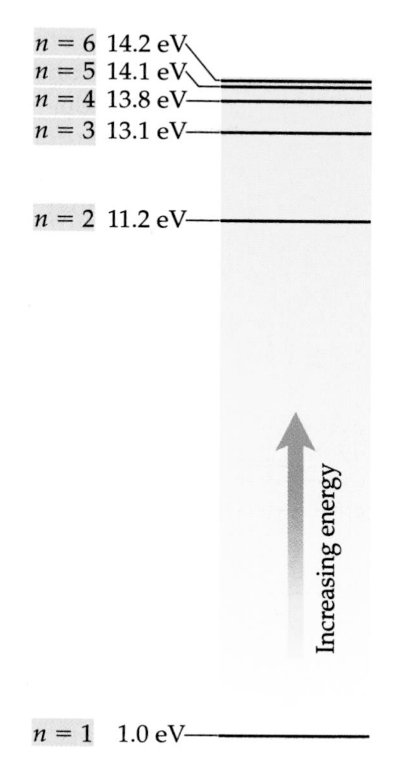

The energies of the Bohr shells in the hydrogen atom scaled so that the energy of the $n = 1$ shell is 1.0 eV.

Examine the energy-level diagram for hydrogen shown above, and answer the following questions:

(a) If an electron is in the $n = 1$ shell, what is the electron's scaled energy in electronvolts?

(b) If an electron is in the $n = 2$ shell, what is the electron's scaled energy in electronvolts?

(c) How much energy would it take to transfer an electron from the $n = 1$ shell to the $n = 2$ shell?

(d) Would it take the same, more, or less energy than your answer to (c) to take an electron from the $n = 2$ shell to the $n = 3$ shell? Why?

Were you able to answer (c) and (d) intuitively? You know the electron's energy when it is in the $n = 1$ shell, and you know its energy when it is in the $n = 2$ shell. The difference between these two energies, therefore, is the energy it will cost you to move the electron from the $n = 1$ to the $n = 2$ shell. This is a very important concept, so make sure you understand it. As for part (d), shell energies get closer together with increasing n. Clearly then, the energy cost will be less to go from $n = 2$ to $n = 3$ than to go from $n = 1$ to $n = 2$.

To predict the line spectrum of hydrogen, Bohr assumed that the atom was in its *ground state*. The **ground state** of an atom is defined as the arrangement of electrons that has the lowest total energy. Thus, it is the configuration where all the electrons are placed in the shells having the lowest energy values. For a hydrogen atom, which has just one electron, the ground state is as shown at the left below (we use a Bohr diagram to show you where the electron is and an energy-level diagram to indicate the energy of the electron).

Under normal conditions, all hydrogen atoms spend most of their time in the ground state. However, when enough energy is added to an atom (for instance, by heating it or passing an electric current through it), its electron can move ("jump") into a higher-energy shell. When that occurs, we say the energy was absorbed by the atom and the atom is now in an *excited state*. In an **excited state**, one or more of the electrons in an atom are located in high-energy shells even though there is still room for those electrons in a lower-energy shell.

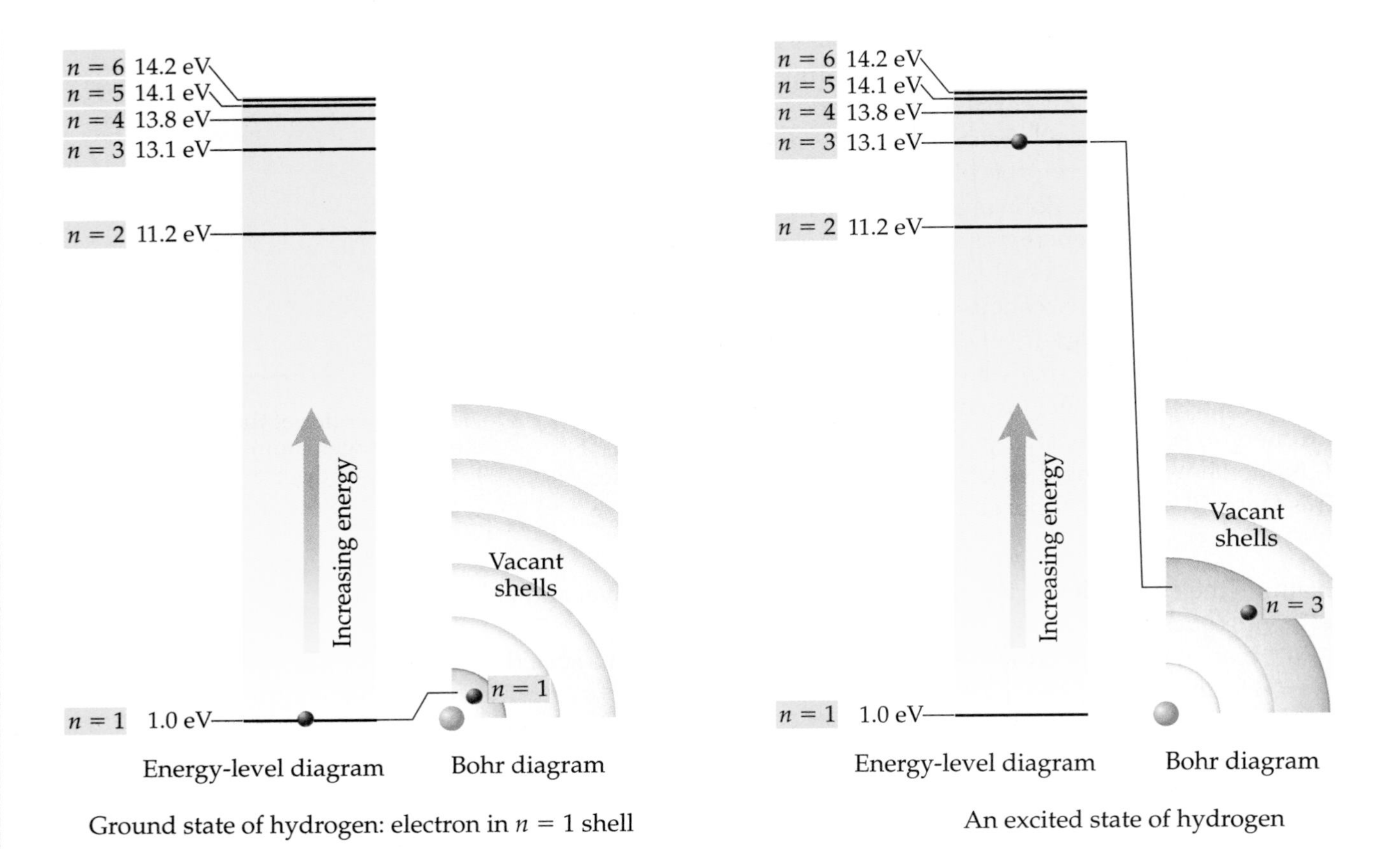

Ground state of hydrogen: electron in $n = 1$ shell

An excited state of hydrogen

One possible excited state for the hydrogen atom is shown on the right above. This energy-level diagram represents an excited state because there are vacant lower-energy shells below the electron (levels have been skipped before they are full). Another possible excited state could have the electron in the $n = 2$ shell, and yet another in the $n = 4$ shell, and so on. We can create any of these excited states by starting with the ground-state hydrogen atom and adding enough energy to make the electron jump from the $n = 1$ shell to any upper shell.

4-3 WORKPATCH What is the lowest-energy excited state of the hydrogen atom?

The WorkPatch points out that all excited states are not created equal. For instance, in the hydrogen excited state with the electron in the $n = 2$ shell, the electron has less energy than in the excited state with the electron in the $n = 3$ level, because it takes more energy to raise the electron to $n = 3$ than to $n = 2$.

Starting from the ground state, how much more energy does it take to create the $n = 3$ excited state of hydrogen than the $n = 2$ excited state?

4.4 WORKPATCH

Check your WorkPatch answer against ours. When you can do the WorkPatch, then you are becoming skilled at understanding and using an energy-level diagram.

When a tube of hydrogen gas is electrified, the electricity supplies the energy to produce various excited states. To account for the line spectrum exhibited by such a gas, we must consider what happens *after* the excited states are created. As it turns out, atoms do not usually remain in an excited state for long. Because there are vacancies at lower energy levels, the electrons can drop back down into them (a process called *relaxation*). Relaxation is desirable from an energy point of view because in general things tend to settle in the lowest energy state possible (as, for instance, when a ball rolls to the bottom of a hill). To create the excited state, the atom absorbs energy, and when the atom returns to the ground state (relaxes), it loses the same amount of energy.

One way an atom can lose energy is to emit it as light. The amount of energy in the light created during relaxation is the same as the amount of energy the electron lost in dropping to a lower energy level. This light has a particular wavelength, and this is where line spectra come from. When an excited hydrogen atom relaxes from the $n = 2$ state to the ground state, the energy of the electron goes from 11.2 eV down to 1.0 eV, meaning the electron loses 10.2 eV of energy in the process. This energy is released as light with a wavelength corresponding to an energy of 10.2 eV. Using Table 4.1, we can see that this particular electron transition results in ultraviolet light being emitted. Because ultraviolet light is outside the visible region, we can't see this particular line in the hydrogen line spectrum.

Table 4.1 Energy and Wavelength Ranges for Ultraviolet and Visible Light

Radiation	Energy (eV)	Wavelength λ (nm)
Ultraviolet	> 3.74	< 380
Indigo/violet	3.34–2.95	380–430
Blue	2.95–2.64	430–480
Green	2.64–2.24	480–565
Yellow	2.24–2.10	565–620
Orange	2.10–1.95	620–650
Red	1.95–1.69	650–750

However, the hydrogen spectrum does contain four lines in the visible region of the electromagnetic spectrum—one red line, one green line, one blue line, and one indigo/violet line. According to Table 4.1, red corresponds to light with an energy in the range from 1.69 to 1.95 eV. A close inspection of the energy-level diagram for hydrogen on page 130 shows that there is one and only one jump that corresponds to energy in this range. An electron in an excited hydrogen atom relaxing from the $n = 3$ level to the $n = 2$ level produces light with exactly 1.9 eV of energy, as you saw in the above WorkPatch.

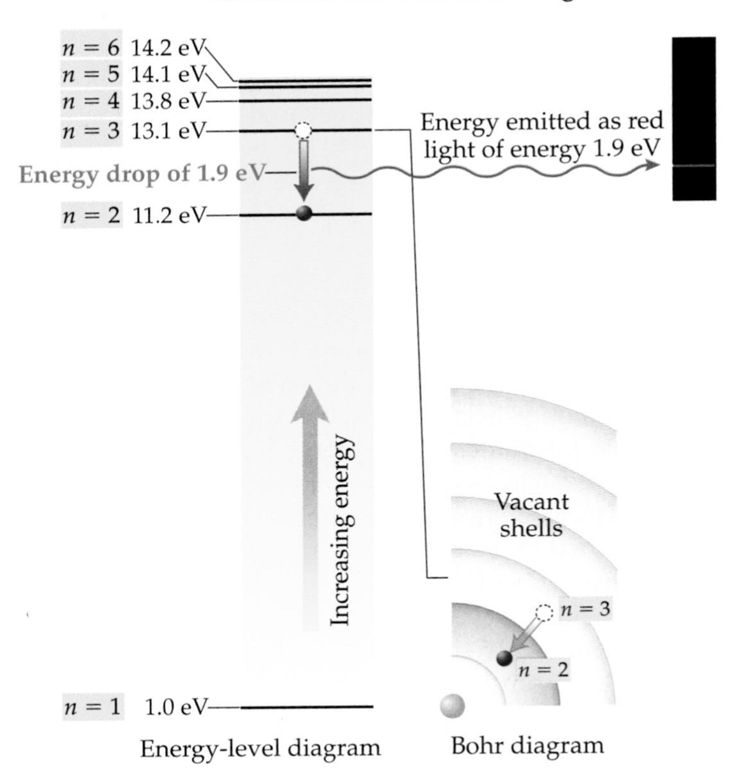

With the electron in the $n = 2$ shell, we are not yet to the ground state. Thus we can conclude that the electron in an excited atom does not necessarily return to its ground state in a single jump. It can also reach the ground state by making a succession of smaller downward jumps.

4.5 WORKPATCH

The hydrogen atom has three other visible lines in its line spectrum: green, blue, and indigo/violet. Use the energy-level diagram for hydrogen and Table 4.1 to determine the electron jumps responsible for these lines.

Were you able to do the WorkPatch? Because indigo/violet has the highest energy of these four colors of visible light, the jump corresponding to the indigo/violet line should be the largest. Check your answer to see that it is.

Bohr's model accounts for the existence of line spectra because the electrons in an atom are allowed only certain energies and never anything in between. Because the energy levels are fixed, the differences between them are also fixed, and only certain wavelengths (energies) of light can be produced. The energy levels for atoms of other elements have different numerical values and hence different line spectra.

The rapid acceptance of Bohr's model was due mainly to its success in exactly reproducing the line spectrum of the hydrogen atom. Not only could Bohr correctly calculate the exact position and energy of the four lines in the visible part of the spectrum, but he could also match exactly the lines detected in other regions of the spectrum (ultraviolet and infrared). Nevertheless, Bohr's model of the atom did have some problems. The lines Bohr predicted would be present in the line spectrum of the helium atom did not match up at all with the real spectrum. In fact, his calculations failed to predict the correct spectrum for any atom with more than one electron. In less than 20 years, Bohr's model would be replaced with the modern model of the atom. However, his postulate that the energy levels of electrons in atoms are quantized and his rules for the maximum number of electrons that each shell can accommodate are still correct. Therefore we can continue to use these aspects of Bohr's model to explain many things in chemistry. Before we do, however, we must look at an important modification to the model.

Practice Problems

4.9 The ground state for the lithium (Li) atom and the scaled energies of its shells are shown below. Draw a Bohr diagram for the lowest-energy excited state of lithium.

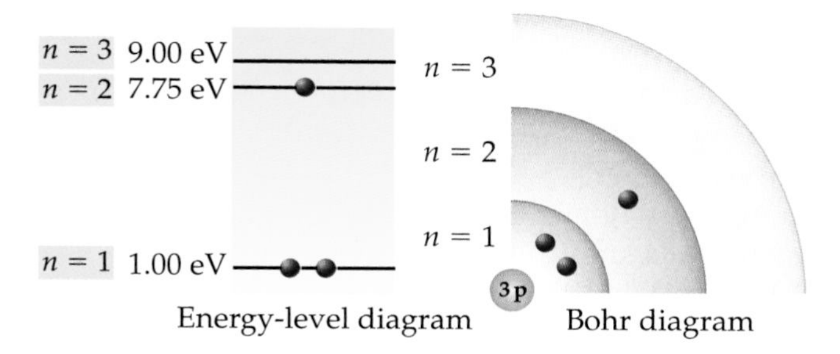

Answer: Two excited states are drawn below, but only (a) correctly answers the question. Remember, both excited states are created from the ground state. To produce (a), an electron must be lifted from the $n = 2$ shell to the $n = 3$ shell. This takes 1.25 eV (1.25 eV of energy must be put into the atom). To produce (b), an electron is lifted from the $n = 1$ shell to the $n = 2$ shell. This takes 6.75 eV. Because it took more energy to create (b), (b) is a higher-energy excited state than (a). Thus, (a) is the lowest-energy excited state that can be prepared from the ground-state lithium atom.

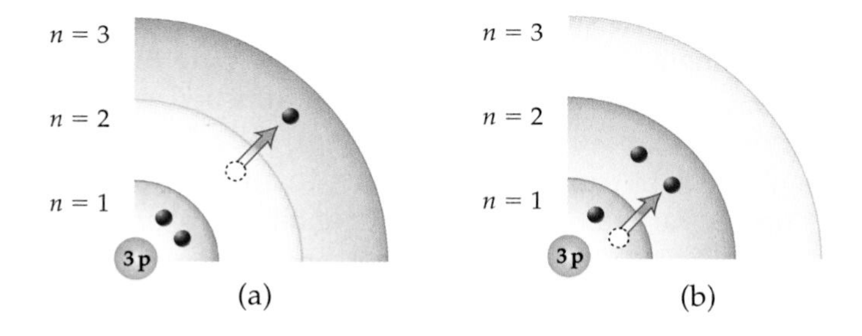

4.10 Draw a Bohr diagram for the highest-energy excited state you can make using the three shells shown in Practice Problem 4.9 and exciting only a single electron.

4.11 Draw a Bohr diagram for a Li^+ cation in its ground state.

4.12 An atom has atomic number 6 and has eight electrons.
(a) Which element is this?
(b) Is this a neutral atom, a cation, or an anion? If it is an ion, what is its charge?
(c) Draw a Bohr diagram for this atom in its ground state.

4.13 An F^- anion has ten electrons. A student draws the following ground state Bohr diagram for this anion. There are two things wrong with the diagram. What are they?

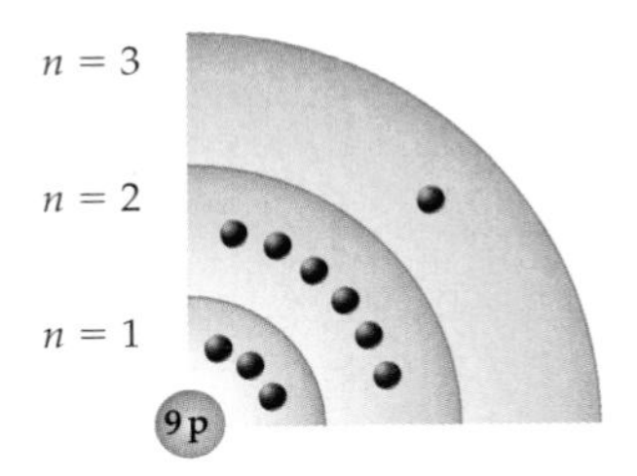

4.5 Subshells and Electron Configuration

Now we need to look at an important refinement to the Bohr model, the essence of which is important in the modern model. During Bohr's time, instruments improved enough to show that the simple line spectra generated by excited atoms were not so simple after all. In many cases, what was originally thought to be a single line turned out to be a number of very closely spaced lines of nearly identical color (energy). Bohr struggled with these findings for several years and tried various modifications to his basic model in an attempt to account for them. What he and others eventually proposed was, in essence, that some of the electron shells in an atom consist of a set of **subshells** that are very closely spaced in energy and size.

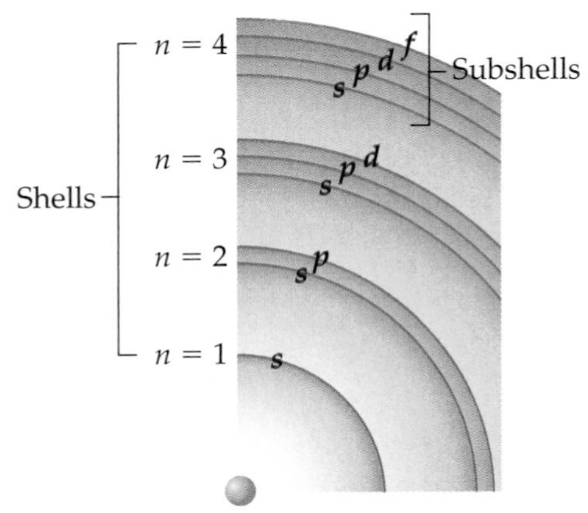

The principal quantum number n tells you the number of subshells in a shell. The $n = 1$ shell has one subshell, the $n = 2$ shell has two subshells, and so on. All subshells in a given shell are close to one another both in size and in energy.

As shown at the left, these subshells are given single-letter designations (s for the first subshell in a shell, p for the second, d for the third, and f for the fourth). Not every shell contains all these subshells. The first shell ($n = 1$) contains just an s subshell. As we move outward from the $n = 1$ shell, each subsequent shell gets an additional subshell, so the $n = 2$ shell consists of two subshells (s and p), the $n = 3$ shell has three subshells (s, p, and d), and so on. In addition, as shown at the top of the next page, each subshell has a different electron capacity. An s subshell can hold a maximum of 2 electrons, a p subshell can hold a maximum of 6 electrons, a d subshell can hold up to 10 electrons, and an f subshell can hold 14 electrons. These subshells also increase in energy and size, with the s subshell in a given shell being the smallest and having the lowest energy. Notice that the electron capacity of each shell is unchanged from the original Bohr model. For instance, the $n = 3$ shell still holds a maximum of 18 electrons.

When we introduced the energy-level diagram for the hydrogen atom, we pointed out that the energies of the shells get closer and closer together as n increases. We also said that this would become important later on. Now is the time to show you why. In part because the energy levels get closer together as n increases, the energies of some subshells can actually cross over with those in an adjacent shell. The diagram below illustrates this.

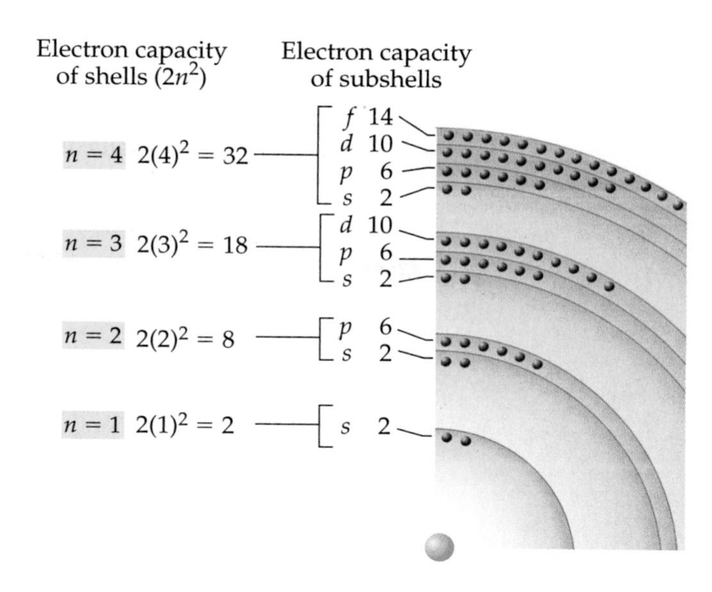

As we've said, the subshells in a shell increase in energy in the order $s \rightarrow p \rightarrow d \rightarrow f$, and all of the $n = 3$ subshells are higher in energy than the $n = 2$ subshells. But look carefully at the 4*s* subshell. It is *lower* in energy than the 3*d* subshell, even though the 4*s* subshell has a larger value of n. Additional crossings occur as we proceed beyond $n = 4$. The consequences of these crossings are extremely important! Remember, Bohr stated that the ground state of an atom is achieved by filling the shells in order of increasing energy, starting with the $n = 1$ shell and working up. The same rule holds when we take subshells into account. This solves a problem that initially puzzled Bohr—what was called the potassium problem.

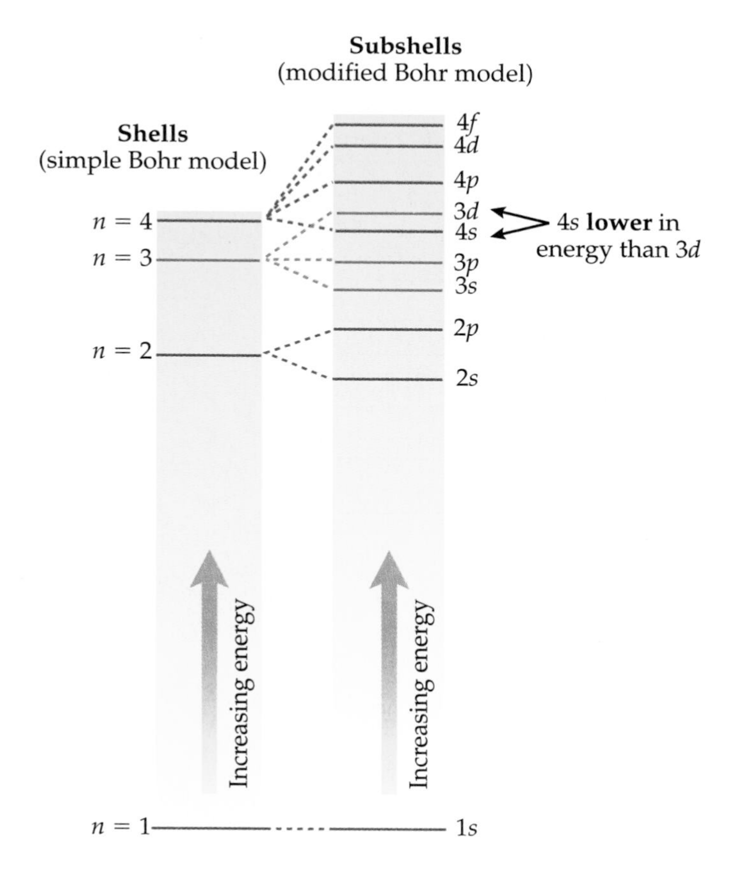

Potassium (K) is a group IA (1) atom possessing 19 electrons. Group IA atoms are supposed to have one valence electron, but that's not what Bohr's initial model gave. Look at what happens when we position potassium's 19 electrons in the shells of the simple version (shells only) of Bohr's model:

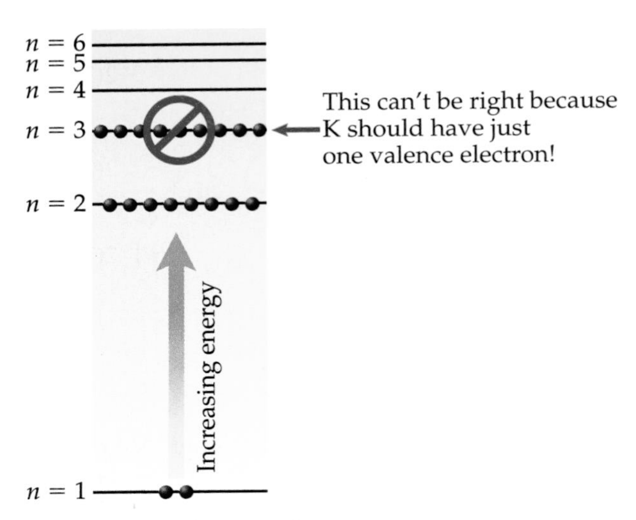

The ground state of K according to the simple Bohr model (shells only)

The $n = 1$ and $n = 2$ shells are filled, as expected. But the valence $n = 3$ shell has nine electrons in it instead of the expected one electron for this group IA atom. The modified subshell model fixed this problem. Demonstrate this for yourself now.

Using dots to represent potassium's 19 electrons, fill the subshells in the diagram below to arrive at the ground-state energy-level diagram for the potassium atom.

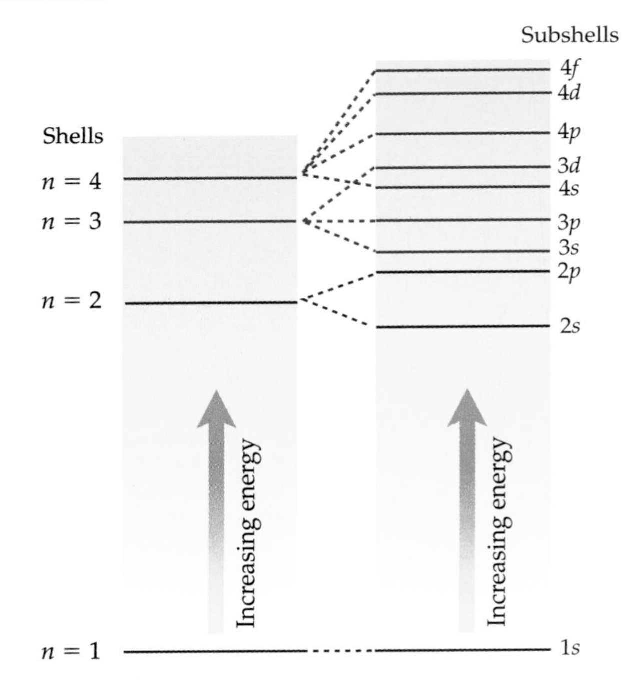

In doing the WorkPatch, you should have arrived at an atom with one electron in the valence 4*s* subshell. You have just solved the potassium problem! We are about to discuss this energy-level diagram in detail, so check your answer against the one at the end of this chapter.

Electron Configuration

The way the electrons are arranged in the subshells of an atom is called the **electron configuration**. An energy-level diagram can accurately depict the electron configuration of an atom; however, it is a tedious way to represent it. Chemists have therefore developed a notation to indicate an atom's electron configuration. Each occupied subshell, starting with the lowest-energy subshell, is indicated with a lowercase italic letter (*s*, *p*, *d*, *f*) preceded by a number indicating the principal quantum number of the shell (n). The number of electrons occupying a subshell is indicated by a superscript. Thus, we can replace the modified Bohr diagram for potassium with the notation

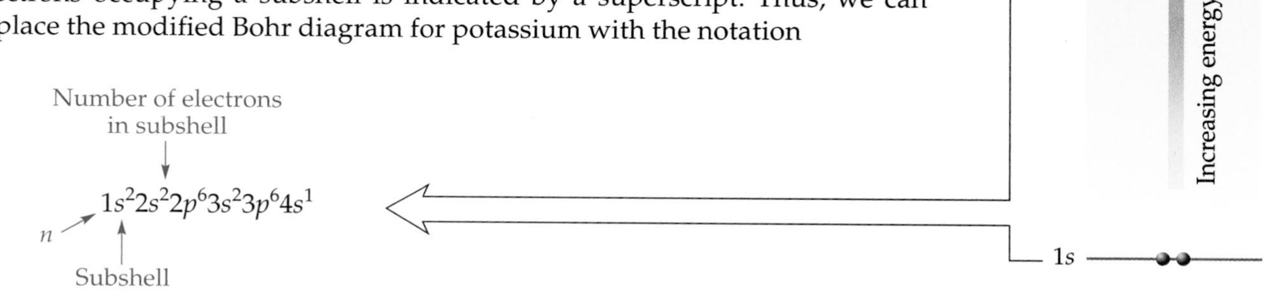

$$1s^2 2s^2 2p^6 3s^2 3p^6 4s^1$$

This electron-configuration notation is a lot easier to write than the energy-level diagram, and it communicates the same information. It tells us how the electrons are arranged in the subshells. In addition, it is an easy matter to identify the valence electrons. In general, the valence electrons are those in the subshells of highest principal quantum number n. For potassium, the highest principal quantum number is 4, and if we sum up the electrons present in the $n = 4$ subshells, we get 1, the correct number of valence electrons for a potassium atom.

Valence electron

$$1s^2 2s^2 2p^6 3s^2 3p^6 \overbrace{4s^1}$$

A slightly more difficult example is bromine (atomic number 35). When this atom is in the ground state, its electron configuration is

$$1s^2 2s^2 2p^6 3s^2 3p^6 \underbrace{4s^2} 3d^{10} \underbrace{4p^5}$$

Valence electrons

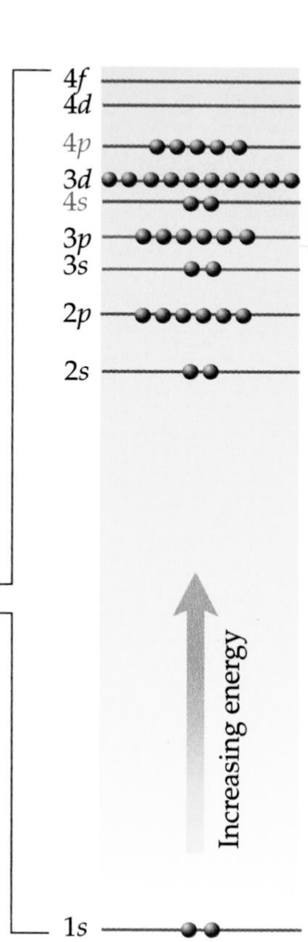

If we sum up the electrons in the subshells having the highest principal quantum number (4), we get seven valence electrons. To emphasize the valence electrons, chemists often write them last:

$$1s^2 2s^2 2p^6 3s^2 3p^6 3d^{10} 4s^2 4p^5$$

Practice Problems

4.14 Use electron-configuration notation to show how electrons are distributed in a ground-state silicon (Si) atom. How many valence electrons does this atom have?

Answer: Silicon has atomic number 14, so it has 14 electrons. We start by putting the first two electrons in the 1s subshell, the next two in the 2s subshell, the next six in the 2p subshell, the next two in the 3s subshell, and the remaining two in the 3p subshell:

$$1s^22s^22p^63s^23p^2$$

The highest principal quantum number is 3, and summing up the electrons in the two $n = 3$ subshells tells us this atom has four valence electrons.

4.15 Use electron-configuration notation to show how electrons are distributed in a ground-state arsenic (As) atom. Does the number of valence electrons in your answer agree with the number predicted by this atom's roman-numeral group number in the periodic table?

4.16 Use electron-configuration notation to show how electrons are distributed in a ground-state scandium (Sc) atom.

All the subshell energy-level diagrams we have presented so far went only to $n = 4$. As we proceed beyond this value of n, additional subshell energy crossings occur. You will be pleased to know that you do not need to memorize any of these crossings. That's because the periodic table can be used as a quick and easy guide for deducing the electron configuration of any element. To do this, we think of the periodic table as being constructed from four blocks—the s block, the p block, the d block, and the f block—with the designation (s, p, d, f) of each block representing the subshell being filled.

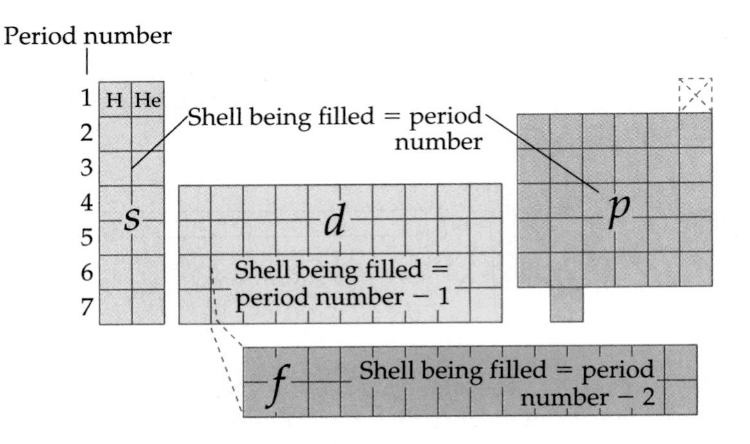

Notice that the s block is two elements wide. That is because an s subshell can hold a maximum of two electrons. The p block is six elements wide because a p subshell can hold up to six electrons. Likewise, the d and f blocks are 10 and 14

elements wide, respectively, indicating the maximum number of electrons those subshells can hold. Also, notice that we moved helium (He) from its usual position at the extreme right of the table to a position in the *s* block, adjacent to hydrogen.

The period (row) number tells you which shell (n) is currently filling, *unless you are in the d or f block*. In the *d* block, the value of n for the *d* subshell being filled is 1 less than the period number. If you are in the *f* block, the value of n for the *f* subshell being filled is 2 less than the period number. This takes into account the subshell crossings. A few examples will help make this clear.

Suppose we want to determine the electron configuration of sulfur (S). Keep glancing at the periodic table below and follow the red arrows as we discuss how to do this. We start at hydrogen (as we always will) and move across the first period from left to right until we reach the end of the period at helium. Each element we encounter on our way adds one more electron to the atom. Helium is the second element, so we have two electrons so far. They are both in the 1*s* subshell because we have moved along a path in the first period and we are in the *s* block. So far we have $1s^2$. Next, we move to the beginning of period 2. Here we move through both the *s* and *p* blocks of the period, gaining $2s^2$ and $2p^6$. Having reached the end of period 2, we move on to period 3, going first through the *s* block ($3s^2$) and then four elements into the *p* block to reach our target, sulfur. Four deep into the *p* block of period 3 gives us four electrons, or $3p^4$. The complete notation for sulfur (S), is therefore $1s^22s^22p^63s^23p^4$. As expected, there are six valence electrons in the $n = 3$ shell for this group VIA element.

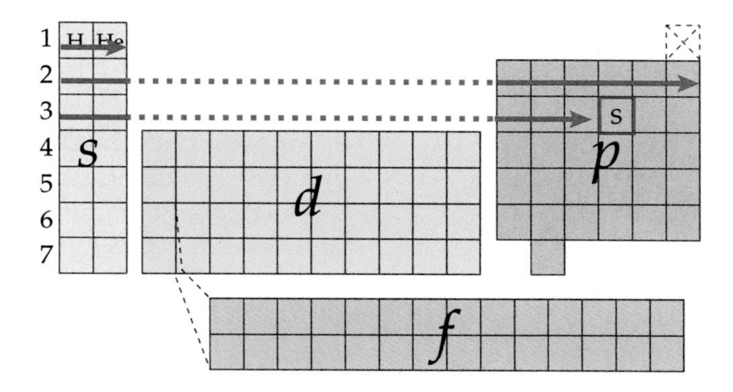

Now let's try iron (Fe), a *d*-block element. Again we start at H. The only catch is to remember that when you plunge into the *d* block in period 4, you must subtract 1 from the period to get the correct n (to account for the subshell crossings). This means that the final six electrons should be listed as $3d^6$, not as $4d^6$. Thus, the electron configuration for iron is $1s^22s^22p^63s^23p^64s^2\,3d^6$.

If we were looking for the electron configuration of selenium (Se), our arrow would continue past iron, out of the *d* block, and into the *p* block. Once we leave the *d* block and enter the *p* block, we must bump n back up by 1. The electron configuration of selenium (Se) therefore is $1s^22s^22p^63s^23p^64s^2\,3d^{10}4p^4$.

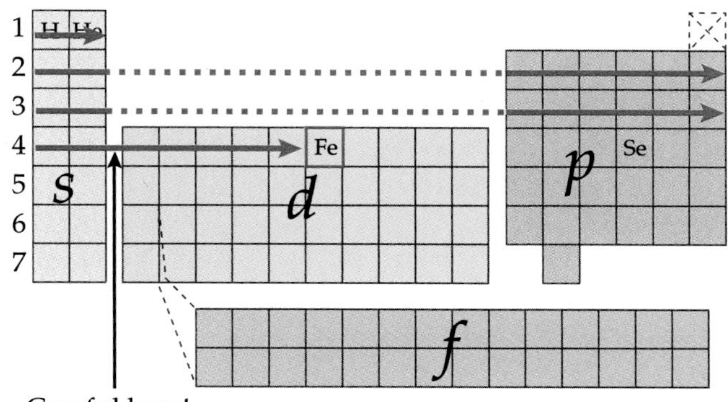

The ground-state electron configuration obtained by using either a subshell energy-level diagram or a four-block periodic table is not always correct because there are exceptions to the rules we have been following. The rules work well for the representative (*s* and *p* block) elements, but slightly more than one-quarter of the transition (*d* block) elements violate the rules, sending one or both of the valence *ns* electrons into the $(n - 1)$ *d* shell. Practice Problem 4.17 shows you one of them.

Practice Problems

4.17 Using a four-block periodic table as a guide, determine the ground-state electron configuration of copper (Cu).

Answer: You should have arrived at $1s^22s^22p^63s^23p^64s^23p^9$*, because copper is nine deep into the d block of period 4. However, copper is one of those exceptions, and its true electron configuration is* $1s^22s^22p^63s^23p^64s^13p^{10}$.

4.18 Using a four-block periodic table as a guide, determine the ground-state electron configuration of krypton (Kr). Explain why it is proper for this element to be in group VIIIA.

4.19 Using a four-block periodic table as a guide, determine the ground-state electron configuration of palladium (Pd).

So far we have been neglecting the *f*-block elements—that is, the elements beyond lanthanum (La) and actinium (Ac). Lanthanum and actinium can be thought of as detour signs to the *f* block. Remember that the value of n in the *f* block is 2 less than the period you are in. Once you get through the *f* block, it's back into the *d* block (where the value of n is only 1 less than the period). For example, let's determine the electron configuration of gadolinium (Gd).

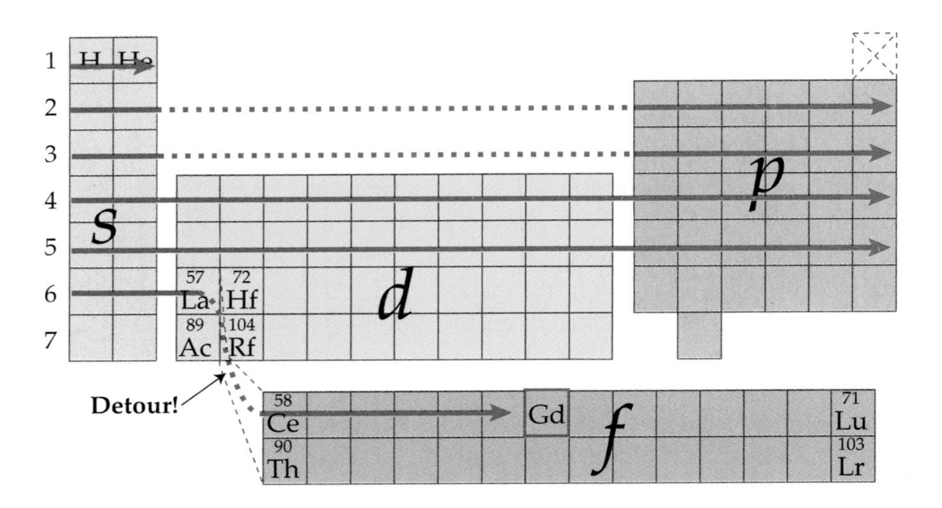

Follow the arrows. We get $1s^2$ from the first row, $2s^22p^6$ from the second row, $3s^23p^6$ from the third row, $4s^23d^{10}4p^6$ from the fourth row, $5s^24d^{10}5p^6$ from the fifth row, and finally $6s^25d^14f^7$ from the sixth row (we get only 1 deep into the *d* block and then must detour and go 7 deep into the *f* block). The result is $1s^22s^22p^63s^23p^64s^23d^{10}4p^65s^24d^{10}5p^66s^25d^14f^7$. We must point out that for the *f*-block elements there are more exceptions to this method of filling the subshells than there are elements that obey it. For most, you must transfer the single *d*

electron into the f subshell that is being filled. This pattern is shown in the abbreviated electron configurations listed here:

Electron Configurations of Highest-Energy Subshells (the Ones Being Filled)

58	59	60	61	62	63	64	65	66	67	68	69	70	71
Ce	**Pr**	**Nd**	**Pm**	**Sm**	**Eu**	**Gd**	**Tb**	**Dy**	**Ho**	**Er**	**Tm**	**Yb**	**Lu**
$6s^24f^15d^1$	$6s^24f^3$	$6s^24f^4$	$6s^24f^5$	$6s^24f^6$	$6s^24f^7$	$6s^24f^75d^1$	$6s^24f^9$	$6s^24f^{10}$	$6s^24f^{11}$	$6s^24f^{12}$	$6s^24f^{13}$	$6s^24f^{14}$	$6s^24f^{14}5d^1$
90	91	92	93	94	95	96	97	98	99	100	101	102	103
Th	**Pa**	**U**	**Np**	**Pu**	**Am**	**Cm**	**Bk**	**Cf**	**Es**	**Fm**	**Md**	**No**	**Lr**
$7s^26d^2$	$7s^25f^26d^1$	$7s^25f^36d^1$	$7s^25f^46d^1$	$7s^25f^6$	$7s^25f^7$	$7s^25f^76d^1$	$7s^25f^9$	$7s^25f^{10}$	$7s^25f^{11}$	$7s^25f^{12}$	$7s^25f^{13}$	$7s^25f^{14}$	$7s^25f^{14}6d^1$

Writing electron configurations for atoms that possess many electrons, such as the f-block elements, can get pretty tedious. Chemists have developed an additional shortcut which recognizes that the electron configuration of the noble gas atom that directly precedes any element is identical to the early part of the electron configuration for that element. For example, the noble gas that directly precedes gadolinium (Gd, atomic number 64) is xenon (Xe, atomic number 54). The first 54 electrons in Gd can thus be replaced with the Xe symbol, and the electron configuration of Gd can be written $[Xe]6s^25d^14f^7$. In this notation, the symbol for the noble gas is always enclosed in square brackets.

Practice Problems

4.20 Write the ground-state electron configuration of chlorine (Cl) and lutetium (Lu), using both the full notation and the noble gas abbreviated notation.

Answer: Cl $1s^22s^22p^63s^23p^5$; $[Ne]3s^23p^5$
Lu $1s^22s^22p^63s^23p^64s^23d^{10}4p^65s^24d^{10}5p^66s^25d^14f^{14}$; $[Xe]6s^25d^14f^{14}$

4.21 Write the ground-state electron configuration of radium (Ra), using both the full notation and the noble gas abbreviated notation.

4.22 Write the ground-state electron configuration of uranium (U), using both the full notation and the noble gas abbreviated notation.

Finally, we can turn this around and ask questions about an atom once we've determined its electron configuration. For example, arsenic (As) has the ground-state electron configuration $1s^22s^22p^63s^23p^64s^23d^{10}4p^3$ (or $[Ar]4s^23d^{10}4p^3$). We could now ask, "Where does this atom belong in the periodic table? Which period is it in? Which group is it in?" You should be able to answer these questions. Because the outermost occupied (valence) shell is $n = 4$, we expect to find As in the fourth period. In addition, As has five valence electrons, which we can emphasize by rewriting the electron configuration as $1s^22s^22p^63s^23p^63d^{10}4s^24p^3$. Five valence electrons puts it in group VA.

Here are a few additional hints. If an element has a partially filled *d* subshell, it is a transition metal (a *d*-block element). If an element has a partially filled *f* subshell, it is an *f*-block element. A partially filled *s* or *p* subshell usually identifies the element as an *s*-block element or a *p*-block element, respectively, and thus as a representative element (there are exceptions, however).

Practice Problems

4.23 In which period and which group is the element that has the ground-state electron configuration $[Ar]4s^2 3d^{10} 4p^5$? Which element is this?

Answer: Rewriting the electron configuration as $[Ar]3d^{10}4s^2 4p^5$ *to emphasize the valence electrons, we see that the highest value of n is 4, so this atom is in the fourth period. There are a total of seven valence electrons, and the p subshell is partially filled, meaning this is a group VIIA element. It is bromine (Br).*

4.24 In which period and which group is the element that has the ground-state electron configuration $1s^2 2s^2 2p^6 3s^2 3p^6 4s^2 3d^3$? Which element is this?

4.25 In which period and which group is the element that has the ground-state electron configuration $[Xe]6s^1$? Which element is this?

4.6 Compound Formation and the Octet Rule

The simple Bohr model (shells only) can be very useful in predicting and understanding the chemical composition of many compounds. For example, it can be used to explain why the compound sodium chloride has the formula NaCl and never Na_2Cl, $NaCl_2$, or any other variation. The one-to-one ratio of sodium to chlorine in this compound can be understood in terms of an early chemical principle known as the *octet rule*. This rule arose from the observation of the unique nature of the elements in group VIIIA (18), the noble gases. As we saw in Chapter 3, the elements in this group are exceptionally unreactive. This lack of "desire" to combine with other elements was interpreted as meaning that the elements in group VIIIA are exceptionally stable. The Bohr model provided another unifying feature for these elements—they all have eight electrons in their valence shell (with the exception of helium, which has two). The obvious implication is that an element is unreactive (unusually stable) if it has eight valence electrons. Thus was born the octet rule.

The Octet Rule: Elements react to form compounds in such a way as to put eight electrons in their outermost occupied (valence) shell, giving them a valence-shell configuration identical to that of a noble gas and making them exceptionally stable.

According to the **octet rule** (from the Latin *octo*, meaning "eight"), the reason the other elements in the periodic table (groups IA–VIIA) participate in chemical reactions is to ultimately end up with eight electrons in their valence shell, thus gaining a stability similar to that of the noble gases.

The octet rule works well for many of the atoms found in biological and organic molecules (in particular, C, H, and N), and it is used to predict how the atoms in these molecules bind together. We can also use this rule for sodium and chlorine to understand why the chemical formula for sodium chloride is NaCl and not something else. Such applications make the octet rule very useful. However, the rule has many exceptions. For example, hydrogen's outermost occupied shell is the $n = 1$ shell, and so the hydrogen atom achieves stability (which means a filled outermost occupied shell) with only two electrons (a duet), so that hydrogen's electron configuration becomes like that of helium. The elements beryllium (Be) and boron (B) often settle for only four and six electrons, respectively, instead of eight. Many atoms in the third period and beyond can end up with more than an octet in some compounds, and the transition metals usually don't obey the octet rule at all. Despite all these exceptions, go ahead and use the rule—it is useful because so much of chemistry involves organic and biological molecules—but remember that it has its limitations.

Let's consider the formation of sodium chloride from sodium and chlorine atoms in terms of the octet rule. We begin with a chlorine atom, which has 17 electrons and the electron configuration shown at right. Because chlorine is a member of group VIIA, chlorine atoms have seven electrons in their outermost occupied shell before the reaction. To understand how chlorine atoms react, focus on the valence shell and apply the octet rule by asking, "How can a chlorine atom get eight electrons in this shell?" Because the valence shell is only a single electron short of an octet, the obvious answer is for each chlorine atom to pick up one electron, which would produce a chloride ion with a -1 charge. But where will this electron come from? To answer this, let's look at a sodium atom, which has 11 electrons and the electron configuration shown at right.

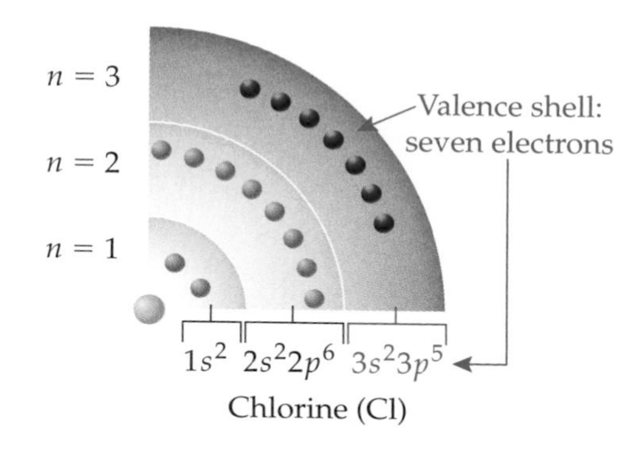

Chlorine (Cl)

Sodium atoms have only a single electron in their valence shell. Again we apply the octet rule and ask, "What is the best way for a sodium atom to end up with eight electrons in its valence shell?" Like chlorine, sodium could do this by picking up more electrons, but each atom would have to gain seven electrons. This would result in a sodium ion with a -7 charge, and that's quite a bit of excess negative charge for the sodium nucleus to hold on to. An easier way for sodium to get its octet is for it to lose its single valence electron from the $n = 3$ shell, leaving this shell empty. The $n = 2$ shell then becomes the valence shell, and this shell already has eight electrons in it! In addition, if sodium loses one electron, then this electron is now available for a chlorine atom.

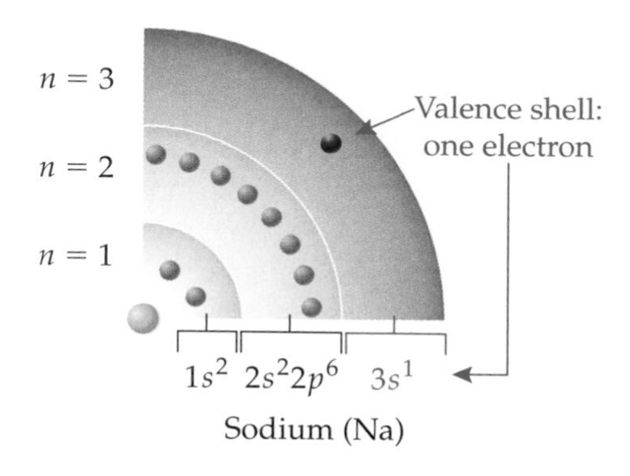

Sodium (Na)

Recall that Cl needs to gain one electron to satisfy the octet rule. Now we know where it might come from. In reacting with chlorine, each sodium atom loses one electron to satisfy the octet rule and becomes a cation with a $+1$ charge. Each chlorine atom gains one electron and becomes an anion with a -1 charge. In doing so, both of the newly formed ions satisfy the octet rule (eight electrons in the valence shell).

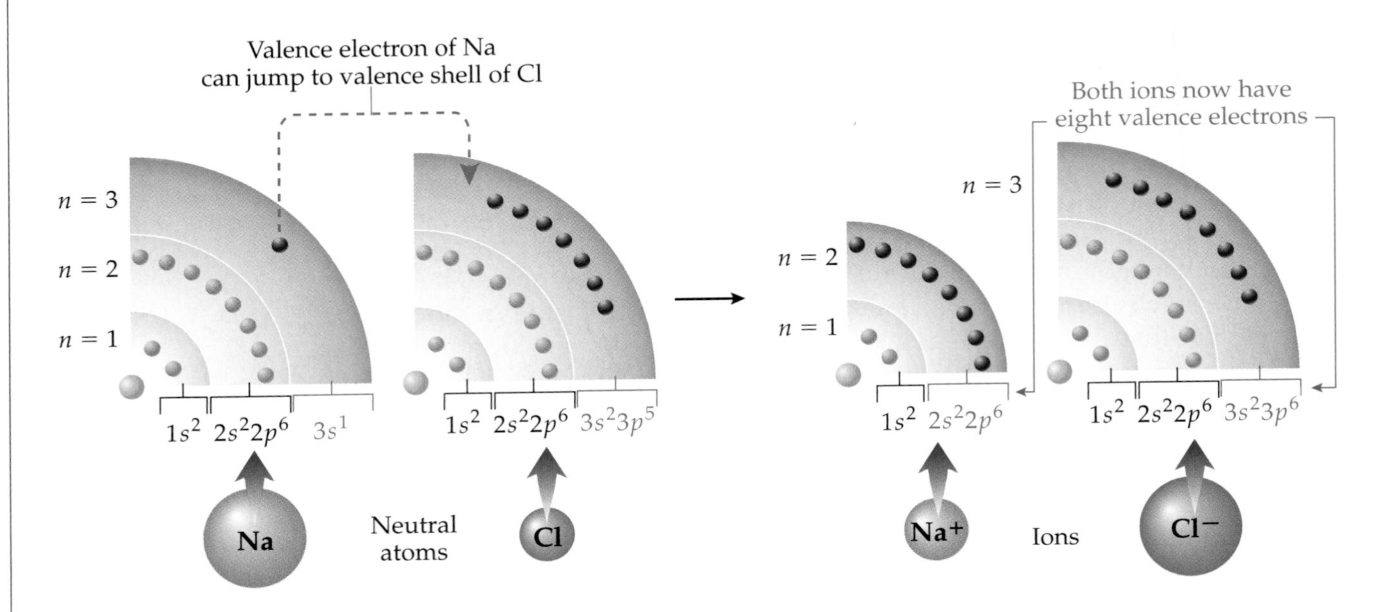

In essence, one electron is exchanged between the atoms. Note that in this exchange the atom that loses the electron, sodium, is a metal. The atom that gains the electron, chlorine, is a nonmetal. This turns out to be typical behavior for metals and nonmetals in general.

We can also now explain why the composition of sodium chloride is NaCl and not, say, $NaCl_2$. Each sodium atom loses one electron, and each chlorine atom picks up one electron. In order to balance this electron exchange and not end up with a deficit or excess of electrons, these elements must combine in a one-to-one ratio.

That is not the case for all elements. Let's try to predict the formula of the compound magnesium fluoride. We start by considering the electron configurations of Mg, a group IIA metal, and F, a group VIIA nonmetal:

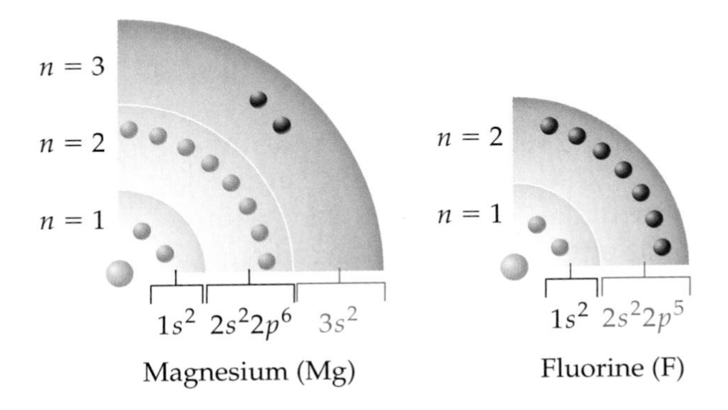

Now consider the best way for each atom to achieve an octet in its valence shell. Like chlorine, each nonmetal fluorine atom needs to gain only one electron, and in doing so will form the anion F^-. Magnesium, a metal atom, can most easily arrive at an octet by losing electrons, but it must lose two electrons and will therefore form a cation with a +2 charge (Mg^{2+}). This has a profound effect on the formula of the compound they form. Because Mg must lose two electrons but each F gains only one, it takes two F atoms to react with one Mg atom and balance the electron exchange. The +2 charge and the two −1 charges of the resulting ions add up to zero, making the compound electrically neutral.

The chemical formula of magnesium fluoride is thus MgF_2. The following diagram summarizes all this. Keep your eyes on the blue valence electrons and you'll see that the octet rule is obeyed.

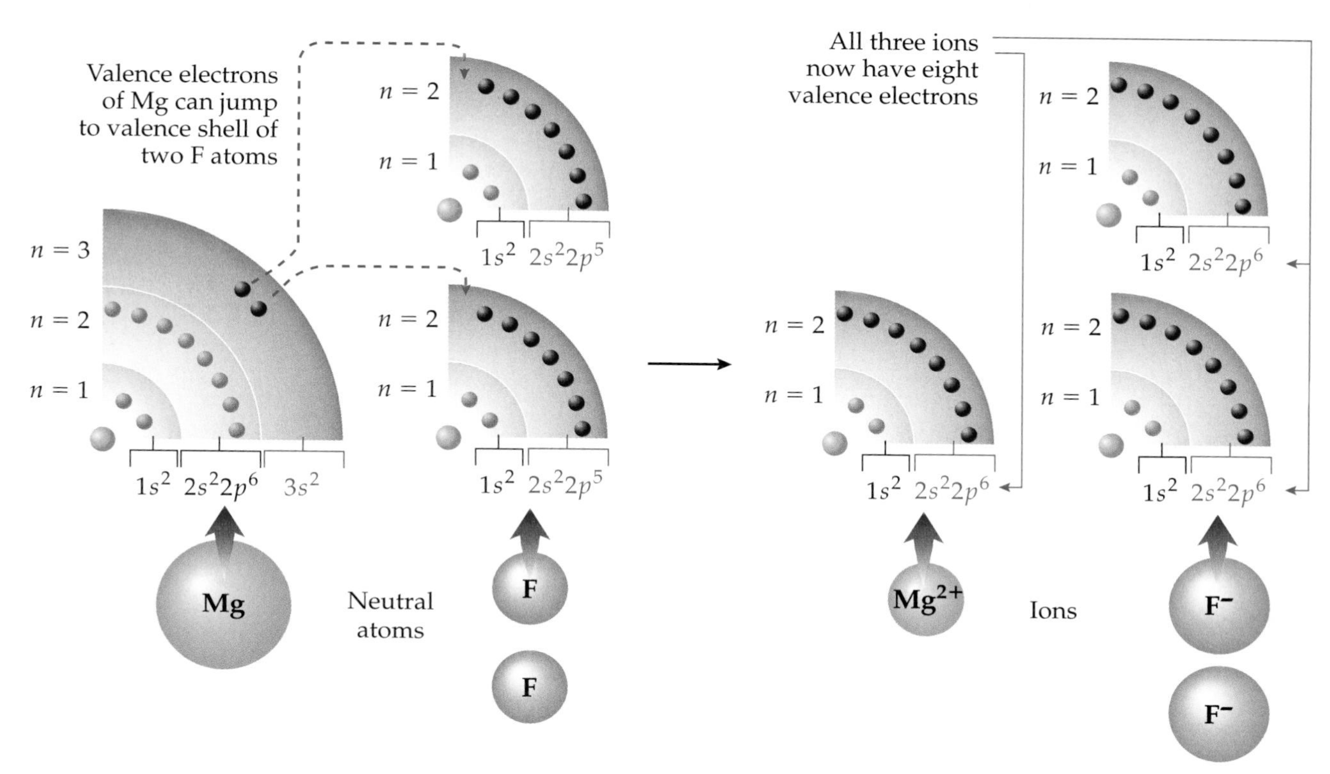

You might realize by now that it is not really necessary to draw the electron configurations of the atoms in order to determine how each achieves its octet. There is a shortcut, at least for the representative (group A) elements. Reconsider the magnesium fluoride example. Magnesium is in group IIA, so we know that its valence shell contains two electrons without resorting to an electron-configuration diagram. We also know that Mg is a metal and that metals tend to lose electrons when they react, so it's a good bet that Mg will lose both its valence electrons to satisfy the octet rule. Fluorine is a group VIIA element and thus has seven electrons in its valence shell. Fluorine is also a nonmetal, and nonmetals tend to gain electrons when they react, so it is likely that F will gain one electron to satisfy the octet rule. Thus, instead of having to draw electron configurations to determine formulas, we just need to consider the roman numeral of the group the atoms are in and whether the elements are metals or nonmetals. Table 4.2 at the top of the next page sums up the behavior of the representative elements for the metals in groups IA–IIIA and the nonmetals in groups VA–VIIA. You can use this table to predict the most likely formula for the compound that results from the reaction between any representative metal and nonmetal.

Notice that the elements of group IVA are not included in this table. We purposely left them out because they can be a bit schizophrenic. Each of these elements has four valence electrons and therefore would need to either gain or lose four electrons to achieve an octet. Neither alternative is very satisfactory, and thus these elements tend to react in an entirely different way. We shall discuss this point in more detail in the next chapter.

Table 4.2 Summary of the Chemistry of the Representative Elements

Group	Valence-shell electron configuration	To achieve octet in valence shell, atom		To become
IA	ns^1	Loses one electron	$\longrightarrow$	+1 cation
IIA	ns^2	Loses two electrons	$\longrightarrow$	+2 cation
IIIA	ns^2np^1	Loses three electrons	$\longrightarrow$	+3 cation
VA	ns^2np^3	Gains three electrons	$\longrightarrow$	−3 anion
VIA	ns^2np^4	Gains two electrons	$\longrightarrow$	−2 anion
VIIA	ns^2np^5	Gains one electron	$\longrightarrow$	−1 anion

Practice Problems

4.26 Predict the chemical formula Na_xO_y of the compound that results from the reaction between the elements Na and O.

Answer: Na is a group IA metal and thus loses one electron to become Na^+.
O is a group VIA nonmetal and so gains two electrons to become O^{2-}.
In order for the compound formed to be electrically neutral, the formula must be Na_2O.

4.27 Draw the reaction above in terms of electron configurations, as was done for MgF_2 in the text.

4.28 Predict the chemical formula of the compound that results from the reaction between the elements Ba and F.

4.29 Predict the chemical formula of the compound that results from the reaction between the elements Al and O.

You can see from these practice problems that the octet rule is a powerful predictive tool.

In Chapter 3 we introduced the terms *metal* and *nonmetal* and defined them according to their observable physical properties and appearance. Now we are in a position to revisit these terms and define them according to their chemical reactions.

Chemical definition of metals and nonmetals

A **metal** is an element that tends to lose its valence electrons in chemical reactions, becoming a cation in the process.

A **nonmetal** is an element that tends to gain valence electrons in chemical reactions, becoming an anion in the process.

To a chemist, this tendency to lose or gain valence electrons in a reaction is the defining property that makes an element a metal or a nonmetal. It is a definition to remember.

4.7 Atomic Size Revisited

In Chapter 3 we discussed the trends in atomic size that occur as we move either across or down the periodic table. These trends are illustrated again in the figure below. As you can see, atoms get larger going down a group and smaller going from left to right across a period.

Relative Atomic Sizes of the Representative Elements

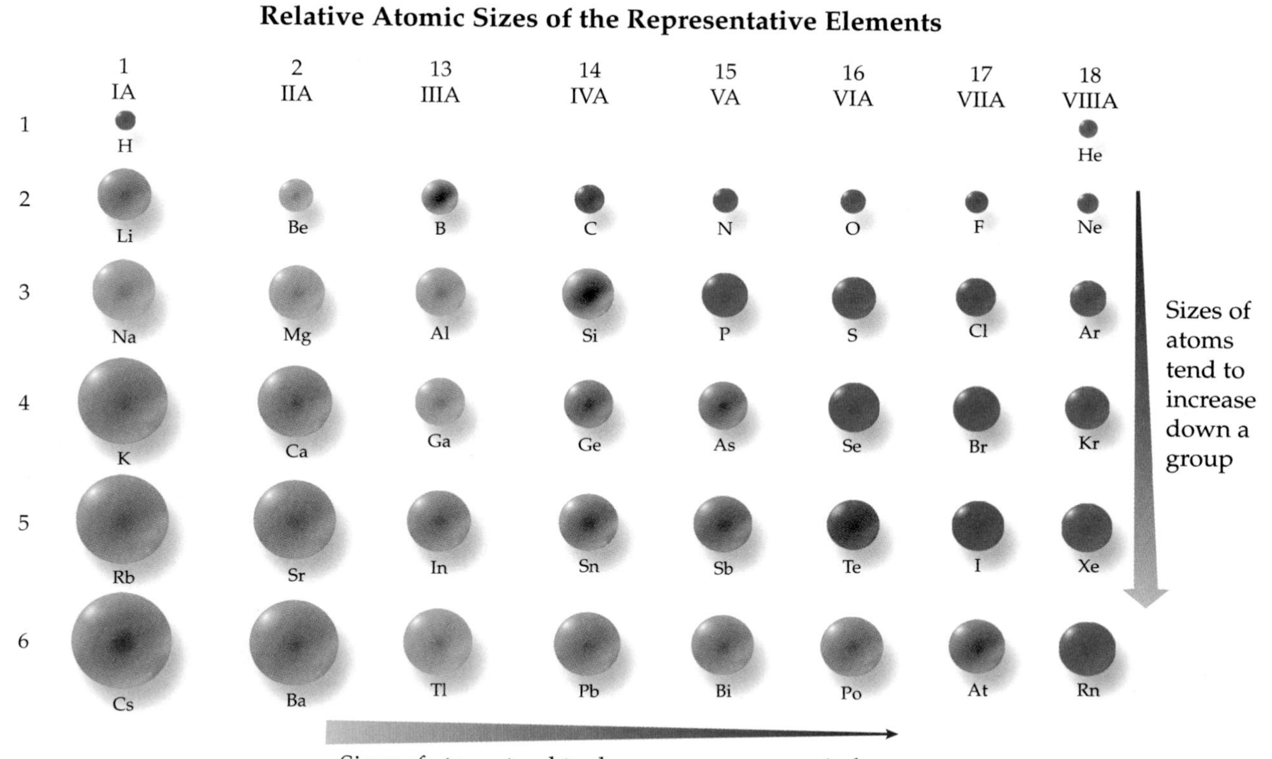

We are now in a position to explore why size varies in this way. Let's start with the trend within a group. The electron configurations for the first three elements in group IA are:

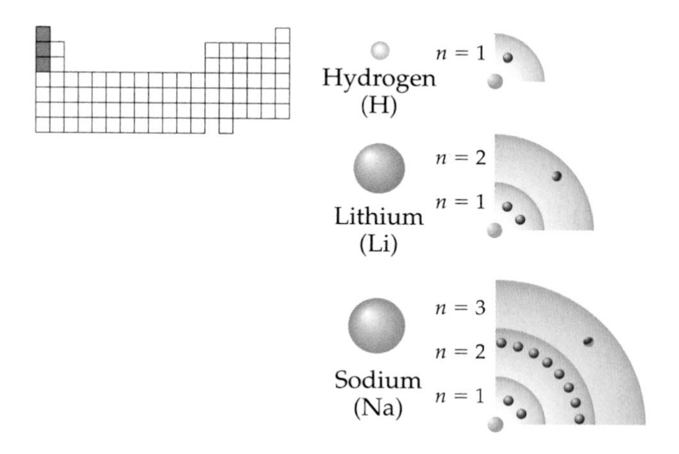

One of the consequences of Bohr's model is that shells get larger as n increases. Therefore a sodium atom, with three occupied shells, is larger than a lithium atom, which has only two occupied shells. Each step down a group adds an additional occupied shell, thus increasing the size of the atom.

To examine the size trend across a period, let's look at the electron configurations for the first four elements in period 2:

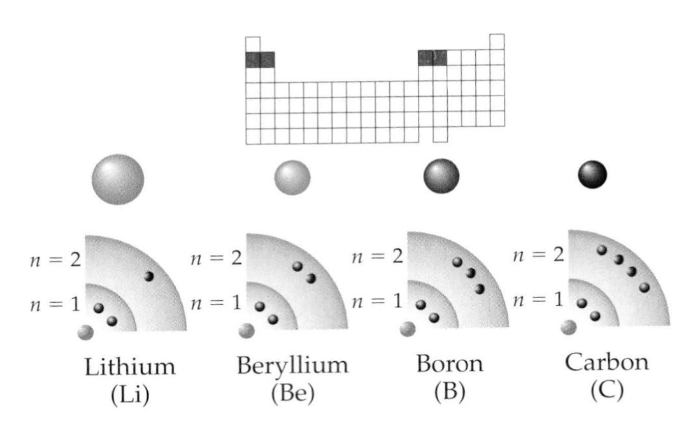

All of these atoms have two occupied shells. Therefore you might expect that the sizes of all the atoms would be roughly the same. But remember, as we go across a period from left to right, each step adds an additional electron to the valence shell and one more proton to the nucleus. The added proton makes the nucleus more positive, which causes an increased attraction for all its electrons, including the valence electrons. This increased positive nuclear charge results in all the electrons being pulled in more tightly and shrinks the shells, thus shrinking the atom. This combination of the same valence shell and an increased nuclear charge is responsible for the observed size trend within a period.

4.7 WORKPATCH

Redraw the shells of Li, Be, B, and C shown in the above drawing to make the overall size of each atom fit the size trend. In each nucleus, write the total number of protons present. Which atom would be most difficult to ionize?

What happens to the relative sizes of the shells in your WorkPatch drawings as you go from left to right? Check your drawings carefully against the answers at the end of the chapter. Good drawings make the answer to the ionization question obvious.

4.8 The Modern Quantum Mechanical Model of the Atom

The Bohr model began with the assumption that the electrons in an atom can possess only certain amounts of energy (in other words, that energies are quantized). It is an understatement to say that many scientists of the day, having been trained in classical physics, had difficulty adjusting to this concept. But there was little time to complain. In less than ten years, quantum physics went even further, with a new description of the electrons in an atom so unusual that even the scientists who gave birth to the field weren't very happy about this new description. Bohr himself said, "Anyone who is not shocked by quantum theory has not understood it." And, as noted at the beginning of this chapter, Erwin Schrödinger, the Austrian physicist who developed the equation on which much of quantum physics is based, has been quoted as saying, "I don't like it, and I'm sorry I ever had anything to do with it." Much of this discomfort was generated not so much by the concept of quantized energy levels but by the other bizarre behaviors that quantum physics demanded of electrons. For example, the equations of quantum physics predict that electrons can exhibit a form of behavior known as *tunneling*. To best understand this behavior, let's assume that a baseball follows the same laws of physics as an electron. The tunneling theory says that if you place a baseball in a solid steel box and weld a cover onto the box, you can come back sometime later and find the baseball sitting outside the still-sealed box, even though the baseball lacks sufficient energy to penetrate the box. No magic or trickery is involved here. It's just possible according to the laws of quantum physics. Baseballs are large objects and therefore follow the laws of classical physics, but electrons are small enough that quantum effects are important, and electron tunneling is experimentally observed. In fact, a number of electronic devices on the market today (for example, the scanning tunneling electron microscope and Josephson junctions, used to make superfast switches for supercomputers) make use of this strictly quantum mechanical behavior. But this was not the worst of it.

This bizarre tunneling behavior where electrons appear in "forbidden" places (on the opposite side of barriers meant to constrain them) is a direct manifestation of what is called the **uncertainty principle**, put forth by Werner Heisenberg in the 1930s. The essence of this principle is that in dealing with particles the size of an electron, it is impossible to know, with any real accuracy, exactly where the electron is or where it is going. In addition, the more carefully and precisely you measure the location of an electron, the less well you will be able to predict where it will be in the next instant. This is exactly the opposite of classical physics, where, for example, careful determination of the present position and velocity of the Moon allows us to know exactly where it will be days or weeks (even centuries) later so that we can aim a rocket and land on it.

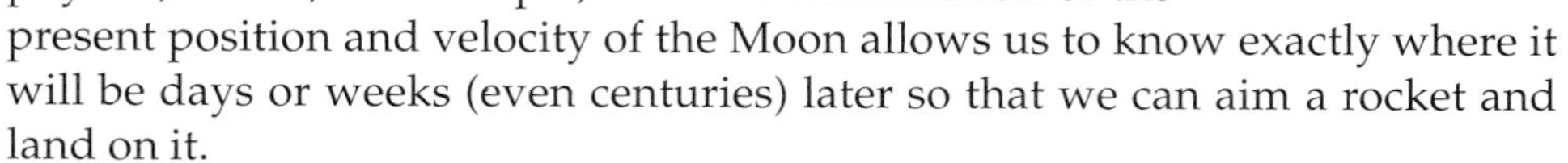

The uncertainty principle has some direct consequences when applied to an electron. Because we can never know an electron's precise location, we cannot describe it as a particle traveling on a well-defined orbit or path because such a particle could be precisely located (in other words, you could point to it). Couple this fact with the experimental observation made in the early 1900s that the electron displays wavelike behavior (*it has a measurable wavelength, just like light!*), and we really have a mess. Exactly how are we supposed to think of an electron in an atom? What picture should come to mind? The answer came from Schrödinger in 1925. He accepted the uncertainty principle and said that, instead of viewing the electron as a particle, we should think of it as a nebulous cloud of negatively charged matter. He went on to show that the density of this electron cloud varies. That is, in some places the cloud is like a thick fog, and in other places it is very thin. Places where the cloud density is high are places where the probability of finding the electron is high, and places where the density is low are places where the probability of finding the electron is low. The cloud might look like this:

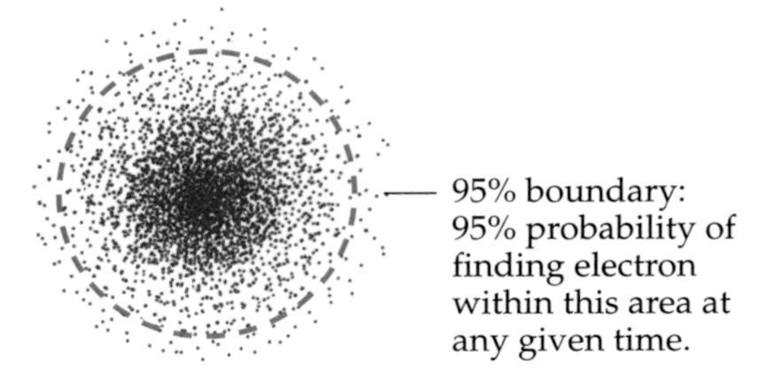
95% boundary:
95% probability of
finding electron
within this area at
any given time.

Electron cloud corresponding to 95% boundary

The left side of the illustration shows a cross section of the cloud for a 1*s* electron (the actual cloud is a three-dimensional sphere). The red dotted line encompasses the volume within which the electron resides 95% of the time. The right side of the illustration shows the spherical shape of the cloud by representing the 95% boundary as a sphere.

QUANTUM
MECHANICS

Schrödinger called this electron cloud an **orbital**. For some physicists, this was too much to accept. Einstein, for instance, refused to believe in this newest quantum mechanical model and its probability-based view of the electron, insisting that "God does not play dice with the universe." Today, however, most physicists and chemists have been forced by the weight of many experimental observations to subscribe to Schrödinger's description of the atom, which is the current, modern, quantum mechanical model of the atom. Of course, some "believers" have had to be dragged, kicking and screaming, to acceptance.

Schrödinger did not stop here. He treated the electron cloud (the orbital) almost like a mound of Jello. Have you ever shaken Jello? If you have, you know that it wiggles back and forth with periodic (wavelike) motion. Well, Schrödinger said that the electron cloud also exhibits wavelike motion and that this motion gives rise to orbitals of various shapes. These shapes are generated by a mathematical equation called a wave equation. The three shapes for orbitals that you should know about are

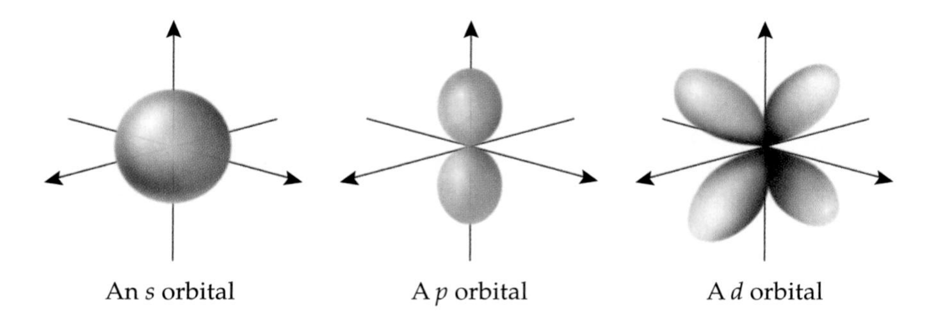
An *s* orbital
A *p* orbital
A *d* orbital

An *s* orbital is spherical, a *p* orbital is dumbbell-shaped, and a *d* orbital looks a bit like a four-leafed clover (there are also *f* orbitals). Notice the *s*, *p*, and *d* designations. These are the same designations we saw in Bohr's subshell model and when we were obtaining electron configurations from the periodic table.

So here is how the modern model works. Let's consider lithium, with its $1s^22s^1$ ground-state electron configuration. Bohr would say that the two 1*s* electrons are in a 1*s* shell and the one 2*s* electron is in a larger 2*s* shell. Schrödinger would say that the two 1*s* electrons exist as 1*s* orbitals and the one 2*s* electron exists as a larger 2*s* orbital.

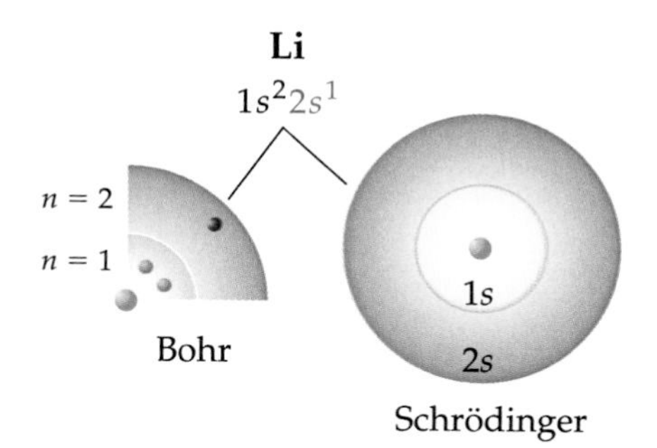

According to Schrödinger, the orbitals get larger as the principal quantum number *n* increases, as shown here with *s* orbitals:

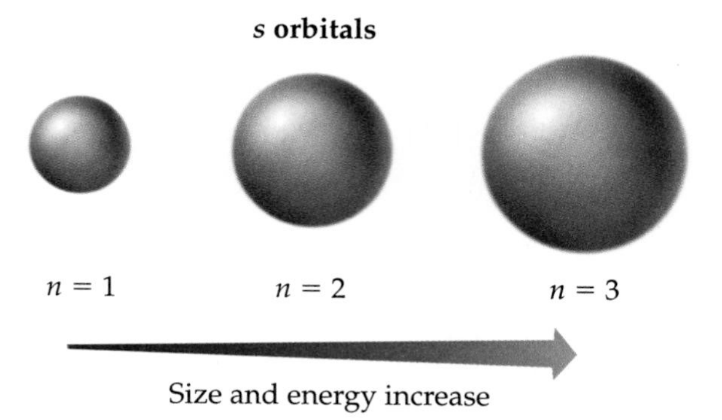

A larger cloud means that the electron spends more time, on average, farther from the nucleus. This means that as *n* increases, the electron gets farther from the nucleus (just as it does in the Bohr model), increasing the electron's energy. Indeed, if we gave sufficient energy to the $1s^22s^1$ ground-state lithium atom so as to produce the excited-state $1s^22p^1$ configuration, Schrödinger's view of what's happening would be

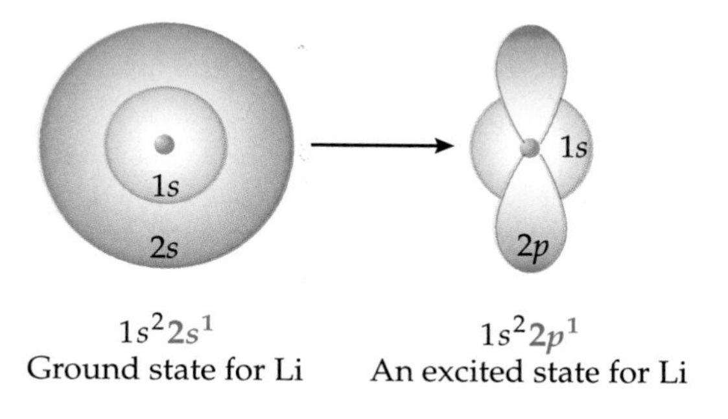

The valence electron has changed from a spherical 2*s* cloud to a dumbbell-shaped 2*p* cloud.

So let's sum up the current model of the atom. It retains Rutherford's tiny, massive, positively charged nucleus and Bohr's proposal about the quantized energy of the electrons, but it replaces the orbiting-particle version of an electron with a nebulous electron cloud (an *orbital*) that has (a) a particular shape resulting from the electron's wavelike motion and (b) a particular size and energy determined by its principal quantum number n. The orbital defines the volume of space about the nucleus where the electron is most likely to be found.

If you continue on with advanced courses in chemistry, you will study the modern quantum mechanical model of the atom in much greater depth. However, what you have learned in this chapter will serve you well for understanding chemical bonding, which comes next.

Have You Learned This?

Wavelength (p. 120)
Electromagnetic radiation (p. 121)
Electromagnetic spectrum (p. 121)
Quantized energy (p. 123)
Classical physics (p. 123)
Quantum physics (p. 123)
Shell (p. 126)
Principal quantum number (n) (p. 126)
Valence shell (p. 127)
Energy-level diagram (p. 129)
Ground state (p. 129)
Excited state (p. 130)
Subshell (p. 134)
Electron configuration (p. 137)
Octet rule (p. 142)
Metal (p. 146)
Nonmetal (p. 146)
Uncertainty principle (p. 149)
Orbital (p. 150)

The Modern Model of the Atom
www.chemistryplace.com

SKILLS TO KNOW

Wavelength

The wavelength of light (symbolized by the Greek letter λ, lambda) is the distance from one wave crest to the next.

The shorter the wavelength of light, the higher the energy of the light. Visible light ranges from low-energy red light to high-energy indigo/violet light.

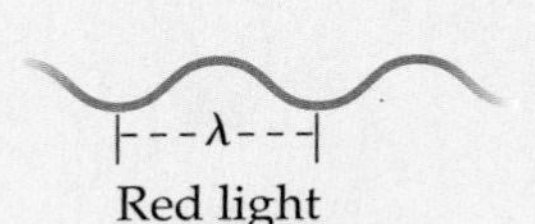

Converting between energy and wavelength

1. To find the energy in joules of light that has a particular wavelength, use the equation at right. But be careful! The wavelength must be in meters. [1 nm = 10^{-9} m]	Planck's constant $h = 6.626 \times 10^{-34}$ J · s; Speed of light $c = 3.00 \times 10^8$ m/s; $E = \frac{hc}{\lambda}$ — Wavelength—*must be in meters*
2. Rearranging the above equation lets you find the wavelength of light that has a given energy. The energy must be in joules, and the equation gives you wavelength in meters.	$\lambda = \frac{hc}{E}$ — Wavelength—*must be in meters*; Energy—*must be in joules*

Example: What is the energy of red light of wavelength 720 nm?

First convert wavelength to meters (as the equation demands):

$$720\ \text{nm} \times \frac{1 \times 10^{-9}\ \text{m}}{1\ \text{nm}} = 7.20 \times 10^{-7}\ \text{m}$$

Then plug in the numerical values and solve:

$$E = \frac{hc}{\lambda} = \frac{(6.26 \times 10^{-34}\ \text{J} \cdot \text{s}) \times (3.00 \times 10^{8}\ \text{m/s})}{7.20 \times 10^{-7}\ \text{m}} = 2.76 \times 10^{-19}\ \text{J}$$

Energy transitions between shells

In order for an electron to move to a higher shell, an atom must absorb an amount of energy equal to the energy difference between the shells.

When an electron relaxes to a lower shell, the atom emits an amount of energy equal to the energy difference between the shells.

The absorbed/emitted energy may be electromagnetic radiation having a wavelength corresponding to that energy.

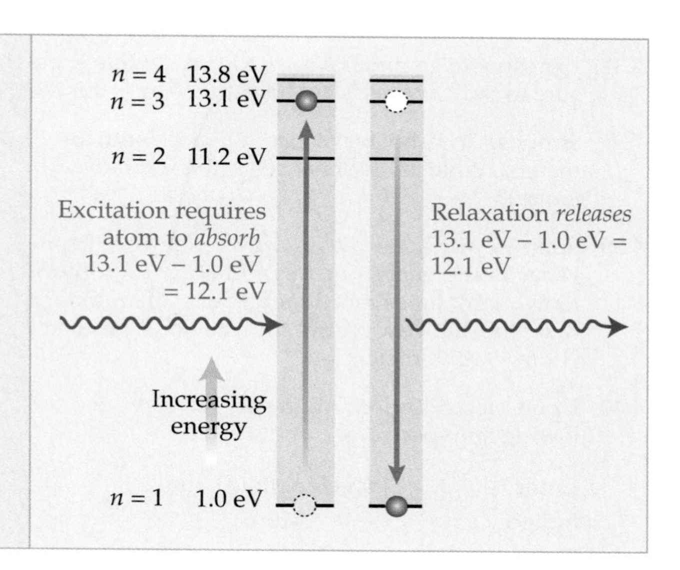

SEEING THE LIGHT—A NEW MODEL OF THE ATOM

4.30 Electromagnetic radiation has a wavelength λ associated with it. Draw two waves of electromagnetic radiation, one representing green light and the other representing X rays, showing in a relative way how the two differ in wavelength.

4.31 X rays have a higher energy than green light. How can you prove this without doing any calculations?

4.32 Why are X rays and gamma rays so dangerous to humans?

4.33 Light travels extremely rapidly ($c = 3.00 \times 10^8$ m/s). Suppose you had to travel 30 miles to work every day. If you traveled at the speed of light for the entire trip, how long would it take you to get to work (in seconds)? [1 mile = 1.61 km]

4.34 The Earth is 9.3×10^7 miles from the Sun. How long does it take the Sun's visible light to reach us (in minutes)? The Sun also emits high-energy gamma rays. How long does it take the gamma rays to reach us (in minutes)? [$c = 3.00 \times 10^8$ m/s; 1 mile = 1.61 km]

4.35 According to the equation for the energy of light, which statement is true? (*1*) The energy of light increases as its wavelength increases. (*2*) The energy of light decreases as its wavelength increases. Explain how the equation for the energy of light tells you which is true.

4.36 The unit of nanometers (nm) is commonly used for the wavelength of visible light. What does 1.00 nm equal in meters? What does it equal in inches? [1 inch = 2.54 cm exactly]

4.37 Exposure to gamma rays can kill you, while exposure to radio waves is not harmful. Why is this so?

4.38 Suppose a radio wave has a wavelength of 10 meters. What is the energy of this radiation (in joules)?

4.39 Suppose an X ray has a wavelength of 10 pm. What is the energy of this radiation (in joules)? How many times more energetic is this radiation than the radio wave in the previous problem? [1 pm = 10^{-12} meter]

4.40 Upon electrification, hydrogen produces the following line spectrum:

Color	Wavelength (λ), nm
Indigo	410.1
Blue	434
Blue-green	486
Red	656.3

What are the energies (in joules) of each of the lines?

ENERGY IS QUANTIZED

4.41 If you exhibited quantum behavior while running in a footrace, describe how you would appear to the people watching the race.

4.42 Why don't we see everyday objects exhibiting quantum behavior?

4.43 What is meant by the term *quantized energy*?

4.44 Which is more general, classical physics or quantum physics? Explain your answer.

4.45 The fact that electrified atoms emit only certain colors of light in sharp lines and not all colors blended together tells us what about an atom?

THE BOHR THEORY OF ATOMIC STRUCTURE

4.46 How does the energy of an object relate to its stability?

4.47 The following drawings represent charged particles. Which situation is more stable? Explain your answer.

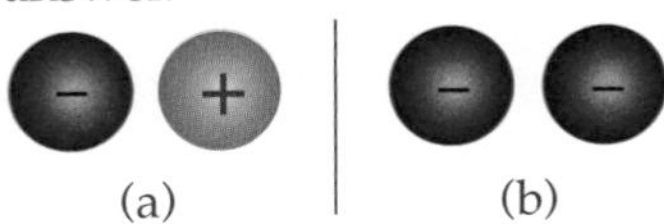

4.48 The following drawings represent charged particles. Which situation is most stable? Which situation is least stable? Explain your answers.

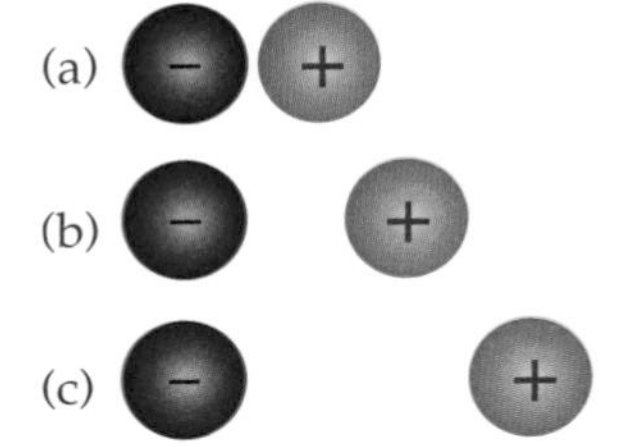

4.49 What happens to both the energy of an electron in an atom and its distance from the nucleus as n increases?

4.50 What type of physics could be used to describe an electron in an atom if the electron could have *any* energy?

4.51 What is another name for a Bohr orbit?

4.52 Is an input of energy required or is energy released when an electron is moved farther away from its nucleus? Explain.

4.53 What are two things that happen to an electron in an atom as the n value of the electron increases?

4.54 What is the maximum number of electrons a Bohr orbit can accommodate?

4.55 Explain why we construct a Bohr model of the atom by first filling a lower shell to capacity before going to an upper shell.

4.56 Does the following Bohr model represent a ground or excited state? Explain your answer.

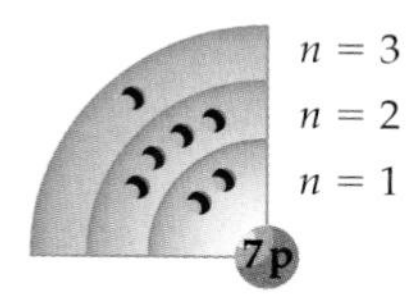

4.57 Using Bohr's rules, draw the lowest-energy configuration for the atom represented in Problem 4.56.

4.58 Why can't an atom's electrons ever be located between orbits?

4.59 Use the Bohr model of the atom to explain why saying that an electron can be only at certain distances from the nucleus is the same thing as saying that the electron can have only certain energies.

PERIODICITY AND LINE SPECTRA EXPLAINED

4.60 What is meant by an atom's valence shell?

4.61 According to Bohr, why do atoms in the same group in the periodic table have similar chemical properties?

4.62 Explain how the Bohr model of the atom accounts for the existence of atomic line spectra.

4.63 What is a ground-state electron configuration? What is an excited-state electron configuration? Explain in words and also by drawing a Bohr model of the Mg atom in both the ground state and an excited state.

4.64 True or false? It is impossible for the H^+ cation to exist in an excited state. Justify your answer.

4.65 Draw Bohr models and use them to explain why phosphorus (P) and arsenic (As) have similar chemical properties.

4.66 (a) How much energy would be necessary to take a hydrogen atom from its ground state to an excited state in which the $n = 3$ orbit is occupied?
(b) What would be the energy of light emitted when the excited atom relaxed back to the ground state?
(c) What kind of electromagnetic radiation would be emitted? (Consult Table 4.1.)

4.67 What would happen to the electron in a ground-state hydrogen atom if the atom were given 5.1 eV of energy?

4.68 In order for hydrogen atoms to give off continuous spectra, what would have to be true?

4.69 For each of the diagrams below do the following:
(a) Determine whether anything is wrong with the diagram. Elaborate.
(b) Give a full atomic symbol for the diagram if there is nothing wrong with it.
(c) For those diagrams that have nothing wrong with them, indicate whether they represent a ground state or an excited state.

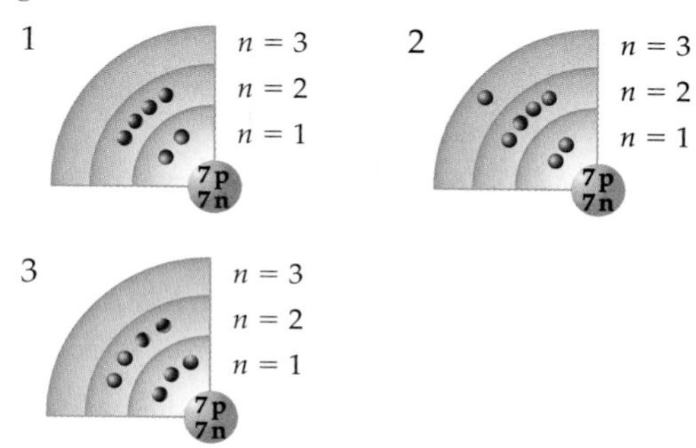

4.70 What is a very useful rule regarding valence electrons for representative elements?

4.71 Referring to Table 4.1, draw a wave of red light and a wave of blue light that show (relatively) how the wavelengths compare. Which light has the higher energy? Without doing any numerical calculations, which light would you expect to be associated with relaxation from the $n = 3$ energy level to the $n = 2$ energy level? Which light would you expect to be associated with relaxation from the $n = 4$ level to the $n = 2$ level?

4.72 Using the energy-level diagram on page 129, calculate the wavelength of the light emitted when the electron in a hydrogen atom falls from the $n = 5$ shell to the $n = 3$ shell. Report your answer in meters and in nanometers. [1 eV $= 1.602 \times 10^{-19}$ J]

4.73 The electron in a hydrogen atom relaxes from the $n = 4$ shell to some lower-energy shell. The light emitted during the relaxation has a wavelength of 1772.6 nm. By calculating the energy of this light, determine the shell to which the electron relaxed. [1 eV $= 1.602 \times 10^{-19}$ J]

SUBSHELLS AND ELECTRON CONFIGURATION

4.74 The simple Bohr model (the model without subshells) works well up to the nineteenth element, potassium (K). Explain how it fails at K.

4.75 What was the experimental evidence that supported the existence of subshells? Explain how this evidence suggested subshells.

4.76 (a) What is the numbering system used to label shells?
(b) What is the lettering system used to label subshells?
(c) How many subshells are there in a given shell?

4.77 How many electrons can each subshell hold before it is considered full?

4.78 Bohr solved the potassium problem by putting its last electron where? How did he justify this?

4.79 You have seen that the 4*s* subshell fills before the 3*d* subshell. This being true, what would be wrong with drawing the Bohr diagram as follows? [*Hint*: What does the diagram drawn this way suggest about relative subshell size that is in fact not true?]

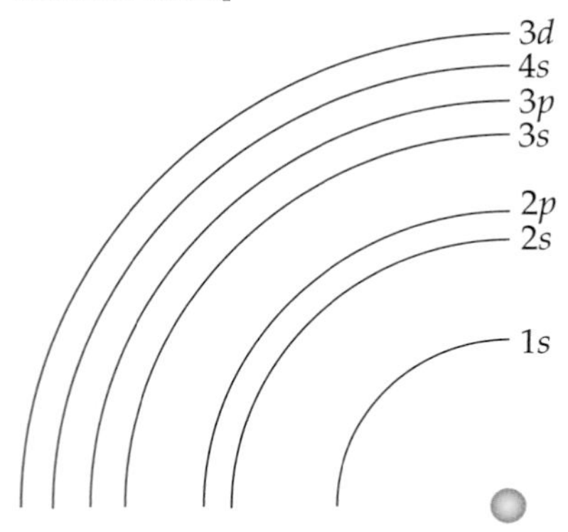

4.80 Write the electron configuration for the following elements without using the noble gas abbreviated form (use the periodic table to assist you).
(a) B (b) Sc (c) Co (d) Se (e) Ru

4.81 Repeat Problem 4.80, but use the noble gas abbreviated form this time.

4.82 Write the electron configuration for the following elements without using the noble gas abbreviated form.
(a) Ba (b) W (c) Pb (d) Pr (e) Pa

4.83 Repeat Problem 4.82, but use the noble gas abbreviated form this time.

4.84 In which period and group in the periodic table are these atoms found?
(a) $1s^22s^22p^3$ (b) $1s^22s^22p^63s^1$
(c) $1s^22s^22p^63s^23p^64s^23d^{10}4p^5$

4.85 A student has written what he thinks are some ground-state electron configurations. Which ones have something wrong with them? What is wrong?
(a) $1s^22p^63s^1$ (b) $1s^22s^63s^23p^64s^24p^6$
(c) $2s^22p^63s^23p^64s^23d^7$ (d) $1s^22s^22p^73s^33p^6$
(e) $1s^22s^22p^53s^1$
(f) $1s^22s^22p^63s^23p^64s^23d^{10}4p^65s^14d^4$ [This is the transition metal niobium, which violates the four-block electron-assignment rules, so be careful.]

4.86 How many valence electrons does each of these atoms have?
(a) $1s^22s^22p^3$ (b) $1s^22s^1$ (c) $1s^22s^22p^63s^23p^64s^23d^7$
(d) $1s^22s^22p^63s^23p^6$ (e) $1s^2$

4.87 What does knowing the period of an atom tell you about the atom's valence electrons?

4.88 Write electron configurations for O, O^{2+}, and O^{2-}. Which form would you expect to find in most compounds of oxygen? Why?

4.89 When using the periodic table to assign electron configurations, what is the rule for the *n* quantum number when you are in the *d* block? When you are in the *f* block?

4.90 Why are the *s*, *p*, *d*, and *f* blocks in the periodic table 2, 6, 10, and 14 blocks wide, respectively?

COMPOUND FORMATION AND THE OCTET RULE

4.91 What is the octet rule, and what is the justification behind it?

4.92 How do metal atoms usually attain an octet in chemical reactions?

4.93 How do nonmetal atoms usually attain an octet in chemical reactions?

4.94 Draw the Bohr model for a Cl atom and for a Cl^- ion. How many electrons are there in the valence shell in each drawing?

4.95 Draw a Bohr model for an Al atom and for an Al^{3+} ion. How many electrons are there in the valence shell in each drawing?

4.96 Predict the formula of the compound that forms when sodium atoms react with sulfur atoms. Completely explain your reasoning.

4.97 Predict the formula of the compound that forms when lithium atoms react with nitrogen atoms. Completely explain your reasoning.

4.98 Why are roman-numeral group numbers for the representative elements useful in predicting how many electrons an atom will gain or lose in a chemical reaction?

4.99 Consider the following Bohr diagrams for two reactants:

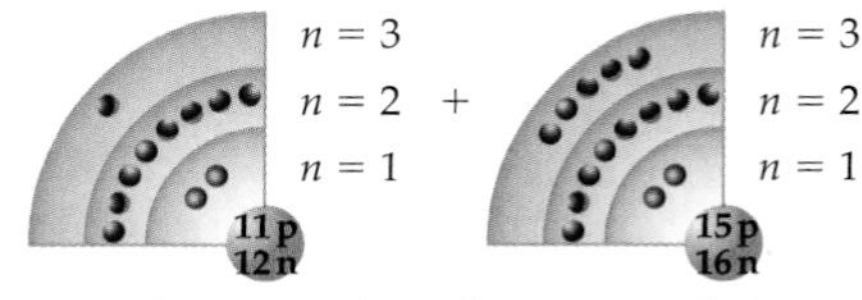

(a) What are the identities of the two elements reacting with each other? Give full atomic symbols.
(b) Which is the metal, and which is the nonmetal?
(c) What is the formula of the resulting compound?
(d) Draw Bohr diagrams for the ions formed.

4.100 When a group IIA element X reacts with a group VIIA element Y, what will be the formula? Why?

4.101 Explain what is meant by this statement: The element hydrogen is an exception to the octet rule and yet it obeys it in principle.

4.102 Aluminum forms compounds with both sulfur and oxygen. Why are the formulas of the resulting compounds Al_2S_3 and Al_2O_3, respectively? Why are the formulas similar?

4.103 True or false? The O^{2-} and F^- anions have identical electron configurations. Justify your answer with Bohr diagrams and electron configuration notation.

4.104 Which part of the following statement is true and which part is false? Mg^{2+} and Na^+ have identical electron configurations, and they also have similar properties. Explain your answer fully.

4.105 How can you tell how many electrons a representative metal is likely to lose? What, in general, will be the charge of the cation it forms?

4.106 How can you tell how many electrons a representative nonmetal is likely to gain? What, in general, will be the charge of the anion it forms?

4.107 When a cation and an anion join to form a compound, how does knowing the charge of the ions help to determine the formula?

ATOMIC SIZE REVISITED

4.108 (a) According to the Bohr model, why might someone expect that atoms would not change in size as you go from left to right across a period?
(b) In fact, atoms do change in size across a period. What is the trend and what is the explanation for it?

4.109 According to the Bohr model, why do atoms get larger as you proceed down a group in the periodic table?

4.110 Which atom has a smaller 1*s* subshell, lithium (Li) or beryllium (Be)? Justify your answer.

4.111 Which atom has a smaller valence shell, lithium (Li) or sodium (Na)? Explain your answer.

4.112 Which atom is larger, lithium (Li) or beryllium (Be)? Explain your answer.

4.113 Which atom is larger, lithium (Li) or sodium (Na)? Explain your answer.

4.114 Rank the following atoms from smallest to largest: Si, Mg, Rb, Na.

THE MODERN QUANTUM MECHANICAL MODEL OF THE ATOM

4.115 What is wrong with Bohr's planetary model of atomic electrons according to modern quantum mechanical theory? [*Hint:* Use Heisenberg's uncertainty principle in your answer.]

4.116 Quantum mechanical tunneling is one consequence of quantum theory. What is quantum mechanical tunneling?

4.117 What is an orbital?

4.118 Draw an *s* orbital, a *p* orbital, and a *d* orbital.

4.119 Suppose you behaved like a quantum mechanical particle. Describe what a picture of you taken by a camera might look like.

4.120 What gave Schrödinger justification to think of an electron in an atom as a nebulous cloud?

ADDITIONAL PROBLEMS

4.121 How many seconds does it take light to travel 3.00×10^3 miles from New York to California? [1 mile = 1.61 km]

4.122 A laser emits a beam of green light that has a wavelength of 5.00×10^{-5} cm. What is the wavelength of this light in nanometers?

4.123 What is the wavelength in nanometers of infrared light for which $\lambda = 2.50 \times 10^{-5}$ m? How many times longer is this wavelength than red light that has a wavelength of 750 nm?

4.124 Electromagnetic radiation emitted by magnesium has a wavelength of 285.2 nm. (a) Is this radiation visible to the eye? (b) What is the energy of this radiation?

4.125 Fill in the table:

Wavelength (m)	Wavelength (nm)	Energy (J)
		5.65×10^{-18} J
	2602.0 nm	
7.85×10^{-12} m		

4.126 Would moving an electron farther from an atom's nucleus give off light energy or require the absorption of light energy? Explain your answer.

4.127 According to Bohr's model, is energy absorbed or released when an electron moves to a shell of lower *n*?

4.128 According to Bohr's model, energy must be put into an atom to move an electron from a low-energy shell to a higher-energy shell. How do you calculate the amount of energy needed for the move?

4.129 Convert this excited boron, B, atom to (a) an excited state of lower energy and (b) the ground state:

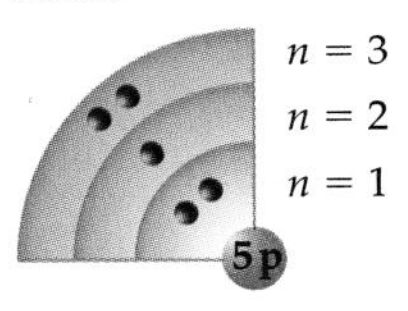

4.130 According to Bohr, what is so special about the valence shell of an atom?

4.131 What are the wavelength in nanometers and energy in joules of the light emitted when a hydrogen electron originally in the $n = 6$ shell relaxes to the ground state? [1 eV = 1.602×10^{-19} J]

4.132 Use electron configuration notation to explain why aluminum, Al, and gallium, Ga, have similar chemical properties even though gallium has *d* electrons but aluminum does not.

4.133 These energy-level diagrams are for iron, Fe. Indicate whether each represents the ground state, an excited state, or is incorrect. Explain your answers.

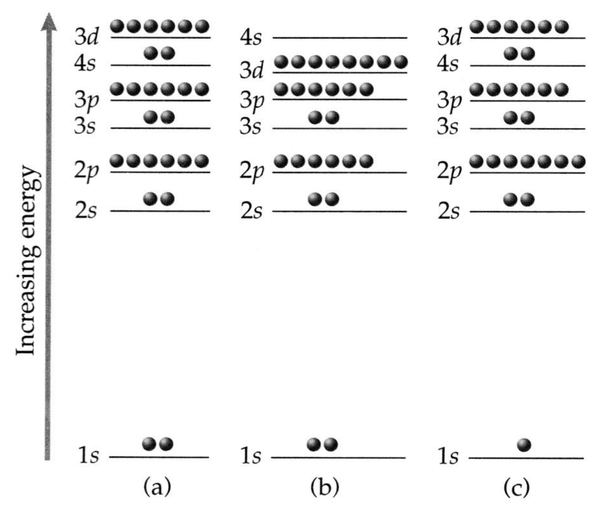

4.134 Name each element and tell how many valence electrons it has:
(a) $1s^2 2s^2 2p^6 3s^2 3p^2$
(b) [Ne]$3s^1$
(c) [Ar]$4s^2 3d^{10} 4p^3$

4.135 Which subshell is being filled in transition metals? What is unique about its order of filling?

4.136 Consider the anion whose charge is 2− and whose electron configuration is identical to that of argon, Ar. What is the symbol for this anion?

4.137 What do F^-, O^{2-}, Na^+, and Mg^{2+} all have in common?

4.138 Why is the formula for calcium sulfide CaS and not Ca_2S?

4.139 Select the element from each pair expected to have the lower first ionization energy. Explain.
(a) N and F (b) Mg and Ba (c) N and Ca

4.140 How does the first ionization energy of the alkali metal in a given period compare with the first ionization energy of the halogen in the same period? What is the result in terms of how these elements tend to react with each other?

4.141 Predict the formula for lithium nitride, made from lithium, Li, and nitrogen, N. Explain your reasoning.

4.142 Suppose the 3*s* valence electron in a ground-state sodium atom is excited to the 4*s* orbital. Why does the electron now have higher energy?

4.143 What is the wavelength in nanometers of electromagnetic waves that have an energy of 1.00×10^{-15} J?

4.144 Are the waves in Problem 4.143 more or less energetic than visible light (λ = 400–750 nm)?

4.145 Which electromagnetic radiation is most dangerous to humans:
(a) X rays (b) Ultraviolet light
(c) Gamma rays (d) Infrared light
(e) Radio waves?

4.146 Write the ground-state electron configuration for each of the following atoms or ions. Which have a valence-shell octet?
(a) Ca^+ (b) Li (c) P^{3-} (d) Ar^+ (e) Si^{2+}

4.147 Halogens are very reactive because
(a) They need to gain only one electron to satisfy the octet rule.
(b) They have seven electrons in their valence shell, and the more electrons an atom has, the more reactive it is.
(c) They are nonmetals, and all nonmetals are reactive.
(d) They can easily lose their seven valence electrons to satisfy the octet rule.

4.148 Write the ground-state electron configuration for each of the following atoms or ions. Which have a valence-shell octet?
(a) Ar (b) Na^+ (c) C^{2-} (d) O^{2-} (e) Ca^{2+}

4.149 Which subshell has the lowest energy:
(a) 4*s* (b) 3*p* (c) 2*p* (d) 3*s* (e) 2*s*

4.150 Which is the correct ground-state electron configuration for antimony, Sb:
(a) $1s^2 2s^2 2p^6 3s^2 3p^6 3d^{10} 4s^2 4p^6 4d^{10} 5s^2 5d^3$
(b) $1s^2 2s^2 2p^6 3s^2 3p^6 3d^{10} 4s^2 4p^6 4d^{10} 5s^2 5p^3$
(c) $1s^2 2s^2 2p^6 3s^2 3p^6 4s^2 3d^{10} 4p^6 5s^2 4d^{10} 5p^4$
(d) $1s^2 2s^2 2p^6 3s^2 3p^6 4s^2 3d^{10} 4p^6 5s^2 5p^3$
(e) $1s^2 2s^2 2p^6 3s^2 3p^6 4s^2 3d^{10} 4p^6 5s^2 4f^3$

4.151 What is the energy in joules of green light that has a wavelength of 500. nm?

4.152 Indicate whether or not the following volumes are quantized:
(a) The volume of water available in 16-oz bottles
(b) The volume of water available from a drinking fountain
(c) The volume of soft drink available from a soda-fountain dispenser
(d) The volume of soft drink available in 12-oz cans

4.153 Predict the formula of the compound formed by the reaction between
(a) Ca and Br (b) K and N (c) Al and S
(d) Na and I (e) Mg and O

4.154 Arrange in order of increasing valence-shell size: Sr, Mg, Ba, Ca, Be.

4.155 The order of the colors in the visible spectrum can be remembered by the acronym ROY G. BIV, which stands for red, orange, yellow, green, blue, indigo, violet. In the acronym, are the colors arranged in order of increasing or decreasing wavelength? In order of increasing or decreasing energy?

4.156 (a) Bohr is the scientist who proposed that the energy of electrons is __________.
(b) The two kinds of physics are __________ physics and __________ physics.

4.157 Write the ground-state electron configuration, both full notation and noble gas abbreviation, and indicate the number of valence electrons for
(a) Al (b) I (c) Rb (d) Ar (e) Mg

4.158 Metals tend to ____________ valence electrons, whereas nonmetals tend to __________ valence electrons.

4.159 What is the valence shell in Mg, Ge, W, Cl, Cs?

4.160 If gamma radiation has a wavelength of 1.00×10^{-12} m, what is the energy of gamma radiation in joules?

4.161 What is the formula for the maximum number of electrons in each shell of the Bohr atom? How many electrons are allowed in the $n = 2$ shell? The $n = 6$ shell?

4.162 What are the group number, period number, and name of the element whose electron configuration is $1s^22s^22p^63s^23p^64s^23d^{10}4p^2$?

4.163 Circle the correct choice to indicate how many electrons each element must gain or lose to form an octet:

(a) Mg	gains, loses	1, 2, 3	electrons
(b) Se	gains, loses	1, 2, 3	electrons
(c) Al	gains, loses	1, 2, 3	electrons
(d) Sr	gains, loses	1, 2, 3	electrons
(e) Br	gains, loses	1, 2, 3	electrons
(f) P	gains, loses	1, 2, 3	electrons

4.164 State the Heisenberg uncertainty principle and what it implies about the structure of an atom.

4.165 Conversion factor fiesta. If a car were able to travel at the speed of light, how many seconds would it take the car to travel 1000 miles? [1 mile = 5280 ft; 1 inch = 2.54 cm exactly]

4.166 In the Bohr model of the atom, are the electrons in shells closer to the nucleus higher or lower in energy than electrons in shells farther from the nucleus? Explain.

4.167 What is the total number of p = subshell electrons in P, Mg, Se, Zn?

4.168 Write the full ground-state electron configuration for Ca^{2+}, S^{2-}, Ar, K^{+}.

4.169 Rank visible light, gamma rays, X rays, radio/television waves, infrared radiation, and ultraviolet light in order of (a) increasing wavelength and (b) increasing energy.

4.170 Draw a simple Bohr model (no subshells) for an oxygen atom. How many electrons are in the valence shell? How many more electrons can be put into the valence shell?

4.171 Taking all configurations to be for the neutral atoms, identify the elements having the following electron configurations:
(a) $1s^22s^22p^63s^23p^63d^{10}4s^24p^5$
(b) $1s^22s^22p^1$
(c) $1s^22s^22p^63s^23p^64s^23d^{10}4p^65s^1$
(d) $1s^22s^22p^63s^23p^64s^23d^{10}4p^65s^24d^{10}5p^66s^24f^{14}5d^{10}6p^67s^26d^15f^3$
(e) $1s^22s^22p^63s^23p^64s^2$

4.172 Explain the major difference between the orbits in the Bohr model of the atom and the orbitals in the quantum mechanical model of the atom.

4.173 What is the highest value of n for the electrons in
(a) Co (b) As (c) Sr (d) Po

4 WORKPATCH SOLUTIONS

4.1 Atom-sized cars on the highway could travel only at certain allowed speeds. Thus, instead of accelerating smoothly from zero to 60 miles/h as it entered traffic, a car stopped at an entrance ramp would have to be at 0 miles/h one instant and, say, 45 miles/h the next instant, and 85 miles/h the next instant. If all the other cars on the highway were not traveling at one of these allowed speeds, think of the major pileup that would occur!

4.2 (a) 1.0 eV
(b) 11.2 eV
(c) The difference between them, 11.2 eV − 1.0 eV = 10.2 eV
(d) Less energy, 13.1 eV − 11.2 eV = 1.9 eV

4.3 The hydrogen atom with the electron in the $n = 2$ shell.

4.4 12.1 eV − 10.2 eV = 1.9 eV more energy

4.5 Green: $n = 4$ to $n = 2$; blue: $n = 5$ to $n = 2$; indigo/violet: $n = 6$ to $n = 2$.

For example, the green line corresponds to light of energy between 2.24 and 2.64 eV (Table 4.1). The only relaxation that produces this energy is from the $n = 4$ level (13.8 eV) to the $n = 2$ level (11.2 eV): 13.8 eV − 11.2 eV = 2.6 eV.

4.6

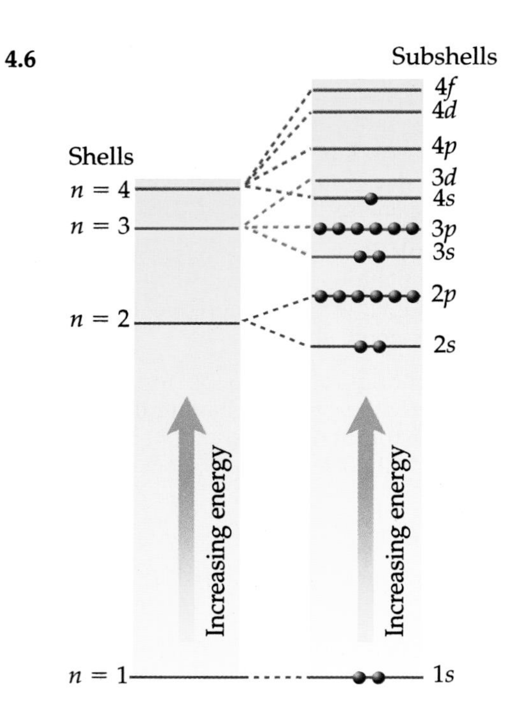

4.7

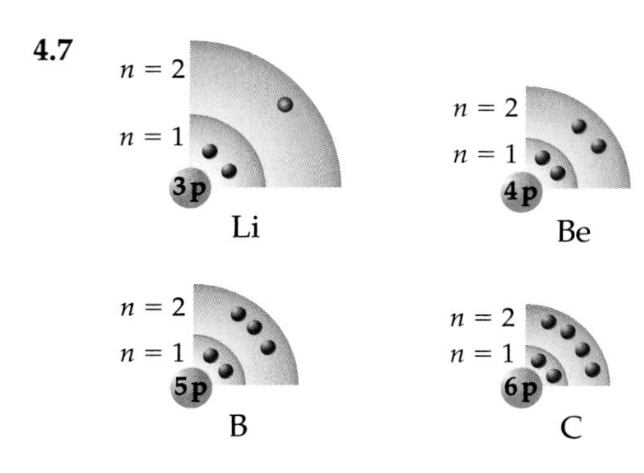

Carbon would be the hardest to ionize because its shells are closest to the nucleus, increasing the force of attraction between the negative electrons and the positive nucleus. The shells are smallest for C because it has the most positive nucleus of the atoms in this period.

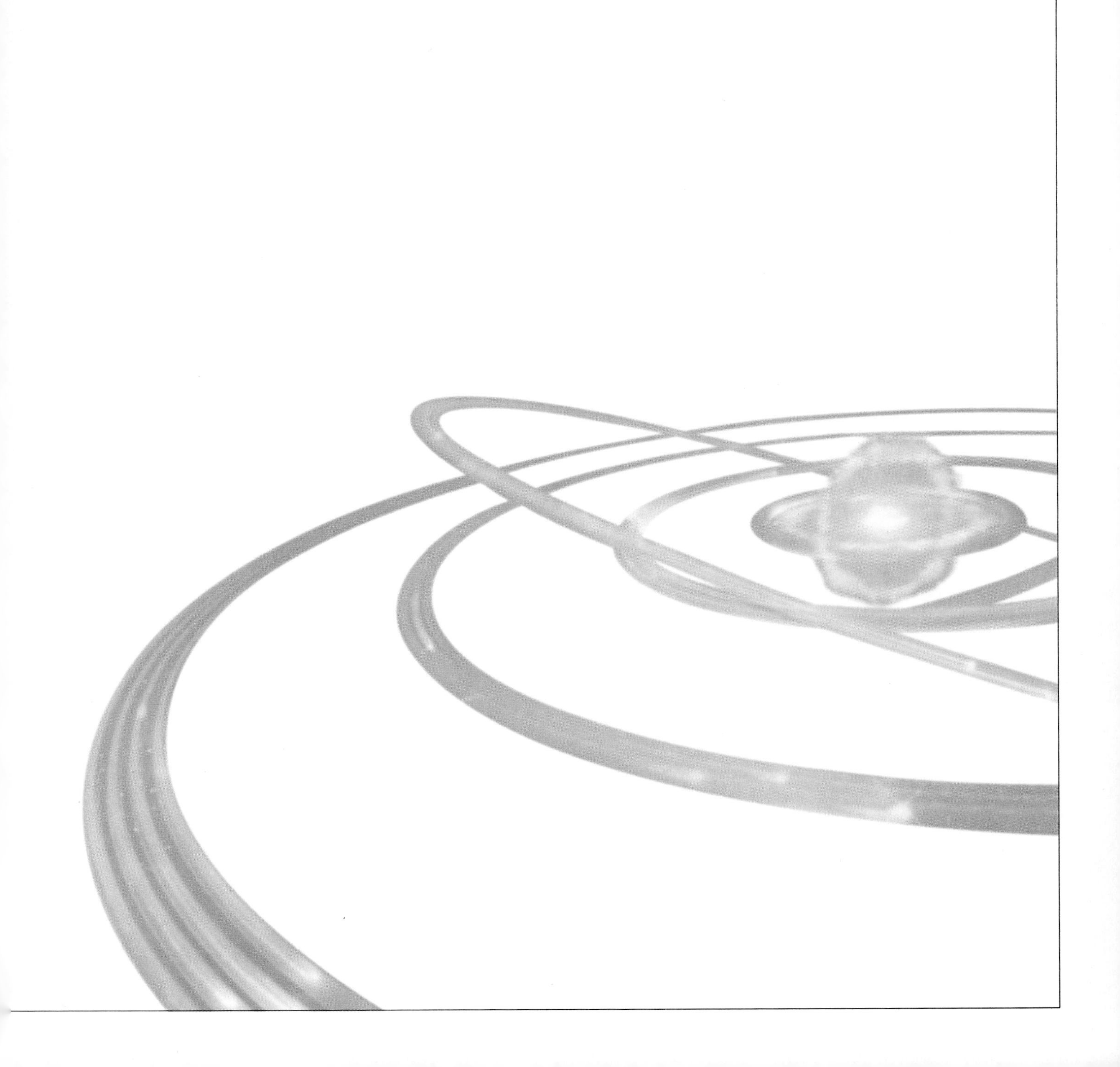

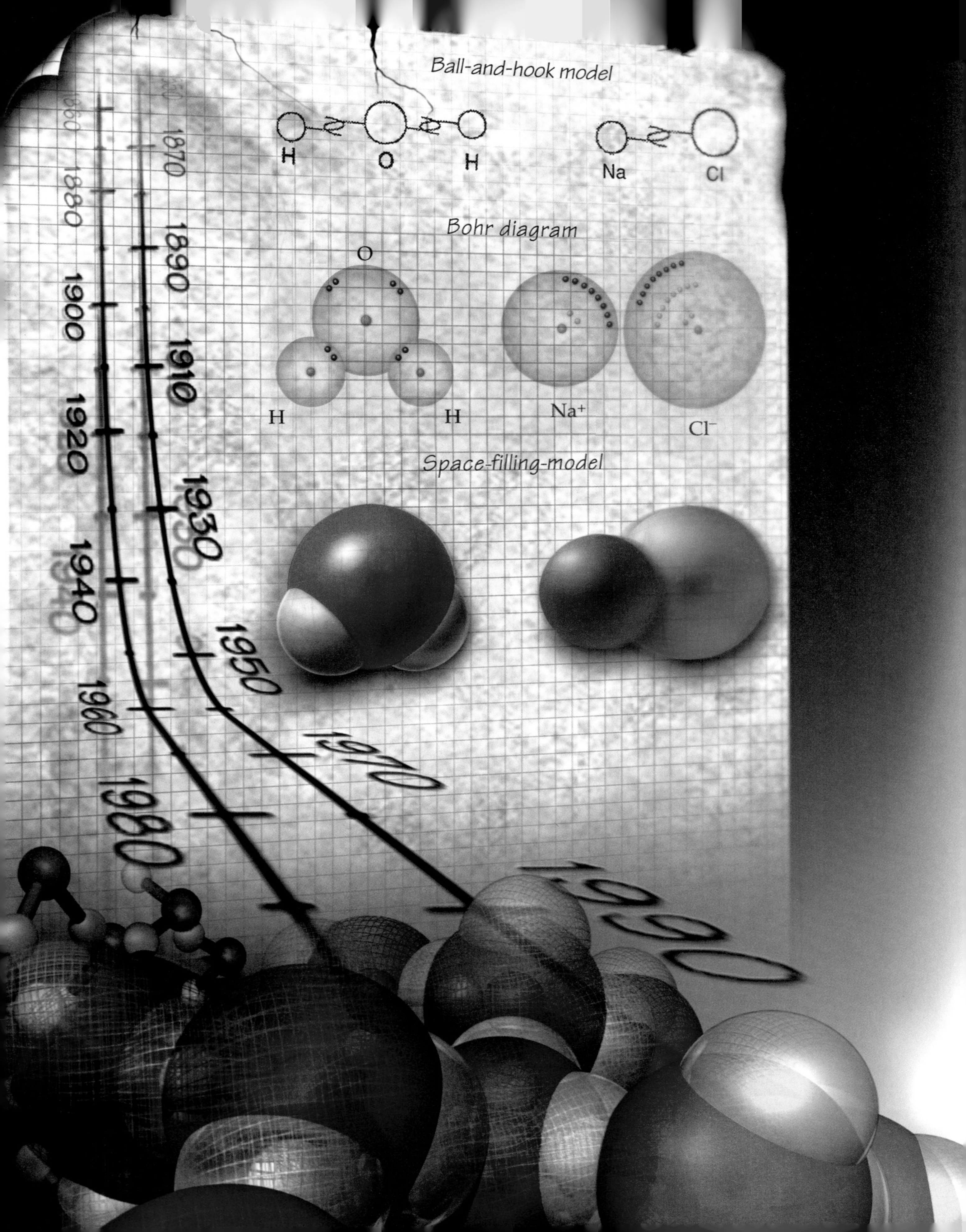

Ball-and-hook model
H
O
H
Na
Cl
Bohr diagram
O
H
H
Na+
Cl−
Space-filling-model
1860
1870
1880
1890
1900
1910
1920
1930
1940
1950
1960
1970
1980
1990

CHAPTER 5

Chemical Bonding and Nomenclature

5.1 Molecules—What Are They? Why Are They?

Take a deep breath. As everyone knows, the air filling your lungs contains oxygen gas, but exactly what is it that you are breathing? The gas you are inhaling is not made up of isolated oxygen (O) atoms. Oxygen atoms are extremely reactive, and your lungs would be instantly destroyed by them. Instead, you are breathing oxygen *molecules* (O_2).

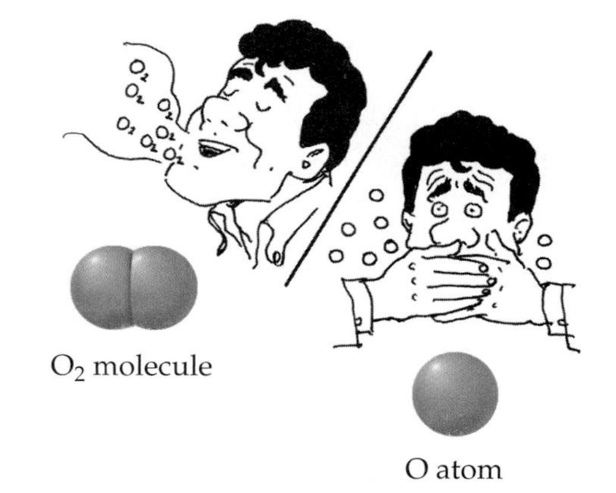

A **molecule** is defined as a collection of atoms bound together. In the case of oxygen, two oxygen atoms are all that is needed to form a molecule. Thus the molecular formula for the oxygen molecule is O_2.

An example of a molecule that is made from more than two atoms is methane, the main component of natural gas, the fuel burned in home furnaces, kitchen stoves, and the bunsen burners in your chemistry laboratory. The methane molecule, which has the molecular formula CH_4, is a collection of five atoms—a single carbon atom and four hydrogen atoms as shown at right.

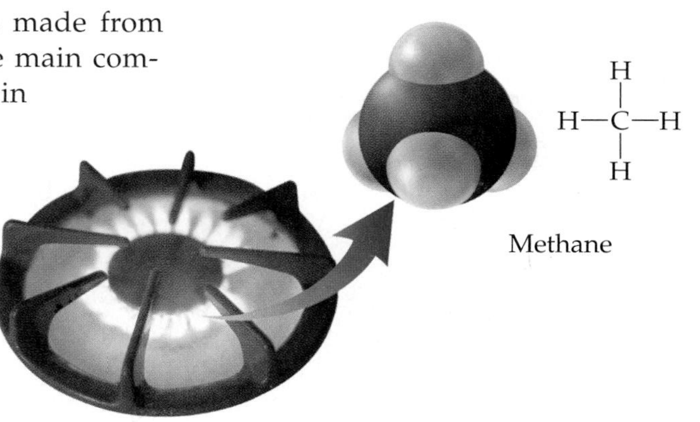

Note that while methane and oxygen both exist as molecules, only CH_4 is classified as a *compound* because it consists of more than one type of element. Molecular O_2 is considered an elemental substance.

The glucose molecule

Natural sugars are broken down to glucose in the body.

One of the things that make molecules so interesting is that they can have chemical and physical properties that are very different from those of the atoms from which they are made. For example, a simple change from individual O atoms to O_2 molecules changes the substance from a lethal to a life-sustaining gas. Another example is the molecule glucose, one of the simplest kinds of sugar. Much of the food you eat either contains glucose or gets converted to glucose during digestion. The glucose molecule is assembled from 24 atoms (6 carbon atoms, 12 hydrogen atoms, and 6 oxygen atoms). Whereas individual carbon, hydrogen, and oxygen atoms have no particular nutritional value, the glucose molecule, $C_6H_{12}O_6$, is a major source of food for the cells in your body.

Huge corporations invest large amounts of money to design and produce new molecules having specific, tailor-made properties. Two examples you are probably familiar with are the compounds aspartame (NutraSweet), which has the property of being sweet without providing significant calories, and Viagra, which has been immensely profitable for its manufacturer. There are literally millions upon millions of different molecules, ranging from simple two- or three-atom combinations, such as oxygen, O_2, and water, H_2O, to huge supermolecules containing thousands or even hundreds of thousands of atoms, such as DNA.

Would you expect these two molecules to have nearly identical properties?

H–O–H (H_2O)   H–O–O–H (H_2O_2)

Did you answer the WorkPatch correctly? The answer (given at the end of the chapter) emphasizes what we have been discussing about the chemical properties of molecules.

5.2 Holding Molecules Together—The Covalent Bond

One of the most fundamental and important questions in chemistry is, "What holds a molecule together?" Atoms aren't sticky, and they don't come with hooks attached. What is it then that prevents molecules from coming apart into separate atoms?

To answer this question, let's consider the element hydrogen. In the periodic table, an atom of hydrogen is represented by the letter H. However, a tank of hydrogen gas contains no individual H atoms. It is filled with H_2 molecules. Like oxygen, hydrogen is more stable as a *diatomic* (two-atom) molecule. Indeed, if you could get your hands on a tank full of H atoms, you would not have it for very long. The gas in the tank would rapidly transform into a collection of H_2 molecules. To discover why hydrogen prefers to exist this way, let's begin by doing a thought experiment. Imagine that you have been shrunk down to a height of 10^{-10} m so that you are about the size of a hydrogen atom. (This length 10^{-10} m is equivalent to 1 angstrom, Å, as noted in Section 3.7). In front of you are two H atoms. Each atom consists of a positively charged nucleus and a single negatively charged valence electron. Each electron is held firmly to its own nucleus by the attraction between their opposite electrical charges. Let's begin with the two H atoms far from each other (say, about 10 Å apart). At this distance, the atoms would just sit there, each unaware that the other one exists.

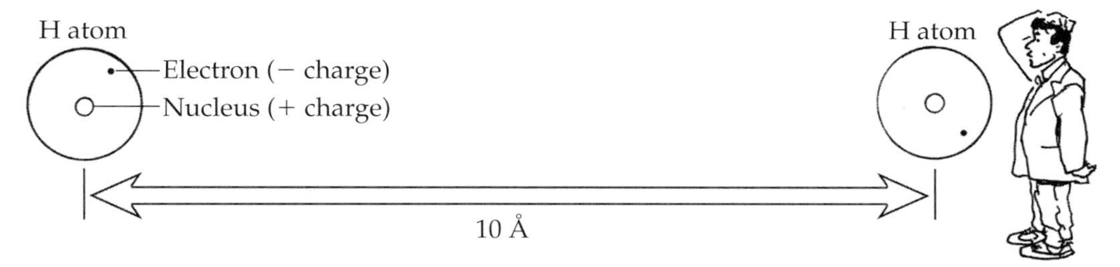

Now imagine moving the atoms closer to each other, a little bit at a time. At 9 Å, they still just sit there. At 8 Å, still no change.

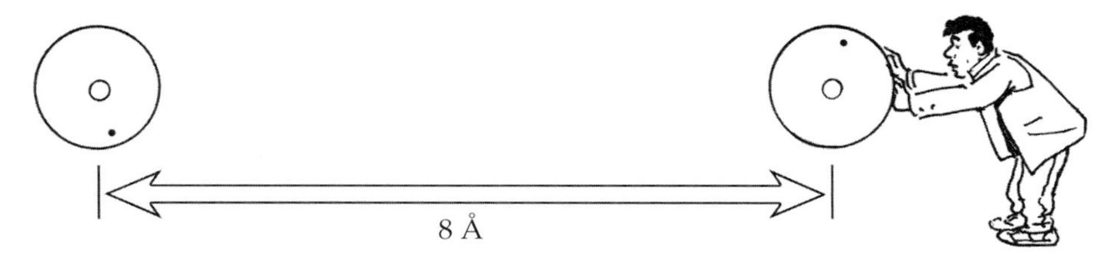

But when the atoms get close enough to each other, say to within 3 Å, something interesting happens. Suddenly, the atoms spontaneously start to move toward each other! Their movement is slow at first, but as they get closer, they begin to move faster. Soon they're racing toward each other!

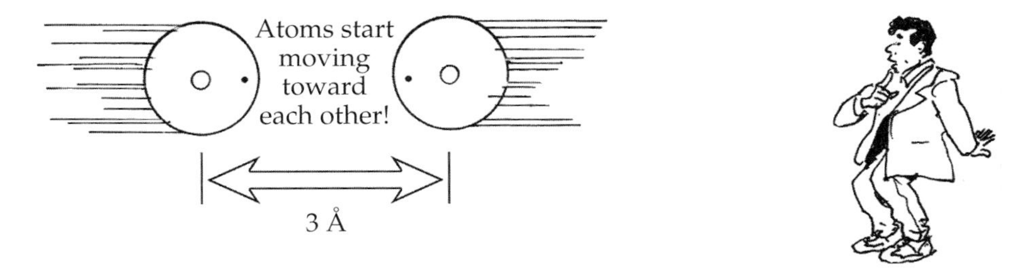

As the two atoms get closer and closer together, you turn away in anticipation of a messy crash.

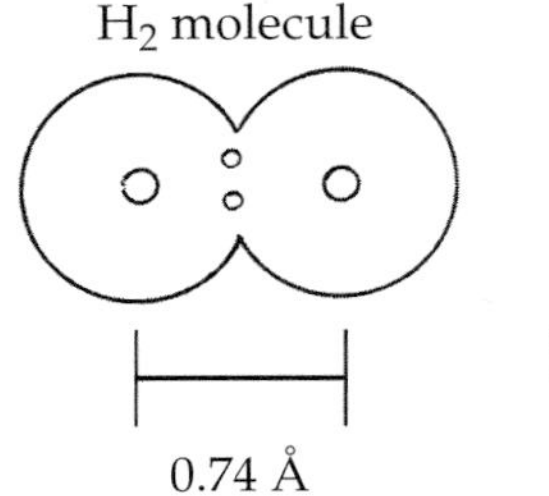
H_2 molecule
0.74 Å

When you turn back, however, you discover that the two atoms are still intact, with their nuclei separated by a distance of 0.74 Å. At the same time, the temperature in the vicinity of the atoms has increased. An H_2 molecule has been created.

What happened? Why did the H atoms start moving toward each other when they got close, and why did they stop when the two nuclei were 0.74 Å apart? And where did the increase in temperature come from?

There is much to explain. Let's start with the first observation—when the atoms were brought to within 3 Å of each other, they began to move toward each other. This spontaneous movement was caused by the same force that holds each electron close to its own nucleus: opposite charges attract. When the two atoms got close enough, the negative electron of one atom began to be attracted to the positive nucleus of the other atom, and vice versa. These attractive forces *between atoms* began to pull the atoms toward each other.

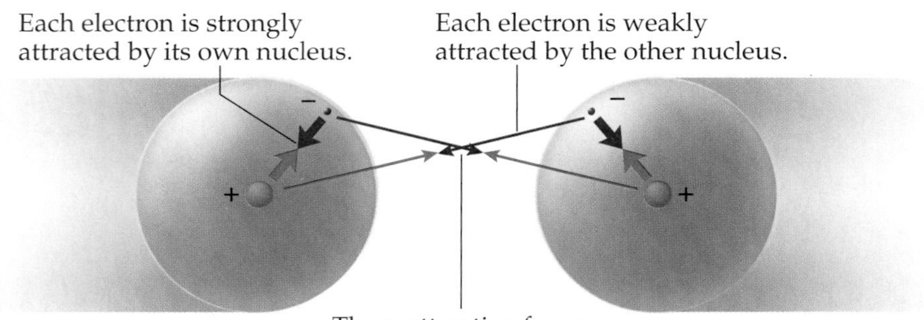
Each electron is strongly attracted by its own nucleus.
Each electron is weakly attracted by the other nucleus.
These attractive forces pull the atoms toward each other.

As the atoms got closer, these attractive forces became stronger, and so the atoms picked up speed. Upon formation of the H_2 molecule, each electron found itself strongly attracted to *two* nuclei. Thus there is a greater total amount of positive–negative attraction in the H_2 molecule than in two isolated H atoms.

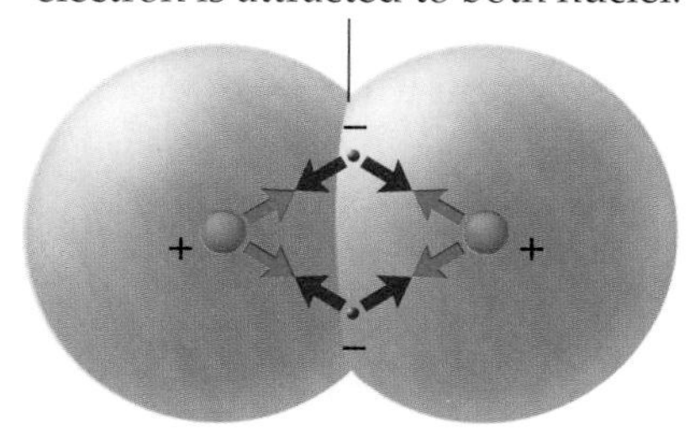
In the H_2 molecule, each electron is attracted to both nuclei.

Because attractive interactions stabilize particles, this greater total amount of positive–negative attraction means that the molecule consisting of two H atoms is more stable than the two isolated H atoms. It is the additional attractive force that holds the molecule together. We call this additional attractive force a *covalent bond*. In order to separate the H_2 molecule into isolated hydrogen atoms, you would have to provide enough energy to overcome this additional attractive force and break the bond.

Why did the atoms stop when the nuclei were 0.74 Å apart? Once again the reason is a force between charges, but this time it's between like charges. Like charges repel each other, and the nuclei of both atoms are positive. When the nuclei are 0.74 Å apart, the repulsion between the two nuclei becomes great enough to keep the H atoms from getting any closer. A covalent bond is therefore a delicate balance between attractive and repulsive forces.

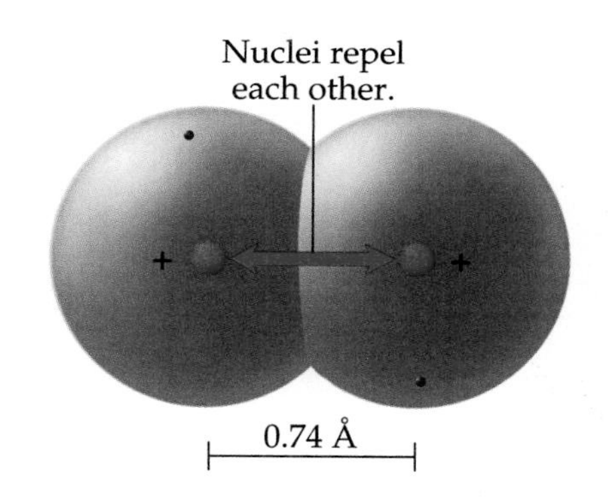
Nuclei repel
each other.
+
+
0.74 Å

Finally, as the two atoms formed a molecule, the temperature increased. Why? We said that the formation of a covalent bond results in the H_2 molecule being more stable than the two isolated H atoms. This gain in stability is the source of the temperature increase because one way things can become more stable is to release energy. (The opposite is also true—feeding energy into something makes it less stable.) The energy released is manifested as heat. You would have to put this same amount of heat (energy) back into the molecule to separate it into isolated atoms—that is, to break the covalent bond.

In the H_2 molecule, both electrons spend their time "buzzing around" both nuclei, but when we draw a representation of the molecule, we usually show the electrons occupying the region between the nuclei. This is a reasonable place to put them because they are attracted equally to both nuclei, and this is in fact where they are located most of the time. Drawn this way, it looks as if the nuclei are sharing the two electrons. In fact, that is exactly what a chemist would say is happening. We can therefore define a covalent bond as follows:

The electrons spend most of their time between the nuclei:

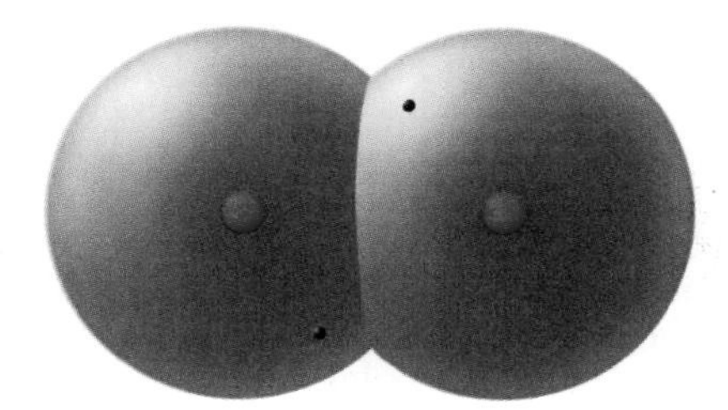

> A **covalent bond** is the force of attraction that results from valence electrons being *shared* between two nuclei. The bond holds molecules together.

although they spend some time in other places:

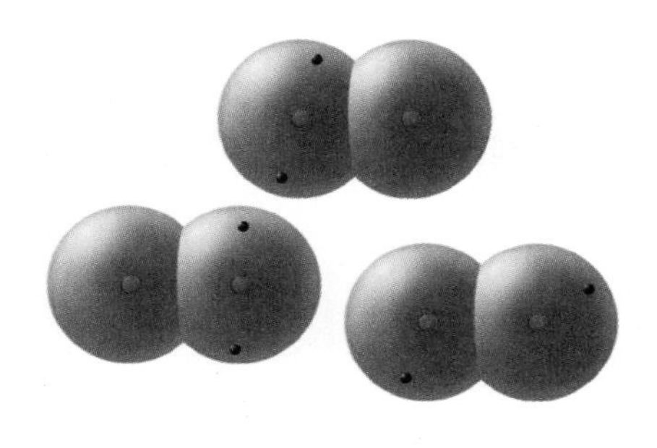

Co- is a prefix that means partner, as in somebody you share with; *valent* refers to the fact that it is the valence electrons that are being shared and are responsible for bonding the H atoms to each other. So the term *covalent* emphasizes a very important point: When atoms come together to form molecules, it is always the valence electrons that are involved. An atom's inner-shell electrons (also called *core electrons*) are almost never involved in bonding because they are too close to their own nucleus and therefore too strongly attracted to it to be shared with another nucleus. We didn't have to worry about this for hydrogen because a hydrogen atom has only one electron.

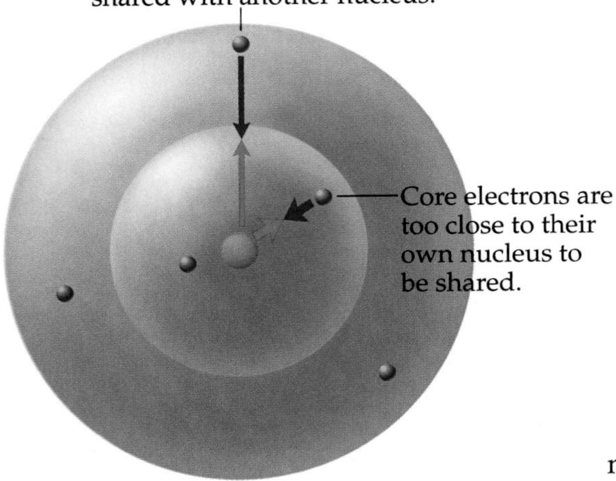
Valence electrons can be shared with another nucleus.
Core electrons are too close to their own nucleus to be shared.

Only valence electrons participate in covalent bonding.

However, as we consider bonding between atoms of other elements, we shall concern ourselves only with the valence electrons.

There are several ways to represent atoms joined by covalent bonds. Three of these ways are

H:H   H··H   H—H

The line drawn between the H atoms stands for the two shared electrons that give rise to the covalent bond. A line is easier to draw than two dots, and it is the representation most often used. Every time you see a line drawn between atoms in a molecule, remember that the line stands for a pair of electrons shared between two nuclei and the covalent bond that results. Take another look at the water and methane molecules we saw earlier. In water there are two covalent bonds (two lines); in methane there are four:

Water, H_2O Methane, CH_4

Though we shall use lines to represent covalent bonds throughout this book, you should remember that they are just a chemist's device. If you could see an individual water molecule, you would not see any lines. You would see only one oxygen atom and two hydrogen atoms that always stay close to one another.

Covalent bonds are quite strong; in fact, energies involved in this type of bonding are among the largest that chemists encounter. The O–H covalent bonds in water molecules are the reason H_2O molecules don't fly apart into atoms even when you boil the water. The steam that rises from a pot of boiling water is made up of intact H_2O molecules. To break the O–H bonds in water would take an amount of energy far greater than the amount required to boil it.

To summarize, *energy is always released* when a covalent bond forms between two atoms. There are *no* exceptions to this statement. The converse is also true: it always takes energy to break a covalent bond.

To further explore covalent bonding, let's return to hydrogen atoms. Recall that when two hydrogen atoms are isolated and far apart, each has only one valence electron. What about after they come together to form a molecule of H_2? Does each hydro-

Covalent bonds are very strong. The amount of energy it takes to boil water is far too small to break them. Thus, steam and water are both made of intact H_2O molecules.

Covalent bonds

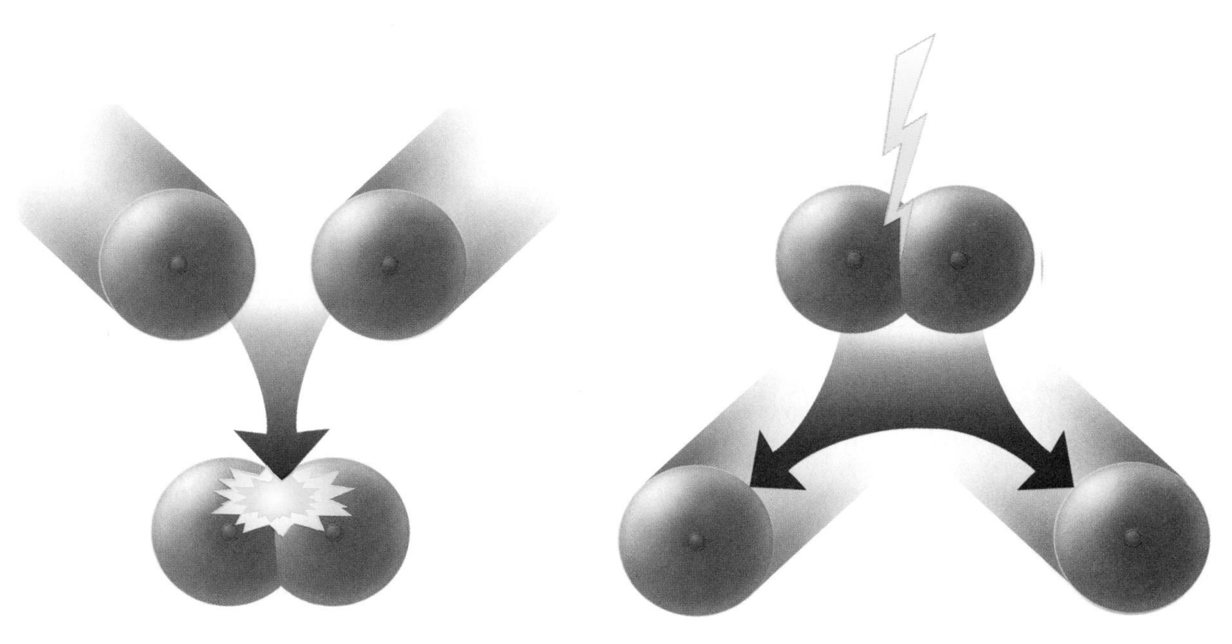

Bond formation ALWAYS releases energy.

Bond breaking ALWAYS consumes energy.

NO EXCEPTIONS!

gen atom in the molecule still have only one valence electron? The answer depends on how we choose to do our electron counting. In the case of a covalent bond between atoms, chemists have decided to double-count the shared electrons. That is, in H_2, each H atom in the molecule is considered to have two electrons.

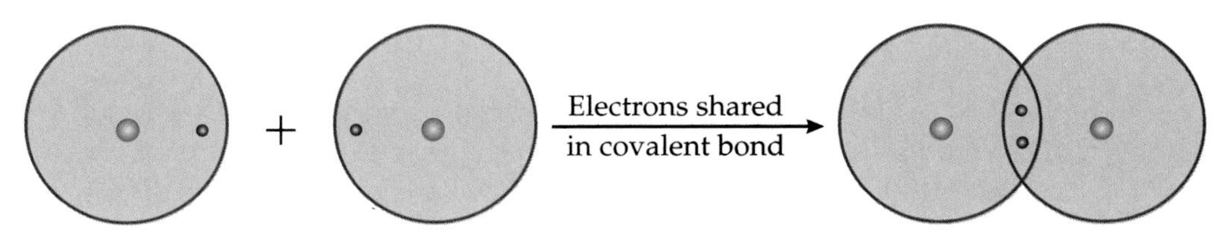

H atoms:
each atom has one electron.

H_2 molecule:
each atom "has" both electrons.

If you had a joint bank account with a friend and you did your accounting this way, you could quickly get into some financial trouble. Obviously, a shared $100 bank account does not give each of you $100 to spend. Nevertheless, this is how we count shared valence electrons. This means that sharing electrons is one way for atoms to get more electrons. The covalent bond in the H–H molecule effectively gives each hydrogen atom one more electron than it started with. Every covalent bond an atom forms increases its number of valence electrons by 1. One justification for this counting method is that it helps us to predict how many atoms will combine when they come together, as we shall see shortly.

How many valence electrons does each atom in the methane molecule have? Double-count shared electrons.

```
    H
    |
H—C—H
    |
    H
```

You should have found an octet on one of the atoms in the WorkPatch if you did your counting correctly.

How many valence electrons does a carbon atom normally have? How many valence electrons does it have in the methane molecule? How many electrons did it gain, and how many covalent bonds to hydrogen did it form?

Are you starting to get an insight into why methane is CH_4 and not CH_5 or CH_6 or anything else other than CH_4?

We are leading up to something important here regarding bonding between atoms, and the best way to introduce you to it is to consider bringing two helium (He) atoms together and attempt to form a helium molecule:

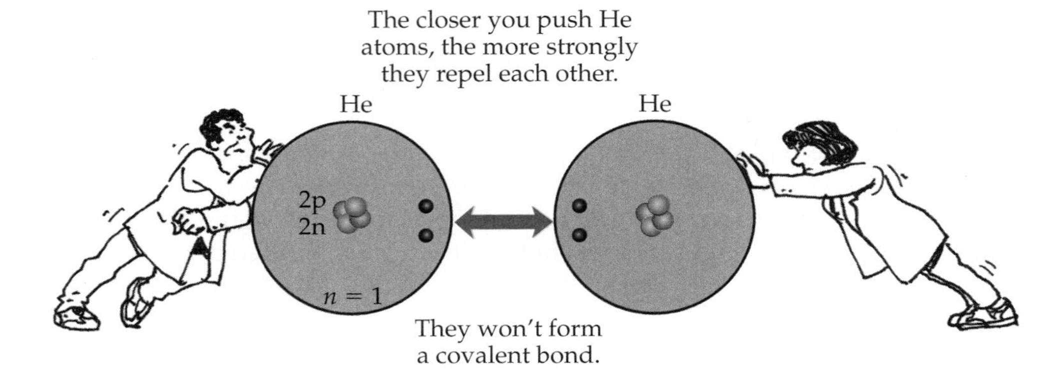

The situation we have now is very different from the situation with two hydrogen atoms. The closer two He atoms get to each other, the more they repel each other. Indeed, they refuse to form a covalent bond. Why the difference between this case and hydrogen? A hint comes from examining the behavior of the other elements in helium's group in the periodic table, group VIIIA (18), the noble gases. All the other elements in this group have eight electrons in their valence shell, and all of them strongly resist forming covalent bonds. Does the octet rule play a role in determining when covalent bonds form and how many of them form? The answer is an emphatic yes.

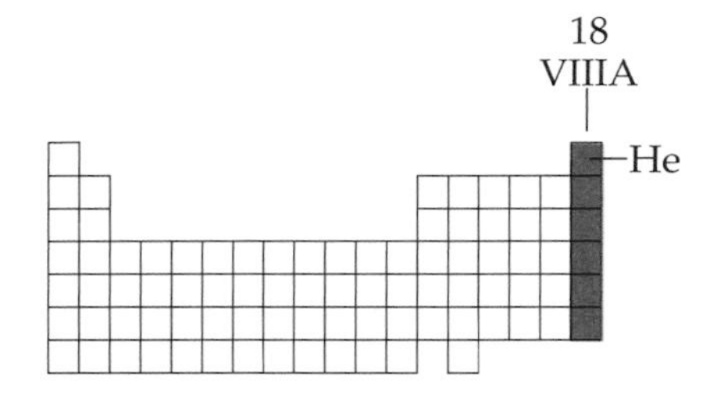

5.3 Molecules, Dot Structures, and the Octet Rule

What determines whether an element forms covalent bonds or not? How many covalent bonds is an element likely to form? Why is a water molecule always H_2O? Why not H_3O or H_4O or HO_2? The answers to these questions come from applying the octet rule, which says that many of the elements "want" to be like group VIIIA atoms and have eight electrons in their valence shell (or two electrons if their valence shell is the $n = 1$ shell). This rule often holds true when atoms are part of a molecule, where they achieve an octet by sharing electrons with other atoms.

By understanding how many electrons an atom wants to share, we can better understand why water has the molecular formula H_2O and not something else, as illustrated in the adjacent figure. The best way to begin is to determine the number of valence electrons a neutral, isolated atom has before it becomes part of a molecule. This is easy. Recall that for a representative element, the roman-numeral group number tells you the number of valence electrons. While the noble gases in group VIIIA have eight valence electrons (He, with two, is the only exception), all the other representative elements have fewer than eight valence electrons—that is, less than an octet. For example, every group IVA element is four electrons short of an octet.

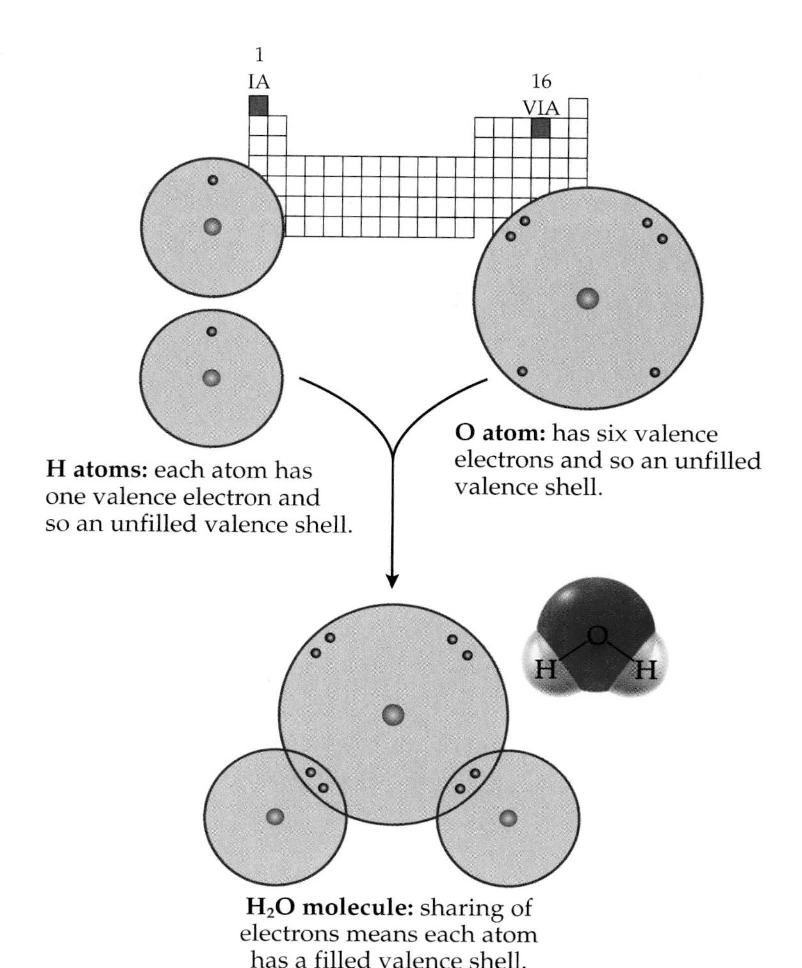
1
IA
16
VIA
H atoms: each atom has one valence electron and so an unfilled valence shell.
O atom: has six valence electrons and so an unfilled valence shell.
H
O
H
H₂O molecule: sharing of electrons means each atom has a filled valence shell.

One way group IVA elements can get eight electrons is to share all four of their valence electrons with other atoms. For example, remember the methane molecule, CH_4, we looked at earlier? Carbon starts with only four valence electrons and needs four more to complete an octet. Therefore, it forms four covalent bonds with other atoms:

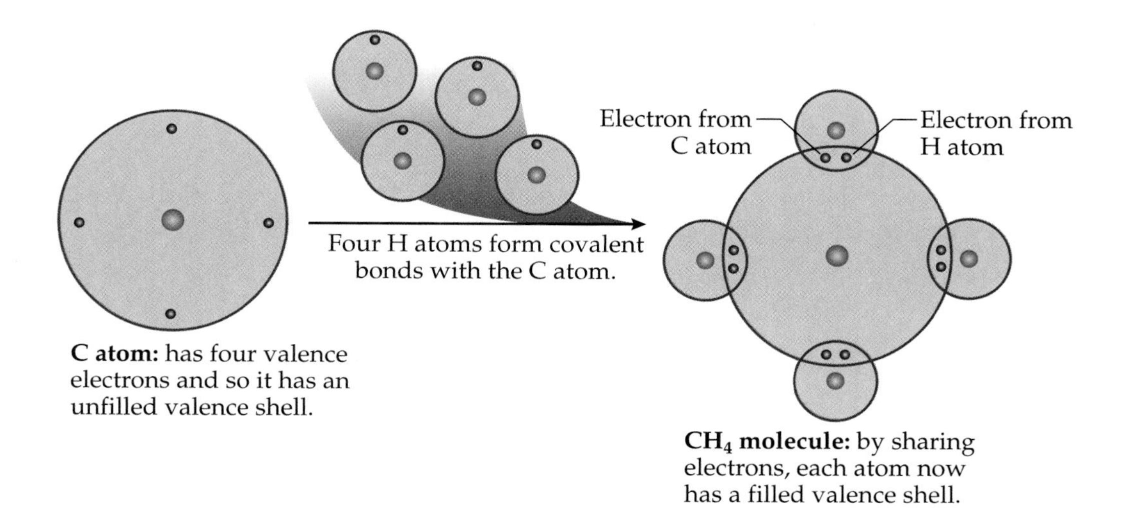
C atom: has four valence electrons and so it has an unfilled valence shell.
Four H atoms form covalent bonds with the C atom.
Electron from C atom
Electron from H atom
CH₄ molecule: by sharing electrons, each atom now has a filled valence shell.

Because the roman-numeral group number tells you the number of valence electrons an atom has, it is a simple matter to figure out how many additional electrons it needs to reach eight. This number is also the number of covalent bonds it must form to get there. Table 5.1 summarizes the requirements for each group of nonmetal atoms (we'll save the metals for later). Examine it closely; then we'll see how it can be used to determine the molecular formula of a molecule.

Table 5.1 Number of Covalent Bonds Nonmetal Atoms Usually Form

Group number	**IA**	**IVA**	**VA**	**VIA**	**VIIA**
Nonmetal element	H	C Si Ge	N P As Sb	O S Se Te	F Cl Br I At
Number of valence electrons	1	4	5	6	7
Number of electrons needed to achieve octet (duet for H)	1	4	3	2	1

This is the number of bonds an atom of the element usually forms.

The number in the last row of each column in Table 5.1 is the number of covalent bonds the atoms in that column must form to achieve an octet. For example, let's look at ammonia, a molecule made from nitrogen and hydrogen. Find nitrogen in the table. It is in group VA and therefore has five valence electrons. It needs three more electrons to achieve an octet. In ammonia, these three electrons come from forming three covalent bonds to H atoms. That is why the formula for ammonia is NH_3—not NH_2 or NH_4.

Practice Problems

5.1 Predict the formula of the compound that forms between carbon and chlorine.

Answer: Table 5.1 tells you that carbon needs to form four bonds and chlorine needs to form one bond. The formula is therefore CCl_4.

5.2 Predict the formula of the compound that forms between phosphorus (P) and hydrogen.

5.3 Predict the formula of the compound that forms between silicon (Si) and bromine (Br).

Table 5.1 is fine, but to build more complex molecules, such as glucose and DNA, we need a system for drawing atoms that can tell us at a glance how many covalent bonds each atom forms. A chemist named G. N. Lewis invented just such a system for molecules back around 1916. His drawings are called **dot diagrams** or sometimes **Lewis dot diagrams**. Dot diagrams are used to show an atom's valence electrons. In a dot diagram, the symbol of an element is surrounded by dots, one dot for each valence electron. For the period 2 elements, for instance, the dot diagrams are

	Li	Be	B	C	N	O	F	Ne
Group ⟶	IA	IIA	IIIA	IVA	VA	VIA	VIIA	VIIIA

In every case, the number of dots is equal to the group number for these representative elements. Notice how the electrons are added. The first four electrons (Li through C) are positioned on the four sides of the symbol. It is not until we have a fifth electron, in N, that we begin to pair the electrons. Thus, carbon is surrounded by four single electrons, whereas nitrogen has three single electrons and one pair. We refer to the pairs as either **electron pairs** or **paired electrons**, and we call the single electrons **unpaired electrons**. Thus, nitrogen has three unpaired electrons and one electron pair. As we shall see, the difference between paired and unpaired electrons is important.

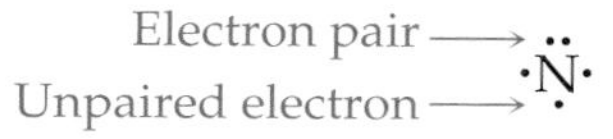

Electron pair ⟶
Unpaired electron ⟶ N

There is some flexibility with respect to how you arrange the unpaired and paired electrons. For example, all three of these dot diagrams for oxygen are correct:

All these forms are equivalent.

O   O   O

Because the number of valence electrons determines what the dot diagram for an atom looks like, all atoms with the same number of valence electrons have identical dot configurations. Thus, the dot diagrams for periods 3–7 of the representative elements are identical to those shown above for period 2. We won't concern ourselves with dot diagrams for the transition metals, lanthanides, or actinides. This leaves just two more elements to describe, hydrogen and helium, both of which are unique. Their dot diagrams are

H·   He:

Note that helium's two valence electrons are always paired.

We now have simple diagrams for all the representative elements. These dot diagrams can help us determine the bonding that can occur between these elements. If you examine all the dot diagrams that we have considered, only the group VIIIA atoms (He, Ne, Ar, Kr, Xe, Rn) have all their valence electrons paired up and possess a complete octet (duet for He). All the other representative elements have unpaired valence electrons and thus less than an octet. The octet rule states that they must have eight valence electrons (duet for H) to achieve stability. One way to get eight valence electrons is for these elements to share their unpaired electrons with other atoms.

Let's see how this works by using fluorine as an example. Fluorine is in group VIIA, so each atom has seven valence electrons, three pairs and one unpaired electron:

Unpaired electron is used for sharing with another atom. ⟶ ·F: ⟵ Electron pairs usually are not used for sharing.

The fluorine atom needs one more valence electron to achieve an octet. If two F atoms come together and share their unpaired electrons, then both atoms can have eight valence electrons. When the atoms share in this way, a single covalent bond forms between them. In this case, no more bonds will form because both atoms now have octets. This is why fluorine exists as the diatomic molecule F_2—not as F, F_3, or any other combination.

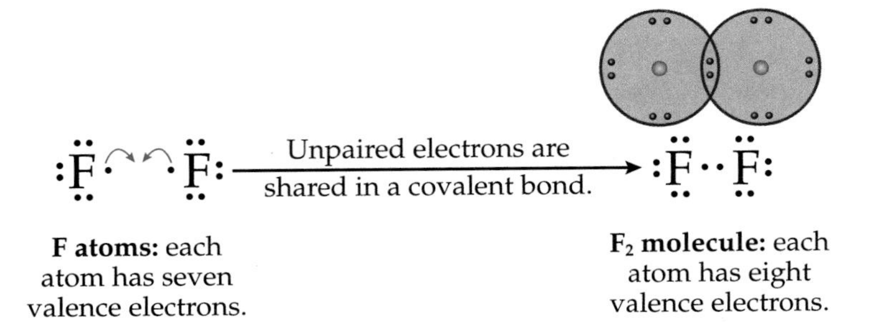
Unpaired electrons are shared in a covalent bond.
F atoms: each atom has seven valence electrons.
F_2 molecule: each atom has eight valence electrons.

WORKPATCH

Why do two He atoms refuse to form the molecule He_2 when brought close to each other?

How did you answer the WorkPatch? You could have said that there were no unpaired electrons to pair up. Or you could have answered in terms of the octet rule. Either way is correct.

Practice Problems

5.4 In forming molecules, atoms can share unpaired electrons in order to achieve an octet in their valence shell. Hydrogen is an exception. What number of electrons does hydrogen "want," and why is this number not eight?

Answer: Hydrogen wants two electrons (a duet instead of an octet) because it is a period 1 atom and so its valence shell is the $n = 1$ shell. This shell is filled when there are two electrons in it.

5.5 What is the molecular formula of the compound that forms between H atoms and F atoms? Justify your answer with dot diagrams.

5.6 Why is water H_2O and not H_3O or something else? Justify your answer with dot diagrams.

The strategy of sharing unpaired electrons and aiming for octets works for all sorts of molecules. For example, the molecules NF_3 and CH_4 are formed from atoms as follows:

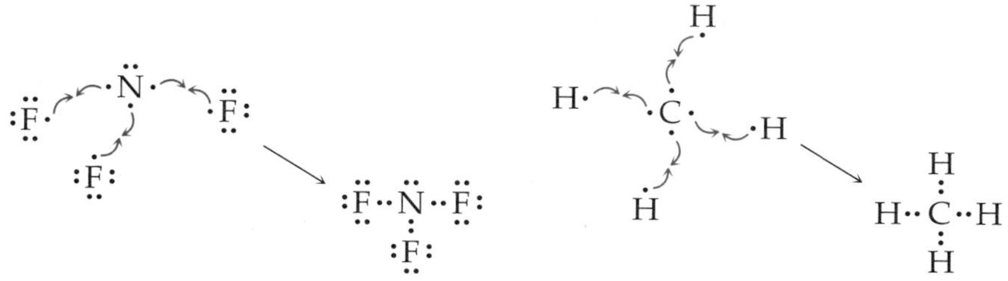

Drawing all these dots gets tedious, so we usually just draw a line for each shared pair of electrons:

As we said previously, a line between two atoms is understood to represent two shared electrons, which constitute a covalent bond. Therefore, shared pairs of electrons are often called **bonding pairs**. Any unshared electron pairs, such as the three pairs on the F atoms and the one pair on the N atom in NF_3, are called **lone pairs** because they belong to a single (lone) atom.

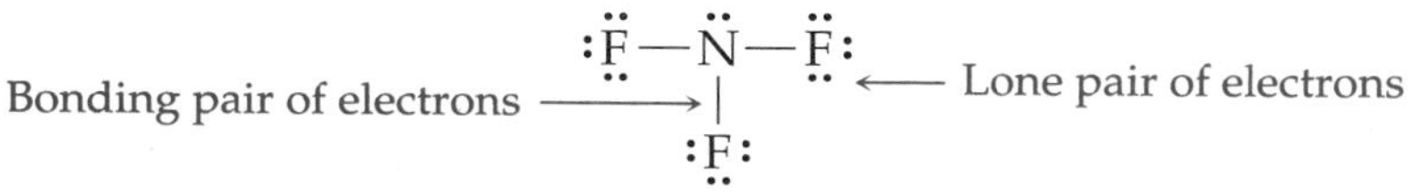

Bonding pair of electrons → ← Lone pair of electrons

Note that not all atoms in a molecular dot diagram necessarily have lone pairs (two examples are the carbon atom in methane and hydrogen atoms in any molecule). Because they are not shared, lone pairs are not directly involved in bonding. Nevertheless, they are important and should never be left out of a molecular dot diagram.

5·5 WORKPATCH

How many valence electrons are present on each atom in the NF_3 molecule?

Did you count the lone pairs once but double-count the shared pairs when doing your electron counting for the WorkPatch? This exercise demonstrates the importance of lone pairs with respect to the octet rule.

With dot diagrams in hand, you are now ready to build more complicated molecules.

5·6 WORKPATCH

Consider ethane, C_2H_6, a minor component of natural gas. What does this molecule look like? Start with the atoms as dot diagrams and try putting them together so that each C ends up with eight valence electrons and each H ends up with two valence electrons. Then replace every pair of shared electrons with a line. When you are done, compare your attempt with the answer at the end of the chapter.

There should be no lone pairs in your dot diagram for this WorkPatch. Remember, hydrogen never has a lone pair in a molecule, because two electrons are all it needs to fill its valence shell, and these will come from the covalent bond. As for carbon, it starts out with four unpaired electrons, so it will end up with no lone pairs when it forms bonds to four other atoms in order to gain four more electrons, which is the case in ethane. It begins to look like carbon will never have lone pairs in a molecule. As we shall see in the next section, lone pairs on carbon are possible but rare.

Practice Problems

5.7 Draw a dot diagram for hydrazine, N_2H_4, sometimes used as a rocket fuel.

Answer: Start with the pieces—individual dot diagrams for all the atoms—and then put them together in a way that satisfies the octet rule (duet for H):

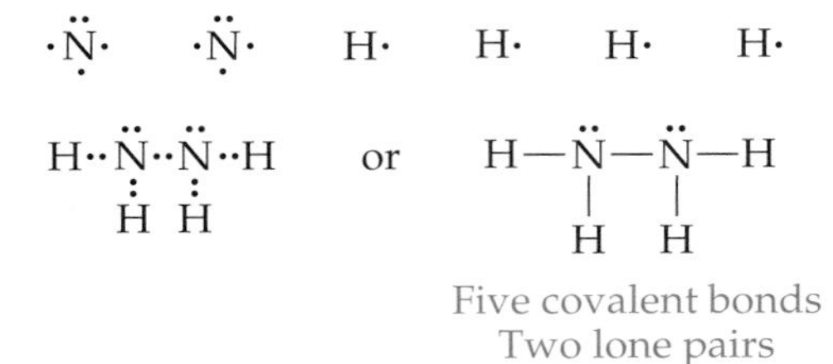

5.8 Draw a dot diagram for hypochlorous acid, HClO.

5.9 Draw a dot diagram for propane, C_3H_8, used as a fuel for heating.

5.4 Multiple Bonds

Sometimes two atoms share more than one pair of electrons with each other. Such a situation is called either a **multiple covalent bond** or simply a **multiple bond**. For example, let's reconsider those diatomic O_2 molecules you are breathing. To understand the bonding in this molecule, we start as usual by drawing a dot diagram for one oxygen atom:

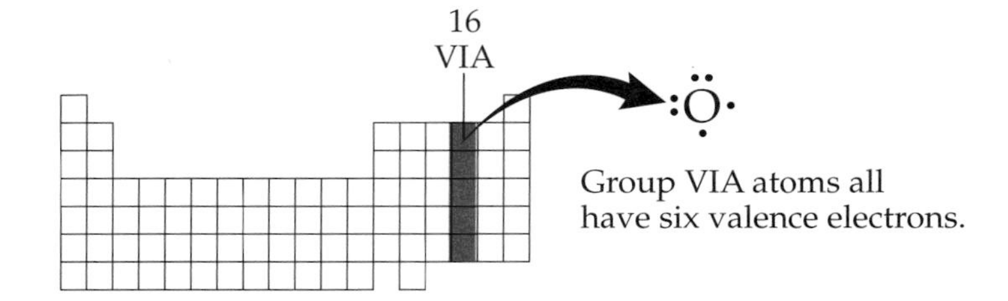

Being a group VIA atom, oxygen has two unpaired valence electrons. As two oxygen atoms get close, they can form a covalent bond, with each atom sharing one of its unpaired electrons:

:Ö· ·Ö: ⟶ :Ö—Ö:

Each oxygen atom has six valence electrons.

Each oxygen atom now has seven valence electrons.

Where do we go from here? Both oxygen atoms now have seven valence electrons, not yet an octet. Each atom still has one unpaired electron left, though, and unpaired electrons are for sharing. So don't stop. Share the remaining unpaired electrons:

$$:\ddot{O}-\ddot{O}: \longrightarrow :\ddot{O}=\ddot{O}:$$

Each oxygen atom has seven valence electrons. Each oxygen atom has eight valence electrons.

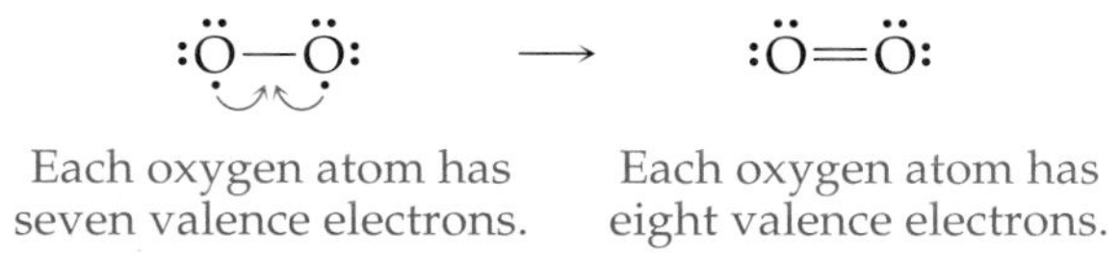

Look what this did for each oxygen atom. Each oxygen is now surrounded by eight valence electrons. The octet rule has been satisfied for both oxygen atoms by the formation of a multiple bond. In this case, there are two pairs of electrons shared between the atoms—in other words, two covalent bonds. We refer to this as a **double bond**. When there is just one covalent bond between two atoms, it is called a **single bond**. As you might expect, a double bond is stronger than a single bond.

The air you breathe contains another diatomic gas, nitrogen, N_2. To draw this molecule, start with dot diagrams for the isolated nitrogen atoms, then bring them together to share electrons until both atoms end up with eight:

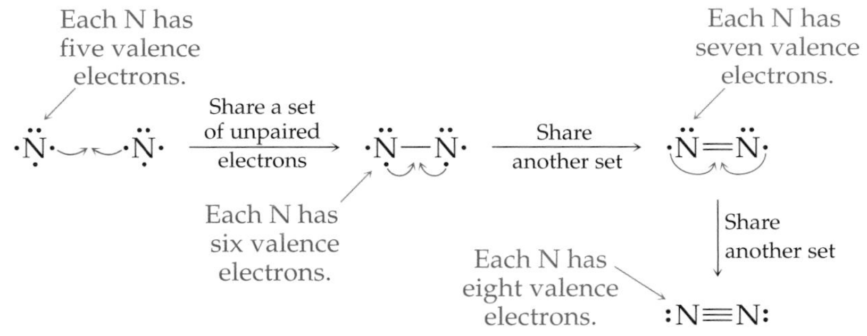

Each N has five valence electrons.

$$\cdot\ddot{N}\cdot \quad \cdot\ddot{N}\cdot \xrightarrow{\text{Share a set of unpaired electrons}} \cdot\ddot{N}-\ddot{N}\cdot \xrightarrow{\text{Share another set}} \cdot\ddot{N}=\ddot{N}\cdot \xrightarrow{\text{Share another set}} :N\equiv N:$$

Each N has six valence electrons.

Each N has seven valence electrons.

Each N has eight valence electrons.

This time we end up with a **triple bond**, three shared pairs between the N atoms, a very strong bond indeed.

Related to the C_2H_6 (ethane) molecule of WorkPatch 5.6 is the acetylene molecule, C_2H_2, used in oxyacetylene torches that can cut through steel. This molecule also contains a multiple bond. Determine which kind in the following WorkPatch.

5.7 WORKPATCH

Draw a dot diagram for the acetylene molecule, C_2H_2.

You should have found that there is only one way to put the acetylene molecule together. If you put both hydrogens on the same carbon, you would end up violating the octet rule and leaving unpaired electrons on the other carbon atom.

Practice Problems

5.10 Draw a dot diagram for carbon dioxide, CO_2.

Answer: $:\ddot{O}\cdot \; \cdot\dot{C}\cdot \; \cdot\ddot{O}: \longrightarrow :\ddot{O}=C=\ddot{O}:$

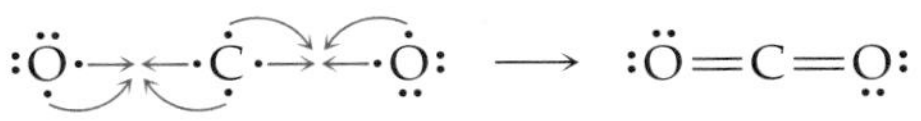

5.11 Draw a dot diagram for propyne, C_3H_4. One carbon has three hydrogens bound to it, one has one hydrogen bound to it, and one has no hydrogens bound to it.

5.12 Draw a dot diagram for hydrogen cyanide, HCN. The hydrogen is attached to the carbon.

5.13 Draw a dot diagram for acetone, C_3H_6O, the active ingredient in nail polish remover. The oxygen is bound to only one carbon, and that carbon has no hydrogens bound to it.

Drawing dot diagrams for molecules is a great way to learn about covalent bond formation, but it is a tedious process and doesn't always work. There is a simpler, four-step method that we want to show you now. We shall demonstrate this method using two examples—carbon monoxide, CO, and carbon dioxide, CO_2.

Step 1: Determine the *total* number of valence electrons (dots) that will be in the final diagram. To do this, simply add up the group numbers of all the atoms in the molecular formula.

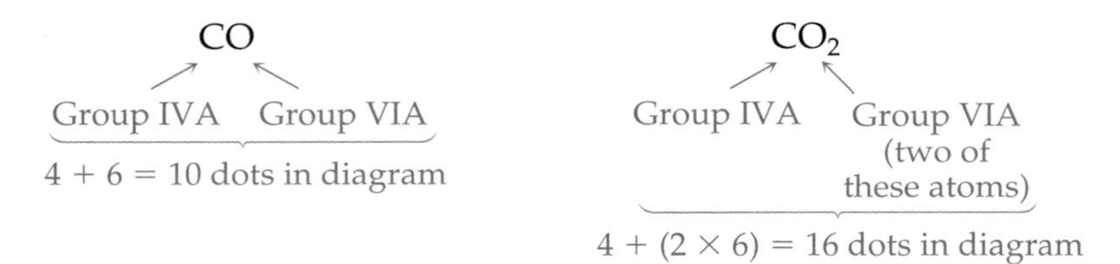

Step 2: Connect the atoms with single bonds. If you are not told how the atoms are connected to one another, it often works to assume that the first nonhydrogen atom in the molecular formula is the central atom. For example, the central atom in phosphorus trichloride, PCl_3, is P, and the central atom in sulfuric acid, H_2SO_4, is S. Also remember that every time you connect two atoms with a single bond, you are using *two* electrons.

C—O
2 dots used

O—C—O
4 dots used

Step 3: Put in the remaining dots two at a time as lone pairs, first on the terminal atoms of your structure and then, if there are any dots left, on the central atoms. Keep adding electrons to an atom until it has an octet before moving on to the next atom. You must stop putting in lone pairs when you run out of electrons. This is very important. Do not use more electrons than you calculated in step 1.

:C—Ö:
All 10 dots now in diagram

:Ö—C—Ö:
All 16 dots now in diagram

If at the end of this step, every atom except hydrogen has eight valence electrons around it and every hydrogen has two electrons, you are done. The molecule will have only single bonds. In our two examples, we are not done because each carbon has only four valence electrons. When this happens, we must go on to step 4.

Step 4: If there are atoms that do not yet have an octet, send lone pairs to the rescue by moving them into a bonding position between the atoms. The lone pairs must come from an *adjacent* atom (one that is attached to the octet-deficient atom).

CO

:C—Ö: ⟶ :C≡O:

CO_2

:Ö—C—Ö: ⟶ :O=C=O:
(a)

or

:Ö—C—Ö: ⟶ :O≡C—Ö:
(b)

or

:Ö—C—Ö: ⟶ :Ö—C≡O:
(c)

This solution reveals that the carbon monoxide dot diagram contains a triple bond between the carbon and oxygen atoms. Each atom also has one lone pair, and each atom has a complete octet. This was accomplished using only the ten valence electrons we started with. By following steps 1–4, we came up with a valid molecular dot diagram. For carbon dioxide, we drew three dot diagrams, all equally valid with respect to the octet rule. Convince yourself of this by examining the diagrams and answering the following WorkPatch.

How did we get three different dot diagrams for carbon dioxide?

Do you understand why we arrived at more than one valid dot diagram for CO_2? If you do, you will know when to expect this situation. But all three dot diagrams for CO_2 can't be correct, can they? Well, actually they are all correct. They are identical in their arrangement of atoms (carbon in the middle with oxygens attached to it). The only difference is the arrangement of electrons. Valid dot diagrams that differ from one another only in the arrangement of electrons are called **resonance forms**. The actual molecule is considered by chemists to be some combination of these three forms, with all the forms not necessarily equally weighted. Experiments and theoretical considerations show that carbon dioxide is most like resonance form (a), with perhaps just a tad of (b) and (c) mixed in.

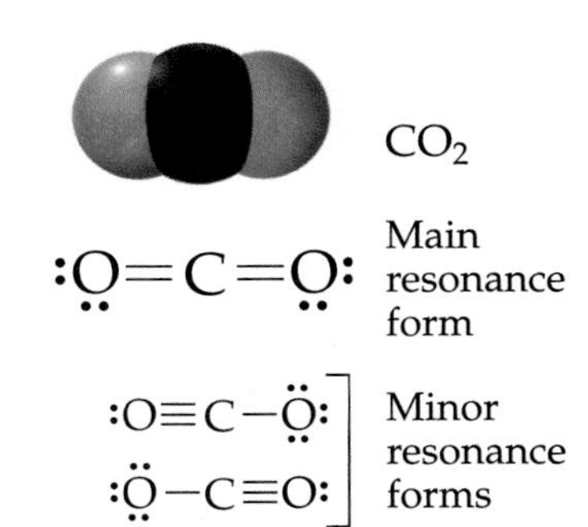

Up to now, we have been generating dot diagrams for electrically neutral (uncharged) molecules. Before we present some practice problems, we want to modify the first step in the above procedure so that it can be extended to cover *polyatomic ions*. **Polyatomic ions** are just what they sound like—ions made from many atoms. Some examples are nitrate, NO_3^-, used to preserve meats; phosphate, PO_4^{3-}, found in soda pop; and ammonium ions, NH_4^+, often found in soaps and detergents. These are all ions because they have an overall electrical charge.

When drawing dot diagrams for ions, we follow steps 1–4 with one small change. Because the charge affects the total number of dots in the diagram, step 1 has to be modified to account for this. We must consider how to convert a neutral molecule to a cation or an anion. An anion is negative because it has a surplus of electrons. The surplus is equal to the charge of the anion. For example, consider the nitrate anion, NO_3^-. Neutral NO_3 has 23 valence electrons (one group VA atom and three group VIA atoms). The anion NO_3^-, with its -1 charge, must therefore have 24 valence electrons. As a second example, consider the phosphate anion, PO_4^{3-}. Neutral PO_4 has 29 valence electrons (one group VA atom and four group VIA atoms). The anion PO_4^{3-}, with its -3 charge, must therefore have 32 valence electrons.

A cation is positive because it has a deficiency of electrons. For example, consider the ammonium ion, NH_4^+. Neutral NH_4 has nine valence electrons (one group VA atom and four group IA atoms). The NH_4^+ cation, with its $+1$ charge, therefore has eight valence electrons.

Modified Step 1: Determine the total number of valence electrons (dots) that will be in the final diagram. To do this, add up the group numbers of all the atoms in the molecular formula. *If you are dealing with an ion, adjust the dot count according to the charge (subtract electrons for cations, add them for anions).*

The method for drawing dot diagrams is summarized at left.

You are now ready for some practice problems. Some will have resonance forms, some will not—it's up to you to decide. When resonance forms exist, draw them all.

Drawing dot diagrams

Step 1
Determine the total number of valence electrons (dots) that will be in the diagram. For a polyatomic ion, adjust the electron count to reflect the charge.

Step 2
Connect the atoms using single covalent bonds. If you do not know how the atoms are arranged, try using the first nonhydrogen atom in the formula as the central atom.

Step 3
Add the remaining electrons as lone pairs, completing octets as you go.

Step 4
If any atom lacks an octet, move lone pair(s) from adjacent atom(s) into bonding positions on the octet-deficient atom.

Practice Problems

5.14 Draw a dot diagram for the carbonate anion, CO_3^{2-}.

Answer:

Step 1: CO_3^{2-} has a total of 24 dots:

C is group IVA = 4 dots
O is group VIA (there are three of them) = 6×3 = 18 dots
-2 charge gives an additional two electrons = 2 dots
24 dots

Step 2: Assume the first atom written is the central atom and all other atoms are attached to it. Connect them with single bonds.

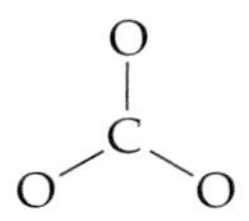

O
C
O O

The three bonds consume 6 electrons, leaving 18 to go.

Step 3: Put in the remaining electrons as lone pairs, completing octets as you go.

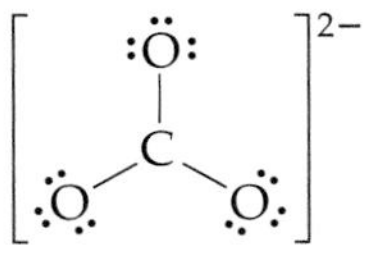

All 24 electrons are in. Are we done? No. Carbon does not have an octet. We must proceed to step 4.

Step 4: If all octets are not completed, move lone pairs from adjacent atoms to the atom needing electrons.

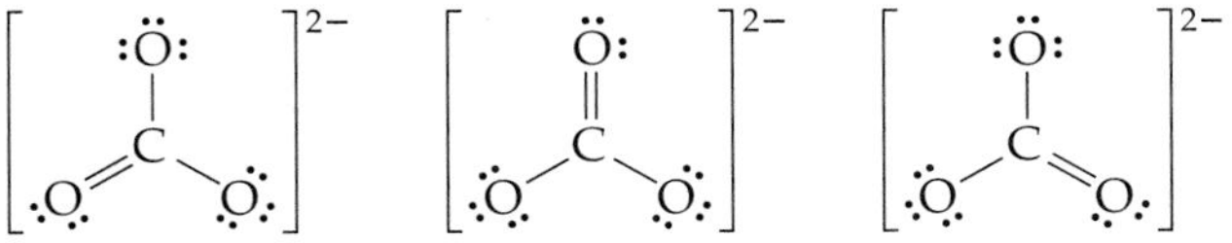

We get three resonance forms, depending on which lone pair we use to create the double bond. Note that if we are dealing with a polyatomic ion, it is traditional to put the ion in square brackets and indicate the overall charge as a superscript.

5.15 Draw a dot diagram for neutral SO_3.

5.16 Draw a dot diagram for SO_3^{2-}.

5.17 Draw a dot diagram for NO^+.

5.18 Are the dot diagrams shown below resonance forms? Explain.

$$\ddot{\underset{..}{O}}=C=\ddot{\underset{..}{O}} \qquad :C\equiv O-\ddot{\underset{..}{O}}:$$

5.5 Ionic Bonding—Bring on the Metals

In our discussion of molecules, we haven't mentioned metal atoms. In general, metal atoms don't share their valence electrons with other atoms because most metal atoms have little ability to attract additional electrons to themselves. In fact, when metal atoms form compounds with nonmetals, the metal atoms tend to completely give up their valence electrons instead of sharing them. We introduced this concept in Section 4.6, but we can now look at it again in terms of the bonding that occurs.

Consider the compound sodium chloride, NaCl, table salt. It is a stable compound in which group IA Na atoms are bound to group VIIA Cl atoms. But this bonding is not the result of shared electrons as was the case in, for instance, H_2. Look at what happens to the metal sodium atom when it wanders close to a nonmetal chlorine atom:

$$Na\cdot \;\; \cdot\ddot{\underset{..}{Cl}}: \longrightarrow Na^+ \;\; :\ddot{\underset{..}{Cl}}:^-$$

There is no electron sharing. Instead, chlorine takes an electron from sodium. We say that the metal's electron has been transferred to the nonmetal. Stripped of its single valence electron, the Na atom is now a Na^+ cation. Having gained the electron, the Cl atom is now a Cl^- anion. This agrees with what we said in Chapter 4, that metals tend to lose electrons and nonmetals tend to gain them. When a metal and a nonmetal get together, this is generally what happens.

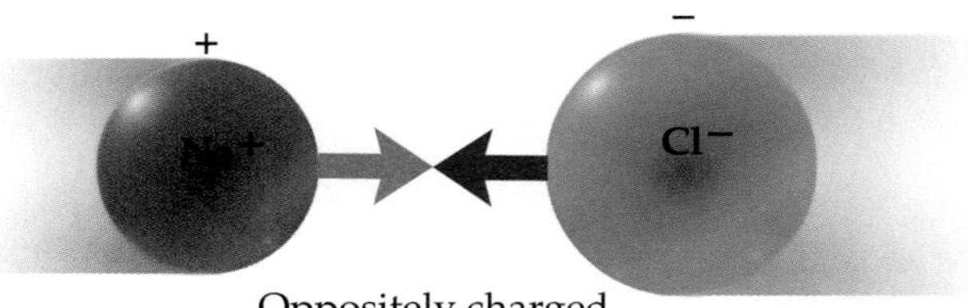

Oppositely charged ions attract each other.

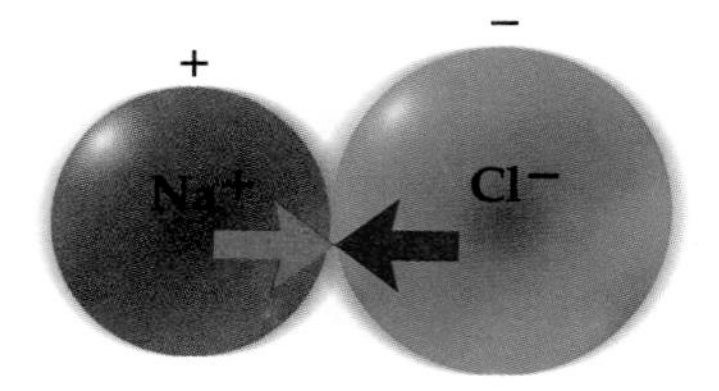

Ions stick together.

If sodium and chlorine don't share electrons, they can't bond together covalently. But clearly, the Na^+ and Cl^- ions in table salt stick together (bond) somehow. What holds these ions together is the fact that they have become oppositely charged. Because opposite charges attract, the Na^+ and Cl^- ions "stick" to each other via a strong electrostatic force of attraction.

This attractive force holding NaCl together is similar to a covalent bond in that it arises from the attraction between something positive and something negative. It is different from a covalent bond in that there is no sharing of electrons. This time, the attractive force arises from oppositely charged ions that were formed because of a transfer of electrons. Since this bond is formed by an attraction between ions, it is called an **ionic bond**.

In general, ionic bonds form between a metal and a nonmetal, and covalent bonds form between two nonmetals. Even though these bonds arise in different ways, they are similar in strength, because both are the result of the attraction between relatively large positive and negative charges—between electrons and nuclei in covalent bonds and between oppositely charged ions in ionic bonds.

Finally, we should point out that compounds like NaCl, made from one nonmetal and one metal, do not exist as molecules. Rather, they exist as a large, ordered, three-dimensional network of positively and negatively charged ions called an **ionic lattice**. Any compound that is such a collection of ions held together by ionic bonds is called an **ionic compound**. In the ionic lattice, the ions are arranged in such a way that positive ions are adjacent to negative ions (resulting in ionic bonds) but like-charged ions, which repel one another, are kept farther apart. A portion of the NaCl lattice is shown here. This entire lattice is held together by ionic bonds between oppositely charged ions.

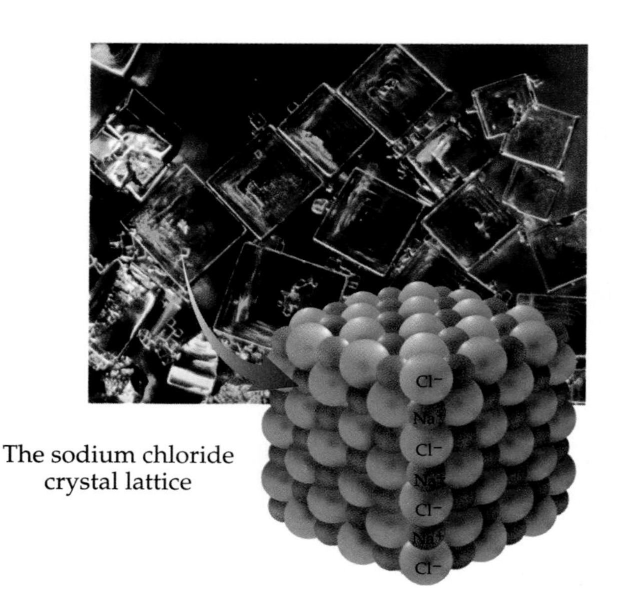

The sodium chloride crystal lattice

With no individual molecules present, NaCl could not be called a molecular compound because a **molecular compound** is defined as a compound made up of individual molecules. Water, which consists of individual H_2O molecules, is one example of a molecular compound.

At this point, you might have the impression that bonds between atoms are either covalent or ionic. Nature, however, is rarely so cut and dried. Many bonds have both covalent and ionic character. Read on.

5.6 Equal Versus Unequal Sharing of Electrons—Electronegativity and the Polar Covalent Bond

When two atoms come close enough to each other, a bond may form between them. The bond occurs because the valence electrons of the two atoms redistribute themselves in such a way as to give rise to an attractive force between the atoms. So far, we have examined redistribution involving either a sharing (covalent bond) or a complete transfer (ionic bond) of electrons. We can represent this pictorially. We begin with two atoms each having a single valence electron as illustrated at right. Two things can happen when these two atoms get close to each other. Either the valence electrons are shared, or they are transferred. Examine the pictures closely:

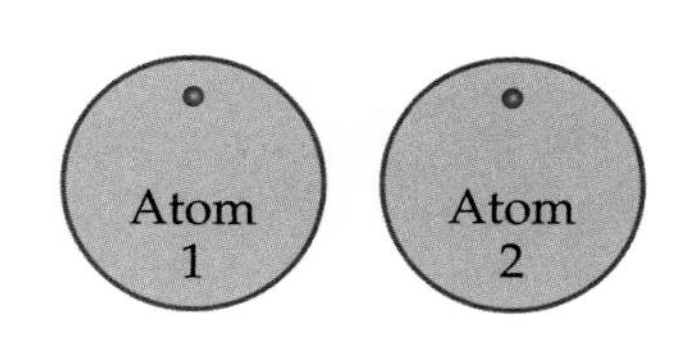

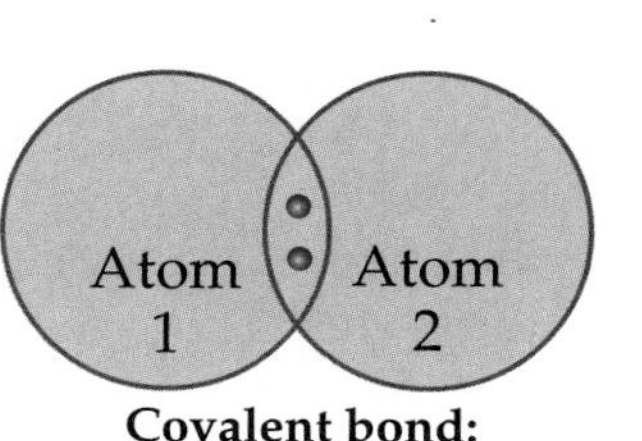

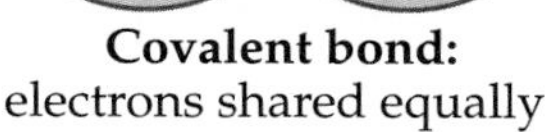

Covalent bond:
electrons shared equally

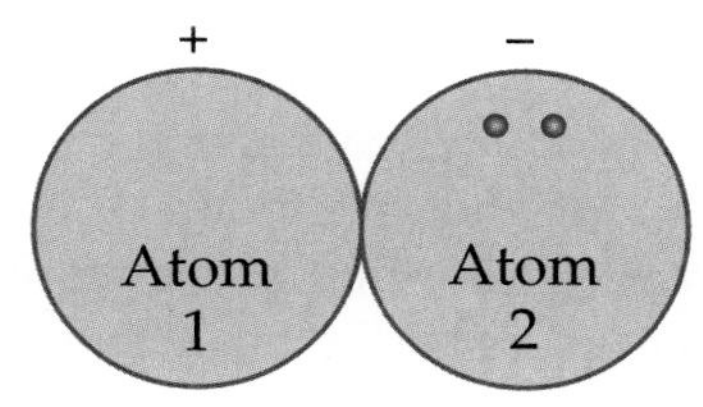

Ionic bond:
electron transferred

Whether the electrons are shared or transferred, the result is the same in that a strong attractive force—a **chemical bond**—arises. But are equal sharing or complete transfer the only possibilities? To gain further insight into this question, we must turn to a defined property of atoms called *electronegativity*. **Electronegativity (EN)** is a numerical rating of an atom's ability to attract to itself the shared electrons in a covalent bond. The chart below lists the electronegativity values of the representative elements.

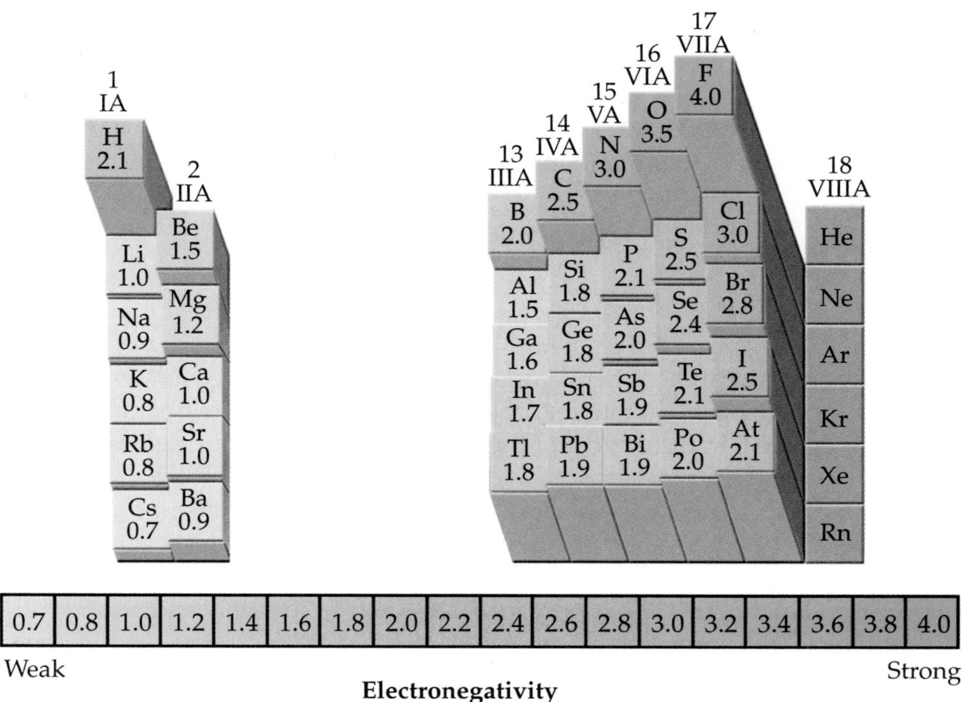

The electronegativity scale ranges from 0.7 to 4.0. The higher the value, the greater an atom's ability to attract shared electrons to itself. Fluorine is the best at attracting shared electrons to itself and is assigned the highest electronegativity value, 4. All other electronegativity values are scaled relative to fluorine = 4.0. Francium, at 0.7, is the least electronegative atom. In general, metal atoms tend to have low electronegativity values and nonmetal atoms tend to have high electronegativity values.

There is a noticeable periodic trend in electronegativity values, which is emphasized by the color shading in the chart on the previous page. The highest values are in the upper right-hand corner (orange), and the lowest values are in the lower left-hand corner (yellow).

Nitrogen, N, oxygen, O, fluorine, F, and chlorine, Cl, are the most electronegative elements in the periodic table. These are the electron "hogs." Because they are the most electronegative elements in the periodic table, it can be very useful to remember both the identity of these hogs and their position in the table. The relative electronegativities of other elements can be estimated by considering how close in the table they are to this group. The closer an element is to these hogs, the more electronegative the element is. This means that as you go either up a group or left to right across a period, electronegativity generally increases.

Notice that the noble gas atoms of group VIIIA are not assigned any electronegativities. This is because they don't typically form bonds to other atoms, so their ability to attract shared electrons is a moot point.

One place where electronegativity is useful is in drawing Lewis dot diagrams. Earlier we said that you can often assume that the first nonhydrogen atom in the chemical formula of a simple molecular compound is the central atom. Thus, for instance, P is the central atom in $POCl_2$:

```
      :O:
       ||
:C̤̈l—P—C̤̈l:
```

However, if someone chooses to write the formula $OPCl_2$ (which is a perfectly correct way to do it), blindly following this rule would lead to an incorrect dot diagram:

Electronegativity can help here. For simple molecules, it is generally true that the least electronegative atom (excepting hydrogen) is the central atom. Thus, we would place P in the middle of the molecule even if the formula were written $OPCl_2$ or Cl_2OP.

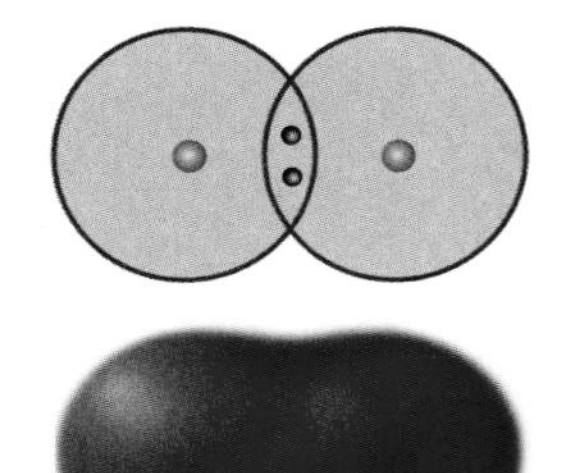

Covalent bond:
electrons shared equally

Another important use of electronegativity is to help understand what type of bonding occurs between two atoms. When two nonmetal atoms of identical electronegativity bond, the result is an equal sharing of electrons, giving rise to a pure covalent bond. This is the case for the bond in such molecules as H_2, Cl_2, O_2, and N_2. If the atoms have different electronegativities, however, the bonding electrons will spend more time near the atom of greater electronegativity. In other words, the more electronegative atom will hog the bonding electrons.

This results in a pileup of the bonding electrons around the more electronegative atom, causing it to develop a partial negative charge (indicated by the symbol δ^-, where the Greek lowercase letter delta, δ, means "partial"). The opposite happens around the less electronegative atom—it develops a partial positive charge δ^+ that is equal in magnitude to the δ^- charge on the more electronegative atom. The result is a **polar covalent bond**, a covalent bond in which the electrons are shared *unequally*. This is the case for the bond or bonds in molecules like HCl, H_2O, and HF.

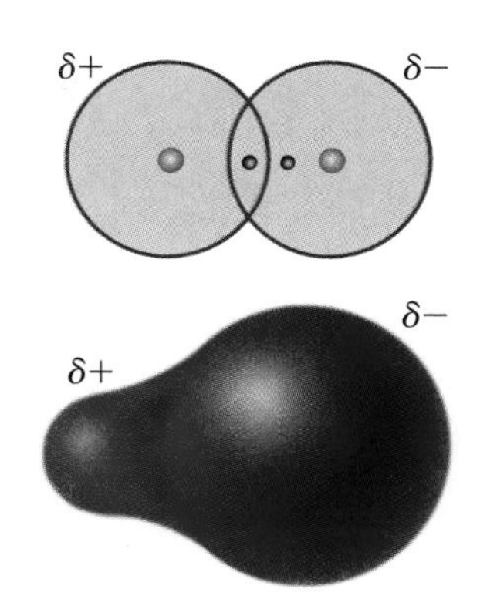

Polar covalent bond: electrons shared unequally

The greater the electronegativity difference ΔEN between the atoms, the greater the partial charges that develop. (The symbol Δ is the Greek uppercase letter delta. In science, this symbol usually stands for a difference in two quantities.) Increasing ΔEN values between atoms involved in a chemical bond are said to increase the polarity of the bond.

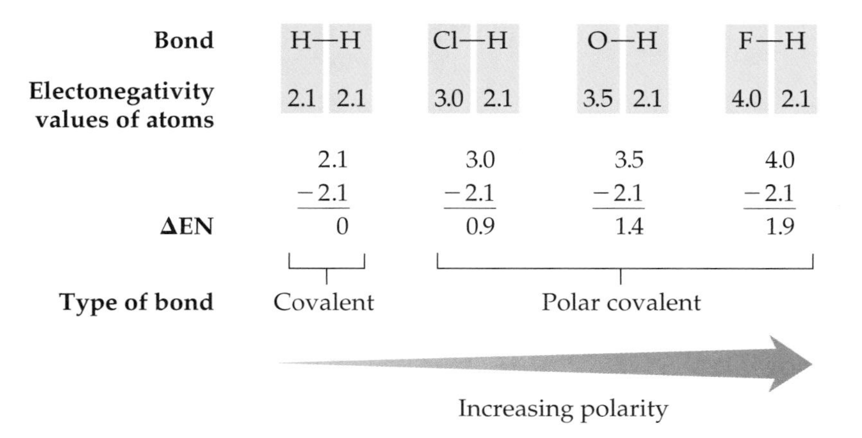

Bond
H—H
Cl—H
O—H
F—H
Electonegativity values of atoms
2.1 2.1
3.0 2.1
3.5 2.1
4.0 2.1
2.1
−2.1
3.0
−2.1
3.5
−2.1
4.0
−2.1
ΔEN
0
0.9
1.4
1.9
Type of bond
Covalent
Polar covalent
Increasing polarity

Do you notice something here? A polar covalent bond with its δ^+ and δ^- atoms is starting to look like an ionic bond. A pure ionic bond, resulting from the complete transfer of electrons, has full positive and negative charges on the ions:

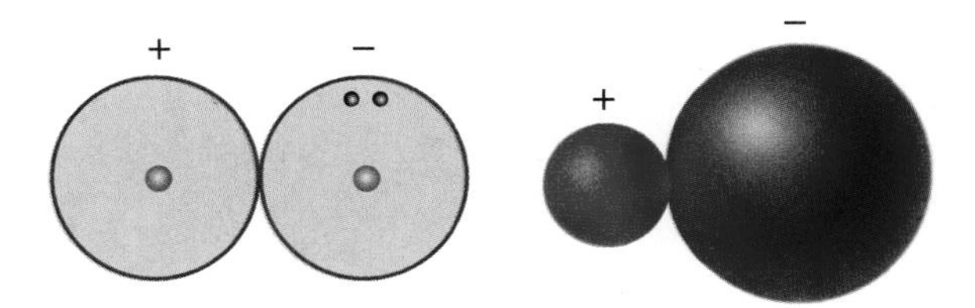

Ionic bond: electron transferred

A polar covalent bond can therefore be thought of as a covalent bond that has some fraction of ionic character. Indeed, the brilliant chemist Linus Pauling, who originated the concept of electronegativity and determined the values listed in this book, developed a simple relationship between electronegativity difference and the percent ionic character in a bond. This relationship is nicely shown in the graph at the right. The graph shows that when ΔEN equals 1.7, a bond can be considered to be 50% ionic (and therefore 50% covalent).

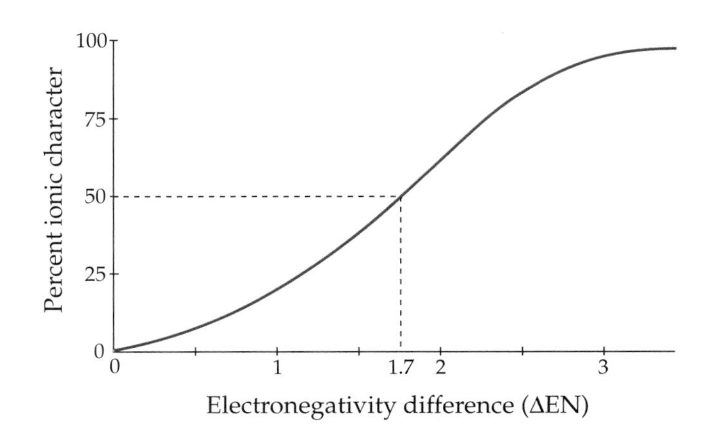

What should you take away from this discussion? Simply that it is often an oversimplification to refer to a bond between two elements as being either ionic or covalent. Except when $\Delta EN = 0$, the bond will have some of the characteristics of both bond types. Ionic and covalent are therefore extremes at the ends of a continuum of bonding types:

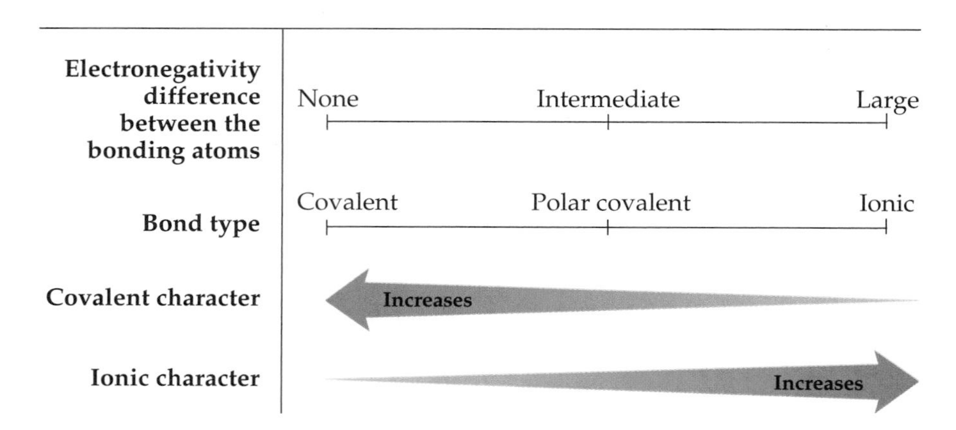

So how do we classify bonds? To answer this question, consider these cases:

F_2	$\Delta EN = 0$
CO_2	$\Delta EN = 1.0$
HF	$\Delta EN = 1.9$
NaBr	$\Delta EN = 1.9$
NaF	$\Delta EN = 3.1$

The bond in the fluorine molecule, F_2, is purely covalent and nonpolar. The bond in NaF is, according to the graph we just looked at, about 90% ionic. What do we call a bond that is 90% ionic? When a metal is bound to a nonmetal (as is the case with NaF) and the percent ionic character is greater than 50%, it is traditional to refer to the compound and the bonds in it as being ionic (even though at 90% ionic, the electron transfer is not actually complete).

What about HF? Here, the electronegativity difference is quite large and leads to a bond which is roughly 60% ionic. Nevertheless, this bond is classified polar covalent and not ionic.

F_2	$\Delta EN = 0$	0% ionic	Covalent
CO_2	$\Delta EN = 1.0$	23% ionic	Polar covalent
HF	$\Delta EN = 1.9$	60% ionic	Polar covalent
NaBr	$\Delta EN = 1.9$	60% ionic	Ionic
NaF	$\Delta EN = 3.1$	90% ionic	Ionic

The HF bond is classified as polar covalent because the term *ionic* is generally reserved for molecules in which a metal bonds to a nonmetal, as in NaBr. Here we have a case where the ΔEN value is again 1.9, just as with polar covalent HF, but we call this bond ionic because a metal, Na, is involved.

There are exceptions to the metal–nonmetal rule. For example, the compound ammonium chloride, NH_4Cl, is best thought of as an ionic compound made up of NH_4^+ cations and Cl^- anions, even though it contains no metals.

Practice Problems

For Practice Problems 5.19–5.21, use the electronegativity values in the chart on page 183 to calculate ΔEN and predict whether the bonds (Ba–Cl, O–O, and Si–O) are covalent, polar covalent, or ionic.

5.19 $BaCl_2$

Answer: EN values: Ba 0.9, Cl 3.0. ΔEN = 2.1.

The bond between Ba and Cl is ionic because the bonding is between a metal (Ba) and a nonmetal (Cl) and this ΔEN value corresponds to about 62% ionic character according to our graph.

5.20 O_3

5.21 SiO_2

5.22 Without knowing electronegativity values, a student claims $BaCl_2$ is more ionic than $BeCl_2$. All she has access to is a periodic table. How does she know she is right?

In summary, substances such as Cl_2 and HCl and NaCl do not fly apart because valence electrons are equally shared, unequally shared, or transferred between one atom and the other. All three situations give rise to an attractive electrostatic force between the atoms, a force we call a chemical bond (covalent, polar covalent, or ionic). Valence electrons truly are the "glue" that bonds atoms to each other.

5.7 Nomenclature—Naming Chemical Compounds

In the early days of chemistry, as compounds were discovered and characterized, they were often given names that had little to do with their molecular formula, if the formula was even known. For example, the compounds H_2O, NH_3, and Hg_2Cl_2 have the *common names* water, ammonia, and calomel, respectively. The problem with these common names is that they give no clue to the molecular formula of the compound. For example, the only way to know that *ammonia* means NH_3 is memorization. Such a naming system is doomed to fail when you consider that there are today literally millions of known compounds. Just imagine trying to memorize a few hundred thousand different names and their associated formulas and you see the problem.

For this reason, it quickly became obvious that chemistry needed a naming system (a *nomenclature* system) that would generate a unique name for each compound based on its chemical formula. Through international agreement, just such a nomenclature system has been developed. Once you understand the rules of this system, just hearing the name of a compound will allow you to write its molecular formula. We shall show you these rules for naming the simplest of compounds—**binary compounds**, which are those made up of only two elements. This category can be further subdivided into *binary ionic compounds* and *binary covalent compounds*. **Binary ionic compounds** are composed of a metal and a nonmetal and are usually ionic. **Binary covalent compounds** are composed of two nonmetals or metalloids and are usually covalent. We'll start with the binary ionic compounds.

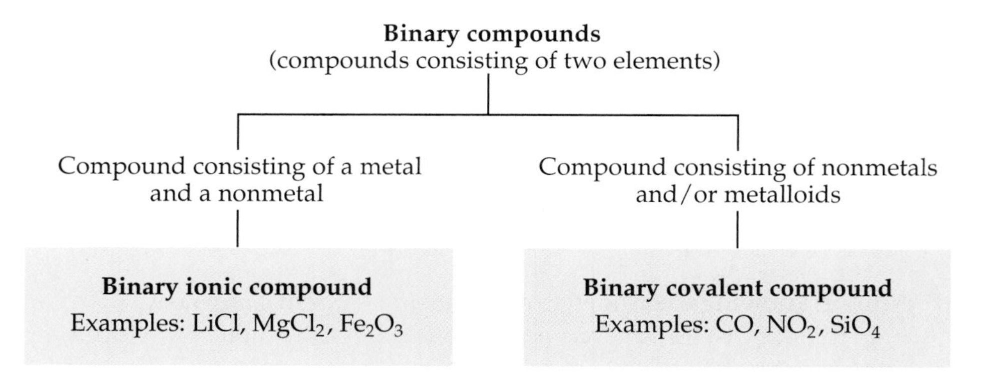

NAMING BINARY IONIC COMPOUNDS

The molecular formula for a binary ionic compound is always written with the metal preceding the nonmetal. For example, the chemical formula for sodium chloride is always written NaCl and never ClNa. When naming the compound, the metal is considered the cation and the nonmetal the anion. Like the formula, the name identifies the cation first, then the anion. The cation is given the name of its element, and the anion has the suffix *-ide* added to its elemental name (*-ide* means "negative"). For example, the ionic compound made from fluorine and lithium has the formula LiF and is called lithium fluor*ide*. Here are a few more examples of binary ionic compounds:

	Cation	**Anion**	Cation / Anion / -*ide* suffix
Na_2S	Na^+	S^{2-}	Sodium sulf*ide*
$MgCl_2$	Mg^{2+}	Cl^-	Magnesium chlor*ide*
Al_2O_3	Al^{3+}	O^{2-}	Aluminum ox*ide*
Li_3N	Li^+	N^{3-}	Lithium nitr*ide*

Notice that each name tells you which elements are present in the compound and even identifies the anion (the one with the *-ide* ending) but does not seem to tell you how many of each ion are in the chemical formula. In fact, the name does tell you how many. Just by looking at the name magnesium chloride, you should be able to determine that the formula is $MgCl_2$. You already have the knowledge necessary to do this. We know that the metal in an ionic compound has lost electrons. Because magnesium is in group IIA, it has lost two electrons to become Mg^{2+}. We also know that nonmetals gain electrons. Because chlorine is in group VIIA, it has gained one electron and become Cl^-.

Finally, in order for magnesium chloride to be electrically neutral, there must be two Cl^- ions for every Mg^{2+} ion, making the molecular formula $MgCl_2$. So, just from seeing the name magnesium chloride, you can deduce the formula. This is the beauty of this naming system.

Thallium, Tl, forms a compound with oxygen called thallium oxide. Based on the location of these elements in the periodic table, what is the expected molecular formula of this compound?

Check your answer to the WorkPatch against that given at the end of the chapter. You should be able to go from chemical formula to name or from name to chemical formula for any binary ionic compound.

Practice Problems

5.23 What is wrong with the molecular formula MgCl?

Answer: To have this formula, the ions would need to have charges that were equal in magnitude but opposite in charge, such as +1 and −1, but Mg forms Mg^{2+} cations and Cl forms Cl^- anions.

5.24 Fill in the blanks:

Formula	Cation	Anion	Name
CaF_2	?	?	?
?	?	?	Cesium bromide
?	Al^{3+}	S^{2-}	?
K_2O	?	?	?

5.25 What do all the molecular formulas of the group IIA bromides have in common?

The transition metals and some of the heavier representative metals throw a monkey wrench into this simple nomenclature system because these elements can form more than one stable cation. For example, unlike sodium, which always exists in compounds as Na^+, iron commonly exists as either Fe^{2+} or Fe^{3+}. This means that *iron chloride* could be the name for either $FeCl_2$ or $FeCl_3$. This is not good. We can't allow ambiguity in our naming system. Fortunately, a simple addition to the system fixes this problem. When dealing with a transition metal or one of the heavier representative metals, a roman numeral in parentheses is included in the name after the cation to indicate the magnitude of the positive charge. For example, iron(II) chloride (pronounced "iron two chloride") implies Fe^{2+} and has the molecular formula $FeCl_2$, whereas iron(III) chloride is $FeCl_3$ (Fe^{3+}). This roman-numeral notation takes care of the problem and allows us to go from name to formula or from formula to name. The table at the top of the next page shows some of the transition metals along with their most common ionic charges.

Table 5.2 The Most Common Ionic Charges of Some Transition Metal Atoms

Sc	Ti	V	Cr	Mn	Fe	Co	Ni	Cu	Zn
3+	4+ 3+	5+ 4+ 3+	6+ 3+	7+ 4+ 3+ 2+	3+ 2+	3+ 2+	2+	2+ 1+	2+
Y	**Zr**	**Nb**	**Mo**					**Ag**	**Cd**
3+	4+	5+	6+					1+	2+
La	**Hf**	**Ta**	**W**				**Pt**	**Au**	**Hg**
3+	4+	5+	6+				4+ 2+	3+ 1+	2+ 1+

Practice Problems

5.26 Name the compound $TiCl_4$.

Answer: Cl is in group VIIA and so has gained one electron to become a Cl^- anion. Because there are four of these anions, titanium, Ti, must have a +4 charge to maintain electrical neutrality. The name is therefore titanium(IV) chloride.

5.27 Name the compound $TiCl_3$.

5.28 Give the molecular formula for tin(II) fluoride.

You should be aware of an older but still commonly used system for naming these types of compounds, a system that does not use roman numerals to indicate ionic charge. Instead, the name of the cation is changed by adding a suffix to indicate charge. For the element iron, the suffix *-ous* is used for Fe^{2+} (ferrous) and the suffix *-ic* is used for Fe^{3+} (ferric). However, the suffixes *-ous* and *-ic* do not necessarily mean +2 and +3, respectively. Rather, *-ous* is used for the ion with lower positive charge and *-ic* is used for the ion with higher charge. For example, tin, Sn, commonly exists in two cationic forms, +2 and +4. The compounds formed between tin and fluorine could be either SnF_2 or SnF_4. In this case, SnF_2, tin(II) fluoride, is stannous fluoride and SnF_4, tin(IV) fluoride, is stannic fluoride. Think about this the next time you are brushing your teeth with fluoride toothpaste whose active ingredient is stannous fluoride.

Table 5.3 contrasts the old versus the modern systematic names for some binary ionic compounds.

Table 5.3 Old Versus Systematic Names for Some Binary Ionic Compounds

Ion	Compound formed with Cl	Old name	Systematic name
Fe^{2+}	$FeCl_2$	Ferrous chloride	Iron(II) chloride
Fe^{3+}	$FeCl_3$	Ferric chloride	Iron(III) chloride
Cu^{+}	$CuCl$	Cuprous chloride	Copper(I) chloride
Cu^{2+}	$CuCl_2$	Cupric chloride	Copper(II) chloride
Co^{2+}	$CoCl_2$	Cobaltous chloride	Cobalt(II) chloride
Co^{3+}	$CoCl_3$	Cobaltic chloride	Cobalt(III) chloride
Hg_2^{2+} (each Hg is +1)	Hg_2Cl_2	Mercurous chloride	Mercury(I) chloride
Hg^{2+}	$HgCl_2$	Mercuric chloride	Mercury(II) chloride
Sn^{2+}	$SnCl_2$	Stannous chloride	Tin(II) chloride
Sn^{4+}	$SnCl_4$	Stannic chloride	Tin(IV) chloride

Naming binary ionic compounds

The name of the compound consists of the names of the two ions, with the metal ion named first.

- *Metal ion name:* based on element name.
- *Nonmetal ion name:* element name with suffix *-ide* (Cl^-, chloride; S^{2-}, sulfide; O^{2-}, oxide).

When the metal has the same ionic charge in all compounds (most representative metals),

the name of the metal ion is simply the name of the metal:

LiCl	Lithium chloride
$MgCl_2$	Magnesium chloride
CaO	Calcium oxide

Only one chemical formula is possible in each case, and so the name is unambiguous.

When the metal has different ionic charges in different compounds (most transition metals and some heavier representative metals),

Systematic names
the charge of the metal ion is indicated by a roman numeral in parentheses after the metal name:

$FeCl_2$	Iron(II) chloride
$FeCl_3$	Iron(III) chloride

or

Older system names
the charge of the metal ion is indicated by a suffix. When a metal commonly occurs as either of two ions, *-ous* is used for the ion of lower charge and *-ic* for the ion of higher charge:

$FeCl_2$	Ferrous chloride
$FeCl_3$	Ferric chloride

NAMING BINARY COVALENT COMPOUNDS

Binary covalent compounds generally consist of two different nonmetals or metalloids and are usually covalent. They are named in much the same way as binary ionic compounds with one slight difference. In the case of ionic compounds, the name magnesium chloride was enough to tell us that the molecular formula was $MgCl_2$ because we could count on the magnesium always being +2 and the chlorine always being −1. With binary covalent compounds, this is no longer the case. When two nonmetallic elements combine, a range of possibilities exists. This means that we need to insert prefixes into the name to tell us how many atoms of each element are present in the molecular formula. The prefixes we use are shown in Table 5.4 and are derived from early Greek names for numbers.

Table 5.4 Greek Prefixes

1	mono-	5	penta-	9	nona-
2	di-	6	hexa-	10	deca-
3	tri-	7	hepta-		
4	tetra-	8	octa-		

Also, because both elements in these compounds are nonmetals, we treat the less electronegative element as if it were a metal, list it first, and use the unmodified name of the element. The more electronegative element is listed second and has the suffix *-ide* appended to its name. For example, the compound HCl is called hydrogen chloride.

Why does it make sense to give the more electronegative element in a binary covalent compound the suffix *-ide*?

Let's consider a few examples. Carbon and oxygen can combine to form two binary compounds, CO and CO_2. These compounds are called carbon *mon*oxide and carbon *di*oxide. We use the prefixes *mon(o)-* and *di-* to indicate that either one or two atoms of oxygen are present in the compound. These prefixes can also be used with the first element. For example, the compounds NO_2 and N_2O_3 are named nitrogen dioxide and dinitrogen trioxide, respectively. Notice that NO_2 is *not* called mononitrogen dioxide. If there is only one atom of the less electronegative element, the prefix *mono-* is generally left out. Some additional examples are

PCl_3	Phosphorus trichloride
P_4O_6	Tetraphosphorus hexoxide*
P_4O_{10}	Tetraphosphorus decoxide*
SF_6	Sulfur hexafluoride
SO_2	Sulfur dioxide
SO_3	Sulfur trioxide
NO_2	Nitrogen dioxide
N_2O_4	Dinitrogen tetroxide*
N_2O_5	Dinitrogen pentoxide*

*To avoid two vowels together, the final *a* on a prefix is dropped when the element name begins with a vowel.

Naming binary covalent compounds

The name of the compound is constructed from the names of the two elements.

- The *less electronegative element* is named first (just as with the metal in a binary ionic compound).
- The *more electronegative element* is named second and takes the suffix *-ide* (just as with the nonmetal in a binary ionic compound).
- *For each element,* a prefix indicates the number of atoms of that element in the formula. If only one atom of the element is present, the prefix is generally omitted:

SF_6	Sulfur hexafluoride
N_2O_4	Dinitrogen tetroxide

Practice Problems

5.29 Name the compound N_2O.

Answer: Dinitrogen monoxide

5.30 Name the compound NO.

5.31 Name the compound PCl_5.

5.32 Give the chemical formula for tetraphosphorus decasulfide.

NOMENCLATURE AND POLYATOMIC IONS

Because polyatomic ions show up in many of the compounds chemists deal with, it pays to become familiar with the names and formulas of these ions. Many of the important (common) ones are listed in Table 5.5. You should commit these names, chemical formulas, and charges to memory. Examine the ions now, and then we'll show you how to name compounds that contain them.

Table 5.5 Some Common Polyatomic Ions

Ion formula	Ion name	Ion formula	Ion name
NH_4^+	Ammonium		
$CH_3CO_2^-$	Acetate	PO_4^{3-}	Phosphate
CN^-	Cyanide	HPO_4^{2-}	Hydrogen phosphate
NO_2^-	Nitrite	$H_2PO_4^-$	Dihydrogen phosphate
NO_3^-	Nitrate	ClO^-	Hypochlorite
CO_3^{2-}	Carbonate	ClO_2^-	Chlorite
HCO_3^-	Hydrogen carbonate (or bicarbonate)	ClO_3^-	Chlorate
SO_3^{2-}	Sulfite	ClO_4^-	Perchlorate
SO_4^{2-}	Sulfate	CrO_4^{2-}	Chromate
HSO_4^-	Hydrogen sulfate (or bisulfate)	$Cr_2O_7^{2-}$	Dichromate
MnO_4^-	Permanganate	O_2^{2-}	Peroxide

You may be familiar with some of these polyatomic ions. Aqueous solutions of the hypochlorite ion are sold as chlorine bleach. Nitrites are used to preserve meats. Aqueous solutions of peroxide ions are found in many medicine cabinets and used as an antiseptic. The cyanide ion is extremely toxic and is the favorite way for Hollywood movie spies to do themselves in. Only one of the ions in Table 5.5 is a cation (ammonium, NH_4^+). Several of the anions consist of a nonmetal plus one or more oxygen atoms, such as sulfate, SO_4^{2-}. These ions are called oxyanions, and they often exist in a series in which the individual members differ only in the number of oxygen atoms that accompany the nonmetal atom. When there are two members in such a series, the ion with the fewer oxygens gets the *-ite* ending, and the ion with the greater number of oxygens gets the *-ate* ending:

SO_3^{2-}	Sulfite
SO_4^{2-}	Sulfate

When there are more than two oxyanions in a series, the prefix *hypo-* is used for the ion with the fewest oxygens, and the prefix *per-* is used for the oxyanion with the most oxygens:

ClO^-	Hypochlorite
ClO_2^-	Chlorite
ClO_3^-	Chlorate
ClO_4^-	Perchlorate

Naming compounds that contain polyatomic ions is easy. You simply use the names from Table 5.5 and then name the compound as you would a binary ionic compound. For example, you call NaCl sodium chloride. Therefore you call NaClO sodium hypochlorite and NH_4Cl ammonium chloride.

And just as for binary ionic compounds, knowledge of the charges of the ions allows you to go from the name of a compound to its chemical formula. As an example, let's consider the compound sodium phosphate. From Table 5.5 we know that the phosphate ion has the composition PO_4 and a −3 charge. Because sodium, a group IA element, forms ions with a charge of +1, the molecular formula must be Na_3PO_4 to achieve charge neutrality:

$$\underbrace{PO_4^{3-}}_{-3} + \underbrace{\begin{matrix} Na^+ \\ Na^+ \\ Na^+ \end{matrix}}_{+3} \longrightarrow \text{Neutral compound, } Na_3PO_4$$

What is the formula for magnesium phosphate? The phosphate ion still has a −3 charge, but now the magnesium cation, a group IIA element, has a +2 charge. To arrive at an electrically neutral compound, we need to combine two PO_4^{3-} ions with three Mg^{2+} ions:

$$\underbrace{\begin{matrix} PO_4^{3-} \\ PO_4^{3-} \end{matrix}}_{+6} + \underbrace{\begin{matrix} Mg^{2+} \\ Mg^{2+} \\ Mg^{2+} \end{matrix}}_{-6} \longrightarrow \text{Neutral compound, } Mg_3(PO_4)_2$$

Naming polyatomic ionic compounds

The compound is named the way a binary ionic compound is, except that the anion name is the name of the polyatomic ion.

- The metal ion is named first, using the rules for binary ionic compounds.
- The polyatomic ion is named second. The names of the common polyatomic ions are given in Table 5.5.

$NaNO_2$	Sodium nitrite
$NaNO_3$	Sodium nitrate
$Cu(NO_3)_2$	Copper(II) nitrate or cupric nitrate

The formula for magnesium phosphate is therefore $Mg_3(PO_4)_2$. Notice that we put parentheses around the phosphate; otherwise, the formula would look like this: Mg_3PO_{42}, which says there are 42 oxygen atoms in the compound. Whenever there is more than one polyatomic ion in the formula for a compound, the formula for the polyatomic ion must be surrounded by parentheses: $CuNO_3$ for copper(I) nitrate but $Cu(NO_3)_2$ for copper(II) nitrate.

Practice Problems

5.33 Write the formula for ammonium phosphate.

Answer: Because ammonium is NH_4^+ *and phosphate is* PO_4^{3-}, *the formula must be* $(NH_4)_3PO_4$ *to obtain charge neutrality.*

5.34 Write the formula for sodium carbonate.

5.35 Write the name for $Ca(ClO)_2$.

5.36 Write the formula for magnesium bicarbonate.

NOMENCLATURE OF ACIDS

Although the subject of acids (and bases) is covered in later chapters, we can say a bit about the nomenclature of acids now. For the purposes of this discussion, any compound that dissolves in water and falls apart (*dissociates*) to produce H^+ ions in the water is an **acid**.

Acids are named by one of two systems, depending on whether or not the acid contains oxygen. Let's first consider acids that do not contain oxygen. When gaseous HBr is dissolved in water, the HBr dissociates to produce H^+ ions. Therefore HBr is an acid. Because this acid does not contain oxygen, it is named by adding the prefix *hydro-* and the suffix *-ic acid* to the root name of the nonhydrogen element (or elements). So HBr is hydrobromic acid. Some other examples are

HF	Hydrofluoric acid
HCl	Hydrochloric acid
HI	Hydroiodic acid
HCN	Hydrocyanic acid
H_2S	Hydrosulfuric acid

These compounds produce acidic solutions when dissolved in water, and thus their aqueous solutions are named as indicated above. However, when they are not dissolved in water, they are named according to the rules for nonacidic compounds. Thus whereas an aqueous solution of HCl gas is called hydrochloric acid, pure HCl gas is called hydrogen chloride.

If the acid contains one or more oxygens, it is called an *oxyacid,* and the rules for naming it are somewhat different. Now the focus is on the name of the polyatomic anion in the compound. If the anion name ends in *-ate,* we name the acid by changing the suffix from *-ate* to *-ic:*

Formula of acid	Formula of ion	Name of ion	Name of acid
H_2SO_4	SO_4^{2-}	Sulf*ate*	Sulfur*ic* acid
HNO_3	NO_3^-	Nitr*ate*	Nitr*ic* acid
$HClO_4$	ClO_4^-	Perchlor*ate*	Perchlor*ic* acid
$HC_2H_3O_2$	$C_2H_3O_2^-$	Acet*ate*	Acet*ic* acid

If the anion name ends in *-ite,* we name the acid by changing the suffix from *-ite* to *-ous:*

Formula of acid	Formula of ion	Name of ion	Name of acid
H_2SO_3	SO_3^{2-}	Sulfite	Sulfurous acid
HNO_2	NO_2^-	Nitrite	Nitrous acid
$HClO$	ClO^-	Hypochlorite	Hypochlorous acid

Naming acids

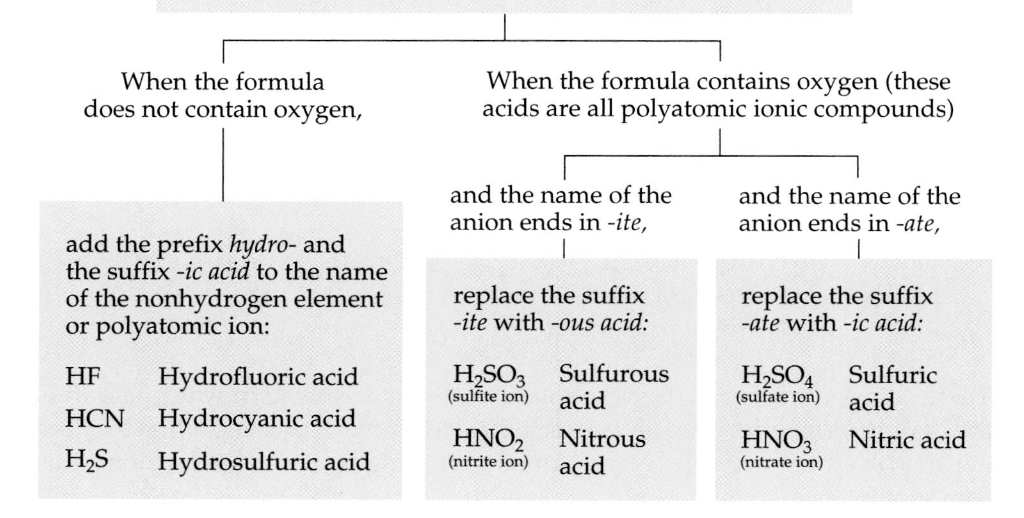

• For now, an acid is any substance that dissociates in water to yield an H+ ion.
• The naming rules given here are used when the acid is dissolved in water.
• When not dissolved in water, acids of this type are either binary compounds or polyatomic ionic compounds and are named by the appropriate rules.
When the formula does not contain oxygen,
add the prefix hydro- and the suffix -ic acid to the name of the nonhydrogen element or polyatomic ion:
HF Hydrofluoric acid
HCN Hydrocyanic acid
H2S Hydrosulfuric acid
When the formula contains oxygen (these acids are all polyatomic ionic compounds)
and the name of the anion ends in -ite,
replace the suffix -ite with -ous acid:
H2SO3 (sulfite ion) Sulfurous acid
HNO2 (nitrite ion) Nitrous acid
and the name of the anion ends in -ate,
replace the suffix -ate with -ic acid:
H2SO4 (sulfate ion) Sulfuric acid
HNO3 (nitrate ion) Nitric acid

We shall learn much more about the nomenclature and properties of acids in Chapter 15.

Finally, some compounds have been around for so long that the common names are the only ones used. For example, NH_3 is always referred to as ammonia and not nitrogen trihydride. And absolutely nobody calls H_2O dihydrogen monoxide! To prove this to yourself, the next time you go to a restaurant, see how clever your waiter thinks you are when you ask for a glass of dihydrogen monoxide.

Have You Learned This?

Molecule (p. 163)
Covalent bond (p. 167)
Dot diagram (p. 173)
Lewis dot diagram (p. 173)
Electron pair (p. 173)
Paired electron (p. 173)
Unpaired electron (p. 173)
Bonding pair (p. 175)
Lone pair (p. 175)
Multiple covalent bond (p. 176)
Multiple bond (p. 176)
Double bond (p. 177)
Single bond (p. 177)
Triple bond (p. 177)
Resonance form (p. 179)
Polyatomic ion (p. 179)
Ionic bond (p. 182)
Ionic lattice (p. 182)
Ionic compound (p. 182)
Molecular compound (p. 182)
Chemical bond (p. 183)
Electronegativity (p. 183)
Polar covalent bond (p. 185)
Binary compound (p. 188)
Binary ionic compound (p. 188)
Binary covalent compound (p. 188)
Acid (p. 195)

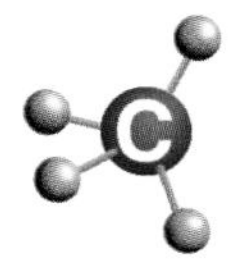

Chemical Bonding and Nomenclature
www.chemistryplace.com

SKILLS TO KNOW

Method for drawing dot diagrams

Example: Draw a dot diagram for HCO_3^-.

1. Determine the total number of valence electrons (dots) that will be in the diagram. For a polyatomic ion, adjust the electron count to reflect the charge.

C	$1 \times 4 =$	4 electrons
O	$3 \times 6 =$	18 electrons
H	$1 \times 1 =$	1 electron
Ionic charge of 1−		1 electron
Total		24 electrons

2. Connect the atoms with single bonds. If you do not know how the atoms are arranged, try using the first nonhydrogen atom in the formula as the central atom.

Choose C as central atom

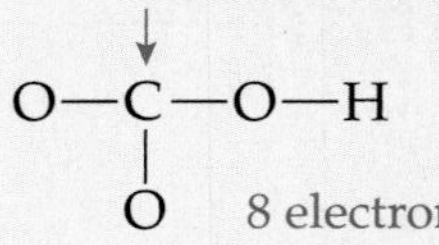

O—C—O—H (with O bonded below C)

8 electrons assigned

3. Add the remaining electrons as lone pairs, satisfying octets as you go. Do terminal atoms first (except H, which needs only two electrons).

Lacks an octet

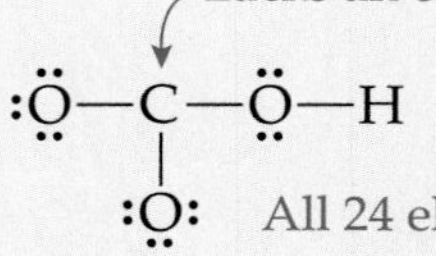

:Ö—C—Ö—H (with :Ö: bonded below C)

All 24 electrons assigned

4. If any atom lacks an octet, move lone pair(s) from adjacent atom(s) into bonding positions on the octet-deficient atom.

Lone pair to the rescue

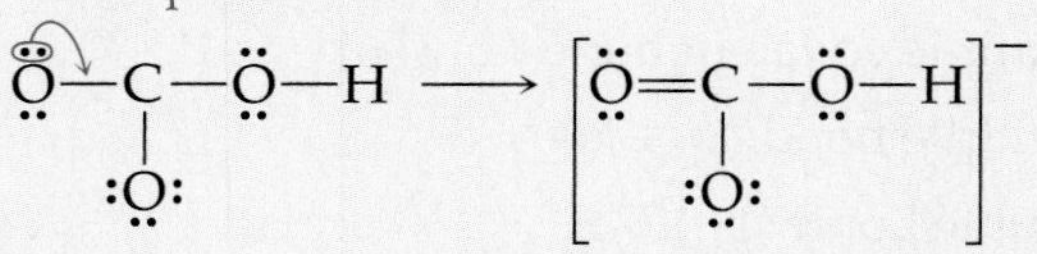

:Ö—C—Ö—H (with :Ö: below C) ⟶ [Ö=C—Ö—H (with :Ö: below C)]⁻

Final dot structure
(one of a resonance set)

Electronegativity and bond polarity

1. The electronegativity, EN, of an element is a number that tells how strongly an atom of the element attracts bonding electrons to itself (how much of an electron "hog" it is). EN values range from 0.7 (the metal francium) to 4.0 (the nonmetal fluorine).

The biggest electron hogs of all

2. The polarity of a bond depends on the difference in electronegativity, ΔEN, between the bonded atoms. ΔEN is calculated by subtracting the smaller from the larger EN.

H—Cl

EN = 2.1 EN = 3.0

$\Delta EN = 3.0 - 2.1 = 0.9$

3. The greater the ΔEN of a bond, the stronger are the δ+ and δ− charges on the bonded atoms and hence the more polar is the bond. When ΔEN = 0, the bonding is pure covalent. When ΔEN is very large, the bonding is ionic. In between, the bonding is polar covalent.

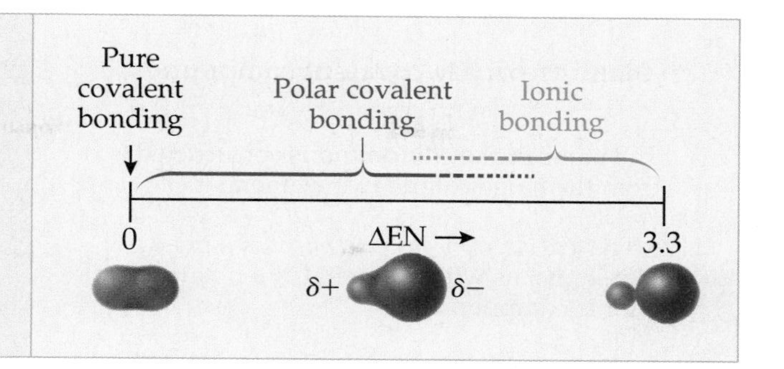

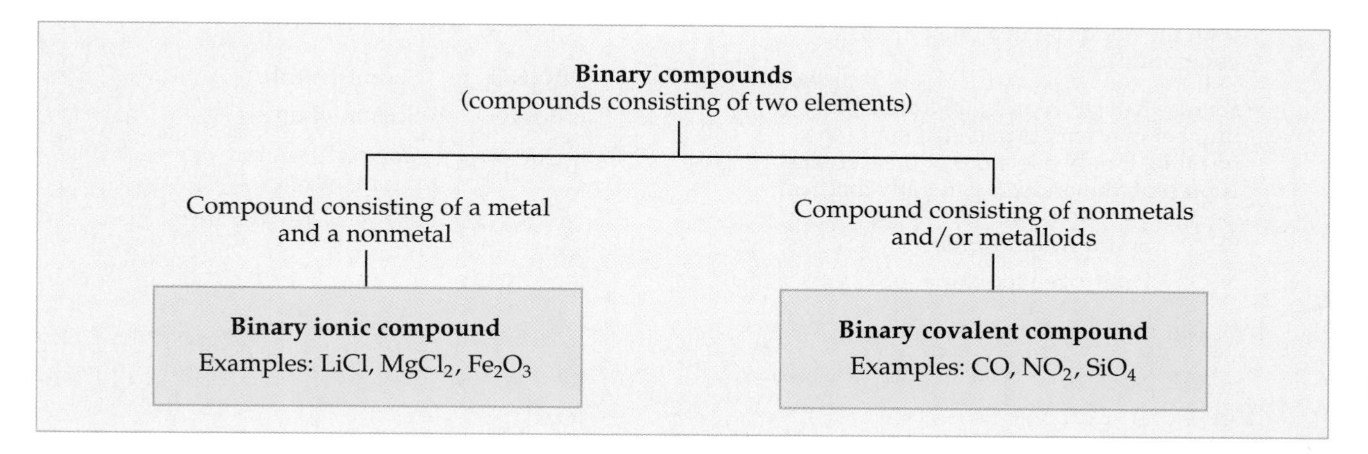

Naming binary ionic compounds

The name of the compound consists of the names of the two ions, with the metal ion named first.

- *Metal ion name:* based on element name.
- *Nonmetal ion name:* element name with suffix *-ide* (Cl^-, chloride; S^{2-}, sulfide; O^{2-}, oxide).

When the metal has the same ionic charge in all compounds (most representative metals),

the name of the metal ion is simply the name of the metal:

LiCl	Lithium chloride
$MgCl_2$	Magnesium chloride
CaO	Calcium oxide

Only one chemical formula is possible in each case, and so the name is unambiguous.

When the metal has different ionic charges in different compounds (most transition metals and some heavier representative metals),

Systematic names
the charge of the metal ion is indicated by a roman numeral in parentheses after the metal name:

$FeCl_2$	Iron(II) chloride
$FeCl_3$	Iron(III) chloride

or

Older system names
the charge of the metal ion is indicated by a suffix. When a metal commonly occurs as either of two ions, *-ous* is used for the ion of lower charge and *-ic* for the ion of higher charge:

$FeCl_2$	Ferrous chloride
$FeCl_3$	Ferric chloride

Naming binary covalent compounds

The name of the compound is constructed from the names of the two elements.

- The *less electronegative element* is named first (just as with the metal in a binary ionic compound).
- The *more electronegative element* is named second and takes the suffix *-ide* (just as with the nonmetal in a binary ionic compound).
- *For each element*, a prefix indicates the number of atoms of that element in the formula. If only one atom of the element is present, the prefix is generally omitted:

SF_6	Sulfur hexafluoride
N_2O_4	Dinitrogen tetroxide

Naming polyatomic ionic compounds

The compound is named the way a binary ionic compound is, except that the anion name is the name of the polyatomic ion.

- The metal ion is named first, using the rules for binary ionic compounds.
- The polyatomic ion is named second. The names of the common polyatomic ions are given in Table 5.5.

$NaNO_2$	Sodium nitrite
$NaNO_3$	Sodium nitrate
$Cu(NO_3)_2$	Copper(II) nitrate or cupric nitrate

Naming acids

- For now, an acid is any substance that dissociates in water to yield an H^+ ion. (You will learn more in Chapter 15.)
- The naming rules given here are used when the acid is dissolved in water.
- When not dissolved in water, acids of this type are either binary compounds or polyatomic ionic compounds and are named by the appropriate rules.

When the formula does not contain oxygen,

add the prefix *hydro-* and the suffix *-ic acid* to the name of the nonhydrogen element or polyatomic ion:

HF	Hydrofluoric acid
HCN	Hydrocyanic acid
H_2S	Hydrosulfuric acid

When the formula contains oxygen (these acids are all polyatomic ionic compounds)

and the name of the anion ends in *-ite*,

replace the suffix *-ite* with *-ous acid*:

H_2SO_3 (sulfite ion)	Sulfurous acid
HNO_2 (nitrite ion)	Nitrous acid

and the name of the anion ends in *-ate*,

replace the suffix *-ate* with *-ic acid*:

H_2SO_4 (sulfate ion)	Sulfuric acid
HNO_3 (nitrate ion)	Nitric acid

MOLECULES

5.37 Give a general definition of a molecule.

5.38 Are all molecules also compounds? Explain.

5.39 What is it about molecules that makes them worth preparing? Give some examples.

5.40 What common diatomic molecules exist in our atmosphere? What triatomic molecules exist in our atmosphere? Give the formulas for all your answers.

5.41 Ethanol (C_2H_6O) is consumed in alcoholic beverages, but another alcohol, methanol (CH_4O), is toxic. What explanation can you give for this?

5.42 Can a molecule also be an elemental substance? If so, give some examples.

THE COVALENT BOND

5.43 Use your own words to describe (define) a covalent bond.

5.44 What is wrong with saying that a covalent bond is just "two shared electrons"?

5.45 Why does the sharing of two electrons between two atoms bond the atoms to each other?

5.46 In terms of energy, why is an H_2 molecule more stable than two isolated H atoms?

5.47 In terms of interaction between the atoms, why is an H_2 molecule more stable than two isolated H atoms?

5.48 The bond distance in an H_2 molecule is 0.74 Å. Why isn't it shorter than this? Why isn't it longer than this?

5.49 Is energy released or absorbed when a covalent bond forms between two atoms?

5.50 Why are an atom's valence electrons the only electrons involved in bonding?

5.51 Suppose one of the electrons from the covalent bond in H_2 suddenly vanished. Why would the bond between the atoms weaken?

5.52 A student decides to boil water to produce hydrogen gas and oxygen gas. Will this work? Explain your answer.

5.53 What do we mean by "double-counting" when it comes to counting the electrons around an atom in a molecule?

5.54 What is the accepted shortcut for drawing a shared pair of electrons (a covalent bond) in a molecular drawing?

5.55 How many electrons does an atom gain for each covalent bond that it forms in a molecule?

MOLECULES, DOT STRUCTURES, AND THE OCTET RULE

5.56 A student draws the following dot diagrams for N, O, and F. What, if anything, is wrong with each?

5.57 For any representative element, what is the relationship between the element's roman-numeral group number and the number of dots in the Lewis dot diagram?

5.58 Draw dot diagrams for the following atoms:
(a) S (b) I (c) He (d) B

5.59 Chlorine (Cl), neon (Ne), and helium (He) all exist as gases, but only one of them is diatomic. Which is it, and why is it diatomic while the others are monatomic?

5.60 Valence electrons in an isolated atom can exist in either of two ways. What are they?

5.61 Why isn't the formula for water HO?

5.62 Hydrogen (H) and sulfur (S) form the toxic compound hydrogen sulfide, a gas that smells like rotten eggs and is spewed from volcanoes. Predict the formula of hydrogen sulfide starting with dot diagrams for the atoms.

5.63 Phosphorus (P) and bromine (Br) form a compound. Predict the formula of this compound starting with dot diagrams for the atoms.

5.64 Nitrogen (N) forms a compound with bromine (Br) similar to that in Problem 5.63. How is it similar? Explain why this is so.

5.65 Ethers are compounds of C, H, and O that are often used as solvents. One particular ether molecule has the formula C_2H_6O. The structure is such that both carbons are attached to the oxygen atom, and there are no O–H bonds. Starting with dot diagrams for the atoms, draw a dot diagram for this ether molecule. How many bonding pairs does the molecule have? How many lone pairs?

5.66 Ethanol, the alcohol found in alcoholic beverages, has the same formula as the ether of Problem 5.65. However, it has a different structure, in which only one carbon atom is bound to the oxygen atom. Starting with dot diagrams for the atoms, draw a dot diagram for the ethanol molecule. Indicate the total number of bonding pairs and lone pairs of electrons. Also comment on what the fact that one form of C_2H_6O is a solvent (poisonous) but another form is drinkable says about the role of structure in determining chemical properties of a molecule.

5.67 The molecule HCl is known, but the molecule HeCl is not. Explain why this is so.

5.68 In the dot diagram for a helium atom, why is it important to draw the two valence electrons as a lone pair?

5.69 Oxygen, in almost all of its compounds, has two bonds to it. Explain why this is so.

5.70 How many bonds will an atom from group VA generally form? Explain why this is so.

MULTIPLE BONDS

For all the problems in this section, you must draw all the resonance forms when they exist.

5.71 Draw a dot diagram for ethylene, C_2H_4.

5.72 Is the bond between the carbon atoms in the ethylene molecule of Problem 5.71 stronger or weaker than the bond in acetylene, C_2H_2 (WorkPatch 5.7)? Explain fully.

5.73 Explain what is meant by resonance forms.

5.74 Draw a dot diagram for the molecule SO_2, sulfur dioxide, a gas that comes from burning coal and is responsible for acid rain. [*Hint:* Sulfur is in the middle.]

5.75 Draw a dot diagram for the molecule O_3, ozone, the molecule in our upper atmosphere that protects us from the Sun's harmful ultraviolet electromagnetic radiation. [*Hint*: Only the middle oxygen atom forms bonds with the two other oxygen atoms.]

5.76 Should there be any similarities between the dot diagrams of Problems 5.74 and 5.75? Explain your answer.

5.77 A student claims that the bonds in ozone are really not double bonds or single bonds, but somewhere between (roughly 1.5 bonds between each oxygen). Justify this statement.

5.78 Experiments show that it takes more energy to break the bond between oxygen atoms in the O_2 molecule than in the O_3 molecule. How can you explain this?

5.79 Is anything wrong with the following dot diagram for carbon monoxide (CO)? If so, explain what it is and fix it.

$$:\underset{\cdot\cdot}{\ddot{C}}-\underset{\cdot\cdot}{\ddot{O}}:$$

5.80 Is anything wrong with the following dot diagram for acetone (C_3H_6O)? If so, then fix it.

```
  H  :Ö:  H
  |   |   |
H—C———C———C—H
  |       |
  H       H
```

5.81 How many covalent bonds do you think would form between the phosphorus atoms in the molecule P_2? Explain your answer and draw a dot diagram.

5.82 Draw a dot diagram for the nitrate (NO_3^-) ion.

5.83 Draw a dot diagram for the hypothetical O_2^{2+} ion.

5.84 Acetaldehyde has the formula C_2H_4O. Draw a dot diagram for it given the following atomic connections.

```
         O
  H      |
   \     C
    C  /   \
H /   \      H
       H
```

5.85 Acetic acid, a major component of vinegar, has the formula $C_2H_4O_2$. Draw a dot diagram for it given the following atomic connections.

```
         O
  H      |
   \     C
    C  /   \
H /   \      O—H
       H
```

5.86 The acetate ion has the formula $C_2H_3O_2^-$. Draw a dot diagram for it given the following atomic connections. [*Hint:* Watch for resonance forms.]

```
         O
  H      |
   \     C
    C  /   \
H /   \      O
       H
```

5.87 Draw a dot diagram for perchloric acid, $HClO_4$. The chlorine is the central atom to which all the oxygens are attached, and the hydrogen is attached to one of the oxygens.

5.88 Draw a dot diagram for the NO_2^+ cation.

5.89 Consider benzene, a pleasant-smelling but carcinogenic (cancer-causing) liquid. Benzene molecules have the formula C_6H_6, with the atoms connected as shown. Complete the dot diagram for this molecule and include any resonance forms.

```
          H
          |
   H      C      H
     \  /   \  /
      C       C
      |       |
      C       C
     /  \   /  \
   H      C      H
          |
          H
```

5.90 How many valence electrons are present on each carbon atom and each hydrogen atom in the benzene molecule of Problem 5.89?

IONIC BONDING

5.91 Draw dot diagrams for the following simple ions:
(a) I^- (b) O^{2-} (c) Cl^- (d) H^+

5.92 What is an ionic bond? How does it differ from a covalent bond, and how is it similar to a covalent bond?

5.93 What must occur between atoms for an ionic bond to form?

5.94 Between what types of elements is ionic bonding most likely to occur?

5.95 Draw dot diagrams for Cl_2 and for $MgCl_2$ (put Mg in the middle) that agree with the type of bonding occurring in each. (That is, show electrons as being shared or transferred, and if they are transferred, show the charges of the resulting ions.)

5.96 Predict the formulas of the ionic compounds that result from combining:
(a) Mg (solid metal) with Br_2 (liquid)
(b) Be (solid metal) with O_2 (gas)
(c) Na (solid metal) with I_2 (solid)

5.97 Predict the formula of the ionic compound formed by each of the following pairs of elements:
(a) Ca, I (b) Ca, O (c) Al, S (d) Ca, Br

5.98 What is wrong with saying that NaCl exists as a molecule, just like H_2O does?

ELECTRONEGATIVITY AND THE POLAR COVALENT BOND

5.99 There is one very important word missing from the following definition of electronegativity. What is it? "Electronegativity is an indication of an atom's ability to attract electrons to itself."

5.100 Why did we call the elements N, O, Cl, and F electron "hogs" in Section 5.6?

5.101 What is the most electronegative element in the periodic table? What is the least electronegative element? What kind of compound would form if these two elements were brought together?

5.102 How do the categories metal and nonmetal relate to electronegativity?

5.103 What are the trends for electronegativity in the periodic table:
(a) Down a group?
(b) Across a period from left to right?
(c) Going from the bottom left corner to the upper right corner?

5.104 What are the guidelines for using electronegativity to predict which type of bond forms between two atoms?

5.105 How does a polar covalent bond differ from a covalent bond? Give examples of diatomic molecules that contain a polar covalent bond and of diatomic molecules that contain a covalent bond.

5.106 Would you classify the C–H bonds in ethane (C_2H_6) as covalent, polar covalent, or ionic? Explain.

5.107 Lithium is a metallic element. Consider the hypothetical species dilithium, Li_2. Predict whether the bonding in such a species would be covalent, polar covalent, or ionic. Explain fully.

5.108 Which molecule has bonds that are the most polar covalent?

H_2 CO H_2S H_2O

5.109 Chemists sometimes think of molecules with polar covalent bonds as being part covalent and part ionic. How can a bond be both covalent and ionic?

5.110 In the following molecule, atoms X and Y are from the same period, meaning they are approximately the same size. The difference in the size of the spheres indicates a difference in the amount of time the shared electrons spend near each atom. Which atom is more electronegative, and how do you know?

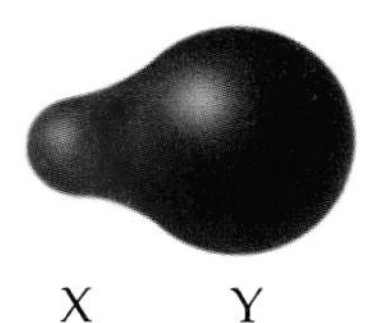

X Y

5.111 In the following molecule AB, one of the atoms is more electronegative than the other. Which is more electronegative, and how do you know?

$\overset{\delta-}{A}—\overset{\delta+}{B}$

5.112 How would you change the electronegativities of the atoms at either end of a polar covalent bond to make the bond ionic? To make the bond covalent?

5.113 Electronegativity is not about measuring an electron's negativeness. Given your knowledge of what electronegativity measures, think of a new word or expression that could replace this misleading term and also communicate what it measures.

5.114 Without knowing the electronegativity values, in what situation can you be absolutely sure that the bonding between two atoms will be purely covalent?

5.115 Classify the bonds in each of the following as ionic, covalent, or polar covalent. Explain each choice.
(a) Br_2 (b) PCl_3 (c) LiCl (d) ClF (e) $MgCl_2$

NOMENCLATURE

5.116 What is a binary compound? Is atmospheric oxygen an example of a binary compound?

5.117 What does the suffix *-ide* mean, and which element in the name of a binary ionic compound gets it?

5.118 Which element in a binary covalent compound gets the suffix *-ide*? Why?

5.119 Any ionic compound has an overall charge of zero. How does this fact help to determine its formula?

5.120 Consider the two binary compounds Al_2O_3 and N_2O_3.
(a) Which is a binary ionic compound and which is a binary covalent compound? Explain.
(b) Al_2O_3 is properly named aluminum oxide, but N_2O_3 is named dinitrogen trioxide. Explain fully why one name uses Greek prefixes and the other does not.

5.121 Name the following binary ionic compounds:
(a) Ca_3N_2 (b) AlF_3 (c) Na_2O (d) CaS

5.122 Give the formulas of the following binary ionic compounds:
(a) Calcium bromide (b) Sodium sulfide
(c) Potassium nitride (d) Lithium oxide

5.123 What "monkey wrench" do transition metals throw into the picture when naming binary compounds? What do we include in the nomenclature system to accommodate them?

5.124 Name the following binary compounds of transition metals using both the modern and older systems.
(a) CuCl and $CuCl_2$
(b) $Fe(OH)_2$ and $Fe(OH)_3$

5.125 Name the following compounds:
(a) Na_2SO_4
(b) $(NH_4)_3PO_4$
(c) KClO
(d) $CaCO_3$
(e) $Al(NO_3)_3$

5.126 Give the formulas for the following compounds:
(a) Ammonium acetate
(b) Ammonium carbonate
(c) Iron(II) nitrate
(d) Ferric hydroxide
(e) Calcium hypochlorite

5.127 Name the following binary covalent compounds:
(a) PCl_3 (b) SO_2 (c) N_2O_4 (d) P_5O_{10}

ADDITIONAL PROBLEMS

5.128 Consider the sulfate, sulfite, nitrate, nitrite, chlorate, and chlorite ions. What information do the *-ate* and *-ite* suffixes communicate?

5.129 The oxide ion is O^{2-}. How does this ion differ from the peroxide ion, O_2^{2-}? Draw dot diagrams for both.

5.130 A polyatomic ion not listed in Table 5.5 is iodate, IO_3^-. What is the formula for the periodate ion, and what is the formula for magnesium periodate? [*Hint*: Look at the oxyanions of chlorine in Table 5.5.]

5.131 Students commonly confuse ammonia with ammonium. Write the formula and draw dot diagrams for both species and describe the difference between them.

5.132 A student writes the formula for magnesium hydroxide as $MgOH_2$. What is wrong with that formula? How should the student fix it?

5.133 A compound consisting of one oxygen atom and two nitrogen atoms is written N_2O and is called dinitrogen oxide. It is *not* written as ON_2, and it is *not* called oxygen dinitride. Explain why.

5.134 Whereas N_2O is called dinitrogen oxide, Na_2O is called sodium oxide, not disodium oxide. Explain why, and also explain how the name sodium oxide still lets us know that there are two sodium atoms in the formula.

5.135 Many acids can be thought of as anions that have an H^+ ion attached to them. Fill in the following table:

Anion	Anion name	Acid formula	Acid name
F^-			
NO_3^-			
Cl^-			
$C_2H_3O_2^-$			
NO_2^-			

5.136 When do you use the *-ic* suffix and when do you use the *-ous* suffix when naming acids?

5.137 What is the name of the polyatomic ion NO_3^-? What are the names of the acids HNO_3 and HNO_2?

5.138 Acetic acid, sulfurous acid, and phosphoric acid can each be thought of as an anion with one or more H^+ ions attached. For each acid, how many H^+ ions have been attached to the anion?

5.139 Give the molecular formula for hypochlorous acid and perchloric acid.

5.140 What is wrong with the name *hypochloric acid*?

5.141 What is a shortcut rule for determining the number of covalent bonds a representative element from group IVA, VA, VIA, VIIA, or VIIIA can form?

5.142 What is incorrect about this statement: Electron–electron electrostatic forces hold molecules together.

5.143 Why are the noble gases monatomic?

5.144 How many lone pairs of electrons are on the P atom of PCl_3?

5.145 Which one of these four molecules contains a triple bond: F_2, O_3, HCN, or H_2CO?

5.146 Which is the correct Lewis dot diagram for carbonyl fluoride, COF_2? What is wrong with the other two?

(a) :F̤—C(—:Ö:)—F̤: (b) :F̤—C(=:O:)—F̤:

(c) :F̤—O(—:C̈:)—F̤:

5.147 Which is the correct Lewis dot diagram for N_2H_2? What is wrong with the other two?

(a) H—N̈—N̈—H (b) H—N̈=N̈—H

(c) H—N≡N—H

5.148 Draw dot diagrams for all resonance forms of the NCO^- ion.

5.149 Complete the dot diagram for this molecule to determine (a) how many double bonds it has and (b) how many lone pairs of electrons:

H—O—C(—O)—C(—O)—H

5.150 Which one of the following is expected to have resonance forms? Explain your choice.
(a) NH_4^+ (b) HCN (c) CO (d) NO_2^-

5.151 How many lone pairs of electrons are there on Br in BrF_2^+?

5.152 Which of the following represents the correct resonance forms for SO_2? Explain what is wrong with the sets you did not choose.

(a) O=S—O and O—S=O

(b) O=S—O and O—S—O

(c) O=S—O and O—S=O

5.153 Which element has the lowest electronegativity, and how can you answer this question without looking up electronegativity values?
(a) Mg (b) Cl (c) Ca (d) Br

5.154 Which one of the following statements is true about BeH_2:
(a) It is ionic with H as the anion.
(b) It is ionic with H as the cation.
(c) It has polar covalent bonds with a partial negative charge on H.
(d) It has polar covalent bonds with a partial positive charge on H.

5.155 The greater the electronegativity difference between two bonded atoms, the greater the percent ______________ character of the bond.

5.156 The phosphorus atom in PCl_3 should have a
(a) $\delta+$ charge
(b) $\delta-$ charge
(c) 3+ charge
(d) 3− charge
Explain.

5.157 Iodine atoms in I_2 should have a
(a) 1− charge
(b) $\delta-$ charge
(c) $\delta+$ charge
(d) No charge

5.158 Arrange in order of increasing ionic character: CsBr, KBr, PBr_3, $MgBr_2$.

5.159 Complete this table of ionic compounds. Give both old and new names where you can.

Name	Formula
	$AgNO_3$
Aluminum selenide	
Lithium oxide	
Ammonium iodide	
	$CuSO_4$
	$KMnO_4$
	$Ca(ClO_3)_2$

5.160 Complete this table of ionic compounds. Give both old and new names where you can.

Name	Formula
Sodium bicarbonate	
Magnesium acetate	
Barium hypochlorite	
	$Fe(NO_3)_3$
	$(NH_4)_2SO_4$
Calcium phosphate	
	$Co_2(CrO_4)_3$

5.161 Complete this table of molecular compounds.

Name	Formula
	CdTe
Nitrogen triiodide	
	SiI_4
Bromine trifluoride	
Hydroiodic acid	
	HI (as a pure gas)
	S_4N_4

5.162 Complete this table of molecular compounds.

Name	Formula
	XeF_4
	XeF_2
Iodine monochloride	
Bromine trichloride	
Diboron hexahydride	
	N_2O
	S_4O_2

5.163 Which of the following substances are ionic and which are molecular? Name each substance. Draw Lewis dot diagrams for each molecular substance and for the anion of each ionic substance.
(a) Cl_2
(b) $HCl(g)$
(c) NaCl
(d) $Mg(ClO_2)_2$
(e) CH_3OH (methanol)
(f) $Fe(NO_3)_3$
(g) $Pb(C_2H_3O_2)_2$

5.164 How many electrons does the nickel atom lose when forming nickel(II) nitrate? When forming nickel(II) sulfate?

5.165 Why is the oxygen atom listed first in OF_2 but last in Br_2O?

5.166 The use of chlorofluorocarbons (CFCs) in air-conditioning and refrigeration systems has been linked to depletion of the ozone layer of the atmosphere. One of the most widely used CFCs, freon-12, has the formula CF_2Cl_2. It is now being replaced with a compound that has the formula CH_2FCH_3. Draw the dot diagrams for CF_2Cl_2 and CH_2FCH_3. [*Hint*: In CH_2FCH_3, the carbons bond to each other and one of them bonds to fluorine.]

5.167 Why is HF not considered an ionic compound even though its percent ionic character is greater than 50%?

5.168 Give the formula for (a) calcium phosphate, (b) potassium hydrogen phosphate, (c) magnesium cyanide, (d) barium chlorate.

5.169 Name these compounds:
(a) $Fe_2(SO_3)_3$
(b) $Au(NO_3)_3$
(c) NaH_2PO_4
(d) $Pb(CH_3CO_2)_2$

5.170 Hydroxylamine is a compound that contains one nitrogen atom and one oxygen atom connected with a single bond. The remaining atoms are hydrogens. Predict the formula for hydroxylamine and draw its dot diagram.

5.171 Using only the periodic table, arrange these sets of atoms in order of increasing electronegativity: (a) Li, Be, K (b) O, Si, S (c) Br, I, Te

5.172 Name each compound and indicate whether it is ionic or polar covalent:
(a) NH_4I (b) Cl_2O_7 (c) $SrCl_2$ (d) $Li_2Cr_2O_7$

5.173 Give the formula for each compound and indicate whether it is ionic or polar covalent:
(a) Copper(I) sulfide (electronegativity of Cu is 1.9)
(b) Aluminum carbide
(c) Diiodine pentoxide
(d) Chlorine trifluoride

5.174 Arrange the following sets of bonds in order of increasing ionic character. Use the symbols $\delta+$ and $\delta-$ to indicate partial charges, if any, in the bonds: O–Cl, C–F, N–Cl, O–H, S–O.

5.175 Hydrogen sulfide and hydrosulfuric acid have the same molecular formula. How are they different?

5.176 Using only the periodic table, determine which bond in each pair is most ionic:
(a) H–F or H–Cl (b) O–F or C–F

5.177 The polyatomic ion MnO_4^- is listed in your book as permanganate. What would be the name of MnO_3^-?

5 WorkPatch Solutions

5.1 The molecules H_2O (water) and H_2O_2 (hydrogen peroxide) have very different chemical properties. For example, you drink water, but H_2O_2 is a harsh antiseptic and bleach. Different formulas mean different compounds with different properties, even when the elements in the formula are the same.

5.2 The carbon atom has eight valence electrons (two from each bond). Each hydrogen atom has two valence electrons (from the bond that attaches it to carbon).

5.3 Carbon normally has four valence electrons (it is a group IVA atom). In methane it has eight valence electrons, so it has gained four electrons. That it forms four bonds to hydrogen tells you it gains one valence electron with each bond it forms.

5.4 Helium atoms have no unpaired electrons available for sharing and therefore cannot form covalent bonds. In addition, helium atoms already have a filled valence shell (a duet) and so have no "desire" to gain more electrons by sharing with atoms of other elements.

5.5 Every atom in the NF_3 molecule has an octet of valence electrons (the bonding pairs must be double-counted).

5.6

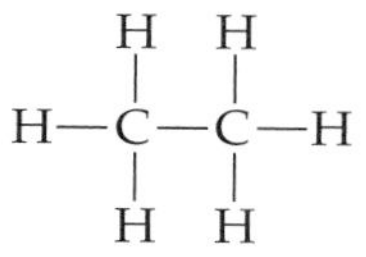

```
    H   H
    |   |
H — C — C — H
    |   |
    H   H
```

5.7 H—C≡C—H

5.8 By moving lone pairs into bonding positions from different atoms.

5.9 Thallium is a metal in group IIIA, and so it loses three electrons to become Tl^{3+}. Oxygen is a nonmetal in group VIA, and so it gains two electrons to become the oxide ion, O^{2-}. To obtain a neutral compound, we must combine two Tl^{3+} ions (for a +6 charge) with three O^{2-} ions (for a −6 charge). The formula is therefore Tl_2O_3.

5.10 The more electronegative element in a binary covalent compound has the partial negative charge, and the suffix *-ide* is associated with the anionic, or negative, part of a compound.

EQUAL
SWEETENER WITH NUTRASWEET
As sweet as 2 tsp sugar
QUADE
Phyllis: We agree
FOR CHAPTER 6
Suggest Highlighting A BLUE
NUTRASWEET ("EQUAL")
(STRUCTURAL,
peptide bond
H2NCH
HOOCCH2
CH2
Asp
Phe

CHAPTER 6

The Shape of Molecules

6.1 Why Is the Shape of a Molecule Important?

In the preceding chapter, we discussed the forces that hold a molecule like methane together and why it has the formula CH_4. Now we'll take a closer look at the way we've been drawing molecules. So far, we've been using dot diagrams to represent molecules and drawing them as flat as a pancake. For example, we represented methane, CH_4, and ammonia, NH_3, as follows:

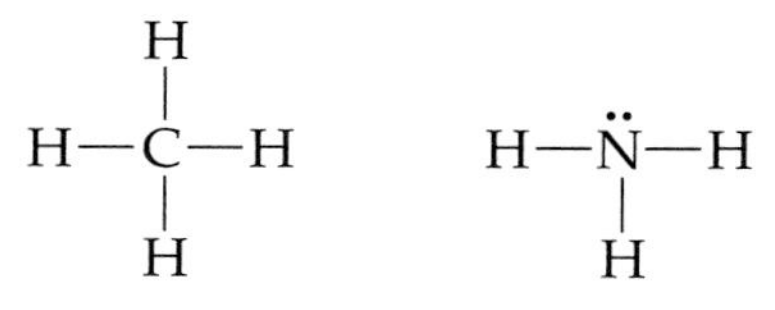

This method was fine because dot diagrams are meant to show only how atoms are connected in a molecule and how the valence electrons are arranged (into bonding pairs and lone pairs). However, neither methane nor ammonia is actually a flat, two-dimensional molecule. Both have much more interesting, three-dimensional shapes.

What do methane and ammonia molecules actually look like? In this chapter, we shall answer this question and learn how to predict the three-dimensional shape of a molecule.

At this point you might be wondering why we should care about the shape of a molecule. After all, we can't see individual molecules. What possible relevance could their shape have? Quite simply, the shape of a molecule could save your life, and most likely it already has. Chemists figured this out shortly before World War II began. In all wars up to that point, the majority of combat deaths were caused not by the wounds inflicted in battle but by the infections that set in shortly thereafter. A soldier with a stomach full of shrapnel would lie in a makeshift field hospital recovering from abdominal surgery. The operation went well, but there was just no way to keep a few microscopic bacteria from entering his bloodstream and reproducing wildly. It was an all too common scenario. The soldier's body would try to fight the infection, but eventually he would die. During World War I, medics could do almost nothing to prevent this, and hundreds of thousands died. But World War II was different. For the first time, a medic could intervene with an effective defense—a drug called sulfanilamide, a trap meant to fool bacteria. The molecular structure of this sulfur-containing compound is similar to that of another molecule found in the bloodstream, *para*-aminobenzoic acid (PABA):

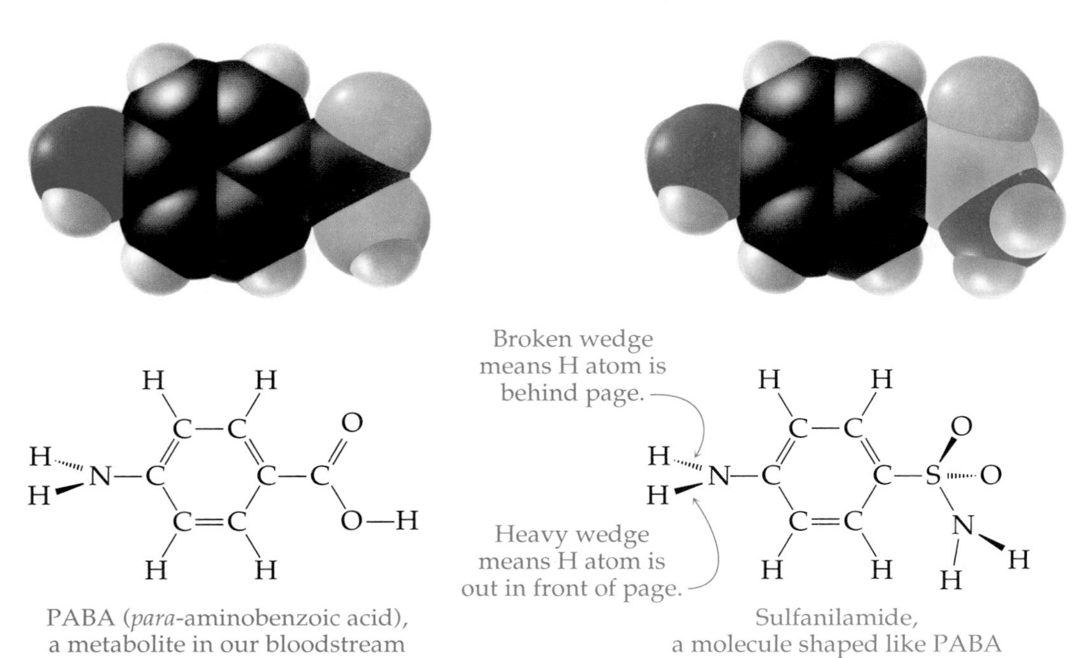
Broken wedge means H atom is behind page.
Heavy wedge means H atom is out in front of page.

PABA (*para*-aminobenzoic acid), a metabolite in our bloodstream

Sulfanilamide, a molecule shaped like PABA

We all have lots of PABA in our bloodstream. PABA is a *metabolite*—that is, a natural product of our body's metabolism. Bacteria can use PABA to their own advantage by combining it with two other molecules also present in our bloodstream. Once connected together, these three molecules form one of the B vitamins—folic acid:

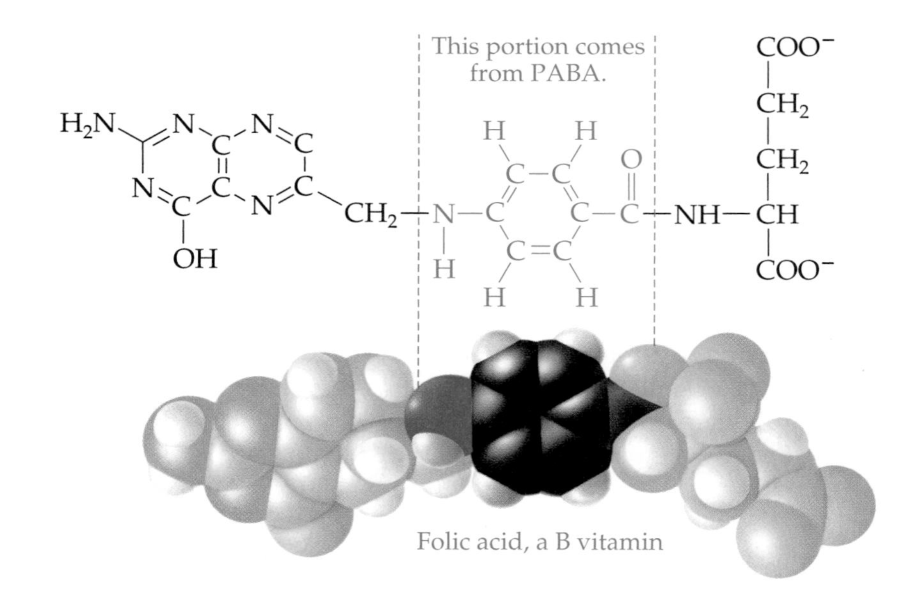
This portion comes from PABA.

Folic acid, a B vitamin

This vitamin is essential for the health and continued functioning of the bacteria. Humans also need this vitamin in order to live, but we have lost the ability to make it ourselves and so must ingest it in the food we eat. It is this difference in the source of folic acid between bacteria and their human hosts that makes sulfanilamide and many other antibiotics the wonder drugs they are (read on).

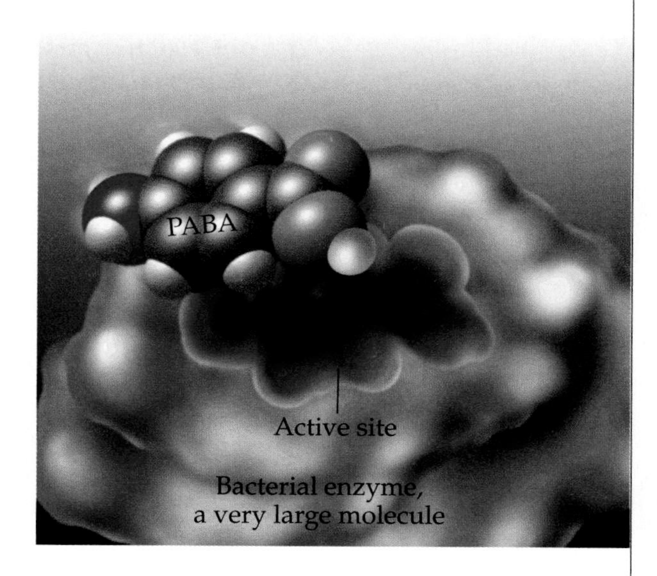

The B vitamin folic acid is created inside the bacterial cell by a specific molecule called an *enzyme*. Enzymes are large protein molecules that control chemical reactions in living systems. They increase the speed of specific reactions that take place in the cells of the body. Enzymes have precisely shaped pockets (called *active sites*) on their surface, pockets that they use to bind with smaller molecules that have complementary shapes. The bacterial enzyme that constructs folic acid from its three precursor molecules has a pocket that is complementary to the PABA molecule.

Efficiently manufacturing their own folic acid, the bacteria can thrive and reproduce in the human bloodstream. But not when sulfanilamide molecules are waiting. Because sulfanilamide has almost the same shape as PABA, the sulfanilamide fits nicely into the pocket of the bacterial enzyme, which is thus tricked into making a defective form of folic acid, one that incorporates sulfanilamide instead of PABA. Starved for the real B vitamin, the bacteria stop replicating, and the infection subsides. Sulfanilamide saved many thousands of lives in World War II and was the precursor to a number of other antibiotics called sulfa drugs.

So, is the shape of a molecule important? Unquestionably! You can be sure that new antibacterial and antiviral agents that have yet to be discovered—including drugs that will one day free humanity of herpes, AIDS, hepatitis, meningitis, and cancer—will have molecular shape at the core of their ability to cure.

6.2 Valence Shell Electron Pair Repulsion (VSEPR) Theory

Of course, no one can build a molecule with a specific shape or even predict what the shape of a molecule will be without first understanding what it is that determines a molecule's shape. **Valence shell electron pair repulsion (VSEPR) theory** is the model most often used to predict molecular shape. It is based on the simple concept that the valence electrons in a molecule repel one another. Let's go back to methane to investigate this. We'll start by drawing a dot diagram for methane:

$$\begin{array}{ccc} & \mathrm{H} & \\ & \vdots & \\ \mathrm{H}\cdot\cdot & \mathrm{C} & \cdot\cdot\mathrm{H} \\ & \vdots & \\ & \mathrm{H} & \end{array} \longrightarrow \begin{array}{ccc} & \mathrm{H} & \\ & | & \\ \mathrm{H}- & \mathrm{C} & -\mathrm{H} \\ & | & \\ & \mathrm{H} & \end{array}$$

All electrons are negatively charged, and so each bonding pair of electrons represents a region of concentrated negative charge. Because like charges repel, the electrons in the four bonds of methane repel one another. To minimize this repulsion, the bonding pairs try to get as far apart from one another as they can:

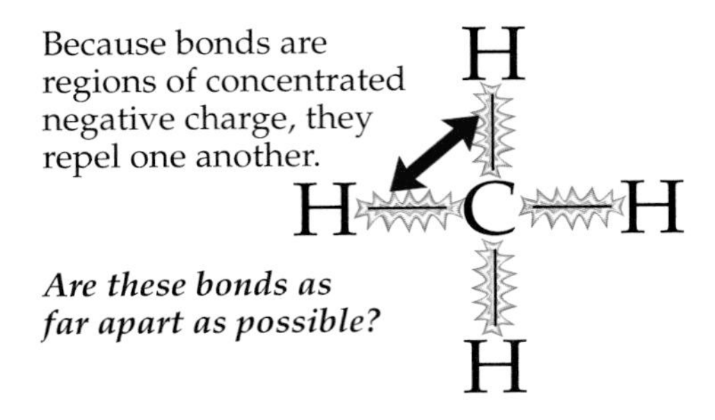

Because bonds are regions of concentrated negative charge, they repel one another.
H
H
C
H
H
Are these bonds as far apart as possible?

Our flat, square version of methane, with its 90° angles, would seem to accomplish that pretty well. Certainly, we can envision other ways of attaching four hydrogens to a central carbon atom that would make the repulsions between the bonds worse. For example, we could clump the hydrogen atoms to make the angles between the bonds 60°:

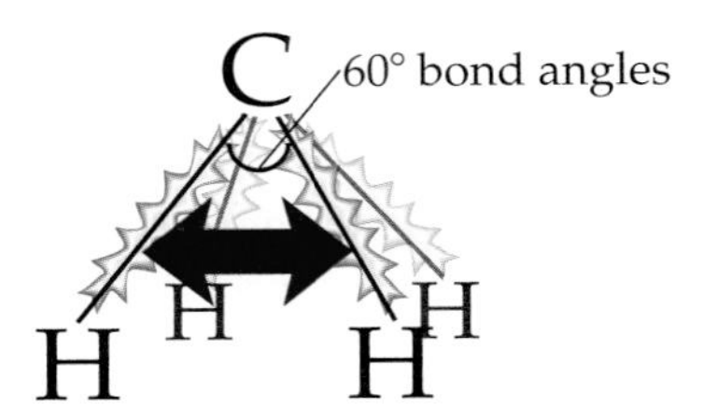

When you cram electron groups together, they REALLY repel one another!

A methane molecule would never assume this shape. With such small bond angles, the bonding pairs of electrons would be practically on top of one another, causing strong repulsion.

While we can easily generate structures that are *worse* than our original square planar (flat) shape, the real question is "Can we do any *better*?" The answer is "Yes, provided we don't insist that our methane molecule be flat." There is a better arrangement, one that opens up the H–C–H bond angles beyond 90° and maximizes the distance between the four bonding pairs of electrons. Here it is:

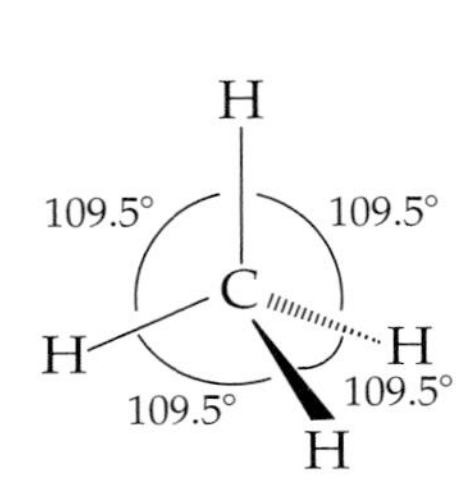

Structure diagram
Bond projects out from page.
Bond goes back behind page.
Bond is in plane of page.

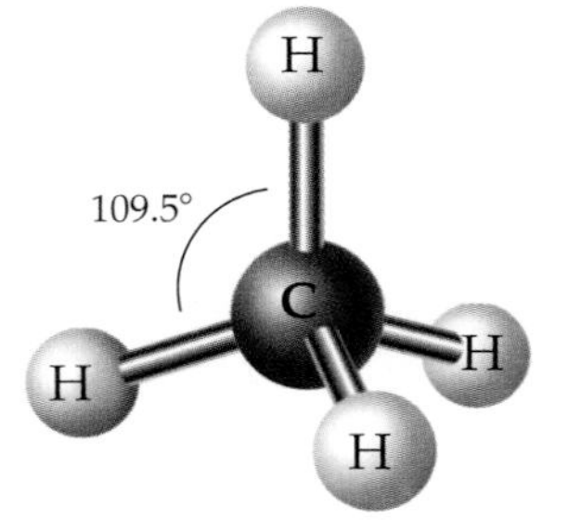

Ball-and-stick model emphasizes bonding pattern.

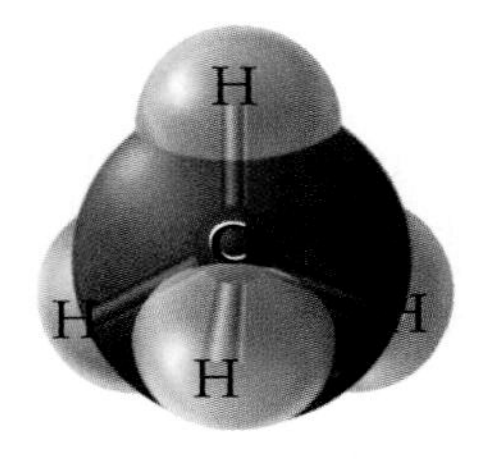

Space-filling model approximates actual molecule shape.

The methane molecule is said to have a *tetrahedral* shape. A **tetrahedron** is a regular four-sided polygon in which the four sides are identical equilateral triangles. The following figure shows a methane molecule inscribed within a tetrahedron:

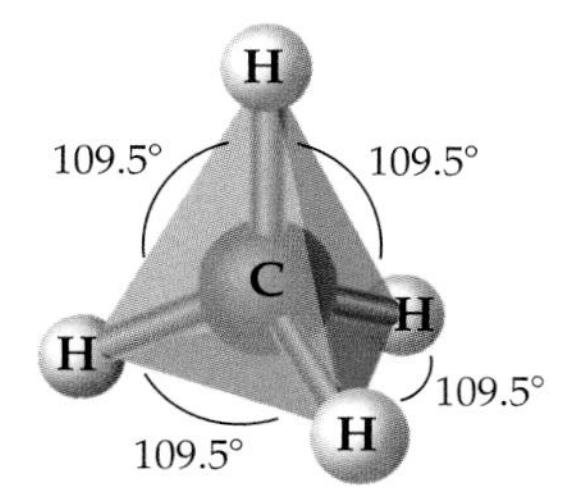

The H–C–H angles are all 109.5°. This is as large as they can be. When methane adopts this nonplanar (not flat) shape, the four bonding pairs are as far apart from one another as possible. This minimizes the repulsion between the electron pairs and thus maximizes the stability of the molecule.

What we have said so far about methane is not specific to that molecule. Any time there are four pairs of electrons around an atom in a molecule, a tetrahedron is the preferred arrangement. This is simply the best way to keep the four groups of electrons as far apart from one another as possible. Once you understand this, you understand one of the most critical factors that determine molecular shape.

Of course, we shall not always be dealing with four pairs of electrons. Let's look at two other organic molecules, formaldehyde (used to preserve biological specimens) and acetylene:

```
   :O:              H   H
    ||              |   |
H—C—H              C≡C
```

Formaldehyde, CH_2O Acetylene, C_2H_2

The dot structures for these molecules have been drawn flat, with arbitrary right angles because, as noted earlier, dot structures are meant to show only the bonds and lone pairs in a molecule, not its actual shape. Once the dot structure of a molecule has been determined, the question of how the bonds and atoms are oriented in space can be answered. We must arrange the electrons in such a way as to minimize the repulsions between them. We start just as we did for methane, by counting the number of bonds around a particular carbon atom, but there is a new twist here. Both of these molecules have multiple bonds (a double bond in formaldehyde, a triple bond in acetylene). The rule for counting multiple bonds for the purpose of VSEPR is to *treat each multiple bond as a single electron group*, just the way we treat a single bond. We do this because all the electrons in a multiple bond occupy roughly the same region of space. This means that there are just three groups of electrons surrounding the carbon atom in formaldehyde, two single bonds and a double bond. Around each carbon atom in acetylene, there are two groups of electrons, one single bond and one triple bond.

Once the counting of electron groups is done, the task of determining the shape of the formaldehyde molecule comes down to answering the question "How can three groups of electrons best arrange themselves around a central (carbon) atom?" In the case of acetylene, the question becomes "How can two groups of electrons best arrange themselves around a central (carbon) atom?" The answer in both situations involves a planar arrangement of the bonds. Can you determine the best angles between these bonds? Try WorkPatch 6.1. If you succeed, you've caught on to VSEPR theory.

6.1 WORKPATCH

Redraw each structure to get the electron groups as far apart as possible. What angles did you use?

(a) 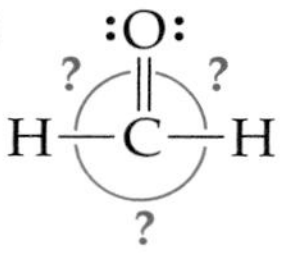 :O: ? ? H—C—H ?   (b) 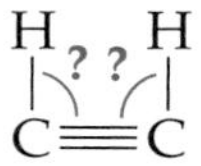 H ? ? H C≡C

Before going on, be sure to compare your answers with the solution given at the end of the chapter.

So far we have looked at three specific molecules and used VSEPR theory to predict their shape. In methane there are four electron groups to accommodate around the carbon atom, in formaldehyde there are three, and in acetylene there are two (counting each multiple bond as a single electron group). But there is nothing special about these molecules. The same results can be expected for any other molecules that contain an atom with these same numbers of electron groups around it, as shown in Table 6.1.

Because determining the shape of a molecule requires knowing the number of electron groups that surround each atom, you really can't get started until you have a good dot diagram in front of you. This means that when you are asked to determine the shape of a molecule, you must first draw a dot diagram if one has not been supplied.

Table 6.1 VSEPR Arrangements of Electron Groups Around an Atom

Number of electron groups	Geometry of electron groups	Corresponding molecular shape	Example
Two	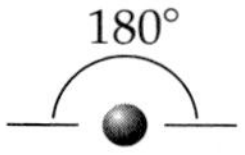 180° Bond angle 180°	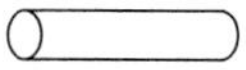 Linear	C=O=C  Carbon dioxide, CO_2
Three	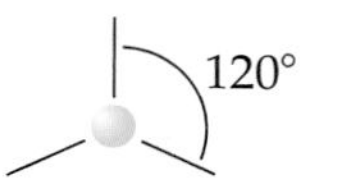 120° Bond angles all 120°	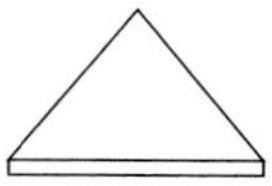 Trigonal planar	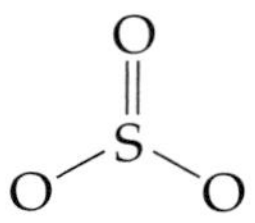 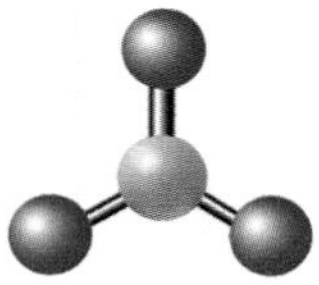 O, S, O, O Sulfur trioxide, SO_3
Four	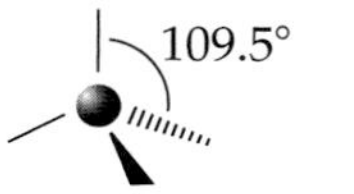 109.5° Bond angles all 109.5°	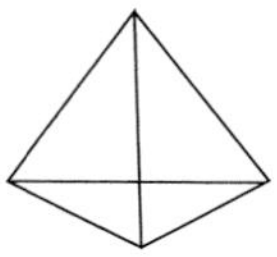 Tetrahedral	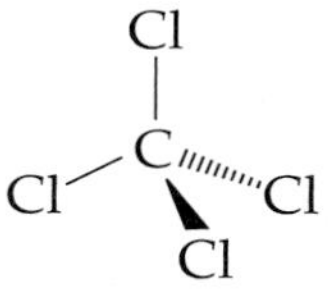 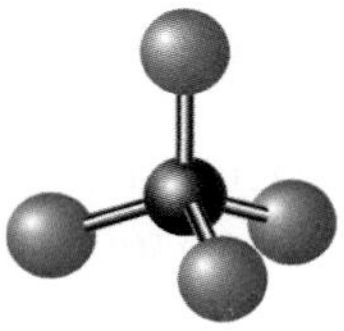 Cl, C, Cl, Cl, Cl Carbon tetrachloride, CCl_4

Practice Problems

For all these problems, describe the geometry of the electron groups and name the molecular shape resulting from that geometry. Also, draw the molecule, and label the size of all bond angles in your drawing.

6.1 CO_2.

Answer: First draw a dot diagram, then examine it via VSEPR:

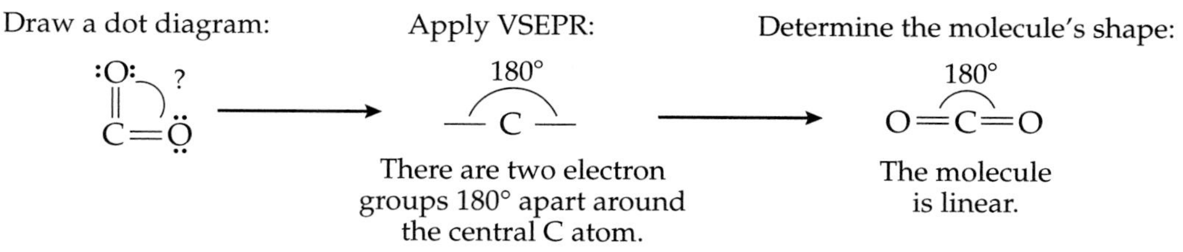

Draw a dot diagram: → Apply VSEPR: → Determine the molecule's shape:

:O: ? C=O (dot diagram) → 180° —C— There are two electron groups 180° apart around the central C atom. → 180° O=C=O The molecule is linear.

Note: Don't worry about the arrangement of the electron pairs about the peripheral (outer) atoms of a molecule. Only the electrons on the inner atoms affect the shape.

6.2 NH_4^+ cation.

6.3 C_2Cl_2, which is connected Cl–C–C–Cl.

6.4 $COCl_2$, in which all atoms are connected to C, and there is a carbon–oxygen double bond.

So far, none of the molecules we have considered had lone pairs of electrons on the central atom. It's now time to consider these molecules. A good example is the ammonia molecule, NH_3. Ammonia is an interesting substance. It exists as a gas at room temperature, but you are probably more familiar with it dissolved in water as a household cleaner. It is the fifth most commonly produced chemical in the United States, used in greater quantities as an agricultural fertilizer than as a household grease-cutter. What is its shape? To answer this question, we have to decide how the electron groups are arranged in space around the central nitrogen atom, so we need a dot diagram to begin:

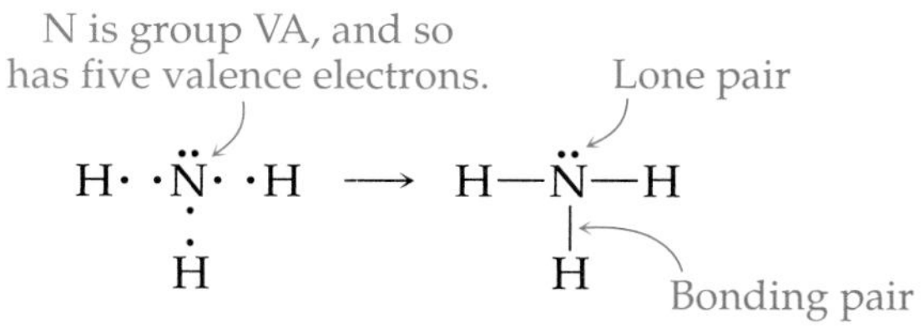

N is group VA, and so has five valence electrons.

H· ·N̈· ·H (with ·H below) → H—N̈—H (with H below)

Lone pair

Bonding pair

As usual, the dot diagram is drawn with a planar shape. Now focus on the central nitrogen atom and count the total number of electron groups that surround it: three bonding pairs plus one lone pair. The lone pair must be included in the total count because it is negatively charged, just like a bonding pair, and thus repels the other electron groups. Therefore, much like the carbon in methane or any other atom with four electron groups around it, the electron groups on the nitrogen adopt a tetrahedral arrangement in order to minimize repulsion:

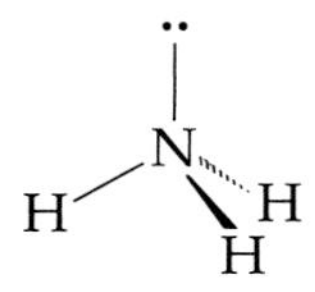

So, the shape of the ammonia molecule is tetrahedral, right? Well, not quite. Ammonia is not considered to be a tetrahedral molecule. The above drawing is perfectly correct, but we describe the shape of a molecule differently when lone pairs are present. The convention is to include lone pairs in our initial count of electron groups in determining the special arrangement of these electron groups (in other words, how they are arranged in space relative to one another), *but when it is time to describe the molecule's shape, we ignore the lone pairs.* This is done because ultimately we want the name of the molecular shape to indicate the arrangement of the *atoms* in a molecule, not the electrons.

Count all electrons when determining the electron-group geometry.

How then do we describe the shape of the ammonia molecule? Ignoring the lone pair of electrons, we see that the molecule is a tetrahedron with the top part removed. What's left looks likes a pyramid, and so the ammonia molecule is said to have a pyramidal shape:

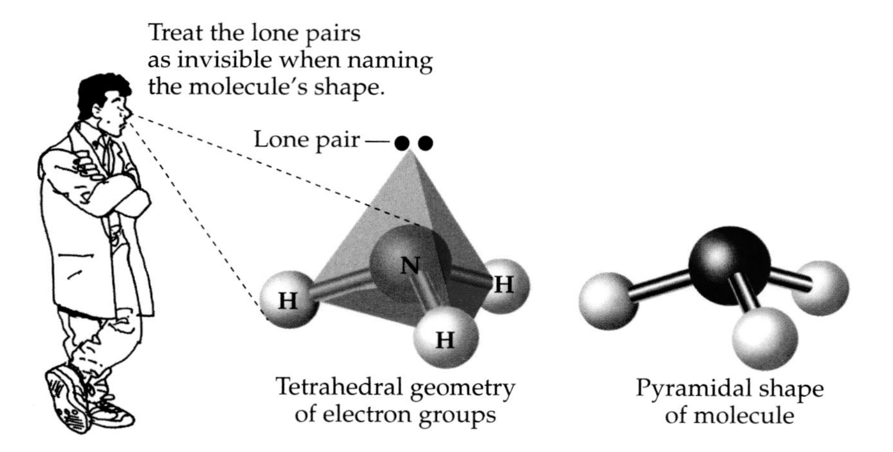

(If you're having trouble seeing the difference between a tetrahedron and a pyramid, remember from our definition of tetrahedron that all four sides are *identical* triangles. When we ignore the lone pair in NH_3, however, we essentially lop off the top of three triangles. Because we've not touched the base triangle, however, the four faces of our polyhedron are no longer identical. In our tetrahedral shape all four faces are identical, but in our pyramidal shape only three of the four faces are identical.)

We can therefore say that, even though the geometry of the electron groups about N is tetrahedral, *the shape of the molecule is not tetrahedral but rather pyramidal.* Don't forget: the shape of a molecule describes how the *atoms* are arranged, not how the electron groups are arranged. Therefore, we use VSEPR theory and *all* of the electron groups (bonding groups plus lone pairs) to predict how the electrons should be arranged relative to one another, but then treat the lone pairs as if they were invisible when describing the shape of the molecule.

Let's take this one step further by asking what the H–N–H bond angles in ammonia are. To answer this question, think about where we started, before we ignored the lone pair. The four electron pairs surround the N atom in a tetrahedral geometry. Therefore, the H–N–H angles in ammonia should be close to the standard 109.5° tetrahedral angle:

All the angles should be close to the 109.5° value expected for a tetrahedral geometry. But are they really 109.5°?

The experimentally determined H–N–H bond angles in ammonia are actually a little bit smaller than the ideal 109.5° tetrahedral angle because the electrons in a lone pair take up more space than the electrons in a bonding pair. This tends to squeeze down on the bonding pairs, compressing the angles. A useful rule of thumb is that each lone pair *on a period 2 atom* compresses the remaining bond angles around that atom by approximately 2°. Thus, we predict the H–N–H bond angles in ammonia to be approximately 107°, which is in good agreement with experiment results.

In summary, follow steps 1–4 to use VSEPR theory to determine the shape of a molecule:

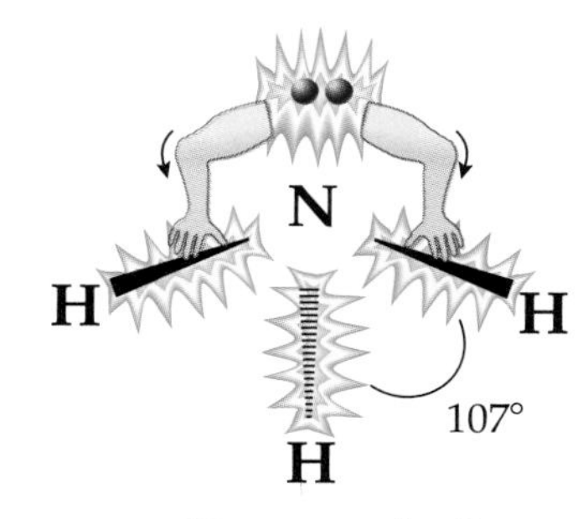

"Squeeze play"

The actual H–N–H bond angles are 107°, slightly smaller than the value for perfect tetrahedral geometry.

Using VSEPR theory to determine molecular shape

Step 1 Draw a dot diagram for the molecule.

Step 2 Count the number of electron groups around the central atom. Include lone pairs, and count each multiple bond as a single electron group.

Step 3 Using the count from step 2, decide on the best arrangement for the electron groups, including lone pairs, around the central atom.

Step 4 Pretend the lone pairs are invisible and describe the resulting shape of the molecule.

Let's see if you've got it. We'll work part way through the determination of the shapes of water and ozone, and then we'll ask some questions. We begin as usual with electron dot diagrams:

H—Ö—H (Water)   Ö=Ö—Ö: (Ozone)

Focus on the central atom in each molecule, and count the total number of electron groups around it (remember that multiple bonds count only once). The oxygen atom in water is surrounded by four electron groups (two bonding pairs and two lone pairs). The central oxygen atom in ozone has three electron groups around it (a double bond, a single bond, and a lone pair). How do these groups arrange themselves in space? VSEPR theory tells us what the arrangement must be:

Water

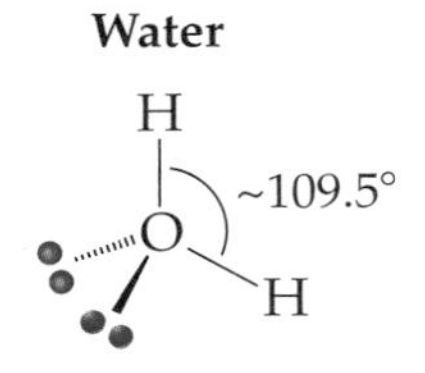

Four electron groups are arranged tetrahedrally around the oxygen atom.

Ozone

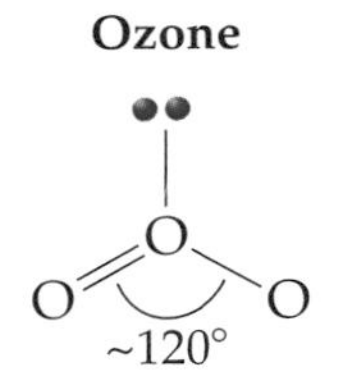

Three electron groups are arranged around the oxygen atom in a planar triangle.

How would you describe the shape of the water and ozone molecules? Follow the convention! Ignore the lone pairs on each central atom. Cover them up and use a word or phrase that best describes the shape created by the atoms.

What did you call the shapes of these molecules? Certainly they are not linear because that would mean all three atoms must lie on a single straight line. They should not be called trigonal planar, our second choice from Table 6.1, because we reserve that term for cases when there are three atoms surrounding a central atom. Here the molecules are again without the lone pairs:

Check the WorkPatch answer at the end of the chapter to see how the shape of these molecules is most commonly described.

Finally, you should have noticed that the water and ozone molecules have essentially the same shape, but there is an important difference between them. The H–O–H angle in water is quite different from the O–O–O angle in ozone. If you can predict these two angles, you are ready to go on.

The drawings preceding WorkPatch 6.2 give the ideal values for the bond angles in the molecules. Recalling that there is roughly a 2° compression from the ideal VSEPR angle for every lone pair on the central atom, give a more accurate bond angle for each molecule.

Applying steps 1–4 for determining the shape of a molecule to all possible cases of four, three, and two electron groups, we obtain Table 6.2.

To make sure you understand how Table 6.2 was generated, try the next WorkPatch and the following practice problems.

We left three possible situations out of Table 6.2:

1. Four electron groups: one bonding group and three lone pairs
2. Three electron groups: one bonding group and two lone pairs
3. Two electron groups: one bonding group and one lone pair

These are considered trivial cases. Why? What shape(s) do these bonding situations give rise to?

Table 6.2 VSEPR Molecular Shape Table

Four electron groups		Three electron groups		Two electron groups	
Electron-group arrangement	**Molecular shape**	**Electron-group arrangement**	**Molecular shape**	**Electron-group arrangement**	**Molecular shape**
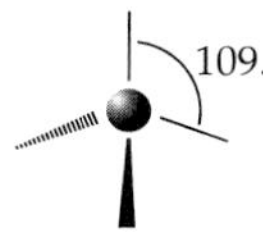 109.5° Four bonding groups No lone pairs Tetrahedral	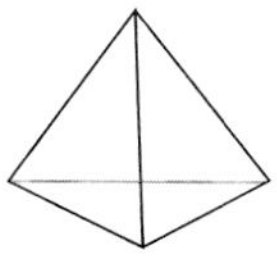 Tetrahedral	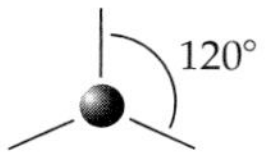 120° Three bonding groups No lone pairs Trigonal planar	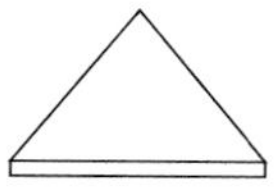 Trigonal planar	 180° Two bonding groups No lone pairs Linear	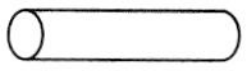 Linear
Bond angles all 109.5°		Bond angles all 120°		Bond angle 180°	
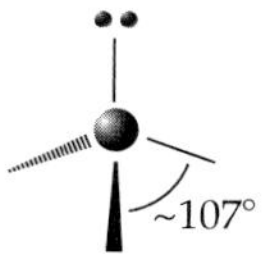 ~107° Three bonding groups One lone pair Tetrahedral	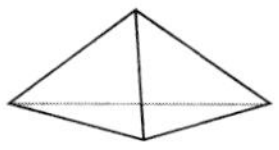 Pyramidal	 ~118° Two bonding groups One lone pair Trigonal planar	 Bent		
Bond angles compressed to ~107° by lone pair*		Bond angle compressed to ~118° by lone pair*			
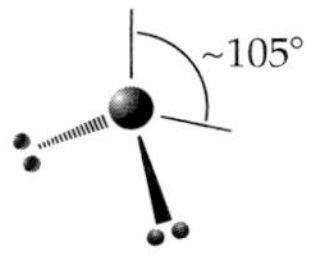 ~105° Two bonding groups Two lone pairs Tetrahedral	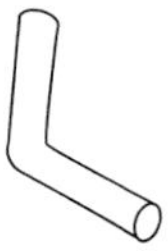 Bent				
Bond angle compressed to ~105° by lone pair*					

Key:
- ● **Central atom**
- **Bonding electron group (single, double, or triple covalent bond)** 
- •• **Lone pair of electrons**

* For elements in period 2, the bond angles around a central atom are compressed by about 2° for every lone pair.

Practice Problems

For each molecule or polyatomic ion in the following five problems, (a) draw the electron-group geometry, (b) name the shape of the molecule, and (c) estimate the size of all the bond angles in the molecule or ion. (Some of the molecules and ions have lone pairs on the central atom.)

6.5 PBr_3.

Answer: First draw a dot diagram (which must show 26 electrons):

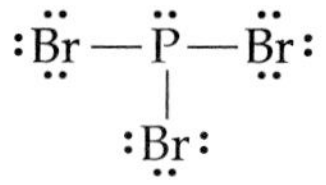

```
 ..    ..   ..
:Br — P — Br:
 ..    |    ..
      :Br:
       ..
```

Next, count electron groups about the central atom. There are four (three single bonds and one lone pair), indicating a tetrahedral geometry of these groups about the P atom. However, because you must ignore the lone pair on P, the molecule is pyramidal in shape, with a predicted Br–P–Br angle of less than 109.5°. (You might predict 107°, but P is not a period 2 element. The actual bond angle is 102°.)

P
Br Br Br

Pyramidal molecule

6.6 SO_4^{2-}.

6.7 HCN, in which hydrogen and nitrogen are attached to carbon.

6.8 Sulfur dioxide, SO_2, a pollutant that comes from burning coal contaminated with sulfur.

6.9 HNO_3, in which hydrogen is attached to one of the oxygens and all oxygens are attached to nitrogen. To simplify, ignore the hydrogen when you name the molecule's shape.

6.3 Polarity of Molecules, or When Does 2 + 2 Not Equal 4?

We have already seen how the shape of a molecule can be used to trick bacteria. Molecular shape is important in many other ways as well. The chemical and physical properties of molecules are often dependent their on shape.

Molecular "stickiness" is one property that is very dependent on molecular shape. Molecules can "stick" to each other so tightly that they cannot easily be pulled apart from each other. Try it! The next time you have a drink with some ice in it, take out an ice cube, grab it with both hands, and try ripping it in half. You can't. The trillions of individual water molecules that make up your ice cube are all stuck to one another as if someone had used superglue. This simple experiment is proof of the existence of a large net attractive force holding together all the water molecules in the ice cube. Such forces between molecules are called **intermolecular attractive forces**, or *intermolecular forces* for short. If the intermolecular forces between water molecules could somehow be turned off, the ice cube would come apart and instantly turn into a gas (water vapor). It is because of these intermolecular forces that molecular substances can exist in condensed phases (liquid and solid). So where do these intermolecular forces come from? What factors determine how strong or weak they are? Molecular shape plays a key role here, so let's look at a few examples.

We'll start with two simple molecules, H_2 and HF. Both have a linear shape (all diatomic molecules are linear). The biggest difference between these two molecules, besides their elemental composition, is in their bonding. The H_2 molecule is held together by a covalent bond (equal sharing of electrons). In

HF, the very different electronegativities of hydrogen and fluorine give rise to a polar covalent bond (unequal sharing of electrons). This unequal sharing means that the bonding electrons spend more time around the F atom and less time around the H atom. That inequality produces a molecule with a partial negative charge ($\delta-$) on the more electronegative F atom and a partial positive charge ($\delta+$) on the less electronegative H atom. If you are not sure about this, stop and review electronegativity and polar covalent bonding in Chapter 5.

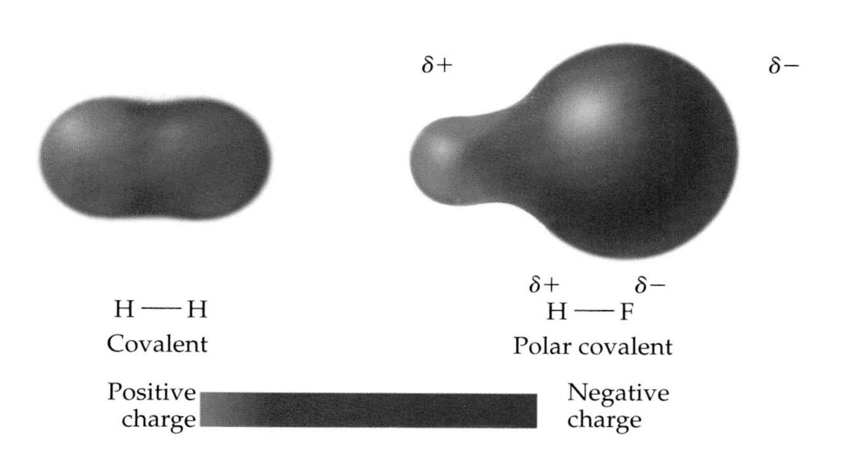

Because of the presence of partial charges, the behavior of HF molecules is in some ways significantly different from the behavior of H_2 molecules. One of these differences can be seen when samples of these molecules are placed between two oppositely charged metal plates. Here is what happens:

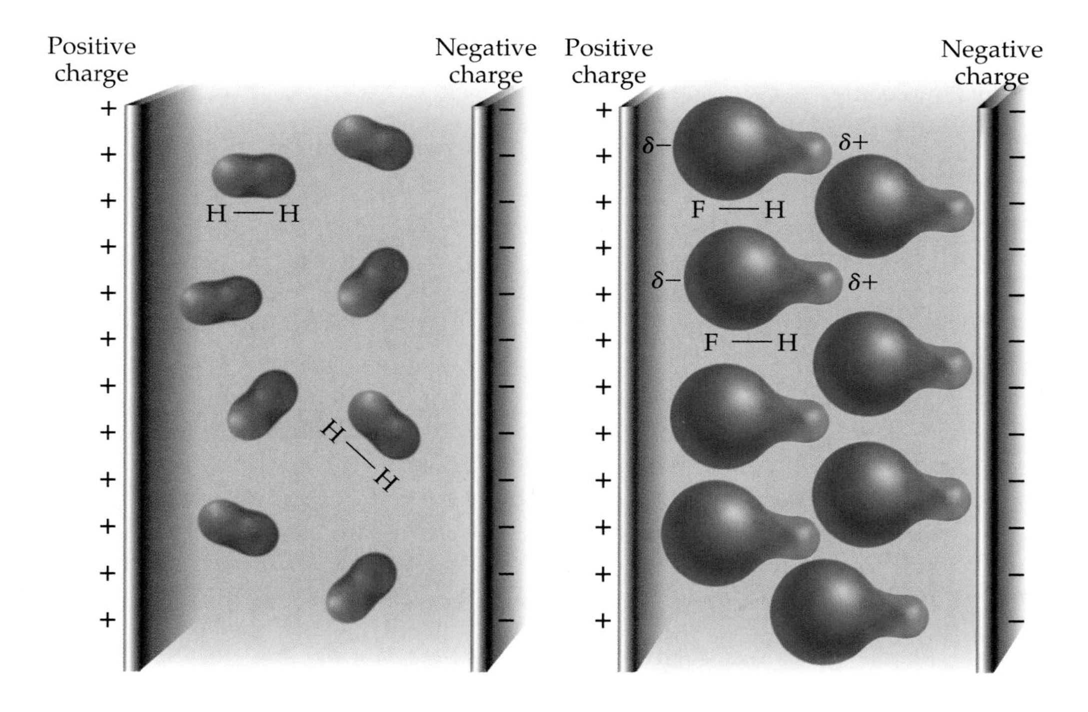

The H_2 molecules are not noticeably affected by the charged plates, and they orient themselves randomly between the plates. The situation is different for the HF molecules. Here the partially negative fluorines are attracted to the positive plate and the partially positive hydrogens are attracted to the negative plate, and the molecules all assume the same orientation. Based on this behavior, HF is called a **polar molecule** and H_2 is called a **nonpolar molecule**. (This is an *operational definition*, which is a definition based on how something behaves in a particular situation.)

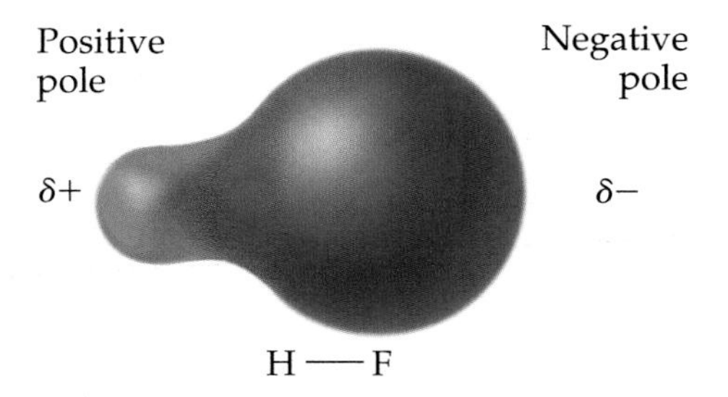

A polar molecule has two charge poles.

Diatomic molecules like N_2 and O_2, the two most abundant molecular substances in our atmosphere, are nonpolar because the bonding electrons are shared equally between the atoms in each molecule. Diatomic molecules like HF and HCl are polar owing to the polar covalent nature of their bonds. Polar molecules are not all equally polar, however. To understand why this is so, look at the electronegativities of the atoms in HF and HCl:

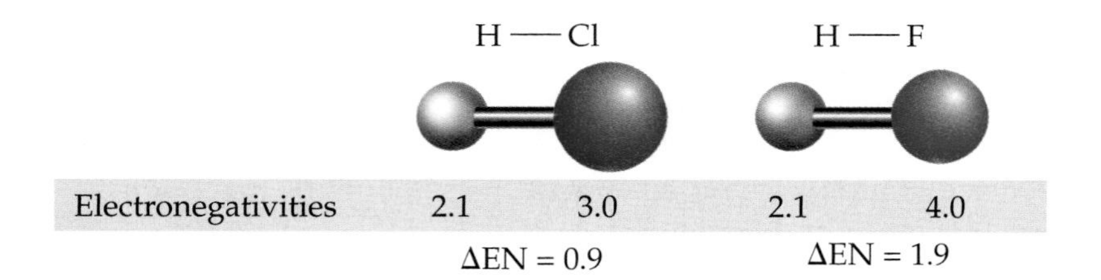

Both these molecules are polar, but HF is more polar than HCl. In both molecules the bonding electrons are shared unequally, but fluorine is more electronegative than chlorine (recall that fluorine is the most electronegative element there is). The greater value of ΔEN for HF tells us that fluorine really hogs those shared electrons. Thus, the bonding electrons are shared more unequally in HF than in HCl, meaning that HF molecules have larger partial charges than do HCl molecules. HF molecules therefore are more polar than HCl molecules.

One way we might show this difference would be to use larger type for the partial charges in HF:

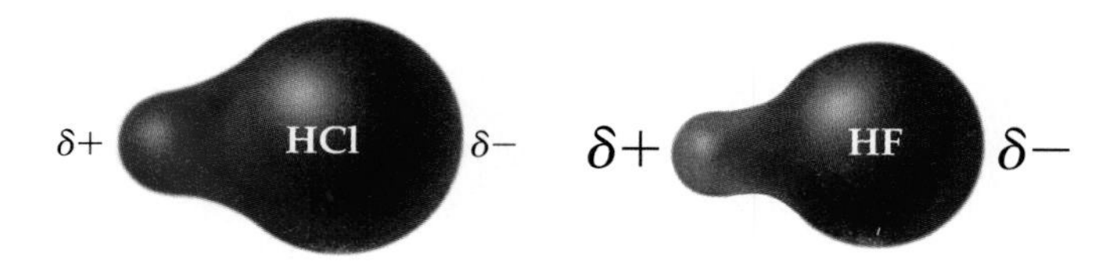

However, we don't do it that way. The convention for indicating the polarity of a bond (and even, as we shall see, of the molecule as a whole) is to use an arrow that points from the δ+ charge to the δ− charge. The arrowhead points to the negative end of the bond, and the arrow is crossed at the positive end:

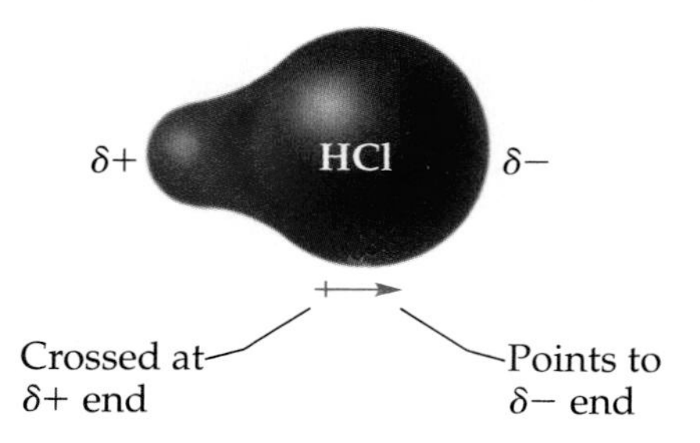

This arrow represents a property of the bond called the **dipole moment**, which is simply a measure of how polar the bond is. The longer the arrow (the larger the dipole moment), the more polar the bond (the larger the δ+ and δ−

charges). Physicists refer to these arrows as *vectors*. (A **vector**, usually represented by an arrow, can best be thought of as a number that has a direction associated with it.)

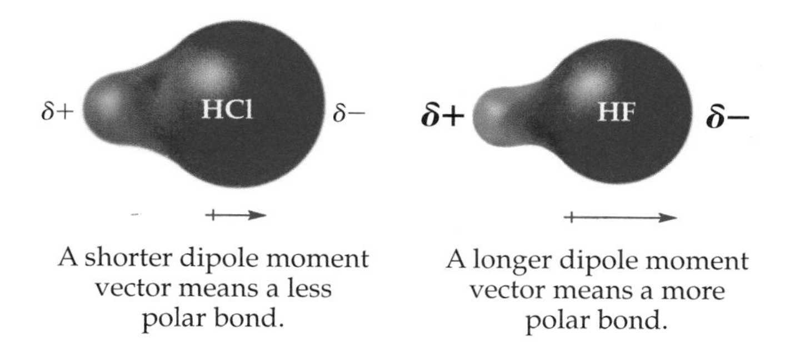

A shorter dipole moment vector means a less polar bond.

A longer dipole moment vector means a more polar bond.

When does a bond have a dipole moment of zero?

Here is a hint for the WorkPatch. The dipole moment will be zero only when the two atoms have no partial charges. When does this happen? What does a dipole moment of zero mean for the bond? Check your answer against ours at the end of the chapter.

Notice that the WorkPatch asked you about the dipole moment (polarity) associated with a bond. There can also be a dipole moment associated with an *entire molecule*. It is called a **molecular dipole moment**, and it points from the δ+ end of the entire molecule to the δ− end. If a molecule is polar, it has a molecular dipole moment. For simple molecules such as H_2, HCl, and HF, considering the polarity of the bond is the same as considering the polarity of the entire molecule because there is only one bond in the molecule. But what about molecules with more than one bond? In these cases, determining whether the entire molecule is polar and therefore has a molecular dipole moment becomes a bit more involved and molecular shape plays a role!

Let's consider two molecules that have more than one polar bond—water, H_2O, and dichloroacetylene, C_2Cl_2:

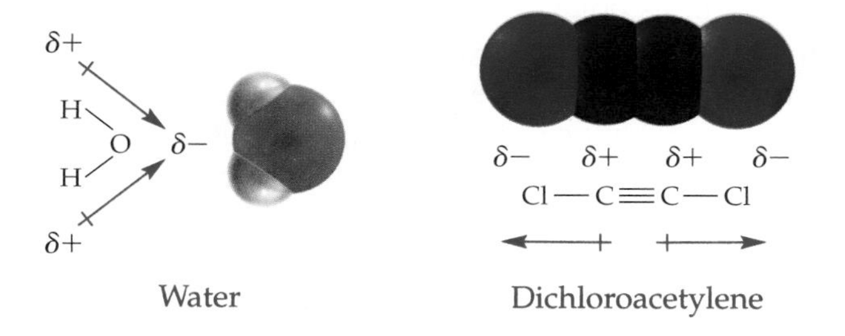

Water

Dichloroacetylene

Notice a few things about these examples. As always, the individual bond dipole moment vectors point from the δ+ to the δ− end of the bond, and the end of the arrow over the δ+ atom is crossed, like a plus sign. There is no bond dipole moment vector over the carbon–carbon triple bond in dichloroacetylene because this bond is nonpolar (electrons in this bond are shared equally). Both molecules contain two polar covalent bonds: the O–H bonds in water and the C–Cl bonds in dichloroacetylene. So, are both molecules polar?

The only way to truly know is to put both molecules between oppositely charged plates and see what happens. Here are the results of just such an experiment:

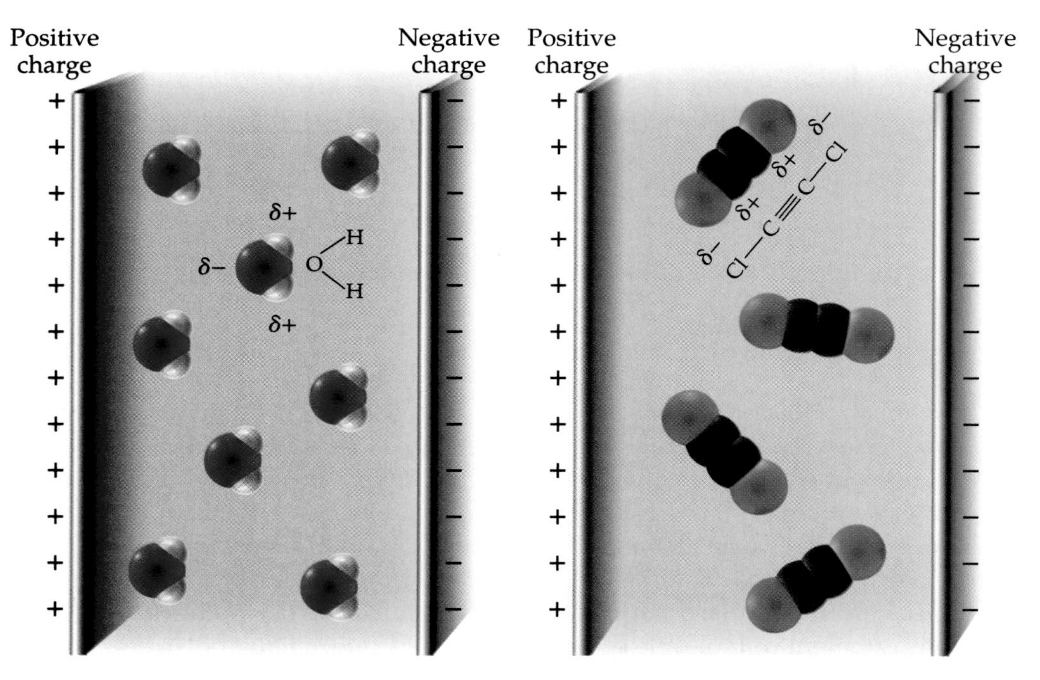

The molecules all assume the same orientation between the charged plates, meaning H_2O must be polar.

The molecules are oriented randomly, meaning C_2Cl_2 must be nonpolar.

Notice how the water molecules all assume the same orientation. The dichloroacetylene molecules, on the other hand, remain randomly oriented. By our operational definition of polar molecules, water is considered a polar molecule but dichloroacetylene is nonpolar. Nonpolar? Can dichloroacetylene be nonpolar even though it contains polar covalent bonds with $\delta+$ and $\delta-$ ends? The answer is yes. A molecule can be nonpolar even though it contains polar bonds, and you must be able to predict when this will occur. Making this prediction requires that you add up all the individual bond dipole moments.

The trick in doing this is to recall that each individual bond dipole moment is a vector quantity, which is a quantity that has both a numeric value and a direction associated with it (the direction of the vector). When you add the numbers 2 + 2, you always get 4. But when you add two vectors, each of length 2, the answer you get depends on which way the two vectors are pointing relative to each other. In vector arithmetic, 2 + 2 can equal anything from 0 to 4:

Vectors pointing in the same direction add:

$$\begin{array}{r} \overrightarrow{2} \\ + \ \overrightarrow{2} \\ \hline \overrightarrow{4} \end{array}$$

Vectors pointing in opposite directions cancel:

$$\begin{array}{r} \overrightarrow{2} \\ + \overleftarrow{2} \\ \hline 0 \end{array}$$

Vectors at an angle to each other add to give a *resultant vector* having a length that can have any value between zero and the sum of the lengths of the two original vectors. This resultant-vector length depends on the angle between the two original vectors:

2
140°
+ 2
0.68

2
90°
+ 2
2.8

You need not worry about how to calculate the length of the resultant vectors. Just note that the length in both cases is between 0 and 4.

If the vectors have the same value but point in opposite directions, they cancel each other and give zero as a result, as shown at the top of this page. This is exactly what happened in dichloroacetylene. The individual bond dipole moment vectors have equal lengths, but because the molecule has a linear shape, they point in opposite directions, as shown below. They therefore completely cancel and give a value of zero. Therefore, even though two of the bonds in the dichloroacetylene molecule are polar, we predict that the molecule as a whole will be nonpolar (which it is).

A case where individual bond dipole moment vectors cancel:

Cl—C≡C—Cl

Nonpolar
The dipole moment vectors add up to zero (they cancel each other).

$$\begin{array}{r} \longrightarrow \\ + \longleftarrow \\ \hline 0 \end{array}$$

In the case of the water molecule, because of its bent shape, the two O–H vectors do not cancel each other. Instead, they add to give a resultant vector that points in the direction shown below—between the two original vectors. This vector represents the molecular dipole moment for the entire molecule. Because this vector is not zero, we predict that the molecule will be polar (which it is), and we call water a polar molecule.

A case where individual bond dipole moment vectors do not cancel:

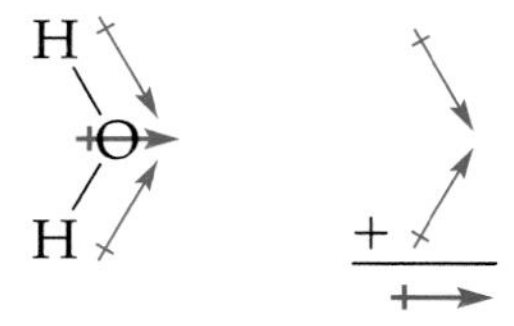

The dipole moment vectors for the O–H bonds add to give the dipole moment vector for the entire molecule.

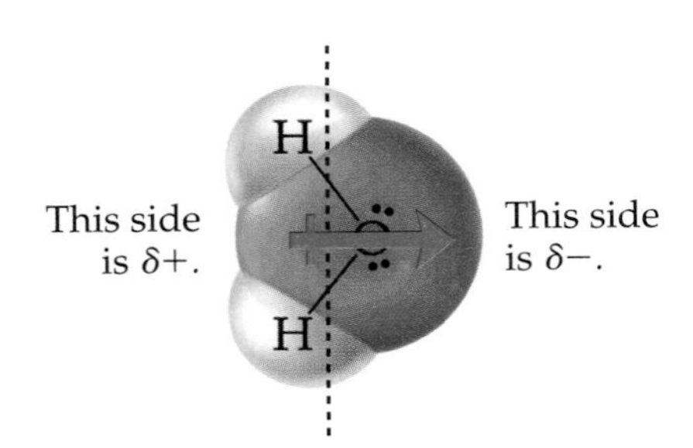

The molecular dipole moment vector points from the δ+ side of the entire molecule to the δ− side of the entire molecule as shown at left.

So molecular shape is important once again. For molecules with more than two atoms, just having polar bonds is not enough to make the entire molecule polar. For these molecules, molecular polarity also depends on molecular shape. Try the following WorkPatch to see if you are catching on to these concepts.

Which molecule is more polar? [Be careful!]

(a) $\ddot{\underset{\cdot\cdot}{O}}=C=\ddot{\underset{\cdot\cdot}{O}}$ (b) $\ddot{\underset{\cdot\cdot}{S}}=C=\ddot{\underset{\cdot\cdot}{S}}$

Which molecule is more polar?

(c) H–Ö–H (bent) (d) H–S̈–H (bent)

The molecules in (a) and (b) of the WorkPatch are a bit tricky. Did you catch it? The molecules in (c) and (d) should have you thinking about the length of the vectors drawn over the bonds.

Adding up all the individual bond dipole moment vectors can be tricky for molecules with more than three atoms. However, there is a quick way to determine whether or not they cancel by doing a "tug-of-war test." To perform this test, imagine you are standing at the central atom. Let each bond dipole moment vector be a rope that tugs on you in the direction of the vector. The longer a vector is, the harder that rope pulls on you. Now consider two molecules, SO_3 and $SOCl_2$. Both have the same shape (trigonal planar), and both have three polar covalent bonds:

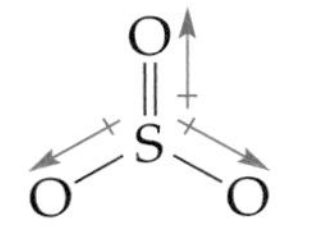

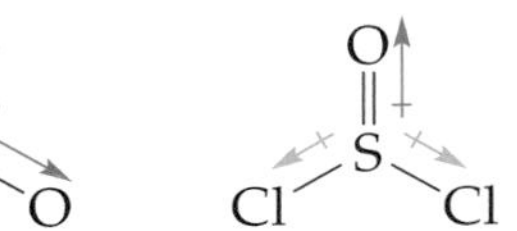

Don't worry about the bonds being single or double. Focus only on the vectors (the "ropes"). As all the ropes pull on you simultaneously, ask the question "Will I move?" If the answer is no, the bond dipoles cancel and the entire molecule is nonpolar. If the answer is yes, the bond dipoles do not cancel and the molecule is polar. For SO_3, the three equal-strength tugs to the corners of an equilateral triangle leave you sore but unmoved. That means the individual bond dipole moments cancel. Therefore SO_3 is a nonpolar molecule even though all of its bonds are polar! However, the tugs for $SOCl_2$ are not of equal strength. The stronger tug along the S–O bond is not canceled by the two weaker tugs along the S–Cl bonds. You move because of this difference. Because the individual bond dipole moments do not cancel, $SOCl_2$ is a polar molecule.

meaning the SO_3 molecule is nonpolar.

meaning the $SOCl_2$ molecule is polar.

Because the length of a bond dipole moment vector determines the strength of the tug for our tug-of-war test, it is critical that you don't forget what it is that determines the vector length.

Why is the upward tug in $SOCl_2$ slightly stronger than either of the downward tugs?

This WorkPatch is extremely important. Don't go on until you understand the answer.

The above examples teach you that three equal-length dipole moment vectors in a trigonal planar molecule cancel to leave the entire molecule nonpolar. What about tetrahedral molecules? To answer this question, let's look at carbon tetrachloride, CCl_4, a tetrahedral molecule:

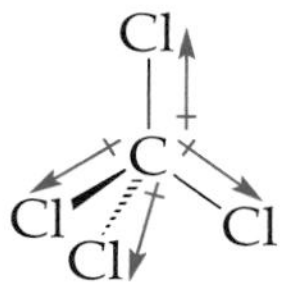

Carbon tetrachloride

The electronegativity difference between C and Cl (remember that Cl is one of our four electron hogs from Chapter 5—see page 184) ensures that all four bonds are polar covalent. In addition, all four bond dipole moment vectors have equal length. Do the dipole moments cancel? To find out, let's look at our molecule in a different orientation:

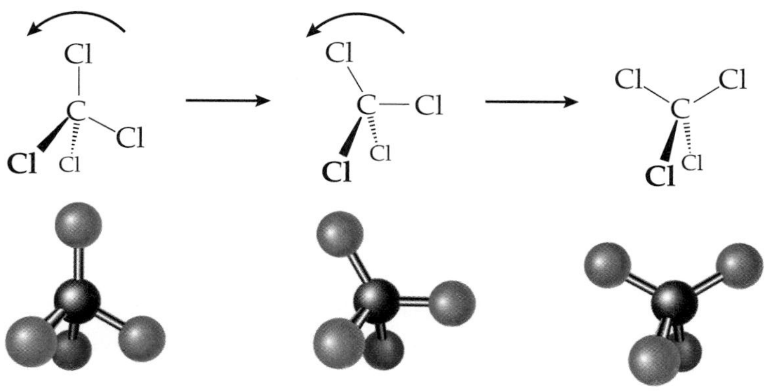

Drawn this way, two tugs are directed upward (to the left and right) and two are directed downward (toward the front and back). Our assistant at the right kindly demonstrates this for you. Once again, such combinations of tugs cancel (our assistant does not move). Carbon tetrachloride therefore is not polar even though its individual bonds are. Remove one of those tugs, however, and the story changes, as you'll see in Practice Problem 6.10.

The tug-of-war test can be very useful in determining the polarity of larger molecules. However, the test is useless unless you can first determine the shape of the molecule and estimate the relative strengths of the different tugs. Try the following practice problems, keeping in mind that for purely covalent bonds, the dipole moment is zero. In addition, you may approximate and treat C–H bonds as nonpolar because C and H have very similar electronegativities.

Practice Problems

6.10 Is phosphorus trichloride, PCl_3, a polar molecule? If it is, draw the dipole moment vector for the entire molecule and show where the $\delta+$ and $\delta-$ regions of the molecule are.

Answer: Yes, it is a polar molecule. PCl_3 *should have 26 electrons.*

VSEPR

The P atom feels three equal downward tugs, and so by the tug-of-war test the molecule is polar.

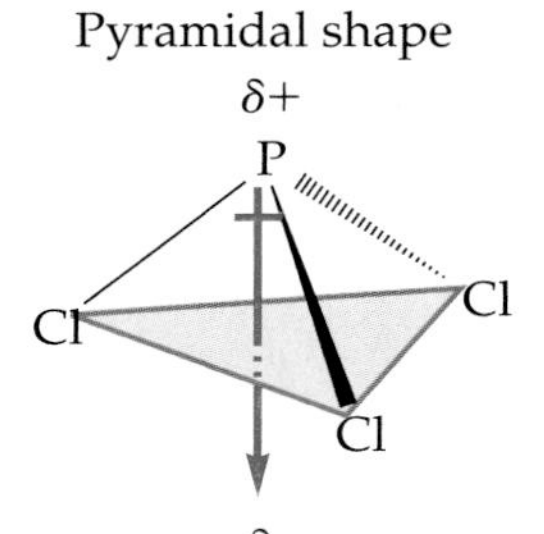

The three individual bond dipole moment vectors add up to give the dipole moment vector for the entire molecule.

6.11 Is chloroform, $CHCl_3$, a polar molecule? If it is, draw the dipole moment vector for the entire molecule and show where the $\delta+$ and $\delta-$ regions of the molecule are.

6.12 Show how the molecules of Problems 6.10 and 6.11 would arrange themselves if they were placed between two plates, one carrying a positive electrical charge and the other carrying a negative electrical charge.

When we started this section, we asked, "What makes molecules stick to one another?" or, to put it another way, "Why can't you rip an ice cube in half?" As we promised, molecular shape plays a key role in answering this question. This is because the way molecules are attracted to one another depends on whether the molecules are polar, and as we have just seen, molecular polarity depends on molecular shape. We are now prepared to take a closer look at intermolecular forces.

6.4 Intermolecular Forces—Dipolar Interactions

How do water molecules stick to one another to form liquid water or solid ice? We now know enough about the water molecule to begin answering this question.

First of all, we know that the water molecule is V-shaped (bent), and because of this shape and its polar covalent bonds, the molecule itself is polar. The dipole moment vector for the molecule shows a partially negative area located at the oxygen atom and a partially positive portion located at the hydrogen atoms. What happens if we bring a group of water molecules together? Because opposite charges attract, the $\delta+$ portion of one molecule is attracted to the $\delta-$ portion of another, causing them to come close to each other:

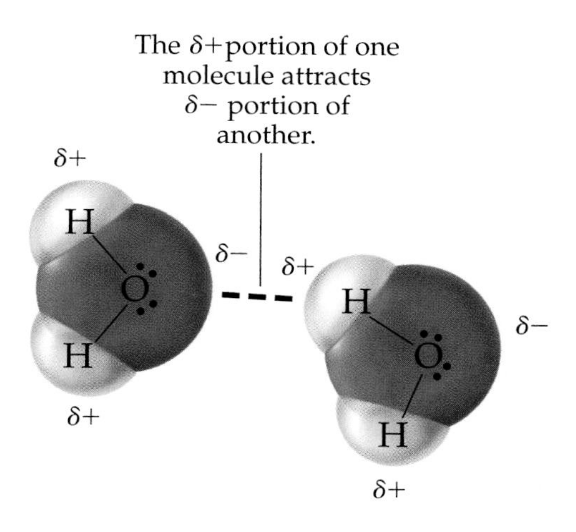

In a piece of ice, billions of water molecules are attracted to one another by dipolar *intermolecular forces*. These forces are the origin of the attractive force between water molecules—the reason these molecules stick together so strongly in ice.

Intermolecular forces between polar molecules are called either **polar forces** or **dipole–dipole forces**, indicating an attraction between dipole moments of different molecules. On a per-atom basis, dipole–dipole forces are among the strongest that exist between molecules. Even so, these forces are nowhere near as strong as the ionic and covalent bonds, sometimes referred to as *intramolecular forces* (*intra-* meaning "within"), we discussed earlier.

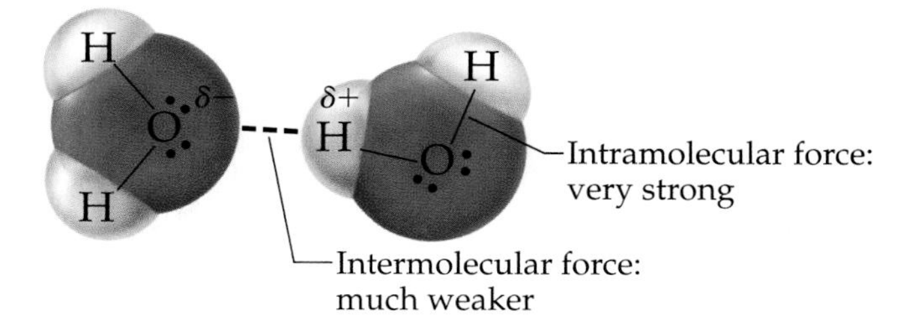

There are other types of intermolecular forces besides dipole–dipole attractions. For example, the hydrogen molecule, H_2, is nonpolar. It has no partial charges, and yet NASA fills fuel tanks on the space shuttle with *liquid* hydrogen. The H_2 molecules must stick to each other to form a liquid, but why? We shall take a more in-depth look at intermolecular forces in Chapter 10.

For now, you shouldn't lose sight of the importance of molecular shape in all of this. If water were a linear molecule instead of bent, it would have no dipole moment (the individual bond dipole moments would cancel). Water molecules would no longer have strong attractive forces between them, and all H_2O on this planet would exist as vapor. Life, at least as we now know it, could not exist if this were so.

You should get comfortable with being able to predict the shape of a molecule from its molecular formula, such as H_2O or PH_3. You can't answer questions about molecular polarity and intermolecular forces unless you know the shape of the molecule. And remember, you can't determine a molecule's shape without a dot diagram.

The process for going from molecular formula to molecular polarity

Formula

↓

1. Determine the total number of valence electrons (use roman-numeral group numbers to help you).
2. Connect the atoms using single bonds.
3. Put in remaining dots as lone pairs to complete octets.
4. If all octets are not complete, create multiple bonds.

Dot diagram

↓

Determine the molecule's shape by applying the concepts of VSEPR to the interior atom(s).

Shape

↓

1. Draw vectors representing all bond dipole moments (consider the electronegativities).
2. Determine the molecular dipole moment, if any.

Molecular polarity
Intermolecular forces

The dot diagram is the key starting point in this process. Start with it, stick to the outline above, and keep practicing.

Have You Learned This?

Valence shell electron pair repulsion (VSEPR) theory (p. 211)

Tetrahedron (p. 213)

Intermolecular attractive force (p. 220)

Polar molecule (p. 221)

Nonpolar molecule (p. 221)

Dipole moment (p. 222)

Vector (p. 223)

Molecular dipole moment (p. 223)

Polar force (p. 229)

Dipole–dipole force (p. 229)

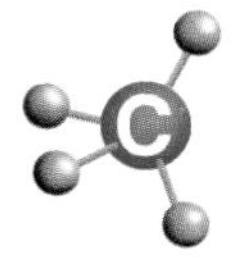

The Shape of Molecules
www.chemistryplace.com

SKILLS TO KNOW

VSEPR

1. For the purposes of VSEPR, a single bond, a multiple bond, and a lone pair each count as a single electron group.

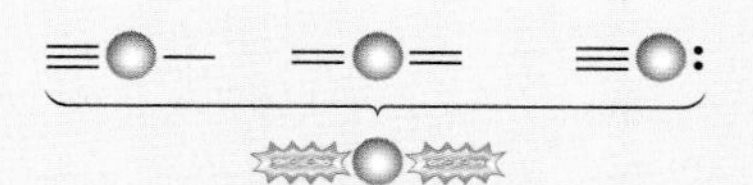

Three ways in which an atom can have two VSEPR electron groups

2. The VSEPR method of determining molecular shape is based on the geometrical arrangment of electron groups around the central atom(s) of a molecule. This geometry depends on the number of electron groups, as shown at right.

   Know your bond angles. As shown at right, the linear bond angle is 180°, the trigonal planar bond angle is 120°, and the tetrahedral bond angle is 109.5°.

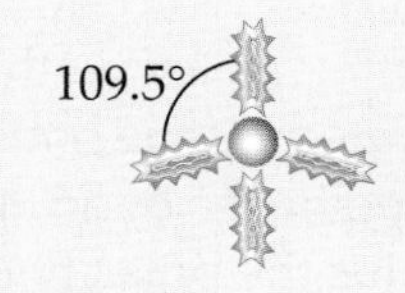

Four electron groups
Tetrahedral electron-group geometry, bond angles all 109.5°

Three electron groups
Trigonal planar electron-group geometry, bond angles all 120°

Two electron groups
Linear electron-group geometry, bond angle 180°

Using VSEPR to find the shape of a molecule

Example: What is the molecular shape of PCl_3?

1. Draw a dot diagram for the molecule.

$$:\ddot{\text{Cl}}-\ddot{\text{P}}-\ddot{\text{Cl}}: \quad (\text{with } :\ddot{\text{Cl}}: \text{ bonded below P})$$

2. Count the number of electron groups around the central atom. Include lone pairs, and count each multiple bond as a single electron group.

$$:\ddot{\text{Cl}}-\ddot{\text{P}}-\ddot{\text{Cl}}: \quad (\text{with } :\ddot{\text{Cl}}: \text{ bonded below P})$$

Four electron groups around P atom

3. Using the count from step 2, use Table 6.2 to determine how the electron groups are arranged around the central atom. Include lone pairs.

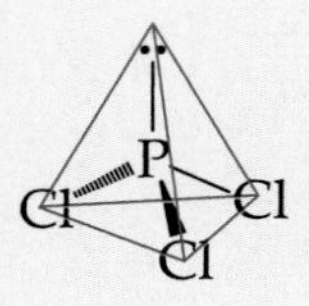

Tetrahedral electron-group geometry

3. Pretend the lone pairs are invisible and describe the resulting shape of the molecule.

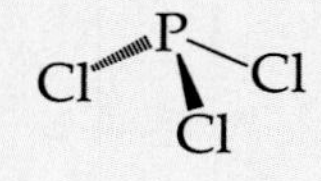

Pyramidal molecular shape

Bond polarity and molecule polarity

1. Every polar covalent bond has a dipole moment. We represent the dipole moment by an arrow that points from the $\delta+$ atom to the $\delta-$ atom of the bond. The longer this arrow, the more polar the bond. The strength of the partial charges, and thus the polarity of the bond, are determined by the size of ΔEN, the difference between the electronegativities of the bonded atoms.

Dipole moment: H—Cl ($\delta+$ → $\delta-$)

2. A molecule that contains polar bonds may have an overall polarity. To decide whether the molecule is polar, do the tug-of-war test. If the individual bond polarities do not cancel each other out, the molecule is polar overall.

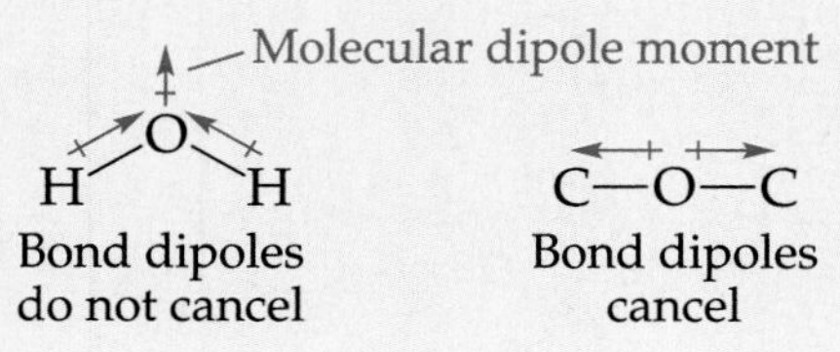

The process for going from molecular formula to polarity

Example: Is PCl_3 polar?

Formula

1. Determine the total number of valence electrons (use roman-numeral group numbers to help you).
2. Connect the atoms using single bonds.
3. Add remaining dots as lone pairs to satisfy octets.
4. If all octets are not satisfied, send adjacent lone pairs to create multiple bonds.

From molecular formula to dot diagram:

PCl_3 :C̈l—P̈—C̈l:
|
Cl

Dot diagram

Determine the molecule's shape by applying the concepts of VSEPR to the interior atom(s).

From dot diagram to molecular shape:

:C̈l—P̈—C̈l:
|
Cl

Cl P Cl
Cl

Pyramidal shape

Shape

1. Draw vectors representing all bond dipole moments (consider the electronegativities).
2. Determine the molecular dipole moment, if any.

Determine molecular dipole moment (if any):

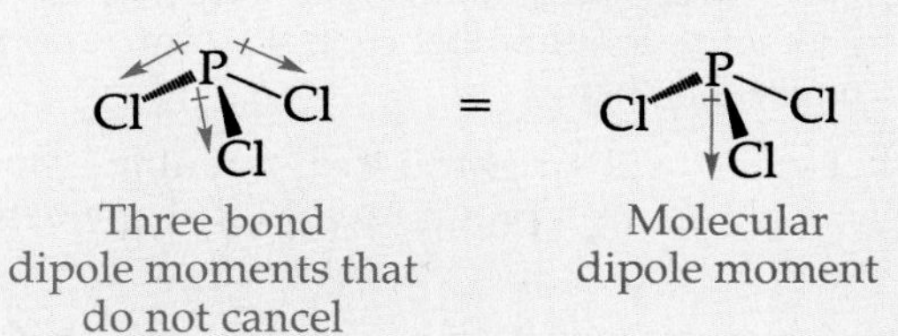

Three bond dipole moments that do not cancel

Molecular dipole moment

Molecular polarity
Intermolecular forces

SHAPES OF MOLECULES AND VSEPR THEORY

6.13 Why is it better for methane to have 109.5° bond angles rather than 90° bond angles?

6.14 How do you think a bacterium species might evolve to develop resistance to a drug like sulfanilamide that is designed to kill it?

6.15 Why is the theory that governs the shape of molecules called VSEPR and not just EPR?

6.16 Draw the methane molecule showing its tetrahedral shape, using lines, solid wedges, and dashed wedges to show three-dimensionality.

6.17 Draw a methane molecule inscribed inside a tetrahedron such that the H atoms touch the vertices of the tetrahedron.

6.18 Shown below are dot diagrams for some simple molecules:

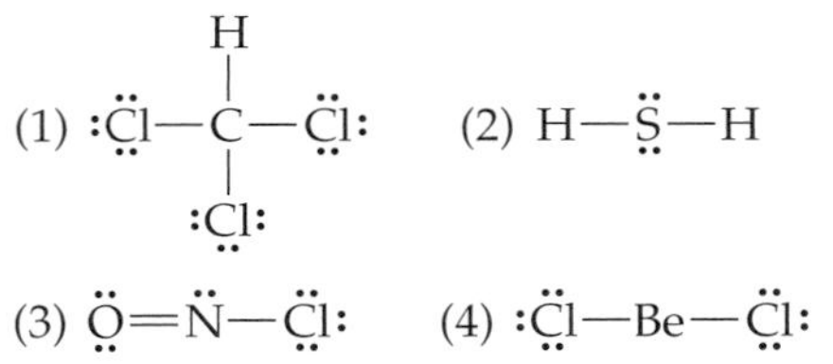

```
        H
        |
(1) :Cl—C—Cl:      (2) H—S—H
        |
       :Cl:

(3) O=N—Cl:      (4) :Cl—Be—Cl:
```

(a) Draw the three-dimensional shape for each molecule. Use lines, solid wedges, and dashed wedges as necessary. Indicate the numeric value of all bond angles.
(b) For each molecule, name both the electron-group geometry around the central atom and the molecular shape.

6.19 Shown below are dot diagrams for some simple molecules and polyatomic ions:

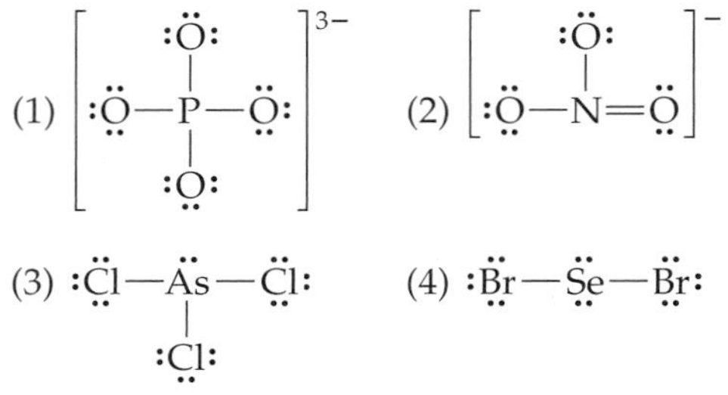

```
       [    :O:    ]3−          [    :O:   ]−
       [     |     ]            [     |    ]
(1)    [ :O—P—O:   ]      (2)   [ :O—N=O   ]
       [     |     ]
       [    :O:    ]

(3) :Cl—As—Cl:      (4) :Br—Se—Br:
         |
        :Cl:
```

(a) Draw the three-dimensional shape for each molecule. Use lines, solid wedges, and dashed wedges as necessary. Indicate the numeric value of all bond angles.
(b) For each species, name both the electron-group geometry around the central atom and the molecular shape.

6.20 What gives us the right to treat a multiple bond as a single electron group in VSEPR shape determination?

6.21 Two molecules may both be correctly described as bent even though one has a bond angle of approximately 118° and the other has a bond angle of approximately 105°. How is this possible?

6.22 Why are the bond angles in ammonia smaller than the expected tetrahedral value?

6.23 Ammonia has four pairs of electrons around the central nitrogen atom, and yet we don't call it a tetrahedral molecule. Why not? What is the shape of this molecule?

6.24 Consider the following molecule. The diagram shows how the atoms are connected, but it is not a complete dot diagram.

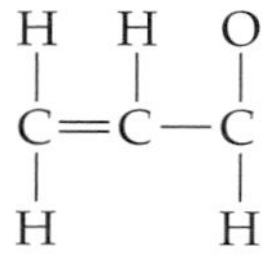

```
H  H  O
|  |  |
C=C—C
|     |
H     H
```

(a) Complete the dot diagram.
(b) Redraw the molecule showing its three-dimensional shape. Use lines, solid wedges, and dashed wedges if necessary. [*Hint*: Work on the C atoms one at a time. If the C you are working on is bonded to an atom that itself is bonded to other atoms, treat all those atoms as a single unit. Each C here, for instance, has three bonding groups, no lone pairs.]
(c) Indicate the numeric value of all bond angles.

6.25 Consider the following molecule. The diagram shows how the atoms are connected, but it is not a complete dot diagram.

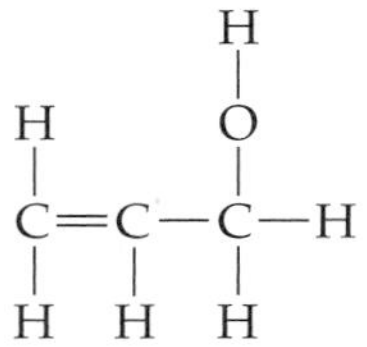

```
         H
         |
H        O
|        |
C=C—C—H
|  |  |
H  H  H
```

(a) Complete the dot diagram.
(b) Redraw the molecule showing its three-dimensional shape. Use lines, solid wedges, and dashed wedges if necessary. [*Hint*: Work on the three C atoms and the O atom one at a time. If the atom you are working on is bonded to an atom that is itself bonded to other atoms, treat all those atoms as a single unit. The O, for instance, has two bonding groups, two lone pairs.]
(c) Indicate the numeric value of all bond angles.

6.26 Why is it necessary to draw a correct Lewis dot diagram before trying to predict the shape of a molecule?

6.27 A hexagon has angles of 120° inside the ring and is flat. The following two molecules are often drawn as flat hexagons, but only one is truly flat. Which one is it and why?

```
            H
            |
      H     C     H
       \  /   \\ /
        C       C
(1)     ||      |
        C       C
       /  \   // \
      H     C     H
            |
            H

          H  H
      H    \/    H
       \   C    /
   H—C       C—H
(2)    |       |
   H—C       C—H
       /   C    \
      H    /\    H
          H  H
```

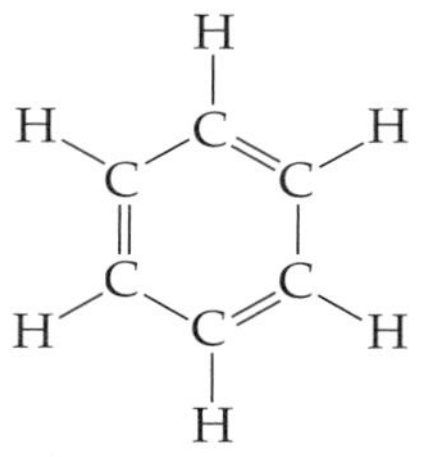

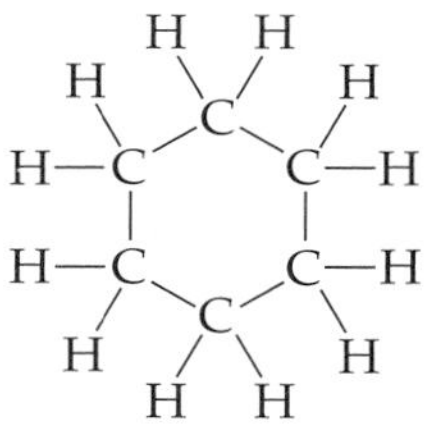

6.28 *Proteins* are long-chain molecules that make up our skin, hair, muscles, and enzymes, among other things. Shown below is a dot diagram of the part of a protein known as a *peptide bond*. Redraw the diagram to show the three-dimensional shape around each carbon atom and around the nitrogen atom, using lines, solid wedges, and dashed wedges as needed. Name the shape around each of these four atoms and give the numeric value of all bond angles.

```
             :O:
   |    ..    ||    |
 —C—N—C—C—
   |    |           |
        H
```

POLARITY OF MOLECULES

6.29 In terms of an operational definition, when is a molecule considered to be polar?

6.30 Is it possible to predict whether a molecule will be polar? Explain completely what you would need to know about the molecule.

6.31 True or false? All molecules that contain polar bonds must be polar. Explain your answer.

6.32 Consider the molecules HCl and HBr.
(a) Which molecule has the larger bond dipole moment? Explain why.
(b) Which molecule is more polar? Explain why.

6.33 What does the magnitude of a bond dipole moment (the length of the arrow) tell you about the bond?

6.34 Why do we use an arrow to represent a bond dipole moment? Why not just use a number?

6.35 What does electronegativity have to do with a bond dipole moment?

6.36 How is a bond dipole moment drawn?

6.37 The molecule CO_2 has very polar bonds, and yet the molecule itself is nonpolar. Explain how this is possible.

6.38 The molecule SO_2 has polar bonds and is polar. Why the difference from CO_2 (Problem 6.37)?

6.39 Consider all the hydrogen halide molecules HX, where X is a group VIIA atom.
(a) Which is the most polar? Why?
(b) Which is the least polar? Why?
(c) Draw all these molecules, showing their relative bond dipole moments.

6.40 Consider the two molecules CO and CO_2. They are both made of the same elements, and yet only one is polar. Why is this so?

6.41 Consider the following molecules. For those that are polar, draw the molecular dipole moment.
(a) $CHBr_3$ (b) CH_3Br (c) H_2S (d) NOCl
(e) C_2Cl_2 (connected Cl–C–C–Cl)

6.42 What is a dipole–dipole force? Give an example.

6.43 Are dipole–dipole forces between molecules as strong as the forces between oppositely charged ions? Explain.

6.44 What do we mean by intermolecular forces? What evidence is there that they exist?

6.45 Draw two HCl molecules and show how they would be attracted to each other. Show the partial charges and dipole moment vectors for both molecules, and orient the molecules properly with respect to each other.

6.46 Draw two ammonia molecules in their three-dimensional shape and show how they would be attracted to each other. Show the partial charges and individual bond dipole moment vectors for both molecules, and orient the molecules properly with respect to each other.

6.47 Why is it important to always show the lone pairs in a Lewis dot diagram?

6.48 The dot diagram $\underset{\cdot\cdot}{\ddot{\mathrm{O}}}=\mathrm{S}=\underset{\cdot\cdot}{\ddot{\mathrm{O}}}$ for SO_2 is incorrect.
(a) Draw the correct dot diagram.
(b) How do the correct and incorrect dot diagrams differ in their prediction of molecular shape and polarity?

6.49 Consider the molecule $SiCl_4$.
(a) Draw the dot diagram.
(b) Draw the molecule's three-dimensional shape, and label the numeric value of all bond angles.
(c) What is the shape of this molecule?
(d) Draw in the individual bond dipole moments.
(e) Is the molecule polar? If yes, draw the molecular dipole moment vector.

6.50 Consider the molecule AsF_3.
(a) Draw the dot diagram.
(b) Draw the molecule's three-dimensional shape, and label the numeric value of all bond angles.
(c) What is the shape of this molecule?
(d) Draw in the individual bond dipole moments.
(e) Is the molecule polar? If yes, draw the molecular dipole moment vector.

6.51 Consider the molecule SO_3.
(a) Draw the dot diagram.
(b) Draw the molecule's three-dimensional shape, and label the numeric value of all bond angles.
(c) What is the shape of this molecule?
(d) Draw in the individual bond dipole moments.
(e) Is the molecule polar? If yes, draw the molecular dipole moment vector.

6.52 The atoms in the molecule HSCN are connected in the order given in the formula.
(a) Draw the dot diagram.
(b) Draw the molecule's three-dimensional shape, and label the numeric value of all bond angles.
(c) Draw in the individual bond dipole moments.
(d) Is the molecule polar? If yes, draw the molecular dipole moment vector.

6.53 Consider the molecule hydrazine, N_2H_4, used in rocket fuel. Each nitrogen is bonded to two hydrogens, and the nitrogens are bonded to each other.
(a) Draw the dot diagram.
(b) Draw the molecule's three-dimensional shape, and label the numeric value of all bond angles.
(c) What is the shape of this molecule?
(d) Draw in the individual bond dipole moments.
(e) Is the molecule polar? If yes, draw the molecular dipole moment vector.

6.54 Consider the molecule HNF_2 (N is the central atom in the molecule).
(a) Draw the dot diagram.
(b) Draw the molecule's three-dimensional shape, and label the numeric value of all bond angles.
(c) What is the shape of this molecule?
(d) Draw in the individual bond dipole moments.
(e) Is the molecule polar? If yes, draw the molecular dipole moment vector.

6.55 Consider the molecule N_2O (connected N–N–O).
(a) Draw the dot diagram.
(b) Draw the molecule's three-dimensional shape, and label the numeric value of all bond angles.
(c) What is the shape of this molecule?
(d) Draw in the individual bond dipole moments.
(e) Is the molecule polar? If yes, draw the molecular dipole moment vector.

6.56 Some molecules pose special challenges to the rules for obtaining the correct Lewis dot diagram. Consider the molecule NO_2 (N is the central atom).
(a) What challenge does it present?
(b) Suppose you were allowed to violate the octet rule for the N atom. What would the dot diagram look like?
(c) Based on your answer to part (b), what is the shape of such a molecule? Is the molecule polar?

6.57 Consider the phosphonium ion, PH_4^+.
(a) Draw the dot diagram.
(b) Draw the ion's three-dimensional shape, and label the numeric value of all bond angles.
(c) What is the shape of this polyatomic ion?
(d) Draw in the individual bond dipole moments.

6.58 Imagine that you could vary the magnitude of the molecular dipole moment of all molecules in a substance. What should happen to the boiling point of the substance as the dipole moment is increased? Explain.

6.59 PH_3 is a nonpolar molecule. Why is this so? [*Hint:* Consider electronegativities.]

ADDITIONAL PROBLEMS

6.60 An atom has no lone pairs of electrons on it and four other atoms bound to it. Why is 109.5° the bond angle adopted by this molecule?

6.61 Using lines, solid wedges, and dashed wedges, draw the three-dimensional shape of ethane, C_2H_6. Indicate the numeric value of all bond angles.

```
    H   H
    |   |
H — C — C — H
    |   |
    H   H
```

6.62 Using lines, solid wedges, and dashed wedges, draw the three-dimensional shape of acetaldehyde, C_2H_4O. Indicate the numeric value of all bond angles.

```
   :O:  H
    ‖   |
H — C — C — H
        |
        H
```

6.63 Using lines, solid wedges, and dashed wedges, draw the three-dimensional shape of diethyl ether, C_2H_6O. Indicate the numeric value of all bond angles.

```
    H       H
    |  ..   |
H — C — O — C — H
    |  ..   |
    H       H
```

6.64 Using lines, solid wedges, and dashed wedges, draw the three-dimensional shape of the hydronium ion, H_3O^+. Indicate the numeric value of all bond angles.

```
     ..      +
[ H — O — H ]
      |
      H
```

6.65 Consider the molecules CH_4, CH_3Cl, CH_2Cl_2, $CHCl_3$, and CCl_4. Which are polar and which are nonpolar?

6.66 Using lines, solid wedges, and dashed wedges, draw the three-dimensional shape of chloromethylacetylene, C_3H_3Cl. Indicate the numeric value of all bond angles.

```
    H
    |             ..
H — C — C ≡ C — Cl:
    |             ..
    H
```

6.67 The connections in $C_2N_2H_2O$ are

```
        O
        |
N — C — C — N — H
            |
            H
```

(a) Put in the remaining valence electron to complete the Lewis dot diagram.

(b) Using lines, solid wedges, and dashed wedges, draw the three-dimensional shape of this molecule. Indicate the numeric value of all bond angles.

6.68 The connections in CH_3N_2O are

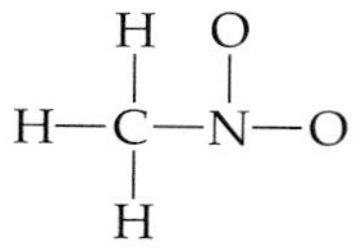

```
    H  O
    |  |
H—C—N—O
    |
    H
```

(a) Put in the remaining valence electron to complete the Lewis dot diagram.
(b) Using lines, solid wedges, and dashed wedges, draw the three-dimensional shape of this molecule. Indicate the numeric value of all bond angles.

6.69 Up until now, you've been drawing first a Lewis dot diagram (two-dimensional, shows lone electron pairs) and then a separate diagram showing molecule shape (three-dimensional, no lone pairs shown). Now begin combining the two, drawing one structure showing both three-dimensional shape and all lone pairs. Draw such a structure for each of the following polyatomic ions. Name each shape, and indicate whether the ion has an overall dipole moment. If so, draw the dipole moment vector. [*Hint*: It's a good idea to continue to draw a regular Lewis diagram first, even though you do not show it in your final answer.]
(a) CN^- (b) ClO_4^- (c) PCl_4^+ (d) NO_2^-

6.70 Draw a combined Lewis dot, molecular-shape diagram for each of the following species. Name each shape, and indicate whether the molecule or ion has an overall dipole moment. If so, draw the dipole moment vector.
(a) Cl_2O (b) NOF (c) PF_3 (d) ICl_2^+
[*Hint*: See problem statement and hint for Problem 6.69. *Hint for (b)*: N=O bonds are common.]

6.71 Draw a combined Lewis dot, molecular-shape diagram for each of the following species. Name each shape, and indicate whether the molecule or ion has a dipole moment. If so, draw the dipole moment vector. [*Hint*: See problem statement and hint for Problem 6.69.]
(a) $CHBr_3$ (b) NF_3 (c) ClO_3^- (d) CS_2

6.72 Draw dot diagrams for the carbonate anion, CO_3^{2-}, and for the sulfite anion, SO_3^{2-}. What is the electron-group configuration around the central atom in each anion? What is the molecular shape of each anion? Is either ion polar? If so, which one? Explain your choice.

6.73 Which, if any, of these molecules do you expect to be polar: CO_2, CS_2, or CSO (carbon is the central atom in all three molecules). Explain your answer.

6.74 Which, if any, of these molecules are polar? For any molecule you classify as polar, show both the individual bond dipole moment vectors and the overall molecular dipole moment vector. Explain your answers. [*Hint*: For the tug-of-war test, place yourself midway between the carbons.]

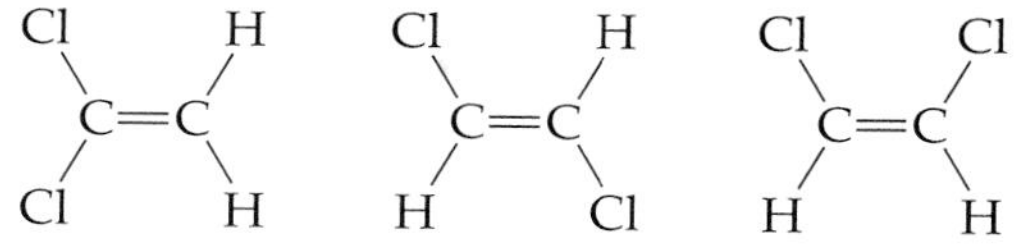

```
Cl     H      Cl     H      Cl     Cl
  \   /         \   /         \   /
   C=C           C=C           C=C
  /   \         /   \         /   \
Cl     H      H      Cl     H      H
```

6.75 Draw the three-dimensional shape of each molecule shown below. Label the numeric value of all bond angles, and indicate whether the molecule is polar or nonpolar. If polar, draw the molecular dipole moment vector.

```
    :F̤̈:         H  :F̤̈:
     |          |   |
  H—C—H         C=C          H—Ö̤—F̤̈:
     |          |   |
     H          H  :F̤̈:
```

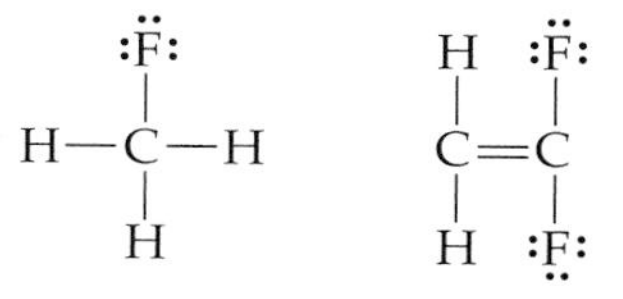

6.76 List the steps you take to decide (a) whether or not a covalent bond is polar and (b) whether or not a molecule is polar.

6.77 Which should have the largest molecular dipole moment: H_2, CO_2, CH_3F, or CH_3I?

6.78 What is the molecular shape of NCl_3? What is the electron-group geometry around N? Is this a polar molecule?

6.79 When asked the shape of NH_2^-, a student begins by drawing two dot diagrams:

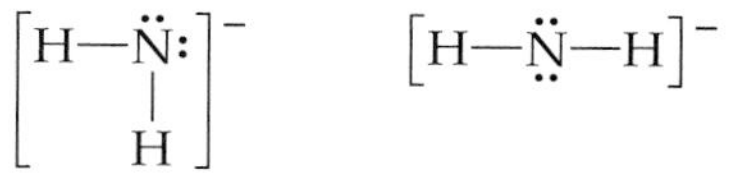

```
[H—N̈:]⁻        [H—N̤̈—H]⁻
[   | ]
[   H ]
```

She then answers that the shape can be either bent or linear depending on the arrangement of the hydrogen atoms. What is wrong with this answer? What is the correct shape?

6.80 What is the difference between intermolecular forces and intramolecular forces? Which are stronger?

6.81 Draw the three-dimensional shape of S_2F_2 (connected F–S–S–F) and N_2F_2 (connected F–N–N–F). Indicate the numeric value of all bond angles.

6.82 Lewis dot diagrams are useful in understanding how the valence electrons in a molecule are arranged, but they do not show what important aspect of the molecule?

6.83 Draw the three-dimensional shape of methanol, CH_3OH. Indicate the numeric value of all bond angles. Is this molecule polar? If so, draw the molecular dipole moment vector.

6.84 Hydrogen peroxonitrite has the formula HNO_3, with the atoms connected O–N–O–O–H.
(a) Draw the dot diagram for this molecule.
(b) Draw the three-dimensional shape of the molecule, showing the numeric value of all bond angles.
(c) Describe the electron-group geometry and molecular shape around the N and around each O to the right of the N.

6.85 Explain the difference between electron-group geometry and molecular shape. How do you use electron-group geometry when deciding what shape a molecule has?

6.86 Draw each ion or molecule showing its three-dimensional shape and valence electrons. For each species, name the electron-group geometry around the central atom and the molecular shape. Indicate whether each species has a molecular dipole moment. If so, draw the dipole moment vector.
(a) NSF (b) CNO^- (c) SF_3^+ (d) SiO_2

6.87 A student forgets that the N in ammonia, NH_3, has a lone pair as well as its three single bonds. After checking Table 6.2, he mistakenly draws the molecule—three bonding groups, no lone pairs—as having a trigonal planar shape. If the ammonia molecule really were trigonal planar, how would the intermolecular forces in an ammonia solution differ from what they actually are?

6.88 Under what conditions is the electron-group geometry for a molecule the same as the molecular shape?

6.89 What is the shape of the N_3^- anion? What is the numeric value of each bond angle in the anion?

6.90 Which should have the largest dipole moment: O_3, H_2O, or OF_2?

6.91 Covalent molecules that contain an O–H, N–H, or F–H bond have very strong intermolecular forces. Explain why.

6.92 Draw the three-dimensional shape of H_2O_3. Label the numeric value of all bond angles. In this molecule, the three oxygen atoms form a chain with a hydrogen atom at each end.

6.93 Draw each ion or molecule showing its three-dimensional shape and valence electrons. For each species, name the electron-group geometry around the central atom (or atoms) and the molecular shape. Indicate whether each species has a molecular dipole moment. If so, draw the dipole moment vector.
(a) NCCN
(b) BF_4^-
(c) $NClO_2$
(d) SeH_2

6.94 White phosphorus, P_4, consists of four phosphorus atoms, with each atom bonded to the other three. Draw the three-dimensional shape of this molecule, showing all valence electrons and bond angles. [*Hint*: Begin by drawing one atom bonded to the three others using VSEPR theory, and then make the remaining connections.]

6.95 BH_3 and PH_3 each contain four atoms, with the three hydrogens surrounding the central atom. Do the two molecules have the same electron-group geometry? Do they have the same molecular shape? Name the geometry and the shape for both molecules. You know from Problem 6.59 that PH_3 is nonpolar. What about BH_3? Is it polar or nonpolar? Explain.

6 WORKPATCH SOLUTIONS

6.1 (a) (b)

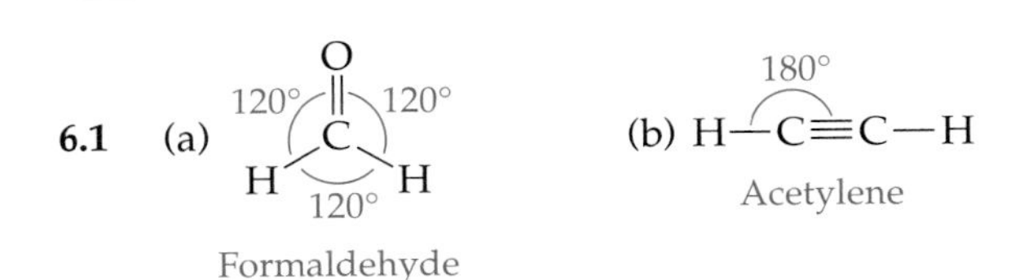

6.2 Both molecules would be called either bent or V-shaped.

6.3 In H_2O, expect a roughly 4° compression from 109.5° because there are two lone pairs on the oxygen:

H–O–H 105.5°

In O_3, expect a roughly 2° compression from 120° because of the one lone pair on the central oxygen:

O=O–O 118°

6.4 These are trivial because they all represent a central atom with only one other atom attached (we know this from the information that there is only one bonding group in each case). When a molecule consists simply of two atoms, the only shape it can have is linear. No other shape is possible.

6.5 A bond is guaranteed to have a zero dipole moment when the electrons forming the bond are shared absolutely equally by the two atoms. This occurs when the two atoms have identical electronegativities (this is assured when the atoms are of the same element).

6.6 (*a*) and (*b*) The electronegativity values are C 2.5, O 3.5, and S 2.5. Thus the individual bonds in CO_2 are more polar because the electronegativity difference between C and O ($\Delta EN = 1$) is greater than the difference between C and S ($\Delta EN = 0$). However, because both molecules are linear, the bond dipole moments in both of them cancel. Therefore, neither molecule is polar.

(a) 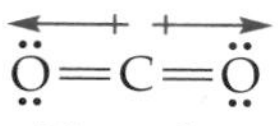 $\ddot{O}=C=\ddot{O}$ Nonpolar

(b) 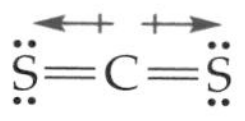 $\ddot{S}=C=\ddot{S}$ Nonpolar

(*c*) and (*d*) The electronegativity of H is 2.1. Therefore the differences are now O–H $\Delta EN = 1.4$ and S–H $\Delta EN = 0.4$. Because the bonds in H_2O are more polar, H_2O is the more polar molecule.

(c) 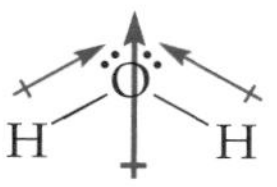 H–O–H More polar

(d) 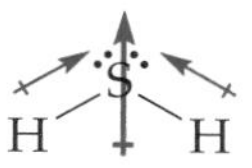 H–S–H Less polar

6.7 The upward tug in the $SOCl_2$ molecule is slightly stronger because the S–O bond is more polar than the S–Cl bonds. This is because the electronegativity difference between S and O ($\Delta EN = 3.5 - 2.5 = 1.0$) is greater than the electronegativity difference between S and Cl ($\Delta EN = 3.0 - 2.5 = 0.5$). If the three tugs were equal, and assuming the bond angles were 120° (trigonal planar), the tugs would cancel and the molecule would be nonpolar.

CHAPTER 7

Chemical Reactions

7.1 What Is a Chemical Reaction?

We touched briefly on the subject of chemical reactions in Section 1.4, but now we want to look at them in more detail.

In the mid-1980s, chemists combined just the right amounts of two metal oxides—yttrium oxide, Y_2O_3, and copper(II) oxide, CuO—with barium carbonate, $BaCO_3$, and heated the combination in a stream of oxygen gas, O_2. The result was a new ceramic compound, $YBa_2Cu_3O_7$, that possessed extraordinary properties. For example, when this compound is cooled to 90 K (−183°C, −298°F), a magnet placed above it floats as if defying gravity!

This new ceramic compound is called a *superconductor*, a class of materials that conduct electricity with zero resistance. In addition, magnetic field lines cannot penetrate a superconductor (as they do ordinary materials), and this is the reason a magnet floats above one. Magnetically levitated (maglev) trains are just one of the uses envisioned for superconductors. Just put some strong magnets on the bottom of a train and replace the steel railroad tracks with tracks made from $YBa_2Cu_3O_7$. Then cool the tracks to 90 K

An experimental Japanese maglev train floats above superconducting tracks.

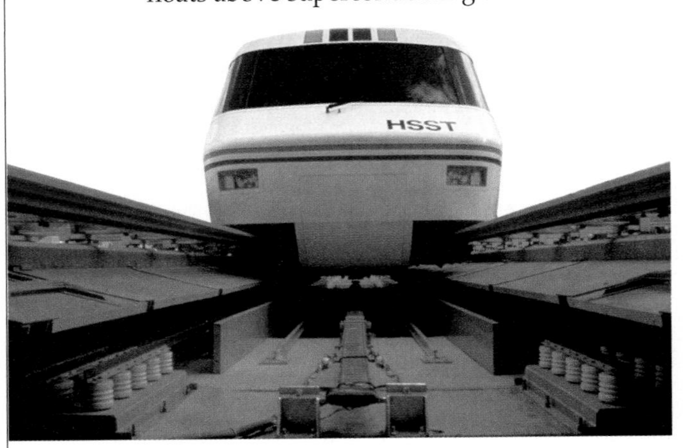

with liquefied nitrogen, a relatively cheap industrial commodity, and the train floats virtually frictionless above the tracks. This is the near future, brought to you by a chemical reaction that turned a few piles of otherwise not very useful metal oxides into a brand-new compound.

It would appear from our example that a chemical transformation has taken place—that is, a chemical reaction produces something new, just as we said in Section 1.4. In a **chemical reaction**, one or more substances are converted into new substances that have compositions and properties that are different from those of the starting substances. We call the starting substances **reactants** and the new substances **products**. We represent a chemical reaction by writing a **chemical equation**, which is something like a mathematical equation in that it shows us two things (or two groups of things) that are related to each other. Instead of numbers, however, a chemical equation uses molecular formulas, and instead of an equals sign, it uses an arrow. We write a chemical equation with the reactants to the left of the arrow and the products to the right:

$$\text{Reactants} \longrightarrow \text{Products}$$

Remember, the arrow means "reacts to form." That is, reactants *react* with each other *to form* products.

It is important that you understand what constitutes a chemical reaction and what doesn't. Try the WorkPatch to see whether or not you do.

Which chemical equations represent chemical reactions:

(a) $H_2O(l) \longrightarrow H_2O(g)$

[Remember, (l) means "liquid" and (g) means "gas."]

(b) $CO_2 + H_2O \longrightarrow H_2CO_3$

(c) $CO_2 + O_2 \longrightarrow O_2 + CO_2$

Did you get the correct answer? There has been a change in (a), true, but a phase change is not a chemical reaction. To be a chemical reaction, at least one new substance must be formed, but how does this happen? How do reactants become products? It is this *how* that we are now prepared to consider.

7.2 How Are Reactants Transformed into Products?

To see how chemical reactions take place, we'll look at a reaction that is much simpler than the one that makes a superconductor. Imagine you have a flask filled with two different diatomic molecules, hydrogen, H_2, and iodine, I_2. Hydrogen is a colorless gas. Iodine is a dark solid at room temperature, but if we heat the flask, we can drive the iodine into the gas phase and produce a purple vapor of iodine molecules. The hydrogen and iodine molecules in the gas phase react with each other to form the colorless gas hydrogen iodide, HI.

The chemical and physical properties of this product are completely different from those of the reactants, as this figure shows:

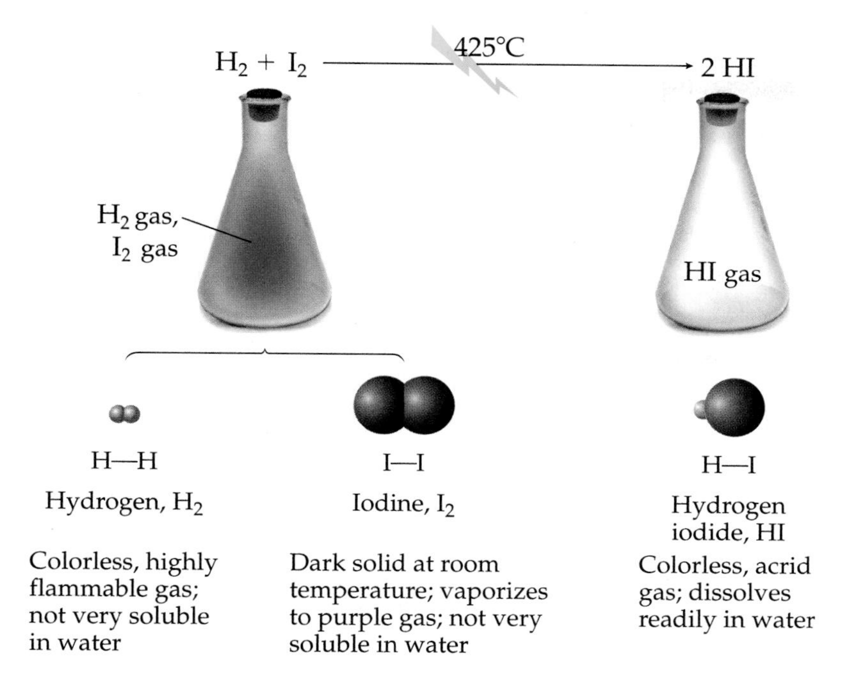
$H_2 + I_2 \xrightarrow{425°C} 2\,HI$
H_2 gas,
I_2 gas
HI gas
H—H
Hydrogen, H_2
Colorless, highly flammable gas; not very soluble in water
I—I
Iodine, I_2
Dark solid at room temperature; vaporizes to purple gas; not very soluble in water
H—I
Hydrogen iodide, HI
Colorless, acrid gas; dissolves readily in water

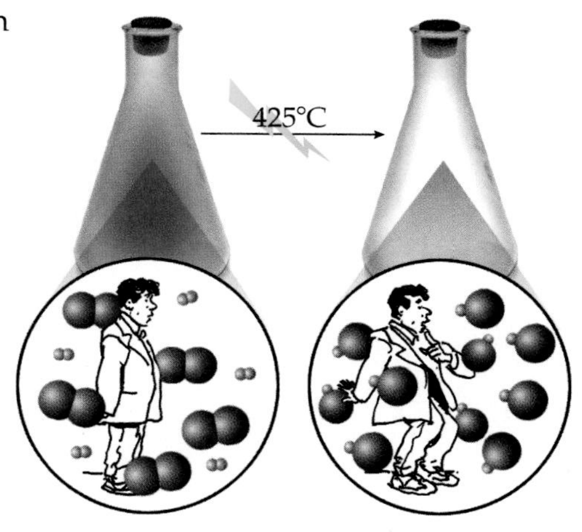
425°C

Because both reactants are gases, we say that the reaction between them is a **gas-phase reaction**. How does a mixture of H_2 and I_2 molecules turn itself into HI molecules? To get some insight, let's change our point of view. Imagine you are so small you can see individual atoms and molecules. At the very beginning of the reaction, you can clearly see the H_2 and I_2 molecules, each with a strong single covalent bond between the atoms. An instant later the reaction is over. In the blink of an eye, all the H–H and I–I covalent bonds have broken and new H–I covalent bonds have formed.

At first you might be puzzled. Why did the H–H and I–I bonds break? It couldn't be that they were weak and ready to break anyway because, as we've seen, covalent bonds are among the strongest chemical forces. And strong bonds certainly would not break for no reason at all. That would be like the legs on a solid oak table suddenly breaking off for no reason. (When was the last time you saw that happen?) You should be just as surprised to see all those H–H and I–I covalent bonds break. To find out why they did, let's change our point of view once again. This time, picture yourself *on* an H_2 or I_2 molecule in the gas-filled flask. You'd better hold on tight because the molecules are not sitting still. At room temperature you would be zipping about at hundreds of miles per hour!

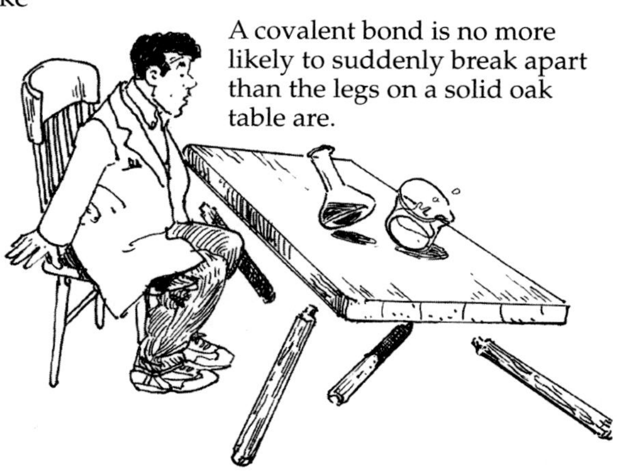
A covalent bond is no more likely to suddenly break apart than the legs on a solid oak table are.

This is true for all small molecules in the gas phase. For example, at room temperature, the oxygen, O_2, and nitrogen, N_2, molecules in the air you are breathing are

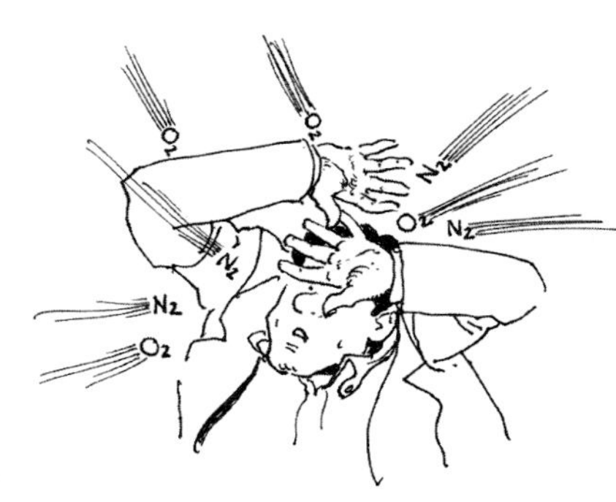

traveling at speeds close to 500 miles per hour. (Even faster if the temperature is increased.) So why don't you feel like you are in a hurricane at this very moment? The answer is that although the molecules in the air are moving rapidly, they are moving in random directions. They are crashing into you at high speeds, each pushing you a little bit, but in different directions. The net result is that all the pushes cancel one another and the air feels still.

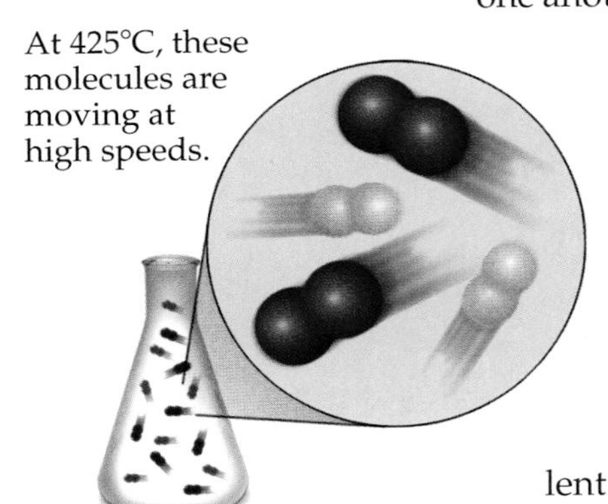

At 425°C, these molecules are moving at high speeds.

Now, let's return to the flask of fast-moving H–H and I–I reactant molecules. Sooner or later, an H_2 and an I_2 molecule will collide with each other. As with two speeding automobiles involved in a collision, the energy associated with their motion (their kinetic energy) gets absorbed by the colliding molecules. We are all familiar with the result of cars' absorbing all that energy. Windows break, tires fly off, and steel crumples. In the case of colliding H_2 and I_2 molecules, the energy absorbed on impact can begin to break their covalent bonds. The fundamental concept here is that

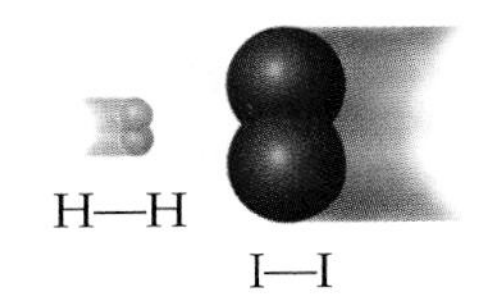

Collision Course!

It always takes energy to break a chemical bond.

There are no exceptions to this rule—ever. And the opposite is also true: energy is always released when a chemical bond forms.

So, for our gas-phase $H_2 + I_2$ reaction, the energy needed to break bonds in the reactant molecules comes primarily from collisions between the fast-moving molecules. Immediately after these collisions, however, the situation differs radically from a crash between automobiles. When automobiles crash and break apart, they stay that way. When H_2 and I_2 molecules collide and break apart, new H–I covalent bonds can form. A new substance is formed. A chemical reaction has occurred.

A Chemical Reaction Taking Place

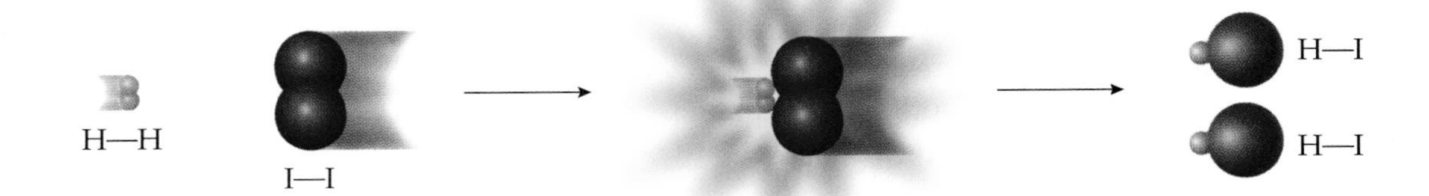

This description of a chemical reaction is a simplified and idealized one. For example, it is possible to re-form the original bonds and restore the H_2 and I_2 reactant molecules. In Chapter 14 we shall deal with this possibility. For now, we shall focus on cases in which collisions between reactant molecules result in the direct formation of product molecules.

Imagine a collision between an oxygen, O_2, molecule and a hydrogen, H_2, molecule to give the product hydrogen peroxide, H_2O_2. Draw Lewis dot diagrams for all the molecules, and then indicate which bonds must be broken and which bonds must form. [*Hint:* The atoms in hydrogen peroxide are connected H–O–O–H.]

Did you notice in the WorkPatch that two reactant molecules collide to give just one molecule of product? This was not the case in the $H_2 + I_2$ reaction. This observation sets us up for what comes next. So far, we have been talking about what a chemical reaction is in terms of bond breaking and bond making, and what causes the bonds to break. When we represent chemical reactions in writing, however, we must always go one step further and *balance* the chemical equation.

7.3 Balancing Chemical Equations

If we combine a certain amount of H_2 with a certain amount of I_2, how much HI will we make? It is easy to answer this question on a molecular level because we can "see" it in the bottom illustration on page 244. If we take one H_2 molecule and bang it into one I_2 molecule with a force sufficient to break the bonds, we can form two HI molecules. We can represent this reaction with a *balanced chemical equation*:

$$H_2 + I_2 \longrightarrow 2\,HI$$

This equation is shorthand for "one molecule of H_2 and one molecule of I_2 react to give two molecules of HI." Note that the number of atoms to the left of the arrow is equal to the number of atoms to the right of the arrow, meaning the equation is *balanced*. A *balanced chemical equation* is one in which the law of conservation of matter is obeyed. If we start with an unbalanced equation, say $H_2 + I_2 \longrightarrow HI$, we have to insert numerical *coefficients* in front of the reactants and/or products to balance the equation. (If no numerical coefficient is written in front of a reactant or product, the coefficient is assumed to be 1.) A chemical equation is balanced when the numbers of atoms of each element are equal on both sides of the arrow:

H_2	+	I_2	$\longrightarrow$	2 HI
H—H		I—I		H—I H—I
Two H atoms		Two I atoms		Two H atoms, two I atoms

On each side of the arrow there are two H atoms and two I atoms. This means we start with two H atoms and two I atoms, and we end up with two H atoms and two I atoms. Nothing is lost or gained. Because any chemical reaction must obey the law of conservation of matter, all chemical equations must be balanced.

Here is another version of the HI reaction. Is this equation balanced?

$$2\,H_2 + 2\,I_2 \longrightarrow 4\,HI$$

It is certainly not the same as the equation above, but it is still balanced. There are now four of each type of atom on each side of the arrow. There is nothing wrong with writing the balanced equation this way, but it is traditional to use the smallest possible whole numbers as coefficients.

What if you come across an equation that is unbalanced, like this one:

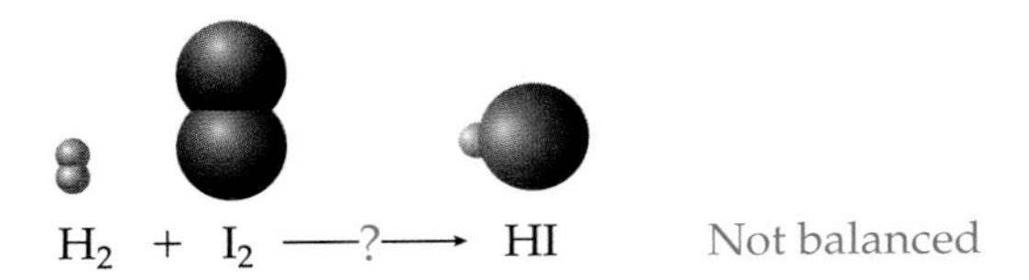

$H_2 + I_2 \xrightarrow{?} HI$ Not balanced

You adjust the coefficients (which are also called *balancing coefficients*) to balance the equation. In this case, you simply insert a balancing coefficient 2 in front of the HI:

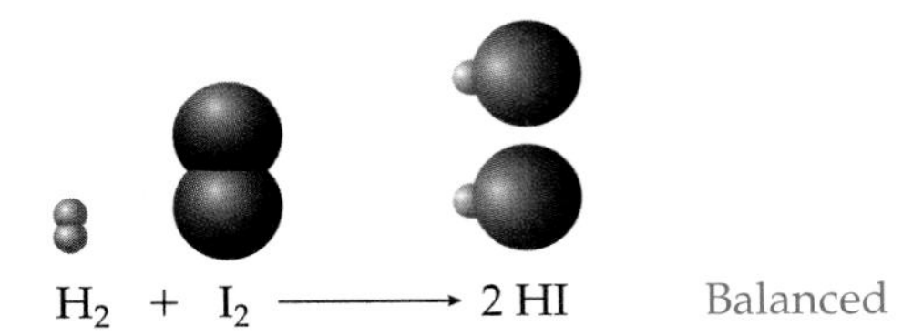

$H_2 + I_2 \longrightarrow 2\,HI$ Balanced

To balance a chemical equation, you can always insert balancing coefficients. What you should never, *never* do is adjust the subscripts in a molecular formula. For example, you should never balance the HI equation this way:

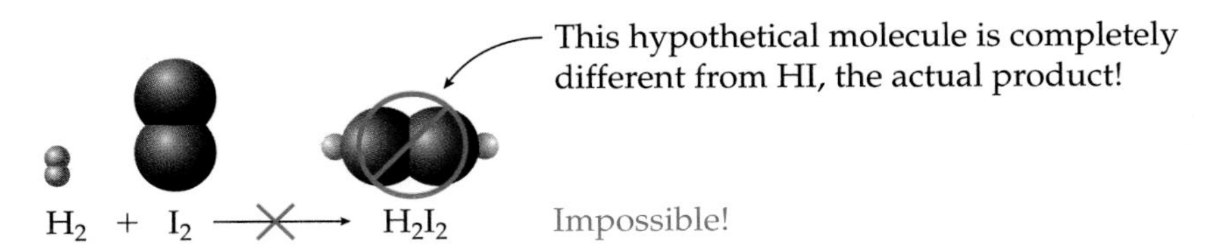

This hypothetical molecule is completely different from HI, the actual product!

$H_2 + I_2 \not\longrightarrow H_2I_2$ Impossible!

Changing the subscripts from HI to H_2I_2 does balance the equation (which now obeys the law of conservation of matter), but what we did is *absolutely wrong* because the reaction of H_2 and I_2 makes HI, not H_2I_2. (The latter, if it existed, which it doesn't, would be a compound completely different from HI, with a whole different set of chemical and physical properties.) One last time, then: when balancing a chemical equation, *never* change the subscripts in a molecular formula. If you do, you will have changed the identity of the substance and the nature of the chemical reaction.

Many of the chemical equations we deal with in this book can be balanced by the method of inspection, which is just a formal way of saying a trial-and-error approach. The best way to proceed is to consider one element at a time, adjusting the coefficient in front of every molecular formula that contains that element so that there are equal numbers of atoms of that element on each side of the arrow. Then move on to another element in the equation and repeat the procedure.

To illustrate this method, let's balance this equation for the production of ammonia from nitrogen and hydrogen:

$N_2 + H_2 \longrightarrow NH_3$ Not balanced

Ammonia

In scanning this equation from left to right, the first element we come to is nitrogen. There are two nitrogen atoms on the left, which means we must have two nitrogen atoms on the right. We can achieve this by putting a coefficient 2 in front of the NH_3:

$$N_2 + H_2 \longrightarrow 2\,NH_3$$

Now the nitrogen atoms are balanced. The next element to balance is hydrogen. There are two H atoms on the reactant (left) side of the equation. How many hydrogen atoms are there on the product (right) side? If you said three, you need to look again. There are three H atoms in each NH_3 molecule, but there are now two NH_3 molecules, meaning a total of six H atoms on the right side. We need to fix this by increasing the total number of hydrogen atoms on the left side to six. To do this, we place a coefficient 3 in front of the H_2:

$$N_2 + 3\,H_2 \longrightarrow 2\,NH_3 \qquad \text{Balanced}$$

Now there are six hydrogen atoms on each side. Because both the nitrogen and the hydrogen are balanced, the overall equation is also balanced.

Try balancing the following unbalanced chemical equations. Here's a helpful hint. If the equation has a pure elemental substance in it, such as H_2, N_2, O_2, or Fe, save this element for last when balancing by inspection. Doing things this way sometimes makes the job of balancing a bit easier.

Practice Problems

Balance the following chemical equations.

7.1 $CH_4 + O_2 \longrightarrow CO_2 + H_2O$

Answer: Because oxygen appears as a pure elemental substance, we save it for last. Starting with carbon (an arbitrary choice), we note that there is one carbon on each side of the equation, meaning C is balanced. There are four hydrogen atoms on the left and only two on the right. To fix this, we put a balancing coefficient 2 in front of the H_2O:

$$CH_4 + O_2 \longrightarrow CO_2 + 2\,H_2O$$

This balances the hydrogen. Now it's time to balance the oxygen. There are two oxygen atoms on the left and four on the right (two from the CO_2 *molecule and two from the two* H_2O *molecules). To balance the equation, we put a 2 in front of the* O_2:

$$CH_4 + 2\,O_2 \longrightarrow CO_2 + 2\,H_2O$$

7.2 $CaH_2 + H_2O \longrightarrow Ca(OH)_2 + H_2$

7.3 $C_6H_{12}O_6 + O_2 \longrightarrow CO_2 + H_2O$

7.4 $HCl + Na_2CO_3 \longrightarrow NaCl + CO_2 + H_2O$

Of course, things don't always go as smoothly as in the preceding practice problems. For example, consider the combustion of liquid octane, C_8H_{18}, a component of gasoline. The unbalanced equation is

$$C_8H_{18}(l) + O_2(g) \longrightarrow CO_2(g) + H_2O(g) \qquad \text{Not balanced}$$

We'll use the trick of saving the elemental substance (O_2) for last. To balance carbon, we need an 8 in front of the CO_2:

$$C_8H_{18}(l) + O_2(g) \longrightarrow 8\,CO_2(g) + H_2O(g) \qquad \text{Not balanced}$$

To balance hydrogen, we need a 9 in front of the H_2O:

$$C_8H_{18}(l) + O_2(g) \longrightarrow 8\,CO_2(g) + 9\,H_2O(g) \qquad \text{Not balanced}$$

That leaves oxygen. The right side has 25 oxygen atoms (16 from the 8 CO_2 molecules plus 9 from the 9 H_2O molecules). How are we going to get 25 oxygen atoms on the left side? Here's the easiest way:

$$C_8H_{18}(l) + \tfrac{25}{2}O_2(g) \longrightarrow 8\,CO_2(g) + 9\,H_2O(g) \qquad \text{Balanced}$$

We now have 25 oxygen atoms on the left because $\left(\frac{25}{2}\right) \times 2 = 25$ (the 2 comes from the molecular formula O_2). Can we do this? Can we use fractional coefficients to balance equations? Absolutely! However, even though our equation is now balanced, many chemists often take one more step to clear out the fractions. To do this, we multiply the coefficients by a number that removes any fractions present. In this case, all we need to do is multiply through by 2:

$$2\,C_8H_{18}(l) + 25\,O_2(g) \longrightarrow 16\,CO_2(g) + 18\,H_2O(g) \qquad \text{Balanced}$$

We must multiply all the coefficients by the same number so as not to change the relationship between them. Try the following practice problems to see if you're getting the hang of this.

Practice Problems

7.5 Balance the chemical equation for the combustion of hexane, $C_6H_{14}(l)$. The unbalanced equation is

$$C_6H_{14}(l) + O_2(g) \longrightarrow CO_2(g) + H_2O(g)$$

Answer:

Balance C: $C_6H_{14}(l) + O_2(g) \longrightarrow 6\,CO_2(g) + H_2O(g)$

Balance H: $C_6H_{14}(l) + O_2(g) \longrightarrow 6\,CO_2(g) + 7\,H_2O(g)$

Balance O: $C_6H_{14}(l) + \tfrac{19}{2}O_2(g) \longrightarrow 6\,CO_2(g) + 7\,H_2O(g)$

Multiply through by 2 to get rid of the fractional coefficient:

$$2\,C_6H_{14}(l) + 19\,O_2(g) \longrightarrow 12\,CO_2(g) + 14\,H_2O(g)$$

7.6 Balance the chemical equation

$$C_6H_{12}(l) + O_2(g) \longrightarrow H_2C_6H_8O_4(l) + H_2O(l)$$

7.7 Balance the chemical equation

$$CH_3NH_2(g) + O_2(g) \longrightarrow CO_2(g) + H_2O(g) + N_2(g)$$

Once a chemical equation is balanced, it is a valuable tool for answering all sorts of "how much?" questions. How much of each reactant must I mix together to make a certain amount of product? How much product will I make if I mix certain amounts of reactants together? Will any of the reactants be left over (unused), and if so, which ones and how much?

We should point out that a number of chemical equations are extremely difficult, if not impossible, to balance by the method of inspection. A variety of methods have been developed to balance such reactions, but these methods are beyond the scope of this book. As we come across these reactions in later chapters, we shall present their equations in balanced form.

7.4 Types of Reactions

Our planet teems with life, from amoebae to zebras. Consider then the plight of the early biologists who set out to understand living organisms. How were these scientists ever going to make sense of it all? Their answer was to create a biological classification scheme that grouped every living thing into one or another kingdom, followed by one or another phylum, then class, order, family, genus, and finally species. Using this scheme, a modern biologist can tell you that the only relationship between koala "bears" (order Marsupialia) and grizzly bears (order Carnivora) is that both are mammals. In other words, the classification scheme became a tool for organizing and understanding life.

Chemists had a similar problem understanding chemical reactions because hundreds of thousands of seemingly unrelated reactions have been studied over the years. By organizing them into various types, chemists could begin to understand them. Questions such as "What is this reaction doing?" and "Are these two reactions related?" could be answered. The goal in this and the next two sections is to introduce you to some of the reaction types chemists have identified so that you can recognize some common features of various chemical reactions as well. For example, suppose you come across a reaction that you have never seen before, such as the one between calcium ions and carbonate ions:

$$Ca^{2+}(aq) + CO_3^{2-}(aq) \longrightarrow CaCO_3(s)$$

[The (*aq*) means "aqueous solution" (the ion is dissolved in water) and serves the same purpose as the (*s*), (*l*), (*g*) designations used to indicate physical state.] This reaction is used to soften "hard" water. How does it soften water? What is the reaction doing? Can other reactions also be used to soften water? You can get some insight into the answers to these questions by assigning this reaction a reaction type. But don't be surprised if a reaction can be assigned more than one reaction type. Reaction types are not meant to be exclusive or exact. Rather, they are meant to help us understand what a reaction is doing, and some reactions do a few things at once. Let's look at some reaction types.

Have you ever tossed a ball to a friend? Some reactions work like this and are called *single-replacement reactions*. Consider the reaction between iron oxide, Fe_2O_3 (rust), and aluminum powder, Al, called the thermite reaction:

$$2\,Al(s) + Fe_2O_3(s) \longrightarrow Al_2O_3(s) + 2\,Fe(l)$$

The iron and aluminum are "playing ball" with the oxygen. As a result, aluminum has replaced iron in the oxygen compound. In a **single-replacement reaction**, one element replaces another element in a compound.

Now consider the reaction

$$Mg(s) + 2\,HCl(aq) \longrightarrow H_2(g) + MgCl_2(aq)$$

At first glance, this reaction seems unrelated to the thermite reaction. Look closely, though, and you'll see that the chlorine compound has replaced its H atom with a Mg atom (hydrogen and magnesium are "playing ball" with the chlorine). Therefore, this is also a single-replacement reaction. By recognizing this, you begin to understand the reaction.

Now let's get fancier. Suppose you have a red ball and a friend has a green ball. You each toss the other your ball. This is what happens in a *double-replacement reaction*. In a **double-replacement reaction**, two reactants AB and CD react to form products AD and CB. For example, consider this reaction between phosphorus trichloride and silver fluoride:

$$PCl_3(l) + 3\,AgF(s) \longrightarrow PF_3(g) + 3\,AgCl(s)$$

Phosphorus, P, "tossed" its chlorine to silver, Ag, and received fluorine in return. Another example of a double-replacement reaction is

$$Ca(NO_3)_2(aq) + 2\,NaF(aq) \longrightarrow CaF_2(s) + 2\,NaNO_3(aq)$$

Next we come to *decomposition reactions*. A **decomposition reaction** is one in which a compound breaks down (decomposes) into two or more simpler substances. The molecules in living organisms decompose after death to give simpler compounds that serve as nutrients for new life. For example, bacteria cause the decomposition of protein molecules that make up skin and muscles into simple amino acids. Another way to decompose compounds is to heat them. Heat the red compound mercury(II) oxide, for instance, and it decomposes into elemental mercury and oxygen:

$$2\,HgO(s) \xrightarrow{\text{Heat}} 2\,Hg(l) + O_2(g)$$

Combination reactions (also called *synthesis reactions*) are those in which two or more substances combine to give another substance. (In other words, combination reactions are the reverse of decomposition reactions.) Some examples are

$$H_2(g) + I_2(g) \longrightarrow 2\,HI(g)$$

$$2\,Na(s) + Cl_2(g) \longrightarrow 2\,NaCl(s)$$

$$4\,Ni(s) + 4\,Al(s) + S_8(s) \longrightarrow 2\,Ni_2Al_2S_4(s)$$

Combination reaction is a common method pharmaceutical companies use to prepare the drugs that have revolutionized modern medicine.

If at this point you are feeling a little lost, take a look at the following visual summary of the reaction types we have covered so far, and then try the practice problems.

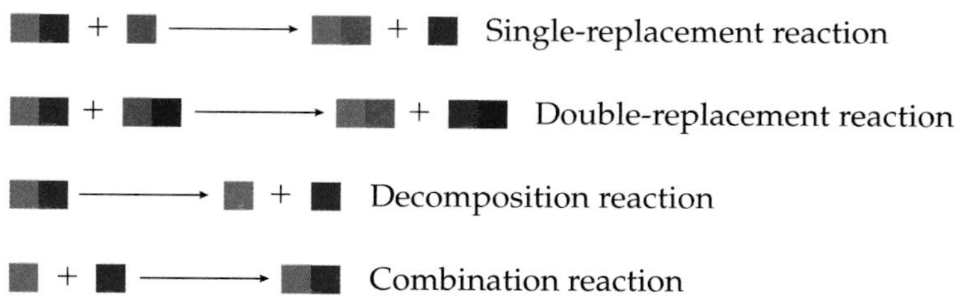

Single-replacement reaction

Double-replacement reaction

Decomposition reaction

Combination reaction

Practice Problems

Balance each equation and classify the reaction as combination, decomposition, single-replacement, or double-replacement.

7.8 $KBr(aq) + Cl_2(aq) \longrightarrow KCl(aq) + Br_2(aq)$

Answer: The balanced equation is

$$2\,KBr(aq) + Cl_2(aq) \longrightarrow 2\,KCl(aq) + Br_2(aq)$$

This is a single-replacement reaction because the potassium compound replaces its bromine with chlorine.

7.9 $Pb(NO_3)_2(s) \xrightarrow{\text{Heat}} PbO(s) + NO(g) + O_2(g)$

7.10 $H_2C{=}CH_2 + Cl_2C{=}CCl_2 \xrightarrow{\text{Catalyst}} H_2C{=}CCl_2$

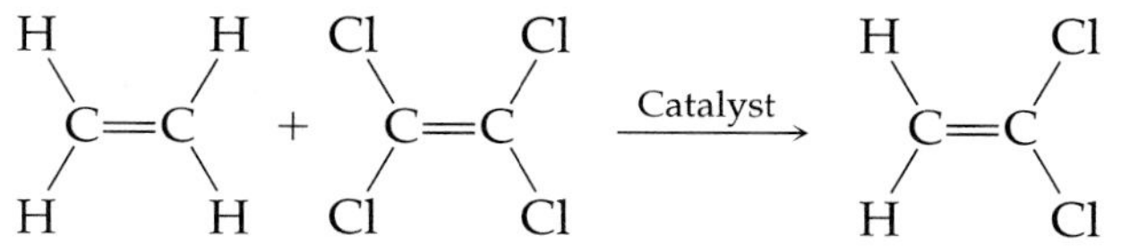

7.5 Solubility and Precipitation Reactions

The ambulance had just pulled up to the hospital's emergency room. The paramedic was shouting out her status report to the doctor: "26-year-old lab technician! Chemical burns to the right arm and both legs. Doused with hydrofluoric acid."

Hydrofluoric acid, an aqueous solution of HF gas, is a good source of fluoride ion, F^-. In less than a minute, the medical team had wheeled the patient into the ICU and begun intravenous fluids. The doctor called out instructions.

"He needs calcium chloride, 10% solution. Set up an IV drip, 1 gram every 10 minutes."

The patient—awake, conscious, and panicked—pummeled the doctor with questions.

Patient: My whole arm is tingling, doctor. Is that bad?
Doctor: Right now we're more worried about the fluoride from the acid. Fluoride ions can remove calcium ions from your cellular fluids, and without calcium ions your heart will stop beating.

The doctor was referring to the Ca^{2+} ions dissolved in the aqueous fluids of our cells. Fluoride ions combine with calcium ions to form solid calcium fluoride, CaF_2, an insoluble salt.

Patient: Is that why you're giving me calcium?
Doctor: Yes, but everything depends on how much fluoride has entered your bloodstream. We'll just have to wait and see.

Within minutes, the patient's electrocardiogram stabilized.

Doctor: Looks like you are going to be all right.
Patient: That's great! Thank you, doctor. You really know your chemistry!
Doctor: Don't thank me. Thank my college chemistry instructor.

Well, what did you expect? This *is* a chemistry text, after all! The reaction occurring in the patient's cellular fluid is called a **precipitation reaction**, defined as one in which an insoluble solid (the **precipitate**) forms when two or more aqueous solutions are mixed. The particular precipitation reaction here is

$$Ca^{2+}(aq) + 2\,F^{-}(aq) \longrightarrow CaF_2(s)$$

The insoluble calcium fluoride, $CaF_2(s)$, is the precipitate. It is an ionic compound, and ionic compounds are often called *salts* (that's why sodium chloride, NaCl, is called table salt). Of course, a big difference between CaF_2 and NaCl is that NaCl is very soluble in water but CaF_2 is not.

SOLUBILITY OF IONIC COMPOUNDS

What happens when an ionic compound dissolves in water? Remember from Section 5.5 that ionic solids consist of a lattice of ions held together by ionic bonds. When an ionic solid dissolves, its lattice comes apart to yield cations and anions in solution, as shown in the figure at the top of the next page. This breaking up of the lattice is called **dissociation** (to dissociate means to separate). When an ionic compound dissolves in water, it dissociates into ions. This is in sharp contrast to most covalent compounds. For example, consider sucrose (table sugar, $C_{12}H_{22}O_{11}$), a covalent compound. When it dissolves in water, you get intact, dissolved sucrose molecules, $C_{12}H_{22}O_{11}(aq)$, with no ions in solution. Keep this in mind because you will often see the molecular formulas for ionic compounds written with an (*aq*) label, as in NaCl(*aq*). The notation NaCl(*aq*) *does not mean there are intact NaCl units dissolved in water.* Rather, it means that $Na^{+}(aq)$ ions and $Cl^{-}(aq)$ ions are present. It is up to you to recognize this.

Of course, not all ionic compounds are soluble in water. Some are incredibly insoluble. We'll discuss why in Chapter 12, but for now it is useful for you to

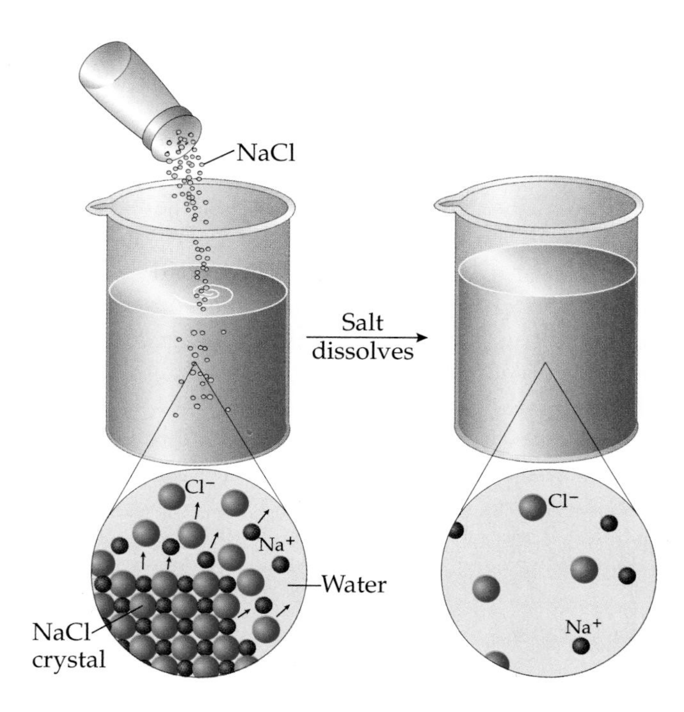

Ions break away from the dissolving crystal.

The solution consists of aqueous Na^+ and Cl^- ions.

know which ionic compounds are soluble and which are not. Thus what we need is a set of solubility rules that gives us this information. Just such a set of rules—all based on observations—has been developed over the years and is shown in Table 7.1. You can see from the table that all sodium salts and all nitrates are soluble in water (this is something worth remembering). Even though this table is not exhaustive (for example, fluorides are not included), you are going to find it very useful for what comes next.

Table 7.1 Solubility Rules

	Soluble in Water	
All sodium, potassium, and ammonium salts		
All acetates and nitrates		
Most halides (chlorides, bromides, iodides)	*except*	Halides of lead(II), silver(I), and mercury(I)
Most sulfates	*except*	Sulfates of calcium, barium lead(II), and strontium
	Insoluble in Water	
Most phosphates, carbonates, and sulfides	*except*	Sodium, potassium, and ammonium salts; calcuim sulfide
Most hydroxides	*except*	Sodium, potassium, calcium, and barium hydroxides

PRECIPITATION REACTIONS—WRITING NET IONIC EQUATIONS

The hydrofluoric acid accident described earlier put the unfortunate laboratory technician in such grave danger because of a precipitation reaction in which CaF_2 was the precipitate. Precipitation reactions can be unwanted, or they can be quite desirable. Until recently, for example, lead salts were used as the coloring pigments in many paints. (Lead compounds are used much less frequently today because of their toxicity.) For example, lead(II) iodide, PbI_2, is a bright yellow powder. Suppose you wanted to make some yellow paint using $PbI_2(s)$ as the pigment. You go to your laboratory to get some PbI_2 but find you have run out. No problem. You'll just make some via a precipitation reaction. On the shelf are bottles of lead(II) nitrate, $Pb(NO_3)_2$, and sodium iodide, NaI, both white powders. After checking the solubility rules and confirming that both these salts are soluble, you prepare two colorless solutions:

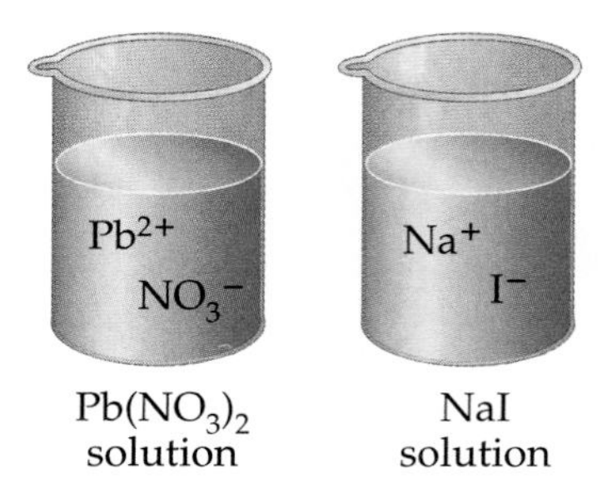

If you combine these two solutions, all of the cations and anions suddenly find themselves in the same solution. This makes possible some new combinations of cations and anions, as indicated here by the crisscross lines:

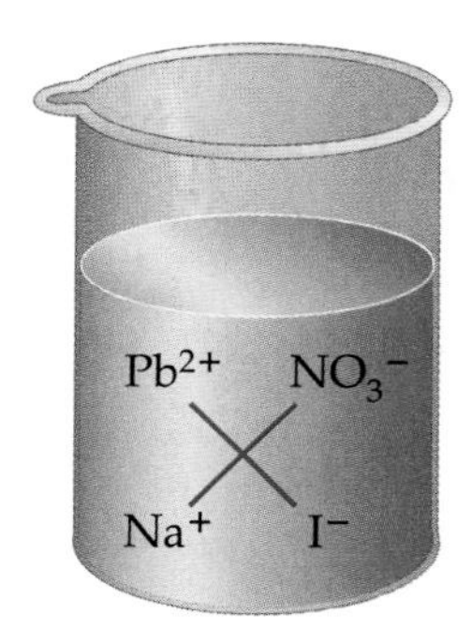

Possible new combinations of cations and anions

This crisscross method of considering new cation–anion combinations tells you which precipitates *might* form. According to the lines, the possible precipitates are lead(II) iodide, PbI_2, and sodium nitrate, $NaNO_3$. *You must be careful with molecular formulas here!* Ionic compounds are always electrically neutral. This means the only precipitate possible when Pb^{2+} is combined with I^- is PbI_2 and *not* PbI. The crisscross method can tell you which ions might form precipitates, but it's up to you to account for charge neutrality to get the correct molecular formula.

So, will you get a precipitate when you combine the two solutions? To predict whether or not you will, simply consult the solubility rules regarding PbI_2 and $NaNO_3$. If both salts are soluble in water, no reaction takes place, but that is not the case here. The solubility rules state that the halides of lead(II) are insoluble, which means yellow PbI_2 precipitates. All you need to do now is isolate the PbI_2 by filtering the solution, and you're ready to mix up a batch of yellow paint. The story is not complete quite yet, however. We know what happened to the $Pb^{2+}(aq)$ ions and $I^-(aq)$ ions, but what happened to the $Na^+(aq)$ ions and $NO_3^-(aq)$ ions? The answer is that absolutely nothing happened to them. These ions started out in solution, and they end up in solution. No $NaNO_3(s)$ precipitates because it is a water-soluble salt.

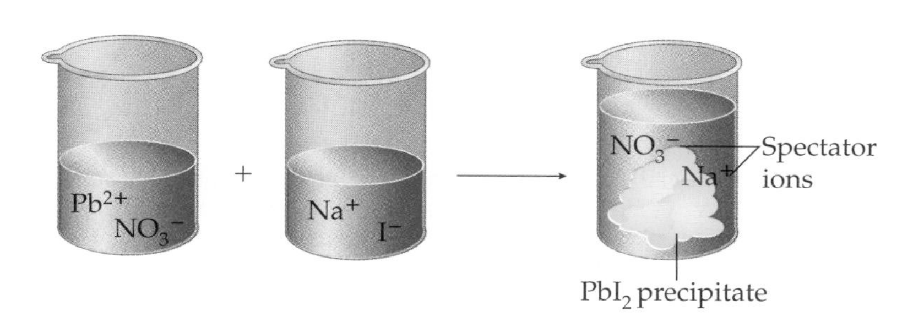

In other words, the $Na^+(aq)$ ions and $NO_3^-(aq)$ ions did not participate in the reaction; they just went along for the ride. We call such ions **spectator ions** because they take no part in the reaction.

All that is left for us to do is to represent what happened with a balanced chemical equation. There are a few ways to do this. We can write it with every species intact rather than dissociated:

$$Pb(NO_3)_2(aq) + 2\,NaI(aq) \longrightarrow PbI_2(s) + 2\,NaNO_3(aq)$$

However, to show the actual species present during the reaction, we can write a **complete ionic equation**, which is one that shows all the ions, both those that take part in the reaction and the spectator ions:

$$Pb^{2+}(aq) + 2\,NO_3^-(aq) + 2\,Na^+(aq) + 2\,I^-(aq) \longrightarrow$$
$$PbI_2(s) + 2\,Na^+(aq) + 2\,NO_3^-(aq)$$

Finally, we can cancel the spectator ions from the two sides of the complete ionic equation. After all, a reaction is defined as a process in which reactants *change* into products, and spectator ions do not change. The result is a chemical equation called a **net ionic equation**:

$$Pb^{2+}(aq) + 2\,I^-(aq) \longrightarrow PbI_2(s)$$

Whichever of the three ways we choose to represent this reaction, we can see that it is a precipitation reaction because an insoluble solid precipitates from solution. However, if we write our equation in the form

$$Pb(NO_3)_2(aq) + 2\,NaI(aq) \longrightarrow PbI_2(s) + 2\,NaNO_3(aq)$$

we can see that it is also a double-replacement reaction, and the net ionic equation form,

$$Pb^{2+}(aq) + 2\,I^-(aq) \longrightarrow PbI_2(s)$$

shows us it is also a combination reaction. As we said earlier, don't be surprised if one reaction can be assigned more than one reaction type.

The following chart summarizes the steps you should use to predict whether a precipitate will form and what it will be.

Identifying the precipitate(s) in a potential precipitation reaction

Step 1: Identify all ions present when the starting solutions are mixed. (These are the ions produced when the starting materials dissolved.)

Step 2: Identify all possible cation–anion combinations that are different from the ones in the starting materials. These combinations represent possible precipitates.

Step 3: Use the solubility rules to decide if any of these new cation–anion combinations represent insoluble salts. Any combination that does will precipitate. If no combinations represent insoluble salts, no precipitation reaction occurs.

Step 4: If a precipitation reaction occurs, write the net ionic equation for it. Make sure the molecular formula you write for the precipitate is electrically neutral.

Now try the following problems. As usual, the first one is worked out for you.

Practice Problems

7.11 A solution of sodium sulfate is combined with a solution of barium chloride. Does a precipitation reaction occur? If so, which salt precipitates?

Answer:

Step 1: Identify the ions present:

$BaCl_2$ solution

Na_2SO_4 solution

Step 2: Identify the possible new combinations of ions (candidates for precipitation):

Mixture

Step 3: Check the solubility rules to find out whether either candidate for precipitation is insoluble in water. The rules tell you NaCl is soluble (all sodium salts are soluble) and $BaSO_4$ is insoluble. Therefore $BaSO_4$ precipitates when the solutions are mixed.

Step 4: Write the net ionic equation for the precipitation reaction:

$$Ba^{2+}(aq) + SO_4^{2-}(aq) \longrightarrow BaSO_4(s)$$

7.12 Aqueous solutions of lead nitrate and sodium sulfate combine to produce a precipitate.
(a) What are the spectator ions?
(b) Write the net ionic equation for this precipitation reaction.

7.13 Aqueous solutions of nickel(II) nitrate and ammonium phosphate combine to produce a precipitate.
(a) What are the spectator ions?
(b) Write the net ionic equation for this precipitation reaction.

Now that you are getting the hang of writing net ionic equations, let's pull out all the stops! Let's try combining *three* solutions of ionic compounds.

Consider three separate aqueous solutions of sodium nitrate, aluminum sulfate, and sodium hydroxide. Draw a beaker for each, and show which ions are present in solution.

Will there be a precipitate when the three solutions of this WorkPatch are combined? To find out, draw a beaker containing all the ions shown in the three beakers of the WorkPatch and apply the crisscross method.

Write the molecular formulas for all potential precipitates in the beaker below.

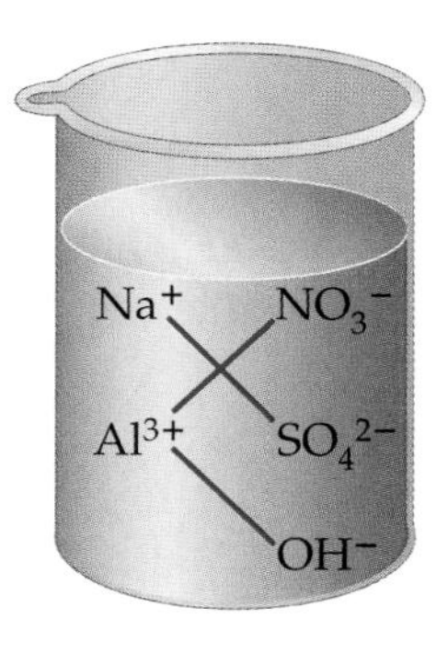

Red lines show only new cation/anion combinations.

Check your formulas against those at the end of the chapter. Of the three precipitates possible, only $Al(OH)_3$ is insoluble in water. Therefore, the net ionic equation for the precipitation reaction is

$$Al^{3+}(aq) + 3\,OH^-(aq) \longrightarrow Al(OH)_3(s)$$

If you feel comfortable with this three-solution problem, you are ready to tackle any other precipitation problem.

7.6 Introduction to Acid–Base Reactions

Do you like the taste of aqueous protons—$H^+(aq)$? Chances are you do. Sour candies, lemons, and other citrus fruits owe their sourness to the $H^+(aq)$ ions they deposit on your tongue. The principal compound responsible for this eye-squinting, mouth-puckering sensation is citric acid:

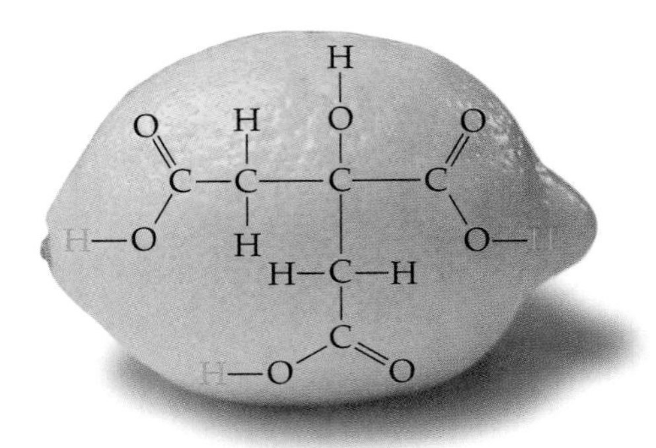

Lemons are loaded with citric acid, $C_3H_4OH(COOH)_3$.

When added to water, a molecule of citric acid dissolves and dissociates to produce a maximum of three $H^+(aq)$ ions:

$$C_3H_4OH(COOH)_3 \xrightarrow{H_2O(l)} C_3H_4OH(COO)_3^{3-}(aq) + 3\,H^+(aq)$$

Any compound that produces $H^+(aq)$ ions when added to water can be called an **acid**. Some other acids present in the food you eat are acetic acid (in vinegar), ascorbic acid (vitamin C), and phosphoric acid (in many soft drinks). There are also many acids you would never eat:

HCl	Hydrochloric acid (also called muriatic acid, used to clean steel)
HNO_3	Nitric acid (used to dissolve many metals)
H_2SO_4	Sulfuric acid (used in car batteries)
HF	Hydrofluoric acid (highly toxic, used to etch glass)

All these compounds are acids because they dissociate to produce $H^+(aq)$ ions when dissolved in water. For instance, if we add HCl gas to water, the HCl dissociates into H^+ and Cl^- ions:

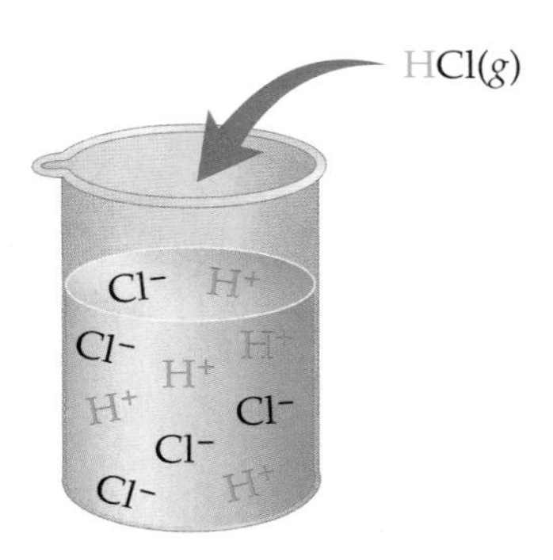

HCl is an acid because it dissociates in water to produce $H^+(aq)$ ions.

You can glance back at the very end of Section 5.7 to see a few more examples of acids. The story doesn't end here, though, because acids have "opposites" called *bases*.

A **base** can be thought of as any substance that produces aqueous hydroxide ions, $OH^-(aq)$, when added to water. A well-known base is sodium hydroxide, NaOH, also called lye.

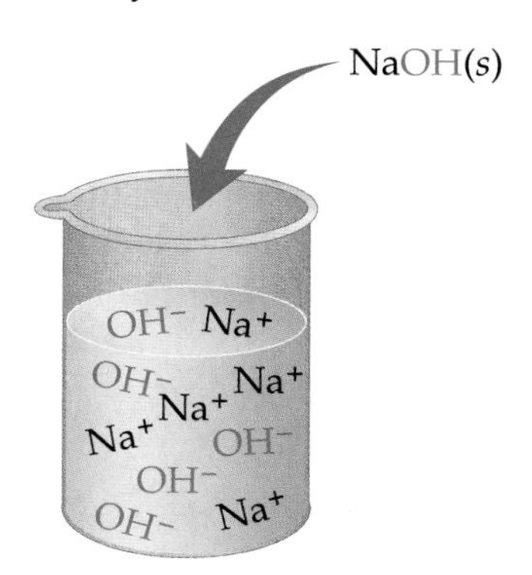

NaOH is a base because it produces $OH^-(aq)$ ions when added to water.

These ionic compounds are some typical bases:

NaOH	Sodium hydroxide
KOH	Potassium hydroxide
$Ca(OH)_2$	Calcium hydroxide

An interesting thing happens when an aqueous solution of an acid meets an aqueous solution of a base, as shown here:

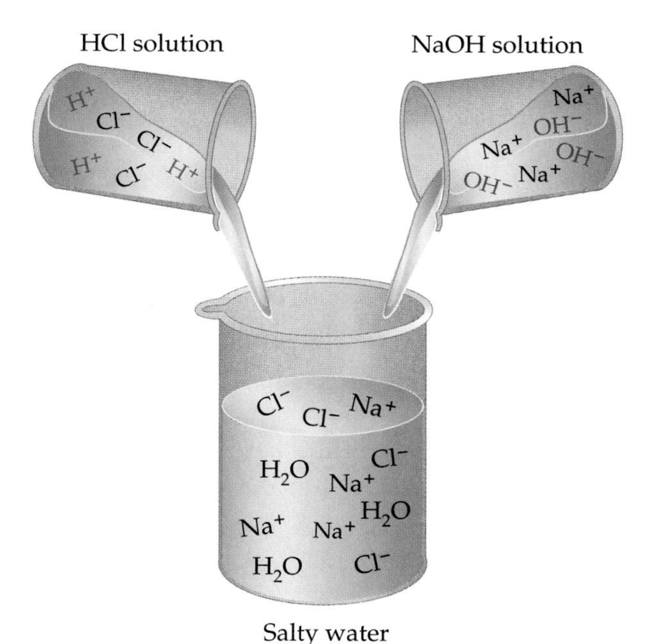

Salty water

We could write the equation for the reaction that occurs with all compounds intact,

$$HCl(aq) + NaOH(aq) \longrightarrow NaCl(aq) + H_2O(l)$$

or we could write the net ionic equation:

$$H^+(aq) + OH^-(aq) \longrightarrow H_2O(l)$$

Either way, the bottom line is that the $H^+(aq)$ ions and $OH^-(aq)$ ions react with each other to form water. We are left with a solution of the two spectator ions Na^+ and Cl^-—in other words, we're left with salty water, just as though we had dissolved some table salt, NaCl, in the beaker.

Any time you react an acidic solution with a basic solution, you end up with a solution of the spectator ions. We call the reaction between an acid and a base an **acid–base neutralization reaction**. The acid *neutralizes* the base by removing the OH^- ions of the base. At the same time, the base *neutralizes* the acid by removing the H^+ ions of the acid. If you were feeling dramatic, you might even say that the $H^+(aq)$ ions and $OH^-(aq)$ ions killed each other. If you combine equal quantities of H^+ ions and OH^- ions, you will end up with a solution that is *neutral* (neither acidic nor basic). This neutralization is why we think of acids and bases as opposites. We shall have much more to say about acids and bases in Chapter 15.

Practice Problems

7.14 (a) Write the intact-compound equation and the net ionic equation for the neutralization reaction that occurs between aqueous solutions of sulfuric acid and potassium hydroxide.
(b) What salt would be isolated if you evaporated the water from this solution after neutralization?

Answer:

(a) Intact-compound equation:

$$H_2SO_4(aq) + 2\,KOH(aq) \longrightarrow K_2SO_4(aq) + 2\,H_2O(l)$$

Net ionic equation:

$$H^+(aq) + OH^-(aq) \longrightarrow H_2O(l)$$

(b) K_2SO_4, *potassium sulfate.*

7.15 It takes 1 mole of NaOH to neutralize 1 mole of HCl, but it takes 2 moles of NaOH to neutralize 1 mole of H_2SO_4. Explain why this is true.

7.16 Write the intact-compound equation and the net ionic equation for the neutralization of aqueous hydrofluoric acid by aqueous calcium hydroxide. What is the name of the salt that forms?

This chapter has given you a feel for what chemical reactions are, how they occur, some of the types that exist, and how to write and balance equations representing them. In the next chapter, we shall look at chemical reactions again but from a quantitative point of view, dealing with such questions as "How many grams of reactants must be combined to get a desired mass of product?" You will learn how to use a balanced chemical equation to answer questions like this. And, as you might have guessed, you will need your calculator.

Have You Learned This?

Chemical reaction (p. 242)
Reactants (p. 242)
Products (p. 242)
Chemical equation (p. 242)
Gas-phase reaction (p. 243)
Single-replacement reaction (p. 250)
Double-replacement reaction (p. 250)
Decomposition reaction (p. 250)
Combination reaction (p. 250)
Precipitation reaction (p. 252)
Precipitate (p. 252)
Dissociation (p. 252)
Spectator ion (p. 255)
Complete ionic equation (p. 255)
Net ionic equation (p. 255)
Acid (p. 258)
Base (p. 258)
Acid–base neutralization reaction (p. 260)

SKILLS TO KNOW

Balancing a chemical equation

1. An equation is balanced when the same numbers of atoms of each element are present on the reactant side and on the product side.	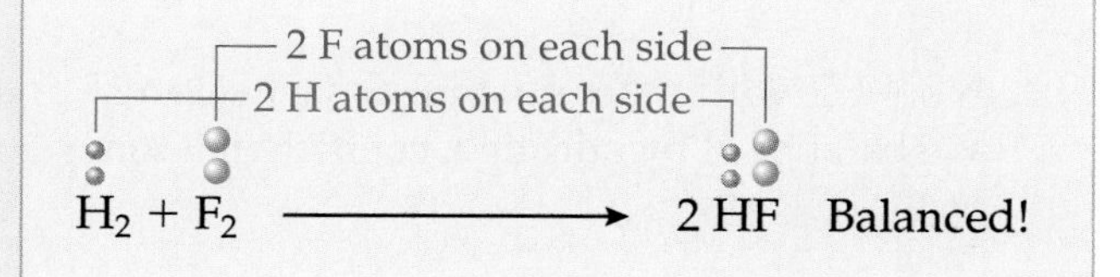
2. Balance an equation by adjusting the balancing coefficients in front of the formulas. Never alter the subscripts in the formulas!	Adjust the balancing coefficients $\blacksquare H_2 + \blacksquare F_2 \longrightarrow \blacksquare HF$ Never change any subscripts!

3. If the equation is complex, balance it one element at a time. First, choose an element. Adjust the coefficients until the equation is balanced for that element. Then, choose another element and repeat the process.	See example below.
4. If the equation contains a pure elemental substance, save that element for last. This sometimes makes the job of balancing easier.	Pure elemental substance, so balance O last in this equation $C_6H_{12}O_6 + O_2 \longrightarrow H_2O + CO_2$
5. If your balanced equation contains a fractional coefficient, it is customary to convert the fraction to a whole number. To do that, multiply all the coefficients on both sides of the equation by the smallest number that will turn the fraction into a whole number.	Balanced equation with fractional coefficient $C_8H_{18} + {}^{25}\!/_2\, O_2 \longrightarrow 9\, H_2O + 8\, CO_2$ Multiply all coefficients by 2: $2\, C_8H_{18} + 2({}^{25}\!/_2)\, O_2 \longrightarrow 2(9)\, H_2O + 2(8)\, CO_2$ $2\, C_8H_{18} + 25\, O_2 \longrightarrow 18\, H_2O + 16\, CO_2$

Example: Balance the equation for the combustion of ethanol, C_2H_5OH:

We shall balance this equation one element at a time, saving O for last because the elemental substance O_2 is present (guideline 4 above).	$C_2H_5OH + O_2 \longrightarrow H_2O + CO_2$
1. We choose to start with C. We can balance it by adding a coefficient 2 for CO_2.	2 C atoms on each side $C_2H_5OH + O_2 \longrightarrow H_2O + 2\, CO_2$
2. Now H. Notice that ethanol has 6 H atoms! We balance H by adding a coefficient 3 for H_2O.	6 H atoms on each side $C_2H_5OH + O_2 \longrightarrow 3\, H_2O + 2\, CO_2$
3. Now O. We add a coefficient 3 to O_2. (We don't alter the coefficient of C_2H_5OH, because that would unbalance C and H.) A check verifies that the equation is balanced.	7 O atoms on each side $C_2H_5OH + 3\, O_2 \longrightarrow 3\, H_2O + 2\, CO_2$ 2 C + 6 H + 7 O atoms on each side **Balanced!**

Solubility rules

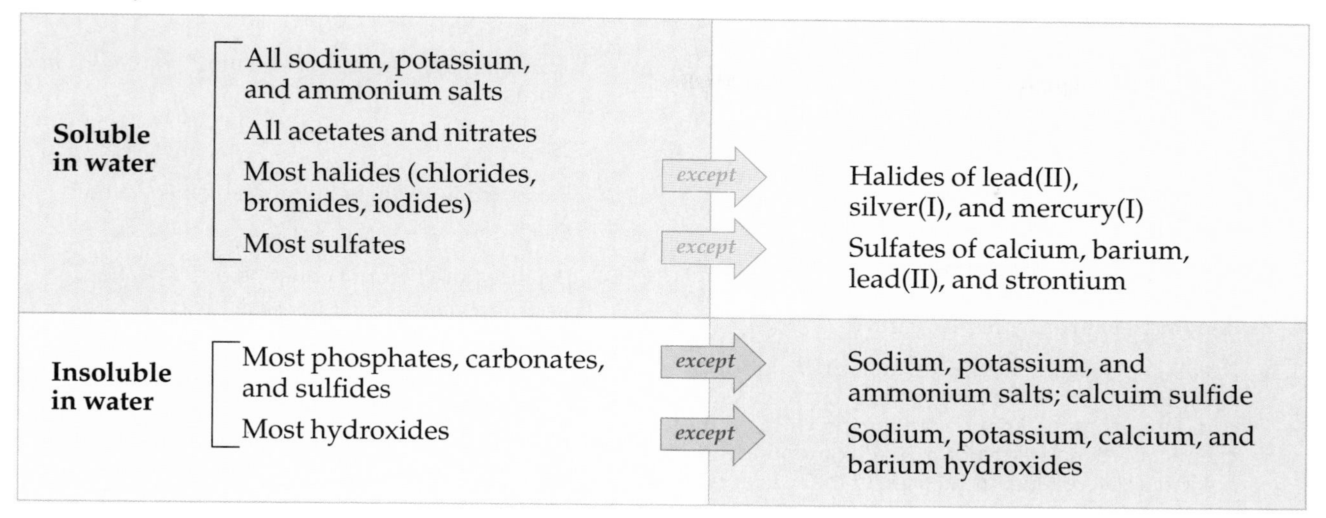

Soluble in water	All sodium, potassium, and ammonium salts		
	All acetates and nitrates		
	Most halides (chlorides, bromides, iodides)	*except*	Halides of lead(II), silver(I), and mercury(I)
	Most sulfates	*except*	Sulfates of calcium, barium, lead(II), and strontium
Insoluble in water	Most phosphates, carbonates, and sulfides	*except*	Sodium, potassium, and ammonium salts; calcuim sulfide
	Most hydroxides	*except*	Sodium, potassium, calcium, and barium hydroxides

Identifying the precipitate(s) in a potential precipitation reaction

Example: A solution of barium chloride is mixed with a solution of sodium sulfate. Will a precipitation reaction occur? If so, what will precipitate?

1. Identify all the ions that will be present when the starting solutions are mixed. (These are the ions produced when the starting materials dissolve.)

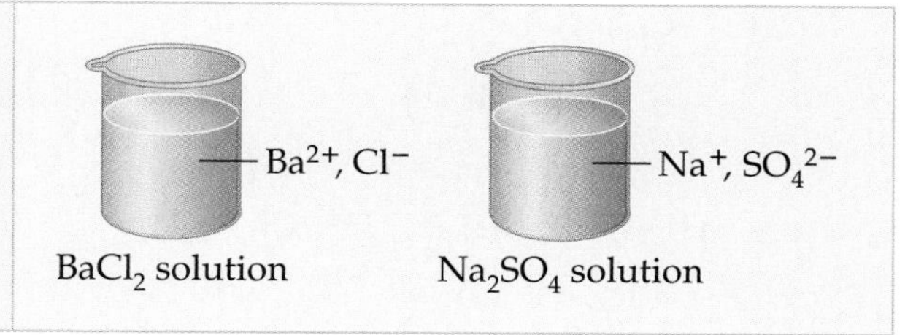

2. Identify all the possible cation/anion combinations that are different from the ones in the starting materials. These combinations represent possible precipitates.

Ba^{2+} Cl^-

Na^+ SO_4^{2-}

Mixture

Possible new combinations:

$BaSO_4$

NaCl

3. Use the solubility rules to decide if any of the ion combinations represents an insoluble salt. Any combination that does will precipitate out of the solution. If none of the combinations represents an insoluble salt, no precipitation reaction will occur.

$BaSO_4$: Insoluble. Will precipitate.

NaCl: Soluble (like all sodium salts). Will not precipitate.

4. If a precipitation reaction occurs, identify it by writing the net ionic equation. Make sure the formula for the precipitate is correct (electrically neutral).

Net ionic equation:

$$Ba^{2+}(aq) + SO_4^{2-}(aq) \longrightarrow BaSO_4(s)$$

CHEMICAL REACTIONS

7.17 Tap water contains dissolved oxygen gas, $O_2(aq)$. Adding heat to tap water causes the dissolved oxygen to leave. Which of the two equations below represents a chemical reaction, and which does not? Defend your answer.

(a) $2\,H_2O_2(l) \longrightarrow 2\,H_2O(l) + O_2(g)$

(b) $H_2O(l) + O_2(aq) \xrightarrow{\text{Heat}} H_2O(l) + O_2(g)$

7.18 Someone claims that a substance has undergone a chemical reaction. What would you have to demonstrate in order to prove this?

7.19 Recall the reaction to make the superconductor described at the beginning of this chapter. What could be done to show that the product of the chemical reaction is truly a new compound and not just a heterogeneous mixture of the reactants?

7.20 Under what circumstances (if any) does it *not* take energy to break a chemical bond?

7.21 Consider the following reaction. Describe it in terms of which bonds must be broken and which bonds must be formed.

$$2\,Cl{-}Cl + \begin{array}{c} \quad H \;\; H \\ \quad | \;\;\; | \\ H{-}C{-}C{-}H \\ \quad | \;\;\; | \\ \quad H \;\; H \end{array} \longrightarrow \begin{array}{c} \quad H \;\; H \\ \quad | \;\;\; | \\ H{-}C{-}C{-}H \\ \quad | \;\;\; | \\ \quad Cl \;\; Cl \end{array} + 2\,H{-}Cl$$

7.22 For most gas-phase chemical reactions, where does the energy come from to break covalent bonds in reactant molecules?

7.23 Cooling a gas-phase reaction mixture slows down the rate at which the reaction proceeds. Cooling it enough can actually stop the reaction. Why do you think this is so?

7.24 Describe how you would explain to someone how a chemical reaction changes one set of compounds into another.

7.25 For most gas-phase reactions, increasing the number of reactant molecules in a flask makes the reaction go faster. Why do you think this is so?

BALANCING CHEMICAL EQUATIONS

7.26 Why must a chemical equation be balanced?

7.27 What is wrong with changing the subscripts in a chemical formula to balance a chemical equation?

7.28 Balance this chemical equation by inspection:

$C_2H_4 + O_2 \longrightarrow CO_2 + H_2O$

7.29 Translate this balanced chemical equation into words:

$\underset{\text{Methane}}{CH_4} + 2\,O_2 \longrightarrow CO_2 + 2\,H_2O$

7.30 Balance this chemical equation by inspection:

$Fe_2O_3 + C \longrightarrow Fe + CO_2$

7.31 Balance this chemical equation by inspection:

$Al_2Cl_6 + H_2O \longrightarrow Al(OH)_3 + HCl$

7.32 Balance this chemical equation by inspection:

$KClO_3 \longrightarrow KCl + O_2$

7.33 Balance this chemical equation by inspection:

$\underset{\text{Acetylene}}{C_2H_2} + O_2 \longrightarrow H_2O + CO_2$

7.34 Translate the balanced chemical equation from Problem 7.33 into words. The compound C_2H_2 is called acetylene and is used as a fuel for torches.

TYPES OF REACTIONS

7.35 Balance this chemical equation and assign it a reaction type:

$N_2O_5(g) \longrightarrow NO_2(g) + O_2(g)$

7.36 Balance this chemical equation and assign it a reaction type:

$CO(g) + NO(g) \longrightarrow CO_2(g) + N_2(g)$

7.37 Balance this chemical equation and assign it a reaction type:

$Fe(NO_3)_3(aq) + Na_2S(aq) \longrightarrow Fe_2S_3(s) + NaNO_3(aq)$

7.38 Balance this chemical equation and assign it a reaction type:

$SO_2(g) + O_2(g) \longrightarrow SO_3(g)$

7.39 Balance this chemical equation and assign it a reaction type:

$Li(s) + N_2(g) \longrightarrow Li_3N(s)$

7.40 Balance this chemical equation:

$C_5H_{10}O_2(l) + O_2(g) \longrightarrow CO_2(g) + H_2O(g)$

7.41 Under the proper conditions, hydrogen peroxide, $H_2O_2(l)$, can be converted to water and oxygen gas. Write a balanced equation for this reaction and assign it a reaction type.

7.42 Compare and contrast combination reactions and decomposition reactions. Give an example of each.

SOLUBILITY AND PRECIPITATION REACTIONS

7.43 Fill in this table:

Name	Formula	Soluble or insoluble in water?
Sodium phosphate		
	$Ba(CH_3CO_2)_2$	
	$(NH_4)_2S$	
Iron(II) carbonate		
Mercury(II) chloride		
	$Co(OH)_2$	
	$HgCl$	

7.44 Draw a beaker that shows the result of dissolving methanol, CH_3OH, in water and a second beaker that shows the result of dissolving sodium phosphate in water. Indicate all species present in solution in each beaker. Describe the difference between how these two substances behave when added to water.

7.45 "Hard" water is hard because it contains $Ca^{2+}(aq)$ and $Mg^{2+}(aq)$ ions. It can be softened by adding borax powder, which is sodium tetraborate, $Na_2B_4O_7$. Adding borax removes the $Ca^{2+}(aq)$ and $Mg^{2+}(aq)$ from solution. Write a net ionic equation for each of these two reactions. What ion replaces the $Ca^{2+}(aq)$ and $Mg^{2+}(aq)$ ions in the water?

7.46 Write a net ionic equation for the precipitation reaction, if any, that occurs when aqueous solutions of the following ionic compounds are mixed:
(a) $Pb(NO_3)_2$ and NaCl
(b) $Ba(NO_3)_2$ and $NiSO_4$
(c) NaCl and KNO_3

7.47 Write a net ionic equation for the precipitation reaction, if any, that occurs when aqueous solutions of the following ionic compounds are mixed:
(a) $Bi(NO_3)_3$ and NaOH
(b) K_2SO_4 and SrI_2
(c) $Cu(CH_3COO)_2$ and Na_3PO_4

7.48 Draw a beaker that shows the result of combining aqueous solutions of Na_2S and nickel(II) nitrate. Indicate which ions are in solution, and write the formula for the precipitate that forms. Write a net ionic equation for the precipitation reaction.

7.49 Aqueous solutions of iron(III) sulfate and barium hydroxide are combined. Does a precipitate form? If yes, write a net ionic equation for the precipitation reaction.

7.50 Aqueous solutions of calcium chloride and potassium carbonate are combined. Does a precipitate form? If yes, write a net ionic equation for the precipitation reaction.

7.51 You need calcium sulfate, but there is none in the lab. However, there are lots of other ionic compounds on the shelves. Propose a synthesis for calcium sulfate.

7.52 You need nickel(II) hydroxide, but there is none in the lab. However, there are lots of other ionic compounds on the shelves. Propose a synthesis for nickel(II) hydroxide.

7.53 You dissolve some silver nitrate in your tap water from home, and the water turns cloudy. What chemical species might be in your tap water?

7.54 When an aqueous solution of ammonium sulfate is added to an aqueous solution of calcium nitrate, a precipitate forms.
(a) Write a net ionic equation for the precipitation.
(b) Write the precipitation reaction in a way that emphasizes that a double-replacement reaction has taken place.

INTRODUCTION TO ACID–BASE REACTIONS

7.55 An aqueous solution of hydrosulfuric acid, $H_2S(aq)$, reacts with an aqueous solution of NaOH.
(a) Write the complete ionic equation for the acid–base neutralization that occurs. [*Hint*: $S^{2-}(aq)$ ion is present in the neutralized solution.]
(b) What salt would remain behind if you evaporated the water from the neutralized solution?

7.56 When an aqueous solution of H_2S is mixed with an aqueous solution of copper(II) sulfate, a black precipitate of CuS forms.
(a) Write an equation that shows how $H_2S(aq)$ supplies sulfide ion, $S^{2-}(aq)$, in solution.
(b) Explain how your answer to (a) demonstrates that $H_2S(aq)$ is an acid.
(c) Write the net ionic equation for the formation of CuS.

7.57 Suppose you had oxalic acid, $H_2C_2O_4$, and you wanted to make sodium oxalate, $Na_2C_2O_4$. How might you do this?

7.58 Write (a) a reaction with all species intact and using (*aq*), (*s*), or (*l*) for the reaction between aqueous solutions of sulfuric acid and calcium hydroxide. (b) What salt is formed as a result of neutralization?

7.59 When aqueous H_2SO_4 is added to an aqueous solution of $Pb(NO_3)_2$, a precipitate forms.
(a) What is the precipitate?
(b) Write a net ionic equation for the precipitation reaction.
(c) Did adding $Pb(NO_3)_2(aq)$ to the sulfuric acid neutralize the acid? Explain.

7.60 Ammonium ion is an acid. Write an equation that shows $NH_4^+(aq)$ behaving as an acid.

7.61 You need to prepare some calcium nitrate.
(a) Write an intact-molecule balanced equation for an acid–base neutralization reaction that could be used to prepare calcium nitrate.
(b) Write the net ionic equation for the neutralization.

7.62 Write the complete ionic equation and the net ionic equation for the reaction that occurs when the following solutions are mixed, and name the salt formed in each case:
(a) $HBr(aq) + NaOH(aq)$
(b) $Ca(OH)_2(aq) + HNO_3(aq)$
(c) $Ca(OH)_2(aq) + HBr(aq)$
(d) $Mg(OH)_2(s) + HCl(aq)$

7.63 The compound CH_3COOCH_3 (methyl acetate) is not an acid, but the compound CH_3COOH (acetic acid) is.
(a) Which of the hydrogen atoms in acetic acid is the acidic hydrogen [the one that becomes $H^+(aq)$ ion]?
(b) Speculate about why none of the hydrogen atoms in methyl acetate is acidic. [*Hint*: Look carefully at what the acidic hydrogen atom is bonded to in acetic acid.]

ADDITIONAL PROBLEMS

7.64 When a precipitate forms in water, the water often becomes warmer. What is the source of the heat energy that warms the water?

7.65 Consider the covalent bond in H_2. Why must energy be added to break this bond (in other words, what is the added energy used for)?

7.66 Consider the ionic bonds between Na^+ ions and Cl^- ions in $NaCl(s)$. Why must energy be added to break these bonds (in other words, what is the added energy used for)?

7.67 Balance each equation and assign each a reaction type:
(a) $SiO_2(s) + C(s) \longrightarrow Si(s) + CO(g)$
(b)

```
    H
    |      ..         ..
H — C — O — H + H — Cl: ⟶
    |      ..         ..
    H

        H
        |  ..          ..
    H — C — Cl: + H — O — H
        |  ..          ..
        H
```

(c) $Al(s) + O_2(g) \longrightarrow Al_2O_3(s)$
(d) $(NH_4)_2Cr_2O_7(s) \longrightarrow$
$N_2(g) + Cr_2O_3(s) + H_2O(g)$

7.68 Aqueous solutions of sodium sulfide and iron(III) nitrate are combined.
(a) Does a precipitation reaction occur? Explain.
(b) If it does, write a net ionic equation for the reaction.

7.69 Aqueous solutions of calcium chloride and potassium carbonate are combined.
(a) Write the formulas for both reactants.
(b) Does a precipitation reaction occur? Explain.
(c) If it does, write a net ionic equation for the reaction.

7.70 Describe what happens when $CaCl_2(s)$ and $NaNO_3(s)$ are added to the same beaker of water. What, if anything, dissolves, and what, if anything, precipitates?

7.71 Suppose you have an aqueous solution containing Cu^{2+} ions and Na^+ ions. What could you add that would remove the Cu^{2+} ions from solution without adding any new type of ions to the solution?

7.72 Suppose you have an aqueous solution of Pb^{2+} and Ba^{2+} ions.
(a) How could you remove just the Pb^{2+} ions from the solution?
(b) How could you remove both the Pb^{2+} and the Ba^{2+} ions from the solution at the same time?

7.73 Nitric acid consists of NO_3^- and a proton, H^+.
(a) Draw a Lewis dot diagram for the nitrate ion.
(b) Draw a Lewis dot diagram for nitric acid. [*Hint*: Nitric acid contains an O–H bond.]
(c) Which species, nitrate ion or nitric acid, is more likely to be found in aqueous solution? Explain.

7.74 Write the intact-molecule equation and the net ionic equation for the neutralization reaction between perchloric acid, $HClO_4(aq)$, and $NaOH(aq)$. What salt is in solution once the reaction is run?

7.75 $Na^+(aq)$, $Pb^{2+}(aq)$, and $Ba^{2+}(aq)$ are all present in the same solution. Fully describe a procedure whereby you could separate these ions from one another.

7.76 $Fe^{3+}(aq)$, $Ca^{2+}(aq)$, and $Ba^{2+}(aq)$ are all present in the same solution. Fully describe a procedure whereby you could separate these ions from one another.

7.77 Heptane, C_7H_{16}, is one of the components of gasoline. Burning heptane in air yields carbon dioxide and water. Balance the equation for this reaction:

$$C_7H_{16}(l) + O_2(g) \longrightarrow CO_2(g) + H_2O(g)$$

7.78 The burning of ethanol, C_2H_5OH, in air yields carbon dioxide and water. Balance the equation for this reaction:

$$C_2H_5OH(l) + O_2(g) \longrightarrow CO_2(g) + H_2O(g)$$

7.79 Sodium chloride reacts with bromine gas, Br_2, to produce sodium bromide plus chlorine gas. Write a balanced equation for this reaction.

7.80 Indicate which bonds are broken and which are formed when each reaction occurs:
(a) $N_2 + 3\,Br_2 \longrightarrow 2\,NBr_3$
(b) $P_4 + 6\,H_2 \longrightarrow 4\,PH_3$
(c) $KF + NaI \longrightarrow KI + NaF$
(d) $CH_4 + Cl_2 \longrightarrow CH_3Cl + HCl$

7.81 Balance the following equation and classify the reaction it represents as combination, decomposition, single-replacement, or double-replacement:

$$Fe + CdCl_2 \longrightarrow FeCl_3 + Cd$$

7.82 Indicate whether each compound is soluble or insoluble in water:

$BaCO_3$ K_2S $CaSO_4$ $MgNO_3$ $(NH_4)_3PO_4$

7.83 Would you expect a reaction in which all reactants are in the gas phase to go faster, slower, or remain at the same rate if the temperature is increased while everything else is held constant? Explain your answer.

7.84 Draw dot diagrams for the reactants and products in each equation, and indicate which bonds are broken and which bonds are formed.
(a) $CH_4 + I_2 \longrightarrow CH_3I + HI$
(b) $C_2H_4 + F_2 \longrightarrow FCH_2CH_2F$

7.85 Which equations represent chemical reactions:
(a) $CO_2(s) \longrightarrow CO_2(g)$
(b) $3\,O_2(g) \longrightarrow 2\,O_3(g)$
(c) $NaCl(s) + H_2O(l) \longrightarrow NaOH(aq) + HCl(aq)$
(d) $H_2O(g) + CH_4(g) \longrightarrow CH_4(g) + H_2O(g)$

7.86 Hydrogen gas is produced when elemental tin reacts with HF to produce hydrogen and tin(II) fluoride. Write the balanced equation for this reaction.

7.87 Classify each compound as acid, base, or salt:

KCl CH_3COOH $Al(OH)_3$ H_3BO_3 $LiC_2H_3O_2$

7.88 Write the balanced chemical equation for the reaction between pentane, C_5H_{12}, and oxygen gas to produce carbon dioxide and water.

7.89 Write the balanced equation, the complete ionic equation, and the net ionic equation for the neutralization reaction that occurs when nitric acid, HNO_3, is mixed with calcium hydroxide, $Ca(OH)_2$.

7.90 Write the balanced chemical equation for the production of P_2O_5 from P and O_2.

7.91 Write the formula of the precipitate formed when these solutions are mixed:
(a) Aluminum nitrate and sodium hydroxide
(b) Potassium phosphate and calcium chloride
(c) Magnesium sulfate and sodium carbonate

7.92 Write the balanced equation for the preparation of iron(III) oxide from iron metal and oxygen gas. What type of reaction is this?

7.93 Balance this chemical equation, and classify it as combination, decomposition, single-replacement, or double-replacement:

$$H_3PO_4 + Ca(OH)_2 \longrightarrow Ca_3(PO_4)_2 + H_2O$$

7.94 Ammonia, NH_3, reacts with chlorine gas, Cl_2, to form ammonium chloride, NH_4Cl, and nitrogen trichloride, NCl_3. Write a balanced equation for this reaction.

7.95 Classify each reaction as combination, decomposition, single-replacement, or double-replacement:
(a) $2\,NaCl \longrightarrow 2\,Na + Cl_2$
(b) $Ba_3N_2 \longrightarrow 3\,Ba + N_2$
(c) $Cl_2 + NaI \longrightarrow NaCl + I_2$
(d) $KNO_3 + AgBr \longrightarrow AgNO_3 + KBr$

7.96 What is the number of moles of NaOH required to neutralize 1 mole of
(a) H_2SO_4
(b) HI
(c) H_3PO_4
(d) HNO_3
(e) CH_3COOH

7.97 Oxygen can be produced by the decomposition of potassium chlorate, $KClO_3$. The products of the reaction are KCl and O_2. Write a balanced equation for the reaction.

7.98 Balance each equation:
(a) $Zn + HCl \longrightarrow ZnCl_2 + H_2$
(b) $Na_2O + H_2O \longrightarrow NaOH$
(c) $CH_4 + H_2S \longrightarrow CS_2 + H_2$
(d) $CO + H_2 \longrightarrow CH_3OH$

7.99 Calcium hydroxide can be used as an antacid to neutralize HCl, the acid found in the stomach. Water and calcium chloride, $CaCl_2$, are the products of the reaction. Write the balanced equation for the reaction.

7.100 Carbon dioxide gas and calcium oxide are produced when calcium carbonate, $CaCO_3$, is heated strongly. Write the balanced equation for this process.

7.101 Balance this chemical equation and classify it as combination, decomposition, single-replacement, or double-replacement:

$$K_2SO_3 + S_8 \longrightarrow K_2S_2O_3$$

7.102 Name the acid and base that combine to form
(a) KNO_3 (b) $Ca_3(PO_4)_2$ (c) Li_2SO_4 (d) NaI

7.103 Mineral compounds containing sulfur are converted to oxygen-containing compounds by a process known as *roasting*, which involves heating the mineral in the presence of oxygen. In addition to the mineral oxide, sulfur dioxide is also produced.

Write the balanced chemical equation for the roasting of zinc sulfide, ZnS, to produce zinc oxide, ZnO, plus sulfur dioxide.

7.104 Chemical reactions occur when new substances are produced as a result of bonds being ________ in the reactants and bonds being ________ in the products.

7.105 When propane gas, C_3H_8, reacts with oxygen gas, carbon dioxide gas and water are produced. Write the balanced equation for the reaction.

7.106 Balance this equation and classify it as combination, decomposition, single-replacement, or double-replacement:

$$K_2CrO_4 + Al(NO_3)_3 \longrightarrow Al_2(CrO_4)_3 + KNO_3$$

7.107 The reaction of carbon monoxide gas with water vapor produces carbon dioxide gas and hydrogen gas. Write the balanced equation for the reaction.

7.108 When an aqueous acid is combined with an aqueous base, the products are a ________ and ________.

7.109 Balance this equation and classify it as combination, decomposition, single-replacement, or double-replacement:

$$Mn + S \longrightarrow MnS_2$$

7.110 Which anions form compounds that are generally soluble in water: CO_3^{2-}, NO_3^-, Br^-, SO_4^{2-}, OH^-, PO_4^{3-}?

7.111 Sulfur dioxide and water are formed when hydrogen sulfide, H_2S, reacts with oxygen gas. Write the balanced equation for the reaction.

7.112 Write the balanced equation for the decomposition of magnesium oxide, MgO, into Mg and O_2.

7.113 A solution containing which anion can be added to separate
(a) $Na^+(aq)$ from $Ag^+(aq)$
(b) $Na^+(aq)$ from $Ca^{2+}(aq)$

7.114 Balance this equation and classify it as combination, decomposition, single-replacement, or double-replacement:

$$H_2O_2 \longrightarrow H_2O + O_2$$

7.115 Write the net ionic equation for the reaction between solutions of
(a) Aluminum nitrate and sodium hydroxide
(b) Potassium phosphate and calcium chloride
(c) Magnesium sulfate and sodium carbonate

7.116 Heating magnesium carbonate results in the production of carbon dioxide and magnesium oxide. Write the balanced equation for this process, and indicate the type of reaction.

7.117 Is the reaction $LiOH + HCN \longrightarrow H_2O + LiCN$ a combination, decomposition, single replacement, or double replacement?

7.118 Indicate whether a precipitate forms when these solutions are mixed, and write the formula of the precipitate if one forms:
(a) Silver nitrate and potassium iodide
(b) Lithium sulfate and silver acetate
(c) Sodium chloride and ammonium sulfate

7.119 What does it mean when the label (*aq*) is attached to the formula for an ionic compound?

7.120 The ____________ ionic equation for a reaction is obtained by canceling the spectator ions in the ____________ ionic equation.

7.121 Write the balanced equation, the complete ionic equation, and the net ionic equation for the reaction between aqueous potassium chloride and lead(II) nitrate to produce lead chloride and potassium nitrate.

7.122 Acids produce _____ ions when dissolved in water, and bases produce ______ ions when dissolved in water.

7.123 The reaction between an acid and a base is called an acid–base ________________ reaction.

7.124 Write the balanced equation for the reaction that occurs when an aqueous solution of carbonic acid, H_2CO_3, is mixed with an aqueous solution of lithium hydroxide, LiOH.

7.125 Of the types of reactions discussed in the chapter, list three ways the reaction $Ca(OH)_2 + H_2SO_4 \longrightarrow CaSO_4 + 2\,H_2O$ can be classified.

7.126 (a) True or false? It is possible for a reaction to be simultaneously a single-replacement reaction and a combination reaction.
(b) True or false? It is possible for a reaction to be simultaneously a double-replacement reaction and a neutralization reaction.
If either statement is false, explain why.

7.127 Would you expect a reaction in which all reactants are in the gas phase to go faster, slower, or remain at the same rate if the size of the container is increased while everything else is held constant? Explain.

7.128 Nitrous oxide, N_2O, is produced when ammonium nitrate, NH_4NO_3, is heated gently. Water is also produced in the reaction. Write the balanced equation for the reaction.

7.129 Write the complete ionic equation and the net ionic equation for the reaction that occurs between aqueous solutions of
(a) Silver nitrate and potassium iodide
(b) Lithium sulfate and silver acetate

7.130 Write a balanced chemical equation for each reaction:

(a) Calcium metal reacts with solid phosphorus, P_4, to form calcium phosphide.
(b) Sodium metal reacts with water to produce aqueous sodium hydroxide and hydrogen gas.
(c) Ammonium nitrate powder explodes to form nitrogen gas, oxygen gas, and water vapor.

7.131 Ammonia gas reacts with fluorine gas to form ammonium fluoride and nitrogen gas.

(a) Write a balanced chemical equation for this reaction.
(b) Draw a picture representing the molecular view of the reactant mixture.
(c) Draw a picture representing the molecular view of the product mixture.

7.132 Write the complete ionic equation and net ionic equation for the neutralization reaction between aqueous hydrobromic acid and barium hydroxide.

7 WORKPATCH SOLUTIONS

7.1 (a) Not a chemical reaction because the same compound exists on both sides of the arrow. This is simply a phase change.
(b) A chemical reaction because a new compound, different from the reactants, is produced.
(c) Not a chemical reaction because the same compounds appear on both sides of the arrow.

7.2 There are two correct answers to this question:

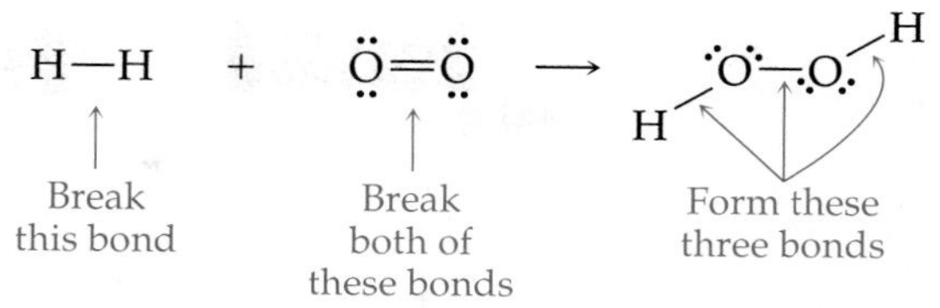

H—H + O=O ⟶ H—O—O—H

Break this bond

Break both of these bonds

Form these three bonds

or

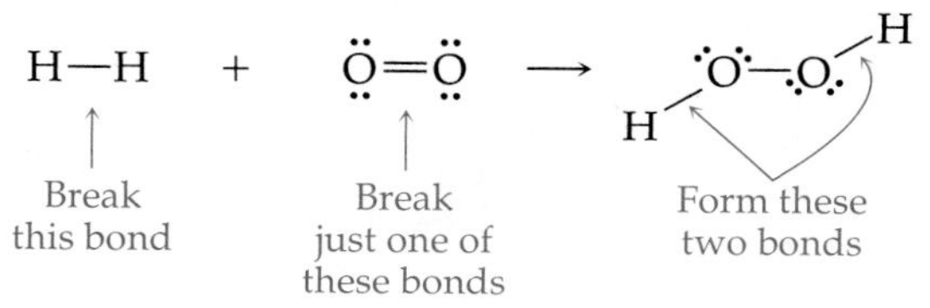

H—H + O=O ⟶ H—O—O—H

Break this bond

Break just one of these bonds

Form these two bonds

7.3

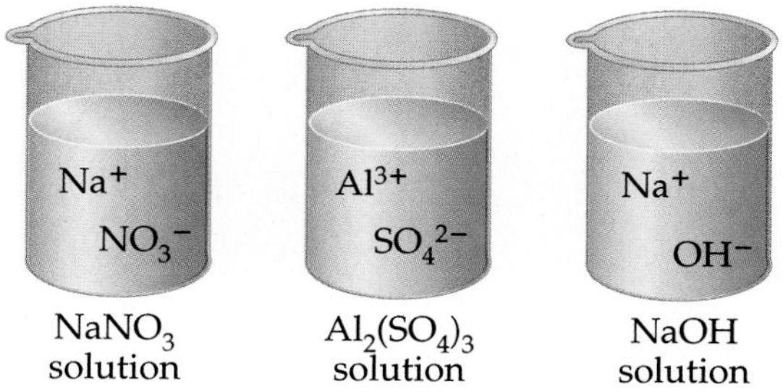

Na^+ NO_3^- — $NaNO_3$ solution

Al^{3+} SO_4^{2-} — $Al_2(SO_4)_3$ solution

Na^+ OH^- — NaOH solution

7.4 $Al(OH)_3$, $Al(NO_3)_3$, Na_2SO_4

$$CH_4 + 2O_2 \longrightarrow CO_2 + 2H_2O$$

CHAPTER 8

Stoichiometry and the Mole

8.1 Stoichiometry—What Is It?

All sorts of people love to cook—to mix ingredients and create a *pièce de résistance*. Indeed, some of the most popular books on the market these days are cookbooks. To use these cookbooks, you must follow their recipes. In this chapter you are going to learn how to follow some unusual recipes. Now, you might think this is a strange topic for a chemistry book, but in fact it is not all that strange because chemists really do like to "cook." Many chemists spend their time mixing "ingredients" (reactants) to create a *pièce de résistance* (products). The recipes they follow come in the form of *balanced chemical equations*, and now you are going to learn how to use these equations—how to read them and how to do calculations based on them.

For example, we could whip up a batch of hydrogen iodide! Here is the recipe:

$$H_2 + I_2 \longrightarrow 2\,HI$$

Balanced chemical equation for formation of hydrogen iodide

Suppose you wanted to go into the laboratory and make 10 g of HI. How much H_2 should you use? How much I_2? How could you do the reaction in a "balanced" way, using just the right amounts of H_2 and I_2 so that when the reaction was over, both reactants were completely gone and the only thing in the reaction vessel was 10 g of HI? These "how much" questions are the subject of **stoichiometry** (stoy-key-OM-e-tree), the study of the quantitative aspects of chemical reactions. The word is derived from two Greek roots,

Cheesecake
3 blocks cream cheese
1 cup sugar
5 eggs

stoicheon (for "element") and *metron* ("to measure"). We use stoichiometry to calculate such things as the amount of reactant needed in order to make a certain amount of product or how much product we can expect from a certain amount of reactant.

This is a good place to stop examining chemical recipes and return to cooking something edible because doing so will show you that you already know how to do stoichiometry calculations. Every time you follow a recipe, any recipe, you are using stoichiometry. For example, consider the recipe for cheesecake at left. It's an absolute diet-buster. According to this recipe, five eggs are required to make one cheesecake. So how many eggs would you need to make two cheesecakes? If you said ten, you have just done a successful stoichiometry calculation in your head. Now try doing a few more.

Practice Problems

Do the following stoichiometry problems based on the cheesecake recipe given above.

8.1 You have plenty of eggs and cream cheese but only 3 cups of sugar. How many cheesecakes can you make?

Answer: Because 1 cup of sugar makes one cake, 3 cups can make three cakes.

8.2 If you use 3 cups of sugar in making your cakes, how many eggs will you need?

8.3 If you use 21 blocks of cream cheese, how many eggs will you need?

You probably did all these problems in your head, but suppose you had to teach someone how to do the calculations on paper. How might you do it? As a budding chemist, you would first write the recipe in a form that makes it look more like a balanced chemical equation:

$$\text{3 blocks cream cheese} + \text{5 eggs} + \text{1 cup sugar} \longrightarrow \text{1 cheesecake}$$

It may look less appetizing this way, but it still says the same thing. Now, on paper, how do we solve the following problem: how many eggs are required to bake two cakes? We saw this type of problem in Chapter 2—it is a unit conversion problem, much like converting feet to meters. In this case, we want to convert from the given unit *cakes* to the unit we need in the answer, *eggs*. We begin just as we did in Chapter 2, by writing down the given information (number of cakes in this case) and then multiplying by the appropriate conversion factor:

$$\underbrace{2\ \text{cakes}}_{\text{Number}} \times \underbrace{\frac{5\ \text{eggs}}{1\ \text{cake}}}_{\text{Conversion factor}} = 10\ \text{eggs}$$

The conversion factor 5 eggs/1 cake came directly from the balanced cheesecake equation (the recipe). Of course, we could have written the conversion factor as 1 cake/5 eggs, but we want the unit *cakes* to cancel. (Remember that identical units cancel only when they are on both top and bottom in an equation.) What is important to realize here is that the conversion factor came from the balancing coefficients of the cheesecake equation.

To get practice in doing this type of conversion calculation, do the stoichiometry practice problems again. This time, though, don't do them in your head.

Practice Problems

Rework Practice Problems 8.1–8.3 by the method of unit conversion. Write out each solution in full, making sure to show your conversion factor and which units cancel.

8.1 You have plenty of eggs and cream cheese but only 3 cups of sugar. How many cheesecakes can you make?

Answer: Because sugar is the ingredient in short supply, it is the amount of sugar available that determines how many cakes you can make. The recipe tells you that one cake requires 1 cup of sugar. Therefore your conversion factors are

$$\frac{1 \text{ cup sugar}}{1 \text{ cake}} \quad \text{and} \quad \frac{1 \text{ cake}}{1 \text{ cup sugar}}$$

Because you want the unit of your answer to be cakes, *use the latter conversion factor:*

$$\underbrace{3 \text{ cups } \cancel{\text{sugar}}}_{\text{Number}} \times \underbrace{\frac{1 \text{ cake}}{1 \text{ cup } \cancel{\text{sugar}}}}_{\text{Conversion factor}} = 3 \text{ cakes}$$

8.2 If you use 3 cups of sugar in making your cakes, how many eggs will you need?

8.3 If you use 21 blocks of cream cheese, how many eggs will you need?

As the answer to Practice Problem 8.1 notes, the balanced cheesecake equation is a source of conversion factors for doing stoichiometry problems. All you have to do to get your answer is write down the number given in the problem and multiply it by the appropriate conversion factor from the balanced equation. That's basically all there is to solving all stoichiometry problems.

Before we apply these techniques to a chemical reaction, let's try one more cheesecake stoichiometry problem. Suppose we are given 24 ounces of cream cheese. How many cakes can we make? Asking this question is throwing a curveball because there is no way to answer it with the information given. The recipe (equation) deals with cream cheese in units of *blocks* and the problem is stated in terms of *ounces*. The only way we can solve this problem is to know (or be given) the relationship between ounces of cream cheese and blocks of cream cheese. For example, if we are told there are 8 ounces of cream cheese in each block, we can convert from ounces to blocks and then solve the problem. Using the conversion factor

$$\frac{\text{1 block cream cheese}}{\text{8 ounces cream cheese}}$$

we can translate the information we've been given into the language of the recipe:

$$\text{24 ounces } \cancel{\text{cream cheese}} \times \frac{\text{1 block cream cheese}}{\text{8 ounces } \cancel{\text{cream cheese}}} = \text{3 blocks cream cheese}$$

Now we can finish the problem. We take the translated amount of cream cheese and multiply it by the appropriate conversion factor from the balanced equation:

$$\text{3 blocks } \cancel{\text{cream cheese}} \times \frac{\text{1 cake}}{\text{3 blocks } \cancel{\text{cream cheese}}} = \text{1 cake}$$

We have our answer. We can make one cheesecake with 24 ounces of cream cheese.

This problem was complicated because we were given the amount of cream cheese in units that were different from the units in the balanced equation. This will always happen when we stop baking cheesecakes and start mixing chemicals. That's because balanced chemical equations *always* speak in terms of numbers of atoms or molecules but our given information will usually be in terms of grams. Therefore, before we can use a balanced chemical equation as a recipe or source of conversion factors for stoichiometry problems, we have to learn how to translate back and forth between *grams* of atoms/molecules and *numbers* of atoms/molecules. This is where we get some help from the unit known as the *mole*.

The mole is used to convert numbers of atoms/molecules to grams of atoms/molecules. No, not this mole.

8.2 The Mole

Let's whip up another batch of hydrogen iodide. Suppose we need to prepare 10 g, following the "recipe"

$$H_2 + I_2 \longrightarrow 2\,HI$$

A very common beginner's mistake is to assume that because the equation shows that H_2 and I_2 react in a one-to-one ratio, all we need to do is combine 5 g of H_2 and 5 g of I_2 to produce 10 g of HI:

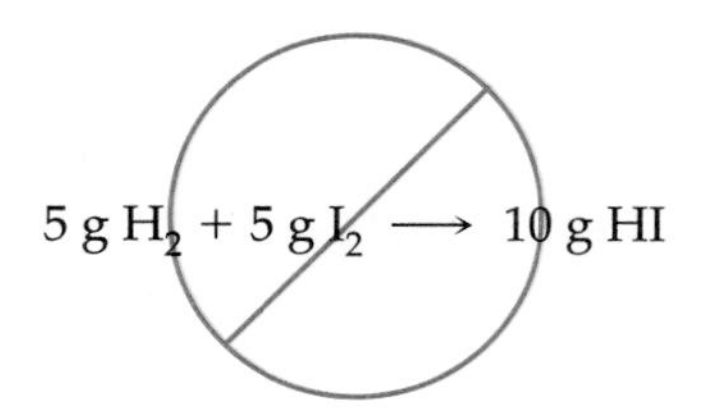

This is not correct, however, because the balancing coefficients in a chemical equation *do not refer to masses*. Instead, they refer to the *number* of atoms or molecules that react and are produced. In other words, the balanced equation does *not* say that 1 g of H_2 reacts with 1 g of I_2 to produce 2 g of HI. What it *does* say is that one *molecule* of H_2 reacts with one *molecule* of I_2 to produce two *molecules* of HI.

Because the balanced equation "speaks" in terms of numbers of molecules, it would be easy to whip up a batch of ten HI molecules. This is just a matter of scaling up the recipe. Our task, however, is to prepare 10 *grams* of HI, and for this task this recipe isn't much help. Plus, a recipe that required us to count out individual molecules would be pretty worthless in any case. Not only can't we see individual H_2 and I_2 molecules, but the numbers we would need to count to prepare 10 g of HI would be astronomical.

. . . one billion and one, one billion and two, . . .

How can we use a balanced chemical equation written in terms of numbers of molecules when we are working with grams? This is where we get assistance from the mole. In concept, a mole is similar to a dozen, which always stands for 12 of anything. A mole is just a whole lot larger: one **mole** (mol) is 6.022×10^{23} of anything. This very large number is called **Avogadro's number**, in honor of Amadeo Avogadro (1776–1856), who first explained how equal volumes of different gases contain the same number of gas molecules. Just as you can have a dozen of something (donuts, dollars, molecules), you can have a mole of something (or Avogadro's number of something). A mole of anything (donuts, dollars, molecules) is always 6.022×10^{23} of that thing. Because this is so, we can always write

$$\frac{1 \text{ mole of something}}{6.022 \times 10^{23} \text{ somethings}} \quad \text{or} \quad \frac{6.022 \times 10^{23} \text{ somethings}}{1 \text{ mole of something}}$$

These are useful conversion factors for doing calculations, as you will soon see.

Whereas a dozen is a conveniently sized number for things like donuts, a mole is a conveniently sized number when dealing with atoms and molecules. Because atoms and molecules are so tiny, you need a tremendous number of them to have a useful and measurable amount. You can get a feel for just how large a mole is by writing Avogadro's number in normal notation:

602,200,000,000,000,000,000,000

Let's put this number—almost a trillion trillion—in perspective. If you counted every grain of sand on every beach and ocean floor on this planet, you would only come close to 1 mole. If you were given 1 mole of dollars the day you were born and then spent a billion dollars a second for your entire life before dying at age 70, you could spend only about 0.001% of your money before you died!

Avogadro's number is huge, but atoms and molecules are the opposite—so tiny that it takes a lot of them to make a dent. One mole of water molecules is barely a swallowful, for instance, and 1 mole of sugar molecules will fit in your cupped hands. It might interest you to know that Avogadro had been in his grave for many years by the time his number was experimentally determined with reasonable accuracy. He never knew the numeric value of the quantity named after him, but you should never forget it.

To see how the mole is going to help us with our problem of mixing grams and molecules, let's look once again at the balanced equation for making HI:

$$H_2 + I_2 \longrightarrow 2\,HI$$

The balancing coefficients in this equation tell us the proportions in which hydrogen and iodine combine, and the key to solving our problem is that these coefficients can be interpreted in more than one way:

H_2	+	I_2	$\longrightarrow$	2 HI
1 molecule		1 molecule		2 molecules
12 molecules (1 dozen molecules)		12 molecules (1 dozen molecules)		24 molecules (2 dozen molecules)
6.022×10^{23} molecules (1 mole)		6.022×10^{23} molecules (1 mole)		$2 \times 6.022 \times 10^{23}$ molecules (2 moles)

So instead of reading the reaction exclusively as "1 H_2 molecule and 1 I_2 molecule react to give 2 HI molecules," we can also read it as "1 mole of H_2 molecules and 1 mole of I_2 molecules react to give 2 moles of HI molecules":

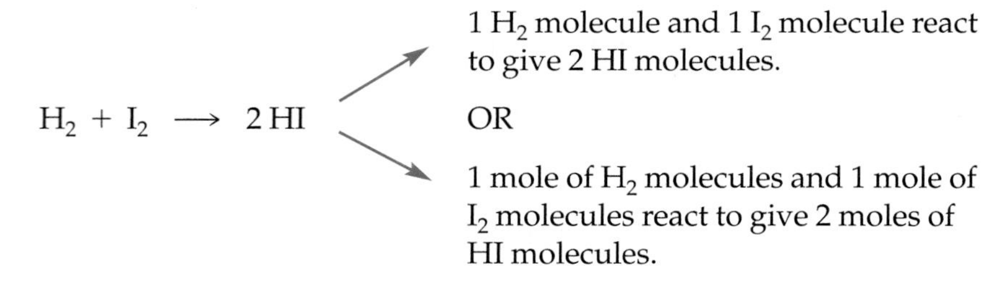

1 H₂ molecule and 1 I₂ molecule react to give 2 HI molecules.
H₂ + I₂ ⟶ 2 HI
OR
1 mole of H₂ molecules and 1 mole of I₂ molecules react to give 2 moles of HI molecules.

This is a fundamental concept in chemistry. Any balanced chemical equation can always be read in terms of moles of molecules instead of individual molecules. This should make sense to you. After all, if the balanced equation says

> 1 H_2 molecule reacts with 1 I_2 molecule to give 2 HI molecules,

it also says

> 1 dozen H_2 molecules react with 1 dozen I_2 molecules to give 2 dozen HI molecules

and

> 1 mole of H_2 molecules react with 1 mole of I_2 molecules to give 2 moles of HI molecules.

I NEED ONE TRILLION HI MOLECULES. PLEASE TELL HER HOW MANY GRAMS OF HI THAT IS.
BUT OF COURSE! THAT'S MY JOB!

This information by itself doesn't help us because we're still speaking in terms of numbers of molecules even though in the laboratory we don't count out molecules directly. Rather, we determine masses. We go to the storage room, find a bottle of iodine, and measure out 10 g of it. If we use 10 g of I_2, how many grams of H_2 do we need? How many grams of HI will we make? To answer these questions, we need to learn one more thing about the mole. Just as we were able to translate from ounces of cream cheese to blocks of cream cheese before using the cheesecake recipe, knowing one more thing about the mole will allow us to translate from grams to numbers of molecules or from numbers of molecules to grams, so that we can use a balanced chemical equation as our recipe.

The mole can do this because it is defined as the number of carbon atoms in exactly 12 g of $^{12}_{6}C$. Experiments over the years have determined that there are 6.022×10^{23} atoms of $^{12}_{6}C$ in 12 g of this carbon isotope. Knowing this fact allows us to "count" the $^{12}_{6}C$ atoms in a sample by determining the mass of the sample. To count out 6.022×10^{23} (1 mole) of $^{12}_{6}C$ atoms, all we have to do is measure out exactly 12 g.

The mole lets us count atoms by determining the mass of a sample.

$$12 \text{ g of } ^{12}_{6}C = 1 \text{ mole of } ^{12}_{6}C = 6.02 \times 10^{23} \text{ atoms of } ^{12}_{6}C$$

Knowing this allows us to do a few things. For example, recall from Chapter 3 that atomic masses of elements are given in atomic mass units and that 1 amu = 1.66056×10^{-24} g. Where did this equivalent value for 1 amu come from? We can use the mole concept to answer this question. By definition, 1 mole of $^{12}_{6}C$ atoms has a mass of exactly 12 g, which means we can write

$$\frac{12 \text{ g}}{1 \text{ mol } ^{12}_{6}C}$$

We can use this conversion factor, along with the knowledge that a mole of anything contains Avogadro's number of that thing, to calculate the mass of one $^{12}_{6}C$ atom:

$$\frac{12 \text{ g}}{1 \text{ mol } ^{12}_{6}C} \times \frac{1 \text{ mol } ^{12}_{6}C}{6.022 \times 10^{23} \text{ atoms } ^{12}_{6}C} = 1.993 \times 10^{-23} \text{ g/atom } ^{12}_{6}C$$

Finally, because 1 amu is defined as 1/12 the mass of a $^{12}_{6}C$ atom, we can say

$$1 \text{ amu} = \frac{1.993 \times 10^{-23} \text{ g}}{12} = 1.661 \times 10^{-24} \text{ g}$$

Now let's get back to how the mole concept allows us to count atoms of any element simply by determining masses. All we need to do is use the atomic masses discussed in Chapter 3. For example, what if we want 1 mole of magnesium, Mg, atoms? The average atomic mass of naturally occurring magnesium is 24.305 amu, as shown in the periodic table. This atomic mass value tells us that a magnesium atom is 24.305 amu/12 amu = 2.025 times more massive than a $^{12}_{6}C$ atom. So, if 1 mole of $^{12}_{6}C$ has a mass of 12 g, then 1 mole of magnesium has a mass 2.025 times that, which is 24.305 g. In other words, to get 1 mole of atoms of any element, the mole concept tells us to take its atomic mass, which is reported in atomic mass units, and change it to units of grams! Consider these elements and their atomic masses:

12
Mg
24.305

Atomic mass (amu) → 1 H 1.0079

1 mole of naturally occurring H atoms has a mass of 1.0079 g.

1 mole of naturally occurring C atoms has a mass of 12.011 g.

1 mole of naturally occurring U atoms has a mass of 238.029 g.

We can generalize the information given for these three elements and say that one mole of any atom has a mass equal to the atom's atomic mass expressed in grams. We call the mass of 1 mole of an atom or molecule its **molar mass**. Therefore, the molar mass of hydrogen is 1.0079 g/mol, the molar mass of carbon is 12.011 g/mol, and the molar mass of uranium is 238.029 g/mol. Once you understand this concept, you can count the number of atoms present in any sample of an element simply by determining the mass of the sample. Said another way, you can convert back and forth between numbers of atoms/molecules and grams of atoms/molecules. Try the following WorkPatch to get some practice.

8.1 WORKPATCH

Consider the element gold, Au, a yellow metal. Suppose you obtained 393.934 g of it.

(a) What is the molar mass of gold?
(b) How many moles of gold do you have?
(c) How many atoms of gold do you have?

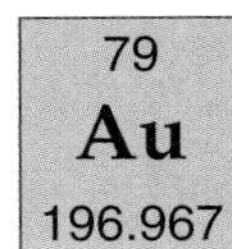

Check your answers against ours at the end of the chapter, and do not go on until you agree with them. Understanding what follows depends on understanding this fundamental mole concept.

Practice Problems

Try to do the following problems in your head.

8.4 Suppose you have 0.5 mole of gold, Au, atoms.
(a) How many gold atoms do you have?
(b) What is the mass in grams of this much gold?

Answer:
(a) Because 1 mole of gold atoms means 6.022×10^{23} atoms, you have 3.011×10^{23} gold atoms.
(b) Because 1 mole of gold has a mass of 196.967 g, 0.5 mole has a mass of 98.484 g.

8.5 Suppose you have 0.10 mole of uranium, U, atoms.
(a) How many uranium atoms do you have?
(b) What is the mass in grams of this much uranium?

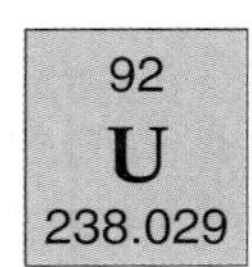

8.6 Suppose you have 120.11 g of carbon atoms.
(a) How many moles of carbon atoms do you have?
(b) How many carbon atoms do you have?

The understanding that 1 mole of atoms of an element has a mass equal to the element's atomic mass expressed in grams is critical to your being able to do chemical stoichiometry problems.

We can calculate the mass of 1 mole of molecules in the same way. For instance, what is the mass in grams of 1 mole of CO_2 molecules? The molecular formula tells us that one carbon dioxide molecule is made from one carbon atom and two oxygen atoms. Following the line of reasoning we developed earlier, it must also be true that 1 mole of CO_2 molecules is made from 1 mole of carbon atoms and 2 moles of oxygen atoms. This is very important! The subscripts in a molecular formula not only say how many atoms there are in one molecule; they also say how many moles of atoms there are in 1 mole of the molecule.

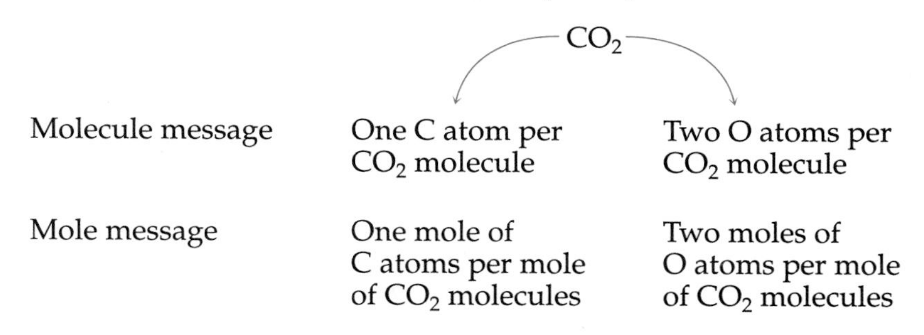

This information contained in formula subscripts is similar to the information contained in the balancing coefficients in a chemical equation, which, as we've already seen, can always be interpreted as moles. To see if you are catching on, try the following WorkPatch. You should be able to do parts (a) through (c) in your head, but a calculator will help for (d).

Suppose you are given 1 mole of methane, CH_4.
(a) How many moles of carbon atoms do you have?
(b) How many moles of hydrogen atoms do you have?
(c) How many atoms of carbon do you have?
(d) How many atoms of hydrogen do you have?

[*Hint for (c) and (d)*: Remember that a mole of anything is 6.022×10^{23} units of it.]

Once you understand the principle behind the answers to the WorkPatch, it is a simple matter to figure out the molar mass of a molecule—that is, the mass in grams of 1 mole. Let's do this for CO_2. We know that 1 mole of CO_2 contains 1 mole of carbon atoms and 2 moles of oxygen atoms, and we also know the mass of 1 mole of each of these atoms. Thus all we have to do is add up the molar masses:

	Molar mass
1 mol C atoms	12.011 g/mol
1 mol O atoms	15.999 g/mol
1 mol O atoms	15.999 g/mol
1 mol CO_2 molecules	44.009 g/mol

Or, because there are 2 moles of O atoms, we could have written

	Molar mass
1 mol C atoms	12.011 g/mol
2 × 1 mol O atoms	2 × 15.999 g/mol
1 mol CO_2 molecules	44.009 g/mol

Either way, the molar mass of CO_2 is 44.009 g/mol. If you measure out 44.009 g of CO_2, you have 1 mole of it, which means 6.022×10^{23} molecules of it. Once again you have counted by measuring masses.

Using the molar masses of the elements from the periodic table, we can calculate the molar mass for any chemical compound so long as we know its molecular formula. Here are some more examples:

Molecular formula	Molar mass
HI	H (1.0079 g/mol) + I (126.905 g/mol) = 127.913 g/mol
H_2	2 × H (1.0079 g/mol) = 2.0158 g/mol
I_2	2 × I (126.905 g/mol) = 253.810 g/mol
H_2O	2 × H (1.0079 g/mol) + O (15.999 g/mol) = 18.015 g/mol

Knowing about moles allows us to go back and forth between numbers of atoms or molecules and their masses in grams.

Practice Problems

8.7 If you have 1 mole of propane, C_3H_8, a gas used in outdoor grills and industrial torches,

(a) How many propane molecules do you have?
(b) How many hydrogen atoms do you have?
(c) What is the mass in grams of the 1 mole of propane?

Answer:

(a) Having 1 mole of propane molecules means you have 6.022×10^{23} propane molecules. (That is the meaning of 1 mole.)

(b) You know that 1 mole of propane contains 8 moles of hydrogen atoms. (That is what the subscript 8 in the formula C_3H_8 tells you.) You also know there are 6.022×10^{23} H atoms per mole of them, which means you have

$$8\ \text{mol H atoms} \times \frac{6.022 \times 10^{23}\ \text{H atoms}}{1\ \text{mol H atoms}} = 4.818 \times 10^{24}\ \text{H atoms}$$

(c) $$\underbrace{(3 \times 12.011\ \text{g})}_{3\ \text{mol C}} + \underbrace{(8 \times 1.0079\ \text{g})}_{8\ \text{mol H}} = \underbrace{44.096\ \text{g}}_{\substack{\text{Molar mass of propane} \\ \text{(mass of 1 mol } C_3H_8)}}$$

8.8 What is the mass in grams of 2 moles of propane?

8.9 How many moles of carbon atoms are there in 2 moles of propane?

8.10 Propane reacts with oxygen to produce CO_2 and H_2O, as shown in the balanced equation

$$C_3H_8 + 5\,O_2 \longrightarrow 3\,CO_2 + 4\,H_2O$$

Write this reaction in words using the word *mole(s)* four times.

Molar mass is very important because it allows you to convert from mass to moles or from moles to mass.

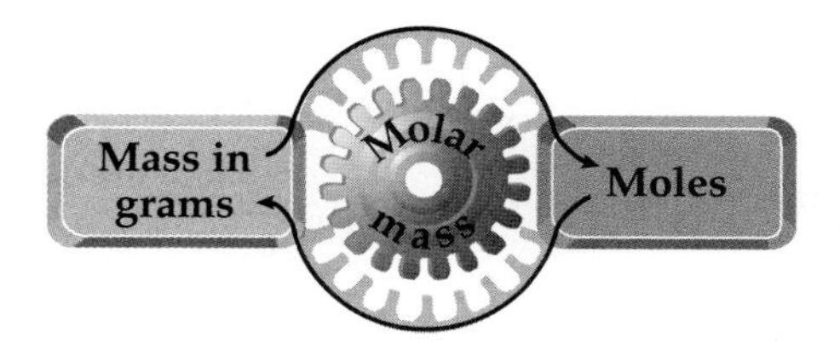

For example, the molar mass of H_2O is 18.015 g/mol; in other words, 1 mole of H_2O has a mass of 18.015 g. Thus, you can write the molar mass of H_2O as if it were a conversion factor:

$$\text{Molar mass } H_2O = \frac{18.015\text{ g } H_2O}{1\text{ mol } H_2O} \qquad \text{and its inverse} \qquad \frac{1\text{ mol } H_2O}{18.015\text{ g } H_2O}$$

You can write the conversion factor either way. Which fraction you use depends on whether you want to convert from moles to grams or from grams to moles. Suppose you had 250.0 g of H_2O. To convert this mass to moles, you simply do the calculation

$$250.0\,\cancel{g}\,H_2O \times \underbrace{\frac{1\text{ mol } H_2O}{18.015\,\cancel{g}\,H_2O}}_{\text{Conversion factor, inverse of molar mass}} = 13.88\text{ mol } H_2O$$

Doing the conversion this way is true to the method of unit analysis (note the cancellation of identical units on top and bottom). However, you will often see mass-to-moles conversions done by dividing the mass by the molar mass rather than multiplying by the inverse of the molar mass:

Mass → Divided by → Molar mass

$$\frac{250.0\text{ g } H_2O}{18.015\text{ g } H_2O/\text{mol}} = 13.88\text{ mol } H_2O$$

You get the same result either way. In other words, dividing something by a conversion factor is identical to multiplying it by the inverse of the conversion factor. We mention this only because many people prefer converting from mass to moles via the latter method.

Now, let's go the other way and convert from moles to mass. Suppose you have 13.88 moles of H_2O. How many grams do you have? Once again, the molar mass takes you to the answer:

$$13.88 \text{ mol } H_2O \times \underbrace{\frac{18.015 \text{ g } H_2O}{1 \text{ mol } H_2O}}_{\text{Conversion factor, molar mass}} = 250.0 \text{ g } H_2O$$

Armed with your knowledge of molar mass and Avogadro's number, you are now ready to convert from mass to moles to numbers of atoms or molecules at will. The following chart summarizes how. Now try some practice problems to see if you are catching on.

Conversions in Stoichiometry Calculations

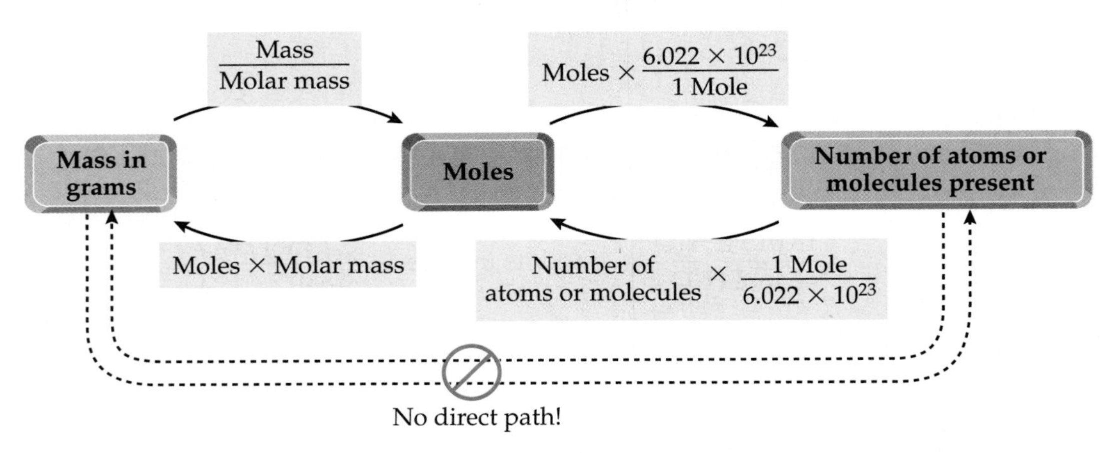

Practice Problems

8.11 Suppose you have 100.0 g of H_2O.
 (a) How many moles of H_2O molecules do you have?
 (b) How many H_2O molecules do you have?

Answer:

(a) $100.0 \text{ g } H_2O \times \frac{1 \text{ mol } H_2O}{18.015 \text{ g } H_2O} = 5.551 \text{ mol } H_2O$

or

$$\frac{100.0 \text{ g } H_2O}{18.015 \text{ g } H_2O/\text{mol}} = 5.551 \text{ mol } H_2O$$

(b) $5.551 \text{ mol } H_2O \times \frac{6.022 \times 10^{23} \text{ molecules } H_2O}{1 \text{ mol } H_2O} = 3.343 \times 10^{24} \text{ molecules } H_2O$

8.12 Suppose you have 0.565 moles of HI.
 (a) How many grams of HI do you have?
 (b) How many HI molecules do you have?

8.13 Suppose you have 5.000×10^{24} molecules of methane, CH_4.
(a) How many moles of methane do you have?
(b) How many grams of methane do you have?

8.3 Reaction Stoichiometry

We are now ready to use the mole in reaction stoichiometry problems. Let's go back to our HI reaction:

$$H_2 + I_2 \longrightarrow 2\,HI$$

Viewed as a recipe, this says "Combine 1 mole of H_2 molecules with 1 mole of I_2 molecules, and they will react to give 2 moles of HI molecules." We can finally go into the laboratory and carry out this reaction because we can now count molecules by determining masses. One mole of H_2 molecules has a mass of 2.0158 g (which we got by doubling 1.0079 g/mol, the molar mass of the H atom), and 1 mole of I_2 molecules has a mass of 253.810 g (twice 126.905 g/mol, the molar mass of the I atom). If we combine these amounts of H_2 and I_2, the reaction tells us how many grams of product to expect. Because the molar mass of HI is 127.912 g/mol, we should produce 2 mol × 127.912 g HI/mol = 255.824 g of HI. This is called the **theoretical yield** of the reaction, the maximum amount of product you can hope to make from a given amount of reactants.

Interestingly, reactions do not always produce the theoretical yield, for a variety of reasons. Many reactions have competing *side reactions* that consume some of the reactants to produce unwanted products, called *side products*. Or, when it is time to recover the desired product, difficulties arise in collecting it or purifying it, causing some of it to be lost. Whatever the reason, it is very common to get something less than the theoretical yield of product. The amount of product you actually end up with is called the **actual yield** of the reaction. Suppose, for example, we were able to isolate only 225.10 g of HI instead of the theoretical yield of 255.824 g. If this were so, the actual yield of the reaction would be 225.10 g of HI.

Actual yield allows us to define yet another quantity, the *percent yield* of the reaction, which is just what it sounds like. The **percent yield** (% yield) of a reaction is the percentage of the theoretical yield we are able to isolate:

$$\%\ \text{yield} = \frac{\text{Actual yield}}{\text{Theoretical yield}} \times 100$$

For example, the calculation for our HI reaction is

$$\%\ \text{yield} = \frac{225.10\ \text{g HI}}{255.824\ \text{g HI}} \times 100 = 87.990\%$$

Our percent yield is almost 88%, which is not bad. Some reactions done in the chemical and pharmaceutical industries have smaller percent yields and are thus quite wasteful of their starting reactants. You end up absorbing this loss in the price you pay for pharmaceuticals and other useful chemicals.

Practice Problems

8.14 (a) Express the balanced chemical equation

$$CH_4 + 2\,O_2 \longrightarrow 2\,H_2O + CO_2$$

in words, using the word *mole(s)* wherever appropriate.

(b) To produce 1 mole of CO_2 from this reaction, how many grams of CH_4 and O_2 must you combine?

(c) What is the theoretical yield in grams of H_2O for this reaction?

(d) If you recover 30.0 g of H_2O, what is the percent yield?

Answer:

(a) One mole of methane molecules and 2 moles of oxygen molecules react to give 2 moles of water molecules and 1 mole of carbon dioxide molecules.

(b) The molar mass of methane is (12.011 g/mol C) + (4 × 1.0079 g/mol H) = 16.043 g/mol CH_4. This is the mass of 1 mole of methane, which is what the reaction calls for. The molar mass of oxygen, O_2, is (2 × 15.999 g/mol O) = 31.998 g/mol. This is the mass of 1 mole of O_2. Because the reaction calls for 2 moles, we multiply by 2 to get 63.996 g of O_2 needed.

(c) The most we can hope to form is 2 moles of water. The molar mass of water is (2 × 1.0079 g/mol H) + (15.999 g/mol O) = 18.015 g/mol H_2O. This is the mass of 1 mole of water. So the theoretical yield is just twice this, which is 36.030 g of H_2O.

(d) % yield = (30.0 g/36.030 g) × 100 = 83.3%

8.15 (a) Balance this unbalanced equation by inspection:

$$\underset{\text{Benzene}}{C_6H_6} + \underset{\text{Hydrogen}}{H_2} \longrightarrow \underset{\text{Cyclohexane}}{C_6H_{12}}$$

(b) Express this reaction in words, using the word *mole(s)* wherever appropriate.

(c) To produce 1 mole of C_6H_{12} from this reaction, how many grams of C_6H_6 and H_2 must you combine?

(d) What is the theoretical yield in grams of C_6H_{12}?

(e) Suppose you recover 24.0 g of C_6H_{12}. What is the percent yield?

8.16 (a) Balance this unbalanced equation by inspection:

$$\underset{\text{Glucose}}{C_6H_{12}O_6} + O_2 \longrightarrow CO_2 + H_2O$$

(b) Express this reaction in words, using the word *mole(s)* wherever appropriate.

(c) To produce 6 moles of H_2O from this reaction, how many grams of glucose and O_2 must you combine?

(d) What is the theoretical yield in grams of CO_2?

(e) Suppose you recover 196.0 g of CO_2. What is the percent yield?

Up to now we have been using a balanced equation like a recipe, following its instructions exactly. But recall that when we were baking cheesecakes, we sometimes wanted to depart from the recipe, doubling it to make two cakes and so forth. The same holds true for chemical reactions. Our balanced equation for hydrogen iodide is $H_2 + I_2 \longrightarrow 2\,HI$, and it tells us how to make 2 moles of HI. But we rarely want to produce exactly 2 moles of HI (255.824 g, twice its molar mass). Suppose we want to make only 10.0 g of HI. How do we cut back on the recipe? The answer comes from a concept we covered in Chapter 2—conversion factors. Both a balanced chemical equation and a molecular formula are sources of conversion factors that are useful for doing stoichiometry problems. We simply read the conversion factors directly from the balanced equation or the formula, like this:

Conversion Factors from Balanced Equation $H_2 + I_2 \longrightarrow 2\,HI$

$\frac{2 \text{ mol HI}}{1 \text{ mol } H_2}$ 2 moles of HI is produced for every 1 mole of H_2 consumed.

$\frac{1 \text{ mol } H_2}{2 \text{ mol HI}}$ 1 mole of H_2 is consumed for every 2 moles of HI produced.

Same, just inverted

$\frac{2 \text{ mol HI}}{1 \text{ mol } I_2}$ 2 moles of HI is produced for every 1 mole of I_2 consumed.

$\frac{1 \text{ mol } I_2}{2 \text{ mol HI}}$ 1 mole of I_2 is consumed for every 2 moles of HI produced.

Same, just inverted

$\frac{1 \text{ mol } I_2}{1 \text{ mol } H_2}$ 1 mole of I_2 is consumed for every 1 mole of H_2 consumed.

$\frac{1 \text{ mol } H_2}{1 \text{ mol } I_2}$ 1 mole of H_2 is consumed for every 1 mole of I_2 consumed.

Same, just inverted

Conversion Factors from Molecular Formulas

$\frac{2 \text{ mol H atoms}}{1 \text{ mol } H_2 \text{ molecules}}$ From formula H_2

$\frac{2 \text{ mol I atoms}}{1 \text{ mol } I_2 \text{ molecules}}$ From formula I_2

$\frac{1 \text{ mol H atoms}}{1 \text{ mol HI molecules}}$ From formula HI

$\frac{1 \text{ mol I atoms}}{1 \text{ mol HI molecules}}$ From formula HI

This concept is so important that you need to practice it now.

(a) Express the balanced equation

$$\underset{\text{Aluminum}}{4\,Al} + \underset{\text{Oxygen}}{3\,O_2} \longrightarrow \underset{\text{Aluminum oxide}}{2\,Al_2O_3}$$

in words, using the word *moles* for every reactant and product.

(b) Write every conversion factor you can think of from the balanced chemical equation.
(c) Write all the conversion factors from the formula Al_2O_3.

Did you get six conversion factors in (b)? Check your answers before going on because being able to get these conversion factors is essential to doing stoichiometry problems.

Before beginning any stoichiometry problem, you need to identify the unknown reactant or product the problem asks about. The easiest way to do this is to write the chemical equation and then, under each known reactant or product, list the information given in the problem statement. Put a question mark under the reactant or product the problem is asking about; this is your *unknown*. In our case, where we want to end up with 10.0 g of HI, we have two unknowns—the H_2 mass and the I_2 mass:

$$\underset{?\,g}{H_2} + \underset{?\,g}{I_2} \longrightarrow \underset{10.0\,g}{2\,HI}$$

Once you have done this preliminary work, the first step is to be sure the equation you wrote is balanced. If it isn't, then balance it.

Now, how do we use the balanced HI equation to determine how many grams of H_2 and I_2 we need to make 10.0 g of HI? Because the reaction speaks in moles, the next step is to translate the 10.0 g of HI into moles of HI. This is the job of the HI molar mass, 127.912 g/mol, used as a conversion factor:

$$10.0\ \cancel{g\ HI} \times \underbrace{\frac{1\ mol\ HI}{127.912\ \cancel{g\ HI}}}_{\text{Inverse of HI molar mass}}$$

By using this conversion factor, we have changed to units of moles of HI (because this is the only unit not crossed out).

The third step is to use the balanced equation to find out how many moles of H_2 and I_2 we need. Let's tackle H_2 first. From the balanced equation, we find the conversion factor that takes us from moles of HI to moles of H_2 and string it into our calculation:

$$10.0\ \cancel{g\ HI} \times \frac{1\ \cancel{mol\ HI}}{127.912\ \cancel{g\ HI}} \times \underbrace{\frac{1\ mol\ H_2}{2\ \cancel{mol\ HI}}}_{\text{From balanced equation}}$$

This step cancels moles of HI and leaves us with moles of H_2. Because we want to know how many grams of H_2 we need, we want our final answer in grams. So, the final step is to use the molar mass of H_2 to convert moles of H_2 to grams of H_2. Notice that the molar mass conversion factor for H_2 is written so that it makes the proper units cancel:

$$10.0\ \cancel{g\ HI} \times \frac{1\ \cancel{mol\ HI}}{127.912\ \cancel{g\ HI}} \times \frac{1\ \cancel{mol\ H_2}}{2\ \cancel{mol\ HI}} \times \underbrace{\frac{2.0158\ g\ H_2}{1\ \cancel{mol\ H_2}}}_{H_2\text{ molar mass}}$$

Surviving unit ($g\ H_2$)

We are done. The only unit not crossed out is the unit of the question we were asked, grams of H_2. All we have to do is perform the math and attach the surviving unit to it. Here is what you would enter into your calculator:

$$10.0 \div 127.912 \div 2 \times 2.0158 = 0.0788 \text{ g } H_2$$

This is the answer we have been looking for. To make 10.0 g of HI, we need 0.0788 g of H_2. We leave the calculation of the amount of I_2 for you to do in WorkPatch 8.4. The following chart and diagram summarize the steps we have just done.

How to solve stoichiometry problems for reactions run with no reactant in short supply

Step 1: Write the balanced chemical equation.

Step 2: For each known reactant or product, convert the given information to moles.

Step 3: Use the balancing coefficients from the equation to find the relationship between the number of moles of the known reactant/product and the number of moles of the unknown reactant/product. Express this relationship as a conversion factor and use it to determine the number of moles of the unknown reactant/product.

Step 4: Convert the molar amount calculated in step 3 to the units asked for in the problem.

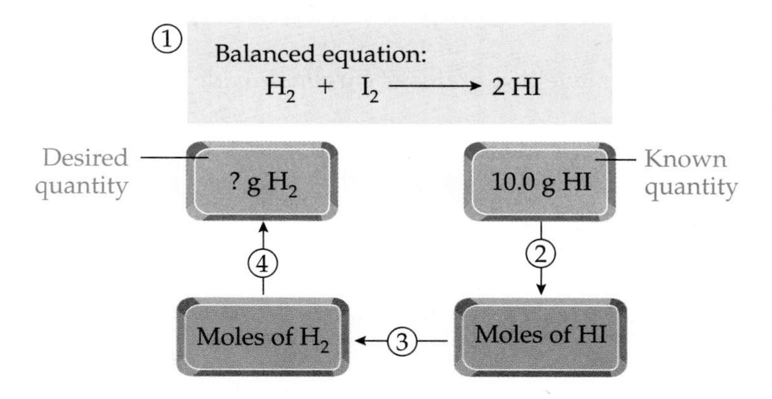

8.4 WORKPATCH

Using the four-step procedure, determine the mass in grams of I_2 required to make 10.0 g of HI. Write out all your steps and cross out all units that cancel.

Steps 1 and 2 in your solution should have been identical to steps 1 and 2 we used when solving for the amount of H_2. Check your work against ours to see how you did.

Practice Problems

In the presence of O_2, glucose "burns" to give CO_2 and H_2O. The balanced equation and molar masses are

	$C_6H_{12}O_6$	+	$6\ O_2$	$\longrightarrow$	$6\ CO_2$	+	$6\ H_2O$
In words →	1 mol of glucose	+	6 mol of oxygen	React to give	6 mol of carbon dioxide	+	6 mol of water
Molar mass →	180.155 g/mol		31.998 g/mol		44.009 g/mol		18.015 g/mol

8.17 How many grams of oxygen does it take to burn 10.0 g of glucose?

Answer:

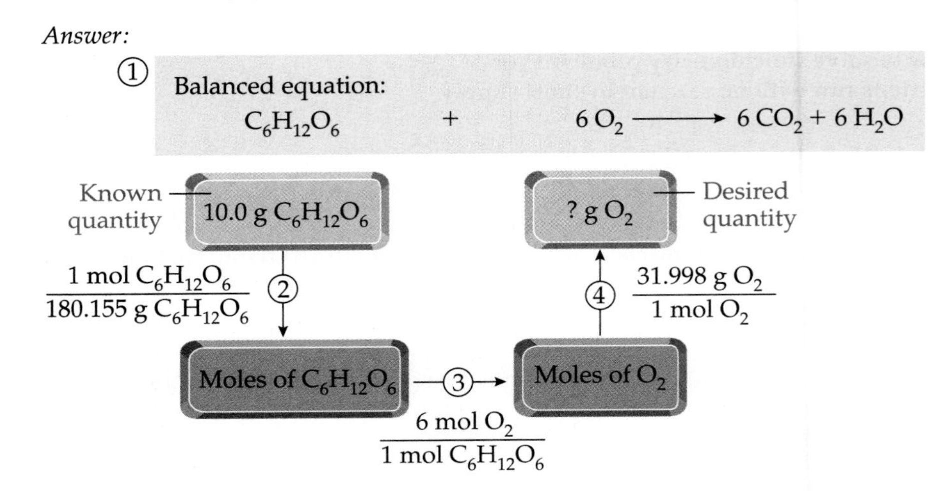

$$\underbrace{10.0\ \cancel{g\ C_6H_{12}O_6} \times \frac{1\ \cancel{mol\ C_6H_{12}O_6}}{180.155\ \cancel{g\ C_6H_{12}O_6}}}_{②} \times \underbrace{\frac{6\ \cancel{mol\ O_2}}{1\ \cancel{mol\ C_6H_{12}O_6}}}_{③} \times \underbrace{\frac{31.998\ g\ O_2}{1\ \cancel{mol\ O_2}} = 10.7\ g\ O_2}_{④}$$

8.18 How many grams of water will be produced by burning 10.0 g of glucose?

8.19 What is the theoretical yield in grams of carbon dioxide if 10.0 g of glucose is burned?

8.20 If 10.0 g of carbon dioxide was produced, how many grams of glucose must have been burned?

Though we have stressed following the four-step method, it can be modified to suit the occasional oddball question. For example, the answer to Practice Problem 8.17 is that it takes 10.7 g of O_2 to burn 10.0 g of glucose. We could have been asked to go further and calculate how many O atoms are in 10.7 g of O_2. Our knowledge of the definition of mole (6.022×10^{23} of anything) and of what the molecular formula O_2 means (each oxygen molecule consists of two oxygen atoms) gives us more conversion factors, which we use to find the answer in Practice Problem 8.21.

Practice Problems

8.21 How many O atoms are there in 10.7 g of oxygen, O_2, molecules?

Answer: Don't be afraid to string out conversion factors. They may come from a balanced equation, a molecular formula, or your knowledge of the mole, depending on the question you are trying to answer.

$$10.7\ \cancel{g\ O_2} \times \underbrace{\frac{1\ \cancel{mol\ O_2}}{31.998\ \cancel{g\ O_2}}}_{\text{Inverse of molar mass}} \times \underbrace{\frac{2\ \cancel{mol\ O\ atoms}}{1\ \cancel{mol\ O_2}}}_{\text{Conversion factor from formula } O_2} \times \underbrace{\frac{6.022 \times 10^{23}\ \text{O atoms}}{1\ \cancel{mol\ O\ atoms}}}_{\text{Conversion factor from knowledge of mole}} = 4.027 \times 10^{23}\ \text{O atoms}$$

8.22 How many aluminum atoms are there in 10.0 g of aluminum oxide, Al_2O_3?

8.23 How many molecules of water are there in 10.0 g of water?

With the proper conversion factors, you can use balanced equations to calculate "how much" when running a reaction at any scale. Because reactions speak in the language of moles, you can bet that for almost every stoichiometry problem, you will be converting to moles. Thus our final advice on the subject is this—if in doubt, convert to moles.

8.4 Dealing with a Limiting Reactant

Let's try one more cheesecake problem. Here is the recipe again:

3 blocks cream cheese + 5 eggs + 1 cup sugar ⟶ 1 cheesecake

Suppose you have 100 cups of sugar, 100 blocks of cream cheese, and 10 eggs. How many cheesecakes can you make? This stoichiometry question is a bit more complicated than our earlier ones because now one of the ingredients—eggs—is in short supply. However, once you realize this, it is easy to determine in your head that the correct answer is two cakes.

This is an example of a *limiting-reactant problem*. For this particular problem, chemists would call the eggs the *limiting reactant* because it is the eggs you'll run out of first, limiting the number of cakes you can make. It does not matter that you have a lot of extra sugar and cream cheese. Once the eggs run out, the cake-making (the reaction) stops. A **limiting reactant** can therefore be defined as any reactant that is present in short supply relative to the other reactants according to the demands of the balanced equation. In our example, the sugar and the cream cheese are referred to as **excess reactants**—reactants that are present in excess relative to the limiting reactant. When a reaction with a limiting reactant is complete, the limiting reactant will be completely gone (used up), replaced by product and unused excess reactant(s). In our example, you would be left with no eggs, two delicious cheesecakes, 94 blocks of cream cheese, and 98 cups of sugar.

Up to now we have done stoichiometry problems for reactions in which there are no excess reactants. That is, we have combined exactly the right amounts of reactants so that, when the reaction is done, all of the reactants are used up and only product remains. Reactions performed in this manner are said to be done in a *stoichiometric*, or *balanced, fashion*. That is, the reactants are present in exactly the mole ratio dictated by the balanced equation. When a reaction is performed in a balanced fashion, it doesn't matter which reactant you use to calculate the theoretical yield of product. In reality, however, chemists often perform reactions so that some reactants are present in excess. In these cases, one reactant is a limiting reactant and the reaction is said to be run in a *limiting fashion*.

When a reaction is performed in a limiting fashion, you must use the limiting reactant to calculate the theoretical yield—for the same reason that only the number of eggs could tell us the cheesecake yield in our example. Therefore, if you are asked to determine the theoretical yield for a reaction performed in a limiting fashion, you must first identify which reactant is limiting.

Unfortunately, to identify the limiting reactant, you must usually do some calculations. In the case of our cheesecake-making, it was pretty obvious that eggs were a limiting reactant. However, when you do real chemical reactions—measuring amounts in grams—typically it is not obvious which reactant, if any, is limiting. The limiting reactant *would* be easy to identify if we always worked in moles instead of grams. To show you why that is so, we'll work with moles first and then switch to grams. Let's work with the simplest kind of reaction, one where all the reactants in the balanced equation are used in the same molar amount. Our HI reaction is a good example.

$$H_2 + I_2 \longrightarrow 2\,HI$$

Reactants used in same molar amount (H_2, I_2)

If we ran this reaction by combining 10 moles of H_2 with 10 moles of I_2, we would be running it in a balanced fashion because both reactants are present in the same molar amount, just what the balanced equation calls for. If we combined 10 moles of H_2 with 11 moles of I_2, we would be running the reaction in a limiting fashion. The H_2 would be the limiting reactant and the I_2 the excess reactant. There would be 1 mole of I_2 left at the end of the reaction. More important, if we wanted to know the theoretical yield of HI, we would have to use the H_2 to find out.

Using the limiting reactant to determine theoretical yield

$$\underbrace{10\ \cancel{\text{mol H}_2}}_{\text{Limiting reactant}} \times \underbrace{\frac{2\ \text{mol HI}}{1\ \cancel{\text{mol H}_2}}}_{\text{From balanced equation}} = \underbrace{20\ \text{mol HI}}_{\text{Theoretical yield}}$$

To determine whether this reaction was run in a stoichiometric fashion or in a limiting fashion and, if the latter, which reactant was limiting, all we had to do was compare moles of H_2 with moles of I_2 and see if one number was smaller than the other. The one present in the smaller molar amount was the limiting reactant. In actuality, you are much more likely to be told reactant amounts in grams rather than moles. If you are told you have 10.0 g of H_2 and 10.0 g of I_2, it is no longer obvious if the reaction is being run in a balanced or limiting fashion because the amounts are in grams but the equation speaks in moles. It's that old language problem again, but all we need to do is to translate grams to moles by using the appropriate conversion factors:

$$10.0\ \cancel{\text{g H}_2} \times \underbrace{\frac{1\ \text{mol H}_2}{2.0158\ \cancel{\text{g H}_2}}}_{\text{Inverse of H}_2\text{ molar mass}} = 4.96\ \text{mol H}_2$$

$$10.0\ \cancel{\text{g I}_2} \times \underbrace{\frac{1\ \text{mol I}_2}{253.81\ \cancel{\text{g I}_2}}}_{\text{Inverse of I}_2\text{ molar mass}} = 0.0394\ \text{mol I}_2$$

Now it's easy to see that the I_2 is present in a far smaller molar amount and is therefore the limiting reactant. The theoretical yield of HI is thus twice the number of moles of I_2.

For reactions where all the reactants appear in the same molar amount in the balanced equation, finding the limiting reactant is just a matter of converting everything to moles and then looking for the reactant present in the smallest molar amount. However, for reactions where the reactants are required in different molar amounts, such as the combustion of propane,

$$C_3H_8 + 5\,O_2 \longrightarrow 3\,CO_2 + 4\,H_2O$$

Reactants used in different molar amounts

this method won't work. For example, suppose we try to run this reaction by combining 1 mole of propane, C_3H_8, with 4 moles of oxygen, O_2. If we blindly did things as we did for the HI reaction, we would notice that propane was present in the smaller molar amount and call it the limiting reactant. That would be wrong, though, because the balanced equation says that each mole of propane requires 5 moles of oxygen. We have 1 mole of propane but only 4 moles of oxygen. Therefore the oxygen is in short supply and will run out first, making oxygen the limiting reactant.

For reactions that have unequal reactant balancing coefficients, you can't just look for which reactant is present in the smaller molar amount to determine which, if any, is limiting. You must also take the balanced equation into account. There is a simple way to do this. After you convert all the amounts of reactants to moles, determine the mole-to-coefficient ratios by dividing each molar amount by its corresponding coefficient from the balanced equation. The reactant that generates the smallest mole-to-coefficient ratio is the limiting reactant. If all the reactants generate the same number, the reaction is being run in a balanced fashion.

Why should this procedure work? We'll show you why with bicycles being assembled by our assistant:

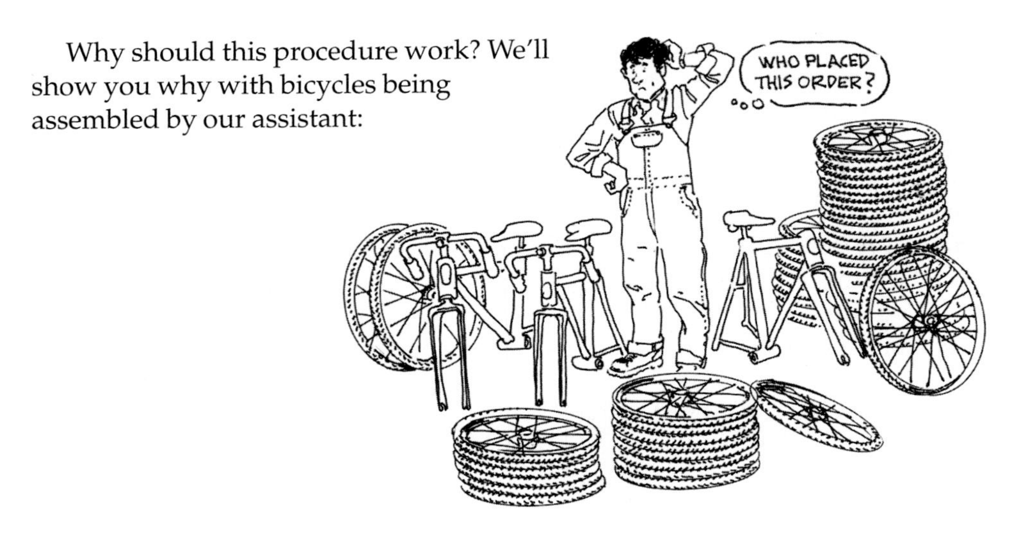

The balanced bicycle equation is

$$1 \text{ frame} + 2 \text{ wheels} \longrightarrow 1 \text{ bicycle}$$

The latest workshop inventory reveals 101 frames but only 200 wheels. Now let's divide the number of each reactant by its balancing coefficient from the balanced bicycle equation to get number-to-coefficient ratios:

Number-to-coefficient ratio

$$\frac{101 \text{ frames}}{1} = 101$$

$$\frac{200 \text{ wheels}}{2} = 100$$ Smaller number-to-coefficient ratio

Unlike the number of wheels and frames, the quotients of these ratios *can* be compared to identify which reactant, if any, is limiting because dividing by the coefficients takes into account the needs of the balanced equation. The smaller number-to-coefficient ratio is associated with the wheels, which makes them the limiting reactant.

This procedure also works when we deal with balanced chemical equations and moles of reactants because a mole of something is also a number of something. For example, let's return to our propane combustion reaction. Suppose we combine 6 moles of propane with 29 moles of oxygen. Are we running the reaction in a balanced or limiting fashion? If limiting, which reactant is the limiting reactant? To find out, we divide each molar amount by its balancing coefficient:

$$C_3H_8 + 5\,O_2 \longrightarrow 3\,CO_2 + 4\,H_2O$$

$$\frac{6 \text{ mol } C_3H_8}{1} = 6 \qquad \frac{29 \text{ mol } O_2}{5} = 5.8$$ ← Smaller mole-to-coefficient ratio, meaning O_2 is limiting reactant

That the ratios are not equal tells you the reaction is being run in a limiting fashion. The smaller mole-to-coefficient ratio is associated with O_2, making it the limiting reactant. This method always works, but be careful! *You must always do this with moles and never with grams.* If you are given the amounts of reactants in grams (which is almost always the case), you must first convert everything to moles before using this method.

The following chart summarizes the method for solving a stoichiometry problem when a limiting reactant is involved. It is essentially the earlier chart with an additional step (between steps 2 and 3) to calculate mole-to-coefficient ratios. Modifications to steps 2 and 3 are emphasized in boldface.

How to solve stoichiometry problems for reactions run in a limiting fashion

Step 1: Write the balanced chemical equation.

Step 2: For **all reactants**, convert the given information to moles.

Step 2a: Determine the mole-to-coefficient ratio for each reactant, using the appropriate coefficients from the balanced equation. The reactant that has the smallest mole-to-coefficient ratio is the limiting reactant.

Step 3: Use the balancing coefficients and **the molar amount of the limiting reactant** to calculate the theoretical yield of the product in moles.

Step 4: Convert the molar amount calculated in step 3 to the units asked for in the problem.

Now try your hand at some limiting-reactant problems.

Practice Problems

In Practice Problems 8.24–8.26, use the method of dividing moles by coefficients to determine whether the reaction is being run in a balanced or limiting fashion. If it is being run in a limiting fashion and you are asked for theoretical yield, remember to use only the limiting reactant in your calculation.

8.24 In the burning of propane,

$$C_3H_8 + 5\,O_2 \longrightarrow 3\,CO_2 + 4\,H_2O$$

suppose 10.0 g of propane is combined with 10.0 g of oxygen. What is the theoretical yield of CO_2, in grams?

Answer: Before we begin, we'll calculate the molar masses of everything because we'll probably need them:

$C_3H_8 = 44.096$ g/mol $\quad O_2 = 31.998$ g/mol
$CO_2 = 44.009$ g/mol $\quad H_2O = 18.015$ g/mol

Step 1: *The equation is already balanced.*

Step 2: *We convert the amounts of reactants to moles:*

$$10.0 \text{ g } C_3H_8 \times \frac{1 \text{ mol } C_3H_8}{44.096 \text{ g } C_3H_8} = 0.227 \text{ mol } C_3H_8$$

$$10.0 \text{ g } O_2 \times \frac{1 \text{ mol } O_2}{31.998 \text{ g } O_2} = 0.312 \text{ mol } O_2$$

Step 2a: *We find the mole-to-coefficient ratios by dividing each molar amount by a balancing coefficient:*

$$\frac{0.227 \text{ mol } C_3H_8}{1} = 0.227 \qquad \frac{0.312 \text{ mol } O_2}{5} = 0.0624$$

Smaller mole-to-coefficient ratio indicates O_2 as limiting reactant.

Steps 3 and 4: *We use the balancing coefficients and the limiting-reactant molar amount to calculate the theoretical yield of CO_2 first in moles and then in grams:*

$$\underbrace{0.312 \text{ mol } O_2 \times \frac{3 \text{ mol } CO_2}{5 \text{ mol } O_2}}_{\text{Step 3}} \times \underbrace{\frac{44.009 \text{ g } CO_2}{1 \text{ mol } CO_2}}_{\text{Step 4}} = 8.24 \text{ g } CO_2$$

The following diagram shows the strategy we used:

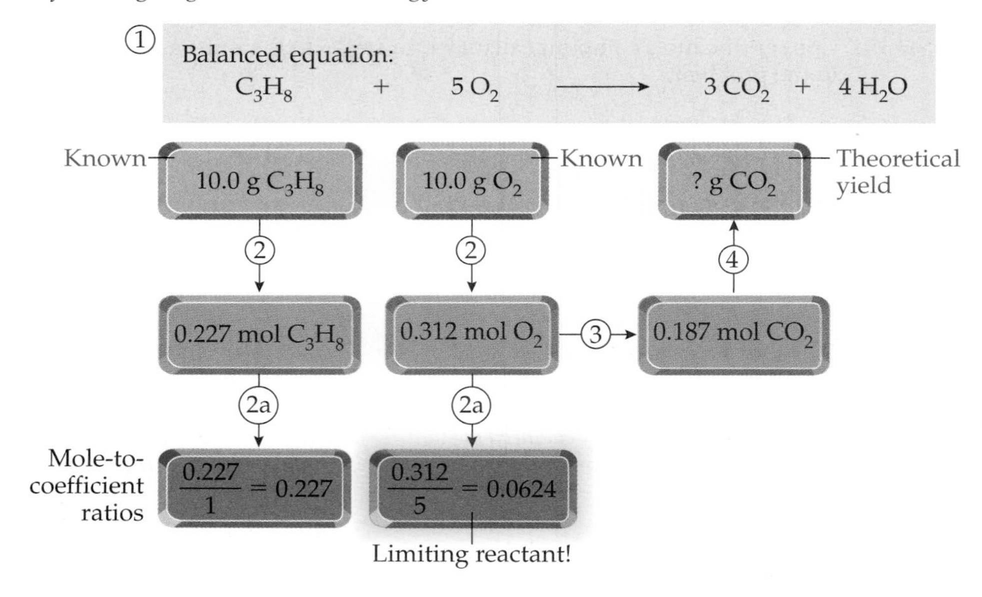

8.25 (a) What mass of oxygen is necessary to burn 100.0 g of propane in a balanced fashion?

(b) What is the theoretical yield of water, in grams?

8.26 Chemical treatment of zinc sulfide, ZnS, with oxygen, O_2, gives zinc oxide, ZnO, and sulfur dioxide gas, SO_2.

(a) Write a balanced equation for the reaction.

(b) If 10.0 g of ZnS is combined with 10.0 g of O_2, what is the theoretical yield of each product, in grams?

(c) How much of the excess reactant is left over, in grams?
(d) Suppose only 7.50 g of ZnO is recovered. What is the percent yield of ZnO?

For many introductory chemistry courses, this is a good stopping point. The next two sections deal with slightly more advanced applications of stoichiometry. Your instructor will decide if they are appropriate for your course. If not, have no fear. We have covered the basics of stoichiometry, and you are well prepared for a next chemistry course or for laboratory work. Just remember that balanced chemical equations "speak" in moles. To use them, you must speak in moles also. If in doubt, convert to moles!

8.5 Combustion Analysis

When we write reactions and work stoichiometry problems on paper, we always know what the products are. In reality, reactions sometimes give totally unexpected products and/or products contaminated with impurities. For example, suppose you have a headache and a friend offers you a tablet, saying, "This is aspirin, I just synthesized it in a chemical reaction. Have some."

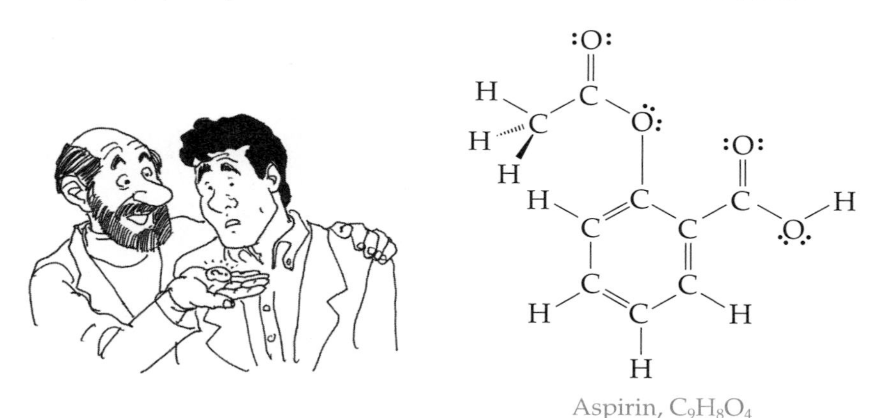

Aspirin, $C_9H_8O_4$

If enough impurities or unexpected products are in the aspirin, it could cure your headache forever! Before it is considered fit for consumption, a chemist would have to confirm both the aspirin's identity and its purity. One good test would be to determine the molecular formula of the product formed in the reaction run by your friend. Hopefully, it's $C_9H_8O_4$, which is aspirin. One of the best ways to determine the molecular formula of an unknown compound is to take a known mass of the compound and—believe it or not—burn it. This procedure is called **combustion analysis**. *Combustion* means "burning," and burning something means reacting it with oxygen. Therefore when you burn a compound, all the elements in it combine with oxygen. For example, the combustion reaction for aspirin is

$$C_9H_8O_4 + 9\,O_2 \longrightarrow 9\,CO_2 + 4\,H_2O$$

All the carbon ends up in CO_2, and all the hydrogen ends up in H_2O. You can think of the carbon dioxide and the water as "garbage cans" where all the carbon and hydrogen end up.

What Combustion Analysis Does

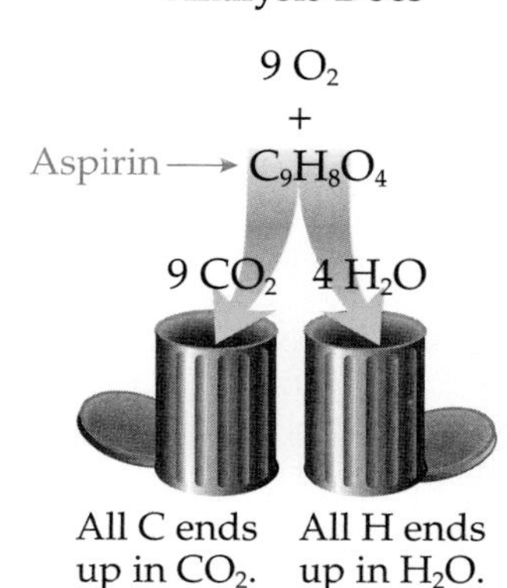

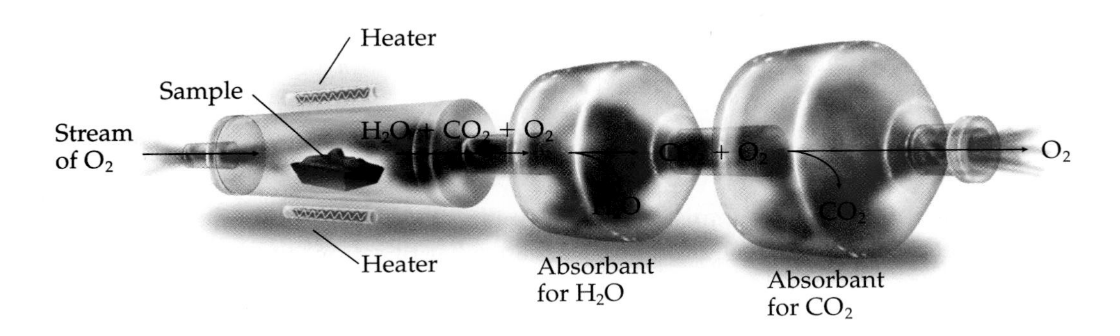

In the laboratory, a combustion analyzer like the one diagrammed above is commonly used to burn unknown substances and determine their molecular formulas. Different materials in the analyzer absorb the H_2O and CO_2 produced during the burning. The masses of the absorbants are determined before and after the combustion. The mass gain of each absorbant is equal to the mass of water or carbon dioxide produced.

Here is the experimental data from the combustion analysis of a 0.100-g sample of your friend's "aspirin":

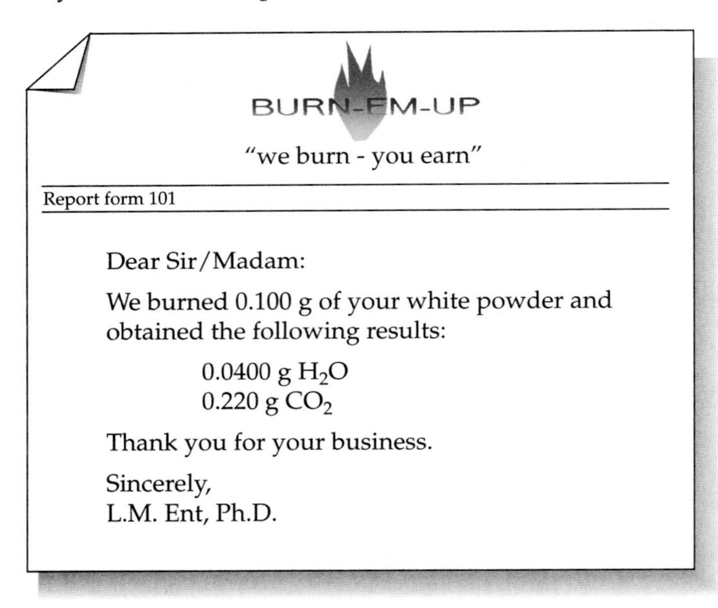

BURN-EM-UP

"we burn - you earn"

Report form 101

Dear Sir/Madam:

We burned 0.100 g of your white powder and obtained the following results:

0.0400 g H_2O
0.220 g CO_2

Thank you for your business.

Sincerely,
L.M. Ent, Ph.D.

Our job now is to turn the data into a molecular formula. Right now, all we know is that the formula of this white powder is $C_xH_yO_z$; we don't know the values of any of the subscripts. We do know, however, that the subscripts can be thought of as numbers of moles. Because all the hydrogen in the burned sample went into the water, if we figure out how many moles of hydrogen there are in the 0.0400 g of water, we will know how many moles of hydrogen there were in the sample. This is just a simple application of stoichiometry and conversion factors:

$$0.0400\ \text{g}\ \cancel{H_2O} \times \underbrace{\frac{1\ \cancel{\text{mol}\ H_2O}}{18.015\ \text{g}\ \cancel{H_2O}}}_{\text{Inverse of molar mass}} \times \underbrace{\frac{2\ \text{mol H atoms}}{1\ \cancel{\text{mol}\ H_2O}}}_{\text{From formula } H_2O} = \underbrace{0.004\ 44\ \text{mol H atoms}}_{\text{Value of } y \text{ in } C_xH_yO_z.}$$

We can also do this for carbon. Because all the carbon in the burned sample went into making carbon dioxide, if we figure out how many moles of carbon there are in the 0.220 g of carbon dioxide, we will know how many moles of carbon there were in the sample:

$$0.220\ \text{g}\ \cancel{CO_2} \times \underbrace{\frac{1\ \cancel{\text{mol}\ CO_2}}{44.009\ \text{g}\ \cancel{CO_2}}}_{\text{Inverse of molar mass}} \times \underbrace{\frac{1\ \text{mol C atoms}}{1\ \cancel{\text{mol}\ CO_2}}}_{\text{From formula } CO_2} = \underbrace{0.005\ 00\ \text{mol C atoms}}_{\text{Value of } x \text{ in } C_xH_yO_z.}$$

So far we have $C_{0.005\ 00}H_{0.004\ 44}O_z$. Don't worry that the subscripts are not whole numbers because we shall take care of that shortly. Now we have to find the value of z. We have a small problem, though. We don't know where the oxygen in the combustion products CO_2 and H_2O came from. Some may have come from the sample, but some may have come from the stream of O_2 used in the combustion analyzer. The only way to find out is to calculate the mass of the 0.005 00 mole of C and 0.004 44 mole of H in our sample, and subtract this mass from the mass of the sample (0.100 g). Anything left must be the mass of oxygen in the sample.

$$0.004\ 44\ \cancel{\text{mol H}} \times \underbrace{\frac{1.0079\ \text{g H}}{1\ \cancel{\text{mol H}}}}_{\text{Molar mass}} = 0.004\ 48\ \text{g H}$$

$$0.005\ 00\ \cancel{\text{mol C}} \times \underbrace{\frac{12.011\ \text{g C}}{1\ \cancel{\text{mol C}}}}_{\text{Molar mass}} = 0.0600\ \text{g C}$$

$$0.004\ 48\ \text{g} + 0.0600\ \text{g} = \underbrace{0.0645\ \text{g}}_{\text{Combined H + C mass}}$$

Of the 0.100-g sample of aspirin, 0.0645 g comes from C and H. If the only elements in the compound are carbon, hydrogen, and oxygen, the rest must be oxygen:

$$\begin{array}{rll} & 0.1000\ \text{g} & \text{Sample} \\ - & 0.0645\ \text{g} & \text{C + H} \\ \hline & 0.0355\ \text{g O} & \end{array}$$

If the result here were zero, we could conclude that the compound had no oxygen in it. Our nonzero result tells us the sample does contain oxygen, however, and therefore let's convert grams of oxygen to moles to find the subscript z:

$$0.0355\ \cancel{\text{g O}} \times \underbrace{\frac{1\ \text{mol O}}{15.999\ \cancel{\text{g O}}}}_{\text{Inverse of molar mass}} = 0.002\ 22\ \text{mol O}$$

Notice that we used the molar mass of O (not O_2) in our calculation. Oxygen is not diatomic in compounds.

We now have the complete formula of the sample we burned. It is $C_{0.005\,00}H_{0.004\,44}O_{0.002\,22}$. OK, so you don't like it. Neither do we. That's because we want whole numbers of moles in our formula. To fix this, we divide all the subscripts by the smallest one (z, in this case, which is 0.002 22):

$$C_{\frac{0.005\,00}{0.002\,22}}H_{\frac{0.004\,44}{0.002\,22}}O_{\frac{0.002\,22}{0.002\,22}} \longrightarrow C_{2.25}H_{2.00}O_{1.00} = C_{2.25}H_2O$$

Well, it almost worked. Many times you will be done at this point, but sometimes not, as is the case here. We are allowed to change to whole numbers any subscripts that are *very close* to whole numbers, but the subscript 2.25 on C is not very close to either 2 or 3. Therefore we need to find a multiplier that makes 2.25 a whole number. After a little thought, we see that with 4 we get $2.25 \times 4 = 9$, a whole number. We must be careful, though. If we multiply 2.25 by 4, we must also multiply all the other subscripts by 4. The rule is, whatever we do to one subscript, we must always do to the others; otherwise we would change the relationships among them. Therefore we now say

$$C_{2.25\times 4}H_{2.00\times 4}O_{1.00\times 4} \longrightarrow C_9H_8O_4$$

We are done. We have gone from the data of a combustion analysis to a chemical formula. The analysis indicates that the sample does indeed have the formula of aspirin (assuming the sample was pure). Of course, this does not prove that the sample *is* aspirin because different compounds can have the same molecular formula. For example, both ethyl ether, CH_3OCH_3, and ethanol, CH_3CH_2OH, have the formula C_2H_6O.

The chart at the top of the next page summarizes the steps we used in our combustion analysis.

The results from a combustion analysis can be used to calculate other things besides a chemical formula. We can also calculate the *mass percent* of each element present in a sample. For example, the human body is roughly 70% by mass water, meaning roughly 70% of your mass is due to water. We could do the same sort of thing for our aspirin sample. What percent of the sample's mass is due to carbon? What percent is due to hydrogen? What percent is due to oxygen? The answers are that aspirin is 60.0% by mass carbon, 4.48% by mass hydrogen, and 35.5% by mass oxygen. We calculate these mass percents from the combustion-analysis results. To do this, we divide the mass of each element in the sample (which we already calculated in finding the formula) by the mass of the sample and then multiply by 100:

$$\%C = \frac{0.0600 \text{ g C}}{0.100 \text{ g aspirin}} \times 100 = 60.0\% \text{ C}$$

$$\%H = \frac{0.004\,48 \text{ g H}}{0.100 \text{ g aspirin}} \times 100 = 4.48\% \text{ C}$$

$$\%O = \frac{0.0355 \text{ g O}}{0.100 \text{ g aspirin}} \times 100 = 35.5\% \text{ O}$$

If we add these mass percents (and obey the rules of significant figures), we should always get 100%. Now try the following practice problems.

Determining a chemical formula from combustion analysis data

In a combustion analysis, the compound whose formula is to be determined is made to react completely with oxygen and the masses of all products are reported. In our example, the products are CO_2 and H_2O, and the preliminary formula is $C_xH_yO_z$.

Step 1: Convert grams of CO_2 to moles of C. This is the subscript x.
Convert grams of H_2O to moles of H. This is the subscript y.

Step 2: Convert moles of C from step 1 to grams of C.
Convert moles of H from step 1 to grams of H.
Add these masses to get grams of C + H.

Step 3: Determine the mass of O:

Grams of sample burned
− Grams of C + H from step 2

Grams of O ←

- If this number is zero, the subscript z in $C_xH_yO_z$ is zero and the compound contains no oxygen.
- If this number is not zero, convert it to moles of O. The result is the subscript z in $C_xH_yO_z$.

Step 4: Divide all subscripts by the smallest one, and if the resulting numbers are very close to whole numbers, round them to whole numbers. If one or more subscripts are still very far from whole numbers after you do the division, find some multiplier that makes them whole. Then multiply all the subscripts by this number.

Practice Problems

8.27 A compound is known to contain carbon and hydrogen. It might also contain oxygen. A 0.250-g sample of the compound is burned to produce 0.561 g of H_2O and 0.686 g of CO_2.
(a) What is the chemical formula for this compound?
(b) What is the mass percent of each element?
(c) Write the balanced combustion reaction (reaction with O_2) for this compound.

Answer (not worked out on purpose—you do it):
(a) CH_4 *(b)* 74.9% C; 25.1% H *(c)* $CH_4 + 2\,O_2 \longrightarrow 2\,H_2O + CO_2$

8.28 A compound is known to contain carbon and hydrogen. It might also contain oxygen. A 0.250-g sample of the compound is burned to produce 0.293 g of H_2O and 0.478 g of CO_2.
(a) What is the chemical formula for this compound?

(b) What is the mass percent of each element?
(c) Write the balanced combustion reaction (reaction with O_2) for this compound.

The combustion-analysis problems we have been doing so far have been chosen to avoid a complication we want to deal with now. Let's look at one more compound, but this time we'll tell you the molecular formula right at the start. When we burn 0.500 g of the simple sugar glucose, $C_6H_{12}O_6$, we get the results shown here. Now we ought to get the formula $C_6H_{12}O_6$ from the report data. Let's see if we can. The problem is worked out below. Follow it through carefully.

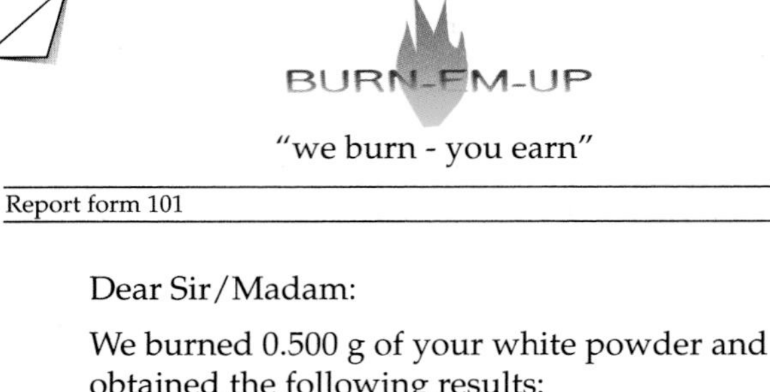

BURN-EM-UP

"we burn - you earn"

Report form 101

Dear Sir/Madam:

We burned 0.500 g of your white powder and obtained the following results:

0.300 g H_2O
0.733 g CO_2

Thank you for your business.

Sincerely,
L.M. Ent, Ph.D.

Step 1: Convert grams of CO_2 to moles of C and grams of H_2O to moles of H:

$$0.733\ \cancel{\text{g CO}_2} \times \frac{1\ \cancel{\text{mol CO}_2}}{44.009\ \cancel{\text{g CO}_2}} \times \frac{1\ \text{mol C}}{1\ \cancel{\text{mol CO}_2}} = 0.0166\ \text{mol C}$$

$$0.300\ \cancel{\text{g H}_2\text{O}} \times \frac{1\ \cancel{\text{mol H}_2\text{O}}}{18.015\ \cancel{\text{g H}_2\text{O}}} \times \frac{2\ \text{mol H}}{1\ \cancel{\text{mol H}_2\text{O}}} = 0.0333\ \text{mol H}$$

The formula so far is therefore $C_{0.0166}H_{0.0333}O_z$.

Step 2: Convert moles of C to grams of C and moles of H to grams of H:

$$0.0166\ \cancel{\text{mol C}} \times \frac{12.011\ \text{g C}}{1\ \cancel{\text{mol C}}} = 0.199\ \text{g C}$$

$$0.0333\ \cancel{\text{mol H}} \times \frac{1.0079\ \text{g H}}{1\ \cancel{\text{mol H}}} = 0.0336\ \text{g H}$$

$$0.199\ \text{g} + 0.0336\ \text{g} = \underset{\text{Combined C + H mass}}{0.233\ \text{g}}$$

Step 3: Determine the mass of O. If this mass is nonzero, convert it to moles.

$$\begin{array}{r l} & 0.500 \text{ g sample} \\ - & 0.233 \text{ g C + H} \\ \hline & 0.267 \text{ g O} \end{array}$$

$$0.267 \cancel{\text{g O}} \times \frac{1 \text{ mol O}}{15.999 \cancel{\text{g O}}} = 0.0167 \text{ mol O}$$

The formula so far is $C_{0.0166}H_{0.0333}O_{0.0167}$.

Step 4: Divide all subscripts by the smallest one:

$$C_{\frac{0.0166}{0.0166}}H_{\frac{0.0333}{0.0166}}O_{\frac{0.0167}{0.0166}} \longrightarrow C_{1.00}H_{2.006}O_{1.006} \longrightarrow \underset{\text{Formula}}{CH_2O}$$

What went wrong? The calculations based on the combustion analysis give us the formula CH_2O, but we know glucose has the formula $C_6H_{12}O_6$. The answer is that nothing went wrong. In general, a combustion analysis does not give the molecular formula of a compound; it gives the *empirical formula.* An **empirical formula** (also called a *simplest formula*) is one that gives only the ratio of the elements present in a compound, with all subscripts being the smallest numbers possible. So how does an empirical formula differ from a *molecular formula,* a term you've been familiar with since Chapter 5? Simple—a molecular formula gives the *actual composition* of the compound. Thus, CH_2O is the empirical formula of glucose. It tells us that there are twice as many H atoms in the compound as C or O atoms and that the C and O atoms are present in equal numbers. If you divide all the subscripts in the glucose molecular formula, $C_6H_{12}O_6$, by the smallest subscript, you get $C_1H_2O_1 = CH_2O$, the glucose empirical formula.

There are cases where the empirical formula of a compound is also the molecular formula (this was the case in all the examples and problems we did up to this point), but you can't count on this. Most often, the empirical formula and the molecular formula are not the same.

That combustion analysis always gives an empirical formula presents a problem. How do you know if what you get is also the molecular formula? And if it is not, how can you figure out what the molecular formula is? To answer these questions, you need to know the molar mass of the substance. But wait a minute—if you don't know the molecular formula in the first place, you can't calculate the molar mass! Therefore, when you need a molar mass to go from an empirical formula to a molecular formula, we shall give it to you (based on values from experimental results).

For glucose, the molar mass is 180.155 g/mol. Now our job is easy. We calculate the molar mass of the empirical formula and compare it with the molar mass of the compound:

Empirical formula CH_2O

$$\text{Molar mass of } CH_2O = \underbrace{12.011 \text{ g/mol}}_{\text{C}} + \underbrace{(2 \times 1.0079 \text{ g/mol})}_{\text{H}} + \underbrace{15.999 \text{ g/mol}}_{\text{O}} = 30.026 \text{ g/mol}$$

Molecular formula $C_6H_{12}O_6$

Molar mass of $C_6H_{12}O_6$ = 180.155 g/mol

Because these numbers are not the same, the empirical formula cannot be the molecular formula.

If the (known) molar mass of the compound and the calculated molar mass of the empirical formula are the same, the empirical formula is the molecular formula. When they are different, divide the molar mass of the empirical formula into the molar mass of the compound. The quotient tells you how many empirical formula units go into making the compound. Then multiply the subscripts in the empirical formula by this number to get the molecular formula. Here is how it works for glucose:

Converting an empirical formula to a molecular formula

Step 1: Divide the molar mass of the compound by the molar mass of the empirical formula:

$$\frac{\text{Molar mass of glucose}}{\text{Molar mass of empirical formula}} = \frac{180.155\ \text{g/mol}}{30.026\ \text{g/mol}} = 6\ \text{(approximately)}$$

Step 2: Multiply all subscripts in the empirical formula by the number calculated in step 1:

$$C_{1\times6}H_{2\times6}O_{1\times6} = C_6H_{12}O_6 \qquad \text{Molecular formula}$$

Now try the following practice problems.

Practice Problems

8.29 A compound is known to contain carbon and hydrogen, and might also contain oxygen. A 0.500-g sample is burned to produce 0.409 g of H_2O and 0.999 g of CO_2.

(a) What is the empirical formula of the compound?

(b) What is the mass percent of each element in the compound?

(c) The molar mass of the compound is 132.159 g/mol. What is the molecular formula?

(d) Write the balanced combustion reaction (reaction with O_2) for the compound.

Answer (not worked out on purpose—you do it):

(a) C_2H_4O

(b) 54.6% C; 9.16% H; 36.2% O

(c) $C_6H_{12}O_3$

(d) To balance the equation, look at C and H first. Then balance the elemental substance (O_2) *last:*

$$\underbrace{C_6H_{12}O_3}_{3\ \text{O}} + \underbrace{O_2}_{\text{Need 15 O here to match 18 on right}} \longrightarrow \underbrace{6\,CO_2 + 6\,H_2O}_{18\ \text{O}}$$

We make O_2 *provide 15 O atoms by multiplying it by 7.5 (that is,* $\frac{15}{2}$*):*

$$C_6H_{12}O_3 + \underbrace{7.5\,O_2}_{7.5\times2\,=\,15\ \text{O atoms}} \longrightarrow 6\,CO_2 + 6\,H_2O$$

This is a perfectly correct balanced equation, but if you prefer the balancing coefficients to be whole numbers, you can multiply all of them by some number that makes them whole (in this case, 2):

$$2\,C_6H_{12}O_3 + 15\,O_2 \longrightarrow 12\,CO_2 + 12\,H_2O$$

8.30 A compound is known to contain carbon and hydrogen, and might also contain oxygen. A 1.000-g sample is burned to produce 1.284 g of H_2O and 3.137 g of CO_2.
(a) What is the empirical formula for the compound?
(b) The molar mass is 28.054 g/mol. What is the molecular formula?
(c) Write the balanced combustion reaction.

8.6 Going Back and Forth Between Formulas and Percent Composition

In this section you are going to exercise your developing stoichiometry skills. You just saw how combustion analysis can give you both the empirical formula and the elemental mass percents for a compound. It is also possible to be given a molecular formula and calculate elemental mass percents or to be given elemental mass percents and calculate an empirical formula. Let's continue to use glucose, $C_6H_{12}O_6$, as our example. To go quickly from this (or any other) molecular formula to the mass percent of each element in the compound, all we have to do is assume we have 1 mole of the compound. The mass of 1 mole of any compound is its molar mass in grams.

If we have 1 mole of glucose, we have 6 moles of C atoms, 12 moles of H atoms, and 6 moles of O atoms (this information comes straight from the molecular formula $C_6H_{12}O_6$). To get the mass percents, we have to figure out the masses of these different moles of atoms, divide each by the total mass of 1 mole of glucose (180.155 g/mol, calculated earlier), and multiply by 100. Remember, the key to getting started is to assume you have 1 mole of the compound.

Going from molecular formula to elemental mass percents

Step 1: Assume you have 1 mole of the compound (glucose in this case). The molecular formula, in this example $C_6H_{12}O_6$, tells you how many moles of each element you have. Figure out all masses in grams:

$$6 \text{ mol C} \times \frac{12.011 \text{ g C}}{1 \text{ mol C}} = 72.066 \text{ g C}$$

$$12 \text{ mol H} \times \frac{1.0079 \text{ g H}}{1 \text{ mol H}} = 12.0948 \text{ g H}$$

$$6 \text{ mol O} \times \frac{15.999 \text{ g O}}{1 \text{ mol O}} = 95.994 \text{ g O}$$

Step 2: Divide the mass of each element by the mass of the 1 mole of compound (in other words, divide by its molar mass) and multiply by 100:

$$\%C = \frac{72.066 \text{ g C}}{180.155 \text{ g}} \times 100 = 40.00\% \text{ C}$$

$$\%H = \frac{12.0948 \text{ g H}}{180.155 \text{ g}} \times 100 = 6.71\% \text{ H}$$

$$\%O = \frac{95.994 \text{ g O}}{180.155 \text{ g}} \times 100 = 53.28\% \text{ O}$$

$$99.99\%$$

Arbitrarily rounded to hundredths place

Mass percents add up to 100 after rounding.

We just went from molecular formula to elemental mass percents. We can also go the other way, only this time we'll end up not with a molecular formula but rather with an empirical formula. The key to getting started is to assume we have 100 g of the compound. This sample size allows us to reinterpret the elemental percents as grams. We can then convert grams of each element to moles for use as subscripts in the empirical formula.

Going from elemental mass percents to empirical formula

Step 1: Assume you have 100 g of the compound (glucose in this case). First convert all percents to grams, and then convert grams to moles:

$$40.00\% \text{ C} \longrightarrow 40.00 \text{ g C} \times \frac{1 \text{ mol C}}{12.011 \text{ g C}} = 3.33 \text{ mol C}$$

$$6.71\% \text{ H} \longrightarrow 6.71 \text{ g H} \times \frac{1 \text{ mol H}}{1.0079 \text{ g H}} = 6.66 \text{ mol H}$$

$$53.28\% \text{ O} \longrightarrow 53.28 \text{ g O} \times \frac{1 \text{ mol O}}{15.999 \text{ g O}} = 3.33 \text{ mol O}$$

Step 2: Use the calculated numbers of moles as formula subscripts and divide by the smallest one to get whole numbers:

$$C_{\frac{3.33}{3.33}}H_{\frac{6.66}{3.33}}O_{\frac{3.33}{3.33}} \longrightarrow CH_2O$$

As noted above, this method gives the empirical formula. If you want the molecular formula, you need to know the molar mass of the compound.

You should now be able to quickly go from molecular formula to elemental mass percents and from elemental mass percents to empirical formula. Just remember what initial assumption to make for each procedure.

Practice Problems

8.31 What is the mass percent of each element in hydrogen peroxide, H_2O_2?

Answer: 5.93% H, 94.07% O

8.32 What is the mass percent of each element in trinitrotoluene (TNT), $C_7H_5N_3O_6$?

8.33 A compound is found to have the following elemental mass percents: Cl = 89.09%, C = 10.06%, H = 0.84%. The molar mass of the compound is 119.378 g/mol. What are the empirical and molecular formulas?

8.34 A compound is known to contain C and H, and might also contain O. It is analyzed for C and H only, yielding the mass percents C = 54.53% and H = 9.15%. The molar mass of the compound is 88.106 g/mol. What are the empirical and molecular formulas?

Stoichiometry problems of the type we have been doing in this chapter are the heart of chemical calculations. "How much?" and "What did I make?" are always important questions, whether you are baking cheesecakes or producing aspirin. Just keep in mind that our recipes—the balanced equations of chemical reactions and the molecular formulas in them—speak in moles. To use these recipes correctly, you must speak in moles also. If in doubt, convert to moles. It's almost always a good first step.

Have You Learned This?

Stoichiometry (p. 258)
Mole (p. 261)
Avogadro's number (p. 261)
Molar mass (p. 264)
Theoretical yield (p. 269)
Actual yield (p. 269)
Percent yield (p. 269)
Limiting reactant (p. 275)
Excess reactant (p. 275)
Combustion analysis (p. 281)
Empirical formula (p. 287)

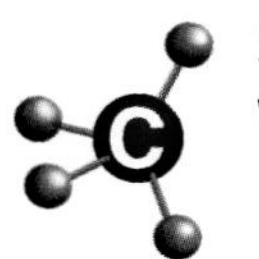

Stoichiometry and the Mole
www.chemistryplace.com

SKILLS TO KNOW

The mole

A mole of anything is 6.022×10^{23} of them. This number is also called Avogadro's number. The definition of a mole gives rise to the useful conversion factors at right.	$\frac{1 \text{ mole of something}}{6.022 \times 10^{23} \text{ somethings}}$ or $\frac{6.022 \times 10^{23} \text{ somethings}}{1 \text{ mole of something}}$

Molar mass

The mass in grams of 1 mole of atoms or molecules is called the molar mass. Molar mass can be viewed as a conversion factor that converts mass to moles or moles to mass, as shown at right.	Moles × Molar mass = Mass ⟶ $\text{Moles} \times \frac{\text{Grams}}{\text{Mole}} = \text{Grams}$ $\frac{\text{Mass}}{\text{Molar mass}} = \text{Moles}$ ⟶ $\text{Grams} \times \frac{\text{Moles}}{\text{Gram}} = \text{Moles}$

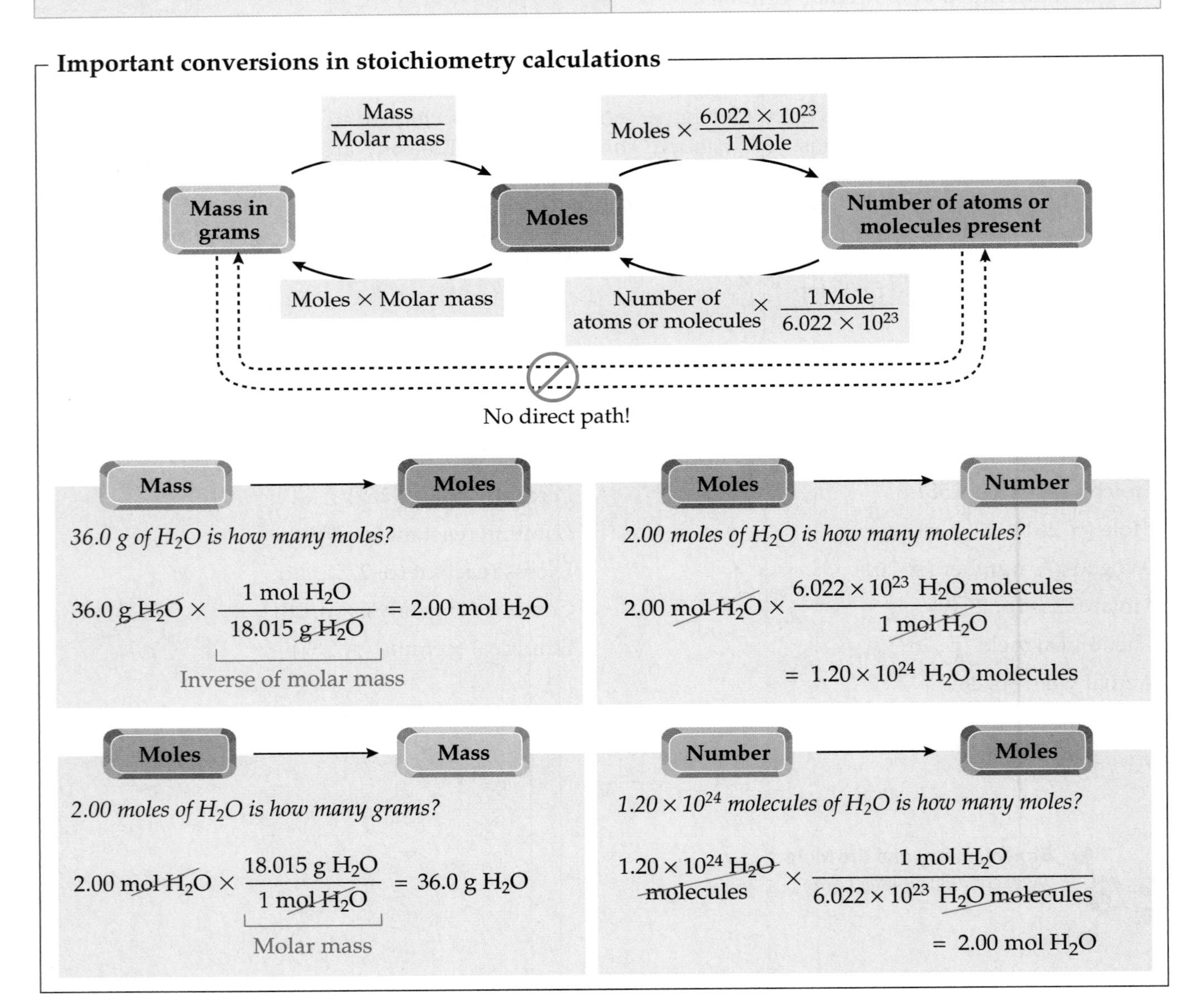

Tips for working stoichiometry problems

1. The coefficients from a balanced chemical equation and the subscripts from a formula can always be interpreted as moles and used to write coversion factors that are needed for doing stoichiometry problems.

$H_2 + I_2 \longrightarrow 2\,HI$ (arrows under H_2, I_2 and $2\,HI$): Can always be interpreted as moles

2. If a reaction is run in a balanced fashion, any reactant can be used to calculate the theoretical yield. If the reaction is performed in a limiting fashion, only the limiting reactant can be used to calculate theoretical yield.

3. For any stoichiometry problem, you will have to work in units of moles. That is because all balanced chemical equations and all formulas are written in terms of moles. If you are not sure how to start a stoichiometry problem, start by converting what you are given to moles.

How to solve a stoichiometry problem

Step 1: Begin with a balanced chemical equation.

Step 2: Identify a reactant or product whose mass you know. Convert its mass to moles by using its molar mass.

Step 3: Use the coefficients from the balanced equation to find the amount in moles of the reactant or product whose amount you need to know.

Step 4: Convert the amount from step 3 from moles to grams, using the molar mass of the reactant or product in question.

Example: You need to produce 10.0 g of HI by reacting H_2 with I_2. How many grams of H_2 will you need? (Assume you get the theoretical yield.)

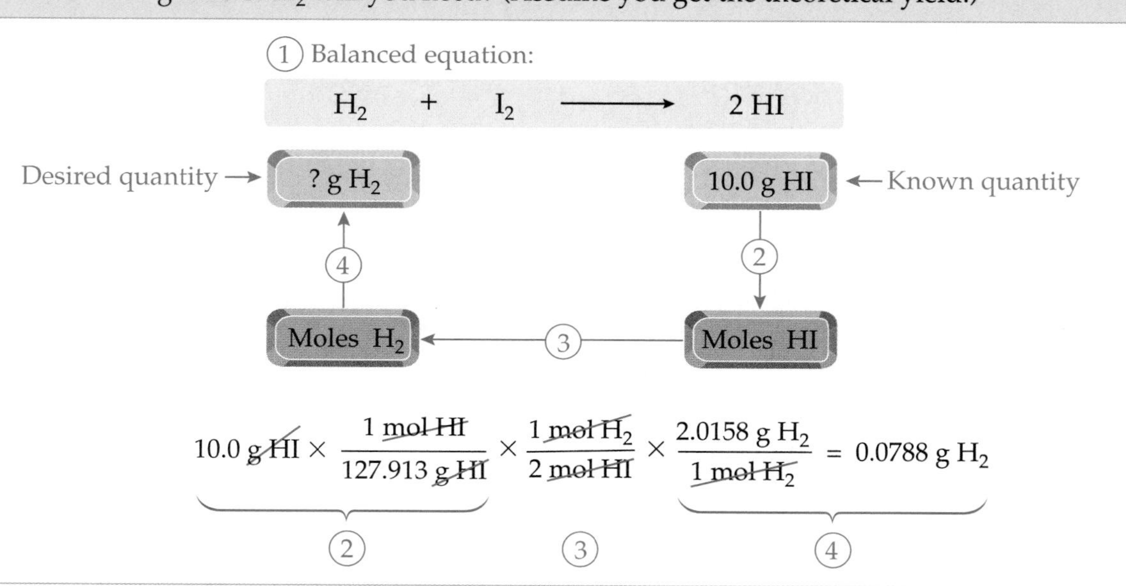

Solving a limiting reactant stoichiometry problem

Example: If you react 5.00 g of methane, CH_4, with 12.00 g of oxygen, O_2, what is the theoretical yield of H_2O?

Step 1: Begin with a balanced chemical equation.	$CH_4 + 2\,O_2 \longrightarrow CO_2 + 2\,H_2O$
Step 2: Convert the given masses of all the reactants to moles, using the molar masses of the reactants.	**CH_4:** $5.00\ \text{g}\ \cancel{CH_4} \times \dfrac{1\ \text{mol}\ CH_4}{16.043\ \text{g}\ \cancel{CH_4}} = 0.312\ \text{mol}\ CH_4$ **O_2:** $12.00\ \text{g}\ \cancel{O_2} \times \dfrac{1\ \text{mol}\ O_2}{31.998\ \text{g}\ \cancel{O_2}} = 0.375\ \text{mol}\ O_2$
Step 2a: Determine the mole-to-coefficient ratio for each of the reactants, using the appropriate coefficients from the balanced equation. The reactant that has the smallest mole-to-coefficient ratio is the limiting reactant.	**CH_4:** $\dfrac{0.312}{1} = 0.312$ **O_2:** $\dfrac{0.375}{2} = 0.188$ ← This is the limiting reactant
Step 3: Use the molar amount of the limiting reactant to calculate the theoretical yield of the desired product in moles. Do this by using the appropriate coefficients from the balanced equation.	**O_2:** $0.375\ \cancel{\text{mol}\ O_2} \times \dfrac{\cancel{2}\ \text{mol}\ H_2O}{\cancel{2}\ \cancel{\text{mol}\ O_2}} = 0.375\ \text{mol}\ H_2O$
Step 4: Convert the theoretical yield from moles to grams, using the molar mass of the desired product.	**H_2O:** $0.375\ \cancel{\text{mol}\ H_2O} \times \dfrac{18.015\ \text{g}\ H_2O}{1\ \cancel{\text{mol}\ H_2O}} = 6.76\ \text{g}\ H_2O$

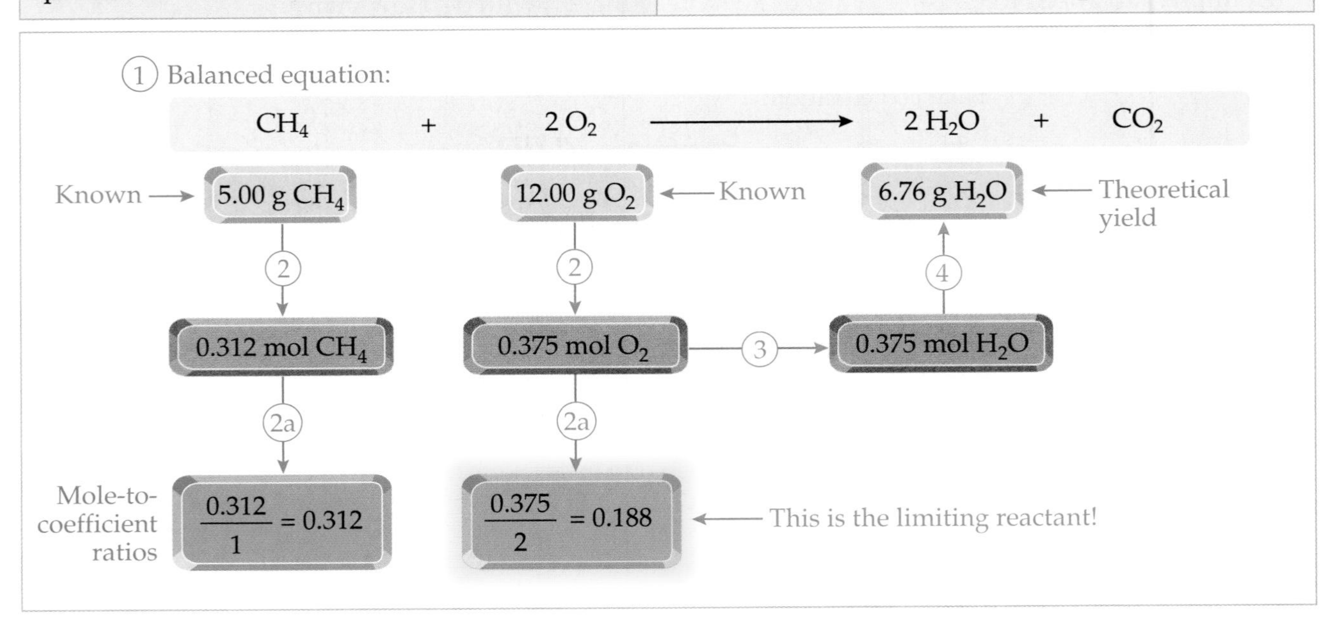

STOICHIOMETRY—WHAT IS IT?

8.35 How is a balanced chemical equation like a recipe?

8.36 Consider this recipe for one dozen cookies:

1 egg
2 cups flour
3 cups sugar
$\frac{1}{4}$ pound butter
1 cup milk

(a) Write at least five conversion factors that might be useful for solving cookie stoichiometry problems.
(b) Using unit analysis, calculate how many cups of flour are needed to make 30 cookies. Write down your method and cross out units that cancel.
(c) Why can't you answer the following question using just the information in the recipe: How many eggs are required to completely use up 1 container of milk?
(d) Given the conversion factor that 1 container of milk = 4 cups of milk, answer the question in part (c). Write down your method and cross out units that cancel.
(e) Write the recipe in chemical equation form.

8.37 One recipe for cheesecake is the following, which makes one cake:

3 blocks of cream cheese
5 eggs
1 cup sugar

(a) To bake 3 cakes, how many eggs do you need? Write down your method and cross out units that cancel.
(b) If you have 63 blocks of cream cheese, how many eggs do you need? Write down your method and cross out units that cancel.

THE MOLE

8.38 How is the concept of "one mole" similar to the concept of "one dozen"?

8.39 When is it allowed to insert the word *mole* into a chemical equation when translating the equation into words?

8.40 The human population of our planet is about 6 billion people. What percentage of a mole is 6 billion?

8.41 How many bicycles are there in 1 mole of bicycles? How many tires are there in 1 mole of bicycles?

8.42 How many O_2 molecules are there in 1 mole of O_2 molecules? How many O atoms are there in 1 mole of O_2 molecules?

8.43 How many pennies are there in 2.5 moles of pennies? How many dollars does this equal? Answer both questions by using conversion factors, and show which units cancel.

8.44 How many years are there in 1 mole of seconds? Use conversion factors, and show which units cancel.

8.45 Translate the following balanced equation into words, first without using the word *moles*, then again with the word *moles*:

$$2\,SO_2 + O_2 \longrightarrow 2\,SO_3$$

8.46 Sometimes using the word *moles* when translating a chemical equation into words can be very helpful. For example, consider the following *correctly balanced* equation. What difficulty do you run into when you try to translate it into words without using the word *moles*? How does using the word *moles* solve the difficulty?

$$C_2H_2 + \tfrac{5}{2}O_2 \longrightarrow 2\,CO_2 + H_2O$$

8.47 In words, what is the mass of 1 mole of atoms of any element?

8.48 In words, what is the mass of 1 mole of any molecule?

8.49 How do you calculate the molar mass of a compound?

8.50 How does molar mass solve the "language problem" when a chemical equation is being used as a recipe?

8.51 How many atoms of the $^{12}_{6}C$ isotope are there in 12 g of the isotope?

8.52 How many carbon atoms are in a sample of naturally occurring carbon that has a mass of 12.011 g?

8.53 If you have 1 mole of glucose ($C_6H_{12}O_6$):
(a) How many moles of carbon atoms do you have?
(b) How many moles of hydrogen atoms do you have?
(c) How many oxygen atoms do you have? (Note that we are not asking for moles here.)

8.54 Consider the ammonia (NH_3) molecule.
(a) If you have 1 mole of ammonia, how many moles of H atoms do you have?
(b) If you have 2 moles of ammonia, how many moles of H atoms do you have?
(c) If you have 2 moles of ammonia, how many N atoms do you have? (We are not asking for moles here.)

REACTION STOICHIOMETRY

8.55 What do we mean by the theoretical yield of a reaction?

8.56 What do we mean by the actual yield of a reaction?

8.57 Why is the actual yield of a reaction often not equal to the theoretical yield?

8.58 What do we mean by the percent yield of a reaction?

8.59 A student runs a reaction to prepare 40.0 g of aspirin and yet recovers only 15.5 g. What is the percent yield?

8.60 Consider the unbalanced chemical equation

$NO + O_2 \longrightarrow NO_2$

(a) Balance the equation.
(b) Translate the equation into words using the word *mole(s)* wherever you can.
(c) To produce 2 moles of NO_2 by the reaction you just wrote, how many grams of NO and O_2 must you combine?
(d) What is the theoretical yield in grams of NO_2?
(e) You carry out the reaction and recover 22.5 g of NO_2. What is the percent yield?

8.61 Consider the unbalanced chemical equation

$HCl + Zn \longrightarrow H_2 + ZnCl_2$

(a) Balance the equation.
(b) Translate the equation into words using the word *mole(s)* wherever you can.
(c) To produce 1 mole of H_2 from the reaction you just wrote, how many grams of HCl and Zn must you combine?
(d) What is the theoretical yield in grams of H_2?
(e) You recover 2.00 g of H_2 after carrying out the reaction. What is the percent yield?

8.62 Consider the unbalanced chemical equation

$Na + Cl_2 \longrightarrow NaCl$

(a) Balance the equation.
(b) Translate the equation into words using the word *mole(s)* wherever you can.
(c) To produce 1 mole of NaCl from the reaction, how many grams of Na and Cl_2 must you combine?
(d) What is the theoretical yield in grams of NaCl from the reaction you ran in part (c)?
(e) You carry out the reaction and recover 45.50 g of NaCl. What is the percent yield?

8.63 The formula $C_6H_{12}O_6$ is a source of many conversion factors. Write at least three of them.

8.64 The balanced chemical equation $2\ AgBr \longrightarrow 2\ Ag + Br_2$ and the formulas of the substances in it are a source of many conversion factors. Write all that are possible using the word *mole(s)*.

8.65 The following chemical equation is unbalanced:

$I_2 + Cl_2 \longrightarrow ICl_3$

(a) Balance the equation.
(b) The balanced equation is a source of many conversion factors containing the two reactants and one product. Write all of them that use the word *mole(s)*.
(c) The balanced equation is also a source of conversion factors involving the individual atoms and molecules. Write all of them that use the word *mole(s)*.

8.66 How many H atoms are there in 2.0158 g of H atoms?

8.67 How many moles of O_2 molecules are there in 24.0 g of O_2 molecules?

8.68 How many moles of O atoms are there in 24.0 g of O_2 molecules?

8.69 Consider sulfuric acid, H_2SO_4, used in car batteries.
(a) What is the molar mass of sulfuric acid?
(b) What is the mass in grams of 1 mole of H_2SO_4?
(c) What is the mass in grams of 2.50 moles of H_2SO_4?
(d) What is the mass in grams of 1000 molecules of H_2SO_4? [*Hint*: Start by writing "1000 molecules H_2SO_4" and then apply conversion factors. Remember our admonition: If in doubt, convert to moles.]

8.70 How many O_2 molecules are there in 1.00 g of O_2 molecules?

8.71 Suppose you wanted one billion (1.00×10^9) water molecules and you didn't have time to sit and count them out. How many grams of water would you need to get one billion water molecules?

8.72 How many grams of glucose ($C_6H_{12}O_6$) would you need to get 5.00×10^{30} carbon atoms?

8.73 Consider the balanced chemical equation

$2\ H_2O_2 \longrightarrow 2\ H_2O + O_2$

(a) Given 20.0 g of H_2O_2 (hydrogen peroxide), how many grams of water will the reaction yield?
(b) How many grams of hydrogen peroxide would you need to get 20.0 g of water?
(c) How many grams of hydrogen peroxide would you need to get 20.0 g of O_2?

8.74 Consider the balanced chemical equation

$SCl_4 + 2\ H_2O \longrightarrow SO_2 + 4\ HCl$

(a) How many grams of H_2O will react with 5.000 g of SCl_4?
(b) How many grams of SO_2 can you make from 10.00 g of H_2O?
(c) Suppose you react 5.000 g of SCl_4 with the amount of H_2O you calculated in part (a). How many grams of HCl will you form?

DEALING WITH A LIMITING REACTANT

8.75 Suppose you run the reaction A + 2 B ⟶ C, and you discover that the product C is contaminated with A. How would you explain this?

8.76 With 3 cups of flour, 3 cups of sugar, and unlimited amounts of everything else, how many cookies can you make from the recipe in Problem 8.36? Write down your method and cross out units that cancel.

8.77 If you have 25 eggs, 9 blocks of cream cheese, and 4 cups of sugar, how many cakes can you make using the recipe of Problem 8.37? Write down your method and cross out units that cancel.

8.78 Consider the following balanced chemical equation:

$$2\,H_2 + O_2 \longrightarrow 2\,H_2O$$

(a) How many grams of water is formed from 5.00 g of H_2 and an excess amount of O_2?
(b) How many grams of O_2 do you need to produce 5.00 g of H_2O?
(c) Given 100.0 g of H_2, how many grams of O_2 is required to run the reaction in a stoichiometric fashion?
(d) What is the theoretical yield in grams of water upon combining 50.0 g of O_2 with an excess amount of H_2?
(e) Express the answer to part (d) in terms of the number of water molecules.

8.79 Consider the following unbalanced chemical equation:

$$P + O_2 \longrightarrow P_2O_5$$

(a) How many grams of phosphorus (P) is required to react completely with 20.0 g of O_2?
(b) What is the theoretical yield in grams if you combine the amounts of reactants in part (a)?

8.80 Consider the following unbalanced chemical equation:

$$H_2 + N_2 \longrightarrow NH_3$$

(a) To run this reaction in a balanced fashion, how much nitrogen is required if you start with 10.0 g of H_2?
(b) How many grams of ammonia (NH_3) will you produce if you run the reaction with the masses calculated in part (a)?
(c) How many molecules of ammonia will you produce?

8.81 A gaseous mixture containing 5.00 moles of H_2 and 7.00 moles of Br_2 reacts to form HBr.
(a) Write a balanced chemical equation for this reaction.
(b) Which reactant is limiting?
(c) What is the theoretical yield for this reaction in moles?
(d) What is the theoretical yield for this reaction in grams?
(e) How many moles of excess reactant is left over at the end of the reaction?
(f) How many grams of excess reactant is left over at the end of the reaction?

8.82 Chlorine (Cl_2) and fluorine (F_2) react to form ClF_3. A reaction vessel is charged with 2.50 moles of Cl_2 and 6.15 moles of F_2.
(a) Write a balanced chemical equation for this reaction.
(b) Which reactant is limiting?
(c) What is the theoretical yield for this reaction in moles?
(d) What is the theoretical yield for this reaction in grams?
(e) How many moles of excess reactant is left over at the end of the reaction?
(f) How many grams of excess reactant is left over at the end of the reaction?

8.83 5.00 g of solid sodium (Na) and 30.0 g of liquid bromine (Br_2) react to form solid NaBr.
(a) Write a balanced chemical equation for this reaction.
(b) Which reactant is limiting?
(c) What is the theoretical yield for this reaction in grams?
(d) How many grams of excess reactant is left over at the end of the reaction?
(e) When this reaction is actually performed, 14.7 g of NaBr is recovered. What is the percent yield of the reaction?

8.84 Chlorine (Cl_2) and fluorine (F_2) react to form ClF_3. A reaction vessel is charged with 10.00 g Cl_2 and 10.00 g F_2. [*Hint*: Refer to Problem 8.82.]
(a) Write a balanced chemical equation for this reaction.
(b) Which reactant is limiting?
(c) What is the theoretical yield for this reaction in grams?
(d) How many grams of excess reactant is left over at the end of the reaction?
(e) When this reaction is actually performed, 12.50 g of ClF_3 is recovered. What is the percent yield of the reaction?

8.85 Sodium (Na) reacts with hydrogen (H_2) to form sodium hydride (NaH). A reaction mixture contains 10.00 g Na and 0.0235 g H_2.
(a) Write a balanced chemical equation for this reaction.
(b) Which reactant is limiting?
(c) What is the theoretical yield for this reaction in grams?
(d) How many grams of excess reactant is left over at the end of the reaction?

(e) When this reaction is actually performed, 0.428 g of NaH is recovered. What is the percent yield of the reaction?

8.86 Butane (C_4H_{10}), used as the fuel in disposable lighters, reacts with oxygen (O_2) to produce CO_2 and H_2O. Suppose 10.00 g butane is combined with 10.00 g O_2.
(a) Write a balanced chemical equation for the combustion of butane.
(b) Which reactant is limiting?
(c) What is the theoretical yield of each product for this reaction in grams?
(d) How many grams of excess reactant is left over at the end of the reaction?
(e) How many additional grams of the limiting reactant is required to run this reaction in a balanced fashion?

COMBUSTION ANALYSIS

8.87 A compound has the empirical formula C_2H_4O. Its molar mass is about 90 g/mol. What is its molecular formula?

8.88 An organic compound of carbon and hydrogen has the empirical formula CH. What is its molecular formula if its molar mass is:
(a) 26 g/mol (b) 52 g/mol
(c) 78 g/mol

8.89 A 1.540-g sample of a liquid burns in oxygen to produce 2.257 g of CO_2 and 0.9241 g of H_2O.
(a) What are the mass percents of the elements present in this sample?
(b) What is the empirical formula for this compound?
(c) The molar mass of this compound is determined to be about 30 g/mol. What is the molecular formula for this compound?

8.90 A 2.230-g sample of a solid burns in oxygen to produce 6.258 g of CO_2 and 1.274 g of H_2O.
(a) What are the mass percents of the elements present in this sample?
(b) What is the empirical formula for this compound?
(c) The molar mass of this compound is determined to be about 94 g/mol. What is the molecular formula for this compound?

8.91 A 1.000-g sample of a liquid burns in oxygen to produce 3.383 g of CO_2 and 0.692 g of H_2O.
(a) What are the mass percents of the elements present in this sample?
(b) What is the empirical formula for this compound?
(c) The molar mass of this compound is determined to be about 78 g/mol. What is the molecular formula for this compound?

8.92 A 1.000-g sample of a liquid that contains only carbon and hydrogen burns in oxygen to produce 1.284 g of H_2O.
(a) What are the mass percents of the elements present in this sample?
(b) What is the empirical formula for this compound?
(c) The molar mass of this compound is determined to be about 71 g/mol. What is the molecular formula for this compound?

8.93 A compound used as an insecticide contains only C, H, and Cl. When a 3.200-g sample is burned in oxygen, 6.162 g of CO_2 and 0.9008 g of H_2O are produced. What are the mass percents of the elements present in this sample?

8.94 A compound used as an insecticide contains only C, H, and Cl. When a 3.000-g sample is burned in oxygen, 2.724 g of CO_2 and 0.5575 g of H_2O are produced.
(a) What are the mass percents of the elements present in this sample?
(b) What is the empirical formula for this compound?

GOING BACK AND FORTH BETWEEN FORMULA AND PERCENT COMPOSITION

8.95 Determine the empirical formula of the compound with the following mass percents of the elements present: 66.63% C; 11.18% H; 22.19% O.

8.96 Determine the empirical formula of the compound with the following mass percents of the elements present: 58.5% C; 4.91% H; 19.5% O; 17.1% N.

8.97 Determine the empirical formula of the compound that is 26.4% by mass Na, 36.8% by mass S, and also contains oxygen.

8.98 Determine the empirical formula of the compound that is 43.2% by mass K, 39.1% by mass Cl, and also contains oxygen.

8.99 Ethanol, the alcohol in beer and wine, has the molecular formula C_2H_6O. Calculate the mass percent of each element in ethanol.

8.100 Ethylene glycol, used for antifreeze, has the molecular formula $C_2H_6O_2$. Calculate the mass percent of each element in ethylene glycol.

8.101 Penicillin G, used as an antibiotic, has the molecular formula $C_{16}H_{18}N_2O_4S$. Calculate the mass percent of each element in penicillin G.

8.102 The hormone thyroxine has the molecular formula $C_{15}H_{11}NO_4I_4$. Calculate the mass percent of each element in thyroxine.

ADDITIONAL PROBLEMS

8.103 (a) What is the molar mass of ribose ($C_5H_{10}O_5$)?
(b) What is the mass of 3.87 moles of ribose?
(c) How many ribose molecules are there in 3.87 moles?
(d) How many oxygen atoms are there in 3.87 moles of ribose?
(e) What is the mass in grams of the oxygen atoms in part (d)?

8.104 Which has more mass:
(a) One atom of sulfur or one molecule of oxygen gas (O_2)?
(b) 0.125 mole of sulfur molecules (S_8) or 0.670 mole of ozone (O_3)?

8.105 Can the actual yield ever be greater than the theoretical yield for a chemical reaction?

8.106 Sodium metal reacts with water to form aqueous sodium hydroxide and hydrogen gas.
(a) Write a balanced chemical equation for this reaction.
(b) Which reactant is limiting if 100.0 g of Na and 4.00 moles of water are used?

8.107 Which has the greatest mass: 1 mole of ethylene gas (C_2H_4), 1 mole of carbon monoxide gas, or 1 mole of nitrogen gas (N_2)?

8.108 What is the maximum possible value of the percent yield of a chemical reaction?

8.109 (a) What is the molar mass of sucrose ($C_{12}H_{22}O_{11}$)?
(b) What is the mass of 1.25 moles of sucrose?
(c) How many sucrose molecules are in 1.25 moles?
(d) How many hydrogen atoms are in 1.25 moles of sucrose?
(e) What is the mass in grams of the hydrogen atoms in part (d)?

8.110 Aluminum metal burns in chlorine gas to form aluminum chloride.
(a) Write a balanced chemical equation for this reaction.
(b) Which reactant is limiting if 100.0 g of Al and 5.00 moles of Cl_2 are used?

8.111 Copper(I) oxide reacts with solid carbon to form copper metal. Carbon dioxide gas is the other product of this reaction.
(a) Write the balanced chemical equation for this reaction.
(b) Coke is a cheap, impure form of solid carbon that is often used industrially. If a sample of coke is 95% C by mass, determine the mass in kilograms of coke needed to react completely with 1.000 ton of copper(I) oxide. [1 ton = 2000 lb; 1 kg = 2.205 lb]

8.112 What is the empirical formula of a compound that is 17.552% Na, 39.696% Cr, and 42.752% O?

8.113 Consider the balanced chemical equation

$$Fe(CO)_5(s) + 2\,PF_3(l) + H_2(g) \longrightarrow Fe(CO)_2(PF_3)_2H_2(s) + 3\,CO(g)$$

(a) How many grams of CO could be produced from 10.0 g of PF_3, excess $Fe(CO)_5$, and excess H_2?
(b) How many grams of CO could be produced from 5.0 moles of $Fe(CO)_5$, 8.0 moles of PF_3, and 6.0 moles of H_2?
(c) How many moles of CO could be produced from 25.0 g of $Fe(CO)_5$, 10.0 g of PF_3, and excess H_2?
(d) The density of hydrogen gas at room temperature and atmospheric pressure is 0.0820 g/L. When 5.00 L of hydrogen gas at room temperature and atmospheric pressure is mixed with excess $Fe(CO)_5$ and excess PF_3, the mass of CO collected is 13.5 g. What are the theoretical yield (in grams) and percent yield of CO?

8.114 (a) Write a balanced chemical equation for the combustion of $CH_4(g)$ with $O_2(g)$ to form $CO_2(g)$ and $H_2O(g)$.
(b) Consider the reaction of three molecules of $CH_4(g)$ with four molecules of $O_2(g)$. The following boxes represent before and after pictures of the tiny, sealed container in which this reaction takes place. Draw pictures to show the contents of this container *before* and *after* the reaction happens:

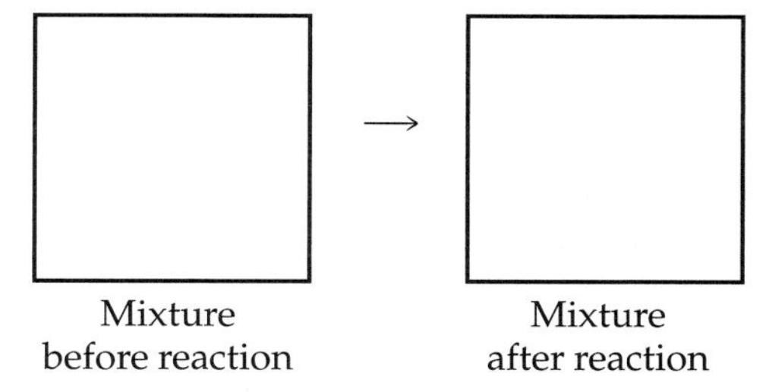

Mixture before reaction — Mixture after reaction

(c) Which is the limiting reactant, $CH_4(g)$ or $O_2(g)$?

8.115 Consider a 5.00-g sample of silver nitrate, $AgNO_3(s)$.
(a) How many moles of $AgNO_3$ are in this sample?
(b) How many moles of O are in this sample?
(c) How many grams of N are in this sample?
(d) How many Ag atoms are in this sample?

8.116 Nitrogen and fluorine react to form nitrogen fluoride according to the balanced chemical equation

$$N_2(g) + 3\,F_2(g) \longrightarrow 2\,NF_3(g)$$

For each of the following reaction mixtures, choose the limiting reactant:
(a) 0.50 mol $N_2(g)$ and 0.50 mol $F_2(g)$
(b) 12.0 mol $N_2(g)$ and 20.0 mol $F_2(g)$
(c) 2.5 mol $N_2(g)$ and 7.5 mol $F_2(g)$
(d) 100 molecules $N_2(g)$ and 500 molecules $F_2(g)$
(e) 5.00 g $N_2(g)$ and 15.0 g $F_2(g)$
(f) 20.0 mg $N_2(g)$ and 70.0 mg $F_2(g)$

8.117 For each part in Problem 8.116, calculate the amount of excess reagent left over at the end of each reaction.

8.118 When you bring your piggy bank of pennies into the bank, instead of counting them individually, the teller measures the mass of a single penny on a scale and then measures the mass of the entire contents of your piggy bank. If a single penny has a mass of 2.49 g and your entire penny collection has a mass of 5789.25 g, how much money in dollars do you receive?

8.119 For the reaction $Ca_3P_2 + H_2O \longrightarrow Ca(OH)_2 + PH_3$
(a) Balance the equation.
(b) How many grams of H_2O would you need to react with 60.0 g of Ca_3P_2?
(c) How many grams of PH_3 could theoretically be produced from the reactant amounts calculated in part (b)?

8.120 If you have 0.262 mole of iron(III) sulfate, how many oxygen atoms do you have?

8.121 Acetaminophen, used in many over-the-counter pain relievers, has a molecular formula of $C_8H_9NO_2$. Calculate the mass percentage of each element in acetaminophen.

8.122 Consider the unbalanced chemical equation

$$KCl + O_2 \longrightarrow KClO_3$$

If you react 42.6 g of KCl and 36.5 g of O_2 and the reaction has a percent yield of 56.0%, how many grams of $KClO_3$ is produced?

8.123 Sucrose has the molecular formula $C_{12}H_{22}O_{11}$. If you were to completely burn 2.00 g of sucrose in a stream of oxygen, how many grams of CO_2 and H_2O would be produced?

8.124 If you have 44.6 g of carbon tetrachloride, how any atoms of chlorine do you have?

8.125 For the reaction $BF_3 + H_2O \longrightarrow H_3BO_3 + HBF_4$
(a) Balance the equation.
(b) If you react 24.2 g of BF_3 with an excess of water and generate 14.8 g of HBF_4, what is your percent yield?

8.126 The ingredients for making 1 dozen cupcakes are

5 tablespoons butter
1 cup sugar
2 eggs
2 cups flour
1 cup milk

If you have only 5 eggs but plenty of all of the other ingredients,
(a) How many cupcakes can you make?
(b) How many cups of sugar do you need to make the number of cupcakes calculated in (a)?

8.127 Chrome yellow, a pigment used in paints, is 64.11% by mass Pb, 16.09% by mass Cr, and 19.80% by mass O. What is the empirical formula of this compound?

8.128 What is the mass of 1.62×10^{25} molecules of hydrogen sulfide?

8.129 Silicon nitride, Si_3N_4, is a ceramic material capable of withstanding high temperatures. It can be produced using the following unbalanced reaction:

$$SiCl_4 + NH_3 \longrightarrow Si_3N_4 + HCl$$

If 64.2 g of $SiCl_4$ is reacted with 20.0 g of NH_3, how many grams of Si_3N_4 is produced if the reaction has a 96.0% yield?

8.130 Vitamin C, also known as ascorbic acid, contains carbon, hydrogen, and possibly oxygen. If 0.160 g of ascorbic acid is burned in oxygen, 0.240 g of CO_2 and 0.0655 g of H_2O are produced. If the molar mass of ascorbic acid is approximately 176 g/mol, what is its molecular formula?

8.131 Consider the unbalanced chemical equation

$$Cr_2O_3 + H_2S \longrightarrow Cr_2S_3 + H_2O$$

(a) How many grams of H_2S is needed to react completely with 13.6 g of Cr_2O_3?
(b) How many grams of Cr_2S_3 could theoretically be produced?

8.132 Determine the mass percent of each element in magnesium phosphate.

8.133 For the reaction $SiCl_4 + H_2O \longrightarrow SiO_2 + HCl$
(a) Balance the equation.
(b) If you want to make 120.0 g of HCl, how many grams of $SiCl_4$ and H_2O do you need?
(c) How many grams of SiO_2 is produced from the quantities calculated in (b)?

8.134 A compound is 91.77% by mass Si and 8.23% by mass H and has a molar mass of approximately 122 g/mol. What is its molecular formula?

8.135 Iron(III) oxide reacts with carbon monoxide to give iron metal and carbon dioxide. If you begin the reaction with 24.0 g of iron(III) oxide and 34.0 g of carbon monoxide, what is the theoretical yield in grams of carbon dioxide?

8.136 The compound ibuprofen, used in some pain relievers, has the molecular formula $C_{13}H_{18}O_2$. If 0.250 g of ibuprofen is burned in oxygen, how many grams of CO_2 and H_2O will be produced?

8.137 If you have 200.0 g of dinitrogen pentoxide, how many atoms of nitrogen do you have? How many atoms of oxygen do you have?

8.138 Potassium nitrate decomposes upon heating to form potassium oxide, nitrogen gas, and oxygen gas. If 18.6 g of potassium nitrate has decomposed, how many molecules of oxygen gas have been formed?

8.139 Tetraphosphorus decoxide, P_4O_{10}, reacts with water to form phosphoric acid. If 52.5 g of P_4O_{10} reacted with 25.0 g of water, how many grams of H_3PO_4 could theoretically be produced?

8.140 Determine the mass percent of each element in aluminum sulfate.

8.141 Consider the unbalanced chemical equation

$$Cl_2O_7 + H_2O \longrightarrow HClO_4$$

The reaction is carried out at 82.0% yield and gives 52.8 g of $HClO_4$.
(a) What is the theoretical yield of $HClO_4$?
(b) How many grams of Cl_2O_7 and H_2O were consumed in the reaction?

8.142 The compound naphthalene, which is used in mothballs, contains carbon, hydrogen, and possibly oxygen. When 0.220 g of naphthalene is combusted in a stream of oxygen gas, it produces 0.755 g of CO_2 and 0.124 g of H_2O. If the molar mass of naphthalene is approximately 128 g/mol, what is its molecular formula?

8.143 Ethylene gas (C_2H_4) reacts with fluorine gas (F_2) to form carbon tetrafluoride gas and hydrogen fluoride gas. If 2.78 g of ethylene reacted with an excess of fluorine, how many grams of each product could theoretically be produced?

8.144 How many grams of sulfur hexafluoride would you need to have 5.25×10^{24} fluorine atoms?

8.145 Consider the unbalanced chemical equation

$$HSbCl_4 + H_2S \longrightarrow Sb_2S_3 + HCl$$

If 118.2 g of $HSbCl_4$ reacts with 47.9 g of H_2S and produces 41.6 g of HCl, what is the percent yield for the reaction?

8.146 Caffeine is made up of 49.48% by mass C, 5.19% by mass H, 28.85% by mass N, and 16.48% by mass O. If its molar mass is approximately 194 g/mol, what is the molecular formula of caffeine?

8.147 For the reaction $NH_3 + NO \longrightarrow N_2 + H_2O$
(a) Balance the equation.
(b) If you react 15.0 g NH_3 with 22.0 g NO and you produce 13.3 g N_2, what is your percent yield?

8.148 Nicotine has the formula $C_{10}H_{13}N_2$. Determine the mass percent of each element.

8.149 Consider the unbalanced chemical equation

$$CaC_2 + CO \longrightarrow C + CaCO_3$$

When the reaction is complete, 135.4 g of $CaCO_3$ is produced and 38.5 g of CaC_2 is left over. Assuming the reaction had a 100% yield, what were the masses of the two reactants at the beginning of the reaction?

8.150 The beryllium mineral beryl is made up of 5.03% by mass Be, 10.04% by mass Al, 31.35% by mass Si, and 53.58% by mass O. If the molar mass is approximately 538 g/mol, what is the molecular formula for beryl?

8.151 Consider the unbalanced chemical equation

$$S + H_2SO_4 \longrightarrow SO_2 + H_2O$$

(a) Balance the equation.
(b) If you react 4.80 g of sulfur with 16.20 g of H_2SO_4, how many grams of SO_2 can theoretically be produced?

8.152 The compound 3′-azido-3′-thymidine (AZT) is used in the treatment of AIDS. The molecular formula of AZT is $C_{10}H_{13}N_5O_5$. What is the mass percent of each element in AZT?

8.153 In a chemical reaction, the mass of the products is equal to the mass of the reactants consumed. Are the moles of product equal to the moles of reactant consumed? Explain your answer.

8.154 Consider the unbalanced chemical equation

$$N_2O_4 + N_2H_4 \longrightarrow N_2 + H_2O$$

(a) Balance the equation.
(b) If 42.32 g of N_2 was produced and the reaction yield was 67.5%, how many grams of N_2O_4 and N_2H_4 were consumed in the reaction?

8.155 Ammonium fluoride reacts with calcium nitrate to produce calcium fluoride, dinitrogen monoxide, and water. If 22.8 g of ammonium fluoride reacts with 38.2 g of calcium nitrate, what is the theoretical yield in grams of each product?

8.156 Some calcium supplements contain calcium citrate, $Ca_3(C_6H_5O_7)_2$, and others contain calcium carbonate, $CaCO_3$. Which compound has the higher mass percent of calcium?

8.157 Lead(II) sulfide reacts with hydrogen peroxide to give lead(II) sulfate as shown in the unbalanced chemical equation

$$PbS + H_2O_2 \longrightarrow PbSO_4 + H_2O$$

If 63.2 g of PbS is reacted with 48.0 g of H_2O_2
(a) Which is the limiting reagent?
(b) How many grams of excess reactant remains after the reaction is complete?

8.158 The hormone estradiol contains carbon, hydrogen, and oxygen and has a molar mass of approximately 272 g/mol. When 1.15 g of estradiol is combusted in a stream of oxygen, it produces 3.34 g of CO_2 and 0.913 g of H_2O. What is its molecular formula?

8.159 Consider the unbalanced chemical equation

$$CaCN_2 + H_2O \longrightarrow CaCO_3 + NH_3$$

If you began the reaction with 5.65 g of $CaCN_2$ and 12.2 g of H_2O, how many grams of NH_3 would be produced if the reaction had an 86.0% yield?

8.160 Saccharin, which was once used as an artificial sweetener, is made up of 45.90% by mass C, 2.75% by mass H, 26.20% by mass O, 7.65% by mass N, and 17.50% by mass S. If the molar mass of saccharin is 183.19 g/mol, what is its molecular formula?

8.161 What is the mass in grams of 5.00×10^2 billion titanium atoms?

8.162 Copper reacts with nitric acid as shown in the balanced chemical equation

$$3\,Cu + 8\,HNO_3 \longrightarrow 3\,Cu(NO_3)_2 + 2\,NO + 4\,H_2O$$

If you react 25.0 g of copper with an excess of nitric acid and produce 7.24 g of NO, what is the percent yield for the reaction?

8.163 The amino acid arginine is made up of 41.37% by mass C, 8.10% by mass H, 32.16% by mass N, and 18.37% by mass O. If the molar mass of arginine is approximately 174 g/mol, what is its molecular formula?

8.164 Consider the unbalanced chemical equation

$$As_4S_6 + O_2 \longrightarrow As_4O_6 + SO_2$$

(a) How many grams of O_2 is needed to react completely with 58.9 g of As_4S_6?
(b) If 41.2 g of SO_2 is produced, what is the percent yield for the reaction?

8.165 Which contains a greater mass of copper, 11.0 g of copper(I) oxide or 12.6 g of copper(I) sulfide?

8.166 Consider the balanced chemical equation

$$2\,A + B \longrightarrow 2\,C + D$$

When 8.0 g of A reacts completely with 6.0 g of B, 10.0 g of C and 4.0 g of D are produced. Assuming the yield is 100%,
(a) Which has a greater molar mass, A or C?
(b) Which has a greater molar mass, A or B?
(c) Which has a greater molar mass, A or D?
(d) If the molar mass of A is 24.0 g/mol, determine the molar mass of B, C, and D.

8.167 A sample of aluminum sulfide contains 4.25×10^{22} atoms of sulfur. What is the mass of the aluminum sulfide sample?

8.168 One of the most important industrial processes is the production of aluminum metal. The overall reaction is aluminum oxide reacting with carbon to give carbon dioxide and aluminum metal. If the reaction is performed at 94.7% yield and produces 284 kg of aluminum, what is the mass in kilograms of each reactant?

8.169 How many molecules of aspirin, $C_9H_8O_4$, would be in a tablet that contained 250 mg of aspirin? How many atoms of carbon would be in the aspirin in that tablet?

8.170 Calcium reacts with nitrogen gas to form calcium nitride. If 33.8 g of calcium reacts with 20.4 g of nitrogen gas
(a) Which reactant is the limiting reagent?
(b) If the reaction has a 72.4% yield, how many grams of calcium nitride is formed?

8.171 The flavoring agent vanillin contains carbon, hydrogen, and possibly oxygen. When 0.450 g of vanillin is combusted in a stream of O_2 gas, it produces 1.04 g of CO_2 and 0.213 g of H_2O. If the molar mass is approximately 152 g/mol, what is the molecular formula of vanillin?

8.172 Calcium hydroxide reacts with phosphoric acid to produce calcium phosphate and water.
(a) How many grams of phosphoric acid is needed to react completely with 34.6 g of calcium hydroxide?
(b) How many grams of calcium phosphate would theoretically be produced with the masses of part (a)?

8.173 Consider the unbalanced chemical equation

$$Cr + S_8 \longrightarrow Cr_2S_3$$

If we need to produce 235.0 g of Cr_2S_3 and the reaction has a 63.80% yield, how many grams of each reactant do we need?

8.174 Consider the unbalanced chemical equation

$$Al_2S_3 + H_2O \longrightarrow Al(OH)_3 + H_2S$$

If 56.0 g of aluminum sulfide reacts with 48.2 g of water
(a) Which is the excess reactant?
(b) What mass in grams of the excess reactant remains after the reaction is complete?

8 WORKPATCH SOLUTIONS

8.1 (a) 196.967 g/mol (which is the atomic mass expressed in grams per mole).

(b) Because 1 mole of gold has a mass of 196.967 g and you have twice this (393.934 g), you have 2 moles of gold.

(c) You know that 1 mole of gold contains 6.022×10^{23} gold atoms, so 2 moles must contain $2 \times (6.022 \times 10^{23})$ atoms $= 1.204 \times 10^{24}$ atoms.

8.2 (a) $$1\ \cancel{\text{mol CH}_4\ \text{molecules}} \times \underbrace{\frac{1\ \text{mol C atoms}}{1\ \cancel{\text{mol CH}_4\ \text{molecules}}}}_{\text{Conversion factor from formula CH}_4} = 1\ \text{mol C atoms}$$

(b) $$1\ \cancel{\text{mol CH}_4\ \text{molecules}} \times \underbrace{\frac{4\ \text{mol H atoms}}{1\ \cancel{\text{mol CH}_4\ \text{molecules}}}}_{\text{Conversion factor from formula CH}_4} = 4\ \text{mol H atoms}$$

(c) $$1\ \text{mol C atoms} \times \frac{6.022 \times 10^{23}\ \text{C atoms}}{1\ \text{mol C atoms}} = 6.022 \times 10^{23}\ \text{C atoms}$$

(d) $$4\ \text{mol H atoms} \times \frac{6.022 \times 10^{23}\ \text{H atoms}}{1\ \text{mol H atoms}} = 2.409 \times 10^{24}\ \text{H atoms}$$

8.3 (a) Four moles of aluminum atoms and 3 moles of oxygen molecules react to give 2 moles of aluminum oxide molecules.

(b) $$\frac{4\ \text{mol Al}}{3\ \text{mol O}_2} \quad \frac{3\ \text{mol O}_2}{4\ \text{mol Al}}$$
$$\frac{4\ \text{mol Al}}{2\ \text{mol Al}_2\text{O}_3} \quad \frac{2\ \text{mol Al}_2\text{O}_3}{4\ \text{mol Al}}$$
$$\frac{3\ \text{mol O}_2}{2\ \text{mol Al}_2\text{O}_3} \quad \frac{2\ \text{mol Al}_2\text{O}_3}{3\ \text{mol O}_2}$$

(c) $$\frac{2\ \text{mol Al}}{1\ \text{mol Al}_2\text{O}_3} \quad \frac{1\ \text{mol Al}_2\text{O}_3}{2\ \text{mol Al}}$$
$$\frac{3\ \text{mol O}}{1\ \text{mol Al}_2\text{O}_3} \quad \frac{1\ \text{mol Al}_2\text{O}_3}{3\ \text{mol O}}$$

8.4 **Preliminary** $$\underset{?\ \text{g}}{H_2 + I_2} \longrightarrow \underset{10.0\ \text{g}}{2\ HI}$$

Step 1 $H_2 + I_2 \longrightarrow 2\ HI$

Step 2 Mass of product HI known; convert to moles:

$$10.0\ \text{g HI} \times \frac{1\ \text{mol HI}}{127.912\ \text{g HI}}$$

Step 3 One mole of I_2 required to make 2 moles of HI:

$$10.0\ \text{g HI} \times \frac{1\ \text{mol HI}}{127.912\ \text{g HI}} \times \frac{1\ \text{mol I}_2}{2\ \text{mol HI}}$$

Step 4 Convert from moles of I_2 to mass of I_2:

$$10.0\ \text{g HI} \times \frac{1\ \text{mol HI}}{127.912\ \text{g HI}} \times \frac{1\ \text{mol I}_2}{2\ \text{mol HI}} \times \frac{253.810\ \text{g I}_2}{1\ \text{mol I}_2} = 9.92\ \text{g I}_2$$

CHAPTER 9

The Transfer of Electrons from One Atom to Another in a Chemical Reaction

9.1 What Is Electricity?

Did you ever accidentally bite down on a piece of aluminum foil? Did it hurt?

The pain is due to a particular type of chemical reaction, one that produces an electric jolt that goes straight to the nerve endings in your teeth. Electricity from a chemical reaction? Yes, some reactions can produce electricity. However, before we study such reactions, let's quickly review what electricity is.

Electricity is simply the flow of electrons. In many ways, electricity is similar to water flowing through a hose. **Voltage** is analogous to water pressure. The higher the voltage, the greater the "pressure," or "push," behind the electrons that causes them to flow. (Remember, we represent an electron by the symbol e^-.)

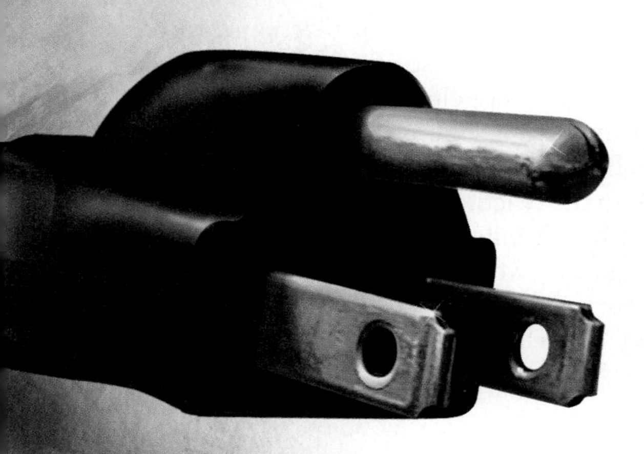

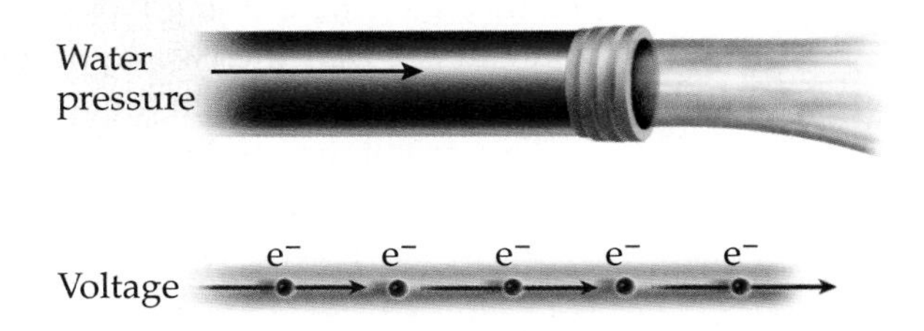

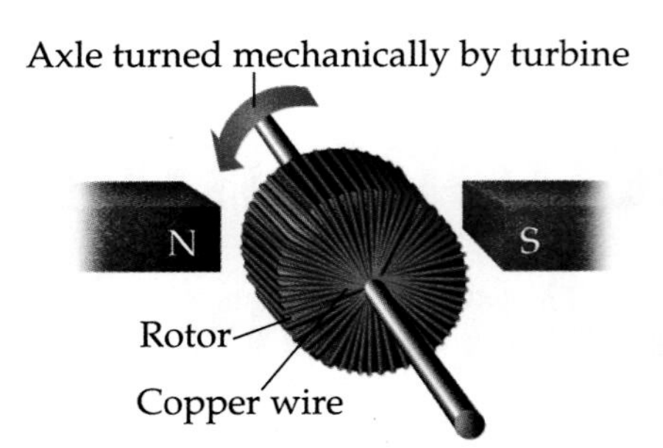

The push begins at a power plant, where electricity is produced mechanically by a generator consisting of copper wire wound on a rotor that is spun between the poles of a magnet. The rotor is spun by a turbine that is usually powered by steam generated as water is vaporized by burning oil, coal, or natural gas or by flowing water in a hydroelectric plant. As the copper wire moves through the magnetic field that exists between the poles of the magnet, electrons in the wire are pushed and made to flow, producing electricity. What you buy from the power company is the right to use their flowing electrons. They flow through your electric devices and do work and then return to the power company to be pushed again.

Electrons can also be made to flow through a wire by chemical means. The chemical reactions that do this often come packaged in containers called *batteries*. In this chapter you will learn how some chemical reactions can generate electricity and how to recognize these reactions.

9.2 Electron Bookkeeping—Oxidation States

All chemical reactions convert reactants into products, but some chemical reactions also transfer one or more electrons from one atom to another as part of the conversion process. For example, the reaction of calcium metal with acid (aqueous H^+ ions) occurs as follows:

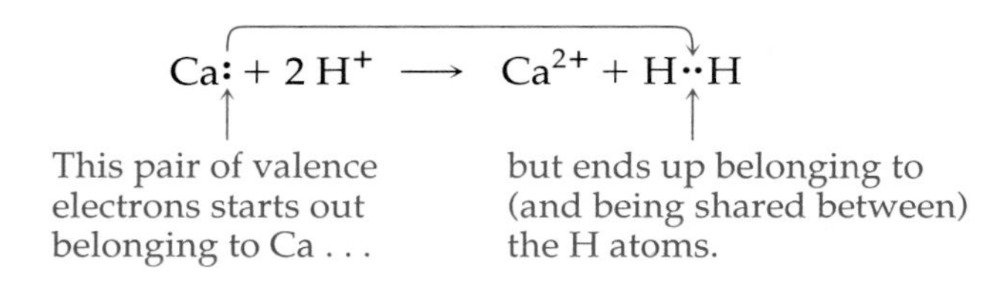

Ca: + 2 H⁺ ⟶ Ca²⁺ + H:H
This pair of valence electrons starts out belonging to Ca . . .
but ends up belonging to (and being shared between) the H atoms.

During this reaction, there must be a transfer of electrons from the calcium atom to the hydrogen ions. Another way to say this is that the calcium atom loses some electrons and the H^+ ions gain them. It is this type of reaction that can be used to push electrons through a wire and thereby produce electricity. These reactions are known as **electron-transfer reactions**. Note that only valence electrons are transferred in these reactions. Core electrons are too tightly held to their respective nuclei to be easily transferred.

Not all chemical reactions involve electron transfer. Some reactions proceed without any atom gaining or losing electrons. Because electron-transfer reactions are the only ones that can be used to produce electricity, it is important to be able to distinguish which reactions involve electron transfer and which do not. This is fairly easy to do if the reaction is written as shown above, with the valence electrons explicitly shown. With the reaction written this way, you can see that the calcium lost two electrons because its lone pair disappeared, and you can see that the two hydrogens gained two electrons because a covalent bond appeared between them.

The problem is that you will rarely see reactions written this way. Rather, you will be presented with just the balanced reaction, which for our example is

$$Ca + 2\,H^+ \longrightarrow Ca^{2+} + H_2$$

Now how can you tell that electrons are transferred as reactants become products? Perhaps you feel you can tell by the fact that charges are changing (for example, calcium goes from a charge of zero to +2). For calcium to change from the neutral atom to the +2 ion, it must lose two electrons, so electrons must have been transferred. But consider the reaction

$$H_2 + Cl_2 \longrightarrow 2\,HCl$$

No apparent changes in charge here. The reactants and product are all neutral molecules. Nevertheless, this reaction is also an electron-transfer reaction.

So, given just a balanced reaction, how are we to tell whether electron transfer is occurring? What we need is information about electron ownership. For example, suppose we wrote the $H_2 + Cl_2$ reaction this way:

$$H_2 + Cl_2 \longrightarrow 2\,HCl$$

Valence-electron ownership information	Each H atom owns one electron.	Each Cl atom owns seven electrons.	H atom owns no electrons.	Cl atom owns eight electrons.

With the ownership information given below the balanced equation, we can now tell that electron transfer has occurred. According to the ownership information, the H atoms in the H_2 molecule each own one valence electron, but the atoms end up with no electrons in the HCl molecules, telling us that each H atom has lost an electron. The ownership information also indicates that each Cl atom has gained one electron. Evidently, electrons moved from H atoms to Cl atoms during this reaction, so it can be classified as an electron-transfer reaction.

Unfortunately, balanced chemical reactions don't come with electron ownership information written below every atom. We need a way, therefore, to determine the number of electrons owned by every atom on both sides of a chemical equation. This can be done in much the same way that we keep track of our personal finances, by doing some bookkeeping. For this, chemists have created an electron-bookkeeping method known as the *oxidation-state method*. Its goal is to assign a number to every atom that tells us how many valence electrons each atom owns. These assigned numbers are called either an *oxidation number* or an *oxidation state*. After assigning oxidation states, we can look at a balanced reaction and tell whether electron transfer occurs.

The best way to understand how oxidation states are assigned to atoms in a molecule or in a polyatomic ion is to use electron dot diagrams. We'll demonstrate with a water molecule:

```
 :Ö:
H   H
```

There are eight valence electrons (dots) in this diagram. Electron bookkeeping means asking the question "How many electrons does each atom in the molecule own?" Because the answer depends on how the bookkeeping is done (in other words, on how we define the word *own*), we need to agree on a few rules. The first rule states that

Lone pairs of electrons are assigned completely to the atom on which they are drawn in the dot diagram.

This makes sense because lone pairs are not shared with another atom.

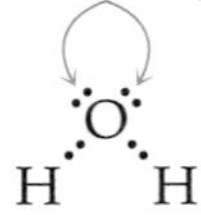

What about the bonding electrons? In our discussion of covalent bonding and the octet rule in Chapter 5, we double-counted the bonding electrons in a dot structure. For example, consider one of the O–H bonds in water. The oxygen atom was said to own both shared electrons, and the hydrogen atom was also said to own both shared electrons. This is how we satisfied the octet rule for each atom (or the duet rule in the case of hydrogen).

In the oxidation-state method of bookkeeping, we count electrons differently. The rule for shared electrons in this method is

In a chemical bond, the more electronegative atom gets complete ownership of both shared electrons.

Recall from Section 5.6 that electronegativity is a measure of an atom's ability to attract shared electrons to itself. It is often useful to remember that the four most electronegative elements in the periodic table are fluorine (EN = 4.0), oxygen (EN = 3.5), chlorine (EN = 3.0), and nitrogen (EN = 3.0), the electron "hogs." Hydrogen has only a moderate electronegativity value (EN = 2.1).

Let's now apply the oxidation-state method of bookkeeping to the water molecule:

Electron counting by the octet rule:
Shared electrons are double-counted.

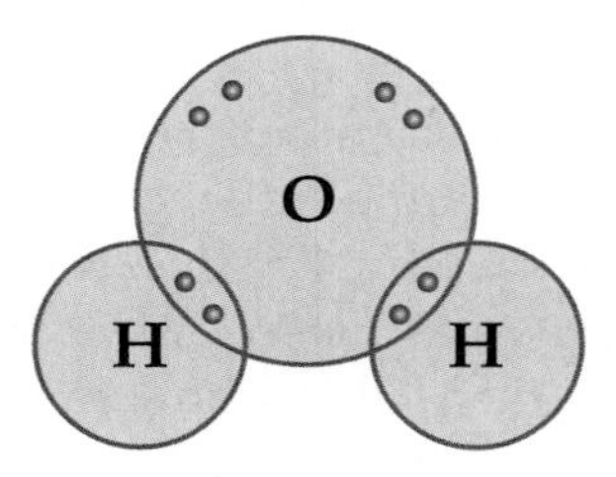

The O has **eight** electrons.
Each H has **two** electrons.

Electron counting by oxidation-state bookkeeping:
Shared electrons are awarded to the more electronegative atom.

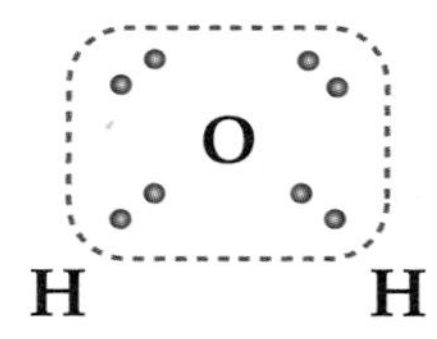

The O has **eight** electrons.
Each H has **zero** electrons.

Because oxygen is more electronegative than hydrogen, oxygen is assigned all the electrons in the O–H bonds. The hydrogen atoms get none of these electrons. Oxygen also gets its own lone pairs. So, by the oxidation-state method of electron bookkeeping, oxygen owns all eight valence electrons in the molecule, and the hydrogens own none.

At least for now, the only way we can do this type of bookkeeping is to have a dot diagram to look at. (In Chapter 6 we saw that it was necessary to have a dot diagram in order to use VSEPR theory to predict molecular shape. You can take it as a general truth that many times in chemistry, you will find that the first step in understanding a problem is to draw a dot diagram.)

9.1 WORKPATCH

Using the oxidation-state method of electron bookkeeping, determine how many electrons each atom owns in NH_2F. [*Hint:* Start by drawing a dot diagram. Nitrogen is the central atom.]

You should have found that only one atom in NH_2F owns an octet by this method of bookkeeping. Was it the most electronegative atom? Check your answer against the one at the end of the chapter.

At this point, you might be puzzled by the existence of two systems for electron bookkeeping. First we saw the double-counting method, which we used to satisfy the octet rule. Now we've introduced the oxidation-state method, which does not double-count shared electrons but instead assigns them to the more electronegative atom. The two counting methods come up with different answers for how many electrons each atom owns. Which method is right? The answer is, both of them. Which one you use depends on what you want to know. If you want to know whether a particular dot diagram is valid with respect to the octet rule, then double-count. If you want to know whether electrons are transferred from one atom to another in a chemical reaction, use the oxidation-state method.

Now that we know how to use the oxidation-state method to assign electrons to atoms in a molecule or polyatomic ion, we are ready to calculate oxidation states for the atoms. We do this by comparing the number of electrons each atom owns (using oxidation-state bookkeeping) with the number of valence electrons a free atom of that element owns. For the representative elements, as we learned in Chapter 4, the number of valence electrons in a free atom is equal to the element's roman-numeral group number in the periodic table.

For the representative elements, the number of valence electrons in a free atom equals the roman-numeral group number.

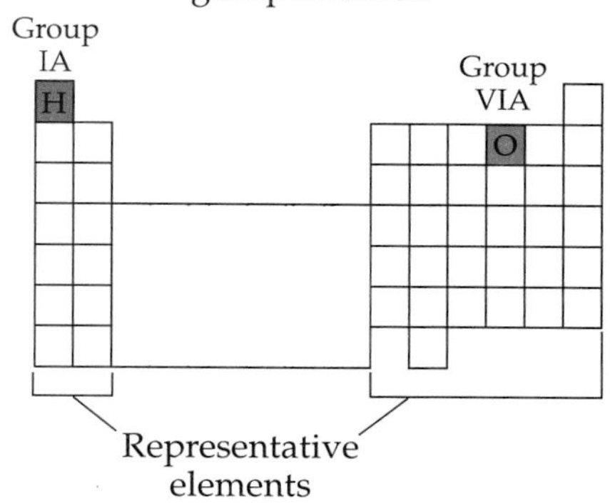

The **oxidation state**, or **oxidation number**,* of an atom in a molecule or polyatomic ion is the difference between the number of valence electrons in a free atom of that element and the number of electrons assigned to the atom by oxidation-state bookkeeping:

	Number of valence electrons in free atom of element
−	Number of electrons assigned to atom by oxidation-state bookkeeping
	Oxidation state

For water,

		Oxygen	Hydrogen	
	Number of valence electrons in free atom of element	6	1	$\cdot\ddot{\text{O}}\colon$ $\dot{\text{H}}$
−	Number of electrons assigned to atom by oxidation-state bookkeeping	− 8	− 0	H :Ö: H (boxed O)
	Oxidation state	−2	+1	

Notice from the above illustration that the oxidation state can be either negative or positive. It is negative if the atom owns more electrons by oxidation-state bookkeeping than it would as a free atom; it is positive if the atom owns fewer electrons than it would as a free atom. An oxidation state of zero means that the

*These two terms are exact synonyms, and we use them interchangeably throughout this text.

number of electrons the atom owns by oxidation-state bookkeeping is the same as the number it owns as a free atom. To summarize:

If the oxidation state of an atom is $+n$, the atom owns n fewer electrons in a molecule or ion than it would as a free atom.

If the oxidation state of an atom is $-n$, the atom owns n more electrons in a molecule or ion than it would as a free atom.

The oxygen atom in water has a -2 oxidation state because, by oxidation-state bookkeeping, it owns two more electrons than it would as a free atom. The hydrogen atoms in water are in the $+1$ oxidation state because they each own one fewer electron than the free atom.

At this point, it is very important to realize that *oxidation states are artificial numbers that result from arbitrary bookkeeping rules.* They do not indicate real charges. For example, even though the oxygen atom in a water molecule is partially negative ($\delta-$) owing to the polar covalent O–H bond, it does *not* carry a full -2 charge.

Using this method of assigning oxidation states to atoms in a molecule or polyatomic ion, it is possible for the same type of atom to have different oxidation states in a molecule. Let's examine this point by considering ozone, O_3.

9.2 WORKPATCH

Assign oxidation states for all the oxygen atoms in ozone, O_3,using this rule:

When two atoms of the same element (and therefore of identical electronegativity) share electrons, the oxidation-state method of electron bookkeeping divides the electrons equally between the atoms.

To do the WorkPatch, you'll need a dot diagram:

$$\ddot{\underset{..}{O}}=\ddot{O}-\ddot{\underset{..}{O}}:$$

This WorkPatch includes our first mention of assigning oxidation states to molecules that contain electrons shared between atoms of identical electronegativity. Using the rule in the WorkPatch, you should have found that two of the oxygen atoms in ozone have nonzero oxidation states.

The chart at the top of the facing page sums up the rules for assigning oxidation states by the oxidation-state bookkeeping method.

Practice Problems

9.1 Consider the NH_2F molecule from WorkPatch 9.1. Assign an oxidation state to each atom in the molecule.

Answer: In WorkPatch 9.1, you should have assigned eight electrons to F, six to N, and none to either H atom, using the oxidation-state method of electron bookkeeping. Now we just subtract from each of these numbers the number of valence electrons in the free atoms:

H_A—N̈—H_B (with :F̤: bonded below N)

	Fluorine	Nitrogen	$Hydrogen_A$	$Hydrogen_B$
Number of valence electrons in free atom of element	7	5	1	1
− Number of electrons assigned to atom by oxidation-state bookkeeping	− 8	− 6	− 0	− 0
Oxidation state	−1	−1	+1	+1

Assigning oxidation states by the electron-bookkeeping method

Step 1
Draw a valid dot diagram for the molecule.

Step 2
Assign every valence electron to an atom using the following rules:
- Assign each lone pair to the atom on which it is drawn in the dot diagram.
- Assign all the electrons of each bond to the atom of higher electronegativity.
- If bonding electrons are shared by atoms of equal electronegativity, divide the electrons evenly between the atoms.

Step 3
Calculate the oxidation state for each atom by subtracting the electron count of the atom (as determined in step 2) from the number of valence electrons in a free atom of that element.

Step 4
Check your answer. The oxidation states for all the atoms should add up to the charge on the molecule or ion.

9.2 In carbon dioxide, CO_2,
- (a) Assign an oxidation state to each atom in the molecule. [*Hint:* Draw a dot diagram first.]
- (b) How many electrons does the C atom own by oxidation-state electron bookkeeping?
- (c) How many more or fewer valence electrons are assigned to the C atom here than are present in a free C atom?
- (d) Is it correct or incorrect to say that the C atom in CO_2 has a complete octet of valence electrons? Explain.

9.3 In methane, CH_4,
- (a) Assign an oxidation state to each atom in the molecule. [*Hint:* Draw a dot diagram first.]
- (b) How many electrons does the C atom own by oxidation-state electron bookkeeping?
- (c) How many more or fewer valence electrons are assigned to the C atom here than are present in a free C atom?

9.4 In chloroform, $CHCl_3$,
- (a) Assign an oxidation state to each atom in the molecule. [*Hint:* Draw a dot diagram first.]
- (b) How many electrons does the C atom own by oxidation-state electron bookkeeping?
- (c) How many more or fewer valence electrons are assigned to the C atom here than are present in a free C atom?
- (d) Is it correct or incorrect to say that the C atom in $CHCl_3$ has a complete octet of valence electrons? Explain.

SHORTCUT METHOD FOR ASSIGNING OXIDATION STATES

The good news about the method we have been using to assign oxidation numbers is that it always works. The bad news is that it requires a dot diagram and drawing one takes time, if it can be done at all. For this reason, chemists have devised a shortcut for assigning oxidation states without the need for a dot diagram or even a table of electronegativities. However, what we gain in time we lose in accuracy because this shortcut method may not always produce a correct answer. Nevertheless, it works reliably for many of the compounds that chemists routinely encounter.

Shortcut rules

Rule 1. The oxidation state of an atom in the most abundant naturally occurring form of an element is assigned as zero.

The atoms in the substances shown below, including atoms in diatomic molecules, are assigned an oxidation state of zero.

Element	Naturally occurring elemental form	Oxidation state of atoms
Hydrogen	H_2	0
Oxygen	O_2	0
Nitrogen	N_2	0
Fluorine	F_2	0
Chlorine	Cl_2	0
Bromine	Br_2	0
Iodine	I_2	0
Helium	He	0
Neon	Ne	0
Phosphorus	P_4	0
Sulfur	S_8	0
Iron	Fe	0
Aluminum	Al	0

The most abundant elemental form for nonmetals is usually (but not always) a diatomic molecule, whereas the noble gases are monatomic. For all metals, the elemental form is represented by the symbol of the element.

Rule 2. The oxidation state of an oxygen atom in a compound is almost always −2.

For example, the oxidation state of the oxygen atoms in the compounds H_2O, CO, CO_2, $C_6H_{12}O_6$ (glucose), and $Fe_2O_3 \cdot H_2O$ (rust) is −2. Molecules called peroxides, which contain an O–O single bond, are exceptions to this rule. Ozone, which also has O–O single bonds, is also an exception (see WorkPatch 9.2). The only other time this rule fails is when oxygen is bound to an element more electronegative than itself. Only fluorine (F) meets this criterion. Except for these situations, it is a safe bet that an oxygen atom in a compound has an oxidation state of −2.

Rule 3. The oxidation state of a hydrogen atom in a compound or polyatomic ion is +1.

For example, the oxidation state of the hydrogen atoms in the compounds H_2O, $C_6H_{12}O_6$ (glucose), CH_4 (methane), and C_2H_2 (acetylene) is +1. The exceptions to this rule, where H is bonded to a less electronegative atom (such compounds are called hydrides) are not considered in this text.

Rule 4. The oxidation state of any group IA (1) atom (alkali metals such as lithium, Li, sodium, Na, and potassium, K) in a compound is always +1.

That's *in a compound*! This means that the oxidation state of the lithium atoms in lithium chloride, LiCl, or lithium oxide, Li_2O, is +1. The oxidation state of lithium metal, Li, its elemental form, is zero by rule 1.

Rule 5. The oxidation state of any group IIA (2) atom (alkaline earth metals such as magnesium, Mg, barium, Ba, and calcium, Ca) in a compound is always +2.

Again, that's *in a compound*! This means that the oxidation state of the magnesium atom in magnesium chloride, $MgCl_2$, is +2. The oxidation state of magnesium metal, Mg, its elemental form, is zero.

Rule 6. The oxidation state of a monatomic (one-atom) ion is equal to the charge of the ion.

Some examples are

Monatomic ion	Oxidation state
Al^{3+}	+3
O^{2-}	−2
H^+	+1
Cl^-	−1
N^{3-}	−3

Finally, we need a way to assign oxidation states to all the atoms in the periodic table not covered by rules 1–6. For this we use an extremely important rule that applies in all situations and to which there are no exceptions.

Rule 7. The sum of the oxidation states of the atoms in a chemical formula must add up to the overall charge of the molecule or ion represented by the formula.

If the formula represents a neutral molecule, such as H_2O, it has no charge and the oxidation states of the atoms must add up to zero. If you are dealing with a polyatomic ion, such as sulfate, SO_4^{2-}, the oxidation states of the atoms must add up to the charge on the ion (for sulfate, −2).

Which of the seven rules to apply and the order of their application can change from example to example, but we recommend that you always try applying rules 2 and 3 first and save rule 7 for last. A few examples using the shortcut rules are worked out below. Follow them through, but realize that we would have arrived at the same result if we had started with a dot diagram.

	Atom		Oxidation state	
H_2O	O		−2	Rule 2: O is −2
	H		+1	Rule 3: H is +1
	H		+1	
		Sum	0	Oxidation states add up to zero, the charge on the water molecule (rule 7).
OH^-	O		−2	Rule 2: O is −2
	H		+1	Rule 3: H is +1
		Sum	−1	Oxidation states add up to −1, the charge on the hydroxide ion (rule 7).

These were straightforward examples. Now we'll make it a bit more difficult by including an atom for which there is no rule. The trick is to start with rules 2 and 3, then apply rules 1, 4, 5, and 6 as needed, remembering to always save rule 7 for last.

	Atom		Oxidation state	
SO_4^{2-}	O		−2	Rule 2: O is −2
	O		−2	
	O		−2	
	O		−2	
	S		?	No rule for S
		Sum	−2	The sum must equal −2, the charge on the sulfate ion (rule 7).

Rule 7 says that all the oxidation numbers in a chemical formula must add up to give the overall charge on the compound, which in this case is −2. We have four oxygen atoms at −2 each, giving a total of −8. What do we have to add to −8 to get the −2 charge of the sulfate ion? The answer is +6. The oxidation number of the sulfur atom must therefore be +6.

We'll work through two more examples, each containing one element for which there is no shortcut rule.

	Atom		Oxidation state	
NH_4^+	H		+1	Rule 3: H is +1
	H		+1	
	H		+1	
	H		+1	
	N		?	No rule for N
		Sum	+1	The sum must equal +1, the charge on the ammonium ion (rule 7).

We have four hydrogen atoms with oxidation states of +1, giving a total of +4. What do we have to add to +4 to get +1, the charge on the ammonium ion, in order to satisfy rule 7? The answer is −3. The oxidation number of the nitrogen atom must therefore be −3.

	Atom	Oxidation state	
$Cr_2O_7^{2-}$	O	−2	Rule 2: O is −2
	O	−2	
	O	−2	
	O	−2	
	O	−2	
	O	−2	
	O	−2	
	Cr	?	No rule for Cr
	Cr	?	
	Sum	−2	The sum must equal −2, the charge on the dichromate ion (rule 7).

We have seven oxygen atoms with oxidation states of −2 each, giving a total of −14. To get −2, the charge on the dichromate ion, so as to satisfy rule 7, we must add +12. Now we have to be careful, however. The unknown oxidation number of Cr is not +12 because there are two Cr atoms in this compound. In cases like this, we can usually assume that the two Cr atoms have the same oxidation state. This means that each chromium atom must have an oxidation state of $+12/2 = +6$.

Before giving you some practice problems, we want to give you one more shortcut rule, which we call the halide rule:

Any halogen atom in a compound most likely has an oxidation state of −1.

Recall that the halogens are the group VIIA (17) atoms (F, Cl, Br, I). So, for example, we can now assign the oxidation states to the atoms in bromoform, $CHBr_3$:

	Atom	Oxidation state	
$CHBr_3$	H	+1	Rule 3: H is +1
	Br	−1	Halide rule: Br is −1
	Br	−1	
	Br	−1	
	C	?	No rule for C
	Sum	0	The sum must equal 0 (rule 7).

We have one hydrogen atom at +1 and three bromine atoms at −1 each, giving a total of −2. What do we have to add to −2 to get 0? The answer is +2. The oxidation number of the carbon atom must therefore be +2.

You have to be careful with the halide rule because it does have exceptions just as rule 2 for oxygen and rule 3 for hydrogen do. Part (c) in WorkPatch 9.3 tests your knowledge of a basic concept that enables you to determine when you can use the halide rule and when you cannot.

Suppose a chlorine atom in a compound has an oxidation state of −1.

(a) Is the number of valence electrons assigned to this atom by oxidation-state bookkeeping more or fewer than the number of valence electrons in a free chlorine atom?

(b) How many more or fewer?

(c) In order for a chlorine atom in a compound to have a −1 oxidation state, what must be true about the electronegativity of the atom to which it is bonded?

Check your answers to the WorkPatch to make sure you understand these basic concepts. Now let's examine a case where the halide rule does not apply.

Consider the chlorate ion, ClO_3^-, in which all three oxygen atoms are bound to a central chlorine atom. The oxidation state of the Cl atom in this ion is not −1.

(a) Why doesn't the halide rule apply for the chlorate ion?
(b) What is the oxidation state of the Cl atom in chlorate?

These last two WorkPatches are very important. They are reminders that in oxidation-state electron bookkeeping, it is the more electronegative atom that gets the shared electrons. More important, though, the WorkPatches tell you when the halogen rule works and when it doesn't. The rule holds only when the atom to which the halogen atom is bonded is less electronegative than the halogen atom. Make sure you understand this principle, and then try the following practice problems using the shortcut rules. For your convenience, the rules are summarized at the top of the opposite page.

Practice Problems

Use the shortcut rules to assign oxidation states to all atoms.

9.5 $COCl_2$ (oxygen and chlorine bonded to central carbon)

Answer: O is −2 (rule 2). Cl is −1 (halide rule applies because Cl is more electronegative than C). Therefore C is +4 (the sum of the oxidation numbers for the O and two Cl's is −4).

9.6 $MgBr_2$

9.7 Fe_2O_3

9.8 HClO (Cl and O bonded to each other)

9.9 CO_3^{2-}

9.10 Cu^{2+}

Chemists frequently use the shortcut rules to assign oxidation states, but when the rules are inadequate (for example, the rules don't cover a molecule like CS_2) or when they present some ambiguity, a dot diagram will always work. A good example of the ambiguity we mentioned is demonstrated by

Assigning oxidation states by the shortcut method

Rule 1
The oxidation state of an atom in the most abundant naturally occuring form of an element is typically zero.

Rule 2
The oxidation state of an oxygen atom in a compound is almost always −2.

Rule 3
The oxidation state of a hydrogen atom in a compound is almost always +1.

Rule 4
The oxidation state of any group IA (1) atom in a compound is +1.

Rule 5
The oxidation state of any group IIA (2) atom in a compound is +2.

Rule 6
The oxidation state of a monatomic ion equals the ion's charge.

Halide rule
A halogen atom in a compound usually has an oxidation state of −1.

Rule 7
The sum of the oxidation states of the atoms in a chemical formula must add up to the overall charge of the molecule or ion represented by the formula.

acetic acid, $C_2H_4O_2$. By the shortcut rules, we would assign the following oxidation states:

	Atom	Oxidation state	
$C_2H_4O_2$	O	−2	Rule 2: O is −2
	O	−2	
	H	+1	Rule 3: H is +1
	H	+1	
	H	+1	
	H	+1	
	C	?	No rule for C
	C	?	
	Sum	0	The sum must equal 0 (rule 7).

The hydrogen and oxygen oxidation numbers sum up to zero. Therefore if we make the oxidation number of both carbons zero, rule 7 will be obeyed. But we can also satisfy rule 7 by making one carbon +1 and the other −1, or +2 and −2, or +3 and −3, and so forth. There is no clear answer here. The only way out is a dot diagram.

9.5 WORKPATCH Using this dot diagram for acetic acid, assign each atom an oxidation number.

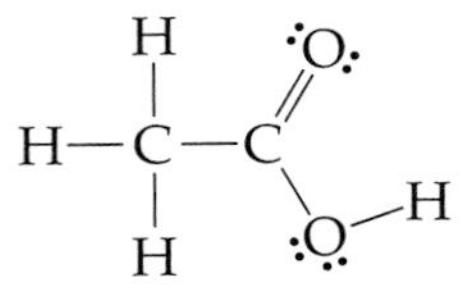

```
     H    :O:
     |    //
  H—C—C
     |    \
     H    :O—H
```

Using the dot diagram and applying the electronegativity rules for oxidation-state electron bookkeeping, you should have arrived at an unambiguous set of oxidation states for the carbon atoms in acetic acid. Check your answer against ours.

So far we have simply been assigning oxidation states to the atoms in a chemical compound. Remember, however, that our original goal was to be able to determine whether, during a chemical reaction, any electrons are transferred from one atom to another. Oxidation numbers were designed to help us do just this, and we are now ready to use them for this purpose.

9.3 Recognizing Electron-Transfer Reactions

Oxidation numbers tell you how many electrons each atom in a compound owns relative to the free atom. Therefore *if the oxidation number of any atom changes during a chemical reaction, a transfer of electrons must have occurred.* Thus, looking for changes in oxidation numbers is a way to determine whether or not electron transfer has occurred during a chemical reaction. To see how this procedure works, look at this chemical reaction, where the oxidation number of each atom is printed above the atom:

$$\overset{+1\,-1}{NaCl} + \overset{+1\,-1}{LiBr} \longrightarrow \overset{+1\,-1}{NaBr} + \overset{+1\,-1}{LiCl}$$

Consider each atom individually. Does the oxidation number of sodium change on going from reactant to product? Does the oxidation number of chlorine change on going from left to right? What about lithium and bromine? In this example, none of the oxidation numbers change during the reaction. This means that even though a chemical reaction takes place, there is no transfer of electrons from one atom to another. Each atom owns the same number of electrons before the reaction as after it.

The situation is quite different, however, in the reaction

$$\overset{-4\,+1}{CH_4} + 2\,\overset{0}{O_2} \longrightarrow \overset{+4\,-2}{CO_2} + 2\,\overset{+1\,-2}{H_2O}$$

Carbon starts with an oxidation state of −4 and ends up as +4. Each oxygen begins as zero and ends up as −2. This means that, according to the oxidation-state method of electron bookkeeping, a transfer of electrons must have taken place. In going from zero to −2, each oxygen atom gained two electrons.

When an atom gains electrons, we say it has been *reduced* (or has *undergone reduction*); **reduction** is defined as a gain of electrons. This makes sense if you think about reduction not in terms of the number of electrons but in terms of the oxidation state of the atom, which is reduced to a lower value.

Reduction

- Gain of electrons (by oxidation-state bookkeeping)
- Reduction in oxidation number (from a higher to a lower value)

O is reduced (gains electrons).

$$\overset{-4\ +1}{CH_4} + 2\,\overset{0}{O_2} \longrightarrow \overset{+4\ -2}{CO_2} + 2\,\overset{+1\ -2}{H_2O}$$

In this reaction, the oxygen atoms gain electrons, so we say the atoms are reduced. In general, an atom is reduced when its oxidation number *becomes smaller or more negative* as the atom goes from reactant to product.

If the oxygen atoms gain electrons in this electron-transfer reaction, where do the electrons come from? The key word here is *transfer*. It implies that, for oxygen to gain electrons, some other atom must lose them. Electron transfer implies *gain* and *loss*—you can never have one without the other. To find the atom that loses electrons, simply look for the one whose oxidation number increases (becomes more positive) during the reaction. In this reaction, it is carbon. To go from an oxidation state of −4 to +4, each carbon atom must lose eight electrons.

When an atom loses electrons, we say it has been *oxidized* (or has *undergone oxidation*); **oxidation** is defined as the loss of electrons. The carbon atoms are oxidized in this reaction.

Oxidation

- Loss of electrons (by oxidation-state bookkeeping)
- Increase in oxidation number (from a lower to a higher value)

$$\overset{-4\ +1}{CH_4} + 2\,\overset{0}{O_2} \longrightarrow \overset{+4\ -2}{CO_2} + 2\,\overset{+1\ -2}{H_2O}$$

C is oxidized (loses electrons).

Now remember, by the octet method of electron bookkeeping, no atom loses or gains electrons. For example, the carbon atom in both methane and carbon dioxide has an octet:

```
    H
    |
H — C — H        :Ö=C=Ö:
    |
    H
```

It is only when we switch to the oxidation-state method of electron bookkeeping that we can talk about the loss and gain of electrons.

Reduction (the gain of electrons) and oxidation (the loss of electrons) go together. One cannot occur in a chemical reaction without the other. We acknowledge this fact by referring to chemical reactions in which electrons are transferred as **redox** (pronounced ree-dox) **reactions**.

The reaction of methane, CH_4, with oxygen in the previous reaction is a redox reaction. The burning of any substance is a redox reaction, including the multistep "burning" of glucose in the cells of your body. Many chemical reactions are redox reactions, but not all of them. In order for a reaction to be classified as redox, electron transfer must take place during the reaction—that is, oxidation numbers must change.

Practice Problems

Indicate whether each reaction is a redox reaction. If it is, which atom gets oxidized and which atom gets reduced? Consult the shortcut rules.

9.11 $P_4 + 6\,Br_2 \longrightarrow 4\,PBr_3$ [*Hint:* Br is more electronegative than P.]

Answer:

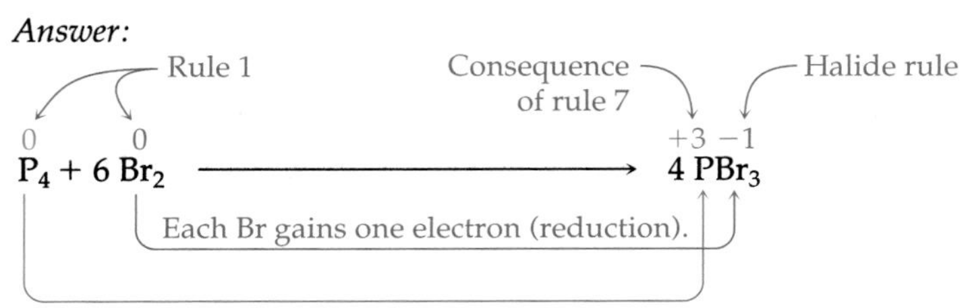

9.12 $4\,Fe + 3\,O_2 \longrightarrow 2\,Fe_2O_3$

9.13 $Ca + 2\,H^+ \longrightarrow Ca^{2+} + H_2$

9.14 $2\,NaBr + MgO \longrightarrow MgBr_2 + Na_2O$

We are getting close to an explanation of how we can use electron-transfer (redox) reactions to produce electricity. But first we need to introduce a bit more terminology. To do this, let's go back to the combustion of methane:

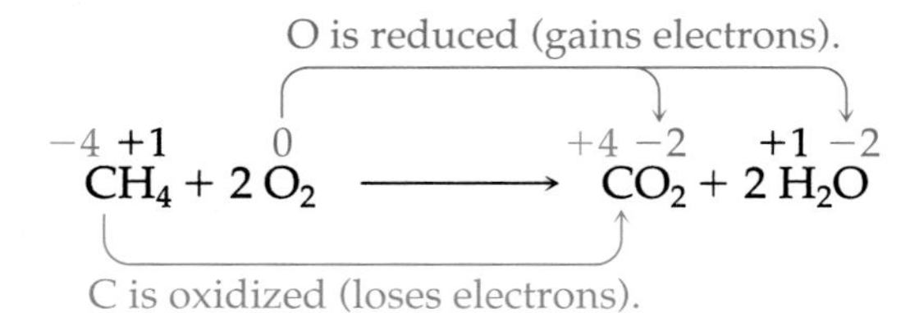

Up to now, we have been indicating the individual atoms that get oxidized or reduced. The C atom in CH_4 is oxidized. The O atoms in O_2 are reduced. While this way of looking at things is completely correct, chemists usually don't think in terms of the oxidation or reduction of individual atoms in a compound. Instead, they say that the entire CH_4 molecule is oxidized and the entire O_2 molecule is reduced. When the situation is described using whole molecules instead of individual atoms, we use the terms *oxidizing agent* and *reducing agent*. An **oxidizing agent** is a substance that oxidizes something else. A **reducing agent** is a substance that reduces something else. In our example, O_2 is an oxidizing agent because it oxidizes something else, CH_4. It does this by gaining the electrons that methane loses. This means that the oxidizing agent, O_2, must be reduced during the reaction.

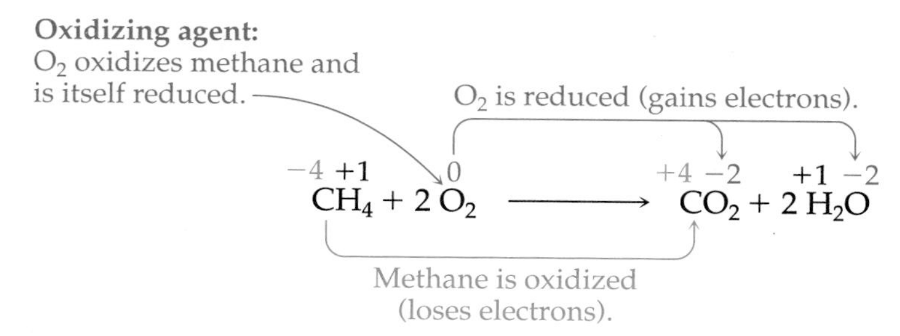

The same kind of logic works in identifying the reducing agent. In our example, CH_4 is a reducing agent because it reduces O_2. It does this by providing the electrons that O_2 gains. This means that the reducing agent, CH_4, must be oxidized during the reaction.

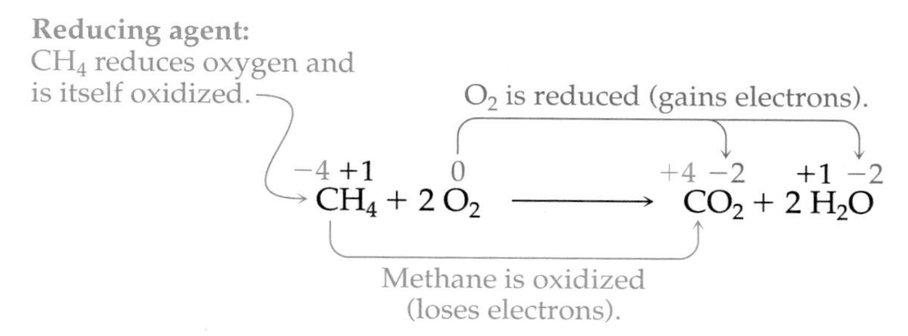

This terminology of oxidizing agent and reducing agent can be confusing, but it is something you will have to become comfortable with, and practice helps. Just think of it this way: Whichever compound (or ion) gets oxidized (whichever compound has an atom whose oxidation number increases) is the reducing agent, and whichever compound gets reduced (whichever compound has an atom whose oxidation number decreases) is the oxidizing agent.

The following are a repeat of Practice Problems 9.11–9.13, but this time you are asked to identify the oxidizing and reducing agents.

Practice Problems

For each redox reaction, indicate which substance is the oxidizing agent and which is the reducing agent.

9.15 $P_4 + 6\,Br_2 \longrightarrow 4\,PBr_3$

Answer:

Reducing agent: reduces Br_2 and is oxidized.

Oxidizing agent: oxidizes P_4 and is reduced.

0 0 +3 −1

$P_4 + 6\,Br_2 \longrightarrow 4\,PBr_3$

Each Br gains one electron (reduction).

Each P loses three electrons (oxidation).

9.16 $4\,Fe + 3\,O_2 \longrightarrow 2\,Fe_2O_3$

9.17 $Ca + 2\,H^+ \longrightarrow Ca^{2+} + H_2$

Now let's carry out a real redox reaction. Put a strip of zinc metal into a beautiful blue aqueous solution of Cu^{2+} ions. Immediately, a chemical reaction begins. Within minutes, the blue color of the liquid begins to fade and the silvery zinc strip becomes plated with copper metal.

A zinc strip is placed in an aqueous solution of blue Cu^{2+} ions.

The reaction is $Zn + Cu^{2+} \longrightarrow Zn^{2+} + Cu$.

By now you should be able to look at the balanced equation and quickly determine that this is indeed a redox reaction. Just to be sure, complete the following WorkPatch.

Step 1: Use rules 1–7 to assign oxidation states to the atoms and ions in the reaction

$$Zn + Cu^{2+} \longrightarrow Zn^{2+} + Cu$$

Step 2: Now answer the following questions:

(a) Is this a redox reaction? How can you tell?
(b) If so, what was oxidized?
(c) If so, what was reduced?
(d) If so, what was the oxidizing agent?
(e) If so, what was the reducing agent?

Without a doubt, electrons are transferred from Zn to Cu^{2+} in this reaction. Now let's harness some electricity from this redox reaction.

9.4 Electricity from Redox Reactions

Here again is the redox reaction we have chosen to harness electricity from:

$$Zn + Cu^{2+} \longrightarrow Zn^{2+} + Cu$$

The zinc metal is oxidized, so the Zn atoms in the zinc strip each lose two electrons, while the copper ions (Cu^{2+}) are reduced, each picking up two electrons. But exactly how do the electrons get from the strip of Zn metal to the aqueous Cu^{2+} ions in solution? Because water is such a poor conductor of electricity, they can't just leap off the zinc strip and swim out to the Cu^{2+} ions. Instead, the Cu^{2+} ions must migrate to the zinc strip and touch it in order for electron transfer from Zn to Cu^{2+} to take place.

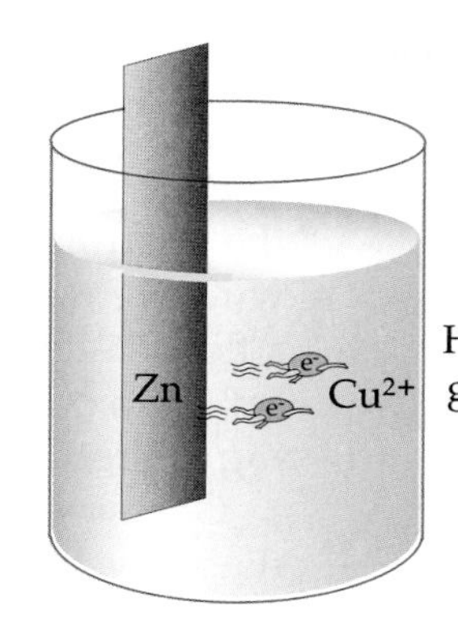

How do the electrons get from Zn to Cu^{2+}? **Not this way!**

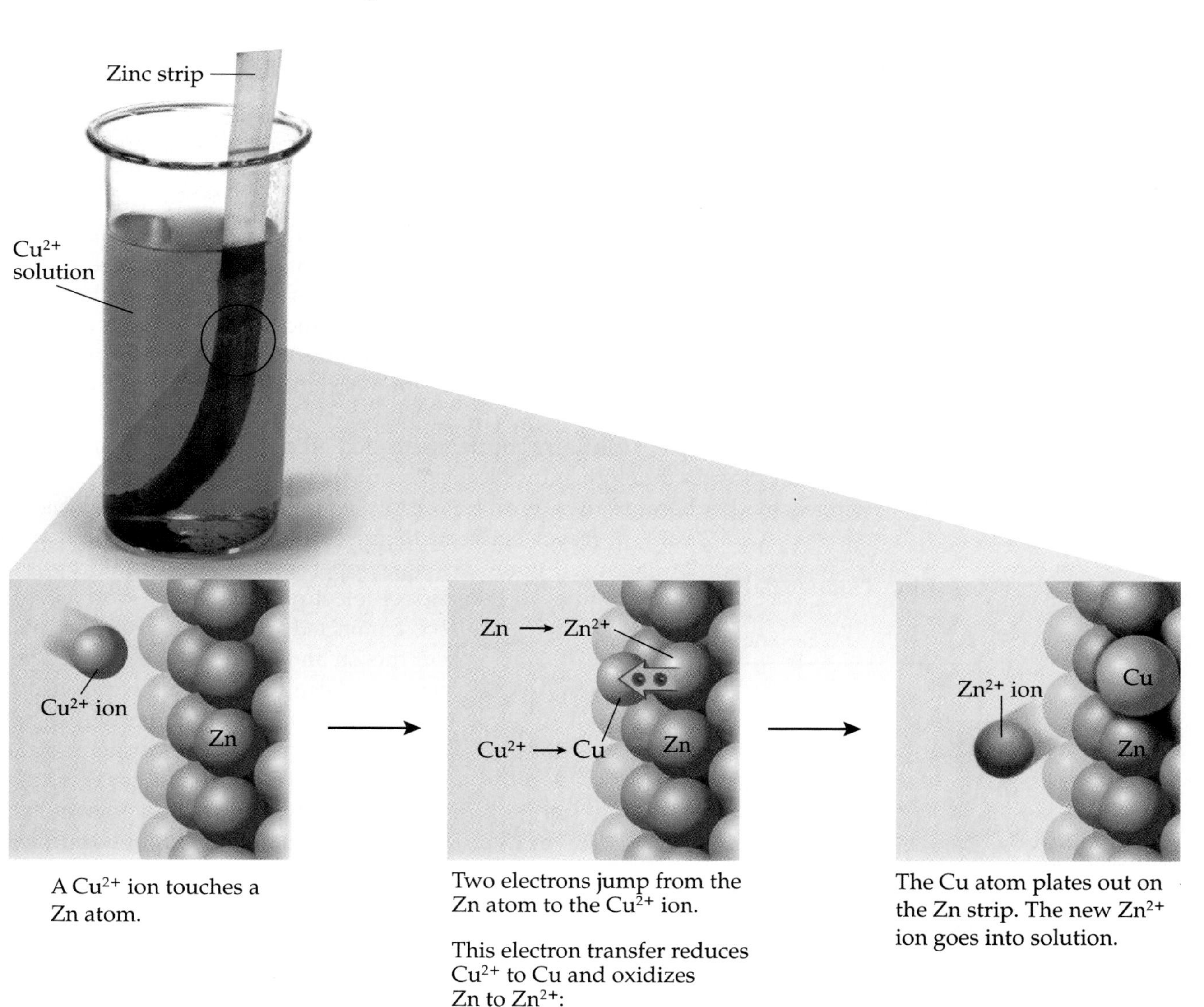

A Cu^{2+} ion touches a Zn atom.

Two electrons jump from the Zn atom to the Cu^{2+} ion.

This electron transfer reduces Cu^{2+} to Cu and oxidizes Zn to Zn^{2+}:

$$Cu^{2+} + Zn \longrightarrow Cu + Zn^{2+}$$

The Cu atom plates out on the Zn strip. The new Zn^{2+} ion goes into solution.

When a Cu^{2+} ion touches the zinc strip, the ion accepts two electrons and is reduced to elemental copper (Cu). This is where the copper metal that plates out on the Zn strip comes from. This also explains why the solution becomes less blue: blue Cu^{2+} ions are being removed from solution and it is the presence of those ions in water that produces the blue color. At the same time, Zn atoms in the zinc strip are being oxidized to colorless Zn^{2+} ions. These newly formed Zn^{2+} ions enter the solution. If we could somehow get the electrons that transfer from the zinc to the copper ions to travel through a wire, we would be creating electricity (recall our description of electricity as a flow of electrons in a wire). This would be electricity produced by a chemical (redox) reaction, and it would last as long as there are reactants converting to products.

One way to accomplish this is to separate the Zn strip from the Cu^{2+} ions so that they are no longer in direct contact with each other, and then provide a wire path for the electrons to flow across. A common laboratory setup for this is shown in the figure below. This setup, called a **galvanic cell** by chemists, is what you would call a **battery**. In this example, a strip of zinc is placed in an aqueous solution of Zn^{2+} ions (created by dissolving zinc sulfate, $ZnSO_4$, in water), and a strip of copper is placed in a solution of Cu^{2+} ions (created by dissolving copper sulfate, $CuSO_4$, in water). In other words, each strip of metal is placed in a solution of its own ions. Both solutions also contain sulfate, SO_4^{2-}, ions. The metal strips are then connected with a wire that can conduct electrons from one strip to the other. An additional component, called a *salt bridge*, is necessary to complete the circuit. This bridge can be as simple as a piece of absorbant paper saturated with a solution of a salt, or it can be a tube like the one shown in the figure. The salt used in the salt bridge should be composed of *spectator ions*—that is, ions that are not directly involved in the redox reaction. A common bridge salt is sodium sulfate, Na_2SO_4, which provides Na^+ and SO_4^{2-} spectator ions to the system. We shall discuss the necessity for the salt bridge shortly.

The metal strips of zinc and copper are called **electrodes**. No longer does a Cu^{2+} ion directly touch the zinc electrode, but they are in "electrical contact" with each other because there is an uninterrupted pathway for the electrons to travel between them. That pathway is through the wire. When a copper ion comes in contact with the copper electrode, the Cu^{2+} ion is, in effect, connected to the Zn electrode (it's as if the Zn and Cu^{2+} are touching). The zinc metal can now give up electrons to the copper ions through the wire. The Cu^{2+} ions are reduced to copper metal, depleting the $CuSO_4$ solution of Cu^{2+} ions. The newly formed copper metal (Cu) plates out on the copper electrode, causing its mass to increase. The Zn atoms in the zinc electrode are oxidized to Zn^{2+}, and these newly formed ions enter the $ZnSO_4$ solution around the electrode, increasing the amount of

A Zinc–Copper Battery

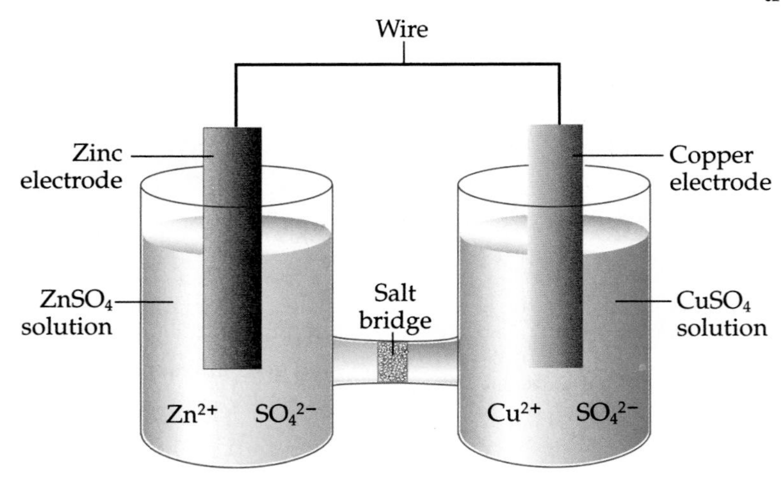

Zn^{2+} ions in the solution. This reaction causes the zinc electrode to be eaten away over time. All of this is shown diagrammatically in the following figure. Study the sequence carefully to see what is happening.

How a Battery Generates Electrical Current

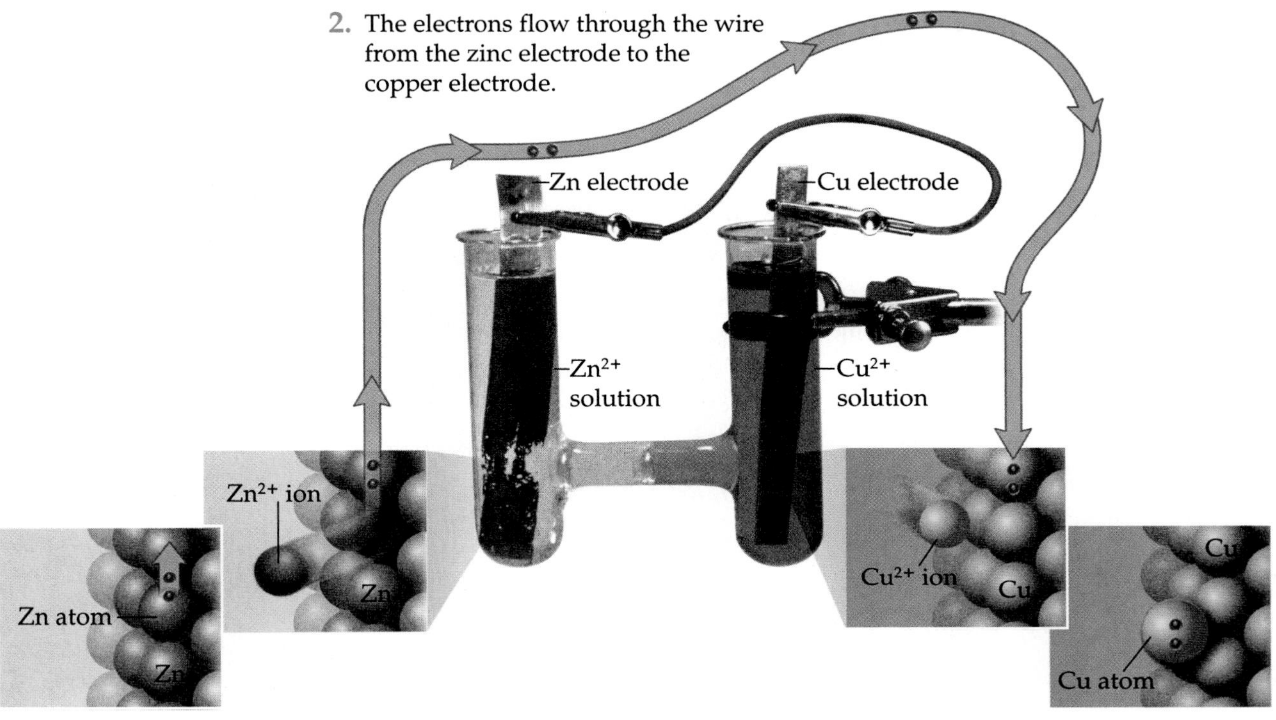

1. A Zn atom gives up two electrons to become a Zn^{2+} ion, which enters the $ZnSO^4$ solution.

Oxidation reaction: $Zn \longrightarrow Zn^{2+} + 2\,e^-$

3. The electrons reduce a Cu^{2+} ion from the solution to a Cu atom. The Cu atom plates out on the electrode.

Reduction reaction: $2\,e^- + Cu^{2+} \longrightarrow Cu$

Now we can better appreciate the presence of the salt bridge. Without it, the zinc side of the galvanic cell becomes more and more positively charged as new Zn^{2+} ions are created. This buildup of positive charge makes it impossible for negatively charged electrons to leave the zinc side. Likewise, the copper side of the cell becomes increasingly negative as the Cu^{2+} ions are reduced and removed from solution, resulting in a relative excess of SO_4^{2-} ions. This excess negative charge repels electrons away from the copper side. This buildup of charge would cause the redox reaction to come to a halt, and electrons no longer flow through the wire. In short, the cell (battery) would not function.

With the salt bridge in place, things are different. The porous plug in the center of the bridge allows ions to move from one side to the other but keeps the two solutions from mixing too quickly with each other. Now as the zinc side becomes more positively charged with Zn^{2+} ions, SO_4^{2-} ions from the salt bridge can move through the porous plug into the $ZnSO_4$ solution and neutralize the excess positive charge. Likewise, Na^+ ions from the salt bridge move into the $CuSO_4$ solution and neutralize the excess negative charge created on the copper side as Cu^{2+} ions plate out on the Cu electrode. In other words, a salt bridge is necessary for this type of reaction in order to maintain a charge balance in the two halves of the system.

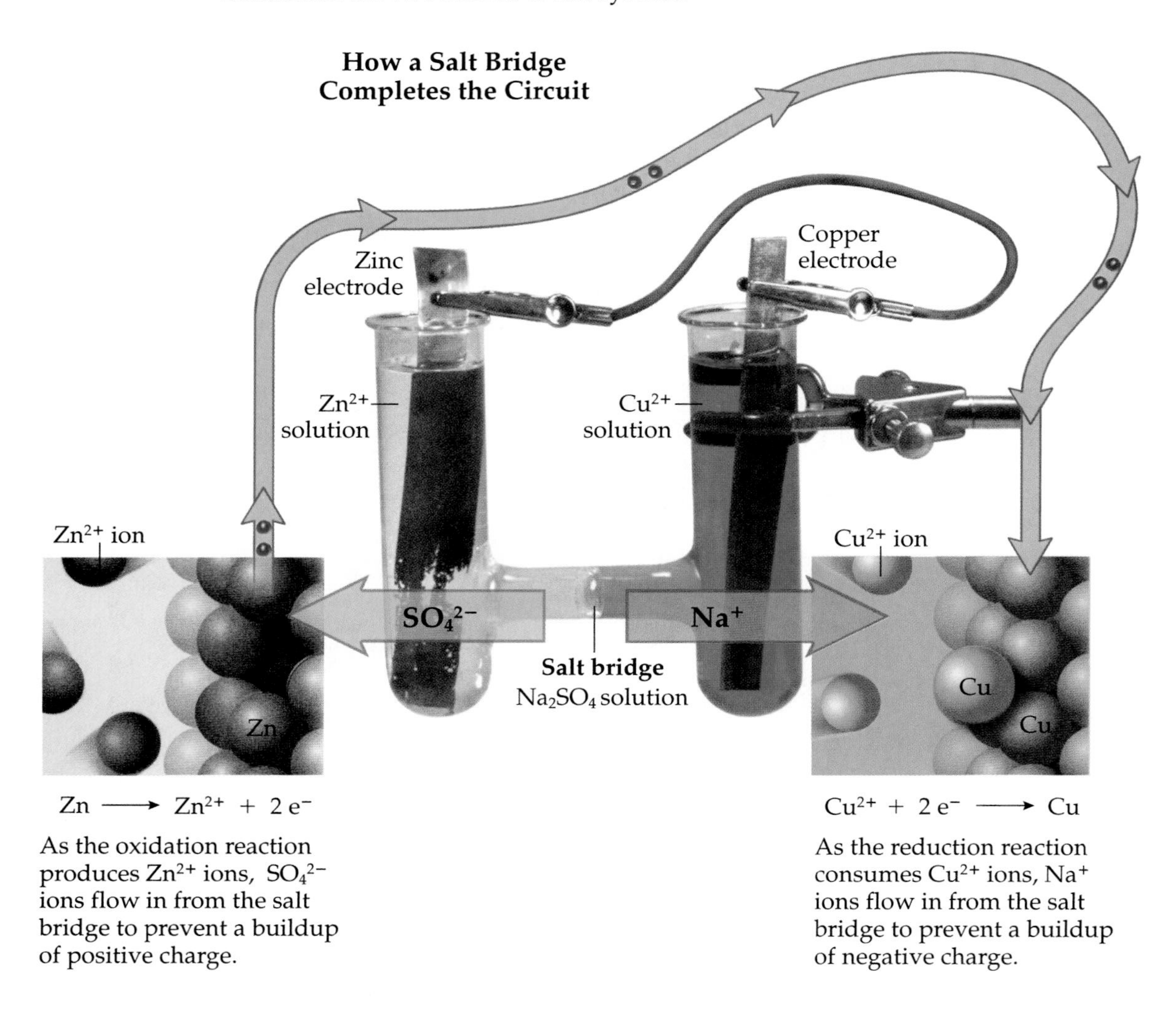

$Zn \longrightarrow Zn^{2+} + 2\,e^-$

As the oxidation reaction produces Zn^{2+} ions, SO_4^{2-} ions flow in from the salt bridge to prevent a buildup of positive charge.

$Cu^{2+} + 2\,e^- \longrightarrow Cu$

As the reduction reaction consumes Cu^{2+} ions, Na^+ ions flow in from the salt bridge to prevent a buildup of negative charge.

Because the electrons in our galvanic cell must flow through a wire to get from zinc atoms to copper ions, we are producing usable electricity. In addition, there is a measurable voltage pushing the electrons along. This zinc–copper battery produces about 0.9 V, enough to light a small bulb or run a motor to do some work. Until one of the reactants (Zn or Cu^{2+}) is used up, this battery will keep on going and going and going.

In principle, any redox reaction can be used to make a battery. All we need to do is separate the oxidizing agent from the reducing agent and then connect them with a wire.

9.7 WORKPATCH

Design and label a battery based on the redox reaction between magnesium metal and silver ions:

$$Mg + 2\,Ag^+ \longrightarrow Mg^{2+} + 2\,Ag$$

Does your battery show electrons flowing from the Mg electrode to the Ag electrode? It should!

Astounding as it may seem, you can make a battery using just electrical wires, strips of zinc and copper, and a lemon or other citrus fruit. You can even use a potato. Here is the lemon battery we built for this book:

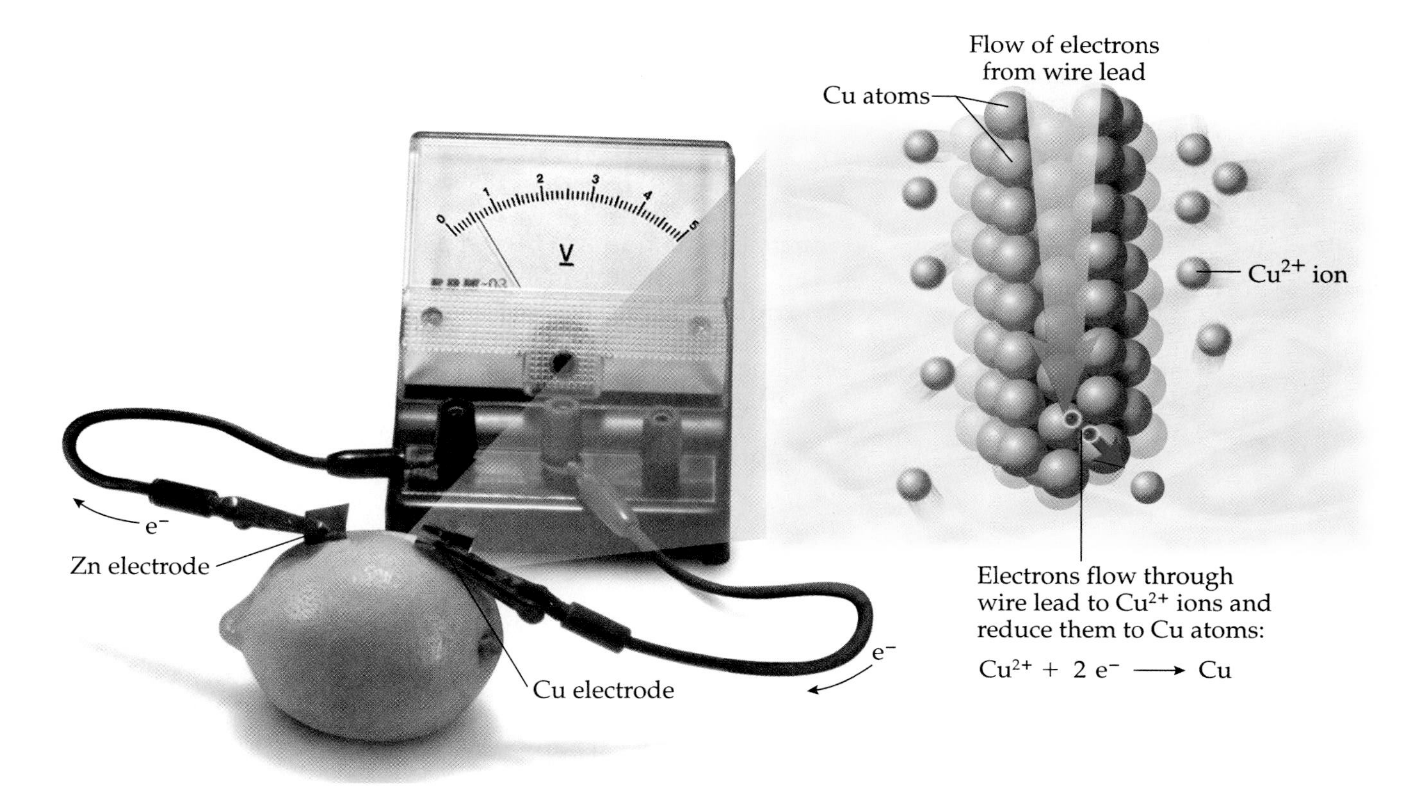

The only significant difference between this lemon battery and the zinc–copper battery we made earlier is that we haven't provided a solution of Cu^{2+} around the Cu electrode. Exactly what happens at the copper electrode is still an area of speculation. One reasonable explanation is that Cu^{2+} ions are already present on the electrode surface as a result of prior corrosion. Another is that the acids in the lemon cause some of the Cu atoms from the copper electrode to oxidize to Cu^{2+} ions. These ions remain in the vicinity of the copper electrode. As soon as the two electrodes are connected by a wire, zinc atoms begin to oxidize, releasing electrons. These electrons travel through the wire and reduce the Cu^{2+} ions back to Cu metal. The lemon serves as a salt bridge.

As we said, any redox reaction can potentially be used to make a battery if you can separate the substance that gets oxidized from the one that gets reduced and then connect the two with a wire. Interestingly, you can simply place a zinc strip and a copper strip in the same beaker of salt water and run a low-current digital clock (a setup like this sits on the desk of one of the authors of this book). This and our lemon battery are a bit unconventional. Perhaps the ones in the photograph look more familiar to you.

The standard dry-cell battery is nothing more than a redox reaction in a can. A cross-section of one type is shown at the top of the next page. The reactants are manganese dioxide, MnO_2, and metallic zinc, Zn. As you can see from the

Cross-Section of a Dry-Cell Battery

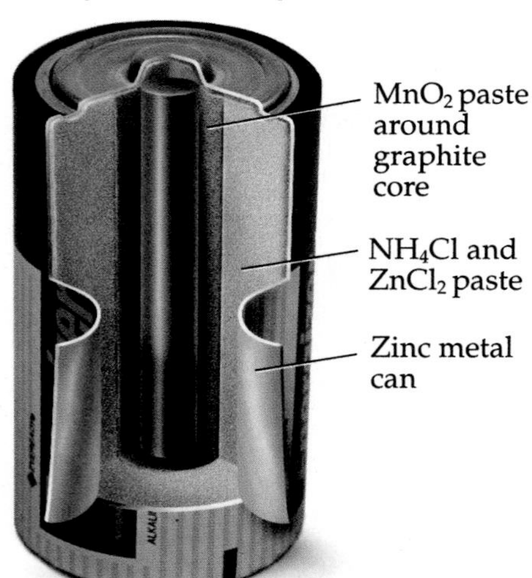

diagram, these reactants are not in direct contact with each other. A semimoist paste containing NH_4Cl (ammonium chloride) that serves as a salt bridge separates them. In this battery, the zinc metal is oxidized (loses two electrons) to become Zn^{2+} ions, so Zn is the reducing agent:

Zinc is oxidized (loses two electrons).
Each Zn atom becomes a Zn^{2+} ion.

The manganese dioxide is changed to dimanganese trioxide, Mn_2O_3, during the reaction. The manganese atom in manganese dioxide is in the +4 oxidation state but becomes +3 in Mn_2O_3. So the manganese is reduced, and MnO_2 is the oxidizing agent:

Manganese is reduced (gains one electron).
Each Mn^{4+} ion in MnO_2 becomes a Mn^{3+} ion in Mn_2O_3.

The overall redox reaction for the battery is

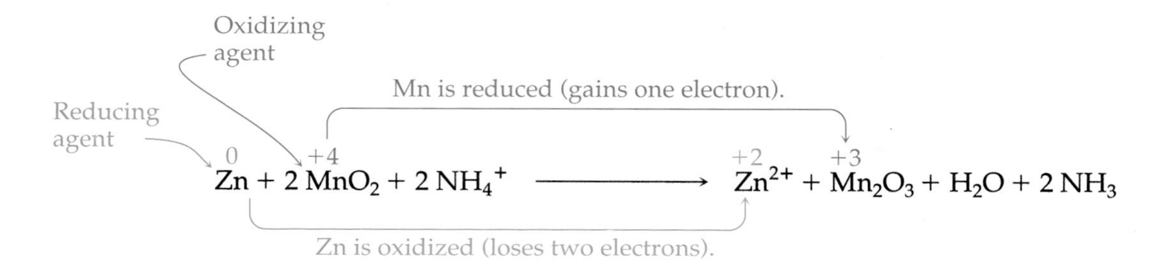

Connect a wire between the two terminals and the reaction begins—electrons flow from Zn to MnO_2. Put a light bulb in the circuit and you've got a flashlight.

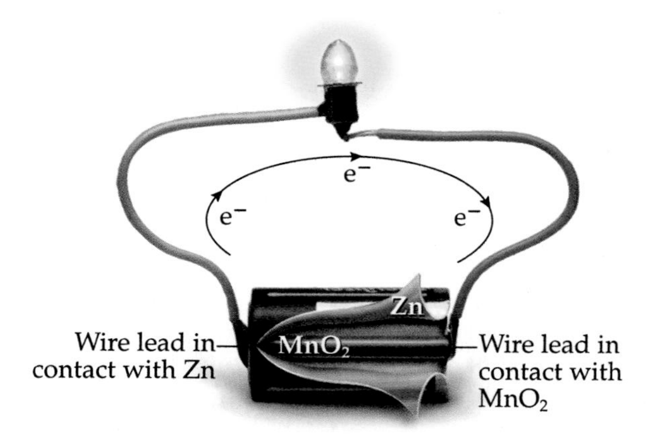

Redox reactions are at the heart of all batteries, from the battery in your car to the tiny battery in a hearing aid. All involve at least one reactant that loses electrons (oxidation) and another reactant that simultaneously gains them (reduction). You have probably noticed the + and − signs used to label the electrodes on a battery. In our zinc–manganese dry cell, which electrode would be marked with which sign? Recall that electrons are negatively charged and are therefore always attracted to anything with a positive charge. In the zinc–manganese dry cell, the electrons flow from the Zn electrode to the MnO_2 electrode. This means that the MnO_2 electrode is assigned the + sign.

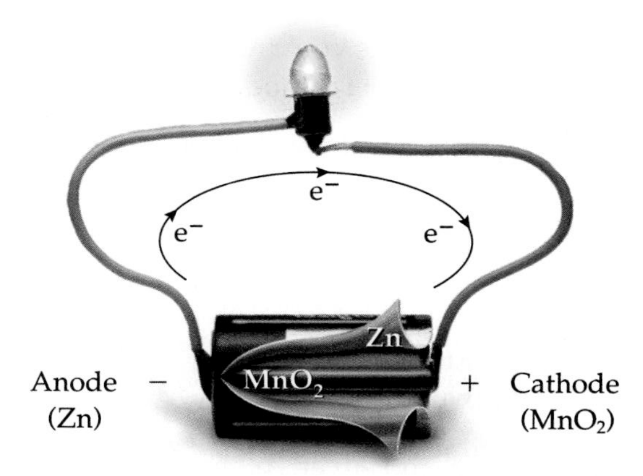

Electrons move from anode (−) to cathode (+).

Not surprisingly, chemists have names for these two electrodes. The positive electrode is called the **cathode**, and the negative electrode is called the **anode**. To help you remember this, just recall the names of positive and negative ions. A positive ion is a *cat*ion, and a negative ion is an *an*ion. For both batteries and ions, the prefixes *cat-* and *an-* mean positive and negative, respectively.

Return to the drawing of the battery you made in WorkPatch 9.7 and add the labels +, −, cathode, and anode.

It's important that you be able to do WorkPatch 9.8 because it leads us to a fundamental question. What ultimately decides which way the electrons flow? For the zinc–copper battery, you were told that zinc is oxidized and feeds electrons to the Cu^{2+} ions. That is, we started with the balanced redox reaction $Zn + Cu^{2+} \longrightarrow Zn^{2+} + Cu$. But what if this were not the case? What if the diagram of the cell were put in front of you and you were asked, "Which way do the electrons flow? What gets oxidized? What gets reduced? Why is the reaction written with electrons flowing from zinc to copper and not the other way?"

Multiple choice: In general, which way do electrons flow in a battery?
(a) From what gets oxidized to what gets reduced.
(b) From what gets reduced to what gets oxidized.
Explain how you made your choice.

This WorkPatch should remind you that the ultimate answer to the question "Which way do the electrons flow?" requires knowing what is oxidized and what is reduced. In the next section, we shall see how you can determine this. For now, try the following practice problems, where the necessary information is given.

Practice Problems

9.18 Construct a battery in which the redox reaction is

$$Ni + Pb^{2+} \longrightarrow Ni^{2+} + Pb$$

To answer, draw a battery similar to the one you drew for WorkPatch 9.7, label which way the electrons flow, and label cathode, anode, +, and −.

Answer: The nickel electrode is the anode (−). The lead electrode is the cathode (+). Electrons flow from the nickel anode to the lead cathode.

9.19 For the nickel–lead battery of Practice Problem 9.18,
(a) What is the oxidizing agent?
(b) What is the reducing agent?
(c) Which electrode gets larger over time?
(d) Which electrode is eaten away over time?

9.20 A battery is constructed from iron, Fe, and silver, Ag, by dipping a strip of each metal into a solution of its ions (Fe^{3+} and Ag^{+}, respectively). As the battery operates, the Fe^{3+} concentration increases while the Ag^{+} concentration decreases.
(a) What is getting oxidized?
(b) What is getting reduced?
(c) Draw a battery similar to the one you drew for WorkPatch 9.7, label which way the electrons flow, and label cathode, anode, +, and −.

9.5 Which Way Do Electrons Flow?—The EMF Series

What Happens Here?

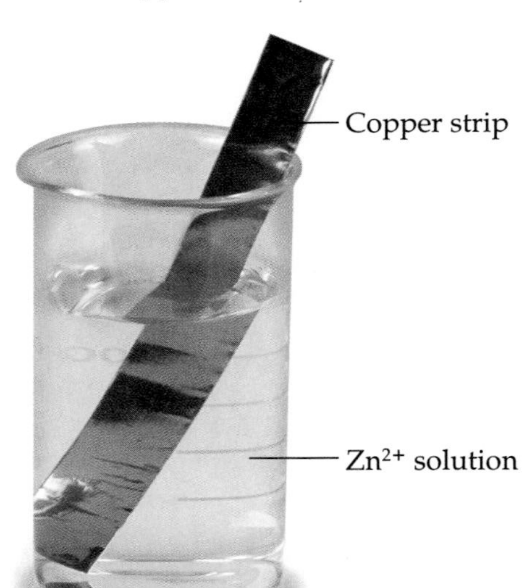

Look back at the illustration on page 336. When we put a strip of zinc into a solution of Cu^{2+} ions, a redox reaction occurred spontaneously. Neutral Zn atoms spontaneously lost electrons to the Cu^{2+} ions in solution, causing copper metal to plate out on the zinc strip. What would happen if we tried the opposite arrangement? Suppose that instead of putting a strip of Zn metal into a solution of Cu^{2+} ions, we put a strip of Cu metal into a solution of Zn^{2+} ions.

Would the Zn^{2+} ions be reduced (gain electrons) and the Cu atoms oxidized (lose electrons)? In other words, would we see zinc metal plate out on the copper strip? The equation for this hypothetical reaction is

$$Zn^{2+} + Cu \xrightarrow{?} Zn + Cu^{2+}$$

This is the opposite (reverse) of the reaction that occurs in a zinc–copper battery:

$$Zn + Cu^{2+} \longrightarrow Zn^{2+} + Cu$$

In fact, our hypothetical reaction doesn't occur; when you immerse a copper strip in a solution of Zn^{2+} ions, nothing happens. By setting up these two experiments, we learn that

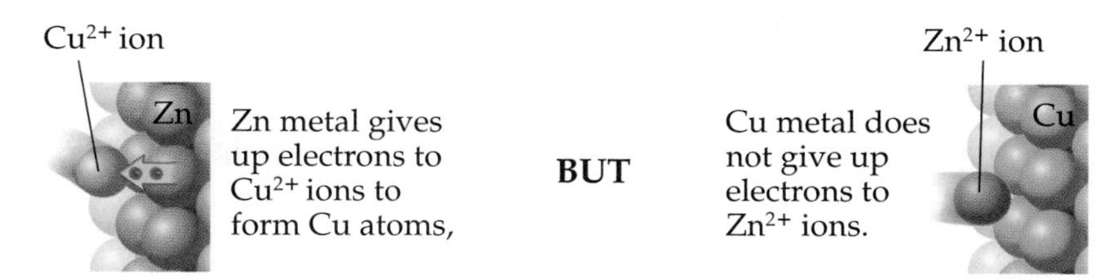
Cu²⁺ ion
Zn
Zn metal gives up electrons to Cu²⁺ ions to form Cu atoms,
BUT
Cu metal does not give up electrons to Zn²⁺ ions.
Zn²⁺ ion
Cu

Why is this so? Recall from Chapter 4 that the chemical definition of a metal is "an element that easily loses its valence electrons." But, given the results summarized above, it must be that all metals are not created equal with respect to this definition. *Some metals lose electrons more readily than others*. In the competition we set up between the two systems Zn/Cu^{2+} and Zn^{2+}/Cu, a spontaneous reaction occurred only in the first case, where Zn metal gave up electrons and copper accepted them. Because of this, we say that zinc is more *active* than copper. An **active metal** is one that easily loses valence electrons. Zinc atoms have a greater tendency than copper atoms to lose valence electrons (in other words, zinc is more easily oxidized than copper), so zinc is the more active metal. Because this is so, zinc metal can "force" Cu^{2+} ions to take its electrons, but copper metal cannot "force" Zn^{2+} ions to take its electrons. The following diagram sums this up. In a zinc–copper system, the electron flow is always from zinc to copper because of zinc's greater tendency to lose electrons.

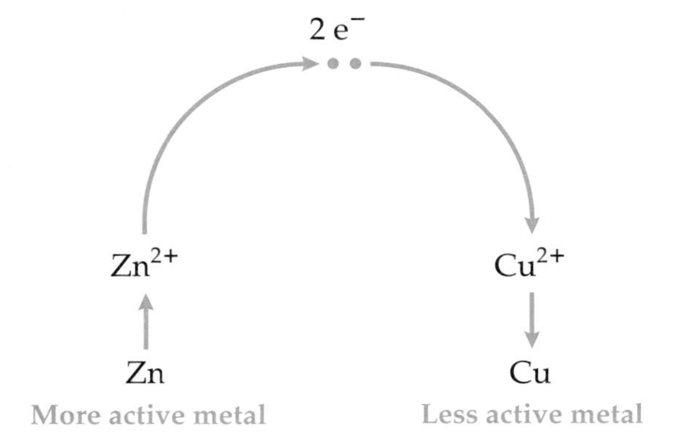
Electrons always flow from the more active metal to the less active metal.
2 e⁻
Zn²⁺
Zn
More active metal
Cu²⁺
Cu
Less active metal

We should point out that, the first time these metals were studied, there was no way to know whether it was the Zn or the Cu that was more active. The result had to be determined by setting up the two experiments described above and noting which one reacted spontaneously.

In principle, we could compare any two metals to determine which one is more active. We could carry out this same experiment using zinc and magnesium, for example, by dipping each metal into a solution of the other's ions and seeing which one reacts spontaneously. This experiment is shown below. We observe that nothing at all happens in the beaker with the zinc strip immersed in a solution of Mg^{2+} ions. However, in the beaker on the right, the situation is quite different. The magnesium strip quickly becomes coated with zinc metal.

Test to Find Out Which Metal Is More Active — Zn or Mg

From these experimental observations, we conclude the following about the relative activities of magnesium and zinc:

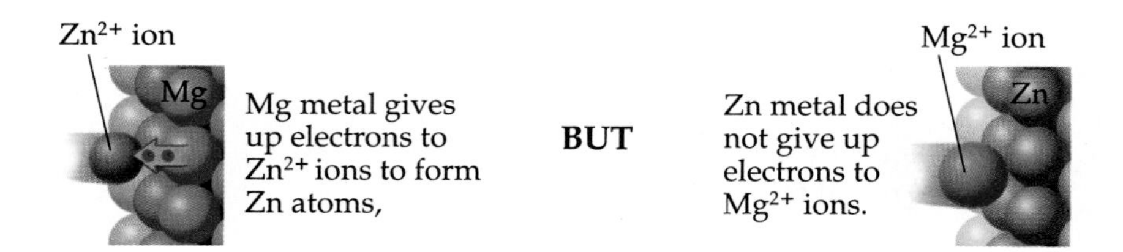

Electrons flow from Mg (more active) to Zn (less active).

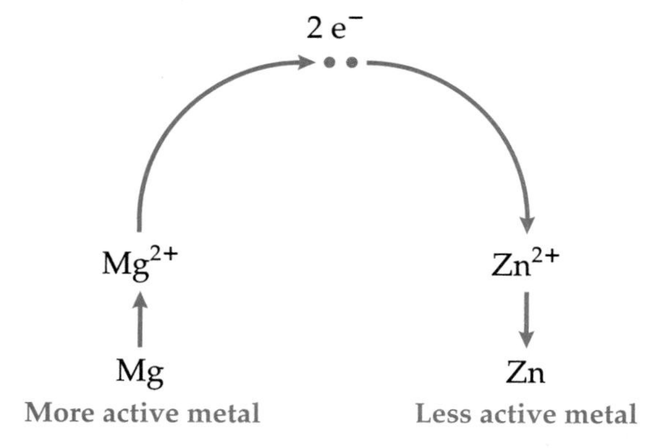

In other words, magnesium is a more active metal than zinc (magnesium's "desire" to lose electrons is greater than zinc's). And because zinc is more active than copper, magnesium must also be more active than copper. This relative ordering of activities is called the **electromotive force series** (or **EMF series**, for short):

An Electromotive Force Series

Mg Most active metal (loses electrons most easily)

Zn Increasing activity

Cu Least active metal (loses electrons least easily)

We could extend this list by testing additional pairs of metals and metal ions, but chemists have already done this for us over the years. A portion of the resulting EMF series is shown in the margin.

Remember that a more active metal can always force electrons onto the ions of a less active metal. Using the EMF series, we can predict which redox reactions will be spontaneous and which way electrons will flow in a battery. For example, if we set up this experiment, in which beaker will a spontaneous redox reaction occur?

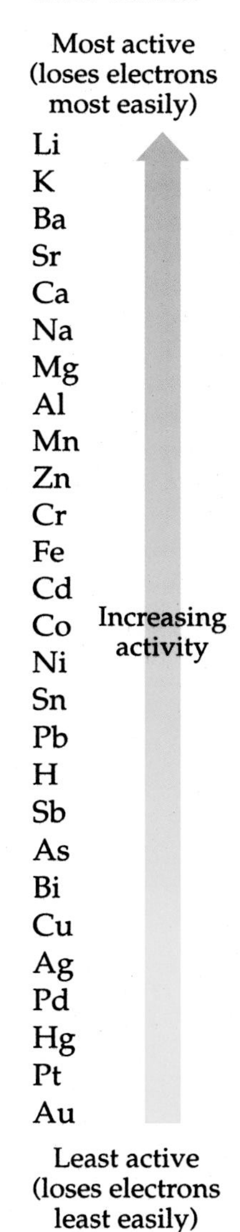

Copper is higher on the EMF list than silver (not by much, but it is higher). This means that copper is more active than silver, and so neutral Cu atoms give electrons to Ag^+ ions. The redox reaction

$$Cu + 2\,Ag^+ \longrightarrow Cu^{2+} + 2\,Ag$$

occurs spontaneously in the beaker on the right. The Ag^+ ions are reduced to elemental silver and plate out on the copper strip (see next page). As for the beaker on the left, absolutely nothing happens in it. The reverse of the spontaneous redox reaction does not happen because silver is less active than copper. In a copper–silver battery, electrons flow from the copper electrode (more active metal) to the silver electrode (less active metal). This means that copper is the anode (−) and silver is the cathode (+). Copper metal is oxidized, and Ag^+ ions are reduced.

Nothing happens. Solution begins to turn blue as Cu atoms from the metal become (blue) Cu^{2+} ions in solution.

All these results can be predicted from a glance at the EMF series. We can even take advantage of this knowledge to make a beautiful Christmas decoration by bending a piece of copper wire into the shape of a Christmas tree and placing it in a solution of Ag^+ ions. The spontaneous redox reaction $Cu + 2\,Ag^+ \longrightarrow Cu^{2+} + 2\,Ag$ occurs, and pure silver metal plates out on the copper tree. In addition, the Cu^{2+} ions formed color the solution a bright blue.

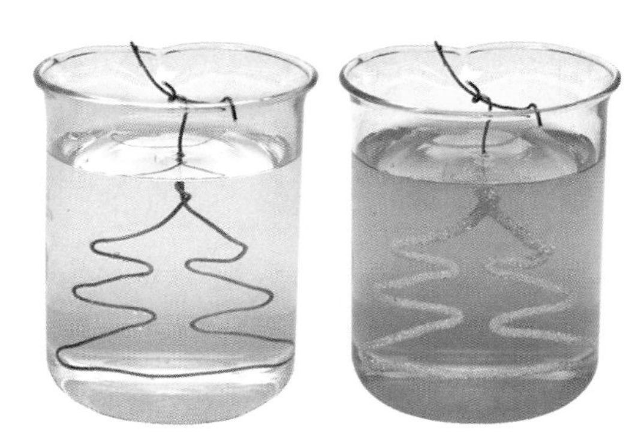

Practice Problems

9.21 What will happen if a piece of Zn metal is placed in a solution of Al^{3+} ions?

Answer: Absolutely nothing; Al is higher on the EMF series and is thus the more active metal. This means that Al could give its valence electrons to Zn^{2+} ions, but Zn does not spontaneously give its electrons to Al^{3+} ions.

9.22 Which of the following reactions is spontaneous?

(a) $3\,Zn^{2+} + 2\,Al \longrightarrow 3\,Zn + 2\,Al^{3+}$

(b) $3\,Zn + 2\,Al^{3+} \longrightarrow 3\,Zn^{2+} + 2\,Al$

9.23 Using a diagram, show how you would construct a zinc–aluminum battery. Label the cathode, the anode, the + electrode, and the − electrode. Indicate the direction of electron flow, and identify the beaker in which oxidation occurs and the beaker in which reduction occurs.

9.24 Consider a battery made from lead, Pb, copper, Cu, along with Pb^{2+} ions and Cu^{2+} ions. In such a battery, what is the oxidizing agent, and what is the reducing agent? [*Hint:* Start by consulting the EMF series, and determine which metal oxidizes and which metal ion gets reduced.]

Understanding the EMF series allows us to predict which redox reactions are spontaneous and how to make batteries. It can also help us combat an enemy that costs us billions of dollars each year, an enemy that weakens our bridges, destroys our infrastructure, and even tried to tear down the Statue of Liberty! Who is this unpatriotic fiend? Read on and find out.

9.6 Another Look at Oxidation: The Corrosion of Metals

Oxidation is defined as the loss of electrons. We are not going to change this definition, but we do want to embellish it a bit. Earlier in this chapter, we talked about the burning of methane, CH_4:

$$\overset{-4+1}{CH_4} + \overset{0}{2\,O_2} \longrightarrow \overset{+4-2}{CO_2} + \overset{+1-2}{2\,H_2O}$$

We say that methane is oxidized because the oxidation state of its carbon atom changes from −4 to +4, signifying a loss of eight electrons. However, there is an older definition of oxidation. One hundred years ago, before chemists knew of the existence of electrons, they still would have said that the methane was oxidized upon burning. At that time, oxidation literally meant "combining with oxygen," which is what happens when something burns. Many elements, including metals, react with oxygen ("burn"). To see an example of this, just go to any junkyard and find a '65 Chevy, or, should we say, what's left of one. Rust! Rust is iron that has combined with oxygen, iron that has "burned":

$$\underset{\text{Iron metal}}{4\,Fe} + 3\,O_2 \longrightarrow \underset{\text{Iron oxide (rust)}}{2\,Fe_2O_3}$$

Of course, most cars don't burst into flames when they rust. This "burning" of metal, this combining with oxygen, occurs slowly. We call this slow burn **corrosion**.

Before we go on, let's get our definitions sorted out. We have the modern definition of oxidation—the loss of electrons. We also have the older definition of oxidation—combining with oxygen. Are these definitions compatible? Or, put another way, does an element lose electrons when it combines with oxygen ("burns")? The answer is yes if the element combining (forming a new bond) with oxygen is less electronegative than oxygen. Because oxygen is more electronegative than every element in the periodic table except fluorine, the answer nearly always is yes. When a metal (or any other element except fluorine) forms new bonds with oxygen, the metal loses the bonding electrons according to the bookkeeping rules of the oxidation-state method. Examine the oxidation numbers of the atoms in the reaction and you will see that iron is indeed oxidized by the modern definition.

$$\overset{0}{4\,Fe} + \overset{0}{3\,O_2} \longrightarrow 2\,\overset{+3}{Fe_2}\overset{-2}{O_3}$$

Iron metal — Iron oxide (rust)

Each iron atom loses three electrons in this reaction. So we see that the new and the old definitions of oxidation are truly one and the same.

Corrosion (rusting, oxidation) of iron or steel (steel is iron combined with small amounts of carbon) is a major economic problem in any modern society because so many things are built with steel. We try to prevent the corrosion of things like cars and bridges by covering the iron with paint or applying sealants, but this ultimately ends up a losing battle. We live in an atmosphere full of oxygen, and sooner or later nature will claim your car. Sometimes, however, we can unwittingly make the situation worse! Such was the case with the Statue of Liberty. She was built with a "time bomb" ticking away inside, an unintentional design flaw involving a redox reaction that almost brought her down.

The statue is made from sculpted sheets of copper metal hung on a steel skeleton that provides structural support. The statue sits on a small island in New York harbor and so is constantly exposed to the wind and spray of salt water. Needless to say, the exposed copper is subject to corrosion. Copper was chosen for the external skin of the statue because when it corrodes it becomes coated with a green material called a patina. The composition of the patina is a mixture of a hydroxy carbonate $Cu_2(OH)_2CO_3$ and a hydroxy sulfate $Cu_2(OH)_2SO_4$. The copper in these compounds is in the +2 oxidation state. This means that, in forming the green patina, the exposed surface of the metallic copper has been oxidized to Cu^{2+}:

$$\overset{0}{Cu} \xrightarrow[\text{air and moisture}]{\text{In the presence of}} \overset{+2}{Cu^{2+}}$$

Cu^{2+} ions in the form of copper hydroxy carbonate and copper hydroxy sulfate

Cu is oxidized (loses two electrons).

By itself, this is not a problem for the statue. Unlike iron oxide (rust), which flakes off and so continuously exposes fresh iron to further corrosion, oxidized copper (the patina) forms a tough, impenetrable coating over the underlying copper, protecting it from contact with oxygen and further corrosion. Given these properties of copper, the Statue of Liberty should not have corroded after the initial patina had formed. For about 100 years, this was true. Then, almost overnight, the statue started to fall apart. Things got so bad that the arm holding the torch was in serious danger of falling off!

The problem was inside the statue—the steel (iron) skeleton was reacting with the copper sheets. Consulting the EMF series on page 347, you can see that iron is a more active metal than copper. If iron metal comes in contact with copper in its oxidized state (Cu^{2+}), the iron will spontaneously oxidize and give electrons to the Cu^{2+} ions, reducing them back to elemental copper.

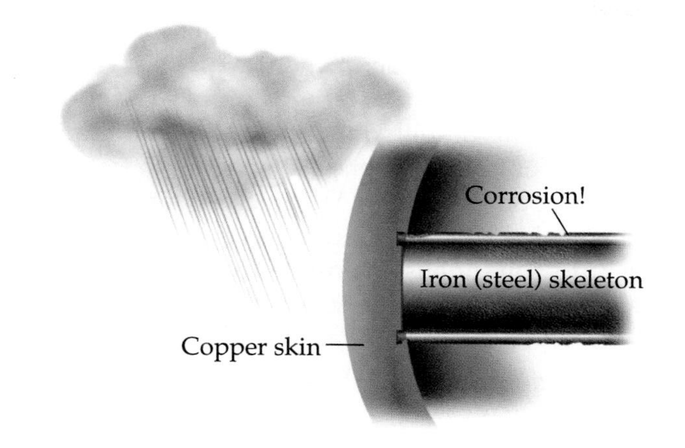

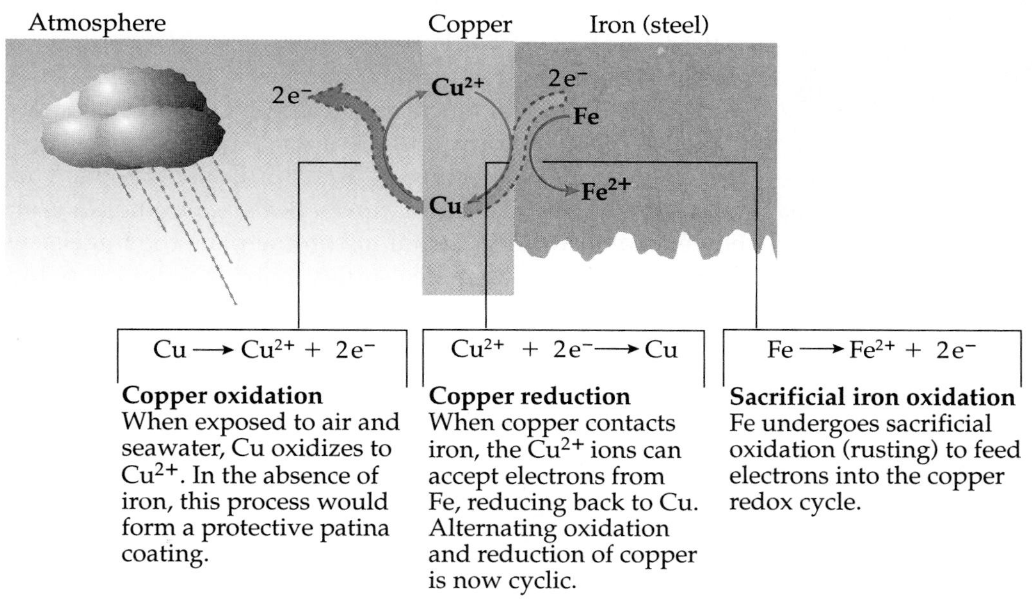

$Cu \longrightarrow Cu^{2+} + 2e^-$	$Cu^{2+} + 2e^- \longrightarrow Cu$	$Fe \longrightarrow Fe^{2+} + 2e^-$
Copper oxidation When exposed to air and seawater, Cu oxidizes to Cu^{2+}. In the absence of iron, this process would form a protective patina coating.	**Copper reduction** When copper contacts iron, the Cu^{2+} ions can accept electrons from Fe, reducing back to Cu. Alternating oxidation and reduction of copper is now cyclic.	**Sacrificial iron oxidation** Fe undergoes sacrificial oxidation (rusting) to feed electrons into the copper redox cycle.

More active metals lose electrons to ions of less active metals. This is great for the copper because the iron "sacrificed" itself to save the copper. As air and moisture re-oxidized the copper, the iron in the skeleton kept losing its electrons to Cu^{2+} ions, reducing them back to elemental Cu. The net result was that the inner steel skeleton of the statue oxidized and began to crumble very quickly.

Actually, the designers of the statue knew that this could happen. To prevent it, they put thick asbestos padding between the steel framework and the copper skin to insulate them from each other. Unfortunately, over the years, the padding rotted away, allowing the copper and iron to come into direct contact. At that point, the statue became one huge redox reaction, with her steel skeleton as one of the reactants.

Scaffolding surrounded the Statue of Liberty as corrosion damage was being repaired.

After ignoring the problem for a number of years, authorities launched a stupendous effort to save the statue. Many millions of dollars later, she now has a reinforced, corrosion-resistant stainless steel skeleton incorporating high-tech plastics to insulate it from the outer copper skin.

An active metal sacrificing itself for a less active one is also responsible for that rather painful sensation you experience when you bite down on a piece of aluminum foil. (Remember, we mentioned this at the beginning of this chapter.) Some of the tin, Sn, or silver, Ag, used in the mercury amalgam of your fillings has been oxidized to Sn^{2+} or Ag^{+} by continuous exposure to air and acids in your mouth. Aluminum, being a far more active metal than either tin or silver (it is far above Sn and Ag in the EMF series), sacrifices itself on contact with your filling. The Al oxidizes to Al^{3+} and feeds electrons to the Sn^{2+} and Ag^{+} ions, reducing them to their elemental forms. This electron transfer gives the nerves in your teeth the equivalent of an electric shock.

The phenomenon of a sacrificial metal can be used in the never-ending battle to prevent corrosion. One way to protect iron (or steel) from corrosion is to coat it with a more active metal, such as chromium, Cr. Many active metals (like Cr) form tough oxide coatings that strongly adhere to the metal underneath, protecting it from further corrosion. This works well for small pieces of iron like car bumpers. Because chromium is an expensive metal, however, you wouldn't want to chrome-plate the huge steel hull of an ocean-going vessel or a large steel utility pole. It would cost too much. What is often done instead is to attach a piece of a more active metal—magnesium, say—to the steel. The more active Mg spontaneously oxidizes to

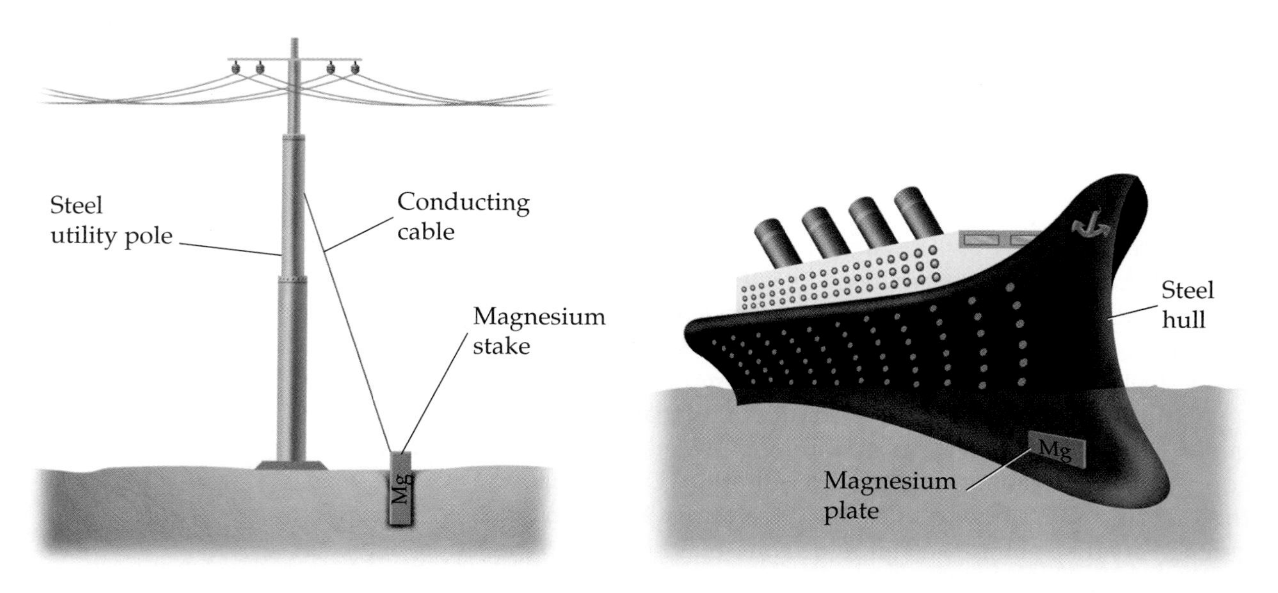

The magnesium sacrifices itself to prevent corrosion of the iron (steel).

Mg^{2+} and feeds its electrons to the iron in the steel, protecting it from oxidation. The only maintenance requirement now is to periodically replace the piece of magnesium, the sacrificial (more active) metal.

Of course, there are some instances, often symbolic, where we want metals to last a lifetime without any signs of corrosion.

If marriage is truly supposed to be "till death do us part," what metal does the EMF series indicate should be used for wedding rings? Explain your answer.

Certainly, a ring made of iron, which corrodes rapidly, would not be appropriate in this case. Do you see why the metals on the bottom of the EMF list are usually the ones used for jewelry and coins?

From the "burning" of glucose in the cells of our bodies to the batteries in our electronic devices to the combustion of methane, chemical reactions that transfer electrons from one atom to another can be found almost everywhere. Not all chemical reactions involve electron transfer, but you now know how to recognize the ones that do.

Have You Learned This?

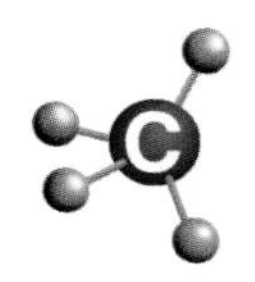

The Transfer of Electrons from One Atom to Another in a Chemical Reaction
www.chemistryplace.com

SKILLS TO KNOW

Assigning oxidation states by the electron-bookkeeping method

This method always works.

Example: Assign an oxidation state to each atom in hydrogen peroxide, H_2O_2.

1. Draw a valid dot diagram for the ion or molecule.

H—Ö—Ö—H

2. Assign every valance electron to an atom.
- Assign each lone pair to the atom on which it is drawn in the dot diagram.
- Assign the electrons of each bond to the bonding atom of higher electronegativity.
- If bonding electrons are shared by atoms of equal electronegativity, divide the electrons evenly between the bonding atoms.

O has higher EN than H, so O gets both shared electrons.

Each O atom gets its own lone-pair electrons.

H··Ö··Ö··H

All O atoms have the same EN, so these electrons are split between the O atoms.

3. Calculate the oxidation state for each atom by subtracting the electron count of the atom (as determined in step 2) from the number of valence electrons on a free atom of that element.

	Each H atom	Each O atom
Number on free atom	1	6
Number assigned	−0	−7
Oxidation state	+1	−1

Check your answer
The oxidation states for all the atoms should add up to the charge on the molecule or ion.

$1 + 1 + (-1) + (-1) = 0$ **OK!**

Example: Assign an oxidation state to each atom in the bicarbonate ion, HCO_3^-.

Step 1: H—Ö—C(=O:)—Ö: Valid dot diagram

Step 2: H··Ö··C··Ö: (with :O: above C) O has higher EN than H or C, so the O atoms get all the shared electrons (plus their own lone pairs).

Step 3:

	H atom	C atom	Each O atom
Number on free atom	1	4	6
Number assigned	−0	−0	−8
Oxidation state	+1	+4	−2

Check answer: $1 + 4 + (-2) + (-2) + (-2) = -1$ **OK!**

Assigning oxidation states by the shortcut method

This method is quick once you know the rules, although it sometimes fails to give a definitive answer.

Example: Determine the oxidation state of each atom in the bicarbonate ion, HCO_3^-.

Rule 1
The oxidation state of an atom in the most abundant naturally occurring form of an element is set to zero.

Rule 2
The oxidation state of an oxygen atom in a compound is almost always −2.

Rule 3
The oxidation state of a hydrogen atom in a compound is +1.

Rule 4
The oxidation state of any group IA atom in a compound is +1.

Rule 5
The oxidation state of any group IIA atom in a compound is +2.

Rule 6
The oxidation state of a monotomic ion equals the ion's charge.

Halide rule
A halogen atom in a compound usually has an oxidation state of −1.

Rule 7
The sum of the oxidation states for every atom in a formula must add up to the formula's overall charge.

Step 1: List the elements and apply the element-specific rules:

Atom	Oxidation state	
H	+1	← Rule 3
C	?	← No rule
O	−2	Rule 2
O	−2	Rule 2
O	−2	Rule 2
Sum:	−5	

Step 2: Use rule 7 to assign an oxidation state to C:

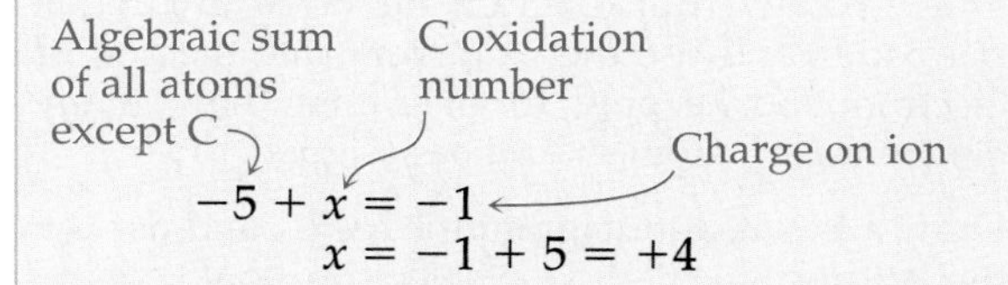

$$-5 + x = -1$$
$$x = -1 + 5 = +4$$

Atom	Oxidation state	
H	+1	
C	+4	← Rule 7 says that the oxidation states must add up to the overall charge (−1 in this case). Therefore, the oxidation state of C must be +4.
O	−2	
O	−2	
O	−2	
Sum:	−1	

If this shortcut method fails to give a definitive answer for all atoms, you must draw a dot diagram and use the oxidation-state bookkeeping method (which always works). For instance, the shortcut rules do not give the correct answer for the H_2O_2 example on the preceding page.

ELECTRICITY AND ELECTRON BOOKKEEPING

9.25 How is electricity similar to water flowing in a hose?

9.26 How is electricity mechanically produced in a power plant?

9.27 What is an electron-transfer reaction?

9.28 Why did chemists find it necessary to invent oxidation states?

9.29 What do we mean by electron bookkeeping?

9.30 Which method of assigning shared electrons is correct, the double-counting method of the octet rule or the oxidation-state method?

9.31 When assigning an oxidation state to an atom, why do we need to know how many valence electrons are present on a free atom of that element?

9.32 According to the oxidation-state method of electron bookkeeping, which atom is assigned the shared electrons in a bond between two atoms?

9.33 Draw a Lewis dot diagram for NF_2H, and use the oxidation-state method of electron bookkeeping to determine how many electrons each atom should be assigned.

9.34 Draw a Lewis dot diagram for H_2O_2 (hydrogen peroxide), and use the oxidation-state method of electron bookkeeping to determine how many electrons each atom should be assigned.

9.35 Draw a Lewis dot diagram for NO_3^-, and use the oxidation-state method of electron bookkeeping to determine how many electrons each atom should be assigned.

9.36 Assign an oxidation state to every atom in the molecules in Problems 9.33–9.35.

9.37 Consider the following molecules. Without using the shortcut rules, determine the oxidation state for each atom in the molecules and show how you calculated it.
(a) H_2 (b) O_2 (c) Cl_2

9.38 What does the answer to Problem 9.37 teach you?

9.39 What must always be true when you add up all the oxidation states for the atoms in a molecule?

9.40 Suppose an atom has an oxidation state of +3.
(a) Would more or fewer electrons be assigned to this atom by oxidation-state bookkeeping than are present on a free atom of that element?
(b) How many more or fewer?

9.41 Suppose an atom has an oxidation state of −2.
(a) Would more or fewer electrons be assigned to this atom by oxidation-state bookkeeping than are present on a free atom of that element?
(b) How many more or fewer?

9.42 Suppose the shortcut rules can determine the oxidation state of every atom in a compound except one. How can you find the oxidation state of the remaining atom?

9.43 Use the shortcut rules to assign an oxidation state to each atom in
(a) PCl_3
(b) H_2S
(c) MnO_4^-
(d) HNO_3
(e) HCOOH
(f) $S_2O_3^{2-}$

9.44 Use the shortcut rules to assign an oxidation state to each atom in
(a) CO
(b) CH_2Cl_2
(c) $HCOO^-$
(d) $PtCl_6^{2-}$

9.45 Use the shortcut rules to assign an oxidation state to each atom in
(a) O^{2-}
(b) Li_3N
(c) $MgSO_4$
(d) MnO_2

9.46 Consider ClO^- and $AlCl_3$. For one of these substances the halide shortcut rule works. For the other it does not.
(a) Which one does it work for, and why?
(b) Why doesn't it work for the other?
(c) Assign oxidation states to all the atoms in both substances.

9.47 Assign an oxidation state to each atom in the molecule IF.

9.48 Consider the organic compound $C_3H_6O_2$, methyl acetate.
(a) What is the problem with using the shortcut method to assign oxidation numbers to the atoms in methyl acetate?
(b) The dot diagram of methyl acetate is shown below. Assign an oxidation state to every atom.

```
   H  :O:       H
   |  ||        |
H—C—C—Ö—C—H
   |            |
   H            H
```

RECOGNIZING ELECTRON-TRANSFER REACTIONS

9.49 What does the term *reduction* mean?

9.50 What does the term *oxidation* mean?

9.51 How are oxidation states useful in determining whether a reaction is a redox reaction?

9.52 What happens to an atom's oxidation state when the atom is reduced?

9.53 What happens to an atom's oxidation state when the atom is oxidized?

9.54 What does the word *transfer* imply about an electron-transfer reaction?

9.55 Why can we always call an electron-transfer reaction a redox reaction?

9.56 Which of the following are redox reactions?
(a) $2\,Na + 2\,H_2O \longrightarrow 2\,NaOH + H_2$
(b) $MgBr_2 + 2\,NaF \longrightarrow MgF_2 + 2\,NaBr$
(c) $2\,CO + O_2 \longrightarrow 2\,CO_2$
(d) $SO_2 + H_2O \longrightarrow H_2SO_3$

9.57 For those reactions in Problem 9.56 that are redox reactions:
(a) Indicate which atoms get oxidized and which atoms get reduced.
(b) Indicate which reactant is the oxidizing agent.
(c) Indicate which reactant is the reducing agent.

9.58 Which of the following are electron-transfer reactions?
(a) $2\,CrO_4^{2-} + 2\,H^+ \longrightarrow Cr_2O_7^{2-} + H_2O$
(b) $Fe + NO_3^- + 4\,H^+ \longrightarrow$
$Fe^{3+} + NO + 2\,H_2O$
(c) $2\,C_2H_6 + 7\,O_2 \longrightarrow 4\,CO_2 + 6\,H_2O$
(d) $2\,AgBr \longrightarrow 2\,Ag + Br_2$

9.59 For those reactions in Problem 9.58 that are electron-transfer reactions:
(a) Indicate which atoms get oxidized and which atoms get reduced.
(b) Indicate which reactant is the oxidizing agent.
(c) Indicate which reactant is the reducing agent.

9.60 The following reaction is responsible for producing electricity in your car battery (often called a lead storage battery):

$$Pb + PbO_2 + 2\,H_2SO_4 \longrightarrow 2\,PbSO_4 + 2\,H_2O$$

(a) Assign an oxidation state to each atom. [*Hint:* For $PbSO_4$, you can get the charge of the Pb if you remember that the charge of the sulfate ion is 2− (SO_4^{2-}). Then use shortcut rule 7 to get the oxidation state of the Pb in $PbSO_4$.]
(b) Identify the atom that gets oxidized and the atom that gets reduced.
(c) Identify the reactant that is the oxidizing agent and the reactant that is the reducing agent.

9.61 Hydrogen gas burns very well in the presence of oxygen to give water:

$$2\,H_2 + O_2 \longrightarrow 2\,H_2O$$

In principle, should it be possible to use this chemical reaction to produce electricity? Explain.

9.62 In the upper atmosphere, sunlight can convert oxygen to ozone:

$$2\,O_2 \longrightarrow \underset{\text{Ozone}}{O_3} + O$$

Is this a redox reaction? Completely justify your answer.

9.63 The hydroquinone molecule can be converted to the quinone molecule as shown below:

2 AgBr + 2 OH⁻ + Hydroquinone ⟶

(Hydroquinone: H–Ö–C ring of C=C, C–C with H atoms and C–Ö–H)

Hydroquinone

2 Ag + 2 H₂O + 2 Br⁻ + Quinone

(Quinone: Ö=C ring of C=C with H atoms and C=Ö)

Quinone

You may want to use a combination of shortcut rules and dot diagrams to assign oxidation numbers before answering the following questions:
(a) Which atom or atoms are oxidized?
(b) Which atom or atoms are reduced?
(c) What is the oxidizing agent?
(d) What is the reducing agent?

ELECTRICITY FROM REDOX REACTIONS

9.64 If you put a piece of iron in an aqueous solution of blue Cu^{2+} ions, the spontaneous redox reaction

$$Fe + Cu^{2+} \longrightarrow Fe^{2+} + Cu$$

will occur. An aqueous solution of Fe^{2+} ions is red-brown.
(a) What is oxidized?
(b) What is reduced?
(c) What is the oxidizing agent?
(d) What is the reducing agent?
(e) What will happen to the piece of iron over time?
(f) Do the electrons move from the oxidizing agent to the reducing agent or from the reducing agent to the oxidizing agent?
(g) How could you use this reaction to make a battery? (Explain and show your battery in a diagram, using a nail and a penny as your source of iron and copper, respectively.)

In Problems 9.65–9.69 you will be given the spontaneous redox reaction (or enough information regarding it) so that you can construct and label a battery.

9.65 Construct a battery from the spontaneous redox reaction

$$Cr + 3\,Ag^+ \longrightarrow Cr^{3+} + 3\,Ag$$

Draw a battery similar to the one you drew for WorkPatch 9.7. Make sure you label which way the electrons flow, and also label the cathode, anode, +, and −.

9.66 For the chromium–silver battery of Problem 9.65:
(a) What is the oxidizing agent?
(b) What is the reducing agent?
(c) Which electrode gains mass with time?
(d) Which electrode is eaten away with time?

9.67 Construct a battery from the spontaneous redox reaction

$$Mg + Ni^{2+} \longrightarrow Mg^{2+} + Ni$$

Draw a battery similar to the one you drew for WorkPatch 9.7. Make sure you label which way the electrons flow, and also label the cathode, anode, +, and −.

9.68 For the magnesium–nickel battery of Problem 9.67:
(a) What is the oxidizing agent?
(b) What is the reducing agent?
(c) Which electrode gains mass with time?
(d) Which electrode is eaten away with time?

9.69 A battery is constructed from tin and copper by dipping strips of each metal into a solution of its ions (Sn^{2+} and Cu^{2+}, respectively). As the battery operates, the Sn^{2+} concentration increases and the Cu^{2+} concentration decreases.
(a) What is getting oxidized?
(b) What is getting reduced?
(c) Draw a battery similar to the one you drew for WorkPatch 9.7. Make sure you label which way the electrons flow, and also label the cathode, anode, +, and −.

9.70 How does knowing which way the electrons flow allow you to assign + and − to the proper electrodes?

9.71 What trick can you use to help you remember the sign (+ or −) of the cathode and anode?

WHICH WAY DO ELECTRONS FLOW?—THE EMF SERIES

9.72 Suppose you have two metals, A and B, and solutions of their ions, A^+ and B^+. Metal B is more active than metal A.
(a) Write the spontaneous redox reaction for this situation.
(b) What is the oxidizing agent?
(c) What is the reducing agent?
(d) Describe which way the electrons flow.

9.73 A zinc electrode and a copper electrode are used to make a battery, as shown at the top of the next column:
(a) Replace the three question marks with appropriate labels.
(b) Indicate which way the electrons flow
(c) Write the spontaneous redox reaction.
(d) Add the labels +, −, anode, and cathode to the drawing.

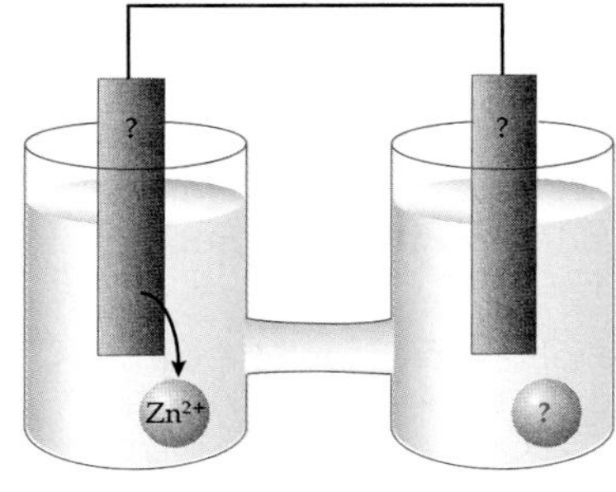

(e) Write "oxidation occurs at this electrode" under the appropriate beaker.
(f) Write "reduction occurs at this electrode" under the appropriate beaker.
(g) Indicate in which beaker the concentration of ions increases with time.
(h) Indicate in which beaker the concentration of ions decreases with time.
(i) Indicate in which beaker the electrode appears eaten away with time.
(j) Indicate in which beaker the electrode gains mass with time.

9.74 Suppose you have three different metals, X, Y, and Z, each of which also exists as a 2+ ion. Design an experiment to place them properly in order of their activity (a mini-EMF series).

9.75 What use is the EMF series?

9.76 What happens when you place an active metal in a solution of ions of a less active metal?

9.77 What happens when you place a less active metal in a solution of ions of a more active metal?

9.78 Using the EMF series on page 347, decide which of the following redox reactions is spontaneous. Explain your answer.
(a) $3\,K + Al^{3+} \longrightarrow Al + 3\,K^+$
(b) $Al + 3\,K^+ \longrightarrow 3\,K + Al^{3+}$

9.79 Using the EMF series on page 347, decide which of the following redox reactions is spontaneous. Explain your answer.
(a) $3\,Ag + Au^{3+} \longrightarrow Au + 3\,Ag^+$
(b) $Au + 3\,Ag^+ \longrightarrow 3\,Ag + Au^{3+}$

9.80 You are trapped on a desert island with plenty of water (both fresh and salt), a drinking glass, some wire, a radio, and no batteries. You do have a tin cup, a tube of toothpaste containing stannous fluoride (SnF_2, a source of Sn^{2+} ions), a silver pendant, and undeveloped black and white film (such film has silver bromide, AgBr, in it, a source of Ag^+ ions).
(a) How would you use the above materials to construct a battery? Show how with a diagram, including an arrow over the wire to show which way the electrons flow. (You can make a salt bridge by soaking a sock in salt water and then dipping one end of the sock in one cell and the other end in the other cell.)
(b) Which metal would be eaten away? Explain.

(c) Which is the oxidizing agent?
(d) Which is the reducing agent?

9.81 Show with a diagram how you would construct a copper–nickel battery. Label the direction of flow of electrons, cathode, anode, +, and −. Also, indicate in which beaker oxidation occurs and in which beaker reduction occurs.

9.82 Recharging a battery means forcing a spontaneous redox reaction to run backwards—in the nonspontaneous direction—once all the reactants have been used up. In Practice Problem 9.24 you considered a battery made from lead and copper.
(a) Write the spontaneous redox reaction for this battery.
(b) Write the recharging reaction for this battery.

9.83 Which two elements from the EMF series on page 347 do you think would combine to give the battery having the highest voltage? Explain your answer.

THE CORROSION OF METALS

9.84 There are two definitions of oxidation, one involving oxygen, the other not.
(a) State the two definitions of oxidation.
(b) Are they compatible? Explain.
(c) When we speak of corrosion of metals, what chemical reaction are we usually talking about (use Fe as an example)?

9.85 How can attaching a piece of Mg to the iron hull of a ship prevent the hull from rusting?

9.86 Suppose gold were not available for a wedding ring, but you still wanted a ring that would last "forever" and not corrode. According to the EMF series on page 347, what would be a good alternative metal to use?

9.87 Explain what almost brought down the Statue of Liberty and what was done to keep the problem from happening again.

9.88 What problem might develop with the casing of the dry-cell battery shown on page 342 if it is not made thick enough? Explain.

ADDITIONAL PROBLEMS

9.89 What is the oxidation state of the iodine atom in $H_2IO_6^{3-}$?

9.90 Identify the oxidizing and reducing agents in the reaction

$$(NH_4)_2Cr_2O_7 \longrightarrow N_2 + Cr_2O_3 + 4\,H_2O$$

9.91 Burning octane, C_8H_{18}, in your car engine forms water and carbon dioxide:

$$2\,C_8H_{18} + 25\,O_2 \longrightarrow 18\,H_2O + 16\,CO_2$$

What gets oxidized and what gets reduced?

9.92 Which metal is most easily oxidized: Pt, Hg, Fe, Mg, Zn?

9.93 Metal strips are immersed in aqueous solutions of various salts. In which combinations do you expect a spontaneous redox reaction:
(a) Silver strip in $CuBr_2(aq)$
(b) Copper strip in $ZnSO_4(aq)$
(c) Zinc strip in $AgNO_3(aq)$
(d) Gold strip in $FeCl_2(aq)$
(e) Copper strip in $Hg(NO_3)_2(aq)$

9.94 Can you protect a steel (predominantly iron) structure from corrosion by connecting it by wire to a small plate made of nickel? What elements (if any) would make better sacrificial metals?

9.95 Which ion is most difficult to reduce: Mn^{2+}, Hg^{2+}, Fe^{3+}, Mg^{2+}, Li^+?

9.96 The blood alcohol level of a person can be detected by reacting a sample of blood plasma with dichromate ion, $Cr_2O_7^{2-}$, which takes part in an electron-transfer reaction with ethanol, C_2H_5OH, in the blood:

$$2\,Cr_2O_7^{2-} + C_2H_5OH + 16\,H^+ \longrightarrow 4\,Cr^{3+} + 2\,CO_2 + 11\,H_2O$$

Assign an oxidation state to each atom in this reaction and indicate the oxidizing agent and the reducing agent.

9.97 A battery was produced using copper metal in a solution of Cu^{2+} ions connected to rhodium metal in a solution of Rh^{3+} ions. Copper is the anode and rhodium is the cathode. Is rhodium higher or lower than copper in the EMF series?

9.98 Assign an oxidation state for each nitrogen atom in N_3^-. [*Hint*: Begin with a dot diagram for the ion.]

9.99 Which of the following are electron-transfer reactions? For those that are, indicate which reactant is the reducing agent and which reactant is the oxidizing agent.
(a) $SeO_3^{2-} + 4\,I^- + 6\,H^+ \longrightarrow Se + 2\,I_2 + 3\,H_2O$
(b) $HI + KOH \longrightarrow KI + H_2O$
(c) $4\,HCl + O_2 \longrightarrow 2\,Cl_2 + 2\,H_2O$
(d) $SiO_2 + H_2O \longrightarrow H_4SiO_4$

9.100 Tin cans are actually iron plated with tin. What is the advantage of the tin coating?

9.101 Fluorouracil is a compound administered to cancer patients as a part of chemotherapy. Assign an oxidation state to every atom:

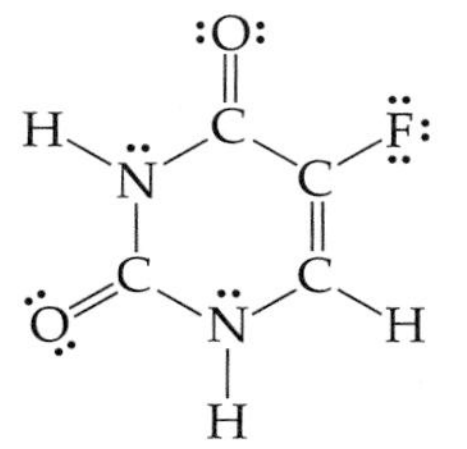

9.102 Which is a better reducing agent, Zn or Pb?

9.103 Assign an oxidation state to each carbon in
(a) H_3CCH_3 (b) H_2CCH_2 (c) HCCH

9.104 The spontaneous redox reaction

$$Mn + Cd^{2+} \longrightarrow Mn^{2+} + Cd$$

takes place in a battery.
(a) What is the oxidizing agent?
(b) What is the reducing agent?
(c) Which metal is the cathode?
(d) Which metal is the anode?

9.105 Assign an oxidation number to each atom and identify the oxidizing agent and reducing agent:

$$3\,Na_2SO_3 + 2\,KMnO_4 + H_2O \longrightarrow 3\,Na_2SO_4 + 2\,MnO_2 + 2\,KOH$$

9.106 You are trying to determine the identity of an unknown metal. You place a strip of it in a solution of Mg^{2+}, and no reaction occurs. You then place a strip of it in a solution of Zn^{2+}, and zinc metal plates out on the strip. Name one possibility for the identity of the unknown metal.

9.107 Use the shortcut rules to assign an oxidation state to each atom:
(a) PF_6^-
(b) $Mo_2O_7^{2-}$
(c) $HPbO_2^-$
(d) $HC_2O_4^-$

9.108 When you turn on an electrical appliance, are you consuming electrons?

9.109 You create a battery using a zinc anode in a solution of Zn^{2+} ions connected to a nickel cathode in a solution of Ni^{2+} ions. The two beakers are also connected with a salt bridge containing K^+ ions and Cl^- ions.
(a) Toward which electrode do K^+ ions flow?
(b) Toward which electrode do Cl^- ions flow?

9.110 Assign an oxidation state to each atom in the amino acid glycine:

```
      H    :O:
      |    //
H—N̈—C—C
   |   |    \
   H   H    :Ö—H
```

9.111 Identify the oxidizing agent and reducing agent in the reaction

$$IO_4^- + 7\,I^- + 8\,H^+ \longrightarrow 4\,I_2 + 4\,H_2O$$

9.112 The black tarnish that forms on silver is silver sulfide, Ag_2S. The tarnish can be removed by wrapping the silver with a piece of aluminum foil and placing it in a solution of $NaHCO_3$ (baking soda) to allow for the flow of ions. Explain how the aluminum foil aids in removing the tarnish.

9.113 Under what circumstances does fluorine *not* have an oxidation state of -1?

9.114 Which of the following are electron-transfer reactions? For those that are, indicate which reactant is the reducing agent and which reactant is the oxidizing agent.
(a) $3\,H_2SO_3 + 2\,HNO_3 \longrightarrow 3\,H_2SO_4 + 2\,NO + H_2O$
(b) $Mg + 2\,H_2O \longrightarrow Mg(OH)_2 + H_2$
(c) $SO_3^{2-} + 2\,H^+ \longrightarrow SO_2 + H_2O$
(d) $PbO + CO \longrightarrow Pb + CO_2$

9.115 Assign an oxidation state to each atom in HCN. [*Hint*: Begin with a dot diagram.]

9.116 A battery is made by connecting strips of lead and palladium and dipping each metal into a solution of its ions (Pb^{2+} and Pd^{2+}, respectively). Over time, the mass of the lead strip decreases and the mass of the palladium strip increases.
(a) Which metal is the anode?
(b) Which metal is the cathode?
(c) Write the spontaneous redox reaction going on in this battery.

9.117 What is the range of oxidation states possible for carbon? [*Hint*: Consider carbon bonded first to atoms that are all higher in electronegativity and then to atoms all lower in electronegativity.]

9.118 Identify the oxidizing agent and reducing agent in the reaction

$$NO_3^- + 4\,Zn + 7\,OH^- + 6\,H_2O \longrightarrow 4\,Zn(OH)_4^{2-} + NH_3$$

9.119 Galvanized steel is steel coated with zinc. Why is galvanized steel less likely to rust than uncoated steel? [*Hint*: Remember, steel is mainly iron.]

9.120 Because of their small size and long life, zinc–mercury batteries were once used in hearing aids, pacemakers, and watches.
(a) Which metal is the anode and which is the cathode?
(b) Which electrode loses mass over time?

9.121 Which of the following are electron-transfer reactions? For those that are, indicate which reactant is the reducing agent and which reactant is the oxidizing agent.
(a) $PF_3 + 3\,H_2O \longrightarrow H_3PO_3 + 3\,HF$
(b) $H_2 + Cl_2 \longrightarrow 2\,HCl$
(c) $2\,Cr_2O_3 + 3\,Si \longrightarrow 4\,Cr + 3\,SiO_2$
(d) $HCl + NaOH \longrightarrow NaCl + H_2O$

9.122 A street vendor sells you a ring he claims is pure gold. However, when you place the ring in a solution of Cu^{2+} ions, copper plates out on the ring. How do you know the ring is a fake? Could it be made of silver?

9.123 Use the shortcut rules to assign an oxidation state to each atom:
(a) $NaPO_3$
(b) $B(OH)_3$
(c) V_2O_5
(d) K_2TiF_6

9.124 The reaction $Cu^{2+} + I_2 \longrightarrow Cu^{+} + 2\,I^{-}$ is not possible as written. Assign an oxidation number to each atom and explain what is wrong with this reaction.

9.125 A battery consists of a strip of titanium metal in a solution of Ti^{2+} ions connected to a strip of zinc metal in a solution of Zn^{2+} ions. Over time, the concentration of Zn^{2+} ions decreases and the concentration of Ti^{2+} ions increases.
(a) Which metal is the anode?
(b) Which metal is the cathode?
(c) Where must titanium be on the EMF scale relative to zinc?
(d) Write an equation that describes the spontaneous electron-transfer reaction occurring in this battery.

9 WORKPATCH SOLUTIONS

9.1 N is assigned six electrons because it is more electronegative than H but less electronegative than F.

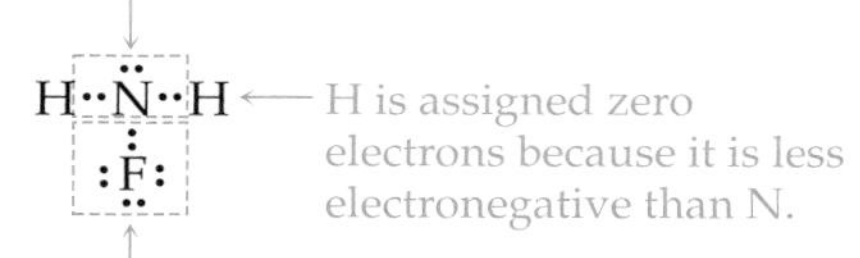
H··N··H
:F:
H is assigned zero electrons because it is less electronegative than N.

F is assigned eight electrons because it is more electronegative than N.

9.2 This O is assigned five electrons. Because a free O atom has six electrons, the oxidation state is $6 - 5 = +1$.

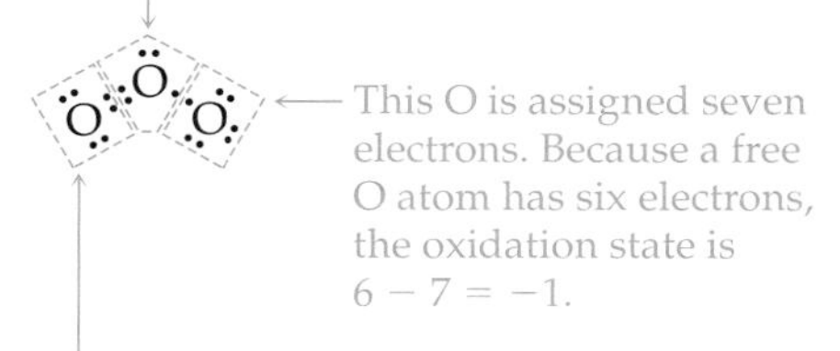
This O is assigned seven electrons. Because a free O atom has six electrons, the oxidation state is $6 - 7 = -1$.

This O is assigned six electrons. Because a free O atom has six electrons, the oxidation state is $6 - 6 = 0$.

9.3 (a) A negative oxidation state means that more valence electrons are assigned to the atom by oxidation-state bookkeeping than are present in a free atom of the element.

(b) An oxidation state of -1 means that oxidation-state bookkeeping assigns that chlorine atom one more valence electron than the number present in a free chlorine atom.

(c) Chlorine has seven valence electrons as a free atom (group VIIA). To have an oxidation state of -1 and therefore eight valence electrons, it must share its unpaired electron with an unpaired electron from an atom that is less electronegative than itself.

9.4 (a) The chlorine atom is bonded to atoms that are more electronegative than it is. Oxidation-state bookkeeping always assigns shared bonding electrons entirely to the more electronegative atom. Therefore, all the bonding electrons in this ion are assigned to the oxygen atoms. The chlorine atom gets none of them, so the halide rule does not apply.

(b) If each O atom has an oxidation state of -2 (rule 2), the oxidation states of the three O atoms add up to -6. Therefore, the Cl atom must have an oxidation state of $+5$ in order for the sum of the oxidation states of all the atoms to add up to the -1 charge on the ClO_3^- ion.

9.5 This C is assigned seven electrons. Because a free C atom has four valence electrons, the oxidation state is $4 - 7 = -3$.

This O is assigned eight electrons. Because a free O atom has six valence electrons, the oxidation state is $6 - 8 = -2$.

All H's in this molecule are assigned zero electrons. Because a free H atom has one valence electron, the oxidation state is $1 - 0 = +1$.

This O is assigned eight electrons. Because a free O atom has six valence electrons, the oxidation state is $6 - 8 = -2$.

This C is assigned one electron. Because a free C atom has four valence electrons, the oxidation state is $4 - 1 = +3$.

9.6 **Step 1:** The oxidation states are Zn 0, Cu^{2+} +2, Zn^{2+} +2, and Cu 0.

Step 2: (a) Yes. You can tell by the fact that the oxidation state of each reactant changed.
(b) Zn was oxidized to Zn^{2+}.
(c) Cu^{2+} was reduced to Cu.
(d) Cu^{2+} was reduced (gained two electrons), so it was the oxidizing agent.
(e) Zn was oxidized (lost two electrons), so it was the reducing agent.

9.7 Electrons flow through wire from Mg to Ag^+.

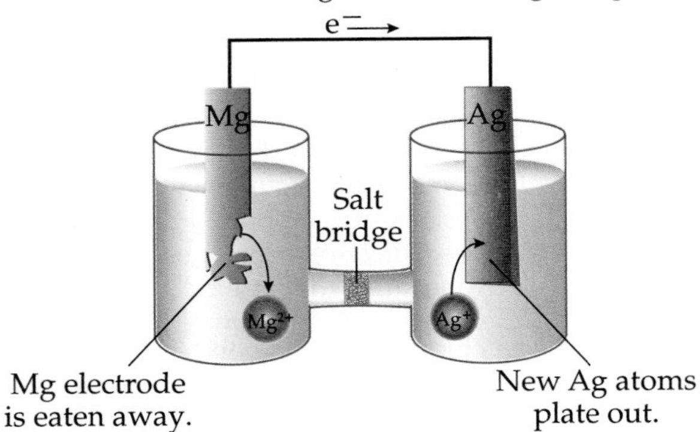
e⁻ →
Mg
Ag
Salt bridge
Mg^{2+}
Ag^+
Mg electrode is eaten away.
New Ag atoms plate out.
Mg gets oxidized to Mg^{2+} ions that enter solution.
Ag^+ ions get reduced to Ag atoms that plate out on the Ag electrode.

9.8 Electrons flow through wire from Mg to Ag^+.

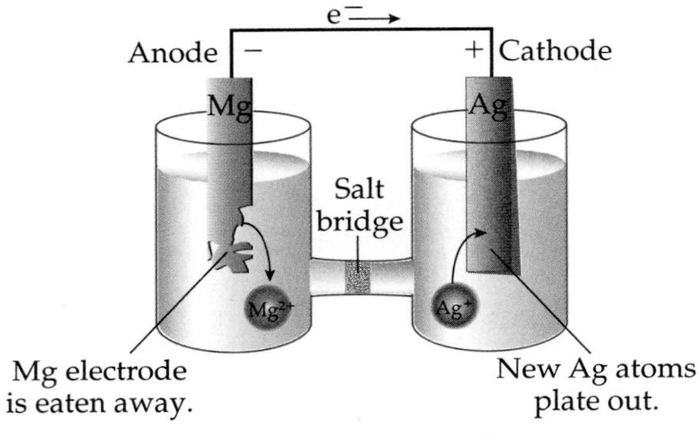
e⁻ →
Anode −
+ Cathode
Mg
Ag
Salt bridge
Mg^{2+}
Ag^+
Mg electrode is eaten away.
New Ag atoms plate out.
Mg gets oxidized to Mg^{2+} ions that enter solution.
Ag^+ ions get reduced to Ag atoms that plate out on the Ag electrode.

9.9 The correct answer is (a). Reduction is the gain of electrons. Oxidation is the loss of electrons. When something is oxidized, it loses electrons, which are then available to be gained by what is being reduced.

9.10 At the bottom of our abbreviated EMF series is gold, Au. It is the least active metal listed and therefore very resistant to oxidation and thus corrosion.

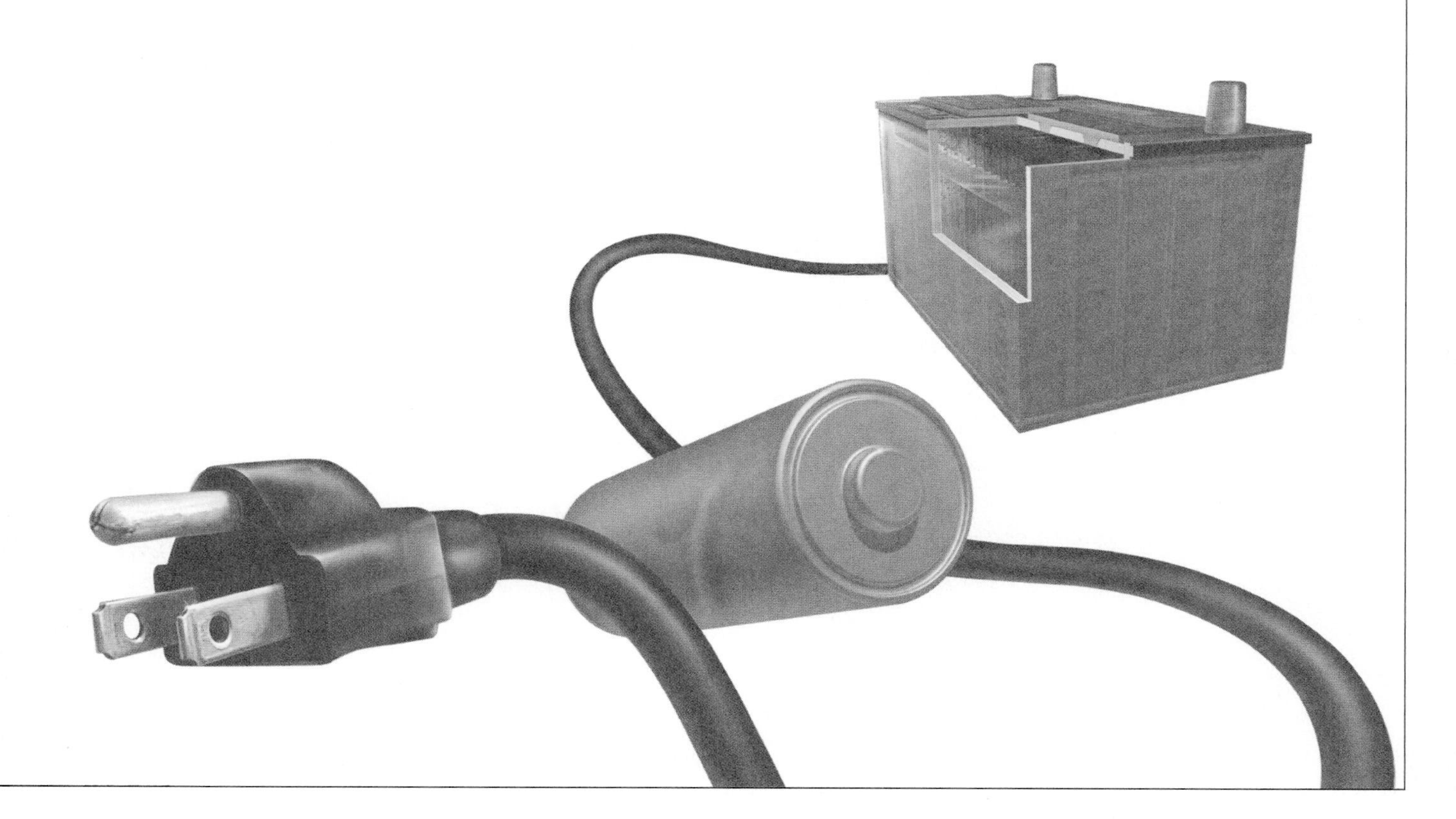

CHAPTER 10

Intermolecular Forces and the Phases of Matter

10.1 Why Does Matter Exist in Different Phases?

It's the middle of July, 34°C, and you feel as if you're covered with a wet blanket. Sound familiar?

These are the sweltering, high-humidity days of summer when there is just too much water in the air. You can't see this water because it exists in the gas, or vapor, phase, but you can certainly feel it. It makes you feel miserable. Today's solution to this problem is of course the air-conditioner, which cools the air by drawing it in with a fan and passing it over cold refrigeration coils. If you were to look at the coils inside an air-conditioner, you would notice they are dripping wet. Cooling humid air causes the water vapor in the air to *condense* to a liquid on the cold coils, and in this way the air is dried as it is "conditioned." This is why air-conditioned air feels so good on a humid summer day—it's not just cooler, it's also drier. By setting an air-conditioner on its highest setting, you can sometimes cause the refrigeration coils to get so cold that the liquid water on them freezes. Thus simply by cooling the air, an air-conditioner can drive atmospheric water through three phases—from the *gas phase* through the *liquid phase* and into the *solid phase*.

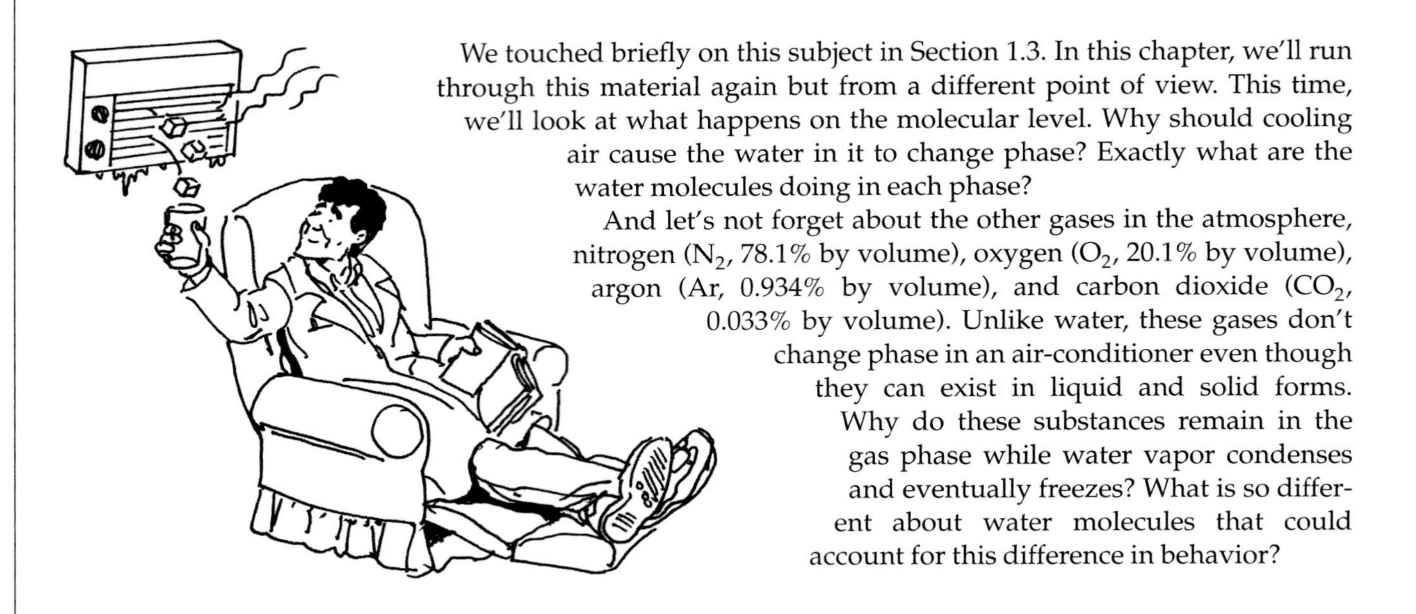

We touched briefly on this subject in Section 1.3. In this chapter, we'll run through this material again but from a different point of view. This time, we'll look at what happens on the molecular level. Why should cooling air cause the water in it to change phase? Exactly what are the water molecules doing in each phase?

And let's not forget about the other gases in the atmosphere, nitrogen (N_2, 78.1% by volume), oxygen (O_2, 20.1% by volume), argon (Ar, 0.934% by volume), and carbon dioxide (CO_2, 0.033% by volume). Unlike water, these gases don't change phase in an air-conditioner even though they can exist in liquid and solid forms. Why do these substances remain in the gas phase while water vapor condenses and eventually freezes? What is so different about water molecules that could account for this difference in behavior?

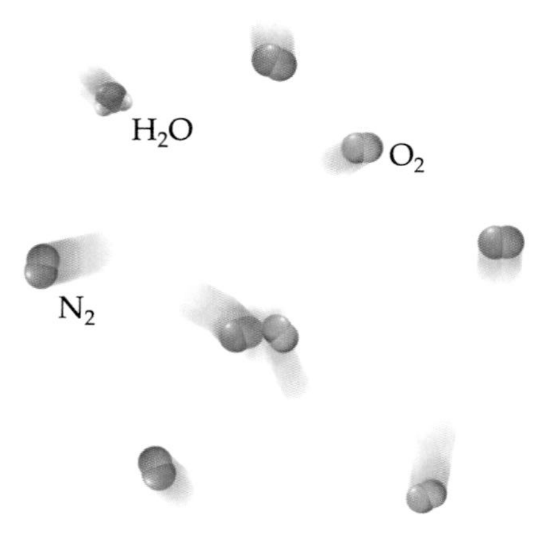

Air consists of fast-moving, independent molecules and atoms.

GASES

If you could see the individual molecules that make up the atmosphere (mostly N_2 and O_2), you would see that they are in constant motion. They travel in all directions, at speeds around 500 miles per hour (as mentioned in Chapter 7). The molecules are either colliding with one another or zipping past one another.

Something very interesting happens when we cool air by passing it over the refrigeration coils of an air-conditioner—the molecules slow down. *This is a very important piece of information*—the average speed of the molecules in a gas depends on temperature. When you heat a gas, the molecules in the gas move faster, on average. When you cool a gas, the molecules in the gas move more slowly, on average.

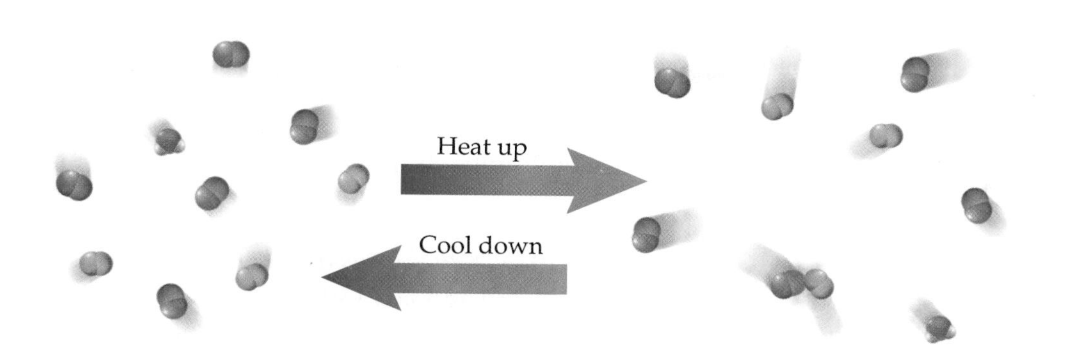

Cold air: molecules move slowly on average.

Hot air: molecules move fast on average.

CONDENSED PHASES—SOLIDS AND LIQUIDS

So maybe the reason water molecules condense out of the gas phase when air is cooled is that they have slowed down. But this cannot be the whole story. If it were, the O_2, N_2, and other gas molecules in the air would also condense because they also slow down as heat is removed.

There must be something else involved. Think back to Chapter 6, where we suggested you try ripping an ice cube in half. It can't be done because the water molecules in the ice cube are "stuck" tightly to one another by the intermolecular forces of attraction between them. The $\delta+$ end of one polar water molecule is attracted to the $\delta-$ end of another. These forces are responsible for water's being able to exist in the solid phase (ice).

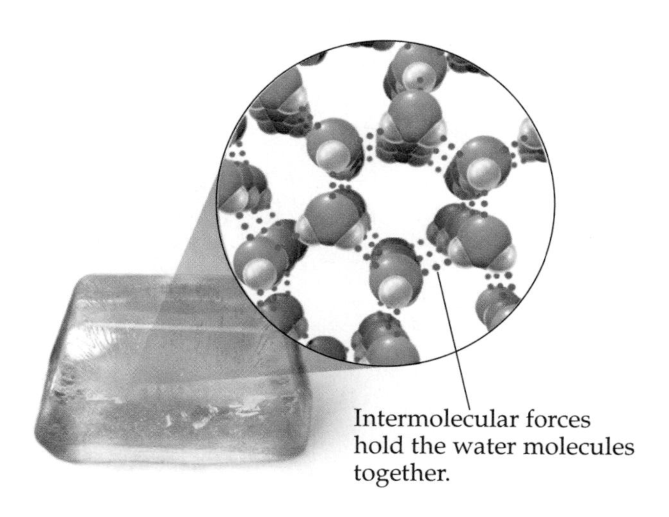

Intermolecular forces hold the water molecules together.

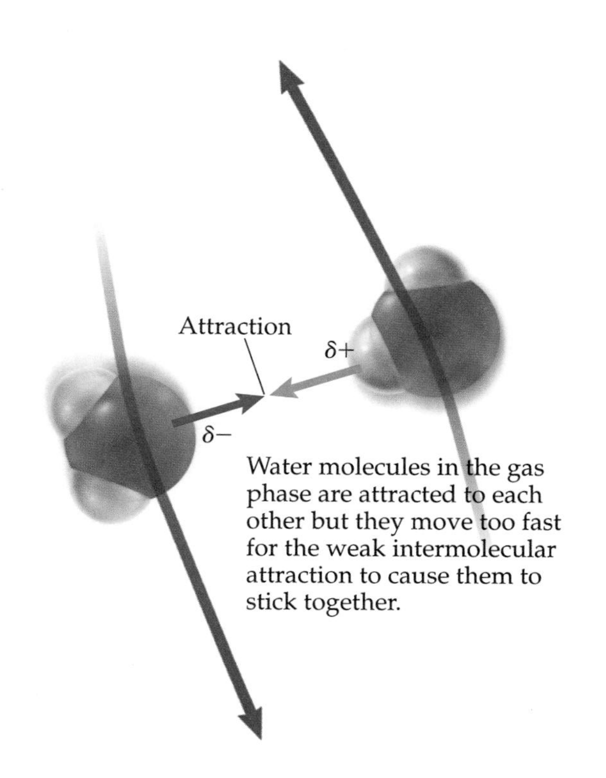

Water molecules in the gas phase are attracted to each other but they move too fast for the weak intermolecular attraction to cause them to stick together.

Water molecules are polar no matter what phase they are in. Even in the gas phase, two H_2O molecules are attracted to each other when they are close. In a gas, however, these attractive forces are too weak to cause the fast-moving molecules to stick together. Their kinetic energy (the energy associated with their motion) is great enough to overcome their mutual attraction, and they shoot on past each other.

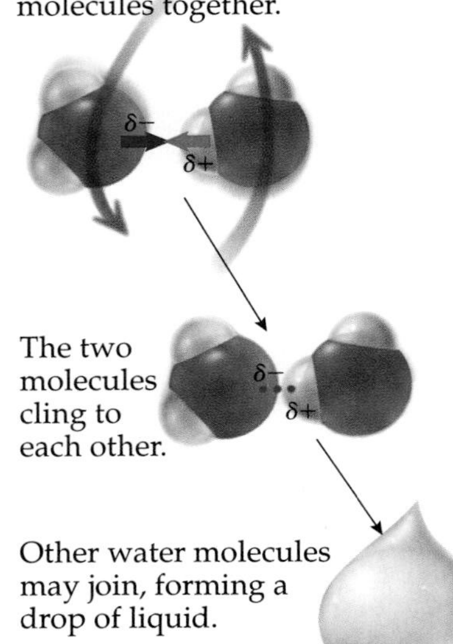

However, as any gas is cooled (energy is removed), the molecules slow down. At some temperature, the molecules are finally moving slowly enough for the attractive intermolecular forces to overcome the molecules' kinetic energy. At that temperature, many of the gas-phase molecules start to stick to one another and form the liquid phase.

When a substance changes from the gas phase to the liquid phase, we say **condensation** has occurred. If you reverse this process—start with a liquid and add heat—the molecules in the liquid move faster and faster until they finally overcome the intermolecular attractions and leap into the gas phase. This reverse process, going from liquid to gas, is called **vaporization**:

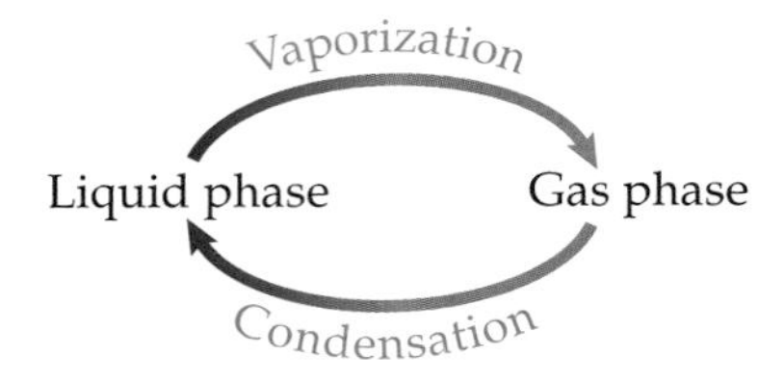

The highest temperature to which a liquid can be heated, and the temperature where it vaporizes most rapidly is called the **boiling point**. Because condensation is just the reverse of vaporization, the boiling point is also the temperature below which a gas condenses.

Continued cooling of a liquid (that is, continued removal of energy) eventually causes the liquid to freeze—in other words, to enter the solid phase. To see why, let's take a close look at liquid water. A molecular view shows that the molecules in the liquid phase are still moving, even though they are all stuck together.

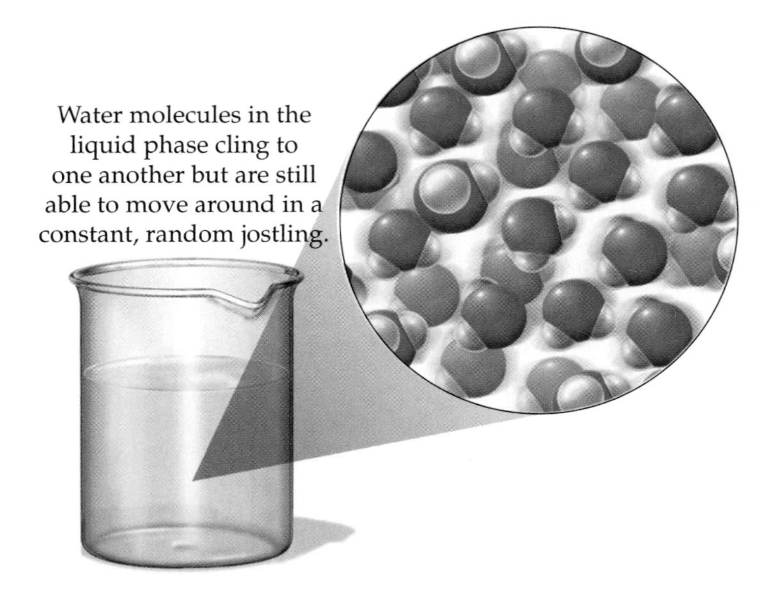

Water molecules in the liquid phase cling to one another but are still able to move around in a constant, random jostling.

Even though the molecules in a liquid are in a much more crowded environment than they would be in a gas, they still have enough kinetic energy to jostle past one another, each molecule continually breaking the attractive force to one molecule and reestablishing it with another. You can come close to experiencing what it's like to be a molecule in the liquid phase by visiting a shopping mall on the day after Thanksgiving. You can continue to move throughout the mall and get your shopping done but all the while getting pushed, shoved, and bumped as you travel.

As the temperature of a liquid drops, the molecules continue to slow down. Soon they are no longer moving fast enough to overcome intermolecular attractive forces and jostle past one another. At that temperature (called the **freezing point**), the molecules are effectively frozen in place, now allowed only small vibrations around their fixed positions. They have arrived at the solid phase.

BIG SALE
THIS FLOOR

This is as close as you'll probably ever come to feeling like a molecule in a liquid.

Structure of Solid Water (Ice)

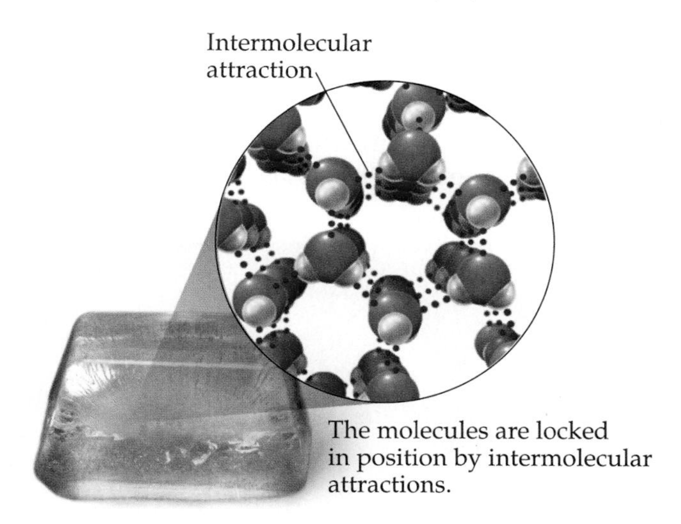
Intermolecular attraction
The molecules are locked in position by intermolecular attractions.

Overview of the Three Phases

A molecular view helps us understand why water exists in three phases and how it changes phases. The key points:

1. Molecules move because they have kinetic energy. Adding heat (energy) increases this kinetic energy. The more heat added, the faster the molecules move and the higher the temperature.
2. Intermolecular attractive forces tend to stick molecules together. If the molecules are moving fast enough, they have sufficient kinetic energy to overcome these forces and remain in the gas phase. Cooling eventually slows them down enough so that they stick together and form the liquid phase. Further cooling slows them down even more, allowing them to form the solid phase.

In the gas phase, the molecules are essentially independent of one another and are, on average, relatively far apart. In the liquid and solid phases, the molecules are quite close to one another (they essentially "touch"). It is for this reason that the liquid and solid phases are often referred to as **condensed phases**.

Now that we understand what happens with water, we can consider the other gases in the atmosphere, the ones that do not change phase when the air is cooled with our air-conditioners. Why don't blocks of, say, solid carbon dioxide clog the coils of our air-conditioners? Why don't pools of liquid nitrogen and liquid oxygen form at our feet? One possibility is that these molecules experience no attractive intermolecular forces. This is wrong, however, because all molecules exert some attraction on one another. They just don't do so with equal strength. The attractive force between two CO_2 molecules (or two N_2 molecules or two O_2 molecules) is much weaker than the attractive force between two H_2O molecules. Thus, in order to get CO_2 or N_2 or O_2 molecules to stick to one another, you have to cool them (slow them down) much more than you have to cool water. It takes a temperature of −78°C (at normal atmospheric pressure) to slow CO_2 molecules to the point where their mutual attraction can overcome their kinetic energy, allowing them to form a solid. (Normal atmospheric pressure is discussed in Section 11.1.) It takes a temperature of −183°C (at normal atmospheric pressure) to slow oxygen molecules to the point where they liquefy, and only when the temperature gets down to −196°C (at normal atmospheric pressure) does nitrogen begin to form a liquid. This is extremely cold, so cold that materials like rubber, plastic, and even bananas lose their flexibility when immersed in liquid nitrogen and become as rigid and fragile as glass.

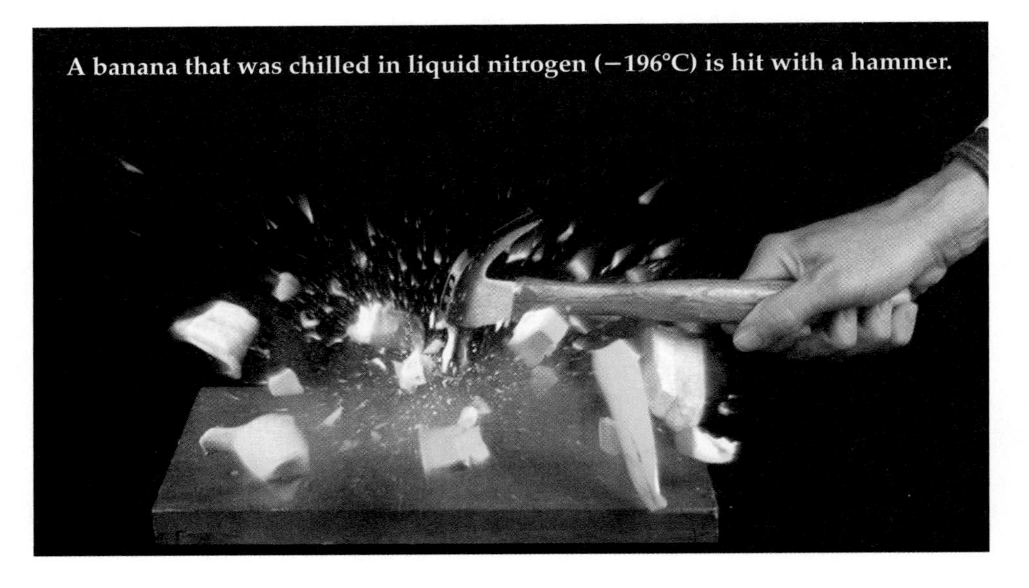
A banana that was chilled in liquid nitrogen (−196°C) is hit with a hammer.

The normal boiling point of ethanol, the alcohol in beer and wine, is 78°C. The normal boiling point of water is 100°C. (*Normal* means that the phase change occurs under the atmospheric pressure usually found at sea level.) What can you say about the strength of the intermolecular forces in ethanol relative to those in water?

The WorkPatch is a reminder that physical properties such as boiling point and freezing point are related to the strength of the attractive forces between molecules.

Practice Problems

10.1 Draw a picture showing how H–Cl molecules are attracted to one another, using dotted lines to represent intermolecular forces.

Answer:

```
δ+  δ−   δ+  δ−
H—Cl···H—Cl
    :    :
   δ+   δ−
   H—Cl
```

Other arrangements of intermolecular forces are possible. Just make sure the $\delta+$ end of one molecule approaches the $\delta-$ end of another molecule.

10.2 Based on the strength of their molecular dipole moments, which compound should have the higher boiling point, HF or HBr? Explain.

10.3 Draw a picture showing how NH_3 molecules attract one another.

10.4 Based on the strength of their molecular dipole moments, which compound should have the higher boiling point, H_2S or H_2O? Explain.

We stated earlier that all molecules are attracted to one another under the right conditions. Even though extremely low temperatures are required, the fact that oxygen and nitrogen can exist as liquids is proof of this statement. For any gas to condense to the liquid phase, there must be some attractive force causing the molecules to stick together. The nature of this force in water is obvious—water molecules are polar, each molecule having $\delta+$ and $\delta-$ ends. The positively charged portion of one molecule is attracted to the negatively charged portion of another molecule. But what about nonpolar molecules? Molecules like CO_2, N_2, and O_2 are completely nonpolar. Since they do not have permanent $\delta+$ and $\delta-$ ends, how can they attract one another?

10.2 Intermolecular Forces

In Chapter 6, we introduced the dipole–dipole forces that exist between polar molecules. However, these are obviously not the only types of intermolecular forces. To begin our study of some other types, let's consider carbon dioxide. In Chapter 6 we determined that although this molecule contains polar bonds, it is linear and therefore nonpolar.

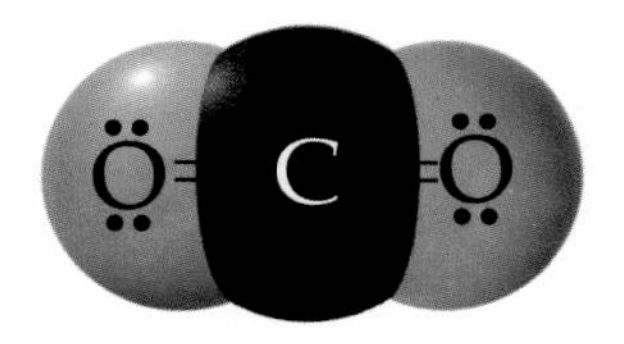

Because the carbon dioxide molecule is nonpolar, it contains no $\delta+$ or $\delta-$ ends.

At −78°C and normal atmospheric pressure, CO_2 molecules go directly from the gas phase to the solid phase. The fact that CO_2 solidifies at a much lower temperature than water (which freezes at 0°C) indicates that the attractive forces between CO_2 molecules are considerably weaker than those between H_2O molecules. But the real question remains "Why are CO_2 molecules attracted to one another if they are truly nonpolar and therefore have no $\delta+$ and $\delta-$ ends?" The answer comes from a model proposed more than 100 years ago by the chemists Fritz London and J. D. van der Waals. Today we refer to the relatively weak attractive forces these men studied as either **London forces** or **dispersion forces**. Although each London force is weak, there can be many such interactions in a large molecule, with the result that the total attraction can be very strong. Nonetheless, on a per-atom basis, it is the weakest intermolecular force there is.

Basically, the original model used to explain London forces treats any atom or molecule as a spherical container of electrons. The electrons inside the sphere are assumed to be in constant, random motion. According to this simplified model, the hydrogen molecule, H_2, is represented by a sphere containing two electrons and the methane molecule, CH_4, by a larger sphere containing ten electrons (six from the carbon and one from each hydrogen):

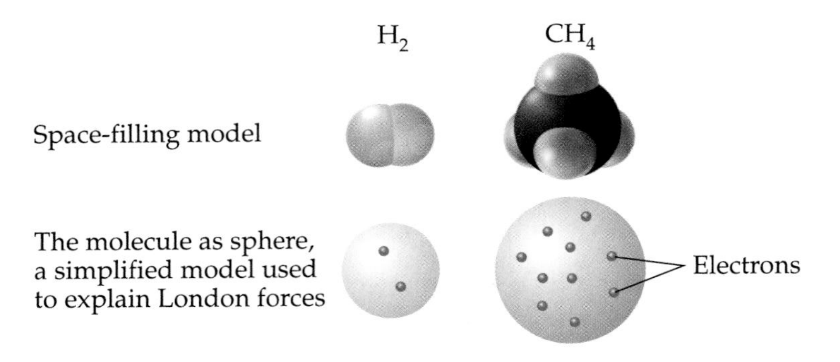

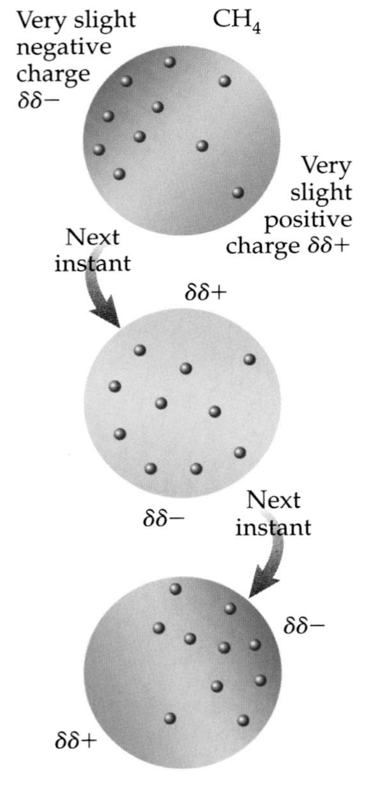

Let's focus on our methane molecule. Remember that its electrons are in constant motion. Imagine that, for a brief instant, this motion happens to put most of the electrons on one side of the molecule, as shown in the top diagram at left. For that instant, the side of the molecule with the excess of electrons has a slight negative charge and the side momentarily drained of electrons is slightly positive. In the next instant, this momentary imbalance disappears, but soon a new imbalance appears randomly in some other part of the molecule.

All atoms and molecules exhibit these brief, constantly shifting electron imbalances. Therefore, even though a nonpolar molecule has no *permanent* dipole moment, it does have weak, randomly oriented dipole moments that appear and disappear from one instant to the next. Because the electron imbalances that produce these dipole moments are typically small (our diagrams exaggerate them), the temporary charges are much weaker than the partial $\delta+$ and $\delta-$ charges in a polar molecule. To signify this extreme weakness, we write these charges as $\delta\delta+$ (partial partial positive) and $\delta\delta-$ (partial partial negative).

This constantly shifting distribution of electrons—and therefore of charges—is what causes nonpolar molecules to attract one another. To really understand this, we must look at more than one molecule at a time. Therefore imagine a row of three methane molecules. Let's assume they happen to have no charge imbalances at the instant we first see them:

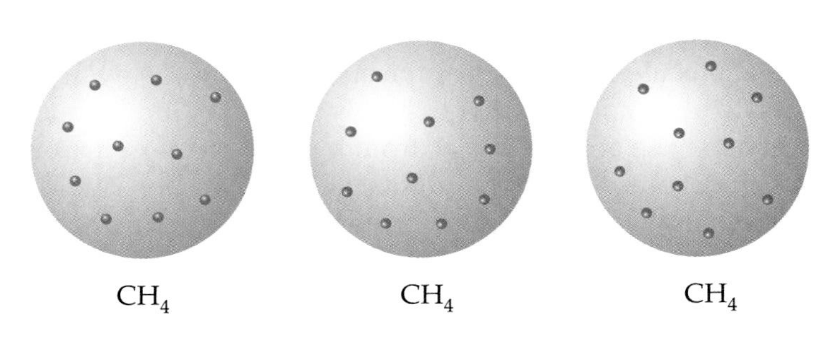

Now suppose the electrons in the center molecule shift left. How do the other molecules respond? A buildup of electrons (and negative charge) on the left side of the center molecule repels the electrons in the atom on the left. Likewise, the partial positive charge on the right side of the center atom attracts the electrons in the atom on the right.

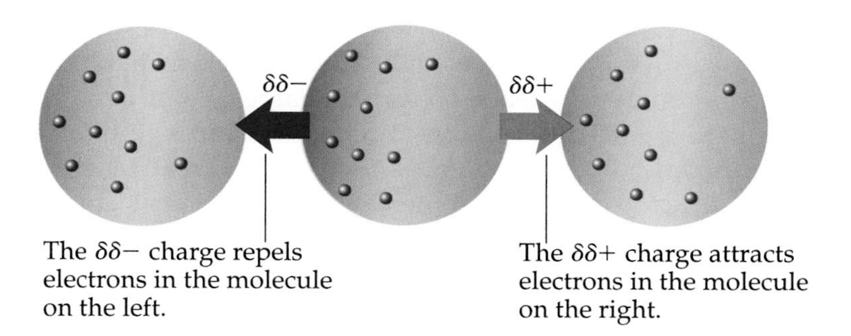
δδ−
δδ+
The δδ− charge repels electrons in the molecule on the left.
The δδ+ charge attracts electrons in the molecule on the right.

The electrons in the left and right molecules shift in response to these forces. We say that the center molecule *induces* dipole moments in the adjacent molecules. These induced dipole moments are oriented in precisely the way that creates an attractive force between adjacent molecules—the $\delta\delta+$ end of each molecule faces the $\delta\delta-$ end of the next:

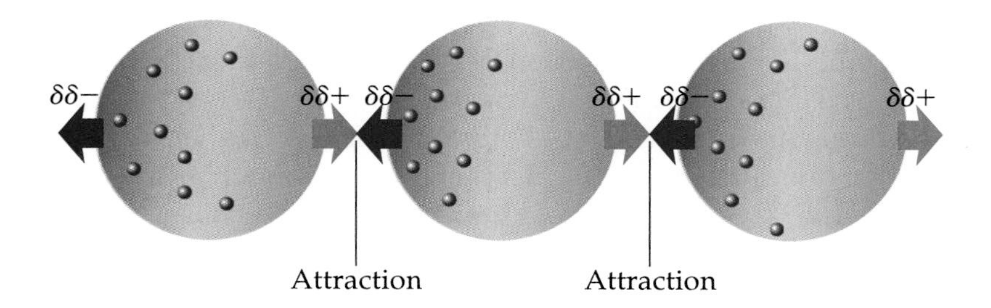
δδ−
δδ+
δδ−
δδ+
δδ−
δδ+
Attraction
Attraction

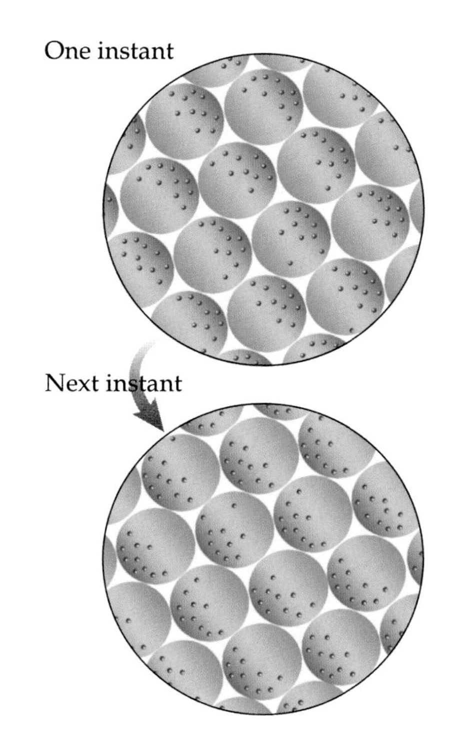
One instant
Next instant

The original temporary dipole moment in the center molecule created, or induced, dipole moments in the adjacent molecules, resulting in attraction. The outcome is that the methane molecules stick together even though they are not polar—that is, even though they don't have permanent $\delta+$ and $\delta-$ ends. Of course, with billions of molecules instead of just the three shown, we get billions of instantaneous induced dipole moments, aligned so as to hold all the methane molecules together. This pattern lasts for only an instant. In the next instant, it is replaced by another aligned pattern that is equally effective at attracting molecules to one another.

London forces exist between any two molecules. Polar molecules experience both dipole intermolecular forces and London intermolecular forces at the same time, whereas nonpolar molecules experience only London intermolecular forces.

How weak (or strong) are London forces? Remember, we have been exaggerating the electron imbalances in our diagrams. In actuality, these imbalances are very slight, and the partial dipole moments that develop are very small. Thus, for molecules containing *only a few electrons* (small molecules made from relatively light atoms, such as N_2, O_2, H_2O, or CH_4), London forces are extremely weak. The dipole–dipole forces that exist between small polar molecules are quite a bit stronger than the London forces that exist between the molecules.

Why are the attractive intermolecular forces between water molecules much stronger than the attractive intermolecular forces between oxygen molecules, O_2? Which do you think would experience stronger London forces, CH_4 or H_2? Explain.

The answer to the WorkPatch gets right to the heart of intermolecular forces: they depend on the existence of partial charges. Even though water molecules have both dipole–dipole forces and London forces going for them, the London forces can be ignored because they are dwarfed by the dipole forces. This is not the case for nonpolar methane, CH_4. All it has attracting its molecules together are weak London forces. For this reason, methane gas must be cooled to −164°C (its normal boiling point) before it liquefies. You really have to slow methane molecules down a lot before the weak London forces between them have any effect. Liquefying hydrogen, H_2, requires even lower temperatures. Because there are fewer electrons in each molecule than in methane molecules, the attractive London forces between these molecules are so weak that almost any motion can overcome them. As a result, you must cool hydrogen gas to −253°C at normal atmospheric pressure before it becomes a liquid.

On the other hand, London forces can be quite substantial for large molecules containing many atoms and therefore many electrons. Thus, although nonpolar methane is a gas at room temperature, nonpolar carbon tetrachloride, CCl_4, with its large chlorine atoms each containing 17 electrons, is a liquid that has a boiling point of 77°C. This is more than 200 C° higher than the boiling point of methane, attesting to the fact that the London forces between CCl_4 molecules are much stronger than those between CH_4 molecules.

Practice Problems

10.5 Carbon tetrabromide, CBr_4, has a melting point of 88–90°C. Explain why this makes sense when compared with the boiling points of methane and carbon tetrachloride.

Answer: Because CBr_4 has a melting point well above room temperature (which is usually about 25°C), CBr_4 is a solid at room temperature. At room temperature, CCl_4 is a liquid and CH_4 is a gas. This makes perfect sense. All three molecules are tetrahedral and nonpolar, which means their only intermolecular forces are London forces. Because CBr_4 has the most electrons (146), it has the strongest London forces of the three molecules.

10.6 The compound ethylene, $H_2C{=}CH_2$, is a gas at room temperature, but the compound polyethylene, made from a large number of ethylene units bonded together, is a solid at room temperature. Explain why.

10.7 (a) Draw a dot diagram for NH_3 and one for PH_3.
(b) Is either molecule polar? (The electronegativities of N, P, and H are 3.0, 2.1, and 2.1, respectively.)

(c) In which substance are the London forces stronger? Explain.
(d) Basing your answer solely on the London forces in the two substances, which substance would you expect to have the higher boiling point? Explain.

10.3 A Closer Look at Dipole Forces —Hydrogen-Bonding

We cannot complete a discussion of intermolecular forces without discussing a special type of attractive force that exists between certain polar molecules. This attractive force is special because it is considerably stronger than the average dipole–dipole attraction. To illustrate this, we shall consider two polar molecules, acetone and water.

Acetone, the main ingredient in many brands of nailpolish remover, has the molecular formula C_3H_6O. Because of the structure of the molecule and the large electronegativity difference between the carbon and oxygen atoms, the acetone molecule has a large permanent dipole moment:

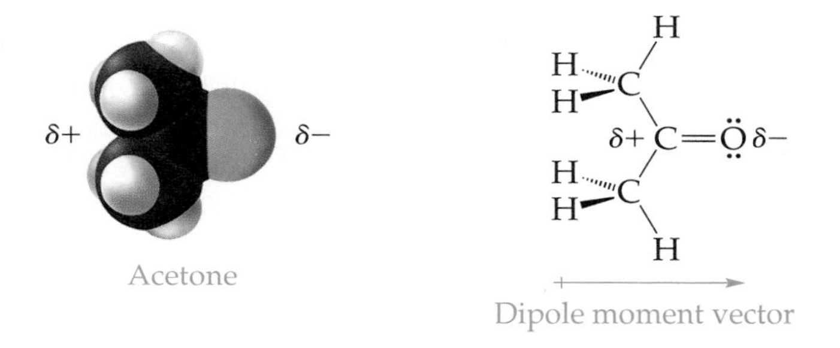

The resulting attractive polar forces between acetone molecules make acetone a liquid at room temperature. At normal atmospheric pressure, it boils at 56°C.

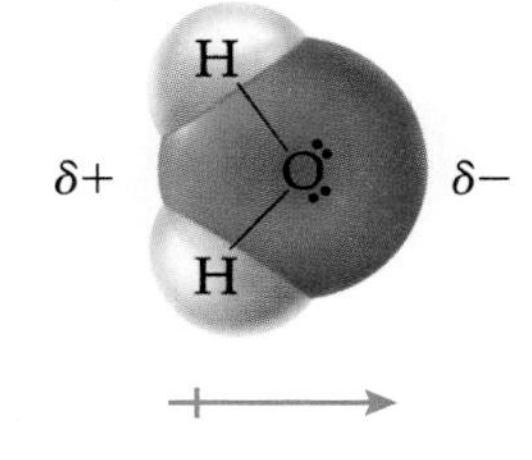

A water molecule is also polar, having a permanent dipole moment that is somewhat smaller than that of acetone. Yet, water boils at 100°C at normal atmospheric pressure, an unusually high boiling point for such a small molecule. If water boiled at the temperature one would expect for a molecule of its size and composition (see Problem 10.38), Earth would be a very different place indeed because virtually all the water on this planet would be in the gas phase, making life as we know it impossible.

The fact that water boils at 100°C and acetone boils at 56°C tells us that the intermolecular attractive forces between water molecules must be stronger than those between acetone molecules. Being that the dipole moment of water is smaller than the dipole moment of acetone, there must be another reason for this difference. The reason has to do with the presence of hydrogen atoms in these molecules. The dipole moment in a water molecule is a result of the polar O–H bonds in the molecule. The dipole moment in an acetone molecule is primarily the result of the polar C=O bond. Even though acetone does contain hydrogen atoms, they are bonded to an almost equally electronegative carbon atom. These hydrogens therefore have no significant partial charges and contribute very little to the overall dipole moment of the acetone molecule:

C–H bonds are nearly nonpolar.

The dipole moment comes from polar O–H bonds.

The dipole moment comes from the C=O bond.

This structural variation makes all the difference in the world to the strength of the polar forces. To understand why, we need to examine the hydrogen and oxygen atoms close up. When a hydrogen atom is covalently bonded to oxygen, the oxygen atom attracts the lion's share of the two shared electrons, and as a result the hydrogen atom develops a sizable $\delta+$ charge. This is not unexpected. Just about any element bound to oxygen develops a $\delta+$ charge because all atoms except fluorine have a lower electronegativity than oxygen. However, there is something unique about the hydrogen atom—it has no other electrons. Thus when the two electrons of the O–H bond in water are pulled toward oxygen, the hydrogen atom looks almost like a naked proton sticking out in space. The picture at left shows what a water molecule looks like to another molecule. The oxygen nucleus is buried deep inside the (blue) cloud of negative charge. Each hydrogen nucleus (a proton) is covered by only a thin veil of negative charge because the shared electrons spend little time on the hydrogen atom.

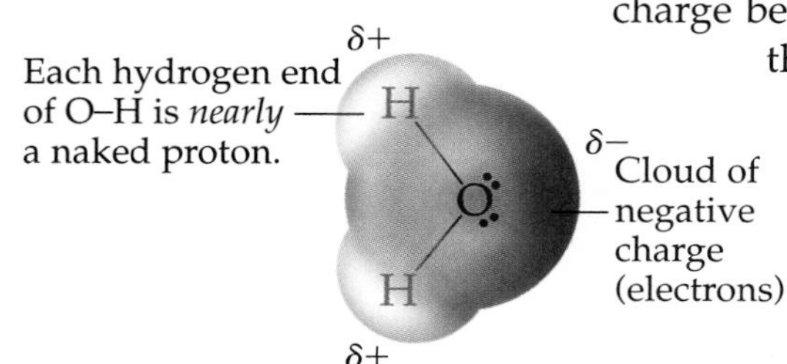

The nearly naked protons have a strong tendency to seek out electrons wherever they can find them. One obvious source is a lone pair of electrons on the oxygen atom in another water molecule. The nearly naked proton can get very close to the lone pair because the proton has no electrons to be repelled by the lone-pair electrons. This combination of the strong $\delta+$ charge on the hydrogen of one water molecule and the short distance between it and an oxygen lone pair in another water molecule lead to an unusually strong attractive force between the two. These polar interactions are given a special name to indicate their uniqueness. They are called **hydrogen bonds** and are usually represented by a dotted line between the atoms involved.

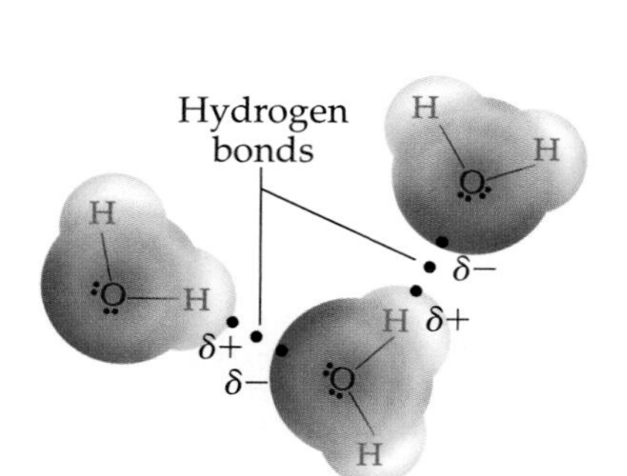

Calling this polar interaction a bond is somewhat misleading because even though this attractive force between water molecules is stronger than a normal dipole–dipole interaction, it is still nowhere near as strong as a covalent bond. The strength of the O–H covalent bond in a water molecule is greater than 400 kJ/mol (that is, it takes more than 400 kJ of energy to break 1 mole of O–H covalent bonds). The hydrogen "bond" between two water molecules is only a fraction as strong—around 16 kJ/mol. This means that it would take roughly 16 kJ to separate 1 mole of water molecules stuck to one another by hydrogen bonds. Despite the dubious choice of the word *bond* in describing this interaction, it is still different enough from normal dipole–dipole attractions (whose strength typically ranges from 4 to 8 kJ/mol) to warrant being treated as a special case.

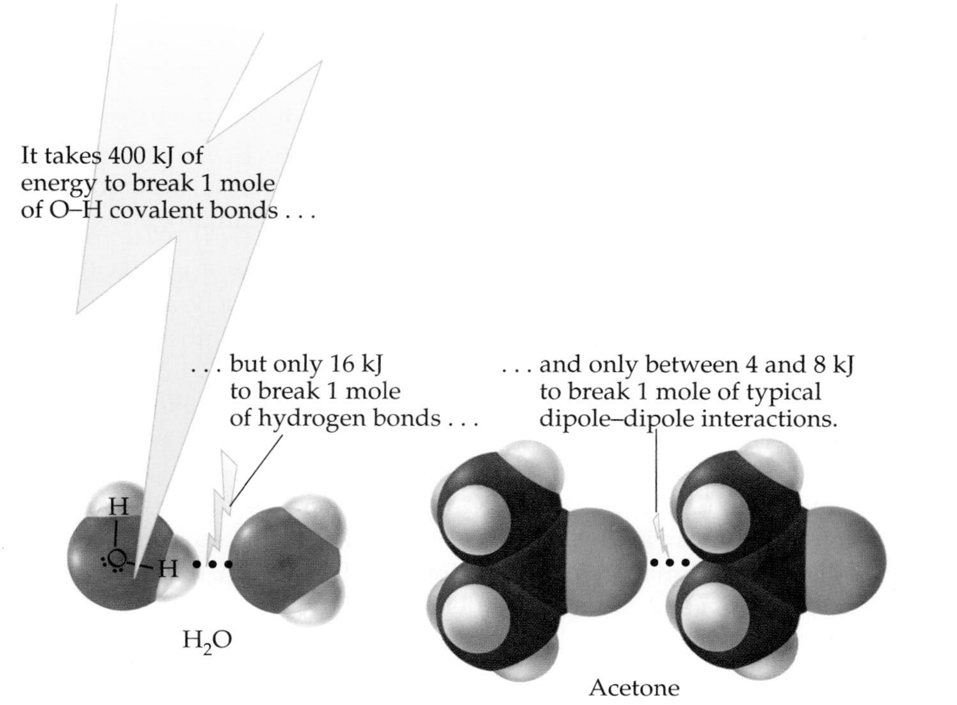

It takes 400 kJ of energy to break 1 mole of O–H covalent bonds . . .

. . . but only 16 kJ to break 1 mole of hydrogen bonds . . .

. . . and only between 4 and 8 kJ to break 1 mole of typical dipole–dipole interactions.

H_2O

Acetone

*Intra*molecular attractions (bonds) are much stronger than *inter*molecular attractions!

Water is an excellent example of hydrogen-bonding, but it doesn't tell the complete story. The fundamental requirement for the existence of a hydrogen bond is a nearly naked proton sticking out in space. This requirement is met whenever a hydrogen atom is directly bonded to a small atom having an electronegativity much higher than that of hydrogen. Only oxygen, nitrogen, and fluorine fill the bill. Thus if there is an O–H, N–H, or F–H covalent bond in a molecule, there can be hydrogen-bonding between the molecules.

To form hydrogen bonds, a molecule must contain at least one of these covalent bonds:

$\cdots H{-}N\langle$ (or $H{-}N{=}$)

$\cdots H{-}O{-}$

$\cdots H{-}F$

Recall that O, N, and F are three of the four electron "hogs" we discussed in Chapter 5 (chlorine is the fourth). These are the four most electronegative elements in the periodic table. So it's not so strange that when a hydrogen atom is bonded to an atom of one of these elements, the hydrogen is practically stripped of its electron and conditions are set for hydrogen-bonding. (Chlorine is usually not included in the set of elements that lead to hydrogen-bonding.)

The boiling point of one of these compounds is 35°C, and the boiling point of the other is 78°C:

(a) H–C(H)(H)–Ö–C(H)(H)–H (b) H–C(H)(H)–C(H)(H)–Ö–H

Which boils at which temperature? Why? If hydrogen bonds are involved in either case, make a drawing using dotted lines to show the hydrogen-bonding.

Both molecules in the WorkPatch have the molecular formula C_2H_6O, which means the difference between their boiling points must be a consequence of their different structures. What is it about their structures that leads to the difference? Make sure you know before going on.

In a very real sense, hydrogen-bonding is responsible for life. DNA—the substance that carries the "blueprint" for your body—depends on hydrogen-bonding to do its work. A DNA molecule is like a very long, twisted zipper made of two strands held together by a ladder of teeth. There are four kinds of teeth, and the information carried by the DNA is encoded in the sequence of teeth. When a cell needs information from a portion of its DNA, it unzips the two DNA strands in that region and reads the sequence of teeth. Then the strands zip back up.

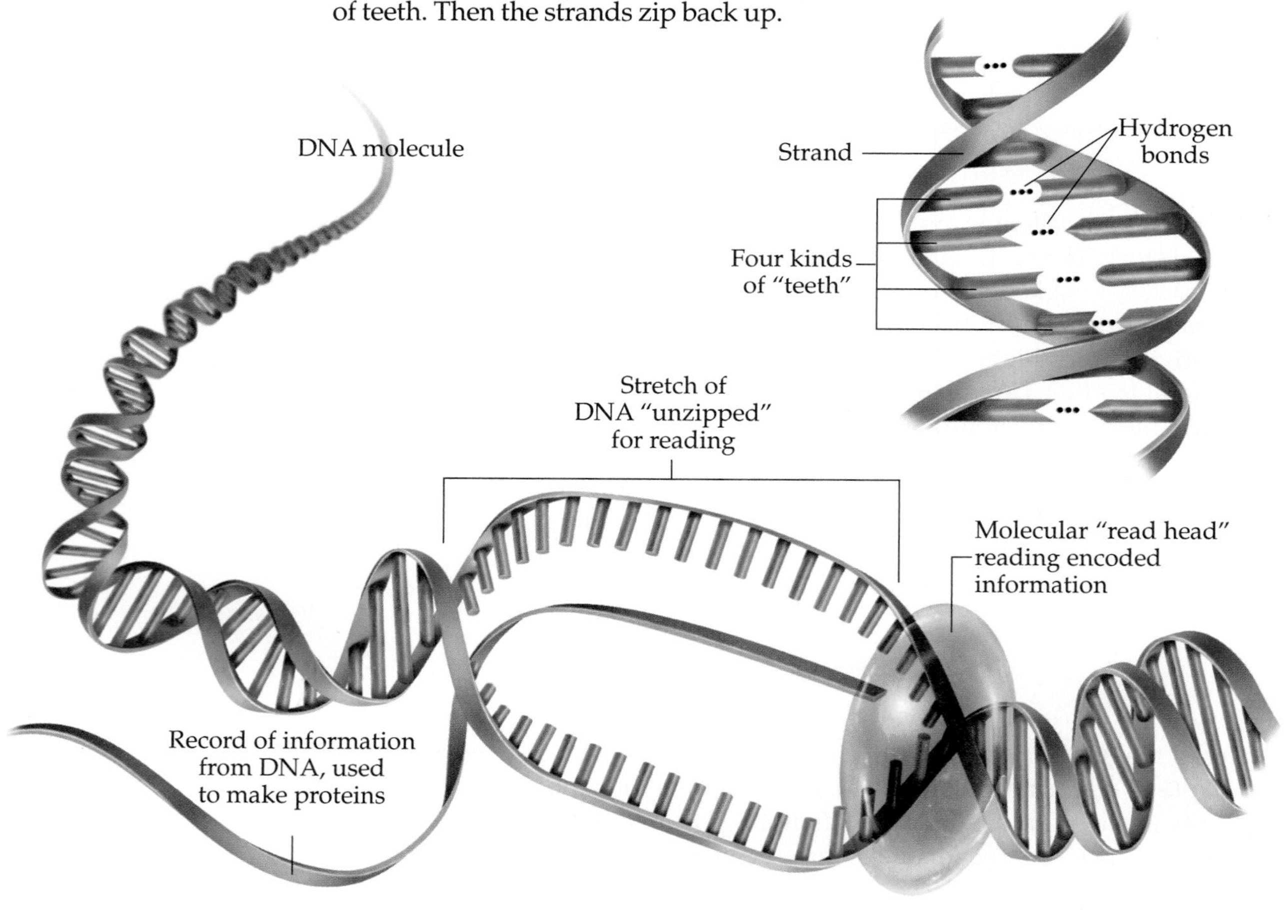

Like a real zipper, a DNA molecule has to be easy to zip and unzip, but when it is zipped, it should stay zipped. Unlike a real zipper, however, the teeth of DNA are held together by chemical interactions. Hydrogen bonds are perfect for this job because covalent bonds would be too strong and London forces would be too weak. The hydrogen bonds are broken when the DNA is unzipped, and they form again when the DNA zips back up.

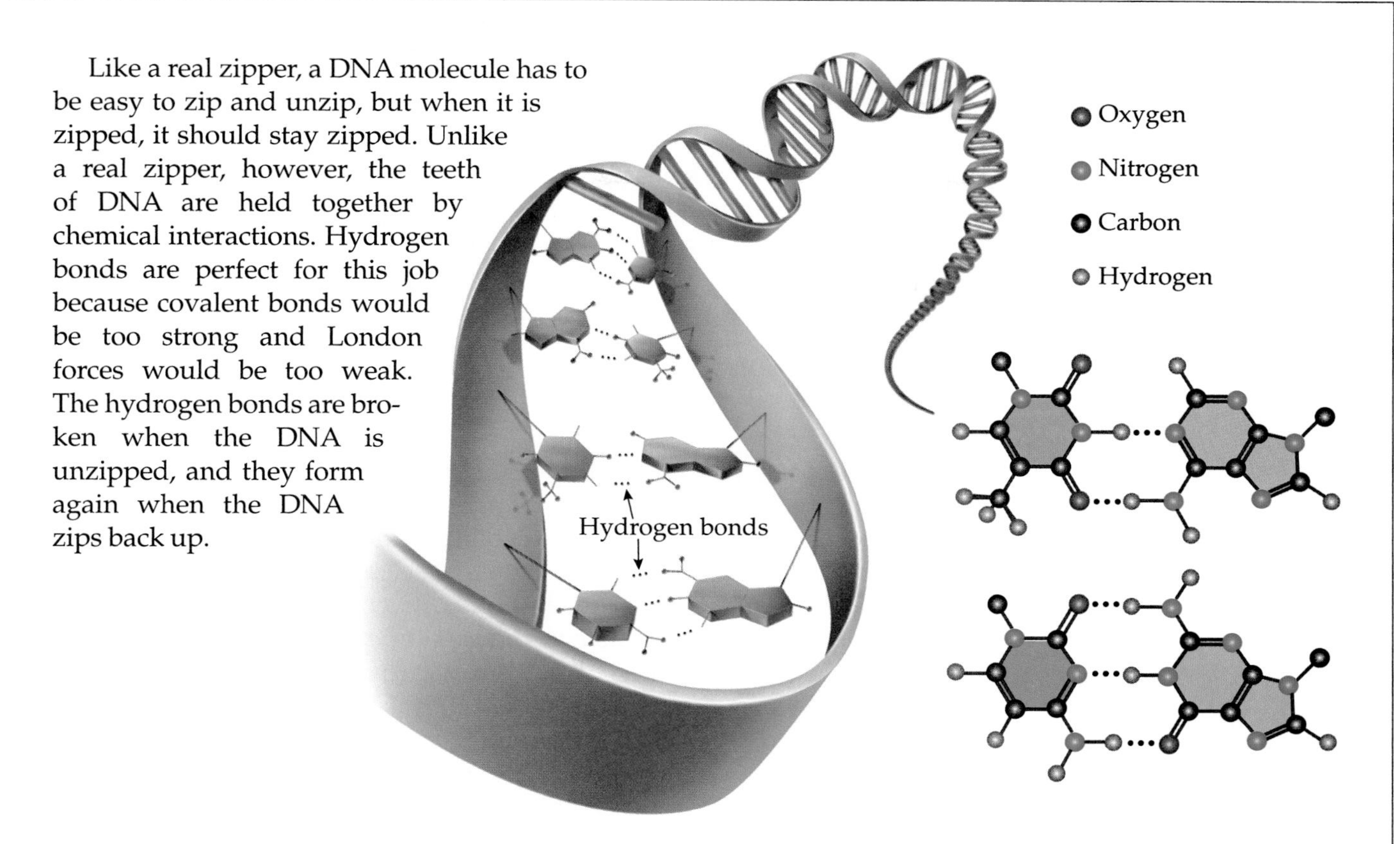

Practice Problems

10.8 In Practice Problem 10.7 you should have predicted that, based solely on London forces, PH_3 (18 electrons) has a higher boiling point than NH_3 (10 electrons). In fact, this is not the case. Explain why.

Answer: Ammonia molecules can hydrogen-bond (H to N). The hydrogen bonds between ammonia molecules, being much stronger than the London forces acting between PH_3 molecules, give ammonia the higher boiling point.

10.9 Draw some molecules of ammonia, NH_3, and use dotted lines to show the hydrogen-bonding between them.

10.10 For the hydrogen halides, the order of boiling points is HF > HI > HBr > HCl.

(a) Why does HF have the highest boiling point?

(b) Why is the boiling point of HI greater than those of HBr and HCl?

We end this discussion of intermolecular forces with a cautionary note. It's easy to get in the habit of thinking that hydrogen bonds are always the strongest intermolecular forces, dipole–dipole forces always the next strongest, and London forces always the weakest. Although this is *generally* true, it is not *always* true. For example, antimony trihydride, SbH_3, is an essentially nonpolar molecule because Sb and H have comparable electronegativities. Nevertheless, SbH_3 has a higher boiling point than does ammonia, NH_3, which forms hydrogen bonds. This seeming discrepancy occurs because antimony is a big atom, containing 51 electrons. This large size results in London forces between SbH_3 molecules that are stronger than the hydrogen bonds between NH_3 molecules. So, as always, respect the general trend but keep an open mind.

10.4 Nonmolecular Substances

So far in this chapter, the solids we have looked at consist of individual molecules held together by intermolecular forces (London forces, dipole–dipole forces, or hydrogen bonds). As you learned in Section 5.5, a solid of this type is called a *molecular solid*. Ice, for instance, is a molecular solid consisting of water molecules held together by hydrogen bonds. If you could shrink yourself down to the size of a molecule and enter an ice cube, you would see an ordered arrangement of water molecules "locked" in place by hydrogen bonds.

Portion of a NaCl Lattice

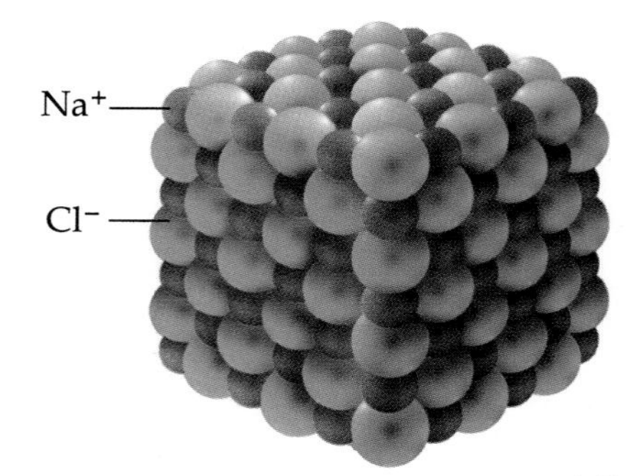

However, not all solids are molecular solids owing their existence to relatively weak intermolecular forces. As you also learned in Chapter 5, some solids contain no molecules at all. Instead, they owe their ability to exist in a condensed phase to far stronger covalent or ionic bonding forces. Sodium chloride is one example of such a *nonmolecular ionic solid*. As we saw in Chapter 5, NaCl is made up of a vast array (lattice) of positive sodium ions, Na^+, and negative chloride ions, Cl^-. In such a lattice, there is nothing you can point to and call a molecule. There is only a repeating pattern of positive and negative ions. Any solid that doesn't consist of discrete molecules is called a **nonmolecular solid**.

All simple ionic compounds are like this. Calcium chloride, $CaCl_2$, sodium bromide, NaBr, and magnesium oxide, MgO, are all lattices of ions. Because there is nothing in these solids that we can identify as being a molecule, it would be incorrect to call the forces holding them together *intermolecular* because this word means "between molecules" and there are no molecules. Instead, these compounds are held together by the much stronger ionic bonds we talked about in Chapter 5. Melting such compounds means breaking these ionic bonds. Because overcoming these strong attractions requires tremendous amounts of energy (heat), many ionic compounds have melting points in excess of 1000°C. Sodium chloride, for instance, has a melting point of 801°C, much higher than the 0°C melting point of water.

10.4 WORKPATCH What is wrong with the reasoning behind this incorrect statement: The fact that water melts at 0°C and NaCl melts at 801°C indicates that ionic bonds are much stronger than covalent bonds.

The incorrect statement in the WorkPatch is a common misconception among chemistry students. After all, water molecules do have covalent bonds in them. If you see why the statement is incorrect, you understand the fundamental difference between melting a molecular solid and melting a nonmolecular solid. Check your answer against the one given at the end of the chapter.

There are nonmolecular solids that are not ionic. Some examples are silicon dioxide (quartz), SiO_2, from which glass is made, and diamond, a form of pure carbon. These compounds are often called either **network solids** or **network covalent substances** because they consist of a large network of atoms held together by covalent bonds.

Two Network Solids

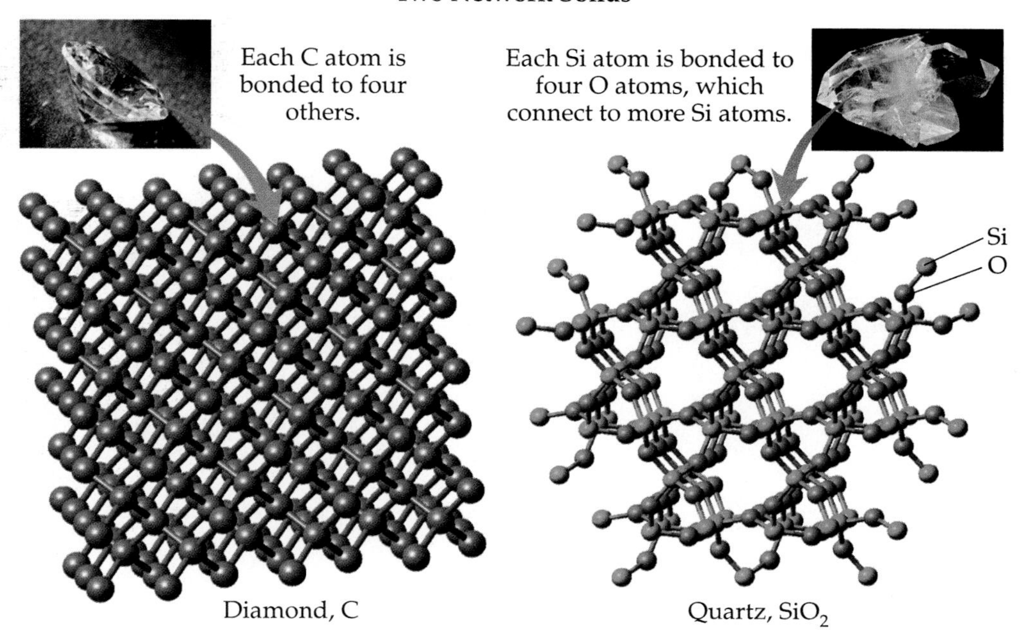
Each C atom is bonded to four others.
Each Si atom is bonded to four O atoms, which connect to more Si atoms.
Si
O
Diamond, C
Quartz, SiO_2

Diamond is a vast network of carbon atoms, with each atom covalently bonded to four other carbon atoms. Quartz is a vast network of silicon atoms, each covalently bonded to four oxygen atoms, which are in turn covalently bonded to more silicon atoms. Just as with an ionic lattice, there is nothing in these network solids that stands out as an individual molecule. What we have now is a lattice of atoms instead of ions, all interconnected by very strong covalent bonds. Melting these compounds involves breaking covalent bonds and consequently requires a lot of heat. Diamond melts above 3550°C, and quartz requires temperatures in excess of 1700°C before it begins to liquefy.

It is not always easy to determine whether a particular solid substance is molecular (like solid H_2O) or nonmolecular (like SiO_2 or NaCl) just by looking at the chemical formula. If the formula includes a group IA (1) or IIA (2) metal plus a nonmetal (like NaCl or $MgBr_2$), it is a good bet the solid is ionic and therefore nonmolecular. For other compounds, such as SiO_2, the formula does not contain enough information to tell you whether the solid is molecular or nonmolecular. If we hadn't told you that quartz is a network solid, for instance, you would have to determine its internal structure by experiment. The melting point of a solid can often provide a good hint, however, because melting point is directly related to the strength of the forces holding the solid together. If a melting point is tremendously high—1000°C or above, say—the solid is most likely nonmolecular. If a melting point is much lower, weaker forces are present, and the solid is more likely molecular.

To melt quartz, a nonmolecular solid, you have to break strong covalent bonds.

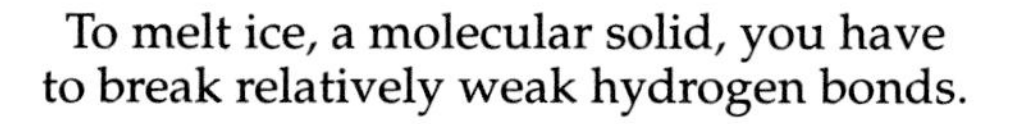
To melt ice, a molecular solid, you have to break relatively weak hydrogen bonds.

Quartz, melting point 1710°C

Ice, melting point 0°C

Another class of solids having high melting points are the pure metals, such as silver, Ag, iron, Fe, copper, Cu, and gold, Au. These nonmolecular solids are made up of a lattice of neutral metal atoms held together by **metallic bonds**. Like covalent bonds, metallic bonds are the result of shared electrons, but there is one important difference. In a piece of metal, the valence electrons on the metal atoms are free to move about and be shared by *all* the nuclei in the piece, giving rise to extra attractive forces between atoms. (This mobility of valence electrons explains why metals conduct electricity and heat so well.)

Like their covalent and ionic counterparts, metallic bonds are true bonds, which means that, in general, they are very strong. As you now know, strong interparticle bonds in a solid usually mean high melting points. Gold, Au, has a melting point of 1064°C, for example, and iron, Fe, has a melting point of 1535°C. These are very high temperatures, reflecting the high strength of metallic bonds.

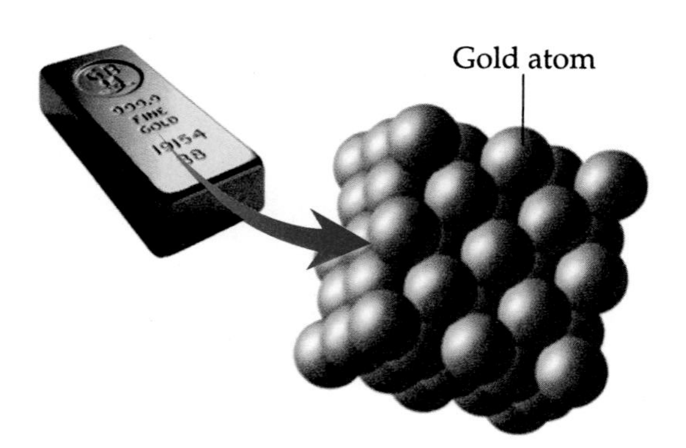

As usual, there are exceptions to this generalization. Sodium metal, Na, has a melting point of only 98°C, and mercury, Hg, is a liquid at room temperature, with a melting point of −39°C. For reasons that go beyond the scope of this book, these are two cases where metallic bonding is quite weak even though these substances are nonmolecular solids.

The following figure summarizes the types of solids we have looked at.

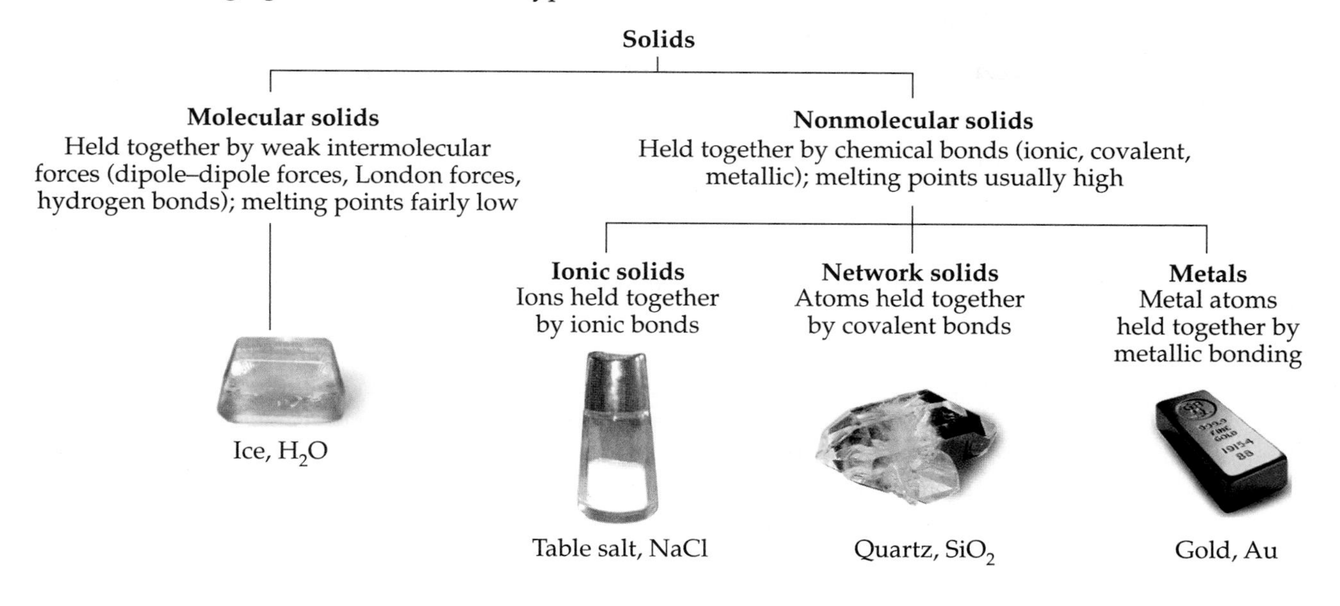
Solids
Molecular solids
Held together by weak intermolecular forces (dipole–dipole forces, London forces, hydrogen bonds); melting points fairly low
Nonmolecular solids
Held together by chemical bonds (ionic, covalent, metallic); melting points usually high
Ionic solids
Ions held together by ionic bonds
Network solids
Atoms held together by covalent bonds
Metals
Metal atoms held together by metallic bonding
Ice, H2O
Table salt, NaCl
Quartz, SiO2
Gold, Au

Practice Problems

10.11 Sodium oxide, Na_2O, is a white-gray powder that sublimes (goes directly from solid to gas) at 1275°C. Is sodium oxide a molecular or nonmolecular solid?

Answer: First, sodium oxide is made from a group IA (1) metal and a nonmetal. This practically ensures that the compound is a nonmolecular ionic solid. Second, this compound sublimes at a tremendously high temperature. This also indicates that the compound is nonmolecular (the high temperature is required to break ionic bonds in the lattice of Na^+ and O^{2-} ions).

10.12 How is a piece of iron similar to a piece of sodium chloride in terms of the forces that must be overcome to melt it?

10.13 Elemental sulfur is a yellow solid that melts at 113°C. Studies show sulfur to be a molecular solid consisting of S_8 molecules held together in a lattice. What forces must be overcome to melt the solid?

10.14 The following are solids at room temperature (about 25°C). Classify them as molecular, ionic, network, or metallic:
(a) Zirconium, Zr, mp = 1852°C
(b) Lead, Pb, mp = 328°C
(c) Calcium nitride, Ca_3N_2, mp = 1195°C
(d) Graphite form of carbon, C, sublimes at 3652°C
(e) Yellow phosphorus, P_4, mp = 44°C

This chapter began by considering the random motion of molecules in the gas phase. Intermolecular forces have little effect on gas behavior because the molecules are too far apart and moving much too fast to feel more than brief, faint touches of attraction and repulsion for one another. We shall have considerably more to say about this state of affairs in the following chapter.

Once we slowed down the gas particles and allowed them to feel various intermolecular forces, we saw a distinct change in both properties and appearance of the collection of molecules. First, the gas phase becomes the liquid phase with its much neater depiction of molecules collected together, touching one another, still moving but much more slowly, sliding over and around one another. And finally the liquid phase gives way to the solid phase, the ultimate in order, where each particle is assigned a distinct location in a lattice and the only allowed motions are slight vibrations about this position. The ordering of the chaotic state of affairs that existed in the gas phase was due to intermolecular forces. To understand the behavior of matter in a condensed phase (solid or liquid), you must be able to accurately describe the interactions between the molecules in that phase. These interactions are difficult to describe mathematically because they depend on many variables. For one thing, they depend on the identities of the molecules involved. Are they polar or nonpolar, large or small? These interactions can also depend on how far apart the molecules are and on how they are oriented relative to one another. In a condensed phase, the interactions even depend on a molecule's location—that is, is the molecule on the surface or in the interior of the condensed phase? Table 10.1 summarizes our discussion.

Table 10.1 Factors Influencing Strength of Intermolecular Forces

Factor	Stronger force	Weaker force
Identity of interacting molecules	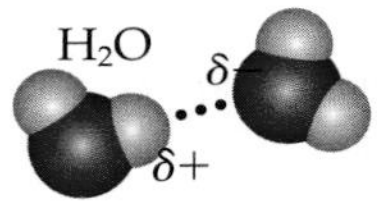 H_2O $\delta-$ $\delta+$ Hydrogen bonding between water molecules	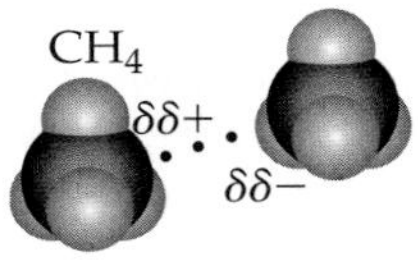 CH_4 $\delta\delta+$ $\delta\delta-$ London forces between methane molecules
Distance between interacting molecules	 Force between close molecules	 Force between distant molecules
Relative orientation of interacting molecules	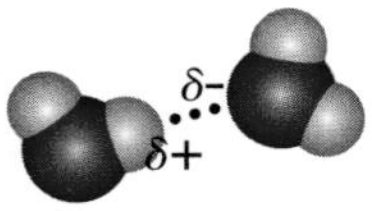 $\delta-$ $\delta+$ Attraction	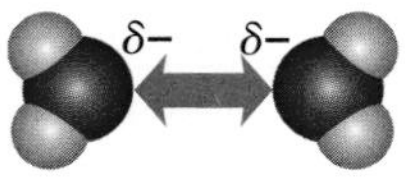 $\delta-$ $\delta-$ Repulsion
Location of molecule in liquid or solid	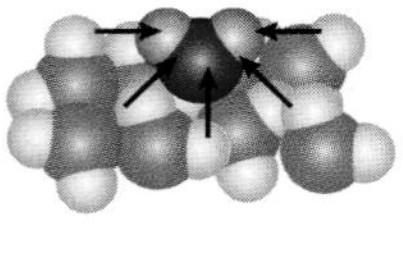 Surface molecule feels forces only from below.	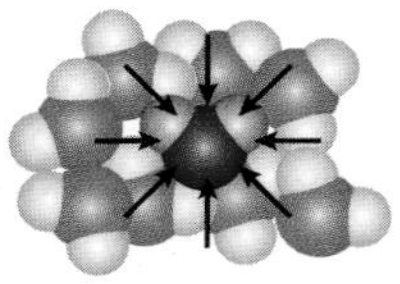 Interior molecule feels forces from all around.

All of this makes understanding and predicting the behavior of condensed phases very complicated. Just imagine, then, how easy it might be to explain the behavior of matter if we could ignore intermolecular forces. For gases we can! Stay tuned for the next chapter.

Have You Learned This?

Condensation (p. 368)
Vaporization (p. 368)
Boiling point (p. 368)
Freezing point (p. 369)
Condensed phase (p. 370)
London force (p. 371)
Dispersion force (p. 371)
Hydrogen bond (p. 376)
Nonmolecular solid (p. 380)
Network solid (p. 381)
Network covalent substance (p. 381)
Metallic bond (p. 382)

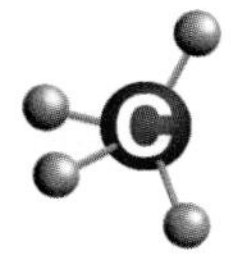

Intermolecular Forces and the Phases of Matter
www.chemistryplace.com

WHY DOES MATTER EXIST IN DIFFERENT PHASES?

10.15 Describe as completely as you can how small molecules like N_2 and O_2 behave at the molecular level in a warm room.

10.16 Explain why cooling a gas should eventually cause it to condense to a liquid.

10.17 What would be true of a gas if there were no such thing as intermolecular forces?

10.18 Is it incorrect to say that molecules are motionless in the liquid phase? Explain.

10.19 Explain in molecular terms how heating causes a liquid to change to the gas phase.

10.20 Water vapor liquefies when cooled below 100°C. Gaseous nitrogen liquefies when cooled below −196°C. What does this information tell you about the relative strengths of the intermolecular forces for these molecules?

10.21 Why are liquids and solids referred to as condensed phases?

10.22 Why does a gas expand to fill the container it is in, but a liquid and a solid do not?

10.23 Describe each phase of matter on the molecular level with respect to the amount of order present.

10.24 For this question you should consider only polarity. Molecule A–B has a smaller dipole moment than molecule C–D.
(a) Which molecule has the higher freezing point? Explain why.
(b) Which molecule has the higher boiling point? Explain why.

10.25 Draw a picture that shows how three polar HBr molecules in the gas phase would attract one another. What kind of intermolecular force is involved?

INTERMOLECULAR FORCES AND HYDROGEN-BONDING

10.26 Explain what gives rise to London forces and when they occur.

10.27 Would you expect CCl_4 or CBr_4 to have the higher boiling point? Explain your answer.

10.28 Chloromethane (CH_3Cl) has a much higher boiling point than methane (CH_4). Give two reasons for this.

10.29 Propane (C_3H_8) is a gas at room temperature, whereas octane (C_8H_{18}) is a liquid. Explain why this is so.

10.30 What is wrong with the statement "London forces are always weaker than dipole–dipole forces"?

10.31 Consider the molecules HF and HCl. The electronegativities of the atoms involved are H, 2.1; Cl, 3.0; F, 4.0.
(a) Which of the two molecules is more polar? Explain your answer.

(b) For which substance are the dipole–dipole attractions between the molecules stronger? Explain your answer.
(c) Which gas would, upon cooling, liquefy first? Explain your answer. Also, indicate which compound would have the higher boiling point based on your answer.
(d) Do both molecules also experience London forces of attraction? If yes, which would have the greater London forces?

10.32 When discussing the intermolecular forces between methanol molecules, chemists usually ignore any London forces between them. Why are they justified in doing this?

H H
H—C—O:
H

Methanol

10.33 At room temperature, Cl_2 exists as a gas, Br_2 exists as a liquid, and I_2 exists as a solid. Explain why.

10.34 Long-chain hydrocarbon molecules of the type $CH_3(CH_2)_{20}CH_3$ are solids and are used for things like waxes. The C–H bonds are essentially nonpolar. Why are waxes solid at room temperature?

10.35 When can hydrogen-bonding occur between molecules?

10.36 What is so special about hydrogen atoms that makes hydrogen-bonding possible?

10.37 Two different compounds have the same elemental composition, C_3H_8O. One has a low boiling point and the other a much higher boiling point. What attractive force must be present in one of these compounds that is not present in the other?

10.38 Consider molecules of the type H_2A, where A is a group VIA atom (O, S, Se, Te). Their relative boiling points are shown below.

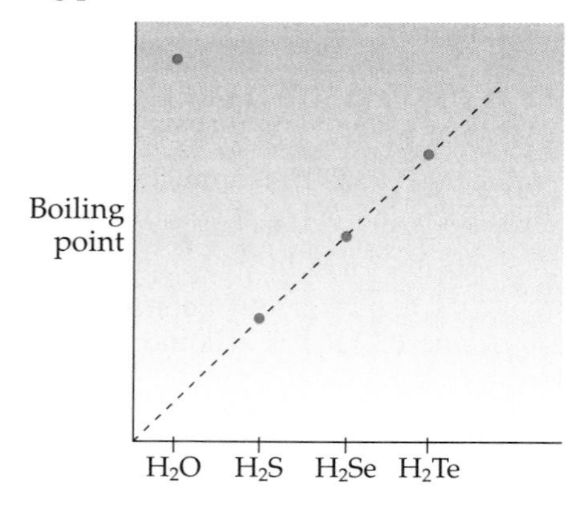

(a) Why is the boiling point trend $H_2S < H_2Se < H_2Te$?
(b) The boiling point of H_2O is well above the line that the others are on. Explain why this is so.

10.39 What do we mean by *induced dipole* when discussing London forces?

10.40 Why do we use dotted lines rather than solid lines to represent hydrogen bonds?

10.41 Examine the diagram of DNA on page 379 to answer this question: Why can hydrogen bonds form in DNA?

10.42 Why would either covalent bonds or London forces be inappropriate for attaching the two strands of DNA to each other?

10.43 How is it possible for a nonpolar molecule to have a higher boiling point than a polar one?

NONMOLECULAR SUBSTANCES

10.44 What is the fundamental difference between a molecular substance and a nonmolecular substance?

10.45 Is it sometimes, always, or never possible to tell from a compound's formula whether it is a molecular or nonmolecular substance?

10.46 What do we mean by a "lattice of ions"?

10.47 In general, nonmolecular solids have much higher melting points than molecular solids. Why is this so?

10.48 What is a network covalent substance? Give an example.

10.49 What evidence is there that metallic bonds can be as strong as ionic or covalent bonds?

10.50 Carbon tetrachloride (CCl_4) is a liquid at room temperature, whereas carbon tetrafluoride (CF_4) is a gas that does not liquefy until $-128°C$ at normal atmospheric pressure. How can you explain this?

10.51 All of the following are solids at room temperature. Classify them as molecular, ionic, network, or metallic.
(a) Potassium (K), mp = 64°C
(b) Potassium chloride (KCl), mp = 770°C
(c) Red phosphorus (P), mp = 590°C
(d) Boron triiodide (BI_3), mp = 50°C

10.52 How does the attraction between two molecules depend on the distance between them?

10.53 Why should the attraction between two water molecules depend on how they are oriented with respect to each other? Use drawings to illustrate your answer.

ADDITIONAL PROBLEMS

10.54 Describe how the molecules in a liquid behave as the temperature of the liquid increases.

10.55 Solid ice is less dense than liquid water. What does this say about how many water molecules there are per unit volume of ice relative to how many water molecules there are per unit volume of liquid water?

10.56 Which of the following best describes a liquid:
(a) The phase of matter in which particles are separated by the least distance.
(b) The phase of matter in which particles are in a fixed, rigid arrangement.
(c) The phase of matter in which particles completely fill the volume of their container.
(d) The phase of matter in which particles are in a loose, changeable arrangement but do not completely fill the volume of their container.

10.57 At 25°C, fluorine, F_2, and chlorine, Cl_2, are gases but bromine, Br_2, is a liquid. What does this say about the intermolecular forces in bromine relative to those in fluorine and chlorine?

10.58 In which of the following are there no dipole–dipole forces between molecules? Justify your answer.
(a) AsH_3
(b) CO_2
(c) H_2O
(d) $SeCl_2$

10.59 Which is more likely to be a gas at room temperature, CH_4 or CH_2Cl_2? Justify your answer.

10.60 Which of the following does not form hydrogen bonds? Justify your choice.
(a) Methyl alcohol, CH_3OH
(b) Hydrofluoric acid, HF
(c) Ammonia, NH_3
(d) Methane, CH_4

10.61 Which of the following would you expect to have the highest boiling point? Justify your choice.
(a) Propane, C_3H_8
(b) Carbon dioxide, CO_2
(c) Ethyl alcohol, CH_3CH_2OH
(d) Methyl fluoride, CH_3F

10.62 Which of the following would most likely be a gas at room temperature? Justify your choice.
(a) NaCl
(b) C_2H_2
(c) Na metal
(d) CH_3F

10.63 (a) Name the strongest intermolecular force in CH_3OH, CH_3Cl, CH_3CH_3, and $CH_3CH_2CH_3$.
(b) Rank these molecules from lowest to highest boiling point.

10.64 Acetic acid,

```
        :O:
         ‖
         C
       /   \
   H3C      Ö:
            /
           H
```

is very soluble in water because of the many hydrogen bonds that can form. Add water molecules to this drawing and show the hydrogen bonds possible.

10.65 Show how two CH_2Cl_2 molecules in the liquid phase are oriented with respect to each other and explain why the molecules align this way.

10.66 (a) Rank these molecules in order of increasing boiling point:

```
     Br                       H
     |                        |
Br — C — Br    Br — Br    H — C — H    Cl — Cl
     |                        |
     Br                       H
```

(b) State your reason for the order you chose in part (a).

10.67 (a) Rank these molecules in order of increasing boiling point:

```
    H                       H
    |                       |
H — C — OH    H — H     H — C — H    H — Cl
    |                       |
    H                       H
```

(b) State your reason for the order you chose in part (a).

10.68 Which compounds would you expect to experience only London forces? Explain your choices.
(a) Carbon tetrabromide, CBr_4
(b) Methyl bromide, CH_3Br
(c) Phosphorous tribromide, PBr_3
(d) Boron tribromide, BBr_3

10.69 Name the strongest intermolecular force expected in
(a) Boron trifluoride, BF_3
(b) 1-Propanol, $CH_3CH_2CH_2OH$
(c) Xenon, Xe
(d) Hydrogen fluoride, HF
(e) Hydrogen iodide, HI

10.70 Which of the following do you expect to be nonmolecular solids?
(a) Sodium hydroxide, NaOH
(b) Solid ethanol, CH_3CH_2OH
(c) Iron, Fe
(d) Solid silane, SiH_4

10.71 SiO_2, the main component of glass, is a solid at room temperature, and CO_2 is a gas at room temperature. How does the structure of these compounds explain this fact?

10.72 Arrange these substances in order of increasing melting point: CH_3OH, SiO_2, C_2H_6, NaCl.

10.73 Ethylene glycol, CH_2OHCH_2OH, and pentane, C_5H_{12}, have approximately the same molar mass. Nevertheless, one of these compounds boils at 198°C, and the other boils at 36°C. Which compound boils at which temperature? Use an argument based on intermolecular forces to justify your choice.

10.74 Sketch the hydrogen bonds present in the liquid 1-propanol, $CH_3CH_2CH_2OH$.

10.75 Both diamond and graphite are network solids consisting solely of carbon atoms:

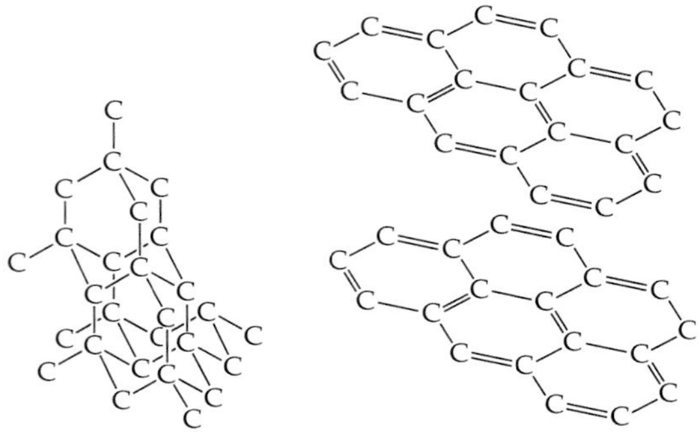

Diamond Graphite

In diamond, every carbon atom is bonded to four other carbon atoms. In graphite, every carbon atom is bonded to three other carbon atoms, creating sheets of atoms that lie on top of one another.

(a) Which type of intermolecular forces exist between the sheets of carbon atoms in graphite?

(b) Diamond is an extremely hard substance, whereas graphite is a soft, slippery substance often used as a lubricant. Use intramolecular and intermolecular forces to explain this difference in physical properties.

10.76 When considering which is stronger, a covalent bond or an ionic bond, why is it fairer to contrast the melting points of diamond and NaCl than to contrast the melting points of ice and NaCl?

10.77 Metallic bonding is often described as a lattice of metal cations in a sea of valence electrons. What is it about metallic bonding that allows it to be described in this fashion?

10.78 As you go down the column of noble gases in the periodic table, boiling point increases. Explain this trend.

10.79 Eicosane, $C_{20}H_{42}$, has a higher melting point and a higher boiling point than water even though water has hydrogen bonds between molecules and eicosane does not. Explain how this can be.

10.80 Most covalent molecular substances have much lower melting points than ionic substances. What does this say about the strength of a covalent bond relative to the strength of an ionic bond? Explain your answer.

10.81 Arrange in order of increasing boiling point: CO_2, SO_2, CH_3CH_2OH, Al.

10 WorkPatch Solutions

10.1 The fact that ethanol has a lower boiling point means it takes less heat (energy) to boil ethanol. Therefore, the intermolecular forces in ethanol must be weaker than those in water.

10.2 Being a nonpolar molecule, O_2 has no permanent $\delta+$ and $\delta-$ charges that can create strong attractive forces between molecules. All it has are the weak, instantaneous $\delta\delta+$ and $\delta\delta-$ partial charges that result in weak London forces between molecules. Being a polar molecule, H_2O has relatively large, permanent $\delta+$ and $\delta-$ charges that can produce relatively strong attractive forces between molecules. The London forces are stronger in CH_4 because there are more electrons involved.

10.3 Compound (b) boils at 78°C, and compound (a) boils at 35°C. The reason can't be London forces because both compounds have the same molecular formula, C_2H_6O, and therefore the same number of electrons to be unbalanced. Both molecules are polar, but only (b) has a hydrogen atom directly attached to an oxygen atom. This H–O combination allows for the formation of hydrogen bonds between molecules, giving compound (b) the higher boiling point:

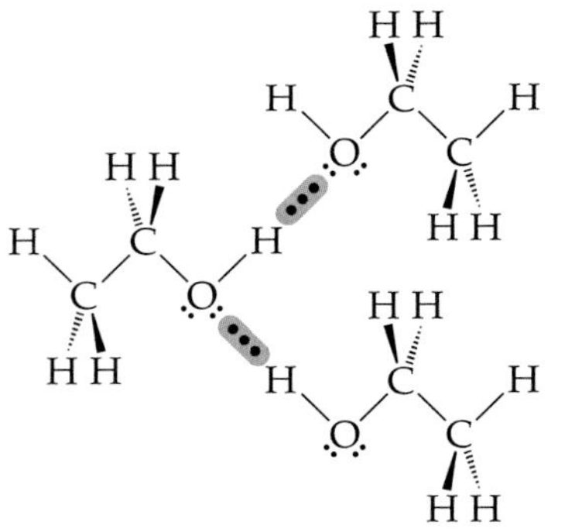

10.4 The reasoning is specious because covalent bonds have nothing to do with melting. Melting ice involves overcoming the hydrogen bonds between water molecules. The covalent bonds in the water molecules remain intact. Thus, comparing the melting points of NaCl and water involves comparing the relative strengths of ionic bonds and hydrogen bonds, not of ionic bonds and covalent bonds.

PV=nRT

CHAPTER 11

What If There Were No Intermolecular Forces? The Ideal Gas

11.1 Describing the Gas Phase—P, V, n, and T

Here is a thousand-year-old recipe discovered by the Chinese, a recipe that changed the world forever. Find some potassium nitrate, KNO_3, some charcoal, and some sulfur. Potassium nitrate, also called saltpeter, can be found as a delicate crust on some rocks and stone walls. Charcoal is a source of carbon. Elemental sulfur can be found in various locations in the form of a yellow mineral sometimes called brimstone. Grind these ingredients *separately* into powders and then make a mixture that is, by mass, 75% potassium nitrate, 15% charcoal, and 10% sulfur. Ignite the mixture with a flame or spark, and you'll see a chemical reaction that rapidly releases great amounts of heat and light—and a large amount of gas:

$$2\,KNO_3(s) + 3\,C(s) + S(s) \longrightarrow N_2(g) + 3\,CO_2(g) + K_2S(s)$$

This reaction is not particularly dangerous unless you do it in an enclosed space, such as the space between the sealed end of a metal tube and a heavy metal ball:

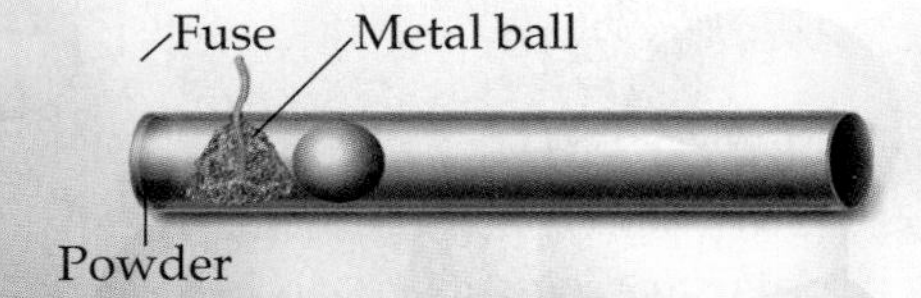

The powder, of course, is gunpowder. The ball becomes a lethal projectile because of all the gas produced by the combustion reaction. Trapped and heated, this gas develops a tremendous pressure that propels the ball out of the tube at high speed, with the gas expanding behind it. A similar phenomenon propels your car. In the engine cylinders, a fine spray of gasoline mixes with oxygen and combusts to produce hot gases at high pressure. The gases push down on a piston that is connected to a crankshaft that is in turn connected to the car's wheels. This combustion reaction is

$$\underset{\text{Octane, a component of gasoline}}{2\,C_8H_{18}(l)} + 25\,O_2(g) \longrightarrow 16\,CO_2(g) + 18\,H_2O(g)$$

Given the usefulness of expanding gases, it's no surprise that, over the centuries, scientists have expended great effort in studying them. After all, if you want a speedier bullet, a more powerful locomotive, or a jetpack the lets you leap tall buildings in a single bound, you had better understand gases. One of the first scientists to contribute to our knowledge about gases was Robert Boyle, who in 1661 studied how gas pressure changes with volume. In 1787, Jacques Alexandre Charles studied how the volume of a gas changes with temperature. Shortly thereafter, Amedeo Avogadro (in whose honor Avogadro's number is named) studied the relationship between the volume of a gas and the number of gas molecules in that volume.

As a result of the work of these and other scientists, it soon became apparent that, under most conditions, all gases behave pretty much the same. For example, when it comes to producing pressure, it does not matter if the gas is O_2, N_2, CO_2, steam, or anything else. In fact, the only factors that typically matter are the number of moles of gas, its temperature, and the volume it occupies. This finding led scientists to develop a model gas, called an *ideal gas,* that captures the essence of gas behavior. An **ideal gas** can be thought of as consisting of pointlike particles that do not attract or repel one another at all. Thus, even though an ideal gas has "molecules" (the particles), the molecules experience no intermolecular forces. The particles of an ideal gas fly around in straight-line paths, changing direction only when they collide with one another or with the walls of their container. When they collide, they rebound elastically.

An Ideal Gas

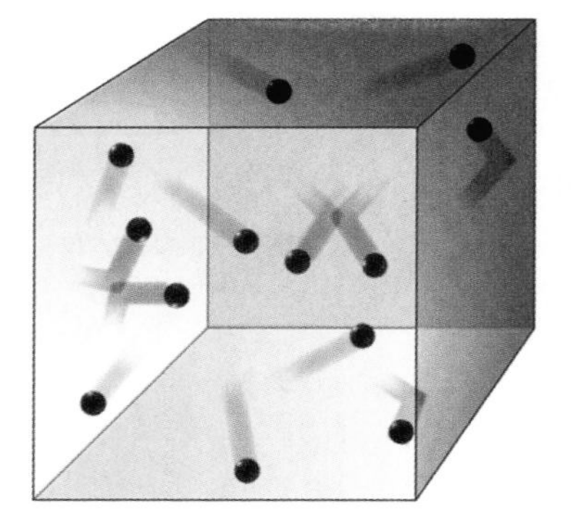

Because there are no forces of attraction between ideal gas particles (no London forces, no dipolar forces, no hydrogen-bond forces—nothing), an ideal gas will never condense to a liquid or solid, no matter how much it is cooled. Clearly no such gas exists, but under non-extreme conditions, real gases come close to behaving this way. That's because real gas molecules travel so fast and are so far apart that they easily overcome the weak attractive forces that *do* exist between them. The only time this isn't true is when a gas is either cooled to near the temperature at which it liquefies or compressed to such a high density that the molecules are quite close to one another. In any other situation, the ideal gas model is appropriate for describing any real gas. For example, the model explains why any real gas always expands to completely fill its container.

Some gas molecules

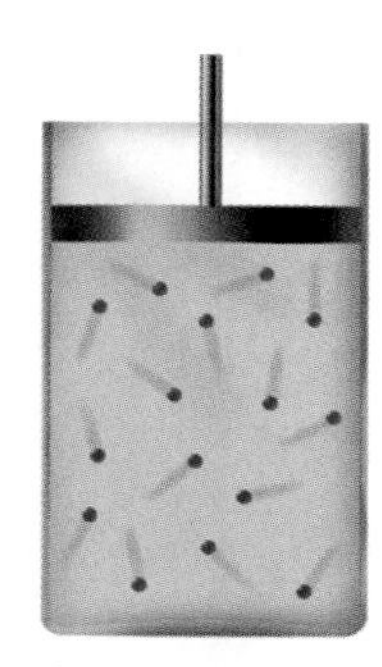
The same group of gas molecules

A gas expands to fill its container.

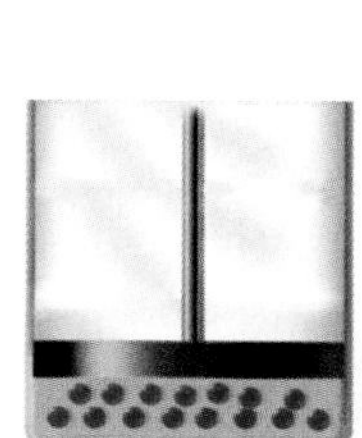
Some liquid molecules

The same group of liquid molecules

A liquid does not.

Because ideal gas molecules don't attract one another, they move freely until they hit something—such as the wall of their container. Thus, a gas expands to fill its container. In a condensed phase, the molecules are not free to wander all over the container because they are attracted to one another via various intermolecular forces.

Why is it never a reasonable approximation to ignore intermolecular forces in liquids and solids?

Your answer to this WorkPatch should remind you of why liquids and solids are referred to as *condensed phases*.

It turns out that we can *completely* describe the behavior of an ideal gas using just four variables:

P, the pressure of the gas
V, the volume the gas occupies
T, the temperature of the gas
n, the number of moles of gas in our sample

In the next section, we'll develop a simple equation that connects these variables to one another. This equation lets us make predictions about real gases that are reasonably accurate under normal circumstances. Before we get to the equation, however, you must become very familiar with the four variables that characterize an ideal gas. Let's start with pressure.

You breathe an atmosphere of pressurized gas. Although you're usually not aware of it, the atmospheric pressure you experience is substantial. Perhaps you've seen the demonstration shown here, in which removing the air from a metal can causes the can to collapse.

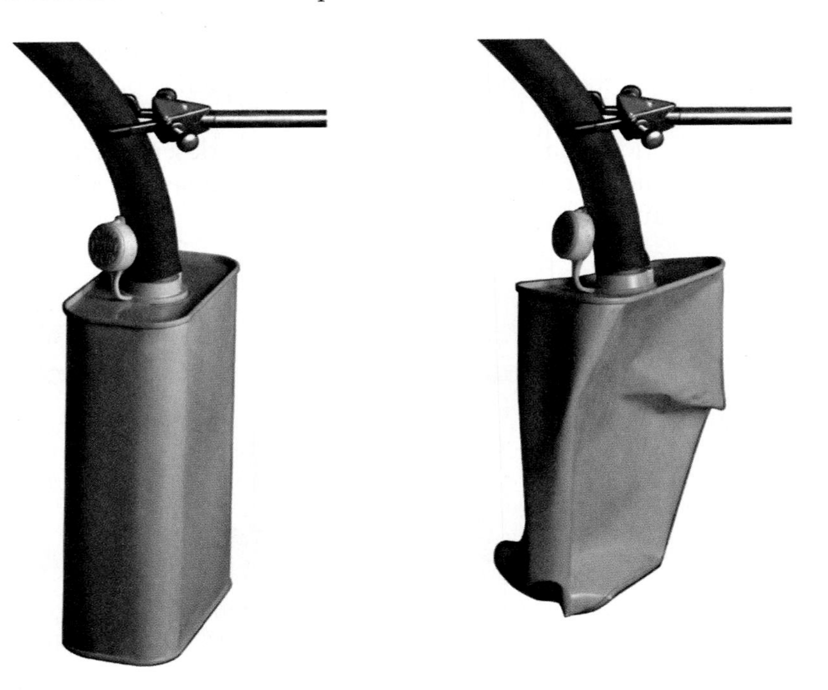

The fact that the can collapses tells you that some external force must have crushed it inward. But what is the source of this force? The source is the billions of fast-moving air molecules in the atmosphere that are colliding with the outside walls of the can. Each collision exerts an inward force on the can. The force from these collisions is felt over the entire surface area of the can and is referred to as either *air pressure* or *atmospheric pressure*.

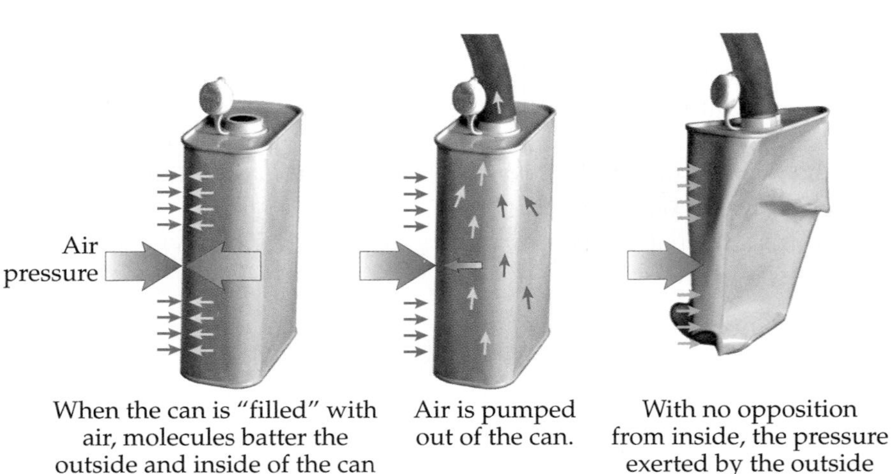

When the can is "filled" with air, molecules batter the outside and inside of the can equally, and as a result the can walls feel no net pressure.

Air is pumped out of the can.

With no opposition from inside, the pressure exerted by the outside air crushes the can.

Pressure (abbreviated P) is defined as force per unit area ($P = F/A$). The same atmospheric pressure that crushed the can is, at this very moment, attempting to crush you. So why don't you collapse? For the same reason the can didn't collapse until the air inside was pumped out. Before they were removed, the air molecules inside the can were colliding with the walls, pushing out on them with a force equal to the force with which the outside molecules were pushing in. The net result was that nothing happened until the inside air was removed.

This same idea can be used to construct a *mercury barometer*, a device used to measure atmospheric pressure. To make a barometer, we immerse one end of a glass tube into a dish of liquid mercury, Hg. Then we attach a vacuum pump to the other end of the tube and remove all the air from the tube. Once the tube is evacuated, the mercury rises into the tube to some specific height and then stops.

A Mercury Barometer

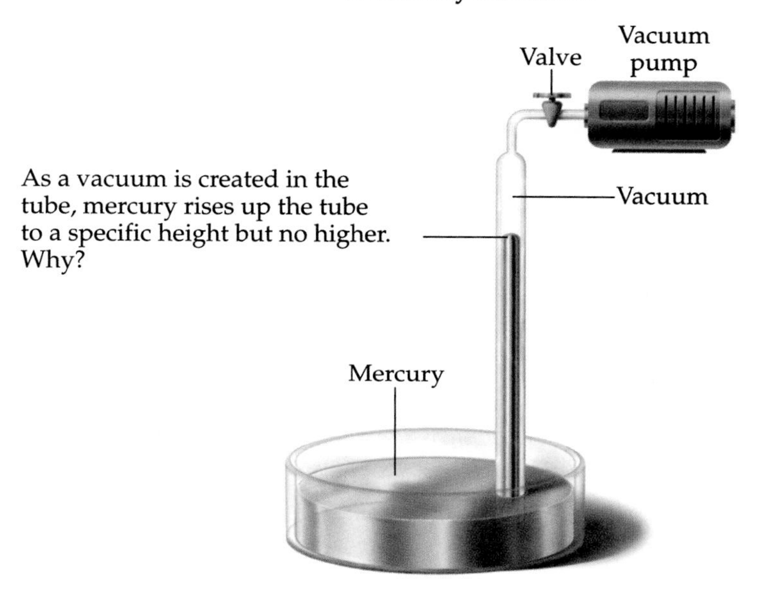

As a vacuum is created in the tube, mercury rises up the tube to a specific height but no higher. Why?

What makes the mercury rise up the tube is the same thing that crushed the can—air pressure. Before we removed the air from the tube, molecules of air both outside and inside the tube collided with the surface of the mercury in the dish, pushing down on the surface with the same force.

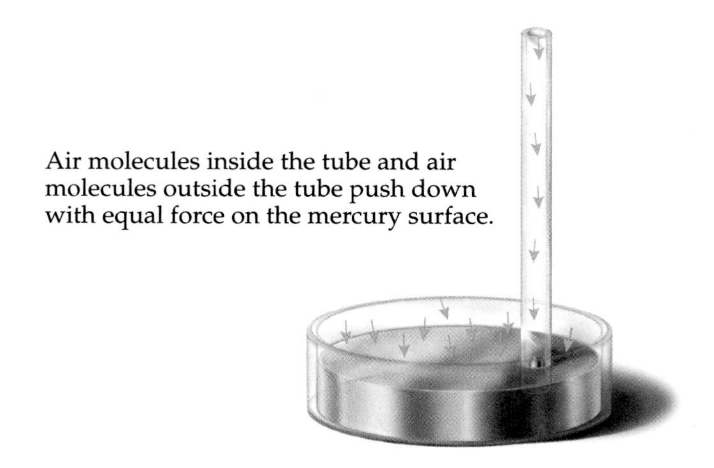

Air molecules inside the tube and air molecules outside the tube push down with equal force on the mercury surface.

When we removed the air from the tube, we eliminated the air pressure inside the tube. Thus, the mercury was pushed up into the tube by the air pressure exerted on the mercury by the air molecules outside the tube.

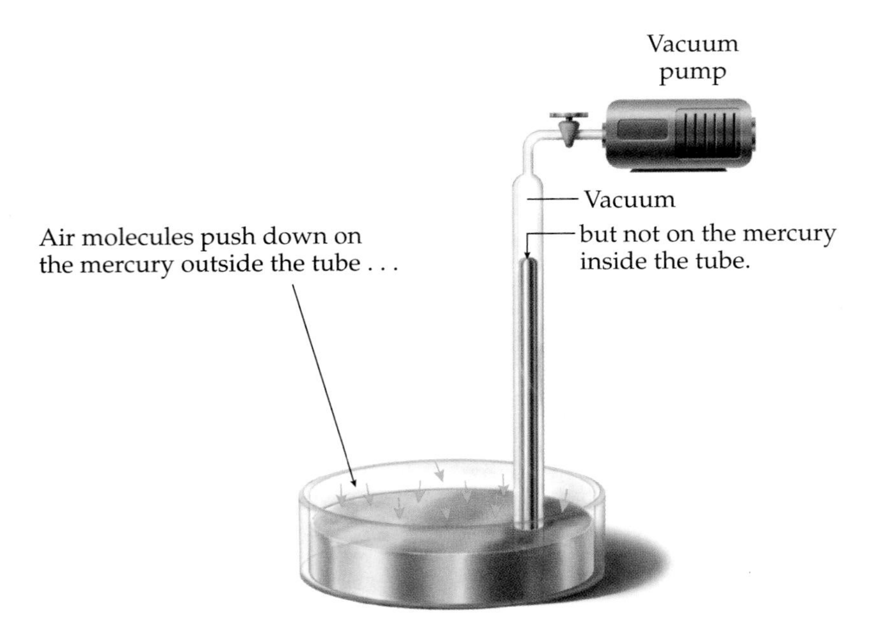
Vacuum pump
Vacuum
Air molecules push down on the mercury outside the tube . . .
but not on the mercury inside the tube.

The mercury is pushed up into the tube by the outside force (atmospheric pressure) until the weight of the mercury column supplies enough opposing force to resist any more movement. On an average day at sea level, this stopping point occurs when the mercury column rises to a height of 760 mm. We therefore define **normal atmospheric pressure** as being equal to 760 mm of mercury (mm Hg) or 1 atmosphere (atm).

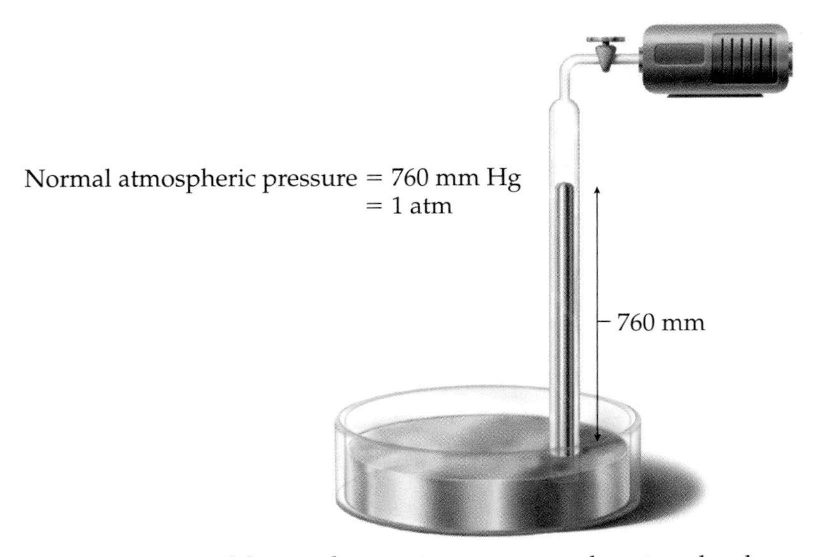
Normal atmospheric pressure = 760 mm Hg
= 1 atm
760 mm

Mercury barometer on average day at sea level

Weather forecasters use barometers to measure atmospheric pressure, although in the United States they usually report the pressure in inches of mercury rather than millimeters (760 mm = 29.9 in.; therefore, normal atmospheric pressure is 29.9 in. of mercury). When the atmospheric pressure drops, the column of mercury falls below its normal 760-mm height. Then it's time to grab your umbrella because low air pressure is often associated with rain.

As we saw earlier, pressure is just one of the variables we can measure to describe an ideal gas. There are three others. We can easily measure the volume V that a sample of gas occupies because the volume of the gas is always equal to the volume of the container confining the gas. We can also easily measure the temperature T of the gas sample. It is conventional to report the temperature of a gas using the Kelvin temperature scale discussed in Chapter 2. This is a bit of an inconvenience because most thermometers are calibrated in °F or °C. However, it is a simple matter to measure a temperature using one scale and then convert the reading to another scale. Recall that if you measure the temperature of a gas in °C, you convert to kelvins by adding 273.15:

$$\text{Temperature in K} = \text{Temperature in °C} + 273.15$$

Along with pressure, volume, and temperature, we also need to know the amount of gas we have. By convention, we specify the amount of gas in moles, and we use the symbol n to represent this quantity. When we say that $n = 1$ for a gas, we mean that we have 1 mol, or 6.02×10^{23} molecules, of the gas.

To summarize:

Measurable variables used to describe a gas

Pressure P, specified in millimeters of mercury or atmospheres

Volume V, specified in liters

Temperature T, specified in kelvins

Amount n, specified as number of moles

Practice Problems

11.1 An "empty" 0.500-gallon milk carton sits in a room where the temperature is 25.20°C. The barometric pressure is exactly 1 atm.

(a) What is the kelvin temperature of the air inside the carton?

(b) What is the pressure in millimeters of mercury of the air inside the carton?

(c) What is the volume in liters of the air inside the carton? [1 gallon = 3.78 L]

Answer:

(a) 25.20°C + 273.15 = 298.35 K

(b) Because the pressure is 1 atm, P = 1 atm = 760 mm Hg.

(c) $0.500 \text{ gallon} \times \dfrac{3.78 \text{ L}}{1 \text{ gallon}} = 1.89 \text{ L}$

11.2 A container is evacuated with a vacuum pump, and its mass is measured. Then it is filled with H_2 gas, and its mass is measured again. If the mass increase is 10.50 g, how many moles of H_2 gas are in the container? [*Hint:* You will need the molar mass of H_2.]

11.3 The gas in a pressurized container is at five times normal atmospheric pressure.
(a) What is the gas pressure in millimeters of mercury?
(b) What is the gas pressure in inches of mercury? [2.54 cm = 1 in.]

11.4 A mercury barometer develops a leak, allowing some air to enter the glass tube. Will such a barometer read too high or too low a pressure? Explain your answer.

11.2 Describing a Gas Mathematically—The Ideal Gas Law

Using the quantities P, V, n, and T, we can characterize the state of any gas. If we phone some friends in another country and give them these values for a particular sample of gas, they can construct a sample identical to ours. In fact, we do not even need to specify all four values—any three will do. That is because P, V, n, and T are not independent of one another. If you change one of these variables, you can predict how the others will change, and if you know three of them, you can figure out the fourth.

In this section, we shall develop the simple mathematical equation that describes how the P, V, n, and T values for an ideal gas are related to one another. Because we use the ideal gas model to develop it, this equation is called either the *ideal gas equation* or the *ideal gas law*. This equation is very useful because we can use it to predict how real gases respond to changes in P, V, n, or T. (Remember that real gases behave almost ideally under most circumstances.) Suppose, for instance, that you leave a tank of air in the sun and notice that it has warmed by 10 C°. What has happened to the pressure in the tank? Could the tank be close to exploding? You can use the ideal gas equation to answer this question.

We can derive the ideal gas equation by thinking about our model for an ideal gas and asking three questions about pressure. Remember that a gas in any container exerts pressure because the fast-moving gas particles (atoms or molecules in the case of a real gas) strike the walls of the container, exerting a force on them. Here then are our three questions about pressure:

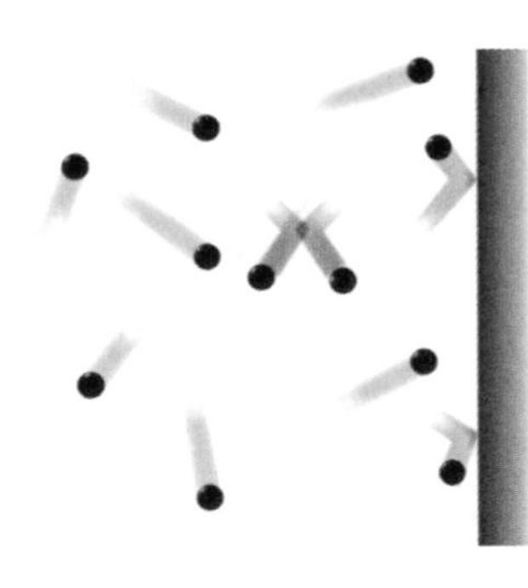

1. What happens to the pressure exerted by a gas confined to a container when we increase or decrease the *volume* of the container?
2. What happens to the pressure exerted by a gas confined to a container when we increase or decrease the *temperature* of the gas?
3. What happens to the pressure exerted by a gas confined to a container when we increase or decrease the *amount* of gas in the container?

To predict the answers to these questions, we need nothing more than our simple model of an ideal gas. Let's begin with the first question. We've already seen that pressure comes about from gas molecules colliding with the container walls. What happens to the pressure when we decrease the volume of the container? You can answer this question graphically, with just two drawings, as the following WorkPatch shows.

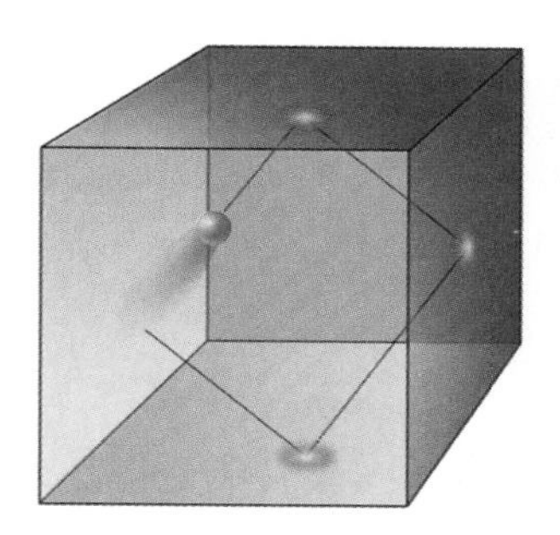

Draw two boxes like this one, making one larger than the other. Put a single fast-moving ideal gas particle inside each box (traveling at the same speed in both boxes). Show the path of the particle over the same time interval for both boxes, and make your interval long enough to allow for several collisions with the walls. Is the pressure in the smaller box the same as, less than, or greater than the pressure in the larger box?

What about the number of collisions with the walls in each box? There should be a difference. Do you see why? Check your answer against the one at the end of the chapter. This simple picture reveals why the pressure of an ideal gas always increases when the volume of its container decreases (and nothing else is changed). Your drawings should have revealed that the number of collisions per unit time increases as the volume of the box decreases. Thus, as the volume of the container gets *smaller*, the pressure exerted by the gas gets *larger*.

Rule 1: The pressure of a gas increases when the volume the gas occupies decreases.

Examine rule 1 very closely. It tells you that pressure and volume behave oppositely. As one increases, the other decreases. If you reexamine WorkPatch 11.2 in terms of what happens to pressure when you increase the size of the box, you'll realize that the second half of rule 1 is that pressure *decreases* as volume *increases*. We can combine these two parts of rule 1 and state our findings mathematically:

Rule 1: P is inversely proportional to V.

These two statements of rule 1 for an ideal gas say the same thing. We can, however, simplify the mathematical version a bit by dropping the word *inversely*. That P is *inversely* proportional to V means that P is *directly* proportional to $1/V$ (the inverse of V). This is a general mathematical fact for any two inversely related quantities.

Rule 1: P is proportional to $\frac{1}{V}$.

This rewording puts the volume in the denominator of the expression. Keep in mind that as V gets smaller, $1/V$ gets bigger. Because P is directly proportional to $1/V$, the same must be true for P: as V gets smaller P gets bigger, which is what our model predicts. All three statements of rule 1 say the same thing—pressure increases as volume decreases and decreases as volume increases. We can even show this graphically. A plot of P versus V for a gas shows that a small volume generates a large pressure whereas a large volume generates a small pressure (at constant n and T).

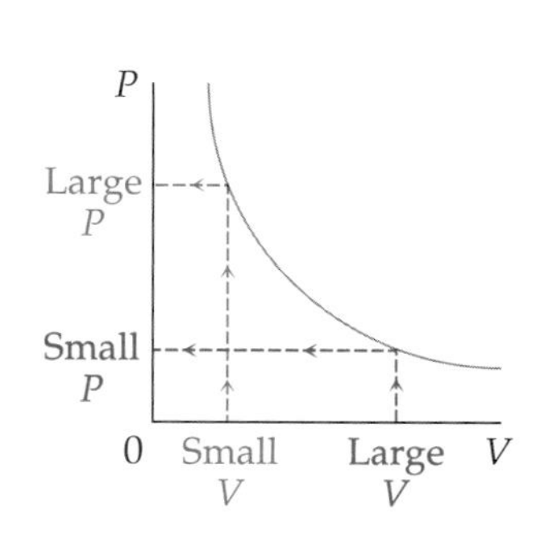

Now let's consider the second question: "What happens to the pressure exerted by a gas confined to a container when we increase or decrease the temperature of the gas?" Once again we use the ideal gas model. As we heat the gas, we are putting energy into it. This added energy causes the gas molecules to move faster, on average. Indeed, heating a gas is like stepping on the accelerator pedal in your car.

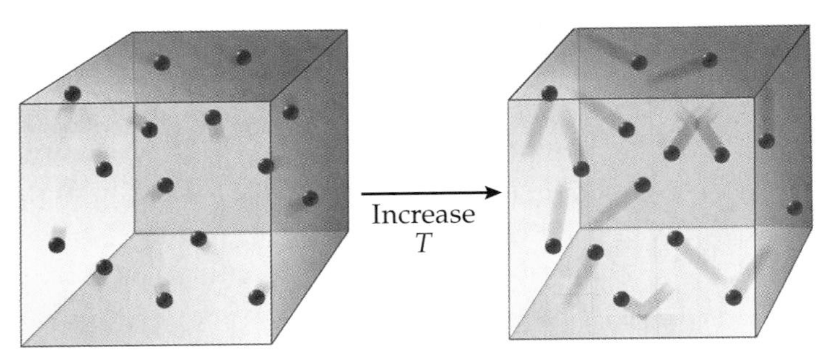

The higher temperature causes the molecules to move faster.

Because the molecules are moving faster, they collide with the walls of the container more often per unit time, and they hit the walls with greater force. Both effects result in more force being exerted on the walls per unit area, and so the pressure increases. Taking this argument in the reverse direction shows that the gas pressure decreases when you lower the temperature. Thus our second rule is

Rule 2: The pressure of a gas increases when its temperature increases and decreases when its temperature decreases.

Look closely at rule 2. Unlike pressure and volume, which we found were inversely related, pressure and temperature are directly related. Whenever the temperature changes, the pressure changes in the same direction. We can once again say the same thing in a mathematical way:

Rule 2: P is proportional to T.

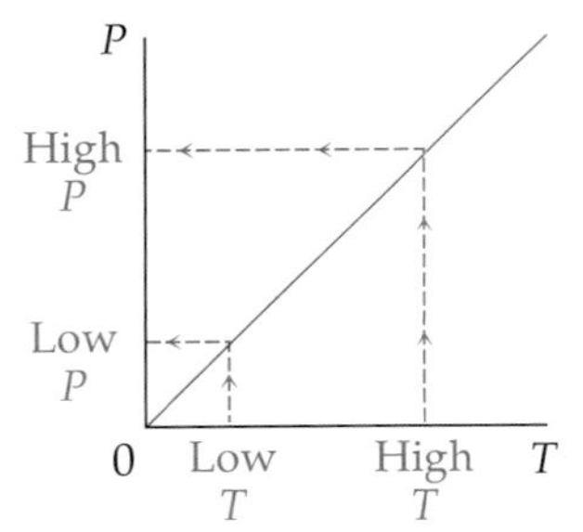

This is shown graphically in the plot of P versus T (at constant n and V) shown at left.

Our third question about pressure is "What happens to the pressure exerted by a gas confined to a container when we increase or decrease the amount of gas in the container?" In what is getting to be a theme, we once again use the model of an ideal gas to answer this question. In the illustration below, the left box has only a few gas molecules in it. The right box has lots more. In which box is the gas pressure higher?

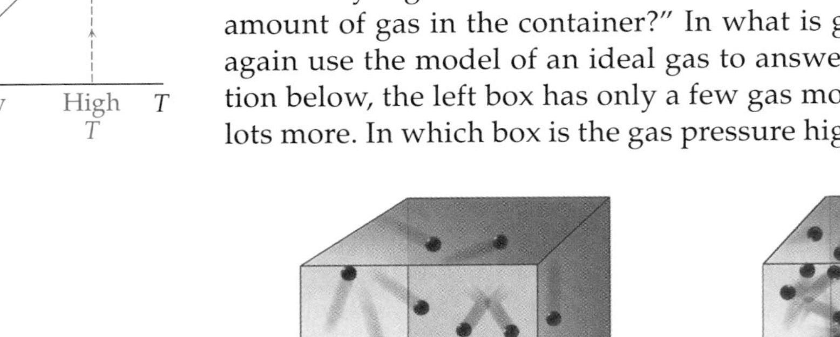

More molecules in the right box mean more collisions with the walls per unit time. More collisions mean more force pushing outward—that is, more pressure. And the reverse is true in the left box—fewer molecules in the same volume mean fewer collisions with the walls and therefore less pressure. This gives us a third rule about the behavior of gases:

Rule 3: The pressure of a gas increases when the number of moles of gas increases and decreases when the number of moles decreases.

Rule 3 states that pressure is directly proportional to the amount of gas present. As the amount increases, so does the pressure. And the same goes for smaller samples: gas pressure decreases as the amount of gas decreases. Mathematically,

Rule 3: P is proportional to n.

Therefore, a plot of P versus n (at constant V and T) looks just like the previous plot.

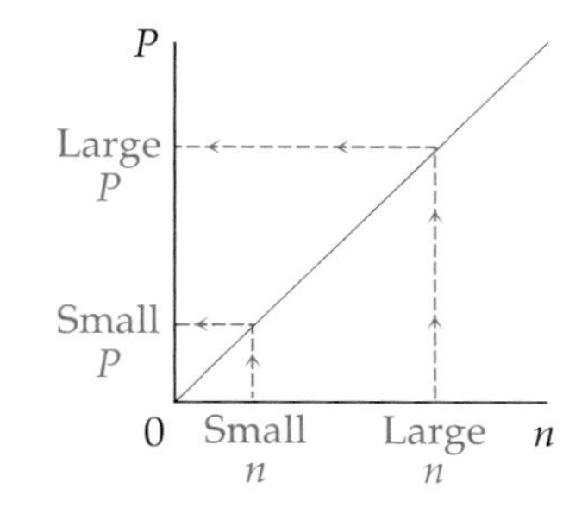

We are done. Here is what we have concluded:

Rule 1: P is proportional to $\frac{1}{V}$.

Rule 2: P is proportional to T.

Rule 3: P is proportional to n.

We can combine these three rules into one summary rule:

P is proportional to $\frac{nT}{V}$.

Notice that n and T are in the numerator (they are directly proportional to P), and V is still in the denominator (it is inversely proportional to P).

We can turn this half-symbols, half-words statement into a full-fledged mathematical equation by replacing the words *proportional to* with some mathematical operator, such as an equals sign. To convert a proportionality to an equality, we need to introduce a constant of proportionality:

$$P = \text{Constant} \times \frac{nT}{V}$$

In principle, this constant is not difficult to determine. All we need to do is take a container of gas (one whose behavior is very close to ideal) and experimentally determine the values of our four variables—P, n, T, and V. Then we can use the above equation to determine a value for the constant. Of course, we don't have to do this at all because others before us have already done it. This constant of proportionality is called the **ideal gas constant** and is represented by the letter R. Its value is

$$\text{Ideal gas constant } R = 0.0821 \text{ L·atm/K·mol}$$

When we include R in the summary rule, we get

$$P = R \times \frac{nT}{V} = \frac{nRT}{V}$$

This is the **ideal gas equation** (sometimes called the **ideal gas law**) we have been seeking, the one that can predict how a gas behaves. This one mathematical equation summarizes everything we know about an ideal gas. It can be algebraically rearranged to be written $PV = nRT$, as it is on the bottle rocket being launched in the chapter opening illustration.

Notice that R has units. *They are important!* When you use this value of R, you *must* express volume in liters, pressure in atmospheres, temperature in kelvins, and the amount of gas in moles. If you use any other units for these variables and attempt to use the ideal gas law with this value of R, you will get a wrong answer.

Now that we have derived the ideal gas law, let's apply it. Suppose we take a closed 2.0-L glass bottle filled with 0.10 mol of air initially at 25°C and heat it while measuring the temperature of the air in the bottle.

At 600°C, the bottle explodes!

Why did this happen? Because heating the air molecules made them move faster, causing them to collide with the bottle walls with more force and more often, creating more pressure. The pressure eventually increased to a point where the bottle could no longer contain it, and so the bottle exploded. Note that during the experiment, we never measured the pressure inside the bottle. The only thing we measured was the temperature, and all we know is that the bottle blew up when the air inside reached 600°C. Even without a barometer, however, we can figure out what the pressure was when the bottle exploded. Knowing the values of n, T, and V, we can use the ideal gas law to calculate the pressure. Here is what we know about the experimental conditions at the moment the bottle blew up:

Amount of gas $n = 0.10$ mol
Volume of gas $V = 2.0$ L
Temperature of gas $T = 600°\text{C}$

But remember! Everything must have the units of the gas constant R. The temperature must therefore be converted to kelvins:

$$\begin{aligned} \text{K} &= °\text{C} + 273.15 \\ &= 600°\text{C} + 273.15 \\ &= 873.15\ \text{K} \end{aligned}$$

Now we can plug the values for n, V, T, and R into our equation and solve for P:

$$P = \frac{nRT}{V}$$

$$P = \frac{0.0821 \frac{\cancel{L} \cdot atm}{\cancel{K} \cdot \cancel{mol}} \times 0.10 \cancel{mol} \times 873.15 \cancel{K}}{2.0 \cancel{L}} = 3.6 \text{ atm}$$

All the units cancel except the units of pressure (atm), which means our answer is expressed in these units. The pressure when the bottle blew up was 3.6 atm, which is 3.6 times greater than normal atmospheric pressure. The ideal gas law has allowed us to determine the pressure of a gas without measuring it directly.

Here is another problem that makes use of the ideal gas law. Suppose you work in a factory filling balloons with helium. One day your boss walks in and accuses you of putting too much helium in each balloon, needlessly cutting into corporate profits. She demands that you tell her how many grams of helium you are putting in each balloon. This is not an easy problem to solve. You can't simply determine the mass of a helium-filled balloon by weighing it because the balloon would float right off the balance!

But then you remember the ideal gas law, $P = nRT/V$. Your boss asked for the number of grams of helium in each balloon. The ideal gas law doesn't have grams in it, but it does have moles (n). You can therefore use the law to calculate the number of moles of helium in a balloon and then convert to grams. Rearranging the ideal gas equation to solve for n gives

$$n = \frac{PV}{RT}$$

Because you know the value of R, the only data you need are the values of P, V, and T for one of the balloons. A thermometer hanging on the wall reads 22°C.

Because the balloon has been in the room for a while, you reason that the gas in the balloon must be at the same temperature. Converting this Celsius temperature to kelvin, you get

$$22°C + 273.15 = 295.15 \text{ K}$$

A barometer in the room contains a column of mercury 748 mm high. The pressure of the gas in the balloon must be the same as the air pressure in the room. (If this were not true, the balloon would expand or shrink until the gas pressure inside was equal to the air pressure outside. Actually, the pressure in the balloon is a little higher than atmospheric pressure because of the elastic force of the rubber balloon, but you can ignore this slight difference.) To use the ideal gas equation, you need the pressure in atmospheres and therefore use the conversion factor 760 mm Hg = 1 atm:

$$748 \cancel{\text{mm Hg}} \times \frac{1 \text{ atm}}{760 \cancel{\text{mm Hg}}} = 0.984 \text{ atm}$$

You now have P and T. That leaves just the volume of the helium to be determined. Because the volume of a gas is always the same as the volume of its container, you realize that all you need to do is determine the volume of the balloon. Remembering that the volume of an object can be determined by water displacement (Section 2.7), you fill a bucket with water to the brim and submerge the balloon in the water. Collecting the overflowing water and pouring it into a graduated cylinder, you find that the volume is 240 mL. To be used in the ideal gas law, the volume must be in liters, not milliliters. There being 1000 mL in 1 L, 240 mL converts to 0.240 L.

You now have the P, V, and T values for the helium in the balloon, all in the appropriate units. Putting them in the ideal gas law and solving for n, the number of moles of helium in the balloon, you get

$$n = \frac{PV}{RT}$$

$$n = \frac{0.984 \cancel{\text{atm}} \times 0.240 \cancel{\text{L}}}{0.0812 \frac{\cancel{\text{L}} \cdot \cancel{\text{atm}}}{\cancel{\text{K}} \cdot \text{mol}} \times 295.15 \cancel{\text{K}}} = 0.010 \text{ mol He}$$

The balloon is filled with 0.010 mol of helium. You are not finished yet, though, because your boss asked for the amount in grams. The periodic table tells you that the atomic mass of helium is 4.003. One mole of helium atoms therefore has a mass of 4.003 g, and you use this information to calculate the grams of helium in each balloon:

$$0.010 \cancel{\text{mol He}} \times \frac{4.003 \text{ g He}}{1 \cancel{\text{mol He}}} = 0.040 \text{ g He}$$

With answer in hand, you proudly march into your boss's office and report that each balloon is being filled with 0.040 g of helium. Unfortunately, you are fired. The company limit is 0.030 g per balloon.

As we stated earlier, the answers we get when we use real-gas data with the ideal gas equation are not exactly correct. In both the exploding-bottle problem, where we solved for pressure, and the helium-balloon problem, where we solved for amount, the calculated answers do not exactly match the answers we would get if we measured these quantities. This is because the ideal gas equation was developed using the model of an *ideal* gas, which assumes that the molecules experience absolutely no attractive intermolecular forces. All real gas molecules do interact somewhat, however, and the small interactions are enough to make the behavior of a real gas deviate from that predicted by the ideal gas law. Usually, though, if we avoid low temperatures and/or high pressures, the deviations are small enough to tolerate in most cases.

Practice Problems

11.5 According to the ideal gas law, what happens to the pressure of a gas in a container when

(a) You double the absolute temperature of the gas?
(b) You double the number of liters the container can hold?
(c) You double the number of moles of the gas?
(d) You double both the absolute temperature of the gas and the number of liters the container can hold?

Answer: $P = \frac{nRT}{V}$

(a) Because T is in the numerator of the ideal gas equation, if it doubles, so does P.
(b) Because V is in the denominator of the ideal gas equation, if it doubles, P is halved.
(c) Because n is in the numerator of the ideal gas equation, if it doubles, so does P.
(d) This question asks what happens when you do (a) and (b) simultaneously. Because (a) doubled P and (b) halved it, the net result is that P does not change.

11.6 Suppose that 3.00 g of gaseous nitrogen, N_2, is placed in a 2.00-L container. The pressure is measured to be 450.5 mm Hg. What is the Celsius temperature of the gas? [*Hint:* Your first step should be solving the ideal gas equation for *T*.]

11.7 An automobile tire is filled with O_2 gas to a total pressure of 40.0 lb/in.^{2}. The temperature is 22.5°C. The inside volume of the inflated tire is 10.5 gallons. How many grams of O_2 are in the tire? (760 mm Hg = 14.696 lb/in.^{2}; 1 gallon = 3.785 L. [*Hint:* Your first step should be solving the ideal gas equation for *n*.]

11.3 Getting the Most from the Ideal Gas Law

With just a little algebraic manipulation and some substitutions, we can get even more from the ideal gas law than *P*, *V*, *n*, and *T* values. For example, our assistant, Albert, has been assigned the task of analyzing an unknown gas leaking from an old, unmarked storage tank. Having learned about combustion analysis back in Chapter 7, he determines the mass of a sample of the gas to be

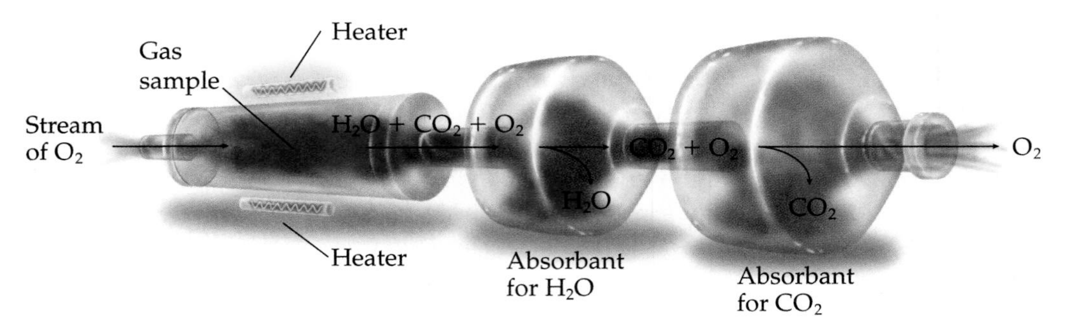

0.125 g and then burns the sample. The products are water and carbon dioxide, and he measures the masses of these products. These masses tell him the gas is 14.37% by mass H and 85.63% by mass C. Using what he learned in Chapter 7 about combustion analysis, he concludes that the empirical formula must be CH_2, but of course this cannot be the molecular formula. (Review Chapter 7 if you've forgotten the relationship between empirical formula and molecular formula.) Albert then generates a table to help him determine the molecular formula:

Molecular formula	Molar mass (g/mol)
C_2H_4	28
C_3H_6	42
C_4H_8	56
⋮	⋮
C_nH_{2n}	$14n$

If Albert knew the molar mass of the gas, he could easily pick the correct molecular formula. But how do you get the molar mass without knowing the molecular formula? From the ideal gas law. In the preceding section, we solved the ideal gas equation for n, the number of moles of gas in a sample:

$$n = \frac{PV}{RT}$$

The key to getting molar mass from this equation is remembering that Albert knows the mass of the sample before it was burned—0.125 g. He uses the symbol m_{sample} to represent the mass of the sample before combustion: $m_{\text{sample}} = 0.125$ g. Any time you have a known mass of some pure substance, you can convert to moles by dividing the mass by the molar mass MM (see Section 7.6 for review):

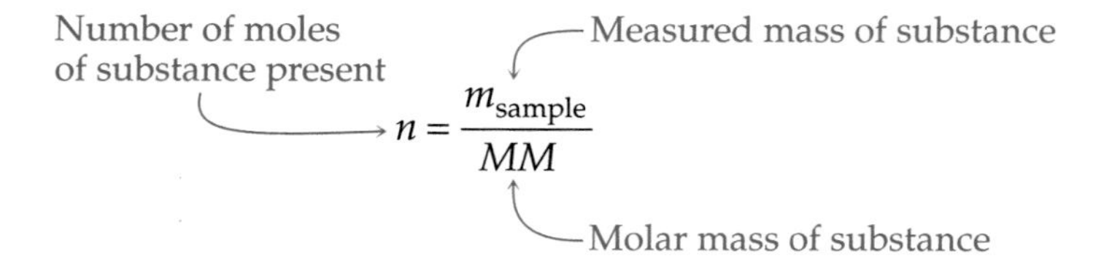

This conversion formula is extremely important. *You should always remember it!*

Because both m_{sample}/MM and PV/RT equal n, these two terms must equal each other:

$$\frac{m_{sample}}{MM} = \frac{PV}{RT}$$

Now all Albert has to do is rearrange this equation to solve for the molar mass:

$$MM = \frac{m_{sample}RT}{PV}$$

This is still a form of the ideal gas law because it still has only R, P, V, T, and n in it, although n is now expressed in terms of m_{sample} and MM. Recall that Albert initially collected $m_{sample} = 0.125$ g of the gas. He collected it in a 0.100-L steel cylinder equipped with a pressure gauge. Here he is with his sample. He is able to measure all the variables (m_{sample}, T, P, V) in the equation for molar mass. He measures the temperature of the room (and thus of the gas) to be 25.0°C (298.15 K). The pressure of the gas is 0.728 atm, and its volume is 0.100 L (the volume of the cylinder). He can just plug these quantities into the equation, along with the gas constant R, and solve for the molar mass of the gas:

0.728 atm

0.5

1

P(atm)

0

$$MM = \frac{(0.125 \text{ g})\left(0.0821 \frac{\cancel{\text{L}} \cdot \cancel{\text{atm}}}{\cancel{\text{K}} \cdot \text{mol}}\right)(298.15 \text{ K})}{(0.728 \cancel{\text{atm}})(0.100 \cancel{\text{L}})} = 42.0 \text{ g/mol}$$

From his table of possible formulas, Albert can now conclude that the molecular formula of the gas is C_3H_6.

The ideal gas equation in terms of molar mass is a powerful tool because it allows you to determine the molar mass of a gas even when you don't know the identity of the gas! All you need to do is determine the mass of a sample of the gas and then measure P, V, and T for the sample. Remember, however, that you must use the correct units. All the variables must have units that agree with those of the gas constant R—L·atm/K·mol. Check out the solution to the following practice problem, where all the variables have been converted to the proper units.

Practice Problems

11.8 A steel cylinder contains 0.013 78 kg of an unknown gas. Combustion analysis indicates that the gas has the empirical formula H_2S. The volume of the cylinder is 2.20×10^3 mL, and the pressure inside the cylinder is 3.42×10^3 mm Hg. The cylinder is stored in a closet at 25.0°C. What is the molecular formula of the gas?

Answer:

$$MM = \frac{m_{\text{sample}}RT}{PV}$$

For gas-law problems you must always convert T to Kelvins—*always*

$$= \frac{\overbrace{(0.013\ 78\ \cancel{\text{kg}})\left(\frac{1000\ \text{g}}{\cancel{\text{kg}}}\right)}^{m_{\text{sample}}}\overbrace{\left(0.0821\frac{\cancel{\text{L}}\cdot\cancel{\text{atm}}}{\cancel{\text{K}}\cdot\text{mol}}\right)}^{R}\overbrace{([25.0+273.15]\ \cancel{\text{K}})}^{T}}{\underbrace{(3.42\times10^3\ \cancel{\text{mm Hg}})\left(\frac{1\ \cancel{\text{atm}}}{760\ \cancel{\text{mm Hg}}}\right)}_{P}\underbrace{(2.20\times10^3\ \cancel{\text{mL}})\left(\frac{1\ \cancel{\text{L}}}{1000\ \cancel{\text{mL}}}\right)}_{V}}$$

$$= 34.1\ \text{g/mol}$$

If you calculate the molar mass of H_2S with atomic mass values from the periodic table, you get 34.082 g/mol. This is essentially identical to the molar mass we just calculated, telling us that the molecular formula in this case is the empirical formula, H_2S.

11.9 Suppose you have a sample of CO_2 gas and want to know its mass without bothering to use a balance. How could you do this?

Answer: Take the form of the ideal gas equation that has molar mass in it and solve for the mass of the sample:

$$MM = \frac{m_{\text{sample}}RT}{PV} \xrightarrow{\text{rearrange}} m_{\text{sample}} = \frac{PV\,MM}{RT}$$

To find the mass of the sample, measure P, V, and T for the gas and plug the numeric values into the rearranged equation along with the value of R and the molar mass of CO_2, which is 44.01 g/mol. Make sure all the variables have units that agree with those in R, and solve for m_{sample}.

11.10 A sample of hydrogen gas, H_2, is in an outdoor 1000.0-gallon tank. It is winter, and the temperature is −4.50°C. A pressure gauge indicates that the pressure inside the tank is 32.6 atm. How many pounds of hydrogen are left in the tank? [*Hint:* Examine the solution to problem 11.9.]

11.11 What do you get when you divide the mass in grams of a sample of a pure substance by the molar mass of the substance? Prove your answer is correct.

Let's not stop here. We can rearrange the molar-mass form of the ideal gas equation once again, this time solving for m_{sample}/V:

$$MM = \frac{m_{\text{sample}}RT}{PV} \xrightarrow[\text{and cross out all variables that cancel.}]{\text{Multiply both sides by } \frac{P}{RT}} \frac{MM\,P}{RT} = \frac{m_{\text{sample}}\cancel{RT}\cancel{P}}{\cancel{P}V\cancel{RT}}$$

Doing so gives

$$\frac{m_{\text{sample}}}{V} = \frac{MM\,P}{RT}$$

We arrive at an equation for m/V, which is the definition of density. This equation says that if you know the molar mass of a gas and measure its pressure and temperature, you can calculate its density. Let's try it. Below are two balloons, one filled with helium and the other with carbon dioxide. Both are in a room that is at 22.0°C, and the pressure inside each balloon is 755.0 mm Hg (slightly less than 1 atm). The densities of both gases are calculated below (notice how we were careful to convert units to those of the gas constant R).

$P = 755.0$ mm Hg
$T = 22.0$°C
$MM_{He} = 4.003$ g/mol

He

$P = 755.0$ mm Hg
$T = 22.0$°C
$MM_{CO_2} = 44.009$ g/mol

CO_2

$$\text{Density} = \frac{m_{\text{sample}}}{V} = \frac{MM\,P}{RT}$$

For helium:

$$\frac{m_{\text{sample}}}{V} = \frac{\left(\dfrac{4.003\text{ g}}{\cancel{\text{mol}}}\right)(755.0\ \cancel{\text{mm Hg}})\left(\dfrac{1\ \cancel{\text{atm}}}{760\ \cancel{\text{mm Hg}}}\right)}{\left(0.0821\dfrac{\text{L}\cdot\cancel{\text{atm}}}{\cancel{\text{K}}\cdot\cancel{\text{mol}}}\right)(295.15\ \cancel{\text{K}})} = 0.164\text{ g/L}$$

For carbon dioxide:

$$\frac{m_{\text{sample}}}{V} = \frac{\left(\dfrac{44.009\text{ g}}{\cancel{\text{mol}}}\right)(755.0\ \cancel{\text{mm Hg}})\left(\dfrac{1\ \cancel{\text{atm}}}{760\ \cancel{\text{mm Hg}}}\right)}{\left(0.0821\dfrac{\text{L}\cdot\cancel{\text{atm}}}{\cancel{\text{K}}\cdot\cancel{\text{mol}}}\right)(295.15\ \cancel{\text{K}})} = 1.80\text{ g/L}$$

Look at the densities we calculated. Under the same conditions, carbon dioxide gas is more than ten times denser than helium gas. Given that the density of air is 1.30 g/L, it should be no mystery why a helium balloon floats in air but a balloon filled with CO_2 would sink.

By now you should be catching on to the idea that rearranging the ideal gas equation can be quite a useful thing to do. For example, try the following WorkPatch, and then we'll discuss your answer.

A steel tank contains 12.992 kg of oxygen gas, O_2. Measurements indicate that the pressure of the gas is 100.0 atm and its temperature is 300.0 K. The internal volume of the tank is 100.0 L. Evaluate the numeric value of PV/nT (include units in your answer), and compare what you get with the accepted value of the gas constant $R = 0.0821$ L·atm/K·mol.

Your WorkPatch answer is extremely important! For any gas that behaves ideally, knowing the number of moles n, the pressure P in atmospheres, the volume V in liters, and the temperature T in kelvins allows you to calculate the gas constant R from the relationship $R = PV/nT$.

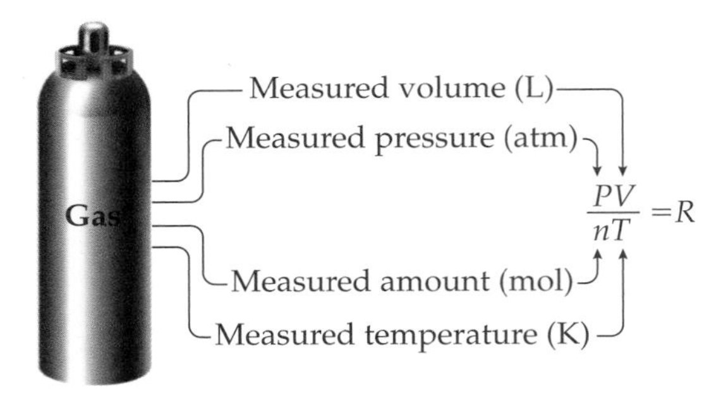

Knowing this makes it much easier to solve a type of problem called an initial-condition, final-condition problem. For example, suppose we have a steel cylinder filled with a gas. The pressure of this gas is 5.00 atm, and its temperature is 200.0 K. The volume of the cylinder is unknown, but it is fixed (unchanging) because the cylinder is made of solid steel. The number of moles of gas in the cylinder is also unknown. So these are our initial (i) conditions for the gas: $P_i = 5.00$ atm, $T_i = 200.0$ K, V_i and n_i both unknown:

Initial Conditions

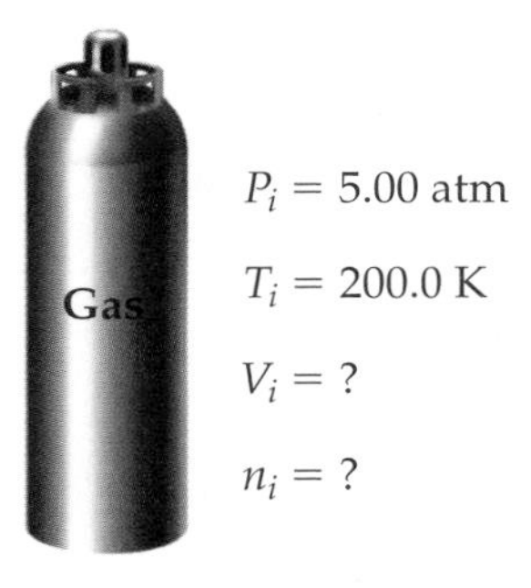

$P_i = 5.00$ atm

$T_i = 200.0$ K

$V_i = ?$

$n_i = ?$

Now let's change the temperature by heating the cylinder to 400.0 K. Because this is the only change we make, our final conditions are

Final Conditions

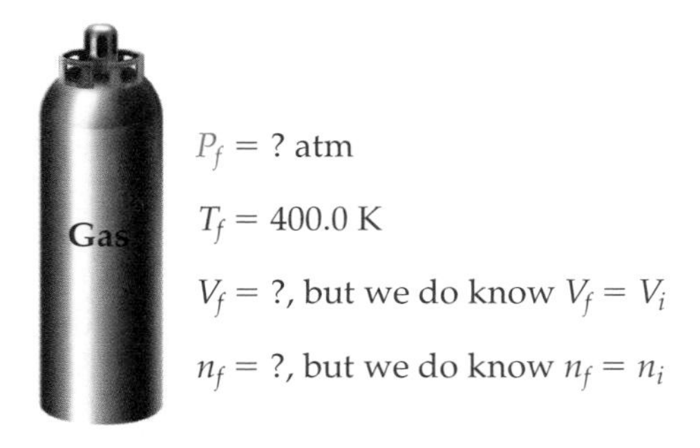

Notice that although we still do not know the volume of the gas or the number of moles of gas present, we *can* say that $V_f = V_i$ and $n_f = n_i$ because we did not change these quantities on going from the initial conditions to the final conditions. The pressure, however, has changed (it increased as the temperature increased because the volume could not change). What is the new pressure P_f inside the cylinder? This is an important question. If a storage cylinder of gas gets too hot, the increase in pressure can cause it to explode.

To determine P_f, just remember that PV/nT is always equal to R. This means that P_iV_i/n_iT_i is equal to R, and P_fV_f/n_fT_f is also equal to R. And of course because these expressions both equal R, they must be equal to each other:

$$\frac{P_iV_i}{n_iT_i} = \frac{P_fV_f}{n_fT_f}$$

This is an extremely useful result. It is the starting point for solving initial-condition, final-condition problems. The second step is to apply the rule of algebra that allows us to cancel *equal* quantities that appear in the same position on both sides of the equation. In our case, V_i on the left side and V_f on the right side are equal, and both appear in the numerator, which means they cancel each other. Likewise, n_i on the left and n_f on the right are equal and both in the denominator, which means they also cancel. These cancellations simplify our equation to

$$\frac{P_i\cancel{V_i}}{\cancel{n_i}T_i} = \frac{P_f\cancel{V_f}}{\cancel{n_f}T_f} \xrightarrow{\text{simplifies to}} \frac{P_i}{T_i} = \frac{P_f}{T_f}$$

All the variables with unknown values have disappeared! The final step is to rearrange the equation to solve for P_f and then plug in the numeric values for P_i, T_f, and T_i:

$$P_f = \frac{P_iT_f}{T_i} = \frac{(5.00 \text{ atm})(400.0 \cancel{\text{K}})}{200.0 \cancel{\text{K}}} = 10.0 \text{ atm}$$

This answer tells us that the pressure doubled when we doubled the temperature while holding the volume constant. Make sure you can do this algebraic rearrangement. (Review Chapter 2 if you can't!)

A summary of the method we just described appears at right.

Try the following initial-condition, final-condition practice problems. The first teaches you a valuable lesson about temperature. The second shows you how a seemingly difficult problem is in fact quite solvable.

Solving initial-condition, final-condition gas problems

Step 1: List P, V, n, and T for the initial and final conditions.

- Write in numeric values for any variables you know.
- Where possible, express the final variable in terms of the initial one. For instance, if volume does not change, you can say $V_f = V_i$.

Step 2: Write the expression

$$\frac{P_iV_i}{n_iT_i} = \frac{P_fV_f}{n_fT_f}$$

- Whenever possible, rewrite this expression to show final variables in terms of initial ones (from step 1).
- Cancel factors that are identical on the two sides of the expression.

Step 3: Solve the equation algebraically for the desired variable. Then plug in numeric values for the quantities you know and do the calculation to get your answer.

Practice Problems

11.12 A steel cylinder is filled with a gas. The initial pressure of this gas is 5.00 atm, and the initial temperature is 200.0°C. The cylinder is heated to a final temperature of 400.0°C. It appears that the temperature has doubled, but the final pressure is not 10.0 atm. Why doesn't the pressure double as it did in the previous example? What is the final pressure? What valuable lesson does this question teach you?

Answer:

Step 1: List P, V, n, and T for the initial and final conditions.

Initial conditions	Final conditions
$P_i = 5.00$ atm	$P_f = ?$
$V_i = ?$	$V_f = V_i$
$n_i = ?$	$n_f = n_i$
$T_i = 200.0°C$ (473.15 K)	$T_f = 400.0°C$ (673.15 K)

Step 2: Write the key expression and cancel identical terms:

$$\frac{P_iV_i}{n_iT_i} = \frac{P_fV_f}{n_fT_f} \xrightarrow{\text{Wherever possible, write final variables in terms of initial variables.}} \frac{P_iV_i}{n_iT_i} = \frac{P_fV_i}{n_iT_f}$$

$$\frac{P_i\cancel{V_i}}{\cancel{n_i}T_i} = \frac{P_f\cancel{V_i}}{\cancel{n_i}T_f} \xrightarrow{\text{Cancel identical factors on the two sides of the equation.}} \frac{P_i}{T_i} = \frac{P_f}{T_f}$$

Step 3: Solve algebraically for the desired variable (P_f in this case), and then plug in the known numeric values:

$$\frac{P_i}{T_i} = \frac{P_f}{T_f} \longrightarrow P_f = \frac{T_fP_i}{T_i}$$

$$P_f = \frac{T_fP_i}{T_i} = \frac{(673.15\text{ K})(5.00\text{ atm})}{473.15\text{ K}} = 7.11\text{ atm}$$

The final pressure is only 7.11 atm, not 10.0 atm. The pressure did not double because the Kelvin temperature did not double! The valuable lesson to be learned is that you can use only the Kelvin scale *for temperature when doing gas-law problems. Never use any other scale.* Never!

11.13 A steel cylinder initially contains some amount of O_2 gas at 500.0 K. Additional O_2 gas is added until the number of moles of gas has doubled. The cylinder is then cooled to 250.0 K. What happened to the pressure inside the cylinder?

Answer:

Step 1: List P, V, n, and T for the initial and final conditions.

Initial conditions	Final conditions	
$P_i = ?$	$P_f = ?$	
$V_i = ?$	$V_f = V_i$	
$n_i = ?$	$n_f = 2n_i$	⟵ We don't know the number of moles of oxygen, but we do know the final number is twice the initial number.
$T_i = 500.0$ K	$T_f = 250.0$ K	

Step 2: Write the key expression and cancel identical terms.

$$\frac{P_iV_i}{n_iT_i} = \frac{P_fV_f}{n_fT_f} \xrightarrow{\text{Wherever possible, write final variables in terms of initial variables.}} \frac{P_iV_i}{n_iT_i} = \frac{P_fV_i}{(2n_i)T_f}$$

$$\frac{P_i\not{V_i}}{\not{n_i}T_i} = \frac{P_f\not{V_i}}{(2\not{n_i})T_f} \xrightarrow{\text{Cancel identical terms on the two sides of the equation.}} \frac{P_i}{T_i} = \frac{P_f}{2T_f}$$

Step 3: Solve algebraically for the desired variable (P_f in this case), and then plug in the known numeric values:

$$\frac{P_i}{T_i} = \frac{P_f}{2T_f} \longrightarrow P_f = \frac{2T_fP_i}{T_i}$$

$$P_f = \frac{2T_fP_i}{T_i} = \frac{2(250.0\text{ K})P_i}{500.0\text{ K}} = \frac{500.0\text{ K}}{500.0\text{ K}} \times P_i = P_i$$

We find that $P_f = P_i$—meaning that the pressure did not change! There's no pressure change because the effect of doubling the amount of gas (a change that doubles the pressure) is counterbalanced by the effect of halving the temperature (a change that halves the pressure).

11.14 A gas initially at $P_i = 2.00$ atm, $T_i = 200.0$ K, and $V_i = 50.0$ mL is in a cylinder that has a movable piston. The cylinder is heated to $T_f = 220.0$ K, and then the piston is pushed down so that the volume of the gas is decreased to $V_f = 25.0$ mL.

(a) Calculate the final pressure P_f of the gas, leaving all volumes in milliliters for your calculation.

(b) Calculate the final pressure P_f of the gas with all volumes expressed in liters.

(c) When you are doing initial-condition, final-condition problems, must volumes always be in liters?

(d) When you are using the ideal gas equation in the form $PV = nRT$, volumes must always be in liters (never milliliters or anything else). Why?

The solution to Practice Problem 11.14 reveals that in initial-condition, final-condition problems, the volumes can be in any units. This is true because R is missing from the equation. The same is true for pressures and amounts of gas. However, *temperature must always be expressed in kelvins.*

We can get other interesting information about gases from the ideal gas equation. For example, we know that real gases act ideally under most circumstances—that is, real gases follow the ideal gas law to a good approximation. We can therefore use our ideal gas model to answer the question "What volume does 1 mol of any gas occupy at standard temperature and pressure?" (**Standard temperature and pressure**, also written **STP**, is defined as 0°C = 273.15 K and 1 atm pressure.) To find the answer, all we need to do is solve the ideal gas equation for volume and plug in the appropriate numeric values:

$$V = \frac{nRT}{P} = \frac{(1\text{ mol})\left(0.0821\frac{\text{L}\cdot\text{atm}}{\text{K}\cdot\text{mol}}\right)(273.15\text{ K})}{1\text{ atm}} = 22.4\text{ L}$$

This is, to a good approximation, the **molar volume** (the volume occupied by 1 mol) of *any* gas at STP. Thus 1 mol of H_2 gas at STP has a volume of 22.4 L. So does 1 mol of CO_2 gas at STP. So does 1 mol of O_2 gas at STP, and so on.

The upshot of all this is that *equal volumes of gas at the same temperature and pressure must contain equal numbers of moles*. For example, suppose we have one tank filled with a gas A and a second tank filled with a gas B. The tanks are the same size (same volume), in the same room (same temperature), and at the same pressure.

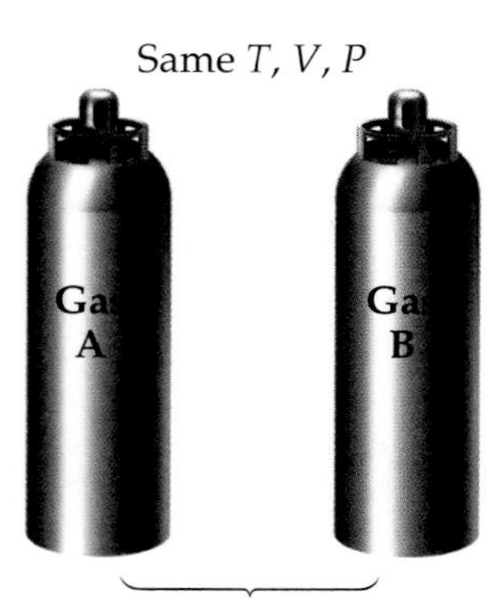

The two tanks must contain the same number of moles regardless of the identities of A and B.

They must therefore contain the same number of moles of gas. (The first person to demonstrate this phenomenon was Avogadro.)

Finally, we can even use the ideal gas equation for stoichiometry problems. This is useful when some of the reactants or products of a reaction are gases. For example, oxygen gas can be produced by heating solid potassium chlorate (a decomposition reaction, Chapter 7):

$$2\ KClO_3(s) \longrightarrow 2\ KCl(s) + 3\ O_2(g)$$

Suppose you are in a spaceship that has an internal volume of 1000.0 m^3. All your oxygen has leaked out, but luckily you have a stash of solid potassium chlorate on board. Your plan is to electrically heat some of it and thereby generate enough oxygen gas to fill the ship at STP. How much $KClO_3$ do you need? This being a stoichiometry problem, you can be sure the first step is to convert to moles, and we do this with the ideal gas equation, converting the volume of 1000.0 m^3 of oxygen gas at STP to moles:

$$n = \frac{PV}{RT} = \frac{(1\ \cancel{\text{atm}})\overbrace{(1000.0\ \cancel{\text{m}^3})\left(\frac{1000\ \text{L}}{\cancel{\text{m}^3}}\right)}^{V \text{ in liters}}}{\left(0.0821\ \frac{\text{L}\cdot\cancel{\text{atm}}}{\cancel{\text{K}}\cdot\text{mol}}\right)(273.15\ \cancel{\text{K}})} = 4.46\times 10^4\ \text{mol } O_2 \text{ gas}$$

Now we can use the balanced chemical equation (which "speaks" in moles) to finish the problem:

$$(4.46\times10^{4}\ \cancel{\text{mol}}\ \text{O}_2)\overbrace{\left(\frac{2\ \cancel{\text{mol}}\ \text{KClO}_3}{3\ \cancel{\text{mol}}\ \text{O}_2}\right)}^{\text{From balanced reaction}}\overbrace{\left(\frac{122.548\ \text{g KClO}_3}{\cancel{\text{mol}}\ \text{KClO}_3}\right)}^{\text{Molar mass of KClO}_3}=3.64\times10^{6}\ \text{g KClO}_3$$

That's over 4 tons of potassium chlorate! Maybe you should look for some spare oxygen tanks instead.

This chapter may seem very mathematical. We certainly gave the ideal gas equation, $PV = nRT$, a healthy algebraic workout. But remember, we derived this equation by visualizing a simple model of an ideal gas while asking three questions about pressure. Everything we did was based on that simple model. Chemists may forget equations and facts, but the underlying pictures and models stay with them forever, and this is where true understanding comes from.

Have You Learned This?

Ideal gas (p. 392)
Pressure (p. 394)
Normal atmospheric pressure (p. 396)
Pressure–volume relationship (p. 399)
Pressure–temperature relationship (p. 400)
Pressure–moles relationship (p. 401)
Ideal gas constant (p. 401)
Ideal gas equation (p. 401)
Ideal gas law (pp. 401)
Standard temperature and pressure (STP) (p. 413)
Molar volume (p. 414)

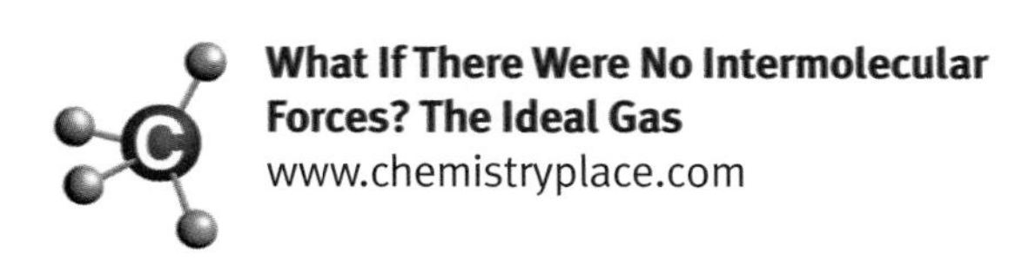

What If There Were No Intermolecular Forces? The Ideal Gas
www.chemistryplace.com

SKILLS TO KNOW

Using the ideal gas law

1. Four variables characterize an ideal gas: pressure (P), volume (V), amount of gas in moles (n), and temperature (T).

2. The relationships among P, V, n, and T can be summarized by the three relations involving pressure:

P is proportional to $\frac{1}{V}$

If you double the volume, the pressure drops by half. (The reverse is also true: If you halve the volume, the pressure doubles.)

P is proportional to T

If you double the absolute temperature, the pressure also doubles. (The reverse is also true: If you halve the absolute temperature, the pressure is halved.)

P is proportional to n

If you double the number of particles, the pressure also doubles. (The reverse is also true: If you halve the number of particles, the pressure is halved.)

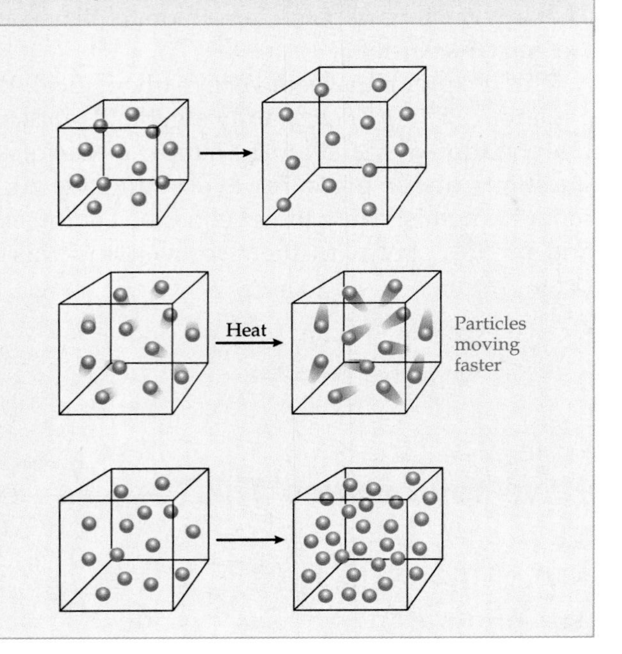

3. The **ideal gas equation** summarizes the above relationships:

$$P = \frac{nRT}{V}$$ where R is the ideal gas constant.

$$R = 0.0821 \frac{\text{L} \cdot \text{atm}}{\text{K} \cdot \text{mol}}$$

Always L and atm

Always K and moles

4. Any time you use a gas equation that contains R, the variables P, V, n, and T must have units that agree with those of R.

For the above value of R:
- P must be in atmospheres
- V must be in liters
- n must be in moles
- T must be in kelvins

5. You should feel comfortable with rearranging the ideal gas equation to solve for any of the variables.

$V = \frac{nRT}{P}$ Solved for V $n = \frac{PV}{RT}$ Solved for n and so on

6. Standard (sea level) atmospheric pressure is 760 mm Hg = 1 atm exactly. This relationship provides conversion factors for changing between atmospheres and millimeters of mercury.

Useful conversion factors:

$$\frac{760 \text{ mm Hg}}{1 \text{ atm}} \qquad \frac{1 \text{ atm}}{760 \text{ mm Hg}}$$

Using the ideal gas equation to identify an unknown gas

You can use the ideal gas equation to identify a pure sample of an unknown gas. First rewrite the equation so you can solve it for molar mass. To do that, replace n in the equation with m/MM, where m is the sample mass and MM is the molar mass.

Now measure P, V, and T for the gas under a given set of conditions, plus the mass m of the sample. Then solve for molar mass and use the molar mass value to figure out what the gas must be. Make sure you use units compatible with the units of R.

For any pure substance,

$$n = \frac{m}{MM}$$ where m is the mass of the sample in grams.

That is, if you divide the mass of the sample by the molar mass of the substance, you get the number of moles present.

Substituting this expression for n in the ideal gas equation and solving for molar mass, you get

$$MM = \frac{mRT}{PV}$$

Solving initial-condition, final-condition gas problems

Example: An air bubble rises through the ocean from a depth of 1000 feet to the surface. As it rises, its pressure decreases from 31.3 atm to 1.00 atm and its volume increases. Its initial volume is 5.00 mL. The bubble's temperature remains constant, and it gains and loses no gas. What is its final volume?

1. List P, V, n, and T for the initial and final conditions.

- Write in numerical values for any variables you know.
- Where possible, express the final variable in terms of the initial one. For instance, if volume does not change, you can say $V_f = V_i$.

Initial conditions	Final conditions
$P_i = 31.3$ atm	$P_f = 1.00$ atm
$V_i = 5.00$mL	$V_f = ?$
$n_i = ?$	$n_f = n_i$
$T_i = ?$	$T_f = T_i$

2. Write the expression

$$\frac{P_iV_i}{n_iT_i} = \frac{P_fV_f}{n_fT_f}$$

- Whenever possible, rewrite this expression to show final variables in terms of initial ones (from step 1).

$$\frac{P_iV_i}{n_iT_i} = \frac{P_fV_f}{n_iT_i}$$

Because $n_f = n_i$ and $T_f = T_i$, we can replace variables on the right side of the equation with those from the left side.

- Cancel quantities that are identical on the two sides of the expression.

$$\frac{P_iV_i}{\cancel{n_i}\cancel{T_i}} = \frac{P_fV_f}{\cancel{n_i}\cancel{T_i}} \longrightarrow P_iV_i = P_fV_f$$

3. Solve the equation algebraically for the desired variable.

$$P_iV_i = P_fV_f \longrightarrow V_f = \frac{P_iV_i}{P_f}$$

Then plug in values for the quantities you know and solve numerically for an answer.

$$V_f = \frac{P_iV_i}{P_f} = \frac{(31.3 \text{ atm})(5.00 \text{ mL})}{1.00 \text{ atm}} = 157 \text{ mL}$$

DESCRIBING THE GAS PHASE

11.15 Why is it proper to think of the gas phase of matter as being more chaotic than either of the condensed phases?

11.16 What assumption is made for an ideal gas, and what gives us the right to make that assumption?

11.17 Consider a container of gas with the pressure inside the container the same as the room pressure outside the container. If a tiny hole is punched in the side of the container, will the gas leak out? Explain your answer.

11.18 How does a gas create pressure?

11.19 Why does liquid rise up a straw when you suck on the liquid through the straw?

11.20 Describe how a mercury barometer works.

11.21 What is the value of normal atmospheric pressure in millimeters of mercury? In atmospheres?

11.22 True or false? 1 atm = 76 cm Hg.

11.23 A weather forecaster reports the barometric pressure as 29.7 inches of mercury. [1 in. = 2.54 cm]
(a) How many millimeters of mercury is this?
(b) How many atmospheres is this?

11.24 The pressure in a tank of oxygen is 2000.5 $lb/in.^2$. [760.00 mm Hg = 14.696 $lb/in.^2$]
(a) How many millimeters of mercury is this?
(b) How many atmospheres is this?

11.25 How do you convert from °C to K? Convert room temperature (22.0°C) to kelvins.

11.26 Convert −100.5°C to kelvins.

11.27 A balloon of methane gas, CH_4, has a temperature of −2.0°C and contains 2.35 g of the gas. What is the temperature of the gas in kelvins? How many moles of the gas does the balloon contain?

11.28 A tank of acetylene gas (C_2H_2) contains 48.5 lb of the gas and is at a pressure of 600.2 $lb/in.^2$. Express the pressure of the gas in atmospheres and the amount of gas in moles. [760.0 mm Hg = 14.696 $lb/in.^2$, 453.6 g = 1 lb]

DESCRIBING A GAS MATHEMATICALLY— THE IDEAL GAS LAW

11.29 What happens to the pressure of a gas when the volume of its container is increased? Explain with pictures and words.

11.30 What happens to the pressure of a gas when its temperature is decreased? Explain with pictures and words.

11.31 What happens to the pressure of a gas when the amount of gas inside the container is increased? Explain with pictures and words.

11.32 Consider the following diagrams representing different gas samples all at the same temperature:

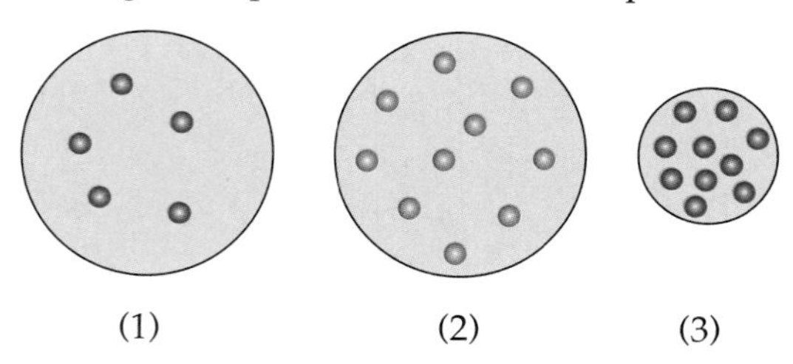

(a) Which gas is at the lowest pressure? Explain.
(b) Which gas is at the highest pressure? Explain.

11.33 Which of the two "gas" samples below is at the lower temperature? Explain, and also tell why we put the word *gas* in quotation marks.

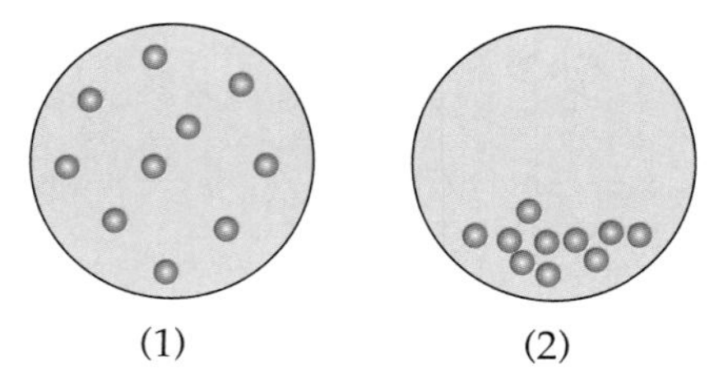

11.34 State how the pressure of a gas depends on its volume.

11.35 State how the pressure of a gas depends on its temperature.

11.36 State how the pressure of a gas depends on the amount of gas present.

11.37 What do we mean by inverse proportionality? By direct proportionality? Give an example of each using the way the pressure of a gas depends on something else.

11.38 Suppose the variable x is proportional to $1/y$. What does this tell you about how the numeric value of x changes as the numeric value of y changes?

11.39 "The older I get, the fewer hairs I have on my head." What kind of relationship (proportion or inverse proportion) exists between this gentleman's age and his hair? Explain your answer.

11.40 What kind of relationship (proportion or inverse proportion) exists between the strength of intermolecular forces and the distance between molecules? Explain your answer. [*Hint*: See Table 10.1.]

11.41 What are the units used in this book for the ideal gas constant R? Why are they important?

11.42 Rewrite the ideal gas law solving for V.

11.43 Rewrite the ideal gas law solving for T.

11.44 Rewrite the ideal gas law solving for n.

11.45 According to the ideal gas law, what should you get if you measure P, V, n, and T for a gas sample and then calculate the quantity PV/nT?

11.46 According to the ideal gas law, what would happen to the pressure of a gas if you doubled the amount of gas in a container while also decreasing the volume of the container to one-half its initial volume? Explain.

11.47 According to the ideal gas law, what would happen to the pressure of a gas if you doubled the amount of gas in a container while also tripling the kelvin temperature of the gas? Explain.

11.48 A student thinks he remembers reading that if you double the temperature of an ideal gas, its pressure doubles. He is given a problem where he has an ideal gas at 25.0°C and 2.5 atm. He is asked what the temperature must be raised to in order to double the pressure to 5.0 atm. He answers, "50.0°C, of course." Why is he wrong? What lesson should he learn about using the ideal gas law? What is the temperature increase in Celsius degrees that will double the pressure?

11.49 The gas inside a balloon is characterized by the following measurements: pressure = 745.5 mm Hg; volume = 250.0 mL; temperature = 25.5°C. What is the number of moles of gas in the balloon?

11.50 A gas is in a container whose volume is variable. The container is in an ice bath at 0.00°C, and there is 2.0 moles of gas in it. What must the volume in liters be adjusted to in order to produce a gas pressure of 2.5 atm?

11.51 What must the Celsius temperature be if 2.0 moles of a gas in a 4.0-L steel container has a measured pressure of 100 atm?

11.52 According to the ideal gas law, what would be the volume of a gas at absolute zero (0 K)? Does the answer make sense? Describe what really happens to any gas as it is cooled toward this temperature.

11.53 An automobile tire at 22°C with an internal volume of 20.0 L is filled with air to a total pressure of 30 psi (pounds per square inch). [1 atm = 14.696 lb/in.^{2}]
(a) What is the amount in moles of air in the tire?
(b) If the air were entirely nitrogen (N_2), how many grams of it would be in the tire? How many pounds of it would be in the tire? [453.6 g = 1 lb]

11.54 Why are the results calculated using the ideal gas law not exactly equal to the "true" results obtained by an experimental measurement?

11.55 Suppose you want to carry out the chemical reaction

$$H_2(g) + Cl_2(g) \longrightarrow 2\ HCl(g)$$

and have 1 mole of Cl_2 gas.
(a) What volume in liters of H_2 gas would you need to have 1 mole of H_2, given that the H_2 pressure and temperature are 1.00 atm and 22.5°C?
(b) What would the volume of the product be if it were collected at 1.00 atm and 22.5°C?

GETTING THE MOST FROM THE IDEAL GAS LAW

11.56 Given some mass m of a known pure substance, what is the quickest way to determine the number of moles n you have of the substance?

11.57 A 7.24-g sample of gas is contained in a 4.00-L flask. Its pressure is 765.0 mm Hg, and its temperature is 25.0°C. What is the molar mass of this gas?

11.58 A 1.56-g sample of gas is contained in a 250.0-mL cylinder. Its pressure is 1255.6 mm Hg, and its temperature is 22.7°C.
(a) What is the molar mass of the gas?
(b) Combustion analysis reveals the empirical formula of this gas to be NO_2. What is the molecular formula?

11.59 A balloon is filled with H_2 gas to a volume of 1610.2 mL. The pressure of the gas in the balloon is 745.4 mm Hg, and the temperature is 22.7°C. What is the mass in grams of the H_2 in the balloon?

11.60 In a room at 24.5°C, a balloon is filled with helium gas until the pressure inside the balloon is 768.0 mm Hg. What is the density of the gas in grams per liter and in grams per milliliter?

11.61 A balloon filled with He gas and another balloon filled with H_2 gas have the same values for P and T.
(a) The density of the He gas is greater than the density of the H_2 gas. How can you prove this using the ideal gas law?
(b) How much more dense than the H_2 gas is the He gas?

11.62 Carbon dioxide and carbon monoxide are very different gases. For example, you exhale CO_2 but CO is extremely toxic. Suppose you have two balloons, one filled with 1.00 mole of CO and the other filled with 1 mole of CO_2. Both gases are at 1.00 atm and 25.0°C.
(a) What is the volume in liters of each balloon?
(b) What did you learn about gases from doing this problem?

11.63 A tank of O_2 gas is stored in a cool basement (T = 10.0°C). Its pressure gauge reads 50.5 lb/in.^{2}. Diver Dan then places the tank in his truck, where the sun warms it to 20.0°C. What does the pressure gauge read at this elevated temperature?

11.64 In Problem 11.63, the temperature doubles but the pressure does not. Explain why.

11.65 In Problem 11.63, to what Celsius temperature would Diver Dan have to warm the tank in order to double the pressure?

11.66 A tank initially at a pressure of 6.70 atm contains 20.0 moles of oxygen gas at 25.0°C. Its temperature is then increased to 45.0°C, and an additional 10.0 moles of gas is added. What is the final pressure in the tank?

11.67 The Kelvin temperature of a gas is doubled. At the same time, its volume is cut in half by compressing it with a piston. What happens to the pressure of the gas? [*Hint*: Set $T_f = 2\,T_i$ and $V_f = 0.5\,V_i$.]

11.68 Consider the reaction

$$2\,H_2(g) + O_2(g) \longrightarrow 2\,H_2O(g)$$

If this reaction were done at STP, what volume in liters of H_2 gas would be required to react completely with 22.4 L of O_2 gas? What volume in liters of $H_2O(g)$ would be produced? Explain how you arrived at your answers.

11.69 Consider the reaction

$$2\,SO_2(g) + O_2(g) \longrightarrow 2\,SO_3(g)$$

(a) What volume in liters of oxygen gas at STP is required to produce 2 moles of sulfur trioxide gas?
(b) What volume in liters of oxygen gas at 25.0°C and 1 atm pressure is required to produce 2 moles of sulfur trioxide gas?

11.70 To produce 500.0 g of gaseous H_2O, what volumes in liters of hydrogen gas and oxygen gas, both at STP, must combine via the reaction

$$2\,H_2(g) + O_2(g) \longrightarrow 2\,H_2O(g)$$

11.71 Zinc metal reacts with hydrochloric acid to form hydrogen gas via the reaction

$$Zn(s) + 2\,HCl(aq) \longrightarrow ZnCl_2(aq) + H_2(g)$$

How many grams of zinc would be required to produce enough hydrogen gas to fill a 50.0-L cylinder to a pressure of 10.0 atm at 25.0°C?

ADDITIONAL PROBLEMS

11.72 Consider two samples of water vapor, one at 101°C and the other at 200°C. Which behaves more ideally and why?

11.73 According to the ideal gas model, why does the chemical identity of a gas not matter when it comes to producing pressure?

11.74 A 5.00-L container is filled with 0.8004 g of helium gas at 107.0°C. What is the pressure of the gas in atmospheres and in millimeters of mercury?

11.75 A chemical reaction produces 100.0 mL of hydrogen gas at 745.5 mm Hg and 24.0°C. What is the amount of H_2 produced in moles and in grams?

11.76 What is the density in grams per liter of nitrogen gas at STP?

11.77 A gaseous compound of carbon and hydrogen contains 80.0% by mass carbon. A 2.00-L sample has a mass of 2.678 g at STP.
(a) Calculate the molar mass of the compound.
(b) What is the empirical formula of the compound?
(c) What is the molecular formula of the compound?

11.78 When it is heated, potassium chlorate decomposes to potassium chloride and oxygen:

$$2\,KClO_3(s) \longrightarrow 2\,KCl(s) + 3\,O_2(g)$$

A sample of solid $KClO_3$ decomposes to produce 150.0 mL of oxygen at 22.0°C and 780.5 mm Hg. How many grams of $KClO_3$ decomposed?

11.79 To what temperature must a gas initially at 20.0°C be heated to double the volume and triple the pressure?

11.80 A sample of argon at 0.0°C and 700.0 mm Hg occupies 15.0 mL. What is its volume at STP?

11.81 A sample of gas at 25.0°C and 655 mm Hg has a density of 2.26×10^{-3} g/mL. What is the molar mass of the compound?

11.82 In the Haber process, nitrogen reacts with hydrogen to produce ammonia:

$$N_2(g) + 3\,H_2(g) \longrightarrow 2\,NH_3(g)$$

(a) Suppose 2.0 L of N_2 gas at STP is combined with 6.0 L of H_2 gas, with the two gases at the same temperature and pressure. Is this reaction being run in a balanced fashion or in a limiting fashion? Explain how you can tell without doing any calculations.
(b) If 50.0 L of N_2 gas at 200.0 lb/in.^{2} and 22.0°C is combined with 100.0 L of H_2 gas at 240.0 lb/in.^{2} and 22.0°C, what mass in grams of ammonia is produced? [14.70 lb/in.^{2} = 760.0 mm Hg]

11.83 How many atoms of nitrogen are there in a 230.0-mL sample of N_2O_4 gas that has a pressure of 745.0 mm Hg at 34.0°C?

11.84 A sample of gas at 25°C and 759.0 mm Hg has a volume of 1.58 L. If the temperature is raised to 35°C but the pressure is held constant at 759.0 mm Hg, will the volume increase or decrease? Explain your answer.

11.85 Consider two identical 1-L containers, both at room temperature (300 K). One of them contains 1 mole of helium gas, and the other contains 1 mole of hydrogen gas. Is the pressure higher in the helium container, higher in the hydrogen container, or the same in the two containers?

11.86 Consider two identical 1-L containers, both at room temperature (300 K). One of them contains 1.0 g of helium gas, and the other contains 1.0 g of hydrogen gas. Is the pressure higher in the helium container, higher in the hydrogen container, or the same in the two containers?

11.87 What volume in milliliters of N_2O_4 gas at 0.996 atm and 25.0°C contains 0.200 g of oxygen?

11.88 Carbon dioxide in a gas cylinder of unchangeable volume is at a pressure of 25.0 atm at 25°C. When placed in the sunlight on a hot summer day, the temperature increases to 40°C. What is the new pressure of the CO_2?

11.89 When a balloon is placed in a freezer, the size of the balloon decreases. Explain why.

11.90 A sample of Ne gas in a 10.0-L container at 25.0°C exerts a pressure of 7.35 atm. What is the number of moles of Ne in the container? What is the mass in grams of the Ne?

11.91 Of the values for P, V, n, and T, which ones remain constant during the following changes?
(a) Air in a sealed glass bulb is heated from 25°C to 32°C.
(b) A He-filled balloon is cooled from 23°C to 10°C.
(c) The amount of CO_2 in a sealed steel cylinder is increased by adding 50.0 g of CO_2.

11.92 Under what two conditions are gas molecules unable to overcome the weak attractive forces that exist between them?

11.93 Arrange these gases, all at STP, in order of increasing density: CO_2, H_2, O_2, CH_4, He.

11.94 If a gas sample occupies 1.80 L at 250°C and 792 mm Hg, what is the number of moles in the sample?

11.95 Which variable is not needed to describe the behavior of an ideal gas: volume, number of moles, temperature, molar mass, or pressure?

11.96 (a) How many liters of O_2 are there in 5.38 moles at STP?
(b) How many moles of NH_3 are there in 859 mL at STP?
(c) How many moles of H_2 are there in 0.518 L at STP?

11.97 Fill in the missing values:

	P	V	n	T
Gas A	757 mm Hg	952 mL	0.300	?°C
Gas B	1.20 atm	? L	5.00	27.0°C
Gas C	800.0 mm Hg	1.20 L	?	298 K
Gas D	? atm	750 mL	0.0875	0.0°C

11.98 Are gases denser or less dense than liquids and solids? Explain.

11.99 If a gas occupies 2.40 L at 1 atm and 22°C, at what Celsius temperature will it occupy 7.20 L at 1 atm?

11.100 A chemistry student realizes she has forgotten the value of R and needs to determine it experimentally. She decides to measure the mass, volume, pressure, and temperature of a sample of carbon dioxide gas. She finds that 4.505 g of the gas occupies 2.50 L at 23°C and 0.9960 atm. Calculate the value of R she determines from these data.

11.101 A gas at 25.0°C occupies a volume of 5.00 gallons and exerts a pressure of 755 mm Hg. Express (a) the temperature of the gas in kelvins, (b) the volume of the gas in liters, and (c) the pressure of the gas in atmospheres.

11.102 In the ideal gas law, pressure must be expressed in units of ___________, volume must be expressed in units of ___________, temperature must be expressed in ___________, and n represents the number of ___________.

11.103 A radial tire has an air pressure of 2.20 atm at 32°F. After hours of high-speed driving, the temperature of the air in the tire reaches 80°F. What is the pressure in the tire at this temperature? Assume that the volume of the tire does not change.

11.104 An 84.0-g sample of an unknown gas occupies 60.8 L at 23°C and 1.2 atm. What is the molar mass of the gas?

11.105 True or false? In order to determine the volume occupied by 3.0 moles of an ideal gas at 28°C and 1.4 atm, the identity of the gas must be known. If the statement is false, explain why.

11.106 What is the density of He gas in a cylinder if the pressure of the gas is 15.2 atm and its temperature is 22°C?

11.107 In the laboratory preparation of oxygen gas, the following reaction is carried out:

$$2\,KClO_3(s) \longrightarrow 2\,KCl(s) + 3\,O_2(g)$$

(a) How many moles of oxygen gas can be produced from the decomposition of 500.0 g of $KClO_3$?
(b) What volume in liters would the oxygen gas produced in part (a) occupy at 24.0°C and 750.0 mm Hg?
(c) What volume in liters would this gas occupy at STP?

11.108 In an ideal gas, the molecules are considered to have no ___________ forces.

11.109 (a) If the temperature of a gas is doubled while the pressure is kept constant, the volume of the gas ___________.

(b) If the pressure of a gas is halved while the temperature is kept constant, the volume of the gas ___________.

11.110 What is the Celsius temperature of a gas if 3.200 moles of it occupies 12.00 L at 8.500 atm?

11.111 A 1.25-g sample of a gas of unknown identity occupies 2.50 L and exerts a pressure of 0.400 atm at 0°C.
(a) What is the molar mass of the gas?
(b) Combustion analysis determines that the empirical formula of the gas is CH_2. What is its molecular formula?

11.112 If a quantity of gas occupies 850 mL at 300 K and 750 mm Hg, what volume in milliliters will the gas occupy at 200 K and 1200 mm Hg?

11.113 Perform the following conversions:
(a) 30.2 in. Hg to millimeters of mercury
(b) 890.0 mm Hg to atmospheres
(c) 300.0 lb/in.^{2} to atmospheres

11.114 If 8.50 moles of He gas occupies a volume of 25.0 L at 28°C, what pressure in atmospheres does the gas exert?

11.115 If 48.3 g of an unknown gas occupies 10.0 L at 40°C and 3.10 atm, what is the molar mass of the gas?

11.116 A gas tank contains CO_2 at a pressure of 6.80 atm. What would the CO_2 pressure be if the container were (a) twice as large and (b) one-fourth as large?

11.117 What is the molar mass of a gas whose density is 1.52 g/L at 0°C and 1 atm pressure?

11.118 Explain why gases always occupy the entire container they are in but solids and liquids do not.

11.119 What happens to the volume of a gas in a cylinder with a movable piston if
(a) The pressure is doubled while the temperature is held constant?
(b) The temperature is doubled while the pressure is held constant?
(c) The pressure and the temperature are both doubled?
(The movable piston means the volume of the cylinder, and therefore of the gas, can change.)

 WORKPATCH SOLUTIONS

11.1 In the condensed phases, the molecules are very close to one another, and as a result the intermolecular forces (no matter what type) are relatively strong and thus can't be ignored. Remember, attractive forces between molecules are stronger when the molecules are closer together.

11.2 The pressure in the smaller box is greater because, in any given time interval, there are more particle–wall collisions in the smaller box than in the larger box.

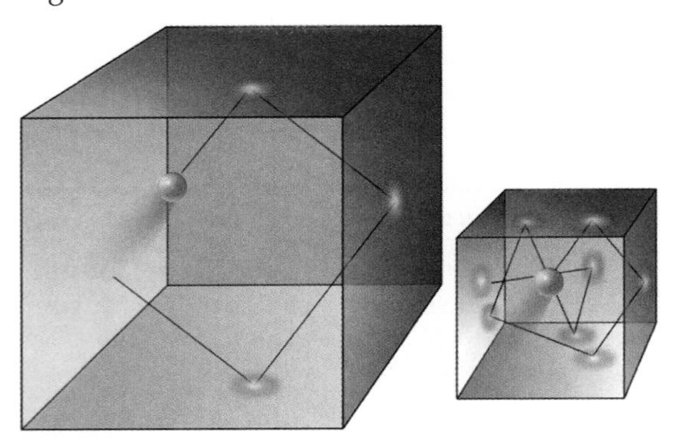

11.3
$$\frac{(100\ \text{atm})(100.0\ \text{L})}{(12.992\times10^3\ \text{g O}_2)\left(\dfrac{1\ \text{mol O}_2}{31.998\ \text{g O}_2}\right)(300.0\ \text{K})}$$

$$=\underbrace{0.08210\ \frac{\text{L}\cdot\text{atm}}{\text{K}\cdot\text{mol}}}_{\text{This equals } R!}$$

solvent
separation
ΔE solute
separation
stem energy
ΔE total
system energy
solutions

CHAPTER 12

Solutions

12.1 What Is a Solution?

The next time you're at the supermarket, take a look at the detergent aisle. You'll find a huge array of products, all claiming that when added to water, they have the muscle to remove dirt and stains. Actually, water by itself does a pretty good job on most household dirt and stains. Consider a puddle of grape juice that has dried on a kitchen counter. Strenuous wiping with a dry cloth will generally not remove the dried juice, but a gentle wipe with a water-dampened cloth does the trick. The purple color on the cloth reveals that the juice has dissolved in the water. It appears that water's stain-removing muscle is due to its ability to dissolve things.

Of course, there are some stains that water alone will not remove. Recall, for instance, that salad-oil stain on your new T-shirt. In this chapter we are going to answer two questions: (*1*) What exactly happens when a substance dissolves? and (2) Why is water so good at dissolving some things and so poor at dissolving others?

A **solution**, as we saw in Chapter 1, is a homogeneous mixture of two or more substances. By *homogeneous* we mean that the composition is the same throughout the solution. This means that when table sugar, for example, is dissolved in water, every tiny volume of the solution contains the same number of water molecules and the same number of sugar molecules. Another way to say this is that the solution has the same *concentration* (the same amount of sugar per amount of water) everywhere.

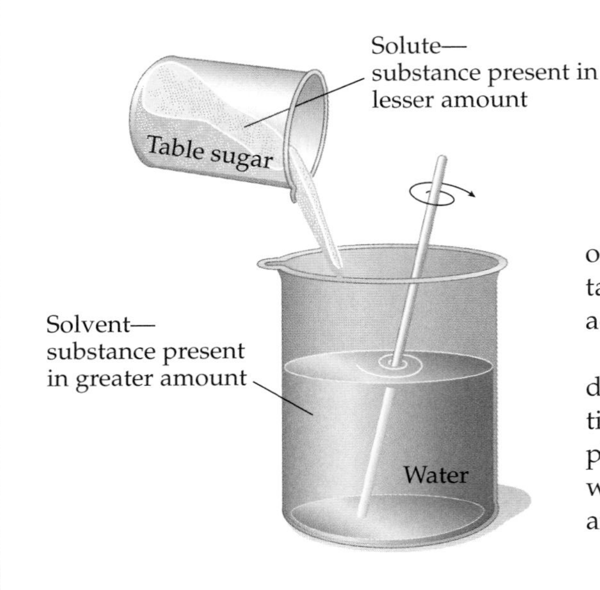

In any solution, the substance present in the greatest amount is called the **solvent** and all other substances in the mixture are called **solutes**. Thus, in our sugar–water example, water is the solvent and sugar the solute. Knowing this, we can now refine our definition to say that a solution is a homogeneous mixture of one or more solutes dissolved in a solvent. It is also important to note that once a solution forms, it will not of its own accord separate back into its components.

The word *solution* often brings to mind one or more solids dissolved in a liquid, but not all solutions are like that. Solutions can also be combinations of liquids. Vinegar, for example, is a solution of the liquid called acetic acid dissolved in water. Because vinegar is mostly water, water is the solvent and acetic acid the solute.

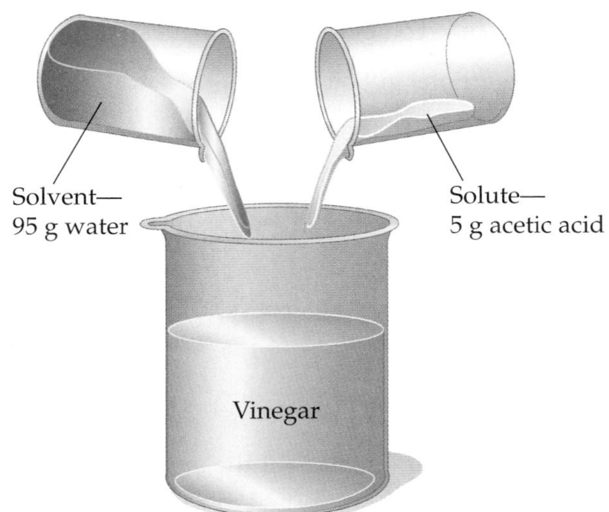

Solution of a liquid in a liquid

Solutions can be more exotic yet. Many metal alloys, for instance, are *solid solutions* consisting of one metal dissolved in another. The solutions are made when the metals are molten (liquid). For example, sterling silver is a solid solution consisting of a small amount of copper (the solute, Cu) dissolved in silver (the solvent, Ag). If the copper atoms are not evenly distributed among the silver atoms, the result is not sterling silver and is not a solution.

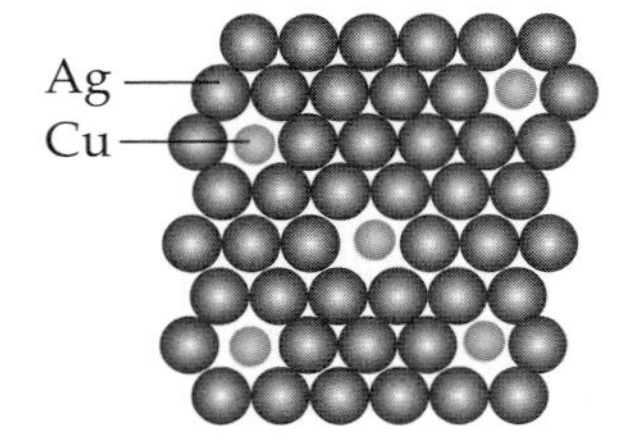

Sterling silver, a solid solution. The Cu atoms are evenly distributed among the Ag atoms.

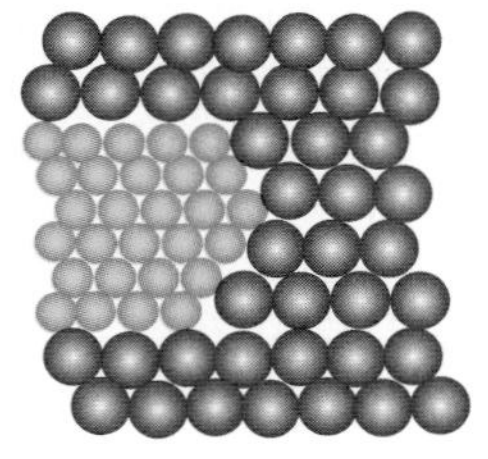

A heterogeneous mixture of copper and silver. This is not a solid solution because the Cu atoms are not evenly distributed among the Ag atoms.

As mentioned in Chapter 1, the air you are breathing is a solution of gases, primarily O_2 dissolved in N_2. Nitrogen, at 78%, is considered the solvent, and oxygen (21%) is the principal solute; other solute gases are also present in smaller amounts.

In most of the solutions we consider in this chapter, water is the solvent. Solutions in which water is the solvent are called **aqueous solutions** (from the Latin *aqua*, "water").

In ending this opening section, we want to look once more at the meaning of *homogeneous* in this context. This topic was covered in Chapter 1, but it is worth another look. You could put silicon dioxide (sand), SiO_2, sodium carbonate, Na_2CO_3, and calcium oxide, CaO, in a mortar and grind the mixture to a fine, pulverized dust. Even if you did this for a thousand years, the result would not be a true solution. It would be a heterogeneous mixture because there would still be tiny grains of pure silicon dioxide, pure sodium carbonate, and pure calcium oxide. Only by melting the components and stirring well would you achieve a molten solution that would become a solid solution (glass) if you allowed it to cool and solidify. The same is true for a mixture of table salt and table sugar. No amount of grinding these two solids together will yield a true solution, but dissolve them both in the same beaker of water and you get a homogeneous mixture—a solution—containing two solutes. Note that for both these examples, we needed a liquid phase to achieve homogeneity. Simply grinding two solids together just won't do the job. (You can also mix different gases together and get a homogeneous mixture—a solution—as we illustrated with air.)

12.1 WORKPATCH

From what you learned about the different phases of matter in Chapter 10, why do you think it is necessary to employ the liquid or gas phase to obtain the homogeneity needed to form a solution?

The models for liquids and gases used in Chapters 10 and 11 should help you understand what it takes to create a solution. The answer to WorkPatch 12.1 illustrates that chemistry works like a jigsaw puzzle, with the concepts and models you learn at different points fitting together to give you a better understanding of the whole.

Practice Problems

12.1 Classify the following as solutions or heterogeneous mixtures:
(a) A hot cup of instant coffee
(b) Chicken vegetable soup
(c) Unfiltered blood
(d) Filtered blood plasma
(e) A chromium-plated steel automobile bumper
(f) A stainless steel automobile bumper (stainless steel is prepared by combining iron, Fe, up to 30% chromium, Cr, and smaller amounts of nickel, Ni, and carbon, C, and heating the mixture until it becomes molten)

Answers: (a), (d), and (f) are solutions. (b), (c), and (e) are heterogeneous mixtures. (Unfiltered blood contains solids that separate out when the mixture is left undisturbed.)

12.2 Identify the solvent and solute or solutes in each solution:
 (a) Nailpolish remover (30% acetone in water)
 (b) Humid air
 (c) Stainless steel (see Practice Problem 12.1)
 (d) Aqueous solution of aspirin

12.3 The "proof" value of any liquor is equal to twice the percentage of alcohol; for example, a 50-proof liquor is 25% alcohol in water. Vodka is normally sold between 80 and 100 proof, but suppose you came across a bottle of 135-proof vodka. Would you be justified in calling the alcohol the solvent and the water the solute? Explain.

12.2 Energy and the Formation of Solutions

Liquid water is an excellent solvent, capable of dissolving many substances (solutes). The human body is approximately 80% water. Blood and cellular liquids are mostly water with small amounts of dissolved solutes (salts, gases, proteins, enzymes, and so forth). So many different substances are soluble in water that it is often referred to as a *universal solvent*.

What is it about water that makes it such a good solvent? To answer this question, let's reexamine the water molecule. We've seen that it is a polar molecule, a consequence of its bent shape and of the large difference in electronegativity between its oxygen and hydrogen atoms:

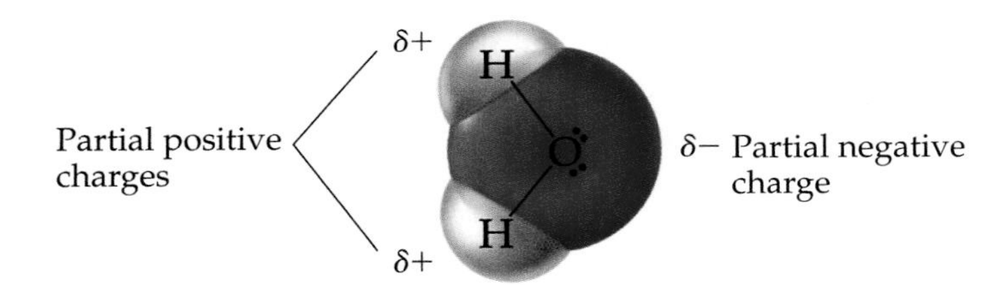

This polarity combined with the almost-naked-proton character of the H atoms leads to formation of the exceptionally strong dipolar attractions called hydrogen bonds (Chapter 10).

We saw in Chapter 10 that water molecules in the liquid state are in constant motion, moving in random directions but always strongly attracted to one another, always touching one another, jostling past one another. Remember the analogy of a crowd during a sale in a shopping mall?

Now let's consider what happens when a solute is dissolved in the water. Because the ionic compound sodium chloride, NaCl, dissolves easily in water, we'll use it as our solute. We've seen that solid sodium chloride crystals are made up of relatively stationary Na^+ and Cl^- ions in a repeating three-dimensional pattern called a lattice. This lattice is held together by ionic bonds. This is why solid NaCl has to be heated to the extraordinarily high temperature of 801°C before it will melt, becoming a liquid of mobile Na^+ and Cl^- ions. Yet, when we put this same compound in water at room temperature, the lattice comes apart easily to produce a solution of mobile Na^+ and Cl^- ions in water.

How can water break the ionic bonds of the crystal at room temperature (22°C) when it takes 801°C to do the same thing without the water? To answer this question, we need to break the dissolving process down into its elementary steps.

Three steps must occur simultaneously in order for NaCl to dissolve in water. (For the sake of clarity, we now talk about these steps one at a time, but always remember—they are taking place *simultaneously*.) First, the NaCl lattice has to be broken apart into individual Na^+ and Cl^- ions. We'll call this the *solute-separation step*. For NaCl, this breaking apart means overcoming the attractive forces (ionic bonds) between the ions in the lattice. Because ionic bonds are strong, this step requires an input of energy.

Second, spaces must be opened up in the solvent to make room for the solute particles. We'll call this the *solvent-separation step*. Recall that the solvent molecules are attracted to one another and touching one another. For a solute particle to dissolve, therefore, the solvent molecules have to move away from one another. Because this moving apart requires overcoming the attractions between solvent molecules, this step also requires an input of energy.

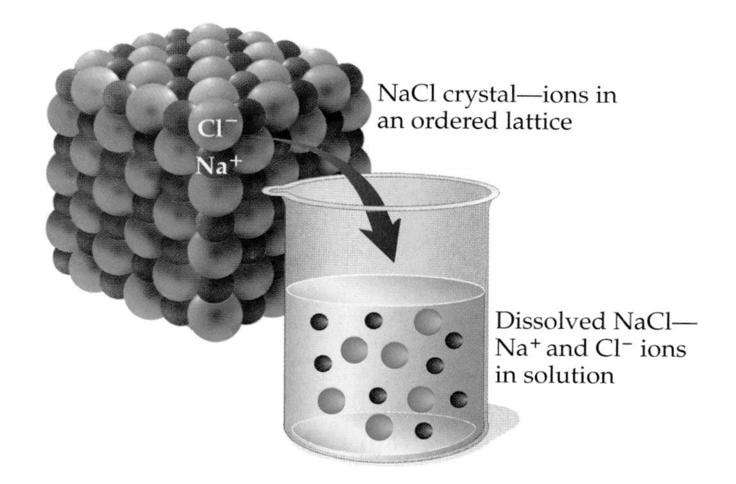
Cl⁻
Na⁺
NaCl crystal—ions in an ordered lattice
Dissolved NaCl— Na⁺ and Cl⁻ ions in solution

Solute-separation step: Freeing ions from the crystal lattice of the solute

Energy must be supplied to break the ionic bonds holding the ions in the lattice.
NaCl
Cl⁻
Na⁺
Na⁺

Solvent-separation step: Making room in the solvent

Energy must be supplied to overcome the attractive interactions between solvent molecules.

The third simultaneous step—called *solvation*—places the individual Na^+ and Cl^- ions into the spaces created in the solvent. Now the charged solute ions can interact with the polar solvent molecules that surround them, resulting in attractive forces called **ion–dipole forces**. Energy is *released* in this step because energy is released anytime an attractive force develops.

Solvation step: Formation of attractive forces between solvent particles and solute particles

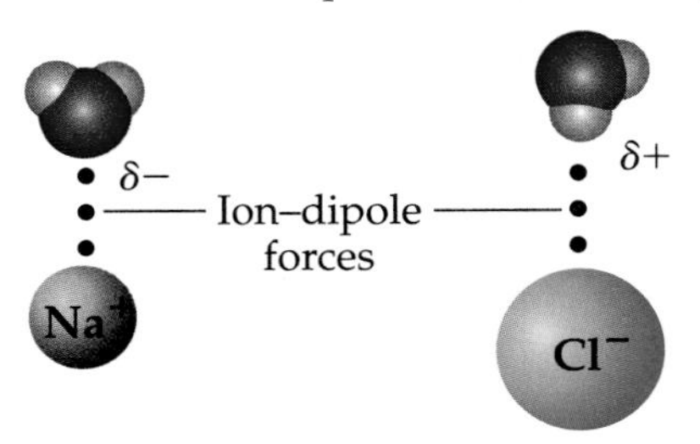

In the above figure, the water molecules are drawn so that an oxygen atom faces the Na^+ ion and a hydrogen atom faces the Cl^- ion. Explain why.

The WorkPatch answer explains why the Na^+ and Cl^- ions are attracted to the water molecules. In fact, there is room for more than one water molecule about each ion. Both Cl^- and Na^+ surround themselves with up to six water molecules, making for many ion–dipole attractions per ion:

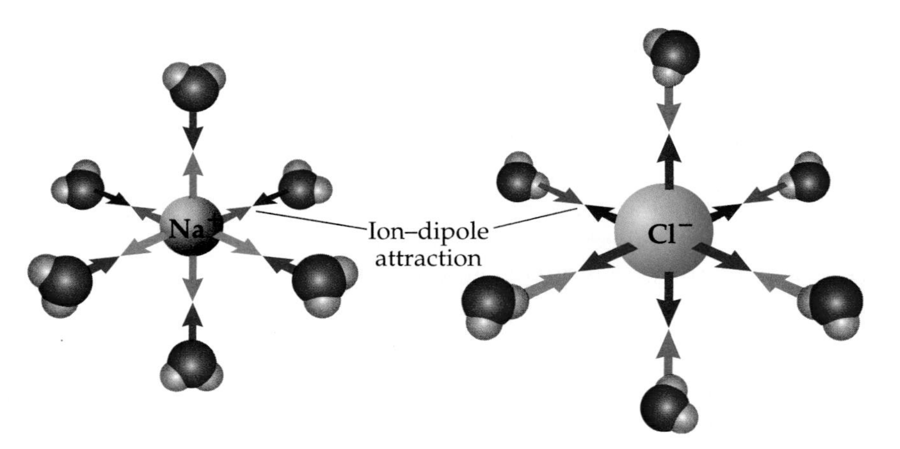

This means that each Cl^- ion and each Na^+ ion is surrounded by a shell of water molecules. The surrounding of solute ions by many solvent molecules along with the development of attractive solvent–solute forces is called **solvation** (hence the name of this step). When the solvent is water, this phenomenon can also be called **hydration**, so this step can be called the *hydration step* as well. When we write the equation

$$NaCl(s) \xrightarrow{H_2O} Na^+(aq) + Cl^-(aq)$$

the symbol (*aq*), standing for "aqueous," reminds us that each dissolved ion is surrounded by and attracted to many water molecules. The ions have been solvated (hydrated).

The hydration step is different from the other two. Solute separation and solvent separation require overcoming attractive forces and thus an energy input. However, energy is released anytime attractive interactions form, which means energy is released as the ions are hydrated. The energy released by this hydration is called—you guessed it!—**hydration energy**. The fact that this step in the dissolving process releases energy is very important because it supplies the bulk of the energy required by the other two steps. However, hydration can supply sufficient energy only if the solvent–solute attractive forces formed are strong enough—that is, only if their formation liberates enough energy to do the job.

We can now finally address the question "When does a solute dissolve in a solvent?" The answer comes from looking at the relative magnitudes of the energies involved in the three steps. The short answer is that the hydration energy released in the third step must be large enough to supply the energy needs of the first two steps. The following problems will give you some practice in determining the magnitude of the hydration energy. As you work them, keep in mind that

When two objects of opposite charge attract each other, the attractive force between them increases as the magnitude of the charges increases and as the distance between the objects decreases.

This fundamental law is called *Coulomb's law* and is very useful for explaining many phenomena, including hydration energy.

Practice Problems

12.4 Consider the ionic compound magnesium chloride, $MgCl_2$. Do you think the hydration energy for this compound is greater than, less than, or about equal to that of NaCl?

Answer: The hydration energy released for $MgCl_2$ is greater than that for NaCl because $MgCl_2$ contains Mg^{2+} ions. The attractive force between oppositely charged particles increases as the magnitude of the charges increases. The +2 charge, which is greater than the +1 charge of the Na^+ ions in NaCl, leads to stronger ion–dipole attractions with water molecules, releasing more hydration energy than for NaCl.

12.5 Imagine you are trying to dissolve NaCl in liquid carbon tetrachloride, CCl_4. Would the energy released in the solvation step be greater than, less than, or about equal to that released when NaCl dissolves in water? Explain.

12.6 Based on the answer to Practice Problem 12.4, is the energy required for solute separation greater for NaCl or $MgCl_2$? Explain.

Very often the combination of a solute (or solutes) and a solvent is referred to as a *system*. In considering the dissolving process for any particular solvent/solute system, what is important is how the energy of the system changes. By examining the change in energy at each step, we can determine whether a solute dissolves in a solvent. The symbol we use to represent the system's energy changes is ΔE. (Recall from Section 5.6 that the Greek letter delta, Δ, means "change in.")

When we feed energy into a system (as we do for the solute-separation and solvent-separation steps), the energy of the system increases, and we represent this increase with a positive ΔE value. When the system releases energy (as it does in the hydration step), the energy of the system decreases, and we represent this decrease with a negative ΔE value. In the following energy analysis for forming a solution, we use a schematic diagram to represent the process, with the overall process represented by two filled circles:

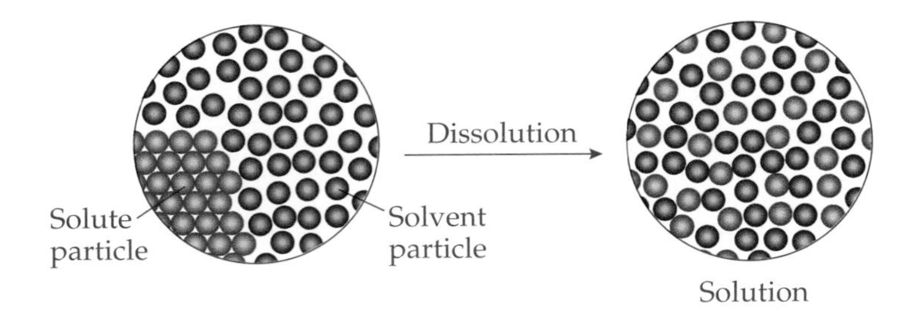

Solute-separation step: We must overcome the solute–solute attractive forces. Because this takes energy, we must feed energy into the system and therefore ΔE is positive. Note that if the solute is a liquid, the energy requirement is much less than for an ionic solid (where the lattice must be separated into ions); for a gas, the energy needed in this step is essentially zero.

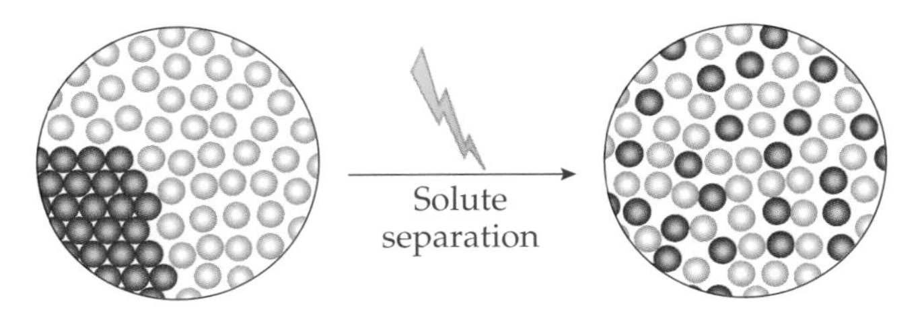

Solvent-separation step: We must overcome the solvent–solvent attractive forces and pry solvent particles apart to make room for solute particles. This takes energy, which means again ΔE is positive.

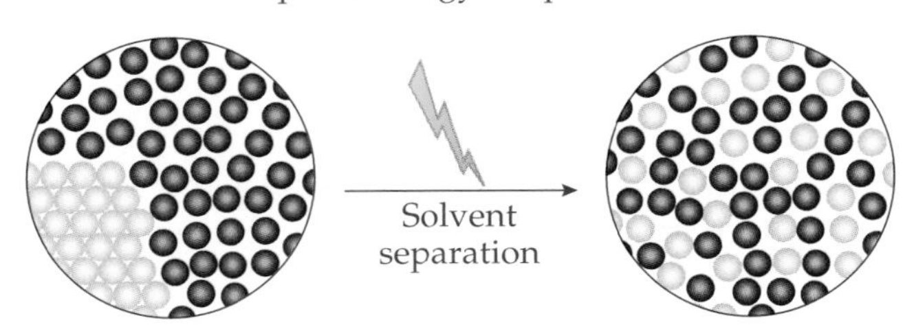

Solvation step: As solute–solvent attractive forces develop, energy is released, making ΔE negative.

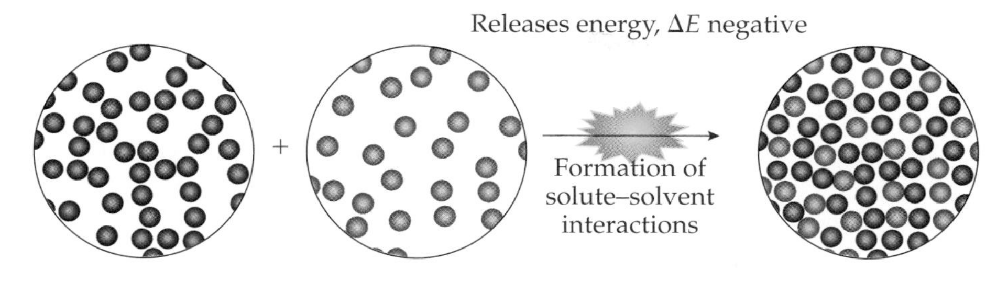

The combined energy changes are represented on an energy diagram:

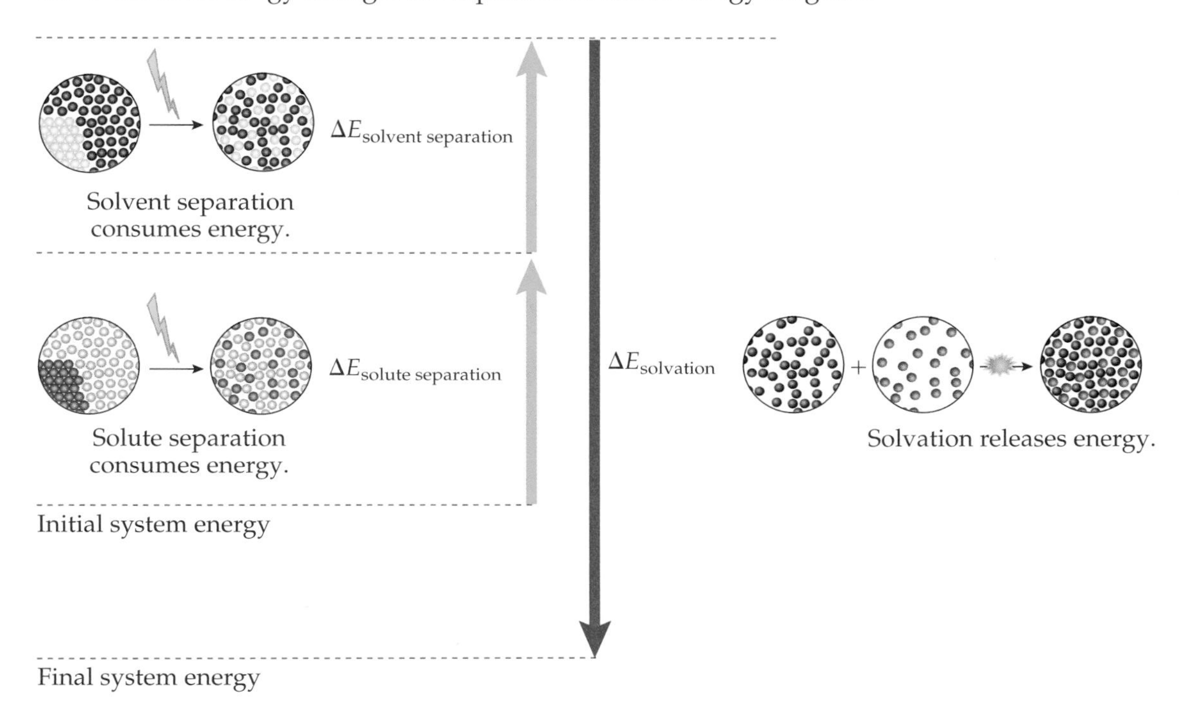

Whether or not a solute dissolves in a solvent depends on the total energy change for the process. This total energy change is the algebraic sum of the energy changes for the three steps:

$$\Delta E_{\text{total}} = \Delta E_{\text{solute separation}} + \Delta E_{\text{solvent separation}} + \Delta E_{\text{solvation}}$$

$\Delta E_{\text{solvent separation}}$

$\Delta E_{\text{solute separation}}$

$\Delta E_{\text{solvation}}$

Initial system energy

ΔE_{total} ← If there is a net release of energy as shown here, the solute dissolves in the solvent.

Final system energy

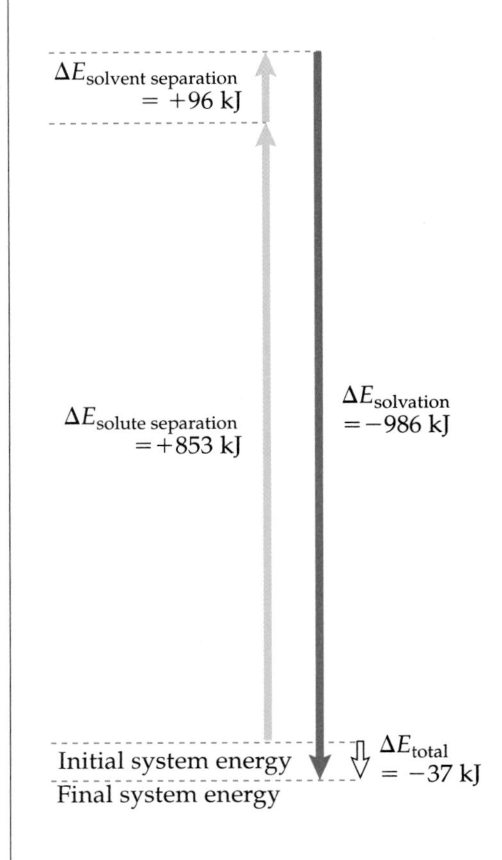

For example, when one mole of LiCl dissolves in water (shown at left),

$$\Delta E_{solute\ separation} = +853 \text{ kJ}$$
$$\Delta E_{solvent\ separation} = +96 \text{ kJ}$$
$$\Delta E_{solvation} = -986 \text{ kJ}$$
$$\Delta E_{total} = 853 \text{ kJ} + 96 \text{ kJ} + (-986 \text{ kJ}) = -37 \text{ kJ}$$

We get a negative value for ΔE_{total}, which means that, overall, energy is released. In other words, the system's energy decrease in the solvation step is more than its energy increase in the solute-separation and solvent-separation steps combined. In this case, the solvation step provides plenty of energy to drive the solute-separation and solvent-separation steps. In fact, solvation provides 37 kJ more energy than is needed. As a general rule, we can say

If $\Delta E_{total} < 0$, the substance will dissolve.

Dissolving LiCl in water results in an energy surplus, but this is not always the case. Sometimes the increase in energy during the solute-separation and solvent-separation steps is greater than the decrease in energy during the solvation step. The result is an energy deficit, which means that solvation can't feed enough energy to the two separation steps to make them occur. The result is that ΔE_{total} is positive (overall, energy is absorbed), and the substance probably does not dissolve. As a general rule, we can say

If $\Delta E_{total} > 0$, the substance will probably not dissolve.

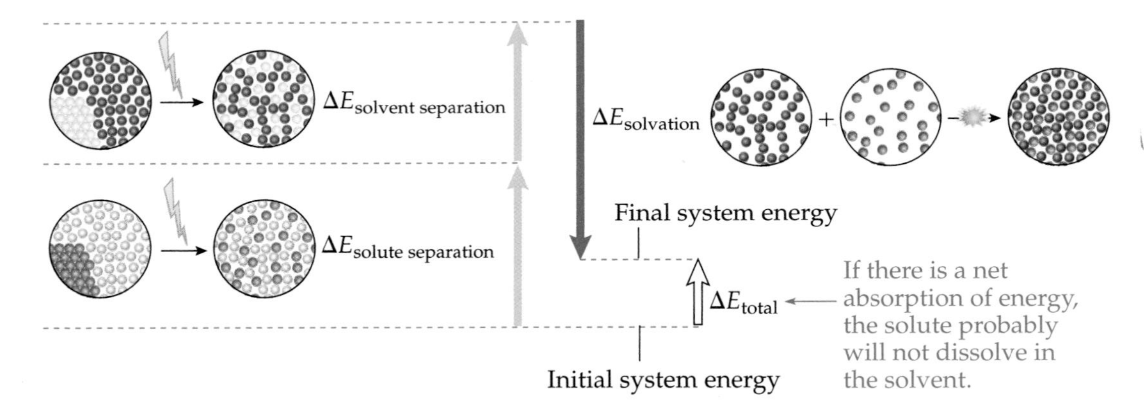

Why "probably"? We shall answer that question in Section 12.4.

Practice Problems

12.7 For a given solute in water, the energy changes are $\Delta E_{solute\ separation} =$ 835 kJ, $\Delta E_{solvent\ separation} = 98$ kJ, and $\Delta E_{solvation} = -805$ kJ. Will this solute dissolve in water? Explain your answer, both numerically and in terms of the three-step model for a solute's dissolving in a solvent.

Answer: $\Delta E_{total} = +835 \text{ kJ} + 98 \text{ kJ} + (-805 \text{ kJ}) = +128$ kJ. *Because* ΔE_{total} *is positive*

(there is an energy deficit), the solute probably will not dissolve. The first two steps take energy (both energies are positive). The third step releases energy but not enough.

12.8 Suppose we want to dissolve a gaseous solute in water. Would you expect $\Delta E_{\text{solute separation}}$ to be larger for the gaseous solute or for a solid ionic solute? Explain.

12.9 The more negative $\Delta E_{\text{solvation}}$ is, the more likely a solute will dissolve. Explain.

12.10 Suppose you want to dissolve some $MgCl_2$ in water.
(a) Why do the magnesium cations have a +2 charge?
(b) $\Delta E_{\text{solute separation}}$ for $MgCl_2$ is much more positive than $\Delta E_{\text{solute separation}}$ for NaCl. Explain what this means and why it might be so.
(c) $MgCl_2$ is more soluble in water than NaCl. Explain how this is possible in light of the information given in (b). [*Hint:* Consider the answer to Practice Problem 12.4.]

A general rule for determining whether a solute will dissolve is often expressed as "like dissolves like," where *like* refers to a similarity in polarity. What this rule means is that polar solutes dissolve best in polar solvents and nonpolar solutes dissolve best in nonpolar solvents. This is reasonable because it is most likely for the solvation step to release enough energy to feed the two separation steps when the solvent particles and the solute particles have similar polarities. For example, NaCl, a very polar substance, dissolves in polar solvents such as water because the ion–dipole interactions of the solvation step help compensate for the energy requirements of the two separation steps. What if we try to dissolve NaCl in a nonpolar solvent like carbon tetrachloride, CCl_4, though? What happens?

For the (attempted) dissolution of NaCl in CCl_4, choose the best answer for the energy required in each step. [*Hint:* Consider your answer to Practice Problem 12.5.]

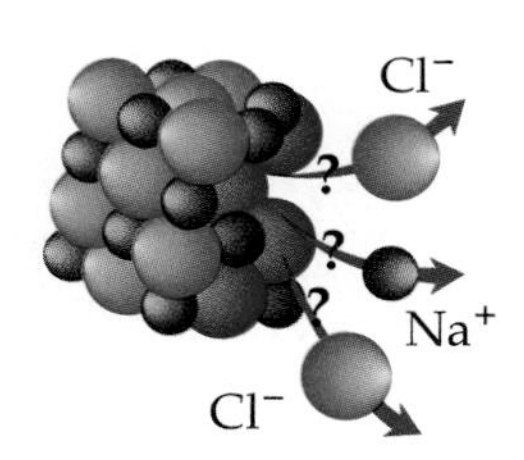

Solute-separation step, overcoming solute–solute interactions. Does it require a

Small amount of energy?
Moderate amount of energy?
Large amount of energy?

Solvent-separation step, overcoming solvent–solvent interactions. Does it require a

Small amount of energy?
Moderate amount of energy?
Large amount of energy?

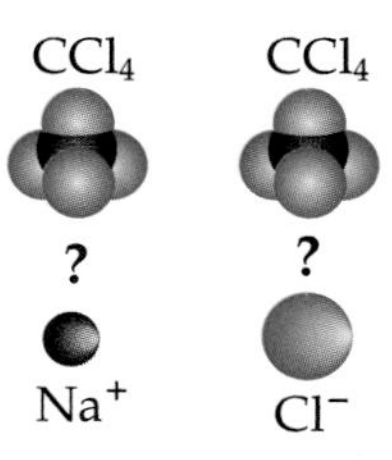

Solvation step, creating solvent–solute interactions. Does it release a

Small amount of energy?
Moderate amount of energy?
Large amount of energy?

You should have answered in the WorkPatch that the solvation step releases a small amount of energy (because there can be no strong ion–dipole forces between the CCl_4 molecules and the Na^+ and Cl^- ions) and that the solute-separation step requires a large amount of energy. Because the energy released during solvation (small amount) is less than the energy required for the two separation steps (large amount + moderate/small amount), this hypothetical solution process has a large energy deficit. Consequently, NaCl does not dissolve in CCl_4.

Now let's look at a different situation. According to our general rule, like dissolves like, nonpolar solutes should dissolve well in nonpolar solvents. Consider the forces involved in the three steps in this case:

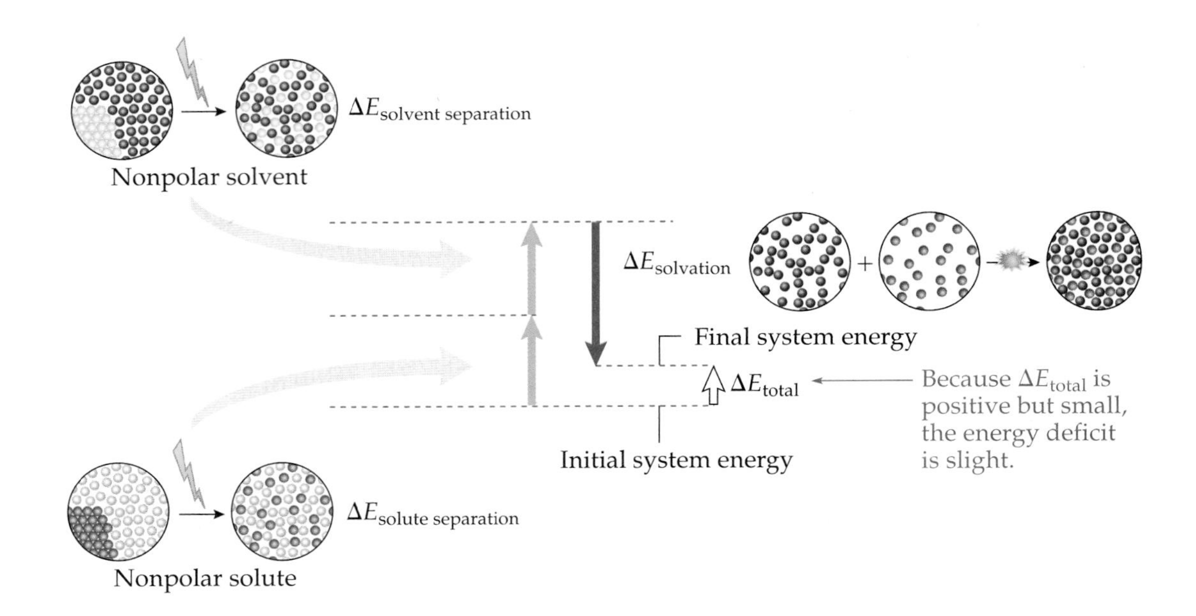

The weak London forces that hold the nonpolar solute molecules together must be broken during solute separation. The weak London forces that hold the solvent molecules together must be broken during solvent separation. The solvent–solute interactions that form during solvation are also London forces. Thus, all three steps involve little energy. Because the energy consumed by solute and solvent separation is small and the energy released by solvation is also small, ΔE_{total} is very small, regardless of whether it's positive or negative.

In fact, even if ΔE_{total} is positive for such a process—that is, even if there's a slight energy deficit—a nonpolar solute is likely to dissolve well in a nonpolar solvent. This is true because there is another factor besides ΔE_{total} that can help a solute dissolve. This factor, called *entropy*, can help tip the balance in favor of dissolution when the energy deficit is small.

12.3 Entropy and the Formation of Solutions

Entropy can be thought of as a measure of the amount of disorder or randomness in a situation. For example, consider your room. When all your clothing is folded and put away, you have a neat and ordered room. This is a low-entropy situation. If you throw all your clothes about the room, you have a messy, disordered room—a high-entropy situation.

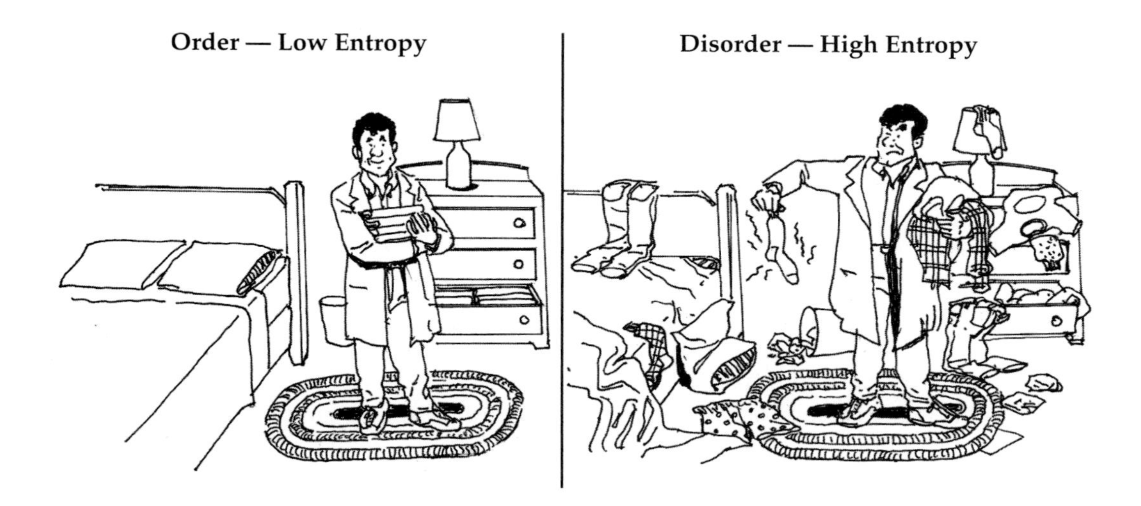

A fundamental law of nature called the **second law of thermodynamics** states that

In order for anything to happen spontaneously anywhere in the universe, the entropy of the universe must increase.

In other words, the universe "prefers" more randomness rather than less. There is a simple reason for this. A random situation is a lot more probable than an ordered one. For example, let's return to your room. Imagine you have just done your laundry and are ready to put things away into the proper drawers. You stand over the open drawers and toss the clothes into the air. Now, there is a chance that the clothes will fall, neatly folded, into the proper drawers. Although this *could* happen, the chances of its happening are minuscule. That's because there are only a limited number of ways to order your clothes neatly and into the proper drawers but an infinite number of ways to get a disordered, chaotic arrangement. The chances of getting a random arrangement are therefore much greater than the chances of getting an ordered arrangement.

A Highly Unlikely Event

Similarly, when solid NaCl dissolves, the ordered rows of Na^+ and Cl^- ions are replaced by a much more random situation in which the ions are scattered throughout the solution. The result is that the entropy of the universe has increased and the second law of thermodynamics has been satisfied.

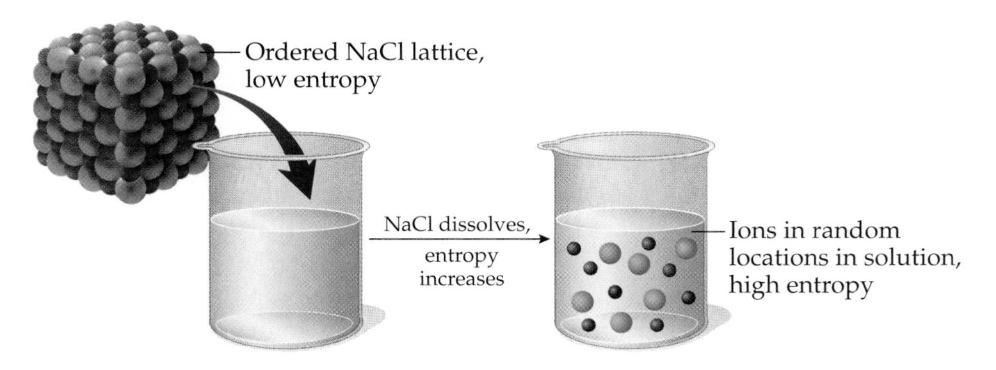

This argument has been simplified for greater impact, and other factors go into determining the entropy change for the universe when a solute dissolves. However, considered by itself, the decomposition of an ordered solute lattice into scattered solute particles always increases the entropy of the universe.

According to the second law, this entropy increase is a factor in the spontaneity of the dissolving process. The significance of entropy to the dissolving process depends on temperature. Entropy becomes more important as temperature increases. This is why some solutes that may not dissolve in cold solvent will more readily dissolve when the solvent is hot. At room temperature however, it is generally a weak factor but can sometimes tip the balance in favor of dissolving. Whether or not it can depends on the size of ΔE_{total}. For example, suppose we have a substance for which dissolving would result in a large energy deficit (ΔE_{total} much greater than zero). Chances are the entropy increase if the solvent were to dissolve is not large enough to make a dent, and the solute does not dissolve. This is the situation we ran into when we tried to dissolve polar NaCl in nonpolar CCl_4. In cases where the energy deficit is not too large, however, the entropy increase can tip the balance in favor of dissolution.

This is what happens when a nonpolar solute dissolves in a nonpolar solvent. In general, the impact of an entropy increase is smaller than the impact of ΔE_{total} and is thus important only when the energy deficit is small. However, dissolving nonpolar solutes in nonpolar solvents almost always results in a small energy deficit. Thus, the entropy increase explains why nonpolar substances are almost always soluble in nonpolar solvents.

An H-shaped tank contains gas A in one arm and gas B in the other arm. When a valve separating the two arms is opened, the gases will spontaneously mix if given enough time. The reverse process—the gases spontaneously separating into all A in one arm and all B in the other—will never happen, no matter how long you wait. Why? (Assume ΔE_{total} for the mixing is zero.)

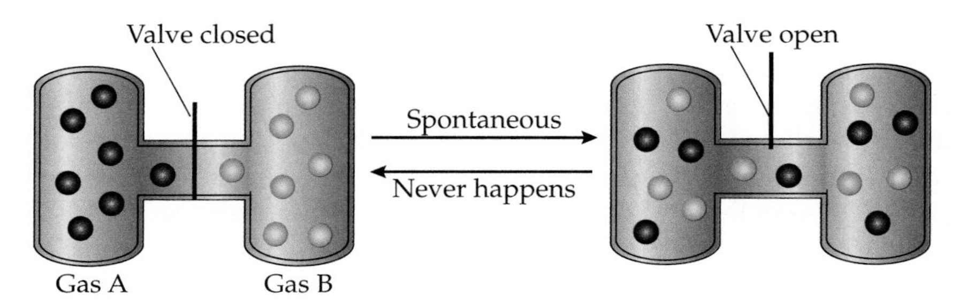

Did you invoke the second law of thermodynamics? If you did, you are beginning to understand entropy.

12.4 Solubility, Temperature, and Pressure

You now know in terms of energies and interactions what must occur for a solute to dissolve in a solvent. However, just knowing that a particular substance dissolves in a solvent is only part of the story. When chemists are working with solutions, they need to know how much solute can be dissolved in a given amount of solvent. The maximum amount of a solute that can be dissolved in a given amount of solvent is called the **solubility** of the solute. Solubility is often reported as the maximum number of grams of solute that will dissolve in 100 g of a solvent. An interesting aspect of solubility is that it can be changed. For example, try to dissolve 200 g of sucrose in 100 g of water at 0°C. Even with lots of stirring, some of the sucrose just sits at the bottom of the glass and refuses to dissolve. Heat the water to 100°C, however, and all the solid dissolves. Evidently, temperature is one variable that can be used to control the solubility of sucrose in water. As the temperature of the water increases, so does the solubility of the sucrose:

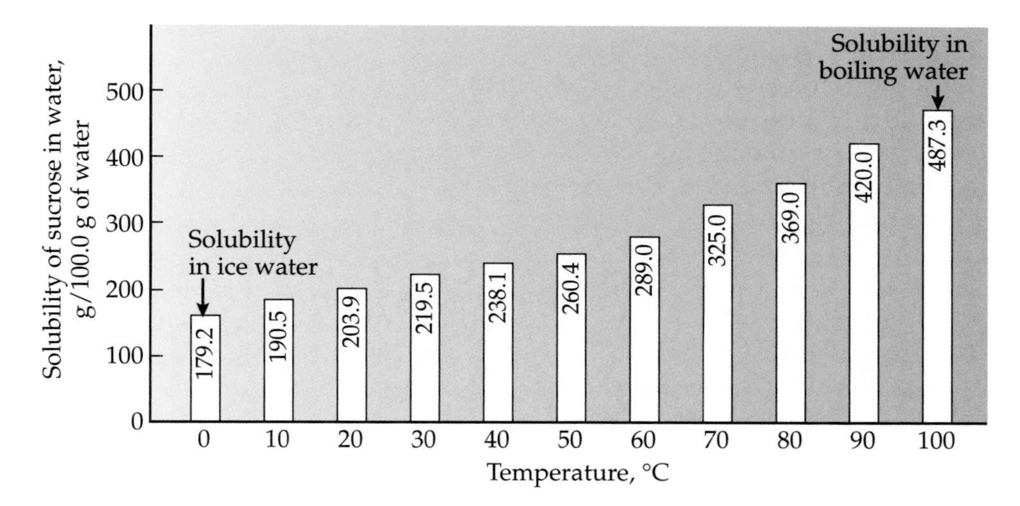

For most molecular solids (such as sucrose), an increase in temperature generally causes an increase in solubility. The same is true for most ionic solids, as you can see from this graph:

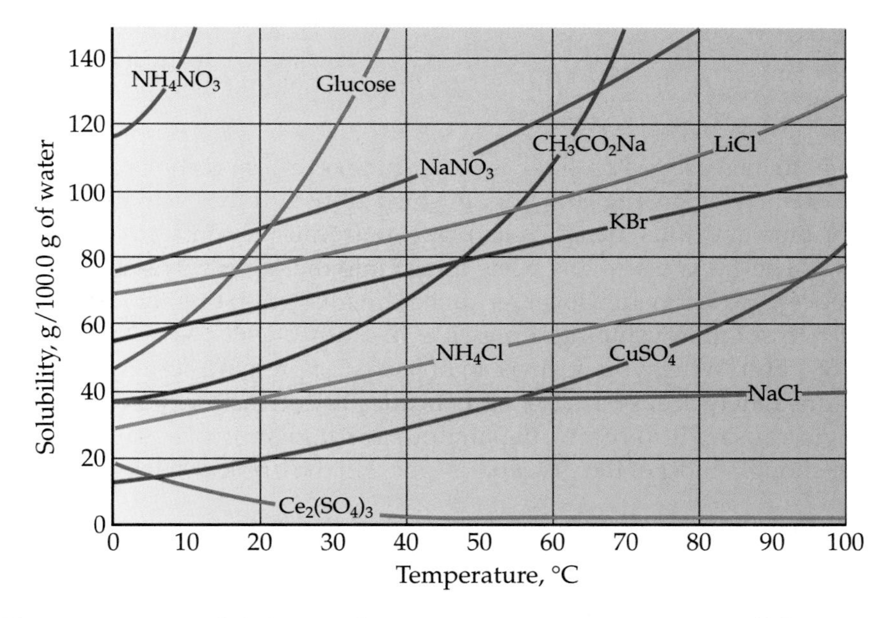

The increase in solubility with increasing temperature is small for some compounds (NaCl) and quite dramatic for others (CH_3CO_2Na, sodium acetate). Note that cerium sulfate, $Ce_2(SO_4)_3$, is an exception and becomes *less* soluble with increasing temperature.

Although it might be tempting to relate the way solubilities change with temperature to the energy changes that take place during the dissolving process, you should not do this. For example, LiCl dissolves with an energy surplus and NH_4NO_3 dissolves with an energy deficit. Yet the solubility of both compounds increases with increasing temperature. The best way to determine how temperature affects solubility is by experiment.

In systems in which a gas dissolves in a liquid, solubility decreases with increasing temperature. Many familiar aqueous solutions fit this profile. The familiar fizz in soft drinks is the result of dissolved CO_2 gas. Raising the temperature of the solvent decreases the solubility of the gas solute. Think of what it's like to drink warm soda pop. It tastes flat because most of the dissolved CO_2 has escaped from the solution.

Another variable that affects the solubility of a gas in a liquid is pressure. Increasing the pressure of the gas in the space above a liquid surface always forces more gas into the liquid—that is, the increase in gas pressure increases the solubility of the gas. This pressure–solubility relationship is known as *Henry's law*.

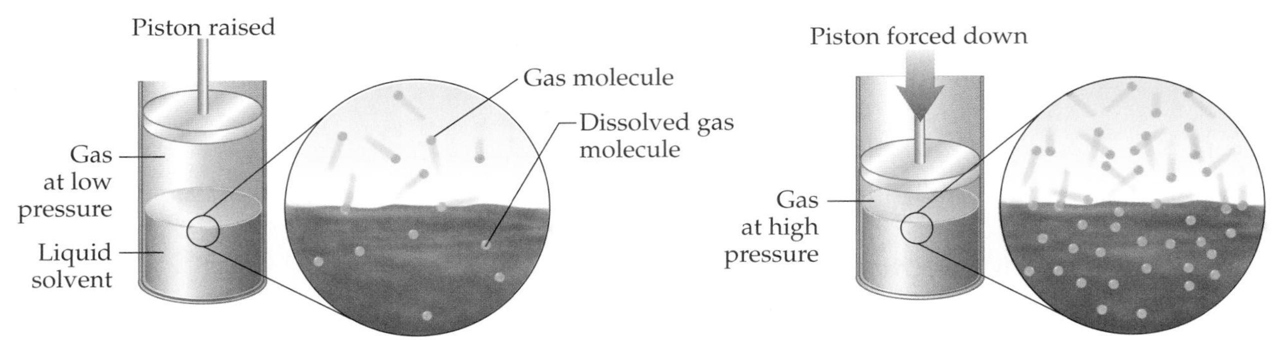

Relatively few gas molecules enter the liquid and dissolve.

When the pressure of the gas is increased, gas molecules enter the liquid more frequently and more of them dissolve.

One way to increase gas pressure is by pushing on a piston above the gas, as shown in the accompanying drawing. Pushing down on the piston decreases the gas volume and thus increases its pressure (recall the ideal gas law from Chapter 11). The gas is essentially being forced into the solvent. This is how carbonated beverages are made. Beverage makers increase the solubility of CO_2 by forcing it into solution using high pressure (2–5 atm), well above the normal pressure of 1 atm. When you open a can of soda, the pressure above the liquid surface immediately drops back to 1 atm, drastically decreasing the solubility of the CO_2 and causing it to come bubbling out of solution. If you leave the can open long enough, most of the CO_2 will escape and the drink will taste flat.

A similar thing happens when divers rise too quickly from a deep-sea dive. Deep under water, the increased pressure squeezes more of the gas the divers are breathing into the blood, increasing the solubility of the gas in the blood. This gas is mostly N_2 if they are breathing compressed air. Coming up too rapidly is like popping the top on a can of soda. The rapid decrease in pressure lowers the solubility of the nitrogen in the blood, and the nitrogen comes out of solution and forms bubbles in the blood. This condition, known as the bends, can be fatal.

The most effective treatment for the bends is to place the afflicted diver as quickly as possible in a high-pressure chamber (called a *hyperbaric chamber*). The pressure in the chamber is raised to a level that causes the gas to redissolve in the blood, temporarily curing the bends. Then the pressure is reduced slowly enough to allow the extra dissolved gas to come out of solution through the diver's lungs instead of forming bubbles in the blood.

Solubility values like those given in the two graphs shown on page 439 represent the maximum amount of solute that dissolves in a given amount of solvent. When the maximum amount of solute is dissolved, we say the solution is a **saturated solution**. Knowing the solubility enables us to calculate how much solute it takes to saturate a given amount of solvent. For example, suppose we have 250.0 g of water and want to use it to prepare a saturated sucrose solution at 20.0°C. How much sucrose do we need? From the sucrose-solubility graph, we find that the solubility of sucrose at 20.0°C is 203.9 g sucrose/100 g water. To answer the question, we use the solubility as a conversion factor and multiply it by the amount of solvent we have:

$$250.0\ \cancel{\text{g water}} \times \frac{203.9\ \text{g sucrose}}{100.0\ \cancel{\text{g water}}} = 509.8\ \text{g sucrose}$$

Remember, solubility values are *maximums*, meaning you can dissolve 509.8 g of sucrose in 250.0 g of water at 20.0°C and no more.

Try the following practice problems to gain some experience using solubility data.

Practice Problems

12.11 How many grams of sucrose will it take to saturate 1 ton of water at 20.0°C? [1 ton = 2000 lb; 1 lb = 453.6 g]

Answer:

$$1.00\ \cancel{\text{ton water}} \times \frac{2000\ \cancel{\text{lb}}}{1\ \cancel{\text{ton}}} \times \frac{453.6\ \cancel{\text{g}}}{1\ \cancel{\text{lb}}} \times \frac{203.9\ \text{g sucrose}}{100.0\ \cancel{\text{g water}}} = 185 \times 10^{6}\ \text{g sucrose}$$

12.12 How many grams of potassium chloride, KCl, will it take to prepare a saturated solution in 500.0 mL of boiling water? [The solubility of KCl at 100°C is 56.7 g/100.0 g water; assume the density of water is 1.000 g/mL.]

12.13 On the basis of the three steps involved in the formation of a solution, what is the biggest difference between dissolving a gas and dissolving a solid?

12.5 Getting Unlikes to Dissolve—Soaps and Detergents

"Dry"-cleaning

We've been talking so far about like dissolving like. However, much of the dirt that accumulates on our bodies and our clothing—such as grease and oil—is nonpolar, which means washing with polar water won't accomplish much in the way of cleaning. In order to dissolve these nonpolar substances, we need a nonpolar solvent. We could use nonpolar carbon tetrachloride, but its expense and toxicity make this approach to cleaning impractical at home. (This is effectively what happens when you send your clothes to be dry-cleaned, however. The clothes are immersed in a nonpolar liquid called perchloroethane in which nonpolar grease and oil dissolve. Dry-cleaning is "dry" only in the sense that no water is used.)

Water—abundant, cheap, nontoxic, and recyclable—is still the solvent of choice for household cleaning, but how do we make nonpolar dirt dissolve in polar water? To find out how, let's go back to the detergent aisle of the supermarket. Getting "unlikes" to dissolve is the job of soaps and detergents. Each soap/detergent molecule consists of a long nonpolar "tail" of carbon and hydrogen (called a *hydrocarbon tail*) and a highly polar "head":

A Typical Soap Molecule

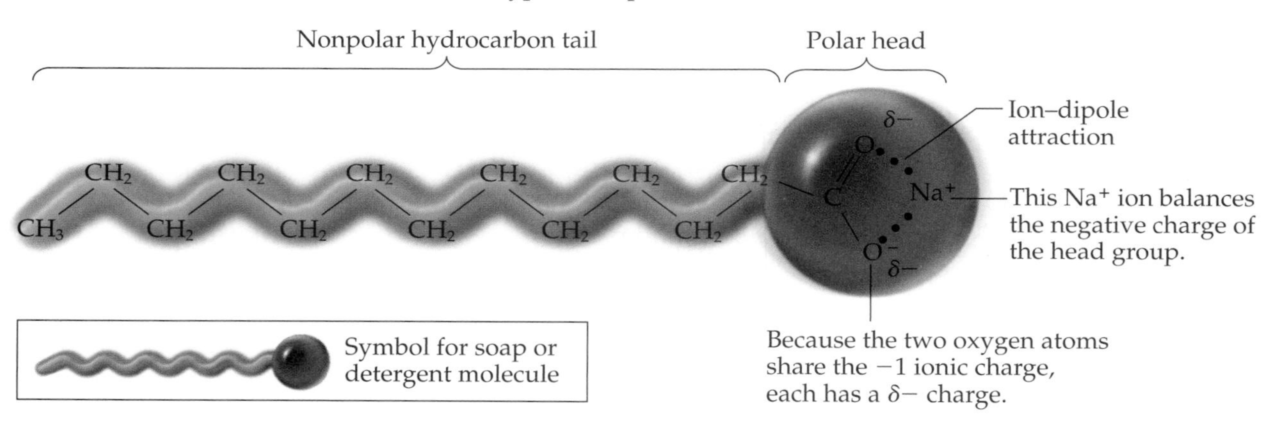

Being polar because of its ionic charges, the head of a soap/detergent molecule is soluble in polar water and is called **hydrophilic** ("water-loving"). The nonpolar hydrocarbon tail is insoluble in water and is referred to as **hydrophobic** ("water-fearing"). (The major difference between soaps and detergents is that soaps contain a carbon–oxygen group at the polar head and detergents contain a sulfur–oxygen group at the polar head. For our purposes in this discussion, you can think of the two as behaving identically in how they clean.)

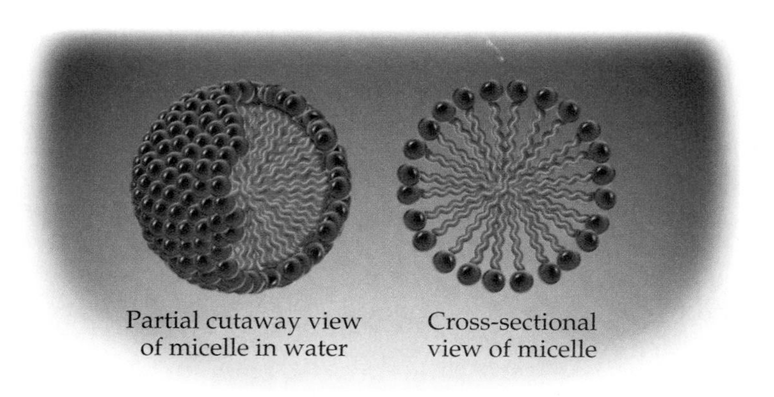

When soap/detergent molecules are added to water, an interesting spherical structure called a *micelle* forms.

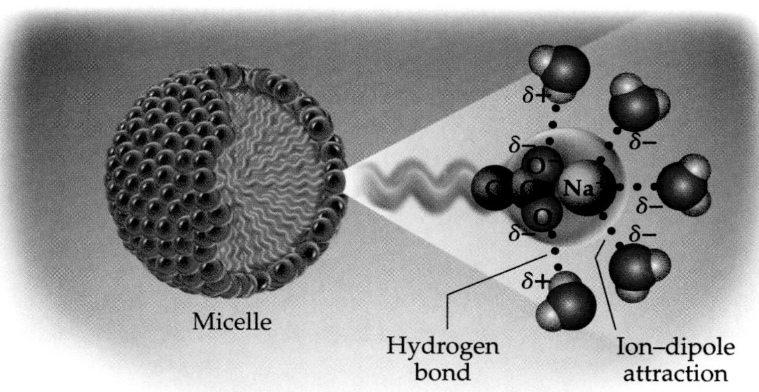

The outside of the micelle is formed from the polar heads and is therefore hydrophilic. The inside is made up of the nonpolar tails, all attracted to one another via London forces, and is therefore hydrophobic. A micelle thus presents a polar surface to the water molecules, which, being polar themselves, are attracted to the micelle via intermolecular forces. The polar heads contain O atoms carrying a partial negative charge ($\delta-$), and because of this charge the O atoms form hydrogen bonds with water molecules. As a result of these attractions to water, the micelle is soluble in water.

It's a different story inside the micelle, where, among the conglomeration of nonpolar tails, the environment is strictly hydrophobic. This is just the right kind of environment for dissolving nonpolar grease and oil molecules (like dissolves like). Grease/oil molecules that encounter the micelle quickly dissolve in its interior. Because the grease/oil dissolves in the micelle and the micelle dissolves in water, the result is that the soap/detergent dissolves the grease/oil in the water—a clever way of getting unlikes to dissolve in one another.

The first soaps were prepared from animal fats cooked with potash (potassium hydroxide, KOH) from the ashes of wood fires. Animal fat and bacon grease provided the hydrocarbon tail, and the potash helped to create the polar head.

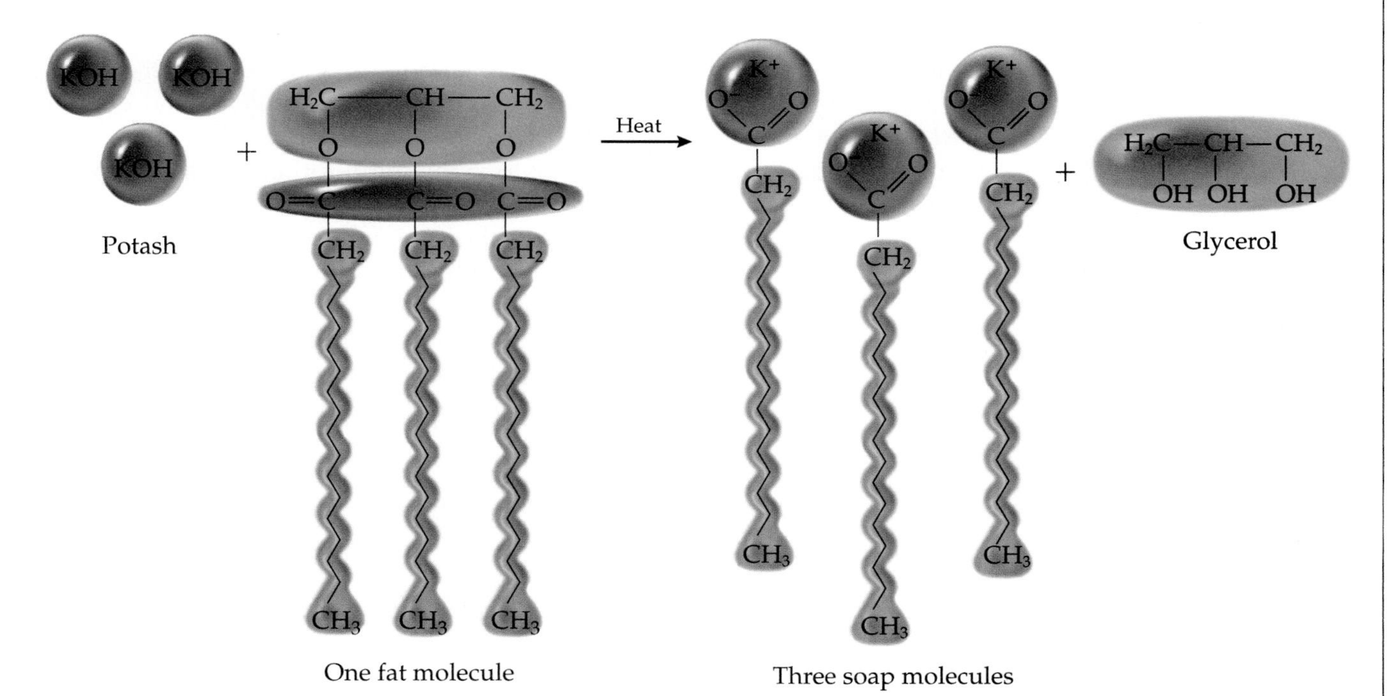

Today, the chemical industry spends a great deal of time trying to improve soaps and detergents, making grease dissolve even better in water, attempting to whiten and brighten your socks until your feet practically glow when you put them on.

Suppose you put detergent in a dry-cleaning machine that uses carbon tetrachloride. Assuming micelle-type structures form,

(a) Draw the micelle-type structures that would form and explain your diagram.
(b) Explain what kind of dirt these structures would be good at removing.

Indeed, there are many dry-cleaning detergents. The micelle you drew to answer the WorkPatch explains how they might work.

12.6 Molarity

The solubility graphs we looked at earlier tell us the maximum amount of a particular solute we can dissolve in a fixed amount of solvent. For example, at 20°C we can dissolve a maximum of 203.9 g of sucrose in 100.0 g of water. This sucrose–water solution is saturated and would be quite sweet. We could also make a less sweet solution by dissolving less than the maximum, say 20 g of sucrose, in 100 g of water. The **concentrations** of these two solutions—which means the amount of solute either per amount of solvent or per amount of solution—are different.

In this and the next section, we look at the quantitative ways of describing the concentration of a solution, but first we'll take a brief look at ways of describing concentration qualitatively. For example, you take a swallow of some watery, light-brown liquid purchased at a snack-bar coffee machine and mutter, "This coffee's pretty weak." Or you sip some dark, expensive cappuccino at a classy restaurant and proclaim, "This is strong!" *Weak* and *strong* are qualitative terms describing how much coffee (solute) is dissolved in the water (solvent). Instead of *weak* and *strong* you could have said *dilute* and *concentrated,* other terms often used to describe concentration. These latter terms are also qualitative because they don't describe the composition of the solution in numeric terms.

There will be times when you need quantitative information about how much solute is dissolved in a given amount of solution. For example, suppose you have been in an automobile accident and find yourself in immediate need of an intravenous saline (salt) solution. Too concentrated a solution will kill you, and too dilute a solution won't help. It's at a time like this that you seriously hope the attending health professional knows something more about concentration than just "weak" or "strong." Considering that your life depends on it, an accurate quantitative value of the concentration would be nice.

Saturated is a quantitative term. As we mentioned earlier, a solution is said to be saturated when the maximum amount of solute is dissolved in it. If you add more solute, it just settles to the bottom of the container and doesn't dissolve. In fact, a good way to prepare a saturated solution is to add an excess of solute so that some doesn't dissolve and then filter the solution.

There are times when knowing the concentration of a solution takes on a whole new level of importance.

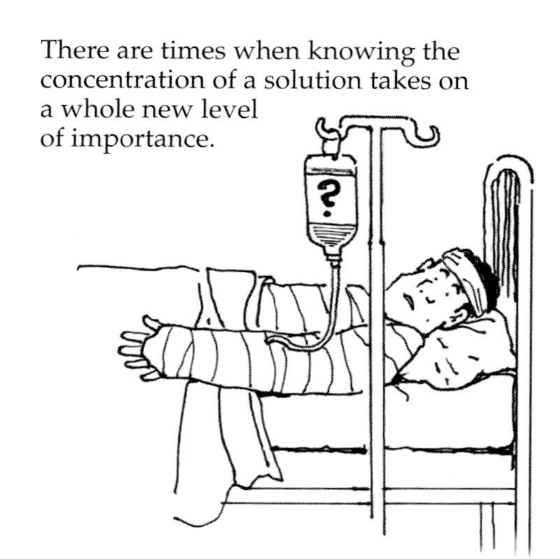

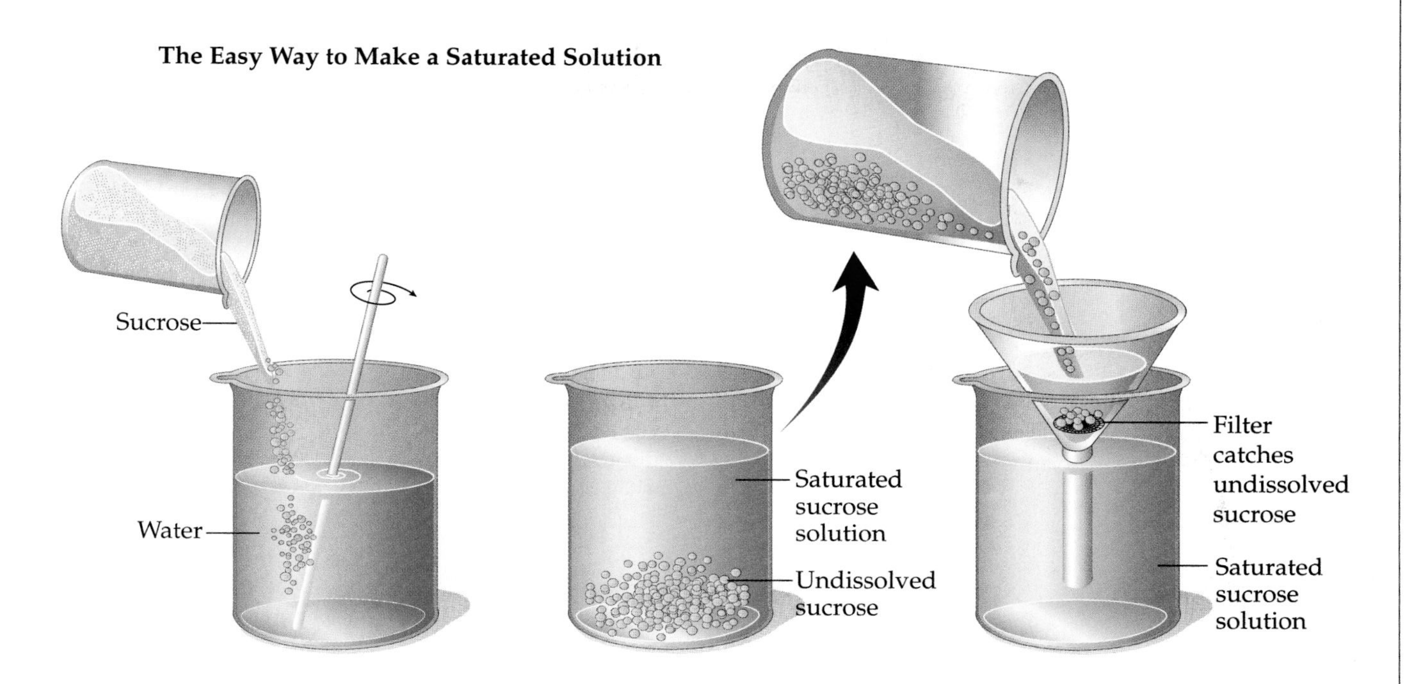

At 0°C, you can dissolve a maximum of 37.5 g of NaCl in 100 g of water. No more will dissolve. The word *saturated* therefore has a quantitative meaning, telling you the solubility of the solute at the temperature of the solution. Unfortunately, just saying the word *saturated* doesn't tell you how much salt is dissolved in a saturated NaCl solution. You would have to look up the solubility in a table or graph. It would be much more useful to have some quantitative description of the concentration that could instantly tell you the amount of solute present in the solvent. One such description commonly used by chemists is called *molarity*. From the name you can probably guess what's coming.

Did you think you were rid of me?

When describing the amount of solute dissolved, chemists most often think in terms of moles and express the concentration in terms of molarity. (If you need to review the concept of the mole, look back at Section 8.2.) The **molarity** of a solution is defined as the number of moles of solute dissolved in 1 L of solution:

$$\text{Molarity} = \frac{\text{Moles of solute}}{\text{1 L of solution}}$$

For example, if you dissolve 1 mole (58.443 g) of NaCl in enough water to give exactly 1 L of solution, you will have a 1-molar (1 M) solution of NaCl:

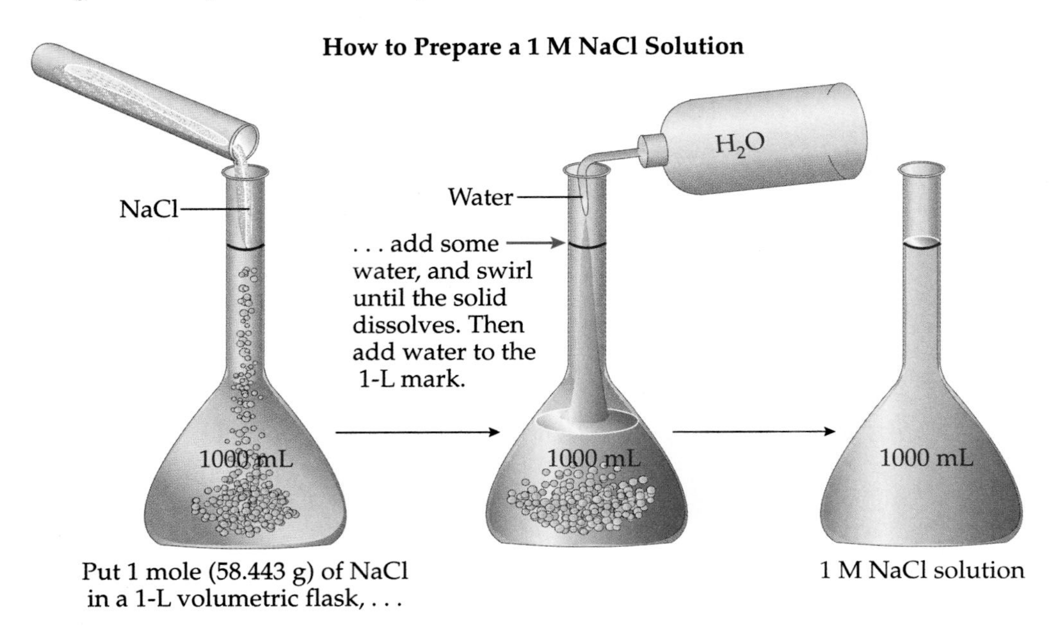
How to Prepare a 1 M NaCl Solution
H_2O
NaCl
Water
. . . add some water, and swirl until the solid dissolves. Then add water to the 1-L mark.
1000 mL
1000 mL
1000 mL
Put 1 mole (58.443 g) of NaCl in a 1-L volumetric flask, . . .
1 M NaCl solution

Now we want to warn you about a mistake that many students make (but you never will). A 1 M NaCl solution is *not* made by putting 1 mole of NaCl in a container and adding 1 L of water to it. It is made by putting 1 mole of NaCl in a container and adding enough water so that you *end up with* 1 L of solution. The denominator in the molarity expression says "L of solution," *not* "L of solvent." Make sure you understand this difference.

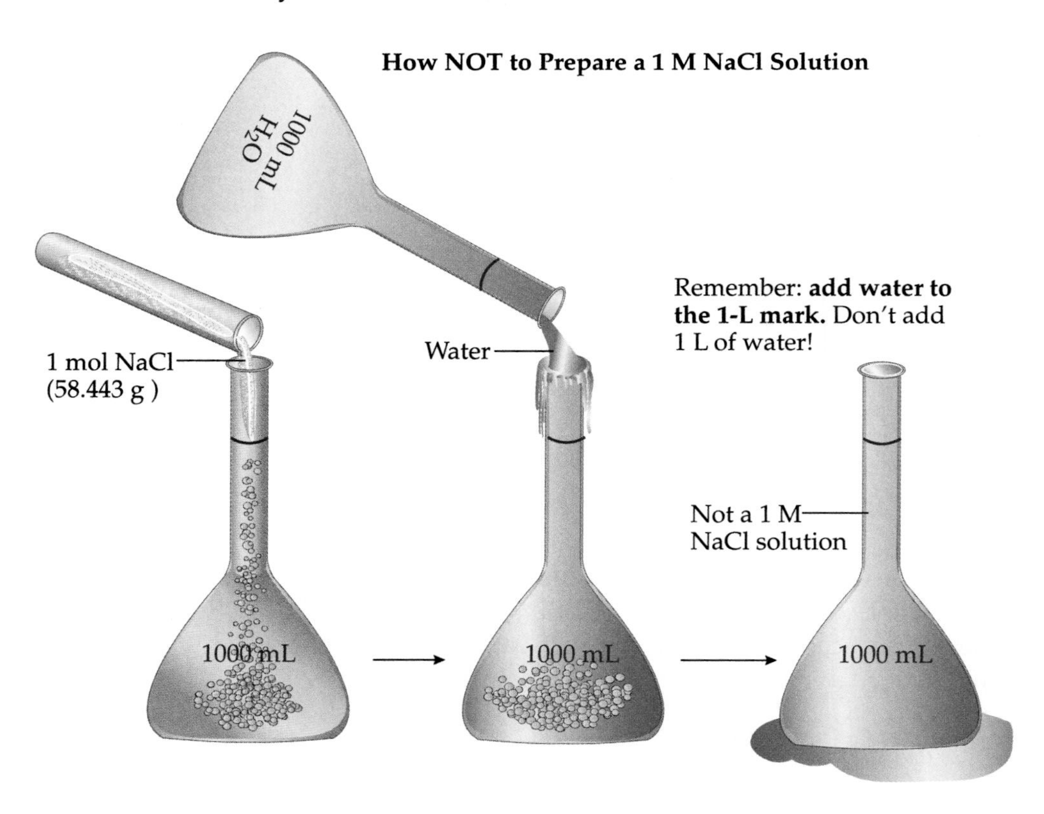
How NOT to Prepare a 1 M NaCl Solution
1000 mL H_2O
1 mol NaCl (58.443 g)
Water
Remember: **add water to the 1-L mark.** Don't add 1 L of water!
Not a 1 M NaCl solution
1000 mL
1000 mL
1000 mL

Molarity gives us what we wanted—a quantitative measure of concentration. If someone hands you a bottle of saline solution labeled 1 M, you quantitatively and absolutely know its concentration. No longer is it weak or strong, saturated or unsaturated. It's 1 M. Every liter of that solution contains exactly 1 mole (58.443 g) of NaCl. A 2 M NaCl solution is twice as concentrated as a 1 M solution, meaning every liter of 2 M solution contains 2 moles (116.886 g) of NaCl. And a 0.5 M solution is half as concentrated as a 1 M solution—every liter of a 0.5 M solution contains 0.5 mole (29.222 g) of NaCl.

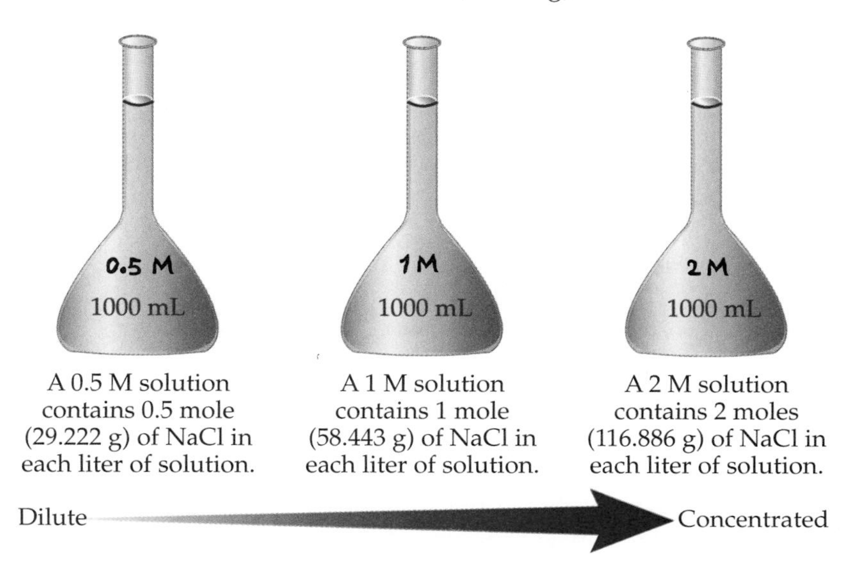
0.5 M
1000 mL
1 M
1000 mL
2 M
1000 mL
A 0.5 M solution contains 0.5 mole (29.222 g) of NaCl in each liter of solution.
A 1 M solution contains 1 mole (58.443 g) of NaCl in each liter of solution.
A 2 M solution contains 2 moles (116.886 g) of NaCl in each liter of solution.
Dilute
Concentrated

In fact, knowing the molarity, we can quickly determine how much solute we have for any volume of solution. We just use molarity as a conversion factor. For example, suppose you have a flask that contains 200.0 mL of a 1.00 M NaCl solution. If all the water evaporated, solid sodium chloride would remain, but how much? In other words, what mass of NaCl is contained in 200.0 mL of a 1.00 M solution?

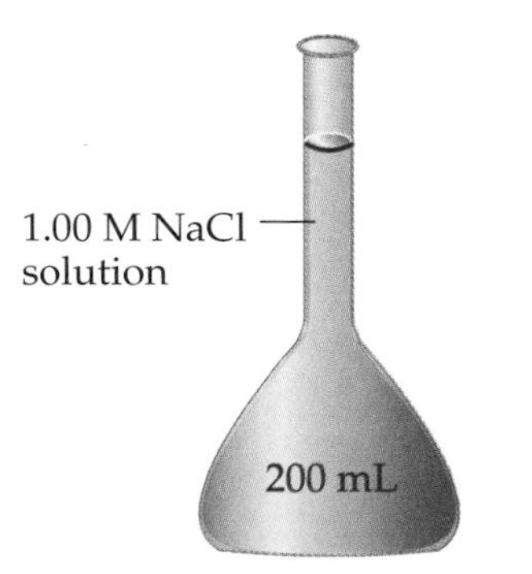
1.00 M NaCl solution
200 mL
How many moles of NaCl are there in the flask? How many grams?

We can answer these questions by using unit analysis (Section 2.8). We can treat the molarity as a conversion factor because it has two units (moles of solute over liters of solution). We treat the given volume, which has just one unit (mL), as the measurement. (This is not to say that the molarity is not a measured quantity. It is, but it can serve as a conversion factor.)

Step 1: Write down the measurement: 200.0 mL solution

Step 2: Multiply it by the molarity conversion factor to make the starting unit (mL solution) cancel. Also, because our conversion factor contains the unit *liter*, we'll convert our measurement unit from milliliters to liters:

$$\underbrace{200.0\ \text{mL solution}}_{\text{Measurement}} \times \underbrace{\frac{1\ \text{L solution}}{1000\ \text{mL solution}}}_{\substack{\text{Conversion factor} \\ \text{to change mL to L}}} \times \underbrace{\frac{1.00\ \text{mol NaCl}}{1\ \text{L solution}}}_{\substack{\text{Molarity} \\ \text{conversion factor}}} = 0.200\ \text{mol NaCl}$$

So 200.0 mL of a 1.00 M NaCl solution contains 0.200 mole of NaCl. To convert the answer to grams, we multiply by the appropriate conversion factor, the molar mass of NaCl:

$$0.200\ \text{mol NaCl} \times \underbrace{\frac{58.443\ \text{g NaCl}}{1\ \text{mol NaCl}}}_{\substack{\text{Molar mass} \\ \text{conversion factor}}} = 11.7\ \text{g NaCl}$$

There is 11.7 g of NaCl in our 200.0 mL of 1.00 M solution. Of course, we did not have to do this last calculation separately. We could have just strung the molar mass conversion factor onto the end of the preceding calculation.

Armed with molarity as a conversion factor, you are now ready to practice some numeric problems. Remember, you can always use the inverse of a conversion factor to make units cancel.

Practice Problems

12.14 How many milliliters of a 1.500 M solution of NaCl do you need to obtain 100.0 g of NaCl?

Answer: As usual, we write the measurement and multiply it by appropriate conversion factors:

$$\underbrace{100.0\ \text{g NaCl}}_{\text{Measurement}} \times \underbrace{\frac{1\ \text{mol NaCl}}{58.443\ \text{g NaCl}}}_{\substack{\text{Conversion factor} \\ \text{to change to moles} \\ \text{(inverse of NaCl} \\ \text{molar mass)}}} \times \underbrace{\frac{1\ \text{L solution}}{1.500\ \text{mol NaCl}}}_{\substack{\text{Conversion factor} \\ \text{(inverse of} \\ \text{molarity)}}} \times \underbrace{\frac{1000\ \text{mL solution}}{1\ \text{L solution}}}_{\substack{\text{Conversion factor} \\ \text{to change L to mL}}} = 1141\ \text{mL solution}$$

Notice that the first thing we did was convert to moles. Remember our general advice for stoichiometry problems in chemistry. Convert to moles as quickly as you can. It's almost always the right thing to do.

12.15 How many milliliters of a 2.55 M solution of glucose, $C_6H_{12}O_6$, molar mass = 180.155 g/mol, do you need to obtain 25.0 g of glucose?

12.16 How many grams of ethanol, C_2H_6O, are there in 200.0 mL of a 2.00 M aqueous solution of ethanol?

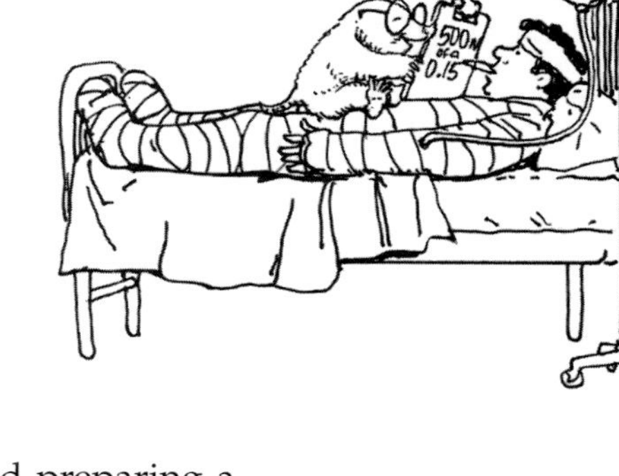

There is one other type of problem involving molarity you should know how to do, a practical problem that asks "How do you prepare a specific amount of a solution that has a specific molarity?" For example, let's go back to you lying in the hospital hooked up to a bottle of intravenous saline solution. Noticing that the bottle is just about empty, your nurse says, "I need to prepare 500.0 mL of a 0.15 M NaCl solution to refill this bottle." Realizing that the nurse studied introductory chemistry quite a while ago whereas you just completed the course, you remind him how to do this. Here's how it's done.

There are two ways to prepare this life-saving solution. One is to measure out the proper amount of NaCl solid and dissolve it in the proper amount of water. We'll call this method preparing a solution from scratch. A second way is to take a more concentrated solution of NaCl off the shelf and dilute it to the proper concentration. Because both ways are commonly used, we'll go over both.

Method 1 Preparing a Solution from Scratch

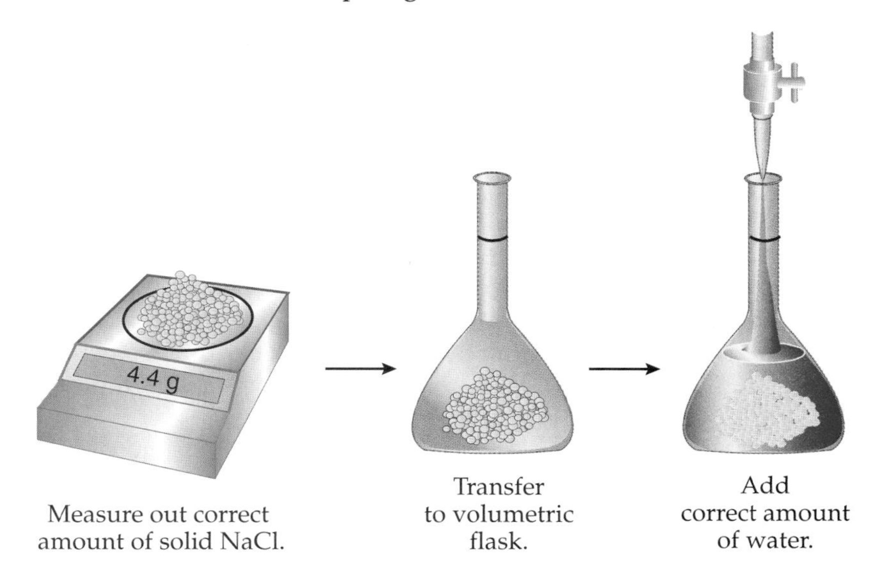
4.4 g
Measure out correct amount of solid NaCl.
Transfer to volumetric flask.
Add correct amount of water.

Method 1: Preparing a solution from scratch. Our task is to prepare 500.0 mL of a 0.15 M NaCl solution. Our first step is to calculate how much NaCl solid we need. Just use the usual method of unit analysis:

$$\underbrace{500.0\ \cancel{\text{mL solution}}}_{\text{Measurement}} \times \underbrace{\frac{1\ \cancel{\text{L solution}}}{1000\ \cancel{\text{mL solution}}}}_{\substack{\text{Change}\\ \text{mL to L}}} \times \underbrace{\frac{0.15\ \text{mol NaCl}}{1\ \cancel{\text{L solution}}}}_{\substack{\text{Molarity}\\ \text{conversion factor}}} = 0.075\ \text{mol NaCl}$$

The answer is in moles, but because we measure amounts of solids in grams, we need to convert to grams:

$$0.075\ \cancel{\text{mol NaCl}} \times \underbrace{\frac{58.443\ \text{g NaCl}}{1\ \cancel{\text{mol NaCl}}}}_{\text{Molar mass conversion factor}} = 4.4\ \text{g NaCl}$$

So to prepare 500.0 mL of a 0.15 M NaCl solution from scratch, place 4.4 g of NaCl in a 500-mL volumetric flask, add some water, swirl to dissolve the solid, and then add enough water to get exactly 500.0 mL of solution. (Remember, do *not* add 500 mL of water!)

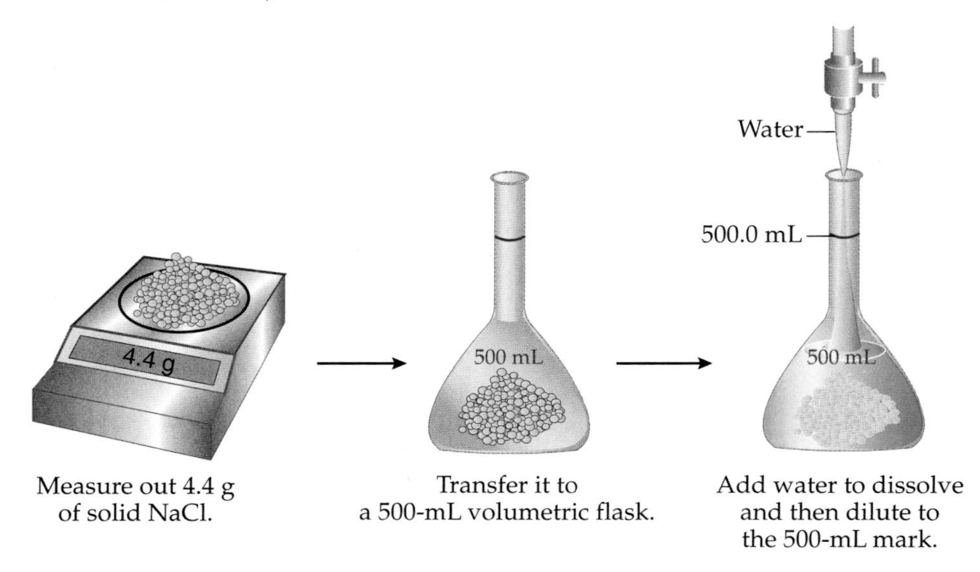
Water
500.0 mL
4.4 g
500 mL
500 mL
Measure out 4.4 g of solid NaCl.
Transfer it to a 500-mL volumetric flask.
Add water to dissolve and then dilute to the 500-mL mark.

Method 2: Preparing a solution by diluting a more concentrated solution. On the shelf is a bottle of 0.50 M NaCl solution. This is our "reservoir" of solution, what is called a *stock solution*. We can use it to prepare the 500.0 mL of 0.15 M solution we need. Because the stock solution is more concentrated than the solution we need, we have to dilute some amount of it with water. How much stock solution should we use, and how much water do we add to dilute it to the desired concentration?

Method 2 Preparing a Solution by Diluting a More Concentrated Solution

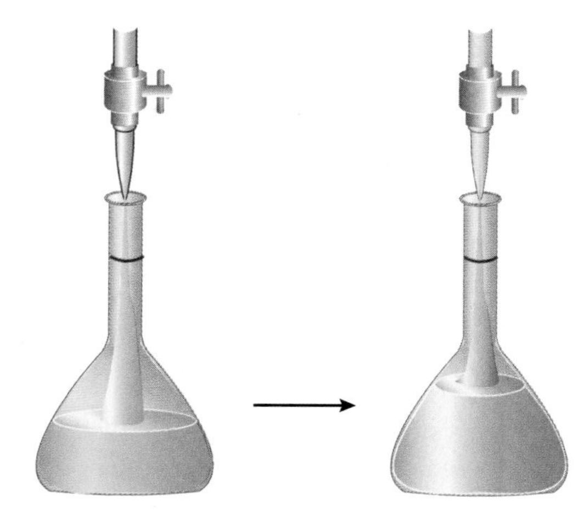
Add correct amount of concentrated solution.
Add correct amount of water.

At first glance, this might seem more difficult than preparing the solution from scratch because now we have to consider two solutions, the stock solution and the one we want to prepare. To start, though, forget about the stock solution. Just worry about the solution we want to make (500.0 mL of a 0.15 M NaCl solution). When we prepared this solution from scratch, the first question we asked was "How much NaCl do we need?" We ask *exactly* the same question here. How much NaCl do we need to prepare 500.0 mL of a 0.15 M solution? Because this is the same question we asked in method 1, we know that the answer is 0.075 mole of NaCl. In method 1 we converted this amount to grams because we were preparing the solution from solid NaCl. Here is where the stock solution comes into play because this time we get our 0.075 mole of NaCl from the stock solution. The question now becomes "What volume of 0.50 M stock solution do we need to get 0.075 mole of NaCl?" Look closely at this question. There is a measurement (0.075 mol NaCl) and a conversion factor (0.50 M). It's time to use unit analysis again:

$$0.075\ \cancel{\text{mol NaCl}} \times \underbrace{\frac{1\ \text{L stock solution}}{0.50\ \cancel{\text{mol NaCl}}}}_{\substack{\text{Conversion factor} \\ \text{(inverse of stock} \\ \text{solution molarity)}}} = 0.15\ \text{L stock solution}$$

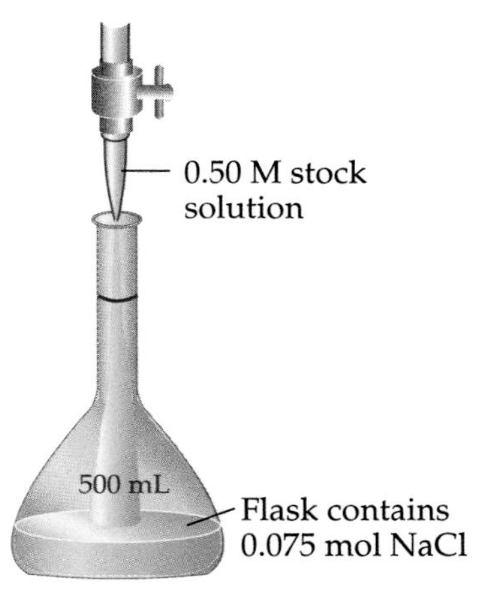

Add 150 mL of stock solution.

This calculation tells us that 0.15 L of stock solution contains 0.075 mole of NaCl. To prepare our solution, we put 0.15 L of stock solution in a 500-mL volumetric flask. This puts 0.075 mole of NaCl in the flask. Then we add enough water to get 500.0 mL of solution. Once again we have 0.075 mole of NaCl dissolved in a total volume of 500.0 mL of solution, which is a 0.15 M NaCl solution.

Method 2 can be considered a dilution problem because we are diluting a stock solution to a desired concentration. An equivalent way to solve such a problem is to use this **dilution equation**:

$$M_{\text{stock solution}} \times V_{\text{stock solution}} = M_{\text{diluted solution}} \times V_{\text{diluted solution}}$$

For example, let's repeat our preparation of 500.0 mL of 0.15 M NaCl from a 0.50 M NaCl stock solution. We are still looking for the same thing, the volume of stock solution we need to use. This is $V_{\text{stock solution}}$. All we need to do is solve the dilution equation for $V_{\text{stock solution}}$:

$$V_{\text{stock solution}} = \frac{M_{\text{diluted solution}} \times V_{\text{diluted solution}}}{M_{\text{stock solution}}}$$

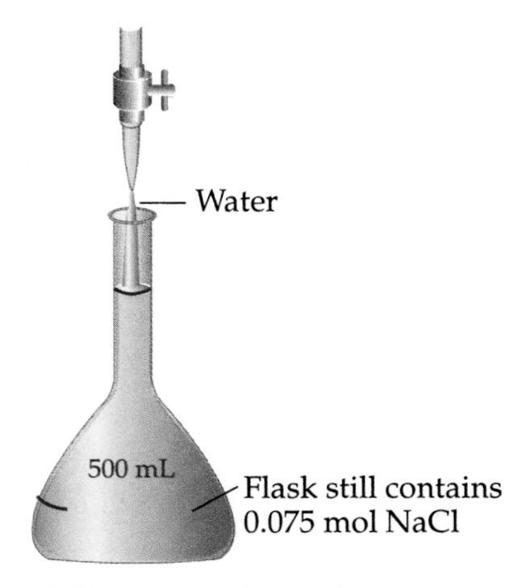

Add water until the flask contains 500.0 mL of solution. The final concentration is 0.075 mol/0.5000 L = 0.15 M.

Now we plug in the numeric values for the three factors on the right and we have a value for $V_{\text{stock solution}}$:

$$V_{\text{stock solution}} = \frac{\overbrace{0.15\ \text{M}}^{M_{\text{diluted solution}}} \times \overbrace{500.0\ \text{mL}}^{V_{\text{diluted solution}}}}{\underbrace{0.50\ \text{M}}_{M_{\text{stock solution}}}} = 150\ \text{mL stock solution}$$

This is the same answer we arrived at a moment ago but now done in one quick step.

The dilution equation does not work because of magic. Rather, it works because it is based on a simple fact—when you dilute some volume of stock solution, the number of moles of solute contained in that volume does not change!

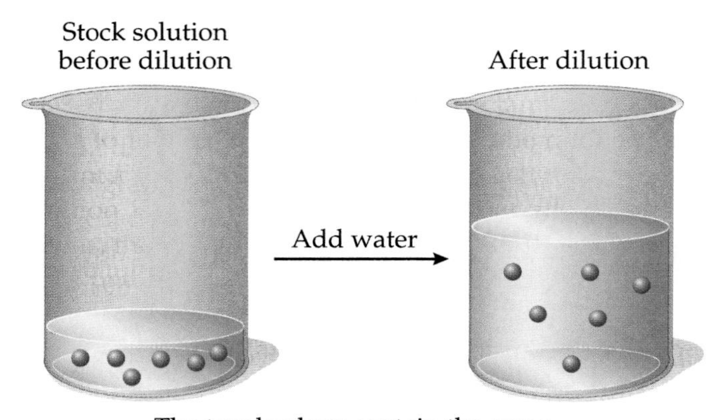

The two beakers contain the same number of moles of solute.

When you multiply the molarity of a solution by its volume, the product is number of moles of solute ($M \times V$ = moles):

$$\overbrace{\frac{\text{Moles of solute}}{\cancel{\text{Liters of solution}}}}^{M} \times \overbrace{\cancel{\text{Liters of solution}}}^{V} = \text{Moles of solute}$$

Therefore the dilution equation is just the mathematical version of the above diagram. $M_{\text{stock solution}} \times V_{\text{stock solution}}$ equals moles of solute in the left beaker, and $M_{\text{diluted solution}} \times V_{\text{diluted solution}}$ equals moles of solute in the right beaker, and these are equal to each other. So feel free to use the dilution equation anytime a dilution problem comes up.

A summary of what we just did appears at the top of the next page.

Preparing various volumes of solutions having particular molarities, both from scratch and by diluting stock solutions, is routine practice in chemical laboratories, medical offices, and hospitals throughout the world. Now it's time to convince yourself you can do it.

How to prepare a solution of desired molarity

Step 1: Determine the number of moles of solute needed to prepare the solution. This quantity equals the desired volume in liters multiplied by the desired molarity:

Moles of solute needed = Volume of solution × Molarity of solution

Step 2: **If preparing the solution from scratch:** Convert number of moles of solute to grams, using the molar mass as a conversion factor. Place this number of grams of solute in the appropriate volumetric flask.

If diluting a stock solution*: Convert number of moles of solute to volume of stock solution, using the stock solution molarity as a conversion factor. (Multiply the number of moles of solute by the inverse of the stock solution molarity.) Place this volume of stock solution in the appropriate volumetric flask.

Step 3: Add enough water to reach the desired volume.

*Alternatively, you can combine steps 1 and 2 and use the dilution equation, as described in the text.

Practice Problems

12.17 What mass of solid NaCl do you need to prepare 400.0 mL of a 2.00 M NaCl solution from scratch? How much water do you add to the NaCl?

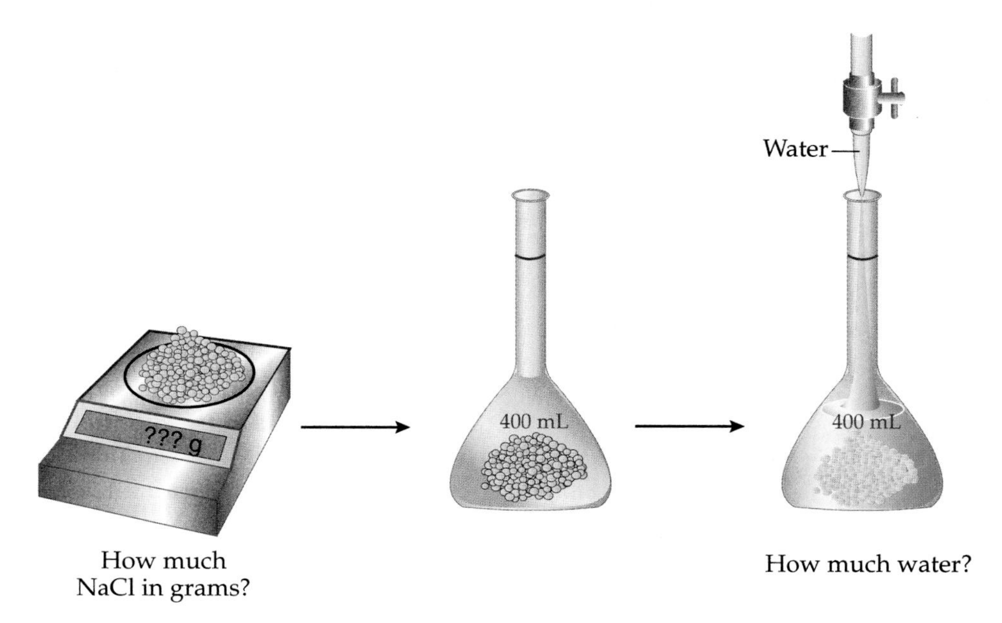

12.18 You have 1 L of a 3.00 M stock solution of NaCl and need to prepare 400.0 mL of a 2.00 M solution. Describe how you would do it.

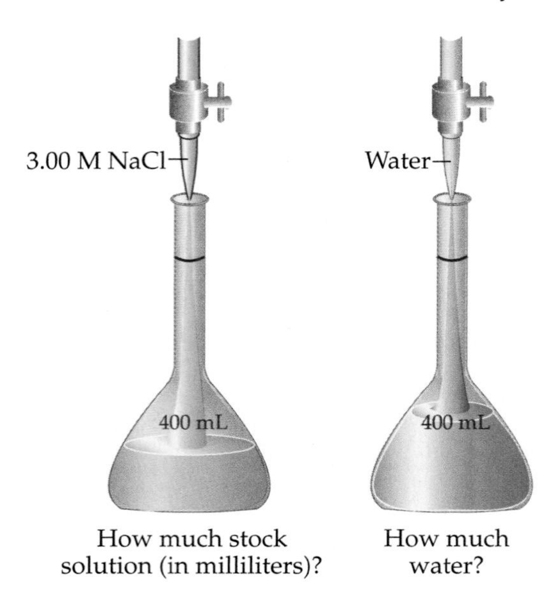

12.7 Percent Composition

Molarity is an excellent way to specify solution concentration quantitatively, and chemists use it routinely. However, solutions encountered in hospitals, doctors' offices, and grocery stores often have concentrations expressed as *percent composition*. In a fundamental sense, percent composition is similar to molarity. In fact, most of the ways of numerically reporting solution concentrations are based on the definition

$$\text{Concentration} = \frac{\text{Amount of solute}}{\text{Amount of solution}}$$

For molarity, the numerator (amount of solute) is expressed in moles and the denominator (amount of solution) is expressed in liters. Percent composition uses the same definition of concentration but with different quantities in the numerator and the denominator.

There are three types of **percent composition**, depending on whether the solute and solution values are given by mass or by volume:

Percent composition by mass	**Percent composition by volume**	**Percent composition by mass/volume**
$\frac{\text{Grams of solute}}{\text{Grams of solution}} \times 100$	$\frac{\text{Volume of solute}}{\text{Volume of solution}} \times 100$	$\frac{\text{Grams of solute}}{\text{Volume of solution}} \times 100$

Percent composition by mass is often used to describe the concentration of a solution prepared by dissolving a solid in a liquid solvent. Let's try one example. If you dissolve 10.0 g of sucrose in enough water to make the final solution have a mass of 100.0 g, the percent-by-mass concentration is 10.0 mass %:

Percent Composition by Mass

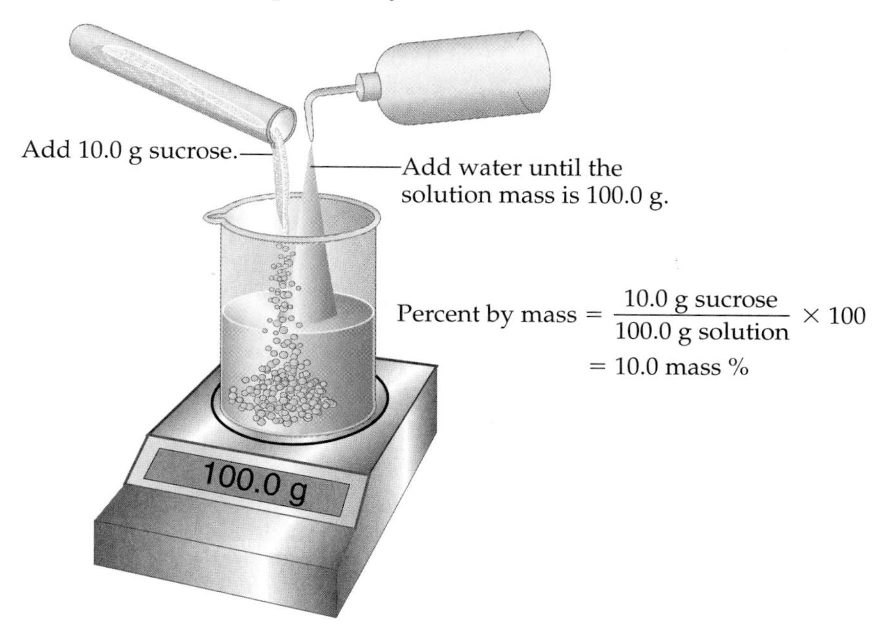
Add 10.0 g sucrose.
Add water until the solution mass is 100.0 g.
Percent by mass = $\frac{10.0 \text{ g sucrose}}{100.0 \text{ g solution}} \times 100$
= 10.0 mass %
100.0 g

When the solute is a liquid, percent composition by volume is often used. For example, dissolving 25.0 mL of the liquid ethanol in enough water to give a total solution volume of 100.0 mL produces a solution that has a concentration of 25.0 vol %:

Percent Composition by Volume

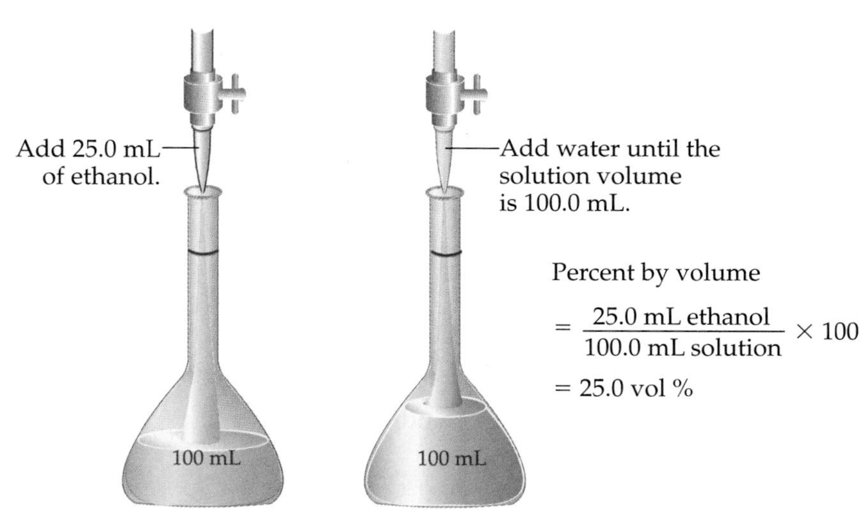
Add 25.0 mL of ethanol.
Add water until the solution volume is 100.0 mL.
Percent by volume
= $\frac{25.0 \text{ mL ethanol}}{100.0 \text{ mL solution}} \times 100$
= 25.0 vol %
100 mL
100 mL

Interestingly, you cannot prepare this solution by adding 75 mL of water to the 25 mL of alcohol because when you mix different liquids together, the volumes are usually not exactly additive. In the case of ethanol and water, a solution made by mixing 25 mL and 75 mL, respectively, has a volume measurably less than 100 mL. That is why the diagram says "Add water until the solution volume is 100 mL" instead of "Add 75 mL of water."

The examples we used to calculate percent compositions are a bit unrealistic because we set the denominator equal to 100.0 to make the calculations easy. In real situations, it's unlikely you will have exactly 100.0 g or 100.0 mL of a solution. For instance, suppose you have 273.0 g of a solution that you know contains 35.0 g of NaCl. What is its percent composition by mass? To find out, you use the same equation we used for our simpler example—percent by mass = (grams of solute/grams of solution) × 100:

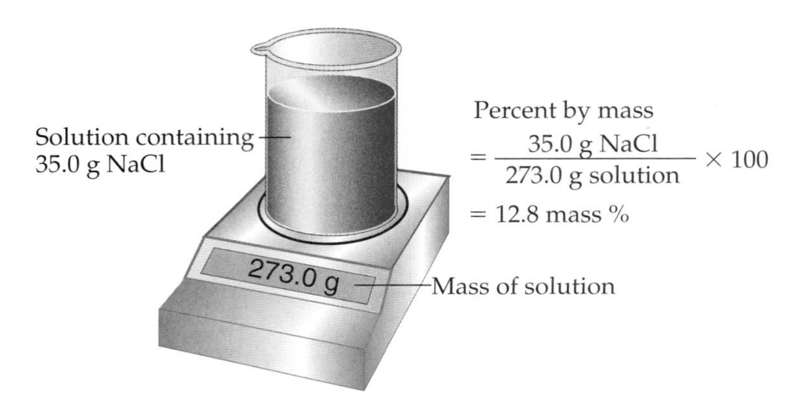

The percent composition for a solution can serve as a conversion factor, just as molarity can. For example, suppose you have 265.5 mL of an aqueous ethanol solution that is 30.0 vol % ethanol. The calculation for determining how much ethanol is in the bottle is

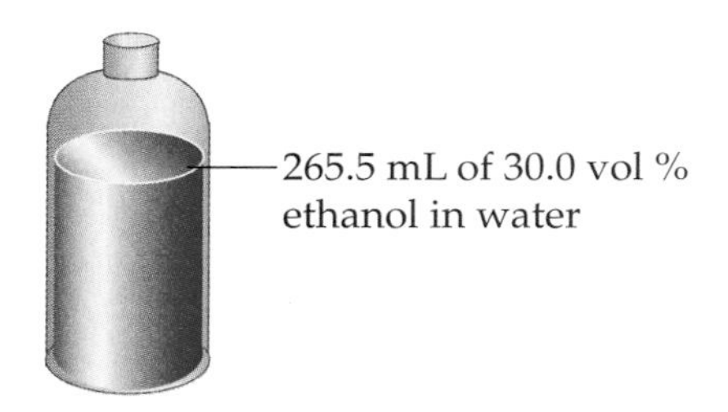

$$\underbrace{265.5 \text{ mL solution}}_{\text{Measurement}} \times \underbrace{\frac{30.0 \text{ mL ethanol}}{100 \text{ mL solution}}}_{\substack{\text{Vol \%} \\ \text{conversion factor}}} = 79.6 \text{ mL ethanol}$$

12.6 WORKPATCH The density of ethanol is 0.7893 g/mL. Use this information to determine the number of grams of ethanol in the 79.6 mL of 30 vol % solution just discussed.

The WorkPatch is one more reminder that many quantities in chemistry can serve as conversion factors.

The third type of percent composition (percent mass/volume) is less frequently seen and uses grams for amount of solute and milliliters for amount of solution.

Because there are three types of percent composition, it is always important to state which definition you are using. For example, a bottle of an ethanol–

water solution labeled 5% ethanol in water does not give you enough information about the solution's composition. If the 5% is by mass, 100 g of the solution contains 5 g of ethanol. If the 5% is by volume, 100 mL of the solution contains 5 mL of ethanol, which is only 3.9 g. This potential for ambiguity is one reason chemists prefer to use molarity.

Practice Problems

12.19 A solution is prepared by dissolving 25.0 g of sucrose in 175.0 g of water. Characterize its concentration by the appropriate percent composition.

Answer: The appropriate percent composition is by mass because both measurements are given in grams.

$$\text{Grams of solution} = \text{Grams of solute} + \text{Grams of solvent} = 25.0\text{ g} + 175.0\text{ g} = 200.0\text{ g}$$

$$\text{Percent composition by mass} = \frac{\text{Grams of solute}}{\text{Grams of solution}} \times 100 = \frac{25.0\text{ g}}{200.0\text{ g}} \times 100 = 12.5\text{ mass }\%$$

12.20 Gasohol is a solution of gasoline and ethanol. Every liter of gasohol contains 90.0 mL of ethanol dissolved in gasoline. Characterize the solute concentration by the appropriate percent composition. [*Hint:* Assume you have 1 liter of gasohol solution.]

12.21 Suppose you have 65.0 g of the solution from Practice Problem 12.19. How many grams of sucrose do you have?

12.8 Reactions in Solution

This chapter has been all about solutions—why they form, how they form, their concentrations—and we've saved the best for last—working with chemical reactions that take place in solution. This is an important topic because so many chemical reactions occur in solution. Your own body is approximately 80% by mass water, for instance, meaning that the biochemical reactions that occur in your cells do so in aqueous solution. How do they occur? What would you see if you could observe the reactants at the molecular level?

To begin answering these questions, let's return to the precipitation reaction that almost killed the laboratory technician in Chapter 7:

$$Ca^{2+}(aq) + 2\,F^-(aq) \longrightarrow CaF_2(s)$$

A molecular view would show you that the ions move about in a random, zigzag manner called *diffusion*. They move in this fashion because the solvent molecules are in constant motion, jostling past one another and constantly buffeting the ions and one another. You would also notice something else—each ion is surrounded by water molecules attracted to it by ion–dipole attractions.

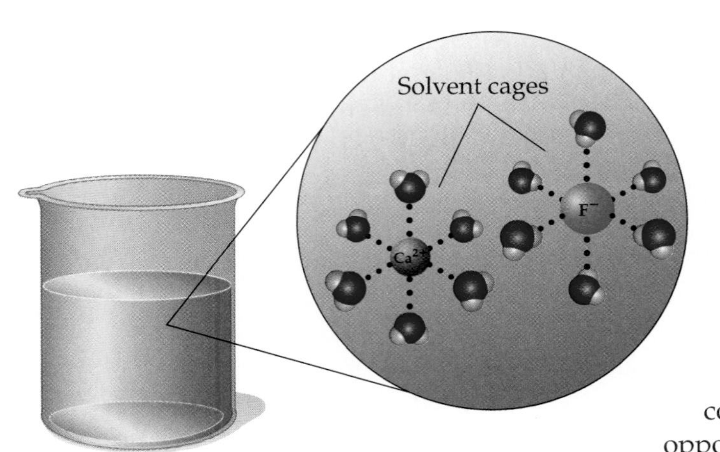

Calcium fluoride solution

The water molecules completely surround an ion, and in three dimensions this means the water molecules form a hollow sphere with the ion at the center. This sphere of water molecules is referred to as a **solvent cage**, and it gets "dragged" along with the ion as the ion diffuses through the solution. If they are to react with one another, the caged $Ca^{2+}(aq)$ and $F^-(aq)$ ions must diffuse toward one another. When they get close enough to allow their solvent cages to make contact, electrostatic attraction between the oppositely charged ions can occur and an ionic bond can form. When enough ions have come together in this way, solid CaF_2 precipitates. This explains how the reaction occurs in solution. Now let's turn to the subject of how much—stoichiometry, our old friend from Chapter 8.

STOICHIOMETRY

You will be pleased to know that the strategy for doing solution-based stoichiometry calculations is the same as what you learned earlier. After all, we are still dealing with balanced chemical equations that "speak" in moles, which means we must convert to moles in order to make use of the equations (*if in doubt, convert to moles!*). When we worked with pure substances, molar mass was the key, serving as a conversion factor between mass and moles. When it comes to solutions, molarity is the key, giving us a way of moving back and forth between a solution volume and the number of moles of reactant/product in that volume. All we need to do is treat molarity like a conversion factor:

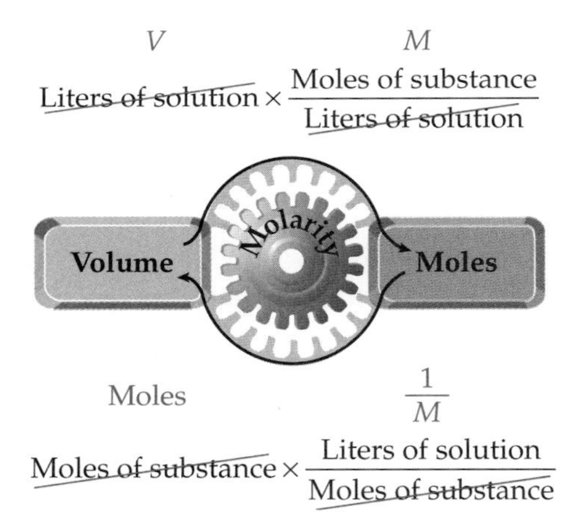

In other words, molarity times volume is moles ($M \times V$ = number of moles) and moles divided by molarity is volume (number of moles/$M = V$). This is very useful stuff when it comes to doing solution stoichiometry calculations, which means you should try exercising your skills now.

Practice Problems

12.22 How many moles of CaF_2 are there in 25.0 mL of 0.350 M $CaF_2(aq)$?

Answer:
$$\overset{V}{0.0250\ \cancel{\text{L solution}}} \times \overset{M}{\frac{0.350\ \text{mol } CaF_2}{1\ \cancel{\text{L solution}}}} = 0.00875\ \text{mol } CaF_2$$

Notice how the units cancel, leaving moles. Also notice that we converted the volume from milliliters to liters before multiplying by molarity. You must do this or else the units won't match! Never multiply molarity by milliliters if you are trying to calculate number of moles!!!

12.23 What volume of 0.350 M CaF_2 solution is required to obtain 0.00875 mole of CaF_2?

Answer: This is just the reverse of Practice Problem 12.22:

$$\overset{\text{Moles}}{0.00875\ \cancel{\text{mol } CaF_2}} \times \overset{\frac{1}{M}}{\frac{1\ \text{L solution}}{0.350\ \cancel{\text{mol } CaF_2}}} = 0.0250\ \text{L solution} = 25.0\ \text{mL}$$

12.24 How many moles of glucose are there in 255.0 mL of a 0.998 M solution of glucose?

12.25 What volume of 0.350 M $BaCl_2$ solution is required to obtain 0.500 mole of $BaCl_2$?

12.26 What volume of 0.350 M $BaCl_2$ solution is required to obtain 0.500 mole of $Cl^-(aq)$?

Did you see the warning in Practice Problem 12.22? Heed it well. Make sure you also do Practice Problems 12.25 and 12.26. The difference between them is subtle but important!

Armed with this new understanding of molarity as a conversion factor, you are ready to do some solution stoichiometry problems. The best way to show you how is to do a few for you. Follow them through.

Suppose we want to synthesize $CaF_2(s)$ by combining these solutions:

Combining them causes insoluble CaF_2 to precipitate (Chapter 7):

$$Ca^{2+}(aq) + 2\ F^-(aq) \longrightarrow CaF_2(s)$$

How much of each solution do we need to produce 1.00 g of $CaF_2(s)$ in a balanced fashion? The first step is to write the balanced equation for the reaction, which we did. The second step is to convert the mass of product we want, 1.00 g of $CaF_2(s)$, to moles (no surprise here).

Step 2: Convert mass of desired product to moles, using product's molar mass as conversion factor:

$$1.00 \text{ g } \cancel{\text{CaF}_2} \times \frac{1 \text{ mol CaF}_2}{78.074 \text{ g } \cancel{\text{CaF}_2}} = 0.0128 \text{ mol CaF}_2$$

Now speaking the right language, we can "ask" the balanced equation how many moles of each reactant we need. To do this, we use the result from step 2 along with conversion factors derived from the balanced equation.

Step 3: Determine number of moles of each reactant, obtaining conversion factors from balanced equation:

$$0.0128 \text{ } \cancel{\text{mol CaF}_2} \times \frac{1 \text{ mol Ca}^{2+}}{1 \text{ } \cancel{\text{mol CaF}_2}} = 0.0128 \text{ mol Ca}^{2+}$$

$$0.0128 \text{ } \cancel{\text{mol CaF}_2} \times \frac{2 \text{ mol F}^-}{1 \text{ } \cancel{\text{mol CaF}_2}} = 0.0256 \text{ mol F}^-$$

The final step is to calculate the volume of each reactant solution required to obtain these molar amounts. This is the job of molarity.

Step 4: Calculate volume of each solution needed, using solute molecular formulas and molarities as conversion factors.

For the $Ca(NO_3)_2$:

$$\underbrace{0.0128 \text{ } \cancel{\text{mol Ca}^{2+}}}_{\text{Moles of reactant needed}} \times \underbrace{\frac{1 \text{ } \cancel{\text{mol Ca(NO}_3)_2}}{1 \text{ } \cancel{\text{mol Ca}^{2+}}}}_{\text{From reactant molecular formula}} \times \underbrace{\frac{1 \text{ L Ca(NO}_3)_2 \text{ solution}}{0.100 \text{ } \cancel{\text{mol Ca(NO}_3)_2}}}_{1/M_{\text{reactant}}} = 0.128 \text{ L Ca(NO}_3)_2 \text{ solution}$$

For the NaF:

$$\underbrace{0.0256 \text{ } \cancel{\text{mol F}^-}}_{\text{Moles of reactant needed}} \times \underbrace{\frac{1 \text{ } \cancel{\text{mol NaF}}}{1 \text{ } \cancel{\text{mol F}^-}}}_{\text{From reactant molecular formula}} \times \underbrace{\frac{1 \text{ L NaF solution}}{0.100 \text{ } \cancel{\text{mol NaF}}}}_{1/M_{\text{reactant}}} = 0.256 \text{ L NaF solution}$$

Combining 128 mL of the calcium nitrate solution with 256 mL of the sodium fluoride solution will yield 1.00 g of $CaF_2(s)$, which we can isolate via filtration. There will be no unused $Ca^{2+}(aq)$ or $F^-(aq)$ because we performed the reaction in a balanced fashion.

We can also turn things around a bit. Instead of specifying how much product we want to make, we'll give you a specified amount of reactant solution and ask you to calculate the theoretical yield. For example, what is the maximum amount of $Ba(OH)_2$ produced by combining 120.0 mL of 0.250 M NaOH with an excess amount of $BaCl_2$? The strategy is always the same—begin by converting to moles.

Step 1: Write the balanced chemical equation:

$$Ba^{2+}(aq) + 2\,OH^-(aq) \longrightarrow Ba(OH)_2(s)$$

Step 2: Convert known amount of NaOH solution to moles:

$$\overset{V}{0.1200\ \text{L solution}} \times \overset{M}{\frac{0.250\ \text{mol NaOH}}{1\ \text{L solution}}} = 0.0300\ \text{mol NaOH}$$

Step 3: Use molecular formulas and the balanced equation to obtain conversion factors for determining number of moles of unknown—$Ba(OH)_2$:

$$0.0300\ \text{mol NaOH} \times \frac{1\ \text{mol OH}^-}{1\ \text{mol NaOH}} \times \frac{1\ \text{mol Ba(OH)}_2}{2\ \text{mol OH}^-} = 0.0150\ \text{mol Ba(OH)}_2$$

Step 4: Convert from moles to appropriate units:

$$0.0150\ \text{mol Ba(OH)}_2 \times \frac{171.341\ \text{g Ba(OH)}_2}{1\ \text{mol Ba(OH)}_2} = 2.57\ \text{g Ba(OH)}_2$$

If you think about it, we are doing solution stoichiometry calculations exactly as we did our stoichiometry calculations in Chapter 8, the only difference being that molarity has joined the party as a conversion factor. Here's the summary chart.

How to solve solution stoichiometry problems

Step 1: Write the balanced chemical equation.

Step 2: For each known reactant or product, convert the given information to moles.

Step 3: Use molar amounts from step 2 and conversion factors derived from molecular formulas and the balanced equation to calculate number of moles of desired reactant or product.

Step 4: Convert molar amounts from step 3 to desired units, using molarities and/or molar masses as conversion factors.

Now try your hand at some practice problems.

Practice Problems

12.27 How would you prepare 9.70 g of $PbCl_2(s)$ from a 0.100 M solution of $Pb(NO_3)_2$ and a 0.200 M solution of $CaCl_2$?

Answer: **Step 1:** Write the balanced chemical equation:

$$Pb^{2+}(aq) + 2\,Cl^-(aq) \longrightarrow PbCl_2(s)$$

Step 2: Convert given information to moles:

$$9.70 \text{ g } \cancel{PbCl_2} \times \frac{1 \text{ mol } PbCl_2}{278.1 \text{ g } \cancel{PbCl_2}} = 0.0349 \text{ mol } PbCl_2$$

Step 3: Determine moles of each reactant required:

$$0.0349 \text{ } \cancel{\text{mol } PbCl_2} \times \frac{1 \text{ mol } Pb^{2+}}{1 \text{ } \cancel{\text{mol } PbCl_2}} = 0.0349 \text{ mol } Pb^{2+}$$

$$0.0349 \text{ } \cancel{\text{mol } PbCl_2} \times \frac{2 \text{ mol } Cl^-}{1 \text{ } \cancel{\text{mol } PbCl_2}} = 0.0698 \text{ mol } Cl^-$$

Step 4: Convert from moles to desired units:

$$0.0349 \text{ } \cancel{\text{mol } Pb^{2+}} \times \frac{1 \text{ } \cancel{\text{mol } Pb(NO_3)_2}}{1 \text{ } \cancel{\text{mol } Pb^{2+}}} \times \frac{1 \text{ L } Pb(NO_3)_2 \text{ solution}}{0.100 \text{ } \cancel{\text{mol } Pb(NO_3)_2}} = 0.349 \text{ L } Pb(NO_3)_2 \text{ solution}$$

$$0.0698 \text{ } \cancel{\text{mol } Cl^-} \times \frac{1 \text{ } \cancel{\text{mol } CaCl_2}}{2 \text{ } \cancel{\text{mol } Cl^-}} \times \frac{1 \text{ L } CaCl_2 \text{ solution}}{0.200 \text{ } \cancel{\text{mol } CaCl_2}} = 0.175 \text{ L } CaCl_2 \text{ solution}$$

Combine 349 mL of the lead nitrate solution with 175 mL of the calcium chloride solution, and then use a funnel and filter paper to isolate the 9.70 g of $PbCl_2$ that forms.

12.28 Suppose you did as required in Practice Problem 12.27 but, because of poor filtering technique, isolated only 8.24 g of $PbCl_2$. What is the percent yield for your synthesis?

12.29 How would you prepare 20.0 g of iron(III) hydroxide from a 0.250 M solution of $Fe(NO_3)_3$ and a 0.150 M solution of $Ba(OH)_2$? How much product would you have isolated if your percent yield were 67.5%?

TITRATIONS

Now we want to tell you about a useful stoichiometric procedure for determining solution concentrations, a procedure you will do many times if your chemistry course includes a laboratory section. The process is called **titration**, and it entails using a known amount of a solution of known concentration to determine the concentration of some other solution.

A common application of titration is determining acid concentrations. Recall that in Chapter 7 we said that acids and bases are compounds that produce $H^+(aq)$ and $OH^-(aq)$ ions in water, respectively. You also learned that an acid and base neutralize each other to form water, in what is called an acid–base neutralization reaction:

$$H^+(aq) + OH^-(aq) \longrightarrow H_2O(l)$$

We can use this reaction to determine an acid concentration via titration. For example, suppose we have a hydrochloric acid solution of unknown concentration. To titrate it, we use a *buret*, which is a calibrated tube with a valve (called a *stopcock*), positioned above a collection flask. The buret is filled with 0.100 M NaOH. *To titrate an acid, it is important to know the exact concentration of the base in the buret!* Exactly 50.00 mL of the hydrochloric acid solution of unknown concentration is in the flask. *To titrate an acid, it is important to know the exact volume of acid in the flask!* We have also added to the flask a few drops of a colorless compound called an *indicator*, the purpose of which is described in a moment.

To do the titration, we open the stopcock and begin adding the base to the acid. Because the buret is calibrated and we fill it exactly to the 0.00-mL mark, we can measure the volume of base we drain from the buret. As it enters the flask, the base reacts with the acid, converting $H^+(aq)$ ions to water. If we add sufficient base, we can convert all of the acid's $H^+(aq)$ ions to water. When this occurs, we say the acid has been neutralized. *The key to determining the molarity of the acid is to determine how many moles of OH^-(aq) base were required to accomplish this neutralization.*

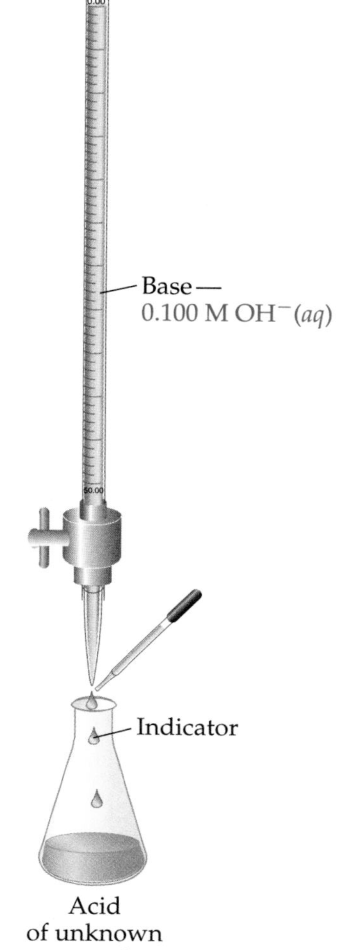

This is where the indicator comes in. We added a common indicator called phenolphthalein, which is colorless in acidic solution but red in basic solution. Therefore, the solution in the flask starts out colorless because the flask contains only our acid of unknown concentration. When we've added enough base to neutralize the acid and then just one more drop to make the solution basic, the phenolphthalein turns red, alerting us that neutralization has been accomplished. At this point, called the *equivalence point* (because at this point the number of moles of base we've added is equivalent to the number of moles of acid in the flask), we record the volume of base added. Once we know this volume, we multiply it by the base molarity to calculate the moles of base required for neutralization:

$$V_{\text{base added}} \times M_{\text{base}} = \text{Moles of base required for neutralization}$$

For simple acids like $HCl(aq)$ that can produce only one mole of $H^+(aq)$ per mole of acid, this is also the number of moles of acid that were in the flask. Remember this!

OK, let's do the titration. The buret is filled with our 0.100 M NaOH to the 0.00-mL mark, and the collection flask contains 50.00 mL of our acid of unknown concentration plus a few drops of phenolphthalein solution. We begin draining base from the buret to neutralize the acid. At a buret reading of 22.50 mL, the solution in the flask changes from colorless to pale red:

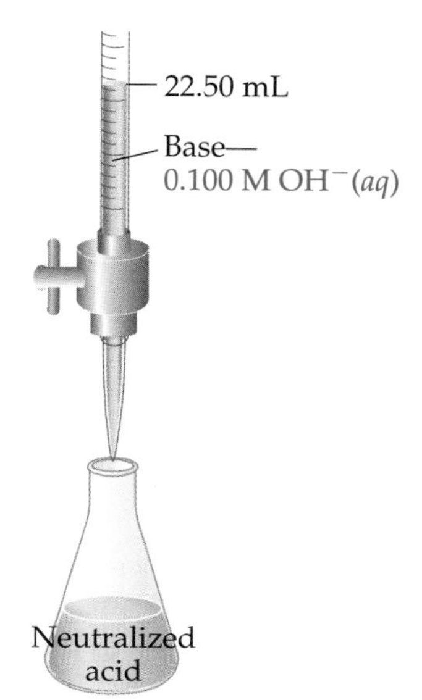

This change in color tells us it took 22.50 mL of base to reach the equivalence point ($V_{\text{base added}} = 22.50$ mL). The calculation we can now do to find the concentration of the acid is just like any other stoichiometry calculation.

Step 1: Write the balanced chemical equation:

$$H^+(aq) + OH^-(aq) \longrightarrow H_2O(l)$$

Step 2: Convert given information to moles. In this case, convert the volume of NaOH solution required to reach the equivalence point to moles of OH^-:

$$\overset{V_{\text{base}}}{0.02250\ \cancel{\text{L NaOH}}} \times \overset{M_{\text{base}}}{\frac{0.100\ \text{mol OH}^-}{1\ \cancel{\text{L NaOH}}}} = 0.002250\ \text{mol OH}^-$$

Step 3: Use moles of OH^- from step 2 and conversion factors from molecular formulas and the balanced equation to find moles of HCl in flask:

$$0.002250\ \cancel{\text{mol OH}^-} \times \underbrace{\frac{1\ \cancel{\text{mol H}^+}}{1\ \cancel{\text{mol OH}^-}}}_{\text{From the balanced equation}} \times \underbrace{\frac{1\ \text{mol HCl}}{1\ \cancel{\text{mol H}^+}}}_{\text{From the formula for HCl}} = 0.002250\ \text{mol HCl}$$

This is how many moles of acid were in the flask!

Step 4: Convert moles from step 3 to desired units. In this case, wanting the acid molarity, we divide moles of HCl by the volume of acid we originally put in the collection flask:

The titration reveals there was 0.002250 moles of HCl(aq) present.

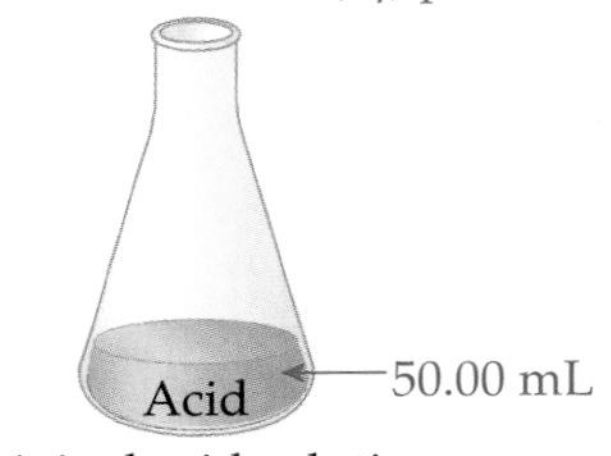

Original acid solution

$$M_{\text{acid}} \times \frac{0.002250\ \text{mol HCl}}{0.05000\ \text{L}} = 0.04500\ \text{M}$$

Try the following practice problems involving titration. The first one takes you through a titration of sulfuric acid, H_2SO_4, which can produce two $H^+(aq)$ ions per molecule of acid. Then Practice Problem 12.31 drills you on this concept.

Practice Problems

12.30 A 30.00-mL volume of aqueous sulfuric acid, H_2SO_4, is titrated using 0.200 M NaOH and an indicator that turns color only after all the sulfuric acid protons have reacted with OH^- ions. The solution turns color when 48.32 mL of base has been added. What is the molar concentration of the sulfuric acid?

Answer:

Step 1: Write the balanced neutralization equation:

$$H^+ + OH^- \longrightarrow H_2O$$

Step 2: Convert given amount of NaOH to moles of OH^-:

$$0.04832 \text{ L NaOH} \times \frac{0.200 \text{ mol } OH^-}{\text{L NaOH}} = 0.00966 \text{ mol } OH^-$$

Step 3: Use moles of OH^- from step 2 to calculate moles of H_2SO_4:

$$0.00966 \text{ mol } OH^- \times \frac{1 \text{ mol } H^+}{1 \text{ mol } OH^-} \times \frac{1 \text{ mol } H_2SO_4}{2 \text{ mol } H^+} = 0.00483 \text{ mol } H_2SO_4$$

Step 4: Convert moles of H_2SO_4 to molarity:

$$\frac{0.00483 \text{ mol } H_2SO_4}{0.0300 \text{ L}} = 0.161 \text{ M } H_2SO_4$$

12.31 A 50.00-mL volume of aqueous phosphoric acid, H_3PO_4, is titrated using 0.100 M NaOH and an indicator that turns color only after all the phosphoric acid protons have reacted with OH^- ions. The solution turns color when 38.60 mL of base has been added. What is the molar concentration of the phosphoric acid?

12.32 A 1.65-g sample of an acid that has one acidic proton per molecule is dissolved in water to give 25.00 mL of solution. It takes 27.48 mL of 1.000 M NaOH to neutralize the acid.

(a) What is the molar concentration of the acid?

(b) What is the molar mass of the acid?

(c) The empirical formula of the acid is CH_2O. What are the molecular formula and name of the acid?

12.9 Colligative Properties of Solutions

We end this chapter with a discussion of some very interesting properties of solutions. To be sure you have the proper background to understand these properties—called *colligative* properties—let's take a look first at some things happening in pure solvents rather than in solutions.

PURE SOLVENTS

You think you know something, and then all of a sudden you find out you don't. For example, suppose you are asked what happens to something when you heat it. This sounds like it could be the easy first question on some television quiz show.

What happens to something when you heat it?

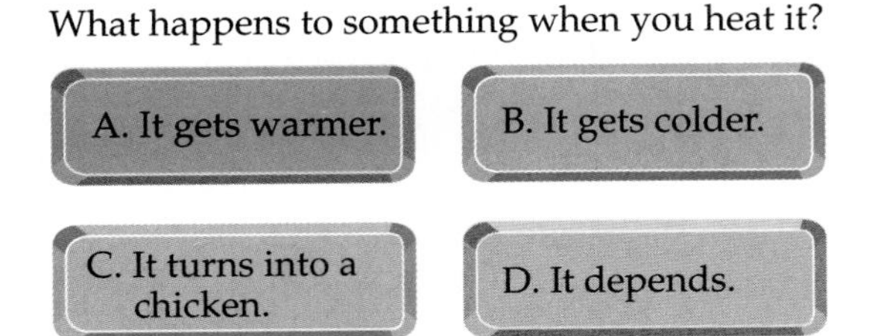

A. It gets warmer.
B. It gets colder.
C. It turns into a chicken.
D. It depends.

"It gets warmer—A. That's my final answer!"

"Oh! I'm sorry! The answer is D, it depends."

It depends? Yes, it depends. To show you why, we need to heat a substance, and to keep things uncomplicated, let's work with one you are familiar with, water. The two physical properties of water we are interested in here are its boiling point and its freezing point. The boiling point of water is 100°C, and the freezing point is 0°C. These numerical values are more properly called the **normal boiling point** and **normal freezing point**, with the word *normal* indicating that these are the freezing and boiling points at standard pressure (760 mm Hg = 1 atm).

What you may be less familiar with is water's behavior at these temperatures. For example, consider a beaker of water initially at room temperature (around 22°C) on a day when atmospheric pressure is exactly 760 mm Hg. If we place the beaker in a 200°C oven, the temperature of the water increases. Certainly this is no surprise. After all, things get warmer when you add heat to them. Or do they? When the water reaches 100°C, its temperature stops increasing even though it is still being heated in the 200° oven.

Water in a 200°C Oven

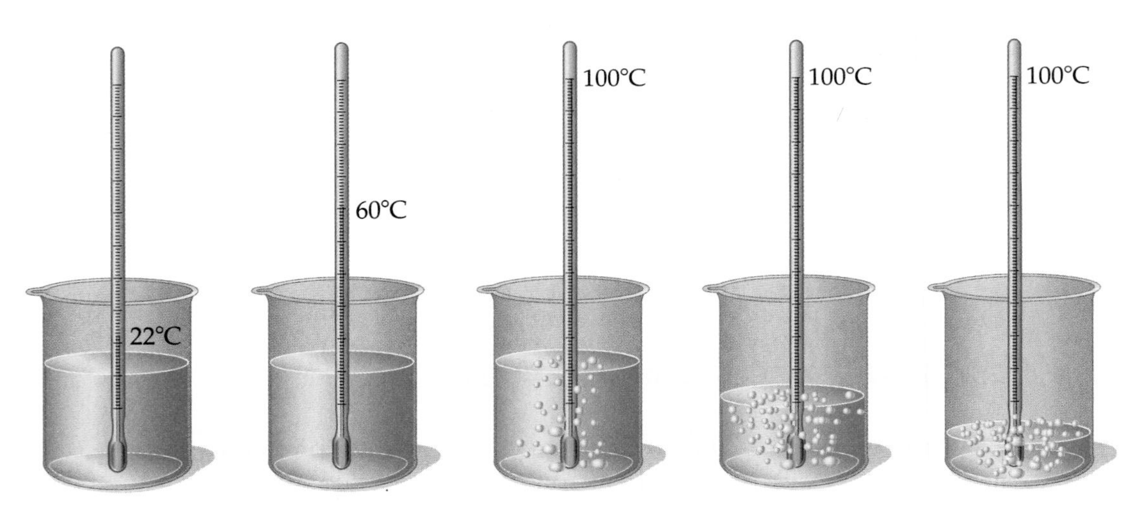

The temperature stays constant at 100°C and the water boils, as evidenced by the formation of bubbles (composed of water vapor). During boiling, the water is undergoing a rapid phase change from liquid to gas. Of course, the water was undergoing the same phase change before its temperature reached 100°C (some of the water was certainly evaporating), but that phase change occurred much more slowly and did not prevent the water's temperature increase. It is only when the liquid reaches its boiling point that additional heating does not increase its temperature as this phase change occurs.

Once the liquid water is all converted to vapor at 100°C, the temperature of the vapor starts to increase, and the vapor is ultimately heated to the temperature of the oven. Why is this so? How is it possible to put heat into liquid water and not have the water get hotter? The answer is that below the boiling point, the heat energy flowing into the water is mostly being used to increase the kinetic energy of the water molecules making them move faster. (As we'll see in Section 13.3, temperature is a measure of the average kinetic energy of molecules.) At the boiling point, however, adding heat no longer increases the molecules' kinetic energy. Now all the heat added is used to overcome the attractive forces (hydrogen bonds) between water molecules and separate them from one another so that they can escape into the gas phase.

We get this same type of temperature behavior in the other direction. If you place liquid water initially at room temperature in a freezer set at −20°C, the temperature of the water drops until it reaches 0°C. At this point the water temperature stops decreasing as the water changes phase from liquid to solid. Only when the phase change is complete does the temperature of the ice begin to decrease again.

This behavior is shown below on a plot of temperature versus heat added, called a *heating curve*.

So if anyone asks you whether or not the temperature of a particular substance increases when you heat the substance, your answer should be, "It depends! Even though a substance is being heated, its temperature remains constant anytime it is changing phase at its freezing (melting) point or at its boiling point. That's my final answer!"

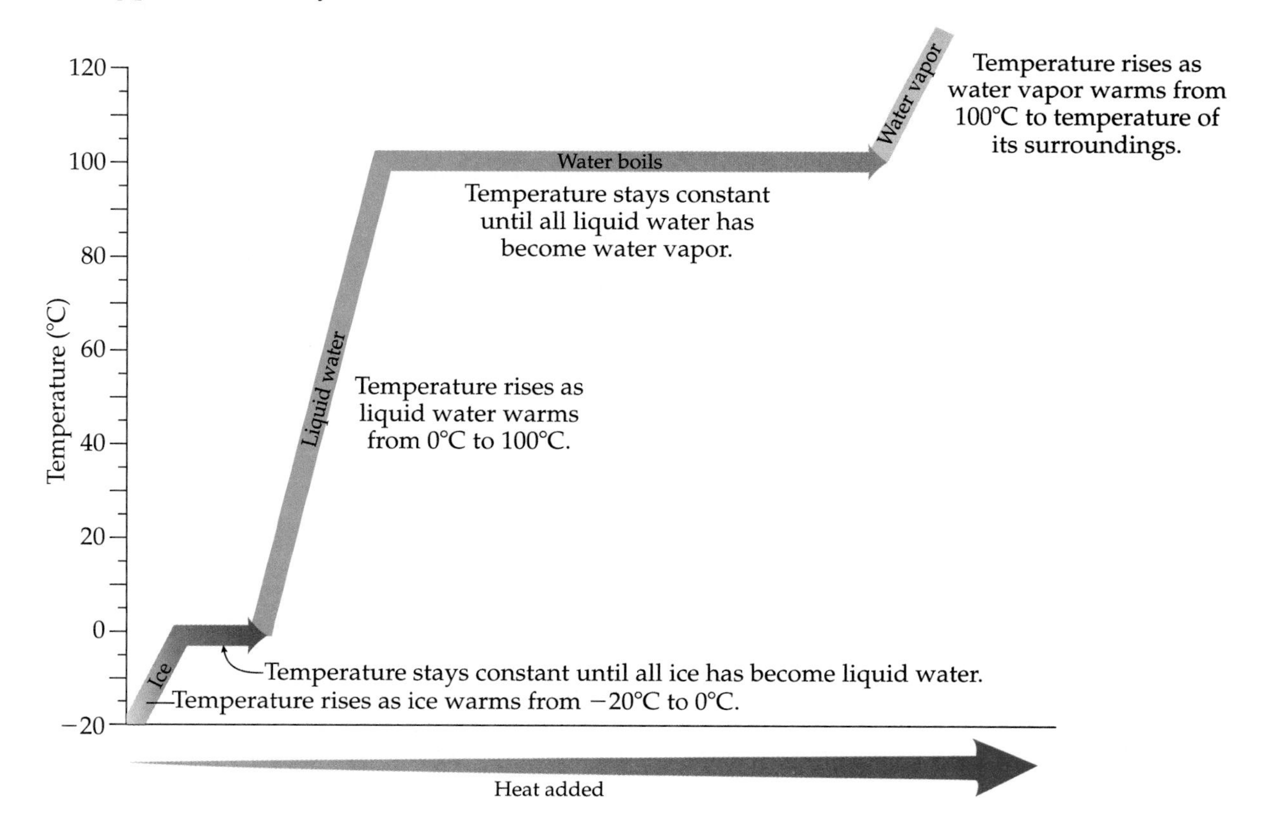

VAPOR PRESSURE

An interesting property of liquids is their "urge" to enter the gas phase even at temperatures well below their normal boiling points. To demonstrate this property, let's again use water and run a simple experiment with it. First we remove all the air from a flask connected to a vacuum pump as shown in the drawing and then close the stopcock.

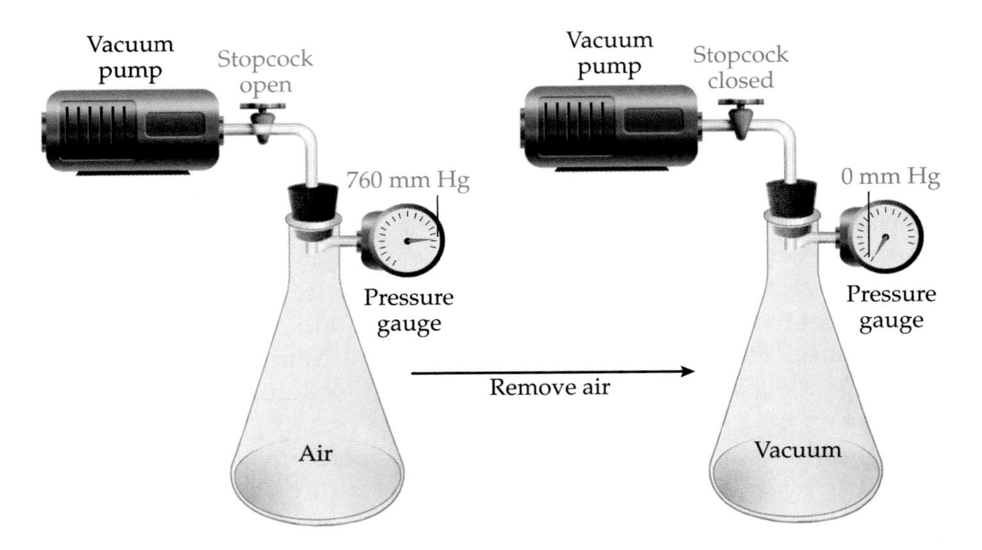

Because we removed all the air, the pressure inside the flask is 0 mm Hg. From this point on, we keep the stopcock closed so that no air can get back into the flask. Next, we use a syringe to place a small amount of water at 25°C in the flask. (The syringe lets no air into the flask.)

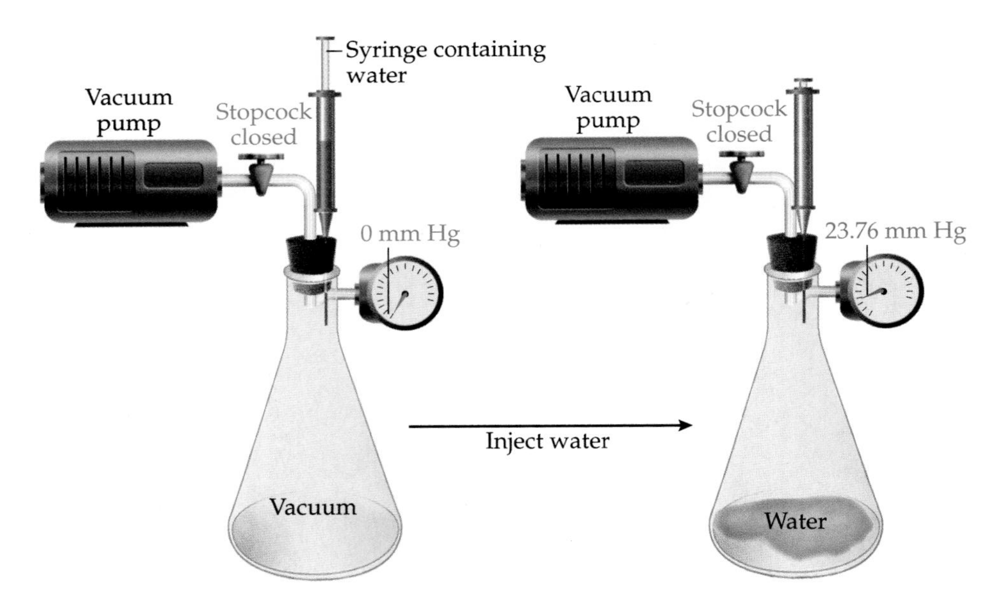

We keep injecting until a small puddle of water has formed at the bottom of the flask. At this point, the pressure inside the flask has increased to 23.76 mm Hg. If we continue to add water, the pressure remains constant at 23.76 mm Hg. The fact that the pressure in the flask is no longer zero tells us there must be some gas in the flask. That gas is water vapor. Evidently, some of the liquid water evaporated into the gas phase, causing the pressure to increase.

Why did the pressure in the flask stop increasing when it reached 23.76 mm Hg? The answer is that evaporation is not the only thing going on. At the same time, some of the gas-phase water molecules collide with the liquid surface and re-enter the liquid phase (condensation). At 23.76 mm Hg, the number of water molecules in the gas phase is enough to make the condensation rate equal to the evaporation rate.

Stopcock closed

23.76 mm Hg

Water

When these two rates are equal, the pressure above the liquid no longer changes. We call such a situation in which the rate of a forward process equals the rate of the reverse process a **dynamic equilibrium** and represent it by a set of two arrows:

$$H_2O(l) \rightleftharpoons H_2O(g)$$

The word *dynamic* reminds us that things are happening (in this case, evaporation and condensation) even though the situation appears static.

With this definition of dynamic equilibrium under our belts, we can see why the pressure in our flask stays constant at 23.76 mm Hg. Because there is no air in the flask, that pressure is exerted solely by water molecules in the gas state. Once the pressure reaches 23.76 mm Hg, which is the pressure at which the rate of the change $g \rightarrow l$ is the same as the rate of the change $l \rightarrow g$, the number of gas H_2O molecules no longer increases and so the pressure no longer increases. When the liquid and gas states of a substance are in dynamic equilibrium in this way, we say that the pressure exerted by the gas molecules is the **vapor pressure** of the liquid. (Strange but true—the pressure is caused by the gas phase but is said to be a characteristic of the liquid phase.)

Vapor pressure is a function of temperature, as Table 12.1 shows. Had we run our experiment at a different temperature, we would have found a different vapor pressure for water.

Table 12.1 Vapor Pressure of Water at Various Temperatures

Temperature (°C)	Vapor pressure (mm Hg)	Temperature (°C)	Vapor pressure (mm Hg)
0	4.58	55	118.0
5	6.54	60	149.4
10	9.21	65	187.5
15	12.79	70	233.7
20	17.54	75	289.1
25	23.76	80	355.1
30	31.82	85	433.6
35	42.2	90	525.8
40	55.3	95	633.9
45	71.9	100	760.0
50	92.5	105	906.1

What happens to the vapor pressure of liquid water as its temperature is increased?

A good way to think about the vapor pressure of a liquid is as a measure of its "urge" to evaporate. The higher the vapor pressure, the greater the tendency for liquid molecules to escape into the gas phase. You can picture the vapor pressure as an arrow that "lives" in the liquid and points up. The length of this arrow is a measure of the vapor pressure of the liquid. Now imagine a second arrow, one above the liquid and pointing down. This arrow represents any external pressure pushing down on the liquid. (For an open container, this external pressure is simply atmospheric pressure.) For example, we can envision the two arrows for an open beaker of water at 25°C as

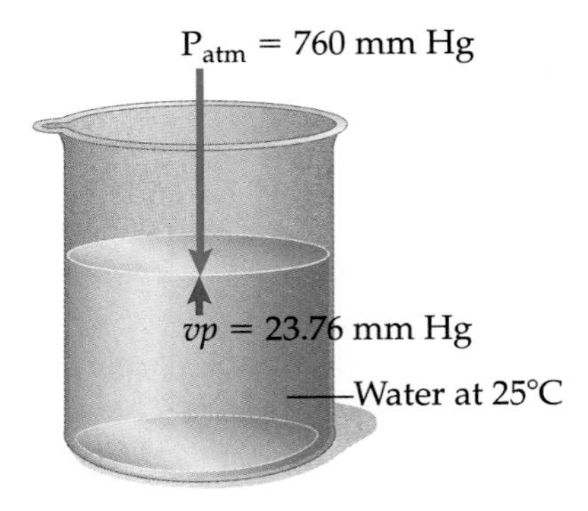

Getting the water to boil is a matter of dueling arrows. As we heat the water, the vapor pressure arrow gets longer until at 100°C, its length indicates a vapor pressure of 760 mm Hg (Table 12.1). This means that at 100°C, the length of the vapor pressure arrow matches the length of the external pressure arrow. It is at this point of equal pressures that the water boils.

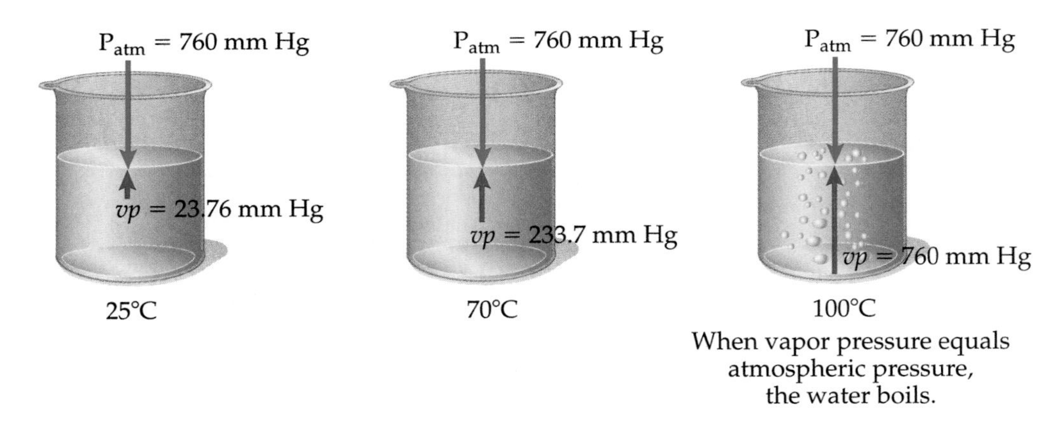

Any liquid boils when its vapor pressure becomes equal to or greater than the external pressure. Knowing this allows us to give another definition of normal boiling point. The **normal boiling point** of a liquid is the temperature at which its vapor pressure equals 760 mm Hg.

It is important to realize that water is changing phase from liquid to gas at all temperatures shown in the above diagram (as well as at all the intermediate temperatures not shown). As it is heated, the water evaporates at a faster and faster rate because its vapor pressure increases. It is said to become more *volatile* as its temperature increases. The word *volatile* means readily vaporizable.) However, only when the water reaches its boiling point and the vapor pressure equals the external pressure does boiling occur.

Once you understand how to make a liquid boil in terms of dueling arrows, you should realize that a liquid can boil at temperatures other than its normal boiling point. All you have to do is vary the external pressure. For example, in Denver, a city 1609 m above sea level, the average atmospheric pressure is only 0.83 atm (631 mm Hg). From Table 12.1 you can see that water has a vapor pressure of 631 mm Hg at approximately 95°C. Thus, you have to heat water to only 95°C in Denver to get it to boil (to get the arrows to be of equal lengths). Being 5°C cooler than they would be in water boiling at sea level, your noodles will take a bit longer to cook.

All liquids have a vapor pressure. For example, chloroform, CH_3Cl, is a useful industrial solvent. At 25°C, it has a vapor pressure of 190 mm Hg, much greater than the vapor pressure of water at the same temperature (23.76 mm Hg). The result is that, at any given temperature, chloroform has a much greater "urge" to evaporate than water, and we say that chloroform is more volatile than water. This fact implies that the boiling point of chloroform is lower than the boiling point of water, and that is indeed the case. The normal boiling point of chloroform is 61.7°C.

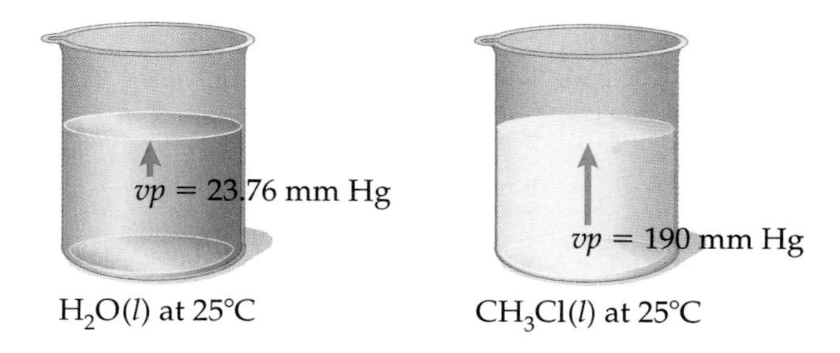

What is the vapor pressure of chloroform at 61.7°C?

Make sure you check your answer against the one at the end of the chapter. If you answered the WorkPatch correctly, you understand what has to happen to the vapor pressure of a liquid in order to make the liquid boil.

Generally speaking, the greater a liquid's vapor pressure at a given temperature, the more volatile the liquid is (the faster it evaporates).

Finally, we can understand this difference between chloroform and water by looking back to the material in Chapter 10 on intermolecular forces. The attractive forces between chloroform molecules are of the dipole–dipole and London type, both types weaker than the hydrogen-bonding attractions between H_2O molecules in liquid water. Therefore, chloroform molecules have an easier time overcoming one another's grasp and escaping into the vapor phase.

SOLUTIONS

It had been slightly below freezing for the past week, and Jim couldn't wait to try out his new ice skates. The freshwater pond in the backyard had frozen solid. It was a very small pond, though, and Jim wished for a much larger expanse of ice. There was a nearby ocean inlet, but it was still liquid. Why, he wondered, hadn't it frozen yet? Luckily, Jim's grandfather was a chemist, and he answered Jim's question by explaining that the ocean is not pure water. Rather, it is a solution of various salts (mostly NaCl).

"Jim," he said, "the physical properties of a liquid change as soon as you dissolve a solute in it. The boiling point goes up (*boiling-point elevation*), the freezing point goes down (*freezing-point depression*), and the vapor pressure decreases (*vapor-pressure lowering*). *Stay awake, Jim!*"

Thanks, Grandpa; we'll take it from here. The amount the boiling point goes up and the freezing point goes down are represented by the symbols ΔT_b and ΔT_f, respectively:

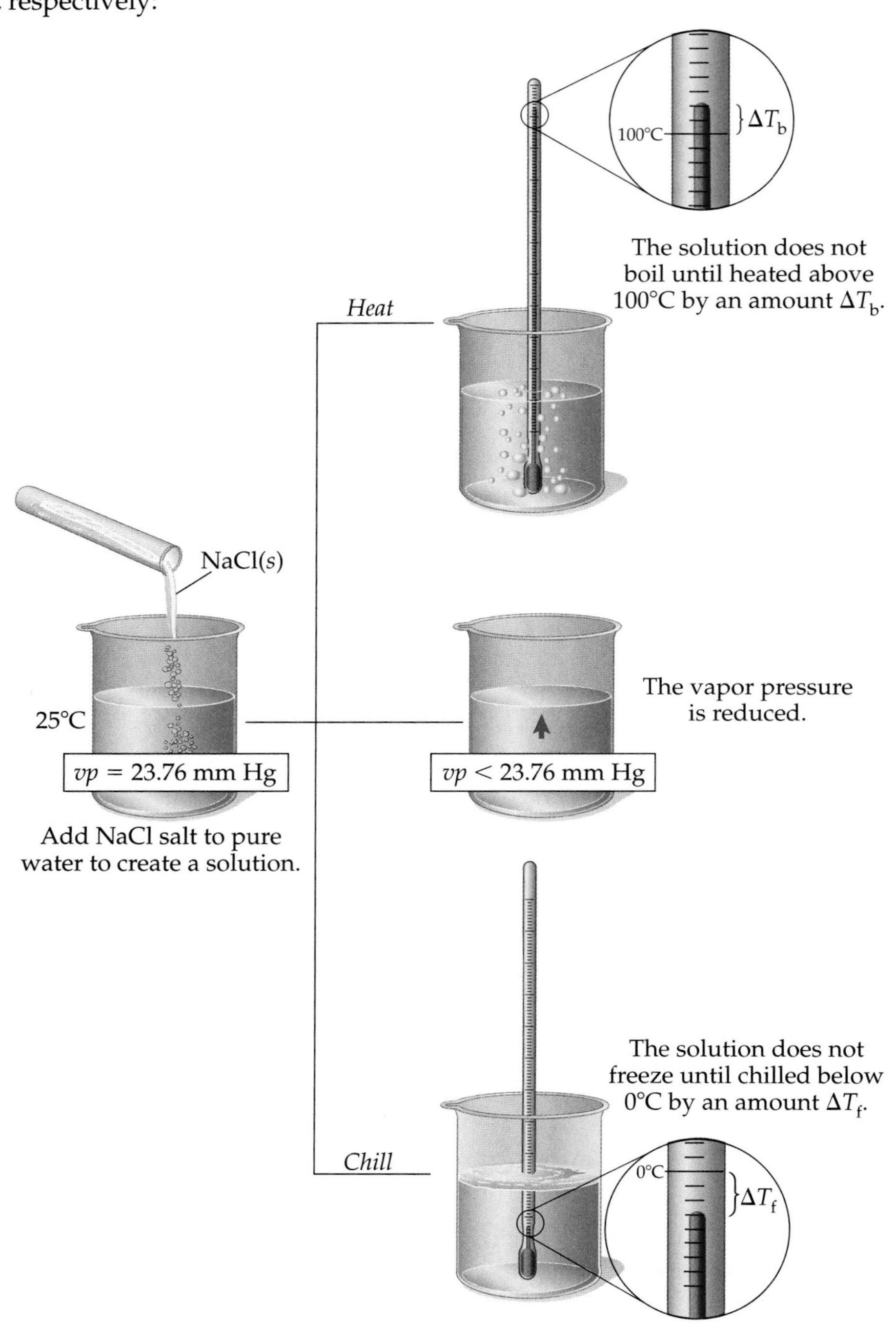

100°C
}ΔT_b
The solution does not boil until heated above 100°C by an amount ΔT_b.
Heat
NaCl(s)
25°C
vp = 23.76 mm Hg
Add NaCl salt to pure water to create a solution.
The vapor pressure is reduced.
vp < 23.76 mm Hg
The solution does not freeze until chilled below 0°C by an amount ΔT_f.
Chill
0°C
}ΔT_f

An aqueous solution is therefore more difficult to boil (you have to heat it to above 100°C) and more difficult to freeze (you have to cool it to below 0°C) than pure water. The elevation of the boiling point, the depression of the freezing point, and the lowering of the vapor pressure are called *colligative properties* of a solution. A **colligative property** is one that depends on the number of solute particles in a solution but not on their identity. *This is an important point!* How much the boiling point goes up, how much the freezing point goes down, and how much the vapor pressure goes down depend *only on the number of dissolved solute particles*, not on what they are. For example, consider two beakers containing equal volumes of water, and pay close attention to the amount of solute added to each:

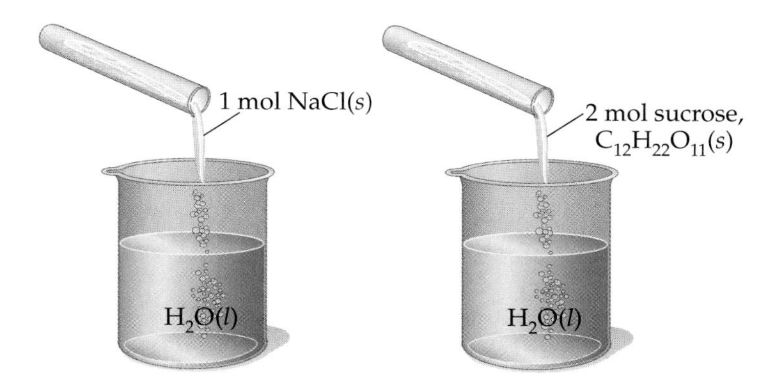

How many moles of dissolved solute particles (molecules or ions) are present in each beaker in the above diagram? [*Hint*: Don't forget that soluble salts dissociate into ions when they dissolve.]

The two solutions contain equal numbers of dissolved solute particles. Therefore, the two solutions experience the same increase ΔT_b in boiling-point elevation and the same decrease ΔT_f in freezing-point depression. The fact that the solute particles are aqueous sugar molecules in one solution and $Na^+(aq)$ and $Cl^-(aq)$ ions in the other is not important. The only thing that is important is the total number of dissolved solute particles in a given volume of solution.

Finally, it is possible to calculate ΔT_b and ΔT_f for a solution by using the equations

$$\Delta T_f = \frac{K_f \times \text{Moles of solute particles}}{\text{Kilograms of solvent}}$$

$$\Delta T_b = \frac{K_b \times \text{Moles of solute particles}}{\text{Kilograms of solvent}}$$

The K_f and K_b are called the *freezing-point constant* and *boiling-point constant*, respectively, and each solvent has its own values for these constants, as Table 12.2 shows. We can use these equations to calculate the freezing and boiling points of ocean water. On average, ocean water contains about 35.0 g (0.599

Table 12.2 Freezing-Point and Boiling-Point Constants for Some Common Solvents

Solvent	K_f $\left(\frac{C° \cdot \text{kg solvent}}{\text{Moles solute}}\right)$	Normal freezing point (°C)	K_b $\left(\frac{C° \cdot \text{kg solvent}}{\text{Moles solute}}\right)$	Normal boiling point (°C)
Acetic acid, CH_3CO_2H	3.9	16.7	3.07	118
Benzene, C_6H_6	5.12	5.53	2.53	80.1
Camphor, $C_{10}H_{16}O$	40	179.5	—	—
Carbon tetrachloride, CCl_4	31.8	−22.96	5.03	76.5
Cyclohexane, C_6H_{12}	20.0	6.47	2.79	80.7
Ethanol, C_2H_5OH	1.99	−115	1.22	78.4
Methanol, CH_3OH	—	−97.8	0.80	64.7
Naphthalene, $C_{10}H_{18}$	6.9	80.2	5.65	218
Phenol, C_6H_6O	7.40	43	3.56	181.8
Tertiary butanol, C_4H_9OH	8.3	25.6	—	82.4
Water, H_2O	1.86	0.00	0.52	100.0

mole) of NaCl per kilogram of water. This means there are 2 × 0.599 mole = 1.20 moles of solute particles per kilogram of water. Knowing this allows us to calculate ΔT_f and ΔT_b:

$$\Delta T_f = \frac{K_f \times \text{Moles of solute particles}}{\text{Kilograms of solvent}}$$

$$= \frac{\frac{1.86\ C° \cdot \cancel{\text{kg solvent}}}{\cancel{\text{mol solute particles}}} \times 1.20\ \cancel{\text{mol solute particles}}}{\cancel{\text{kg solvent}}} = 2.23\ C°$$

$$\Delta T_b = \frac{K_b \times \text{Moles of solute particles}}{\text{Kilograms of solvent}}$$

$$= \frac{\frac{0.52\ C° \cdot \cancel{\text{kg solvent}}}{\cancel{\text{mol solute particles}}} \times 1.20\ \cancel{\text{mol solute particles}}}{\cancel{\text{kg solvent}}} = 0.62\ C°$$

These are the changes to the normal freezing and boiling points. To find the depressed freezing point, we must subtract ΔT_f from the normal freezing point of pure water:

$$\text{Depressed freezing point} = 0.00°C - 0.62\ C° = -0.62°C$$

To find the elevated boiling point, we must add ΔT_b to the normal boiling point of pure water:

$$\text{Elevated boiling point} = 100.00°C + 2.23\ C° = 102.23°C$$

Try the following practice problems. The first two are worked out for you, and the second one is very interesting. It shows you how to calculate the molar mass of a solute without knowing its molecular formula (similar to what we did for gases using the ideal gas law in Chapter 11). This is an important use of colligative properties.

Practice Problems

12.33 Automotive antifreeze is essentially the compound ethylene glycol, $C_2H_6O_2$, molar mass 62.07 g/mol. Suppose you fill your car radiator with 6.00 L of water and then add 4.00 kg of ethylene glycol. At what temperature will this mixture freeze? (The density of water is 1.00 g/mL.)

Answer:

$$\Delta T_f = \frac{\overbrace{\frac{1.86\ C° \cdot \cancel{kg\ H_2O}}{\cancel{mol\ C_2H_6O_2}}}^{K_f} \times \overbrace{4.00\times10^3\ \cancel{g\ C_2H_6O_2} \times \frac{1\ \cancel{mol\ C_2H_6O_2}}{62.07\ \cancel{g\ C_2H_6O_2}}}^{\text{Moles of solute}}}{\underbrace{6.00\ \cancel{L\ H_2O} \times \frac{1000\ \cancel{mL\ H_2O}}{1\ \cancel{L\ H_2O}} \times \frac{1.0\ \cancel{g\ H_2O}}{1\ \cancel{mL\ H_2O}} \times \frac{1\ \cancel{kg\ H_2O}}{1000\ \cancel{g\ H_2O}}}_{\text{Kilograms of solvent}}} = 20.0\ C°$$

Therefore the new freezing point = 0°C − 20 C° = −20°C

12.34 A white powder is analyzed via combustion analysis. Results indicate its empirical formula is CH_2O. To determine its molecular formula, a chemist dissolves 117.5 g into 1.175 kg of water, and the resulting solution freezes at −1.03°C. What is the molecular formula?

Answer: The molar mass of the empirical formula CH_2O is 30.026 g/mol. If we can determine the molar mass of the compound, we can divide it by 30.026 g/mol to find out how many empirical units there are in one molecular formula (Section 8.5). To do this, all we need do is solve the freezing-point-depression equation for moles of solute (we know everything else):

$$\text{Moles of solute} = \frac{\Delta T_f \times \text{kg solvent}}{K_f} = \frac{1.03\ \cancel{C°} \times 1.175\ \cancel{kg\ H_2O}}{1.86\ \cancel{C°} \cdot \cancel{kg\ H_2O}/\text{mol solute}} = 0.651\ \text{mol solute}$$

Thus we added 0.651 mole of solute to the water, but we also know that the mass of this number of moles of the compound is 117.5 g. To find the molar mass of the compound, we divide this mass by the number of moles:

$$\text{Molar mass} = \frac{117.5\ \text{g}}{0.651\ \text{mol}} = 180\ \text{g/mol}$$

Finally, 180 g/mol ÷ 30.026 g/mol = 6. Therefore, the molecular formula consists of six empirical formulas, which means the molecular formula must be $C_6H_{12}O_6$ (it's glucose!).

12.35 A solution is prepared at sea level (external pressure = 1 atm) by dissolving 100.0 g of calcium nitrate in 450.0 g of water. At what temperature will this solution have a vapor pressure of 760 mm Hg? [*Hint*: What does an aqueous solution at sea level do when its vapor pressure is 760 mm Hg?]

12.36 A student dissolves 45.0 g of an unknown solid in 225.0 g of cyclohexane. It is known to dissolve without dissociating. She cools the solution and finds that the temperature remains at 2.70°C while the solution changes phase from liquid to solid. What is the molar mass of the solid?

We have spent most of this chapter examining solutions—what they are, why solutes dissolve, how to quantify the amount of solute dissolved, how the properties of solvents change when solutes are added, and how to do stoichiometry calculations for reactions that occur in solution. This emphasis on solutions is certainly understandable given that we are aqueous beings living on a planet covered with water, one of the best all-purpose solvents known. So the next time you dissolve some sweetener in your iced tea, add some antifreeze to a car radiator, or wipe a stain clean with a wet cloth, think about all that is happening as solute and solvent mingle, and you will have a newfound respect for solutions.

Have You Learned This?

SKILLS TO KNOW

Energetics of dissolving

A substance will disolve in a given solute if the energy released by solvation ($\Delta E_{solvation}$) is greater than the energy required for solute separation and solvent separation. If $\Delta E_{solvation}$ is slightly less than the amount required, the solution is likely still to dissolve, owing to the effect of entropy.

These results can be summed up as "like dissolves like."

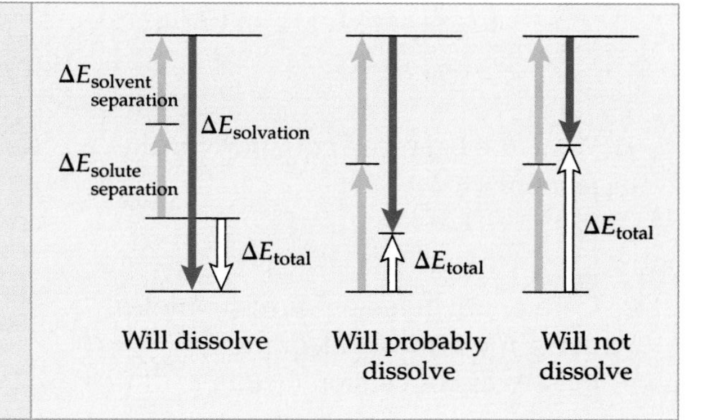

Molarity

Molarity (M) is one of the most common ways to characterize a solution's concentration.	$\text{Molarity} = \frac{\text{Moles of solute}}{\text{Liters of solution}}$
Molarity times volume always gives the moles of solute present.	Molarity × Volume = Moles of solute

How to prepare a solution of desired molarity

Example: Prepare 250.0 mL of a 1.50 M solution of NaCl.

1. Determine the moles of solute that you need in order to prepare the solution. This quantity equals the desired volume in liters times the desired molarity: $\text{Moles of solute needed} = \text{Volume of solution} \times \text{Molarity of solution}$	$0.2500\ \cancel{L} \times \frac{1.50\ \text{mol NaCl}}{\cancel{L}} = 0.375\ \text{mol NaCl}$
2. From scratch: Convert the moles of solute to grams by using the molar mass as a conversion factor.	$0.375\ \cancel{\text{mol NaCl}} \times \frac{58.443\ \text{g NaCl}}{1\ \cancel{\text{mol NaCl}}} = 21.9\ \text{g NaCl}$
By diluting a stock solution*: Convert moles of solute to volume of stock solution by using the molarity of the stock solution as a conversion factor. (Multiply the moles of solute by the inverse of the stock solution's molarity.	Using 2.00 M NaCl stock solution: $0.375\ \cancel{\text{mol NaCl}} \times \frac{1\ \text{L}}{2.00\ \cancel{\text{mol NaCl}}} = 0.188\ \text{L stock solution} = 188\ \text{mL}$
3. Add enough water to reach the desired volume.	Add water to make 250.0 mL of solution.

* Alternatively, you can combine steps 1 and 2 and use the dilution equation, as described in the text.

Solving solution stoichiometry problems

Example: Given 0.100 M solutions of $Ca(NO_3)_2$ and NaF, how much of each solution would you need to prepare 1.00 g of $CaF_2(s)$?

1. Write the balanced equation for the reaction.	Net ionic reaction: $Ca^{2+}(aq) + 2\,F^{-}(aq) \longrightarrow CaF_2(s)$
2. Convert all given amounts to moles, using molarities and/or molar masses as conversion factors.	$1.00\ \text{g CaF}_2 \times \frac{1\ \text{mol CaF}_2}{78.074\ \text{g CaF}_2} = 0.0128\ \text{mol CaF}_2$
3. Use the known molar amounts from step 2 to calculate the molar amounts of the substances the question asks for. Use conversion factors derived from the formulas and the balanced equation.	$0.0128\ \text{mol CaF}_2 \times \frac{1\ \text{mol Ca}^{2+}}{1\ \text{mol CaF}_2} = 0.0128\ \text{mol Ca}^{2+}$ $0.0128\ \text{mol CaF}_2 \times \frac{2\ \text{mol F}^-}{1\ \text{mol CaF}_2} = 0.0256\ \text{mol F}^-$
4. Convert the molar amounts from step 3 to the desired units, using molarities and/or molar masses as conversion factors.	$\underbrace{0.0128\ \text{mol Ca}^{2+}}_{\text{Moles of reactant needed}} \times \underbrace{\frac{1\ \text{mol Ca(NO}_3)_2}{1\ \text{mol Ca}^{2+}}}_{\text{From formula of reactant}} \times \underbrace{\frac{1\ \text{L solution}}{0.100\ \text{mol Ca(NO}_3)_2}}_{1/M_{reactant}}$ $= 0.128$ L solution $\underbrace{0.0256\ \text{mol F}^-}_{\text{Moles of reactant needed}} \times \underbrace{\frac{1\ \text{mol NaF}}{1\ \text{mol F}^-}}_{\text{From formula of reactant}} \times \underbrace{\frac{1\ \text{L solution}}{0.100\ \text{mol NaF}}}_{1/M_{reactant}} = 0.256\ \text{L solution}$

Colligative effects

Adding a solute to a solvent has the following colligative effects on the solvent's physical properties: • The freezing point is depressed by an amount ΔT_f. • The boiling point is elevated by an amount ΔT_b. • The vapor pressure is decreased.	$\Delta T_f = \frac{K_f \times \text{Moles of solute particles}}{\text{kg solvent}}$ $\Delta T_b = \frac{K_b \times \text{Moles of solute particles}}{\text{kg solvent}}$ where ΔT_f and ΔT_b are in C°, and K_f and K_b are constants for the given solvent and have units of (C° • kg solvent)/(moles of solute particles).

WHAT IS A SOLUTION?

12.37 Define the terms *solute*, *solvent*, and *solution*.

12.38 Suppose you mixed a small amount of table sugar with a large amount of flour and then spent hours grinding the mixture to a very fine powder. Is this mixture a solution? Explain.

12.39 Is it appropriate to call a soft drink an aqueous solution? Justify your answer.

12.40 Is it appropriate to call the atmosphere we breathe a solution? Justify your answer.

12.41 Vinegar is a common household solution that we consume. What is the solvent and what is the solute in vinegar?

12.42 What is a solid solution? Give some examples.

12.43 Classify the following as solutions or heterogeneous mixtures:
(a) 14 karat gold (prepared by mixing 10 parts molten copper with 14 parts molten gold and then allowing the substance to cool until it solidifies)
(b) Filtered ocean water
(c) A piece of wood
(d) Your exhaled breath
(e) A bottle of salad oil and vinegar shaken extremely well

12.44 Suppose you had a 50:50 homogeneous mixture of oxygen gas in helium gas. Which would you call the solvent and which would you call the solute?

12.45 A solution of table salt and table sugar in water is allowed to evaporate, leaving behind the two solid solutes. Is what remains a solution or a heterogeneous mixture? Explain your answer. [*Hint:* As solids, both solutes exist as lattices that would exclude each other.]

ENERGY AND THE FORMATION OF SOLUTIONS

12.46 Is there anything wrong with the following diagram? Explain your answer.

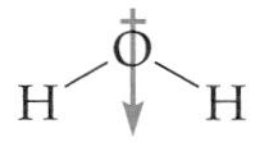

12.47 What evidence is there in your everyday experience to indicate that intermolecular attractive forces must exist between water molecules?

12.48 If there were no attractive forces between water molecules, what phase or phases of water would you expect to exist? Explain your answer.

12.49 In what way is melting a piece of NaCl similar to dissolving it in water?

12.50 Why does it take such a high temperature to melt NaCl but a much lower temperature to dissolve it in water?

12.51 When an ionic substance such as NaCl dissolves, the crystal lattice has to break apart to release the individual ions into the solution. Does this part of the dissolving process absorb energy or release energy? Explain your answer.

12.52 In dissolving any solute, room must be made in the solvent to accommodate solute particles. Does making this room absorb energy or release energy? Explain why.

12.53 What is *hydration*?

12.54 Does hydration absorb energy or release energy? Explain your answer.

12.55 List the three steps that occur as NaCl dissolves in water and illustrate each with a diagram.

12.56 Solid sucrose consists of individual sucrose molecules fixed in a lattice. The molecules are hydrogen-bonded to one another via OH groups. Knowing this, comment on which takes more energy, breaking up an NaCl lattice or breaking up a sucrose lattice. Justify your answer.

12.57 When NaCl dissolves, what helps keep the dissolved Na^+ and Cl^- ions from coming back together and reforming the lattice, precipitating the solid?

12.58 What is the name of the attractive force between dissolved Na^+ ions and water molecules? Diagram this force, showing how a water molecule would approach an Na^+ ion. Do the same for a Cl^- ion.

12.59 Some alcohols are quite soluble in water. For example, isopropyl alcohol, shown below, is sold as an aqueous solution we call rubbing alcohol. Show how a water molecule would be attracted to isopropyl alcohol, and name the strongest intermolecular force involved. Would you call this interaction hydration?

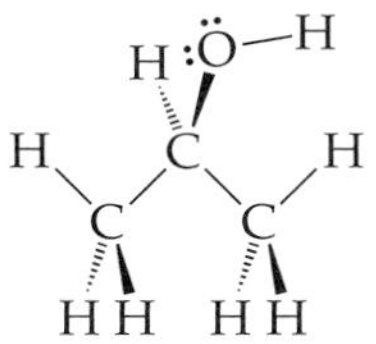

12.60 Why is the amount of energy associated with the solvation step of the dissolving process so critical to whether or not a solute will dissolve?

12.61 Consider an ionic solid dissolving in a liquid. Suppose the energy released as solute–solvent interactions take place is substantially less than the energy required to break up the lattice and to make room in the solvent. Would the solid be very soluble in this liquid? Explain your answer fully.

12.62 When some salts dissolve in water, the water heats up, even when no external heat is added. How is this possible?

12.63 When some salts dissolve in water, the water gets colder. How is this possible?

12.64 Suppose the hydration energy for an ionic compound is much less than the energy required to pull apart the lattice. Is such a compound likely to be soluble or insoluble in water? Explain your answer.

12.65 Sodium chloride is very soluble in water but insoluble in liquid hexane, C_6H_{14}. Why is this so?

12.66 Even though NaCl does not dissolve in hexane, C_6H_{14}, if we imagine the dissolving process for this system, there is one step that would take less energy than the corresponding step for NaCl dissolving in water. Which step is it? Why would it require less energy?

12.67 Which has the greater hydration energy, $AlCl_3$ or NaCl? Explain your answer.

12.68 In theory, it is possible for one ionic compound to have a greater hydration energy than another and still be the less soluble of the two. What would have to be true for this to be so?

12.69 What is meant by the rule of thumb "like dissolves like"?

12.70 What do we mean by the total energy change ΔE_{total} for the dissolving process, and why is it important to know about this energy change?

12.71 Consider the following diagram and answer the questions below:

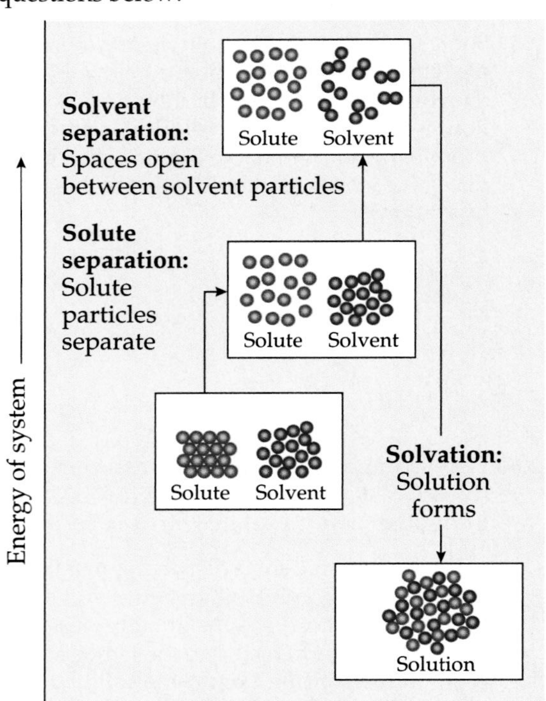

(a) Why does solute separation require energy?
(b) Why does solvent separation require energy?
(c) Why does solvation release energy?
(d) On the basis of this diagram, would you expect the solute to be soluble or insoluble in the solvent? Explain.
(e) Assuming the solute does dissolve, would you expect the solution to get warmer or colder? Explain.

12.72 When we write (*aq*) for a dissolved ion, such as $Na^+(aq)$, exactly what does it mean? Answer both in words and with a diagram.

12.73 Suppose you have two ionic compounds, A and B. The only significant difference between them is that the ionic bonds that hold the lattice together in compound A are stronger than those in compound B. Which compound would you expect to be more soluble in water? Explain your answer.

12.74 When a gaseous solute dissolves in water, which step in the dissolving process is essentially skipped? Explain why.

12.75 When a liquid solute dissolves in water, there is still a solute-separation step that absorbs energy, but the step doesn't require breaking up a crystal lattice as for a typical solid solute. What happens in this step with a liquid solute, and why does the step absorb energy?

12.76 These three substances are all liquids at room temperature:

CH_3Cl $\qquad$ CCl_4 $\qquad$ CH_3OH

Which do you expect to be least soluble in water? Most soluble in water? Explain your answers fully.

12.77 Why is oil insoluble in water? (Don't just say one is polar and one is nonpolar; give an answer based on energy considerations.)

ENTROPY AND THE FORMATION OF SOLUTIONS

12.78 What is entropy?

12.79 What does the second law of thermodynamics say?

12.80 What is the justification for the second law of thermodynamics?

12.81 When you release a drop of blue food dye into a beaker of water, the drop eventually dissolves to give a homogeneously light blue solution. No matter how long you wait, the dye molecules will never regroup to form the original drop. Why not?

12.82 In terms of total energy change, when is entropy an important factor in determining whether or not a solute dissolves in a solvent? When is it not an important factor?

12.83 True or false? Whenever something spontaneously dissolves in water, the entropy of the universe has increased. Justify your answer.

12.84 Certainly if NaCl dissolved in liquid hexane, C_6H_{14}, the entropy associated with the Na^+ and Cl^- ions would increase. Nevertheless, NaCl does not dissolve in hexane. Why not?

12.85 If the second law of thermodynamics is indeed true, what is the ultimate fate of the universe?

SOLUBILITY, TEMPERATURE, AND PRESSURE

12.86 How does increasing temperature affect the solubility in water (a) of most solids and (b) of gases?

12.87 What effect does increasing pressure have on the solubility of gaseous solutes? Explain your answer.

12.88 How is the medical condition known as the bends related to solubility?

12.89 Examine the bottom graph on page 439 showing solubility in water as a function of temperature. What is the trend for most of the ionic substances shown?

12.90 One ionic compound in the bottom graph on page 439 shows almost no temperature dependence, and one clearly violates the general trend. Identify these two ionic compounds.

12.91 Aquatic life is often damaged when hot water is discharged from power stations into rivers and lakes. What might this have to do with gas solubility in water?

12.92 Why does a can of soda pop go flat when left open to the atmosphere?

12.93 What is the mass in grams of sucrose necessary to saturate 100.0 lb of water at 90°C? [Use the top graph on page 439; 1 lb = 453.6 g.]

12.94 (a) Use these data to plot solubility as a function of temperature for KCl and Li_2SO_4:

	Solubility (g/100 g water)	
Temp (°C)	**KCl**	**Li_2SO_4**
0	27.6	35.3
10	31.0	35.0
20	34.0	34.2
30	37.0	33.5
40	40.0	32.7
50	42.6	32.5
100	56.7	29.9

(b) Using the plot, estimate the solubility of both compounds in water at 70°C.

(c) How much of each compound can be dissolved in a beaker containing 75 g of water at 70°C?

12.95 How many more grams of Li_2SO_4 can you dissolve in 250.0 g of water at 10.0°C than at 50.0°C? [See Problem 12.94.]

12.96 A saturated solution of KCl was prepared at 40.0°C using 5.00 lb of water. How many grams of KCl were required to prepare this solution? [See Problem 12.94; 1 lb = 453.6 g.]

12.97 A saturated solution of KCl was prepared at 50.0°C using 2.00 L of water. How many grams of KCl were required to prepare this solution? [See Problem 12.94; assume the density of water is 1.00 g/mL.]

GETTING UNLIKES TO DISSOLVE— SOAPS AND DETERGENTS

12.98 Soap molecules have two portions with very different properties. Draw a typical soap molecule and discuss the properties of the two parts.

12.99 What do the terms *hydrophobic* and *hydrophilic* mean?

12.100 Micelles are spherical, although they are usually drawn as a flat cross-section (as shown on page 442). Why wouldn't micelles exist in water as flat, two-dimensional structures?

12.101 Soap molecules have a hydrophobic portion and yet they dissolve in water. Explain how they accomplish this.

12.102 Explain how soaps allow water to wash oily, nonpolar dirt off clothes and skin.

12.103 A common molecule used in detergents is sodium lauryl sulfate:

$$CH_3(CH_2)_{11}O{-}\overset{\displaystyle O}{\overset{\|}{\underset{\underset{\displaystyle O}{\|}}{S}}}{-}O^- \quad Na^+$$

Identify the hydrophobic and hydrophilic portions of this molecule.

MOLARITY

12.104 Define *molarity*.

12.105 Give precise instructions to your laboratory assistant as to how to prepare 1.00 L of a 1.00 M aqueous solution of $CaCl_2$. Remember that your assistant will be measuring out the $CaCl_2$ in grams. She has available a 1-L volumetric flask.

12.106 Give precise instructions to your laboratory assistant as to how to prepare 1.00 L of a 1.00 M aqueous solution of sucrose, $C_{12}H_{22}O_{11}$. Remember that she will be measuring out the sucrose in grams. She has available a 1-L volumetric flask.

12.107 Give precise instructions to your laboratory assistant as to how to prepare 1.00 L of a 0.250 M aqueous solution of sucrose, $C_{12}H_{22}O_{11}$. Remember that she will be measuring out the sucrose in grams. She has available a 1-L volumetric flask.

12.108 Give precise instructions to your laboratory assistant as to how to prepare 0.500 L of a 1.50 M aqueous solution of sucrose, $C_{12}H_{22}O_{11}$. Remember that she will be measuring out the sucrose in grams. She has available a 0.50-L volumetric flask.

12.109 Your assistant tells you she measured out 2.50 moles of NaCl and then added enough water to get 500.0 mL of solution to prepare a 5.00 M solution of NaCl.
(a) What was the mass of the NaCl in grams?
(b) Did she successfully prepare a 2.5 M solution? Prove your answer.

12.110 Your assistant tells you he measured out 116.886 g of NaCl and then added exactly 1.00 L of water to it to prepare a 2.00 M solution of NaCl. Do you fire him or give him a promotion? Explain.

12.111 How many grams of NaCl are there in 2500.0 mL of a 0.250 M solution?

12.112 How many grams of sucrose, $C_{12}H_{22}O_{11}$, are there in 45.0 mL of a 0.250 M solution?

12.113 How many milliliters of a 1.00 M solution of NaCl are required to obtain 5.00 g of NaCl?

12.114 How many milliliters of a 0.250 M solution of glucose, $C_6H_{12}O_6$, are required to obtain 100.0 g of glucose?

12.115 There is a bottle of 0.500 M sucrose stock solution in the laboratory. Give precise instructions to your assistant on how to use the stock solution to prepare 250.0 mL of a 0.348 M sucrose solution.

12.116 There is a bottle of 4.50 M NaCl solution in the laboratory. Give precise instructions to your assistant on how to use the stock solution to prepare 100.0 mL of a 4.00 M NaCl solution.

12.117 What do you always get when you multiply molarity times volume in liters?

PERCENT COMPOSITION

12.118 What are the three types of percent composition? Give names and definitions.

12.119 In Practice Problem 12.3 you learned that "proof" for an alcoholic drink equals twice the percentage of alcohol in the drink. The complete definition of proof is that it is twice the percentage *by volume* of alcohol. Knowing this, exactly what does it mean to have a 90-proof drink?

12.120 A solution of ethanol is prepared by combining 22.5 g of ethanol with 49.6 g of water. What is the percent composition by mass of alcohol in this solution?

12.121 A solution of a particular solid solute in water has a concentration of 25.0 mass %.
(a) Given 100.0 g of this solution, how many grams of solute do you have?
(b) Given 48.0 g of this solution, how many grams of solute do you have?
(c) How many grams of this solution do you need to obtain 56.5 g of solute?

12.122 How would you prepare 2.00 kg of an NaCl solution that is 30.0 mass % NaCl?

12.123 How would you prepare 1.00 L of an alcohol–water solution that is 5.00 vol % alcohol?

12.124 An alcohol–water solution is 35.00 vol % alcohol. How much solution is required to obtain 200.0 mL of alcohol?

12.125 An aqueous solution is 25.0 mass % NaCl. The density of the solution is 1.05 g/mL. What is the molarity of the solution? [*Hint:* Assume you have 100.0 g of solution.]

REACTIONS IN SOLUTION

12.126 What is a solvent cage?

12.127 What do we mean when we say that solute particles *diffuse* through a solution?

12.128 You have two solutions, one 1.50 M sodium sulfide and the other 1.00 M $Pb(NO_3)_2$.
(a) Write a net ionic equation for the precipitation that occurs when these solutions are combined.
(b) How many milliliters of the two solutions must be combined to prepare 10.00 g of precipitate?
(c) Suppose you filter off the precipitate and find that your percent yield is 50.0%. What volumes of the solutions should you have combined to isolate 10.00 g of precipitate?

12.129 You have two solutions, one 0.755 M barium nitrate and the other 1.250 M calcium hydroxide.
(a) Write a net ionic equation for the precipitation that occurs when these solutions are combined.
(b) How many milliliters of the two solutions must be combined to prepare 5.00 g of precipitate?
(c) Suppose you filter off the precipitate and find that your percent yield is 85.0%. What volumes of the solutions should you have combined to isolate 5.00 g of precipitate?

12.130 You have two solutions, one 0.650 M iron(III) nitrate and the other 1.500 M ammonium carbonate.
(a) Write a net ionic equation for the precipitation that occurs when these solutions are combined.
(b) If you pour 200.0 mL of each solution into the same flask, what is the theoretical yield of the precipitate in grams?
(c) What is the molar concentration of the excess reactant ion?

12.131 You have two soltuions, one 0.800 M sodium phosphate and the other 0.800 M lead(II) acetate.
(a) Write a net ionic equation for the precipitation that occurs when these solutions are combined.
(b) If you pour 100.0 mL of the sodium phosphate solution and 50.0 mL of the lead(II) acetate solution into the same flask, what is the theoretical yield of the precipitate in grams?
(c) What is the molar concentration of the excess reactant ion?

12.132 A 25.00-mL sample of aqueous hydrobromic acid of unknown concentration is neutralized by 43.28 mL of 0.1001 M NaOH(*aq*).
(a) Write a net ionic equation for this acid–base neutralization reaction.
(b) What is the molar concentration of the hydrobromic acid?

12.133 A 25.00-mL sample of aqueous sulfuric acid of unknown concentration is neutralized by 27.55 mL of 1.0002 M NaOH(*aq*).
(a) Write a net ionic equation for this acid–base neutralization reaction.
(b) What is the molar concentration of the sulfuric acid?

12.134 Proprionic acid has one acidic proton per molecule. A solution is prepared by dissolving 0.273 g of proprionic acid in enough water to yield 100.0 mL of solution. This solution is neutralized by 36.82 mL of 0.1001 M NaOH(*aq*).
(a) What is the molar concentration of the proprionic acid?
(b) What is the molar mass of proprionic acid?

COLLIGATIVE PROPERTIES OF SOLUTIONS

12.135 A substance has the following heating curve:

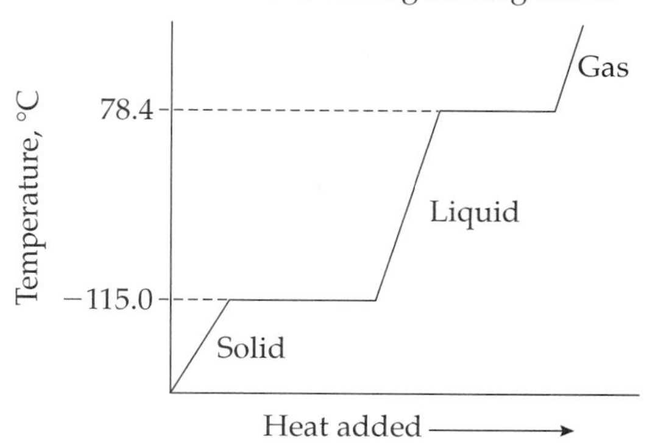

(a) What is the melting point of the substance?
(b) What is its freezing point?
(c) What is its boiling point?
(d) At what temperature does the vapor condense?
(e) Why isn't the temperature increasing as the substance is being heated at the horizontal portions of the curve?
(f) According to Table 12.2, what is this substance?

12.136 When it is warmed, a piece of plastic changes from a solid phase in which its molecules are positioned in an ordered arrangement to a solid phase in which its molecules are arranged randomly. How could you determine the temperature at which this phase change occurs?

12.137 When water boils, its temperature does not increase even though it is being heated. If the heat energy isn't being used to increase the temperature of the water, what is it being used for?

12.138 Dynamic equilibrium between liquid water and water vapor occurs when the ________________ rate equals the ________________ rate.

12.139 Define *vapor pressure* and include the word *equilibrium* in your definition.

12.140 Hexane, which is a liquid at room temperature, is more volatile than water.
(a) What does this mean?
(b) Which liquid has the higher vapor pressure at a given temperature?
(c) Which liquid most likely has the higher boiling point?
(d) Why is hexane more volatile than water?

12.141 How could you make water boil at 20°C (approximately room temperature)? [*Hint*: Use information from Table 12.1.]

12.142 Use the dueling-arrows model to explain what heating does to a liquid that eventually causes the liquid to boil.

12.143 What is the vapor pressure of benzene at 80.1°C? [*Hint*: Consult Table 12.2.]

12.144 What is a colligative property? Give three examples of colligative properties.

12.145 A student dissolves 28.7 g of NaCl(*s*) in 2.00 kg of water. How many grams of sucrose, $C_{12}H_{22}O_{11}$, would she have to dissolve in 2.00 kg of water to make the sucrose solution have the same freezing point as the NaCl solution?

12.146 A student dissolves 36.9 g of calcium nitrate in 500.0 g of water. How many grams of glucose, $C_6H_{12}O_6$, would he have to dissolve in 500.0 g of water to make the glucose solution have the same boiling point as the calcium nitrate solution?

12.147 A student dissolves 45.6 g of sodium sulfate in 550.0 g of water at a given temperature. How many grams of sodium chloride would she have to dissolve in 550.0 g of water at the same temperature to make the chloride solution have the same vapor pressure as the sulfate solution?

12.148 What are (a) the boiling point and (b) the freezing point of a solution made by dissolving 38.60 g of sodium phosphate in 200.0 mL of water? [The density of water is 1.000 g/mL.]

12.149 What are (a) the boiling point and (b) the freezing point of a solution made by dissolving 5.75 g of solid cetyl alcohol, $C_{16}H_{34}O$, in 100.0 mL of benzene? [The density of benzene is 0.874 g/mL.]

12.150 A solution is prepared by dissolving 5.00 g of caffeine in 100.0 g of carbon tetrachloride. The solution is cooled and the temperature plotted over time:

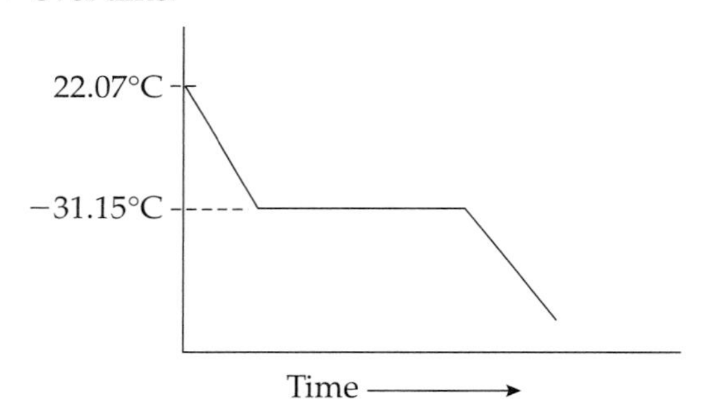

(a) What is the molar mass of caffeine?
(b) Combusiton analysis reveals that the empirical formula of caffeine is $C_4H_5N_2O$. What is the molecular formula?

12.151 When 1.56 g of cholesterol is dissolved in 50.0 mL of cyclohexane, the resulting solution freezes at 4.40°C. What is the molar mass of cholesterol? [The density of cyclohexane is 0.779 g/mL.]

ADDITIONAL PROBLEMS

12.152 Which would you expect to release the most hydration energy when dissolved in water: KCl(*s*), $Mg(OH)_2$(*s*), or CO_2(*g*)? Which would you expect to release the least hydration energy? Explain your answers.

12.153 Which would you expect to release the most hydration energy when dissolved in water: CH_3CH_2OH(*l*), CH_3Cl(*l*), or C_8H_{18}(*l*)? Which would you expect to release the least hydration energy? Explain your answers.

12.154 The ionic lattice in which compound requires the most energy to break: KCl, $Mg(OH)_2$, or $NaNO_3$? Explain your answer.

12.155 Which one of the following steps in the dissolving process must have a negative value for ΔE? Explain your answer.
(a) The physical separation of solute particles.
(b) The formation of solvent–solute interactions.
(c) The physical separation of solvent particles.
(d) None of the above.

12.156 Which one of the following statements is true?
(a) Gases are more soluble in liquids when the pressure is decreased.
(b) Pressure has no effect on how soluble a gas is in a liquid.
(c) Gases are more soluble in liquids when the pressure is increased.
(d) Gases are less soluble in liquids when the pressure is increased.

12.157 How many moles of potassium permanganate, $KMnO_4$, are there in 28.86 mL of a 5.20×10^{-3} M solution of $KMnO_4$?

12.158 (a) How many grams of NaOH are needed to prepare 500.0 mL of a 0.300 M NaOH solution?
(b) Describe how you would make this solution, including the equipment needed.

12.159 How would you prepare 250.0 mL of a 0.350 M NaOH solution from a 6.00 M NaOH stock solution?

12.160 Use the bottom graph on page 439 to determine what mass of water is required to dissolve 50.0 g of $NaNO_3$ at 40.0°C.

12.161 Calculate the number of moles of each ion present in 2.00×10^2 cm^3 of (a) 0.200 M NaCl, (b) 0.350 M K_3PO_4, (c) 1.44 M $Al(NO_3)_3$.

12.162 Complete the table:

Solute	Solute mass (g)	Solution volume (L)	Molarity (mol/L)
NaBr	3.96	0.150	
$Ba(OH)_2$	2.58		0.0800
$(NH_4)_2SO_4$	8.65	2.40	
NH_4Cl		4.20	0.420

12.163 4.70 g of $CuSO_4$ is added to enough water to make 150.0 cm^3 of solution.
(a) What is the molarity of the solution?
(b) How many moles of $CuSO_4$ are there in 1.00 mL of this solution?
(c) What is the percent by mass of $CuSO_4$ of this solution? [The density of the solution is 1.01 g/mL].

12.164 Calculate the freezing point and boiling point of each solution:
(a) 18.4 g of glucose, $C_6H_{12}O_6$, in 95.5 g of water.
(b) 15.00 g of urea, N_2H_4CO, in 75.0 g of water.

12.165 Combustion analysis reveals vitamin C to be 40.9% by mass C and 4.58% by mass H. The only other element present is oxygen. A solution of 19.40 g of vitamin C in 100.0 g of water freezes at −2.05°C. What is the molecular formula of vitamin C?

12.166 The freezing point of a solution prepared by dissolving 0.200 mole of HF(*g*) in 2.00 kg of water is −0.19°C. Is HF primarily intact in solution, existing as HF(*aq*), or has it dissociated to H^+(*aq*) and F^-(*aq*) ions? Does this mean HF is a weak or a strong acid?

12.167 How many gallons of 24-proof wine would you have to drink to consume 0.100 gallon of alcohol?

12.168 Draw pictures illustrating the solute–solvent interactions and the solvent–solvent interactions for methanol, CH_3OH, in water. Is this a case of like dissolves like? Explain.

12.169 After 125.0 mL of water is added to 50.0 mL of a 0.250 M solution of ammonium phosphate, (a) what is the molar concentration of ammonium phosphate in the diluted solution?

Once in solution, the ammonium phosphate exists not as intact ammonium phosphate but rather as ammonium ions and phosphate ions. What are the molar concentrations (b) of ammonium ion in the diluted solution and (c) of phosphate ion in the diluted solution?

12.170 After 50.0 mL of a 0.250 M solution of calcium nitrate is combined with 100.0 mL of a 0.835 M solution of calcium nitrate, (a) what is the molar concentration of $Ca(NO_3)_2$(*aq*) in the combined solution?

Once in solution, the calcium nitrate exists not as intact calcium nitrate but rather as calcium ions and nitrate ions. What are the molar concentrations (b) of Ca^{2+}(*aq*) in the combined solution and (c) of NO_2^-(*aq*) in the combined solution?

12.171 Suppose 250.0 mL of a 0.600 M solution of barium nitrate is combined with enough of 0.500 M sodium sulfate solution to precipitate the maximum amount of barium sulfate. What volume of the sodium sulfate solution is required?

12.172 A bottle contains 1.00 L of a stock solution of $Fe(NO_3)_3$ of unknown concentration. A lab technician dilutes 5.00 mL of the stock solution to 100.0 mL with water. He then determines the $Fe(NO_3)_3$ concentration of this solution to be 0.0478 M. What is the $Fe(NO_3)_3$ concentration of the stock solution?

12.173 What volume of a 0.245 M solution of NaI would you need to add to 100.0 mL of a 0.300 M solution of lead acetate to precipitate out all the lead? Which ions are spectators?

12.174 Citric acid can produce 3 H^+(*aq*) ions per molecule. A solution of citric acid is prepared by dissolving 0.177 g of solid citric acid in enough water to yield 100.0 mL of solution. When this solution is titrated with 0.1001 M NaOH(*aq*), the indicator turns color after 27.55 mL of NaOH(*aq*) has been added.
(a) What is the molar concentration of citric acid?
(b) What is the molar mass of citric acid?

12.175 A 25.00-mL sample of a hydrochloric acid solution of unknown concentration is titrated with 0.1004 M NaOH. However, before the acid is titrated, 27.65 mL of water is added to it. The phenolphthalein indicator turns red after 28.70 mL of NaOH has been added. What is the concentration of the 25.00-mL sample of hydrochloric acid?

12.176 Identify the solvent and the solute in each solution:
(a) Brass, which is 60 to 80% Cu and 18 to 40% Zn.
(b) Household ammonia cleaner, which is 1% by mass ammonia gas in water.
(c) 2.59 g of sucrose in 1.00 g of water at 50°C.

12.177 Carbon dioxide, CO_2, is a compound of carbon and oxygen. Does this mean that a sample of pure carbon dioxide can be considered to be a solution of carbon dissolved in oxygen? Explain your answer.

12.178 Describe at the molecular level what is happening when solid iodine, I_2(*s*), dissolves in carbon tetrachloride liquid, CCl_4(*l*).

12.179 Would you expect solid iodine, I_2(*s*), to be more or less soluble in CCl_4(*l*) than in H_2O(*l*)? Explain your answer in terms of energy.

12.180 What is the difference between hydration and solvation? Does one process always release more energy than the other? Explain.

12.181 How would you explain the fact that Na_3PO_4 is quite soluble in water but $AlPO_4$ is essentially insoluble?

12.182 Consider dissolving these two molecules in water:

Diethyl ether, C_2H_6O

Ethanol, C_2H_6O

Even though the solute-separation step for ethanol requires more energy than does the solute-separation step for diethyl ether, ethanol is more soluble in water. Explain why.

12.183 Whether or not a solute dissolves in a solvent is decided not only by changes in energy but also by changes in ________________.

12.184 Correct the two errors in this statement: A solute can dissolve even when the energy absorbed by the solute–solvent interactions is less than the energy absorbed by the solute–solute and solvent–solvent interactions because the entropy of the system decreases.

12.185 If the dissolution of a particular solute in water is endothermic, what must be true about (a) ΔE_{total} and (b) the change in entropy for the universe as a result of the dissolution?

12.186 Henry's law tells us that the solubility s of a gas in a liquid increases as the pressure P of the gas increases. Which of the following mathematical expressions of Henry's law is correct? Explain your choice and explain why the other two expressions are incorrect. [*Hint*: The k is just a constant of proportionality and can be ignored.]
(a) $s = kP$ (b) $s = k/P$ (c) $s = kP/P$

12.187 Could 32 g of NaCl be dissolved in 75 g of water at 20.0°C? [*Hint*: Refer to the bottom graph on page 439.]

12.188 Could 75 g of sodium nitrate be dissolved in 90.0 g of water at 60.0°C? [*Hint*: Refer to the bottom graph on page 439.]

12.189 What is the mass in grams of glucose dissolved in 60.0 g of water at 20.0°C if the solution is saturated? [*Hint*: Refer to the bottom graph on page 439.]

12.190 Why do most solids become more soluble in water with increasing temperature? [*Hint*: Think about what happens to the water molecules as temperature increases.]

12.191 Phospholipids are naturally occurring soaplike molecules present in the membranes of living cells. A typical phospholipid structure is

$$
\begin{array}{l}
CH_2-O-\overset{\overset{\large O}{\|}}{C}-(CH_2)_{16}CH_3 \\
| \\
CH-O-\overset{\overset{\large O}{\|}}{C}-(CH_2)_7-CH=CH-(CH_2)_7-CH_3 \\
| \\
CH_2-O-\underset{\underset{\large O}{\|}}{\overset{\overset{\large O}{\|}}{P}}-O-(CH_2)_2-\overset{+}{N}(CH_3)_3
\end{array}
$$

Identify the hydrophobic and hydrophilic portion(s) of this molecule.

12.192 When a small amount of soap is added to a beaker of water, the soap molecules end up positioning themselves at the surface (hence soaps are often called *surfactants*):

Explain why the soap molecules migrate to the surface and why they orient themselves with their hydrocarbon tails sticking out of the water.

12.193 Draw the structure of the fat molecule that, when it reacts with KOH, forms glycerol and a soap molecule containing 14 carbon atoms.

12.194 Consider the three soap molecule shown on page 443 and the ionic compound potassium acetate.
(a) Draw a dot diagram for the acetate ion, and then place a K^+ ion next to the O bonded to C by a single bond.
(b) What is the similarity between potassium acetate and a soap molecule? How do they differ?
(c) Hard water often contains calcium ions, which react with soap molecules to form a solid soap scum. Draw a soap scum molecule, using a soap molecule containing ten C atoms. What is the molecular formula of this soap scum molecule.

12.195 Soap molecules not only form spherical micelles in water, they also form spherical vesicles, which you can picture as thick-walled hollow spheres. Here is a cross-section of such a vesicle, with the blue regions representing water:

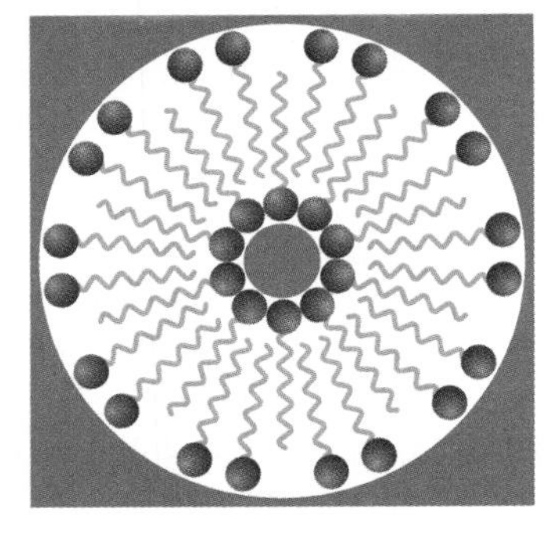

Unlike a micelle, a vesicle traps water in its interior. What gives a vesicle this ability? [If you are having trouble with the difference between micelles and vesicles, think of a baseball and a hollow rubber ball. The baseball, with no empty space inside, is analogous to the micelle, and the hollow ball is analogous to the vesicle.]

12.196 A student dilutes 75.0 mL of a 2.00 M solution of iron(III) nitrate with sufficient water to prepare 2.00 L of solution. (a) What is the molar concentration of iron(III) nitrate in the diluted solution?

Once in solution, the iron(III) nitrate exists not intact but rather as dissociated ions. What are the molar concentrations (b) of Fe^{3+}(*aq*) in the diluted solution and (c) of NO^{3-}(*aq*) in the diluted solution?

12.197 A student dilutes 45.0 mL of a 0.500 M solution of aluminum sulfate with sufficient water to prepare 1.50 L of solution. (a) What is the molar concentration of aluminum sulfate in the diluted solution?

Once in solution, the aluminum sulfate exists not intact but rather as dissociated ions. What are the molar concentrations (b) of Al^{3+}(*aq*) in the diluted solution and (c) of SO_4^{2-}(*aq*) in the diluted solution?

12.198 A student combines 60.0 mL of 0.250 M NaOH with 60.0 mL of 0.125 M NaOH. What is the NaOH molar concentration in the resulting solution?

12.199 A student combines 60.0 mL of 0.250 M NaOH with 60.0 mL of 0.125 M $Ba(OH)_2$. What is the hydroxide ion molar concentration in the resulting solution?

12.200 What is the mass in grams of the nitrogen atoms in 100.0 mL of 1.00 M $Ca(NO_3)_2$ solution?

12.201 How many grams of NaOH are needed to prepare 500.0 mL of 0.300 M NaOH solution?

12.202 What is the molarity of 3.69 mL of solution containing 0.0025 mole of calcium chloride?

12.203 How many moles of potassium permanganate, $KMnO_4$, are there in 28.68 mL of a 5.20×10^{-3} M solution of $KMnO_4$?

12.204 What is the volume in milliliters of 0.0150 M NaOH solution required to neutralize 50.0 mL of 0.0100 M HNO_3(*aq*)?

12.205 What is the volume in milliliters of 0.0150 M NaOH solution required to neutralize 50.0 mL of 0.0100 M H_2SO_4(*aq*)?

12.206 A 50.00-mL sample of hydrochloric acid of unknown concentration was neutralized by 47.35 mL of 0.01020 M sodium hydroxide solution. Calculate the concentration of the original HCl solution.

12.207 A solution is prepared by dissolving 5.00 g of sucrose, $C_{12}H_{22}O_{11}$, in 1.00 L of water. What is the percent by mass concentration of sucrose? [The density of water is 1.00 g/mL].

12.208 A solution is prepared by combining 4.00 mL of hexane with a sufficient volume of ethanol to obtain 250.0 mL of solution. What is the percent by volume concentration of hexane?

12.209 Air is approximately 21% by volume oxygen, O_2. How many grams of oxygen are present in 200.0 L of air at 1.00 atm and 25°C? [*Hint*: Use the ideal gas equation from Chapter 11.]

12.210 When 5.00 g of a solute is dissolved in sufficient solvent to produce 100.0 mL of solution,

(a) What is the percent by mass/volume concentration of the solute?

(b) What additional information do you need to calculate the percent by mass concentration of the solute, and how would you calculate that concentration once you had the missing information?

12.211 A sample of stainless steel, which is an alloy of iron, chromium, and minor amounts of other components, is 11.5% by mass chromium. How many grams of chromium are there in 250.0 lb of stainless steel?

12.212 Your boss says, "Prepare 0.5000 kg of a 1.00% solution of hexane (C_6H_{14}, a liquid) in dichloromethane (CH_2Cl_2, also a liquid)."

(a) What is wrong with his request?

(b) How would you prepare this solution if he meant a 1.00% by mass solution?

12.213 When 26.5 mL of 0.100 M $Ca(NO_3)_2$ is combined with 49.8 mL of 0.100 M NaF,

(a) What is the theoretical yield of CaF_2(*s*) in grams?

(b) What is the molar concentration of Ca^{2+} (excess reactant) in the combined solution?

12.214 Suppose 200.0 mL of a 2.50 M solution of sodium hydroxide is combined with 100.0 mL of a 1.50 M solution of iron(III) nitrate.

(a) Write a net ionic equation for the formation of the expected precipitate.

(b) What is the theoretical yield of precipitate in grams?

(c) Suppose only 10.95 g of precipitate is isolated. What is the percent yield for the reaction?

(d) What is the molar concentration of the excess reactant in the combined solution?

12.215 Suppose 20.0 g of NaBr(*s*) is added to 50.0 mL of a 2.00 M silver nitrate solution.

(a) Write a net ionic equation for the formation of the expected precipitate.

(b) What is the theoretical yield of precipitate in grams?

(c) Suppose only 15.0 g of precipitate is isolated. What is the percent yield for the reaction?

(d) What is the molar concentration of the excess reactant in the combined solution?

12 WORKPATCH SOLUTIONS

12.1 Homogeneity can occur only when the molecules are free to move about randomly, and such motion is possible only in the liquid or gas phase.

12.2 The water molecule is polar, with its partial negative charge located on the oxygen and its partial positive charges located on the hydrogens. Because opposite charges attract, the $\delta-$ oxygen atoms are attracted to Na^+ ions and the $\delta+$ hydrogen atoms are attracted to Cl^- ions.

12.3 The solute-separation step requires a large amount of energy because ionic bonds are very strong.

The solvent-separation step requires a moderate or small amount of energy because these are nonpolar solvent molecules, meaning that the only forces to be overcome are weak London forces.

The solvation step releases a small amount of energy. This is a case of very different particles trying to attract one another. CCl_4 is nonpolar, but Na^+ and Cl^- ions are charged. The only force that can exist between a CCl_4 molecule and either ion is a weak London force. Development of this weak attractive force releases only a small amount of energy.

12.4 Because ΔE_{total} is zero, any mixing is due solely to entropy. In the spontaneous direction (from left to right in the drawing), we go from an ordered situation (pure A and pure B) to a chaotic situation (A and B mixed). The latter is a very probable state, and it contributes to an entropy increase for the universe. The second law of thermodynamics therefore predicts that mixing should be spontaneous. The reverse direction goes from a disordered situation to an ordered situation. It contributes to an entropy *decrease* for the universe and is far less probable than the mixed state because there is only one way to achieve the ordered arrangement and many more ways to achieve the mixed arrangement. The spontaneous unmixing of A and B is in violation of the second law and therefore does not happen.

12.5 (a) The detergent molecules would arrange themselves with the polar heads pointing inward, exactly the opposite of the arrangement shown earlier for a micelle forming in water. This way, the nonpolar tails are presented to the nonpolar solvent, leading to a maximum attractive force. The polar heads buried inside avoid the nonpolar solvent that surrounds the structure.

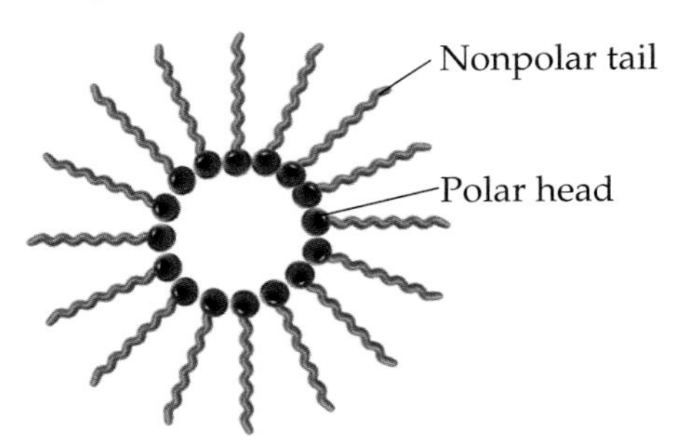

(b) This structure would be good at removing polar dirt from clothes being dry-cleaned.

12.6 Use the density as a conversion factor:

$$79.6 \text{ mL ethanol} \times \frac{0.7983 \text{ g ethanol}}{1 \text{ mL ethanol}} = 62.8 \text{ g ethanol}$$

12.7 The vapor pressure increases, as Table 12.1 shows.

12.8 The definition of normal boiling point is the temperature at which a liquid's vapor pressure is 760 mm Hg. Thus chloroform's vapor pressure must be 760 mm Hg at 61.7°C because, as you were told in the text, this temperature is the normal boiling point of chloroform.

12.9 There are 2 moles of solute particles in each beaker—1 mole of $Na^+(aq)$ ions plus 1 mole of $Cl^-(aq)$ ions in the left beaker and 2 moles of $C_{12}H_{22}O_{11}(aq)$ molecules in the right beaker.

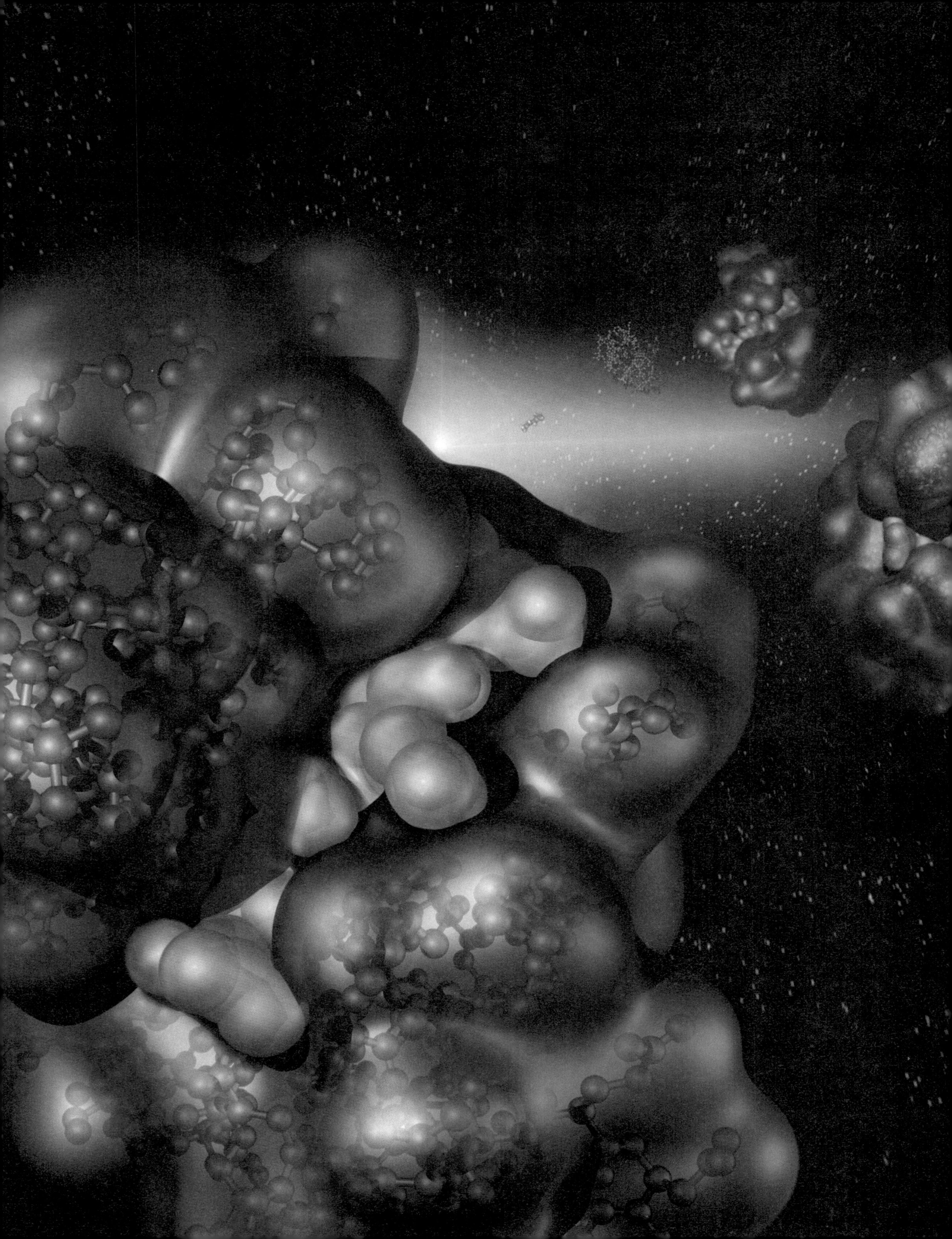

CHAPTER 13

When Reactants Turn into Products

13.1 Chemical Kinetics

So far we have examined atoms, molecules, forces within molecules (bonds, also known as intramolecular forces), and forces between molecules (intermolecular forces). All of these appetizers have prepared us to take a closer look at the main course of chemistry—chemical reactions.

Methane, CH_4, is one of the molecules we have seen many times already. Now consider what happens in the combustion reaction between methane and oxygen:

$$CH_4 + 2\,O_2 \longrightarrow CO_2 + 2\,H_2O$$

According to this balanced equation, one methane molecule reacts with two molecules of oxygen to produce one molecule of carbon dioxide and two molecules of water. After the reaction is over, the carbon and hydrogen atoms that were bonded to each other in methane end up bonded to oxygen. For this to happen, all four carbon–hydrogen single bonds and both oxygen–oxygen double bonds must have been broken. This is not a bad way to think about what happens. During most chemical reactions, atoms exchange partners to form new compounds.

But how do molecules do this? To put this question another way, what process does the arrow drawn between reactants and products represent? In this chapter, we answer this question by examining what happens to reactants as they are converted to products. It is usually not simply reactants banging into one another and forming products instantaneously. In the reaction between methane and oxygen, for instance, the CH_4 and O_2 molecules have to collide with one another sufficiently hard to begin breaking the strong covalent bonds in each molecule. But do the molecules have to collide so hard that they are blasted completely into separate atoms before being reassembled into products?

Such a scenario is very unlikely. First of all, it requires a simultaneous collision of *three* molecules—one methane and two oxygens. A simultaneous collision of three molecules has a very low probability of occurrence (much lower than the likely probability of a collision between two molecules). Second, at normal reaction temperatures, most collisions would not supply nearly enough energy to split molecules into atoms. Instead, in most reactions, the reactants are converted to products via a series of one-molecule or two-molecule steps, each of which requires less energy than the one-big-crash route. The sequence of steps molecules go through on changing from reactants to products is called a **reaction mechanism**.

13.1 WORKPATCH

Give a third reason a single-step mechanism of product formation is unlikely. [*Hint:* Focus on the products. The reaction of methane and oxygen produces CO_2 and H_2O. There are certainly many other possible combinations of carbon, hydrogen, and oxygen, however, such as CH_2O (formaldehyde), CO (carbon monoxide), H_2O_2 (hydrogen peroxide), and CH_3OH (methanol).]

Could you answer the WorkPatch? Check your answer against ours.

To deduce a reaction mechanism, we must study the reaction while it is occurring. Unfortunately, this is often very difficult to do. Many reactions, like the combustion of methane, are essentially instantaneous, occurring within a few millionths of a second. Until recently, this was just too fast to allow a direct probe of what was going on. Other reactions, such as the rusting of steel, are inconvenient to study for the opposite reason—they are too slow.

A number of reactions, however, occur on a time scale suitable for study, and these examples have allowed chemists to determine what happens during a chemical reaction. But how exactly do we determine the mechanism of chemical transformation? Because we cannot see individual molecules or atoms, we cannot directly observe a reaction's mechanism, which means we have to examine it indirectly. One way to do this is to study the speed (rate) of the reaction and see how that speed changes as we vary the temperature and/or the concentration of one or more reactants. The study of how chemical reaction rates change is a field called **chemical kinetics**, and chemists called *kineticists* devote much of their careers to obtaining these measurements, which provide a "window" through which we can "look at" the mechanisms of reactions.

One reaction that has been studied to the point where most chemists feel confident they know exactly how reactants become products is the substitution of an OH^- group for the Br atom in methyl bromide:

$$OH^- + CH_3Br \longrightarrow CH_3OH + Br^-$$

In this type of reaction, called a **substitution reaction**, one atom or group of atoms replaces (in other words, substitutes for) another atom or group of atoms in a molecule. In this case, a hydroxyl group replaces a bromine atom.

We shall be referring to this reaction throughout the chapter. Substitution reactions are a very important class of reactions, frequently used to modify molecules in the preparation of pharmaceutical compounds. Substitution reactions are one of the major ways in which specific genes are activated or repressed during the development of an organism.

Why is it necessary to understand reaction mechanisms? A major reason is that, if we understand on the molecular level exactly how reactants turn into products, we can do things to influence the course of the reaction. For example, rust, Fe_2O_3, is produced in large quantities for use as the magnetic medium in audiotapes, videotapes, and computer disks. What if you decide to go into the business of producing rust? You know you can't just set out a pile of iron nails and hope to have rust anytime soon. It would be helpful, therefore, to understand the mechanism of rusting well enough so that you could do something to speed up the reaction.

To take another example, consider the reaction of hydrogen and oxygen to produce water. This reaction is generally so fast and releases so much energy so quickly that it is explosive. If you knew enough about the mechanism of the reaction, you might be able to slow it down enough to

$$2\,H_2 + O_2 \xrightarrow{\text{Explosively}} 2\,H_2O + \text{Heat}$$

The *Hindenburg* dirigible was one of the last filled with hydrogen, for obvious reasons. Today, nonflammable helium is used.

make it useful. This is precisely what is done inside a fuel cell, where hydrogen and oxygen are slowly combined to supply spacecraft not only with water but also with electricity.

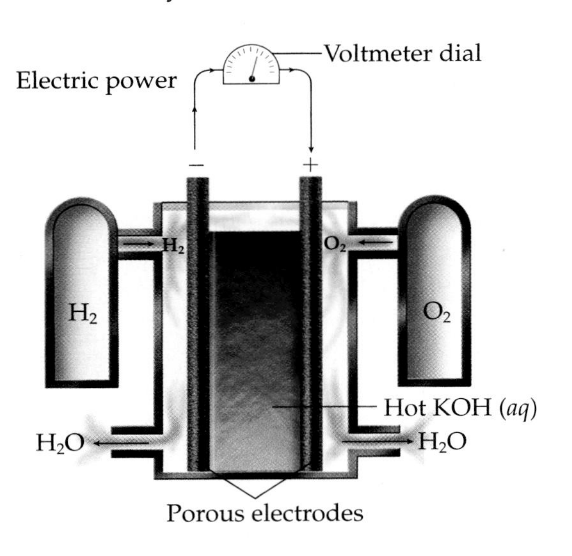

A Fuel Cell

$$2\,H_2 + O_2 \xrightarrow{\text{Slowly}} 2\,H_2O + \text{Electricity}$$

So deducing a reaction mechanism is important, and we get insight into the mechanism by studying how the reaction rate depends on external factors like temperature and concentration. The next question is "How do we get this insight?" Suppose we study a reaction and successfully determine how its rate depends on temperature and concentration. Now what? What is its mechanism? To answer this question, we must first understand the energy changes that take place during a reaction.

13.2 Energy Changes and Chemical Reactions

The key to understanding the mechanism of chemical change lies in knowing something about the changes in energy that take place during a reaction. Most chemical reactions occur with some net change in energy; that is, the combined energy of the products is different from the combined energy of the reactants. If the product molecules in a reaction contain more energy than the reactants, the reaction must have absorbed energy from the surroundings as it occurred. Alternatively, a reaction can result in products that contain less energy than the reactants. In this case, energy must have been released into the surroundings (typically in the form of light or heat) as the reaction occurred. The detonation of nitroglycerine is just such a reaction:

$$\underset{\text{Nitroglycerine}}{4\,C_3H_5N_3O_9} \longrightarrow 6\,N_2 + 10\,H_2O + 12\,CO_2 + O_2 + \textbf{5720 kJ}$$

Each nitroglycerine molecule contains a lot of energy, stored mostly in its chemical bonds. The combined energy of the products of the detonation—N_2, H_2O, CO_2, and O_2—is lower than the energy of the nitroglycerine molecule. The difference in energy between reactant and products is the energy released when the nitroglycerine explodes.

Reactions that release energy into the surroundings as they occur are called **exothermic** (from Greek roots meaning "heat out"). These are the reactions usually associated with the stereotype of the bumbling and careless "mad scientist" seen on television and in the movies. The detonation of nitroglycerine is an extremely exothermic reaction, for instance. Combustion reactions, such as the burning of natural gas or methane in your stove or furnace, are also very exothermic, and we make direct use of the energy released for cooking or heating.

Less familiar are reactions that absorb energy (usually in the form of heat) from their surroundings. These reactions are called **endothermic** (from Greek roots meaning "heat in"). A flask in which such a reaction is occurring feels cold to the touch. This is because heat is being drawn out of your hand to drive the reaction. If the reaction is very endothermic, such as the one shown in the photograph below, you could end up with the flask frozen to your skin.

A Strongly Endothermic Reaction

The beaker of reactants was placed in a puddle of water on the block of wood. The reaction froze the water solid.

$$Ba(OH)_2 \cdot 8\,H_2O + 2\,NH_4SCN \longrightarrow Ba^{2+} + 2\,SCN^- + 10\,H_2O + 2\,NH_3$$

Reactions that neither absorb energy from the surroundings nor release energy to the surroundings are called **energy-neutral reactions**. In such reactions, the combined energy of the product molecules is equal to the combined energy of the reactant molecules.

To sum up:

- If a reaction releases energy in some form (usually as heat or light), the reactants have more energy than the products and the reaction is called exothermic.
- If a reaction absorbs energy from its surroundings, the reactants contain less energy than the products and the reaction is called endothermic.
- If a reaction neither releases nor absorbs energy, the reactants and products contain the same amount of energy and the reaction is called energy-neutral. This type of reaction is much less common than exothermic and endothermic reactions.

Chemists often use a **reaction-energy profile** to represent the energy changes that occur during a chemical reaction. Such a profile is a graph that plots the relative energies of the reactants and products. For example, let's consider the hypothetical exothermic reaction R $\longrightarrow$ P, where R represents the reactants and P the products. Because it is exothermic, the reaction releases energy. Its reaction-energy profile looks like this:

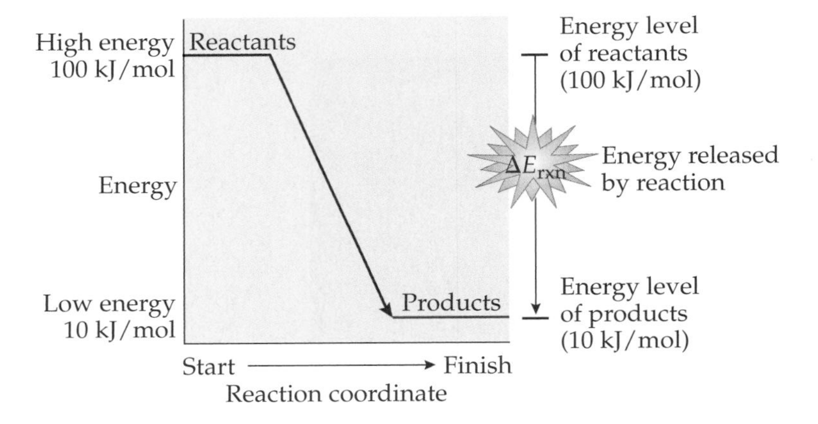

Energy is plotted on the vertical axis. The horizontal axis, called the *reaction coordinate*, traces the progress of the reaction from start to finish. Notice that the reactants are higher up on the energy axis than the products. These relative positions indicate that the reactant molecules contain more energy than the product molecules. The difference in energy between reactants and products is called the **net energy change** for the reaction and is indicated on the diagram by $\boldsymbol{\Delta E_{rxn}}$. (Remember, the Greek letter delta, Δ, stands for either difference or change; the subscript *rxn* is an abbreviation for reaction.) ΔE_{rxn} is the exact amount of energy that is released to the surroundings when the higher-energy reactants are converted to the lower-energy products.

We can calculate the amount of energy released because the energy per mole of R used or per mole of P formed is indicated on the reaction-energy profile. We subtract the energy of the reactants from that of the products:

$$\Delta E_{rxn} = E_{products} - E_{reactants}$$

If we apply this formula to our hypothetical reaction $R \longrightarrow P$, we get

$$\Delta E_{rxn} = 10 \text{ kJ/mol} - 100 \text{ kJ/mol} = -90 \text{ kJ/mol}$$

That is, 90 kJ of energy is released per mole of P formed (or per mole of R consumed).

Note the minus sign in our calculated value of ΔE_{rxn}. Because of the way ΔE_{rxn} is defined ($E_{products} - E_{reactants}$), exothermic reactions always have a negative ΔE_{rxn} value. In our hypothetical reaction, 90 kJ of energy is released for every mole of reactant that becomes product. You should get used to this sign convention. Any reaction with a negative ΔE_{rxn} value releases energy. Because exothermic reactions go from higher-energy reactants to lower-energy products, they are often said to go *downhill* in energy.

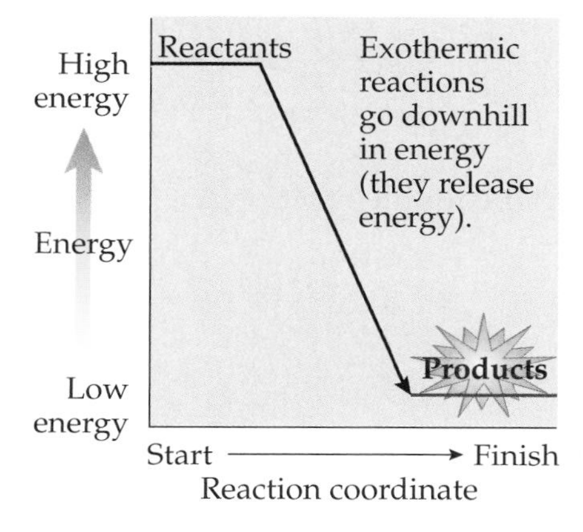

An endothermic reaction is just the opposite. It must absorb energy to proceed because the products contain more energy than the reactants. The path from reactants to products is now *uphill* in energy.

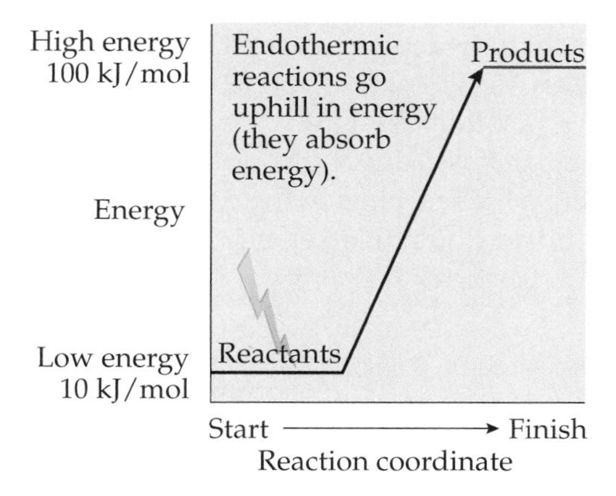

When we calculate ΔE_{rxn} for an endothermic reaction, the number we get is the amount of energy that must be *absorbed* from the surroundings as reactants go to products. For instance, if we have a reaction for which $E_{products} = 100$ kJ/mol and $E_{reactants} = 10$ kJ/mol, the energy change for the reaction is

$$\Delta E_{rxn} = E_{products} - E_{reactants} = 100 \text{ kJ/mol} - 10 \text{ kJ/mol} = +90 \text{ kJ/mol}$$

This value for ΔE_{rxn} tells us that 90 kJ of energy is absorbed from the surroundings for every mole of reactants converted to products. This time the ΔE_{rxn} value is positive, which is always the case for an endothermic reaction. Reactions that absorb energy always have a ΔE_{rxn} value greater than zero.

Because it is important to keep straight the meaning of the sign of ΔE_{rxn}, let's summarize what we've said:

For exothermic (energy-releasing) reactions, ΔE_{rxn} is negative: $\Delta E_{rxn} < 0$.

For endothermic (energy-absorbing) reactions, ΔE_{rxn} is positive: $\Delta E_{rxn} > 0$.

Finally, $\Delta E_{rxn} = 0$ kJ/mol for energy-neutral reactions, which means there is no net energy change. Overall, energy is neither released nor absorbed because the reactants and products have the same energy.

Practice Problems

13.1 Consider this reaction-energy profile for the reaction $A \longrightarrow B$:

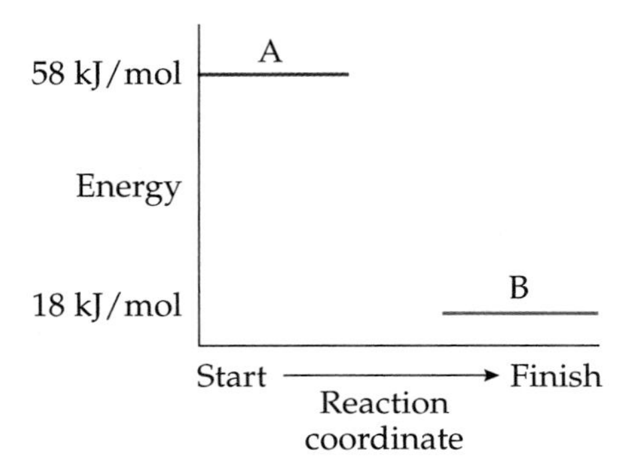

(a) Is this reaction exothermic or endothermic?
(b) Will the container in which this reaction is carried out feel hot or cold to the touch? Explain.
(c) What is the value of ΔE_{rxn} (include the sign)?
(d) What does the sign of ΔE_{rxn} tell you about this reaction?
(e) Which contain more energy, the reactants or the products?
(f) Does this reaction go uphill or downhill in energy?

Answer:
(a) Exothermic (energy is released).
(b) Hot because the reaction is exothermic.
(c) $\Delta E_{rxn} = E_{products} - E_{reactants} = 18$ kJ/mol $- 58$ kJ/mol $= -40$ kJ per mole of product formed.
(d) The fact that ΔE_{rxn} is negative tells you the reaction releases energy.
(e) The reactants have more energy (they are higher on the plot).
(f) Downhill.

13.2 A reaction is endothermic by 65 kJ/mol (that is, $\Delta E_{rxn} = +65$ kJ for every mole of product formed).
(a) Draw a reaction-energy profile for this reaction, assuming the energy of the reactants to be 100 kJ/mol.
(b) Does this reaction go uphill or downhill in energy?
(c) What is the value of ΔE_{rxn}?

(d) Will the container in which this reaction is carried out feel hot or cold to the touch? Explain.

13.3 A reaction has a ΔE_{rxn} of -450 kJ/mol. The products have an energy of 20 kJ/mol.

(a) Is the energy of the reactants higher or lower than that of the products? Explain.
(b) What is the energy of the reactants?
(c) Is the reaction exothermic or endothermic? How do you know?
(d) Does the reaction go uphill or downhill in energy?
(e) Draw a reaction-energy profile for the reaction.

Now let's add a bit of a twist. So far we have been looking at reactions proceeding in one direction, from reactants to products, the so-called *forward* direction. This makes sense. After all, that's the way the arrow points, and that's the way the reaction goes. But we are always free to consider the reverse reaction in which the reaction proceeds from products to reactants. Even if a reaction doesn't proceed in the reverse direction, there is nothing to keep us from thinking about its doing so. So let's return to Practice Problem 13.1, where we considered the exothermic reaction $A \longrightarrow B$ ($\Delta E_{rxn} = -40$ kJ/mol). What about the reverse reaction, where B turns into A? What is ΔE_{rxn} for it? To find out, try the following WorkPatch.

Here is the reaction-energy profile from Practice Problem 13.1 for the exothermic reaction $A \longrightarrow B$, for which $\Delta E_{rxn} = -40$ kJ/mol. What is ΔE_{rxn} for the reverse reaction $A \longleftarrow B$?

58 kJ/mol — A

Energy

18 kJ/mol — B

Start ⟵ Finish
Reaction coordinate

We've included this WorkPatch before explaining the energy change for a reverse reaction because we wanted you to answer it intuitively. Because the forward reaction $A \longrightarrow B$ goes downhill in energy (it's exothermic; $\Delta E_{rxn} = -40$ kJ/mol), the reverse reaction $A \longleftarrow B$ goes uphill in energy by exactly the same amount (it's endothermic; $\Delta E_{\text{reverse rxn}} = +40$ kJ/mol). A reaction that releases energy as it proceeds in one direction must absorb the same amount of energy when it proceeds in the reverse direction. Another way to say this is that the magnitude (size) of ΔE_{rxn} for the reverse reaction is the same as it is for the forward reaction, but the sign has changed. We can express this with an equation:

$$\Delta E_{\text{reverse rxn}} = -\Delta E_{\text{forward rxn}}$$

This principle is always true. So remember that when you know the ΔE_{rxn} value for any reaction, you also know the ΔE_{rxn} value for the reverse reaction. Just change the sign.

Practice Problems

13.4 The energy released by the reaction $A \longrightarrow B$ is 400 kJ/mol.
(a) What is $\Delta E_{forward\ rxn}$? (b) What is $\Delta E_{reverse\ rxn}$?

Answers:
(a) $\Delta E_{forward\ rxn} = -400$ kJ/mol *(b)* $\Delta E_{reverse\ rxn} = +400$ kJ/mol

13.5 As the reaction $A \longrightarrow B$ proceeds, the container in which it is run feels cold to the touch. Measurements show a net energy change of 250 kJ/mol.
(a) What is $\Delta E_{forward\ rxn}$? (b) What is $\Delta E_{reverse\ rxn}$?

13.6 As the reaction $A \longrightarrow B$ proceeds, the container in which it is run feels hot to the touch. Measurements show a net energy change of 250 kJ/mol.
(a) What is $\Delta E_{forward\ rxn}$? (b) What is $\Delta E_{reverse\ rxn}$?

Some questions logically arise at this point. *Why* are there energy changes in a chemical reaction? Where does the energy released to the surroundings during an exothermic reaction come from? And in an endothermic reaction, where is the energy absorbed from the surroundings stored? Molecules don't have pockets where they can store energy, but they do have bonds. The answers to all these questions have to do with the fact that, as reactants become products, there are usually changes in the number and type of chemical bonds in the reacting molecules. For example, consider the reaction that produces water from hydrogen molecules and oxygen molecules (the combustion of hydrogen). The large negative value for ΔE_{rxn} indicates that this reaction is highly exothermic:

$$2\,H_2(g) + O_2(g) \longrightarrow 2\,H_2O(g) \qquad \Delta E_{rxn} = -479\ \text{kJ}$$

Note that this is a real reaction rather than the hypothetical "1 mol of reactant gives 1 mol of product" ($R \longrightarrow P$, $A \longrightarrow B$) reactions we have been considering. For real reactions, ΔE_{rxn} is interpreted as being the energy released or absorbed *for the balanced reaction as written*. In other words, in this reaction 479 kJ of energy is released for every 2 mol of $H_2O(g)$ formed *or* for every 2 mol of $H_2(g)$ consumed *or* for every 1 mol of $O_2(g)$ consumed. That is why we write -479 kJ and not -479 kJ/mol. It is up to you to interpret ΔE_{rxn} properly for a real reaction by examining the balanced equation.

Let's imagine we can see the hydrogen-combustion reaction at the molecular level. At the start of the reaction, we see H_2 molecules with their single H–H covalent bonds and O_2 molecules with their double O=O covalent bonds. To understand the source of the energy released during the reaction, we can postulate a simple reaction mechanism wherein the reactant molecules break apart into atoms. Although this is not the actual mechanism for this reaction, it does take us from H_2 and O_2 to H_2O, meaning we can use this model to understand where the released energy comes from. In this case, we can break the H–H and O=O bonds to give individual H and O atoms. Of course, because covalent bonds are quite strong, it takes energy to break them.

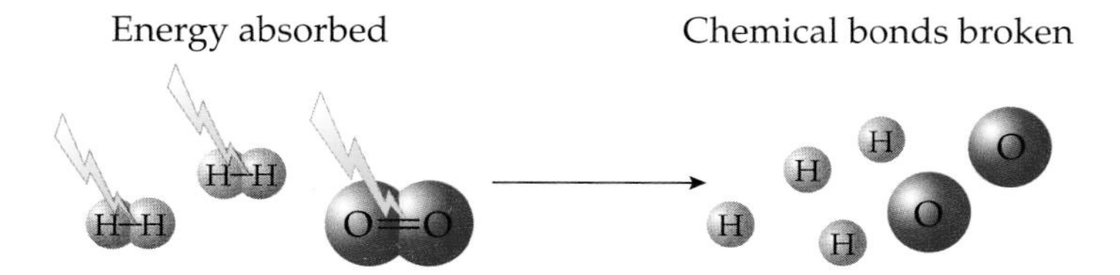

Energy must *always* be put *into* a molecule to break a chemical bond—no exceptions, ever! So can we assume that the reaction between hydrogen and oxygen to form water must absorb energy and must therefore be endothermic? Not necessarily. Breaking the covalent bonds in the reactant molecules is only half the story. After we've broken the bonds, we must bring the atoms together in a new arrangement, forming the new bonds of the products. This step releases energy. Energy is *always released* when a chemical bond is formed—no exceptions, ever!

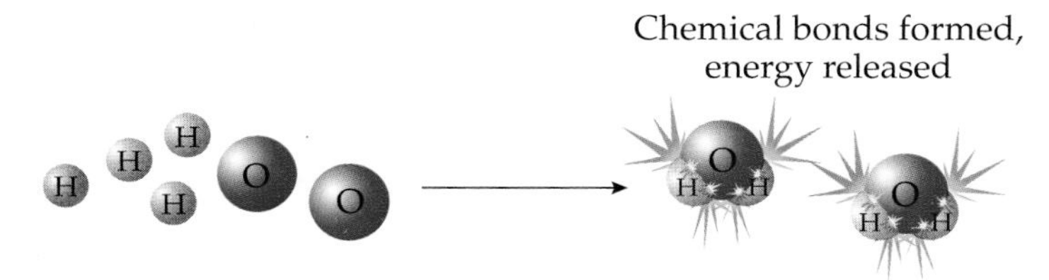

Energy is absorbed in breaking reactant bonds, and energy is released in forming product bonds. *Whether the overall reaction releases or absorbs energy depends on the relative magnitudes of these bond-breaking and bond-forming steps.* In principle, we have two possibilities (shown here using generic reactions):

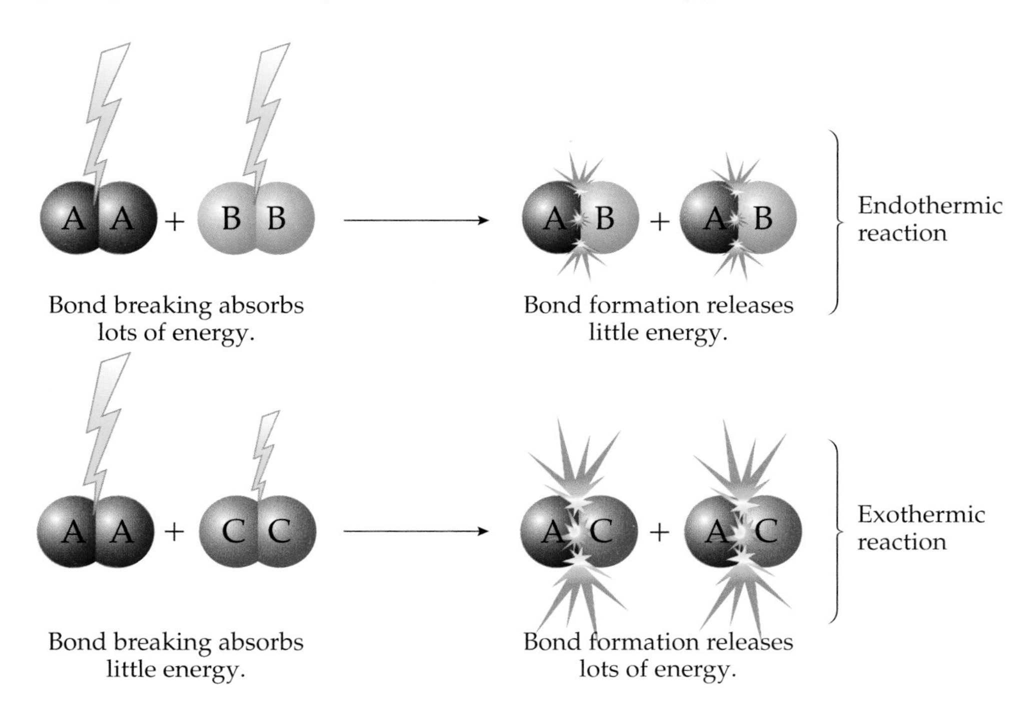

Simply compare the total energy absorbed to break all the reactant bonds with the total energy released as product bonds are formed. Whichever is larger in magnitude wins and determines the energetics of the overall reaction. That is, ΔE_{rxn} is the difference between the energy released and the energy absorbed; it is a positive number if the energy absorbed has the larger magnitude and a negative number if the energy released has the larger magnitude.

When we run our hydrogen–oxygen reaction—that is, when we burn hydrogen—we observe that a tremendous amount of heat is released (remember the *Hindenburg*). Therefore, the amount of energy released as O–H bonds form must be greater than the amount of energy absorbed as H–H and O=O reactant bonds break. The exothermic scheme shown on the previous page therefore holds true for this reaction.

Practice Problems

13.7 Consider the hydrogen combustion reaction, where $\Delta E_{rxn} = -479$ kJ. If the energy absorbed in breaking the reactant bonds is 1370 kJ, how much energy is released as the product bonds form?

Answer: To break the reactant bonds, the reactants must absorb 1370 kJ of energy. However, the fact that this reaction is exothermic (ΔE_{rxn} is negative) tells you that more than this amount of energy (479 kJ more, to be exact) must be released. Therefore, the amount of energy released as the O–H bonds form is 1370 kJ + 479 kJ = 1849 kJ.

13.8 Consider the reaction $A_2 + B_2 \longrightarrow 2\,AB$. Breaking 1 mol of A–A bonds and 1 mol of B–B bonds requires 2200 kJ. Forming 1 mol of A–B bonds releases 1000 kJ.

(a) Is this reaction exothermic or endothermic? Explain.

(b) What is the value of ΔE_{rxn}? (Get the sign right.) How is it to be interpreted?

Answer:

A–A + B–B ⟶ 2 A–B

A–A + B–B: 2200 kJ required to break these bonds.

2 A–B: Formation of each mole of A–B bonds releases 1000 kJ, but we are forming 2 mol of A–B bonds, which means that 2000 kJ is released.

(a) Endothermic. More energy is absorbed as the reactant bonds break than is released as the product bonds form, meaning the overall reaction absorbs energy.

(b) Because 2200 kJ is absorbed but only 2000 kJ is released, the overall reaction absorbs 200 kJ. Thus $\Delta E_{rxn} = +200$ kJ (the positive sign means endothermic). This can be interpreted as 200 kJ absorbed per 2 mol of AB formed OR 200 kJ absorbed per 1 mol of A_2 consumed OR 200 kJ absorbed per 1 mol of B_2 consumed.

13.9 Consider the reaction $A_2 + B_2 \longrightarrow 2\,AB$, for which $\Delta E_{rxn} = -100$ kJ. Forming 1 mol of A–B bonds releases 150 kJ. How much energy does it take to break the reactant bonds?

13.10 Draw reaction-energy profiles for the reactions in Practice Problems 13.8 and 13.9. For both, label the gap that is ΔE_{rxn}.

Let's not lose sight of our original goal—gaining insight into the mechanism of a reaction by studying how temperature and concentration affect the reaction rate. Remember, we said that to do this, we first had to know something about the energy changes that take place during a reaction. So far we have looked at the initial total energy of the reactants, the final total energy of the products, and the difference in energy between these two values (ΔE_{rxn}). To reach our goal, we also need to look at the energy changes that take place during the course of the reaction.

13.3 Reaction Rates and Activation Energy—Getting over the Hill

Of the three reaction types—exothermic, endothermic, and energy-neutral—which do you think is fastest? Which is slowest? Because exothermic reactions are downhill in energy and endothermic reactions are uphill, you might be tempted to declare that exothermic reactions are faster. In fact, this was a trick question because there is no single answer. There are many examples of endothermic reactions that proceed faster than some exothermic ones. The total energy change ΔE_{rxn} does not determine the *rate* of a chemical reaction. Rather, the rate is determined by the energy changes that occur as reactants change to products.

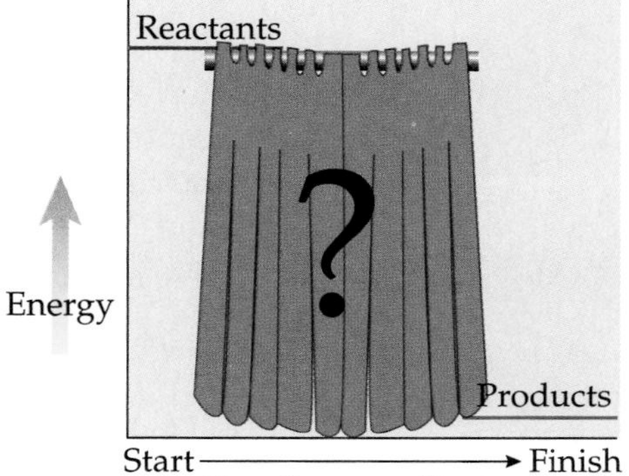

Reaction rate depends on what is going on behind the curtain.

In general, in order for reactants to convert to products, some bonds in the reacting molecules must break, and as you now know, this step always absorbs energy. In other words, almost all reactions must begin by going uphill and absorbing energy, even if the overall reaction is downhill (exothermic). This creates an energy barrier between reactants and products. The barrier indicates how hard reactant molecules must collide (in other words, how much energy the collision must supply) in order for a reaction to occur (bonds break). If two reactant molecules collide with less energy than this, they just bounce off each other unchanged. This special amount of energy is called the **activation energy E_a** of the reaction. It is the amount of energy that must be available to the reactants before they can become products. It is indicated graphically by the difference in energy between the initial energy of the reactants and the top of the energy barrier.

Top of energy barrier
Reactant molecules must attain at least this amount of energy before they can proceed downhill to product molecules.
E_a
Reactants
Energy
Products
Start
Finish
Reaction coordinate

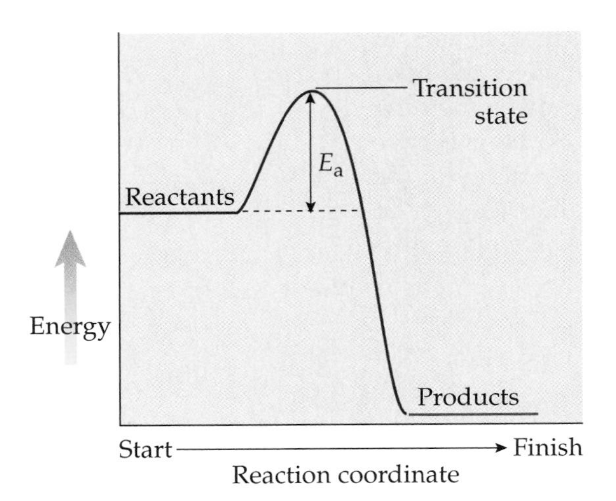

Each reaction has its own value of activation energy. Some reactions have tremendously large barriers to overcome, while other reactions have smaller barriers. At the top of the barrier, the reactant molecules are in a special, high-energy, unstable condition called the *transition state*. We can't usually observe what the reactants look like in the transition state, but we can describe the state for a typical one-step reaction. In the **transition state**, the amount of energy E_a that has been absorbed by the reactants is just enough to stretch to the breaking point—but not break—all the bonds destined to break so that the reaction can proceed. In addition, the new bonds that will be present in the product have begun to form. This transition-state condition occurs right at the top of the energy barrier. It represents molecules in transition, reactants on the way to becoming products.

We shall talk a bit more about the transition state shortly, but for now let's keep focused on the activation energy. Try the following WorkPatch to see if you can plot some reaction-energy profiles.

Here are data for three hypothetical reactions:

Reaction	Energy of reactants (kJ/mol)	Energy of products (kJ/mol)	Energy of transition state (kJ/mol)
1	50	60	70
2	20	10	90
3	10	20	90

For each reaction,

(a) Sketch the reaction-energy profile.
(b) Indicate whether the reaction is endothermic or exothermic.
(c) Determine the value of ΔE_{rxn}.
(d) Determine the value of E_a.

WorkPatch 13.3 is extremely critical; everything we are about to discuss is based on your being able to answer it. Don't go on until you get it right.

Your WorkPatch plots reveal what ultimately determines the rate of a reaction. In order for a chemical reaction to occur, the reacting molecules must absorb enough energy E_a from collisions to reach the top of the energy barrier and form the transition state. It is the size of this energy barrier that determines the reaction rate:

The higher the activation energy barrier, the slower the reaction.

This is one of the most important fundamental concepts in chemistry. You should also note that it is an inverse relationship. As E_a increases, the reaction rate decreases, and vice versa. Thus,

A large E_a value gives rise to a slow reaction.

A small E_a value gives rise to a fast reaction.

Given this inverse relationship, we can tell which of the three reactions plotted in WorkPatch 13.3 is fastest. If all three reactions are carried out at comparable temperatures and reactant concentrations, reaction 1 is fastest because it has the smallest E_a value (20 kJ/mol).

It makes intuitive sense that reactions that must pass over large barriers proceed more slowly than reactions that must pass over smaller barriers, but we can use our model of reactant molecules colliding with one another to go even further.

Practice Problems

13.11 Draw two reaction-energy profiles, one for an endothermic reaction A and one for an exothermic reaction B. Make the profile for reaction A represent a faster reaction than reaction B by drawing the two E_a barriers to scale.

Answer:

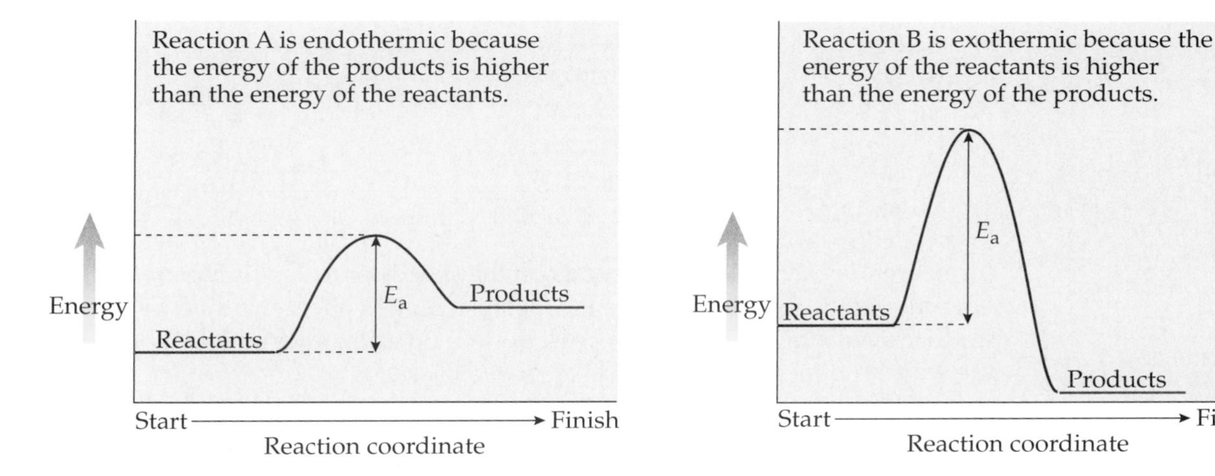

Reaction A is faster than reaction B even though it is endothermic because A has a smaller activation energy E_a.

13.12 Draw two reaction-energy profiles, one for an endothermic reaction A and one for an exothermic reaction B. Make the profile for B represent the faster reaction by drawing the two E_a barriers to scale.

13.13 True or false? If reaction X is more exothermic than reaction Y, reaction X must be faster than reaction Y. Whatever your answer, back it up with reaction-energy profiles.

TEMPERATURE AND THE COLLISION THEORY OF REACTION RATES

Every chemical reaction has its own characteristic activation energy. This energy depends to a large extent on the number and types of bonds that are broken and formed in the transition state. If many strong bonds are breaking but only a few weak bonds are forming in the transition state, the activation energy for the reaction is very high and the reaction takes place very slowly. Fortunately, with the understanding we now have as to why such a reaction proceeds only slowly, we can also suggest some things that can be done to speed it up.

One of the factors that can always be used to manipulate the speed of a chemical reaction is temperature. Temperature is a measure of the average kinetic energy of a collection of atoms or molecules. As such, it is also a measure of the average speed of the atoms or molecules in the collection.

We have said that in order for two reactant molecules to react with each other, they must collide and the energy of the collision must be equal to or greater than the activation energy E_a. For example, ozone and nitrogen oxide molecules do not react unless the collision between them has an energy equal to or greater than E_a for the reaction:

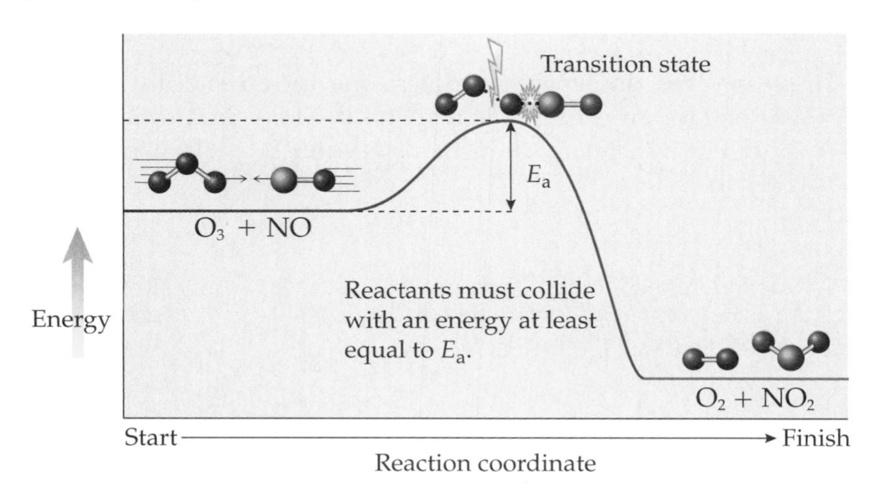

The energy of each collision depends on the speeds of the colliding molecules. Slow-moving molecules that result in collisions less energetic than E_a do not produce a reaction. The molecules just bounce off each other and go their separate ways, chemically unchanged.

In a gas or in a liquid, molecules are traveling at different speeds, some very fast and others more slowly. Thus, at any given temperature, some molecules are moving fast enough to lead to energetically sufficient collisions. If we now increase the temperature of the system, the average speed of the molecules increases and the molecules are, on average, traveling faster than before. This means that more of the molecules are now capable of collisions with sufficient energy to react. At a higher temperature, collisions with energy greater than E_a are more frequent than at a lower temperature.

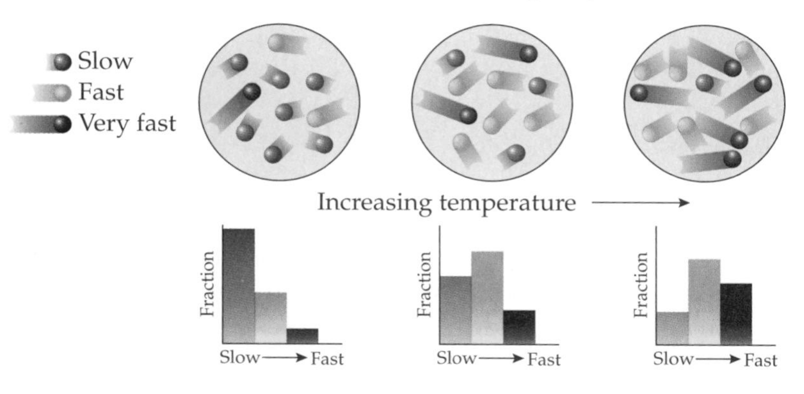

This is why an increase in temperature always increases the rate of a chemical reaction (or, conversely, why cooling always slows it down). In fact, a general rule of thumb is that the rate of a reaction doubles for every 10 C° increase in temperature. Like all other rules of thumb, this one is approximate; the actual factor by which a reaction rate increases for a fixed change in temperature depends on the value of E_a for that reaction.

Another factor that influences reaction rate is the relative orientation of the colliding molecules. In many reactions, two reactant molecules must collide not only with sufficient energy but also in an appropriate orientation with respect to each other. To see why, let's look again at the substitution reaction in which an OH^- ion substitutes for the Br atom in methyl bromide:

$$OH^- + CH_3Br \longrightarrow CH_3OH + Br^-$$

In the transition state, the C–Br bond is partially broken and at the same time the C–OH bond is partially formed:

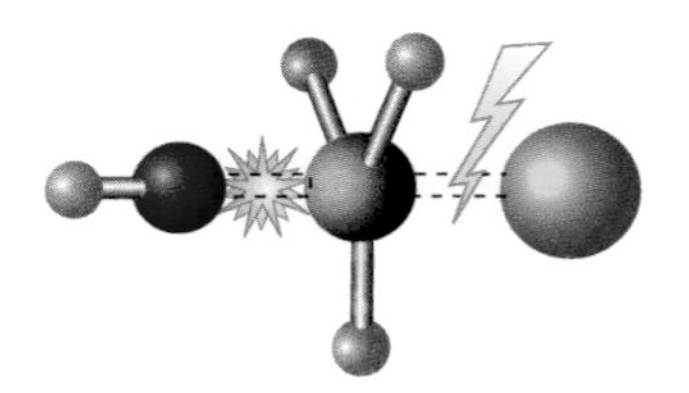

$OH \cdots CH_3 \cdots Br$

The Transition State

For the reaction to occur, it is not enough that the OH^- and CH_3Br collide with sufficient energy; they must also come together with the correct geometry. Many different experiments have shown that the OH^- must approach the methyl bromide from the side opposite the Br, usually referred to as the back side. Any other approach generally does not lead to formation of a product molecule even if the energy of the collision is sufficient to break the C–Br bond.

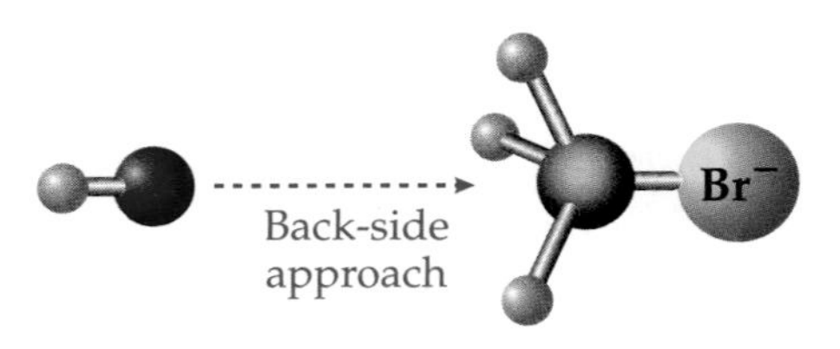

So we now see that the rate of a reaction depends not only on the number of collisions (per unit time) that occur with sufficient energy but also on the fraction of those collisions in which the molecules are in the proper orientation. Any collision that is sufficiently energetic and properly oriented such that it leads to the formation of product is called an **effective collision**. We can summarize this concept as a mathematical expression:

$$\text{Reaction rate} = \underbrace{\begin{array}{c}\text{Number of collisions per}\\ \text{unit time with energy equal}\\ \text{to or greater than } E_a\end{array}}_{\text{Energy factor}} \times \underbrace{\begin{array}{c}\text{Fraction of collisions in which}\\ \text{molecules have proper orientation}\end{array}}_{\text{Orientation factor}} \qquad \textbf{(13.1)}$$

Basically, this equation says that the reaction rate is equal to the number of effective collisions per unit time. The *orientation factor* is a number between 0 and 1—that is, a fraction. An orientation factor of 1 means that orientation doesn't matter, and in this case any sufficiently energetic collision between reactant molecules results in formation of a product molecule. For most reactions, the orientation factor is quite small (less than 0.1, meaning that in fewer than 10% of the collisions are the molecules oriented properly). Because molecules come in a variety of shapes and sizes, each reaction has its own orientation requirements.

We now have enough information to understand the accepted mechanism for the substitution reaction we have been studying. It is a one-step mechanism in which an OH^- ion in solution collides with a CH_3Br molecule in a back-side approach to form a transition state that then goes on to form products:

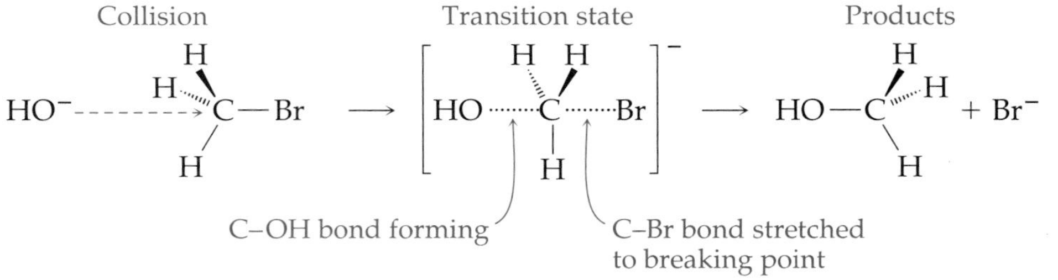

Collision: $HO^- \cdots\cdots \rightarrow CH_3$—Br $\longrightarrow$ Transition state: $[HO\cdots CH_3 \cdots Br]^-$ $\longrightarrow$ Products: HO—CH_3 + Br^-

C–OH bond forming

C–Br bond stretched to breaking point

Examine the transition state. Notice the dotted lines that represent the two bonds in the process of breaking and forming, and also notice how the hydrogen atoms have moved. This transition state exists at the very peak of the activation-energy hill. Furthermore, this mechanism reflects the principle that collisions are rarely energetic enough to blast molecules entirely to pieces. The hydrogen atoms remain attached to the carbon atom throughout the reaction. Only one bond breaks, and only one bond forms.

Practice Problem

13.14 Consider our substitution reaction between OH^- and CH_3Br. Imagine it is occurring in a solution where there are 1000 collisions every second between OH^- ions and CH_3Br molecules. Suppose only 10% of these collisions are sufficiently energetic to lead to products. Also, assume that the orientation factor is 0.2.

(a) What does an orientation factor of 0.2 mean?

(b) What is the rate of this reaction? Give your answer in number of CH_3OH molecules formed per second.

Answer:

(a) An orientation factor of 0.2 means that colliding molecules have the proper orientation in only 20% of the collisions.

(b) The reaction rate expression (Equation 13.1) says that, for this reaction,

$$\text{Reaction rate} = (100 \text{ sufficiently energetic collisions/s}) \times 0.2$$
$$= 20 \text{ effective collisions/s}$$

Because every effective collision results in the formation of product, the rate of product formation is 20 CH_3OH molecules per second.

Notice from Practice Problem 13.14 that rate is expressed in units of molecules formed per second (or, more typically, moles per second). Rate is always expressed in units of something over time; for example, the legal rate of travel on many highways is 55 miles/h. So thinking about how fast a reaction goes (its rate) is no different from thinking about how fast anything else goes. Now try the remaining practice problems for this section.

Practice Problems

13.15 Suppose the temperature of the reaction analyzed in Practice Problem 13.14 is increased by 10 C°.

(a) What should happen to the number of effective collisions per second? Estimate the new value based on the rule of thumb given in the text.

(b) Why does the number of effective collisions per second change when the temperature is increased?

(c) What is an approximate value for the new rate of this reaction?

13.16 Repeat Practice Problem 13.14 for an orientation factor of 0.1.

CATALYSTS

As we mentioned earlier, a reaction that has a high activation energy is a very slow reaction. When confronted with such a reaction, there are a number of things you might do to increase its rate. One is to raise the temperature. But there are limits to just how high a temperature you can use in many circumstances. Sometimes the reactants are sensitive to temperature, and if the temperature gets too high, they decompose before they can react. Often, a large increase in temperature causes a number of *side reactions* to take place, giving rise to undesired products (called *by-products*). So temperature control is not the complete answer to speeding up slow reactions. To examine what else we might do, let's continue with our substitution reaction. Here is the energy profile for this reaction:

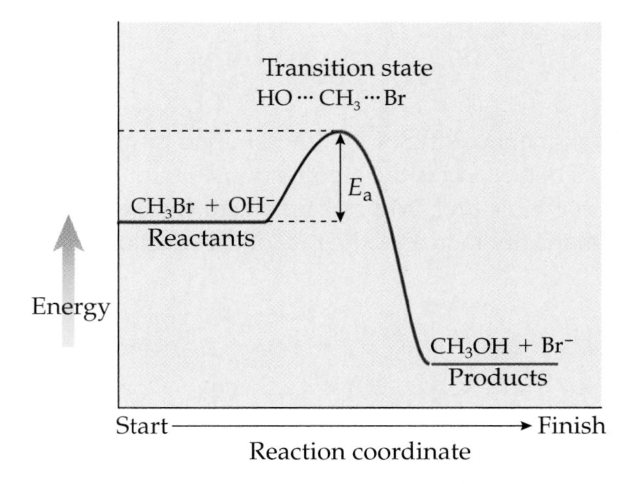

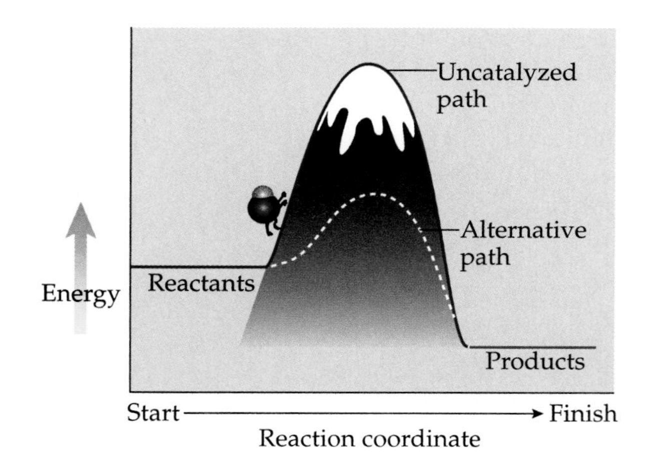

What if you were a reactant molecule and had to make the trip over the E_a hill to become a product molecule? Faced with having to climb the hill, you might instead look for an alternative, lower-energy path. In fact, this is sometimes possible in a chemical reaction. Creating such a new path essentially involves changing the way in which the reaction occurs—that is, changing its mechanism. One way to do this is with a **catalyst**, something that, when added to the container in which a reaction takes place, increases the rate of the reaction by lowering its activation energy. In addition, a catalyst does this without being used up in the process and for this reason usually needs to be present in only trace amounts. A catalyst works by providing a new mechanism—a different way for the reactants to combine and form products. Most important, the new path made possible by the catalyst has a lower E_a value than the path reactant molecules must take in the absence of the catalyst.

How can the addition of an extraneous substance (one not needed in the balanced equation for the reaction) change the mechanism? Another substitution reaction, the conversion of 2-propanol, $(CH_3)_2CHOH$, to 2-chloropropane, $(CH_3)_2CHCl$, helps us answer this question. If you try to run this reaction by combining the 2-propanol with Cl^-, expecting a back-side attack and substitution the way we saw earlier, the reaction barely proceeds; its rate is practically zero:

$$(CH_3)_2CHOH + Cl^- \not\longrightarrow (CH_3)_2CHCl + OH^-$$

This substitution reaction is a no-go because the mechanism for it is different from the mechanism for the methyl bromide reaction and has a much larger E_a value. (Recall that each reaction has its own unique E_a.) The uncatalyzed version of this reaction takes place in two steps:

Step 1: $$CH_3{-}\overset{CH_3}{\underset{H}{C}}{-}O{-}H \xrightarrow{\text{Very, very slow}} CH_3{-}\overset{CH_3}{\underset{H}{C^+}} + {}^-O{-}H$$

Step 2: $$CH_3{-}\overset{CH_3}{\underset{H}{C^+}} + Cl^- \xrightarrow{\text{Fast}} CH_3{-}\overset{CH_3}{\underset{H}{C}}{-}Cl$$

The first step involves just the breaking of the C–OH bond, without simultaneous formation of the C–Cl bond, and is the step responsible for the high activation energy of the reaction. However, adding some acid (H^+) and a trace of Zn^{2+} ions as a catalyst dramatically increases the rate of the reaction. Here is the alternative mechanism:

Step 1: $$CH_3{-}\overset{CH_3}{\underset{H}{C}}{-}\ddot{\underset{..}{O}}{-}H \;(\curvearrowright Zn^{2+}) \xrightarrow{\text{Fast}} CH_3{-}\overset{CH_3}{\underset{H}{C}}{-}\overset{Zn^+}{\dot{O}^+}{-}H \xrightarrow[\text{but much faster than step 1 above}]{\text{Slow}} CH_3{-}\overset{CH_3}{\underset{H}{C^+}} + ZnOH^+$$

$$\textbf{Step 2:}\quad CH_3{-}\overset{\overset{\displaystyle CH_3}{|}}{\underset{\underset{\displaystyle H}{|}}{C^+}} + Cl^- \xrightarrow{\text{Fast}} CH_3{-}\overset{\overset{\displaystyle CH_3}{|}}{\underset{\underset{\displaystyle H}{|}}{C}}{-}Cl$$

$$\textbf{Step 3:}\quad ZnOH^+ + H^+ \xrightarrow{\text{Fast}} Zn^{2+} + HOH$$

In the first step, the Zn^{2+} ion bonds with the –OH portion of the propanol molecule. This addition of Zn^{2+} to the oxygen of the OH group makes it easier to break the C–OH bond, which in turn makes it easier (faster) to form the **carbocation** (pronounced car-bo-CAT-ion, the ion with the C^+ in it). Once formed, the carbocation rapidly bonds with the Cl^- ion to give the final product, as step 2 shows.

The trace amount of Zn^{2+} catalyst provides a new, lower-E_a mechanism for turning reactants into products by making use of zinc's ability to remove the OH^- group from the propanol in step 1. Notice that in step 3, the Zn^{2+} is regenerated when an H^+ ion from the added acid removes the OH^- group from $ZnOH^+$. This regeneration means that the Zn^{2+} is again available to be used over and over. This is why only a trace amount of Zn^{2+} is necessary to speed up the reaction.

Catalysts are tremendously valuable to the chemical industry because, by using them, high-E_a reactions that would normally require great amounts of energy (and therefore dollars) can be run for less money and more profit. Even more important, though, is the role of catalysts in biological systems. In short, you would not be alive without them because, at normal body temperature, many of the biochemical reactions responsible for keeping you alive would run too slowly.

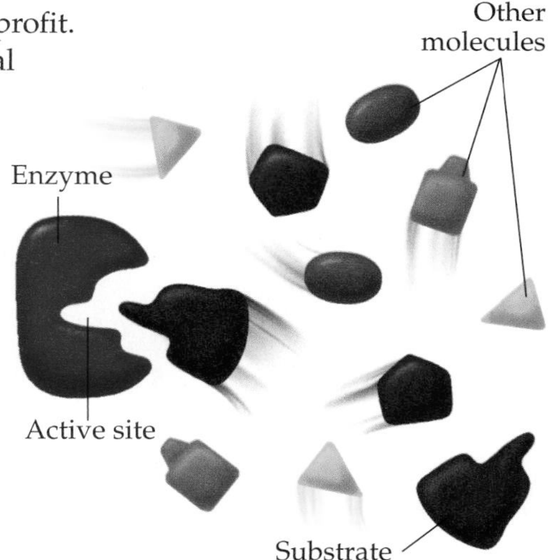

Very large molecules called **enzymes** are the major catalysts in living systems. These large biomolecules speed up chemical reactions, but in a highly selective fashion—each enzyme molecule is designed to interact with only certain specific reactants. Enzymes are selective because of their shape, as we saw in our discussion of sulfanilamide in Section 6.1. Each enzyme molecule contains on its surface a specifically shaped pocket that fits only a specific reactant molecule. A reactant molecule that fits into this pocket and undergoes the reaction is called a **substrate**. The pocket itself is known as the enzyme's **active site**. How an enzyme interacts with its substrate is often referred to as a *lock-and-key mechanism*, where the active site is the "lock" and the substrate molecule is the only "key" that fits it. Once the key is inserted into the lock, intermolecular interactions between the enzyme and the substrate weaken certain substrate bonds and bring the substrate into the proper orientation to react. The E_a value for the catalyzed reaction is always lower than the E_a value of the uncatalyzed reaction.

For example, the enzyme sucrase has an active site that conforms to the shape of sucrose, the sugar you sprinkle on cereal or put into coffee. Once a molecule of sucrose is attached to the sucrase active site, this enzyme catalyzes the conversion of sucrose to two simpler sugars, glucose and fructose. This transformation is necessary to keep us alive because it is glucose, not the sucrose we eat, that our cells use to produce the energy needed to carry out other life-sustaining reactions.

Interactions between the sucrase and sucrose molecules weaken one of the C–C bonds in sucrose, thereby lowering the E_a for the reaction

sucrose $\longrightarrow$ glucose + fructose.

Sucrose in active site

Weakened bond

Sucrase

Glucose part of molecule

Bond that gets broken

Fructose part of molecule

$C_{12}H_{22}O_{11}$ + H_2O $\longrightarrow$ $C_6H_{12}O_6$ + $C_6H_{12}O_6$

Sucrose, Glucose, Fructose

Many diseases can be traced to hereditary defects that lead to the production of enzymes that have malformed active sites, incapable of accommodating the substrate molecules for which they were meant.

WORKPATCH

In Equation 13.1, which factor or factors do you think a catalyst would alter?

Our answer to the WorkPatch, shown at the end of the chapter, tells you how thorough nature can be when it has to solve a problem.

13.4 How Concentration Affects Reaction Rate

Up to this point, if we wanted to discuss a reaction mechanism, we simply presented it to you. Now we want to show you one of the primary experimental methods used in determining a reaction mechanism: varying the concentration of the reactants and observing how the rate changes. To show you how this works, we begin by discussing why varying reactant concentrations changes a reaction rate.

So far, all the factors we have examined that influence reaction rate are factors that can't be changed (without help from a catalyst) because they are properties of the reacting molecules, properties such as activation energy and orientation. We repeat the relationship we stated earlier:

$$\text{Reaction rate} = \underbrace{\begin{matrix}\text{Number of collisions per}\\ \text{unit time with energy equal}\\ \text{to or greater than } E_a\end{matrix}}_{\text{Energy factor}} \times \underbrace{\begin{matrix}\text{Fraction of collisions in which}\\ \text{molecules have proper orientation}\end{matrix}}_{\text{Orientation factor}} \qquad (13.1)$$

Now let's modify this expression a bit by separating the first factor into two parts:

$$\text{Reaction rate} = \underbrace{\begin{matrix}\text{Number of collisions}\\ \text{per unit time}\end{matrix} \times \begin{matrix}\text{Fraction of collisions}\\ \text{with energy equal to}\\ \text{or greater than } E_a\end{matrix}}_{\text{Energy factor}} \times \underbrace{\begin{matrix}\text{Fraction of collisions}\\ \text{in which molecules have}\\ \text{proper orientation}\end{matrix}}_{\text{Orientation factor}} \qquad \textbf{(13.2)}$$

This expression says exactly the same thing as Equation 13.1 because the two new factors are equal to the single energy factor in Equation 13.1. So why did we make this modification? Because Equation 13.2 highlights a very important fact—in addition to depending on the activation energy, E_a, and molecule orientation, the rate of a reaction also depends on the total number of collisions per unit time.

For example, suppose that when a certain reaction is run at a fixed temperature, only 1% of all collisions lead to product formation. Let's run the reaction in two equal-sized boxes that are at the same temperature but have one important difference. In box 1 there are 20 collisions between reactant molecules every second, and in box 2 there are 20,000 collisions every second. Don't forget, in both boxes only 1% of the collisions are effective.

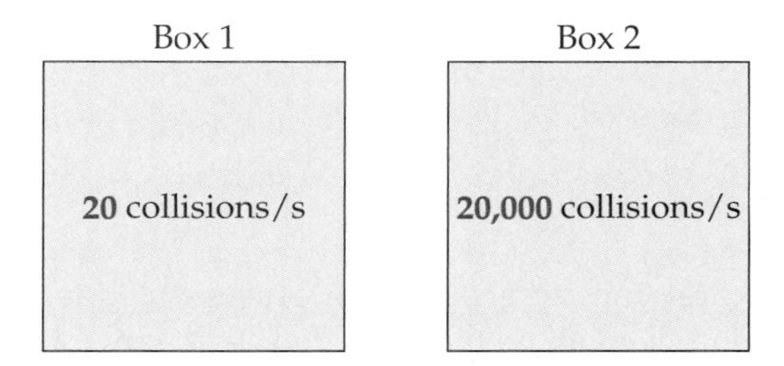

In which box does product form faster? Obviously, in box 2 because, even though only 1% of the collisions are effective in both boxes, there are 1000 times more collisions each second in box 2.

The next question is "How can we arrange to have 1000 times more collisions per second in box 2?" You can see the answer by thinking about rush-hour traffic. More cars on a highway result in more collisions. To get 1000 times more collisions per second in box 2 without changing either the size of the box or the temperature, we make sure the number of reactant molecules in box 2 is much greater than the number in box 1.

Because the two boxes are the same size (volume), putting more reactant molecules in box 2 means that box 2 has more reactant molecules per unit volume than box 1. Remember from our discussion of molarity in Chapter 12 that number of molecules per unit volume is a unit of concentration. And so we arrive at how reactant concentration affects rate:

The greater the concentration of reactants, the greater the total number of collisions per second and thus the higher the reaction rate.

Both boxes are at the same temperature; the same reaction is occurring in both boxes; only 1% of collisions are effective.

Box 1

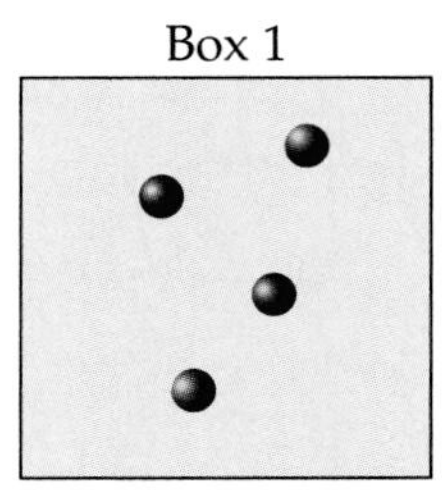

20 collisions/s

Box 2

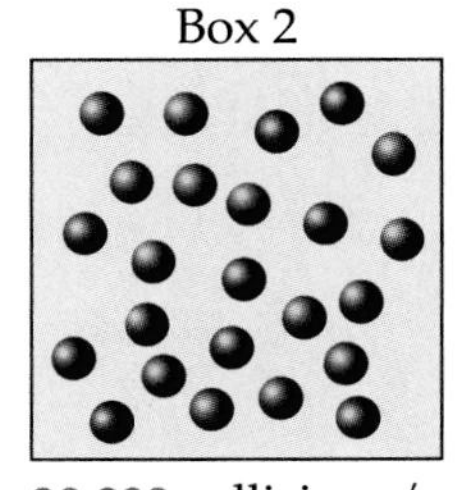

20,000 collisions/s

Another way to increase reactant concentration is to decrease the volume of the container. For example, imagine a gas-phase reaction in which $A + B \longrightarrow$ Products. There are five A molecules and five B molecules in each of two reaction vessels, one larger than the other. In which vessel does the reaction go faster?

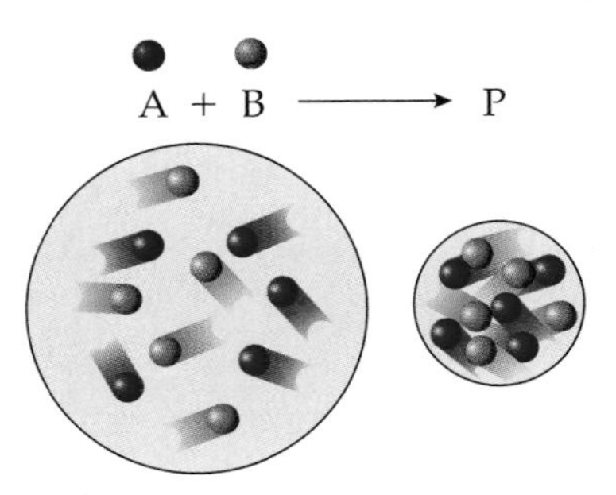

The reaction in the smaller vessel has the higher rate because here the concentration of reactants is greater (there are more molecules per unit volume). Being crowded into a smaller volume, the molecules travel shorter distances, on average, before bumping into one another. Because the molecules are moving at the same average speed in the two vessels (the temperature is the same), this means there are more collisions per second in the smaller vessel.

However you do it, whether by injecting more molecules into the system or by squeezing the same number of molecules into a smaller volume, increasing reactant concentration speeds up the reaction by increasing the number of collisions per unit time. The bottom line is that the magnitude of the first factor in Equation 13.2 depends on reactant concentration:

$$\text{Reaction rate} = \underbrace{\begin{array}{c}\text{Number of collisions}\\\text{per unit time}\end{array}}_{\substack{\text{This factor depends on}\\\text{reactant concentration and}\\\text{so can be manipulated. When}\\\text{concentration increases, so}\\\text{does this factor.}}} \times \underbrace{\begin{array}{c}\text{Fraction of collisions}\\\text{with energy equal to}\\\text{or greater than } E_a\end{array} \times \begin{array}{c}\text{Fraction of collisions}\\\text{in which molecules have}\\\text{proper orientation}\end{array}}_{\substack{\text{These factors are inherent. They depend}\\\text{on bond strength and molecular shape}\\\text{(\textit{not} on concentration).}}} \qquad (13.3)$$

From this form of our rate expression, we see that there are two kinds of factors influencing reaction rate: one that can be manipulated (by changing reactant concentration) and two *inherent factors*, which are factors that depend on unchanging characteristics of the reactant molecules, such as bond strength and molecular shape. Chemists like to simplify equations. In this case, we replace the two inherent factors by the symbol k:

$$k = \underbrace{\begin{array}{c}\text{Fraction of collisions}\\\text{with energy equal to}\\\text{or greater than } E_a\end{array} \times \begin{array}{c}\text{Fraction of collisions}\\\text{in which molecules have}\\\text{proper orientation}\end{array}}_{\text{Inherent rate factors}} \qquad (13.4)$$

The k is called the **rate constant** for a reaction. Each reaction has its own value of k. If k is large, the reaction is fast. If k is small, the reaction is slow. And, as Equation 13.4 clearly shows, the value of k depends on both the E_a value of a reaction and the shapes (orientations) of the reactant molecules.

Suppose a reaction has a large k value.

(a) What does this say about E_a for the reaction?
(b) What does this say about the orientation requirements of the reaction?
(c) Is such a reaction intrinsically fast or slow?

Check your answer to the WorkPatch against ours. If you got it right, you are catching on as to how all the aspects of a reaction—rate, k, E_a, orientation, concentration—are related.

To simplify Equation 13.3, we use the definition given for k in Equation 13.4 and let k replace the two inherent factors in Equation 13.3:

$$\text{Reaction rate} = k \times \begin{array}{c}\text{Number of collisions} \\ \text{per unit time}\end{array} \qquad \textbf{(13.5)}$$

A final word about the rate constant k. It is called a constant because the things it depends on—the E_a value for a reaction and the shape of the reacting molecules—do not change. However, there are two things that can change the value of a reaction's rate constant. One is adding a catalyst. A catalyst provides a new mechanism that has a different and, by definition, lower E_a value and a different set of orientation requirements. As a result, the value of k would be larger for the catalyzed reaction than for the uncatalyzed reaction.

The second thing that can change the value of k is temperature. As temperature increases, so does k (and vice versa). Why is this so? You already know the answer. Look at Equation 13.4. Will either factor get larger as the temperature increases? Yes, the first factor will. As temperature increases, molecules move faster, meaning that there will be a larger fraction of collisions between molecules having energy equal to or greater than E_a. In addition, there will be an increase in the number of collisions per unit time. As a result, the reaction speeds up. Remember the rule of thumb that a reaction rate approximately doubles for every 10 C° increase in temperature? This is true because k nearly doubles.

Calculating reaction rates is looking pretty simple now. If we know the value of the rate constant k for a reaction, we can plug it into the rate expression, Equation 13.5, and calculate the rate. All we need to do is multiply k by the number of collisions per unit time. Now we have a problem, however. After all, you can't just look at a reaction and know how many collisions occur each second. You can't see molecules, and even if you could, you would be hard-pressed to count the billions of collisions occurring each second. But don't lose hope. We have seen that the total number of collisions per unit time depends on reactant concentration, and reactant concentration is something we *can* measure.

What comes next is extremely important. In any chemical reaction at a fixed temperature, the number of collisions per unit time is proportional to the product of the concentrations of the reactants, each concentration value raised to some power (exponent) called an **order**. For the general reaction $A + B \longrightarrow P$, we express this relationship in the form

$$\begin{array}{c}\text{Number of} \\ \text{collisions between A and B} \\ \text{per unit time}\end{array} \propto [A]^x[B]^y \qquad \textbf{(13.6)}$$

Orders

This symbol means "is proportional to."

Square brackets mean "concentration of."

Don't get lost in the details. Just keep in mind the basic concept—increasing reactant concentrations results in more collisions between reactant molecules. The orders just fine-tune this idea, and we shall discuss them more thoroughly soon. For the moment, the proportionality in Expression 13.6 allows us to replace the factor for number of collisions per unit time in Equation 13.5 with measurable concentrations. For the general reaction $A + B \longrightarrow P$, this substitution gives us

$$\text{Reaction rate} = k[A]^x[B]^y \tag{13.7}$$

This equation is called the **rate law** for the reaction. The rate constant k embodies the inherent rate factors (E_a, orientation requirements), and the concentrations $[A]^x[B]^y$ embody the number of collisions. Now, go back and look at Equation 13.2 to see where we started. The rate law in Equation 13.7 is a symbolic expression of Equation 13.2, with k representing the second and third terms of Equation 13.2 and $[A]^x[B]^y$ representing the first term.

In principle, it is easy to write the rate law for any chemical reaction. For example, consider the general reaction

$$a\text{A} + b\text{B} + c\text{C} \longrightarrow \text{P}$$

where a, b, and c represent the coefficients in the balanced equation. The rate law for this reaction is

$$\text{Reaction rate} = k[A]^x[B]^y[C]^z \tag{13.8}$$

Note that the concentrations of all the reactants are included—[A], [B], [C]—each raised to its respective order (x, y, z). The rate law for any reaction is the rate constant k times the concentration of each reactant raised to some power. This equation is the promised window that allows us to "look at" reaction mechanisms, the reason we started looking at kinetics in the first place.

There is one more thing, however. In order to look through this window, you have to understand what the orders x, y, z are all about and how we find them. We'll turn to them after you do the following practice problems.

Practice Problems

13.17 Write a general rate law for the reaction

$$2\,NO + O_2 \longrightarrow 2\,NO_2$$

using x and y as orders.

Answer: Rate = $k[NO]^x[O_2]^y$
Note that only the reactants show up in the rate law, not the products. (Rate laws with products in them go beyond the scope of this book.)

13.18 Write a general rate law for the reaction

$$H_2O_2(aq) + 3\,I^-(aq) + 2\,H^+(aq) \longrightarrow I_3^-(aq) + 2\,H_2O(l)$$

using x, y, z as orders.

13.19 Suppose that when the two reactions described in Practice Problems 13.17 and 13.18 are run at the same temperature and the same concentrations, the NO reaction goes much, much faster than the H_2O_2 reaction.

(a) What can you say about the relative sizes of the k values for the reactions?

(b) What can you say about the relative sizes of the E_a values for the reactions?

(c) What are three ways you might speed up the slow reaction to make it faster than the fast reaction?

13.20 Why should the number of collisions per second between reactant molecules have anything to do with their concentration?

13.5 Reaction Order

There's still one important part of the rate law we haven't explained—a part that seems to have appeared out of thin air. What do the orders associated with reactant concentrations mean? And where do they come from?

Orders

$$\text{Rate} = k[\text{A}]^x[\text{B}]^y[\text{C}]^z$$

To answer this question, we start by telling you where they *do not* come from. In general, the orders do *not* come from the coefficients a, b, c of the balanced equation:

$$a\,\text{A} + b\,\text{B} + c\,\text{C} \longrightarrow \text{P}$$

So, for example, for the reaction

$$H_2O_2(aq) + 3\,I^-(aq) + 2\,H^+(aq) \longrightarrow I_3^-(aq) + 2\,H_2O(l)$$

from Practice Problem 13.18, you would be incorrect if you wrote Rate = $k[H_2O_2]^1[I^-]^3[H^+]^2$.

The correct way to find the orders in a rate law is by doing rate experiments, often called *kinetics experiments*. In a **kinetics experiment**, we vary the concentration of one reactant and observe what happens to the reaction rate. To see how this is done in principle, let's again consider the hypothetical gas-phase reaction A + B $\longrightarrow$ Products. Suppose we start with a 1-L vessel filled with 1.0 mole of A molecules and 1.0 mole of B molecules. This means that the starting concentration of both A and B is 1.0 M. We run the reaction for 1 min and then measure the number of product molecules formed. Suppose that after 1 min, 0.1 mole of product has formed. Keep in mind that as time passes, the reactants get used up (their concentrations decrease) as product forms.

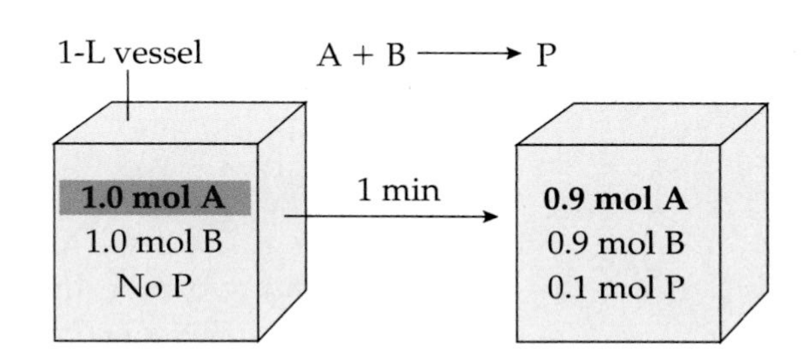

What is the rate of this reaction? The data from the kinetics experiment let us answer this question. Remember, the rate is always something over time. In this case, we want to know how fast the concentration of product is changing during a certain time interval, which means our "something" is the change in product concentration at the end of the time interval:

$$\text{Rate} = \frac{\text{Change in product concentration}}{\text{Time}}$$

At the end of 1 min, 0.1 mole of product has formed. Because the reaction takes place in a 1-L vessel, the molar concentration of the product after 1 min is (0.1 mol P)/(1 L) = 0.1 M. (The nice thing about working with a 1-L vessel is that the number of moles of each substance present is equal to the molarity of the substance.) At the start of the reaction, the product concentration was 0.0 M, which means the change in product concentration is the difference between the two concentrations, or 0.1 M. Thus, for the first minute of this reaction, the rate is 0.1 M/min.

Units of molarity/time may look strange to you, but they are the common units of reaction rate, just as miles/h are common units for car speeds. Practice will help you get used to them. For example, suppose we repeat the kinetics experiment, this time doubling the starting concentration of A to 2.0 mol/L (2.0 M). We'll again measure the rate by determining how much product forms in the first minute of the reaction.

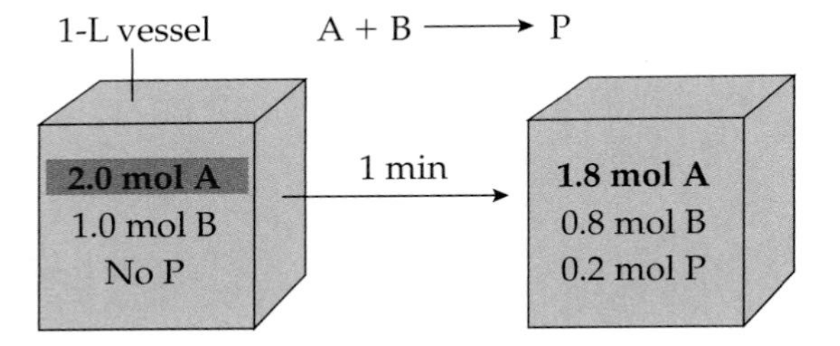

When we measure the amounts of everything in the vessel after 1 min has passed, we find that 0.2 mole of P has formed. Therefore, the change in the product concentration is 0.2 M, and the rate is 0.2 M/min (twice the original value). So, when we double the concentration of A, the rate of product formation also doubles (in other words, the reaction runs twice as fast). It thus appears that there is a one-to-one correspondence between concentration of A and reaction rate (as the concentration of A doubles, so does the rate). For any reactant where this is the case, we assign a value of 1 for its order. *An order of 1 for a reactant means there is a one-to-one correspondence between reactant concentration and rate*:

$$\text{Rate} = k[\text{A}]^1[\text{B}]^y$$

An order of 1 means that if you double the concentration of A, the rate also doubles.

So now we have found the value of x for reactant A. To find this order, we ran two experiments and never changed the concentration of B; we changed only the concentration of A. Doing things this way ensured that any change in rate was due to A only.

Now it's time to look at B. We run the reaction with 1.0 mole of A and 2.0 moles of B, which means we've doubled the concentration of B but left A alone:

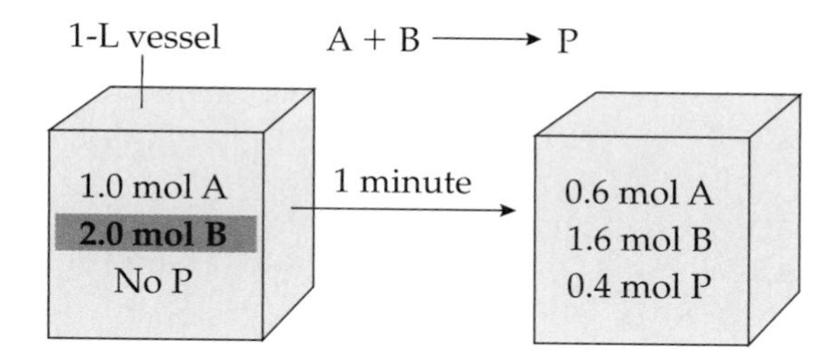

The kinetics data show that after 1 min, 0.4 mole of product has formed. The rate of product formation is thus 0.4 M/min, four times the rate we observed when we used only 1.0 mole of B. Doubling the B concentration quadruples the rate, quite a different result from doubling the concentration of A. How do we get this rate quadrupling effect of B into the rate law? The answer is by assigning B an order of 2. *An order of 2 for a reactant means the rate quadruples when the concentration of that reactant is doubled*:

$$\text{Rate} = k[\text{A}]^1[\text{B}]^2$$

An order of 2 means that if you double the concentration of B, the rate quadruples.

To see the quadrupling effect of the exponent 2, just consider the effect of this exponent on the concentration of B when that concentration is doubled:

Concentration of B → $[1]^2 = 1$

Concentration of B doubled → $[2]^2 = 4$

An exponent of 2 on a concentration causes quadrupling when the concentration is doubled.

When we double the concentration of B from 1 M to 2 M, the quantity $[\text{B}]^2$ increases by a factor of 4. Thus, the term $[\text{B}]^2$ in the rate expression tells us that the rate quadruples when the concentration of B is doubled.

Orders besides 1 and 2 are often observed in rate laws. For some reactions, a reactant may have an order of zero, which means that changing the concentration of that reactant has no effect on the rate; that is, the rate does not depend on the concentration of that reactant. Such reactants usually are not included in the rate law for the reaction. It is even possible for reactants to have fractional or negative orders in some reactions, although we will not deal with such situations in this book. The important thing to remember is that orders must be determined by doing kinetics experiments.

Now let's switch from the hypothetical reaction A + B ⟶ P to a real reaction and examine some real rate data from kinetics experiments. Consider this reaction between nitrogen oxide and hydrogen to give nitrogen and water:

$$2\,\text{NO}(g) + 2\,\text{H}_2(g) \longrightarrow \text{N}_2(g) + 2\,\text{H}_2\text{O}(g) \qquad (13.9)$$

The general rate law for this reaction can be written

$$\text{Rate} = k[\text{NO}]^x[\text{H}_2]^y \qquad (13.10)$$

To find the orders, a kineticist runs this reaction three times, each time measuring the rate of formation of the product N_2 in the first second of the reaction. The data are presented in Table 13.1.

Table 13.1 Rate Data for Reaction 13.9 at 904°C

Experiment	Initial concentration of NO (mol/L)	Initial concentration of H_2 (mol/L)	Rate of N_2 formation (M/s)
1	0.210	0.122	0.0339
2	0.210	0.244	0.0678
3	0.420	0.122	0.1356

First look at the data for experiments 1 and 2. For these two, the initial concentration of NO is unchanged but the concentration of H_2 is doubled. Notice that this change causes the rate to double. This means that the order (y) of H_2 must be 1. Now let's find the order for the other reactant, NO. To do this, examine the data for experiments 1 and 3. Why 1 and 3? Because we are looking for two experiments where the concentration of NO changes but the concentration of H_2 remains unchanged. For experiments 1 and 3, the concentration of NO is doubled and the rate quadruples. This means the order (x) of NO must be 2. The complete rate law is thus

$$\text{Rate} = k[\text{NO}]^2[\text{H}_2]$$

Notice that when the order is 1, we usually don't show the exponent.

Because the order of H_2 is 1, we say that the reaction is *first-order with respect to hydrogen*. Because the order of NO is 2, we say that the reaction is *second-order with respect to NO*. Chemists also often quote the **overall order of the reaction**, which is obtained by adding together all the exponents in the rate expression. For this example, the overall order of the reaction is 3 and we call it a *third-order reaction*.

Practice Problems

13.21 Suppose the rate did not change when we ran experiments 1 and 3 in Table 13.1. Write the rate law. What is the overall order of the reaction?

Answer: If the rate didn't change when we doubled the NO concentration, the order of this reactant is 0, meaning the rate does not depend on NO concentration. The rate law is thus Rate $= k[\text{H}_2]$, *and the overall reaction order is 1.*

13.22 For the reaction

$$\text{BrO}_3^- + 5\,\text{Br}^- + 6\,\text{H}^+ \longrightarrow 3\,\text{Br}_2 + 3\,\text{H}_2\text{O}$$

the experimentally determined rate law is

$$\text{Rate} = k\,[\text{BrO}_3^-][\text{Br}^-][\text{H}^+]^2$$

(a) What is the order of this reaction with respect to Br^-?
(b) What is the order of this reaction with respect to H^+?
(c) What is the overall order of the reaction?
(d) What happens to the rate of this reaction when you double the H^+ concentration?

13.23 Use the given kinetics data to write the rate law for the reaction

$$2\,NO + O_2 \longrightarrow 2\,NO_2$$

Experiment	Initial [NO]	Initial $[O_2]$	Rate of NO_2 formation (M/s)
1	0.015 M	0.015 M	0.048
2	0.030 M	0.015 M	0.192
3	0.015 M	0.030 M	0.096
4	0.030 M	0.030 M	0.384

There is one topic left to cover in this chapter. What determines the values of these orders? Why are orders sometimes 1, sometimes 2, sometimes something else? Once we've answered this question, you will understand how kinetics provides a window into reaction mechanisms.

13.6 Why Reaction Orders Have the Values They Do—Mechanisms

It would be wonderful if we could assign the orders in a rate law directly from the stoichiometric coefficients in the balanced equation instead of having to do kinetics experiments. Unfortunately, this is generally not possible because the balanced chemical equation does not show you how the various reactants come together. For example, consider the balanced equation for the combustion of glucose:

$$\underset{\text{Glucose}}{C_6H_{12}O_6} + 6\,O_2 \longrightarrow 6\,CO_2 + 6\,H_2O$$

Judging from the stoichiometric coefficients, the equation seems to imply that one glucose molecule and six oxygen molecules collide all at once to form the combustion products CO_2 and H_2O. If this were how the reaction actually proceeded, the rate law would be

$$\text{Rate} = k[C_6H_{12}O_6][O_2]^6$$

but this is *not* correct because this is *not* what happens.

We said earlier that the chance of just three molecules colliding in the same spot is remote. The chance of *seven* molecules doing this is essentially zero. Most reactions occur in a series of simpler steps—called **elementary steps**—that involve collisions of just two molecules at a time. A reaction mechanism is therefore a series of elementary steps that show exactly how reactant molecules are converted to product molecules. Each step in a mechanism has its own rate law. In addition, in each step, the balancing coefficients give directly the orders of the reactants in the rate law. Unfortunately, reactions don't come with their mechanism printed below them, and we can't observe the effective collisions responsible for the reaction. So how do we determine the elementary steps and the overall mechanism? The answer is, we work backward. That is, first we determine the rate law for the overall reaction (get the orders) by doing kinetics experiments, and then we postulate a mechanism that would give rise to this rate law. When postulating a mechanism, we always have to be sure to use only elementary steps that involve effective two-molecule collisions. If we postulate collisions involving three or more molecules, our postulated mechanism is not very likely to be correct. Let's work through a few examples of this backward process.

We start by considering the hypothetical reaction $A_2 + B_2 \longrightarrow 2\ AB$. Suppose we do the appropriate kinetics experiments and find that the rate law is

$$\text{Rate} = k[A_2]^2$$

That is, the reaction is second-order with respect to A_2 and zero-order with respect to B_2, which means second-order for the overall reaction. (At this point, you should know what experiments we had to do to determine this.) We call this rate law the **experimental rate law** because we found it by doing kinetics experiments. Now we are going to postulate a couple of mechanisms and see if either of them generates this rate law.

Suppose this reaction occurs in one elementary step involving a two-molecule collision between an A_2 molecule and a B_2 molecule to give two AB molecules. We'll call this mechanism I:

Mechanism I, one step:

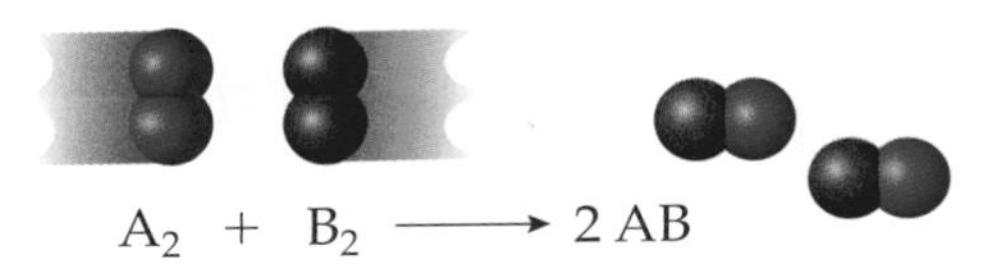

It is important to realize that this chemical equation represents an elementary step. It tells you exactly how the reaction occurs—that is, it reveals the actual mechanism of the reaction. If $A_2 + B_2 \longrightarrow 2\ AB$ is an elementary step, the reaction must occur by a direct collision between an A_2 molecule and a B_2 molecule.

Because we are assuming that our reaction proceeds through mechanism I—a single elementary step—we can write the rate law for the reaction from the balanced equation. In the balanced equation, the stoichiometric coefficients for the reactants are both 1, and therefore the rate law predicted by this mechanism is

$$\text{Rate} = k[A_2]^1[B_2]^1 = k[A_2][B_2]$$

Let's call this our **predicted rate law**, the one generated by the postulated reaction mechanism. Now we need to ask some questions. Is mechanism I correct? Does the reaction occur this way? To answer these questions, we compare the predicted rate law with the experimentally determined one:

Experimental rate law: $\text{Rate} = k[A_2]^2$
Predicted rate law: $\text{Rate} = k[A_2][B_2]$

The predicted rate law is different from the experimental one, which means our postulated mechanism I cannot be correct. We must try again.

This time, we postulate a multistep mechanism involving three elementary bimolecular (two-molecule) steps. We'll call this mechanism II:

Mechanism II, three steps

Step 1 $A_2 + A_2 \longrightarrow 2\,A + A_2$

Step 2 $A + B_2 \longrightarrow AB_2$

Step 3 $AB_2 + A \longrightarrow 2\,AB$

Multistep mechanisms must fulfill an important requirement—the individual elementary steps must add up to give the overall balanced reaction. If they don't, the postulated mechanism can't be correct. So let's add up the three elementary steps of mechanism II. When we add elementary steps, we cancel anything on the left side of one step that appears *in exactly the same form* on the right side of any other step:

Adding the three elementary steps yields the overall net balanced reaction.

Step 1	$A_2 + \cancel{A_2}$	$\longrightarrow$	$\cancel{2A} + \cancel{A_2}$
Step 2	$\cancel{A} + B_2$	$\longrightarrow$	$\cancel{AB_2}$
Step 3	$\cancel{AB_2} + \cancel{A}$	$\longrightarrow$	$2\,AB$
Overall balanced reaction	$A_2 + B_2$	$\longrightarrow$	$2\,AB$

Be careful!
There are two A's here!
They cancel the two separate A's on the left, one in step 2 and one in step 3.

For any reaction with a mechanism that involves two or more steps, each step in the mechanism is an elementary step but the overall balanced reaction (the sum of the steps) is not. Most of the chemical equations you have seen in this book represent reaction mechanisms that involve more than one elementary step.

As you can see, when we add our three elementary steps, we do indeed get the correct balanced overall reaction. This does not mean, however, that this mechanism is necessarily the correct one. To be acceptable, it had to do at least this, but it must also do something else. It must generate the experimental rate law. So we now want to compare the rate law given by this mechanism with the experimental rate law and see if the two agree. But mechanism II has three steps from which we can write three different rate laws! Which one do we use?

A reasonable choice is to use the rate law for the slowest elementary step. Why? Because the overall reaction can't go any faster than the slowest step in its mechanism. The slowest step in any reaction mechanism is the bottleneck in the whole process and is referred to as the **rate-determining step**. The drawing on the next page shows how this concept works. When we experimentally measure the rate of a reaction, it is really the rate of the slowest step we are measuring. Therefore, the experimental rate law should agree with the rate law derived from the slowest elementary step. Because we are postulating the

An Hourglass Illustrating the Concept of Rate-Determining Step

Overall reaction R ⟶ P

Mechanism R ⟶ A ⟶ B ⟶ P

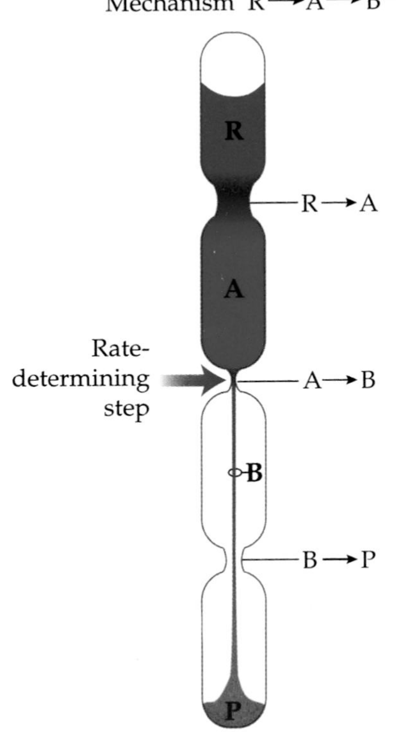

The amount of P formed per second depends on how fast the rate-determining step (A ⟶ B) is.

mechanism, it's up to us to choose the rate-determining step. We do know what the experimental rate law looks like (Rate = $k[A_2]^2$), however, and we can use that knowledge to help us make an intelligent choice. Look closely at the first step:

$$A_2 + A_2 \longrightarrow 2\,A + A_2$$

It involves a collision between two A_2 molecules in which the bond in one molecule is broken to give two A atoms while the other A_2 molecule remains intact. The rate law for this elementary step can be written directly from the balanced reaction:

$$\text{Rate} = k[A_2][A_2] = k[A_2]^2$$

This step is therefore second-order in A_2 and zero-order in B_2. If we assume that this step is the slowest of the three, the speed of this step is what governs the overall speed of the reaction, which means the rate law for this step becomes the predicted rate law for the overall reaction:

Experimental rate law: Rate = $k[A_2]^2$

Rate law predicted by assuming step 1 to be rate-determining step: Rate = $k[A_2]^2$

They match! Does this mean mechanism II is correct? Maybe. However, there are many other mechanisms we could postulate that would also predict a rate law that agrees with the experimentally determined one. Doing kinetics experiments to get the experimental rate law allows us to rule out incorrect mechanisms, but it can never alone prove that a particular mechanism is correct. All it can do is provide support for the validity of a possible mechanism.

Although we can never truly know the mechanism for a chemical reaction from kinetics data alone, kineticists can do additional experiments to further support the validity of a particular mechanism. For example, in our mechanism II, step 1 produces A atoms, which are not among the products of the overall reaction (AB molecules are the only product of our hypothetical reaction). In addition, step 2 produces an AB_2 molecule, which is also not a product of the overall reaction. These species, A and AB_2, are called *reaction intermediates*. A **reaction intermediate** is a species that is produced during one step of a reaction and then consumed in a subsequent step on the way toward making the final product. A chemist can try to observe or detect the presence of a particular reaction intermediate as it first appears and then disappears during the course of a reaction. This is usually a difficult task because intermediates are often present only in very low concentrations and only for very short periods of time. However, if intermediates can be detected, and if they match those predicted by the postulated mechanism, the chemist has additional support for its validity.

Let's try this again using the substitution reaction we introduced at the beginning of the chapter:

$$OH^- + CH_3Br \longrightarrow CH_3OH + Br^-$$

Here are two possible mechanisms for this reaction:

Mechanism I, two steps

$$CH_3Br \xrightarrow{\text{Slow first step}} CH_3^+ + Br^-$$

$$CH_3^+ + OH^- \xrightarrow{\text{Fast second step}} CH_3OH$$

$$\text{Rate} = k[CH_3Br]^1 = k[CH_3Br]$$

Mechanism II, one step

$$CH_3Br + OH^- \longrightarrow CH_3OH + Br^-$$

$$\text{Rate} = k[CH_3Br]^1[OH^-]^1 = k[CH_3Br][OH^-]$$

Both mechanisms have a rate-determining step. In mechanism I, the rate-determining step is the first step because it is the slower step. In mechanism II, there is only one step, which means it must be rate-determining.

Why is the CH_3^+ cation in mechanism I properly considered a reaction intermediate?

Because mechanism I consists of more than one step, the steps must add up to give the proper balanced reaction before we can even consider this as the correct mechanism.

13.7 WorkPatch

Prove that when you add up the steps in mechanism I, you arrive at the proper balanced equation:

$$OH^- + CH_3Br \longrightarrow CH_3OH + Br^-$$

Now comes the important question. Which of these two mechanisms is incorrect? Which might possibly be correct? We need to compare the rate law predicted from each mechanism with the experimental rate law. The experimental data from a series of kinetics experiments are presented in WorkPatch 13.8. See what you can find out.

13.8 WorkPatch

Rate data for the reaction $CH_3Br + OH^- \longrightarrow CH_3OH + Br^-$:

Experiment	Initial $[CH_3Br]$	Initial $[OH^-]$	Rate of CH_3OH production (M/min)
1	0.200 M	0.200 M	0.015
2	0.400 M	0.200 M	0.030
3	0.400 M	0.400 M	0.060

Use these data to find the experimental rate law. In other words, determine the values of x and y in the rate equation

$$\text{Rate} = k[CH_3Br]^x[OH^-]^y$$

Then decide whether mechanism I or mechanism II can possibly be the correct mechanism for the reaction.

You should have found that mechanism I is wrong and that only mechanism II is a possibility.

13.9 WorkPatch

Why is mechanism I wrong?

Why is mechanism II still in the running?

Check your answers against ours. Once you understand the principles behind the answers, you will understand how the field of kinetics is indeed one of our best windows into reaction mechanisms. The time spent doing kinetics experiments and determining rate laws is time well spent. The hard part for even advanced students of chemistry is postulating the mechanisms because doing this takes practice and experience. We have given you a few guidelines. Let's review them here:

1. Any elementary step should never go beyond a two-molecule collision.
2. For a multistep mechanism, the elementary steps must add up to give the balanced equation for the overall reaction.
3. Whatever mechanism you postulate must allow you to predict a rate law that matches the experimental rate law.
4. The only thing a match between a predicted rate law and an experimental rate law can do is support the postulated mechanism. The match does not prove a mechanism is correct because other mechanisms may also generate the same rate law. To prove any mechanism beyond a doubt, chemists must do many more experiments, which is one of the most exciting challenges of modern chemistry.

Practice Problems

13.24 Suppose the experimental rate law for the reaction $X_2 + Y_2 \longrightarrow 2\,XY$ is

$$\text{Rate} = k[X_2][Y_2]$$

A student postulates a multistep mechanism in which the first step is the rate-determining step. For this step, he postulates $2\,Y_2 \longrightarrow Y_3 + Y$. Is it possible for this to be the rate-determining step?

Answer: No. In any multistep mechanism, the rate law for the rate-determining step is also the rate law for the overall reaction. The rate law for this postulated rate-determining step is Rate $= k[Y_2]^2$, *which does not match the experimental rate law.*

13.25 Suppose the experimental rate law for the reaction in Practice Problem 13.24 is

$$\text{Rate} = k[Y_2]^2$$

Use the step postulated in Practice Problem 13.24 as the first step in a possibly correct mechanism for the reaction. Don't be afraid to be creative with your steps, but keep your collisions to two molecules. In addition,

(a) Show that the steps of your proposed mechanism add to give the proper overall reaction.
(b) Circle all reaction intermediates.
(c) What might you do to prove this mechanism beyond a reasonable doubt?

13.26 Suppose the experimental rate law for the reaction in Practice Problem 13.24 is

$$\text{Rate} = k[Y_2]$$

Suppose also that the first step in the mechanism is the rate-determining step. What might this first step look like? [*Hint:* Because it is the rate-determining step, it must generate the experimental rate law. Consider the first step in mechanism I on page 522.]

We began this chapter by saying we would take a close look at what's hidden behind the arrow in any chemical equation—in other words, at the mechanism of the reaction. We've seen that this simple arrow represents a complex, sometimes multistep transformation process. The payoff for all the hard work of doing kinetics experiments is the insight we get into reaction mechanisms. Today chemists have come to realize more than ever that such insight is important because essentially all the diseases that afflict us, from viral to genetic, operate at the molecular level. An inherited gene with a defect causes the production of a malformed enzyme that no longer interacts with the proper substrate. A virus produces a compound that allows the virus to attach to and penetrate our cell membranes. If we don't understand the mechanisms of the chemical reactions that viruses and defective genes depend on, we can never hope to cure the diseases they cause.

Have You Learned This?

Reaction mechanism (p. 492)
Chemical kinetics (p. 493)
Substitution reaction (p. 493)
Exothermic reaction (p. 495)
Endothermic reaction (p. 495)
Energy-neutral reaction (p. 495)
Reaction-energy profile (p. 496)
Net energy change ΔE_{rxn} (p. 496)
Activation energy (E_a) (p. 503)
Transition state (p. 504)
Effective collision (p. 507)
Catalyst (p. 510)
Carbocation (p. 511)
Enzyme (p. 511)
Substrate (p. 511)
Active site (p. 511)
Rate constant k (p. 514)
Order (p. 515)
Rate law (p. 516)
Kinetics experiment (p. 517)
Overall order of the reaction (p. 520)
Elementary step (p. 521)
Experimental rate law (p. 522)
Predicted rate law (p. 522)
Rate-determining step (p. 523)
Reaction intermediate (p. 524)

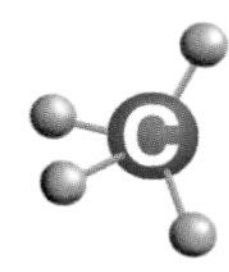

When Reactants Turn into Products
www.chemistryplace.com

SKILLS TO KNOW

ΔE_{rxn}, E_a, and k for a reaction

For any reaction, the amount of energy in the reactants is different from the amount of energy in the products. The difference between these energies is the energy change of the reaction, ΔE_{rxn}.

$$\Delta E_{rxn} = E_{products} - E_{reactants}$$

If $\Delta E_{rxn} < 0$, the reaction is exothermic (it releases energy).

If $\Delta E_{rxn} > 0$, the reaction is endothermic (it absorbs energy).

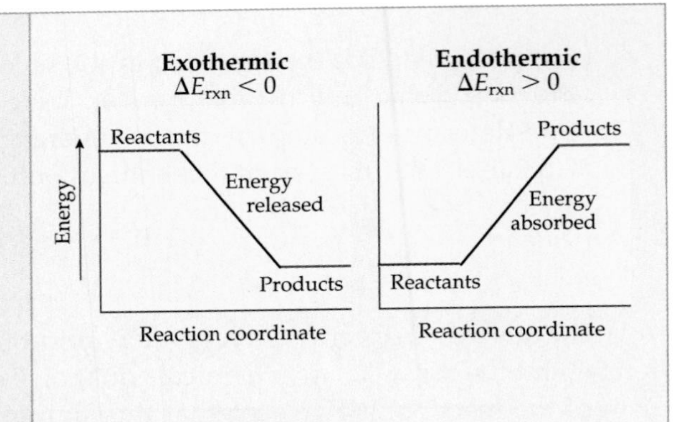

The activation energy E_a of a reaction is the amount of energy that must be supplied to the reactants to enable them to form the transition state.

In general, a larger E_a means a slower reaction rate. (The higher the E_a barrier, the fewer reactant molecules will have the energy to surmount the barrier per unit time.)

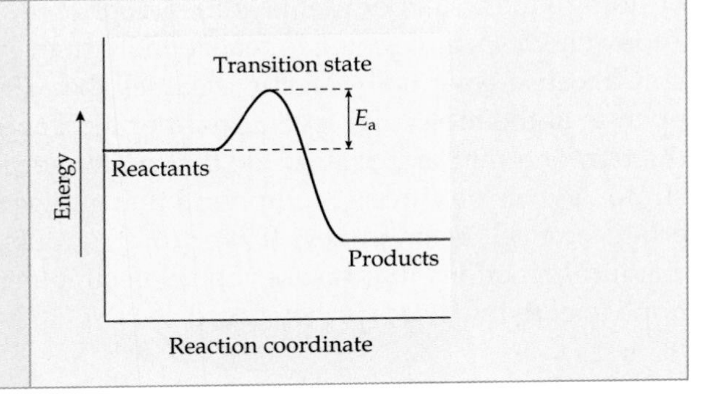

Of the factors that determine the rate of a reaction, those that are inherent to the reacting substances (including E_a) are represented by the rate constant k. Every reaction has a characteristic k.

- If k is large, the reaction is fast.
- If k is small, the reaction is slow.

The value of k depends on temperature; increasing the temperature increases the value of k.

k = Fraction of collisions with energy equal to or greater than E_a × Fraction of collisions in which molecules have proper orientation

The rate law for a reaction

The rate law for a general reaction

$$aA + bB + cC \longrightarrow \text{Products}$$

can be written

$$\text{Rate} = k[A]^x[B]^y[C]^z$$

where k is the rate constant,

[A], [B], and [C] are concentrations, and

x, y, and z are reaction orders, the values of which must be determined by kinetics experiments.

If a given reactant has a reaction order of

0 then changing the concentration of the reactant has no effect on the reaction rate.

1 then doubling the concentration of the reactant doubles the reaction rate.

2 then doubling the concentration of the reactant quadruples the reaction rate.

CHEMICAL KINETICS

13.27 Describe the following reaction in terms of which bonds must be broken and which bonds must be formed.

$$H_2C{=}CH_2 + 3\,O_2 \longrightarrow 2\,CO_2 + 2\,H_2O$$

13.28 Is it likely that a single collision leads to the breaking of all the bonds that you named in Problem 13.27? Explain your answer.

13.29 Is the reaction

$$CH_3I + Cl^- \longrightarrow CH_3Cl + I^-$$

an example of a substitution reaction? Explain.

13.30 Consider the following reaction. Use labeled arrows to indicate which bonds must be broken and which bonds must be formed.

```
    H  H            Cl    H
    |  |              \   |
 H—C—C—O—H  +         C—C—H   ——>
    |  |             //   |
    H  H            O     H

        H  H           H
        |  |           |
     H—C—C—O—C—C—H  +  HCl
        |  |     ||    |
        H  H     O     H
```

13.31 What is meant by the *mechanism* of a chemical reaction?

13.32 Which branch of chemistry concerns itself with the study of reaction rates and the factors that affect rates?

13.33 What do chemists typically do to indirectly "see" a reaction mechanism?

13.34 What is one benefit of understanding a reaction's mechanism?

ENERGY CHANGES AND CHEMICAL REACTIONS

13.35 Compound A converts to compound B; ΔE_{rxn} is -100 kJ/mol. Is compound B at a higher or lower energy level than compound A? By how much?

13.36 Compound A has half as much energy in it as compound B. If compound A converts to B, will this reaction release energy into the surroundings or absorb energy from the surroundings? Explain your answer.

13.37 In a chemical reaction, compound A is converted to compound B. In the process, energy is absorbed from the surroundings. Which compound is at a higher energy level? Explain your answer.

13.38 In a chemical reaction, compound C is converted to compound D. In the process, energy is released into the surroundings. Which compound is at a higher energy level? Explain your answer.

13.39 Referring to Problems 13.37 and 13.38, which reaction is exothermic and which is endothermic? Justify your answer, and describe which reaction could be used to supply heat and which could be used to "supply cold" (which actually means to remove heat).

13.40 A reaction occurs in which 1 mole of A is converted to 1 mole of B. If 1 mole of A has an energy content of 20 kJ and 1 mole of B has an energy content of 60 kJ, is this reaction exothermic or endothermic? Calculate ΔE_{rxn}.

13.41 What does it mean when ΔE_{rxn} for a reaction is negative?

13.42 The reaction of Problem 13.40 is run in the reverse direction. Is it exothermic or endothermic? Calculate ΔE_{rxn} for the reverse reaction.

13.43 What do we mean by an energy-uphill reaction? What do we mean by an energy-downhill reaction? Include reaction-energy profiles with your explanations.

13.44 Judging from the following reaction-energy profile, is the reaction endothermic or exothermic? What is the value of ΔE_{rxn}?

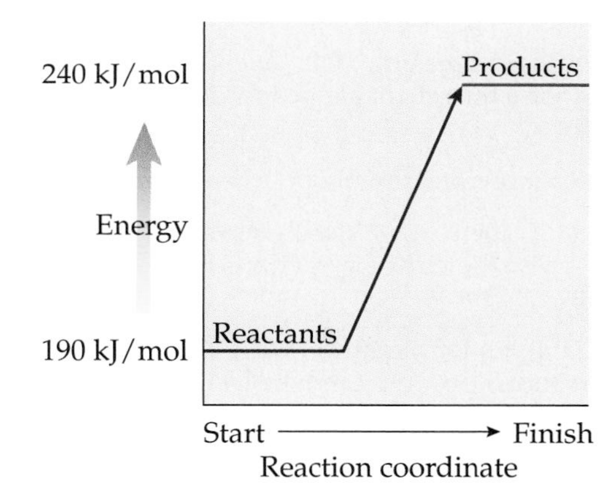

13.45 According to the reaction-energy profile in Problem 13.44, the products are higher in energy than the reactants. What must occur during the course of the reaction for this to occur? [*Hint:* The word *surroundings* must appear in your answer.]

13.46 The value of ΔE_{rxn} for exothermic reactions is always negative. Why is this so?

13.47 For a particular chemical reaction, the absorbed energy is 800 kJ to break old bonds, and 400 kJ is released on forming new bonds. Calculate ΔE_{rxn} and comment on whether this reaction is exothermic or endothermic. Explain why.

13.48 For a particular reaction, the absorbed energy is 800 kJ to break old bonds, and ΔE_{rxn} is equal to -800 kJ. How much energy is released into the surroundings as the product bonds are formed?

REACTION RATES AND ACTIVATION ENERGY

13.49 What do we mean by activation energy?

13.50 What is the relationship between the rate of a reaction and the value of E_a for the reaction?

13.51 True or false? An energy-downhill reaction can always be expected to be faster than an energy-uphill reaction. Explain your answer.

13.52 Use reaction-energy profiles, complete with E_a distances labeled, to show that your answer to Problem 13.51 is correct.

13.53 A reaction is exothermic, with $\Delta E_{rxn} = -40$ kJ, and the transition state is 20 kJ higher in energy than the reactants. Sketch a reaction-energy profile consistent with this information, complete with labels for the distances representing ΔE_{rxn} and E_a.

13.54 Define transition state for a chemical reaction.

13.55 For a particular reaction, the reactants are at 30 kJ, the products are at 60 kJ, and the transition state is at 100 kJ. Sketch a reaction-energy profile showing both ΔE_{rxn} and E_a. Also, calculate the value of ΔE_{rxn} and E_a, and state whether this reaction is endothermic or exothermic.

13.56 Would decreasing the size of E_a increase or decrease the rate of a reaction? Explain your choice fully.

13.57 Is reaction rate directly or inversely related to E_a?

13.58 Would increasing the temperature increase or decrease the rate of a reaction? Explain your choice fully.

13.59 At a given temperature, which reactant molecules can become product molecules?

13.60 Using reaction-energy profiles, plot two exothermic reactions that have the same ΔE_{rxn}, but make one reaction substantially faster than the other. Label the plots "fast" and "slow," and explain why you labeled them as you did.

13.61 What does changing the temperature at which a reaction is run do to the size of E_a?

13.62 Why might one reaction have a much larger E_a than another reaction?

13.63 What is the rule of thumb for how reaction changes as the temperature changes?

13.64 How can the orientation of reactant molecules as they collide play a role in substitution reactions?

13.65 If there were no orientation requirement for collisions, would reactions be faster or slower than they are? Explain your answer.

13.66 What is a catalyst?

13.67 In general, how does a catalyst increase the rate of a chemical reaction?

13.68 In the substitution reaction of Cl^- for OH^- in 2-propanol, explain how Zn^{2+} acts as a catalyst to increase the reaction rate.

13.69 What is the general name given to biological catalysts?

13.70 What is meant by the term *substrate*?

13.71 Explain what is meant by the term *lock-and-key mechanism*.

13.72 The opening of Chapter 6 told of the development of the first antibiotics. Was this accomplished by making modified versions of the lock or the key? Explain your answer.

13.73 Why are catalysts important to industrial chemical processes? Why are they important to biological chemical processes?

HOW CONCENTRATION AFFECTS REACTION RATE

13.74 In general, increasing the concentration of a reactant will usually increase the rate of a reaction. Why is this true?

13.75 Why would decreasing the volume of a container in which a gas-phase reaction is taking place speed up the reaction?

13.76 The product

$$\begin{pmatrix}\text{Fraction of collisions} \\ \text{with energy equal to} \\ \text{or greater than } E_a\end{pmatrix} \times \begin{pmatrix}\text{Fraction of} \\ \text{collisions in which} \\ \text{molecules have} \\ \text{proper orientation}\end{pmatrix}$$

is often called the "inherent rate" of a reaction. Why?

13.77 Does the rate constant k increase, decrease, or stay the same when the temperature of a reaction is increased? Explain your choice fully.

13.78 Does the rate constant k increase, decrease, or stay the same when you add a catalyst? Explain your choice fully.

13.79 The rate of a reaction depends both on inherent factors and on concentration. The rate constant k is associated with the inherent factors. What are they?

13.80 A student says that an exothermic reaction will always have a larger rate constant k than an endothermic one, and will thus always be faster. What is wrong with her line of reasoning?

REACTION ORDER

13.81 Given the general form of the rate law,

$$\text{Rate} = k[\text{Reactant 1}]^x[\text{Reactant 2}]^y$$

answer the following questions:

(a) Which part of the rate law reflects the inherent factors of the reaction?
(b) What is the general name for the exponents x and y?
(c) How do we calculate the overall order of a chemical reaction?
(d) Suppose the reaction is second-order with respect to reactant 1 and first-order with respect to reactant 2. What are the values of x and y, and what is the overall order of a reaction with only these two reactants?
(e) Suppose reactant 1 does not appear in the rate law. What does this say about the value of its order? What is the meaning of the value of its order?

13.82 True or false? The orders x and y in a rate law are written directly from the appropriate balancing coefficients in the overall balanced equation for the reaction.

13.83 How do we go about determining the orders in a rate law?

13.84 If a reaction rate has a first-order dependence on a given reactant concentration, what will happen to the rate when the concentration of that reactant is doubled?

13.85 If a reaction rate has a second-order dependence on a given reactant concentration, what will happen to the rate when the concentration of that reactant is doubled?

13.86 If a reaction rate has a zero-order dependence on a given reactant concentration, what will happen to the rate when the concentration of that reactant is doubled?

13.87 A reaction $A + B \longrightarrow \text{Product}$ is run in a balloon. (Both A and B are gases.) The balloon has a volume of 1 L and is initially loaded with 1 mole of A and 1 mole of B. The reaction has the rate law

$$\text{Rate} = k[A]$$

The reaction is run again using the same amount of reactants, but this time in a balloon that has a volume of 0.5 L. How much faster will the reaction proceed in the smaller balloon? Explain your answer.

13.88 Repeat Problem 13.87 for a reaction that has the rate law

$$\text{Rate} = k[A]^2$$

13.89 Repeat Problem 13.87 for a reaction that has the rate law

$$\text{Rate} = k[A][B]^2$$

13.90 Repeat Problem 13.87 for a reaction that has the rate law

$$\text{Rate} = k[A][B]$$

13.91 What do we mean by a kinetics experiment, and how is it tied to the experimental rate law?

13.92 Given the rate data below from a series of kinetics experiments, determine the orders for the following reaction, and state the overall order of the reaction.

$$H_2O_2(aq) + 3\,I^-(aq) + 2\,H^+(aq) \longrightarrow I_3^-(aq) + 2\,H_2O(l)$$

Experiment	$[H_2O_2]$	$[I^-]$	$[H^+]$	Rate (M/s)
1	0.010 M	0.010 M	0.000 50 M	1.15×10^{-6}
2	0.020 M	0.010 M	0.000 50 M	2.30×10^{-6}
3	0.010 M	0.020 M	0.000 50 M	2.30×10^{-6}
4	0.010 M	0.010 M	0.001 00 M	1.15×10^{-6}

13.93 In a kinetic study of the reaction

$$2\,A(g) + B(g) \longrightarrow P(g)$$

the following rate data were obtained. Write the rate law with proper orders. Give the overall order of the reaction. Finally, state what this problem confirms about the relationship between reactant orders and coefficients in the balanced equation.

Experiment	[A]	[B]	Rate of disappearance of A (M/s)
1	0.0125 M	0.0253 M	0.0281
2	0.0250 M	0.0253 M	0.0562
3	0.0125 M	0.0506 M	0.1124

13.94 In a kinetic study of the reaction

$$2\,ClO_2(aq) + 2\,OH^-(aq) \longrightarrow ClO_3^-(aq) + ClO_2^-(aq) + H_2O$$

the following rate data were obtained. Write a rate law complete with proper values for the orders. What is the overall order of the reaction?

Experiment	$[ClO_2]$	$[OH^-]$	Rate (M/s)
1	0.060 M	0.030 M	0.024 84
2	0.020 M	0.030 M	0.002 76
3	0.020 M	0.090 M	0.008 28

MECHANISMS

13.95 Why is it unlikely that the reaction A + 2 B + C $\longrightarrow$ P occurs in one step?

13.96 True or false? The orders in a rate law are equal to the balancing coefficients in the slowest elementary step in a mechanism.

13.97 What is an elementary step?

13.98 Suppose the reaction in Problem 13.95 did occur in one step. What rate law would predict such a mechanism?

13.99 What is meant by the term *rate-determining step*?

13.100 Why can we ignore other steps and use only the rate-determining step in a mechanism to write the predicted rate law?

13.101 Is it wise to postulate a three-molecule collision as an elementary step in a reaction mechanism? Explain your answer.

13.102 Suppose a postulated reaction mechanism generates a rate law that does not agree with the experimentally determined rate law. What does this say about the postulated mechanism?

13.103 Suppose a postulated mechanism does generate the experimental rate law, but the elementary steps, when added together, do not generate the balanced equation for the overall reaction. What can you say about the postulated mechanism?

13.104 Suppose a postulated mechanism does generate the experimental rate law, and, when the elementary steps are added together, the balanced equation for the overall reaction is generated. What can you say about the postulated mechanism?

13.105 Consider the reaction

$$(CH_3)_3C\text{—}Br + H_2O \longrightarrow (CH_3)_3C\text{—}OH + HBr$$

Kinetics studies reveal a first-order rate dependence on the concentration of the $(CH_3)_3C$–Br and a zero-order dependence on the concentration of H_2O.

(a) What happens to the reaction rate as the $(CH_3)_3C$–Br concentration is changed? What happens to the reaction rate as the H_2O concentration is changed?

(b) Two mechanisms for this reaction are offered below. Can you rule out either of them? Is either mechanism plausible given the overall balanced equation and kinetic data? Explain your answer fully.

Mechanism I

$$(CH_3)_3C\text{—}Br \xrightarrow{\text{Rate-determining step}} (CH_3)_3C^+ + Br^-$$

$$(CH_3)_3C^+ + H_2O \xrightarrow{\text{Fast}} (CH_3)_3C\text{—}OH + H^+$$

$$H^+ + Br^- \xrightarrow{\text{Fast}} HBr$$

Mechanism II

$$(CH_3)_3C\text{—}Br + H_2O \longrightarrow (CH_3)_3C\text{—}OH + HBr$$

13.106 Write the overall balanced chemical equation that goes along with the mechanism

Elementary step 1: $Cl_2 \longrightarrow 2\,Cl$

Elementary step 2: $Cl + CHCl_3 \longrightarrow HCl + CCl_3$

Elementary step 3: $Cl + CCl_3 \longrightarrow CCl_4$

13.107 Suppose the first step in the reaction of Problem 13.106 was the rate-determining step. Would the rate law be $k[Cl_2][CHCl_3]$, $k[Cl_2]$, $k[CHCl_3]$, or $k[Cl_2]^2$? Explain.

ADDITIONAL PROBLEMS

13.108 Does the following reaction-energy profile represent an endothermic or exothermic reaction in the forward direction? In the reverse direction?

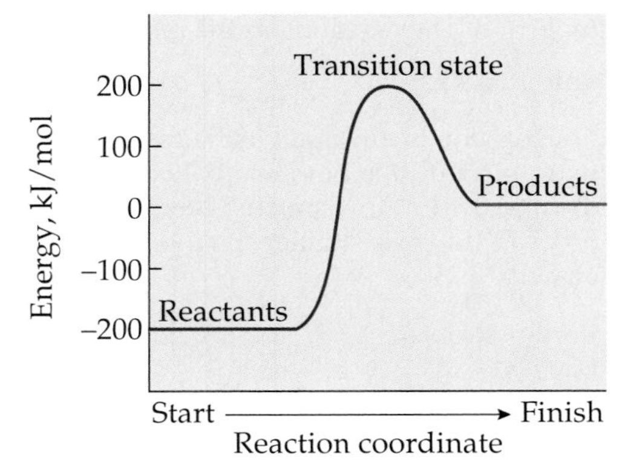

13.109 What is $\Delta E_{\text{forward rxn}}$ for the reaction in Problem 13.108? What is $\Delta E_{\text{reverse rxn}}$?

13.110 What is the activation energy for the forward reaction in Problem 13.108?

13.111 What is the activiation energy for the reverse reaction in Problem 13.108?

13.112 Which of the following will slow down a chemical reaction?
(a) Increase concentration of reactants.
(b) Add catalyst.
(c) Decrease temperature.
(d) All of the above.

13.113 From the reaction-energy profiles, determine whether reactions A and B are exothermic or endothermic in the forward direction:

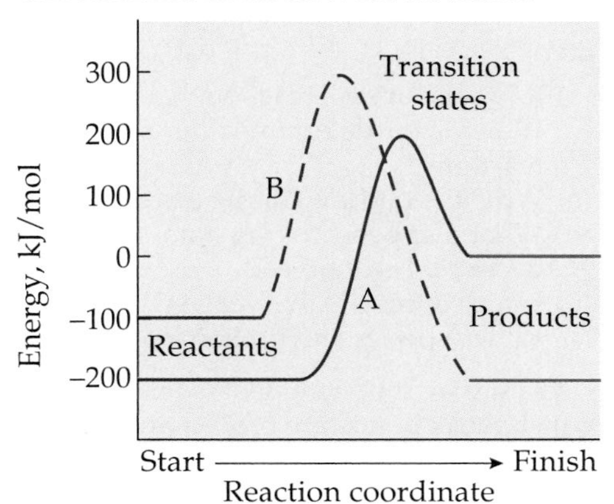

13.114 What are $\Delta E_{\text{forward rxn}}$ for reactions A and B in Problem 13.113?

13.115 What are the forward-direction activation energies for reactions A and B in Problem 13.113? What are the reverse-direction activation energies?

13.116 Assuming the reaction conditions (temperature, concentrations, and so on) are the same, compare the forward-direction rates for reactions A and B in Problem 13.113.

13.117 The rate constant k of a chemical reaction can be changed by
(a) Changing the temperature at which the reaction is run.
(b) Changing the concentration of reactants.
(c) Adding a catalyst.
(d) All of the above.
(e) Only (a) and (c).

13.118 The mechanism for the endothermic reaction

$A + B \longrightarrow C + X$

is

Step 1: $A + A \longrightarrow C + D$ (slow)

Step 2: $B + D \longrightarrow X + A$ (fast)

(a) Draw the reaction-energy profile for this reaction and label reactants, products, reaction intermediates, transition states, activation energies, and ΔE_{rxn}. [*Hint:* First draw a profile for step 1. Make it a very endothermic reaction, and remember that a slow reaction has a large value for E_a. Then draw a profile for step 2, using the line representing the step 1 products as the reactants line for step 2. Remember that a fast reaction has a small value for E_a.]
(b) What is the rate law for this reaction?
(c) If you wanted to quadruple the rate of this reaction, by what factor would you have to increase the concentration of A?

13.119 In each reaction, indicate which bonds are broken and which bonds are formed:
(a) $N_2 + 3\,H_2 \longrightarrow 2\,NH_3$
(b) $PCl_5 \longrightarrow PCl_3 + Cl_2$
(c) $H_2 + 2\,ICl \longrightarrow 2\,HCl + I_2$
(d) $4\,HBr + O_2 \longrightarrow H_2O + 2\,Br_2$

13.120 Indicate whether each reaction is endothermic or exothermic:
(a) $CO_2 + 2\,H_2O \longrightarrow CH_4 + 2\,O_2$
$\Delta E_{\text{rxn}} = +890$ kJ
(b) $CH_4 + 2\,O_2 \longrightarrow CO_2 + 2\,H_2O$
$\Delta E_{\text{rxn}} = -890$ kJ
(c) $S + O_2 \longrightarrow SO_2$ + Heat
(d) $N_2 + 3\,H_2 \longrightarrow 2\,NH_3 \quad \Delta E_{\text{rxn}} = -92$ kJ
(e) Heat + $NH_4NO_3 \xrightarrow{H_2O} NH_4^+(aq) + NO_3^-(aq)$
(f) $2\,H_2 + O_2 \longrightarrow 2\,H_2O \quad \Delta E_{\text{rxn}} = -479$ kJ

13.121 When a chemical reaction is in the ________ ________, the reactant bonds are just ready to break and the product bonds are just ready to form.

13.122 Determine the value of k for a reaction for which
(a) The fraction of collisions having energy $\geqslant E_a$ is 0.42 and the fraction of collisions having the proper orientation is 0.15.
(b) The fraction of collisions having energy $\geqslant E_a$ is 0.42 and the fraction of collisions having the proper orientation is 0.30.
(c) The fraction of collisions having energy $\geqslant E_a$ is 0.84 and the fraction of collisions having the proper orientation is 0.15.
(d) The fraction of collisions having energy $\geqslant E_a$ is 0.84 and the fraction of collisions having the proper orientation is 0.30.

13.123 The rate law for the reaction

$2\,NO + Br_2 \longrightarrow 2\,NOBr$

is Rate $= k[NO]^2\,[Br_2]$. How will the rate change when
(a) [NO] is doubled?
(b) $[Br_2]$ is tripled?
(c) [NO] is tripled?
(d) $[Br_2]$ is quadrupled?
(e) [NO] is doubled and $[Br_2]$ is tripled?

13.124 The slowest step in a reaction mechanism is called the __________-__________ step.

13.125 The experimental rate law for the reaction

$A + A \longrightarrow A_2$

is

Rate $= k[A][BC]$

Two mechanisms have been proposed for the reaction:

Step 1: $A + B \longrightarrow AB$ (slow)

Step 2: $AB + A \longrightarrow A_2 + B$ (fast)

and

Step 1: $A + BC \longrightarrow AB + C$ (slow)

Step 2: $A + AB \longrightarrow B + A_2$ (fast)

Step 3: $B + C \longrightarrow BC$ (fast)

(a) Show that each mechanism results in the correct overall reaction.
(b) Which mechanism is consistent with the rate law?
(c) Why does BC appear in the rate law but not in the overall reaction?

13.126 For a reaction mechanism to be valid, the ______ rate law must agree with the ________ rate law.

13.127 ΔE_{rxn} for the reaction $X \longrightarrow Y$ is +30 kJ.
(a) Is the reaction endothermic or exothermic?
(b) Rewrite the reaction showing heat as either a reactant or a product.
(c) What is the value of ΔE_{rxn} for the reverse reaction $Y \longrightarrow X$?
(d) Is the reverse reaction endothermic or exothermic?
(e) Will the container in which the forward reaction $X \longrightarrow Y$ occurs feel hot or cold to the touch? Why?

13.128 Determine the value of E_a and ΔE_{rxn} for each case below. Also indicate whether each reaction is endothermic or exothermic.

	Energy of reactants, kJ	**Energy of transition state, kJ**	**Energy of products, kJ**
(a)	100	150	130
(b)	100	150	70
(c)	50	175	130
(d)	20	40	10

13.129 The reaction $N_2 + 3\ H_2 \longrightarrow 2\ NH_3$ is exothermic. Draw a reaction-energy profile for the reaction. Label the gap that represents ΔE_{rxn}.

13.130 True or false? A catalyst in a reaction decreases the energy gap between reactants and products. If the statement is false, explain why.

13.131 Which of the following are substitution reactions:
(a) $CH_3Br + I^- \longrightarrow CH_3I + Br^-$
(b) $H_2 + Br_2 \longrightarrow 2\ HBr$
(c) $CH_2{=}CH_2 + H_2 \longrightarrow CH_3{-}CH_3$
(d) $CH_3CH_2Cl + OH^- \longrightarrow CH_3CH_2OH + Cl^-$

13.132 The rate law for a reaction involving A(*g*) as the only reactant is

Rate $= k[A]^2$

What happens to the rate when
(a) The volume of the reaction container is halved?
(b) The concentration of A is tripled?

13.133 The mechanism for the reaction of A_2 with B is

Step 1: $A_2 + Y \longrightarrow AY + A$ (slow)

Step 2: $A + B \longrightarrow AB$ (fast)

Step 3: $AY + AB \longrightarrow Y + A_2B$ (fast)

(a) Write the overall reaction that is occurring.
(b) Which step determines the rate law for the reaction?
(c) Write the rate law for the reaction.
(d) What happens to the rate of the reaction when $[A_2]$ is doubled?
(e) Which species is the catalyst?
(f) Which species are reaction intermediates?

13.134 Explain why this statement is false: A reaction in which weak bonds are broken and strong bonds are formed is an endothermic reaction.

13.135 Indicate whether each statement is true or false. Rewrite each false statement to make it true.
(a) Raising the temperature of a reaction mixture increases the number of reactant molecules that have energy equal to or greater than E_a.
(b) Lowering the temperature of a reaction mixture speeds up the reaction.
(c) The rate of a reaction depends only on the number of collisions per unit time having energy equal to or greater than E_a.
(d) A reaction with an orientation factor of 1 means that all reacting molecules are oriented properly.

13.136 Which reaction occurs fastest, one with $E_a = 20$ kJ, one with $E_a = 50$ kJ, or one with $E_a = 75$ kJ?

13.137 The general rate law for a reaction is

Rate $= k[H_2]^x[Cl_2]^y$

Write a reaction for which this general rate law is correct.

13.138 A student says to you, "Catalysts are not used up in chemical reactions because they are not involved in the reactions." Is this statement true or false? Why?

13.139 Explain the difference between a reaction intermediate and a catalyst in terms of the order in which each appears in the various steps of a reaction mechanism.

13.140 Explain the difference between the energy factor and the orientation factor in the equation for reaction rate (Equation 13.1).

13.141 Write the general rate law for each reaction, using x and y exponents as orders:
(a) $2\ NO + O_2 \longrightarrow 2\ NO_2$
(b) $2\ H_2O_2 \longrightarrow 2\ H_2O + O_2$

13.142 Chemical companies invest a considerable amount of time and energy in search of better catalysts for their chemical processes. Explain how this investment might pay off.

13.143 A reaction releases 900 kJ of energy.
(a) Is the reaction endothermic or exothermic?
(b) Which are higher in the reaction-energy profile, reactants or products? Explain.
(c) Does this reaction go uphill or downhill in energy?
(d) Draw the reaction-energy profile.

13.144 Explain why the value of k gets larger as the temperature of a reaction mixture is increased.

13.145 Indicate whether each statement is true or false. Rewrite each false statement to make it true.
(a) The exponents in the rate law for a reaction that has a one-step mechanism can be determined from the balanced equation for the reaction.
(b) A rate law can be used to prove that a proposed mechanism is correct.
(c) The step $A + X + Y \longrightarrow AX + Y$ in a reaction mechanism is plausible.
(d) A reaction intermediate appears first as a reactant and then as a product in a reaction mechanism.

13 WORKPATCH SOLUTIONS

13.1 If the mechanism were that reactant molecules were blasted apart into atoms, there is no reason for the atoms to recombine to form only CO_2 and H_2O. Instead, all the other molecules listed in the hint could also be produced. (Chemists describe this situation as a *loss of product selectivity*.) We know, however, that the combustion of methane is very selective, producing predominantly CO_2 and H_2O, which tells us there must be something about the reaction mechanism that steers toward the formation of these products.

13.2 If ΔE_{rxn} for the forward reaction is downhill and equal to -40 kJ/mol, ΔE_{rxn} for the reverse reaction is uphill and equal to $+40$ kJ/mol.

13.3 (a) **Reaction 1**

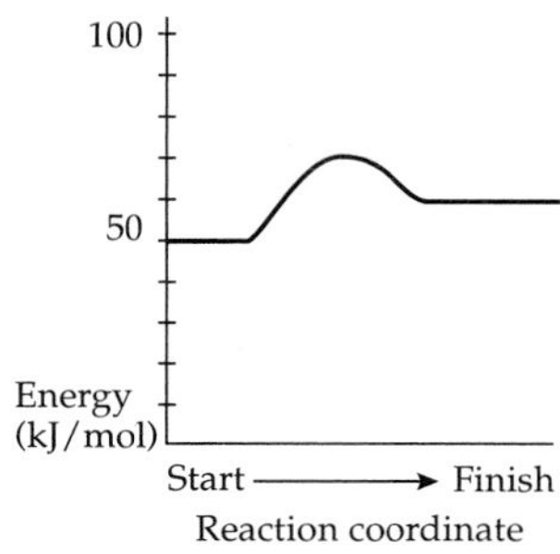

Reaction 2

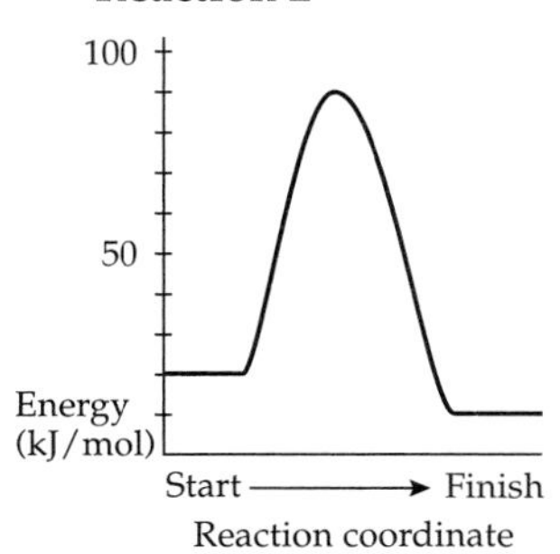

Reaction 3

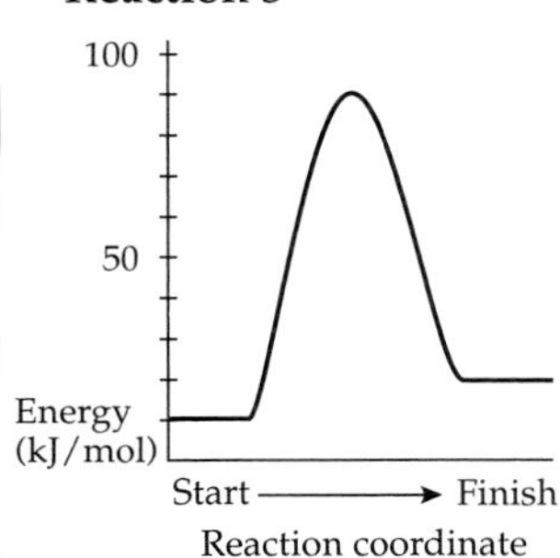

(b) Reaction 1 endothermic; reaction 2 exothermic; reaction 3 endothermic.
(c) Reaction 1: $\Delta E_{rxn} = +10$ kJ/mol; reaction 2: $\Delta E_{rxn} = -10$ kJ/mol; reaction 3: $\Delta E_{rxn} = +10$ kJ/mol.
(d) Reaction 1: $E_a = 20$ kJ/mol; reaction 2: $E_a = 70$ kJ/mol; reaction 3: $E_a = 80$ kJ/mol.

13.4 A catalyst would likely alter both the energy factor and the orientation factor. The number of collisions (per unit time) in which the energy is equal to or greater than E_a would increase because a catalyst lowers the E_a value for the reaction. The fraction of collisions between molecules having the proper orientation would also likely change because, according to the lock-and-key model, each reactant molecule is forced into one specific orientation as it attaches to the enzyme molecule. (It's no stretch to assume that that orientation is one that favors the reaction.)

13.5 (a) A reaction with a large rate constant k has a small E_a (these two constants are inversely related).

(b) A large k for a reaction is a good indication that how reactant molecules are oriented during collisions probably does not matter much. If many different orientations result in product formation, the reaction goes quickly, making k large.

(c) A reaction with a large k is intrinsically fast. The larger k is, the faster the reaction.

13.6 The CH_3^+ cation is a reaction intermediate for two reasons. First, it is not one of the final products of the reaction, and second, it is produced in one step and then consumed in a subsequent step.

13.7

$$\begin{array}{rcl} CH_3Br & \longrightarrow & \cancel{CH_3^+} + Br^- \\ + \ \cancel{CH_3^+} + OH^- & \longrightarrow & CH_3OH \\ \hline CH_3Br + OH^- & \longrightarrow & CH_3OH + Br^- \end{array}$$

13.8 Experiments 1 and 2 tell you that doubling the CH_3Br concentration doubles the rate. Thus, $x = 1$. Experiments 2 and 3 tell you that doubling the OH^- concentration doubles the rate. Thus, $y = 1$. This information gives an experimental rate law of Rate $= k[CH_3Br][OH^-]$. Only mechanism II predicts this rate law, meaning that mechanism II is your only valid choice.

13.9 Mechanism I is wrong because it predicts a rate law that doesn't match the experimental rate law.

13.10 Mechanism II is possibly correct because its predicted rate law matches the experimental rate law. Of course, other mechanisms might also do this, which means you can't be sure mechanism II is correct. All you can say is that it *may* be correct.

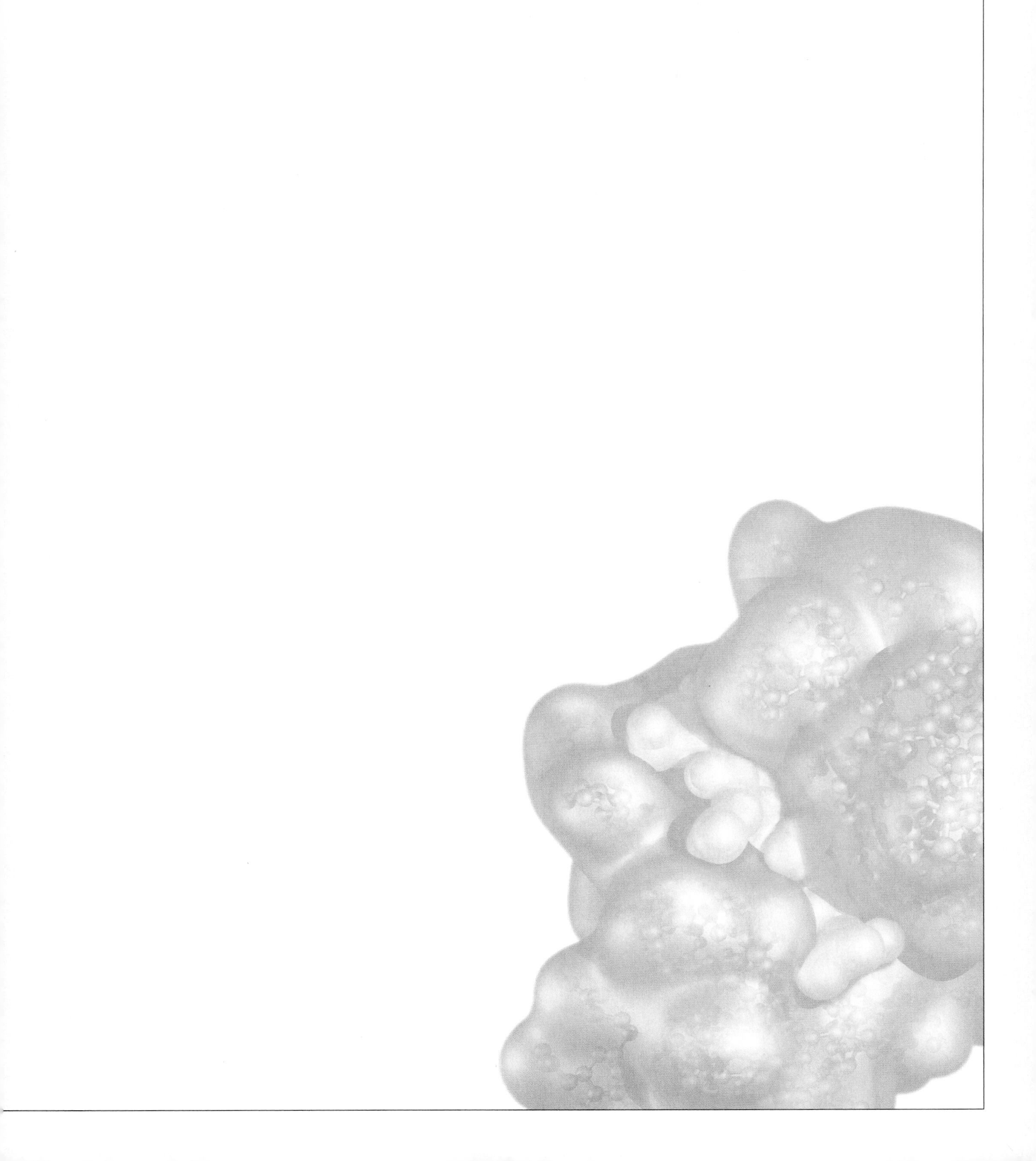

DYNAMIC
EQUILIBRIUM

CHAPTER 14

Chemical Equilibrium

14.1 Dynamic Equilibrium—My Reaction Seems To Have Stopped!

From an economic viewpoint, the reaction that produces sulfuric acid, H_2SO_4, is one of the most important chemical reactions. Millions of tons of sulfuric acid are produced by the chemical industry each year. In fact, no other chemical is produced in greater amounts, the main reason being that one use of sulfuric acid is in the manufacture of many of the agricultural fertilizers used worldwide.

Sulfer burns in air with a distinctive blue flame, forming SO_2 gas.

Suppose you decided to go into the sulfuric acid production business. The first thing you would need to do is learn how it is made. A common method, called the *contact process*, involves three reactions. The first reaction is the partial burning of elemental sulfur to make sulfur dioxide gas:

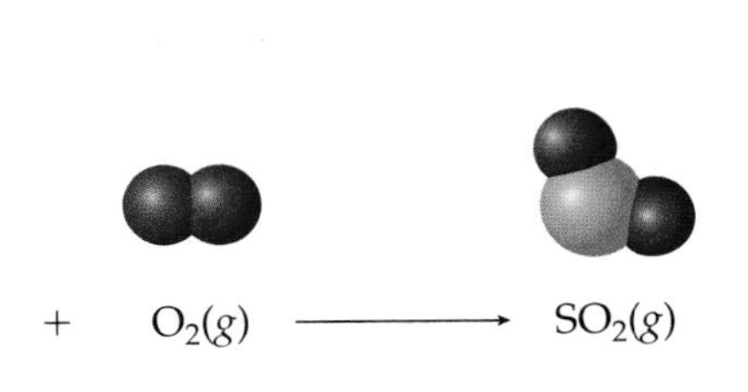

Reaction 1 $\quad S(s) \quad + \quad O_2(g) \longrightarrow SO_2(g)$

This reaction is called a *partial* burning because the sulfur dioxide combines with additional oxygen to form sulfur trioxide gas in the second reaction of the contact process:

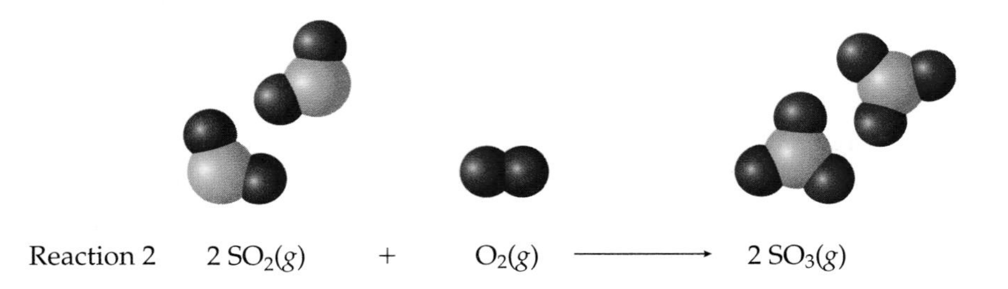

Reaction 2 $2\ SO_2(g) + O_2(g) \longrightarrow 2\ SO_3(g)$

To speed things up, the reaction is usually run over the surface of a metal catalyst, such as platinum. Finally, the SO_3 gas is combined with water and sulfuric acid is produced:

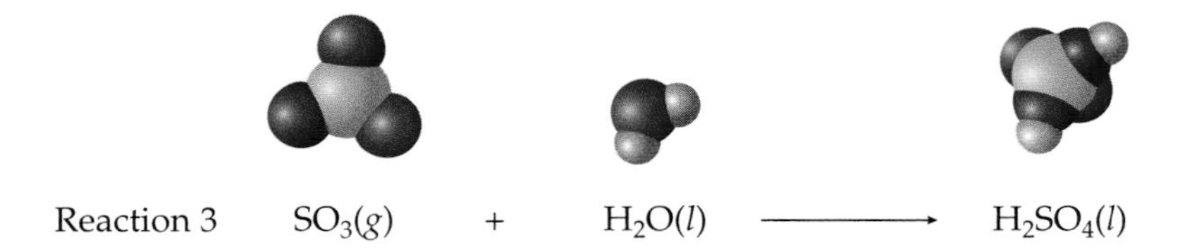

Reaction 3 $SO_3(g) + H_2O(l) \longrightarrow H_2SO_4(l)$

Imagine getting into this business and attempting to turn your knowledge of this chemistry into personal profit. First you need to find someone willing to invest money in a production facility. Having accomplished this, you start with reaction 1, burning sulfur to produce SO_2 gas. Then it is on to reaction 2, converting the SO_2 to SO_3. Following the instructions of the balanced reaction, $2\ SO_2(g) + O_2(g) \longrightarrow 2\ SO_3(g)$, you combine 2 moles of SO_2 gas with 1 mole of O_2 gas and expect to get 2 moles of SO_3 gas. You even throw in some platinum to help the reaction run faster. Time is money, after all.

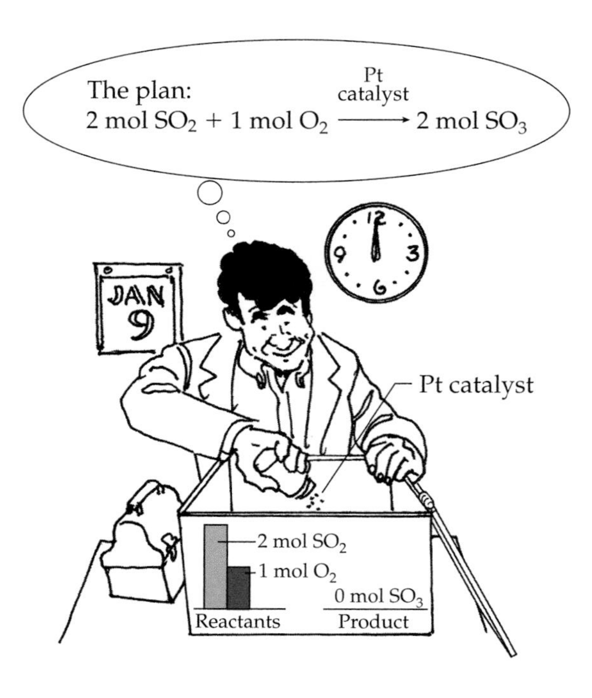

You start the reaction, step out for lunch, and return one hour later in eager anticipation of finding 2 moles of profit-yielding SO_3 gas. What you find instead is disturbing. After a full hour, only 1.8 moles of SO_3 has formed, and 0.2 mole of SO_2 and 0.1 mole of O_2 remain unreacted.

It appears that the reaction is not yet complete. Perhaps the reaction just doesn't run as fast as you thought it would. Maybe it needs more time to reach completion. With the weekend coming up, you decide to check again on Monday. Imagine your horror when you return on Monday, accompanied by your investor, to find absolutely no change. Your attempts to get things going again by shaking the reaction vessel do little to impress either the reaction or your investor. What went wrong?

"My reaction seems to be stuck!"

Your first mistake was going into business before reading this chapter. You see, your reaction hasn't really stopped. It has reached what chemists call *dynamic equilibrium*. In a chemical reaction, **dynamic equilibrium** occurs when the rate of the forward reaction becomes equal to the rate of the reverse reaction.

We introduced the concept of a reaction running in the reverse direction briefly in Chapter 13. In principle, all chemical reactions can proceed in both the forward direction and the reverse direction simultaneously. In chemical reactions where the reverse reaction makes a significant contribution, the reaction is often referred to as a **reversible reaction**. We indicate this property by using two arrows, one pointing to the right (forward direction) and the other pointing to the left (reverse direction):

$$2\,SO_2(g) + O_2(g) \underset{\text{Reverse}}{\overset{\text{Forward}}{\rightleftharpoons}} 2\,SO_3(g)$$

When the forward rate (the rate at which reactants become products) equals the reverse rate (the rate at which products become reactants), the reaction appears to have stopped because the amounts of reactant and product present no longer change with time. Hence *equi-* in the word *equilibrium*, meaning "equal rates." We use the word *dynamic* to describe the equilibrium because this adjective reminds us that, although the reaction *appears* to have stopped, it is really still going on. It's just that the reverse reaction is now proceeding at the same rate as the forward reaction:

$$2\,SO_2(g) + O_2(g) \underset{\text{Reverse}}{\overset{\text{Forward}}{\rightleftharpoons}} 2\,SO_3(g)$$

The two reaction rates are the same at equilibrium.

Dynamic Equilibrium

Therefore, even though your SO_3 reaction seemed to go almost (90%) to completion and then "stop," you now know that it really didn't stop at all. At the 90% point on the reaction coordinate, the reaction reached dynamic equilibrium—or equilibrium, for short. Because the forward and reverse reactions are now occurring at the same rate, the total number of reactant and product molecules present no longer changes. Every time two SO_2 molecules and an O_2 molecule react, producing two SO_3 molecules, two other SO_3 molecules react in the reverse direction to form two "replacement" SO_2 molecules and one O_2 molecule.

Of course, not all reactions reach equilibrium at the 90% point on the reaction coordinate. For example, consider the reaction that produces nitrogen monoxide from nitrogen and oxygen:

$$N_2(g) + O_2(g) \rightleftharpoons 2\,NO(g)$$

If we load a reaction vessel with 1 mole each of N_2 and O_2 according to the balanced equation and adjust the temperature to 2027°C, we find that the reaction appears to stop after only a small amount of product has formed:

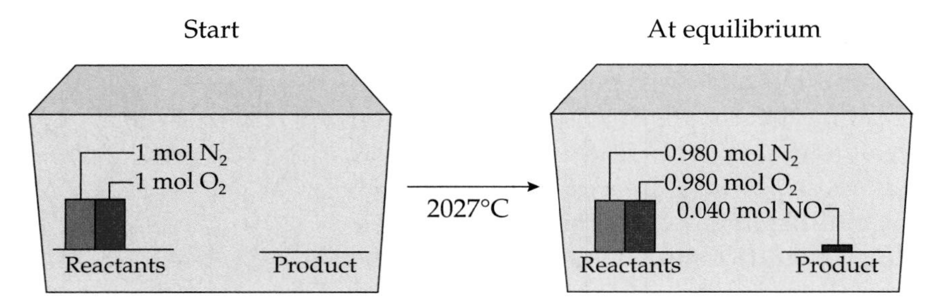

At equilibrium, the vessel contains mostly reactants. You could wait forever and this situation would not change because this reaction reaches equilibrium at only 2% completion.

As our examples demonstrate, some reactions reach dynamic equilibrium near completion (to the right side of the reaction, toward products) and some reach it close to the beginning (to the left side of the reaction, toward reactants). In addition, some reactions reach equilibrium near the middle.

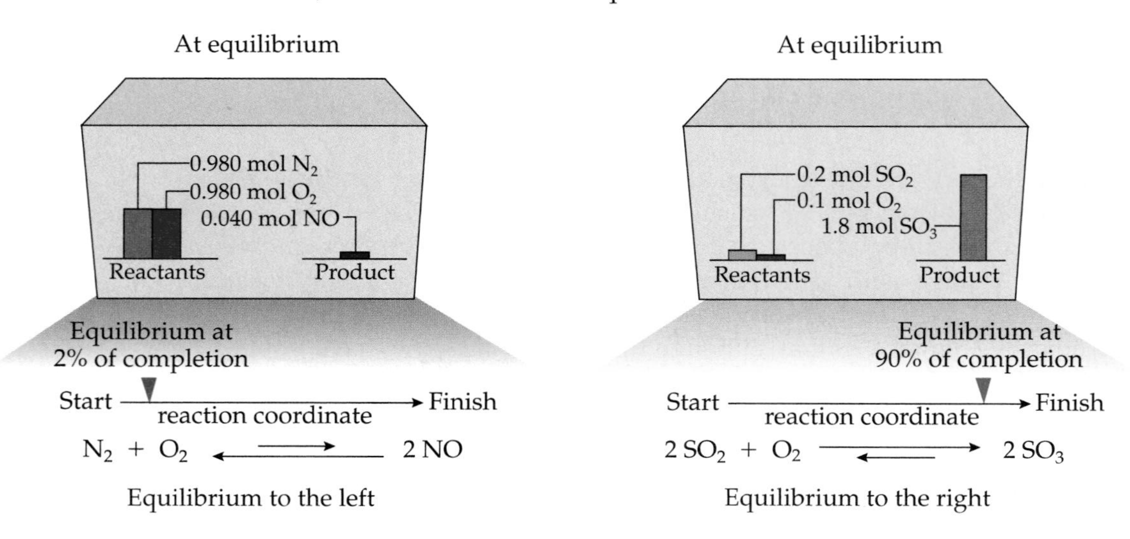

There are, however, two special cases. First, some reactions reach equilibrium so close to the beginning that practically no product is formed. These reactions appear to never even get started before they "stop." Because practically no product is formed at equilibrium, we say that the reaction essentially does not occur:

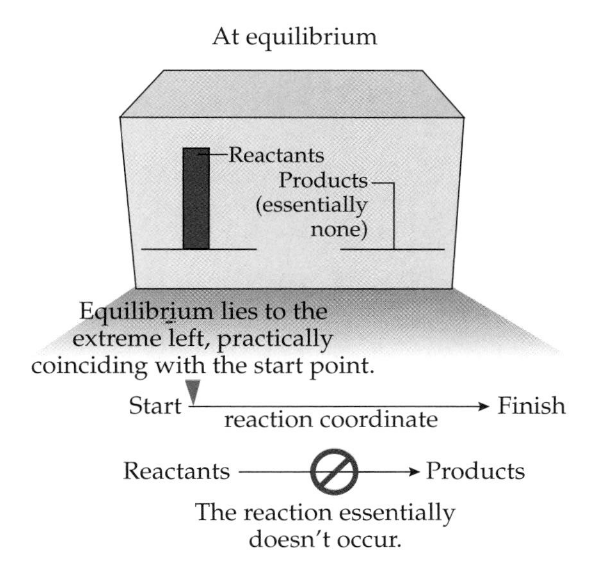

The other special case is just the opposite—reactions that reach equilibrium so close to completion that practically no reactant remains. For this special case, we use only one arrow pointing from left to right instead of two arrows and say that the reaction goes to completion:

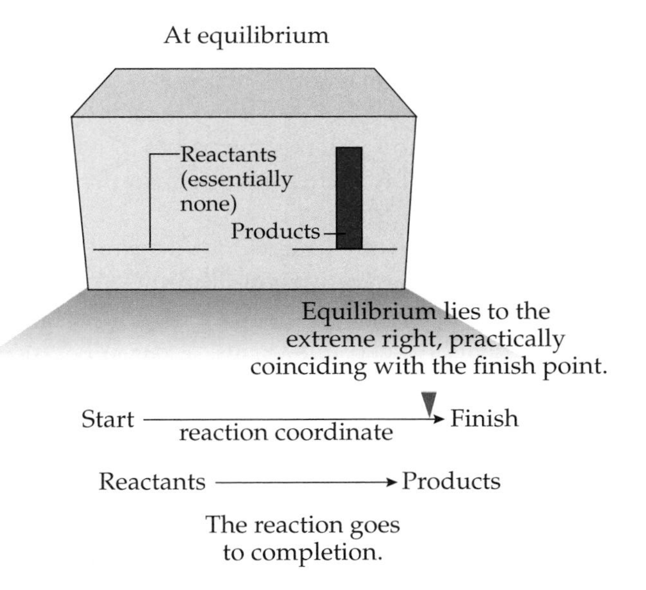

It is important to realize that our treatment of these two special cases as "one-way" reactions is oversimplified. All reactions, given the correct conditions, reach equilibrium. It's just that for these special cases, the equilibrium lies so far to the left or right that we can ignore the concept of equilibrium because it is of no practical importance.

Here are five possibilities for the hypothetical reaction A ⟶ B:

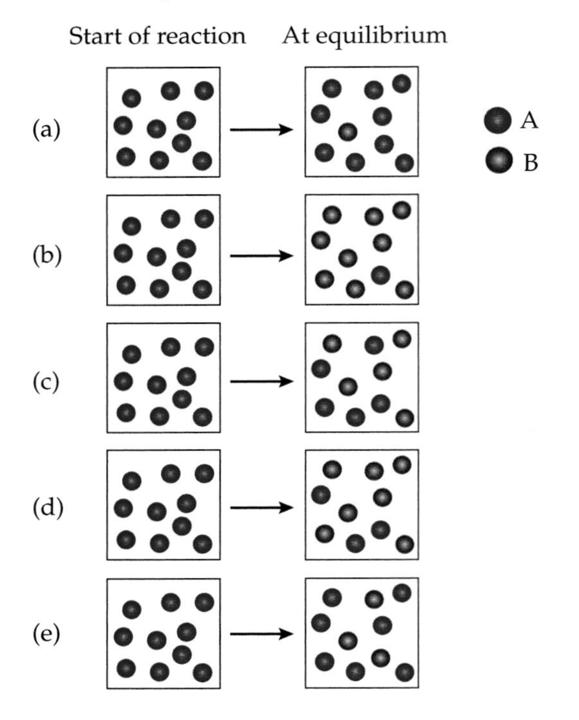

Which sentence best describes each possibility:

1. The equilibrium lies to the left.
2. The equilibrium lies to the right.
3. The equilibrium lies in the middle.
4. The reaction essentially doesn't occur.
5. The reaction essentially goes to completion.

Now, let's return to your SO_3 production facility and your worried investor, who's concerned that the reaction stopped before all the reactants were used up. You calmly explain that it has not stopped. It just appears to have stopped because the rates of the forward reaction and reverse reaction have become equal. The reaction has reached equilibrium.

"Great!" he exclaims, "what about the other reactions you plan to run? Will they reach equilibrium? Where will they reach equilibrium? Can't you do something about all this!!??"

He's got some good questions that need answering (especially since he's holding the checkbook), so let's take a look at them.

Questions about equilibrium

1. Why do reactions reach equilibrium?
2. What determines where a reaction reaches equilibrium?
3. Is there a way to shift the position of a reaction's equilibrium?

You must answer these questions if you are going to stay in business, so we'll look at them one by one. Before we do, though, try these practice problems on dynamic equilibrium.

Practice Problems

14.1 Consider a point during a chemical reaction at which the rate of the forward reaction is less than the rate of the reverse reaction.
(a) Is the reaction at equilibrium at that point?
(b) Which way does the overall reaction appear to be running?

Answer:
(a) The reaction is not at equilibrium because the forward and reverse rates are not equal.
(b) The reaction appears to be running to the left (in reverse, converting products to reactants).

14.2 From a practical point of view, why would you want a reaction equilibrium to lie very far to the right?

14.3 Why can we ignore equilibrium for reactions that go to completion?

14.2 Why Do Chemical Reactions Reach Equilibrium?

We have defined dynamic equilibrium as the point during a chemical reaction at which the forward and reverse rates become equal. Why does this happen? To understand why, we must look at what factors influence reaction rates. We discussed the effects of temperature and concentration in Chapter 13, where we saw that the rate of a reaction can be expressed by a rate law:

$$\text{Reactants} \xrightarrow{\text{Forward}} \text{Products}$$

$$\text{Rate}_{\text{forward rxn}} = k_{\text{forward rxn}}[\text{Reactants}]^{\text{order}}$$

Assume we are running a reaction at some constant temperature. Because the rate constant doesn't change at a given temperature, the only way for the reaction rate to change is for the concentration of one or more of the reactants to change. But this is exactly what happens during a chemical reaction. As a reaction proceeds in the forward direction, reactants are consumed, and their concentrations decrease. Consequently, the rate of the forward reaction also decreases as the reaction proceeds. Our reaction "speedometers," or rate meters, illustrate this behavior, which is the behavior we find in most reactions:

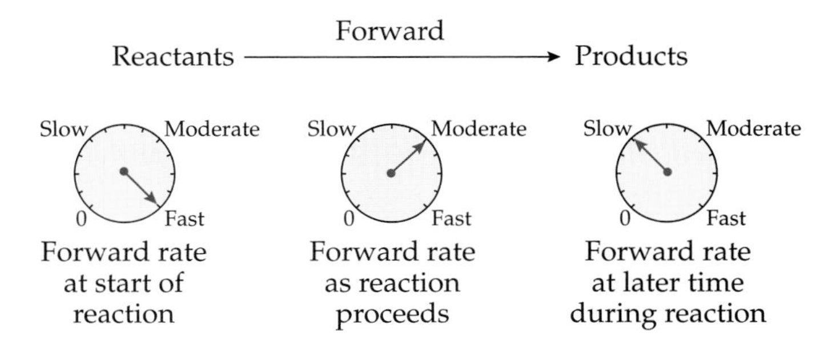

If we now consider our generic reaction to be reversible, we must take into account the reverse reaction, which has its own rate law and its own rate constant $k_{\text{reverse rxn}}$, which is different from $k_{\text{forward rxn}}$:

$$\text{Reactants} \xleftarrow{\text{Reverse}} \text{Products}$$

$$\text{Rate}_{\text{reverse rxn}} = k_{\text{reverse rxn}}[\text{Products}]^{\text{order}}$$

Looking at this rate law, we see that the rate of the reverse reaction depends on the concentration of the products raised to some power. But at the very beginning of a reaction, there are no products, which means the product concentration is zero. Therefore, the reverse reaction starts off with a zero rate. As the forward reaction progresses and products form, the product concentration increases and the speed of the reverse reaction begins to increase. Whereas the forward reaction starts fast and slows down, the reverse reaction starts slow and speeds up. At some point, the two rates become equal. At that point, dynamic equilibrium is established and there are no further changes in concentrations.

Let's see what happened in your SO_3 reaction. Return to the start of the reaction, when 2 moles of SO_2 and 1 mole of O_2 are combined. We shall call the very beginning of the reaction *time zero*. At time zero, no SO_3 has been produced; only reactants are present. Now, consider both possible reactions, the forward and the reverse:

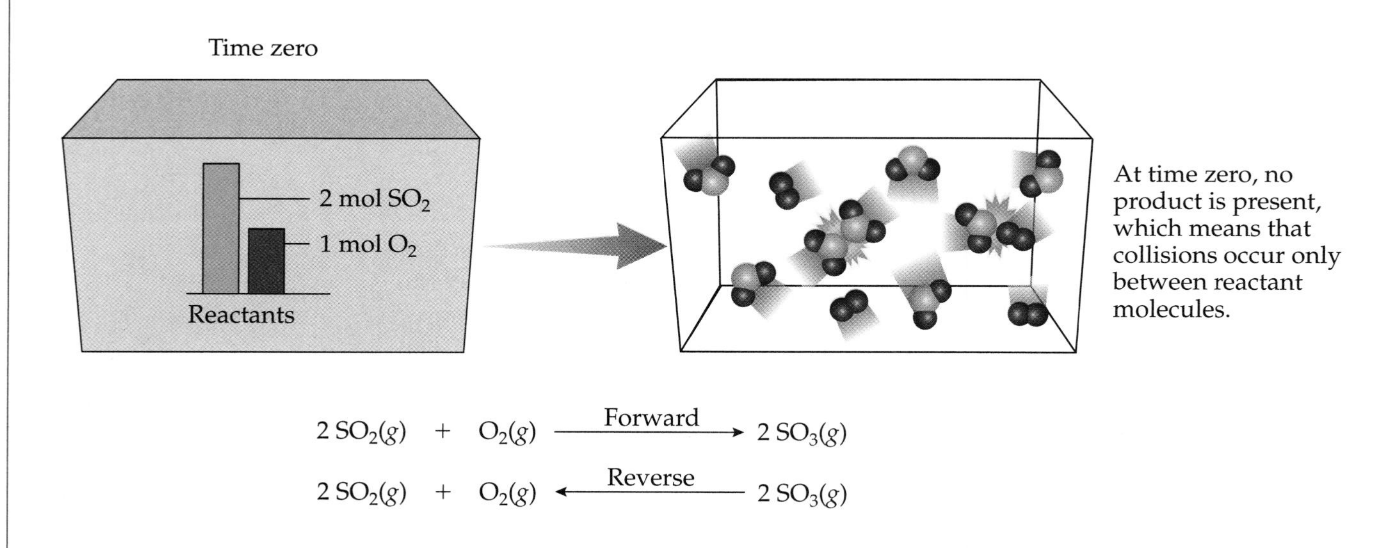

At time zero, no product is present, which means that collisions occur only between reactant molecules.

$$2\,SO_2(g) + O_2(g) \xrightarrow{\text{Forward}} 2\,SO_3(g)$$

$$2\,SO_2(g) + O_2(g) \xleftarrow{\text{Reverse}} 2\,SO_3(g)$$

At time zero, only collisions between reactant molecules occur, meaning the forward reaction is relatively fast at time zero. For the reverse reaction to occur, two SO_3 molecules must collide, but there are no SO_3 molecules present at time zero. Therefore the reverse reaction has a zero rate at time zero.

As time passes, some of the SO_2 and O_2 reactants get used up, decreasing their concentration and thus decreasing the rate of the forward reaction. At the same time, SO_3 product is formed, building up its concentration. The presence of SO_3 enables the reverse reaction to run, and as the SO_3 concentration builds up, more and more collisions occur between SO_3 molecules, increasing the rate of the reverse reaction:

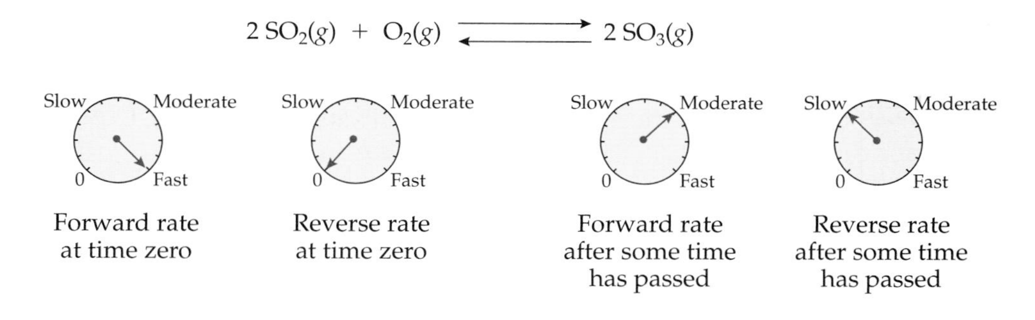

As time passes, the forward reaction keeps slowing down and the reverse reaction keeps speeding up. Eventually, a point is reached where the two reaction rates become equal. At that moment, reactants are being turned into product and product is being turned into reactants at the same rate. The reaction appears to have stopped, and dynamic equilibrium has been reached:

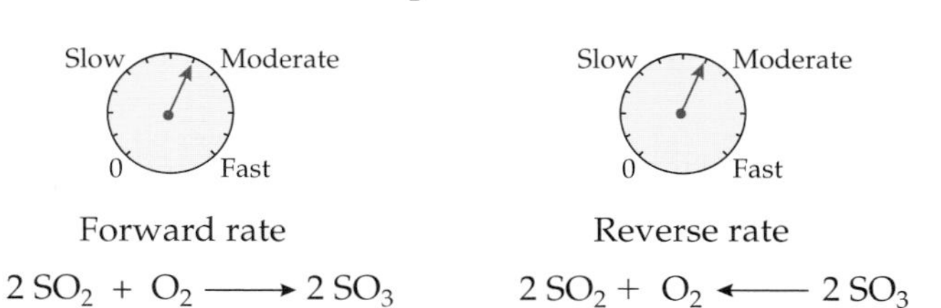

You should realize that we could just as easily have approached this reaction from the other direction. That is, we could have filled the reaction vessel with just product (2 moles of SO_3). At time zero now, the reverse reaction is quite fast and the forward reaction rate is zero:

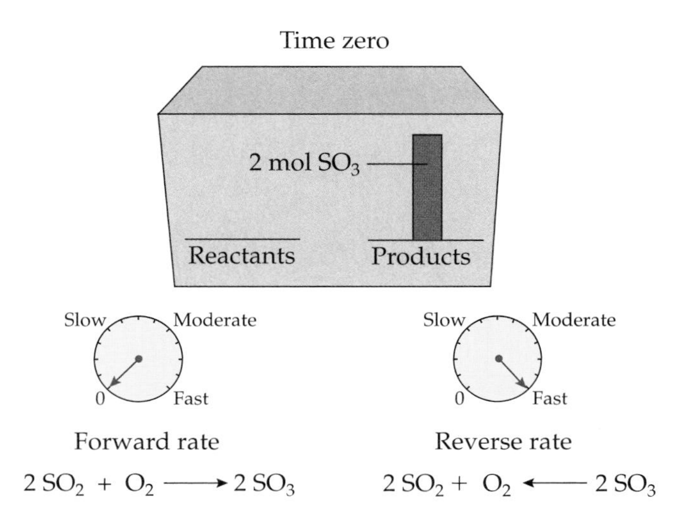

Given enough time, the reverse reaction slows down and the forward reaction speeds up until the two rates are equal. Thus, because the forward and reverse reaction rates change with time, we are guaranteed that at some point, the two rates will be equal.

Now see if you can answer the following WorkPatch.

Given a reaction vessel with the initial conditions shown below, which set of rate meters best describes the time-zero and equilibrium conditions? Explain why your choice is correct and why the other choices are incorrect.

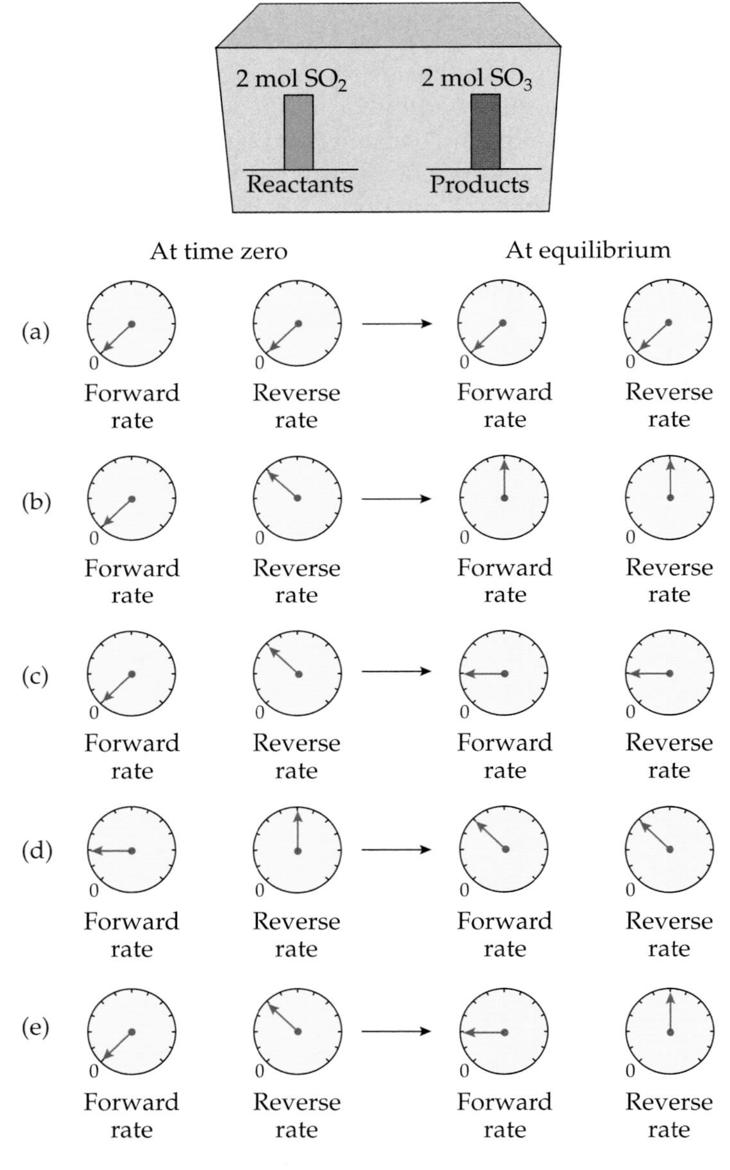

One choice in the WorkPatch can be immediately ruled out. Did you catch it? That choice is (e) because at equilibrium, the forward and reverse rates must be equal—by definition.

Under the proper conditions, all chemical reactions come to equilibrium. The question now is when equilibrium does occur, where does it occur—toward the left? The right? The middle? Is there something that can tell us?

14.3 The Position of Equilibrium—The Equilibrium Constant K_{eq}

Up to now we have described the position of equilibrium in qualitative terms, saying such things as the equilibrium lies to the left or to the right or the reaction goes practically to completion. Such qualitative descriptions are fine until you need to worry about the exact amounts of reactants and products in your reaction vessel when equilibrium is reached. To quantify the position at which a particular reaction comes to equilibrium, chemists use a quantity known as an *equilibrium constant*, K_{eq}. Stated plainly, the value of K_{eq} will tell us where the equilibrium for a reaction lies.

The equilibrium constant for a reaction can be derived by considering the reaction rates for the forward and reverse reactions. This is easy to show only for the simplest reactions, those that occur in one elementary step. For this reason, we shall derive K_{eq} for a hypothetical one-step reaction $A \rightleftharpoons B$ and then present the general result for more complex reactions.

The general definition of dynamic equilibrium in a chemical reaction is

$$\text{Forward rate} = \text{Reverse rate}$$

Therefore, when our simple one-step reaction is at equilibrium, we may write

$$k_f[A] = k_r[B]$$

where k_f represents $k_{\text{forward rxn}}$ and k_r represents $k_{\text{reverse rxn}}$.

Now try your hand at a little algebraic manipulation (Chapter 2) by answering this WorkPatch.

Rearrange the equation $k_f[A] = k_r[B]$ algebraically to solve for the ratio k_f/k_r.

Were you able to do this algebra? (If not, see the solution at the end of the chapter.) The correct result is

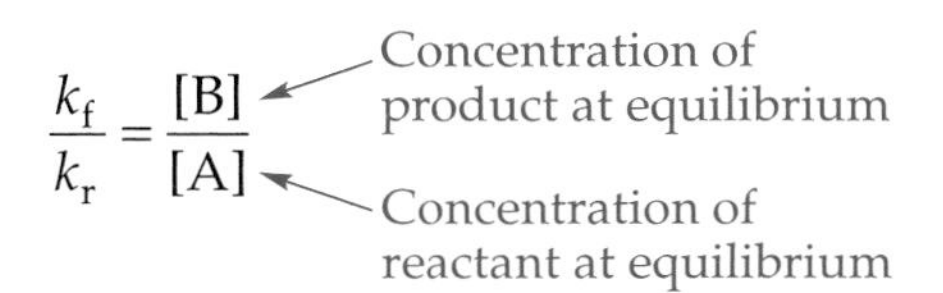

$$\frac{k_f}{k_r} = \frac{[B]}{[A]}$$

During the course of the reaction, the concentrations of A and B keep changing until equilibrium is reached. At a fixed temperature, however, the values of k_f and k_r *never* change (because they are rate *constants*). Thus your rearrangement of the basic equilibrium equation has collected all the constant terms (the rate constants) on one side of the equals sign and all the variable terms (the concentrations) on the other side.

Because k_f and k_r are constants for a given reaction at a set temperature, the ratio k_f/k_r is also a constant for the reaction. Chemists therefore replace the ratio k_f/k_r with a new symbol, K_{eq}, the **equilibrium constant**:

$$K_{eq} = \frac{k_f}{k_r} = \frac{[B]}{[A]}$$

Concentration of product at equilibrium ([B])
Concentration of reactant at equilibrium ([A])

We have arrived at a mathematical expression for the equilibrium constant K_{eq} for a one-step reaction. It is the ratio of forward rate constant to reverse rate constant. It is also the ratio of product concentration to the reactant concentration at equilibrium. These must be equilibrium concentrations because we got to our K_{eq} equation by insisting that the reaction be at equilibrium (remember, we started by saying $k_f[A] = k_r[B]$). Thus, in our expression for K_{eq}, [B] is the molar concentration of product B at equilibrium and [A] is the molar concentration of reactant A at equilibrium.

This equation suggests two ways to determine the value of K_{eq} for a one-step reaction. One way is to determine the values of k_f and k_r for the reaction and then calculate K_{eq} by dividing k_f by k_r. The second way is to go into the laboratory, let the reaction $A \rightleftharpoons B$ come to equilibrium, measure the molar concentrations of A and B, and then calculate [B]/[A]. You will get the same result for K_{eq} either way.

Deriving the equilibrium expression for a multistep reaction is beyond the scope of this text. However, we'll give you the result of such a derivation because you will be needing it. For a multistep reaction that has the general form

$$a\,A + b\,B \rightleftharpoons c\,C + d\,D$$

where a, b, c, and d stand for the stoichiometric coefficients in the balanced equation and the two arrows indicate that the reaction should be treated as an equilibrium reaction, the equilibrium constant is

$$K_{eq} = \frac{[C]^c \times [D]^d}{[A]^a \times [B]^b}$$

Products OVER Reactants

This general definition of the equilibrium constant, which applies to multistep as well as one-step reactions, is clearly related to the definition we derived for a one-step reaction. In both cases, terms for product concentrations are divided by terms for reactant concentrations. The product concentrations are always in the numerator, and the reactant concentrations are always in the denominator. When there is more than one reactant or product, their concentrations are multiplied by one another. In addition, each concentration is raised to a power equal to its stoichiometric coefficient in the balanced equation for the reaction. For example, our balanced SO_3 reaction and its K_{eq} expression are

$$2\,SO_2(g) + O_2(g) \rightleftharpoons 2\,SO_3(g)$$

$$K_{eq} = \frac{[SO_3]^2}{[SO_2]^2 \times [O_2]}$$

Practice Problems

14.4 Write the equilibrium-constant expression for the reaction

$$CH_4(g) + 2\,H_2S(g) \rightleftharpoons CS_2(g) + 4\,H_2(g).$$

Answer: $K_{eq} = \dfrac{[CS_2] \times [H_2]^4}{[CH_4] \times [H_2S]^2}$

14.5 Write the equilibrium-constant expression for the reaction

$$H_2(g) + I_2(g) \rightleftharpoons 2\,HI(g).$$

14.6 Write the equilibrium-constant expression for the reaction

$$Fe^{3+}(aq) + SCN^-(aq) \rightleftharpoons Fe(SCN)^{2+}(aq).$$

In practice, it is easy to write the equilibrium-constant expression for a reaction. And, as we said earlier, it is relatively simple to find the value of K_{eq} for a reaction. All you have to do is go into the laboratory, determine the concentrations of all the reactants and products at equilibrium, plug them into the K_{eq} expression, and evaluate it. For example, let's go back to our SO_3 reaction. We'll start this time with a 1.0-L box containing 2.0 moles of SO_2 and 1.0 mole of O_2. When the reaction reaches equilibrium, the box contains 0.2 mole of SO_2, 0.1 mole of O_2, and 1.8 moles of SO_3. Thus, at equilibrium, the concentrations of SO_2, O_2, and SO_3 are 0.2 M, 0.1 M, and 1.8 M, respectively.

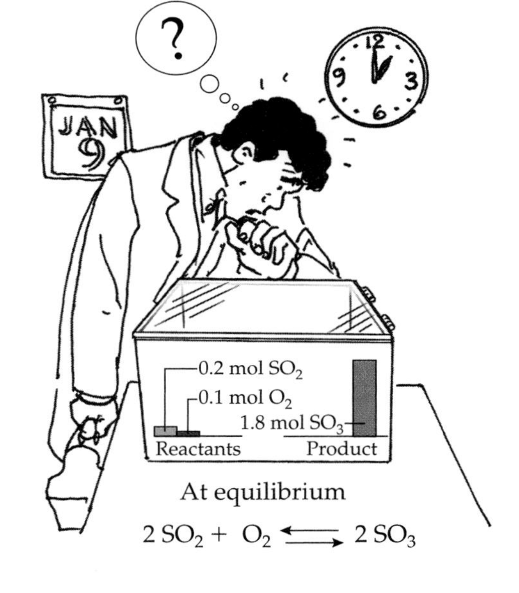

To arrive at a value for K_{eq}, we write the equilibrium expression for the reaction, plug in the equilibrium concentrations, and do the required arithmetic:

$$K_{eq} = \frac{[SO_3]^2}{[SO_2]^2 \times [O_2]} = \frac{(1.8\text{ M})^2}{(0.2\text{ M})^2 \times 0.1\text{ M}} = 810$$

Remember that this numeric value of K_{eq} applies to the reaction *only as long as the temperature doesn't change.* Let's say for the sake of argument that the temperature at which we ran our reaction was 400°C. Then, anytime we come across this reaction at equilibrium at this temperature, we can be sure that no matter what kind of vessel is used and no matter how much of the reactants was initially put into the vessel, we shall get 810 if we measure all concentrations and plug them into the K_{eq} expression. This number is a fundamental constant for this reaction run at this temperature. For example, what if we changed the starting amounts of reactants to 5 moles of SO_2 and 3 moles of O_2? Remember, our box has a volume of 1 L, making the concentrations at time zero: 5 M for SO_2 and 3 M for O_2. If you did this experiment, you would get this equilibrium situation:

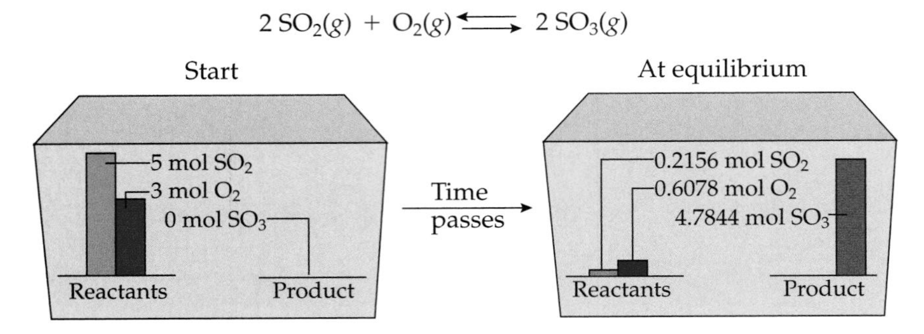

Changing the starting concentrations has given rise to a different set of equilibrium concentrations. Let's calculate the equilibrium constant for this second case, using our new equilibrium concentrations:

$$K_{eq} = \frac{[SO_3]^2}{[SO_2]^2 \times [O_2]} = \frac{(4.7844\ M)^2}{(0.2156\ M)^2 \times 0.6078\ M} = 810$$

So we see that, because we have not changed the temperature, K_{eq} is still equal to 810. The value of K_{eq} for a reaction is a constant unless the temperature changes.

Practice Problems

14.7 At the start of the reaction

$$H_2(g) + I_2(g) \rightleftharpoons 2\,HI(g)$$

the concentrations are $[H_2] = 0.100$ M, $[I_2] = 0.100$ M, $[HI] = 0.000$ M. At 427°C, the equilibrium concentrations are $[HI] = 0.158$ M, $[H_2] = 0.021$ M, $[I_2] = 0.021$ M. Calculate K_{eq} for this reaction.

Answer: 57

14.8 Given the concentrations in Practice Problem 14.7, does the equilibrium for the HI reaction lie to the left or to the right? Explain your choice.

14.9 The equilibrium concentrations for the reaction

$$CH_4(g) + 2\,H_2S(g) \rightleftharpoons CS_2(g) + 4\,H_2(g)$$

are $[CS_2] = 6.10 \times 10^{-3}$ M, $[H_2] = 1.17 \times 10^{-3}$ M, $[CH_4] = 2.35 \times 10^{-3}$ M, $[H_2S] = 2.93 \times 10^{-3}$ M. Calculate K_{eq} for this reaction.

Now that we know how to determine the value of K_{eq} for a reaction, we can ask the question "What exactly does the equilibrium constant tell us about an equilibrium?" For any reaction, if we neglect the stoichiometric balancing coefficients that appear as exponents, the equilibrium constant is essentially a ratio of equilibrium product concentrations over equilibrium reactant concentrations:

$$K_{eq} = \frac{[\text{Products}]}{[\text{Reactants}]}$$

Because K_{eq} represents a ratio, its value is large when the numerator is much larger than the denominator. Suppose, for instance, that at equilibrium the numerator of the K_{eq} expression is much larger than the denominator, making $K_{eq} > 10^3$. For this reaction, there is so much more product than reactant present at equilibrium that the reaction proceeds almost to completion. Looking back at our SO_3 example, you can see that this is almost so. We found that K_{eq} is large (810) and that the reaction reaches equilibrium close to completion (90%). In general, we say that if K_{eq} is close to or larger than 10^3, the equilibrium lies to the right. The larger the value of K_{eq} for a reaction, the farther the equilibrium lies to the right. A reaction having a K_{eq} greater than 10^5 or 100,000 goes essentially to completion.

If a reaction has a small K_{eq} value, the opposite is true. For reactions where $K_{eq} < 10^{-3}$, there is so much more reactant than product present at equilibrium that the reaction hardly proceeds at all. In general, we say that if K_{eq} is on the order of 10^{-3} or less, the equilibrium lies to the left. The smaller K_{eq} is, the farther the equilibrium lies to the left. A reaction with a K_{eq} that is very small essentially does not proceed. The $N_2 + O_2 \rightleftharpoons 2\ NO$ reaction we looked at in Section 14.1 has such a left-lying equilibrium point, and therefore should have a small value for K_{eq}, as you'll see in the WorkPatch.

Calculate the value of K_{eq} for the NO reaction, given the following data:

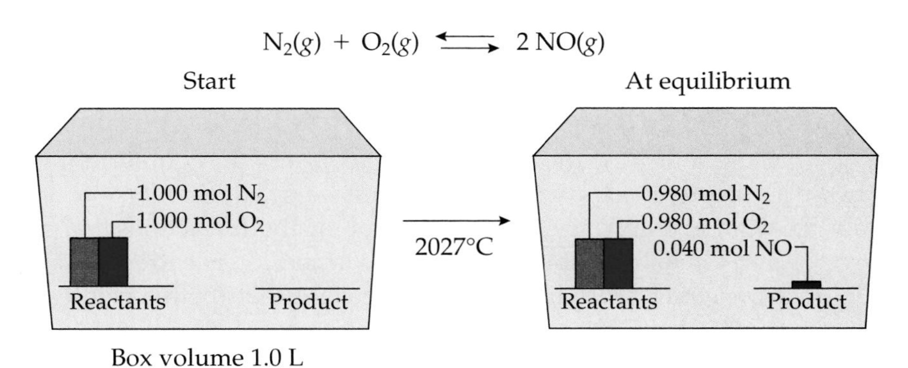

Looking at the equilibrium box in the WorkPatch leaves no question that this equilibrium lies to the left because mostly reactants are present. What did you get for the value of K_{eq}? Was it around 10^{-3}? You may remember that we found earlier that this reaction goes to only 2% of completion.

Finally, there are those reactions that reach equilibrium at a point where appreciable amounts of both reactants and products are present. For these reactions, K_{eq} has values somewhere between 10^{-3} and 10^3. (It is this third type of equilibrium that your H_2SO_4 production business is currently wrestling with.)

Interpreting K_{eq} values

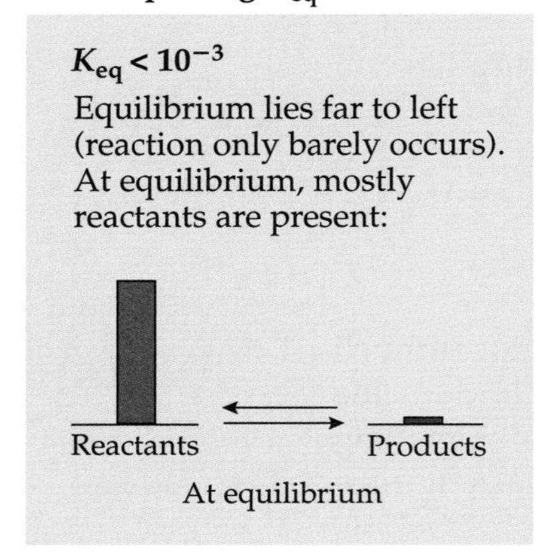

$K_{eq} < 10^{-3}$
Equilibrium lies far to left (reaction only barely occurs). At equilibrium, mostly reactants are present:
Reactants
Products
At equilibrium

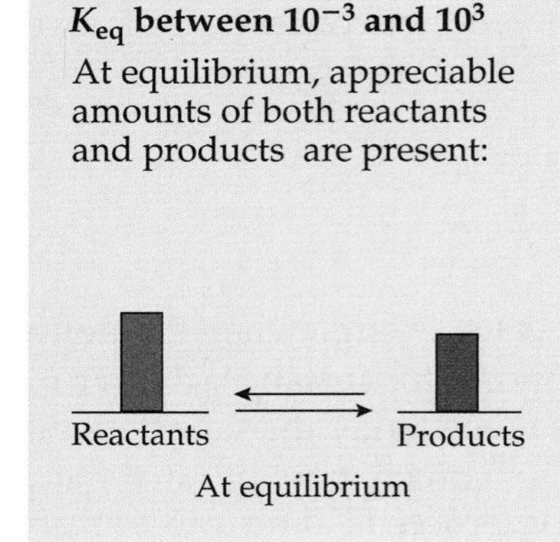

K_{eq} between 10^{-3} and 10^{3}
At equilibrium, appreciable amounts of both reactants and products are present:
Reactants
Products
At equilibrium

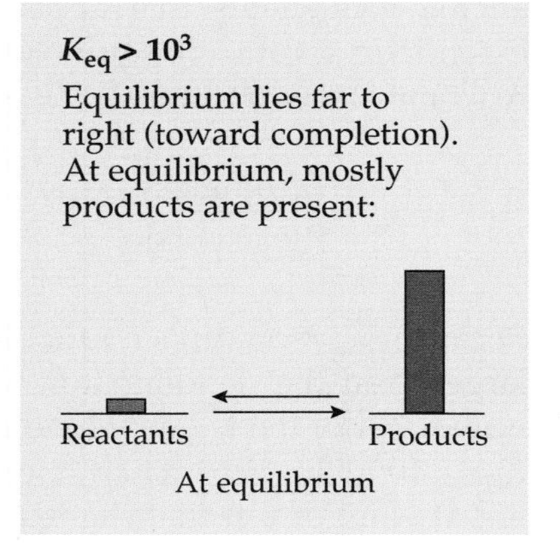

$K_{eq} > 10^{3}$
Equilibrium lies far to right (toward completion). At equilibrium, mostly products are present:
Reactants
Products
At equilibrium

Note that K_{eq} values can never be negative.

Chemists have run many reactions, allowed them to reach equilibrium, measured the concentrations of products and reactants, and then calculated equilibrium constants. Consequently, tables of K_{eq} values for tens of thousands of reactions are recorded in reference books. Some of these reactions are presented in the following WorkPatch. See if you can match the reactions with their measured equilibrium constants.

For each reaction, write the K_{eq} expression.

(a) $HCl(aq) \longrightarrow H^+(aq) + Cl^-(aq)$ Reaction goes to completion

(b) $CH_3COOH(aq) \rightleftharpoons H^+(aq) + CH_3COO^-(aq)$ Equilibrium lies far to the left

(c) $PCl_5(g) \rightleftharpoons PCl_3(g) + Cl_2(g)$ Equilibrium lies somewhat to the left

(d) $H_2O(g) + CH_4(g) \rightleftharpoons CO(g) + 3\,H_2(g)$ Equilibrium lies near middle

Then decide which of the following equilibrium constants goes with each reaction: (1) $K_{eq} = 1.0 \times 10^{-5}$, (2) $K_{eq} = 2.2 \times 10^{-2}$, (3) $K_{eq} = 4.6$, (4) $K_{eq} > 10^6$.

Are you getting a feel for the meaning of the size of an equilibrium constant? To see if you are, check your WorkPatch answers against ours.

Finally, we want to emphasize an important point about reactions that have very large or very small equilibrium constants. When K_{eq} is extremely large, we know that the equilibrium lies so far to the right that the reaction essentially goes to completion. This usually means we can neglect equilibrium. One such reaction is $2\,CO + O_2 \rightleftharpoons 2\,CO_2$, for which $K_{eq} = 1.9 \times 10^{11}$. For this reaction, if you react 2 moles of CO with 1 mole of O_2, you will get, according to the stoichiometry of the balanced equation, 2 moles of CO_2 and essentially no unreacted starting materials.

When K_{eq} is extremely small, the equilibrium lies so far to the left that the reaction essentially never starts. Once again, this means we can ignore equilibrium. One such reaction is $N_2 + O_2 \rightleftharpoons 2\,NO$, for which $K_{eq} = 2.3 \times 10^{-9}$ at 25°C. If you mix 1 mole of N_2 with 1 mole of O_2 at 25°C, you will get essentially no product because essentially no reaction occurs. (Be thankful this is the case, as these gases make up most of the atmosphere you breathe. If K_{eq} for this reaction were large, there would be a substantial amount of NO in the atmosphere at equilibrium. The NO would damage your lungs, and you would not survive for very long because NO is a powerful oxidizing agent.)

If we reverse the reaction for making NO, we have a reaction for the decomposition of NO:

$$2\,NO \rightleftharpoons N_2 + O_2$$

(a) Write the K_{eq} expression for this reaction and for the reaction to form NO ($N_2 + O_2 \rightleftharpoons 2\,NO$).
(b) How is the K_{eq} expression for the NO-formation reaction related to the K_{eq} expression for the NO-decomposition reaction?
(c) The value of K_{eq} for NO formation at 25°C is 2.3×10^{-9}. What is K_{eq} for NO decomposition at 25°C? To which side does the equilibrium lie for the decomposition reaction?

Did you discover something about how the equilibrium constants for the forward and reverse versions of a reaction are related? Be sure to check your answer against the one given at the end of the chapter.

Finally, knowing the value of K_{eq} can do more than tell you where a reaction's equilibrium lies. As we said at the beginning of this section, knowing the value of K_{eq} quantifies the position of equilibrium. The equilibrium constant for a reaction can be used to calculate the actual concentrations of the reactants and products present at equilibrium. Take your sulfuric acid production business as an example. If you had researched the SO_3 reaction adequately, you would have discovered that its K_{eq} value is 810. Then, using this information, you could have calculated that you would end up with 0.1 mole O_2 and 0.2 mole SO_2 left over. Instead, you found out the hard way. You'll see how to use K_{eq} to calculate equilibrium concentrations in the final section of this chapter.

14.4 Disturbing a Reaction Already at Equilibrium—Le Chatelier's Principle

Let's return to our SO_3 production reaction. Last time we looked, you were panicked because the reaction would not go to completion.

What can you do about it? Suddenly, you have an idea! You'll simply suck all the leftover SO_2 and O_2 out of the box with a device designed to remove just these gases. Then you'll have pure SO_3 in the box—right?

You quickly do this and then retrieve your investor and proudly show him the box. But this is what he sees inside the box:

A Disturbed Equilibrium Fights Back

What happened? A cruel joke? No! A lesson learned. Before the SO_2 and O_2 were sucked out, the reaction was at equilibrium. The rate of the forward reaction equaled the rate of the reverse reaction. When the SO_2 and O_2 were removed, the rate of the forward reaction dropped to zero. At that moment (time zero), only the reverse reaction was running, merrily turning SO_3 back into SO_2 and O_2:

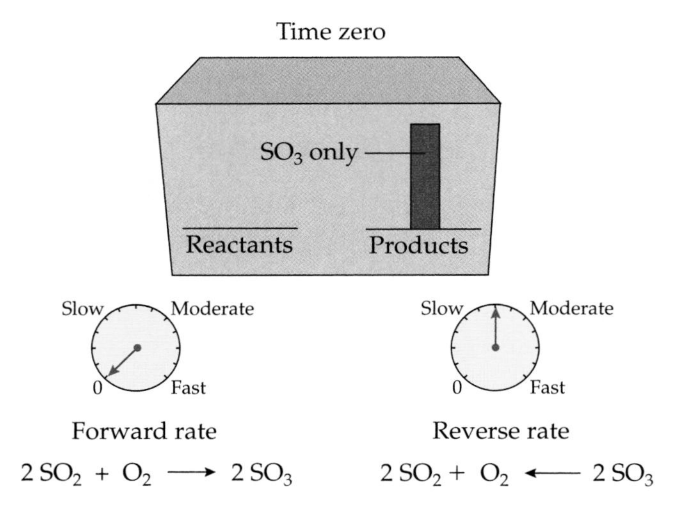

The rates of the forward and reverse reactions are no longer equal, meaning this is a nonequilibrium situation. Of course, as SO_2 and O_2 are produced, the forward reaction begins to pick up speed. And as SO_3 gets used up, the initially faster reverse reaction begins to slow down. Eventually, the forward and reverse rates again become equal, and equilibrium is reestablished.

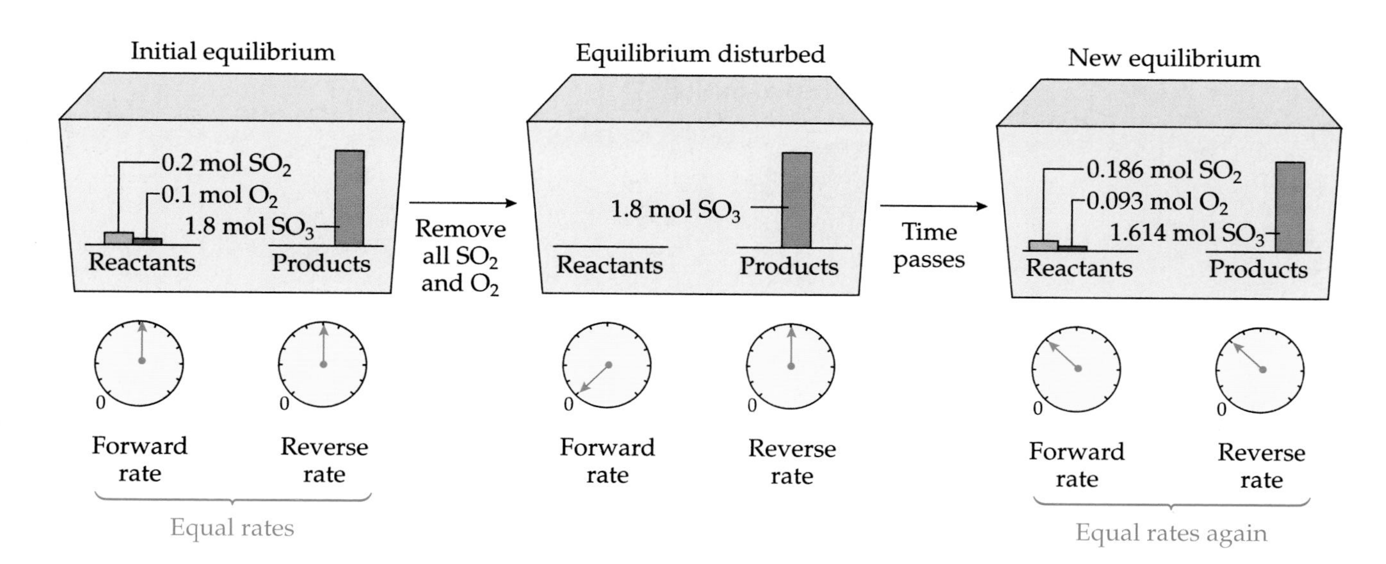

Notice that the amounts of reactants and products for the initial and reestablished equilibria are not the same. Whenever an equilibrium is disturbed by the addition or removal of reactant or product, the reestablished equilibrium concentrations are always slightly different from the initial equilibrium concentrations. However, there is something that does remain the same.

Calculate K_{eq} for both SO_3 equilibrium situations (before and after the disturbance). Remember that the box volume is 1 L.

As we stated previously, the value of K_{eq} does not change unless the temperature changes. The two equilibrium situations described above have the same value of K_{eq}, as you found out in the WorkPatch.

This exercise leads us to a general result—whenever you disturb a reaction at equilibrium by changing the concentration of one or more of the reactants or products, the reaction returns to equilibrium by driving the altered concentrations partially back toward what they were. This principle was summed up many years ago (in 1884, to be exact) by the French chemist Henri Louis Le Chatelier:

Le Chatelier's principle: If you disturb a reaction at equilibrium, the reaction returns to equilibrium by shifting in such a direction as to partially undo the disturbance.

Another way to say this is that anytime you disturb an equilibrium, the reaction attempts to undo what you did. If you remove a reactant or product, the reaction shifts in the direction that produces more of what was removed. If you add additional reactant or product, the reaction shifts in the direction that removes some of what you added.

In our scenario, you removed both SO_2 and O_2. The reaction responded by shifting to make more SO_2 and O_2, partly replacing what you removed.

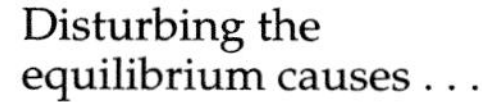
Disturbing the equilibrium causes . . .

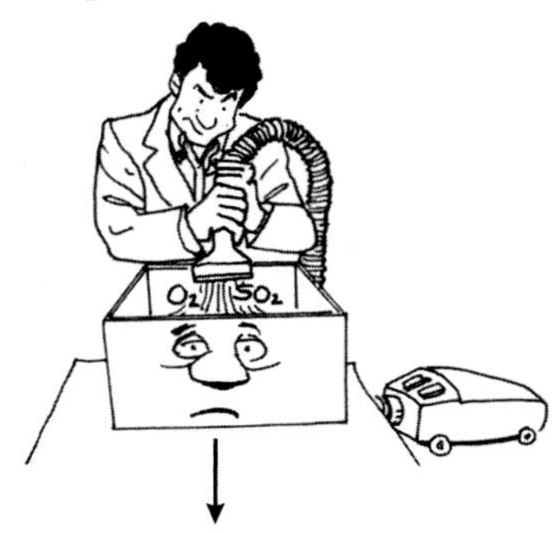

. . . the reaction to shift to restore equilibrium:

$2\,SO_2 + O_2 \longleftarrow 2\,SO_3$

Because SO_2 and O_2 are on the left side of the chemical equation, we say that the reaction shifted to the left.

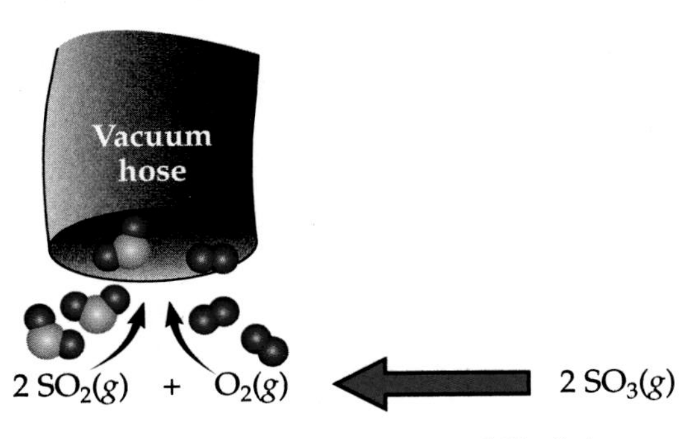

Now try the following problem concerning the $2 SO_2 + O_2 \rightleftharpoons 2 SO_3$ reaction.

14.8 WORKPATCH

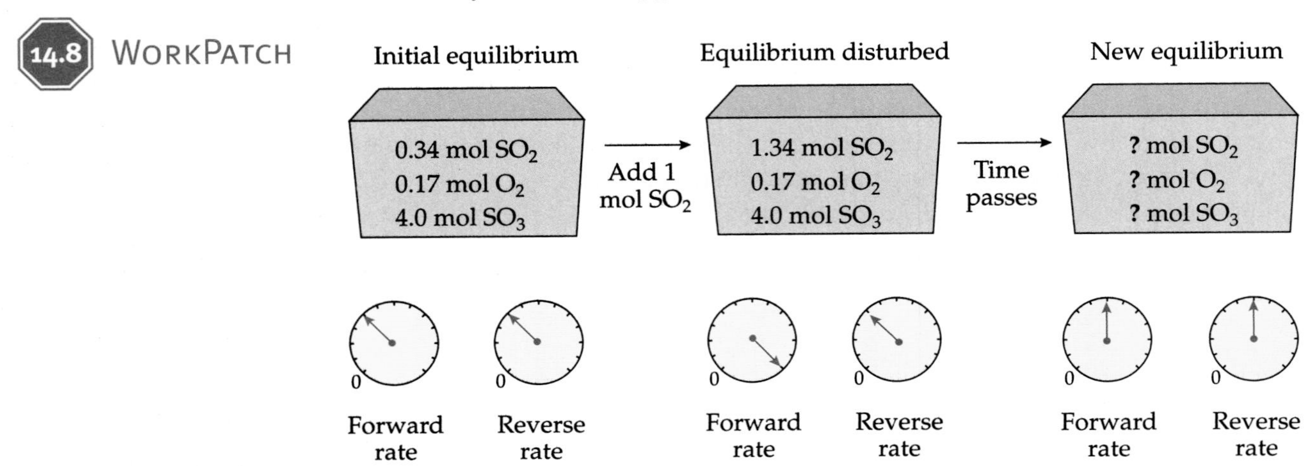

Which is the best choice for the amounts at the new equilibrium:

(a) 0.34 mol SO_2
0.17 mol O_2
4.0 mol SO_3

(b) 1.04 mol SO_2
0.02 mol O_2
4.3 mol SO_3

(c) 1.34 mol SO_2
0.17 mol O_2
4.0 mol SO_3

(d) 1.68 mol SO_2
0.02 mol O_2
4.3 mol SO_3

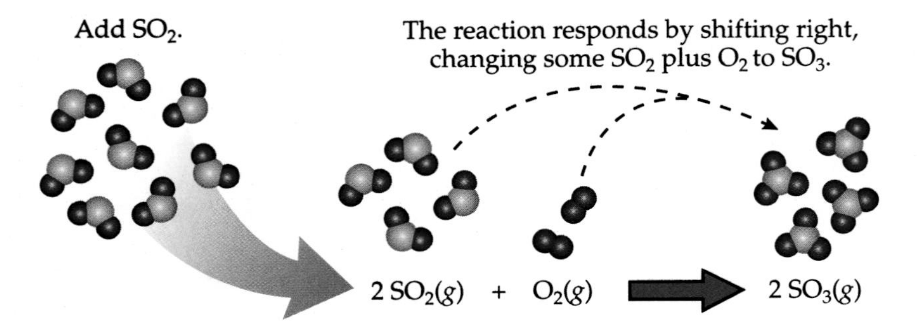

Your answer to the WorkPatch should be consistent with a shift to the right (which rules out choices a or c). Because we added SO_2, a reactant, the reaction shifts to consume the added SO_2 and produce more SO_3. (So which is correct, b or d?) In the process, some O_2 is also consumed. In other words, if you disturb an equilibrium by adding something to the left side, the reaction responds by shifting to the right to partially remove what was added. The addition of SO_2 speeds up the forward reaction while having no effect on the rate of the reverse reaction. The net result is consumption of some of the added SO_2 plus some O_2. As these reactants are consumed, the forward reaction slows. As more SO_3 is produced, the reverse reaction speeds up. Eventually, the forward and reverse reaction rates once again become equal. Equilibrium is reestablished.

Practice Problems

14.10 In WorkPatch 14.8, why are the rates of the forward and reverse reactions greater at the new equilibrium than at the initial equilibrium?

Answer: In both cases, the forward rate equals the reverse rate. This must be true because both situations are at equilibrium. However, when SO_2 is added, some of it is converted to SO_3, meaning that both these substances are present at higher concentration than before. Therefore, collisions between molecules are more frequent. In general, increased collision frequency increases the rate of a reaction.

14.11 If you added SO_3 to a vessel in which the reaction $2\ SO_2 + O_2 \rightleftharpoons 2\ SO_3$ is at equilibrium, which way would the reaction shift?

14.12 Suppose you added N_2 rather than SO_3 to the vessel in Practice Problem 14.11. Which way would the reaction shift?

14.5 How Equilibrium Responds to Temperature Changes

Up to now, we have looked at only one way of disturbing the equilibrium position of a reaction—changing the concentration of one or more substances present in the reaction vessel. In these cases, as stressed above, the reason for the equilibrium shift is to maintain a constant value of K_{eq}. Another way to disturb an equilibrium is to change the temperature. Because the equilibrium constant for any reaction is temperature-dependent, changing the temperature causes the equilibrium to shift because the numeric value of K_{eq} changes. Fortunately, we need not worry about *why* an equilibrium shifts because we can use Le Chatelier's principle in either case to predict the direction of the shift. To understand why temperature affects equilibrium, you must recall some things we discussed previously.

In Chapter 13 you learned that some reactions give off heat (exothermic, ΔE_{rxn} negative),

$$\text{Reactants} \rightleftharpoons \text{Products} + \text{Heat} \qquad \text{Exothermic reaction, } \Delta E_{rxn} \text{ negative}$$

and other reactions absorb heat (endothermic, ΔE_{rxn} positive),

$$\text{Heat} + \text{Reactants} \rightleftharpoons \text{Products} \qquad \text{Endothermic reaction, } \Delta E_{rxn} \text{ positive}$$

Notice that in both cases we wrote the word *heat* as part of the reaction. In an exothermic reaction, heat is a product (heat is released) and therefore it is written on the right side of the balanced equation. In an endothermic reaction, heat is consumed (absorbed), just as a reactant is, and therefore is written on the left side. We normally don't include heat in a chemical equation because the sign of ΔE_{rxn} tells us whether heat is released or absorbed. Nevertheless, when deciding how a reaction at equilibrium responds to changes in temperature, we recommend that you include heat on the appropriate side to make it easier to apply Le Chatelier's principle. As you are about to see, how a reaction at equilibrium responds to a temperature disturbance depends on whether the reaction is exothermic or endothermic.

A good example is our reaction to make SO_3. Being that $\Delta E_{rxn} = -197$ kJ for this reaction, the reaction is exothermic, and we write heat on the right:

$$2\,SO_2(g) + O_2(g) \rightleftharpoons 2\,SO_3(g) + \text{Heat} \qquad \text{Exothermic reaction, } \Delta E_{rxn} = -197 \text{ kJ}$$

If you were to now revisit your SO_3 production facility, you'd see that it is quite cold inside because your investor has stopped paying the bills, including the heating bill. Shivering in the cold, you can hardly believe your eyes. The box is now almost completely filled with SO_3 and contains almost no SO_2 and O_2. After your efforts to purify the product first by shaking the box and then by removing the unreacted SO_2 and O_2 failed, the reaction fixed itself! Does this reaction have a mind of its own?

You forgot about Le Chatelier's principle. As the room cooled, so did the reaction vessel. This cooling removed heat from the reaction vessel. As far as the reaction at equilibrium was concerned, it was being disturbed by this removal of heat and so fought back. To partially undo the disturbance, the reaction shifted in the direction that produces heat. For an exothermic reaction like this one, this means shifting right to produce (release) more heat:

Removing Heat from an Exothermic Reaction

. . . the reaction shifts right . . .

$$2\,SO_2(g) + O_2(g) \longrightarrow 2\,SO_3(g) + \text{Heat}$$

. . . and thereby generates "replacement heat."

The opposite is also true. If you add heat to an exothermic reaction at equilibrium, the equilibrium shifts in the direction that consumes some of the added heat. This means shifting to the left. (An exothermic reaction running in reverse is endothermic.)

Adding Heat to an Exothermic Reaction

If you add heat . . .

. . . the reaction shifts left . . .

$$2\ SO_2(g) + O_2(g) \longleftarrow 2\ SO_3(g) + \text{Heat}$$

. . . and thereby consumes some of the added heat.

Suppose you have a vessel containing an endothermic reaction at equilibrium:

$$\text{Heat} + \text{Reactants} \rightleftharpoons \text{Products} \qquad \text{Endothermic reaction}$$

When you put the vessel in a freezer, which way does the reaction shift? Why?

Did you get the WorkPatch right? Cooling this reaction has the opposite effect of cooling our SO_3 reaction. Only reactions that are neither exothermic nor endothermic ($\Delta E_{rxn} = 0$) establish equilibria that do not shift as the temperature changes.

Practice Problems

14.13 Ethyl acetate, a solvent used as nailpolish remover, is produced by the reaction

$$CH_3COOH + C_2H_5OH \rightleftharpoons \underset{\text{Ethyl acetate}}{CH_3COOC_2H_5} + H_2O \qquad \Delta E = +180.5\ \text{kJ}$$

(a) Rewrite this reaction with the word *heat* in it.
(b) Does the amount of ethyl acetate in the equilibrium mixture increase or decrease when the temperature is raised?

Answer:
(a) Heat + $CH_3COOH + C_2H_5OH \rightleftharpoons CH_3COOC_2H_5 + H_2O$ *(b) It increases.*

14.14 The reaction

$$\underset{\text{Colorless}}{N_2O_4(g)} \rightleftharpoons \underset{\text{Brown}}{2\ NO_2(g)} \qquad \Delta E = +57.2\ \text{kJ}$$

is run at one temperature in flask (a) and at a different temperature in flask (b). Which flask is at the lower temperature and which is at the higher temperature? Explain your answer.

(a)

(b)

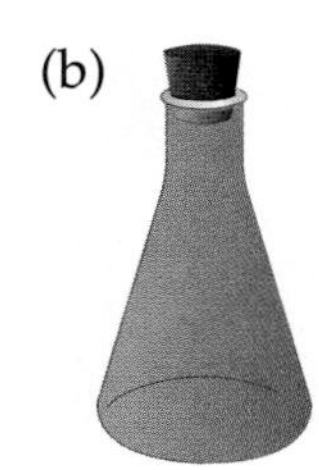

14.15 (a) Rewrite this reaction with the word *heat* in it:

$$2\,CO(g) + O_2(g) \rightleftharpoons 2\,CO_2(g) \qquad \Delta E = -563.5\text{ kJ}$$

(b) Which way does the reaction shift when the temperature is raised? Explain your answer.

Prior knowledge of Le Chatelier's principle would have benefited your relationship with your investor. The exothermic SO_3 reaction should have been run in the cold to drive it further to the right. Of course, there would have been a trade-off. In Chapter 13 we saw that lowering the temperature always slows down a chemical reaction. Thus, running your reaction in the cold would cause it to take longer to reach equilibrium, and time is money in the business world.

A catalyst could help get around this problem. In Chapter 13 we saw that a catalyst can be used to speed up a chemical reaction. The use of a catalyst does not change the position at which the equilibrium occurs because the catalyst speeds up both the forward and reverse reactions by exactly the same amount.

In addition to temperature, there are other ways to handle reactions that have an undesirable equilibrium position, and chemical engineers spend much of their time applying these methods to industrial chemical processes. The moral is, that even in the business world, it pays to know some of the fundamental concepts of chemistry.

14.6 Equilibria for Heterogeneous Reactions, Solubility, and Equilibrium Calculations

Up to now, we have examined only **homogeneous reactions**—those in which reactants and products are in the same phase. Here are two examples:

$$2\,SO_2(g) + O_2(g) \rightleftharpoons 2\,SO_3(g) \quad \text{Everything in gas phase}$$

$$Fe^{3+}(aq) + SCN^-(aq) \rightleftharpoons Fe(SCN)^{2+}(aq) \quad \text{Everything dissolved in water (aqueous phase)}$$

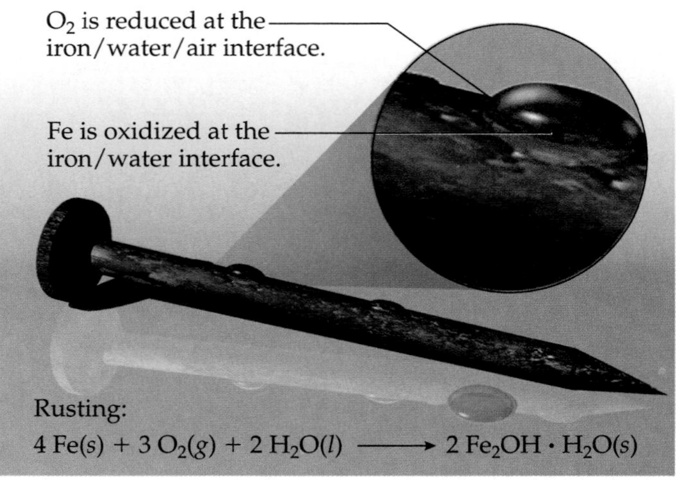
O_2 is reduced at the iron/water/air interface.
Fe is oxidized at the iron/water interface.
Rusting:
$4\,Fe(s) + 3\,O_2(g) + 2\,H_2O(l) \longrightarrow 2\,Fe_2OH \cdot H_2O(s)$

However, many chemical reactions take place at the **interface** between two or more phases—that is, at the surface where different phases are in contact with one another. Reactions that occur at an interface are called **heterogeneous reactions**. The rusting of a nail exposed to water and air is an example. You can recognize heterogeneous reactions by examining the phase designations of the reactants and products. If they are not all the same, the reaction is heterogeneous.

Heterogeneous reactions are common. Almost all reactions involving solids are heterogeneous, and even chemistry that occurs in the atmosphere can be heterogeneous. For instance, some of the reactions that lead to ozone depletion over the Antarctic take place on the surface of stratospheric ice particles.

THE EQUILIBRIUM EXPRESSION FOR HETEROGENEOUS REACTIONS

When we wrote the equation for the rusting of a nail, we used a single arrow because the rusting of iron is a reaction that goes to completion. An example of a heterogeneous reaction that does not go to completion is

$$CO_2(g) + C(s) \rightleftharpoons 2\,CO(g)$$

in which gaseous carbon dioxide reacts with solid carbon in the form of graphite powder to produce carbon monoxide. The reverse reaction also occurs once enough CO accumulates. If we run this reaction in a closed vessel at 1100 K, it achieves an equilibrium for which $K_{eq} = 0.114$. (This value of K_{eq} tells us that substantial amounts of both reactants and product are present at equilibrium. The closed vessel is necessary to keep the gases from escaping so that they may achieve their equilibrium concentrations.) If you write the K_{eq} expression for this reaction in the usual way on an exam, you might be surprised when it's marked incorrect!

$$K_{eq} = \frac{[CO]^2}{[CO_2] \times [C]}$$

Your surprise would be short lived once you read the following section, however. *Anytime we deal with a heterogeneous reaction, we exclude pure liquids and solids from the right side of the K_{eq} expression.* This means that the correct K_{eq} expression for our reaction is actually written

$$K_{eq} = \frac{[CO]^2}{[CO_2]}$$

Can we really toss things out of the equilibrium expression just because they are solids or liquids? The answer is "We haven't really tossed them out." To show you what we mean, let's start again. We'll write the equilibrium expression for our heterogeneous reaction and include everything. We'll call this constant K'_{eq}:

$$CO_2(g) + C(s) \rightleftharpoons 2\,CO(g)$$

$$K'_{eq} = \frac{[CO]^2}{[CO_2] \times [C]}$$

Now we can make use of the fact that the concentration of any solid or liquid remains constant during the course of a heterogeneous reaction (we'll show you why in a moment). This means that the concentration of carbon is constant.

Now let's take our K'_{eq} equilibrium expression and rearrange it algebraically to group all the constant terms on the left:

$$K'_{eq} = \frac{[CO]^2}{[CO_2] \times [C]} \xrightarrow{\text{Multiply both sides by [C]}} K'_{eq}[C] = \frac{[CO]^2 \times \cancel{[C]}}{[CO_2] \times \cancel{[C]}}$$

$$\underbrace{K'_{eq}[C]}_{\text{All constants on left}} = \underbrace{\frac{[CO]^2}{[CO_2]}}_{\text{All variable concentrations on right}}$$

The product K'_{eq}[C] is a constant because both K'_{eq} and [C] are constants. We can therefore simply relabel K'_{eq}[C] and call it K_{eq}:

$$K_{eq} = K'_{eq}[C]$$

Thus we really didn't toss the [C] out. Rather, we've included it in the equilibrium constant.

Practice Problems

14.16 Write the equilibrium-constant expression for the reaction $CaCO_3(s) \rightleftharpoons CaO(s) + CO_2(g)$.

Answer: Because this is a heterogeneous reaction, leave solids and liquids out of the K_{eq} expression: $K_{eq} = [CO_2]$.

14.17 The process of photosynthesis in plants converts carbon dioxide and water to glucose and oxygen:

$$6\,CO_2(g) + 6\,H_2O(l) \rightleftharpoons C_6H_{12}O_6(s) + 6\,O_2(g) \qquad \Delta E_{rxn} = 2801\text{ kJ}$$

(a) Write the equilibrium-constant expression for this conversion.
(b) How would the equilibrium be affected if $CO_2(g)$ were added?
(c) How would the equilibrium be affected if $H_2O(l)$ were added?
(d) How would the equilibrium be affected if the reaction vessel were warmed?
(e) How would the equilibrium be affected if a catalyst were added?

14.18 (a) Write the equilibrium-constant expression for the reaction

$$PbI_2(s) \rightleftharpoons Pb^{2+}(aq) + 2\,I^-(aq)$$

(b) How would the equilibrium be affected if $PbI_2(s)$ were added?
(c) How would the equilibrium be affected if $Pb(NO_3)_2(s)$ were added? [*Hint*: Don't forget that $Pb(NO_3)_2$ is a water-soluble salt.]

Now we must explain our earlier statement that the concentrations of pure liquids and solids remain constant during a heterogeneous reaction. Let's continue with the equilibrium among $CO_2(g)$, $C(s)$, and $CO(g)$ at 1100 K:

$$CO_2(g) \times C(s) \rightleftharpoons 2\,CO(g)$$

$$K_{eq} = \frac{[CO]^2}{[CO_2]} = 0.114 \text{ at } 1100 \text{ K}$$

Remember that the reaction occurs at the interface of the two phases:

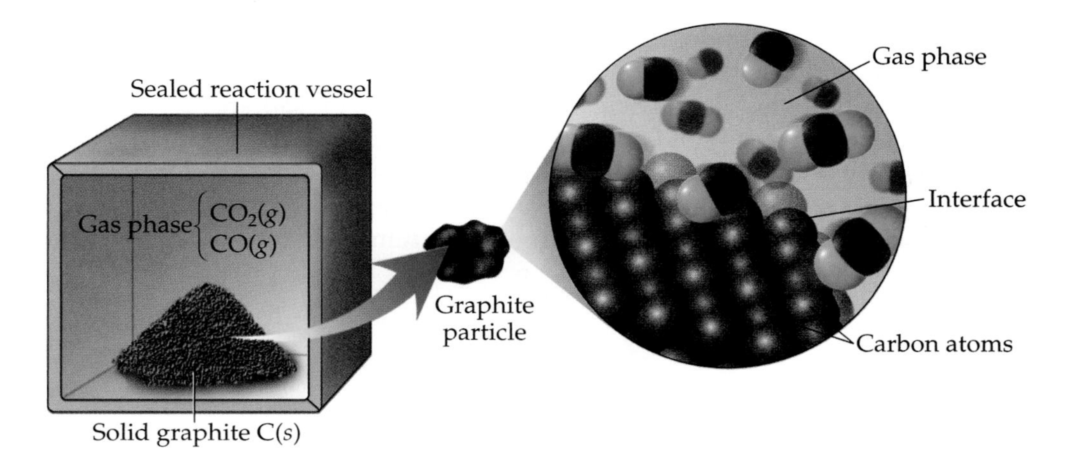

Suppose we start this reaction with only reactants in the vessel. As the reaction approaches equilibrium, the amount of $CO_2(g)$ decreases and the amount of $CO(g)$ increases. Therefore the molar concentrations of these gases change during the approach to equilibrium. These concentrations continue to change until the quantity $[CO]^2/[CO_2]$ equals 0.114. But what about the concentration of $C(s)$? To answer this question, we need to consider the density of this solid. In a quick-and-dirty approach, let us use the density of graphite at 30°C—2.25 g/mL—as our density at 1100 K. Like molarity, density is a form of concentration because both have units of amount of substance divided by volume:

$$\text{Concentration} = \frac{\text{Amount of substance}}{\text{Volume}} \quad \begin{cases} \text{Molarity} = \dfrac{\text{Moles of substance}}{\text{L}} \\[2ex] \text{Density} = \dfrac{\text{Grams of substance}}{\text{mL}} \end{cases}$$

Unlike the density of the gases in our reaction, the density of the graphite does not change during the reaction. Unless the graphite completely disappears, its density (and therefore its concentration) remain a constant 2.25 g/mL no matter the size of the graphite particles. Do you see the fundamental concept behind why this is true? The density of a condensed phase, either solid or liquid, is constant because the substance does not expand to fill its container. Gases, on the other hand, expand to fill their container. This means that as a gas forms, its density increases, and as it is consumed, its density decreases.

SOLUBILITY EQUILIBRIA

Now let's look at a particular type of equilibrium that can have a lot of practical importance—an equilibrium that involves "insoluble" ionic solids. You can determine which ionic solids are soluble or insoluble in water by consulting Table 7.1. One example of an "insoluble" ionic solid is iron(II) sulfide (FeS). The reason we put quotes around the word *insoluble* is that the meaning of this word depends on its context. In Chapter 7 we were dealing with precipitation reactions. We called FeS insoluble because it precipitates when we combine solutions of iron(II) nitrate and sodium sulfide. Now we have a different context. We'll put some solid FeS in water and ask "Does *any* of the FeS dissolve?" The answer is "Yes, but not much." Indeed, all the salts listed as insoluble in Table 7.1 are in fact *sparingly soluble*, meaning a tiny amount does dissolve. FeS(*s*) dissolves very slightly in water to produce $Fe^{2+}(aq)$ and $S^{2-}(aq)$ ions (recall that when ionic solids dissolve, they also dissociate). However, unlike the case with a soluble salt, such as NaCl, it takes only a very small concentration of $Fe^{2+}(aq)$ and $S^{2-}(aq)$ ions before the reverse process (precipitation) occurs with equal speed and a dynamic equilibrium is established:

$$FeS(s) \rightleftharpoons Fe^{2+}(aq) + S^{2-}(aq)$$

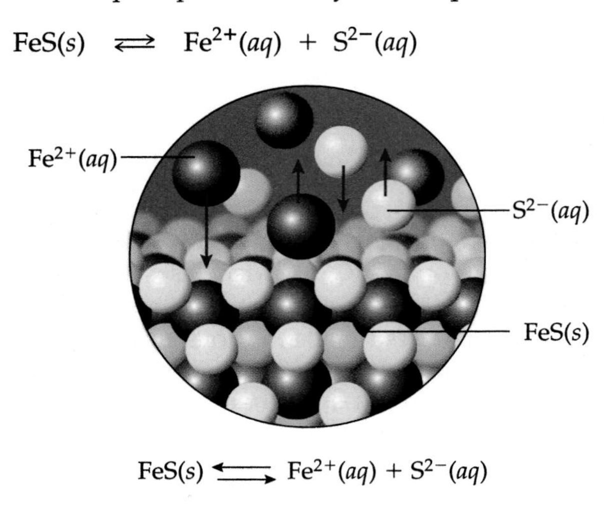

$$FeS(s) \rightleftharpoons Fe^{2+}(aq) + S^{2-}(aq)$$

At equilibrium, ions dissolve and precipitate at the same rate, with the result that the concentrations of the dissolved ions are constant.

If you allow solid FeS to come to equilibrium with water at room temperature, you will find that the concentrations of $Fe^{2+}(aq)$ and $S^{2-}(aq)$ are both 7.746×10^{-10} M.

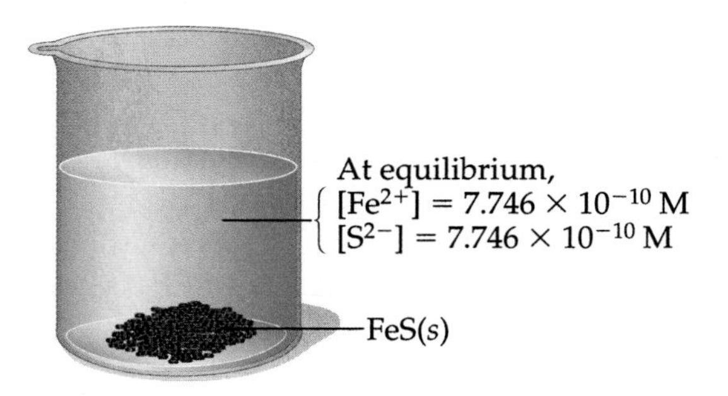

At this point, the solution is saturated. Even though these are extremely small concentrations, they are the highest concentrations of Fe^{2+}(*aq*) and S^{2-}(*aq*) possible when FeS(*s*) is dissolved in water at room temperature. Therefore in the context of dissolving "insoluble" ionic compounds, we instead refer to them as being sparingly soluble.

Perhaps at this point you are wondering why all the fuss? After all, isn't a concentration as tiny as 10^{-10} M so negligible that it's not worth worrying about? Not necessarily! For example, those of us who live in cities or highly populated suburbs drink purified "city water." However, folks in rural settings often get their water from a well drilled on their property. Oftentimes this well water is contaminated with trace amounts of ions that come from contact with sparingly soluble minerals found in the ground. Aqueous sulfide and iron ions are common contaminants, for instance. Concentrations of 10^{-10} M for these ions are not harmful, but they can cause some annoying problems. Living under ground are all sorts of bacteria, including *sulfur bacteria*. Some sulfur bacteria convert aqueous sulfide ions to elemental sulfur, which then combines with bacterial filaments to clog pipes. Other sulfur bacteria generate hydrogen sulfide gas, which is corrosive to pipes and has the foul odor of rotten eggs.

The equilibrium for the dissolving of sparingly soluble FeS is an example of a heterogeneous equilibrium reaction:

$$FeS(s) \rightleftharpoons Fe^{2+}(aq) + S^{2-}(aq)$$

This means that when we write the equilibrium expression, we leave out the solid FeS:

$$K_{eq} = [Fe^{2+}] \times [S^{2-}]$$

Therefore the equilibrium constant is simply the *product* of the ion concentrations at equilibrium (remember that the square brackets [ ] mean molar concentration at equilibrium). For this reason, K_{eq} for sparingly soluble salts is often called the **solubility product** and relabeled K_{sp}:

$$K_{eq} = K_{sp} = [Fe^{2+}] \times [S^{2-}]$$

Determining the value of K_{sp} for FeS(*s*) is easy in principle. All you have to do is add some solid FeS to water, stir awhile to let the solution reach equilibrium, and then measure the concentration of the dissolved ions. We found earlier that the equilibrium concentration for both Fe^{2+}(*aq*) and S^{2-}(*aq*) is 7.746×10^{-10} M. So

$$K_{sp} = [Fe^{2+}] \times [S^{2-}] = (7.746 \times 10^{-10}\text{ M}) \times (7.746 \times 10^{-10}\text{ M})$$

$$K_{sp} = 6.000 \times 10^{-19}$$

An equilibrium constant of 6.000×10^{-19} is very small, indicating that this equilibrium lies very far to the left (the equilibrium favors undissolved FeS). By contrast, a highly soluble salt like NaCl has a very large K_{sp}, reflecting the fact that the concentrations of Na^+ and Cl^- ions in water can be quite large.

Chemists have determined K_{sp} values for many sparingly soluble salts, as Table 14.1 shows.

Table 14.1 Solubility Products of Some Ionic Compounds at 25°C

Compound	Molecular formula	K_{sp}	Compound	Molecular formula	K_{sp}
Aluminum hydroxide	$Al(OH)_3$	1.8×10^{-33}	Lead(II) iodide	PbI_2	1.4×10^{-8}
Barium carbonate	$BaCO_3$	8.1×10^{-9}	Lead(II) sulfide	PbS	3.4×10^{-28}
Barium fluoride	BaF_2	1.7×10^{-6}	Magnesium carbonate	$MgCO_3$	4.0×10^{-5}
Barium sulfate	$BaSO_4$	1.1×10^{-10}	Magnesium hydroxide	$Mg(OH)_2$	1.2×10^{-11}
Bismuth sulfide	Bi_2S_3	1.6×10^{-72}	Manganese(II) sulfide	MnS	3.0×10^{-14}
Cadmium sulfide	CdS	8.0×10^{-28}	Mercury(I) chloride	Hg_2Cl_2	3.5×10^{-18}
Calcium carbonate	$CaCO_3$	8.7×10^{-9}	Mercury(II) hydroxide	$Hg(OH)_2$	3.1×10^{-26}
Calcium fluoride	CaF_2	4.0×10^{-11}	Mercury(II) sulfide	HgS	4.0×10^{-54}
Calcium hydroxide	$Ca(OH)_2$	8.0×10^{-6}	Nickel(II) hydroxide	$Ni(OH)_2$	5.5×10^{-16}
Calcium phosphate	$Ca_3(PO_4)_2$	1.2×10^{-26}	Nickel(II) sulfide	NiS	1.4×10^{-24}
Chromium(III) hydroxide	$Cr(OH)_3$	3.0×10^{-29}	Silver bromide	AgBr	7.7×10^{-13}
Cobalt(II) sulfide	CoS	4.0×10^{-21}	Silver carbonate	Ag_2CO_3	8.1×10^{-12}
Copper(I) bromide	CuBr	4.2×10^{-8}	Silver chloride	AgCl	1.6×10^{-10}
Copper(I) iodide	CuI	5.1×10^{-12}	Silver iodide	AgI	8.3×10^{-17}
Copper(II) hydroxide	$Cu(OH)_2$	2.2×10^{-20}	Silver sulfate	Ag_2SO_4	1.4×10^{-5}
Copper(II) sulfide	CuS	6.0×10^{-37}	Silver sulfide	Ag_2S	6.0×10^{-51}
Iron(II) hydroxide	$Fe(OH)_2$	1.6×10^{-14}	Strontium carbonate	$SrCO_3$	1.6×10^{-9}
Iron(III) hydroxide	$Fe(OH)_3$	1.1×10^{-36}	Strontium sulfate	$SrSO_4$	3.8×10^{-7}
Iron(II) sulfide	FeS	6.0×10^{-19}	Tin(II) hydroxide	$Sn(OH)_2$	5.4×10^{-27}
Lead(II) carbonate	$PbCO_3$	3.3×10^{-14}	Tin(II) sulfide	SnS	1.0×10^{-26}
Lead(II) chloride	$PbCl_2$	2.4×10^{-4}	Zinc hydroxide	$Zn(OH)_2$	1.8×10^{-14}
Lead(II) chromate	$PbCrO_4$	2.0×10^{-14}	Zinc sulfide	ZnS	3.0×10^{-23}
Lead(II) fluoride	PbF_2	4.1×10^{-8}			

For salts that produce the same number of ions per mole, for example, all 1:2 and 2:1 salts such as $CaCl_2$, Na_2S, and $MgBr_2$, there is a direct relationship between K_{sp} and solubility. The smaller the K_{sp}, the lower the solubility of the ionic compound in water.

Practice Problems

14.19 Sparingly soluble $PbCl_2$ dissolves in water to yield an equilibrium $Pb^{2+}(aq)$ concentration of 0.039 M.

(a) Write the balanced equilibrium equation for $PbCl_2(s)$ dissolving in water.

(b) Write the K_{sp} expression for $PbCl_2$.

(c) What is the equilibrium concentration of chloride ion?

(d) Calculate the value of K_{sp} for $PbCl_2$ (show your calculation).

Answer:

(*a*) $PbCl_2(s) \rightleftharpoons Pb^{2+}(aq) + 2\,Cl^-(aq)$

(*b*) $K_{sp} = [Pb^{2+}] \times [Cl^-]^2$

(*c*) *Twice the* $Pb^{2+}(aq)$ *concentration, or 0.078 M.*

(*d*) $K_{sp} = (0.039\ M) \times (0.078\ M)^2 = 2.4 \times 10^{-4}$

14.20 Sparingly soluble calcium phosphate dissolves in water to yield an equilibrium calcium ion concentration of 7.8×10^{-6} M.

(a) Write the balanced equilibrium equation for calcium phosphate dissolving in water.

(b) Write the K_{sp} expression for calcium phosphate.

(c) What is the equilibrium concentration of phosphate ion?

(d) Calculate the value of K_{sp} for calcium phosphate (show your calculation).

Answer:

(*a*) $Ca_3(PO_4)_2(s) \rightleftharpoons 3\,Ca^{2+}(aq) + 2\,PO_4^{3-}(aq)$

(*b*) $K_{sp} = [Ca^{2+}]^3 \times [PO_4^{3-}]^2$

(*c*) *This is a little tougher than the previous problem because now one concentration is not simply twice the other. Use the balanced equation to get a conversion factor:*

$$7.8 \times 10^{-6}\ M\,\cancel{Ca^{2+}} \left(\frac{2\ PO_4^{3-}}{3\ \cancel{Ca^{2+}}} \right) = 5.2 \times 10^{-6}\ M\ PO_4^{3-}$$

(*d*) $K_{sp} = (7.8 \times 10^{-6}\ M)^3 \times (5.2 \times 10^{-6}\ M)^2 = 1.3 \times 10^{-26}$

14.21 Sparingly soluble magnesium hydroxide dissolves in water to yield an equilibrium magnesium ion concentration of 1.44×10^{-4} M.

(a) Write the balanced equilibrium equation for magnesium hydroxide dissolving in water.

(b) Write the K_{sp} expression for magnesium hydroxide.

(c) What is the equilibrium concentration of hydroxide ion?

(d) Calculate the value of K_{sp} for magnesium hydroxide (show your calculation).

14.22 Sparingly soluble aluminum hydroxide dissolves in water to yield an equilibrium hydroxide ion concentration of 8.58×10^{-9} M.

(a) Write the balanced equilibrium equation for aluminum hydroxide dissolving in water.

(b) Write the K_{sp} expression for aluminum hydroxide.

(c) What is the equilibrium concentration of aluminum ion?

(d) Calculate the value of K_{sp} for aluminum hydroxide (show your calculation).

USING K_{SP} TO CALCULATE SOLUBILITIES AND ION CONCENTRATIONS

Now let's switch direction. Instead of using equilibrium ion concentrations to calculate K_{sp} values, let's use K_{sp} values from Table 14.1 to calculate saturation (equilibrium) ion concentrations. There are practical reasons for doing this. For example, suppose your uncle visits you because he knows you've studied chemistry and he has a problem with foul-smelling well water. A geologist has told him his house sits over a deposit of pyrrhotite, FeS, a common mineral. He has two questions for you: "How soluble is FeS?" and "What concentrations of Fe^{2+} and S^{2-} might be in my water?" The first question is easy. A glance at Table 14.1 reveals a small K_{sp} value for FeS (6.0×10^{-19}), which means that this compound is only sparingly soluble in water. To answer his second question, you need to calculate the *solubility* of FeS. The **solubility** s of a substance is the concentration of the substance in a saturated solution. We now show you how to use the K_{sp} value for FeS to calculate the molar solubility of this salt at 25°C. (All solubilities calculated from Table 14.1 values are at 25°C because all the values in the table were determined at that temperature.)

The first step is to write the equilibrium concentration of the aqueous ions in terms of the salt's solubility. To do this, we need to consider the equation that represents the dissolving of the salt in water. For example, the dissolution equations for FeS and $Mg(OH)_2$ are

$$FeS(s) \rightleftharpoons Fe^{2+}(aq) + S^{2-}(aq)$$

$$Mg(OH)_2(s) \rightleftharpoons Mg^{2+}(aq) + 2\,OH^-(aq)$$

The salt FeS is an example of a 1:1 (one cation to one anion) salt. The equation tells you that when 1 mole of FeS dissolves, 1 mole of cation and 1 mole of anion are produced. For this reason, *the cation concentration and anion concentration are both equal to the solubility s*. The salt $Mg(OH)_2$ is an example of a 1:2 (one cation to two anions) salt. When 1 mole of $Mg(OH)_2$ dissolves, 1 mole of cation and 2 moles of anion are produced. For this reason, the cation concentration is equal to s but the anion concentration is equal to $2s$. *That's right!* The concentration of OH^- ion is twice as large as the solubility of $Mg(OH)_2$. The general rule is that if the ion has a coefficient 1 in the dissolution equation, its equilibrium concentration is s; if it has a coefficient 2, its equilibrium concentration is $2s$; and so forth.

Table 14.2 Ion Solubilities in Terms of s for Various Salts

Salt	Solubility	Cation concentration	Anion concentration
FeS	s	s	s
$Mg(OH)_2$	s	s	$2s$
$Al(OH)_3$	s	s	$3s$
$Ca_3(PO_4)_2$	s	$3s$	$2s$

The second step is to plug into the K_{sp} expression the equilibrium ion concentrations expressed in terms of s:

$$[Fe^{2+}] = s$$

$$[S^{2-}] = s$$

$$K_{sp} = 6.0 \times 10^{-19} = [Fe^{2+}][S^{2-}] = s \times s = s^2$$

We now have a simple relationship between K_{sp} and s for FeS: $K_{sp} = s^2$. The third and final step is to solve for s and then do the math:

$$K_{sp} = s^2$$
$$\sqrt{K_{sp}} = \sqrt{s^2}$$
$$s = \sqrt{K_{sp}} = \sqrt{6.0 \times 10^{-19}}$$

Taking the square root of the K_{sp} value on your calculator tells you that $s = 7.7 \times 10^{-10}$ M. This is the saturation solubility of FeS at 25°C, and it is also the equilibrium concentration of $Fe^{2+}(aq)$ and $S^{2-}(aq)$. So show your uncle your calculations, and be thankful he was interested in a 1:1 salt and not a 1:2 salt like $Mg(OH)_2$. For $Mg(OH)_2$, $K_{sp} = 1.2 \times 10^{-11} = [Mg^{2+}][OH^-]^2 = s \times (2s)^2 = 4s^3$. This would have complicated the math and required taking a cube root to solve for s.

Practice Problems

14.23 What is the maximum solubility of PbS in water at 25°C? What are the individual ion concentrations?

Answer: $s = \sqrt{K_{sp}} = \sqrt{3.4 \times 10^{-28}} = 1.8 \times 10^{-14}$ M

The $Pb^{2+}(aq)$ *and* $S^{2-}(aq)$ *concentrations are also* 1.8×10^{-14} *M each.*

14.24 How can you quickly determine the saturation solubility of a sparingly soluble 1:1 salt at 25°C?

14.25 The saturation solubility of Ag_2S at 25°C is 1.14×10^{-17} M. What are the equilibrium concentrations of the cation and anion?

Finally, you should realize that there is nothing special about sparingly soluble salts when it comes to calculating equilibrium concentrations from equilibrium constants. We can do this for any chemical equilibrium. For example, consider one last time your beleaguered $SO_3(g)$ production facility. Suppose you did things in such a way that the reaction came to equilibrium with the concentrations shown below. Let's also suppose that you were unable to measure the equilibrium concentration of $O_2(g)$.

You can use K_{eq} to solve for $[O_2]$, the equilibrium concentration of O_2:

$$K_{eq} = \frac{[SO_3]^2}{[SO_2]^2 \times [O_2]}$$
$$[O_2] = \frac{[SO_3]^2}{[SO_2]^2 \times K_{eq}}$$
$$[O_2] = \frac{(3.50)^2}{(0.50)^2 \times 810} = 0.060 \text{ M}$$

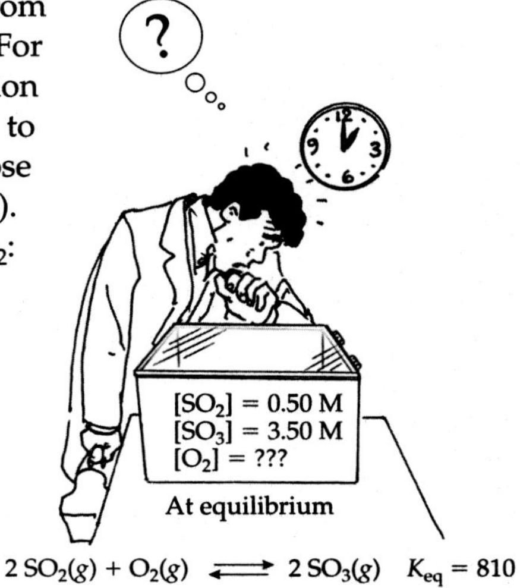

Finding one missing equilibrium concentration is the simplest kind of calculation we can do. Equilibrium constants can be used to do more complicated calculations, such as finding some or all of the equilibrium concentrations. These types of calculations are beyond the scope of this book. However, we are not done with the concept of equilibrium yet. You'll see it again in the next chapter.

Have You Learned This?

Dynamic equilibrium (p. 541)
Reversible reaction (p. 541)
Equilibrium constant K_{eq} (p. 550)
Le Chatelier's principle (p. 557)
Homogeneous chemical reaction (p. 562)
Interface (p. 562)
Heterogeneous chemical reaction (p. 562)
Solubility product K_{sp} (p. 567)
Solubility s (p. 570)

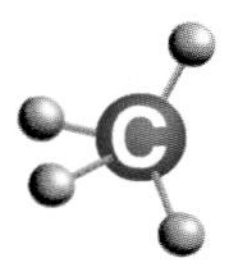

Chemical Equilibrium
www.chemistryplace.com

SKILLS TO KNOW

The equilibrium constant K_{eq}

For a given reaction at a given temperature, the equilibrium constant K_{eq} is a number that tells you the relative concentrations in which reactants and products are present at equilibrium. For a homogeneous reaction

$$aA + bB \rightleftharpoons cC + dD$$

where a, b, c, and d are stoichiometric (balancing) coefficients, K_{eq} is defined as

$$K_{eq} = \frac{[C]^c[D]^d}{[A]^a[B]^b} \quad \frac{\text{Products}}{\text{Reactants}}$$

where [A], [B], [C], and [D] are the molar concentrations at equilibrium.

K_{eq} quantifies the balance between reactant and product concentrations at equilibrium.

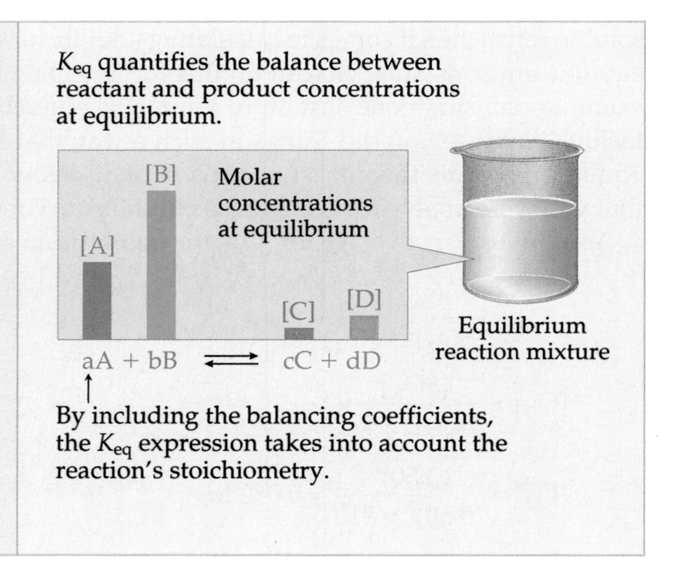

By including the balancing coefficients, the K_{eq} expression takes into account the reaction's stoichiometry.

Interpreting K_{eq} values

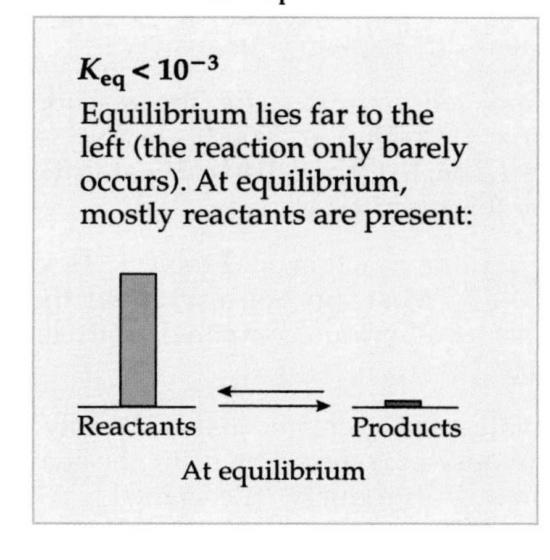

$K_{eq} < 10^{-3}$
Equilibrium lies far to the left (the reaction only barely occurs). At equilibrium, mostly reactants are present:
Reactants
Products
At equilibrium

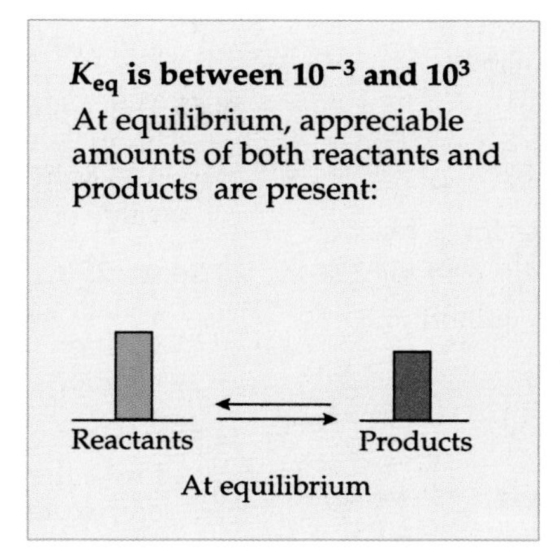

K_{eq} is between 10^{-3} and 10^{3}
At equilibrium, appreciable amounts of both reactants and products are present:
Reactants
Products
At equilibrium

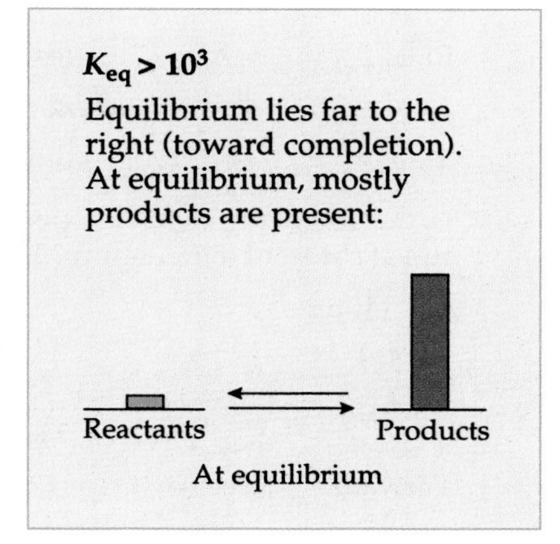

$K_{eq} > 10^{3}$
Equilibrium lies far to the right (toward completion). At equilibrium, mostly products are present:
Reactants
Products
At equilibrium

K_{eq} for heterogeneous reactions

For a heterogeneous reaction (one that occurs at the interface between two phases), the K_{eq} expression *excludes* any reactant or product present in the form of a pure solid or pure liquid. Concentrations of substances in these phases can be omitted from the expression because they remain essentially constant during the reaction.

$$aA(aq) + bB(l) \rightleftharpoons cC(aq) + dD(aq)$$

$$K_{eq} = \frac{[C]^c[D]^d}{[A]^a}$$

Liquid phase — Pure B
Aqueous phase — A, C, D (dissolved)

DYNAMIC EQUILIBRIUM

14.26 Define *equilibrium* in terms of forward and reverse reaction rates.

14.27 What do we mean when we say that a reaction is reversible?

14.28 When we said that your SO_3 reaction was "stuck," we meant that it had stopped. Did it really stop, and how does your answer help explain why we can also describe the reaction as being in a state of dynamic equilibrium?

14.29 What do we mean by the position of a reaction's equilibrium?

14.30 Where on the reaction coordinate is the equilibrium point for a reaction that goes to completion?

14.31 Where on the reaction coordinate is the equilibrium point for a reaction that appears not to occur?

14.32 Sometimes reactions are written with two arrows pointing in opposite directions instead of a single arrow going from reactants to products. What do the two arrows mean?

14.33 What surprise might be encountered if a chemist ran a reaction without ever worrying about equilibrium?

WHY DO CHEMICAL REACTIONS REACH EQUILIBRIUM?

14.34 If a reaction vessel is loaded with just reactants, the forward reaction is fast at first but gradually slows down. Explain why this is so.

14.35 When a reaction vessel is loaded with just reactants, the reverse reaction initially has a rate of zero. Explain why this is so.

14.36 Once a reaction begins, the rate of the reverse reaction gradually speeds up. Explain why this is so.

14.37 Suppose a reaction vessel is loaded only with the products of a reaction. Which would be faster at the moment after loading, the forward reaction or the reverse reaction? Explain your answer.

14.38 Is it possible for the reaction in Problem 14.37 to attain equilibrium? Explain your answer.

14.39 The water in a beaker of water left in a room will slowly evaporate until the beaker is dry. However, place that same beaker in a sealed box and the water level in the beaker will drop a bit but then remain constant. Is the latter case an example of equilibrium? Explain your answer.

14.40 Write the definition of equilibrium in terms of the general rate laws

$$\text{Rate}_{\text{forward rxn}} = k_{\text{forward rxn}}[\text{Reactants}]^{\text{order}}$$

and

$$\text{Rate}_{\text{reverse rxn}} = k_{\text{reverse rxn}}[\text{Products}]^{\text{order}}$$

14.41 Is the following behavior possible for a reaction run at constant temperature? Explain your answer.

At time zero

0 Forward rate

0 Reverse rate

At equilibrium

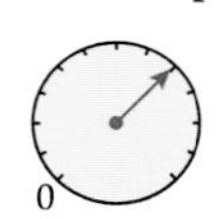

Forward rate

Reverse rate

14.42 Adjust the rate meters of Problem 14.41 to make them show a reaction behavior that is possible.

THE POSITION OF EQUILIBRIUM—THE EQUILIBRIUM CONSTANT K_{eq}

14.43 Using the definition of equilibrium, show how k_f/k_r for the one-step reaction R $\rightleftharpoons$ P is equal to the ratio [P]/[R].

14.44 At a given temperature, why is the ratio k_f/k_r constant for a given reaction?

14.45 What symbol and name are used to replace the ratio k_f/k_r for a reaction?

14.46 If $k_f > k_r$, will K_{eq} be less than 1 or greater than 1? Explain your answer.

14.47 Suppose you have a reaction with many reactants. When you write the equilibrium expression for the reaction, do the reactant concentrations all go in the numerator or in the denominator? Do you add, subtract, multiply, or divide those concentrations by one another?

14.48 Write the equilibrium-constant expression for the gas-state reaction 2 A + 3 B $\rightleftharpoons$ C + D.

14.49 Write the equilibrium-constant expression for the gas-state reaction C + D $\rightleftharpoons$ 2 A + 3 B.

14.50 Compare your answers to Problems 14.48 and 14.49. What conclusion can you draw from the comparison?

14.51 Write the expression for K_{eq} for the reaction $N_2O_4(g) \rightleftharpoons 2\,NO_2(g)$.

14.52 Write the expression for K_{eq} for the reaction $4\,NH_3(g) + 5\,O_2(g) \rightleftharpoons 4\,NO(g) + 6\,H_2O(g)$.

14.53 Suppose you wanted the value of K_{eq} for the reaction of Problem 14.51. What would you do in the laboratory to obtain it?

14.54 Would the value you obtain for K_{eq} for a reaction depend on the initial concentrations of reactants and products you use? Explain your answer.

14.55 The equilibrium concentrations for the reaction $N_2(g) + O_2(g) \rightleftharpoons 2\,NO(g)$ at 2000°C are $[N_2] = 0.25$ M; $[O_2] = 1.2$ M; [NO] = 0.011 M. What is the value of K_{eq} for this reaction?

14.56 At 25°C, K_{eq} for the reaction in Problem 14.55 equals 2.3×10^{-9}. What can you say about the position of this equilibrium compared with its position at 2000°C?

14.57 What does a value of K_{eq} greater than 10^3 imply? Prove that your answer is correct by using the general expression K_{eq} = [Products]/[Reactants].

14.58 Suppose a reaction has a K_{eq} value of 2.05. When we write the reaction, can we use a single arrow to the right instead of a double set of equilibrium arrows? Explain your answer.

14.59 What does a value of K_{eq} less than 10^{-3} imply? Prove that your answer is correct by using the general expression K_{eq} = [Products]/[Reactants].

14.60 A certain reaction has a K_{eq} value of 1.5×10^{-6}. Would this be a practical reaction from which to isolate pure product? Explain your answer.

14.61 The reaction

$$CO(g) + 3\,H_2(g) \rightleftharpoons CH_4(g) + H_2O(g)$$

is run in a 10.0-L vessel. The vessel is loaded with 1 mole of CO and 3 moles of H_2. At equilibrium, the amounts are 0.613 mole of CO, 1.839 moles of H_2, 0.387 mole of CH_4, and 0.387 mole of H_2O. What is the value of the equilibrium constant for this reaction? Describe the position of the equilibrium.

14.62 The reaction in Problem 14.61 is run in the same vessel and at the same temperature, but this time the vessel is loaded with 2 moles of CO and 3 moles of H_2. When the reaction reaches equilibrium, what is the value of K_{eq}?

14.63 Why is the equilibrium constant called a constant?

14.64 For the reaction

$$CH_4(g) + 2\,H_2S(g) \rightleftharpoons CS_2(g) + 4\,H_2(g)$$

$K_{eq} = 3.59$ at 900°C. After the reaction has run for 10 min at 900°C, the concentrations are $[CH_4] = 1.15$ M; $[H_2S] = 1.20$ M; $[CS_2] = 1.51$ M; $[H_2] = 1.08$ M. Is this reaction at equilibrium?

14.65 On the basis of K_{eq} values, which reaction goes essentially to completion? How would you describe the other reaction?

(a) $2\,H_2(g) + O_2(g) \rightleftharpoons 2\,H_2O(g)$; $K_{eq} = 3 \times 10^{81}$

(b) $2\,HF(g) \rightleftharpoons H_2(g) + F_2(g)$; $K_{eq} = 1 \times 10^{-95}$

14.66 Write the balanced chemical equation for the reaction that goes with the equilibrium constant

$$K_{eq} = \frac{[H_2O]^2 \times [Cl_2]^2}{[HCl]^4 \times [O_2]}$$

14.67 An 8.00-L reaction vessel at 491°C contains 0.650 mole of H_2, 0.275 mole of I_2, and 3.00 moles of HI. Assuming that the reaction is at equilibrium, determine the value of K_{eq} and comment on where the equilibrium lies. The reaction is

$$H_2(g) + I_2(g) \rightleftharpoons 2\,HI(g)$$

14.68 How would the value of the equilibrium constant for a one-step reaction calculated as k_f/k_r compare with the value calculated from the concentrations of all substances present at equilibrium?

DISTURBING A REACTION ALREADY AT EQUILIBRIUM—LE CHATELIER'S PRINCIPLE

14.69 Suppose a reaction is at equilibrium and you then disturb the equilibrium by adding reactants. What happens to the value of K_{eq}? Explain your answer.

14.70 State Le Chatelier's principle using the words *undo* and *partially*.

14.71 Suppose we have an equilibrium mixture of reactants and products for the reaction

$$PCl_3(g) + Cl_2(g) \rightleftharpoons PCl_5(g)$$

Predict the direction in which the reaction will shift when

(a) Chlorine (Cl_2) gas is added.
(b) Chlorine is removed.
(c) PCl_5 is added.
(d) PCl_3 is removed.
(e) H_2 gas is added. (Assume the H_2 does not react with any reactant or product.)

14.72 Why can't you simply remove unreacted reactants from an equilibrium mixture to get pure product? Assume that K_{eq} is larger than 10^3 but not all that much larger.

14.73 When a reaction at equilibrium is disturbed by the addition of products,

(a) Which way will the reaction shift?
(b) After the reaction is done shifting, will the product concentration be the same as before the disturbance, greater than before the disturbance, or less than before the disturbance? Explain your answer.
(c) Repeat part (b) but for the reactant concentration.

14.74 Consider the reaction

$$PCl_3(g) + Cl_2(g) \rightleftharpoons PCl_5(g)$$

Use forward and reverse rate meters to represent the forward and reverse reaction rates for

(a) The initial equilibrium.
(b) Moments after you disturb the equilibrium by adding PCl_5.
(c) The restored equilibrium.
(d) Which way did the reaction shift to get from the initial equilibrium, part (a), to the restored equilibrium, part (c)?

14.75 One way of preparing hydrogen is by the decomposition of water:

$$2\,H_2O(g) \rightleftharpoons 2\,H_2(g) + O_2(g) \qquad \Delta E_{rxn} = 484\text{ kJ}$$

Would you expect the decomposition to be more complete when it is run at a high temperature or when it is run at a low temperature? Explain.

14.76 Suppose you are making ammonia (NH_3) by the Haber reaction:

$$3\,H_2(g) + N_2(g) \rightleftharpoons 2\,NH_3(g) \qquad K_{eq} = 0.105 \text{ at } 472°C$$

(a) Describe qualitatively where the equilibrium lies for this reaction.
(b) On the face of it, would this reaction be a good one for isolating pure ammonia?
(c) What would happen if you could keep feeding H_2 and N_2 into the reaction vessel while at the same time removing NH_3?

HOW EQUILIBRIUM RESPONDS TO TEMPERATURE CHANGES

14.77 The equilibrium constant for the synthesis of methanol,

$$CO(g) + 2\,H_2(g) \rightleftharpoons \underset{\text{Methanol}}{CH_3OH(g)}$$

is 4.3 at 250°C and 1.8 at 275°C.

(a) Does this reaction shift to the left or to the right when the reaction mixture is heated? Explain how you know.
(b) Is this reaction endothermic or exothermic? Explain how you know.
(c) Rewrite the equation for the reaction, including heat on the appropriate side.

14.78 The amount of nitrogen dioxide formed by dissociation of dinitrogen tetroxide,

$$N_2O_4(g) \rightleftharpoons 2\,NO_2(g)$$

increases as the temperature rises.

(a) Is the reaction exothermic or endothermic? Explain how you know.

(b) Does K_{eq} increase or decrease as the temperature rises?

(c) Rewrite the equation for the reaction, including heat on the appropriate side.

14.79 Diamond and graphite are two forms of elemental carbon. Under the appropriate conditions they will be in equilibrium with each other:

$$C_{diamond} \rightleftharpoons C_{graphite}$$

If graphite is subjected to very high pressure and temperature, it will convert to the diamond form. Is the above equilibrium reaction exothermic or endothermic? Explain how you know.

14.80 Suppose you have an endothermic reaction with K_{eq} approximately equal to 1×10^3. How could you adjust the temperature of this reaction to drive it to completion? Explain your answer.

14.81 Will K_{eq} for an exothermic reaction increase or decrease when the reaction mixture is (a) heated and (b) cooled? Explain your answer.

14.82 Will K_{eq} for an endothermic reaction increase or decrease when the reaction mixture is (a) heated and (b) cooled? Explain your answer.

14.83 ΔE_{rxn} for the reaction given in Problem 14.77 is -90.8 kJ. Would the fraction of methanol present at equilibrium be increased or decreased by raising the temperature? Explain your answer.

14.84 Cooling an exothermic reaction for which K_{eq} is very low shifts the reaction to the right, so that more product is formed, but there is a trade-off. What is the downside of cooling such a reaction, as far as forming product is concerned?

14.85 What effect does a catalyst have on the position of equilibrium for a reaction and on the value of the equilibrium constant?

14.86 What does a catalyst do to the time it takes for a reaction to reach equilibrium? Explain how it does this.

14.87 As noted in the chapter, the value of K_{eq} for the reaction $N_2(g) + O_2(g) \rightleftharpoons 2\ NO(g)$ is 0.0017 at 2027°C and 2.3×10^{-9} at 25°C.

(a) Judging from the values of K_{eq}, does this reaction shift to the left or to the right when the reaction mixture is heated? Explain your answer.

(b) Is this reaction endothermic or exothermic? Explain your answer.

EQUILIBRIA FOR HETEROGENEOUS REACTIONS, SOLUBILITY, AND EQUILIBRIUM CALCULATIONS

14.88 What is a heterogeneous chemical reaction? Where does a heterogeneous reaction occur?

14.89 Write the equilibrium-constant expression for

(a) $2\ FeCl_3(s) + 3\ H_2O(g) \rightleftharpoons Fe_2O_3(s) + 6\ HCl(g)$

(b) $Fe_2O_3(s) + 3\ CO(g) \rightleftharpoons 2\ Fe(l) + 3\ CO_2(g)$

(c) $PbSO_4(s) \rightleftharpoons Pb^{2+}(aq) + SO_4^{2-}(aq)$

(d) $CO_2(g) + H_2O(l) \rightleftharpoons H_2CO_3(aq)$

14.90 Write the equilibrium-constant expression for

(a) $SnO_2(s) + 2\ H_2(g) \rightleftharpoons Sn(s) + 2\ H_2O(l)$

(b) $H_3PO_4(aq) + 3\ H_2O(l) \rightleftharpoons PO_4^{3-}(aq) + 3\ H_3O^+(aq)$

(c) $Pb^{2+}(aq) + 2\ I^-(aq) \rightleftharpoons PbI_2(s)$

(d) $Ca^{2+}(aq) + 3\ H_2O(l) + CO_2(g) \rightleftharpoons CaCO_3(s) + 2\ H_3O^+(aq)$

14.91 What allows us to incorporate the concentrations of pure solids and liquids into K_{eq} instead of writing these concentrations explicitly in the equilibrium-constant expression?

14.92 Consider a saturated aqueous solution of AgCl, a salt that is only sparingly soluble in water. What happens to this solution if a saturated solution of NaCl (a water-soluble salt) is added to it? [*Hint*: If $[Ag^+(aq)] \times [Cl^-(aq)] > K_{sp}$, precipitation will occur.]

14.93 Consider the reaction

$$SnO_2(s) + 2\ H_2(g) \rightleftharpoons Sn(s) + 2\ H_2O(l)$$

run in an explosion-proof sealed vessel.

(a) Running the reaction in a sealed vessel allows equilibrium to be established. Explain why.

(b) Express the concentration of $H_2(g)$ in terms of K_{eq}.

(c) The equilibrium constant for this reaction decreases as the reaction mixture is heated. Which way does the equilibrium shift?

(d) Is the reaction exothermic or endothermic? Explain.

14.94 What is "dynamic" about the equilibrium that is established when a sparingly soluble salt is added to water?

14.95 At 25°C, the solubility of $Al(OH)_3$ in water is 2.86×10^{-9} M. What are the equilibrium concentrations of the cation and the anion?

14.96 At 25°C, the solubility of $Ca_3(PO_4)_2$ in water is 2.60×10^{-6} M. What are the equilibrium concentrations of the cation and the anion?

14.97 At 25°C, the solubility in water of the moderately soluble salt silver acetate, $AgC_2H_3O_2$, is 10.6 g/L.

(a) Write the chemical equation for the dissolving of silver acetate in water.

(b) Write the K_{sp} expression for silver acetate.
(c) Calculate the value of K_{sp} (show your work).

14.98 At 25°C, the solubility of calcium oxalate, CaC_2O_4, in water is 6.1 mg/L.
(a) What are the equilibrium molar concentrations of $Ca^{2+}(aq)$ and $C_2O_4^{2-}(aq)$?
(b) Calculate K_{sp} for calcium oxalate.

14.99 What is the most soluble 1:2 salt in Table 14.1? Explain how you know.

14.100 At 25°C, the solubility of ferric hydroxide in water is 4.49×10^{-10} M.
(a) What is the solubility in grams per liter?
(b) What is the molar equilibrium concentration of each ion?
(c) How many grams of ferric hydroxide could you dissolve in a 20,000-gallon swimming pool?

14.101 At 25°C, what is the solubility of lead(II) carbonate in water
(a) In moles per liter?
(b) In grams per liter?
(c) What is the molar equilibrium concentration of each ion?
(d) How many grams of lead(II) carbonate could you dissolve in 1.000×10^6 gallons of water?

14.102 Tin(II) sulfide, SnS, is dissolved in an aqueous solution in which the Sn^{2+} concentration is fixed at 2.00×10^{-1} M. What is the sulfide concentration?

14.103 Phosphorus pentachloride gas decomposes to $PCl_3(g)$ and $Cl_2(g)$. At equilibrium, the concentrations of the decomposition products are 5.50×10^{-3} M for $PCl_3(g)$ and 0.125 M for $Cl_2(g)$. What is the equilibrium concentration of PCl_5? The equilibrium constant for this reaction is 7.50×10^{-2}.

14.104 In the Haber process, nitrogen gas reacts with hydrogen gas to form gaseous ammonia (see Problem 14.76).
(a) Write the equilibrium expression for the reaction.
(b) The equilibrium constant at a certain temperature is 1.5×10^{-2}. If the equilibrium concentrations of hydrogen and nitrogen are both 0.20 M at this temperature, what is the equilibrium concentration of ammonia?

ADDITIONAL PROBLEMS

14.105 Suppose the reaction $A_2 + B_2 \rightleftharpoons 2\,AB$ proceeds via a one-step mechanism involving a collision between one A_2 molecule and one B_2 molecule. Suppose also that this reaction is reversible and that the forward reaction is inherently much faster than the reverse reaction.
(a) Does the equilibrium lie to the left or to the right? Explain your choice in terms of the reactant and product concentrations necessary to establish equal forward and reverse rates.
(b) Does an analysis in terms of the relationship $K_{eq} = k_f/k_r$ yield the same answer as in (a)? Explain.

14.106 Suppose the reaction in Problem 14.105 goes to completion. What must be true about E_a for the forward reaction relative to E_a for the reverse reaction?

14.107 When the reaction $N_2(g) + O_2(g) \rightleftharpoons 2\,NO(g)$ is run at 2000°C, appreciable amounts of reactants and product are present at equilibrium.
(a) A sealed 2.00-L container at 2000°C is filled with 1.00 mole of NO(g) and nothing else. At that moment, which reaction is faster, forward or reverse? Justify your answer.
(b) At equilibrium, the concentration of NO(g) is 0.0683 M and the concentration of $N_2(g)$ is 0.2159 M. What is the value of K_{eq} at 2000°C?

14.108 For the reaction

$$2\,NO_2(g) \rightleftharpoons N_2O_4(g)$$

$K_{eq} = 0.500$. What is the equilibrium molar concentration of $NO_2(g)$ if $[N_2O_4] = 0.248$ M?

14.109 For an endothermic reaction, will the equilibrium constant increase, decrease, or stay the same as the temperature of the reaction mixture increases? Explain your answer.

14.110 Write the equilibrium-constant expression for
(a) $SiCl_4(l) + 2\,H_2O(g) \rightleftharpoons SiO_2(s) + 4\,HCl(g)$
(b) $H_2(g) + CO_2(g) \rightleftharpoons CO(g) + H_2O(l)$
(c) $MnO_2(s) + 4\,H^+(aq) + 2\,Cl^-(aq) \rightleftharpoons Mn^{2+}(aq) + Cl_2(g) + 2\,H_2O(l)$
(d) $I_2(s) \rightleftharpoons I_2(g)$
(e) $TiCl_4(g) + 2\,Mg(s) \rightleftharpoons Ti(s) + 2\,MgCl_2(s)$
(f) $Ni(OH)_2(s) \rightleftharpoons Ni^{2+}(aq) + 2\,OH^-(aq)$

14.111 After the reaction $N_2(g) + 3\,H_2(g) \rightleftharpoons 2\,NH_3(g)$ is run, an equilibrium mixture at 300°C is 0.25 M in $N_2(g)$, 0.15 M in $H_2(g)$, and 0.090 M in $NH_3(g)$.
(a) What is the value of K_{eq}?
(b) Which way does the equilibrium shift when $H_2(g)$ is added?
(c) What happens to the value of K_{eq} when $H_2(g)$ is added?
(d) Suppose we write this reaction as

$$2\,N_2(g) + 6\,H_2(g) \rightleftharpoons 4\,NH_3(g)$$

Now what is the value of K_{eq}?
(e) The equilibrium shifts to the right when the reaction mixture is cooled. Is this reaction

exothermic or endothermic? Justify your choice.

14.112 Hydrochloric acid is added to an aqueous solution of silver nitrate.
(a) What precipitate forms?
(b) The equilibrium concentration of $Cl^-(aq)$ is 2.0×10^{-3} M. What is the equilibrium concentration of $Ag^+(aq)$?

14.113 The solubility of silver acetate in water at 20°C is 7.51 g per liter of solution. Calculate K_{sp} for silver acetate.

14.114 How many moles of $BaSO_4$ will dissolve in 200.0 mL of water at 25°C? [*Hint*: Begin with a K_{sp} value from Table 14.1.]

14.115 The solubility of PbI_2 in water at 25°C is 1.52×10^{-3} M. How many grams of PbS will dissolve in 2.50×10^6 gallons of water at 25°C?

14.116 (a) How would you prepare a saturated aqueous solution of copper(I) iodide at 25°C?
(b) What is the mass in milligrams of CuI in 400.0 mL of the saturated solution? [*Hint*: Begin with a K_{sp} value from Table 14.1.]
(c) Suppose you add some CuI* to this saturated solution, where I* is a radioactive form of iodide ion. A student says, "Because the solution is already saturated, the added CuI* won't dissolve and there's no danger of getting any radioactive iodide ion in solution." What is wrong with his thinking?

14.117 For the reaction

$$4\,NO_2(g) + O_2(g) \rightleftharpoons 2\,N_2O_5(g)$$

at 25°C, $K_{eq} = 0.150$. What is the equilibrium concentration of $NO_2(g)$ if $[N_2O_5] = 0.300$ M and $[O_2] = 1.20$ M?

14.118 Which of the following reactions is described by the equilibrium-constant expression

$$K_{eq} = \frac{[A]^2 \times [B]^3}{[C]^3 \times [D]^2}$$

(a) $A_2 + B_3 \rightleftharpoons C_3 + D_2$
(b) $2\,A + 3\,B \rightleftharpoons 3\,C + 2\,D$
(c) $3\,C + 2\,D \rightleftharpoons 2\,A + 3\,B$
(d) $A^2 + B^3 \rightleftharpoons C^3 + D^2$
(e) $2\,C + 3\,D \rightleftharpoons 3\,A + 2\,B$

14.119 Would the solubility of $PbI_2(s)$ be greater in water or in an aqueous solution of NaI? Explain your answer. [*Hint*: If $[Pb^{2+}] \times [I^-]^2 > K_{sp}$, precipitation will occur.]

14.120 In which direction does the reaction

$$CaCO_3(s) + H_2O(l) + CO_2(g) \rightleftharpoons Ca^{2+}(aq) + 2\,HCO_3^-(aq)$$

shift when
(a) $[CO_2]$ is increased?
(b) The volume of the reaction vessel is decreased?
(c) $Ca^{2+}(aq)$ is added?
(d) $CaCO_3(s)$ is removed?

14.121 $K_{eq} = 1.4 \times 10^{-10}$ for the dissolution of calcium fluoride in water:

$$CaF_2(s) \rightleftharpoons Ca^{2+}(aq) + 2\,F^-(aq)$$

(a) What is another name for K_{eq} for this reaction?
(b) If the equilibrium calcium ion concentration in a saturated aqueous solution of calcium fluoride is 3.3×10^{-4} M, what is the equilibrium fluoride ion concentration?
(c) Which is larger, the rate constant for the forward reaction or the rate constant for the reverse reaction?
(d) Which is larger, E_a for the forward reaction or E_a for the reverse reaction?
(e) Which is larger, the rate of the forward reaction or the rate of the reverse reaction?
(f) For lithium carbonate, $K_{sp} = 0.0011$. Write the balanced chemical equation and the equilibrium expression for the dissolution of Li_2CO_3 in water.
(g) Which is more soluble in water, calcium fluoride or lithium carbonate?

14 WorkPatch Solutions

14.1 (a) 4
(b 5
(c) 3
(d) 2
(e) 1

14.2 (c) is correct. Because one of the reactants (O_2) is missing at time zero, the forward rate must be zero. Because there is a large concentration of product (SO_3) at time zero, the reverse rate is fast. As time progresses, the forward reaction speeds up and the reverse reaction slows down, meaning the equilibrium rate meters shown in (c) make sense.
(a) is incorrect because the reverse rate must be fast at time zero because the vessel is full of SO_3.
(b) is incorrect because although there is nothing wrong with the time-zero meters, the reverse reaction slows down as SO_3 is consumed, meaning the equilibrium reverse meter is wrong.

(d) is incorrect because the initial absence of the reactant O_2 means the time-zero forward meter must read zero.

(e) is incorrect because the equilibrium meters do not show the same rate.

14.3 First, divide both sides by k_r and cross out anything identical appearing on the top and bottom of a side:

$$\frac{k_f[A]}{k_r} = \frac{\not{k_r}[B]}{\not{k_r}}$$

$$\frac{k_f[A]}{k_r} = [B]$$

The left side now has k_f/k_r, but this ratio is multiplied by [A]. To get rid of the [A], divide both sides by [A], crossing out anything identical appearing on the top and bottom of a side:

$$\frac{k_f[\not{A}]}{k_r[\not{A}]} = \frac{[B]}{[A]}$$

$$\frac{k_f}{k_r} = \frac{[B]}{[A]}$$

14.4 $$K_{eq} = \frac{[NO]^2}{[N_2][O_2]} = \frac{(0.040\text{ M})^2}{(0.980\text{ M})(0.980\text{ M})} = 0.0017$$

14.5 (a) $\dfrac{[H^+]\times[Cl^-]}{[HCl]}$ (4) $K_{eq} > 10^6$

(b) $\dfrac{[H^+]\times[CH_3COO^-]}{[CH_3COOH]}$ (1) $K_{eq} = 1.0\times10^{-5}$

(c) $\dfrac{[PCl_3]\times[Cl_2]}{[PCl_5]}$ (2) $K_{eq} = 2.2\times10^{-2}$

(d) $\dfrac{[CO]\times[H_2]^3}{[H_2O]\times[CH_4]}$ (3) $K_{eq} = 4.6$

14.6 (a) $$K_{eq(\text{decompose NO})} = \frac{[N_2]\times[O_2]}{[NO]^2}$$

$$K_{eq(\text{form NO})} = \frac{[NO]^2}{[N_2]\times[O_2]}$$

(b) One is the inverse of the other:

$$K_{eq(\text{decompose NO})} = \frac{1}{K_{eq(\text{form NO})}}$$

(c) $$K_{eq(\text{decompose NO})} = \frac{1}{K_{eq(\text{form NO})}} = \frac{1}{2.3\times10^{-9}} = 4.3\times10^8$$

Because this is a very large number, the equilibrium lies far to the right, meaning NO essentially decomposes completely to N_2 and O_2.

14.7 Initial equilibrium:

$$K_{eq} = \frac{(1.8\text{ M})^2}{(0.2\text{ M})^2\times0.1\text{ M}} = 810$$

New equilibrium:

$$K_{eq} = \frac{(1.614\text{ M})^2}{(0.186\text{ M})^2\times0.093\text{ M}} = 810$$

14.8 (b) is the best choice. Remember, Le Chatelier's principle says that a reaction shifts to partially undo a disturbance to an equilibrium. Here we increased the amount of SO_2 from 0.34 mole to 1.34 moles. The reaction therefore shifts to the right to decrease the amount of SO_2 to some extent (but not all the way back to 0.34 mole). A shift to the right (from the amount present in the center box) also causes a decrease in the amount of O_2 and an increase in the amount of SO_3. Only (b) is consistent with these results.

14.9 To the left, to replace the "reactant" heat that's being removed. Why it shifts this way is explained by Le Chatelier's principle. Because you removed heat, the reaction shifts in the direction that produces heat, attempting to partially undo the disturbance. The only way an endothermic reaction can produce heat is to run in reverse (toward reactants, including heat).

BASES

ACIDS

CHAPTER 15

Electrolytes, Acids, and Bases

15.1 Electrolytes and Nonelectrolytes

We have seen that chemistry attempts to understand matter in terms of the atoms and molecules from which all things are made. But chemistry had its beginnings long before the widespread acceptance of the existence of atoms and molecules. As a result, many early chemists concerned themselves with finding, examining, and categorizing various "chemical" substances. Much as the biologist uses the categories kingdom, phylum, class, and so forth to group together living things that have similar traits and behaviors, early chemists did the same sort of thing by creating the two categories *acids* and *bases*.

A chemical substance was considered to be an **acid** if it had these three properties:

1. A sour taste. Early chemists used all five senses to probe the nature of matter. Many substances were tasted to determine whether they had a sour taste and might therefore be classified as acids. (The word *acid* comes from the Latin *acidus*, meaning "sour.") Of course, no one does such taste testing without some risk. We can only imagine the fate of the pioneering chemist who determined that hydrogen cyanide, HCN, a fast-acting and lethal poison, is an acid.

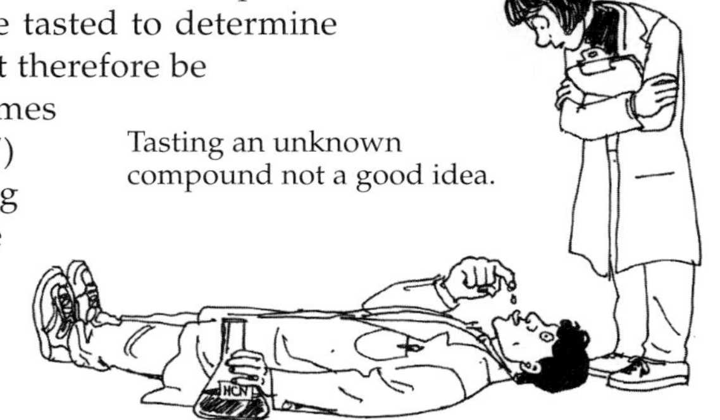

Tasting an unknown compound not a good idea.

Acid
Acids react with active metals, producing hydrogen bubbles.
Zinc
An acid causes an aqueous solution of litmus to turn red.
Acid solution

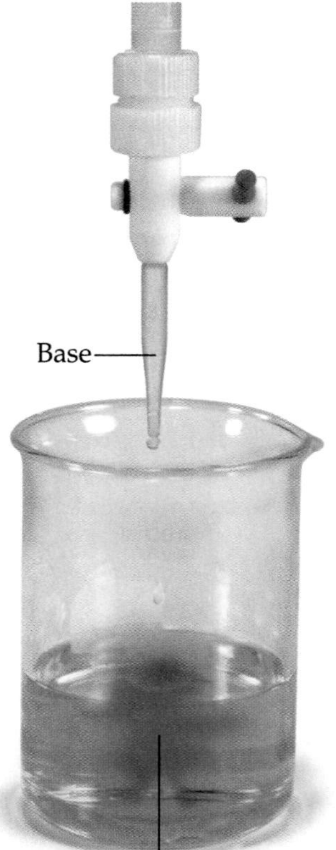
Base
A base causes an aqueous solution of litmus to turn blue.

Not all acids are toxic. After all, vitamin C is an acid (ascorbic acid), and citric acid is found in all kinds of fruits, vegetables, sour candies, and so on. Clearly, though, other tests for acids were needed if the science of chemistry, not to mention the chemists themselves, was to survive. This need led to the other two properties subsequently used to classify compounds as acids.

2. Turns the plant dye *litmus* (extracted from certain mosses) red.
3. Eats away active metals by reacting with them to produce hydrogen gas.

A chemical substance was classified as a **base** if it

1. Had a bitter taste.
2. Turned an aqueous solution of litmus dark blue.
3. Produced aqueous solutions that felt slippery to the touch.

Compounds such as litmus are called *indicators* because they indicate by their color whether an acid or base is present. As early chemists tested many of the substances found in nature, they discovered that some were acids, some were bases, and some were neither. Substances that gave no signs of being either acidic or basic were called **neutral substances**. Testing and categorizing numerous compounds in this way was an important first step in understanding the chemical behavior of many common substances.

The categories acid and base proved so fundamental and useful that they remain with us today. However, modern chemistry goes further, explaining at the molecular level why acids and bases have the properties they do. The key to understanding these properties involves yet another classification scheme, that of *electrolyte* versus *nonelectrolyte*. This was one of the first classification schemes for compounds to be successfully explained on the molecular level, and it led directly to a clearer understanding of acids and bases. Therefore, we shall now spend a little time discussing electrolytes and nonelectrolytes before considering acids and bases.

A substance is considered to be an **electrolyte** if an aqueous solution of the substance conducts an electric current. If the solution does not conduct current, the substance is a **nonelectrolyte**. It is easy to determine whether a substance is an electrolyte or a nonelectrolyte. All we need is a sample of the substance dissolved in pure water, a light bulb attached to a plug and cord, and an electrical outlet. When we plug the cord into the outlet, electrons flowing through the circuit create an electric current, and the bulb lights:

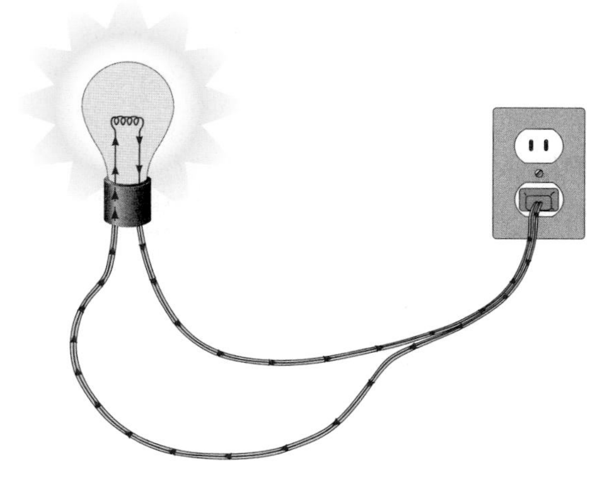

To use this apparatus as an electrolyte tester, all we need to do is unplug the cord and then cut one of its wires. A store-bought tester is the same thing:

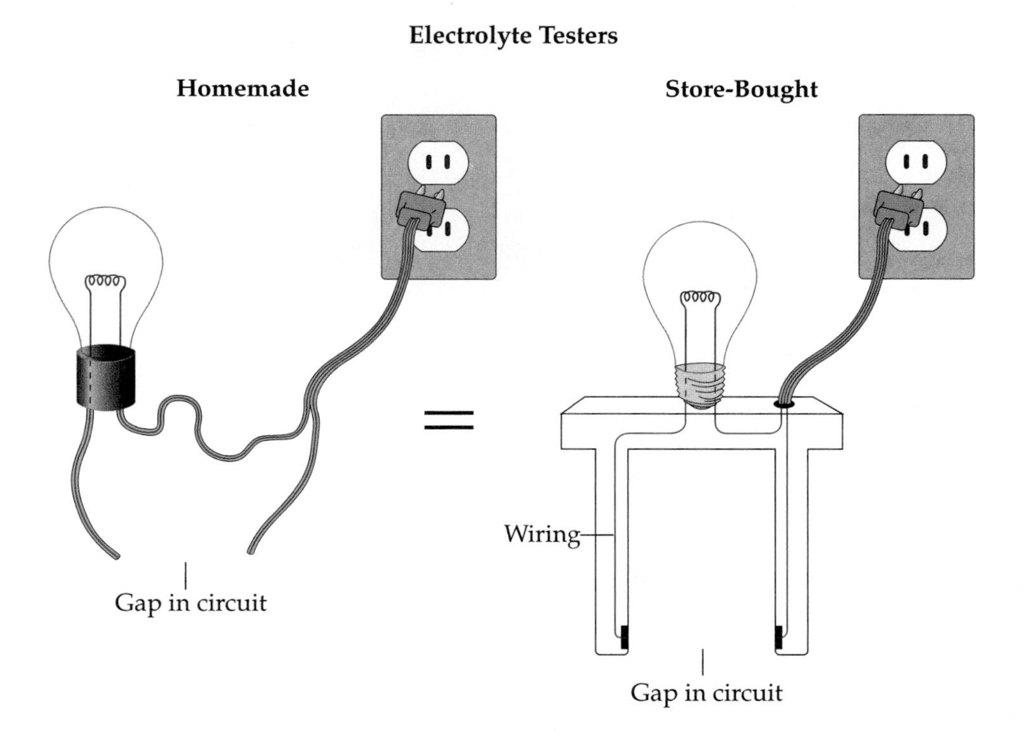

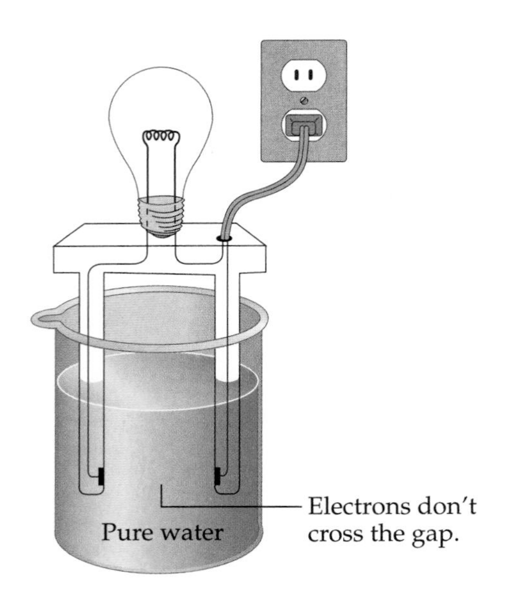
Pure water
Electrons don't cross the gap.

When we plug either tester into the wall outlet, the bulb does not light because electrons cannot flow across the gap in the circuit. Now, let's dip the prongs of the store-bought tester into a beaker of pure water. If electrons could somehow swim through the water from one prong to the other, they would be creating a complete electric circuit, and the bulb would light. The bulb doesn't light, however, because pure water is a poor conductor of electricity.

Now let's test two other familiar substances: table salt, NaCl, and table sugar, sucrose. We dissolve both substances in pure water in separate beakers and insert a tester in each as shown below. The bulb in the NaCl circuit lights up, but the bulb in the sucrose circuit remains dark. Clearly, only the aqueous solution of NaCl conducts electricity. Therefore NaCl is an electrolyte and sucrose is a nonelectrolyte. An electric current doesn't travel through a sucrose solution any better than it does through pure water.

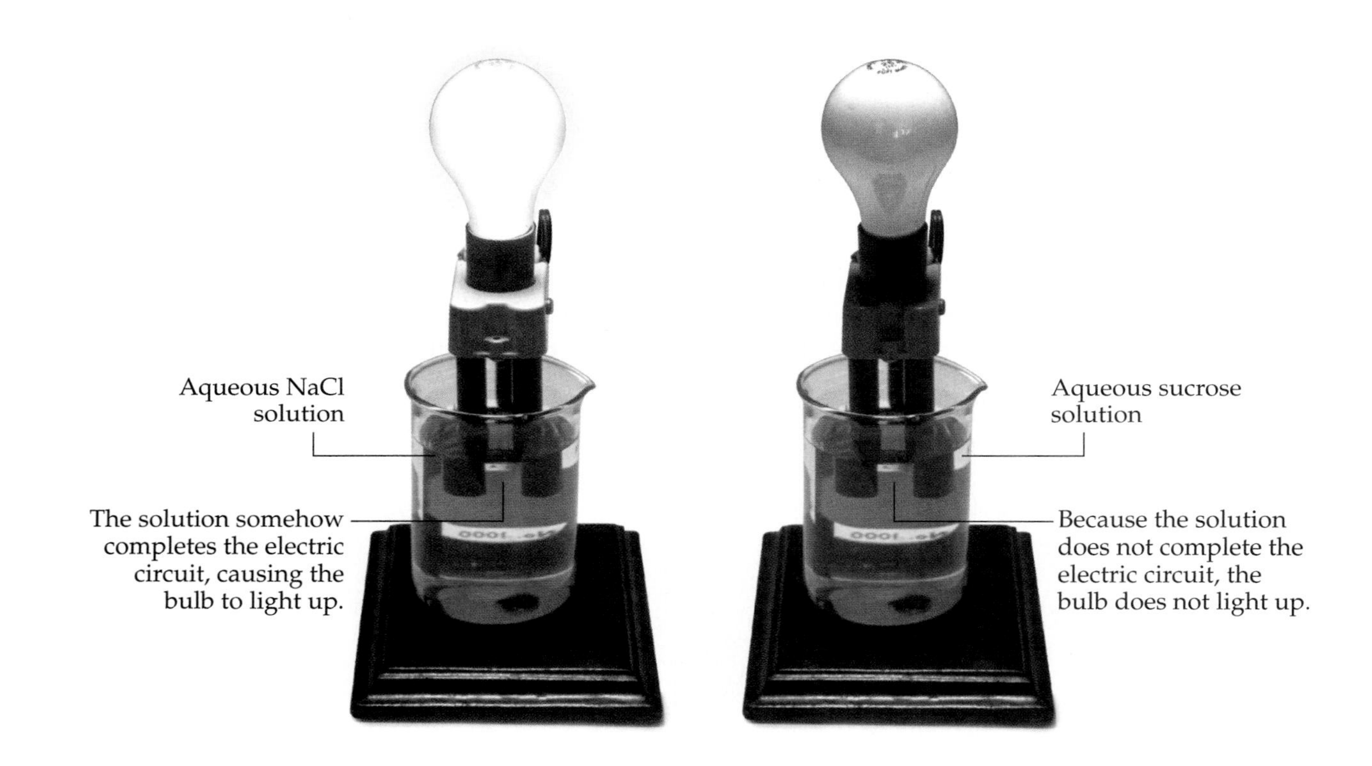
Aqueous NaCl solution
The solution somehow completes the electric circuit, causing the bulb to light up.
Aqueous sucrose solution
Because the solution does not complete the electric circuit, the bulb does not light up.

Using methods like these, chemists have compiled lists of electrolytes and nonelectrolytes. Table 15.1 lists some typical substances from each category.

Until about 200 years ago, chemists had no idea *why* some aqueous solutions conduct electricity but others do not. Michael Faraday, an English physicist who in the early 1800s did much of the basic work on the nature of electricity, was the first to shed some light on what was going on. Faraday proposed that in order for an electric current to pass through an aqueous solution, some form of mobile charged particles must be present in the solu-

Table 15.1 Some Well-Known Electrolytes and Nonelectrolytes

Electrolytes	Nonelectrolytes
Acetic acid, CH_3COOH (vinegar)	Acetone, C_3H_6O (nailpolish remover)
Hydrogen chloride, HCl	Carbon monoxide, CO
Sodium bicarbonate, $NaHCO_3$ (baking soda)	Ethanol, CH_3CH_2OH (drinking alcohol)
Sodium chloride, NaCl	Methane, CH_4
Sodium hydroxide, NaOH (lye)	Oxygen, O_2
Sodium sulfate, Na_2SO_4	Sucrose, $C_{12}H_{22}O_{11}$
Sulfuric acid, H_2SO_4	Turpentine (hydrocarbons like $C_{10}H_{22}$)

tion. He called these particles *ions*. Today we know that ions are either charged single atoms (Na^+, Cl^-) or charged groups of atoms (SO_4^{2-}, NH_4^+). Faraday had no real concept of what these ions were, however, and he incorrectly hypothesized that they were created in the water by the electricity. He was correct, though, about ions being mobile charged particles. He was also correct when he said that electricity can be carried through water only when these charged particles are present.

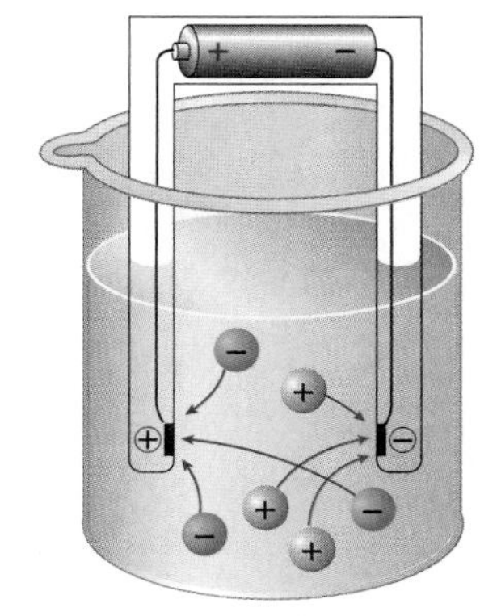

Faraday's mobile charged particles that carry electric current. He called them ions.

In 1884, after years of painstaking measurements on the electrical conductivity of hundreds of aqueous solutions, the Swedish chemist Svante Arrhenius offered the first comprehensive (and correct) explanation for the behavior of electrolytes. In his Ph.D. thesis, Arrhenius hypothesized that Faraday's ions came from the dissociation of the solute. In other words, when an electrolyte is dissolved in water, the solute breaks up into mobile charged particles (ions), and these particles are present in the solution whether or not an electric current is applied.

The response of Arrhenius's thesis advisor was essentially "You have a new theory? Very interesting. Please close the door on your way out." Arrhenius was awarded his degree grudgingly because his work was considered a "radical and ridiculous" theory of ionization. Years later, the world caught up to Arrhenius, and in 1903 he received the Nobel Prize in Chemistry for his work.

Today we realize that a solution of NaCl conducts electricity because the Na^+ and Cl^- ions act as the mobile charge carriers that Faraday proposed, each ion migrating to the wire of opposite electrical charge. Electrons cannot flow through water, but ions can. These mobile charge carriers take the place of electrons, keeping the flow of charge (electricity) going.

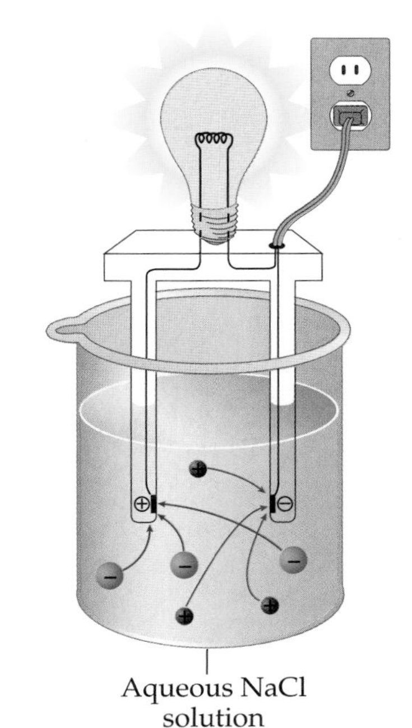

Being positive, cations (Na^+) are attracted to the negative wire. Being negative, anions (Cl^-) are attracted to the positive wire.

The mobile charge carriers exist because the ionic compound NaCl not only dissolves in water but also breaks up into ions. We call this breaking apart to produce ions **dissociation**. Soluble ionic compounds dissolve and dissociate. The alternative would be for a substance to dissolve without dissociation. For example, molecular substances, such as sucrose and ethanol, dissolve without dissociating into ions and therefore are nonelectrolytes. Their solutions consist of hydrated, intact, neutral molecules. Because no ions are present, no electricity can pass through these solutions.

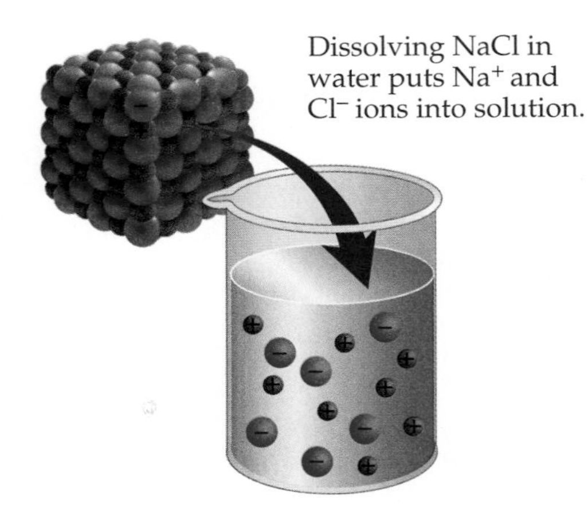

Dissolving ethanol, CH3CH2OH, in
water puts intact ethanol molecules
into solution. No ions are present.

What you want to remember here is this difference between how ionic and molecular solutes dissolve—the former with dissociation, the latter without. This makes sense when you stop to think about it. Molecular solids are made up of discrete, whole molecules, and ionic solids are made up of ions packed into a lattice. Dissolving an ionic solid simply puts into solution the particles that make up the lattice. We can now use the word *ion* to extend our definitions of electrolyte and nonelectrolyte:

> An **electrolyte** is a solute that dissolves in water and dissociates into ions, yielding a solution that conducts electricity.
>
> A **nonelectrolyte** is a solute that dissolves in water without producing ions, yielding a solution that does not conduct electricity.

You should be able to determine from its chemical formula whether a compound is an electrolyte or not. Water-soluble ionic substances are electrolytes. Most ionic compounds consist of a metal combined with either a nonmetal or a group of nonmetals. Therefore, when you see a formula like $NaCl$ or $CaCO_3$, you can be reasonably confident you are looking at a water-soluble ionic compound and therefore an electrolyte. Ionic compounds that do not contain a metal are also electrolytes. For example, compounds containing the polyatomic ion NH_4^+, such as NH_4Cl and $(NH_4)_2SO_4$, are water-soluble ionic compounds and thus electrolytes.

Water-soluble molecular substances generally consist entirely of nonmetals and are usually nonelectrolytes. Therefore, when you see a formula that includes no metals, such as CO_2 or $C_{12}H_{22}O_{11}$ (sucrose), you can be reasonably confident you are looking at a molecular compound and therefore a nonelectrolyte. Of course, there are exceptions here as well. One that you should be aware of pertains to compounds having the formula HX, where X = Cl, Br, or I. Even though hydrogen chloride, HCl, hydrogen bromide, HBr, and hydrogen iodide, HI, are all polar covalent molecules rather than ionic substances, they do dissociate into ions when dissolved in water and therefore are electrolytes.

There are other molecular compounds that dissolve in water and also dissociate to give ions, making them electrolytes. We shall show you some of these shortly. Most molecular substances, however, are nonelectrolytes.

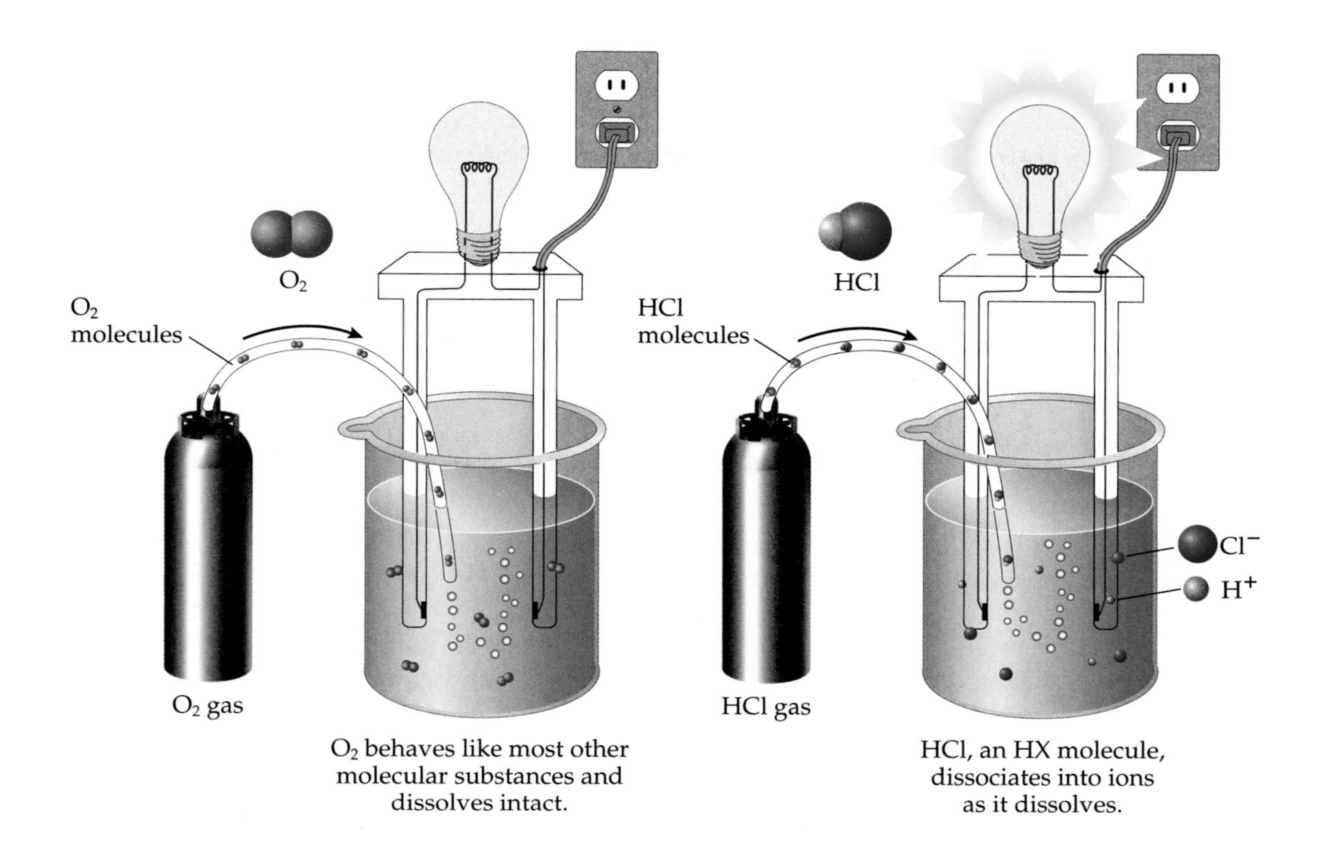

O_2 behaves like most other molecular substances and dissolves intact.

HCl, an HX molecule, dissociates into ions as it dissolves.

Practice Problems

15.1 (a) Is the compound H_2SO_4 molecular or ionic?
(b) How did you decide on your answer to part (a)?
(c) Based on your answer to part (a), would you call H_2SO_4 an electrolyte or a nonelectrolyte?
(d) An aqueous solution of H_2SO_4 causes a light bulb to light. What does this tell you about H_2SO_4?

Answer:
(a) Molecular.
(b) The formula has no metals in it, just nonmetals. This is a sign that the compound is probably molecular.
(c) Most molecular compounds are nonelectrolytes, so the safe bet would be to call H_2SO_4 a nonelectrolyte.
(d) That the bulb lights up tells you H_2SO_4 must be an electrolyte. That H_2SO_4 is an electrolyte tells you two things: (1) just like HCl, H_2SO_4 is an exception to the rule that most molecular compounds are not electrolytes, and (2) H_2SO_4 must dissociate into ions when it dissolves in water.

15.2 Indicate whether each compound is an electrolyte or a nonelectrolyte in water:
(a) $Al(NO_3)_3$ (b) $(CH_3)_2O$ (c) $(NH_4)_2SO_4$ (d) CH_3OH
(e) $CuSO_4$ (f) KBr (g) HBr

15.3 (a) Draw a beaker containing water and show what ions, if any, are present when ammonium bromide, NH_4Br, dissolves.
(b) Does the NH_4Br dissociate? What does the fact that it does or does not dissociate tell you regarding electrolyte/nonelectrolyte classification for this compound?
(c) Add to your drawing from part (a) a light bulb attached to positive and negative wires. Indicate which way the dissolved species move.

15.2 Electrolytes Weak and Strong

Let's look a bit more closely at molecular solutes that are also electrolytes. Both HCl and HF dissociate in water to give H^+ ions and halide ions. However, an HCl solution makes a light bulb glow brightly, whereas an HF solution of the same concentration barely causes the bulb to glow. Why the difference? We can find the answer by considering how efficient each molecule is at dissociating into ions once it dissolves.

Hydrogen chloride is particularly good at dissociating in water. Were you ever unfortunate enough to be swimming around in an HCl solution, you would practically never come across an undissociated HCl molecule. All you would ever find are H^+ and Cl^- ions:

$$HCl(g) \xrightarrow[\text{in } H_2O]{\text{Dissolves}} H^+(aq) + Cl^-(aq) \qquad \text{100\% dissociation}$$

When we write $H^+(aq)$, we are really using an abbreviation. To be absolutely accurate, we should write $H_3O^+(aq)$ to indicate a hydronium ion because "naked" H^+ ions do not exist in water. Instead, they are picked up by water molecules to yield hydronium ions because H^+, being a bare proton with no electrons, is too reactive to exist by itself in water. Thus the more exact equation is

$$HCl(g) + H_2O(l) \longrightarrow H_3O^+(aq) + Cl^-(aq)$$

To form H_3O^+, the incoming H^+ shares one of the lone pair of electrons on the oxygen in the water molecule. This covalent bond fills the valence electron shell of the H^+ and results in the more stable hydronium ion:

$$H^+ + H_2O \longrightarrow H_3O^+$$

Hydronium ion

$$H^+ + H\text{—}\ddot{\underset{\cdot\cdot}{O}}\text{—}H \longrightarrow \left[H\text{—}\ddot{O}(\text{—}H)\text{—}H\right]^+$$

Brackets mean the charge applies to the whole ion.

The situation is quite different for HF. If you examined an aqueous solution of HF, you would discover some H_3O^+ and F^- ions, but you would also come across a far greater number of undissociated HF molecules. The dissociation reaction can be written as an equilibrium that lies to the left, which we indicate by exaggerating the length of the left-pointing arrow:

$$HF(g) + H_2O(l) \;\xrightleftharpoons[\longleftarrow]{}\; H_3O^+(aq) + F^-(aq)$$

This equilibrium puts fewer ions into solution, resulting in a decreased ability to conduct electricity and a light bulb that barely lights. Only a small percentage of the dissolved HF molecules dissociate (about 3% in a 1.0 M solution), as opposed to 100% dissociation of HCl. For this reason, we call HCl a *strong electrolyte* and HF a *weak electrolyte*.

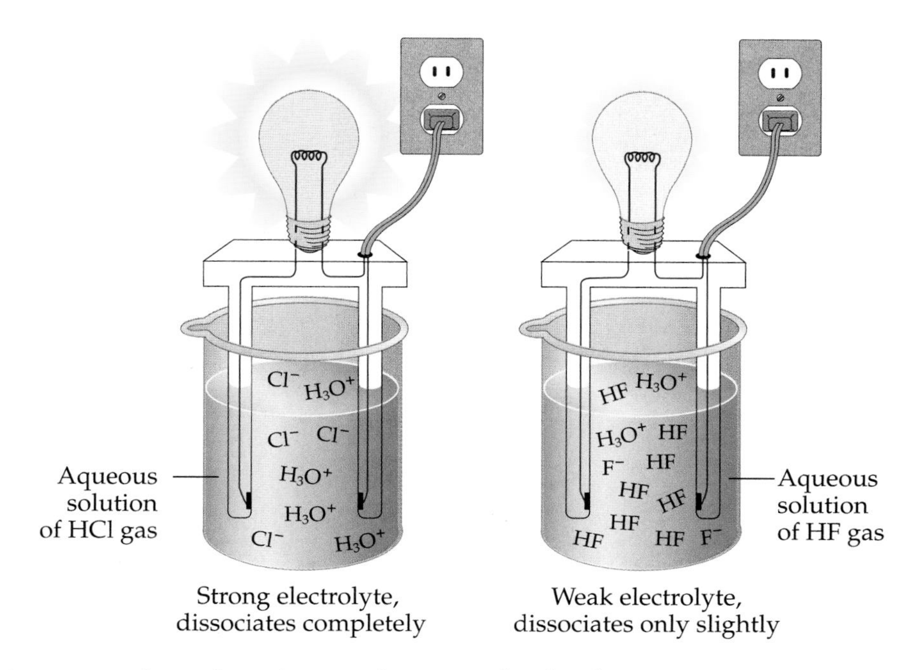

A **strong electrolyte** is one that completely dissociates into ions upon dissolving in water.

A **weak electrolyte** is one that only partially dissociates into ions upon dissolving in water.

All of this relates back to what we learned in Chapter 14 about equilibrium. Both the HCl and the HF dissociation reactions, like all other reactions in principle, are equilibrium reactions:

$$HCl(g) + H_2O(l) \rightleftharpoons H_3O^+(aq) + Cl^-(aq)$$

$$HF(g) + H_2O(l) \rightleftharpoons H_3O^+(aq) + F^-(aq)$$

In the case of HCl, the equilibrium lies so far to the right that we say the reaction goes to completion and drop the double arrows:

$$HCl(g) + H_2O(l) \longrightarrow H_3O^+(aq) + Cl^-(aq) \qquad K_{eq} \gg 10^3$$

In the case of HF, the equilibrium lies far to the left:

$$HF(g) + H_2O(l) \;\xrightleftharpoons[\longleftarrow]{}\; H_3O^+(aq) + F^-(aq) \qquad K_{eq} = 3.5 \times 10^{-4} (< 10^3)$$

Almost all the dissolved HF remains undissociated.

15.1 WORKPATCH

Of the three solutions shown here,

(a) Which is a weak electrolyte, which is a strong electrolyte, and which is a nonelectrolyte? How can you tell?

(b) Which of these compounds is/are molecular? Which is/are ionic?

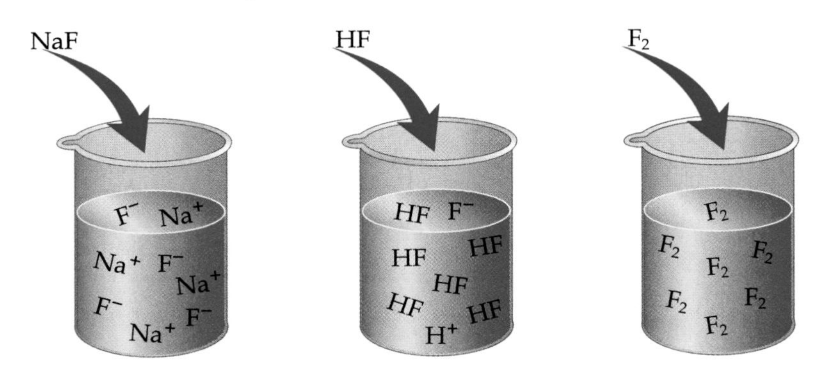

Were you able to tell which electrolyte was weak or strong just from examining the species in solution? Once you understand what an electrolyte is, you are ready to go back to acids and bases, which, as you will see, are particular kinds of electrolytes.

15.3 Acids Weak and Strong

The molecular electrolytes we have considered so far, HCl and HF, were known more than 150 years ago. Indeed, quite a few molecular electrolytes were known at that time, including HBr, HI, HNO_3, H_2SO_4, and HClO. Little was understood about them, but some similarities had been discovered. For example, they were all known to dissolve in water to give aqueous solutions that tasted sour and turned litmus red. They also all reacted with active metals to liberate H_2 gas. As we saw at the beginning of the chapter, this behavior classifies them as acids.

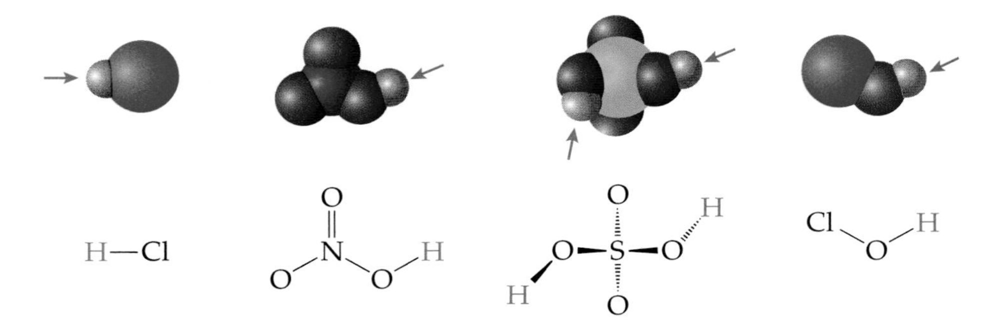

Why very different molecules like HCl, HNO_3, and H_2SO_4 should all give rise to aqueous solutions having similar acidic properties was a mystery, though one clue came from the fact that all these compounds are electrolytes. It was Arrhenius who in 1884 put it all together, postulating that all these compounds dissociate in water to produce H^+ ions. That became the first modern definition of an acid. Chemists have modified this definition a bit to recognize that aqueous H^+ ions really exist as H_3O^+ ions, but otherwise we still use Arrhenius's definition of an acid.

Arrhenius definition of acid—any electrolyte that contains one or more hydrogen atoms and produces H^+ (actually H_3O^+) ions when dissolved in water.

Molecular substances like HCl, HBr, HI, HNO_3, and H_2SO_4 are all strong electrolytes, dissociating extensively to give lots of H_3O^+ ions. This makes their aqueous solutions very acidic, and we refer to these compounds as strong acids.

A **strong acid** is a water-soluble compound that dissociates extensively in water to give large numbers of H_3O^+ ions.

On the other hand, HF is a weak electrolyte and therefore dissociates only partially, producing far fewer H_3O^+ ions. Because intact HF molecules do not make a solution acidic (only H_3O^+ ions do), an aqueous solution of HF is only weakly acidic. We therefore refer to HF as a weak acid.

A **weak acid** is a water-soluble compound that dissociates only partially in water to produce only a few H_3O^+ ions.

The figure below shows the difference in behavior between a strong acid and a weak acid. When you dissolve 1 mol of HCl in water, you get 1 mole of H_3O^+ ions, making the solution strongly acidic. By contrast, 1 mole of HF dissolved in water produces far fewer H_3O^+ ions, making the solution only weakly acidic.

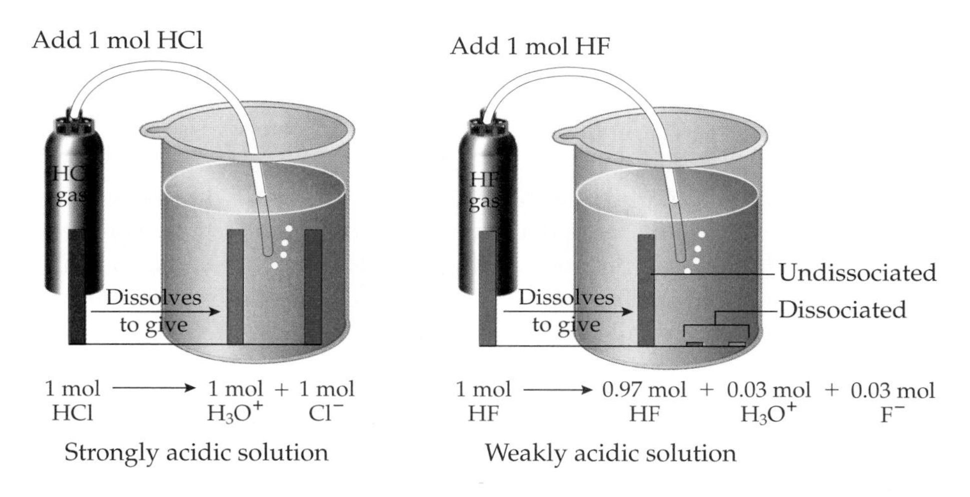

Strong or weak, all these molecular electrolytes produce acidic solutions when dissolved in water. To acknowledge this fact, we change the names of these electrolytes slightly when they are dissolved in water, as we saw already in Chapter 5. An aqueous solution of HCl is called hydrochloric acid, and aqueous solutions of HF, HNO_3, and H_2SO_4 are called hydrofluoric acid, nitric acid, and sulfuric acid, respectively. All these molecular compounds produce H_3O^+ ions when they dissociate.

We can now see why all these electrolytes have similar acidic properties. It is the H_3O^+ ion that is responsible for the sour taste, it is the H_3O^+ ion that turns litmus red, and it is the H_3O^+ ion that reacts with active metals to yield hydrogen gas.

Tables 15.2A and 15.2B list some common strong and weak acids.

Table 15.2A Strong Acids

Molecular formula	Name	Maximum number of H_3O^+ ions produced per molecule of acid
HCl	Hydrochloric acid	1
HBr	Hydrobromic acid	1
HI	Hydroiodic acid	1
HNO_3	Nitric acid	1
H_2SO_4	Sulfuric acid	2

Table 15.2B Weak Acids

Molecular formula	Name	Maximum number of H_3O^+ ions produced per molecule of acid
HF	Hydrofluoric acid	1
HClO	Hypochlorous acid	1
CH_3COOH	Acetic acid	1
H_2CO_3	Carbonic acid	2
H_3PO_4	Phosphoric acid	3

You may be familiar with some of these acids. Sulfuric acid, H_2SO_4, is the acid used in car batteries. Vinegar is a dilute aqueous solution of acetic acid, CH_3COOH. Carbonic acid is present in carbonated soft drinks, as is phosphoric acid, which gives soda pop its ability to completely dissolve a tooth left soaking in it for just a few days.

The rightmost column in Tables 15.2A and 15.2B tells you that some acids are capable of producing more than one H_3O^+ ion per molecule. For example, sulfuric acid, H_2SO_4, can produce a maximum of two H_3O^+ ions. Such an acid is referred to as a **diprotic acid**. Acids that give just one hydronium ion per molecule (HCl, HF) are called **monoprotic acids**. **Triprotic acids**, such as phosphoric acid, H_3PO_4, can produce a maximum of three H_3O^+ ions:

$$H_3PO_4 + 3\,H_2O \longrightarrow 3\,H_3O^+ + PO_4^{3-} \qquad (15.1)$$

Sulfuric acid, H_2SO_4, is a diprotic acid. Write a balanced equation for sulfuric acid dissolving in water, showing the maximum number of H_3O^+ ions it can produce.

The reaction you just wrote for H_2SO_4 and the reaction shown above for H_3PO_4 show the *maximum* number of protons these acids can produce in water. In fact, though, these reactions do not go to completion when these acids are added to water. Read on to find out what really happens.

A common misconception is that diprotic and triprotic acids are stronger than monoprotic acids because diprotic and triprotic acids dissociate to give more hydronium ions. This is not necessarily true. For example, even though carbonic acid, H_2CO_3, is diprotic, it is a much weaker acid than monoprotic HCl. This is because carbonic acid barely dissociates in water, which means that even though each molecule is capable of producing two H_3O^+ ions, most of the dissolved H_2CO_3 molecules prefer to stay intact. So don't confuse the terms *strong* and *weak* with whether an acid is mono-, di-, or triprotic. The two classification schemes are not related to each other. The only thing the terms *strong* and *weak* can tell you is to which side the dissociation equilibrium in water lies—to the right, toward complete dissociation, for strong acids and to the left, toward nondissociation, for weak acids.

The dissociation equilibrium constants for strong acids are 10^3 or greater, so we can say the dissociation goes to completion. For weak acids, knowing the value of K_{eq} for the dissociation reaction can help us gauge just how weak a particular acid is. For example, consider the three dissociation steps for triprotic phosphoric acid:

Loss of first H^+

$$H_3PO_4 + H_2O \rightleftharpoons H_3O^+ + H_2PO_4^- \qquad K_{eq} = 7.5 \times 10^{-3}$$

Loss of second H^+

$$H_2PO_4^- + H_2O \rightleftharpoons H_3O^+ + HPO_4^{2-} \qquad K_{eq} = 6.2 \times 10^{-8} \qquad (15.2)$$

Loss of third H^+

$$HPO_4^{2-} + H_2O \rightleftharpoons H_3O^+ + PO_4^{3-} \qquad K_{eq} = 4.2 \times 10^{-13}$$

All these equilibria have a K_{eq} of approximately 10^{-3} or less, which means that all three equilibria lie to the left. The acid H_3PO_4 gives a proton (H^+) to water to become $H_2PO_4^-$. The $H_2PO_4^-$ is also an acid because it gives up a proton to water in the second dissociation equilibrium, becoming HPO_4^{2-}. And HPO_4^{2-} is yet another acid because it gives up a proton to water in the third dissociation equilibrium. Of these three weak acids, which is the weakest? The values of K_{eq} tell us:

Weak acid	Dissociation K_{eq}
H_3PO_4	7.5×10^{-3}
$H_2PO_4^-$	6.2×10^{-8}
HPO_4^{2-}	4.2×10^{-13}

Even though all three acids are weak, HPO_4^{2-} is clearly the weakest because it has by far the smallest value of K_{eq} for its dissociation reaction.

Acetic acid, CH_3COOH, has a dissociation K_{eq} of 1.8×10^{-5}. This means that it is a weaker acid than H_3PO_4 but a stronger acid than $H_2PO_4^-$. The value of K_{eq} can even be used to calculate what percentage of an acid dissociates. For example (without going into the details), based on its K_{eq} value, we can calculate that 99.58% of the acetic acid in a 1.0 M solution is in the form of intact CH_3COOH molecules. Only 0.42% dissociates to CH_3COO^- and hydronium ions.

CH_3COOH ● H_3O^+ ● CH_3COO^- ●

$$H_3C{-}C(=O){-}O{-}H + H_2O \rightleftharpoons H_3C{-}C(=O){-}O^- + H_3O^+$$

Acetic acid — Acetate ion

$K_{eq} = 1.8 \times 10^{-5}$

Dissociation K_{eq} values can reveal some interesting behavior. For example, sulfuric acid is listed as a strong acid in Table 15.2A, but take a closer look at the dissociation equilibria for this diprotic acid:

Loss of first H$^+$

$$H_2SO_4 + H_2O \longrightarrow H_3O^+ + HSO_4^- \qquad K_{eq} > 1.0 \times 10^3$$

Loss of second H$^+$

$$HSO_4^- + H_2O \rightleftharpoons H_3O^+ + SO_4^{2-} \qquad K_{eq} = 1.2 \times 10^{-2}$$

The equilibrium constant for the first dissociation is quite large, indicating sulfuric acid is indeed a very strong acid. This is true only for the first proton dissociation, however; the second proton is only partially dissociated. So in a sense sulfuric acid is both a strong acid and a weak acid. In other words, even though sulfuric acid is diprotic, it does not lose both protons with equal ease. However, because one dissociation goes to completion, this acid is considered to be strong.

Practice Problems

15.4 True or false? If a compound is a molecular substance and an acid in water, it must also be an electrolyte.

Answer: True. An acid must dissociate to produce H_3O^+ ions in water. Anything that dissociates in water is an electrolyte.

15.5 Write a balanced dissociation equation for carbonic acid in water that shows the maximum number of H_3O^+ ions the acid can yield.

15.6 Your answer to Practice Problem 15.5 shows the dissociation yielding 2 moles of protons per mole of acid, but this reaction actually produces fewer protons. Write the two balanced equilibrium equations for the dissociation of carbonic acid in water.

15.4 Bases—The Opposites of Acids

Many substances discovered in the early days of chemistry could be classified as acids. In addition, early workers discovered a large number of compounds that behaved similarly to one another but were not acidic. These compounds tasted bitter instead of sour, and they turned litmus dark blue instead of red. These compounds were classified as *bases*.

The most interesting property of bases was that, when added in the proper amount to an acidic solution, they destroyed the acidic properties of the solution and left the water neutral (neither acidic nor basic) and with a salty taste. The bases apparently "killed," or somehow *neutralized*, the H_3O^+ ions responsible for the acidity. It was this observation, along with his definition of acids, that led Arrhenius to his definition of bases.

Arrhenius reasoned that a base must remove the H_3O^+ ions produced by acids. One way to do this would be to turn the H_3O^+ ions back to neutral water molecules. The question "What is a base?" then boiled down to the question "What could react with H_3O^+ ions to produce neutral water molecules?" Arrhenius's answer was the hydroxide ion, OH^-. The reaction between an H_3O^+ ion supplied by an acid and an OH^- ion supplied by a base is called an **acid–base neutralization reaction.** A hydronium ion and a hydroxide ion react with each other to give two molecules of water:

$$\underset{\text{Acid}}{H_3O^+} + \underset{\text{Base}}{OH^-} \xrightarrow[\text{other to give}]{\text{Neutralize each}} 2\,H_2O \qquad (15.3)$$

We can understand this reaction at the molecular level by examining dot diagrams. A basic hydroxide ion converts an acidic hydronium ion to two water molecules by accepting a proton from the H_3O^+:

$$[H_3O]^+ + [:\ddot{O}-H]^- \longrightarrow H_2O + H_2O \qquad (15.4)$$

There was another hint that it was the hydroxide ion that gave basic compounds their properties. Most of the bases known at the time were ionic compounds with such formulas as NaOH, LiOH, CaO_2H_2, and MgO_2H_2. No doubt, Arrhenius looked at those last two formulas and rewrote them in his mind as $Ca(OH)_2$ and $Mg(OH)_2$. Today we call these basic compounds *metal hydroxides* and represent them by the generic formula $M(OH)_n$, where M stands for a metal atom. Table 15.3 lists some common metal hydroxides.

Table 15.3 Some Common Metal Hydroxide Bases

LiOH	Lithium hydroxide
NaOH	Sodium hydroxide
KOH	Potassium hydroxide
$Mg(OH)_2$	Magnesium hydroxide
$Ca(OH)_2$	Calcium hydroxide
$Ba(OH)_2$	Barium hydroxide

All these bases are water-soluble to various degrees. They are ionic compounds, and they dissociate to yield OH^- ions in aqueous solution. Arrhenius put all this information together and defined a base as anything that dissolves in water to give hydroxide ions.

Arrhenius definition of base—any electrolyte that contains a metal ion and hydroxide group and produces hydroxide ions when dissolved in water.

As with acids, some bases are strong electrolytes and others are weak electrolytes. And again as with acids, bases that are strong electrolytes are also strong bases and those that are weak electrolytes are also weak bases. The group IA (1) metal hydroxides in Table 15.3—LiOH, NaOH, and KOH—are strong electrolytes that are highly soluble in water and dissociate completely to produce lots of OH^- ions. This makes them **strong bases** because their dissociation reactions go essentially to completion:

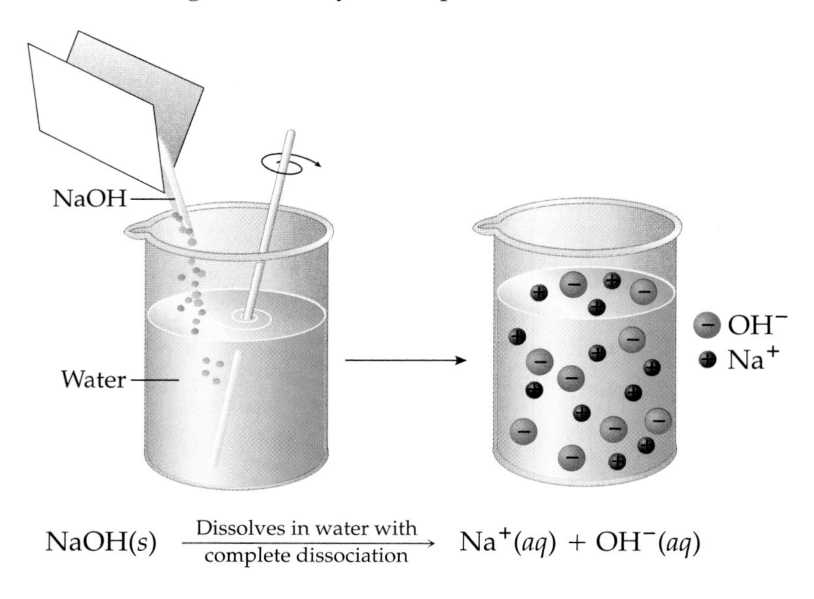

$$NaOH(s) \xrightarrow[\text{complete dissociation}]{\text{Dissolves in water with}} Na^+(aq) + OH^-(aq)$$

We'll take a close look at weak bases later in the chapter.

Seeing bases as providers of hydroxide ions, Arrhenius could even explain the fact that after an acid–base neutralization, the resulting solution is salty. For example, consider what happens when we combine a solution of hydrochloric acid with a solution of sodium hydroxide. The solution of HCl contains Cl^- ions as well as hydronium ions, and the solution of NaOH contains Na^+ ions as well as hydroxide ions. After the OH^- ions have neutralized the H_3O^+ ions, the Na^+ and Cl^- ions are still present. We are left with an aqueous solution of NaCl, a salty solution.

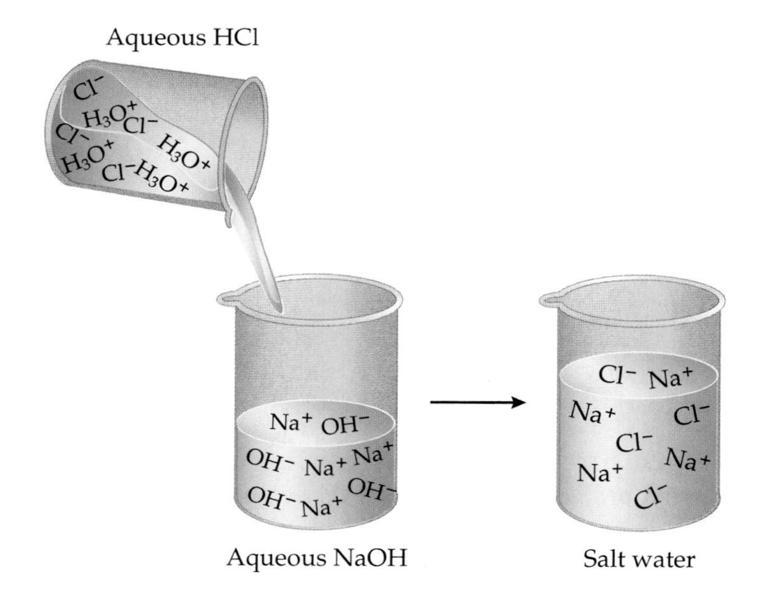

(a) What species are present in each beaker?

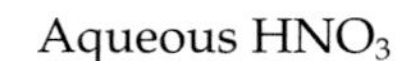

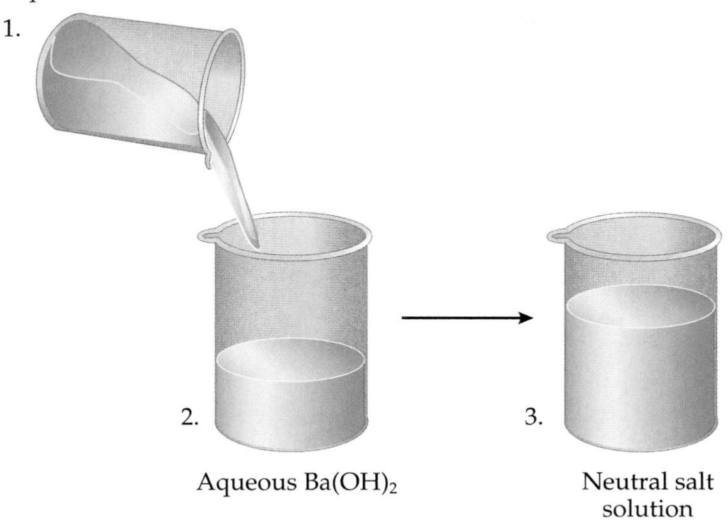

(b) Ionic compounds are often called *salts*. What are the chemical formula and name of the salt that gives rise to the salt solution in beaker 3?

Did you get the correct salt for the WorkPatch? Now try the following practice problems.

Practice Problems

15.7 What salt forms when sulfuric acid completely neutralizes sodium hydroxide? Give both name and formula for the salt.

Answer: Sodium sulfate, Na_2SO_4.

15.8 How many moles of NaOH would it take to neutralize 1 mole of hydrochloric acid? How many moles of NaOH would it take to neutralize 1 mole of sulfuric acid? Why aren't the answers to these two questions the same?

15.9 How many moles of $Ba(OH)_2$ would it take to neutralize 0.1 mole of hydrochloric acid?

15.10 Draw three beakers as shown in WorkPatch 15.3 but change the labels to make them reflect the reaction mentioned in Practice Problem 15.9. Then list all species present in each beaker.

Arrhenius's definition of a base as anything that dissolves in water to give hydroxide ions is still used today. However, it is not the only definition.

15.5 Help! I Need Another Definition of Acid and Base

Arrhenius's definitions are not completely satisfactory. For example, a solution of ammonia dissolved in water turns litmus blue and can be used to neutralize a solution of hydrochloric acid:

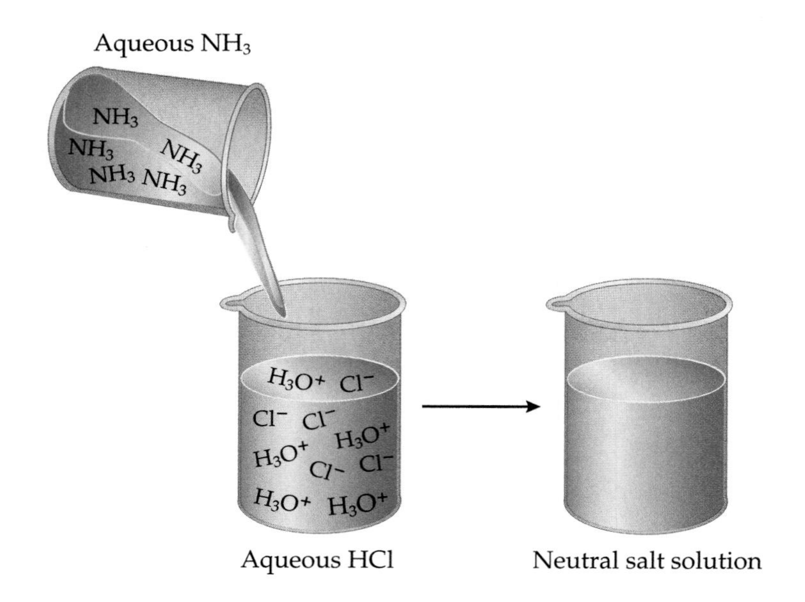

Because ammonia has the properties of a base, it must be a base. We have a problem, however. By the Arrhenius definition, a base is a compound that dissociates in water to produce hydroxide ions. This is not a problem for ionic compounds like $NaOH$ or $Ba(OH)_2$, which are lattices packed with hydroxide ions. The Arrhenius definition of a base fits these compounds like a glove, but Arrhenius would have been hard-pressed to explain the basicity of NH_3. How can this molecule produce OH^- ions in water when it doesn't even have an oxygen atom in it, let alone hydroxide ions? We need a broader definition.

In 1923, working independently of each other, the Danish chemist J. N. Brønsted and the English chemist T. M. Lowry both came up with the answer. Instead of thinking of a base as something that produces hydroxide ions, they defined a base more generally as *any substance that removes H_3O^+ ions from solution*. But they didn't stop there. They went on to detail exactly how this process occurs at the molecular level. To remove an H_3O^+ ion from solution, a molecule of base has to accept a proton (H^+) from the hydronium ion. Remember, a hydronium ion is just a water molecule with a proton attached to it. When a base steals this proton from an H_3O^+, the H_3O^+ is converted back to water, and once again we have an acid–base neutralization reaction:

$$H_3O^+ + NH_3 \longrightarrow H_2O + NH_4^+ \qquad (15.5)$$

Take a very close look at this reaction. When the H moves from H_3O^+ to NH_3, it moves as an H^+ ion (a proton). This means that it leaves behind the two electrons in the O–H bond that initially held it to the oxygen atom in H_3O^+. These two electrons form the second lone pair on oxygen in the resulting H_2O molecule. What comes next is very important. In order for the H^+ to be accepted by a base molecule, the base molecule must have a lone pair of electrons somewhere in it. Remember, it takes two electrons to form a covalent bond between two nuclei. Because H^+ ions have no electrons, a base molecule must therefore have two electrons available with which to form the bond. As you can tell from its Lewis dot diagram, NH_3 does indeed have a lone pair of electrons, on the N atom, and this is why NH_3 is a base.

Which of these three molecules is the only one that can serve as a base according to Brønsted and Lowry's definition? Explain your answer.

(a) CH_4 (b) PH_3 (c) BH_3

From what Equation 15.5 shows—a base removes H_3O^+ ions from a solution—we can reword the Brønsted–Lowry definition of a base to *anything that can accept a proton*. This definition is wonderfully general. It explains why species other than hydroxide ions can be bases, and it even explains why hydroxide ions act as base.

Go back and examine Equation 15.4, where a hydroxide ion reacts with a hydronium ion, and then answer these questions:

(a) Use the Brønsted–Lowry definition of a base to explain why it is proper to call a hydroxide ion a base.
(b) Why is the hydroxide ion able to do what Brønsted and Lowry say a base should do?

Once Brønsted and Lowry defined a base as a proton acceptor, their definition of an acid was fixed. Because an acid is the opposite of a base, an acid must be a proton donor. Certainly the hydronium ion qualifies, as we saw in Reaction 15.5, where H_3O^+ donates a proton to ammonia, and in Reaction 15.4, where H_3O^+ donates a proton to hydroxide. So we now have a more general definition of acids and bases:

Brønsted–Lowry definition of acid—anything that *donates* a proton.
Brønsted–Lowry definition of base—anything that *accepts* a proton.

Anything that **donates** a proton is called an acid.

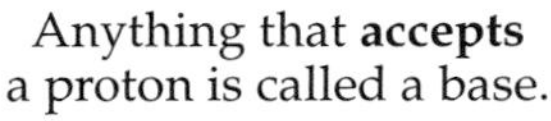
Anything that **accepts** a proton is called a base.

Indeed, these definitions are general enough to explain the concept of acid and base in the absence of water, OH^- ions, and H_3O^+ ions. For example, HCl molecules react with NH_3 molecules in the gas phase (in the absence of water) to give solid ammonium chloride:

$$\underset{\text{Proton donor: acid}}{HCl(g)} + \underset{\text{Proton acceptor: base}}{NH_3(g)} \xrightarrow{} NH_4^+\,Cl^-(s) \qquad (15.6)$$

(Curved arrow labeled H^+ from HCl to NH_3.)

Arrhenius would have a tough time explaining this reaction, but not Brønsted and Lowry. The HCl molecule is the acid because it donates a proton to the NH_3 molecule. The NH_3 molecule is the base because it accepts the proton. In fact, the Brønsted–Lowry definitions allow us to classify as acids and bases substances that we normally don't think of as being either. Water is an example. We usually think of water as being neutral, but when we put HCl into water, the HCl acts as an acid by donating a proton to water. By accepting the proton, water is acting as a base. In this situation, therefore, according to the Brønsted–Lowry definition, water is a base:

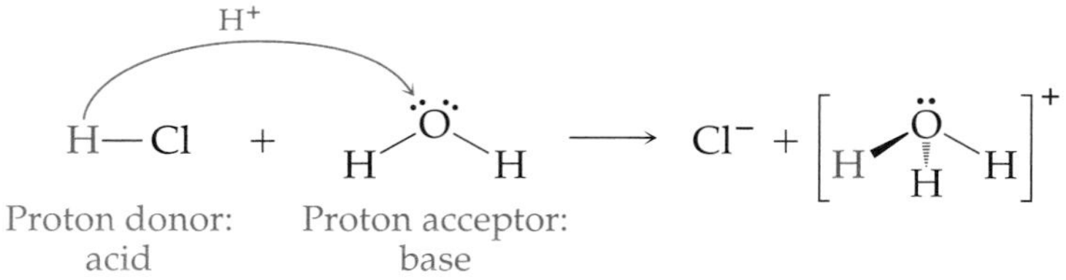

$$\underset{\text{Proton donor: acid}}{H{-}Cl} + \underset{\text{Proton acceptor: base}}{H{-}\ddot{\underset{..}{O}}{-}H} \longrightarrow Cl^- + \left[H_3O\right]^+ \qquad (15.7)$$

(Curved arrow labeled H^+ from H—Cl to the oxygen of water.)

Even chemical reactions that at first appear strange to you can be understood in terms of the Brønsted–Lowry definitions. For example, an important reaction in organic chemistry is that between the amide ion, NH_2^-, and acetylene, C_2H_2, the gas used in high-temperature oxyacetylene torches to cut through steel:

$$\underset{\text{Acetylene}}{C_2H_2} + \underset{\text{Amide ion}}{NH_2^-} \longrightarrow \underset{\text{Acetylide ion}}{C_2H^-} + \underset{\text{Ammonia}}{NH_3}$$

Dot diagrams emphasize what is happening in this reaction:

$$\underset{\text{Acetylene, proton donor: acid}}{H{-}C{\equiv}C{-}H} + \underset{\text{Amide ion, proton acceptor: base}}{H{-}\ddot{N}{:}^-\,(-H)} \longrightarrow \underset{\text{Acetylide ion}}{H{-}C{\equiv}C{:}^-} + \underset{\text{Ammonia}}{H{-}\ddot{N}(H){-}H} \qquad (15.8)$$

(Curved arrow labeled H^+ from acetylene H to the N of the amide ion.)

By accepting a proton from acetylene, the amide ion is acting as a base. By donating the proton, acetylene is acting as an acid. The Brønsted–Lowry definitions allow us to make sense of this unfamiliar reaction by recognizing it as a reaction between an acid and a base. Achieving such an understanding is always one of the most important goals of chemistry.

Practice Problems

15.11 In the reaction

$$NH_3 + PH_3 \longrightarrow NH_2^- + PH_4^+$$

which reactant is the Brønsted–Lowry acid and which is the Brønsted–Lowry base?

Answer: Because it accepts a proton, PH_3 is the base; because it donates the proton, NH_3 is the acid.

15.12 Rewrite the reaction of Practice Problem 15.11 using dot diagrams, and use an arrow to show the proton transfer.

15.13 In Reaction 15.7, water is shown acting as a base. How is water behaving in the following reaction?

$$NH_2^- + H_2O \longrightarrow NH_3 + OH^-$$

Explain your answer.

15.6 Weak Bases

One thing the Brønsted–Lowry definitions do not require is that acids and bases be electrolytes. Recall that Arrhenius said that an acid produces H_3O^+ ions in water and a base produces OH^- ions in water. To produce ions in water, a substance must be an electrolyte, which means that as far as Arrhenius was concerned, acids and bases had to be electrolytes. Neither water nor electrolytes are necessary for a compound to be an acid or a base by the more general Brønsted–Lowry definitions. For instance, Reaction 15.8 does not occur in water, and even if it did, acetylene is a nonelectrolyte. Nevertheless, the Brønsted–Lowry definition has no problem classifying acetylene as an acid in the reaction.

Does this mean we should throw out the Arrhenius definitions? Absolutely not. Much of the chemistry that is done on a daily basis—in a laboratory or in our own bodies—occurs in water, where the Arrhenius definitions are still quite applicable. So, let's go back to aqueous solutions, where a compound has to produce H_3O^+ ions in order to be an acid and has to produce OH^- ions in order to be a base. In other words, a compound has to be an electrolyte.

We have discussed strong acids that are strong electrolytes, such as HCl and H_2SO_4. We have discussed weak acids that are weak electrolytes, such as HF and CH_3COOH. And we have discussed strong bases that are strong electrolytes, such as the ionic compounds NaOH and $Ba(OH)_2$. Are there compounds that are **weak bases** in water—that is, compounds that are weak electrolytes and so produce only a low concentration of OH^- ions in water? The answer is yes. And just like weak acids, weak bases tend to be molecular compounds that produce only a low concentration of ions in water.

One of the most frequently encountered weak bases is ammonia, NH_3. Its aqueous solutions turn litmus blue, a sign of basicity. In addition, ammonia is a weak electrolyte; its aqueous solutions cause the light bulb in an electrolyte tester to glow dimly. By the Arrhenius definition, NH_3 must be producing OH^- ions in solution. Indeed, close examination of an aqueous solution of NH_3 reveals not only dissolved NH_3 molecules but also some OH^- ions. But this brings us back to our earlier problem—ammonia has no OH^- ions in it, so where did they come from? We can help out the Arrhenius interpretation by invoking the Brønsted–Lowry definition. The ammonia molecule can accept a proton from water, thereby converting the H_2O to a hydroxide ion:

$$\underset{\text{Proton donor: water}}{H-\ddot{O}-H} + \underset{\text{Proton acceptor: ammonia}}{:NH_3} \rightleftharpoons \underset{\text{Hydroxide ion}}{[:\ddot{O}-H]^-} + \underset{\text{Ammonium ion}}{[NH_4]^+} \qquad (15.9)$$

(H^+ transferred from water to ammonia)

Because NH_3 produces ions in water, we are justified in calling it an electrolyte. However, the equilibrium lies far to the left ($K_{eq} = 1.8 \times 10^{-5}$). Thus, NH_3 is a weak electrolyte and therefore a weak base. If you place 1 mole of NaOH in 1 L of water in one beaker and 1 mole of NH_3 in 1 L of water in another beaker, the NaOH solution will be much more basic. That's because the beaker of aqueous NaOH contains 1 mole of OH^- ions but the beaker of aqueous NH_3 contains only 0.004 mole of OH^- (and NH_4^+) and almost 1 mole (actually $1.000 - 0.004 = 0.996$ mole) of intact NH_3 molecules. Because ammonia puts much less OH^- into the water than does an equal molar amount of NaOH, ammonia is referred to as a weak base. (As a result of the small amounts of OH^- and NH_4^+ ions present in aqueous solutions of NH_3, we sometimes refer to them as ammonium hydroxide solutions.)

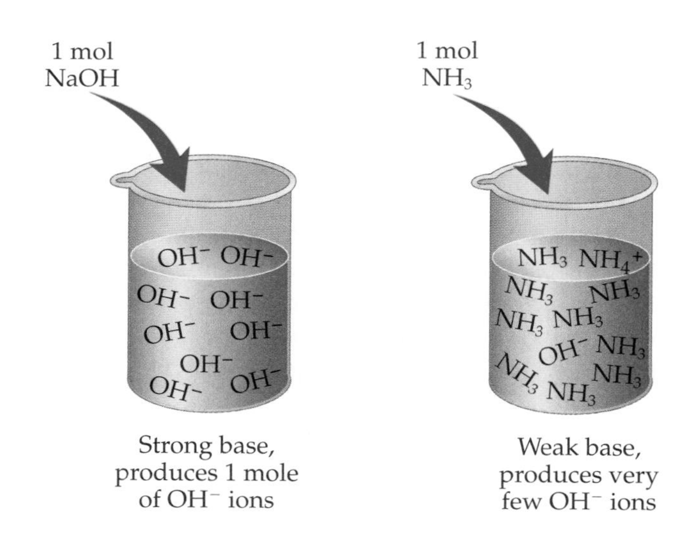

Strong base, produces 1 mole of OH^- ions

Weak base, produces very few OH^- ions

There are other weak bases. Two examples are the carbonate ion, CO_3^{2-}, and the acetate ion, CH_3COO^-:

Carbonate ion, CO_3^{2-} Acetate ion, CH_3COO^-

Like ammonia, carbonate and acetate ions produce small amounts of OH^- in aqueous solution without having any OH^- ions themselves. Demonstrate that you understand why these two ions are bases in water by doing the following WorkPatch.

Write equilibria similar to Reaction 15.9 that show how CO_3^{2-} and CH_3COO^- ions produce small amounts of OH^- ions in aqueous solution.

Were you able to do this WorkPatch? Check your equilibria against ours at the end of the chapter.

Here are the equilibrium constants for the three weak bases we have discussed:

Weak base	K_{eq}
Ammonia, NH_3	1.8×10^{-5}
Acetate ion, CH_3COO^-	5.6×10^{-10}
Carbonate ion, CO_3^{2-}	2.1×10^{-4}

These K_{eq} values go along with the two equilibria you just wrote for WorkPatch 15.6 and with the one we wrote for ammonia in Reaction 15.9. Just as we did for the weak acids, we can judge which of these weak bases is weakest and which is strongest from the K_{eq} values. Because carbonate ion has the largest equilibrium constant, it is the strongest of these three weak bases. Next comes ammonia, and the weakest of the weak—the one whose equilibrium lies farthest to the left—is acetate ion.

You may come across equilibrium constants for weak acids and weak bases represented by the symbols $\boldsymbol{K_a}$ and $\boldsymbol{K_b}$, respectively. For example, the K_b for NH_3 is 1.8×10^{-5}. Whenever you see such a symbol, remember that K_b is just the equilibrium constant for the reaction between a base and water, in which a molecule accepts a proton from water and thereby produces a small amount of OH^-. The same is true for weak acids, where K_a is the equilibrium constant for a reaction in which an acid donates a proton to water, converting a small amount of it to H_3O^+.

Practice Problems

15.14 Aniline, C_6H_7N, is a molecular compound that is a weak base. The molecule has a lone pair of electrons on the N atom.

(a) Why can aniline act as a base?

(b) Write the equilibrium expression that shows how aniline in water makes the water basic. (Be careful about the relative lengths of the two arrows you show in the expression.)

Answer:

(a) Aniline can act as a base because of the lone pair of electrons on the N atom. This lone pair gives the N the ability to accept a proton from an acid:

```
       H       H                                      H       H          +
        \     /                 H+                     \     /         ]
         C—C      H          ⌐                          C—C       H
        //   \\   |         /                          //   \\    |
    H—C        C—N: <--              ——>          H—C        C—N—H
        \\   /    |                                    \\   /     |
         C=C      H                                     C=C       H
        /   \                                          /   \
       H     H                                        H     H
```

(b) $C_6H_7N + H_2O \rightleftharpoons C_6H_7NH^+ + OH^-$

15.15 In the presence of water, the bicarbonate ion, HCO_3^-, can be either a Brønsted–Lowry acid or a Brønsted–Lowry base. Write both equilibrium equations.

15.16 For HCO_3^-, $K_a = 4.8 \times 10^{-11}$ and $K_b = 2.4 \times 10^{-8}$.

(a) Which of the two reactions you wrote in Practice Problem 15.15 has an equilibrium that lies farther to the right? Explain how you know.

(b) Based on your answer to part (a), do you think an aqueous solution of bicarbonate ions would be acidic or basic? Explain your answer. [*Hint:* It will be acidic if the solution contains more H_3O^+ than OH^- and basic if the solution contains more OH^- than H_3O^+.]

15.7 Is this Solution Acidic or Basic? Understanding Water, Autodissociation, and K_w

We have spent a good deal of this chapter talking about what acids and bases are and how they make solutions acidic or basic. What we have yet to do is discuss a way to describe just how acidic or basic a particular solution is. An aqueous solution is described as being **acidic** when it contains more H_3O^+ ions than OH^- ions and **basic** when OH^- ions outnumber H_3O^+ ions.

An **acidic solution** is one in which the concentration of hydronium ions is higher than the concentration of hydroxide ions: $[H_3O^+] > [OH^-]$.

A **basic solution** is one in which the concentration of hydroxide ions is higher than the concentration of hydronium ions: $[OH^-] > [H_3O^+]$.

When we ask how acidic or basic an aqueous solution is, we are really asking how many H_3O^+ or OH^- ions are in the solution. By now you know that when it comes to asking how much of something, chemists answer in moles.

So when you ask how acidic a solution is, a chemist might answer you by reporting the molarity of hydronium ion, meaning how many moles of H_3O^+ ions there are per liter of solution. But that's not enough to let you know if the solution is acidic unless you first know something about neutral solutions.

When we say that a solution is **neutral**, we mean that it is neither acidic nor basic. Pure distilled water is neutral, but does this mean that there are absolutely no H_3O^+ or OH^- ions in the water? The answer is no. In fact, it is impossible to have water that is completely free of these two ions, even if the water is absolutely pure. The purest of water always contains some H_3O^+ and OH^- ions, although not many. For instance, 1 L of pure water at 25°C contains 1.0×10^{-7} mole (0.000 000 10 mole) of H_3O^+ ions and exactly the same amount of OH^- ions. This, in fact, is what it means to be neutral—not an *absence* of acidic H_3O^+ ions and basic OH^- ions but rather an equal number of each. And not just any number, but 1.0×10^{-7} mole of each ion per liter, meaning the concentration of each is 1.0×10^{-7} M.

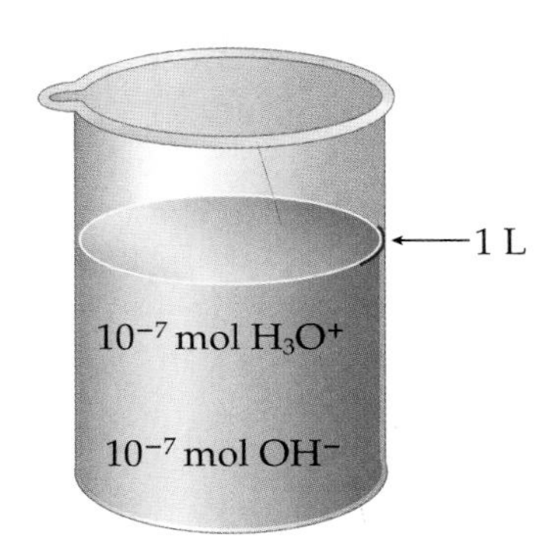

Pure water is neutral, which means it contains equal concentrations of hydronium and hydroxide ions—1×10^{-7} M at 25°C.

Notice that in the drawing we have written 10^{-7} mol instead of 1.0×10^{-7} mol. The number $1.0 \times 10^{\text{power}}$ is the same as the number 10^{power}. (Be warned, though, that we can use this short form only when the number being multiplied by 10^{power} is 1.)

Where do the H_3O^+ and OH^- ions in pure neutral water come from, and why is it impossible to have water without small amounts of these ions present? The answer is that the water molecules dissociate. That is, one water molecule donates a proton to another water molecule. The molecule that donates the proton turns into an OH^- ion, and the molecule that accepts the proton becomes an H_3O^+ ion. The equilibrium equation for this dissociation, often referred to as either the **autodissociation** of water or the **autoionization** of water, is

$$\underset{\text{Proton donor: acid}}{H{-}\ddot{\underset{\cdot\cdot}{O}}{-}H} + \underset{\text{Proton acceptor: base}}{H{-}\ddot{\underset{\cdot\cdot}{O}}{-}H} \rightleftharpoons [:\ddot{\underset{\cdot\cdot}{O}}{-}H]^- + \left[H{-}\ddot{O}(H){-}H\right]^+ \quad (15.10)$$

(H⁺ transferred from the proton donor to the proton acceptor)

$$2\,H_2O \rightleftharpoons OH^- + H_3O^+$$

This equilibrium lies far, far to the left, with only one out of every 555 million water molecules, on average, dissociated at any given time! This is why the concentrations of H_3O^+ and OH^- in pure water are so low. Notice also that water acts as both the acid (proton donor) and the base (proton acceptor) in this reaction. The result is the production of an equal number of H_3O^+ and OH^- ions (10^{-7} mole of each at 25°C) in each liter of neutral pure water.

The fact that the autoionization equilibrium lies so far to the left means that the equilibrium constant has a value much less than 10^{-3}. We can determine the value of this constant because we know that the equilibrium concentrations of H_3O^+ and OH^- in pure water are both 10^{-7} M. First, we write the K_{eq} expression for the balanced equilibrium reaction in the usual way, as products over reactants, each raised to its own stoichiometric coefficient (note that there are two water molecules on the left side of the equilibrium):

$$\underbrace{H_2O + H_2O}_{\text{Reactants}} \rightleftharpoons \underbrace{OH^- + H_3O^+}_{\text{Products}} \qquad K_{eq} = \frac{[OH^-] \times [H_3O^+]}{[H_2O]^2} \qquad \textbf{(15.11)}$$

Next, we simplify the expression. Because this is an aqueous solution, the amount of water present is much higher than the amount of anything else. Pure water is 55.5 M—that is, there are 55.5 moles of water in 1 L of water. This is a number so large that it is not going to change much when we allow a few water molecules to dissociate. So although we can change the concentration of H_3O^+ and OH^- ions in water by adding acid or base, the concentration of H_2O is essentially constant no matter what we do. Therefore, chemists combine the constant $[H_2O]^2$ factor in Equation 15.11 with the equilibrium constant K_{eq} to give a new equilibrium constant $\boldsymbol{K_w}$ (the w stands for water):

$$\underbrace{H_2O + H_2O}_{\text{Reactants}} \rightleftharpoons \underbrace{OH^- + H_3O^+}_{\text{Products}} \qquad \begin{aligned} K_w &= K_{eq} \times [H_2O]^2 \\ &= [OH^-] \times [H_3O^+] \end{aligned}$$

This gives us the result $K_w = [OH^-] \times [H_3O^+]$, an exceptionally important mathematical relationship for water. You should always remember it. In a moment we'll show you why it is so useful, but first we need the numerical value of K_w. To find it, we simply plug in the equilibrium concentrations of H_3O^+ and OH^- in pure water:

$$\begin{aligned} K_w &= [H_3O^+] \times [OH^-] = 10^{-7}\text{ M} \times 10^{-7}\text{ M} \\ &= 10^{-14} \qquad K_w \text{ for autoionization of water at 25°C} \end{aligned}$$

This value of K_w is much less than 10^{-3}, a result that agrees with what we said before, that the autodissociation equilibrium lies far to the left.

Now we are ready to answer the questions "How acidic?" and "How basic?" because we have something to compare our particular solution to—neutral water. Because the OH^- and H_3O^+ concentrations are both 10^{-7} M in neutral water, a solution in which the H_3O^+ concentration is greater than 10^{-7} M is acidic and a solution in which the OH^- concentration is greater than 10^{-7} M is basic:

An **acidic solution** is one in which $[H_3O^+] > 10^{-7}$ M.
A **basic solution** is one in which $[OH^-] > 10^{-7}$ M.

At this point you might be wondering about our earlier definitions of acidic and basic solutions—acidic $[H_3O^+] > [OH^-]$, basic $[OH^-] > [H_3O^+]$. Which definitions are right? They both are because the OH^- and the H_3O^+ concentrations in an aqueous solution are not independent of each other. When one goes up, the other must go down. They can never be equal except for the special case of neutrality, when both are equal to 10^{-7} M. Why? Because of the autodissociation equilibrium and its K_w expression. This equilibrium, which is always occurring, tells us that for any aqueous solution, the H_3O^+ concentration times the OH^- concentration must always equal 10^{-14}. The relationship $K_w = [H_3O^+] \times [OH^-] = 10^{-14}$ is an equilibrium constant for water. It *always* holds true.

For example, suppose we add 1 mole of the strong acid HCl to enough water to make 1 L of aqueous solution. The HCl completely dissociates, producing 1 mole of H_3O^+ ions and making the H_3O^+ concentration 1 M:

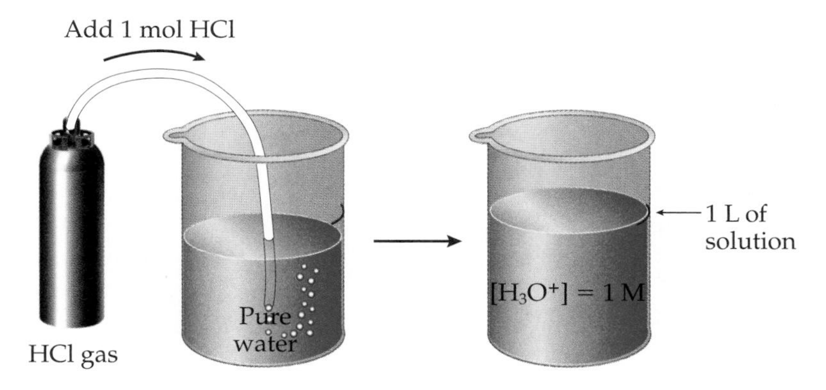

This solution is very acidic because the H_3O^+ concentration is much larger than 10^{-7} M. But is there any OH^- present? Absolutely. There *always* is, no matter how acidic we make the solution, because of the autodissociation equilibrium. How much OH^- is present? The K_w expression tells us.

Determining $[OH^-]$ when you know $[H_3O^+]$

Step 1: Write the K_w expression:
$K_w = 10^{-14} = [H_3O^+] \times [OH^-]$

Step 2: Solve the expression algebraically for $[OH^-]$:

$$[OH^-] = \frac{10^{-14}}{[H_3O^+]}$$

Step 3: Plug in the known concentration of H_3O^+ and do the math. For instance, if $[H_3O^+] = 10^{-3}$, you have

$$[OH^-] = \frac{10^{-14}}{10^{-3}} = 10^{-11}\text{ M}$$

Because $[H_3O^+] = 1$ M in our example and because $[H_3O^+] \times [OH^-]$ must always equal 10^{-14}, the OH^- concentration must be 10^{-14} M in our beaker of acidic solution, as step 3 of the chart shows:

Step 3: Plug in the known concentration of H_3O^+ and do the math:

$$[OH^-] = \frac{10^{-14}}{[H_3O^+]} = \frac{10^{-14}}{1\text{ M}} = 10^{-14}\text{ M}$$

The K_w expression is extremely useful for aqueous solutions. As soon as you know either $[H_3O^+]$ or $[OH^-]$, you can immediately figure out the other one. The solution will be acidic or basic depending on which one is greater than 10^{-7} M. You try it now.

15.7 WORKPATCH

For each solution, determine $[OH^-]$ and decide whether the solution is acidic, basic, or neutral. Arrange your results in tabular form.

Solution	$[H_3O^+]$
1	10^0 (same as 1.0×10^0 M = 1.0 M)
2	10^{-1}
3	10^{-2}
4	10^{-3}
5	10^{-4}
6	10^{-5}
7	10^{-6}
8	10^{-7}
9	10^{-8}
10	10^{-9}
11	10^{-10}
12	10^{-11}
13	10^{-12}
14	10^{-13}
15	10^{-14}

When you did the WorkPatch, were you able to see that as one concentration goes up, the other goes down? The autodissociation equilibrium and the K_w expression are so important for water that we now present the expression one more time along with its important aspects:

$$K_w = [H_3O^+] \times [OH^-] = 10^{-14}$$

- *Both* H_3O^+ ions and OH^- ions are *always* present in any aqueous solution.
- The product of their molar concentrations *always* equals 10^{-14} at 25°C.

Of course, you should be able to solve the K_w expression for concentrations that are not simple. For example, suppose the H_3O^+ concentration in a solution is 2.56 M. What is the OH^- concentration? Is the solution acidic? The answer to the second question is easy. Yes, it's acidic because the H_3O^+ concentration is greater than 10^{-7} M. To find the OH^- concentration, you just plug this H_3O^+ concentration into the equation of step 3:

Step 3: Plug in the known concentration of H_3O^+ and do the math:

$$[OH^-] = \frac{10^{-14}}{[H_3O^+]} = \frac{10^{-14}}{2.56 \text{ M}} = 3.91 \times 10^{-15} \text{ M}$$

If your calculator gave the exponent −14 instead of −15, remember what's implicit in the notation 10^{-14}: it's really 1×10^{-14}, meaning the sequence you should have entered is 1 EXP 14 +/−, not 10 EXP 14 +/−. WorkPatch 15.8 requires you to use the K_w expression to solve for $[H_3O^+]$ and $[OH^-]$. Make sure you can do both.

Determine the unknown concentration in each solution, and then decide whether the solution is acidic or basic. Why did you decide as you did?

(a)

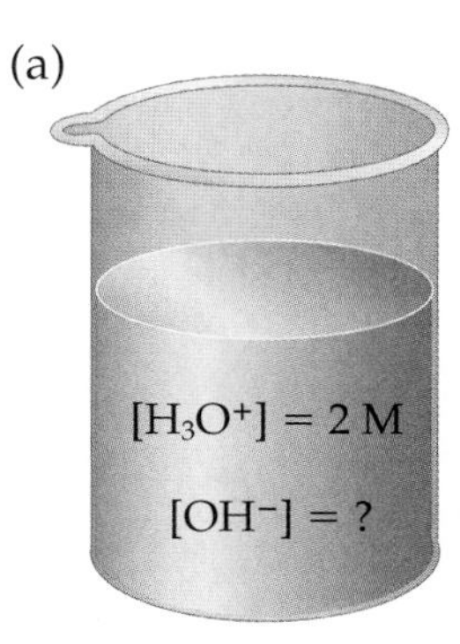

(b)

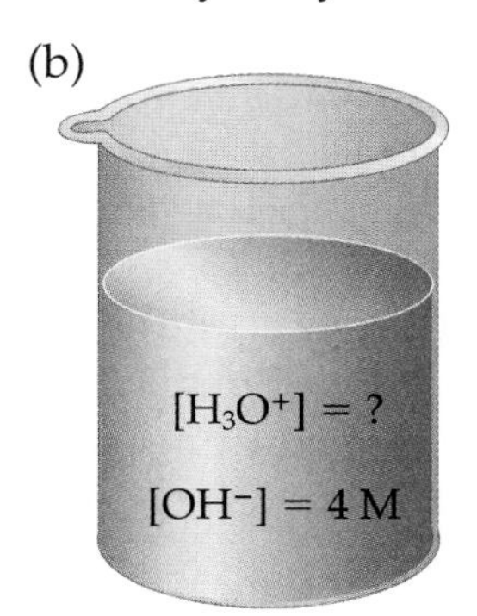

Practice Problems

15.17 In the example we worked above, the concentration of H_3O^+ was 2.56 M and that of OH^- was 3.91×10^{-15} M. Suppose you must multiply these values together but are forbidden to use a calculator or a pencil. What is their product?

Answer: It must be 10^{-14}. The product of $[H_3O^+] \times [OH^-]$ for any aqueous solution is always 10^{-14}.

15.18 The OH^- concentration of an aqueous solution is 0.000 155 M.
(a) Is the solution acidic or basic? Explain how you know.
(b) What is the molar H_3O^+ concentration?

15.19 Suppose 1.00 mole of HCl is dissolved in enough water to give 500.0 mL of solution.
(a) What is the H_3O^+ concentration? [*Hint:* When 1 mole of HCl dissociates, we get 1 mole of H_3O^+ ions because the acid is strong and monoprotic. Your answer should be in moles of H_3O^+ per liter of solution.]
(b) What is the OH^- concentration? [*Hint:* Use the K_w relationship.]

15.20 Suppose 0.0100 mole of $Ba(OH)_2$ is dissolved in enough water to give 500.0 mL of solution.
(a) What is the OH^- concentration? [*Hint:* Remember that every time 1 mole of $Ba(OH)_2$ dissociates, we get 2 moles of OH^- ions.]
(b) What is the H_3O^+ concentration? [*Hint:* Use the K_w relationship.]

15.8 The pH Scale

Let's take stock of where we are. Arrhenius taught us that in water, H_3O^+ is the acidic species and OH^- is the basic species. The fact that water undergoes an autodissociation reaction taught us that both species are always present in water, their concentrations tied to each other via the K_w expression. Only when the water is neutral are the concentrations of H_3O^+ and OH^- the same, 10^{-7} M. If you add acid to neutral water, the H_3O^+ concentration becomes larger than 10^{-7} M and the solution is acidic. If you add base to neutral water, the OH^- concentration becomes larger than 10^{-7} M and the solution is basic. The concentration 10^{-7} M therefore takes on a special significance. It is the *neutral value* against which we must compare the H_3O^+ and OH^- molar concentrations to determine whether a solution is acidic or basic.

As much as chemists are used to molar concentrations, they have devised a special way to indicate how acidic or basic a solution is. Rather than report the molar concentration of H_3O^+ ions, they report the **pH** of a solution:

$$\text{pH} = -\log\,[H_3O^+]$$

Because all scientific calculators have a **log** key, it is easy to punch in a solution's molar H_3O^+ concentration, hit the log key, and then multiply the result by −1. When you do this, you arrive at the solution's pH value. Thus, to get a pH, you do not even have to understand what a log is. You should understand, though, and a brief explanation makes it clear.

The word *log* stands for *logarithm to the base 10*, and we speak of "taking the log of a number." Given any number x, you can enter it into your calculator and press the log key. Your calculator then displays some new number y that is the base-10 logarithm of x. We write this $\log x = y$. Try it for yourself. Enter the number 100 into your calculator, then hit the log key and see what you get. Your calculator should report that the base-10 logarithm of 100 is 2; that is, $\log 100 = 2$.

What does it mean? A base-10 logarithm is an exponent, or power, of 10. In other words, when you take the logarithm of 100, you are really asking your calculator the question "To what power must I raise 10 to get the number 100?" The answer is 2 because $10^2 = 100$. If we ask you for the logarithm of 1000, we are really asking "To what power must you raise 10 to get the number 1000?" You shouldn't even need a calculator for this. You know that $10^3 = 1000$, which means the logarithm of 1000 is 3. Anytime you ask for the base-10 logarithm of a number, you are really asking "To what power must I raise 10 to get that number?" That power is the logarithm of the number.

Practice Problems

Do not use a calculator for the following problems.

15.21 What is the base-10 logarithm of 0.0010?

Answer: We are really asking "To what power must you raise 10 to get 0.0010?" The number 0.0010 can be written as 1.0×10^{-3}. Thus, the logarithm of 0.0010 is −3.

15.22 What is the logarithm of 10,000?

15.23 What is the logarithm of 10? The logarithm of 10^1 is the same. Why?

15.24 What is the logarithm of 0.01? The logarithm of 10^{-2} is the same. Why?

Are you getting a feel for what a logarithm is? It's easy to find the logarithm of numbers that are whole powers of 10, such as the numbers in Practice Problems 15.21–15.24. If such numbers are written in scientific notation, it's even easier. For example, the number 0.000 000 10 can be written either 1.0×10^{-7} or 10^{-7}. The logarithm of 10^{-7} is just -7. After all, when you ask for the logarithm of 10^{-7}, you are really asking "To what power must you raise 10 to get 10^{-7}?"

Practice Problems

Do not use a calculator for the following problems.

15.25 What is the base-10 logarithm of 10^3?

Answer: 3

15.26 What is the logarithm of 1.0×10^{-11} and of 10^{-11}?

15.27 What is the negative logarithm, $-\log$, of 10^{-7}? [*Hint:* The minus sign in $-log$ means you must put a minus sign in front of the number you give for the logarithm.]

You will need to use a calculator when asked for the logarithm of a number that is not a whole power of 10. For example, what is the logarithm of 558? This number being between 100 and 1000, the answer must be a number between 2 and 3 because the logarithm of 100 is 2 ($10^2 = 100$) and that of 1000 is 3 ($10^3 = 1000$). A calculator reveals that the logarithm of 558 is 2.75 (that is, $10^{2.75} = 558$).

Now let's go back to pH. Table 15.4 at the top of the next page lists the results you obtained in WorkPatch 15.7 plus a new column, the pH of the solution. Each row represents a solution having a different H_3O^+ concentration, from highly acidic (1 M H_3O^+) at the top of the table, through neutral, to highly basic (1 M OH^-) at the bottom. The pH column was filled in by taking the negative of the logarithm of the number in the $[H_3O^+]$ column. For example, $-\log 10^0 = -(0) = 0$, $-\log 10^{-1} = -(-1) = 1$, and so on.

Examine the pH column in Table 15.4, and you will learn how pH varies as a solution goes from very acidic, through neutral, to very basic. A neutral solution has pH $= 7$ because its H_3O^+ concentration is 10^{-7} M. A lower pH means

Table 15.4 The pH Scale

$[H_3O^+]$	$[OH^-]$		pH
10^0	10^{-14}	Very acidic	0
10^{-1}	10^{-13}		1
10^{-2}	10^{-12}		2
10^{-3}	10^{-11}		3
		Acidic	
10^{-4}	10^{-10}		4
10^{-5}	10^{-9}		5
10^{-6}	10^{-8}		6
10^{-7}	10^{-7}	◀ Neutral	7
10^{-8}	10^{-6}		8
10^{-9}	10^{-5}		9
10^{-10}	10^{-4}		10
		Basic	
10^{-11}	10^{-3}		11
10^{-12}	10^{-2}		12
10^{-13}	10^{-1}		13
10^{-14}	10^0	Very basic	14

*Remember that $10^0 = 1.0 \times 10^0$ M = 1.0 M, and so forth.

increased acidity, and the pH has to be lower than 7 for a solution to be considered acidic ($[H_3O^+] > [OH^-]$). As the solution becomes increasingly basic, the pH goes up, and the pH has to be higher than 7 for a solution to be considered basic ($[OH^-] > [H_3O^+]$). In addition, because it is a logarithmic scale, each onefold decrease in pH represents a *tenfold* increase in acidity, and each onefold increase in pH represents a *tenfold* decrease in acidity. For example, when the pH decreases from 1 to 0, the H_3O^+ concentration increases from 0.1 M to 1.0 M, a tenfold increase. A solution whose pH = 5 is 100 times more acidic than a solution whose pH = 7 (two pH units = $10 \times 10 = 100$). A solution whose pH = 10 is 1000 times less acidic than a solution whose pH = 7 (three pH units = $10 \times 10 \times 10 = 1000$).

Practice Problems

15.28 Which solution is less acidic, solution A with pH 2 or solution B with pH 6, and by how much?

Answer: Solution B, with the higher pH, is less acidic by 10,000 times. The difference between pH = 6 and pH = 2 is 4, and four pH units = 10^4 = 10,000.

15.29 Basic solution A has pH = 9. Basic solution B is ten times more basic than A. What is the pH of solution B?

15.30 Consider a solution ten times more acidic than the most acidic solution in Table 15.4.
(a) What is the pH of this solution?
(b) What is the molar concentration of H_3O^+ in this solution?

The pH of a solution can be quite important. For example, the pH of your blood is approximately 7.4, just slightly on the basic side of neutral. If the pH should fall below 7.4, a condition called *acidosis* (too much acid) exists. The blood pH of diabetic patients can drop to as low as 6.8. Such severe acidosis can cause coma and death. You wouldn't think that a decrease of just 0.6 of a pH unit could kill you until you remember that the pH scale is logarithmic so that a drop of one pH unit represents a tenfold increase in acidity.

Another area where pH is extremely important is in our environment. Natural rainwater has a pH close to 5.6, acidic because it contains the weak acid carbonic acid, H_2CO_3, formed from atmospheric CO_2. However, the atmospheric pollutants $SO_2(g)$ and $SO_3(g)$, which come from cars and power plants burning the sulfur impurities in gasoline and coal, respectively, form sulfurous acid, H_2SO_3, and sulfuric acid, H_2SO_4. When rain falls in areas with such pollutants in the air, the result is called *acid rain*. In many areas of the industrialized world, scientists have measured rain having a pH value as low as 3! A drop in pH from 5.6 to 3 is almost three pH units, meaning that acid rain is almost 1000 times more acidic than normal rain. The result is often the devastation of forests and the eradication of fish from lakes and streams—often many miles from the source of the pollutants. Environmental agencies devote much time to monitoring the pH of rain, soil, and lakes with a simple device known as a pH meter.

A weakened forest due to acid rain

Once you know the pH of a solution, it is a simple matter to convert that number to molar concentration of H_3O^+. After all, a pH is a logarithm (actually, the negative of a logarithm), and a logarithm is an exponent of 10. Therefore, to convert pH to molar concentration, we simply put a minus sign in front of the pH value and make it an exponent of 10:

The formula for converting pH to molar H_3O^+ concentration:

$$[H_3O^+] = 10^{-(pH)}$$

You can do this with a scientific calculator. Simply enter the pH, multiply it by −1, and then press the 10^x key. For those calculators that do not have this key, depress the INV (inverse) key followed by the log key. For example, the molar H_3O^+ concentration of normal blood is $10^{-7.4}$ (because the pH is 7.4), which gives $[H_3O^+] = 4.00 \times 10^{-8}$ M when entered correctly into your calculator.

Practice Problems

15.31 A basic solution has a pH of 9.8. What is its molar H_3O^+ concentration?

Answer: $[H_3O^+] = 10^{-9.8} = 1.58 \times 10^{-10}$ M

15.32 What is the H_3O^+ concentration in a solution that is 100 times less acidic than one having a pH of 2.56?

15.33 What is the OH^- concentration in a solution having a pH of 5.55? [*Hint:* Use the K_w expression.]

The pH of some common substances are shown below. Note the acidity of your stomach acid!

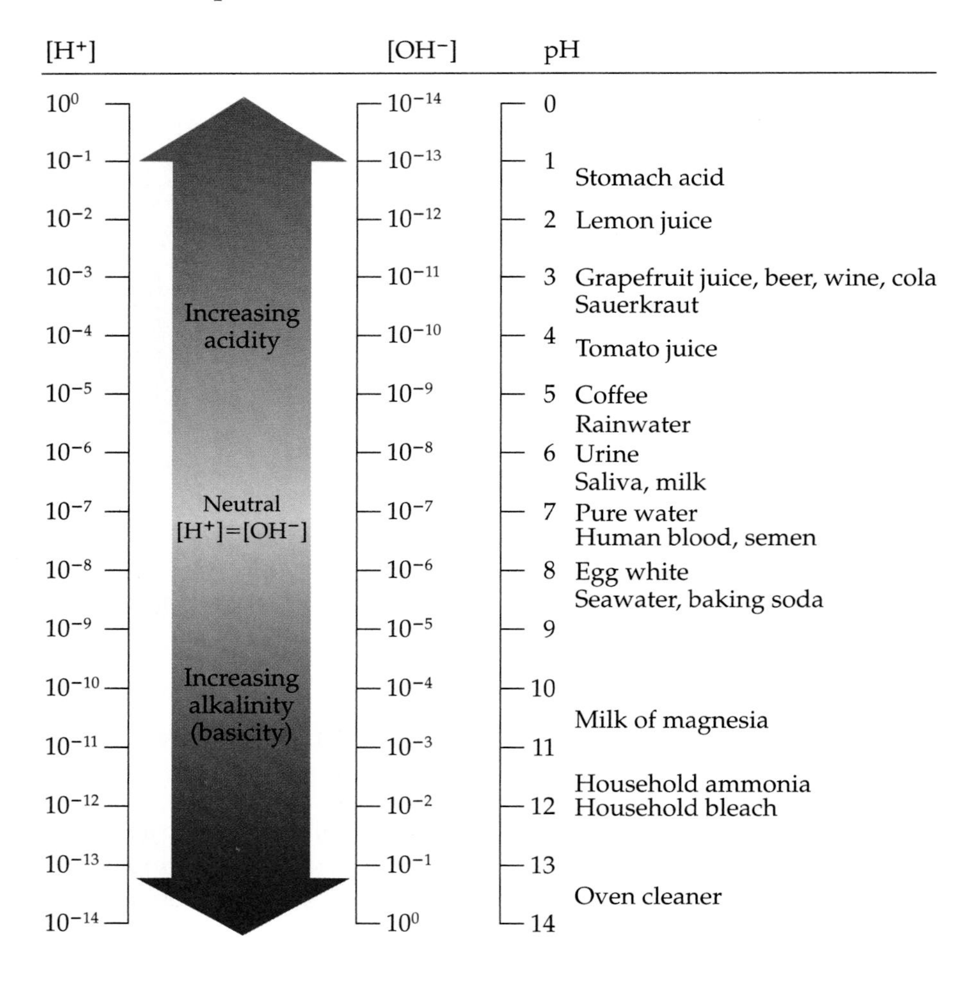

15.9 Resisting pH Changes—Buffers

We just saw that a decrease in blood pH from the normal value of 7.4 down to 6.8 could be fatal. How much acid would it take to cause this drop in pH, and where might that acid come from? Diabetes was mentioned as one source. Another source is your stomach, which produces strong gastric acids having pH values as low as 1.3. Your stomach is lined with specialized acid-resistant cells to prevent this acid from entering your bloodstream. But suppose you should develop a bleeding ulcer or take an aspirin (acetylsalicylic acid) for a headache or catch a stomach flu, all of which could cause some bleeding in your stomach lining. Now your blood is in contact with your stomach acid.

How much of this acid would have to get into your bloodstream to lower your blood pH to a fatal level? Assuming you have 5 L of blood (a typical adult blood volume), a quick calculation shows that it would take only 0.01 mL of stomach acid to do the job. A milliliter is about 10 drops from an eyedropper, meaning that 0.01 mL is only a fraction of a drop! Only a fraction of a drop of your own stomach acid getting into your bloodstream could lower your blood pH from 7.4 to 6.8. So what has been keeping you alive all these years?

The answer is that your blood has the extremely important ability to resist changes to its pH, even upon addition of reasonably large amounts of acids or bases. The ability of a solution to resist changes to its pH is called its **buffering ability**, and such a solution is called either a **buffered solution** or simply a **buffer**. The buffering ability of your blood makes it quite unlike pure water, which has no buffering ability. For example, whereas that fraction of a drop of stomach acid in 5 L of your blood causes almost no change in pH, the same amount added to 5 L of neutral water would decrease its pH from 7.4 to 6.8. If your blood behaved like water, you wouldn't be reading this now.

Chemists have discovered where blood gets its buffering ability from, and in fact, we can buffer any aqueous solution to resist pH changes. In order to understand how a buffer works, you must learn one more thing about acids and bases—the concept of *conjugates*.

If you look up *conjugate* in a dictionary, you will see synonyms such as *couple* and *marry*. Every acid and every base has a conjugate, a "mate" to which it is forever tied. Every acid has a **conjugate base**, and every base has a **conjugate acid**. An acid and its conjugate base make up what is called a **conjugate pair**, and they are called conjugates of each other. A base and its conjugate acid also make up a conjugate pair and are called conjugates of each other.

Conjugate pair

Acid and its **Conjugate base**

Conjugate pair

Base and its **Conjugate acid**

The only difference between the conjugates in a conjugate pair is a proton, H^+. The acid member of the pair has the proton, the basic member does not. This means that an acid turns into its conjugate base by losing a proton, and a base turns into its conjugate acid by gaining a proton. The examples listed in Table 15.5 should make this clear. Notice that the only difference between each acid in the left column and its conjugate base in the right column is a proton. The acid has it, the conjugate base does not.

Table 15.5 Some Weak Acids and Their Conjugate Bases

Weak acid	Conjugate base
Acetic acid, CH_3COOH	Acetate ion, CH_3COO^-
Carbonic acid, H_2CO_3	Bicarbonate ion, HCO_3^-
Hydrofluoric acid, HF	Fluoride ion, F^-
Ammonium ion, NH_4^+	Ammonia, NH_3
Phosphoric acid, H_3PO_4	Dihydrogen phosphate ion, $H_2PO_4^-$
Hypochlorous acid, HClO	Hypochlorite ion, ClO^-

Remove H^+

Remove H^+

What is the conjugate base of citric acid, $C_5H_7O_6OH$? [*Hint:* The last H in the formula is the one donated by this acid.]

Did you get both the formula and the charge of the conjugate base correct? Remember, if a molecule or ion loses something that has a +1 charge (such as an H^+), it must become more negative by −1 so that charge is conserved. Try the next WorkPatch to make sure you understand the difference between conjugates.

Carbonic acid, H_2CO_3, is a weak acid. Carbonate ion, CO_3^{2-}, is a weak base. A student claims that these are conjugates of each other. Is she right or wrong? Explain.

Here's a hint: she is absolutely wrong. Make sure you know why. Check your answer against ours at the end of the chapter.

Table 15.6 lists some weak bases and their conjugate acids. Take a moment to compare Tables 15.5 and 15.6. Do you notice anything similar about them? We hope so. Except for the column headings, the entries are identical. The point is that in any conjugate pair, it is arbitrary which we call the conjugate. Either the acid or the base can be considered the conjugate because they are conjugates of each other. It's up to you.

Table 15.6 Some Weak Bases and Their Conjugate Acids

Weak base	Conjugate acid
Acetate ion, CH_3COO^-	Acetic acid, CH_3COOH
Bicarbonate ion, HCO_3^-	Carbonic acid, H_2CO_3
Fluoride ion, F^-	Hydrofluoric acid, HF
Ammonia, NH_3	Ammonium ion, NH_4^+
Dihydrogen phosphate ion, $H_2PO_4^-$	Phosphoric acid, H_3PO_4
Hypochlorite ion, ClO^-	Hypochlorous acid, HClO

Practice Problems

15.34 (a) Show how HCO_3^- can act as a weak acid.
(b) Show how HCO_3^- can act as a weak base.
(c) Being that HCO_3^- can act as either an acid or a base, can it be its own conjugate?

Answer:
(a) $HCO_3^- + H_2O \rightleftharpoons CO_3^{2-} + H_3O^+$ *(donates proton to water)*
(b) $HCO_3^- + H_2O \rightleftharpoons H_2CO_3 + OH^-$ *(accepts proton from water)*
(c) No, because conjugates must differ by one proton.

15.35 If HCO_3^- is considered a weak acid, what is its conjugate base?

15.36 If HCO_3^- is considered a weak base, what is its conjugate acid?

There is one more thing we need to tell you about conjugates before we go on to how buffers work. If you compare Table 15.5 or 15.6 with Table 15.2B, you will notice that all the acids in Tables 15.5 and 15.6 are weak acids. In addition, there are no strong bases in Tables 15.5 and 15.6, only weak ones. Why are there no strong acids, such as HCl, or strong bases, like NaOH, in these tables? Doesn't hydrochloric acid have a conjugate base? Doesn't sodium hydroxide have a conjugate acid? On paper, yes, they do, but in practice, no, they don't. Consider HCl, a very strong acid. Its conjugate base is just HCl minus a proton, which is the chloride ion:

Strong acid	HCl	They differ by a proton.
Its conjugate base	Cl^-	

So on paper, HCl has a conjugate base. The problem is that chloride ions don't act like a base in water. Solutions of chloride ions are not bitter, they are not slippery to the touch, they do not turn litmus blue, and they contain no excess of hydroxide ions. This behavior is very different from the behavior of the conjugate base of a weak acid, such as acetic acid:

Weak acid	CH_3COOH	They differ by a proton.
Its conjugate base	CH_3COO^-	

If you dissolve acetate ions in water, the water gets moderately basic, telling you that acetate ion is a weak base.

The rule then is as follows: weak acids have conjugates that are bases (and vice versa). However, strong acids and strong bases have conjugates that are so weak they are not bases and acids at all in an Arrhenius sense. Chloride ion, for example, refuses to accept a proton from water. Therefore, no OH^- ions are produced, and a solution of chloride ions is not basic.

$$\underset{\text{Conjugate base of HCl}}{Cl^-} + H_2O \not\rightarrow \underset{\text{Strong acid}}{HCl} + OH^-$$ This reaction does not occur.

This should not surprise you. After all, HCl is a very strong acid, which means it has a tremendous urge to donate its proton. Therefore, the opposite must also be true: Cl^- must have no urge at all to get that proton back.

The same can be said of H_2O, the conjugate acid of OH^-. Hydroxide ion is an extremely strong base, with a tremendous urge to accept a proton. Therefore its conjugate acid, H_2O, must have no real tendency to donate that proton. Thus, although on paper H_2O is the conjugate acid of OH^-, in practice pure water is not acidic at all. So we usually limit the concept of conjugates to weak acids and weak bases. The weak acid acetic acid has a conjugate base that behaves like a base; hydrochloric acid does not. The weak base acetate ion has a conjugate acid that behaves like an acid; the strong base NaOH does not.

Suppose you have a bottle of acetic acid and a bottle of sodium acetate, NaO_2CCH_3, and you add some of each to a beaker of water. Consult Tables 15.5 and 15.6 to help you answer the following questions.

(a) Is there a weak acid in the beaker? If so, what is it?
(b) Is there a weak base in the beaker? If so, what is it?
(c) Is there a conjugate pair in the beaker? If so, what is it?

If you answered the WorkPatch correctly, you are closer than you think to understanding buffers and their ability to resist changes in pH, because the solution described in the WorkPatch is a buffered solution.

When we combine a weak acid and its conjugate base in water, the resulting aqueous solution is a buffered solution. In WorkPatch 15.11, the acetic acid is the weak acid and the acetate ion is the weak base. Of course, they are not just *any* acid and base; they are conjugates of each other. The source of the acetate ions is the sodium acetate, NaO_2CCH_3, an ionic compound that completely dissociates when it dissolves. Because ionic compounds are often referred to as *salts*, we can say that this buffered solution was prepared by dissolving a weak acid and one of its salts. This is a very common way to prepare a buffered solution.

The Na^+ ions in the buffered solution shown here play no role in the solution's buffering ability and are thus often referred to as *spectator ions*, a term we first met in Chapter 7 and again when discussing salt bridges in Section 9.4. It is the presence of the weak acid and its conjugate weak base that gives the solution its buffering ability. Without having a good amount of both, you don't have a buffer. (This last piece of information is the reason an aqueous solution of just acetic acid by itself is not a buffer. The few CH_3COO^- ions created by the dissociation of this weak acid are not enough to give the solution any buffering ability.)

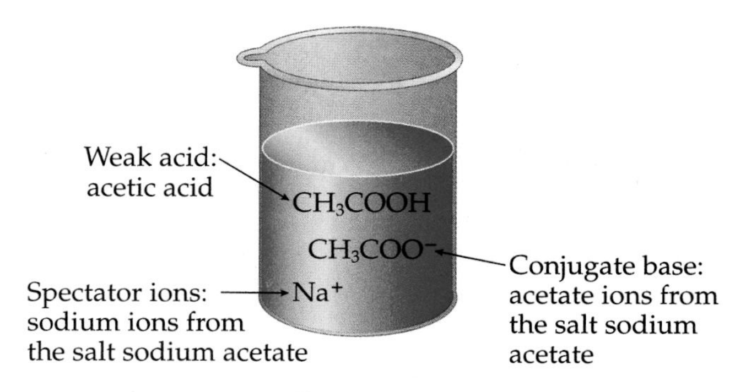

A Buffered Solution

What would you dissolve in water along with the weak acid HF in order to prepare a buffered solution?

Now that you know how to prepare a buffer, the question is, when you add a strong acid like HCl or a strong base like NaOH to a buffered solution, why doesn't the pH of the solution change much? In other words, how does a buffer work? The answer is actually quite simple. A buffer in effect removes the added strong acid or strong base by converting it to a weak acid or weak base, respectively. As a result, the added acid or base has much less effect on the pH of the solution.

A buffer replaces added strong acid with weak acid or added strong base with weak base by sending the appropriate member of its conjugate pair to do battle. Remember, acids and bases react with each other. Lurking in a buffered solution like sharks are a weak acid waiting to react with any added strong base and its conjugate base waiting to react with any added strong acid.

In this acetate buffer, the weak acid acetic acid, CH_3COOH, goes after any added OH^- ions and the weak base acetate ion, CH_3COO^-, goes after any added H_3O^+ ions.

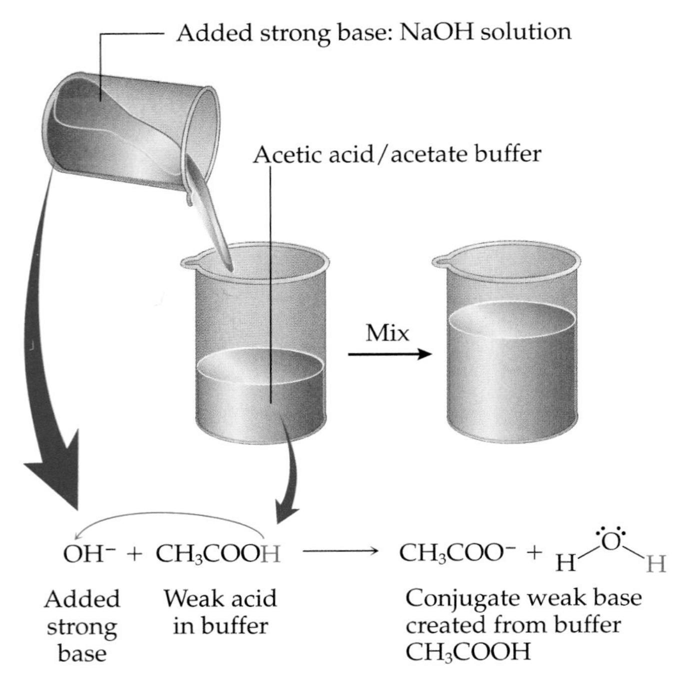

This buffer replaces a strong base with a weak base at the expense of some buffering weak acid.

The acid–base reactions that occur are shown in the drawings. To get rid of added strong base, the buffer sends in its weak acid to donate a proton to the added base. This "kills" the OH^- (turns it into water) and converts the weak acid to its conjugate weak base. The net result is that the strong base OH^- gets replaced by the much weaker base acetate ion. Because OH^- is a strong base, this reaction goes to completion, even though acetic acid is a weak acid. As long as one of the reactants in an acid–base reaction is strong, the reaction goes to completion.

To get rid of added strong acid, the buffer sends in its base to accept a proton from the added acid. This "kills" the H_3O^+ (turns it into water) and converts the base to its conjugate weak acid. The net result is that the strong acid H_3O^+ gets replaced by the much weaker acid acetic acid. This reaction goes to completion even though acetate ion is a weak base because H_3O^+ is a strong acid.

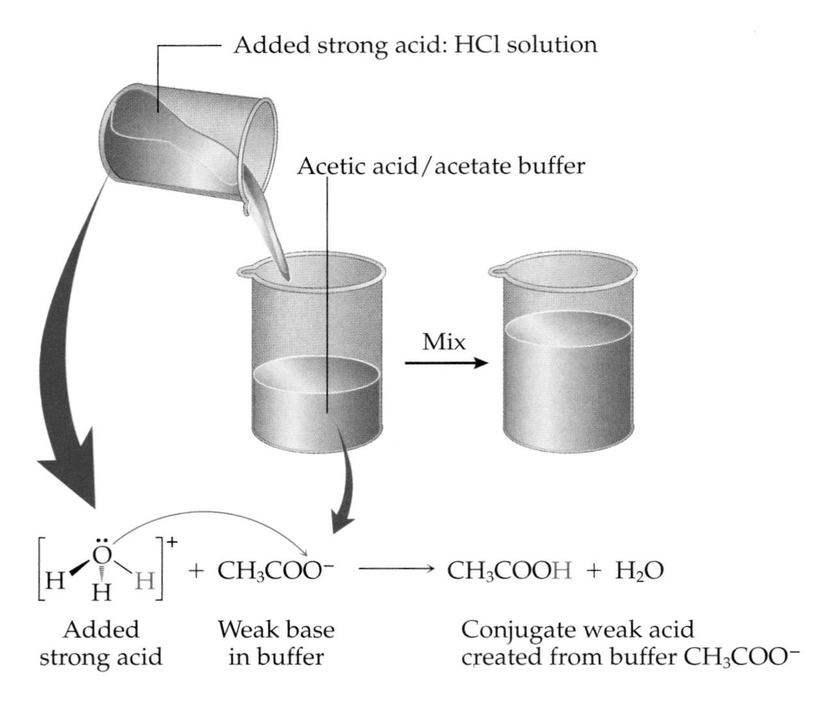

This buffer replaces a strong acid with a weak acid at the expense of some buffering weak base.

(a) How does a weak acid in a buffer eliminate added OH^- ions? To what does it convert these ions? What happens to the weak acid?
(b) How does a weak base in a buffer eliminate added H_3O^+ ions? To what does it convert those ions? What happens to the weak base?

It is essential that you be able to answer this WorkPatch. Examine the whimsical shark acetate buffer drawing two pages back to help you. Each buffer species (shark) always attacks its opposite. Buffer acid goes after added strong base. Buffer base goes after added strong acid. Let's see if you are catching on.

15.14 WORKPATCH

Imagine a buffer made by combining an aqueous solution of the weak acid HF with its sodium salt, NaF.

(a) Write the reaction responsible for replacing any added strong base (OH^-) by a weak base.
(b) Write the reaction responsible for replacing any added strong acid (H_3O^+) by a weak acid.

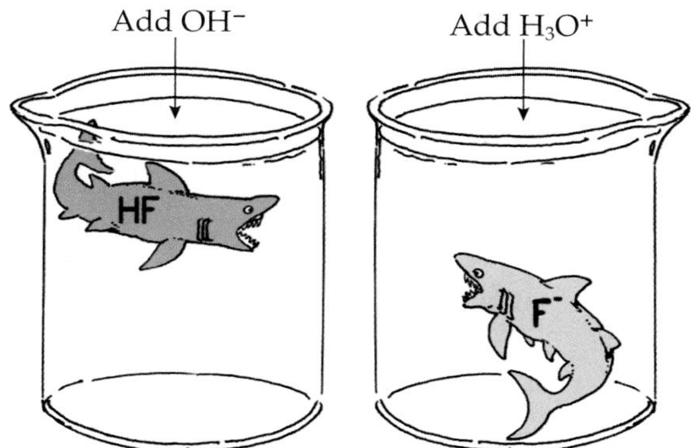

Compare your answers to the WorkPatch against ours, and make sure you understand this WorkPatch completely.

The main buffer in our blood is a HCO_3^-/CO_3^{2-} system that keeps the blood pH at around 7.4. Buffers like this one or the ones discussed above, (CH_3COOH/NaO_2CCH_3 and HF/NaF, keep working until one of their components, either the weak acid or the conjugate weak base, gets used up. If you add enough strong acid to a buffered solution, you will consume all the weak base. If you add enough strong base, you will consume all the weak acid. Once either event occurs, we say either that the buffer *is exhausted* or that its *capacity has been reached*. At that point, the buffer starts acting like plain old water. Therefore, when you prepare a buffered solution, you need to make sure that it contains sufficient amounts of both the weak acid and the salt of the conjugate base.

Finally, make sure you understand that even the best of buffers only *resists* changes in pH when strong acids or bases are added; no buffer entirely cancels out their effect. The pH still goes up when strong base is added, and it still goes down when strong acid is added. It just changes a lot less than it would in pure water. That the pH still changes somewhat should not surprise you. After all, if all the added H_3O^+ were replaced by acetic acid in an acetic acid/acetate buffer, some of the newly produced acetic acid would dissociate to produce H_3O^+ ions, decreasing the pH. Likewise, adding OH^- forms more acetate ion, a weak base, which slightly raises the pH.

Practice Problems

15.37 Ammonium ion, NH_4^+, is a weak acid. Write the equation for its reaction with water.

Answer: $NH_4^+ + H_2O \rightleftharpoons NH_3 + H_3O^+$

Being that the acid is weak, the equilibrium lies far to the left.

15.38 Ammonia, NH_3, is a weak base. Write the equation for its reaction with water.

15.39 What do you call an aqueous solution that contains a large amount of dissolved NH_4Cl and NH_3? Explain.

15.40 Write a reaction to show how the solution in Practice Problem 15.39 resists pH change when OH^- ions are added.

15.41 Write a reaction to show how the solution in Practice Problem 15.39 resists pH change when H_3O^+ ions are added.

If all of this buffer material has given you a headache, you might consider taking a buffered aspirin. Some manufacturers buffer their aspirin (acetylsalicylic acid, a weak acid) in an effort to keep it from damaging your stomach lining. It is debatable whether or not this works, but at least now you can understand the chemistry involved.

Have You Learned This?

Acid (p. 581)
Base (p. 582)
Neutral substance (p. 582)
Electrolyte (pp. 583, 586)
Nonelectrolyte (pp. 583, 586)
Dissociation (p. 585)
Strong electrolyte (p. 589)
Weak electrolyte (p. 589)
Arrhenius definition of acid (p. 591)
Strong acid (p. 591)
Weak acid (p. 591)
Diprotic acid (p. 592)
Monoprotic acid (p. 592)
Triprotic acid (p. 592)
Acid–base neutralization reaction (p. 595)
Arrhenius definition of base (p. 595)
Strong base (p. 596)
Brønsted–Lowry definition of acid (p. 599)
Brønsted–Lowry definition of base (p. 599)
Weak base (p. 602)
K_a and K_b (p. 603)
Acidic solution (pp. 604, 606)
Basic solution (pp. 604, 606)
Neutral solution (p. 605)
Autodissociation (p. 605)
Autoionization (p. 605)
K_w (p. 606)
pH (p. 610)
Buffering ability (p. 615)
Buffered solution (p. 615)
Buffer (p. 615)
Conjugate base (p. 615)
Conjugate acid (p. 615)
Conjugate pair (p. 615)

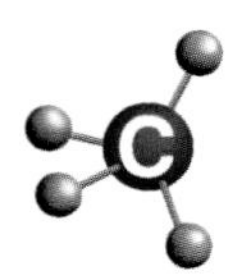

Electrolytes, Acids, and Bases
www.chemistryplace.com

SKILLS TO KNOW

The autodissociation equilibrium constant K_w

In water, autodissociation (autoionization) always occurs:

$$H_2O(l) + H_2O(l) \rightleftharpoons H_3O^+(aq) + OH^-(aq)$$

The equilbrium constant for this reaction, called K_w, has a value of 10^{-14} at 25°C.

The constant K_w provides us with a way to calculate the concentration of H_3O^+ and OH^- both in pure water and in aqueous solutions. First we write the equilibrium expression in the usual way:

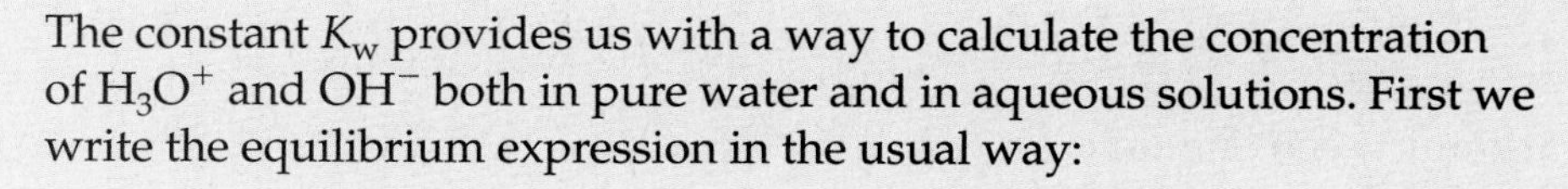

$$K_{eq} = \frac{[H_3O^+] \times [OH^-]}{[H_2O]^2} = 10^{-14}$$

Because the concentration of water, $[H_2O]$, is essentially constant (55.5 M) in any dilute aqueous solution, we can simplify this expression to

$$\underbrace{K_{eq} \times [H_2O]^2}_{K_w} = [H_3O^+] \times [OH^-] = 10^{-14}$$

Therefore, in pure water and in dilute acidic and basic solutions, the *product* of $[H_3O^+]$ and $[OH^-]$ always equals K_w. At 25°C, the numeric value of this product is 10^{-14}:

$$[H_3O^+] \times [OH^-] = K_w = 10^{-14}$$

Using K_w to calculate H_3O^+ and OH^- molar concentrations

The K_w expression allows us to calculate $[H_3O^+]$ if we know $[OH^-]$ and vice versa:

$$[H_3O^+] = \frac{K_w}{[OH^-]} \quad \text{and} \quad [OH^-] = \frac{K_w}{[H_3O^+]}$$

Example: If an aqueous solution has a hydronium ion concentration of 5.00 M, what is its hydroxide ion concentration?

$$[OH^-] = \frac{K_w}{[H_3O^+]} = \frac{10^{-14}}{5.00\ \text{M}} = 2.00 \times 10^{-15}\ \text{M}$$

Acidic solutions, basic solutions, and the pH scale

An aqueous solution is

acidic if $[H_3O^+] > 10^{-7}$ M.

neutral if $[H_3O^+] = [OH^-] = 10^{-7}$ M.

basic if $[H_3O^+] < 10^{-7}$ M.

The pH scale measures how acidic or basic a solution is, and pH is defined in relation to the concentration of H_3O^+:

$$pH = -\log[H_3O^+]$$

Because the pH scale is logarithmic, each one-unit decrease in pH is equal to a tenfold increase in H_3O^+ concentration.

The definition of pH means you can calculate the hydronium ion concentration if you know the pH of a solution:

$$pH = -\log[H_3O^+]$$

so $[H_3O^+] = 10^{-(pH)}$

The hydroxide ion concentration can then be found from the autodissociation equilibrium expression:

$$[H_3O^+] \times [OH^-] = K_w$$

$$[OH^-] = \frac{K_w}{[H_3O^+]}$$

Molar Concentration at 25°C		
$[H_3O^+]$	$[OH^-]$	pH
10^{0}	10^{-14}	0
10^{-1}	10^{-13}	1
10^{-2}	10^{-12}	2
10^{-3}	10^{-11}	3
10^{-4}	10^{-10}	4
10^{-5}	10^{-9}	5
10^{-6}	10^{-8}	6
10^{-7}	10^{-7}	7
10^{-8}	10^{-6}	8
10^{-9}	10^{-5}	9
10^{-10}	10^{-4}	10
10^{-11}	10^{-3}	11
10^{-12}	10^{-2}	12
10^{-13}	10^{-1}	13
10^{-14}	10^{0}	14

Logarithmically increasing acidity

Neutral (pH 7)

Logarithmically increasing basicity

Example: The gastric acid secreted by your stomach has a pH of about 1.5. What are the molar concentrations of H_3O^+ and OH^- in gastric juice at 25°C?

$$[H_3O^+] = 10^{-(pH)}\ M = 10^{-1.5}\ M = 3.2 \times 10^{-2}\ M$$

(Use the 10^x or y^x key on your calculator to compute the answer.)

At 25°C, $K_w = 10^{-14}$. Therefore

$$[OH^-] = \frac{K_w}{[H_3O^+]} = \frac{10^{-14}}{3.2 \times 10^{-2}} = 3.1 \times 10^{-13}\ M$$

Checking the pH chart above, we see that these values for $[H_3O^+]$ and $[OH^-]$ look correct because they correspond to a pH between 1 and 2.

ELECTROLYTES AND NONELECTROLYTES

15.42 List the three criteria early chemists used to classify a compound as an acid.

15.43 List the three criteria early chemists used to classify a compound as a base.

15.44 What is an indicator? Give an example of one.

15.45 What is an electrolyte? What is a nonelectrolyte? Give some examples of each.

15.46 Describe an experimental setup to determine whether a compound is an electrolyte or a nonelectrolyte. What would you look for?

15.47 Ethanol, C_2H_5OH, dissolves in water. So does magnesium chloride, $MgCl_2$. Yet there is a fundamental difference in the way these two substances dissolve. What is this fundamental difference?

15.48 What is meant by the term *dissociation*? Give an example of this process.

15.49 If an ionic compound is water-soluble, it is an electrolyte. Explain why.

15.50 True or false? Because the solid phase of a molecular compound does not consist of ions, the compound cannot dissociate into ions when it dissolves in water. Back up your answer with an explanation and an example.

15.51 What must be present in an aqueous solution in order for it to conduct electricity?

15.52 The molecular compound HCl is an electrolyte.
(a) What do we mean when we say HCl is a molecular compound?
(b) Is it incorrect to call HCl an ionic compound?
(c) What must a molecular compound like HCl do when dissolved in aqueous solution in order to function as an electrolyte?

15.53 The compound NaCl is an electrolyte.
(a) Is it incorrect to call NaCl a molecular compound? Explain.
(b) Why is NaCl expected to be an electrolyte?

15.54 Explain why this statement is true: Water-soluble compounds that consist of one or more metal atoms combined with one or more nonmetal atoms are electrolytes.

15.55 Characterize the following as electrolytes or nonelectrolytes:
(a) Calcium bromide, $CaBr_2$
(b) Bromine, Br_2
(c) Hydrogen, H_2
(d) Hydrogen bromide, HBr [*Hint:* Look at the other elements in group VIIA (17).]
(e) Ammonium fluoride, NH_4F
(f) Potassium chlorate, $KClO_3$
(g) Sucrose, $C_{12}H_{22}O_{11}$

15.56 Write a balanced equation to show what happens to Na_2SO_4 when it dissolves in water. Use the (*aq*) symbol when necessary.

15.57 Write the chemical formula, complete with charge, for the ammonium ion.

15.58 Write a balanced equation to show what happens when $CaBr_2$ dissolves in water. Use the (*aq*) symbol when necessary.

15.59 What is wrong with this instruction: Write a balanced equation showing the sugar glucose, $C_6H_{12}O_6$, dissociating in water.

15.60 What do we mean by saying that some molecular compounds dissolve in water and also dissociate? Give an example of such a compound.

ELECTROLYTES WEAK AND STRONG

15.61 Two beakers each contain 1 L of water. Then 0.1 mole of HCl gas is dissolved in one beaker, and 0.1 mole of HF gas is dissolved in the other beaker. The HCl solution causes a light bulb to glow intensely, and the HF solution causes the same bulb to glow dimly. Explain these observations.

15.62 In which solution of Problem 15.61 is the halide ion concentration 0.1 M, and in which solution is the halide ion concentration much less than 0.1 M? Explain how you know.

15.63 To which side does the dissociation equilibrium lie for a strong electrolyte? For a weak electrolyte?

15.64 What do we mean by the term *partially dissociates*? Give an example of a compound that partially dissociates in water, and draw a picture of the solution showing relative amounts of the various species in the solution.

15.65 What one ion do aqueous solutions of HCl, CH_3COOH, and H_2SO_4 have in common?

15.66 Is it more correct to write $H^+(aq)$ or $H_3O^+(aq)$? Explain.

15.67 A particular molecular electrolyte has a dissociation equilibrium constant of 8.2×10^{-6}. Is this electrolyte strong or weak? Explain.

15.68 A particular molecular electrolyte has a dissociation equilibrium constant of approximately 10^8. Is it appropriate to ignore equilibrium for this electrolyte? Explain.

15.69 Write the formulas and names of two strong molecular electrolytes and two weak molecular electrolytes. For all four, write a balanced dissociation equation.

ACIDS WEAK AND STRONG

15.70 Name these acids:
(a) HCl (b) HNO_3 (c) H_2SO_4
(d) HF (e) CH_3COOH

15.71 Which of the compounds in Problem 15.70 are molecular and which are ionic? Explain how you know.

15.72 Which of the acids in Problem 15.70 are molecular electrolytes, and what common ion do they produce in water?

15.73 According to the Arrhenius definition, why are all the compounds in Problem 15.70 acids?

15.74 True or false? The acidity of a given volume of 1.0 M hydrofluoric acid is the same as that of the same volume of 1.0 M hydrochloric acid. Explain your answer.

15.75 Which of the acids in Problem 15.70 are weak? Which are strong? Suppose you have five beakers, each containing a 1.0 M solution of one the acids of Problem 15.70. What do the terms *weak acid* and *strong acid* tell you about the relative acidities of these solutions?

15.76 What does *diprotic* mean when applied to an acid? Give an example and show both dissociation equilibrium equations.

15.77 Give an example of a triprotic acid, and write down all the dissociation equilibrium equations.

15.78 Other than water, which species would you expect to find in highest concentration in an aqueous solution of CH_3COOH? Explain.

15.79 Other than water, which species would you expect to find in highest concentration in an aqueous solution of HNO_3? Explain.

15.80 Imagine that the eight identical molecules shown below are of a weak molecular acid. Draw a picture showing all the expected ions and/or molecules present after the molecules dissolve in the water in the beaker and equilibrium is reached.

Weak acid

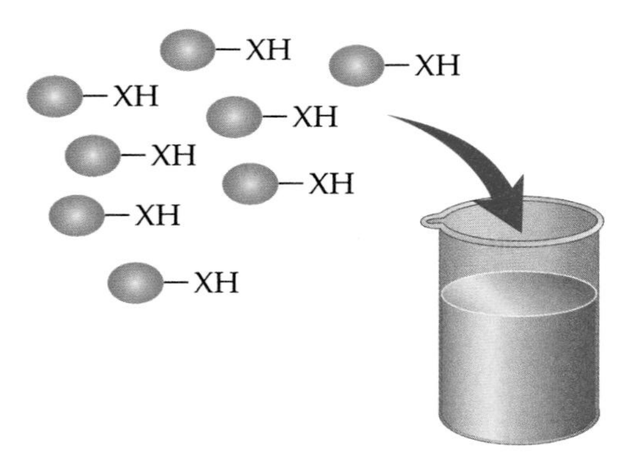

15.81 Answer Problem 15.80 for the case where the eight molecules are of a strong acid.

Strong acid

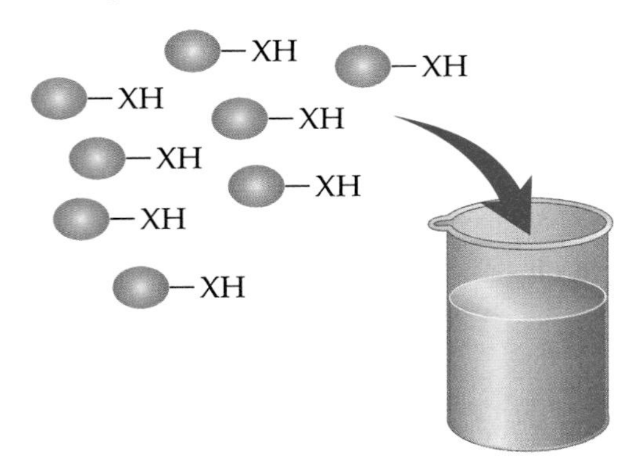

15.82 Draw a dot diagram for acetic acid and indicate which proton is the one that dissociates.

15.83 What is the value of the equilibrium constant for the dissociation of acetic acid's proton? What does this value tell you?

15.84 What is the value of the equilibrium constant for the dissociation of the first proton of sulfuric acid? What does this value tell you?

15.85 True or false? Because carbonic acid, H_2CO_3, is diprotic, it is a stronger acid than HCl, which is only monoprotic. Explain your answer.

BASES—THE OPPOSITE OF ACIDS

15.86 What is the Arrhenius definition of a base?

15.87 What is meant by acid–base neutralization? Write a net ionic equation to go along with your answer.

15.88 What kinds of compounds are typically strong bases?

15.89 When calcium hydroxide, $Ca(OH)_2$, reacts with HCl in a neutralization reaction, the resulting solution is found to conduct electricity very well. Explain.

15.90 Write a complete ionic equation for the reaction between potassium hydroxide and aqueous hydrochloric acid.

15.91 How many moles of $Ba(OH)_2$ does it take to neutralize 0.50 mole of H_2SO_4? Why does the resulting solution after the neutralization no longer conduct electricity? Why does the solution then conduct electricity after an excess of $Ba(OH)_2$ is added?

15.92 How many moles of $Ba(OH)_2$ does it take to neutralize 2.5 moles of HNO_3?

15.93 What salts are formed in the neutralization reactions of Problems 15.91 and 15.92?

15.94 Draw three-beaker sets similar to those in Work-Patch 15.3 for Problems 15.91 and 15.92. Fill in the species present in each beaker. Assume the H_2SO_4 dissociates completely to give two H_3O^+ ions (although it actually doesn't do this).

ANOTHER DEFINITION OF ACID AND BASE

15.95 Write a balanced equation for what happens when gaseous NH_3 is dissolved in water.

15.96 The equilibrium constant for the balanced equation in Problem 15.95 is 1.8×10^{-5}. What does this indicate about which way the equilibrium lies? What are the predominant species in solution?

15.97 Why would Arrhenius have a problem explaining why NH_3 is a base?

15.98 Use the Brønsted–Lowry definition to explain why NH_3 is a base in water.

15.99 According to the Brønsted–Lowry definition, why is it proper to say that water is acting as an acid when ammonia is added to it?

15.100 Solid ammonium chloride, NH_4Cl, reacts with solid sodium hydroxide to produce ammonia gas, water, and sodium chloride, NaCl.
(a) Write a balanced equation for this reaction.
(b) According to the Brønsted–Lowry definition, which species is the acid and which species is the base? Explain.

15.101 The equilibrium constant for the reaction

$$NH_4^+(aq) + H_2O(l) \rightleftharpoons NH_3(aq) + H_3O^+(aq)$$

is 5.6×10^{-10}.
(a) Is a solution of ammonium ion very acidic or only slightly acidic?
(b) Is water acting as an acid or a base according to the Brønsted–Lowry definition? Explain.

15.102 The carbonate ion, CO_3^{2-}, is a weak base.
(a) Write the equilibrium equation that shows how this ion makes water basic.
(b) Which species is the acid and which species is the base according to the Brønsted–Lowry definition?
(c) The equilibrium constant for the reaction between CO_3^{2-} and water is 2.1×10^{-4}. What does this tell you about the strength of CO_3^{2-} as a base?

15.103 Hydride ion, H^-, is an exceptionally strong base, reacting with water to produce lots of hydroxide ion and H_2 gas. The K_{eq} for this reaction is huge.
(a) Write the balanced equation for the reaction between hydride and water.
(b) Explain why H_2 gas forms. [*Hint:* Use the Brønsted–Lowry definition of base.]

15.104 Rewrite the equation of Problem 15.103 using dot diagrams, and show with an arrow the proton being transferred. Note that H^- is an H with a lone pair of electrons on it.

15.105 What is it about the hydride ion that allows it to accept a proton and thereby act as a base?

15.106 If NH_3 can act as a base, why can't CH_4? [*Hint:* Draw dot diagrams.]

15.107 Ethanol, C_2H_5OH, and hydride ion, H^-, react to produce H_2 gas and the $C_2H_5O^-$ anion. Arrhenius could not tell you what is going on, but Brønsted and Lowry would have no trouble. How would they explain this reaction?

WEAK BASES

15.108 How would Brønsted–Lowry define a weak base?

15.109 Why is it proper to call ammonia a weak base?

15.110 Why aren't ionic compounds like NaOH and $Mg(OH)_2$ weak bases?

15.111 Are a 1 M solution of ammonia, NH_3, and a 1 M solution of lithium hydroxide, LiOH, equally basic, or is one solution more basic than the other? Explain.

15.112 Draw a set of beakers similar to those in Work-Patch 15.3 for the solutions in Problem 15.111. Fill in the species present in each beaker.

15.113 In water, the bisulfate ion, HSO_4^-, can act both as a weak base and as a weak acid.
(a) Write an equilibrium equation that shows bisulfate acting as a weak acid in water.
(b) Write an equilibrium equation that shows bisulfate acting as a weak base in water.
(c) What information would you need to help you determine whether a solution of bisulfate was going to be slightly acidic or slightly basic? How would you use that information to find out?

15.114 Amines are organic compounds that contain an NH_2 group, and water-soluble amines are weak bases in water. For example, the compound methylamine, $H_3C–NH_2$, is a weak base.
(a) Draw a dot diagram for methylamine.
(b) Using dot diagrams, show the equilibrium reaction between methylamine and water.
(c) To which side does the equilibrium in part (b) lie? What did we tell you that allowed you to figure out the answer?
(d) The similar compound ethane, $H_3C–CH_3$, does not act as a weak base. Why can methylamine act as a weak base but $H_3C–CH_3$ can't? [*Hint:* Draw a dot diagram for $H_3C–CH_3$.]
(e) Is it appropriate to call methylamine an electrolyte? If so, is it weak or strong? Explain.

15.115 To be a weak base in water, a molecular compound must also be a weak electrolyte. What must be one of the ions it produces in water?

15.116 (a) List all the weak bases phosphoric acid, H_3PO_4, can produce via successive losses of its protons.
(b) Of the bases you listed, which has no ability to serve as a weak acid?

WATER, AUTODISSOCIATION, AND K_w

15.117 What do we mean by the autoionization of water? Write a balanced equation to go along with your explanation.

15.118 An alternate name for the autoionization of water is *autodissociation*. Exactly how is water dissociating?

15.119 Does the equilibrium for water autoionization lie to the left or to the right? What constant verifies your answer?

15.120 Write the mathematical expression that allows you to solve for the OH^- concentration in water when you know only the H_3O^+ concentration.

15.121 True or false? A liter of pure water contains no ions whatsoever. Explain your answer.

15.122 What is the molar concentration of hydronium ion and hydroxide ion in pure water at 25°C?

15.123 If pure water has both hydronium ions (acid) and hydroxide ions (base) in it, how can it be neutral?

15.124 Based solely on concentrations, when is an aqueous solution judged to be acidic? Give two answers to this question.

15.125 Based solely on concentrations, when is an aqueous solution judged to be basic? Give two answers to this question.

15.126 True or false? Even in a very basic aqueous solution, there are some H_3O^+ ions present. Explain your answer.

15.127 True or false? As the H_3O^+ concentration in an aqueous solution increases, the OH^- concentration must decrease. Explain your answer.

15.128 True or false? In an aqueous solution at 25°C, you will always get the same number when you multiply the equilibrium H_3O^+ concentration by the equilibrium OH^- concentration. Explain your answer.

15.129 An aqueous solution has an H_3O^+ concentration of 1.0 M. What is the OH^- concentration? Is this solution acidic or basic? Justify your answer.

15.130 An aqueous solution has an OH^- concentration of 1.0×10^{-11} M. What is the H_3O^+ concentration? Is this solution acidic or basic? Justify your answer.

15.131 A solution is prepared by dissolving 2.50 moles of LiOH in enough water to get 4.00 L of solution. What are the OH^- the H_3O^+ molar concentrations?

15.132 A solution is prepared by dissolving 0.250 mole of $Ba(OH)_2$ in enough water to get 4.00 L of solution. What are the OH^- the H_3O^+ molar concentrations?

15.133 A solution is prepared by dissolving 2.40 g of $Mg(OH)_2$ in enough water to get 4.00 L of solution. What are the OH^- and H_3O^+ molar concentrations? [*Hint:* You need to calculate the molar mass of $Mg(OH)_2$.]

15.134 A solution is prepared by dissolving 2.00 moles of HNO_3 in enough water to get 800.0 mL of solution. What are the H_3O^+ the OH^- molar concentrations?

15.135 A solution has an OH^- concentration of 10^{-11} M.
(a) Is this solution basic or acidic? Explain how you know.
(b) Without using a calculator, explain how you could quickly determine the H_3O^+ concentration of this solution. What is that concentration?

THE pH SCALE

15.136 What is a base-10 logarithm?

15.137 Without using a calculator, what is the base-10 logarithm of 10^{-34}?

15.138 Without using a calculator, what is the base-10 logarithm of 10^{13}?

15.139 Without using a calculator, what is the base-10 logarithm of 10^0? Of 1?

15.140 The base-10 logarithm of 60 is a number
(a) Between −2 and −1 (b) Between −1 and 0
(c) Between 0 and 1 (d) Between 1 and 2
(e) Between 2 and 3

15.141 Explain how you arrived at your answer to Problem 15.140.

15.142 The base-10 logarithm of 0.73 is a number
(a) Between −2 and −1 (b) Between −1 and 0
(c) Between 0 and 1 (d) Between 1 and 2
(e) Between 2 and 3

15.143 Explain how you arrived at your answer to Problem 15.142.

15.144 True or false? As a solution's acidity increases, its pH decreases.

15.145 Write the mathematical definition of pH.

15.146 Why is a pH of 7 equal to neutrality?

15.147 What is the pH of a solution whose hydronium ion concentration is 0.0010 M? Is the solution acidic or basic?

15.148 What is the pH of a solution whose H_3O^+ concentration is 10^{-6} M? Is the solution acidic or basic?

15.149 What is the pH of a solution whose H_3O^+ concentration is 6.40×10^{-9} M? Is the solution acidic or basic?

15.150 What is the pH of a solution whose OH^- concentration is 10^{-14} M? Is the solution acidic or basic?

15.151 What is the pH of a solution whose OH^- concentration is 2.0×10^{-3} M? Is the solution acidic or basic?

15.152 Solution A has a pH of 3. Solution B has a pH of 6. Which solution is more acidic, and by how much?

15.153 What is the pH of a solution whose H_3O^+ concentration is 10.0 M? Is this solution acidic or basic?

15.154 The pH of a solution is 4. What is the H_3O^+ concentration? Is the solution acidic or basic?

15.155 The pH of a solution is 8. What is the OH^- concentration? Is the solution acidic or basic?

15.156 The pH of a solution is -1. What are the H_3O^+ and OH^- concentrations? Is the solution acidic or basic?

15.157 Two students each dissolve 1 mole of acid in enough water to get 1 L of solution. Student A uses acetic acid, and student B uses hydrochloric acid. Do the solutions have the same pH? If not, which has the lower pH? Why should the two pH values be different?

RESISTING pH CHANGES—BUFFERS

15.158 Aniline, $C_6H_5NH_2$, is a weak base, possessing a lone pair of electrons on the nitrogen atom.
(a) According to Brønsted and Lowry, what must aniline do to act as a base?
(b) Why can aniline act as a base?
(c) What are the molecular formula and charge of the conjugate acid of aniline?

15.159 Can a weak acid and its conjugate base ever have the same charge? Explain.

15.160 Knowing that aniline (Problem 15.158) is a weak base, is its conjugate acid a weak acid or a strong acid?

15.161 Nitric acid, HNO_3, is a very strong acid. Solutions of sodium nitrate, $NaNO_3$, contain lots of nitrate ions. Would you expect such a solution to be acidic, basic, or neutral? Explain. [*Hint:* Think about the conjugate base of HNO_3.]

15.162 Why is the conjugate base of a weak acid like acetic acid often referred to as a salt of the acid?

15.163 If Cl^- is the conjugate base of HCl, why isn't an aqueous solution of NaCl acidic?

15.164 What do we mean when we say that a solution has buffering ability?

15.165 Does pure liquid water have any buffering ability? Explain.

15.166 What is the general recipe for making a buffer? Explain the function of each ingredient.

15.167 Name one biological system in which control of pH (buffering ability) is important, and name the buffer that is in control.

15.168 Can a buffer resist pH changes for any added amount of strong acid or base? Explain.

15.169 Does a mixture of carbonic acid, H_2CO_3, a weak acid, and sodium bicarbonate, $NaHCO_3$, in water constitute a buffer? If no, explain why. If yes, explain why and use chemical equations to show what happens when either OH^- or H_3O^+ is added to the solution.

15.170 When a strong acid is added to a buffer, the pH changes a little bit.
(a) Does the pH increase or decrease?
(b) Why does the pH change at all? Why doesn't the buffer hold the pH constant?

15.171 One way to make an acetic acid buffer is to mix substantial amounts of acetic acid and its salt, sodium acetate, in water. Another way to make the same buffer is to add a substantial amount of acetic acid to water and then add half as much NaOH to the water. Explain how and why this latter procedure makes a buffer.

15.172 How would you make a buffer based on ammonia? [*Hint:* Refer to Table 15.5.]

15.173 How would you make a buffer based on hypochlorous acid? [*Hint:* Refer to Table 15.5.]

15.174 It is possible to make two completely different buffers using the dihydrogen phosphate ion, $H_2PO_4^-$.
(a) In one buffer, $H_2PO_4^-$ serves as the weak acid. What is the conjugate weak base?

(b) Write the equations that show how the buffer in part (a) works when either H_3O^+ or OH^- is added.
(c) In the other buffer, $H_2PO_4^-$ serves as the weak base. What is the conjugate weak acid?
(d) Write the equations that show how the buffer in part (c) works when either H_3O^+ or OH^- is added.

15.175 Suppose 2.0 moles of sodium acetate, NaO_2CCH_3, is dissolved in some water and then 1.0 L of a 1.0 M HCl solution is added.
(a) Write the chemical reaction that occurs.
(b) After the reaction, what are the predominant species in solution? How many moles of each species is there?
(c) Is the resulting solution a buffer? If yes, explain why.

15.176 A buffer works by replacing added strong acid with weak acid. Explain how.

15.177 A buffer works by replacing added strong base with weak base. Explain how.

15.178 How does a buffer "kill" added strong base?

15.179 How does a buffer "kill" added strong acid?

15.180 Write the equations that show how a hypochlorous acid buffer defends against added strong acid and base.

ADDITIONAL PROBLEMS

15.181 Perchloric acid, $HClO_4$, is a strong acid.
(a) Write the chemical equation for the reaction between perchloric acid and water.
(b) List all species present in an aqueous solution of perchloric acid in order of concentration, highest to lowest.
(c) Calculate the concentration of all species (except H_2O) present in a 0.100 M $HClO_4$ solution.
(d) What is the pH of this solution?

15.182 Pyridine, C_6H_5N, is a weak base.

Pyridine

(a) Write the chemical equation for the reaction between pyridine and water.
(b) List all species present in an aqueous solution of pyridine in order of concentration, highest to lowest.
(c) In a 0.100 M pyridine solution, 3.2% of the pyridine has reacted with water to form products. Calculate the concentration of all species present (except water) in 1 L of a 0.100 M pyridine solution.
(d) What is the pH of this solution?

15.183 In each of the following pairs, which is the stronger acid?
(a) HPO_4^{2-} and $H_2PO_4^-$
(b) H_2O and H_3O^+
(c) HCN ($K_{eq} = 6.2 \times 10^{-10}$) and HCO_2H ($K_{eq} = 1.8 \times 10^{-4}$)
(d) HI and HF

15.184 How many milliliters of 0.015 M NaOH is needed to neutralize 50.0 mL of 0.010 M $HNO_3(aq)$? Draw a beaker and show all species present after the neutralization. Be sure to include several water molecules in your drawing.

15.185 A solution is formed by mixing 50.0 mL of 0.015 M NaOH(*aq*) with 50.0 mL of 0.010 M $HNO_3(aq)$.
(a) Is the solution acidic or basic?
(b) Draw a beaker and show all species present in the solution.
(c) What is the pH of the solution?

15.186 If 100 mL of an aqueous solution of nitric acid contains 0.030 mole of HNO_3, what is the pH of the solution?

15.187 Write a chemical equation for the reaction between each pair of reactants, using single or double arrows as appropriate:
(a) HNO_3 and OH^-
(b) HF and OH^-
(c) NH_3 and H_2O
(d) HCO_3^- and H_2O

15.188 A 0.20 M aqueous solution of monoprotic acid HX has a pH of 2.14.
(a) Is HX a strong acid or a weak acid?
(b) Calculate K_{eq} for the reaction of HX with water. [*Hint:* What are the equilibrium concentrations of H_3O^+, X^-, and HX?]

15.189 Complete the table:

Substance	Name	Type of electrolyte (strong, weak or nonelectrolyte)	Type of acid/base (strong or weak)	Reaction(s) in water
NaCl			—	(1)
HNO_3				(1)
$Mg(NO_3)_2$			—	(1)
HF				(1)
NaF				(1)
				(2)
NH_4Cl				(1)
				(2)

15.190 Without calculating pH values, list these solutions in order of acidity, from lowest pH to highest pH:
(a) 0.10 M $LiNO_3$, (b) 0.10 M NaF,
(c) 0.10 M KOH, (d) 0.10 M HCN,
(e) 0.10 M HNO_3, (f) 0.20 M HCl.

15.191 How many moles of nitric acid do you need to prepare 200 mL of an aqueous solution that has a pH of 2.0?

15.192 Other than water, what would you expect to find in highest concentration in an aqueous solution of $Ca(OH)_2$? Explain.

15.193 Write a chemical equation for the reaction between each pair of reactants, using single or double arrows as appropriate:
(a) HCO_3^- and OH^- (b) HCl and F^-
(c) H_2CO_3 and OH^- (d) HCN and H_2O

15.194 Other than water, what would you expect to find in highest concentration in an aqueous solution of KOH? Explain.

15.195 What is the pH of these aqueous solutions:
(a) 1.0 M HCl, (b) 0.1 M HCl,
(c) 0.001 M HCl, (d) 1.0×10^{-5} M HCl,
(e) 1.10×10^{-7} M HCl?

15.196 How many moles of $Ca(OH)_2$ would it take to *completely* neutralize 0.4 mole of phosphoric acid, H_3PO_4?

15.197 If 2.00 L of an aqueous solution of sodium hydroxide contain 20.0 g of NaOH, what is the pH of the solution?

15.198 Sulfuric acid, H_2SO_4, is a diprotic acid with dissociation equilibrium constants of $K_{eq} > 1.0 \times 10^3$ and $K_{eq} = 1.2 \times 10^{-2}$. Write the two dissociation equilibrium equations and match the proper K_{eq} to each. Which species is the weak acid?

15.199 What is the pH of a 0.010 M aqueous solution of NaOH?

15.200 How many grams of NaOH do you need to prepare 1.0 L of an aqueous solution that has a pH of 11?

15.201 Which acid and base react to to give an aqueous solution of K_2SO_4? Write a balanced equation for this neutralization reaction.

15.202 The molar concentration of concentrated hydrochloric acid is 12.1 M. What are the pH and OH^- ion molar concentration?

15.203 Rewrite this acid–base reaction in dot-diagram form, then use an arrow to show the proton being transferred and label the Brønsted–Lowry acid and base:

$$F^- + NH_4^+ \rightleftarrows HF + NH_3$$

15.204 Indicate whether each compound is an electrolyte or a nonelectrolyte in water:
(a) N_2 (b) NH_4ClO_4 (c) HI
(d) $CH_3CH_2CH_3$ (propane) (e) Li_3PO_4
(f) $AgNO_3$ (g) CCl_4

15.205 What is wrong with this question: How many grams of NaOH must be added to pure water to give a solution having a pH of 4.20?

15.206 Rearrange the acids in this chart so that they are listed with the weakest acid first and the strongest last:

Name	Formula	K_{eq}
Hypochlorous acid	HClO	2.9×10^{-8}
Nitrous acid	HNO_2	7.2×10^{-4}
Hydrocyanic acid	HCN	6.2×10^{-10}
Formic acid	HCOOH	1.8×10^{-4}

15.207 Hydrazine, H_2NNH_2, is a weak base that can accept two protons. Draw the dot diagram for hydrazine and write equations showing how this compound accepts protons from water in two steps.

15.208 Acrylic acid, C_2H_3COOH, is a weak monoprotic acid. Write a balanced equation for the reaction that occurs when 25.0 mL of 0.200 M KOH is added to 50.0 mL of 0.200 M of C_2H_3COOH. Besides water, what species are present in the solution? Is this solution a buffer? Why or why not?

15.209 How many grams of LiOH are there in 750 mL of an aqueous LiOH solution having an H_3O^+ concentration of 2.30×10^{-13} M?

15.210 Citric acid, $H_3C_6H_5O_7$, is a weak triprotic acid. Write the equations for the three equilibrium reactions for the stepwise dissociation of citric acid in water.

15.211 What is the molar hydroxide ion concentration in a solution that is 1000 times more acidic than a solution that has a pH of 9.20?

15.212 Using dot diagrams, write a reaction showing how the phenoxide ion, $C_6H_5O^-$,

:Ö:⁻
H, C, H
C, C
C, C
H, C, H
H

can act as a weak base in water.

15.213 For each acid–base pair, write a balanced equation for the neutralization reaction:
(a) Lithium hydroxide and hydroiodic acid
(b) Acetic acid and sodium hydroxide

(c) Hydrobromic acid and calcium hydroxide
(d) Potassium hydroxide and phosphoric acid (assume all acidic protons react)

15.214 Which acid and base react to give an aqueous solution of magnesium nitrate? Write a balanced equation for the neutralization reaction.

15.215 Is a solution that contains 0.10 M HBr and 0.10 M NaBr a buffered solution? Why or why not?

15.216 Formic acid, $HCHO_2$, has a K_{eq} of 1.8×10^{-4}, and propionic acid, $HC_3H_5O_2$, has a K_{eq} of 1.3×10^{-5}. Which has a higher pH, a 1 M solution of formic acid or a 1 M solution of propionic acid?

15.217 How many grams of HCl gas are dissolved in 7.50 L of an aqueous HCl solution that has a pH of 2.40?

15.218 Classify each substance as strong electrolyte, weak electrolyte, or nonelectrolyte:
(a) CH_3COOH (b) KCH_3COO
(c) H_2SO_4 (d) CH_2Cl_2 (e) NH_3
(f) H_3PO_4 (g) $ZnSO_4$

15.219 $HClO_4$ is a strong acid, and $HClO_2$ is a weak acid. If you had a 1.0 M solution of $NaClO_4$ and a 1.0 M solution of $NaClO_2$, which would have the higher pH? Explain.

15.220 If you dissolve 0.250 g of $Ba(OH)_2$ in 3.00 L of water, what is the pH of the solution?

15.221 An aqueous solution containing CN^- ions turns litmus blue. Write an equation showing how CN^- makes water basic.

15.222 How many moles of HI are there in 2.75 L of an aqueous HI solution having an OH^- concentration of 8.20×10^{-12} M?

15.223 You mix 500 mL of 1.00 M NaOCl with 500 mL of 0.500 M HNO_3. Write an equation for the reaction that occurs. Besides water, what species are in the solution after reaction? Is this solution a buffer?

15.224 If 0.378 g of HBr dissolved in enough water to make 1.25 L of solution, what is the H_3O^+ concentration? What is the OH^- concentration?

15.225 Rewrite this acid–base reaction in dot-diagram form, then use an arrow to show the proton being transferred and label the Brønsted–Lowry acid and base:

$$HS^- + HOCl \rightleftharpoons H_2S + OCl^-$$

15.226 What is the OH^- concentration of a solution that has a pH of 9.66? Is this solution acidic or basic?

15.227 An ionic compound with the formula NaX (X^- is an unknown anion) is dissolved in water, and the resulting solution is basic. Is HX a strong acid or a weak acid? Explain.

WORKPATCH SOLUTIONS

15.1 (a) NaF is a strong electrolyte because it dissociates completely. HF is a weak electrolyte because very little of it dissociates. F_2 is a nonelectrolyte because it produces no ions when it dissolves.
(b) HF and F_2 are molecular compounds. HF is a polar covalent molecule, and F_2 is a nonpolar covalent molecule. Only NaF is ionic (metal plus nonmetal).

15.2 $H_2SO_4 + 2\,H_2O \longrightarrow 2\,H_3O^+ + SO_4^{2-}$

15.3 (a) Beaker 1: H_3O^+, NO_3^-; beaker 2: Ba^{2+}, OH^-; beaker 3: Ba^{2+}, NO_3^-.
(b) $Ba(NO_3)_2$, barium nitrate. Note that it takes two NO_3^- ions to balance the charge of one Ba^{2+} ion; hence the formula has two nitrates per barium ion.

15.4 PH_3 (phosphine), because it is the only molecule that contains an atom that has a lone pair of electrons.

15.5 (a) It is proper to call OH^- a Brønsted–Lowry base because it accepts a proton from the hydronium ion.
(b) Because it has lone pairs of electrons on the oxygen atom.

15.6 $CO_3^{2-} + H_2O \rightleftharpoons HCO_3^- + OH^-$

$CH_3COO^- + H_2O \rightleftharpoons CH_3COOH + OH^-$

Because both ions are weak bases, these equilibria lie far to the left.

15.7

Solution	$[H_3O^+]$	$[OH^-]$	Acidic, basic, or neutral
1	10^0	10^{-14}	Most acidic
2	10^{-1}	10^{-13}	Acidic
3	10^{-2}	10^{-12}	Acidic
4	10^{-3}	10^{-11}	Acidic
5	10^{-4}	10^{-10}	Acidic
6	10^{-5}	10^{-9}	Acidic
7	10^{-6}	10^{-8}	Acidic
8	10^{-7}	10^{-7}	Neutral
9	10^{-8}	10^{-6}	Basic
10	10^{-9}	10^{-5}	Basic
11	10^{-10}	10^{-4}	Basic
12	10^{-11}	10^{-3}	Basic
13	10^{-12}	10^{-2}	Basic
14	10^{-13}	10^{-1}	Basic
15	10^{-14}	10^0 (1.0 M)	Most basic

15.8 (a) $[OH^-] = \frac{K_w}{[H_3O^+]} = \frac{10^{-14}}{2\ M} = 5 \times 10^{-15}\ M$

This solution is acidic because $[H_3O^+] > 10^{-7}$ M.

(b) $[H_3O^+] = \frac{K_w}{[OH^-]} = \frac{10^{-14}}{4\ M} = 2.5 \times 10^{-15}\ M$

This solution is basic because $[OH^-] > 10^{-7}$ M.

15.9 The conjugate base is citric acid minus a proton: $C_5H_7O_6O^-$, the citrate ion.

15.10 She is wrong. A weak acid and its conjugate base differ by a *single* proton. The difference between H_2CO_3 and CO_3^{2-} is *two* protons:

$H_2CO_3 \longrightarrow 2\,H^+ + CO_3^{2-}$

Thus these two species are not conjugates of each other. (The conjugate base of H_2CO_3 is HCO_3^-.)

15.11 (a) Yes, acetic acid, CH_3COOH.
(b) Yes, acetate ion, CH_3COO^-.
(c) Yes, acetic acid and its conjugate base, the acetate ion.

15.12 Any ionic fluoride salt, such as NaF, LiF, or MgF_2—anything that puts F^- ions into solution.

15.13 (a) The weak acid donates H^+ ions to the added OH^- ions, forming water. The weak acid is converted to its conjugate base.
(b) The weak base takes H^+ ions from the added H_3O^+ ions, forming water. The weak base is converted to its conjugate acid.

15.14 (a) $HF + OH^- \longrightarrow F^- + H_2O$; F^- is a weak base.
(b) $F^- + H_3O^+ \longrightarrow HF + H_2O$; HF is weak acid.

CHAPTER 16

Nuclear Chemistry

16.1 The Case of the Missing Mass—Mass Defect and the Stability of the Nucleus

My name is Otto Bismarck Wahn, and I'm a private investigator. The name is a mouthful, so I go by O. B. on the street. This case was unusual from the start. A woman—she claimed to be a nuclear chemist—showed up at my office late one night. She went on and on about some theft that had apparently occurred in her lab. Seems she and her lab partners had noticed some missing mass while studying the carbon-12 isotope.

"One mole of carbon-12 has a mass of exactly 12 grams," she said.

"Sure, everybody knows that," I said, "so what's your point?" She was starting to get on my nerves.

"I calculate that a mole of carbon-12 should have a mass greater than 12 grams," she said, pointing out that one mole of carbon-12 atoms was made from six moles of protons, six moles of neutrons, and six moles of electrons. I was reaching the limit of my patience.

"Cut to the chase, lady!" She got huffy.

"Don't you get it?" she said. "If we add up the masses of six moles of protons, six moles of neutrons, and six moles of electrons, we get more than 12 grams!"

She proceeded to show me:

What the mass of 1 mole of $^{12}_{6}C$ should be:

$$6 \text{ mol protons} \times \frac{1.007\ 30 \text{ g}}{1 \text{ mol protons}} = 6.043\ 80 \text{ g}$$

$$6 \text{ mol neutrons} \times \frac{1.008\ 70 \text{ g}}{1 \text{ mol neutrons}} = 6.052\ 20 \text{ g}$$

$$6 \text{ mol electrons} \times \frac{0.000\ 55 \text{ g}}{1 \text{ mol electrons}} = 0.003\ 3 \text{ g}$$

$$12.099\ 3 \text{ g}$$

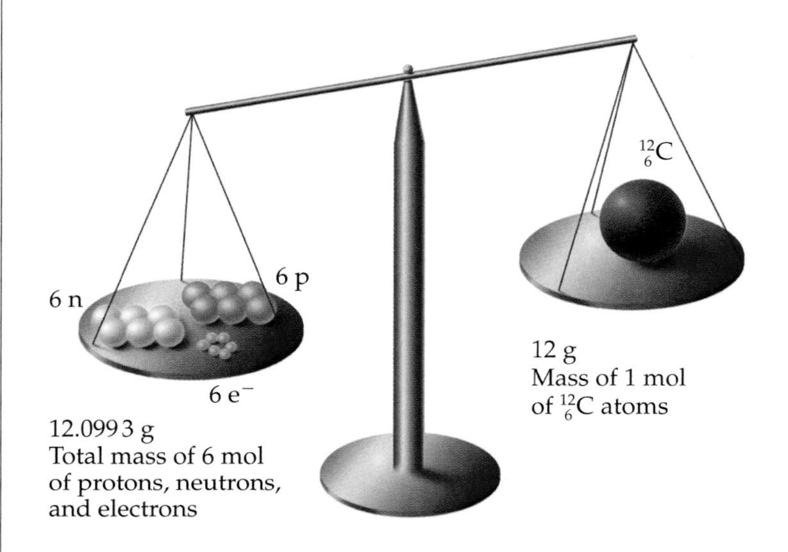

"Do you understand now? I calculate that one mole of carbon-12 atoms should have a mass of 12.0993 grams, but in actuality its mass is only 12 grams. So carbon-12 is lighter than it's supposed to be! Someone has stolen 0.0993 gram from each mole of carbon-12!" She was stumped and desperate. "Help me, O. B. Wahn, you're my only hope!"

Ouch! How could I turn her down? I went to my Rolodex and looked up the number for Professor Albert Einstein. I had used him as a source before. It was no surprise to me that he had already figured this out. The answer involved his famous equation $E = mc^2$, where m is mass and c^2 is the speed of light squared. He said that the missing mass was called the **mass defect** and that all atoms exhibit it to some degree. All atoms are a little bit lighter than they're supposed to be. According to Einstein, the carbon-12 mass defect of 0.0993 gram per mole is converted to energy and released when one mole of the atoms is formed from its subatomic pieces. He even calculated the released energy for me by plugging the values for the molar mass defect and the speed of light into his equation and solving for the energy released:

$$\underbrace{\frac{0.0993 \text{ g}}{\text{mol}} \times \frac{1 \text{ kg}}{1000 \text{ g}}}_{\text{Mass defect in kilograms}} \times \underbrace{\left(\frac{3.00 \times 10^8 \text{ m}}{\text{s}}\right)^2}_{c^2} \times \frac{1 \text{ kJ}}{1000 \text{ J}} = 8.94 \times 10^9 \text{ kJ}$$ of energy released for each mole of carbon-12 atoms formed

Grams cancel, giving kg • m²/s², a unit equivalent to the joule.

"Al, that's incredible!" I said. I realized that 10^9 kilojoules is a billion kilojoules, so he was saying that almost 9 billion kilojoules of energy is released when one mole of carbon-12 is formed from its subatomic particles. Now, I know something about chemistry. Your most highly exothermic chemical reaction might release a few thousand kilojoules per mole of reactant, but he was talking billions of kilojoules!

"Al, do you have a name for all this energy?" I asked, and of course he did.

"The energy released when an atom forms from its subatomic particles is called the atom's **binding energy**," he said. "Think of it as the energy that holds the nucleus together. You would have to put 8.94×10^9 kJ of energy, the amount I just calculated, back *into* a mole of carbon-12 atoms to break all the nuclei apart into separate protons and neutrons. The more binding energy a nucleus has, the more stable it is."

A light went on in my head. "Hold on, Al," I said. I quickly did a similar calculation for carbon-13, the carbon isotope containing six protons, six electrons, and seven neutrons.

What the mass of 1 mole of $^{13}_{6}\text{C}$ should be:

$$6 \text{ mol protons} \times \frac{1.007\,30 \text{ g}}{1 \text{ mol protons}} = 6.043\,80 \text{ g}$$

$$7 \text{ mol neutrons} \times \frac{1.008\,70 \text{ g}}{1 \text{ mol neutrons}} = 7.060\,90 \text{ g}$$

$$6 \text{ mol electrons} \times \frac{0.000\,55 \text{ g}}{1 \text{ mol electrons}} = 0.003\,3 \text{ g}$$

$$13.108\,0 \text{ g}$$

What the mass of 1 mole of $^{13}_{6}\text{C}$ actually is:

13.003 35 g

Mass defect:

13.108 0 g	← What the mass of 1 mole of $^{13}_{6}\text{C}$ should be
−13.003 35 g	← What the mass of 1 mole of $^{13}_{6}\text{C}$ actually is
0.104 65 g	← Missing mass per mole

Binding energy per mole:

$$\frac{0.104\,65 \text{ g}}{1 \text{ mol}} \times \frac{1 \text{ kg}}{1000 \text{ g}} \times \left(\frac{3.00 \times 10^8 \text{ m}}{\text{s}}\right)^2 \times \frac{1 \text{ kJ}}{1000 \text{ J}} = 9.42 \times 10^9 \text{ kJ/mol}$$

"Al, I just calculated the mass defect and binding energy for one mole of carbon-13 and got even bigger numbers than you did for carbon-12. Does that mean the carbon-13 nucleus is even more stable than the carbon-12 nucleus?"

Einstein gave me one of those low chuckles that always got my goat. "O. B. Wahn, my friend, it's not the binding energy per mole of *atoms* you need to look at. The number you need to consider is binding energy *per mole of nucleons*."

He went on to explain that a **nucleon** is the term used to refer to either a proton or a neutron. The lighter carbon-12 atom has 12 nucleons in its nucleus (6 protons and 6 neutrons), whereas a carbon-13 atom has 13 nucleons (6 protons and 7 neutrons). He instructed me to take the total binding energy for one mole of each isotope and divide it by the number of moles of nucleons in the nuclei:

For $^{12}_{6}C$:

$$\left(\frac{8.94\times10^{9}\text{ kJ}}{\text{mol }^{12}\text{C atoms}}\right)\left(\frac{1\text{ mol }^{12}\text{C atoms}}{12\text{ mol nucleons}}\right)=7.45\times10^{8}\text{ kJ of binding energy per mole of nucleons}$$

For $^{13}_{6}C$:

$$\left(\frac{9.42\times10^{9}\text{ kJ}}{\text{mol }^{13}\text{C atoms}}\right)\left(\frac{1\text{ mol }^{13}\text{C atoms}}{13\text{ mol nucleons}}\right)=7.25\times10^{8}\text{ kJ of binding energy per mole of nucleons}$$

"Well, I'll be! It looks like the carbon-12 nucleus is more stable than the carbon-13 nucleus after all because the carbon-12 has more binding energy *per nucleon*. Thanks, professor!" I hung up before I had to listen to another one of his chuckles and turned to my new client. I was about to explain that no one had stolen mass from her carbon-12, but she was already out the door. She had left without saying a word, but she did leave behind a graph on which she had scrawled "Thanks! This makes sense now." Her graph was a plot of binding energy per mole of nucleons versus the number of nucleons in a nucleus:

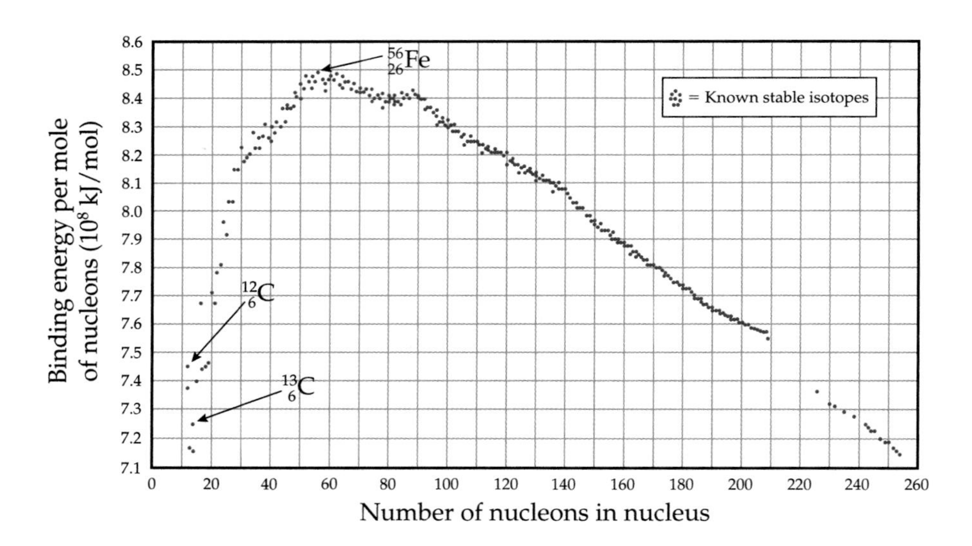

(a) What does the plot say about the stability of the nucleus of the iron isotope $^{56}_{26}Fe$?
(b) What is the total binding energy for the $^{56}_{26}Fe$ nucleus in kilojoules per mole of atoms?

The plot made it clear that while all nuclei have a binding energy, one nucleus is the most stable, and now my client, you, and I know which one that is. Case closed.

The investigation into mass defect and binding energy also solves another mystery—the failure of medieval alchemists to transform base metals like lead

into gold by chemical means. For example, an alchemist might have tried to react lead carbonate, $PbCO_3$, with acid to form carbon dioxide, water, and aqueous gold ions:

$$PbCO_3(s) + 2\,H_3O^+ \not\longrightarrow 3\,H_2O(l) + CO_2(g) + Au^{2+}(aq)$$

Forming gold ions from lead in this manner would be quite a trick. Unfortunately, no chemical reaction has ever accomplished this. If you start with lead, it stays lead. That goes for the hydrogen, oxygen, and carbon as well:

$$PbCO_3(s) + 2\,H_3O^+ \longrightarrow 3\,H_2O(l) + CO_2(g) + Pb^{2+}(aq)$$

In a chemical reaction, the atoms may change partners, but they don't change elemental identity. A chemical reaction simply cannot transform one element into another. The reason for this has to do with energy. Lead has an atomic number of 82, and gold has an atomic number of 79. To turn lead into gold would thus require removing three protons from lead's nucleus. A chemical reaction involves the valence electrons of atoms, redistributing them among atoms, breaking old bonds and forming new ones. The energies associated with redistributing valence electrons are large but not tremendous. That is, the bonds in lead carbonate are strong, but they can be broken with the addition of a reasonable amount of heat. On the other hand, nuclear binding energies are huge—billions of kilojoules per mole of atoms. There is just no way a chemical reaction can produce enough energy to overcome this binding energy and affect a nucleus.

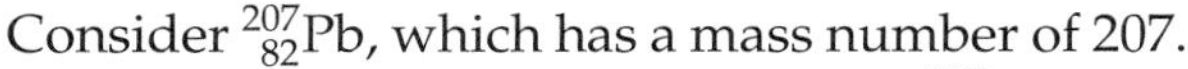

Consider $^{207}_{82}Pb$, which has a mass number of 207.

(a) How many nucleons are in the $^{207}_{82}Pb$ nucleus?

(b) Reading the plot of binding energy per mole of nucleon as best you can, calculate how many kilojoules it would take to remove 3 moles of protons from the atoms in 1 mole of lead-207.

The lead carbonate plus acid reaction releases 1.1 kJ of heat energy per mole of $PbCO_3$. What is the difference between this number and the number you just calculated in the WorkPatch? That is how much the reaction falls short of being able to turn lead into gold. No chemical reaction produces remotely enough energy to cause changes in a nucleus. You might as well try to blow a brick wall down with your breath.

Having said all this, we must now say that some atoms do undergo spontaneous change—without the input of any energy at all. These atoms eject pieces of their nuclei, transforming themselves into new elements. We call these atoms *radioactive*. A **radioactive atom** is one that possesses a nucleus that undergoes spontaneous change. The reasons for spontaneous nuclear change are well beyond the scope of this text—it is not simply a matter of insufficient binding energy. However, scientists have made some empirical observations that can help us predict which nuclei are radioactive and the changes they undergo. Try the following WorkPatch to get ready for what comes next. It reviews some key concepts from Chapter 3.

Three isotopes of carbon have mass numbers 12, 13, and 14.

(a) Write the full atomic symbols for these three isotopes, including mass number and atomic number.

(b) For each isotope, indicate the number of neutrons and protons in the nucleus.

(c) For each isotope, determine the total number of nucleons in the nucleus.

(d) How can all three isotopes be carbon if they have different mass numbers?

Of the three carbon isotopes mentioned in the WorkPatch, only the heaviest one, $^{14}_{6}C$, is radioactive. What kind of change does its nucleus undergo? We shall show you how to answer this question, but don't go on until you feel confident about this WorkPatch, otherwise what comes next will not make sense.

16.2 Half-Life and the Band of Stability

The nuclei of atoms are collections of positively charged protons (p) and uncharged neutrons (n). Nuclei are also tiny. This brings up an interesting question: "Why don't nuclei blow apart?" After all, they have all those positively charged protons jammed into a tiny space, and like charges repel one another. The answer is that the nucleons in a nucleus are glued together by the **strong force**, the strongest force known in nature. This force exists only between pairs of nucleons, and it develops fully only when they are touching. Any time a bunch of nucleons touch one another, the strong force kicks in and gives rise to an attractive force among them. The electrostatic repulsion between pairs of protons still exists, a force that is extremely large when the protons touch each other. The presence of neutrons in an atomic nucleus increases the volume of the nucleus and increases the distance between protons, helping decrease the electrical repulsion among them. For a nucleus to exist at all, the attractive forces must be greater than the repulsive forces. The difference between these two values is the binding energy of the nucleus, which holds the nucleus together.

However, it is possible to have too many or too few neutrons in a nucleus. Here is where empirical observations can help us predict whether a nucleus is radioactive and, if so, how it changes. While studying the composition of the nuclei of all known isotopes, nuclear physicists noticed something about the relative numbers of neutrons and protons in a nucleus. All the known isotopes of the lighter atoms, those with an atomic number between 1 and 20, have nuclei in which the number of neutrons tends to equal the number of protons. For these nuclei, the **neutron-to-proton ratio** is either exactly 1 or close to 1.

Things change for the elements with atomic numbers greater than 20. For these elements, the known isotopes have more neutrons than protons. For example, the most abundant isotope of iron, $^{56}_{26}Fe$, has 26 protons and 30 neutrons, giving it a neutron-to-proton ratio of 1.15. By the time we get to the $^{209}_{83}Bi$ isotope of bismuth, this ratio has increased to 1.52 (126 neutrons/83 protons). In other words, as atoms get heavier, the n/p ratio increases. Evidently, by the time a nucleus has more than 20 protons, it has become so positive that additional neutrons are required to help keep it together. By the time we get to bismuth, with its 83 protons, 1.5 times as many neutrons as protons are required to do the job.

Scientists have plotted the number of neutrons versus the number of protons for all known isotopes. On this plot, as we'll see shortly, the known isotopes form a compact band called the **band of stability**. This n/p plot can be used in predicting whether and how a nucleus changes. To show you how the plot is constructed, we shall use the three carbon isotopes from WorkPatch 16.3:

	$^{12}_{6}C$	$^{13}_{6}C$	$^{14}_{6}C$
Number of protons	6	6	6
Number of neutrons	6	7	8
n/p ratio	1.00	1.17	1.33

To construct our plot, we label a horizontal axis "number of protons" and a vertical axis "number of neutrons," as shown in the figure. First we move along the horizontal axis until we get to 6 because all three of our isotopes have six protons. Then, working on the isotopes one at a time, we move up parallel to the vertical axis until we are alongside the number on the vertical axis that is equal to the number of neutrons in the isotope and mark this point. Any data point that lies on the thick red diagonal line in the graph represents a nucleus with an n/p ratio of 1. The farther a point is above this line, the greater the n/p ratio of the isotope it represents. In other words, points above this line represent nuclei containing more neutrons than protons (these nuclei are *neutron-rich*).

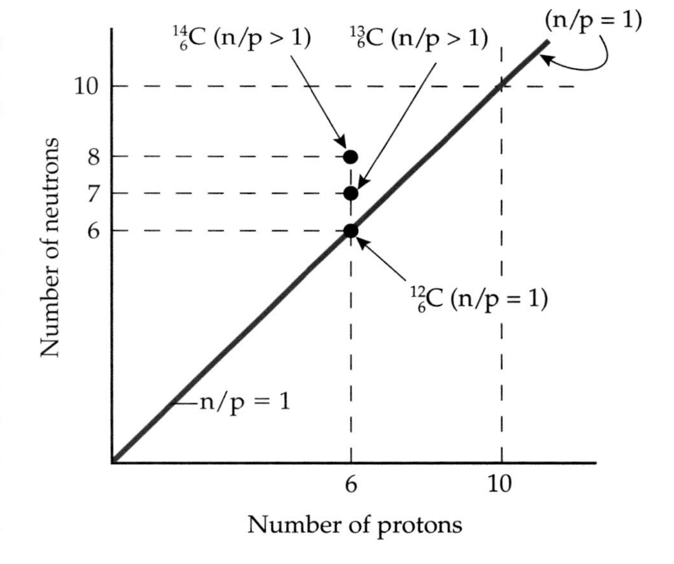

Note that we included $^{14}_{6}C$ in this plot even though its nucleus is radioactive. For the purposes of an n/p plot, the word *stable* includes both nonradioactive isotopes and radioactive isotopes that have a measurable *half-life*. **Half-life** is defined as the time it takes for the amount of radioactive nuclei in a sample of a given isotope to drop to half the initial amount as a result of spontaneous nuclear change. (What the nuclei change into is covered in the next section.) For example, the radioactive isotope of iodine $^{123}_{53}I$ has a half-life of 13.1 h. This means that if you initially have a 10-g sample of $^{123}_{53}I$, you will have only 5 g ($\frac{1}{2} \times 10$ g) left after 13.1 h. After another 13.1 h (two half-lives), only 2.5 g ($\frac{1}{2} \times \frac{1}{2} \times 10$ g) will be left, and so on. Some radioactive isotopes and their half-lives are listed in Table 16.1 at the top of the next page.

Table 16.1 Some Radioactive Isotopes

Element	Radioactive isotope	Half-life
Iodine	$^{123}_{53}I$	13.1 h
Phosphorus	$^{32}_{15}P$	14.28 days
Carbon	$^{14}_{6}C$	5715 years
Uranium	$^{235}_{92}U$	7.04×10^8 years

When scientists plotted all known isotopes the way we just did for carbon-12, carbon-13, and carbon-14, the interesting pattern referred to earlier as the *band of stability* emerged. Any isotope lying in this band is stable. All the isotopes listed in Table 16.1, for instance, lie in the band of stability.

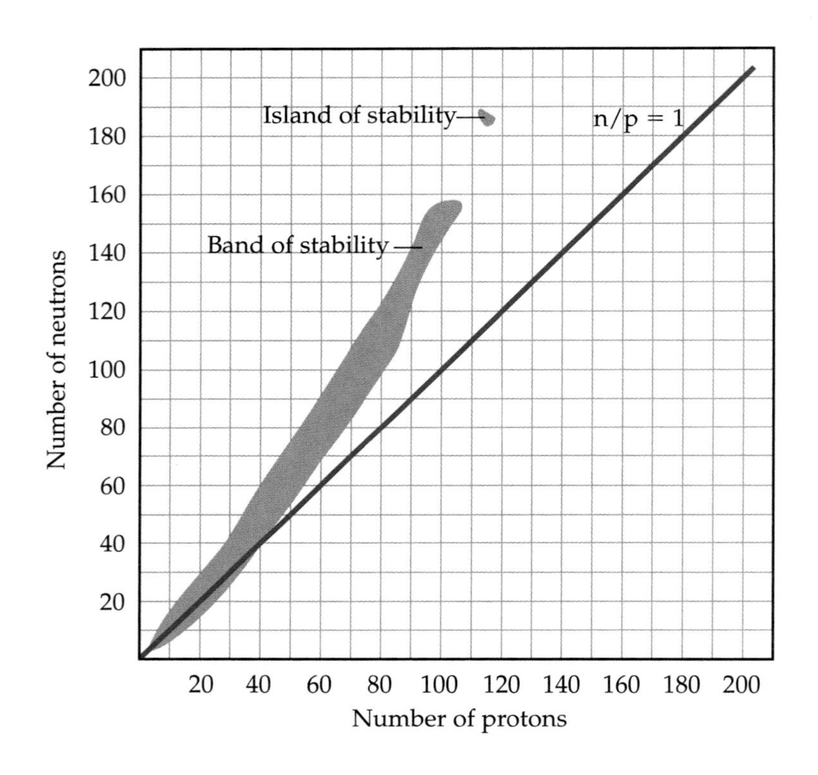

Remember that all the isotopes that lie within the stability band in an n/p plot are either nonradioactive (their nuclei are stable indefinitely) or radioactive with a measurable half-life (their nuclei undergo spontaneous change). The little island at the top of the plot shown here corresponds to superheavy elements that are not found in nature but have been predicted to be stable. Amazingly, in the last few years, scientists actually produced the first element that lies on this island, having 114 protons. The half-life of of this element is less than a millisecond! All points on this plot that lie outside either the band of stability or the island of stability can be thought of as an "ocean of instability." A nucleus composed of the number of neutrons and protons represented by any point in this ocean would be so unstable that its half-life would be immeasurably short.

Practice Problems

16.1 The radioactive isotope of iodine $^{123}_{53}I$ is used to treat thyroid disease. Suppose a patient is given a 30-μg dose. How much will be left in the patient after 39.3 h? (See Table 16.1.)

Answer: Because 39.3 h is three half-lives, there will be 3.75 μg left:

$$30\ \mu g \times \frac{1}{2} \times \frac{1}{2} \times \frac{1}{2} = 3.75\ \mu g$$

16.2 What do we call atoms in which the nuclei undergo spontaneous change?

16.3 Calculate the binding energy per mole (in kilojoules) for $^{4}_{2}He$.

16.4 In an n/p plot, the band of stability curves up above the n/p = 1 line. Explain why.

You are now ready to learn (1) the types of spontaneous changes that radioactive nuclei undergo and (2) how the band of stability can help us predict which type of change occurs in a given nucleus.

16.3 Spontaneous Nuclear Changes—Radioactivity

Let's take a moment to summarize what we have covered so far. All nuclei have a mass defect and considerable binding energy. Nevertheless, only some isotopes have nuclei that are unstable. We call these isotopes radioactive, and their nuclei undergo some sort of spontaneous change at a rate characterized by each isotope's half-life. Although the reason for nuclear instability is beyond the scope of this book, there are some empirical observations involving the neutron-to-proton ratio that can help us predict whether a nucleus is unstable and, if so, how it changes. For the lighter elements, the n/p ratio is 1, but for heavier elements, the n/p ratio increases to 1.5 and above because extra neutrons are needed to hold the nucleus together.

As noted earlier, all existing isotopes of every atom, both nonradioactive isotopes and radioactive ones, lie inside the band of stability. Let us now refine our n/p plot so that we can talk about one particular region inside the band of stability. As the drawing to the right shows, through the interior of the band lies a narrower band that represents all the nonradioactive nuclei. The story is different in the light brown regions above and below this central dark brown region and beyond atomic number 83 (past the "bismuth buoy"). All nuclei that lie within the band of stability but outside the central dark brown region are radioactive. Nuclei that lie in the light brown region above the dark brown band have an n/p ratio that is too large (these nuclei have too many neutrons). One way they could move toward the nonradioactive (dark brown) band would be to convert one or more of their neutrons to protons. Nuclei that lie in the light brown region below the dark brown band have an n/p ratio that is too small (these nuclei have too few neutrons). One way they could move toward the nonradioactive

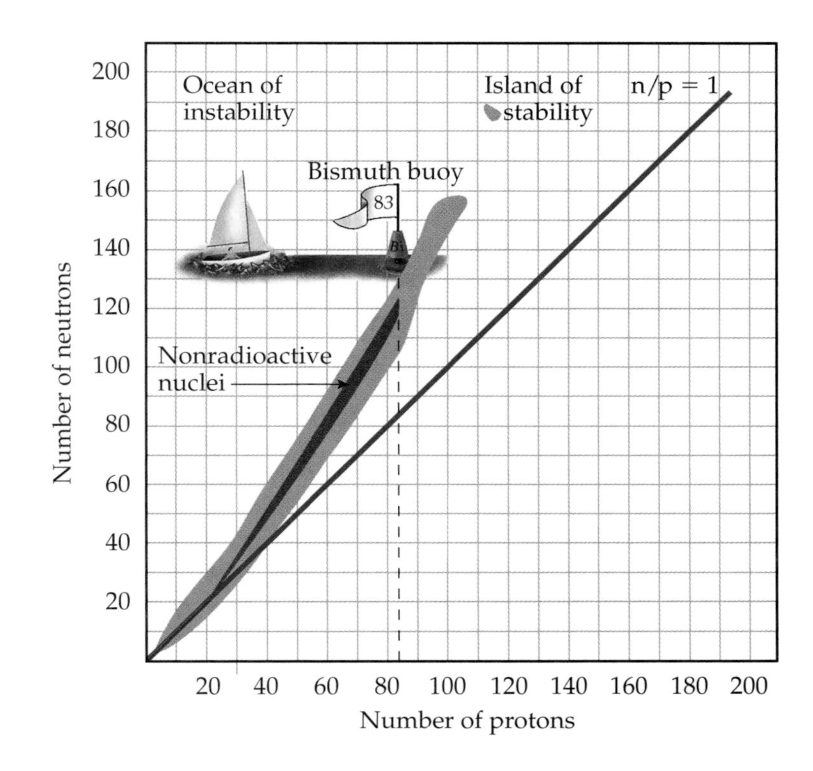

band would be to convert one or more of their protons to neutrons. The far-lying nuclei beyond bismuth are just too big. Evidently, a nucleus with more than 83 protons, if it exists, will be radioactive no matter how many neutrons it has.

Neutrons changing into protons? Protons changing into neutrons? Nuclei losing protons? Believe it or not, these are exactly the processes that spontaneously occur in radioactive nuclei. We call these and other spontaneous nuclear changes **radioactive decay**. The best way to understand these decay processes is to start by reviewing the subatomic particles.

KNOWING YOUR SUBATOMIC PARTICLES

To understand radioactive decay, you need to be familiar with the subatomic particles involved and how they are related to one another. We begin with the neutron and proton, and from this point on, we shall use full symbols for these particles. In other words, we shall include the mass number and atomic number with each symbol, just as we do for atoms:

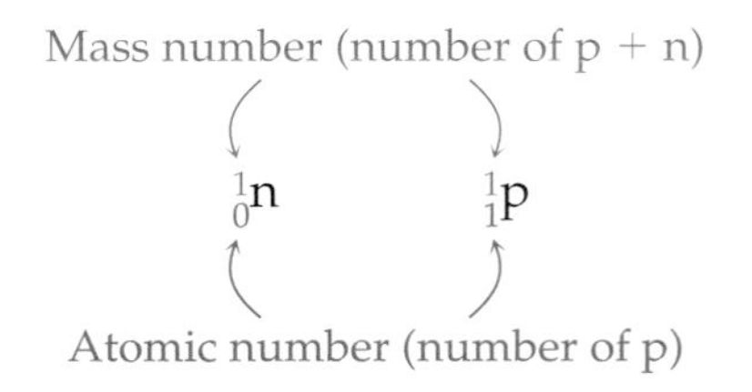

Demonstrate that you understand the full symbols for the neutron and proton by answering the following questions.

(a) Why do neutrons and protons have the same mass number?
(b) Why is the atomic number for the neutron 0?
(c) Why is the atomic number for the proton 1?

Make sure you understand the superscripts and subscripts for these subatomic particles because understanding them is absolutely crucial for understanding nuclear decay processes.

In order to understand how a neutron can turn into a proton or vice versa, we need to consider two additional subatomic particles—the *electron* and the *positron*. You are already familiar with the electron (e^-). Its charge is -1, and it is almost 2000 times lighter than either a proton or a neutron. A **positron** is, in essence, an electron that has a positive ($+1$) charge. Their full symbols are

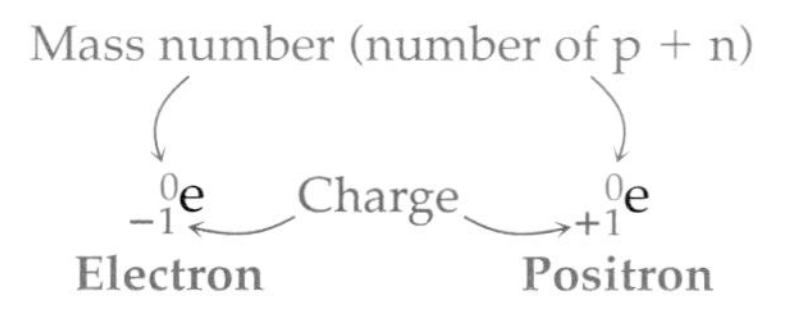

The superscripts (mass numbers) of 0 for both particles should make sense to you, being that both particles contain no protons or neutrons. The subscripts, however, take on a new meaning for these particles. Instead of being the number of protons, the subscripts now represent the electrical charge of the particles.

Positrons are examples of what physicists refer to as **antimatter**, the oppositely charged version of matter. A positron is the antimatter of an electron—we could call the positron an antielectron—identical in size but opposite in charge. If an electron and positron meet, they instantly annihilate each other. This means that their combined mass disappears from the universe and is replaced by its equivalent in energy ($E = mc^2$). Positrons and electrons are the key to understanding how radioactive nuclei convert neutrons to protons and protons to neutrons. As we shall see, large radioactive nuclei undergo these conversions in an attempt to move into the nonradioactive (stable) band.

16.5 WorkPatch

Each of the following symbols has one mistake in it. Identify the mistake, explain why it is a mistake, and fix it:

$$^{0}_{0}\text{n} \qquad ^{1}_{0}\text{p} \qquad ^{1}_{1}\text{e}$$

Is the third symbol in the WorkPatch meant to represent an electron or a positron? How did you know? Check our answer to find out.

Conversion of a Neutron to a Proton—Radioactive Decay via Beta Emission

A neutron changes into a proton when a nucleus ejects an electron. Now, you might ask how a nucleus can eject an electron when the nucleus contains no electrons in the first place. The electron comes into being when the neutron-to-proton conversion occurs. In other words, a neutron turns into a proton *plus* an electron, and the latter is ejected from the nucleus. This conversion is summarized by the equation

$$^{1}_{0}\text{n} \longrightarrow {}^{1}_{1}\text{p} + {}^{0}_{-1}\text{e}$$

We call the ejected electron a **beta particle**, and its symbol is β^-, the Greek letter beta followed by a minus sign superscript. We call the ejection a **beta emission**. (The symbols β^- and $^{0}_{-1}\text{e}$ can be used interchangeably.) Note that charge is conserved in this conversion. That is, the total charge on the right side of the arrow (sum of the subscripts) is equal to the charge (subscript) on the left side. (The same is true for the superscripts because mass is also conserved.) You can use this fact to help you remember that when a neutron converts to a proton, an electron must also be formed.

Because a neutron is turned into a proton, the net effect of beta emission is to increase the atomic number by 1 but leave the mass number unchanged. An example of beta emission is the radioactive decay of $^{14}_{6}\text{C}$. Look at the figure at right and you will see a $^{14}_{6}\text{C}$ nucleus undergoing beta decay. We can summarize this event by the equation

$$^{14}_{6}\text{C} \longrightarrow {}^{14}_{7}\text{N} + {}^{0}_{-1}\text{e}$$

Beta Emission

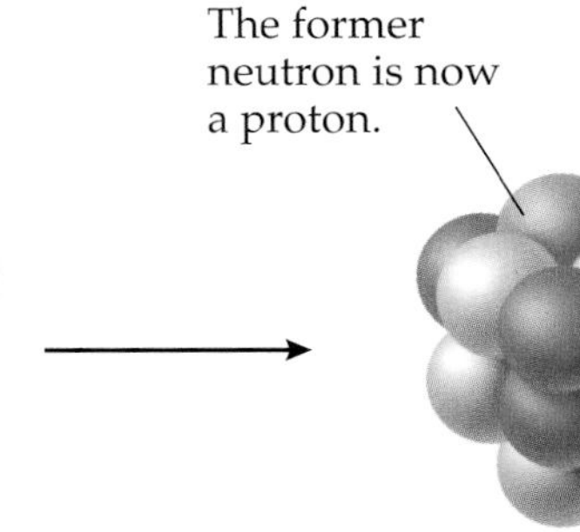
An electron (beta particle) flies out of nucleus.
$^{0}_{-1}\text{e}$
Watch this neutron.
The former neutron is now a proton.
The $^{14}_{6}\text{C}$ nucleus is neutron-rich and radioactive.
The $^{14}_{7}\text{N}$ nucleus has one fewer neutron and one more proton and is nonradioactive.

This equation is called a **nuclear reaction** because it represents a change in the atom's nucleus. The $^{14}_{7}N$ formed is called a **daughter isotope** (or **daughter nucleus**), defined as one that results from a nuclear decay process. The $^{14}_{6}C$ is referred to as the **parent isotope** because it "gave birth" to the daughter. In this case, the $^{14}_{6}C$ parent lies on the upper (radioactive) part of the light brown band of stability and the $^{14}_{7}N$ daughter lies in the dark brown nonradioactive central region.

Before going on, we want to show you an easy way to use the superscripts and subscripts to check that a nuclear reaction you've written is balanced. The sum of the superscripts on the left side of the reaction must equal the sum of the superscripts on the right side of the reaction. The same is true for the subscripts. Notice how this works for our $^{14}_{6}C$ beta emission:

$$^{14}_{6}C \longrightarrow {}^{14}_{7}N + {}^{0}_{-1}e$$

$$14 + 0 = 14$$

$$7 + (-1) = 6$$

By using this method of balancing, you can predict the product of a nuclear reaction. Try it now in the next WorkPatch. First, determine the value of the superscript and subscript question marks. Once you have determined the subscript (the atomic number), you'll know which elemental symbol to use.

Balance this nuclear reaction and predict the product:

$$^{214}_{82}Pb \longrightarrow {}^{?}_{?}? + \underset{\beta^- \text{ particle}}{{}^{0}_{-1}e}$$

Do you understand why the mass number didn't change but the elemental identity did? Remember that beta emission tends to occur in nuclei that have too many neutrons (the light brown part of the band of stability lying above the dark brown central region).

Practice Problems

16.5 What would the answer to WorkPatch 16.6 be if there were another beta emission?

Answer: $^{214}_{83}Bi \longrightarrow {}^{214}_{84}Po + {}^{0}_{-1}e$

16.6 Suppose $^{35}_{14}Si$ undergoes beta emission. Write a nuclear reaction for this spontaneous change.

16.7 Suppose $^{40}_{20}Ca$ is the product of a beta-emission nuclear reaction. What was the parent isotope?

RADIOACTIVE DECAY VIA POSITRON EMISSION AND ELECTRON CAPTURE

Now let's consider radioactive isotopes that require additional neutrons to become stable (these are found in the part of the light brown band of stability that lies below the dark brown central region). These isotopes tend to decay by converting a proton to a neutron. They can do this in one of two ways: *positron emission* or *electron capture*.

A proton becomes a neutron when a nucleus ejects a positron via a process known as **positron emission**.

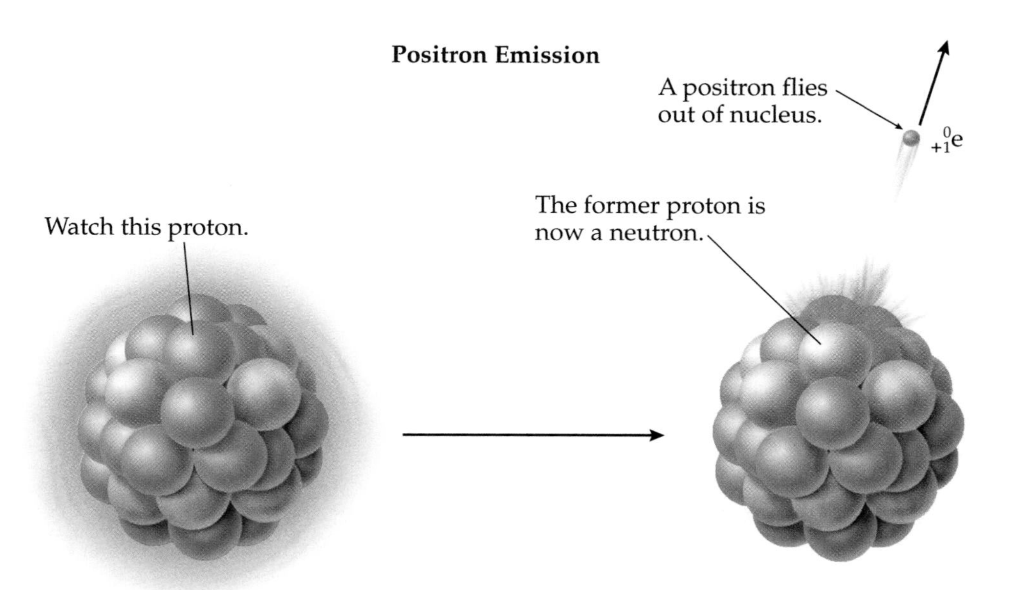
Positron Emission
A positron flies out of nucleus.
$_{+1}^{0}e$
Watch this proton.
The former proton is now a neutron.

The parent nucleus is neutron-poor and radioactive.

The daughter nucleus has one more neutron and one fewer proton.

The positron comes into being when the proton-to-neutron conversion occurs. In other words, a proton turns into a neutron *plus* a positron, and the latter is ejected from the nucleus. This conversion is summarized by the nuclear reaction

$$_{1}^{1}p \longrightarrow {}_{0}^{1}n + {}_{+1}^{0}e$$

Once again, conservation of mass and charge can help you remember this decay process. If a proton turns into a neutron, a particle with positive charge, the positron, must be formed.

Another way to represent a positron is β^{+}, although we shall use the full $_{+1}^{0}e$ symbol when writing nuclear reactions. An example of positron emission is shown here. Notice how the reaction is properly balanced.

$$40 + 0 = 40$$

$$_{19}^{40}K \longrightarrow {}_{18}^{40}Ar + {}_{+1}^{0}e$$

$$18 + 1 = 19$$

Besides positron emission, a proton can convert to a neutron via the radioactive decay process known as *electron capture*. In **electron capture**, the nucleus of an atom absorbs, or captures, one of the atom's inner-shell electrons. This captured electron then combines with one of the protons in the nucleus to create a neutron:

$$_{-1}^{0}e + {}_{+1}^{1}p \longrightarrow {}_{0}^{1}n$$

Electron Capture

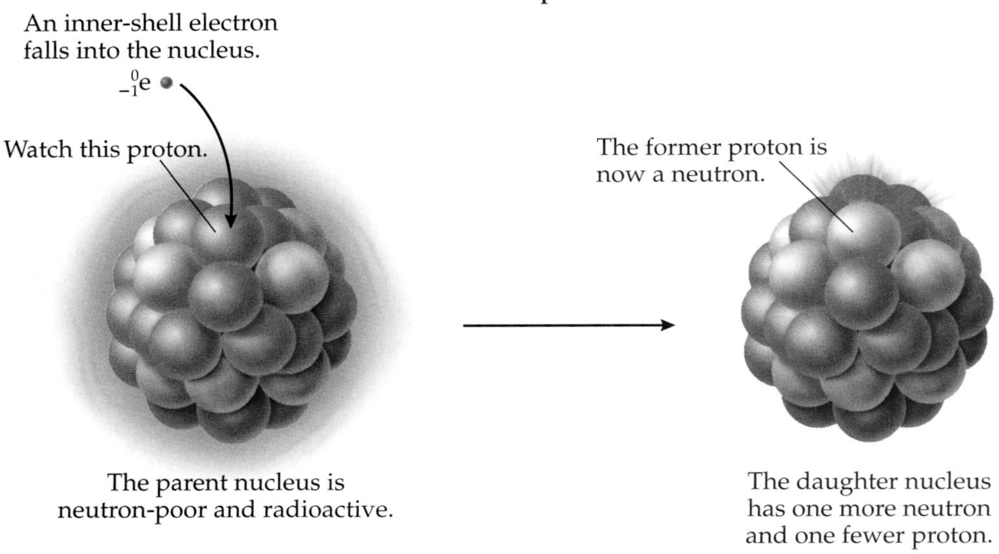

An inner-shell electron falls into the nucleus.

$_{-1}^{0}e$

Watch this proton.

The former proton is now a neutron.

The parent nucleus is neutron-poor and radioactive.

The daughter nucleus has one more neutron and one fewer proton.

Unlike proton-to-neutron conversion, which occurs by positron emission, this time a particle is captured by the nucleus instead of ejected from it. This conversion is summarized by the equation

$$_{1}^{1}p + _{-1}^{0}e \longrightarrow _{0}^{1}n$$

Once again, check the superscripts and subscripts to see that this nuclear reaction is balanced. Electron capture has the effect of decreasing the atomic number by 1 while leaving the mass number unchanged. Note that this is the same result as in positron emission. An example of electron capture is

$$197 + 0 = 197$$

$$_{80}^{197}Hg + _{-1}^{0}e \longrightarrow _{79}^{197}Au$$

$$80 + (-1) = 79$$

Both positron emission and electron capture tend to occur in radioactive isotopes that need to convert a proton to a neutron (on the radioactive lower part of the band of stability). Which decay mode occurs depends on the particular radioactive isotope.

Practice Problems

16.8 The fluorine isotope $_{9}^{17}F$ undergoes positron emission.
(a) Write a nuclear reaction for this emission.
(b) Where on the band of stability would you expect to find $_{9}^{17}F$?

Answer:
(a) $_{9}^{17}F \longrightarrow _{+1}^{0}e + _{8}^{17}O$
(b) Below the dark brown central band.

16.9 The argon isotope $^{37}_{18}Ar$ undergoes electron capture.
 (a) Write a nuclear reaction for this process.
 (b) Where on the band of stability would you expect to find $^{37}_{18}Ar$?
 (c) What is the daughter nucleus produced?

16.10 The magnesium isotope $^{25}_{12}Mg$ is the daughter isotope created when a radioactive parent isotope undergoes electron capture. What is the full symbol for the parent isotope?

16.11 The magnesium isotope $^{25}_{12}Mg$ is the daughter isotope created when a radioactive parent isotope undergoes positron emission. What is the full symbol for the parent isotope?

16.12 Write full nuclear reactions for Practice Problems 16.10 and 16.11.

RADIOACTIVE DECAY VIA ALPHA-PARTICLE EMISSION

Go back and look at the n/p plot on page 643, where we indicated the nonradioactive dark brown central band. Notice how this band ends at the bismuth buoy. That's because all isotopes in the band of stability that have more than 83 protons are radioactive. The nuclei in these isotopes are just too big to be nonradioactive. They need to be smaller if they are to produce daughters nearer to or on the nonradioactive dark brown band. They tend to eject a sizable chunk of their nuclei via a process called **alpha decay**, the emission of an *alpha (α) particle* from the nucleus. As mentioned in Section 3.3 when we were examining Rutherford's gold-foil experiment, an **alpha particle** is a small piece of the nucleus that consists of two protons, two neutrons, and no electrons. This combination of particles gives the alpha particle a charge of +2. An alpha particle is therefore equivalent to the nucleus of a helium atom, and for this reason the full symbol for an alpha particle makes use of the helium symbol, He:

Alpha Particle

Even though the alpha particle has a +2 charge, we usually don't include the charge in the full symbol.

By emitting an alpha particle, a large nucleus can get rid of a reasonably large chunk of itself all at once. When such an emission takes place, the atomic number decreases by 2 and the mass number decreases by 4.

An example of alpha emission is the conversion of uranium to thorium (note that the nuclear reaction is balanced):

$$\overbrace{}^{234 + 4 = 238}$$
$$^{238}_{92}U \longrightarrow {}^{234}_{90}Th + {}^{4}_{2}He$$
$$\underbrace{}_{90 + 2 = 92}$$

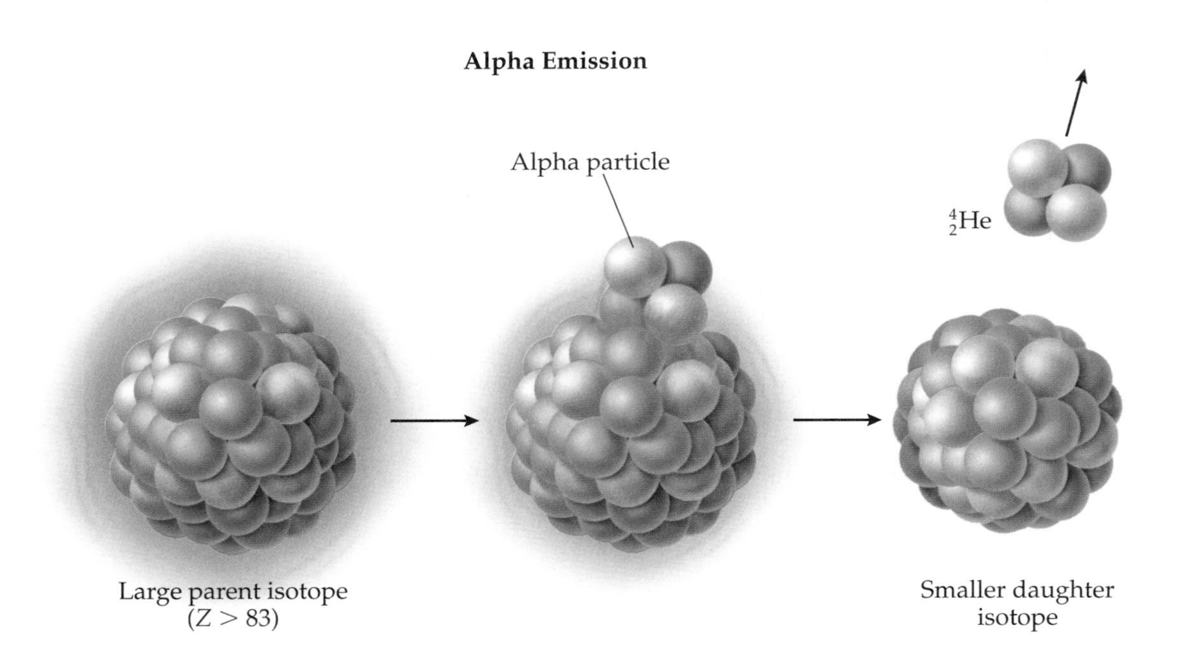

The thorium daughter isotope is also radioactive, but at atomic number 90 it is closer to the nonradioactive dark brown band than its uranium parent.

Test yourself now to see if you are getting the hang of alpha decay.

Balance this nuclear reaction and predict the product:

$$^{209}_{84}Po \longrightarrow \; ^{?}_{?}? + \; ^{4}_{2}He$$

α **particle**

Have the last few WorkPatches made you realize something? We hope so. These different decay modes are, in effect, moving us around on the periodic table in predictable ways. For example, suppose you start with some radioactive isotope of an element somewhere on the periodic table. Let's represent this element with the symbol $^{m}_{Z}E$, where m represents the mass number and Z represents the atomic number. When this parent isotope undergoes beta emission, the daughter isotope is one element to the right. When this parent isotope undergoes either positron emission or electron capture, the daughter isotope is one element to the left. And when this parent isotope undergoes alpha emission, the daughter isotope is two elements to the left.

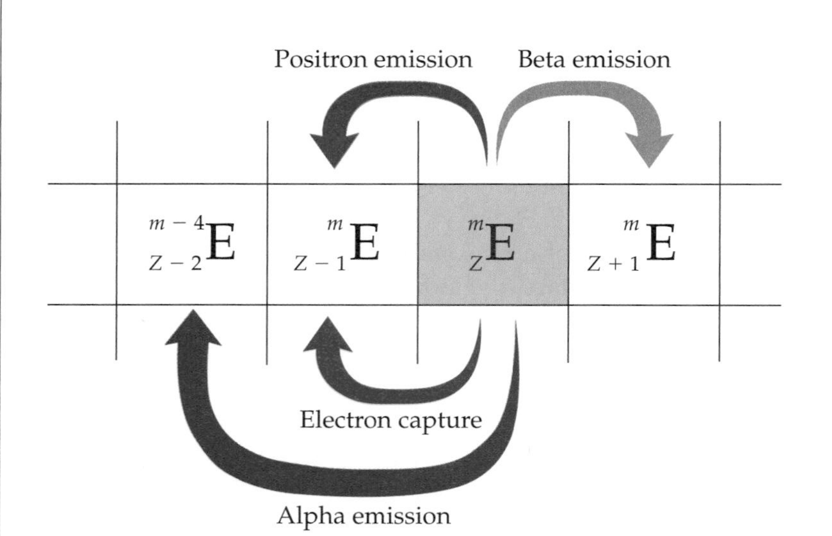

For example, here is the answer to WorkPatch 16.7, an answer arrived at by looking at the periodic table and knowing in advance which way and how much to move a parent isotope undergoing alpha decay. Just follow the red arrow:

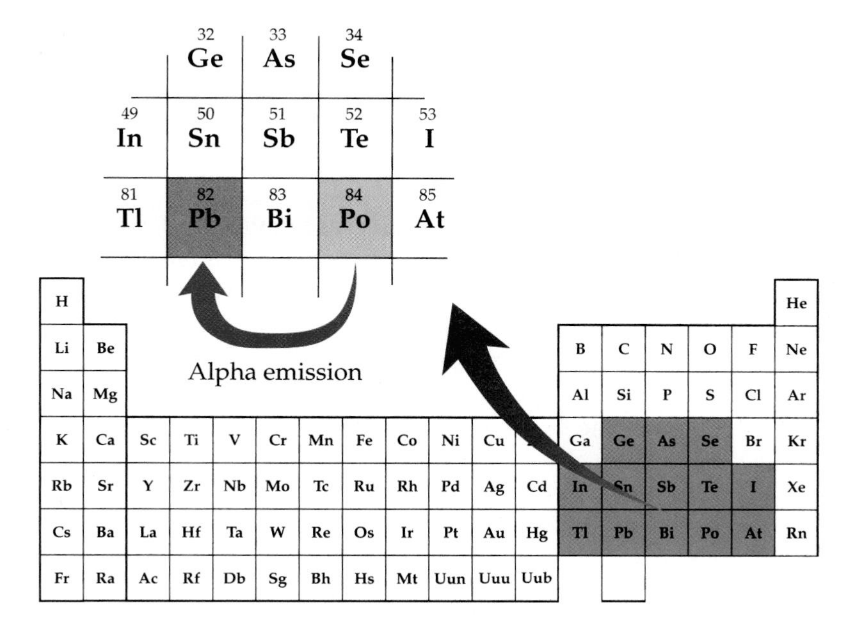

The products of the various decay processes can also be summarized in table form, as in the table we have started for you in the following WorkPatch.

Replace each question mark with the correct number and sign:

	Particle	Change in atomic number	Change in mass number	Change in number of neutrons
Alpha emission (α)	${}^{4}_{2}\text{He}$	-2	-4	-2
Beta emission (β^-)	${}^{0}_{-1}\text{e}$	?	?	?
Positron emission (β^+)	${}^{0}_{+1}\text{e}$	?	?	?
Electron capture (${}^{0}_{-1}\text{e}$)	${}^{0}_{-1}\text{e}$	?	?	?

According to your table, which two decay processes give identical results? Does this result agree with the lower figure on the facing page, which shows how the various decay modes move us around the periodic table? It should.

We'll end this discussion of alpha emission by highlighting one of the most famous (or infamous, depending on your point of view) elements of the twentieth century. Uranium is used to power nuclear reactors that generate electricity. It was also used in the form of a bomb to destroy the Japanese cities of Hiroshima and Nagasaki during World War II. Uranium has an atomic number of 92. This is well beyond our bismuth buoy, so all isotopes of uranium are radioactive. Most of the uranium in naturally occurring uranium ore is the isotope $^{238}_{92}U$, which spontaneously decays to become, ultimately, the nonradioactive lead isotope $^{206}_{82}Pb$:

$$\underset{\text{Radioactive}}{^{238}_{92}U} \xrightarrow{\text{Many steps}} \underset{\text{Nonradioactive}}{^{206}_{82}Pb}$$

Mass number decreases by 32,
atomic number decreases by 10.

As you can see, the atomic number must decrease by 10 and the mass number by 32. None of the simple decay processes we have discussed can accomplish this in one step. The decay of $^{238}_{92}U$ to $^{206}_{82}Pb$ is a multistep process involving eight alpha emissions and six beta emissions, as shown below. All the isotopes in the process are radioactive and decay spontaneously over time, except for the lead isotope $^{206}_{82}Pb$, which is on the nonradioactive dark brown central region in the band of stability.

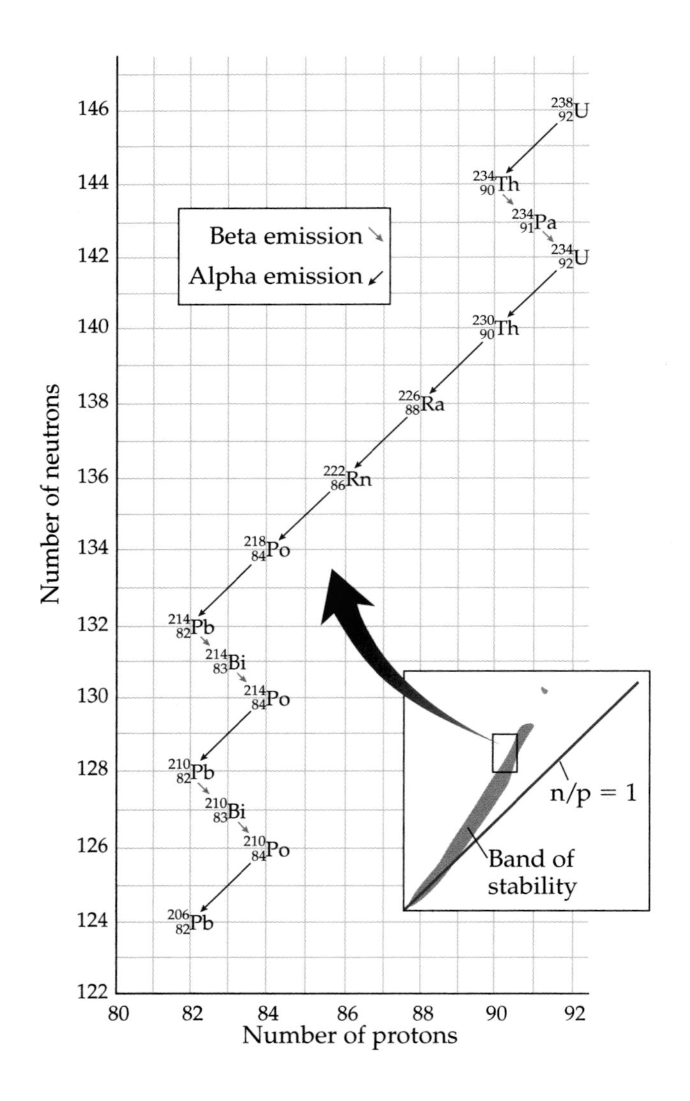

Practice Problems

16.13 Positron emission moves us one step to the left in the periodic table, and alpha emission moves us two steps to the left. Does this mean that, for a given parent isotope, the daughter isotope resulting from two successive positron emissions is the same as the daughter isotope resulting from one alpha emission?

Answer: No. Starting from the same parent isotope, two positron emissions result in the same daughter element *as one alpha emission (the two daughters have the same atomic number, Z), but the daughters are different isotopes of that element. The daughter of two successive positron emissions has a mass number that is 4 larger than the mass number of the daughter resulting from one alpha emission.*

16.14 Starting from $^{238}_{92}U$, demonstrate that the answer to Practice Problem 16.13 is correct.

16.15 If a radioactive element undergoes a single decay process and transforms into an element two spaces away on the periodic table, which decay process must have taken place? Explain why it could not be any other decay process.

16.16 If a radioactive element undergoes a single decay process and transforms into an element one step to the right in the periodic table, did a proton turn into a neutron or did a neutron turn into a proton? What do we call this type of decay?

GAMMA RADIATION

When a nucleus undergoes radioactive decay, it often releases a great deal of energy. For nuclei that decay via particle emission (α, β^-, and β^+ emission), the emitted particle can carry away some of this energy in the form of kinetic energy. That is, ejected particles generally move very fast. However, many radioactive nuclei also release energy in the form of electromagnetic radiation, most usually **gamma rays**, for which the symbol is γ, the Greek lowercase letter gamma. Gamma rays are more energetic than X rays (see the electromagnetic spectrum shown in Section 4.1) and thus can be very harmful to living organisms.

Unlike alpha, beta, and positron emission and electron capture, emission of a gamma ray from a nucleus causes no change in either mass number or atomic number. That's because electromagnetic radiation has no mass. Therefore, if the nucleus of a radioactive isotope does nothing more than emit a gamma ray, the elemental identity of the isotope remains unchanged. However,

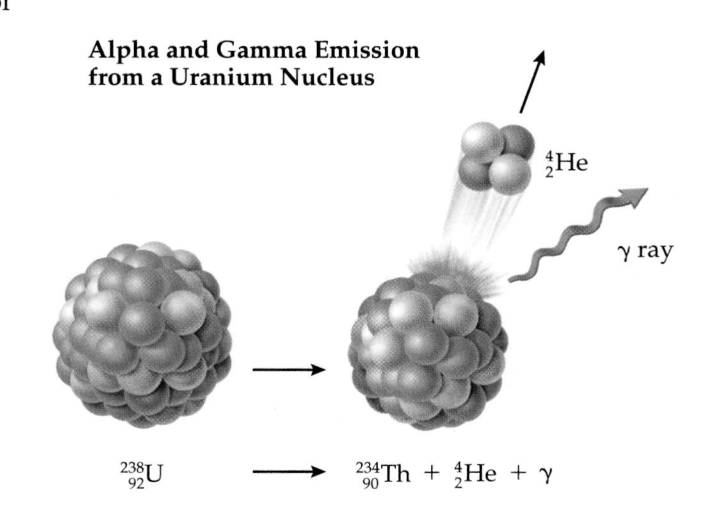

gamma-ray emission usually accompanies the other decay modes we have discussed. For example, the equation we gave earlier for $^{238}_{92}U$ undergoing alpha decay should have included the accompanying release of a gamma ray. Both the alpha particle and the gamma ray carry energy away from the nucleus.

16.4 Using Radioactive Isotopes to Date Objects

We defined half-life as the time it takes for the amount of a particular radioactive isotope in a sample to drop to half its initial amount as a result of spontaneous nuclear change. Now we will see how the measured half-lives of various radioactive isotopes can be used to determine the age of objects. The theory behind this process is simple. All we have to do is determine how much of some radioactive isotope remains in an object. Radioactive isotopes in an object? "There's no radioactive isotope in me!" you say. Actually, there are many. One in particular is the radioactive carbon isotope $^{14}_{6}C$, which, as we saw earlier, is a beta emitter. All living things contain a roughly constant amount of $^{14}_{6}C$ because $^{14}_{6}C$ is constantly being produced in the upper atmosphere as neutrons produced by cosmic radiation bombard nitrogen nuclei:

$$^{14}_{7}N + ^{1}_{0}n \longrightarrow ^{14}_{6}C + ^{1}_{1}H$$

This radioactive carbon isotope enters the food chain and behaves chemically like the nonradioactive carbon isotopes. This means that we metabolize it and excrete it as CO_2 at the same rate as we metabolize and excrete the nonradioactive isotopes, maintaining a constant amount of $^{14}_{6}C$ in our bodies. However, once we die, we no longer take in any $^{14}_{6}C$. The $^{14}_{6}C$ present at the time of death begins to decay at a predictable rate characterized by its half-life, which is 5715 years.

Now consider Table 16.2, which was generated by taking 100% (the original amount of $^{14}_{6}C$ present) and repeatedly multiplying it by $\frac{1}{2}$.

Table 16.2

Number of half-lives that have passed	Isotope remaining, as percent of initial amount
0	100
1	50
2	25
3	12.5
4	6.25
5	3.13
6	1.56
7	0.78
8	0.39
9	0.19
10	0.098

Suppose someone digs up your bones a long time from now, analyzes them, and finds that they contain 75% of the $^{14}_{6}C$ contained in the bones of people living at the time of the analysis. According to Table 16.2, that 75% value puts the age of your bones at somewhere between 0 and 1 half-lives of carbon-14. In other words, your bones are between 0 and 5715 years old. You might be thinking that since 75% is exactly halfway between 100% and 50%, your bones are 5715/2 = 2858 years old. But it's not that simple. To find the actual age, we must use the formula:

$$\text{Age} = \frac{-2.303 \times \log\left(\dfrac{\text{Percent } ^{14}_{6}\text{C remaining}}{100}\right) \times \text{Half-life}}{0.693}$$

$$= \frac{-2.303 \times \log\left(\dfrac{75\%}{100}\right) \times 5715 \text{ years}}{0.693}$$

$$= 2.4 \times 10^3 \text{ years (2400 years old)}$$

Of course, there is some uncertainty in this number, an uncertainty that depends on the accuracy to which the $^{14}_{6}C$ remaining was measured. This calculation also assumes that there is a constant amount of $^{14}_{6}C$ in the food chain over the years—that is, there hasn't been any change in the amount of $^{14}_{6}C$ found in living humans from the time you died until your bones were analyzed. This may not be true. Still, this method works pretty well even with its limitations.

Now suppose an animal's bones are dug up 1 million years after death.

What would be wrong with using $^{14}_{6}C$ to determine the age of fossils that are 1 million years old?

Do you see the problem? One million years is almost 175 half-lives for $^{14}_{6}C$. Check your answer against ours to be sure you understand. The solution to this problem would be to choose a different radioactive isotope present in the fossils—one that has a half-life that is closer to the age of the fossils. For instance, Earth has been dated by determining the amount of the radioactive potassium isotope $^{40}_{19}K$ in rocks, and this isotope was chosen because it has a half-life of 1.26×10^9 years. (Measurements indicate that our planet is approximately 4.5 billion years old.)

Practice Problems

16.17 A rock from an asteroid contains 2.57 g of $^{238}_{92}U$ and 3.83 g of $^{206}_{82}Pb$. The molar mass of $^{206}_{82}Pb$ is 205.974 46 g/mol, the molar mass of $^{238}_{92}U$ is 238.029 g/mol, and the half-life of $^{238}_{92}U$ is 4.46×10^9 years. Assume that all the $^{206}_{82}Pb$ came from the radioactive decay of the $^{238}_{92}U$.

(a) How many atoms of each isotope are present in the rock?

(b) How many atoms of $^{238}_{92}U$ were in the rock when it formed?

(c) What is the percent of $^{238}_{92}U$ atoms remaining in the rock compared to when it was first formed?

(d) How old is the asteroid?

Answer:

(a) $$2.57 \text{ g } {}^{238}_{92}\text{U} \times \frac{1 \text{ mol } {}^{238}_{92}\text{U}}{238.029 \text{ g } {}^{238}_{92}\text{U}} \times \frac{6.02 \times 10^{23} \text{ atoms } {}^{238}_{92}\text{U}}{1 \text{ mol } {}^{238}_{92}\text{U}} = 6.50 \times 10^{21} \text{ atoms } {}^{238}_{92}\text{U}$$

$$3.83 \text{ g } {}^{206}_{82}\text{Pb} \times \frac{1 \text{ mol } {}^{206}_{82}\text{Pb}}{205.974\ 46 \text{ g } {}^{206}_{82}\text{Pb}} \times \frac{6.02 \times 10^{23} \text{ atoms } {}^{206}_{82}\text{Pb}}{1 \text{ mol } {}^{206}_{82}\text{Pb}} = 1.12 \times 10^{22} \text{ atoms } {}^{206}_{82}\text{Pb}$$

(b) *Just add the number of ${}^{238}_{92}U$ atoms and ${}^{206}_{82}Pb$ atoms present in the rock today:*

$$6.50 \times 10^{21} + 1.12 \times 10^{22} = 1.77 \times 10^{22} \text{ atoms of } {}^{238}_{92}\text{U initially present}$$

(c) $$\% \ {}^{238}_{92}\text{U remaining in rock} = \frac{\text{Number of } {}^{238}_{92}\text{U atoms present now}}{\text{Number of } {}^{238}_{92}\text{U atoms intially present}} \times 100$$

$$= \frac{6.50 \times 10^{21} \text{ atoms}}{1.77 \times 10^{22} \text{ atoms}} \times 100$$

$$= 36.7\% \text{ remaining}$$

(d) $$\text{Age} = \frac{-2.303 \times \log\left(\dfrac{\text{Percent } {}^{238}_{92}\text{U remaining}}{100}\right) \times \text{Half-life}}{0.693}$$

$$= \frac{-2.303 \times \log\left(\dfrac{36.7\%}{100}\right) \times (4.46 \times 10^{9} \text{ years})}{0.693}$$

$$= 6.45 \times 10^{9} \text{ years} = 6.45 \text{ billion years}$$

This is a very old rock indeed, predating our solar system!

16.18 In part (b) of Practice Problem 16.17, why is the number of ${}^{238}_{92}U$ atoms initially present in the rock equal to the sum of the number of ${}^{238}_{92}U$ atoms and ${}^{206}_{82}Pb$ atoms present in the rock today?

16.19 An even older asteroid is found. A rock from it yields 1.82 g of ${}^{238}_{92}U$ and 4.02 g of ${}^{206}_{82}Pb$. How old is this asteroid?

16.5 Nuclear Energy—Fission and Fusion

The human race has come a long way since the early alchemists. Indeed, now we can convert one element to another by inducing nuclear reactions, but quite often the energy cost is tremendous. As a result, no one is getting rich turning lead into gold. There are two types of induced nuclear reactions, however, that can be used to release tremendous amounts of energy—*nuclear fission* and *nuclear fusion*.

Nuclear fission is the breaking up of a large nucleus into smaller ones, and **nuclear fusion** is the combining of small nuclei into larger ones. Both processes release energy so long as the combined mass defect of the products is greater than the combined mass defect of the reactants (in other words, as long as mass is lost as reactants are converted to products). Not much mass needs to be lost to generate a tremendous amount of energy. As we saw earlier, this is because the energy equivalent of mass is $E = mc^2$, where $c = 3.0 \times 10^8$ m/s.

The value of c^2 is such a large number that when it is multiplied by even a tiny *m* value, a lot of energy is released. We have the ability to carry out these nuclear reactions either quickly in an uncontrolled manner or slowly in a controlled manner. When the energy is released all at once, we have a nuclear bomb. When it is released slowly in a nuclear reactor, the energy can be used to generate steam for turning turbines in an electric power plant. Let's look now at some examples of induced nuclear fission and fusion.

FISSION

In nuclear reactors, fission is induced by firing neutrons at a heavy radioactive isotope, such as $^{235}_{92}U$, causing the uranium atoms to split into smaller atoms. One of the fission reactions that occurs is

$$^{235}_{92}U + ^{1}_{0}n \longrightarrow ^{142}_{56}Ba + ^{91}_{36}Kr + 3\,^{1}_{0}n$$

Mass lost per mole of $^{235}_{92}U$: 0.1868 g
Energy released per mole of $^{235}_{92}U$: 16,800,000,000 kJ

There are two important things to notice here. First, a tremendous amount of energy is released in the fission reaction of 1 mol (235.04 g) of $^{235}_{92}U$. You would have to burn 132,000 gallons of gasoline to get the same amount of energy. Second, this fission reaction produces more neutrons than it consumes. The three neutrons produced can cause fission of three more $^{235}_{92}U$ atoms, which each produce three neutrons, which can go on to cause fission of nine more $^{235}_{92}U$ atoms, which each produce three neutrons, which can go on to cause fission of 27 more $^{235}_{92}U$ atoms, and on and on. Things can get out of hand pretty quickly.

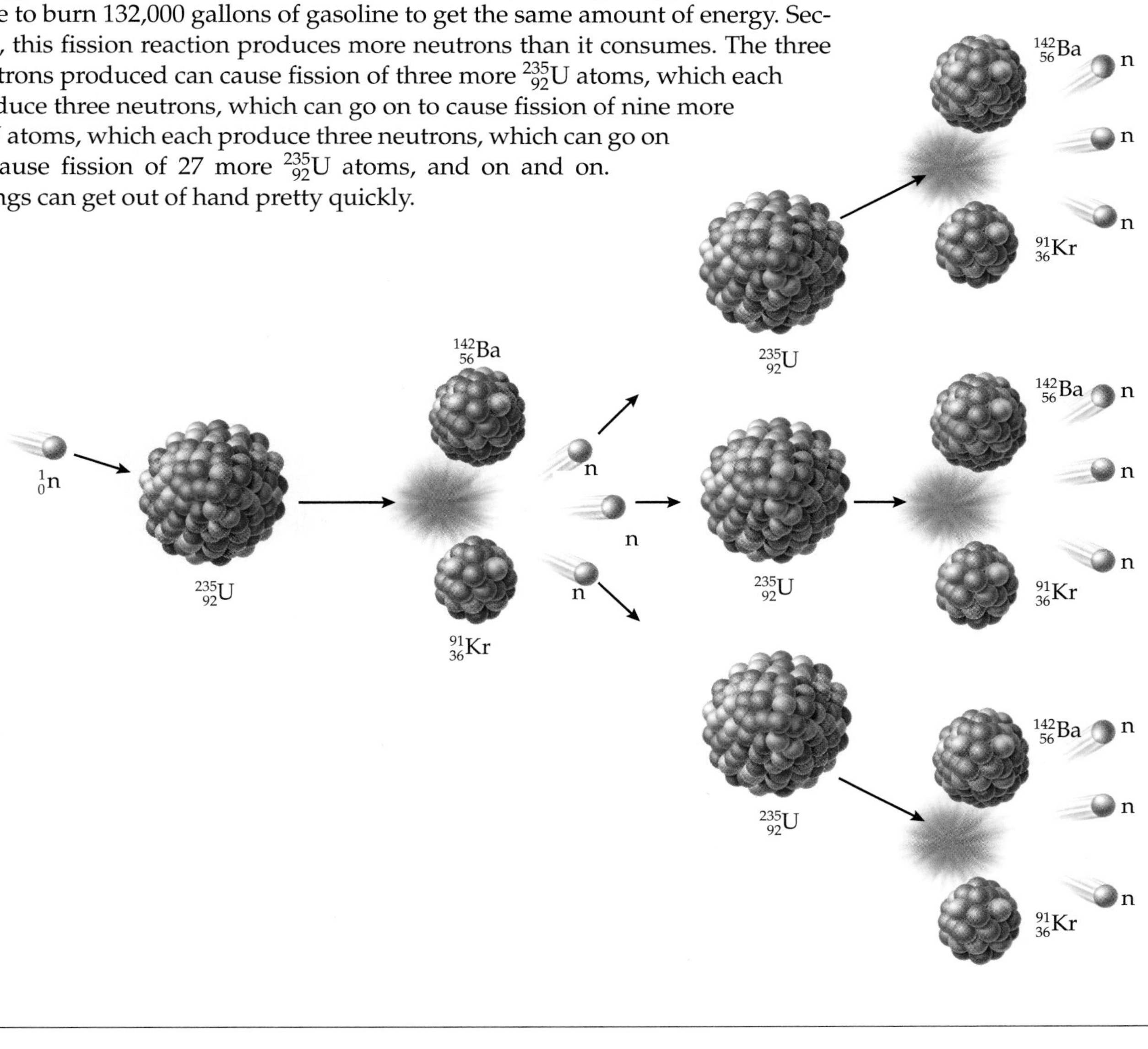

$^{1}_{0}n$ $^{235}_{92}U$ $^{142}_{56}Ba$ $^{91}_{36}Kr$ n n n $^{235}_{92}U$ $^{235}_{92}U$ $^{235}_{92}U$ $^{142}_{56}Ba$ n n n $^{91}_{36}Kr$ $^{142}_{56}Ba$ n n n $^{91}_{36}Kr$ $^{142}_{56}Ba$ n n n $^{91}_{36}Kr$

The third generation of fission of $^{235}_{92}U$ produces 27 neutrons. How many neutrons would the tenth generation of fission produce?

This series of fissions is called a **chain reaction**, which is a sequence of identical reactions in which each reaction event gives rise to more than one subsequent reaction event. In a nuclear reactor, the chain reaction is regulated with control rods containing neutron-absorbing material, such as boron or cadmium.

Even without control rods, there is no danger of a nuclear power plant exploding like an atom bomb. A nuclear explosion can occur only if the $^{235}_{92}U$ is of the proper mass and shape so that most of the neutrons produced by the chain reaction stay within the uranium sample rather than escaping before they can initiate more fission. This mass necessary for explosion during any chain reaction is called the **critical mass**. For $^{235}_{92}U$, the critical mass is 56 kg, but this value can be reduced to 15 kg by reflecting escaped neutrons back into the uranium mass. No nuclear reactor ever designed allows 56 kg of pure $^{235}_{92}U$ to come together. Although they are capable of producing large amounts of energy from relatively little fuel, nuclear reactors account for only 22% of the total electricity generated in the United States. This is because even though there is no chance a fission reactor could ever explode like an atom bomb, there are other problems. There is the potential for accidents that release radioactive materials into the environment. And there is the problem of what to do with the nuclear waste. Spent uranium fuel consists of highly radioactive fission products, some with half-lives of thousands of years. The United States

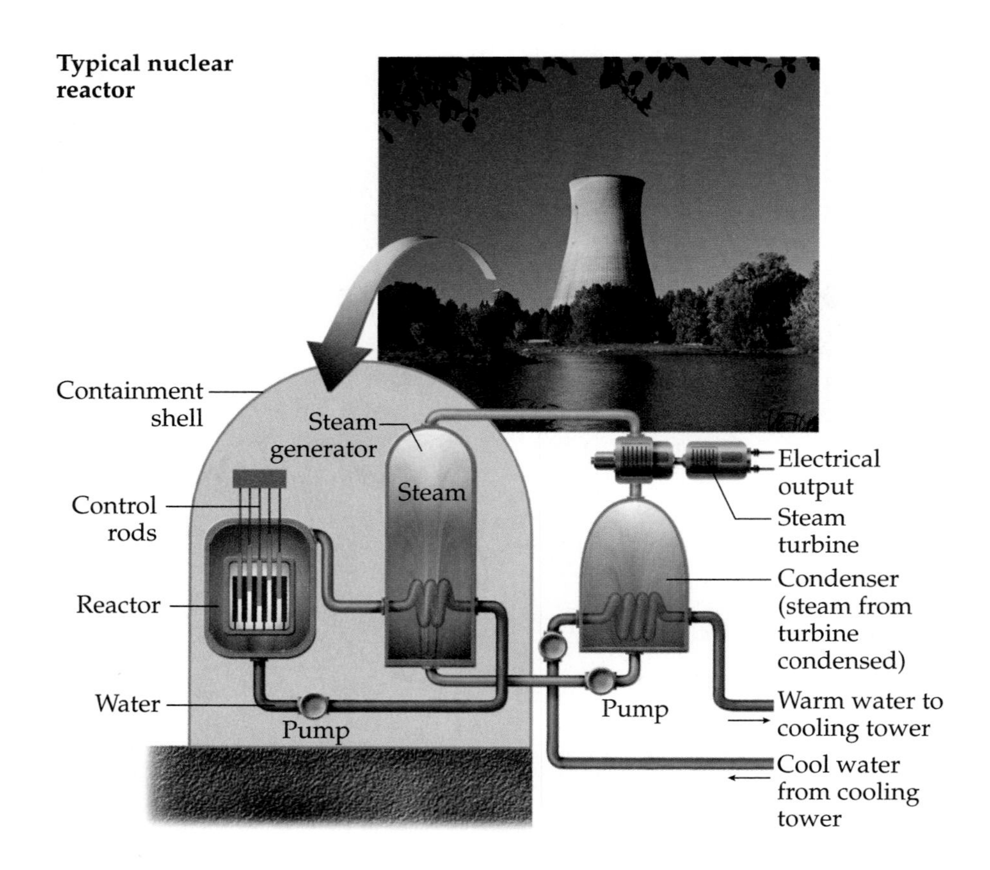

has built an underground nuclear waste storage facility deep under the Nevada desert, where the government plans to bury nuclear waste for thousands of years. (Needless to say, not everyone is pleased with this solution.) And yet another problem with nuclear power plants is that they are tremendously expensive to build and maintain.

Of course, there are benefits to nuclear power generation. It lessens our dependence on fossil fuels, for one thing. Currently, coal and petroleum are the main fuels used in the United States to produce steam for turning electric turbines. As a by-product, these fuels produce millions of tons of atmospheric pollution each year, contributing to acid rain and smog. In addition, mining coal and transporting petroleum have their own drawbacks. Nevertheless, the general fear of anything nuclear will probably prevent nuclear power generation via fission from becoming the predominant source of electrical energy in the United States. This is not true everywhere, however. For example, France produces 73% of its electricity via nuclear fission.

FUSION

Nuclear fusion is the process responsible for the production of energy in all stars, including our Sun. The Sun is essentially a gigantic ball of hydrogen (approximately 80%) and helium (approximately 20%). Inside its core, hydrogen atoms are fusing to form helium atoms. In the process, mass is converted to energy. Scientists calculate that the Sun converts 4×10^{12} g of matter to energy every second. That's 4 trillion grams per second! One sequence of fusion reactions thought to be occurring in our Sun is as follows:

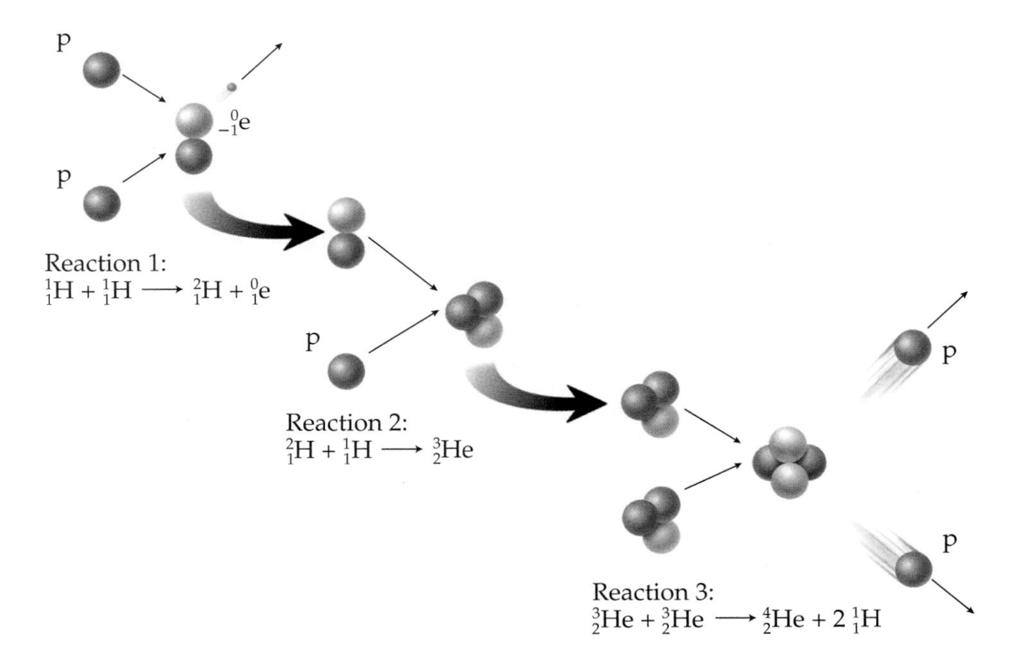

There are no fusion-based nuclear reactors producing electricity here on Earth. The problem with fusion is to get two nuclei to combine. Nuclei repel each other as they get close because all nuclei are positive and like charges repel. In the core of the Sun, hydrogen is heated to incredibly high temperatures (estimated to be 15,000,000 K). At these high temperatures, nuclei are moving so fast that collisions between them overcome their mutual repulsion, and fusion can occur. Here on Earth, generating the temperatures required to sustain fusion is a major problem. This is not a problem for the Sun. When the Sun formed, gravitational collapse provided the energy to heat the gas until fusion started. Now the reaction is self-sustaining; the energy to overcome the repulsion between nuclei comes from the fusion itself. The Sun's immense inward gravitational pull keeps the solar core dense, so that fusion can continue.

On Earth we have yet to figure out a way to heat hydrogen to 15,000,000 K and keep it contained in a dense-enough form to sustain fusion. Such heat would instantly vaporize the thickest concrete or steel vessel that we could build to contain the reaction, so most experimental fusion reactors operating today use magnetic fields to confine the reaction. Although we do have the ability to produce fusion in reactors, we have yet to achieve the *break-even point*, which is the point at which we would get as much energy back from the reaction as we put in to make it happen. Beyond the break-even point, a fusion reactor would become a power source—that is, it would give us back more energy than we feed into it.

For the past 45 years, nuclear physicists have been predicting that practical nuclear fusion is just 20 years away. Today we are closer than ever but still probably at least 20 years away. In addition, the cost of attaining this goal will be astronomical. However, the benefits of generating electricity via fusion instead of fission will be tremendous. Heavy isotopes of hydrogen, the likely fuel, are present in ocean water and therefore plentiful. In addition, there would be no long-lived radioactive waste products. Fusion reactors may one day help us meet our energy requirements.

Meanwhile, scientists have perfected ways to perform uncontrolled nuclear fusion in the form of the hydrogen bomb. In a hydrogen bomb, isotopes of hydrogen surround a nuclear fission bomb. The temperature achieved when the fission bomb goes off is sufficient to induce fusion in the surrounding hydrogen. With no attempt being made to contain the energy released, the result is powerful, to say the least. For comparison, the Hiroshima $^{235}_{92}U$ fission bomb destroyed almost everything within a 1-mile radius of the explosion, but hydrogen bombs have been produced that can obliterate everything within a radius of between 15 and 20 miles.

Practice Problems

16.20 Complete this nuclear reaction, and state whether it is fission or fusion.

$$^{239}_{94}Pu + ^{1}_{0}n \longrightarrow ^{90}_{38}Sr + ? + 3\,^{1}_{0}n$$

Answer: It is fission—you figure out the missing element.

16.21 Could the nuclear reaction of Practice Problem 16.20 be used to produce a chain reaction? Fully explain your answer.

16.22 Explain how you would determine whether the nuclear reaction of Practice Problem 16.20 is exothermic or endothermic.

16.6 Biological Effects and Medical Applications of Radioactivity

Many people have an almost irrational fear of anything nuclear. This is understandable, given the media's preoccupation with the lethal effects of the nuclear radiation that would be unleashed by an atomic war or the meltdown of a nuclear reactor core. Much less front-page coverage is given to the useful applications of nuclear radiation. Consider, for example, a technique called *magnetic resonance imaging (MRI)*, used in hospitals for the last decade or so. In this technique, a person is placed in a magnetic field and radio waves are used to probe certain nuclei in the body. The results are spectacular images of the insides of the body, allowing doctors to locate tumors and other internal problems without dangerous exploratory surgery. This technique is completely safe and involves no nuclear radiation whatsoever. Nevertheless, the original name for the technique, *nuclear magnetic resonance (NMR)*, had to be scrapped and replaced by the name *magnetic resonance imaging* simply because people so feared the word *nuclear* that they often refused to be examined by the instrument.

"I'm not climbing into that nuclear contraption!"

Nuclear radiation is any high-energy particle or electromagnetic radiation emitted by a nucleus during nuclear change. Thus alpha and beta particles and gamma rays are forms of nuclear radiation. Certainly it is true that sufficient exposure to nuclear radiation can cause harm ranging from simple burns to death because the high energy associated with radiation can cook living tissue as well as any stove. Another way radiation energy can harm is by *ionizing* biological molecules in your body. **Ionizing radiation** is defined as any radiation that imparts to a biological molecule enough energy to knock an electron out of the molecule, converting the molecule to a cation:

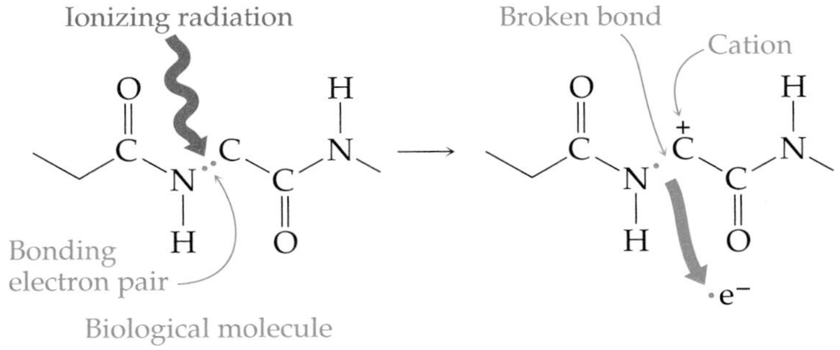

Because biological molecules, such as proteins, carbohydrates, and enzymes, are held together by covalent bonds, ionizing these molecules can cause irreparable damage in the form of weakened and broken bonds. Damage enough biological molecules in an organism, and the organism dies. However, the different types of radiation have different biological effects.

For a given amount of radiation energy, alpha particles are less dangerous than beta particles when the radioactive source is outside the body. This is because alpha particles are much more massive than beta particles and their larger mass causes alpha particles to move more slowly than beta particles. This lower speed decreases the ability of alpha particles to penetrate matter. As a result, normal clothing is enough to protect the skin from alpha radiation. Lighter beta particles of similar energy move much more rapidly, but they can be stopped by a block of wood or by heavy protective clothing. With proper precautions to protect the skin, alpha and beta particles are relatively safe, but we must qualify this statement. If an alpha or beta emitter is ingested, the radiation is then in intimate contact with body tissues and therefore extremely hazardous. Indeed, alpha particles are more dangerous than beta particles when the emitter is inside the body.

Gamma rays are the most dangerous type of external radiation. They have 1000 times the penetrating power of alpha particles and can penetrate deep into the body and damage internal organs. It takes a lead block several inches thick to stop gamma rays.

We don't want to leave you with the impression that the only thing radiation can do in the body is cause harm. Quite the opposite is true, and the medical field depends on radioactive substances for a variety of therapeutic procedures. For example, the gamma rays given off by the radioactive isotope of cobalt $^{60}_{27}Co$ are directed at tumors in a procedure called *external radiation therapy*. Because gamma rays can penetrate into the deepest cells in the body and kill them by breaking covalent bonds in proteins and DNA, carefully controlled and focused doses of gamma radiation can destroy cancer cells. The trick is to expose the cancer cells to the gamma rays while not exposing too much of the surrounding healthy tissue.

Another medical technique, *radiation imaging*, is very useful in diagnosis. For instance, a healthy thyroid gland absorbs iodine, but a hyperactive thyroid gland absorbs too much and a hypoactive gland absorbs too little. A patient is injected with radioactive $^{131}_{53}I$, a beta emitter, and then either a piece of film or a radiation detector is placed close to the patient's neck. The thyroid absorbs the $^{131}_{53}I$ and becomes a "hot" organ, emitting both beta and gamma radiation. This radiation either develops the film or causes the detector to produce a signal. An estimate of thyroid function can be made from the amount of radiation detected. The dose of radioactive iodine is not sufficient to cause harm, and the half-life of $^{131}_{53}I$ is only eight days, meaning it is mostly gone from the body after a month.

Procedures called *radioimmunoassays* are done outside the body on samples of a patient's blood. In these procedures, radioactive substances are bound to substances in the blood in an effort to determine their concentration.

As our final application example of the many uses of radiation, consider irradiated foods. Each year, thousands of people are sickened and many die from food poisoning. In 1999, more than 20 people died in the Midwest from exposure to *E. coli* bacteria in meat contaminated at one processing plant. Salmonella in poultry is also a large problem. A proven way to kill harmful bacteria in meats, vegetables, and fruits is to briefly irradiate the foods with ionizing radiation. In 1986 the U.S. Food and Drug Administration approved the irradiation of fruits, vegetables, herbs, spices, and pork. In 1990 they approved the irradiation of poultry. Indeed, astronauts and cosmonauts eat irradiated foods while in space to protect themselves from food poisoning. However, public fear regarding anything nuclear has limited the commercial success of irradiated food. More education on this subject may eventually overcome this fear.

Have You Learned This?

Mass defect (p. 636)
Binding energy (p. 637)
Nucleon (p. 637)
Radioactive atom (p. 639)
Strong force (p. 640)
Neutron-to-proton ratio (p. 640)
Band of stability (p. 641)
Half-life (p. 641)
Radioactive decay (p. 644)
Positron (p. 644)
Antimatter (p. 645)
Beta particle (p. 645)
Beta emission (p. 645)
Nuclear reaction (p. 646)
Daughter isotope (p. 646)
Parent isotope (p. 646)
Positron emission (p. 647)
Electron capture (p. 647)
Alpha decay (p. 649)
Alpha particle (p. 649)
Gamma ray (p. 653)
Nuclear fission (p. 656)
Nuclear fusion (p. 656)
Chain reaction (p. 657)
Critical mass (p. 657)
Nuclear radiation (p. 661)
Ionizing radiation (p. 661)

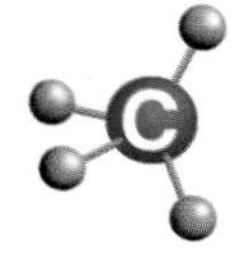

Nuclear Chemistry
www.chemistryplace.com

MASS DEFECT AND THE STABILITY OF THE NUCLEUS

16.23 What do we mean when we say that an atom has a mass defect?

16.24 What do we mean by the binding energy of an atom?

16.25 True or false? The number of nucleons in an atom is equal to its mass number.

16.26 Consider these facts regarding two hypothetical nuclei, one heavy and one light: (*1*) The heavy nucleus has a greater binding energy than the light nucleus. (*2*) The light nucleus is more stable than the heavy nucleus. Explain how both facts can be true.

16.27 Explain why a chemical reaction can never cause changes to an atom's nucleus.

16.28 Of all the isotopes of the elements, which has the greatest mass defect? What does this mean for that isotope?

16.29 What in Einstein's energy equation ensures that a tiny mass defect results in a tremendous amount of energy? Explain.

16.30 The mass of 1 mole of radioactive $^{14}_{6}C$ is 14.003 24 g. Calculate its binding energy in kilojoules per mole of nucleons.

16.31 Examine the plot of binding energy per mole of nucleons versus number of nucleons that appears on page 638. Where does your answer to Problem 16.30 put $^{14}_{6}C$ relative to $^{12}_{6}C$ and $^{13}_{6}C$? What does this imply about the stability of the $^{14}_{6}C$ nucleus relative to that of the lighter isotopes?

16.32 What is a radioactive atom?

16.33 True or false? Radioactive atoms have no mass defect or binding energy. Explain.

HALF-LIFE AND THE BAND OF STABILITY

16.34 Why are neutrons thought to be important for making a nucleus stable?

16.35 As we go from light atoms to heavier ones,
(a) What happens to the neutron-to-proton ratio?
(b) Why does the answer to part (a) make sense?

16.36 Write the full symbols for the isotopes of oxygen having 8, 9, and 11 neutrons.

16.37 Calculate the n/p ratios for the isotopes in Problem 16.36.

16.38 What is the band of stability?

16.39 Define half-life.

16.40 How many half-lives does it take a 10-g sample of $^{123}_{53}\text{I}$ to drop to 0.039 g? What length of time is this? [The half-life of $^{123}_{53}\text{I}$ is 13.1 h.]

16.41 According to the band-of-stability graph on page 642, when would an atom with 60 protons in its nucleus be unstable? (Read the plot as best you can.)

16.42 What is the ocean of instability?

16.43 Why do radioactive isotopes appear on the band of stability?

SPONTANEOUS NUCLEAR CHANGES—RADIOACTIVITY

16.44 What can we say about the n/p ratios of nuclei that lie in the light brown band of stability above the dark brown center region representing nonradioactive nuclei?

16.45 What can we say about the n/p ratios of nuclei that lie in the light brown band of stability below the dark brown center region?

16.46 What is true of all atoms that have an atomic number greater than 83?

16.47 What is meant by the term *radioactive decay*?

16.48 Write the full symbol for a neutron, a proton, an electron, and a positron.

16.49 Why is a positron referred to as antimatter?

16.50 How do we interpret the subscripts for the full symbols of an electron and a positron?

16.51 What kind of nucleus would ever want to convert a proton to a neutron, and why would it want to do that?

16.52 What kind of nucleus would ever want to convert a neutron to a proton, and why would it want to do that?

16.53 True or false? Radioactive decay ends up changing the elemental identity of the isotope undergoing decay. Explain.

16.54 What happens to an atom's nucleus when it undergoes beta emission?

16.55 How is it possible for a nucleus to eject an electron when it contains no electrons?

16.56 The tantalum isotope $^{186}_{73}\text{Ta}$ is radioactive and decays by converting a neutron to a proton.
(a) Where is this atom likely to lie in the band of stability?
(b) Write a nuclear reaction for this decay process.
(c) Which type of decay is this?

16.57 What happens to an atom's nucleus when it undergoes positron emission?

16.58 What happens to an atom's nucleus when it undergoes electron capture?

16.59 The tungsten isotope $^{162}_{74}\text{W}$ is radioactive and decays by converting a proton to a neutron.
(a) Where is this atom likely to lie in the band of stability?
(b) Write two nuclear reactions that describe this decay process.
(c) Which type of decay is represented by each reaction you wrote in part (b)?

16.60 What kind of nucleus would ever want to eject two neutrons and two protons?

16.61 What happens to an atom's atomic number and mass number when it undergoes alpha emission?

16.62 The thorium isotope $^{232}_{90}\text{Th}$ is radioactive and decays by ejecting two protons and two neutrons from its nucleus.
(a) Where is this atom likely to lie in the band of stability?
(b) Write a nuclear reaction for this decay process.
(c) Which type of decay is this?

16.63 Name two forms in which energy can be carried away from a nucleus undergoing radioactive decay.

16.64 In principle, which of Problems 16.56, 16.59, and 16.62 could occur along with gamma-ray emission?

16.65 How do you check to see if a nuclear reaction is balanced?

16.66 Pockets of trapped helium gas are found near some deposits of radioactive ores. How can you explain this?

16.67 What happens to the mass number of an atom when its nucleus
(a) Ejects a beta particle
(b) Ejects a positron
(c) Undergoes electron capture
(d) Ejects an alpha particle

16.68 What happens to the atomic number of an atom when its nucleus
(a) Ejects a beta particle
(b) Ejects a positron
(c) Undergoes electron capture
(d) Ejects an alpha particle

16.69 Explain how simply examining the periodic table allows you to predict the daughter isotope in
(a) Alpha decay
(b) Beta decay
(c) Positron emission
(d) Electron capture

16.70 Why doesn't gamma emission change the elemental identity of a nucleus?

16.71 Postulate a sequence of radioactive decays that converts the lead isotope $^{207}_{82}Pb$ to an isotope of gold.

16.72 Complete this nuclear reaction, and name the decay process:

$$^{8}_{4}Be + ? \longrightarrow {}^{8}_{3}Li$$

16.73 Complete this nuclear reaction, and name the decay process:

$$^{47}_{20}Ca \longrightarrow ? + {}^{47}_{21}Sc$$

16.74 Complete this nuclear reaction, and name the decay process:

$$^{235}_{92}U \longrightarrow {}^{4}_{2}He + ?$$

16.75 Complete this nuclear reaction, and name the decay process:

$$? \longrightarrow {}^{11}_{5}B + {}^{0}_{-1}e$$

16.76 Complete this nuclear reaction, and name the decay process:

$$^{0}_{-1}e + ? \longrightarrow {}^{40}_{18}Ar$$

USING RADIOACTIVE ISOTOPES TO DATE OBJECTS

16.77 Suppose you have 100 g of $^{123}_{53}I$. How much of it will be left after 26.2 h? After 39.3 h? [The half-life of $^{123}_{53}I$ is 13.1 h.]

16.78 Would $^{14}_{6}C$ be useful in dating a fossil that is 120 million years old? Explain.

16.79 Given the half-life of $^{14}_{6}C$, why is any of it present in the environment?

16.80 Measurements show that the percentage of $^{14}_{6}C$ in a particular artifact today is 22.8%. What is the age of the artifact in years?

16.81 Measurements show that a sample of rock contains 14.90 g of $^{238}_{92}U$ and 26.50 g of $^{206}_{82}Pb$. [See Practice Problem 16.17 for molar masses.]
(a) To date this rock, what assumption must be made about where the $^{206}_{82}Pb$ came from?
(b) How many atoms of each isotope are present in the rock?
(c) How many atoms of $^{238}_{92}U$ were present in the rock when it was new?
(d) What percentage of the original amount of $^{238}_{92}U$ remains in the rock today?
(e) How old is the rock in years? [The half-life of $^{238}_{92}U$ is 4.46×10^9 years.]

NUCLEAR ENERGY—FISSION AND FUSION

16.82 For a fission or fusion reaction to be exothermic, what must be true about the mass defect? Explain fully.

16.83 What is nuclear fission?

16.84 What is nuclear fusion?

16.85 Why does nuclear fission often proceed as a chain reaction?

16.86 Complete this fission reaction:

$$^{239}_{94}Pu + {}^{1}_{0}n \longrightarrow ? + {}^{140}_{54}Xe$$

16.87 Complete this fusion reaction:

$$^{8}_{4}Be + {}^{4}_{2}He \longrightarrow ? + \gamma$$

16.88 The hydrogen in our Sun is undergoing fusion and turning into helium. As the hydrogen begins to run out billions of years from now, the helium atoms will fuse, forming heavier atoms. Eventually, these heavier atoms will also fuse. Interestingly, when astronomers examine the remnants of burnt-out stars, they find the remnants to be extremely rich in iron. Explain why this is so.

16.89 What is the definition of critical mass?

16.90 In a nuclear power plant, what is the job of the heat produced in the fission reactions?

16.91 Discuss the benefits and problems associated with using nuclear fission to produce electricity.

16.92 Why can't a nuclear reactor explode like a nuclear bomb?

16.93 Why are such high temperatures needed to initiate nuclear fusion?

16.94 Why are there as yet no fusion reactors operating on Earth to generate power?

16.95 What are the advantages of fusion reactors over fission reactors?

BIOLOGICAL EFFECTS AND MEDICAL APPLICATIONS OF RADIOACTIVITY

16.96 Of the types of radioactive decay studied in this chapter, which is least likely to damage you upon external exposure? Which is most likely? Explain fully.

16.97 How does radiation damage living organisms?

16.98 Radioactivity is often called *ionizing radiation*. Why?

16.99 Briefly discuss the medical uses of radioactivity presented in this chapter.

ADDITIONAL PROBLEMS

16.100 Which has the larger binding energy per mole of nucleons, ${}^{4}_{2}He$ (molar mass 4.00150 g/mol) or ${}^{6}_{3}Li$ (molar mass 6.01348 g/mol)? [Useful masses: proton, 1.00730 g/mol; neutron, 1.00870 g/mol; electron, 0.00055 g/mol]

16.101 Would fusing two ${}^{56}Fe$ atoms together to produce ${}^{112}Te$ be an exothermic reaction or an endothermic reaction? Justify your answer. [*Hint*: Don't do a calculation, just consult the plot of binding energy versus number of nucleons in nucleus.]

16.102 Consider the radioactive decay of radium to radon:

$$ {}^{226}_{88}Ra \longrightarrow {}^{222}_{86}Rn + ? $$

(a) Write the complete equation.
(b) What type of decay is this?
(c) Explain why radium-226 is likely to undergo the type of decay you named in part (b).
(d) How much energy is released, in kilojoules, when 1 mole of ${}^{226}_{88}Ra$ decays? [Molar masses: ${}^{226}_{88}Ra$, 226.0254 g/mol; ${}^{222}_{86}Rn$, 222.0175 g/mol; ${}^{4}_{2}He$, 4.0026036 g/mol]
(e) How much energy is released, in kilojoules, when 1 g of ${}^{226}_{88}Ra$ decays?

16.103 Polonium-210 is an alpha-emitter and has a half-life of 138 days.
(a) Write the equation for the radioactive decay of polonium-210.
(b) How long will it take before only 5.00% of the original amount of ${}^{210}Po$ in a sample remains?

16.104 Rubidium-87, a beta emitter, is a product of positron emission.
(a) Identify the parent nucleus of ${}^{87}Rb$.
(b) When the parent nucleus named in part (a) decays, does the n/p ratio increase or decrease?

16.105 Complete these equations representing nuclear reactions:
(a) ${}^{121}_{51}Sb + {}^{4}_{2}He \longrightarrow ? + {}^{1}_{1}H$
(b) ${}^{27}_{13}Al + {}^{4}_{2}He \longrightarrow ? + {}^{1}_{0}n$
(c) ${}^{238}_{92}U + {}^{1}_{0}n \longrightarrow ? + {}^{0}_{-1}e$

16.106 Which isotope in each pair is more likely to decay by electron capture:
(a) ${}^{13}B$ or ${}^{8}B$
(b) ${}^{209}Bi$ or ${}^{194}Bi$?

16.107 The isotopes ${}^{17}F$, ${}^{20}F$, and ${}^{21}F$ are all radioactive, decaying either by beta emission or by positron emission. Name the decay process for each isotope.

16.108 Suppose you discovered a new radioactive decay mode for which the daughter had a mass number seven lower than the parent and was three places to the left of the parent in the periodic table. What particle would the parent nucleus have to eject to accomplish such a decay?

16.109 A painting supposedly by Rembrandt (1609–1669) was found to contain 96.1% of the amount of ${}^{14}C$ found in a living plant. Could this painting have been done by Rembrandt? [The half-life of ${}^{14}C$ is 5715 years.]

16.110 Consider the fission reaction

$$ {}^{1}_{0}n + {}^{235}_{92}U \longrightarrow {}^{89}_{37}Rb + 3\,{}^{0}_{-1}e + 3\,{}^{1}_{0}n + ? $$

(a) What is the missing fission product represented by the question mark?
(b) What is it about this reaction that allows for a chain reaction?
(c) How much energy, in kilojoules, is released per gram of ${}^{235}U$?
[Molar masses: ${}^{235}U$, 235.0439 g/mol; ${}^{89}Rb$, 88.8913 g/mol; missing product, 143.8817 g/mol; ${}^{1}_{0}n$, 1.00870 g/mol; ${}^{0}_{-1}e$, 0.00055 g/mol]
(d) How many kilograms of TNT must be detonated to produce the same amount of energy if the energy released per gram of TNT detonated is 2.76 kJ?

16.111 Thorium-232 undergoes the following decays successively: parent ${}^{232}Th$ decays to six alpha particles plus daughter 1, then daughter 1 decays to four beta particles plus daughter 2. Identify daughter 2.

16.112 In the ore of what metal would you look for thorium? Explain.

16.113 Two possible mechanisms for a certain reaction are

$$H_3C{-}C({=}O){-}O{-}H + H{-}O{-}CH_3 \longrightarrow H_3C{-}C({=}O){-}O{-}CH_3 + H{-}O{-}H$$

$$H_3C{-}C({=}O){-}O{-}H + H{-}O{-}CH_3 \longrightarrow H_3C{-}C({=}O){-}O{-}CH_3 + H{-}O{-}H$$

The red oxygen is the nonradioactive isotope oxygen-18, and analysis of the products reveals that all of the ^{18}O ends up in the water.

(a) Which mechanism is correct, and which bonds in the reactants must be broken to form the products?

(b) If your laboratory assignment is to determine which mechanism is the right one, how would it help if ^{18}O were radioactive?

16.114 (a) Write the reaction for the beta decay of tritium.

(b) Like ^{14}C, tritium is formed by nuclear reactions in the upper atmosphere. What is the missing product here:

$$^{14}_{7}N + ^{1}_{0}n \longrightarrow ^{3}_{1}H + ?$$

(c) The half-life of tritium is only 12.26 years, and yet there is always some present on Earth (about 1 atom in every 10^{18} H atoms is tritium). How can this be?

16.115 Polonium-210, an alpha-emitter, has a half-life of 138.4 days. Suppose you were to collect the helium gas originating from the alpha particles. How many milliliters of helium gas at standard temperature and pressure would you collect from 1.000 g of polonium dioxide, PoO_2, in a period of 138.4 days? [Assume all the polonium in the sample is ^{210}Po, molar mass 209.98287 g/mol. Alpha emission from polonium-210 yields the nonradioactive isotope lead-206; see Problem 16.103.]

16 WorkPatch Solutions

16.1 (a) Being at the highest point on the plot, the $^{56}_{26}Fe$ nucleus has the most binding energy per mole of nucleons, making it the most stable isotope of all the elements.

(b) Reading the binding energy off the plot, we get 8.48×10^8 kJ per mole of nucleons. Multiplying this value by the conversion factor (56 mol of nucleons/mol of $^{56}_{26}Fe$ atoms), we see that the total binding energy is 4.75×10^{10} kJ per mole of $^{56}_{26}Fe$ atoms.

16.2 (a) There are 207 nucleons (the mass number).

(b) Going to 207 nucleons on the plot of binding energy, we read 7.57×10^8 kJ per mole of nucleons. To remove 3 mol of protons would thus take $3 \times 7.57 \times 10^8\ kJ = 2.27 \times 10^9$ kJ.

16.3 (a) $^{12}_{6}C$, $^{13}_{6}C$, $^{14}_{6}C$

(b) Six protons and six neutrons in $^{12}_{6}C$, six protons and seven neutrons in $^{13}_{6}C$, six protons and eight neutrons in $^{14}_{6}C$.

(c) 12 nucleons in $^{12}_{6}C$, 13 nucleons in $^{13}_{6}C$, 14 nucleons in $^{14}_{6}C$.

(d) They are all carbon because they all have the same atomic number, 6, which means they all have six protons.

16.4 (a) Neutrons and protons have the same mass number (1) because they both consist of one nucleon. Mass number tells you the number of protons plus neutrons in a nucleus.

(b) The atomic number of a neutron is 0 because the atomic number tells you the number of protons in a particle and a neutron has no protons in it.

(c) The atomic number of a proton is 1 because a proton "has" one proton (it *is* a proton).

16.5

$^{0}_{0}n$ — This should be 1. The superscript is the mass number, the sum of protons (0) + neutrons (1).

$^{1}_{0}p$ — This should be 1. For neutrons and protons (as well as for atoms), the subscript in the particle symbol gives the number of protons.

$^{1}_{1}e$ — This should be 0. The superscript is the mass number, the sum of protons (0) + neutrons (0), and either particle having the symbol e has neither protons nor neutrons.

You were told there is only one mistake per symbol, and so this subscript number must be correct. A subscript of +1 means this symbol must represent a positron. Remember, the subscript is +1 for a positron and −1 for an electron.

16.6 $^{214}_{82}\text{Pb} \longrightarrow {}^{214}_{83}\text{Bi} + {}^{0}_{-1}\text{e}$

16.7 $^{209}_{84}\text{Po} \longrightarrow {}^{205}_{82}\text{Pb} + {}^{4}_{2}\text{He}$

16.8

	Particle	Change in atomic number	Change in mass number	Change in number of neutrons
Alpha emission (α)	${}^{4}_{2}\text{He}$	−2	−4	−2
Beta emission (β^-)	${}^{0}_{-1}\text{e}$	+1	0	−1
Positron emission (β^+)	${}^{0}_{+1}\text{e}$	−1	0	+1
Electron capture (${}^{0}_{-1}\text{e}$)	${}^{0}_{-1}\text{e}$	−1	0	+1

16.9 After 1 million years, which is 175 half-lives, the percentage of ${}^{14}_{6}\text{C}$ isotope remaining in the fossil would be too tiny to detect, let alone measure.

16.10 $3^{10} = 59\,049$ neutrons.

CARBON
Royal Sovereign
838

CHAPTER 17

The Chemistry of Carbon

17.1 Carbon—A Unique Element

By applying the concepts covered in the preceding chapters, we can begin to understand some of the mysteries of nature. So, what should we look at? The glow of a firefly? The production of acid rain? The elasticity of rubber? Why limit ourselves? Let's examine the biggest subject of all—life itself. Knowledge of the fundamental concepts of chemistry has brought humanity to the brink of some astounding possibilities regarding life, from the possibility of conquering viruses and cancer to the manipulation of our own genetic code and cloning. If we are to explore the chemistry of life using what we have learned, we must start with the one element upon which all life as we know it is based—carbon.

Carbon is the element with atomic number 6. Its most abundant isotope, $^{12}_{6}C$, has six protons and six neutrons in the nucleus and six electrons outside the nucleus. Of those six electrons, four are valence electrons. When a carbon atom forms covalent bonds to another atom, it is these four valence electrons that are involved:

Carbon bonds via its four valence electrons. $\longrightarrow \cdot\dot{\underset{\cdot}{C}}\cdot$

One of the things carbon atoms are very good at is bonding to other carbon atoms. The atoms bond to one another by sharing valence electrons, forming long chains, rings, and a large variety of other carbon frameworks, a few of which are shown on the next page.

Some chains, rings, and other structures formed by carbon atoms (attached hydrogen atoms not shown):

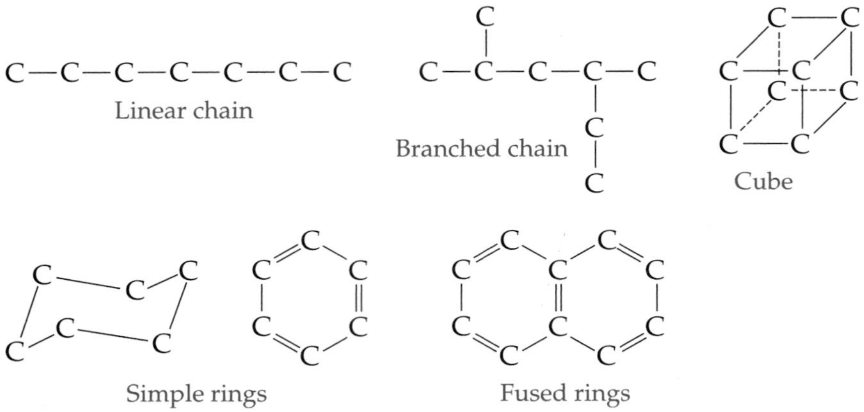

C—C—C—C—C—C—C

Linear chain

Branched chain

Cube

Simple rings

Fused rings

This ability of an atom to link together with like atoms is called **catenation.** No other element is able to catenate as well as carbon. For example, two oxygen atoms can bond to each other to form the diatomic molecule O_2. Also known but much more reactive is the ozone molecule, O_3, a bent chain of three covalently bonded oxygen atoms. But that is it; longer chains of oxygen are unknown. Nitrogen, which falls between carbon and oxygen in the periodic table, can also bond to other atoms of itself but again does not form long chains. For example, you are currently breathing N_2 molecules, but the N_3^- azide ion (three nitrogen atoms in a chain) is so unstable that it explodes when heated. Thus, carbon is unique.

Because of its ability to catenate, carbon gives rise to a larger number of compounds than any other element. Indeed, a whole branch of chemistry, called **organic chemistry**, is devoted to the study of carbon-based molecules. The name *organic* comes from the fact that carbon-based molecules make up the majority of molecules present in living organisms. No other branch of chemistry is based on just one element, an indication of how important carbon is. This chapter and the next introduce you to the rich variety of molecules that carbon forms.

Elemental carbon comes in the three structural forms (called *allotropes*) shown on the next page: diamond, graphite, and fullerene, a form of carbon found in soot that has only recently been discovered. In each allotrope, the carbon atoms are covalently bonded to other carbon atoms. In diamond, each carbon atom is attached by single bonds to four other carbon atoms forming a tetrahedron, as we expect from VSEPR theory. In graphite and fullerene, each carbon atom is connected, by both single and double bonds, to three other carbon atoms, forming a flat triangular structure also expected from VSEPR theory. In other words, the geometry about a carbon atom depends on how many other atoms are bonded to it.

These seemingly small differences give rise to substances having drastically different properties. Diamond has all its carbons locked into a three-dimensional structure by strong covalent bonds. The result is the hardest known natural substance, often used to coat saw blades that can cut through virtually any other material. Graphite, on the other hand, consists of flat sheets of carbon atoms attracted to one another by weak London forces. This weak attraction allows the layers to slide over one another and results in a soft, crumbly material that can be used as a lubricant.

Allotropes

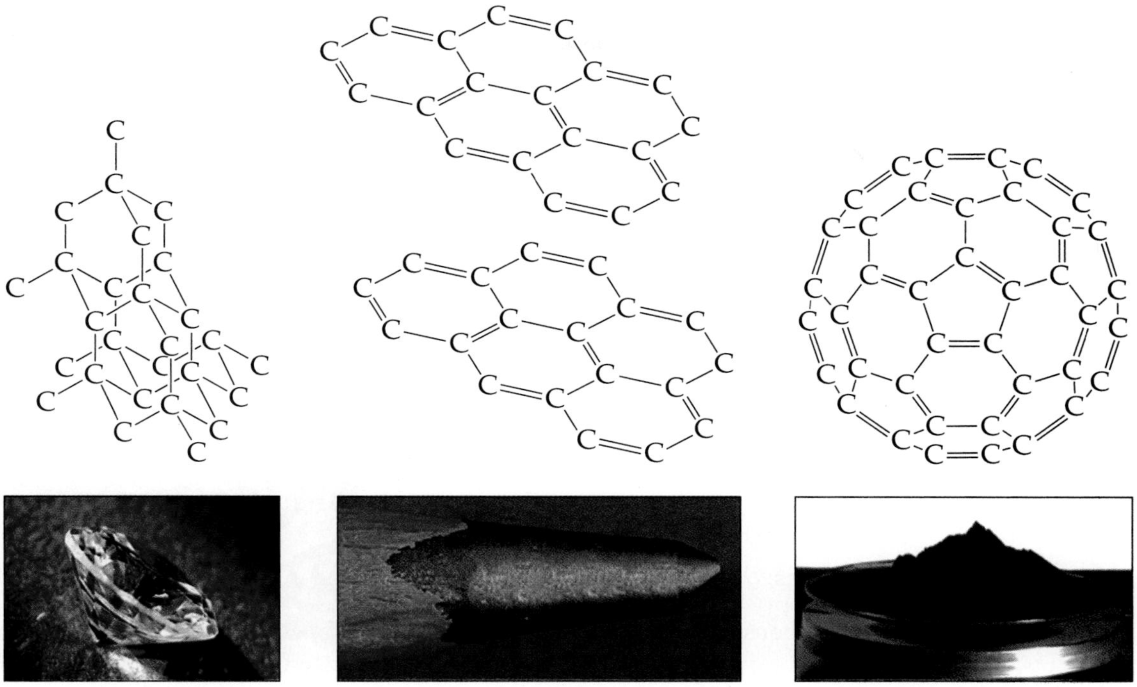

Diamond

Graphite

Soot containing fullerene

Graphite: Each carbon bonded to three others

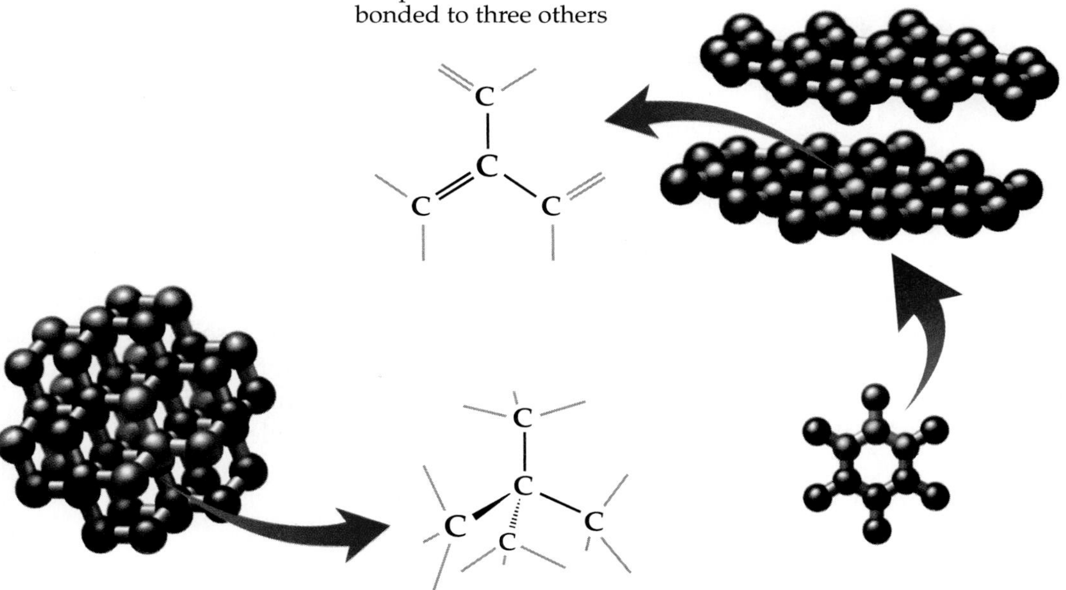

Diamond: Each carbon bonded to four others

Take another look at the structures of the diamond, graphite, and fullerene forms of carbon shown at the top of this page. Can you see what they have in common? It is not the number of attached carbon atoms. Each carbon atom in diamond bonds with four other carbon atoms, but in fullerene and graphite, each carbon atom bonds with only three other carbon atoms. The common feature is that in all three allotropes, every carbon atom forms four covalent bonds:

Four bonds
(diamond)

Four bonds
(graphite and fullerene)

Because carbon has four valence electrons, it always "wants" to form four covalent bonds to arrive at an octet of electrons. This can be accomplished with just single bonds or with a combination of single and multiple bonds:

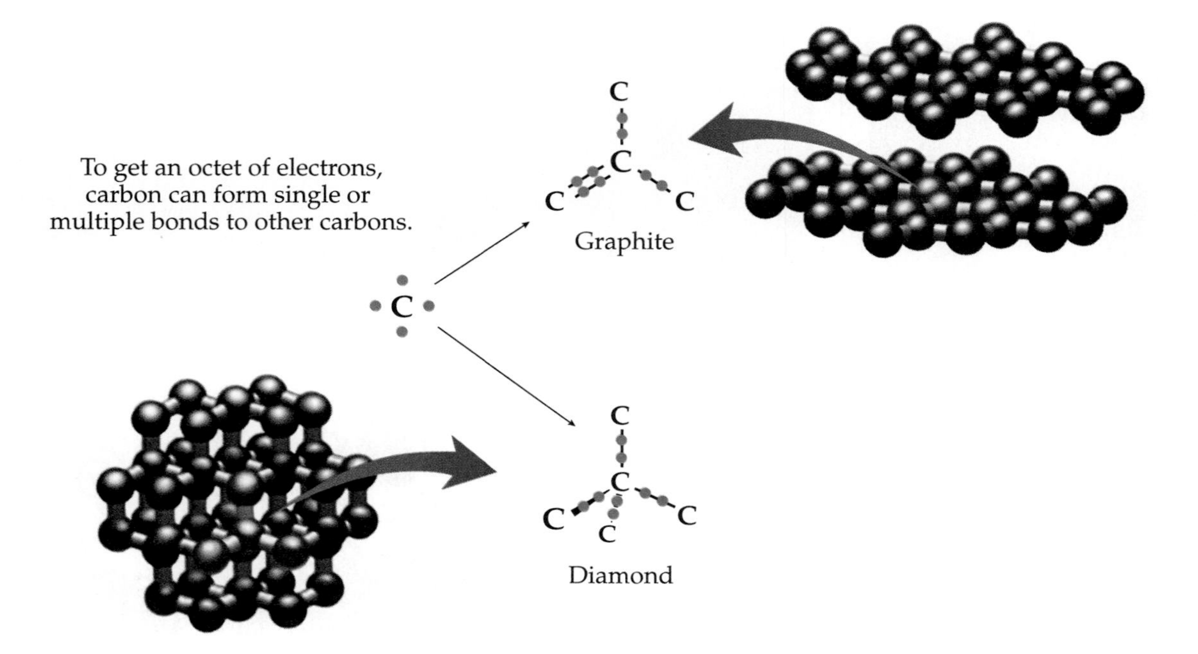
To get an octet of electrons, carbon can form single or multiple bonds to other carbons.
Graphite
Diamond

In one case, carbon is attached to four other carbon atoms. In the other case, carbon is attached to three other carbon atoms. In both cases, however, every carbon atom has four covalent bonds. This fact is often referred to as the cardinal rule of organic chemistry: carbon always forms four covalent bonds.

Another way carbon can form four bonds is with triple bonds:

$$\mathrm{C{-}C{\equiv}C}$$

Single bond Triple bond
Four bonds total

Here, one single bond and one triple bond result in a total of four bonds and a complete octet for the central carbon. Although triple bonds do not occur in the three allotropic forms of elemental carbon, they do occur in compounds of carbon.

17.1 WORKPATCH

There is a problem with at least one carbon atom in each of these three molecules. Identify the problem carbon(s) and then fix the problem by either removing or adding one (and *only* one) hydrogen atom or by removing or adding a covalent bond.

```
              H
              |
       H     C      H
         \ //  \   /
          C      C
(a)       |      ||
          C      C
        /  \   /   \
       H     C      H
             |
             H

      H
        \
(b)      C≡C—H
        /
      H

                      :O:
                       ||
(c) H                  C
      \              /   \
        C                  H
        |
        H
```

17.2 Naturally Occurring Compounds of Carbon and Hydrogen—Hydrocarbons

Anyone who lives in an industrialized society has heard of crude oil, also known as petroleum. The majority of the chemical compounds in petroleum are **hydrocarbons**, molecules made up entirely of carbon and hydrogen. The carbon atoms in a hydrocarbon molecule bond to one another to form the chains and rings described in the preceding section, and the hydrogen atoms bond to the carbon atoms. Although any two carbon atoms in a hydrocarbon may bond to each other by single, double, or even triple bonds, the hydrogen atoms always bond to the carbon atoms by single bonds. A typical hydrocarbon molecule is octane, which has an eight-carbon chain:

```
    H   H   H   H   H   H   H   H
    |   |   |   |   |   |   |   |
H—C—C—C—C—C—C—C—C—H
    |   |   |   |   |   |   |   |
    H   H   H   H   H   H   H   H
```

The chief use of petroleum is as a fuel source. Hydrocarbons burn well, and the heat released can be used to warm your house, drive the wheels of your automobile, push a jet aircraft through the sky, or spin the turbines in an electric power plant. Petroleum is also used as a source of chemicals for the production of fertilizers, insecticides, plastics, food preservatives, paints, inks, lubricants, detergents, solvents, and medicines. In other words, petroleum is the raw material for a large fraction of the things we use each and every day.

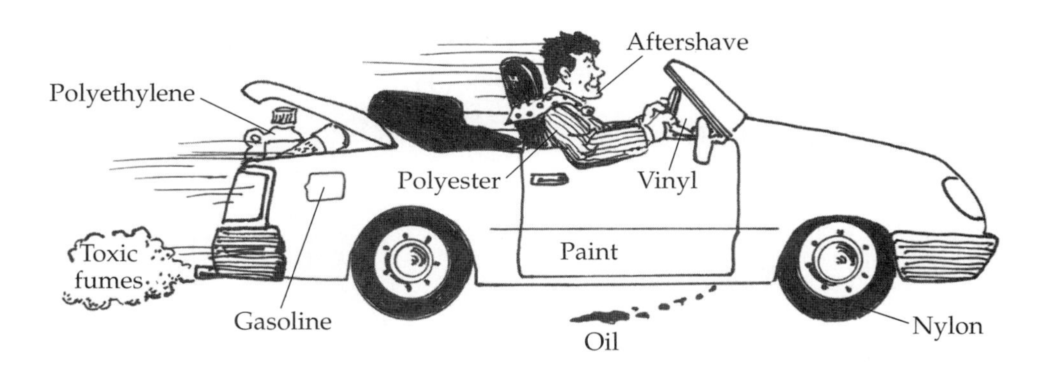

This leads to an ethical dilemma, for as we burn cheap petroleum to support our energy-intensive lifestyle, we are also polluting the environment and using up a limited resource. The most optimistic estimates are that, at our present rate of consumption, we shall have burned up most of the planet's petroleum reserves within 200 years. Research to develop alternative energy sources (such as nuclear power and conversion of sunlight, wind, and oceanic wave motion to electricity) will help extend our petroleum reserves. The pace of research and the deployment of these alternatives will increase as petroleum becomes increasingly scarce and we are forced to pay ever-higher prices for fuel, plastics, detergents, and the hundreds of other everyday items made from petroleum.

CHAINS OF CARBON

Because hydrocarbon molecules consist of chains of carbon atoms, chemists often describe them in terms of their *chain length*. The **chain length** is the number of carbon atoms in the longest continuous carbon chain in a hydrocarbon molecule. Our earlier hydrocarbon example, octane, has a chain length of 8:

```
    H   H   H   H   H   H   H   H
    |   |   |   |   |   |   |   |
H — C — C — C — C — C — C — C — C — H
    |   |   |   |   |   |   |   |
    H   H   H   H   H   H   H   H
```

Octane has a *linear* chain of carbon atoms. A **linear hydrocarbon** is one in which all but the two end carbons (usually called the *terminal carbons*) are attached to two other carbon atoms. We have to be careful with the term *linear* because the actual shape of the octane molecule is anything but a straight line. Recall from Chapter 6 that VSEPR theory predicts tetrahedral bond angles of 109.5° around each carbon atom. Therefore, the real shape of octane is more like

```
          H H     H H     H H     H H
           \ /     \ /     \ /     \ /
  H         C       C       C       C
   \       / \     / \     / \     / \
    C     C   C   C   C   C   C   C   H
   / \  109.5° / \     / \     / \
  H   H     H   H   H   H   H   H
```

All bond angles are tetrahedral.

Linear therefore refers to how the carbon atoms are connected, not to the actual shape of the molecule.

Hydrocarbons can be *branched* as well as linear:

```
    H   H   H   H   H   H   H   H
    |   |   |   |   |   |   |   |
H — C — C — C — C — C — C — C — C — H
    |   |   |   |   |   |   |   |
    H   H   H   H   |   H   H   H
                    |
                H — C — H
                    |
                H — C — H
                    |
                    H
```

A branched hydrocarbon

A **branched hydrocarbon** has at least one carbon atom attached to more than two other carbon atoms. You can also think of a branched hydrocarbon as having a backbone with a smaller branch growing off this backbone. The backbone—called the **main chain**—is the longest continuous chain in the molecule (not necessarily drawn in a straight line). The shorter chain or chains are the branches. In our example, the main chain is eight carbons long, which means its chain length is 8, and the branch has a chain length of 2:

```
    H   H   H   H   H   H   H   H
    |   |   |   |   |   |   |   |     <- Eight-carbon main chain
H — C — C — C — C — C — C — C — C — H
    |   |   |   |   |   |   |   |
    H   H   H   H   |   H   H   H
                    |
                H — C — H  )
                    |      } Two-carbon branch
                H — C — H  )
                    |
                    H
```

Consider the branched hydrocarbon

```
    H   H   H   H   H
    |   |   |   |   |
H — C — C — C — C — C — H
    |   |   |   |   |
    H   H   |   H   H
            |
        H — C — H
            |
        H — C — H
            |
        H — C — H
            |
            H
```

17.2 WORKPATCH

(a) What is the length of the main chain?
(b) What is the length of the branch?

Did you answer that the main chain length is 5 and the branch length is 3? If you did, you were fooled. Remember, the main chain is the *longest continuous* chain of carbon atoms in the molecule, but it does not have to be drawn in a straight line. Check your answer against ours.

One of the most useful rules for understanding the structure of hydrocarbon molecules is the cardinal rule of organic chemistry mentioned earlier: carbon always forms four covalent bonds in a hydrocarbon molecule. This rule is so reliable that chemists have taken advantage of it to develop a shorthand way of representing hydrocarbon molecules with what is called a **line drawing**. In this representation, we use short lines to show just the C–C bonds without any letters. A linked sequence of these lines represents the carbon "skeleton" of the molecule. The line drawing for octane looks like this:

```
/\/\/\/
```

A C atom is implicit wherever two lines connect.

A terminal C atom is implicit at the end of this line.

Each line stands for one C–C bond.

Each line in this zigzag represents a C–C bond, so there is a carbon atom implied at each point where two lines connect and at the two ends of the zigzag. The hydrogen atoms and C–H bonds are not shown. Here is how the line drawing of octane corresponds to the structural formula:

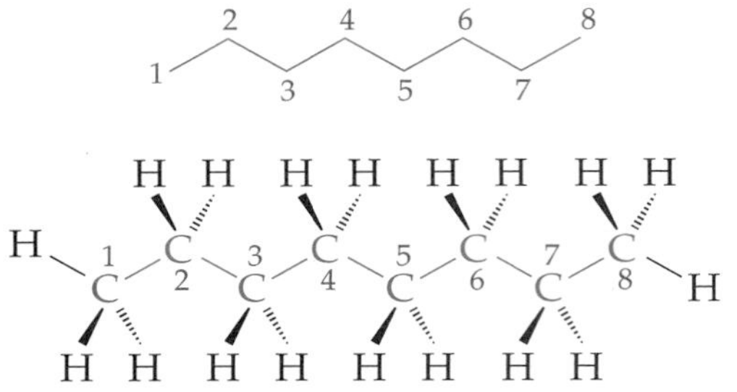

The C–H bonds are not shown in the line drawing because you can fill them in automatically by using the four-bond rule. For instance, look at the carbon at position 1 (called C1). The four-bond rule tells you that this carbon must have four bonds. The line drawing shows only one bond to this carbon (the line running from 1 to 2), so the other three bonds must be C–H bonds. The more detailed structural formula confirms this. The line drawing shows two bonds for each of the carbons at positions 2 through 7 (C2 through C7), so the additional two bonds required by the four-bond rule must be C–H bonds. The terminal carbon at position 8 (C8) is like C1, so it must have three attached hydrogens. We can also figure out the molecular formula from the line drawing: C_8H_{18}.

Double or triple bonds between carbons reduce the number of hydrogens in the molecule. For example, consider this eight-carbon molecule:

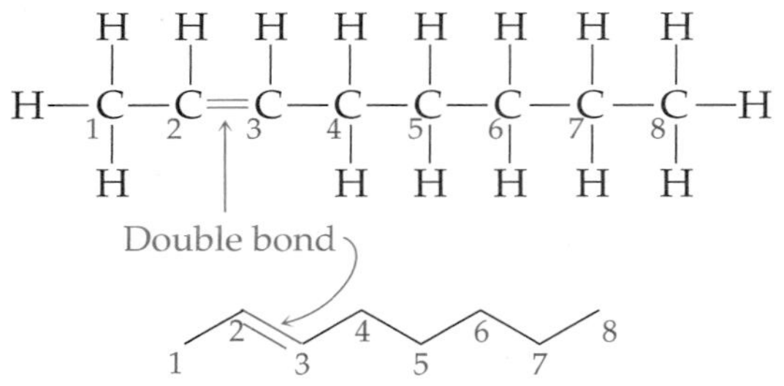

Double bond

Now C2 and C3 are connected by a double bond, so each atom needs only one hydrogen to satisfy the four-bond rule. The molecular formula for this molecule is therefore C_8H_{16}, two hydrogens fewer than in the octane molecule.

With triple bonds, there is one minor change in the way we show a line drawing. Because the C≡C triple bond is linear, it would be misleading to show one as part of a zigzag pattern. Instead, we show the three bonds between the carbons with three stacked lines and leave a space on either side when drawing in the single bond. The stack of three lines represents the C≡C bond, and you need to realize that there is a carbon atom on each end of the short lines:

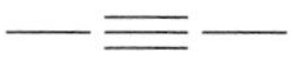

—≡— means C—C≡C—C

Thus for, say, 2-pentyne we draw

$$H_3C-C\equiv C-CH_2-CH_3$$

When interpreting what this line drawing represents, what you should see in your mind's eye is

```
C—C≡C—C
        \
         C
```

A chemist must be able to deduce the molecular formula for a hydrocarbon molecule from its line drawing, as well as be able to turn a line drawing into a structural formula and vice versa. The following practice problems give you a chance to do this.

Practice Problems

17.1 Give the molecular formula for each hydrocarbon molecule:

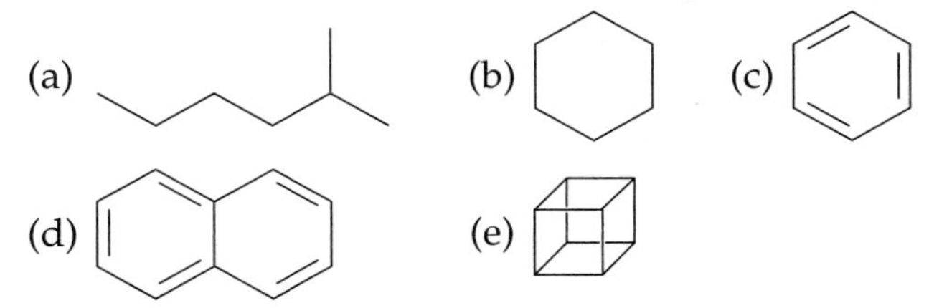

Answer: (a) C_7H_{16} *(b)* C_6H_{12} *(c)* C_6H_6 *(d)* $C_{10}H_8$ *(e)* C_8H_8

17.2 Draw structural formulas for the hydrocarbons in Practice Problem 17.1, showing all carbons, hydrogens, and bonds.

17.3 Create a line drawing representing the hydrocarbon whose structural formula is

```
    H  H        H  H
    |  |        |  |
 H—C—C—C≡C—C—C—H
    |  |        |  |
    H  H        H  H
```

17.4 Consider the hydrocarbon molecule

(a) Is it linear or branched? Explain how you can tell.
(b) What is the length of the main chain?

(c) What is the length of the branch chain?
(d) What is the molecular formula of this hydrocarbon?

SATURATED VERSUS UNSATURATED

You are probably familiar with some of the hydrocarbon molecules found in petroleum:

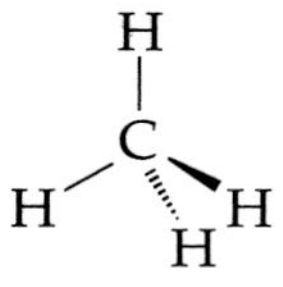

Methane, CH_4 | Ethane, C_2H_6 | Propane, C_3H_8 | Butane, C_4H_{10}

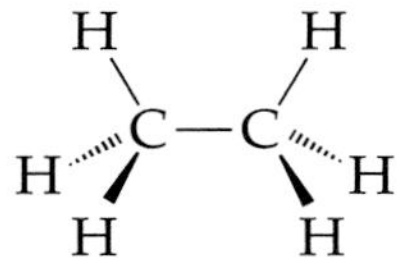

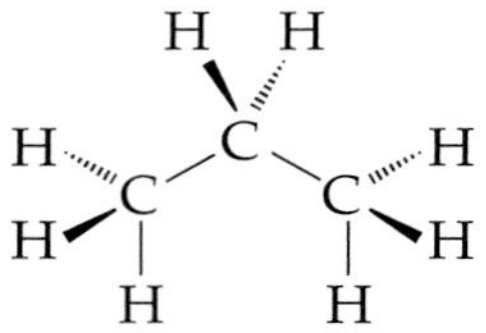

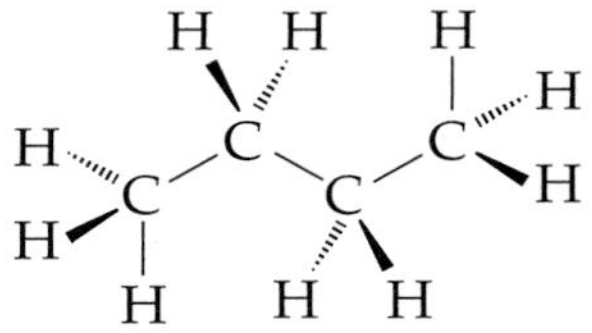

Methane and ethane are the smallest hydrocarbon molecules. They are gases at room temperature and are the principal components of the natural gas burned in home stoves and furnaces. Natural gas is relatively inexpensive, burns cleanly, and is easy to transport through pipes. Propane, also a gas at room temperature, is often used in torches. Butane, at chain length 4, is also a gas at room temperature, but it can be easily compressed to a liquid. This is the fuel found in disposable lighters.

These simple hydrocarbons all have single bonds between the carbon atoms, but as we have seen, hydrocarbons can also contain double and triple bonds. For example, there are three ways to make a hydrocarbon that has a chain length of 2:

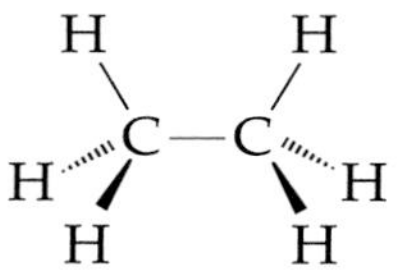

Ethane, C_2H_6 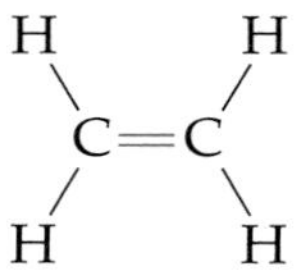 | Ethene, C_2H_4 | $H-C\equiv C-H$ Ethyne, C_2H_2

In all three molecules, each carbon atom has four covalent bonds. However, the molecules have different numbers of hydrogen atoms. The hydrocarbon molecule with all single bonds has the most hydrogens (C_2H_6). The hydrocarbon molecule with the triple bond has the fewest hydrogens (C_2H_2). To talk about these differences, we use the terms *saturated* and *unsaturated*. A hydrocarbon molecule that contains the maximum possible number of hydrogen atoms is a **saturated hydrocarbon**. A hydrocarbon molecule with the same number of carbons but fewer hydrogens is an **unsaturated hydrocarbon**. Unsaturated molecules have fewer hydrogens because they have double or triple bonds. In the three structures we are looking at here, ethane is saturated and ethene and ethyne are unsaturated.

You have probably heard the terms *saturated* and *unsaturated* used in reference to fats and oils. Unsaturated oils contain double bonds in their linear hydrocarbon portions. These oils may be healthier to eat than saturated fats because the body doesn't metabolize the oils into artery-clogging cholesterol (although recent evidence suggests things are not quite this simple).

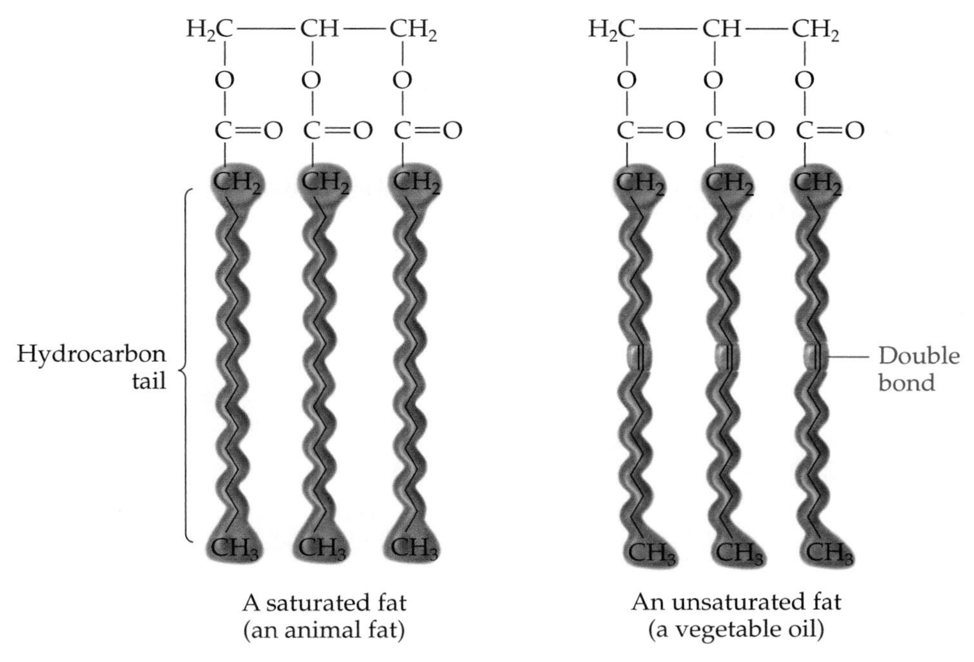

A saturated fat (an animal fat) — An unsaturated fat (a vegetable oil)

It is important to remember that the multiple bonds in hydrocarbon molecules that are responsible for unsaturation occur *only between carbon atoms*. The carbon–hydrogen bonds are *always* single bonds.

Practice Problems

17.5 A particular linear hydrocarbon molecule has six carbons and ten hydrogens. Is it unsaturated or saturated?

Answer: Unsaturated. The best way to discover this is to make a line drawing for a saturated six-carbon hydrocarbon (which means you give the molecule all C–C single bonds) and then count the number of hydrogens:

1 3 5
2 4 6

Carbons 1 and 6 have three hydrogens each. Carbons 2 through 5 have two hydrogens each. The total number of hydrogens is therefore 14. This is the maximum number of hydrogens a six-carbon chain can have. The hydrocarbon you are asked about here has only ten hydrogens, so it is unsaturated.

17.6 Draw all possible line drawings for the unsaturated hydrocarbon molecule in Practice Problem 17.5.

17.7 Which hydrocarbon molecule is least unsaturated? Explain your answer.

(a) (b) (c)

17.3 Naming Hydrocarbons

As they discovered more and more hydrocarbon molecules, chemists ran into a problem. Because carbon atoms are so good at forming chains of various lengths and are so good at forming single, double, and triple bonds, the possible number of different hydrocarbon molecules is astronomical. There can even be many different hydrocarbon molecules that have the same molecular formula. Consider, for example, the formula C_5H_{12}. Three different hydrocarbon molecules have this formula:

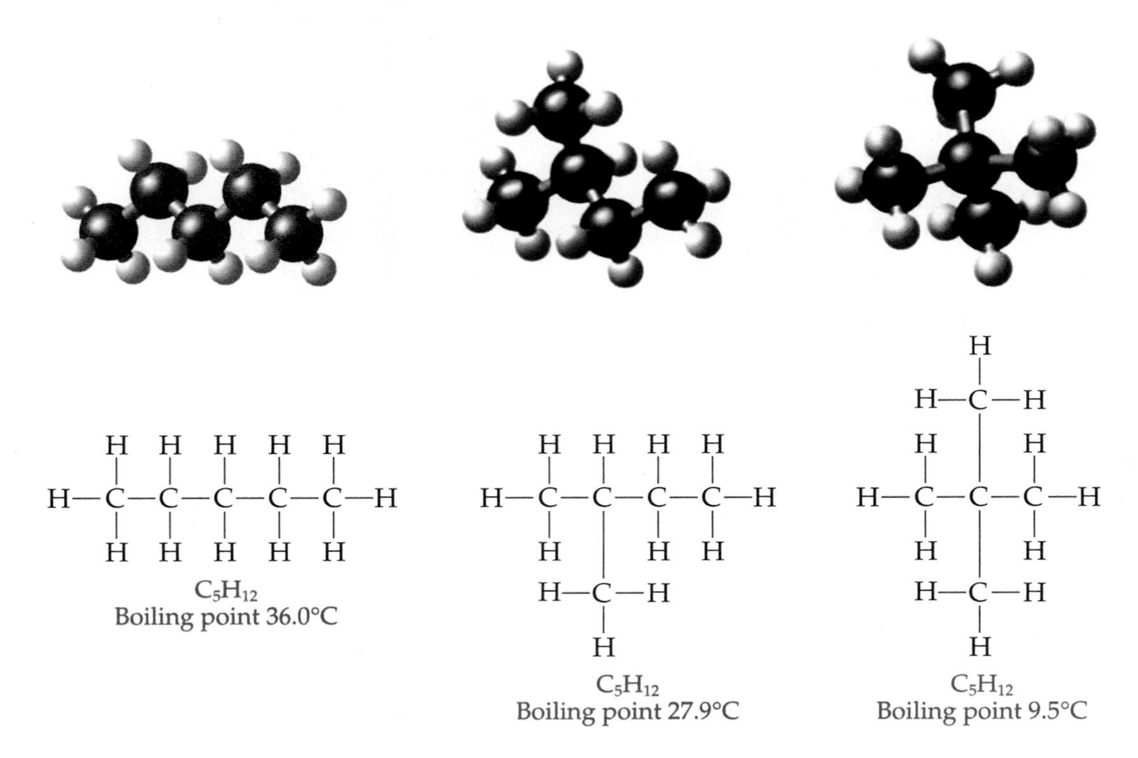

These three hydrocarbons are all saturated, and all have the same molecular formula, but they have different structures, as you can see. Because these molecules are put together differently, they are different compounds, possessing different physical and chemical properties. For example, their boiling points are different, as indicated below the structures. We call these molecules *isomers* of one another. **Isomers** are molecules that have the same molecular formula but different structural formulas.

How shall we name all these molecules? We could use common names, but they wouldn't give us any clues as to the molecular and structural formulas of the compounds. For example, water is the common name for the molecule H_2O, but this name tells us nothing about the molecule; you must simply memorize that water means H_2O. Even if we wanted to make up a different common name for every different hydrocarbon molecule, that approach would quickly get out of hand. While there are only 3 isomers with the formula C_5H_{12}, there are 5 with the formula C_6H_{14}, 75 with the formula $C_{10}H_{22}$, and more than 4 billion with the formula $C_{30}H_{62}$. What we need is a naming system that never runs out of names and somehow instantly paints in our minds a picture of the structure of the molecule.

In Chapter 5 you learned about a better naming system. For example, in this system H_2O is called dihydrogen oxide, a name that clearly indicates the molecular formula. This naming system is known as the **IUPAC nomenclature system**. It does what we want, although the rules for naming organic compounds are a bit more involved than those for naming simple compounds like water. We shall now introduce you to the basics of this system. The best way to start is with linear hydrocarbons.

The IUPAC nomenclature system for organic molecules begins by developing a general name for all simple hydrocarbons. Years ago, organic chemists were busy studying the structure of fats and oils. The Greek word for fat is *aleiphas*. Because, as we have seen, fats and oils contain long chains of carbon atoms, organic chemists began referring to long-chain hydrocarbons as *aliphatic hydrocarbons*. The root *alk-* later came into general use in the names of these molecules. As a result, simple hydrocarbons that have only C–C single bonds are called **alk*anes***. Hydrocarbons that have at least one C=C double bond are called **alk*enes***, and those that have at least one C≡C triple bond are called **alk*ynes*** (pronounced alk-EYE-ns).

```
    H  H  H             H  H  H                      H
    |  |  |             |  |  |                      |
 H—C—C—C—H          H—C=C—C—H          H—C≡C—C—H
    |  |  |                   |                      |
    H  H  H                   H                      H
```

An alk*ane* An alk*ene* An alk*yne*

Thus, the suffixes *-ane*, *-ene*, and *-yne* have special significance. If you see *-ene* in the name of a hydrocarbon, you automatically know the molecule has at least one C=C double bond. However, the words *alkane*, *alkene*, and *alkyne* are general terms and represent a chain of any length. We want a name that also tells us the length of the main chain for a given molecule. To do this, we replace the *alk-* in these three words with a root name that indicates the length of the main chain. Table 17.1 on page 684 lists the names for main chains having between one and ten carbon atoms.

From Table 17.1 you can see that the root *meth-*, as in *meth*ane, means that the hydrocarbon chain contains only one carbon. For this reason, there can be no methene or methyne. The root *eth-* (as in *eth*ane, *eth*ene, and *eth*yne) means the hydrocarbon chain contains two carbons, *prop-* (pronounced pr-oh-p) means three carbons, and *but-* (pronounced b-you-t) means four carbons. For five carbons and beyond, we use Greek number prefixes: *pent-* (5), *hex-* (6), *hept-* (7), *oct-* (8), *non-* (9, pronounced n-oh-n), and *dec-* (10), as indicated in Table 17.2.

As Table 17.1 illustrates, hexane is a linear six-carbon molecule with all single bonds. Hexene and hexyne are the same except that they have one C=C double bond and one C≡C triple bond, respectively. Even though we are not done showing you the basics of this nomenclature system, you can begin to see how the name tells you what the molecule looks like.

The compounds listed in the three columns of Table 17.1 represent three *homologous series*. A **homologous series** is one in which all the compounds in the series are similar. For example, the compounds from methane, CH_4, to

Table 17.1 Hydrocarbons of Increasing Chain Length

Chain length	Alkanes	Alkenes	Alkynes
1	H H—C—H H Methane	—	—
2	H H H—C—C—H H H Ethane	H H C=C H H Ethene	H—C≡C—H Ethyne
3	H H H H—C—C—C—H H H H Propane	H H H C=C—C—H H H Propene	H H—C≡C—C—H H Propyne
4	H H H H H—C—C—C—C—H H H H H Butane	H H H H C=C—C—C—H H H H Butene	H H H—C≡C—C—C—H H H Butyne
5	H H H H H H—C—C—C—C—C—H H H H H H Pentane	H H H H H C=C—C—C—C—H H H H H Pentene	H H H H—C≡C—C—C—C—H H H H Pentyne
6	H H H H H H H—C—C—C—C—C—C—H H H H H H H Hexane	H H H H H H C=C—C—C—C—C—H H H H H H Hexene	H H H H H—C≡C—C—C—C—C—H H H H H Hexyne
7	H H H H H H H H—C—C—C—C—C—C—C—H H H H H H H H Heptane	H H H H H H H C=C—C—C—C—C—C—H H H H H H H Heptene	H H H H H H—C≡C—C—C—C—C—C—H H H H H H Heptyne
8	H H H H H H H H H—C—C—C—C—C—C—C—C—H H H H H H H H H Octane	H H H H H H H H C=C—C—C—C—C—C—C—H H H H H H H H Octene	H H H H H H H—C≡C—C—C—C—C—C—C—H H H H H H H Octyne
9	H H H H H H H H H H—C—C—C—C—C—C—C—C—C—H H H H H H H H H H Nonane	H H H H H H H H H C=C—C—C—C—C—C—C—C—H H H H H H H H H Nonene	H H H H H H H H—C≡C—C—C—C—C—C—C—C—H H H H H H H H Nonyne
10	H H H H H H H H H H H—C—C—C—C—C—C—C—C—C—C—H H H H H H H H H H H Decane	H H H H H H H H H H C=C—C—C—C—C—C—C—C—C—H H H H H H H H H H Decene	H H H H H H H H H—C≡C—C—C—C—C—C—C—C—C—H H H H H H H H H Decyne

Table 17.2 Names for Indicating Chain Length in Hydrocarbons

Chain length	Root name	Alkane example	Chain length	Root name	Alkane example
1	meth-	*meth*ane	8	oct-	*oct*ane
2	eth-	*eth*ane	9	non-	*non*ane
3	prop-	*prop*ane	10	dec-	*dec*ane
4	but-	*but*ane	11	undec-	*undec*ane
5	pent-	*pent*ane	12	dodec-	*dodec*ane
6	hex-	*hex*ane	20	eicos-	*eicos*ane
7	hept-	*hept*ane	30	triacont-	*triacont*ane

decane, $C_{10}H_{22}$, and beyond make up one homologous series of compounds. They are all alkanes because they have only C–C single bonds. In addition, they all have the same general formula C_nH_{2n+2}, where n is the number of carbon atoms in the molecule. Table 17.3 shows the beginning of this homologous series.

Table 17.3 The Alkanes: A Homologous Series of Hydrocarbons Having the General Formula C_nH_{2n+2}

Chain length n

```
            H
            |
1       H—C—H   Methane, CH4
            |
            H

            H   H
            |   |
2       H—C—C—H   Ethane, C2H6
            |   |
            H   H

            H   H   H
            |   |   |
3       H—C—C—C—H   Propane, C3H8
            |   |   |
            H   H   H

            H   H   H   H
            |   |   |   |
4       H—C—C—C—C—H   Butane, C4H10
            |   |   |   |
            H   H   H   H

            H   H   H   H   H
            |   |   |   |   |
5       H—C—C—C—C—C—H   Pentane, C5H12
            |   |   |   |   |
            H   H   H   H   H
```

All the compounds listed under alkenes in Table 17.1 make up another homologous series. All these alkenes have one double bond. Because they have only one double bond, we also refer to them as **monoalkenes**. They all have the general formula C_nH_{2n}, where again n is the number of carbons in the molecule.

Table 17.4 The Monoalkenes: A Homologous Series of Hydrocarbons Having the General Formula C_nH_{2n}

Chain length *n*	
2	$\begin{array}{ccc} H & & H \\ \vert & & \vert \\ C & = & C \\ \vert & & \vert \\ H & & H \end{array}$ Ethene, C_2H_4
3	$\begin{array}{ccccccc} H & & H & & H & & \\ \vert & & \vert & & \vert & & \\ C & = & C & - & C & - & H \\ \vert & & & & \vert & & \\ H & & & & H & & \end{array}$ Propene, C_3H_6
4	$\begin{array}{ccccccccc} H & & H & & H & & H & & \\ \vert & & \vert & & \vert & & \vert & & \\ C & = & C & - & C & - & C & - & H \\ \vert & & & & \vert & & \vert & & \\ H & & & & H & & H & & \end{array}$ Butene, C_4H_8
5	$\begin{array}{ccccccccccc} H & & H & & H & & H & & H & & \\ \vert & & \vert & & \vert & & \vert & & \vert & & \\ C & = & C & - & C & - & C & - & C & - & H \\ \vert & & & & \vert & & \vert & & \vert & & \\ H & & & & H & & H & & H & & \end{array}$ Pentene, C_5H_{10}

Note that the general formula for a monoalkene, C_nH_{2n}, has two fewer hydrogen atoms than the general formula for an alkane, C_nH_{2n+2}. As we have seen, every double bond in a chain of *n* carbon atoms reduces by two the number of hydrogen atoms found in the alkane containing *n* carbon atoms.

Finally, all the compounds listed under alkynes in Table 17.1 make up yet another homologous series of hydrocarbon molecules. All these alkynes have a C≡C triple bond. Because they have only one triple bond, we also call them **monoalkynes**, and they all have the general formula C_nH_{2n-2}. With two fewer hydrogens than an alkene of the same chain length, alkynes are more unsaturated than alkenes.

Table 17.5 The Monoalkynes: A Homologous Series of Hydrocarbons Having the General Formula C_nH_{2n-2}

Chain length *n*	
2	$H-C\equiv C-H$ Ethyne, C_2H_2
3	$\begin{array}{ccccccc} & & & & H & & \\ & & & & \vert & & \\ H & - & C\equiv C & - & C & - & H \\ & & & & \vert & & \\ & & & & H & & \end{array}$ Propyne, C_3H_4
4	$\begin{array}{ccccccccc} & & & & H & & H & & \\ & & & & \vert & & \vert & & \\ H & - & C\equiv C & - & C & - & C & - & H \\ & & & & \vert & & \vert & & \\ & & & & H & & H & & \end{array}$ Butyne, C_4H_6
5	$\begin{array}{ccccccccccc} & & & & H & & H & & H & & \\ & & & & \vert & & \vert & & \vert & & \\ H & - & C\equiv C & - & C & - & C & - & C & - & H \\ & & & & \vert & & \vert & & \vert & & \\ & & & & H & & H & & H & & \end{array}$ Pentyne, C_5H_8

The IUPAC nomenclature system as we have described it so far works fine for alkanes, but we have to do more for alkenes and alkynes. For example, consider the molecule butene. The *but-* tells you there are four carbon atoms in the main chain. The *-ene* tells you that there is a C=C double bond. Put it all together, and the formula must be C_4H_8 (C_nH_{2n}, with $n = 4$). But there are three ways to draw this molecular formula:

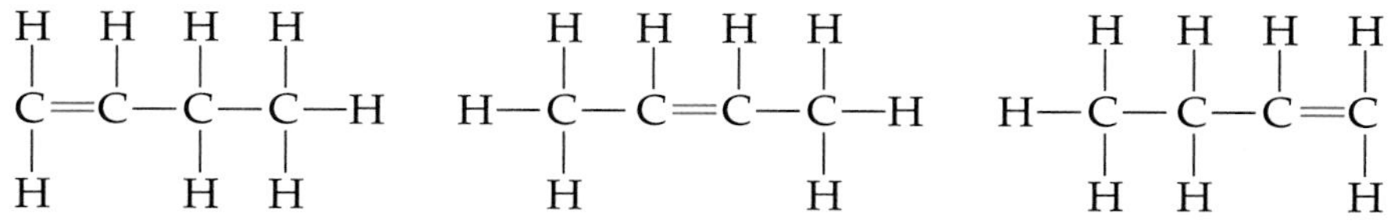

```
H  H  H  H              H  H  H  H                H  H  H  H
|  |  |  |              |  |  |  |                |  |  |  |
C=C—C—C—H           H—C—C=C—C—H           H—C—C—C=C
|     |  |              |        |                |  |     |
H     H  H              H        H                H  H     H
   C4H8                    C4H8                      C4H8
```

Where should we put the C=C double bond? Does it matter where we put it? Let's consider the left and right versions first. These are really the same structure because if we spin the left version about the dashed-line axis shown here, we generate the right version:

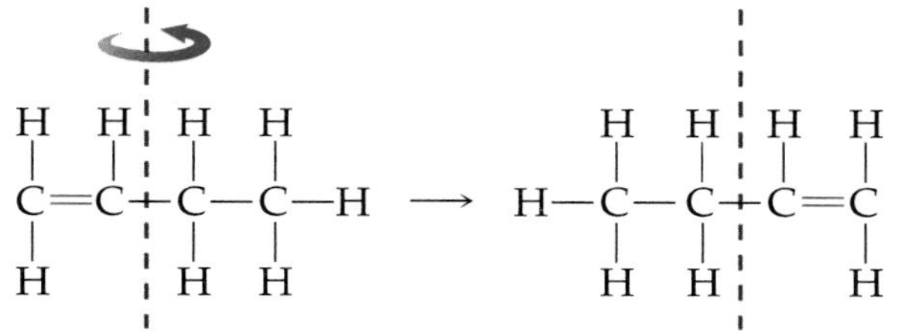

```
H  H ¦ H  H                 H  H ¦ H  H
|  | ¦ |  |                 |  | ¦ |  |
C=C-+-C—C—H   ⟶   H—C—C-+-C=C
|    ¦ |  |                 |  | ¦    |
H    ¦ H  H                 H  H ¦    H
```

Anytime you can convert one molecule to another by simply reorienting one of them in space, the molecules are identical.

However, the butene with the double bond in the middle is unique. No amount of rotating or flipping will make this molecule the same as the others. In fact, these two butenes have different physical properties:

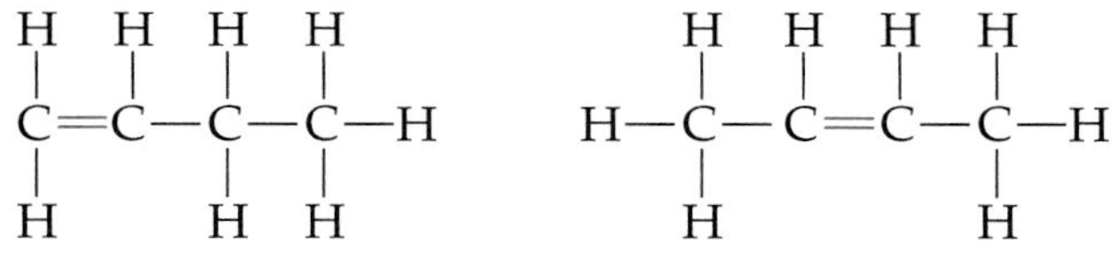

```
H  H  H  H                 H  H  H  H
|  |  |  |                 |  |  |  |
C=C—C—C—H              H—C—C=C—C—H
|     |  |                 |        |
H     H  H                 H        H
Melting point −185°C       Melting point −139°C
Boiling point −6.3°C       Boiling point 3.7°C
```

What we have here are two compounds that have the same molecular formula, C_4H_8, but different structural formulas. They are isomers, just as the linear and branched versions of C_5H_{12} we saw earlier are isomers. Because these two butene molecules are different compounds, they need different names. The IUPAC nomenclature system takes care of this by numbering the carbon atoms sequentially. This numbering allows us to create two names:

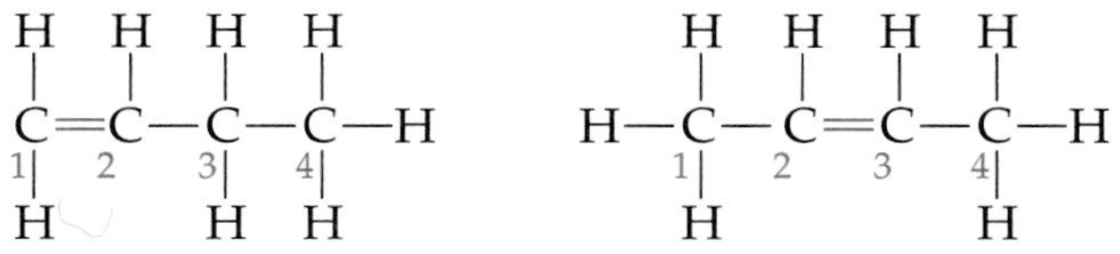

```
H  H  H  H                 H  H  H  H
|  |  |  |                 |  |  |  |
C=C—C—C—H              H—C—C=C—C—H
1| 2  3| 4|                1| 2  3  4|
H     H  H                 H        H
  1-Butene                  2-Butene
```

The isomer with the double bond between C1 and C2 is called 1-butene. The number in this name indicates the lower-numbered carbon sharing the double bond. The isomer with the double bond between C2 and C3 is called 2-butene.

You have to be careful with this numbering system when assigning numbers to the carbon atoms. For example, we began numbering the carbons from the left, but why not number them from the right? That would not change the name of the butene isomer with the double bond in the middle. In this particular case, both ways of numbering the carbon atoms give the name 2-butene:

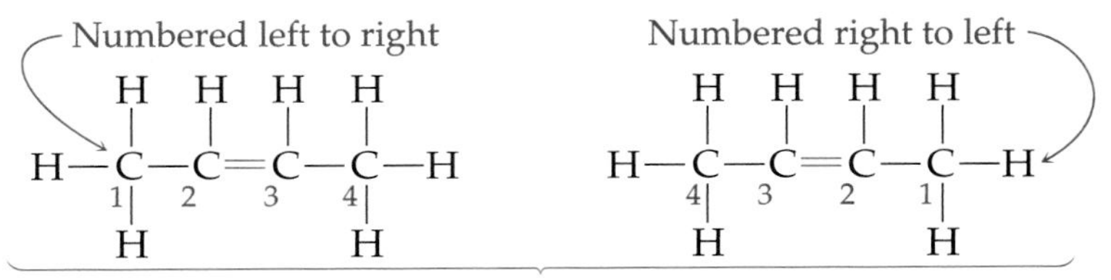

2-Butene—the name is the same either way.

However, the direction of numbering does affect the name of the other butene isomer:

```
Numbered left to right          Numbered right to left
 H  H  H  H                      H  H  H  H
 |  |  |  |                      |  |  |  |
 C==C--C--C--H                   C==C--C--C--H
1|  2 3|  4|                    4|  3 2|  1|
 H     H  H                      H     H  H
   1-Butene                        3-Butene
```

The IUPAC system solves this problem by insisting that you number carbon atoms so that the double-bonded carbons are assigned the *lowest* possible numbers:

```
Start numbering here
 H  H  H  H
 |  |  |  |
 C==C--C--C--H
1| ↑2 3|  4|
 H  \  H  H
     Double bond
```

Therefore, 1-butene is the only correct name for this compound.

The same rule applies to a chain containing a triple bond: number the chain so that the triple-bonded carbons get the lowest possible numbers. Try your hand at this now.

What is the IUPAC name of the hydrocarbon

```
   H  H  H  H  H  H  H
   |  |  |  |  |  |  |
H--C--C--C--C--C==C--C--H
   |  |  |  |        |
   H  H  H  H        H
```

Did you consult Table 17.1 for the proper root name? Did you number the chain starting at the end nearer the double bond? Check your answer against ours.

All the hydrocarbons we have named so far have been linear molecules. But more numerous are the branched hydrocarbons, which have one or more branches attached to a main chain. Let's begin learning how to name these molecules with the branched alkane

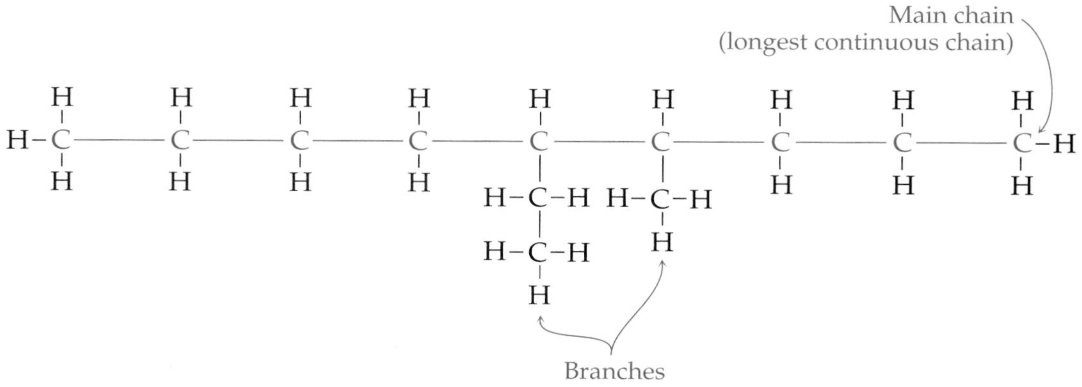

Main chain (longest continuous chain)

Branches

Naming a branched hydrocarbon takes three steps:

Step 1 Find and name the main chain, which is the longest continuous chain in the molecule.

Step 2 Name the branches that run off the main chain.

Step 3 Give the location of each branch on the main chain.

Let's follow these steps to name the above alkane:

Step 1 The longest continuous chain is nine carbons long. There are no carbon–carbon multiple bonds, so you can see from Table 17.2 that this main chain is called nonane.

Step 2 Hanging off the main chain are a one-carbon branch and a two-carbon branch. The one-carbon branch is CH_3, which looks a lot like methane, CH_4, except that one hydrogen is missing. So we call the CH_3 branch a *methyl* branch. The two-carbon branch is CH_3CH_2, which looks a lot like ethane, CH_3CH_3, except that, once again, a single hydrogen is missing. So we call the CH_3CH_2 branch an *ethyl* branch.

Indeed, any alkane can serve as a branch by losing a single hydrogen atom so that the alkane can bond to the main chain. The branch is then referred to as an *alkyl* branch, with the *-yl* suffix replacing the *-ane* in *alkane*:

Chain (alk*ane*)	**Branch (alk*yl*)**
Methane, CH_4	Methyl, $-CH_3$
Ethane, CH_3CH_3	Ethyl, $-CH_2CH_3$
Propane, $CH_3CH_2CH_3$	Propyl, $-CH_2CH_2CH_3$

So our hydrocarbon molecule is a nonane with a methyl branch and an ethyl branch. In the IUPAC system, the branch names precede the name of the main chain and are listed in alphabetical order. Therefore, we could call this molecule ethylmethylnonane.

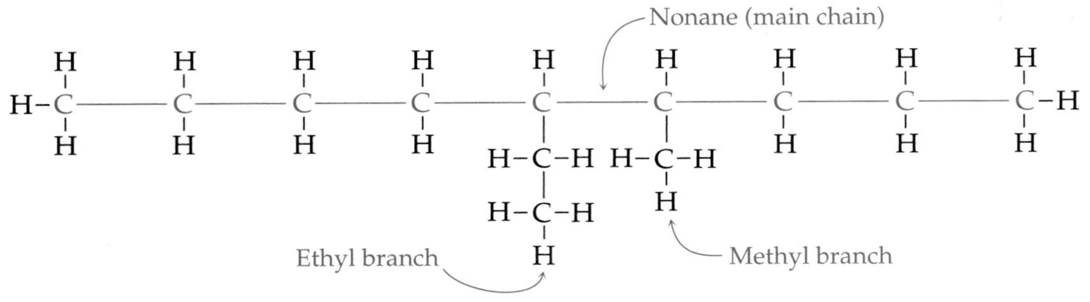

Nonane (main chain)

Ethyl branch

Methyl branch

Ethylmethylnonane

Step 3 All that is left is to give the location of each branch on the main chain. To do this, we number the main chain, but once again, we can number either left to right or right to left:

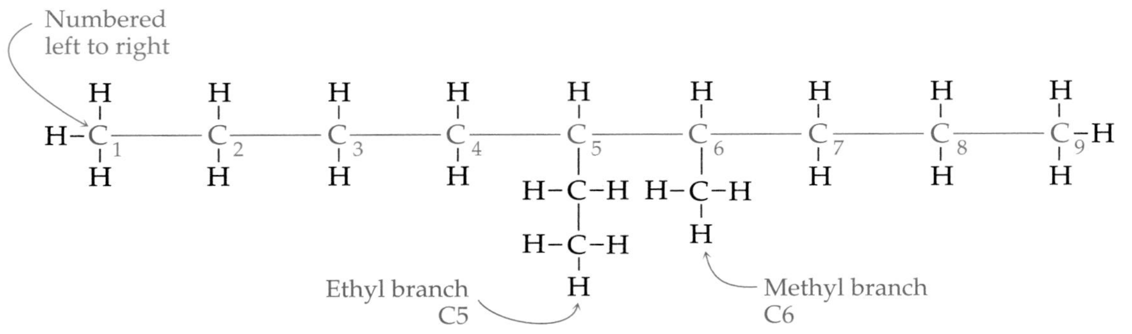

Numbered left to right

Ethyl branch C5

Methyl branch C6

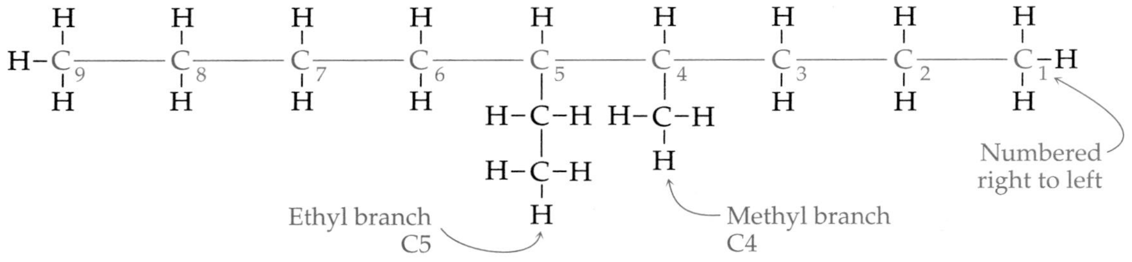

Ethyl branch C5

Methyl branch C4

Numbered right to left

Which is correct? The answer is the same as for multiple bonds: number the chain in the direction that gives the lowest numbers to the branches. Look again at our example, and you will see that the main chain should be numbered from the right because that end is closer to a branch. The correct name of our compound is therefore 5-ethyl-4-methylnonane.

There are two things to notice about this name. First, the IUPAC rules say that the position number for each branch precedes the name of the branch and is set off by hyphens. Second, in accordance with the alphabetical rule, the branches are listed in alphabetical order (*not* in numeric order).

There is one more possibility to deal with. What do you do if a molecule has two or more of the same branch? For example, suppose our nonane had two methyl branches:

$$\begin{array}{ccccccccccccccccc}
9 & & 8 & & 7 & & 6 & & 5 & & 4 & & 3 & & 2 & & 1 \\
CH_3 & - & CH_2 & - & CH_2 & - & CH_2 & - & CH & - & CH & - & CH_2 & - & CH_2 & - & CH_3 \\
 & & & & & & & & | & & | & & & & & & \\
 & & & & & & & & CH_3 & & CH_3 & & & & & &
\end{array}$$

Two branches of the same type

This molecule is *not* named 4-methyl-5-methylnonane. Instead, it is named 4,5-dimethylnonane. The numbers representing branches of the same type get lumped together in the name, separated from one another by commas and from the first letter of the name by a hyphen. We then use the Greek prefixes *di-*, *tri-*, *tetra-*, and so on, to indicate how many of that type of branch are in the molecule.

It's time for you to try your hand at this. Perhaps the most challenging part of naming a branched hydrocarbon is finding the main chain, so your first task is always to identify the longest continuous chain in the molecule. Also, remember to number the hydrocarbon from the end nearer a branch.

Name these branched hydrocarbons:

$$\text{(a)}\quad \begin{array}{ccccc}
 & & CH_3 & & \\
 & & | & & \\
CH_3 & - & C & - & CH_3 \\
 & & | & & \\
 & & CH_3 & &
\end{array}
\qquad
\text{(b)}\quad \begin{array}{ccccccccc}
 & & & & & & CH_3 & & \\
 & & & & & & | & & \\
CH_3 & - & CH_2 & - & CH_2 & - & C & - & CH_3 \\
 & & & & & & | & & \\
 & & & & & & CH_2 & & \\
 & & & & & & | & & \\
 & & & & & & CH_3 & &
\end{array}$$

$$\text{(c)}\quad \begin{array}{ccccccccc}
 & & CH_3 & & & & CH_3 & & \\
 & & | & & & & | & & \\
 & & CH_2 & & & & CH_2 & & \\
 & & | & & & & | & & \\
CH_3 & - & CH & - & CH_2 & - & CH & - & CH_3
\end{array}$$

Make sure you can do WorkPatch 17.4, because we are going to make things a little more complicated now. Were you able to see that the main chain in (c) is *seven* carbons long? Remember, there's no rule that says you must start counting at either the leftmost or rightmost carbon in a drawing. In this molecule, the longest chain is in the shape of the letter U.

The final level of naming complexity for this book is to name a branched unsaturated hydrocarbon, such as

$$\begin{array}{ccccccccccccccccc}
 & & & & CH_2 & & & & & & & & & & & & \\
 & & & & \| & & & & & & & & & & & & \\
CH_3 & - & CH_2 & - & C & - & CH_2 & - & CH & - & CH & - & CH_2 & - & CH_2 & - & CH_3 \\
 & & & & & & & & | & & | & & & & & & \\
 & & & & & & & & CH_2 & & CH_2 & & & & & & \\
 & & & & & & & & | & & | & & & & & & \\
 & & & & & & & & CH_2 & & CH_2 & & & & & & \\
 & & & & & & & & | & & | & & & & & & \\
 & & & & & & & & CH_3 & & CH_3 & & & & & &
\end{array}$$

Double bond (unsaturation)

As always, we start by finding the main chain, but now there is a catch. If the hydrocarbon is unsaturated, the main chain *must* contain the multiple bond.

This means that in some molecules the main chain won't be the longest chain. Such is the case in our example:

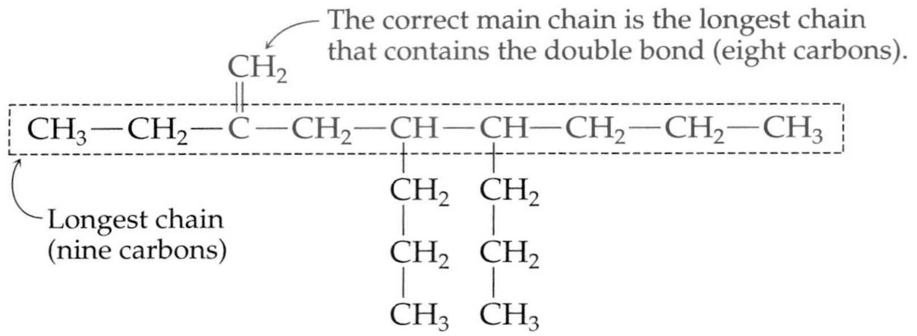

The correct main chain is the longest chain that contains the double bond (eight carbons).

$$\begin{array}{ccccccccc} & & CH_2 & & & & & & \\ & & \| & & & & & & \\ CH_3— & CH_2— & C— & CH_2— & CH— & CH— & CH_2— & CH_2— & CH_3 \\ & & & & | & | & & & \\ & & & & CH_2 & CH_2 & & & \\ & & & & | & | & & & \\ & & & & CH_2 & CH_2 & & & \\ & & & & | & | & & & \\ & & & & CH_3 & CH_3 & & & \end{array}$$

Longest chain (nine carbons)

Thus this molecule is an octene with one ethyl and two propyl branches. Now all we need to do is number the main chain to locate the positions of the branches and of the double bond. We're supposed to number the main chain from the end nearer the double bond—but also from the end nearer a branch point. Which gets priority? In this molecule, we don't have to worry about priority because one of the branches and the double bond are at the same end of the chain. That convenient circumstance means our numbering is

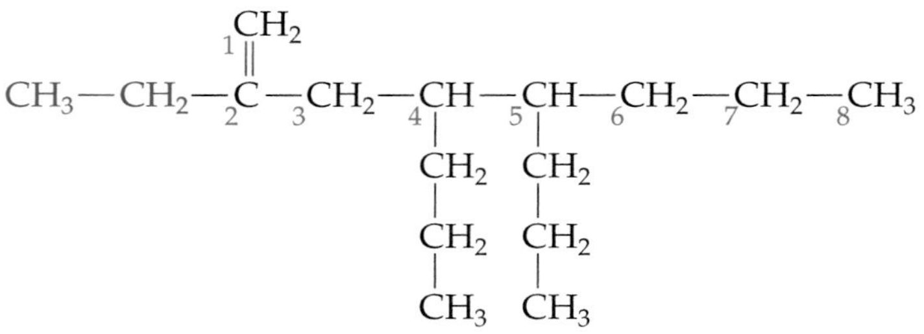

$$\begin{array}{ccccccccc} & & {}_1CH_2 & & & & & & \\ & & \| & & & & & & \\ CH_3— & CH_2— & {}_2C— & {}_3CH_2— & {}_4CH— & {}_5CH— & {}_6CH_2— & {}_7CH_2— & {}_8CH_3 \\ & & & & | & | & & & \\ & & & & CH_2 & CH_2 & & & \\ & & & & | & | & & & \\ & & & & CH_2 & CH_2 & & & \\ & & & & | & | & & & \\ & & & & CH_3 & CH_3 & & & \end{array}$$

Now we have all the information we need to name this molecule: 2-ethyl-4,5-dipropyl-1-octene. Notice that we've put *dipropyl* after *ethyl*, seeming to break our rule about alphabetical order. The trick is that when arranging the branch names alphabetically, we ignore the Greek prefixes *di-*, *tri-*, and so on. Here, therefore, our alphabetizing means putting *ethyl* before *propyl*.

Before you try your hand at all this in a WorkPatch, let's look at a molecule where you do have to make a choice about which end you'll choose for carbon 1. Suppose we are naming the molecule shown at the top of the facing page. Remember, the IUPAC naming system gives priority to multiple bonds, meaning the multiple bond must get the lowest number possible. Thus, this molecule is correctly numbered left to right and named 6,7-dimethyl-3-propyl-1-octene.

The beauty of the IUPAC naming system is not only that it gives a unique name for each hydrocarbon but also that the name can paint a picture of that molecule in your mind. Try it now.

17.5 WORKPATCH

Draw the structural formula for 2-methyl-1-butene. [*Hint:* Start by drawing the main chain, number the carbons, and then position the branches and the multiple bond.] Now draw the line drawing for this molecule.

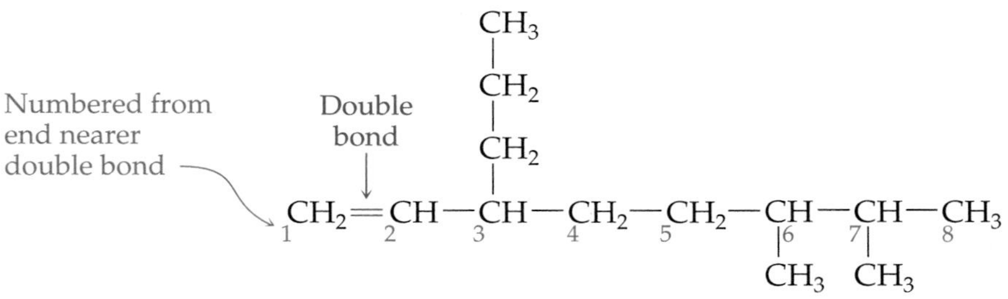

Numbered from end nearer double bond

Double bond

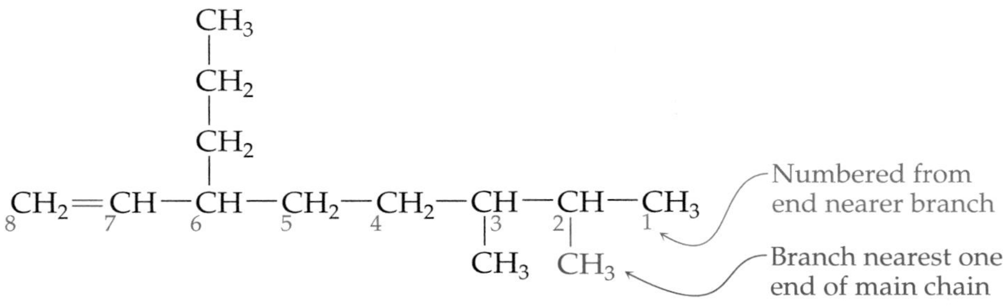

$$CH_2{=}CH{-}CH(CH_2CH_2CH_3){-}CH_2{-}CH_2{-}CH(CH_3){-}CH(CH_3){-}CH_3$$

(numbered 1 2 3 4 5 6 7 8 from left)

$$CH_2{=}CH{-}CH(CH_2CH_2CH_3){-}CH_2{-}CH_2{-}CH(CH_3){-}CH(CH_3){-}CH_3$$

(numbered 8 7 6 5 4 3 2 1 from left)

Numbered from end nearer branch

Branch nearest one end of main chain

When you can do this WorkPatch correctly, you will know that you understand the basics of the IUPAC nomenclature system. The system can also tackle molecules with more than one double or triple bond and combinations of doubles and triples, although we shall not consider such cases in this book.

The next time you put on some deodorant or shampoo your hair, stop and read the product label. You will see IUPAC names of hydrocarbon molecules—names that will now have some meaning for you.

Practice Problems

17.8 Give the IUPAC names for

(a) 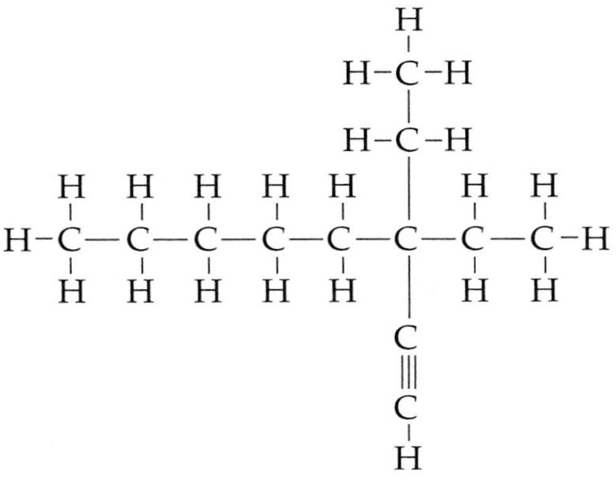 $H{-}CH_2{-}CH_2{-}CH_2{-}CH_2{-}CH_2{-}C(CH_2CH_3)(C{\equiv}C{-}H){-}CH_2{-}CH_2{-}H$

[*Hint:* It's an alkyne with a main chain of length 8.]

(b) 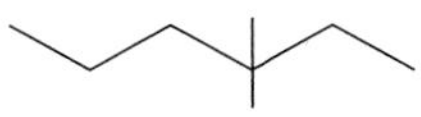 (c)

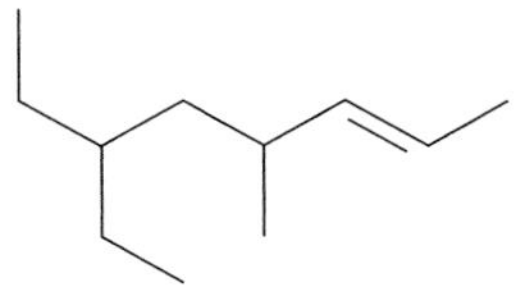

Answers:
(a) 3,3-Diethyl-1-octyne *(b) 3,3-Dimethylhexane* *(c) 6-Ethyl-4-methyl-2-octene*

17.9 Draw the structural formula, showing all C and H atoms, for 6-methyl-4-propyl-2-octene.

17.10 This line drawing is numbered and named incorrectly. First correct the numbering error, and then give the correct name for the molecule.

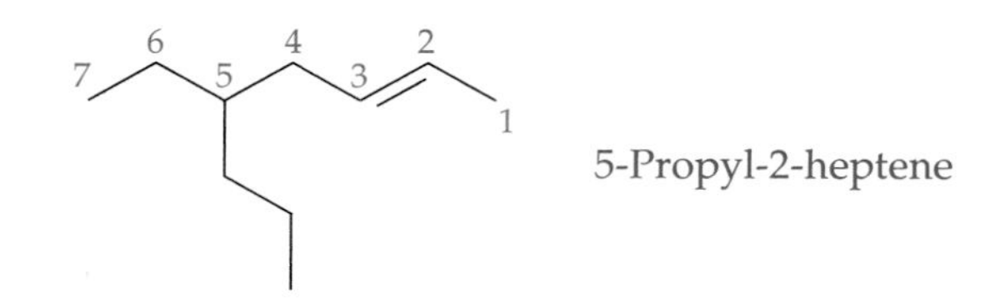

17.4 Properties of Hydrocarbons

As a class, hydrocarbons are essentially nonpolar and are thus quite insoluble in water (which, of course, is polar). However, because like dissolves like, hydrocarbons that are liquids at room temperature are good solvents for fats, oils, and many other nonpolar organic substances. As mentioned in Section 17.2, the four smallest linear alkanes are all gases at room temperature: methane, CH_4; ethane, CH_3CH_3; propane, $CH_3CH_2CH_3$; and butane, $CH_3CH_2CH_2CH_3$. At a chain length of 5 (pentane, $CH_3CH_2CH_2CH_2CH_3$), hydrocarbons become liquids at room temperature. As the chains increase in length, boiling point increases until at chain lengths of 20 and above, hydrocarbons become solids at room temperature. You know some of these hydrocarbons as *waxes* and *asphalt*. Look at Table 17.6, and you will see how the boiling points of linear alkanes steadily increase as chain length increases.

By now you should be able to explain why boiling point increases as the molecules get bigger. A higher boiling point means stronger intermolecular forces. Therefore, because boiling point increases with size, the intermolecular

Table 17.6 Boiling Points for Some Linear Alkanes (at 1 atm pressure)

Chain length n	Alkane	Boiling point (°C)	
1	Methane, CH_4	−164	Gas at 25°C
2	Ethane, CH_3CH_3	−88.6	
3	Propane, $CH_3CH_2CH_3$	−42.1	
4	Butane, $CH_3CH_2CH_2CH_3$	−0.5	
5	Pentane, $CH_3CH_2CH_2CH_2CH_3$	36.1	Liquid at 25°C
6	Hexane, $CH_3CH_2CH_2CH_2CH_2CH_3$	69.0	
7	Heptane, $CH_3CH_2CH_2CH_2CH_2CH_2CH_3$	98.4	
8	Octane, $CH_3CH_2CH_2CH_2CH_2CH_2CH_2CH_3$	124.7	
9	Nonane, $CH_3CH_2CH_2CH_2CH_2CH_2CH_2CH_2CH_3$	150.8	
10	Decane, $CH_3CH_2CH_2CH_2CH_2CH_2CH_2CH_2CH_2CH_3$	174.1	
20	Eicosane, $CH_3CH_2CH_2CH_2CH_2CH_2CH_2CH_2CH_2CH_2CH_2CH_2CH_2CH_2CH_2CH_2CH_2CH_2CH_2CH_3$	343	Solid at 25°C

forces must also be increasing with size. Why should that be so? Remember that hydrocarbons are essentially nonpolar, so the intermolecular forces can't be dipolar forces or the result of hydrogen-bonding. This leaves just London forces (Chapter 10), which are the result of momentary imbalances in the electron distributions in the molecules. The larger the hydrocarbon molecule, the more electrons there are to be unbalanced and the greater the molecular surface area, allowing for greater contact between molecules. These factors lead to more London attractive forces between molecules as hydrocarbon size increases and to the steady increase in boiling point observed in Table 17.6.

17.5 Functionalized Hydrocarbons—Bring On the Heteroatoms

Organic chemists refer to atoms other than C and H as **heteroatoms**. The prefix *hetero-* means "different"; in this case, the atoms are different from C atoms and H atoms. Typical heteroatoms found in many organic compounds are oxygen, nitrogen, sulfur, phosphorus, and the halogens.

Alkanes are important as fuels, but otherwise they are unreactive and not very exciting chemically. However, when heteroatoms are added to an otherwise boring hydrocarbon, we obtain what are called **functionalized hydrocarbons**. The heteroatoms often give functionalized hydrocarbons new and exciting properties and increased reactivity. The number of functionalized hydrocarbons that chemists have prepared and/or characterized is absolutely mind-boggling; literally millions of them are known. A few familiar ones are shown at the top of the next page.

Because there are so many different kinds of functionalized hydrocarbons, each with its own unique properties and reactivities, organizing them in a way that makes sense and allows us to understand their behavior as well as name them is a difficult task. The IUPAC nomenclature system can do all this for us. The portion of a functionalized hydrocarbon molecule that gives the molecule its unique properties is called a **functional group**. There are many types of functional groups, each with its own name. For example, let's start with the simplest hydrocarbon, methane, and remove one hydrogen atom. In its place we put something other than a carbon or hydrogen atom:

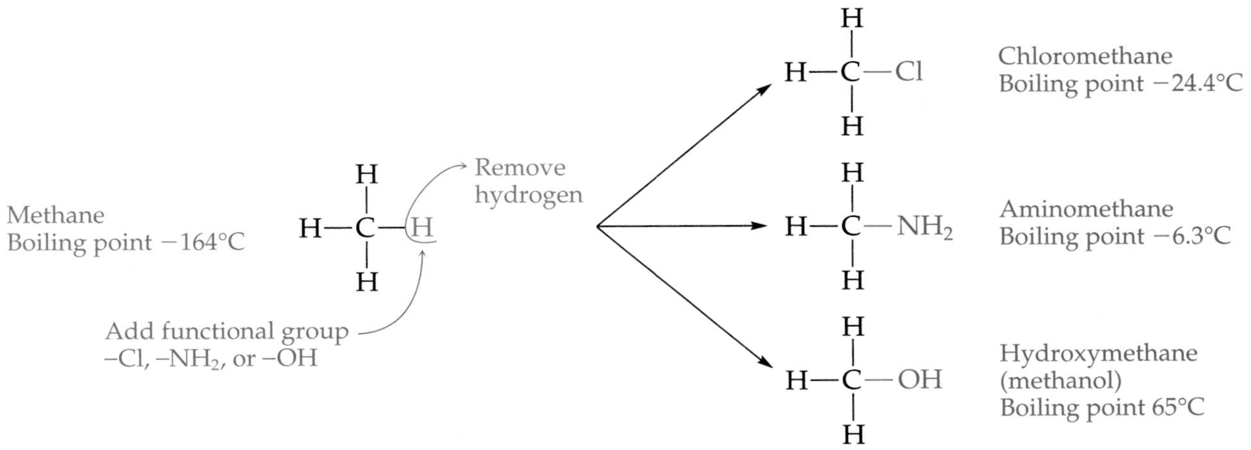

Methane
Boiling point −164°C

H—C(H)(H)—H

Remove hydrogen

Add functional group
$-Cl$, $-NH_2$, or $-OH$

H—C(H)(H)—Cl
Chloromethane
Boiling point −24.4°C

H—C(H)(H)—NH_2
Aminomethane
Boiling point −6.3°C

H—C(H)(H)—OH
Hydroxymethane
(methanol)
Boiling point 65°C

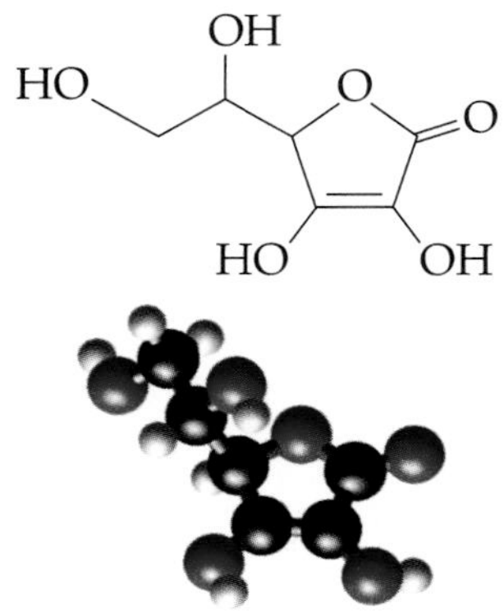

Vitamin C

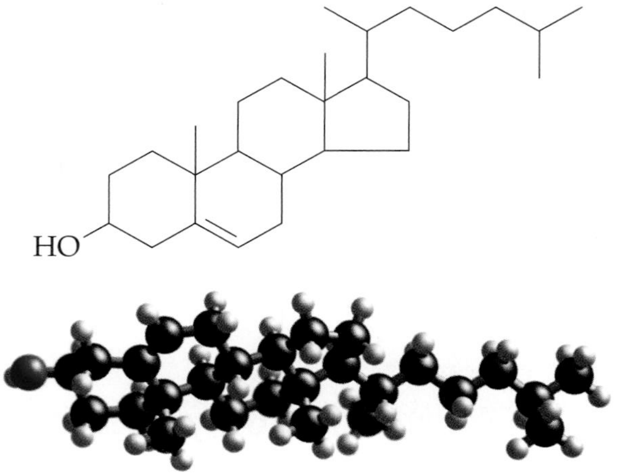

Cholesterol

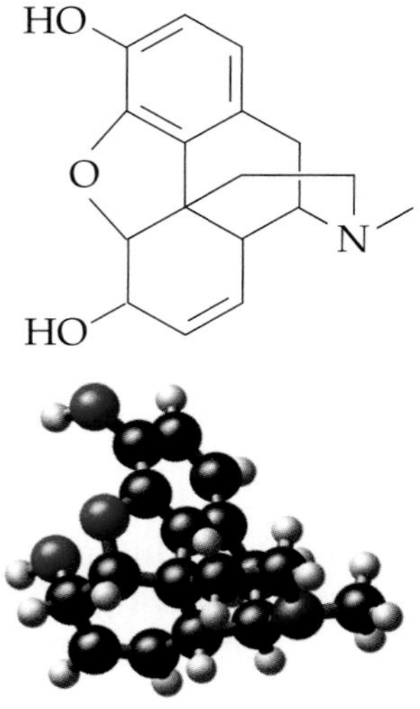

Morphine

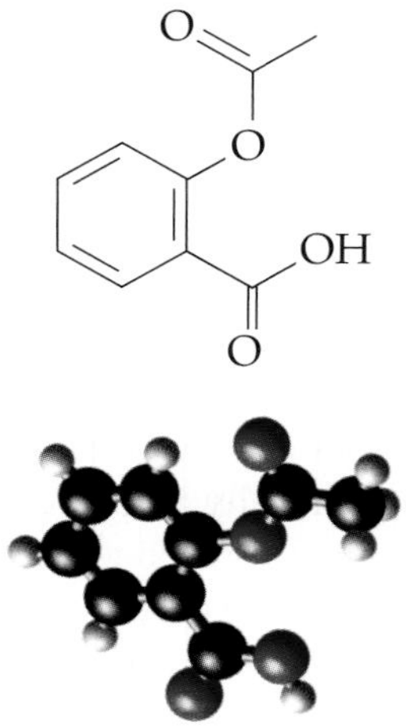

Aspirin

The chemical and physical properties of the three molecules we've just made are drastically different from those of the parent hydrocarbon, methane. They boil at different temperatures, they have different odors, they have different solubilities in water, and they have different chemical reactivities.

To name a hydrocarbon that contains a halogen atom (fluorine, chlorine, bromine, or iodine), we replace the *-ine* ending of the halogen's elemental name with *-o* and then put the name we've created in front of the name of the parent hydrocarbon. For example, the molecule CH_3Cl shown on page 695 is called *chloro*methane. For hydrocarbons made of three or more carbon atoms, we number the position of the halogen, just as we did for branches and multiple bonds. We number the main chain starting at the end nearer the halogen:

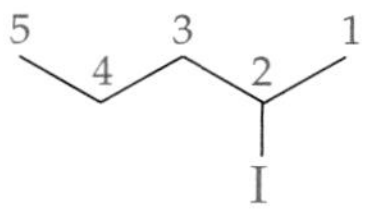

2-Iodopentane

The molecule CH_3NH_2 shown on page 695 is called aminomethane. To name a hydrocarbon with an NH_2 group on it, we put the prefix *amino-* before the name of the parent hydrocarbon and (when the chain has more than two carbons) number the carbon chain from the end nearer the amino group. The NH_2 group in aminomethane gives this functionalized hydrocarbon properties similar to those of NH_3 (ammonia).

Finally, the molecule CH_3OH shown on page 695 has an OH group and is therefore a member of the class of functionalized hydrocarbons known as *alcohols*. Although CH_3OH might be called hydroxymethane, the way we name most alcohols is different from the way we name halogen or amino functionalized hydrocarbons. Instead of putting a prefix in front of the name of the parent alkane, we use the parent name and replace the final *-e* with *-ol*. Thus, the name methan*e* becomes methan*ol* for CH_3OH. Once again, for chains containing more than two carbons, we number the chain from the end nearer the OH group, as in

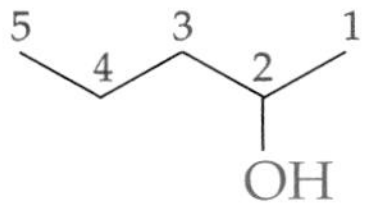

5 3 1
4 2
OH

2-Pentanol

Practice Problems

Name the following functionalized hydrocarbons:

17.11 OH

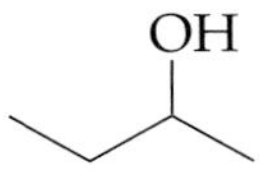

Answer: 2-Butanol

17.12 OH

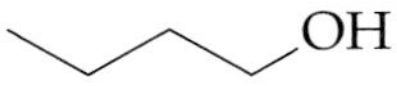

17.13 NH_2

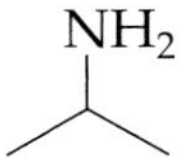

17.14 Br

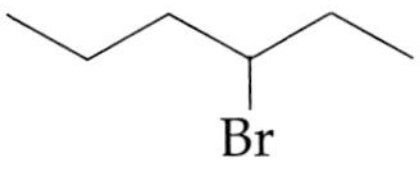

17.15 Draw a line drawing for 3-methyl-2-pentanol.

We have given you just a very brief look at the IUPAC nomenclature system for functionalized hydrocarbons. The halogen, OH, and NH_2 functional groups are just three among an entire zoo of functional groups. If you take a course in organic chemistry, you will spend a lot of time learning the details of this system. For the rest of this chapter, however, we would rather discuss the *properties* of the different types of functionalized hydrocarbons, for it is their properties that make them so interesting and useful. As we proceed, keep in mind that a functional group has a large effect on an organic molecule, giving the molecule properties that are quite different from those of the parent hydrocarbon. Indeed, the effect of a functional group is generally so large that all organic molecules that contain the same functional group tend to have similar chemical and physical properties.

We shall start with a functional group that frequently appears in the environmental headlines—halogens.

HALOGEN-FUNCTIONALIZED HYDROCARBONS

Depending on whom you talk to, halogens are either evil or wonderful functional groups. Halogen-functionalized hydrocarbons (called *alkyl halides*) do not occur naturally. They are prepared from hydrocarbons, often by reacting the hydrocarbons with halogens:

$$CH_4 + Cl_2 \xrightarrow[\text{light}]{\text{UV}} CH_3Cl + CH_2Cl_2 + CHCl_3 + CCl_4$$

Common name: Chloromethane, Dichloromethane, Chloroform, Carbon tetrachloride

This reaction is not balanced—its only point here is to show that many products may be obtained when an alkane is reacted with a halogen. Industrial chemical producers often add chlorine (and to lesser extents fluorine, bromine, and iodine) to hydrocarbon chains to produce compounds that have specific properties. Some examples are as follows:

CCl_3F
Trichlorofluoromethane (Freon 11, refrigerant)

CCl_2F_2
Dichlorodifluoromethane (Freon 12, refrigerant)

$CHCl_3$
Trichloromethane (chloroform, solvent; formerly used as anesthetic)

$CHCl_2{-}CHCl_2$
1,1,2,2-Tetrachloroethane (dry-cleaning fluid)

Keeps going for many carbons $\cdots{-}CF_2{-}CF_2{-}CF_2{-}CF_2{-}CF_2{-}CF_2{-}CF_2{-}CF_2{-}\cdots$ Keeps going for many carbons

Teflon (nonstick coating)

These are interesting compounds. Trichlorofluoromethane, CCl_3F, and dichlorodifluoromethane, CCl_2F_2, are often referred to as *chlorofluorocarbons*. They used to be the principal refrigerant fluid in refrigerators and air-conditioners and, until recently, as propellants in aerosol cans. They are relatively unreactive, nonflammable, and nontoxic—just what's needed if you are going to spray them toward your body or use them in a refrigerator to cool your food. Unfortunately, once released into the atmosphere, their unreactivity allows them to survive and reach the high-altitude ozone layer that protects us from the Sun's harmful ultraviolet radiation. Once there, the chlorofluorocarbons are activated by sunlight and begin to destroy the ozone layer. One chlorofluorocarbon molecule can destroy tens of thousands of ozone molecules by acting as a catalyst for the decomposition of ozone to oxygen. For this reason, chlorofluorocarbons are now banned in most of the industrialized world for use as refrigerants and aerosol propellants.

Halogen-functionalized hydrocarbons can be made to undergo a variety of reactions, allowing them to serve as starting materials for attaching other functional groups to hydrocarbons. For example, although there is no efficient way to turn methane, CH_4, into methanol, CH_3OH, directly, chemists can convert methane to methanol by reacting chloromethane, CH_3Cl, with a strong base:

$$H{-}\overset{\displaystyle H}{\underset{\displaystyle H}{\overset{|}{\underset{|}{C}}}}{-}Cl + OH^- \longrightarrow H{-}\overset{\displaystyle H}{\underset{\displaystyle H}{\overset{|}{\underset{|}{C}}}}{-}OH + Cl^-$$

Before proceeding any further with our tour of functional groups, we want to introduce a shorthand notation that chemists commonly use when drawing functionalized hydrocarbons. Instead of writing the entire hydrocarbon chain attached to a functional group, chemists often simply write the capital letter R. In other words, R represents the hydrocarbon portion of a molecule. Thus, for example, R–Cl stands for all monochlorinated hydrocarbons:

$$H{-}\overset{\displaystyle H}{\underset{\displaystyle H}{\overset{|}{\underset{|}{C}}}}{-}Cl \quad \text{becomes} \quad R{-}Cl$$

R stands for methyl group

Chloromethane

$$H{-}\overset{\displaystyle H}{\underset{\displaystyle H}{\overset{|}{\underset{|}{C}}}}{-}\overset{\displaystyle H}{\underset{\displaystyle H}{\overset{|}{\underset{|}{C}}}}{-}\overset{\displaystyle H}{\underset{\displaystyle H}{\overset{|}{\underset{|}{C}}}}{-}Cl \quad \text{becomes} \quad R{-}Cl$$

R stands for propyl group

1-Chloropropane

This notation allows chemists to quickly draw an abbreviated version of a molecule that emphasizes the functional group.

HYDROXY-FUNCTIONALIZED HYDROCARBONS—ALCOHOLS (R–OH)

As noted earlier, a hydrocarbon with an OH group is called an **alcohol**. By our shorthand notation, all alcohols have the general formula R–OH.

The alcohol CH_3CH_2OH, called ethanol, is consumed in the form of wine, beer, whiskey, or champagne, depending on the occasion. All these drinks are solutions of ethanol in water, with other substances present for flavor. Ethanol acts as a depressant on the central nervous system. High concentrations in the blood lead to impaired physical ability and mental judgment. Pure ethanol (100%, or 200 proof), which is called *absolute ethanol*, is available only to licensed institutions for research and industrial applications. The general public can purchase only a *denatured* form of absolute ethanol, to which poisons have been added to make it undrinkable. Still, 200-proof ethanol is nothing you would want to drink. It is extremely hygroscopic, which means that it readily absorbs water. A drink of it would severely dehydrate the tissues of your tongue and throat, damaging them as you swallowed.

Although ethanol can be produced from ethane, its main source in beer, wine, and champagne is the fermentation of various grains and fruits. For this reason, ethanol is sometimes called *grain alcohol*. Enzymes found in naturally occurring yeasts change the sugars and carbohydrates in grains and fruits into ethanol and carbon dioxide:

$$\underset{\text{Glucose (sugar)}}{C_6H_{12}O_6} \xrightarrow{\text{Enzymes}} \underset{\text{Ethanol}}{2\ CH_3CH_2OH(aq)} + \underset{\text{Carbon dioxide}}{2\ CO_2(g)}$$

The carbon dioxide is what supplies the fizz in champagne and beer.

Other alcohols are very toxic to humans. Only a few milliliters of methanol, CH_3OH, can cause nausea and blindness, and larger quantities can be fatal. Why is there such a large difference between methanol and ethanol, two seemingly similar molecules? The answer is that your body metabolizes methanol and ethanol differently. Enzymes in the human liver oxidize methanol to formaldehyde but convert ethanol to acetaldehyde. Acetaldehyde is produced in the body in other metabolic cycles, so your body knows how to deal with it. Formaldehyde is not normally found in your body, however. Until recently, it was used to pickle and preserve dead animals. In your body, it quickly poisons and pickles your cells, resulting in blindness and death.

Most methanol is produced from methane, but it can also be isolated from the fermentation of wood. For this reason, methanol is often called *wood alcohol*. In certain animals, the liver lacks an enzyme needed to metabolize wood alcohol to formaldehyde. Horses, for example, can ingest large quantities of wood alcohol with no apparent ill effects. (This is why extending animal toxicity tests to humans must be done very carefully and with an understanding of the biochemistry involved.) It might surprise you to know that the treatment for methanol poisoning in humans is, believe it or not, large doses of ethanol! The ethanol keeps the liver enzymes busy, and the methanol, instead of being metabolized, gets flushed out of the body.

As just two more examples of common alcohols, we mention the very poisonous 2-propanol, also known as *rubbing alcohol*, and the dialcohol (or *diol*) 1,2-ethanediol, commonly called ethylene glycol. The latter is the major component of automobile antifreeze and is also highly toxic.

2-Propanol (rubbing alcohol)   1,2-Ethanediol (ethylene glycol; antifreeze)

Interestingly, large alcohol molecules containing many carbons tend to be much less toxic to humans. Small alcohol molecules like the ones we have discussed so far tend to be soluble in water because the OH group is polar and capable of hydrogen-bonding. Because blood is largely water, alcohol molecules having between one and five carbons have no trouble getting into the bloodstream. Larger alcohol molecules are much less water-soluble because the larger nonpolar R portion of the molecule "outweighs" the smaller polar OH portion. This makes the larger alcohols less toxic because they are less soluble in aqueous body fluids. (Another factor that makes larger alcohols less toxic to humans is the body's reflex to regurgitate them.)

With all this talk of toxicity, we don't want you to get the impression that alcohols are not important and useful. Quite the contrary. They are used as solvents for a variety of substances—paints and glues, to name just two. Also, the high reactivity of the OH functional group makes alcohols useful as starting materials in industrial chemical processes, much as halogenated hydrocarbons are used to make alcohols.

ETHERS (R–O–R)

An **ether** has the general formula R–O–R. Here are some examples:

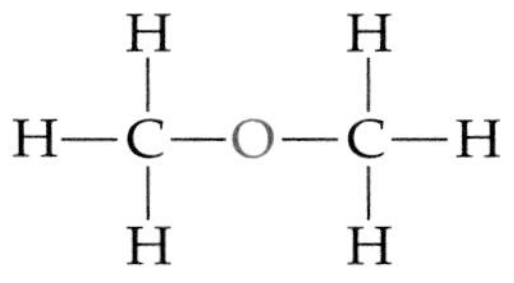

```
    H       H
    |       |
H—C—O—C—H
    |       |
    H       H
```

Dimethyl ether
(both R groups methyl)

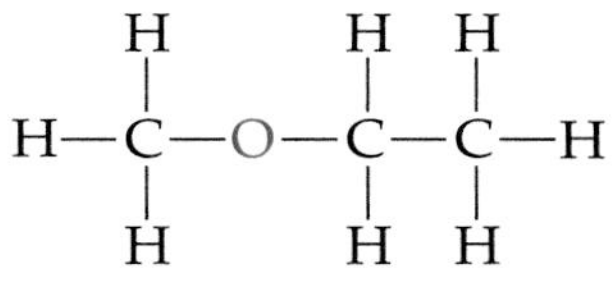

```
    H       H   H
    |       |   |
H—C—O—C—C—H
    |       |   |
    H       H   H
```

Ethyl methyl ether
(one R group methyl,
the other ethyl)

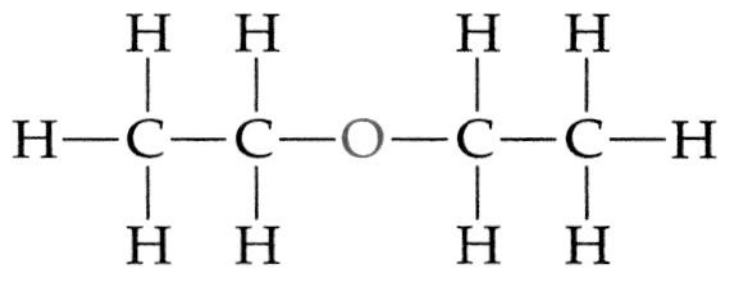

```
    H   H       H   H
    |   |       |   |
H—C—C—O—C—C—H
    |   |       |   |
    H   H       H   H
```

Diethyl ether
(both R groups ethyl)

Ethers are commonly used as solvents for organic solutes. Diethyl ether is a very volatile liquid, readily vaporizing at room temperature. This is the "ether" that once was used as an anesthetic. (One of the authors of this book remembers having a mask put over his face so that he could breathe diethyl ether vapors when he was seven years old. This put him right to sleep, only to awaken later without tonsils.) Today, diethyl ether is no longer used as an anesthetic because the vapors are highly flammable. An errant spark from nearby electrical equipment could easily cause an explosion.

ORGANIC ACIDS—CARBOXYLIC ACIDS (R–COOH)

A hydrocarbon with a COOH group on it is called a **carboxylic acid**:

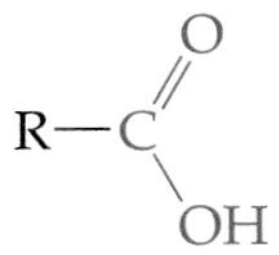

```
      O
     //
R—C
     \
      OH
```

A carboxylic acid

As you can see, this functional group has both a C–O single bond and a C=O double bond. All hydrocarbons that have this group act as weak acids owing to the dissociation equilibrium

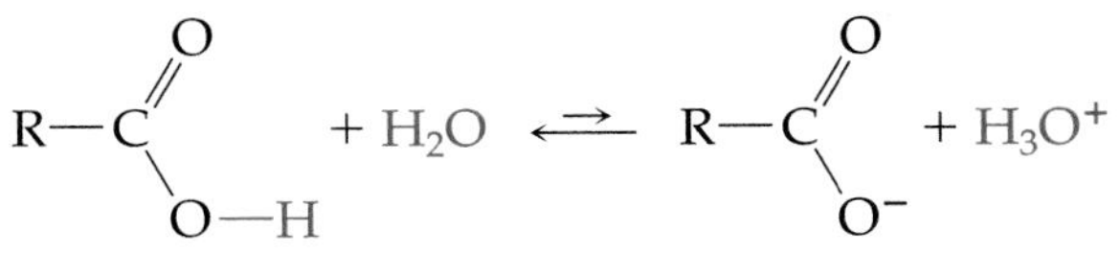

```
      O                          O
     //                         //
R—C       + H2O  ⇌  R—C       + H3O+
     \                          \
      O—H                        O−
```

The equilibrium constant K_a for a carboxylic acid dissociation tends to be much smaller than 1. This means that the equilibrium lies to the left, so carboxylic acids are weak acids. In a carboxylic acid, only the H on the oxygen is acidic.

Acetic acid, which we saw in earlier chapters, is a carboxylic acid with R = CH_3:

$$CH_3-C(=O)OH$$

Acetic acid

Vinegar is an approximately 5% by mass solution of acetic acid in water. It is naturally produced by the bacteria *Acetobacter*, which, in the presence of oxygen, oxidizes ethanol to acetic acid. Wine spoils when *Acetobacter* oxidizes some of the ethanol to acetic acid.

Acetic acid consists of a two-carbon chain. Its IUPAC name is thus ethanoic acid. The ending *-oic* indicates a carboxylic acid, just as *-ol* indicates an alcohol. Acetic acid is the common, and more frequently used, name for ethanoic acid.

The smallest carboxylic acid is methanoic acid, commonly called formic acid. This latter name comes from the fact that the acid is excreted by some ants as a defense against threats, and *formica* is Latin for "ant." Try drawing this compound now.

Draw a dot diagram for methanoic acid, showing all hydrogen, carbon, and oxygen atoms.

Your drawing should have just one carbon atom in it.

ORGANIC BASES—AMINES (H_2NR, HNR_2, NR_3)

Hydrocarbons containing a nitrogen atom attached by a single bond to carbon are called **amines**. They are related to ammonia, NH_3, with one, two, or three of the H atoms replaced by hydrocarbon (R) chains. Like ammonia, amines all have a single lone pair of electrons on the nitrogen atom:

$$\ddot{N}H_3$$

Ammonia

$$R\ddot{N}H_2 \qquad R_2\ddot{N}H \qquad R_3\ddot{N}$$

Primary amine — Secondary amine — Tertiary amine

All amines are Brønsted–Lowry bases because they can accept a proton (H^+). They bond to the proton via the lone pair of electrons on the nitrogen. The result is a positively charged ammonium ion:

$$R_2\ddot{N}H + H^+ \longrightarrow [R_2NH_2]^+$$

Amine (base) Acid Ammonium ion

The smaller amines are soluble in water and act as weak bases, producing small amounts of hydroxide ions in water, just as ammonia does:

$$R_2\ddot{N}H + H_2O \rightleftharpoons [R_2NH_2]^+ + OH^-$$

The equilibrium constant K_b for a water-soluble amine is less than 10^{-3}. This means the reaction equilibrium lies to the left, producing only small amounts of hydroxide ion.

Ammonia has a characteristic sharp odor that jolts your sense of smell. Spirits of ammonia, also called smelling salts, is sometimes used to revive someone who has fainted. Small amines like methyl amine, CH_3NH_2, have a similar odor. Larger amines simply stink, as these two common names suggest:

$$H_2N{-}CH_2{-}CH_2{-}CH_2{-}CH_2{-}NH_2$$

Putrescine (1,4-diaminobutane)

$$H_2N{-}CH_2{-}CH_2{-}CH_2{-}CH_2{-}CH_2{-}NH_2$$

Cadaverine (1,5-diaminopentane)

You can guess where cadaverine is likely to be formed. Putrescine is formed in rotting fish. Knowing something about chemistry, you can deal effectively with putrescine if you ever purchase some old fish. Simply rub the fish with lemon, which is a source of carboxylic acids, such as citric acid.

$$HOOC{-}CH_2{-}C(OH)(COOH){-}CH_2{-}COOH$$

Citric acid has three carboxylic acid groups.

You know that when an acid and a base meet, they neutralize each other. When citric acid meets putrescine (an amine base), a proton from one of the COOH groups is transferred and an ammonium salt of putrescine is formed:

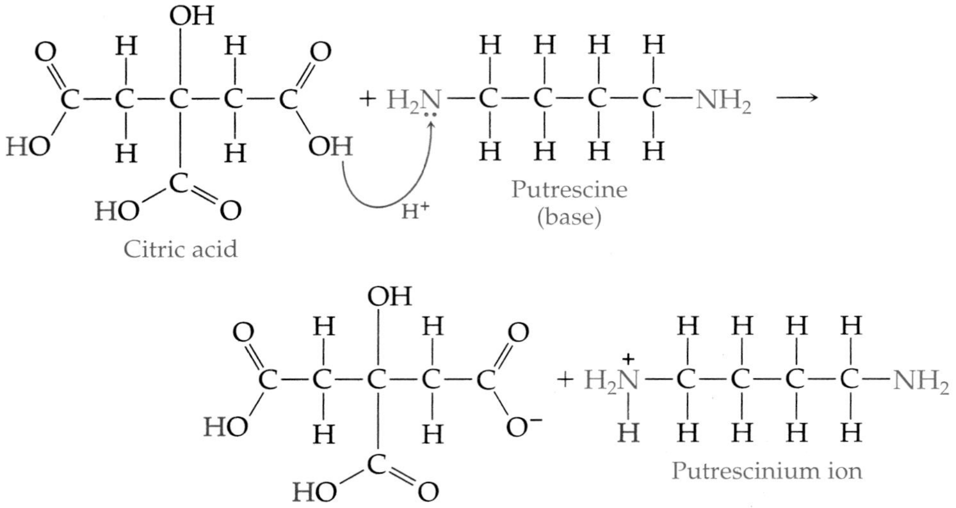

$$\text{Citric acid} + H_2\ddot{N}-C-C-C-C-NH_2 \ (\text{Putrescine (base)}) \xrightarrow{H^+} \text{citrate} \ (C-O^-) + H_2\overset{+}{N}H-C-C-C-C-NH_2 \ (\text{Putrescinium ion})$$

Citric acid

Putrescine (base)

Putrescinium ion

The ammonium salt doesn't have a bad smell the way the free base does, so you've fixed the smell problem. Of course, the fish is still bad, so you shouldn't have bought it in the first place.

KETONES ($R_2C{=}O$) AND ALDEHYDES (RHC=O)

So far we have seen the heteroatom oxygen in alcohols (R–OH), ethers (R–O–R), and carboxylic acids (R–COOH). In the alcohol functional group and the ether functional group, each bond between C and O is a single bond. In the carboxylic acid functional group, there is one C–O single bond and one C=O double bond:

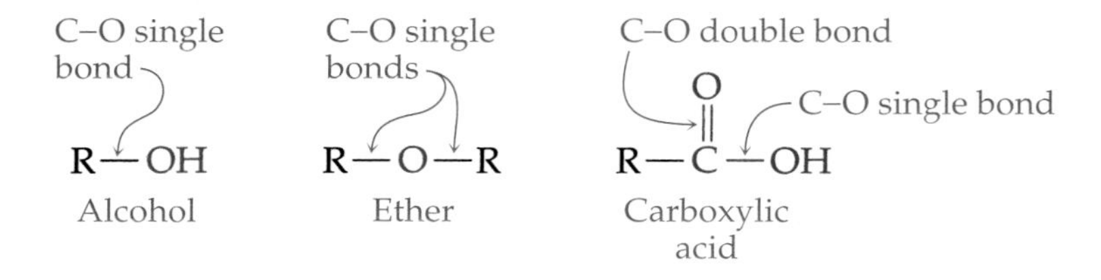

The C=O unit is referred to as a **carbonyl**. Of the three classes of molecules just mentioned—alcohols, ethers, and carboxylic acids—only the carboxylic acids include a carbonyl. The compounds known as **aldehydes** and **ketones** also contain a carbonyl. They differ from carboxylic acids in that they lack the OH portion of the COOH group.

The shorthand representations for aldehydes and ketones are

$$R-\overset{\overset{\displaystyle O}{\|}}{C}-R \qquad\qquad H-\overset{\overset{\displaystyle O}{\|}}{C}-R$$

Ketone   Aldehyde

From these structures, we see how to name the functional groups in aldehydes and ketones. When the carbonyl is attached to one hydrogen atom and one carbon atom (or two hydrogen atoms), it is an *aldehyde functional group*. When the carbonyl is attached to two carbon atoms, it is called the *ketone functional group*.

For an even shorter representation, a ketone can be written either R(CO)R or R_2CO and an aldehyde either R(CO)H or RHCO. When we put the CO in parentheses, we are emphasizing that it is a carbonyl and that the atoms to the right and left of the parentheses are attached to the carbonyl carbon.

Ketones and aldehydes are closely related. In a ketone, two hydrocarbon chains (R) must be attached to the carbonyl. In an aldehyde, at least one hydrogen must be attached to the carbonyl. The IUPAC names for ketones end with the suffix *-one* (pronounced own), and those for aldehydes end with the suffix *-al*. For example, a three-carbon ketone is called propanone, and a three-carbon aldehyde is called propanal:

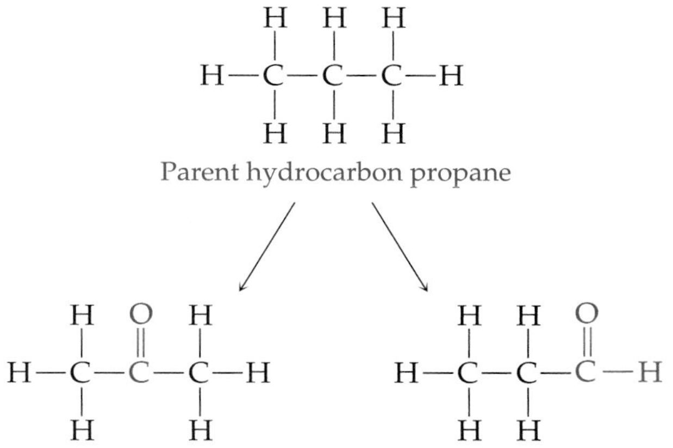

```
    H   H   H
    |   |   |
H — C — C — C — H
    |   |   |
    H   H   H
```

Parent hydrocarbon propane

```
    H   O   H                 H   H   O
    |   ‖   |                 |   |   ‖
H — C — C — C — H         H — C — C — C — H
    |       |                 |   |
    H       H                 H   H
```

The ketone propanone — The aldehyde propanal

Some ketones and aldehydes are already familiar to you. Propanone has the common name of acetone and is a widely used industrial solvent capable of dissolving many glues and paints. In addition, it is sold as a solution in water as nailpolish remover. The simplest aldehyde, which contains just the carbonyl carbon, is called methanal, and its common name is formaldehyde:

```
    O
    ‖
H — C — H
```

Methanal
(formaldehyde)

Practice Problems

17.16 (a) What functional group does the molecule ethanal contain? How can you tell?
(b) Draw a dot diagram for the molecule.

Answer:
(a) The ending -al *tells you this molecule is an aldehyde. It therefore contains the aldehyde functional group.*

(b)
```
      H  :O:
      |   ||
  H — C — C — H
      |
      H
```

17.17 There is no molecule called ethanone. Explain why not.

17.18 Circle and name the functional group in each molecule:

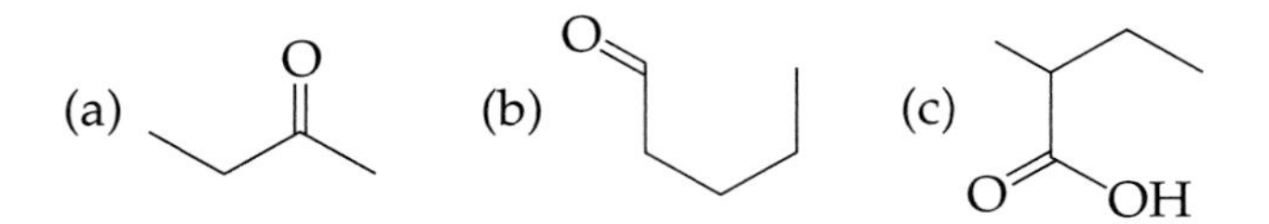

17.19 Give the IUPAC name for each molecule in Practice Problem 17.18.

The following chart summarizes the classes of organic compounds we have looked at in this chapter.

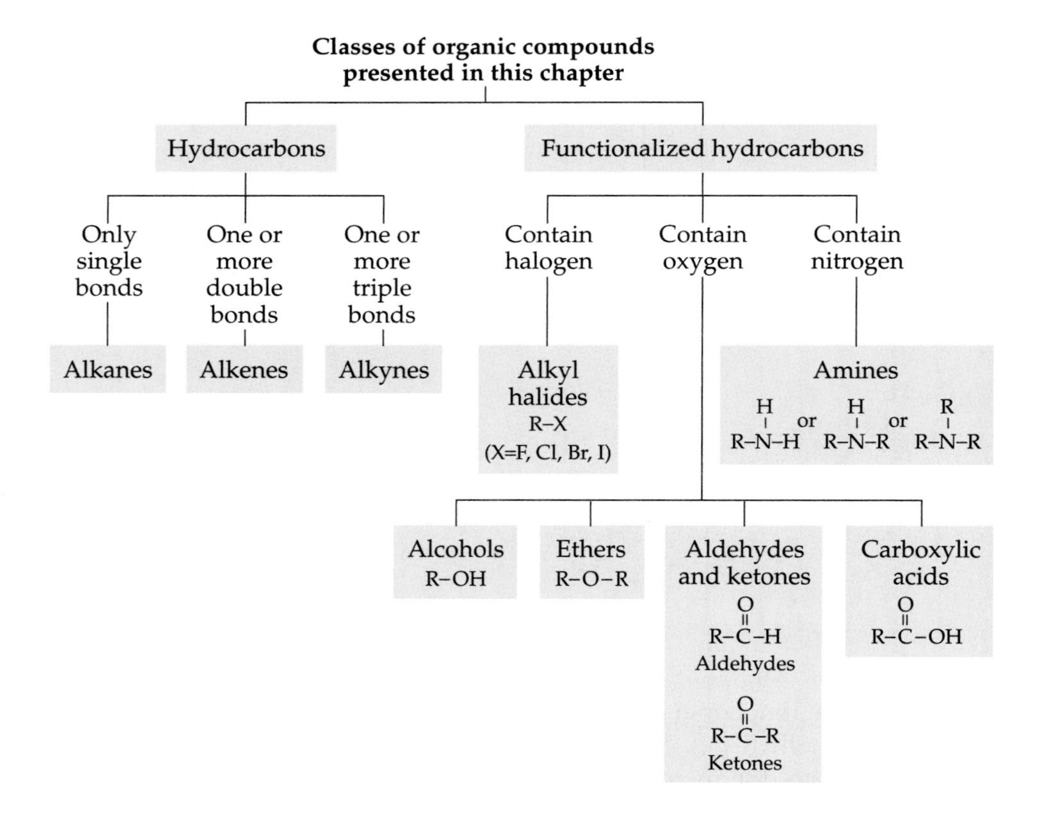

We have only scratched the surface of organic chemistry. A full course in the subject would examine additional kinds of functional groups, the properties they bestow on hydrocarbons, and the reactions those functionalized hydrocarbons undergo. It is a fascinating area of chemistry to study because when we examine functionalized hydrocarbons we examine ourselves. We are made of carbon in the form of functionalized hydrocarbons. Functional groups, such as the NH_2 of amines and the COOH of carboxylic acids, are key to building the proteins from which our skin, hair, blood, cells, and enzymes are made. This realization has lead to chemistry's current and perhaps greatest challenge—the understanding and manipulation of the molecules of life in living organisms. We shall take a glimpse at this subject in the final chapter of this book.

Have You Learned This?

Catenation (p. 672)
Organic chemistry (p. 672)
Hydrocarbon (p. 675)
Chain length (p. 676)
Linear hydrocarbon (p. 676)
Branched hydrocarbon (p. 677)
Main chain (p. 677)
Line drawing (p. 677)
Saturated hydrocarbon (p. 680)
Unsaturated hydrocarbon (p. 680)
Isomer (p. 682)
IUPAC nomenclature system (p. 683)
Alkane (p. 683)
Alkene (p. 683)
Alkyne (p. 683)
Homologous series (p. 683)
Monoalkene (p. 685)
Monoalkyne (p. 686)
Heteroatom (p. 695)
Functionalized hydrocarbon (p. 695)
Functional group (p. 695)
Alcohol (p. 699)
Ether (p. 701)
Carboxylic acid (p. 701)
Amine (p. 702)
Carbonyl (p. 704)
Ketone (p. 704)
Aldehyde (p. 704)

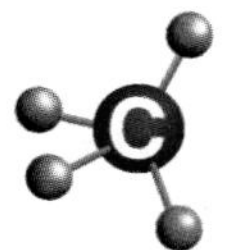

The Chemistry of Carbon
www.chemistryplace.com

CARBON—A UNIQUE ELEMENT

17.20 What is catenation?

17.21 What are the three known allotropes of carbon? How do their structures differ from one another, and how are they similar?

17.22 Explain why a carbon atom always forms four bonds in a covalent substance.

17.23 Consider this incomplete drawing of a three-carbon hydrocarbon:

```
            H
            |
H—C—C—C—H
            |
            H
```

(a) Complete the drawing by adding hydrogen atoms.
(b) Complete the drawing by changing one C–C single bond to a C=C double bond and then adding hydrogen atoms.
(c) Complete the drawing by changing one C–C single bond to a C≡C triple bond and then adding hydrogen atoms.

17.24 A hydrocarbon molecule consists of a chain of four carbon atoms. In this chain there are one C=C double bond and two C–C single bonds. What is the molecular formula? Draw a possible structure.

17.25 The geometry about each carbon in a particular hydrocarbon molecule is tetrahedral. What does this say about the possibility of there being C=C or C≡C bonds in the molecule? Explain.

17.26 This molecule has four bonds to every carbon, yet it doesn't exist. Why not?

```
H     H     H
 \    |    //
  C=C—C
 /         \
H           H
```

17.27 What is the name of the branch of chemistry devoted to carbon-based molecules? Why is it called this?

17.28 Why are there more compounds of carbon than of any other element?

HYDROCARBONS

17.29 Define the term *hydrocarbon*.

17.30 Draw the structural formula for a linear six-carbon hydrocarbon that contains only C–C single bonds. Show all H and C atoms.

17.31 Draw as many different structural formulas as you can of a branched six-carbon hydrocarbon that contains only C–C single bonds. Show all H and C atoms.

17.32 How do the various C–C bonds in a hydrocarbon tell you whether the hydrocarbon is branched or linear?

17.33 Why is it more correct to draw a linear hydrocarbon that contains only C–C single bonds as a zigzag rather than a straight line?

17.34 Is it more proper to draw this hydrocarbon as a straight line or as a zigzag? Explain.

```
H—C≡C—C≡C—H
```

17.35 What is the source of most hydrocarbons?

17.36 What ethical issues are involved in burning hydrocarbons as fuel?

17.37 Consider the hydrocarbon

```
                 H
                 |
               H—C—H
                 |
               H—C—H
                 |
               H—C—H
    H     H      |      H
    |     |      |      |
 H—C——C——C——C—H
    |     |      |      |
    |     H      H      |
 H—C—H              H—C—H
    |                   |
    H                   H
```

(a) What is the length of the main chain?
(b) What is the length of the branch?

17.38 Hydrocarbons are often represented by line drawings. How is it possible to look at a line drawing and deduce the molecular formula of the compound?

17.39 Convert each line drawing to a structural formula, then give the molecular formula and describe the hydrocarbon as being linear or branched:

(a) (b)

(c)

17.40 Make a line drawing for the hydrocarbon in Problem 17.37.

17.41 Make a line drawing for

```
     H     H
      \   /
(a)    C=C
      /   \
     H     H
```

```
    H  H  H  H  H  H
    |  |  |  |  |  |
(b) H—C—C—C—C—C—C—H
    |  |  |  |  |  |
    H  H  H  H  H  H

     H   H
      \ /
(c) H—C—C—H
       \ /
        C
       / \
      H   H
```

17.42 Define the terms *saturated* and *unsaturated* as applied to hydrocarbons.

17.43 Which molecule in Problem 17.39 is unsaturated?

17.44 Which molecule is more unsaturated and why:

(a) (b)

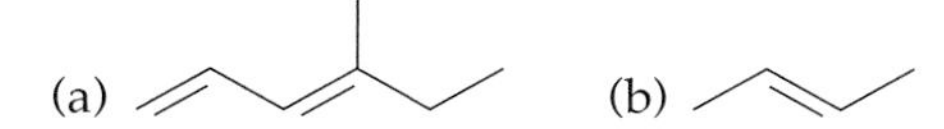

17.45 Suppose you are asked to make molecule (a) of Problem 17.44 even more unsaturated by removing H atoms to form a third double bond.
(a) Why can't this be the molecule you form:

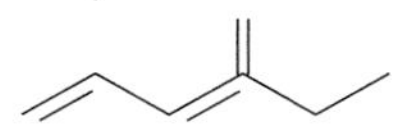

(b) Show a line drawing of the molecule you form.

17.46 A molecule contains 5 carbon atoms and 12 hydrogen atoms. Is it saturated or unsaturated? Explain how you know.

17.47 Is the compound C_7H_{14} unsaturated or saturated? How can you tell?

NAMING HYDROCARBONS

17.48 Which of these hydrocarbons are isomers?

(a) (b) (c) (d)

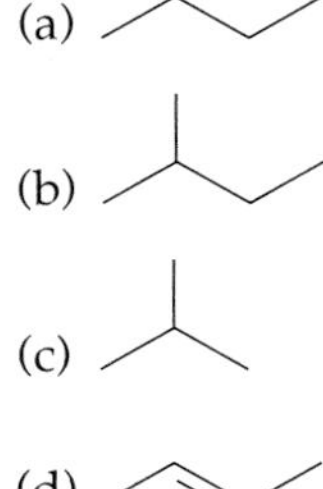

17.49 Make line drawings for all the isomers of C_6H_{14}.

17.50 In order for two molecules to be isomers, what must be true about their molecular formulas, their structural formulas, and their properties?

17.51 Are H_2O and H_2O_2 isomers? Explain.

17.52 What is wrong with using common names for hydrocarbon molecules?

17.53 *Cyclic hydrocarbons* contain at least one ring of carbon atoms. Of the cyclic hydrocarbons

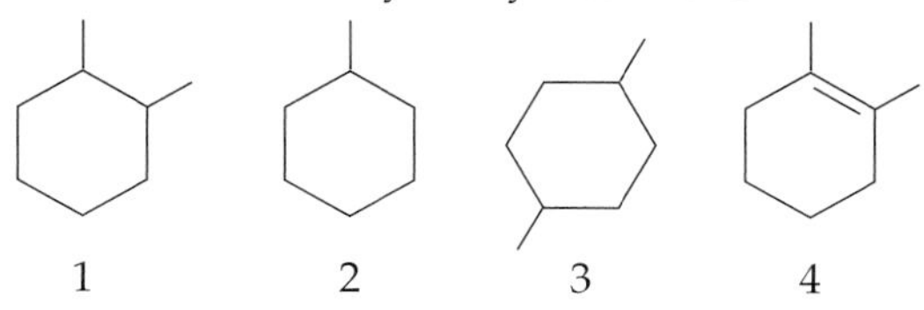

which are isomers? Explain why.

17.54 Relate the words *saturated* and *unsaturated* to the general formulas C_nH_{2n+2}, C_nH_{2n}, and C_nH_{2n-2}.

17.55 What kind of hydrocarbons are called aliphatic hydrocarbons, and where does the word *aliphatic* come from?

17.56 What endings are added to the root *alk-* to name hydrocarbons containing all C–C single bonds, hydrocarbons containing one or more C=C double bonds, and hydrocarbons containing one or more C≡C triple bonds?

17.57 What do the letters in the acronym IUPAC stand for?

17.58 What is meant by the expression *homologous series*? Give an example of one.

17.59 How is the general molecular formula for alkenes different from that for alkanes? Explain the basis of this difference.

17.60 Name the alkanes that do not use Greek prefixes in their names, and draw the structural formula for each.

17.61 The common name for ethyne is acetylene. Draw a dot diagram and a line drawing for this compound, and indicate whether it is more or less unsaturated than ethene.

17.62 Define the prefixes *meth-*, *eth-*, *prop-*, and *but-*.

17.63 Is $C_{22}H_{46}$ an alkane, an alkene, or an alkyne? How can you tell?

17.64 The IUPAC name for C_3H_6 needs no number in it, but the IUPAC name for C_4H_8 does. Explain.

17.65 Do propane, propene, and propyne belong to a homologous series? Explain.

17.66 Draw the structural formula and line drawing for 2-hexene.

17.67 What is the relationship between the molecules 2-hexene and 3-hexene? Justify your answer.

17.68 Explain why the molecules 1-hexene, 2-hexene, and 3-hexene exist but the molecules 4-hexene and 5-hexene do not.

17.69 True or false? We always number hydrocarbons from left to right because that is how we read. Explain your answer, and include an example to back up your explanation.

17.70 What should be your first task in naming any hydrocarbon molecule?

17.71 Name these linear hydrocarbon molecules:

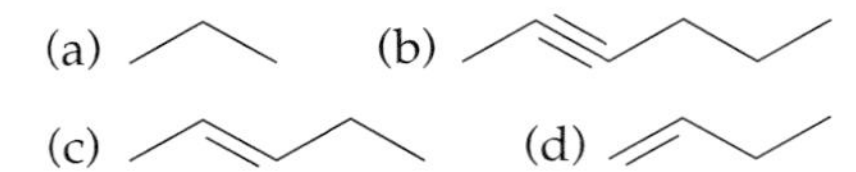

17.72 Give the molecular formula and line drawing for
(a) Propene (b) 1-Octene (c) Hexane
(d) 2-Butyne (e) 1-Butyne

17.73 Which molecules in Problem 17.72 represent a pair of isomers?

17.74 Where do we start numbering carbons on a main chain when both branches and C=C or C≡C bonds are present?

17.75 What is wrong with the name 2-methyl-4-hexene? What is the correct name?

17.76 Why can't the molecule 2,2-dimethyl-1-butene exist?

17.77 Give the line drawing and molecular formula for
(a) 3-Methylpentane
(b) 2,3-Dimethyl-2-butene
(c) 3-Methyl-1-pentene
(d) 5-Methyl-1-hexyne

17.78 Give the IUPAC name for

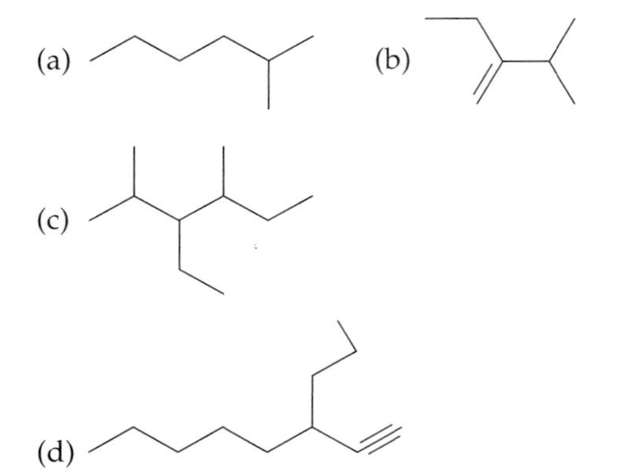

17.79 How many branches does this hydrocarbon have? What are they?

17.80 Make a line drawing of 5,5-dimethyl-1-hexene. What would be wrong with calling this compound 2,2-dimethyl-5-hexene?

17.81 True or false? The longest continuous chain of carbons in a hydrocarbon is always the chain to use in naming a hydrocarbon. Explain your answer.

17.82 Although it was not discussed in the chapter, it is possible for a hydrocarbon molecule to have more than one site of unsaturation. One such molecule is 1,3-butadiene; the *di-* tells us there are two double bonds in the molecule. Give the molecular formula and line drawing for this molecule.

PROPERTIES OF HYDROCARBONS

17.83 The alkanes containing one, two, three, and four carbons are gases at room temperature. However, as alkanes get larger they become liquids and then solids at room temperature. Why?

FUNCTIONALIZED HYDROCARBONS

17.84 What do we mean by the expression *functional group*?

17.85 Explain how halogenated hydrocarbons are named.

17.86 Explain how hydrocarbons containing an NH_2 group are named.

17.87 Explain how hydrocarbons containing an OH group are named.

17.88 Name the molecules

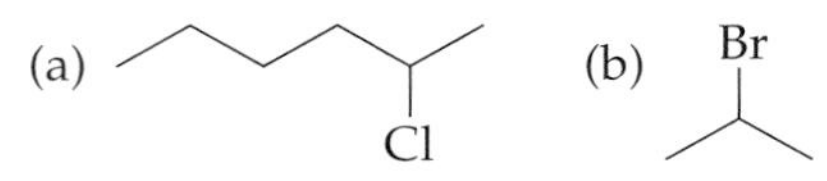

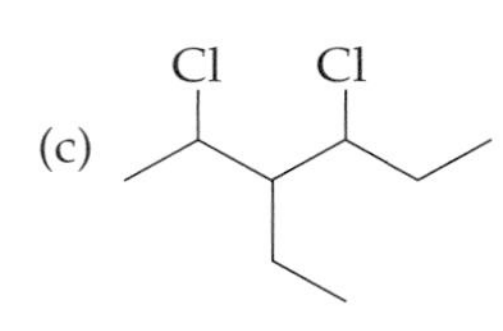

[*Hint:* Halogen and alkyl branches have the same priority; just make sure you come up with the lowest set of position numbers for the molecule.]

17.89 Draw the Teflon molecule.

17.90 Name the molecules

(a) OH (b) NH_2

(c) OH

17.91 Why is chlorine an important functional group?

17.92 Why have Freons been banned?

17.93 What class of compounds does the general formula R–OH represent? What does the R represent?

17.94 How can you convert R–Cl to R–OH?

17.95 A student claims there are three isomers of propanol: 1-propanol, 2-propanol, and 3-propanol. Is he correct? Explain.

17.96 Draw all the isomers of the alcohols having the molecular formula $C_4H_{10}O$.

17.97 Why is wood alcohol toxic to humans but not to horses?

17.98 Why is wood alcohol poisoning treated by having the patient consume ethanol?

17.99 2-Propanol is quite soluble in water, but 2-decanol is insoluble in water. Draw both structural formulas and explain why one compound is soluble but the other isn't.

17.100 This line drawing may look like a flying bat, but it is really a functionalized hydrocarbon. Give its molecular formula and name.

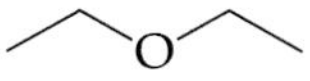

O

17.101 Name the ethers

(a)  O

(b) 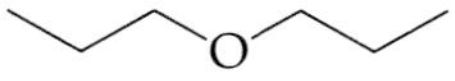 O

(c) 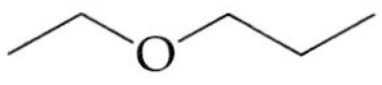 O

17.102 Draw the structural formula of butyl propyl ether and give its molecular formula.

17.103 Give the common name and line drawing for ethanoic acid.

17.104 Write the equilibrium reaction associated with molecules of the type R–COOH. To which side does the equilibrium lie? What does this imply?

17.105 Show how a carboxylic acid, R–COOH, reacts with a tertiary amine, NR_3.

17.106 Write an equilibrium to show how a primary amine, H_2NR, makes water basic. To which side does the equilibrium lie? What does this imply?

17.107 Diethylamine is a secondary amine having the formula $C_4H_{11}N$. Draw a dot diagram for this compound.

17.108 Make a line drawing for 2-aminohexane. Is this a primary, secondary, or tertiary amine?

17.109 The molecule methanal has the common name formaldehyde. Draw a dot diagram for it.

17.110 Draw the structural formulas for 1-chloropropane, 1-propanol, 1-aminopropane, 2-propanone, propanoic acid, and propanal. Then label each molecule to indicate which type of functionalized hydrocarbon it is (alcohol, ketone, aldehyde, and so on).

17.111 Make a line drawing for 2-propanone and give its common name.

17.112 Where are you likely to run across acetic acid outside a chemistry laboratory?

17.113 Circle the portion of each molecule that is the functional group and name the molecule:

(a) 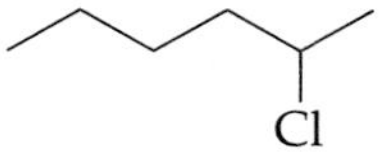 Cl (b) 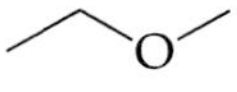 O

(c) 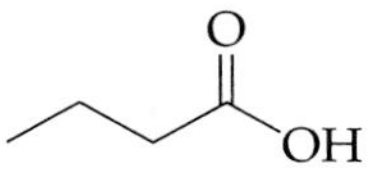 O OH (d) 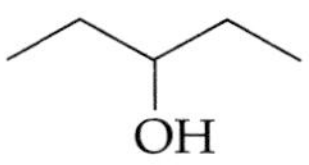 OH

17.114 What classes of molecules contain a carbonyl, C=O?

17.115 This compound is sold as Prozac, used for the treatment of depression:

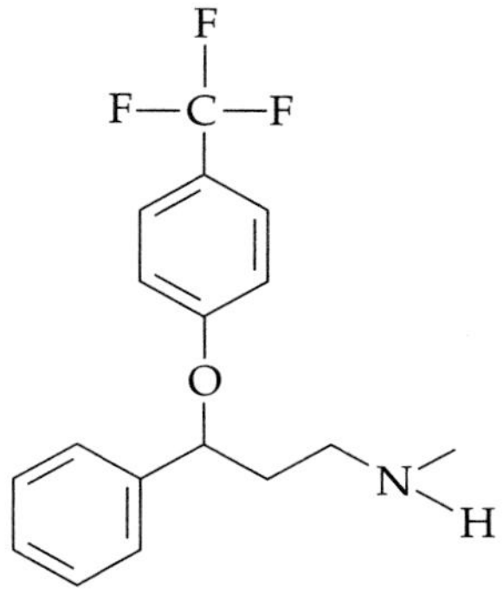

F
F—C—F
O
N H

(a) Circle all the functional groups and name them.
(b) What is the molecular formula of Prozac?

ADDITIONAL PROBLEMS

17.116 Hydrocarbons are insoluble in water, but some alcohols, such as methanol, CH_3OH, dissolve readily in water. Why?

17.117 A solution of methyl alcohol in water does not conduct electricity. Why not?

17.118 Which do you expect to be most water-soluble:
(a) CH_3CH_2OH
(b) $CH_3CH_2CH_3$
(c) $CH_3CH_2CH_2CH_2CH_2CH_2OH$
Explain your answer.

17.119 There are three isomers that have the molecular formula C_5H_{12}. Draw the structural formula for each one. Which do you expect to have the lowest boiling point? Explain.

17.120 There are four isomers that have the molecular formula C_4H_9Br. Draw the structural formula for each one. Name each compound using IUPAC rules.

17.121 Write the molecular formula for each molecule:

(a) Vanillan

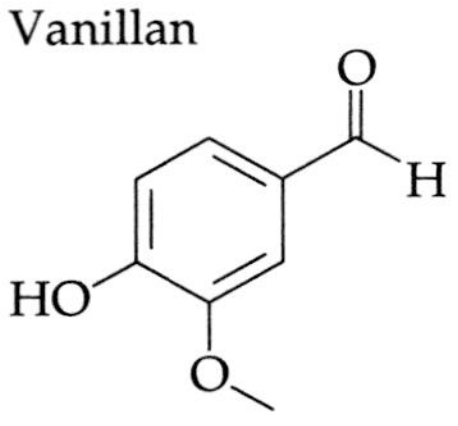

(b) Cholesterol

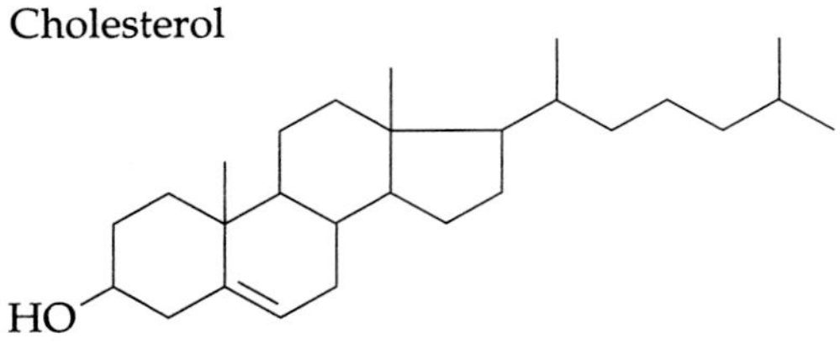

(c) Vitamin C

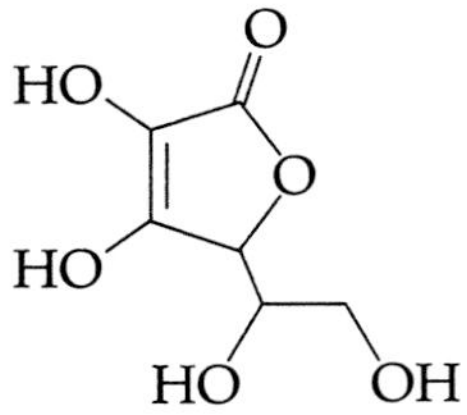

(d) Cocaine

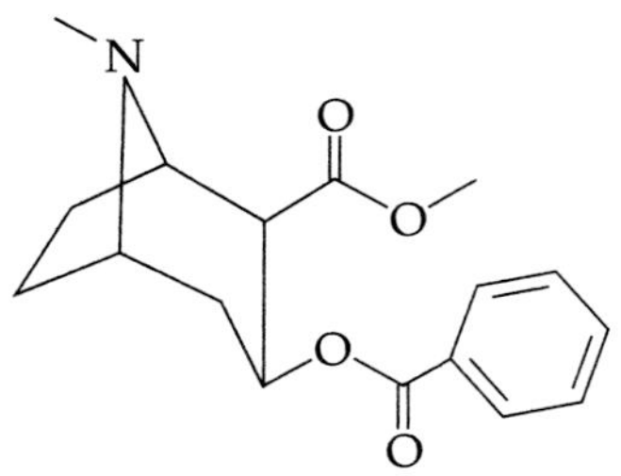

17.122 For each pair of molecules, indicate whether the two are identical, isomers, or completely unrelated.

```
    H  H  H           H  H  H
    |  |  |           |  |  |
(a) H—C—C—C—OH       H—C—C—C—H
    |  |  |           |  |  |
    H  H  H           H  OH H

        H  H  H           H  H  OH
        |  |  |           |  |  |
(b) HO—C—C—C—H       H—C—C—C—H
        |  |  |           |  |  |
        H  H  H           H  H  H

              H  H        H   H  H   H
              |  |         \  |  |  /
(c) H—C≡C—C—C—H           C=C—C=C
              |  |         /        \
              H  H        H          H
```

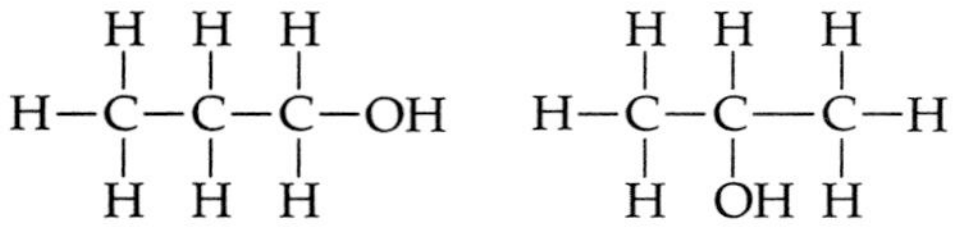

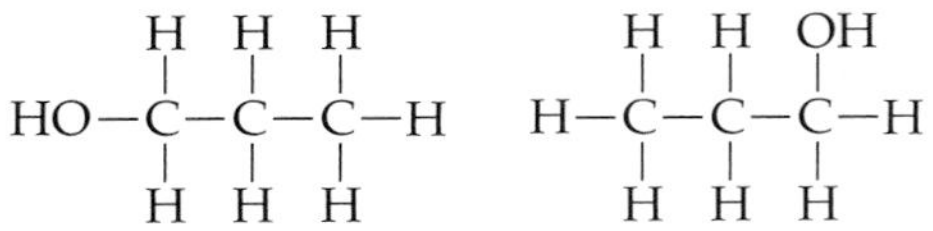

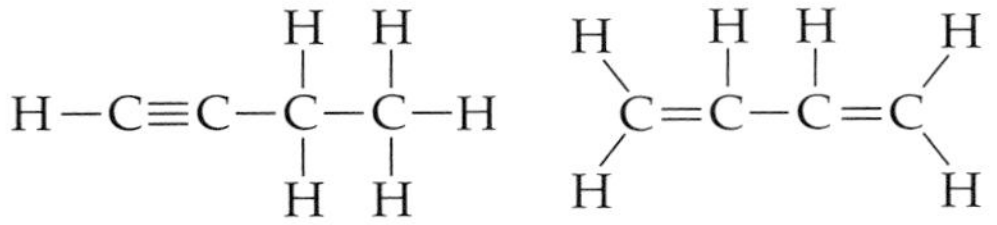

(d) 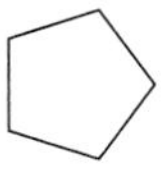 $CH_3CH_2CH_2CH_2CH_3$

(e)

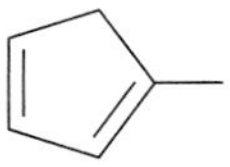

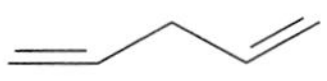

17.123 Using the molecular formula $C_2H_4Cl_2$, draw the structural formula of (a) a molecule that is polar and (b) a molecule that is nonpolar.

17.124 Using the molecular formula $C_4H_6Cl_2$, draw the structural formula of (a) a molecule that is polar, (b) a molecule that is nonpolar, and (c) a molecule that contains only C–C single bonds.

17.125 Benzene is a hydrocarbon that has the molecular formula C_6H_6. The benzene molecule contains a six-carbon ring and three double bonds. Draw its structural formula.

17.126 Benzene (see Problem 17.125) is a planar molecule. How many isomers of benzene are there that

(a) Have one hydrogen atom replaced by a chlorine atom?

(b) Have two hydrogen atoms replaced by chlorine atoms?

(c) Have four hydrogen atoms replaced by chlorine atoms?

17.127 Name all the functional groups possible for molecules that have the molecular formulas (a) $C_4H_{10}O$, (b) C_4H_8O, (c) $C_4H_8O_2$. Draw a structural formula for each functionalized hydrocarbon.

17 WORKPATCH SOLUTIONS

17.1 (a)

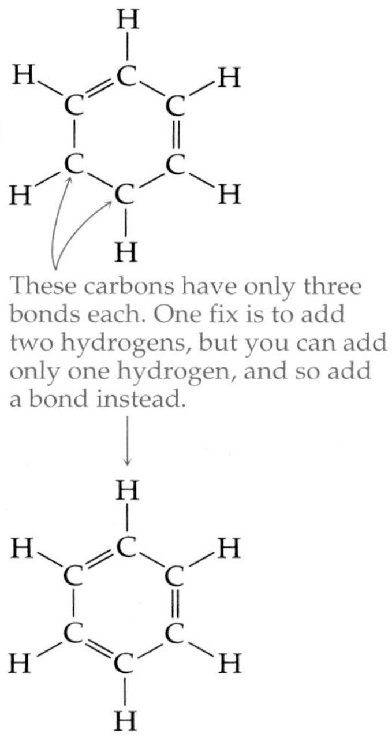

These carbons have only three bonds each. One fix is to add two hydrogens, but you can add only one hydrogen, and so add a bond instead.

(b) H—C≡C—H

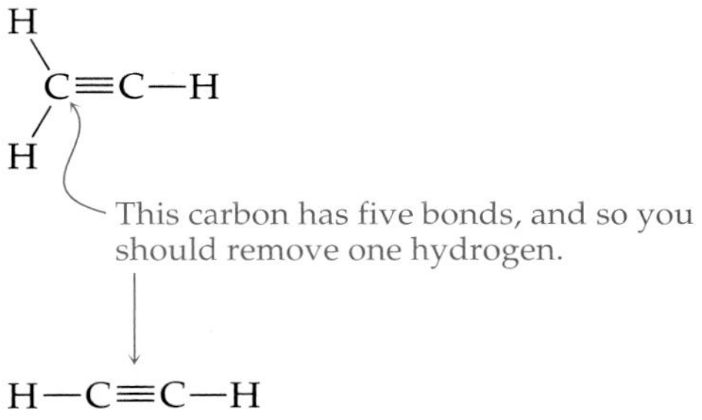

This carbon has five bonds, and so you should remove one hydrogen.

H—C≡C—H

(c)

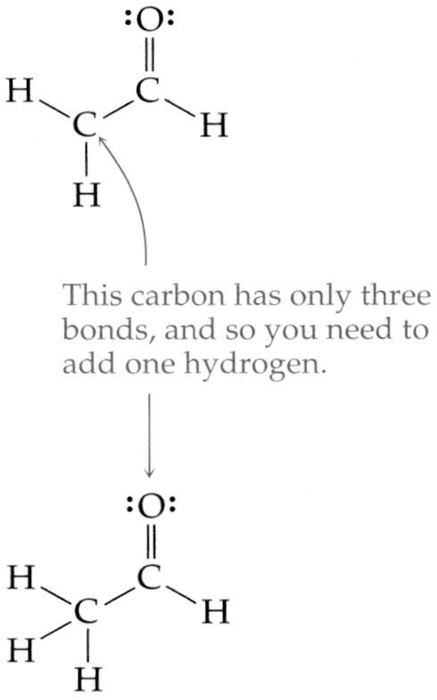

This carbon has only three bonds, and so you need to add one hydrogen.

17.2

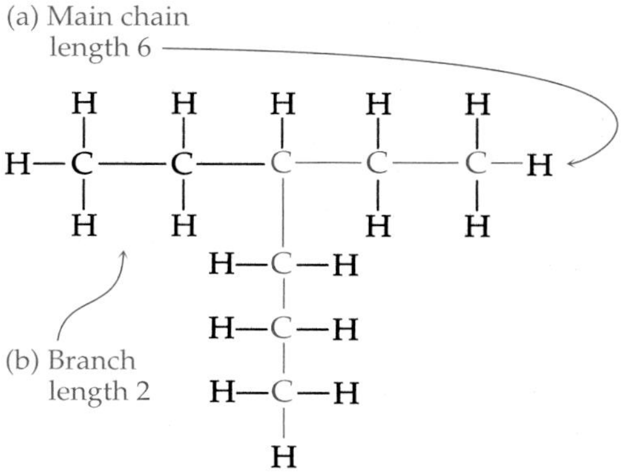

(a) Main chain length 6

(b) Branch length 2

17.3 2-Heptene

17.4 (a) 2,2-Dimethylpropane
(b) 3,3-Dimethylhexane
(c) 3,5-Dimethylheptane

17.5 or

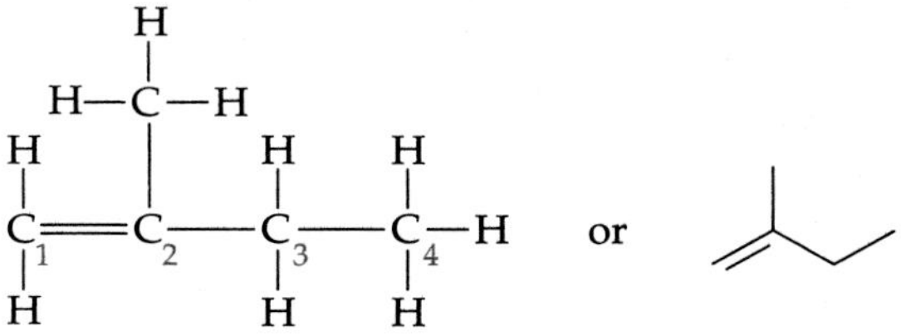

17.6

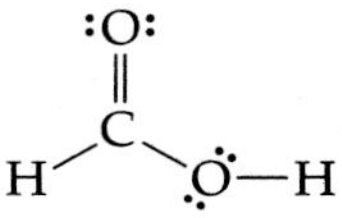

-8.194 1.00 57.58 1HBB1320
8.451 1.00 55.51 1HBB1321
7.611 1.00 55.94 1HBB1322
-8.912 1.00 58.92 1HBB1323
-10.186 1.00 59.11 1HBB1324
-9.707 1.00 59.45 1HBB1325
-9.640 1.00 53.55 1HBB1326
19.934 -10.030 1.00 52.74 1HBB1327
-18.651 -9.234 1.00 49.33
19.018 -19.719 -11.522
19.705 -20.803 -12.358
20.918 -21.496 -11.825
21.968 -20.811 1HBB1332
20.953 -22.631 -11.355 1HBB1333
17.594 -18.464 -8.653 1HBB1334
17.334 -17.291 1HBB1335
17.870 -17.589 1.00 40.32 1HBB1336
1.00 40.33 1HBB1337
1.00 40.21 1HBB1338
-7.708 1.00 30.42 1HBB1339
-8.924 1.00 32.35 1HBB1340
-8.818 1.00 33.93 1HBB1341
13.118 -16.654 -8.233 1.00 34.14 1HBB1342
13.443 -14.754 -9.275 1.00 39.84 1HBB1343
-6.024 1.00 38.38 1HBB1344
19.594 -19.449 -4.727 1.00 37.24 1HBB1345
20.195 -19.278 -4.627 1.00 35.54 1HBB1346
17.453 -20.749 -3.547 1.00 37.44 1HBB1347
17.638 -21.461 -4.522 1.00 44.41 1HBB1348
16.376 -21.990 -3.206 1.00 60.83 1HBB1349
15.450 -22.824 -2.563 1.00 72.76
14.142 -23.091 -3.404 1.00 75.03
-2.735 1.00 66.43

CHAPTER 18

Synthetic and Biological Polymers

18.1 Building Polymers

Believe it or not, you are similar to a collection of plastic sandwich bags filled with water. This may not be a very flattering view, but it is essentially correct. Your body is about 70% by weight water and is enclosed by your skin, the biological equivalent of a plastic bag. If we are ever visited by some nonaqueous lifeform from space, our visitors will likely see us as living bags of water.

The analogy between your skin and a plastic bag goes even further. Both are made of large molecules known as *polymers*, a word that means "many parts" (*poly-* meaning "many" and *-mer* from the Greek *meros,* meaning "part"). A **polymer** is a very large molecule made through the linking together of small molecules called **monomers**:

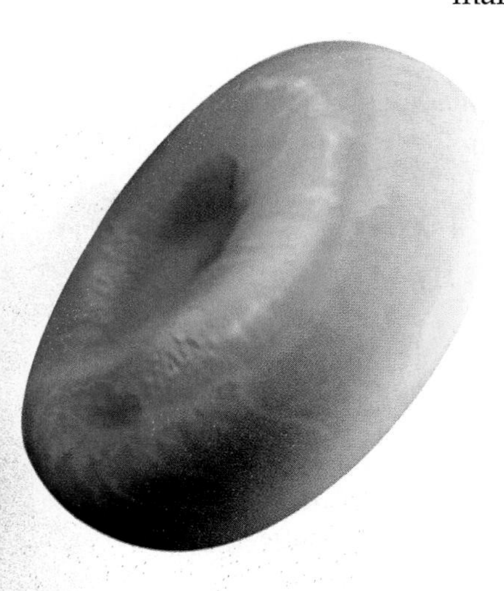

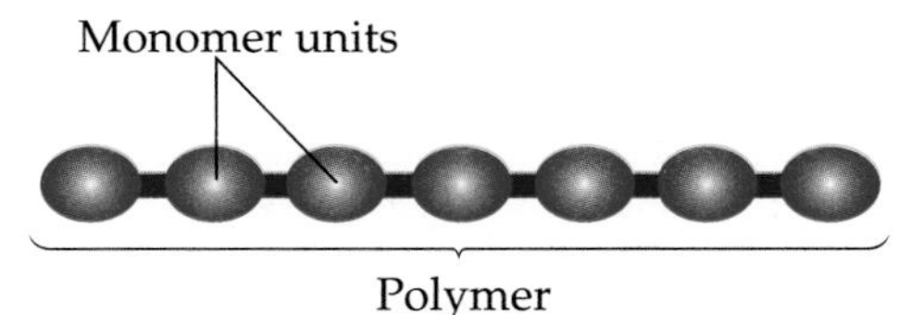

You can think of monomers as the building blocks from which polymers are made. Some polymers are made of only one kind of monomer; other polymers contain more than one kind.

Notice the distinction between the terms *monomer* and *monomer unit*. *Monomer* refers to the small molecule (or molecules) used as the building blocks in the synthesis of a polymer. **Monomer unit** refers to a monomer once it has been made part of a polymer. Because of the rearrangement of electrons and/or atoms required to bond monomers together, there is always a difference between a monomer and the corresponding monomer unit. Watch for the upcoming example called polyethylene, where the molecule CH_2CH_2 is the monomer, and the fragment $-CH_2CH_2-$ is the monomer unit.

In Chapter 17, we mentioned the molecules eicosane and triacontane, which are linear hydrocarbon chains containing 20 and 30 carbon atoms, respectively. Strictly speaking, these molecules may be considered polymers of the monomer unit $-CH_2-$. The term *polymer*, however, is typically used to describe molecules much larger than triacontane, molecules hundreds or thousands of atoms long and having molar masses ranging from thousands to hundreds of thousands of grams per mole. Because of their enormous size, polymer molecules are often referred to as **macromolecules** (*macro-* meaning "large"). The best way to start examining the subject of polymers is with the simplest ones. So let's start with a real sandwich bag and work our way up to your skin.

18.2 Polyethylene and Its Relatives

The common name for ethene is ethylene. Ethylene is the simplest alkene and is very important to the chemical industry.

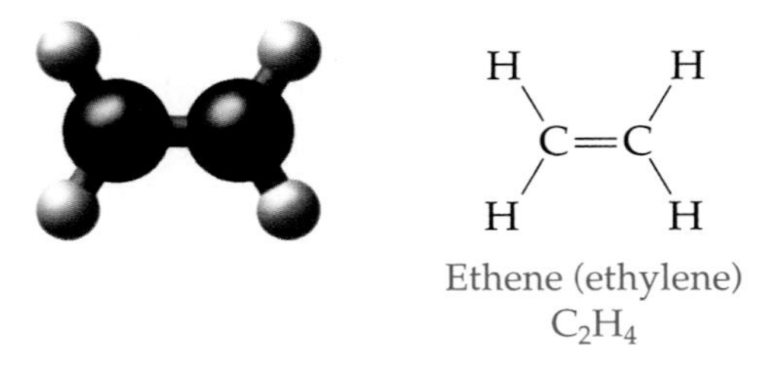

Ethene (ethylene)
C_2H_4

Thirteen billion kilograms of ethylene are produced annually by the U.S. chemical industry. When ethane from natural gas is heated to a high temperature in the absence of air over a metal catalyst, the saturated ethane molecule loses hydrogen and becomes unsaturated ethylene:

$$H_3C{-}CH_3 \xrightarrow[\text{palladium}]{\text{Heat over}} H_2C{=}CH_2 + H_2$$

Ethane from natural gas — Ethylene — Hydrogen gas

When ethylene gas is either heated under pressure in the presence of oxygen or exposed to ultraviolet light, huge polymer molecules form, molecules having molar masses of up to 50,000 g/mol. These macromolecules consist of thousands of ethylene molecules linked together:

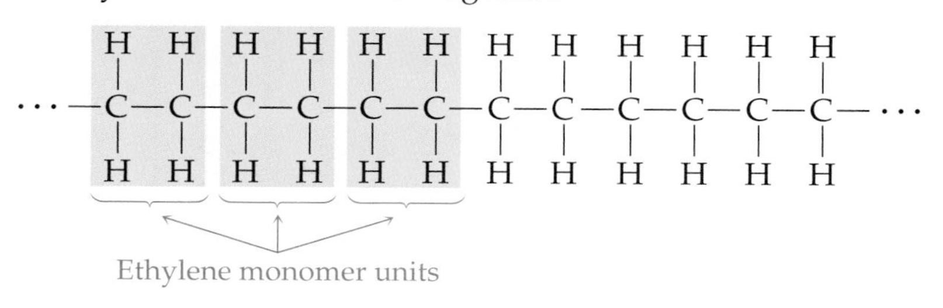
Ethylene monomer units

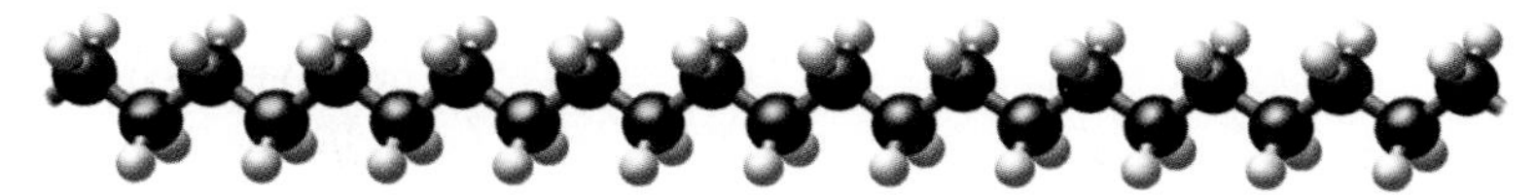

Polyethylene

The polymer is called *polyethylene* because it is made from many ethylene molecules. You might have noticed that there are no C=C double bonds in polyethylene. Nevertheless, it is still called polyethylene and not polyethane. We keep the *-ene* ending to indicate that this polymer was made from ethylene monomers. Sandwich bags, plastic bottles, and bottle caps are made from polyethylene.

Double bonds are missing from polyethylene because one of the electron pairs from the C=C double bond in each monomer is used to bond with another monomer:

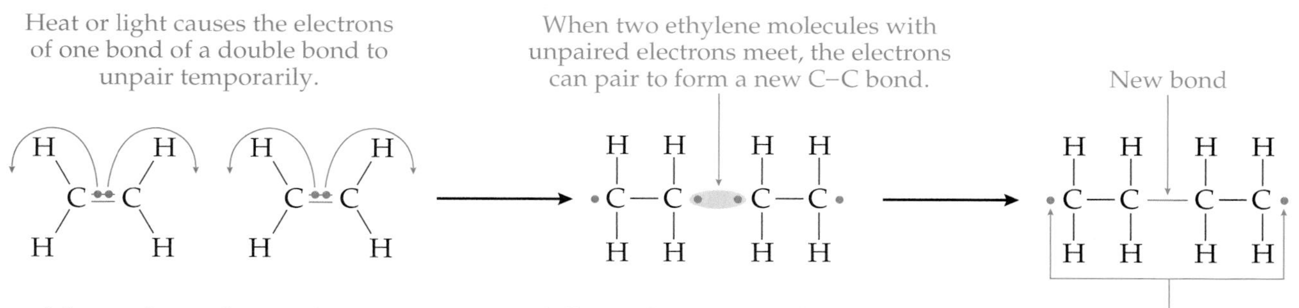
Heat or light causes the electrons of one bond of a double bond to unpair temporarily.
When two ethylene molecules with unpaired electrons meet, the electrons can pair to form a new C–C bond.
New bond
Monomers can continue to hook onto the ends of the new molecule, allowing the polymer to grow.

Many other polymers having properties different from those of polyethylene can be produced by using monomers that are close relatives of ethylene. For example, consider chloroethene, commonly called vinyl chloride. The polymer made from vinyl chloride is called poly(vinyl chloride), often abbreviated PVC. This polymer is most familiar to you as plastic sandwich wrap, and it is also used to make floor tiles, raincoats, and pipe for carrying water and sewage.

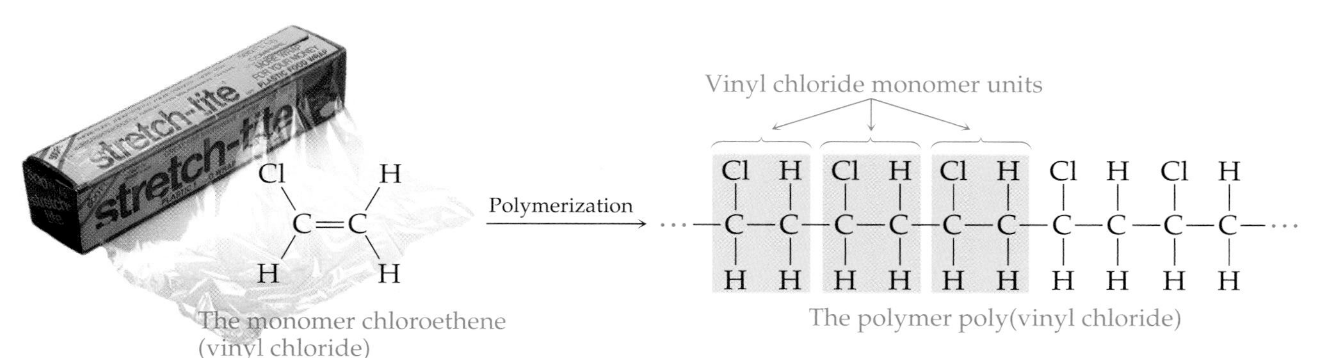
stretch-tite
Polymerization
Vinyl chloride monomer units

The monomer chloroethene (vinyl chloride)

The polymer poly(vinyl chloride)

Teflon, the familiar nonstick coating used on pots and pans, is made by polymerizing tetrafluoroethene:

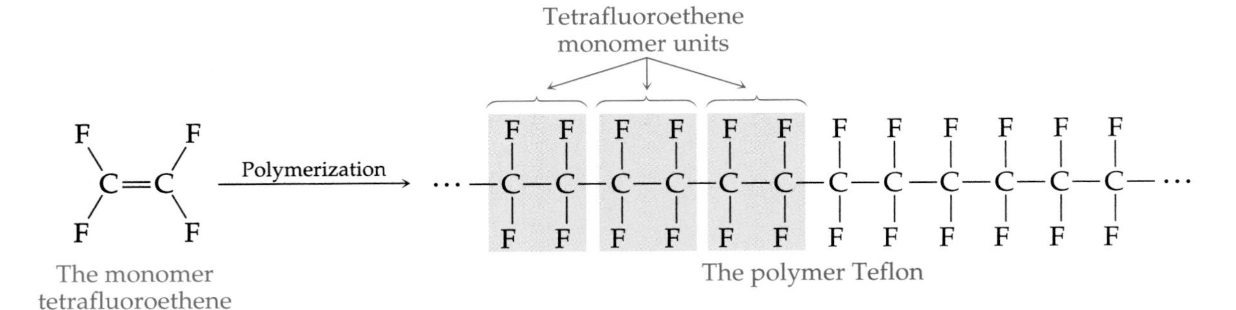

Practice Problems

18.1 (a) Draw the monomer 1,2-dibromoethene.
(b) Draw the polymer formed from this monomer, making it at least three monomer units long.

Answer:

(a) 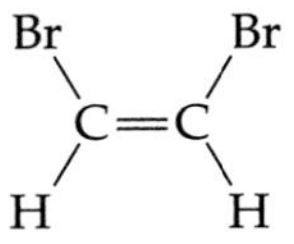 BrHC=CHBr

(b) 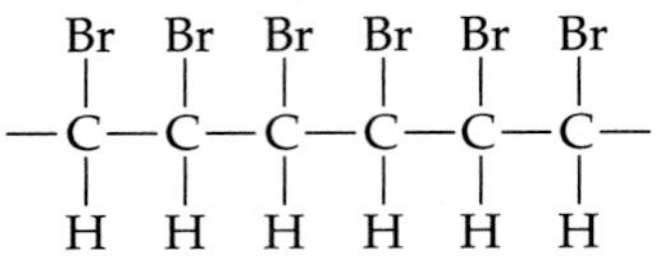 —CHBr—CHBr—CHBr—CHBr—CHBr—CHBr—

18.2 (a) Draw the monomer 1-butene.
(b) Draw the polymer formed from this monomer, making it at least three monomer units long. [*Hint:* There are no double bonds in the polymer.]

18.3 (a) Draw the monomer ethyne (acetylene).
(b) Draw the polymer formed from this monomer, making it at least three monomer units long. [*Hint:* The polymer contains double bonds, not triple bonds.]

So far we have been showing you a monomer first and then the polymer it forms. In order to better understand the relationship between a polymer and its monomer, it is useful to do a reverse analysis. For example, consider the polymer poly(propylene oxide):

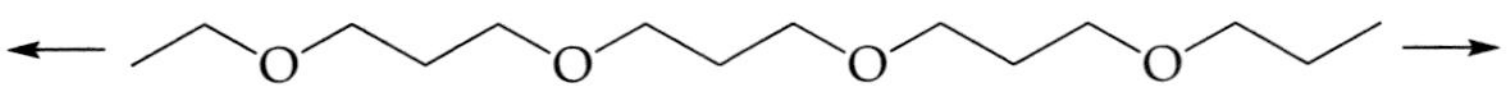

Poly(propylene oxide)

What is the monomer unit? Remember that the monomer unit is the small piece that gets repeated over and over again in the polymer. Here are two candidates:

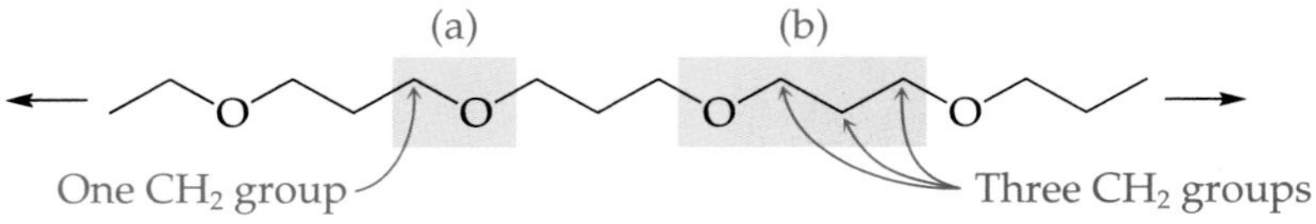

(a) (b)

One CH_2 group

Three CH_2 groups

Both candidates contain one oxygen atom. Choice (a) also contains a CH_2 group, so the formula for this candidate is CH_2O. The formula for candidate (b) is $CH_2CH_2CH_2O$. So which is the monomer unit? To find out, all we have to do is repeat them and see what we get:

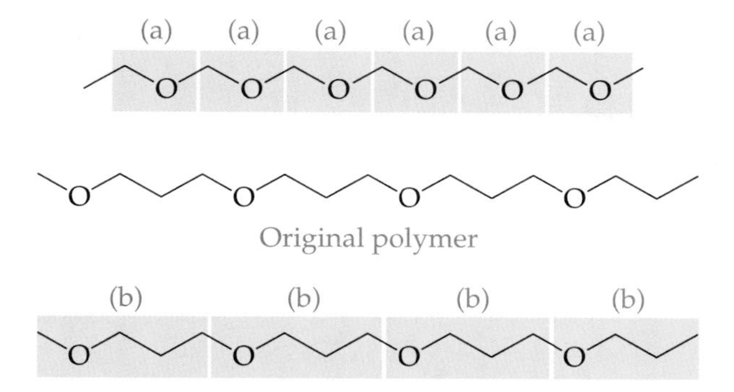

You can see that repeating (a) doesn't reproduce the original polymer, meaning it can't be the monomer unit. The correct choice is (b).

In the remainder of this chapter, we shall examine some very interesting polymers. Being able to look at them and identify the monomer unit or units will help you understand them. Try your skill at this now.

This polymer is an example of a *polyester* and is often used to make clothing. The monomer unit is called an *ester*. Which unit—(a), (b), or (c)—is the correct monomer unit? Explain your choice.

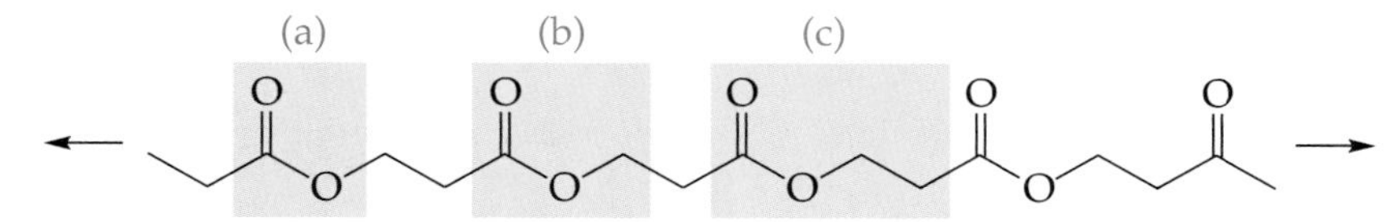

18.3 Nylon—A Polymer You Can Wear

Unless you insist on wearing clothes made only of 100% cotton or wool, chances are that your shirt, blouse, or slacks are made from some petroleum-based synthetic fabric. Polyester, nylon, and Dacron are woven into fibers made from synthetic polymers. One type of nylon polymer has the structure shown below:

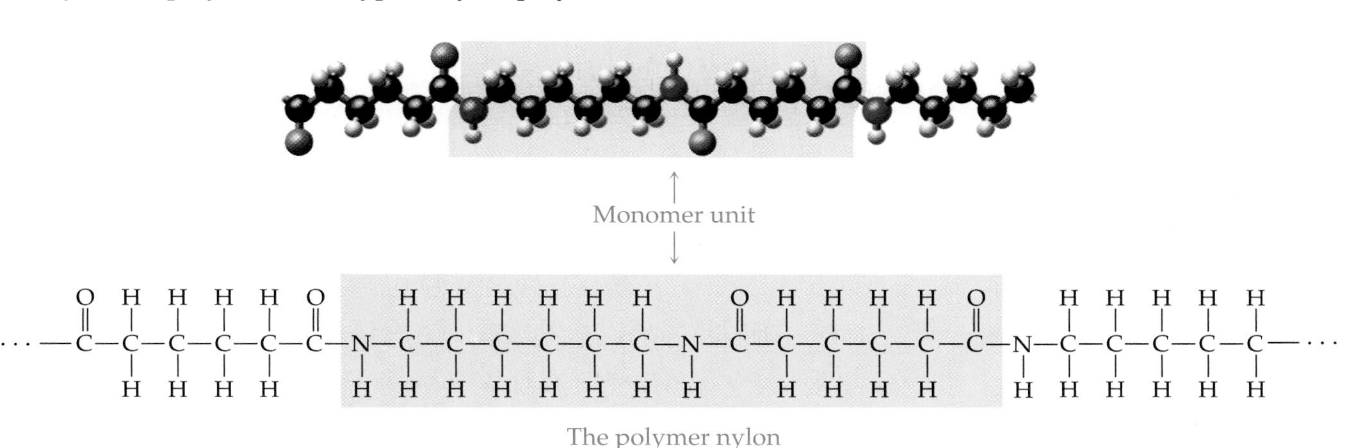

The polymer nylon

Nylon fibers wound around one another are so strong that they are used to make ropes for tying up ships. Much of this strength is due to the hydrogen bonds that form between adjacent polymer chains:

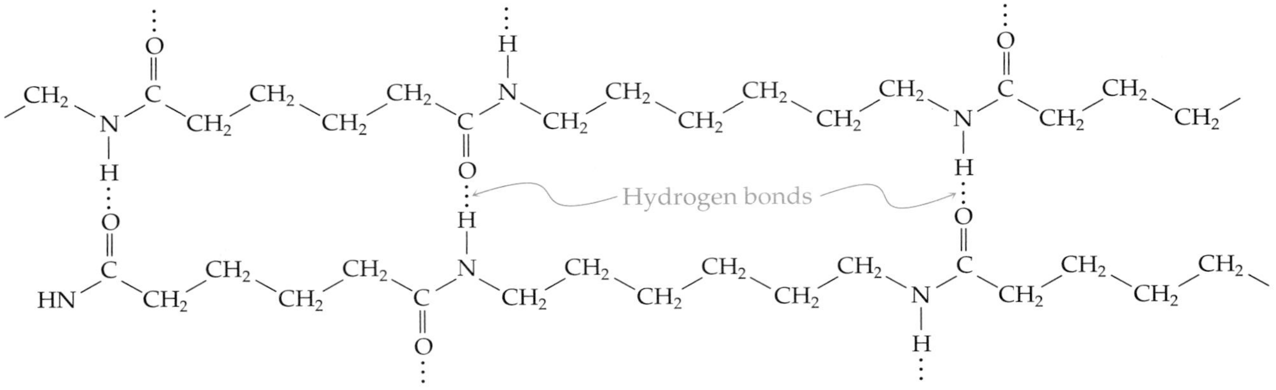

Hydrogen bonds

Nylon polymers are made by starting with two different monomers. We'll show how this works using the nylon shown above, which is called nylon 66. To make this nylon, we start with the following two monomers, each based on hexane with functional groups at both ends:

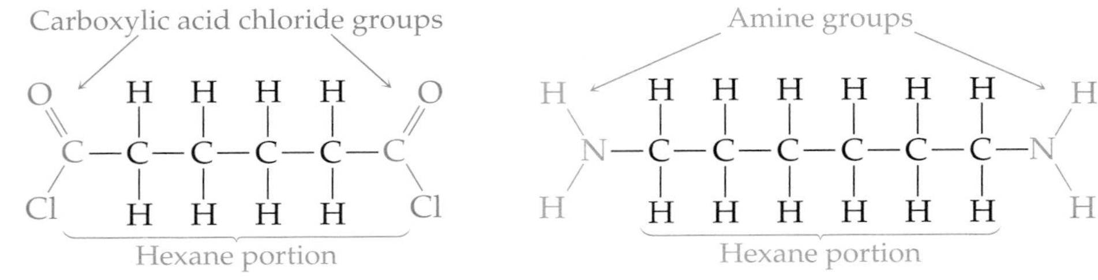

The monomer hexanedioic acid chloride

The monomer 1,6-diaminohexane

In hexanedioic acid chloride, the functional groups are carboxylic acid chloride groups—each essentially a carboxylic acid group, COOH, with the –OH replaced by –Cl. In 1,6-diaminohexane, the functional groups are amines.

The functional groups on these two monomers can react with each other, forming a bond that links the two monomers together. Specifically, the Cl from a carboxylic acid chloride combines with an H from an amine group to form HCl. The HCl departs, leaving the two monomers linked by what is called an **amide bond** (a bond between an amine nitrogen and a carbonyl carbon):

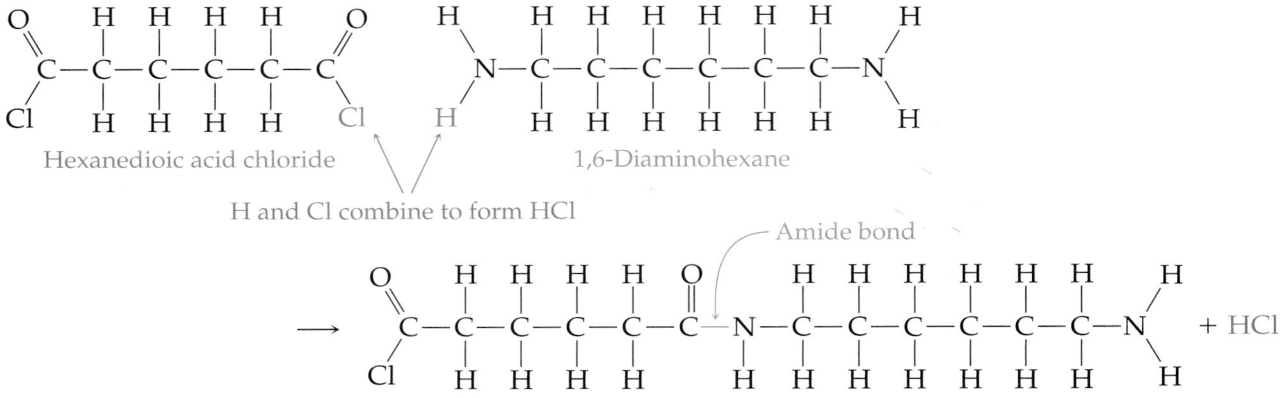

Hexanedioic acid chloride

1,6-Diaminohexane

H and Cl combine to form HCl

Amide bond

$\longrightarrow$ + HCl

Because *both* ends of each monomer carry a functional group, monomer molecules can add onto both ends of this short chain. The reactions are the same as shown above, and the result is a nylon chain held together by amide bonds:

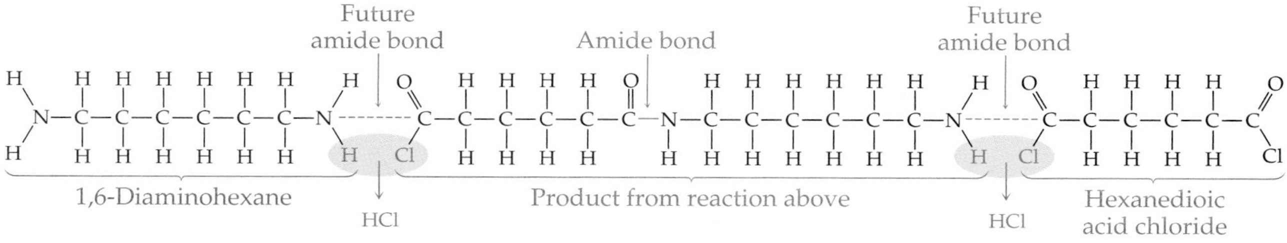

Future amide bond — Amide bond — Future amide bond

1,6-Diaminohexane + Product from reaction above + Hexanedioic acid chloride → HCl, HCl

The chains can be grown long enough to make fibers that can be woven into stockings, shirts, and ropes.

We have chosen to highlight nylon for a reason. The same amide bonds that are in nylon are also part of your skin, muscles, enzymes, hormones, blood cells, and so on, in the form of *proteins*. **Proteins** are biological polymers in which amide bonds link monomer units together, just as they do in nylon. This makes recognizing amide bonds and understanding amide bond formation very important. Try the next WorkPatch to make sure you understand the amide bond.

18.2 WORKPATCH

Show how the amine and the carboxylic acid chloride in these molecules react to form an amide bond. Draw the product, and indicate the new amide bond formed.

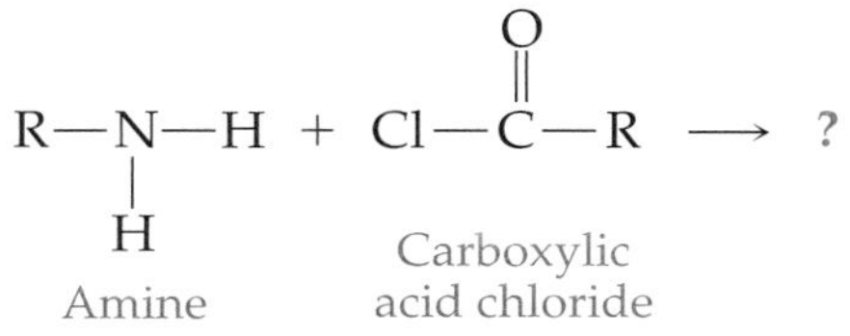

R—NH—H (Amine) + Cl—C(=O)—R (Carboxylic acid chloride) ⟶ ?

Compare your drawing to ours (at the end of the chapter) to make sure you understand the amide bond.

Nature uses a variety of polymers—called **biopolymers**—in constructing living organisms. To give you an introduction to this topic, we now consider two of the most important classes of biopolymers, polysaccharides and proteins.

18.4 Polysaccharides and Carbohydrates

Billions of years ago, when the first single-cell form of plant life was developing, something was needed to form a boundary separating the contents of the cell from its surroundings and protecting the life processes occurring inside. This boundary, called the *cell wall*, is made from a biological polymer known as a *polysaccharide*. The particular polysaccharide that makes up cell walls is *cellulose*. You are probably more familiar with other polysaccharides, such as *starch*, an important food source. The exoskeletons of insects also consist largely of a type of polysaccharide. The word *saccharide* comes from *saccharine*, which means "sugar" or "sweet." Sugar molecules are the monomers in polysaccharides. **Polysaccharides** are therefore polymers in which tens, hundreds, or even thousands of sugar molecules are linked together.

Let's begin our examination of polysaccharides by looking at the monomer, a sugar molecule. Sugars belong to a class of molecules known as *carbohydrates*. **Carbohydrates** are generally aldehydes or ketones that also have many OH groups in them. The most common sugar molecule is glucose:

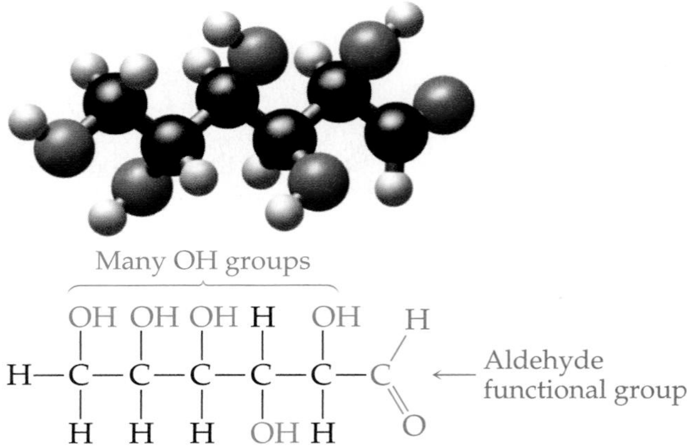

Many OH groups

Aldehyde functional group

Glucose, a simple sugar
$C_6H_{12}O_6$

Notice the aldehyde group and the five OH groups. Glucose is the principal sugar in your blood. It is consumed by your cells to obtain the energy necessary for sustaining life processes.

The linear structure shown above for glucose is not its most common form. Usually, glucose exists in a ring, or cyclic, structure:

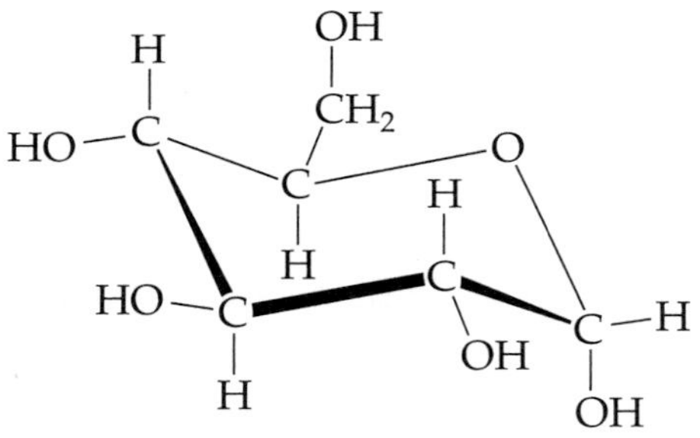

The formula is the same, but the structure is different. The cyclic form exists in equilibrium with the linear form. In the cyclic form, glucose no longer has an aldehyde functional group, but it is still called a carbohydrate. One reason we can call this cyclic sugar a carbohydrate is obvious from the name *carbohydrate*. The *carbo-* comes from carbon, and *-hydrate* means "water." Thus, *carbohydrate* literally means hydrated carbon, or carbon with water attached. Clearly, a cyclic glucose molecule is not carbon with water molecules attached. However, most of the carbons do have the makings of water, an H atom and an OH group, bound to them.

Glucose is a **monosaccharide** because it consists of only one sugar ring. The sugar that you eat on your cereal and in sweets is sucrose, a **disaccharide**. Sucrose consists of two rings: a glucose ring joined to a ring of another monosaccharide called fructose:

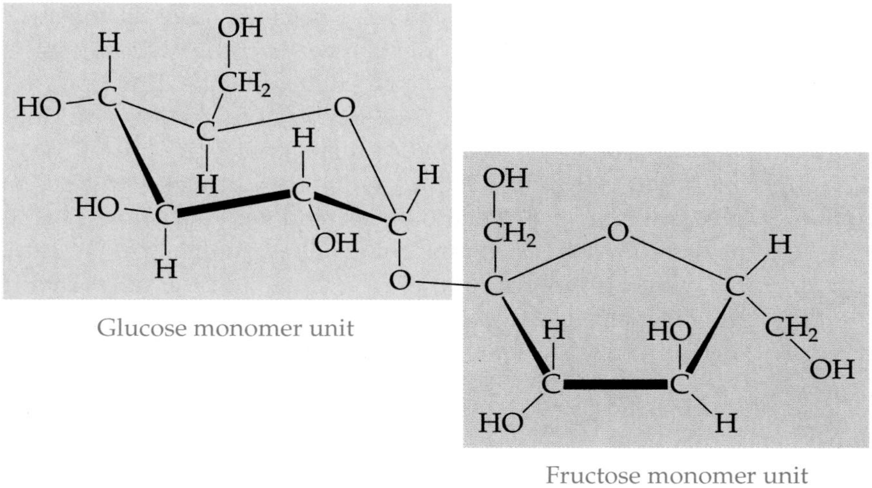

Glucose monomer unit

Fructose monomer unit

Sucrose (table sugar), a disaccharide
$C_{12}H_{22}O_{11}$

Glucose can also form long chains with itself. Two polysaccharides made from glucose monomers are starch and cellulose:

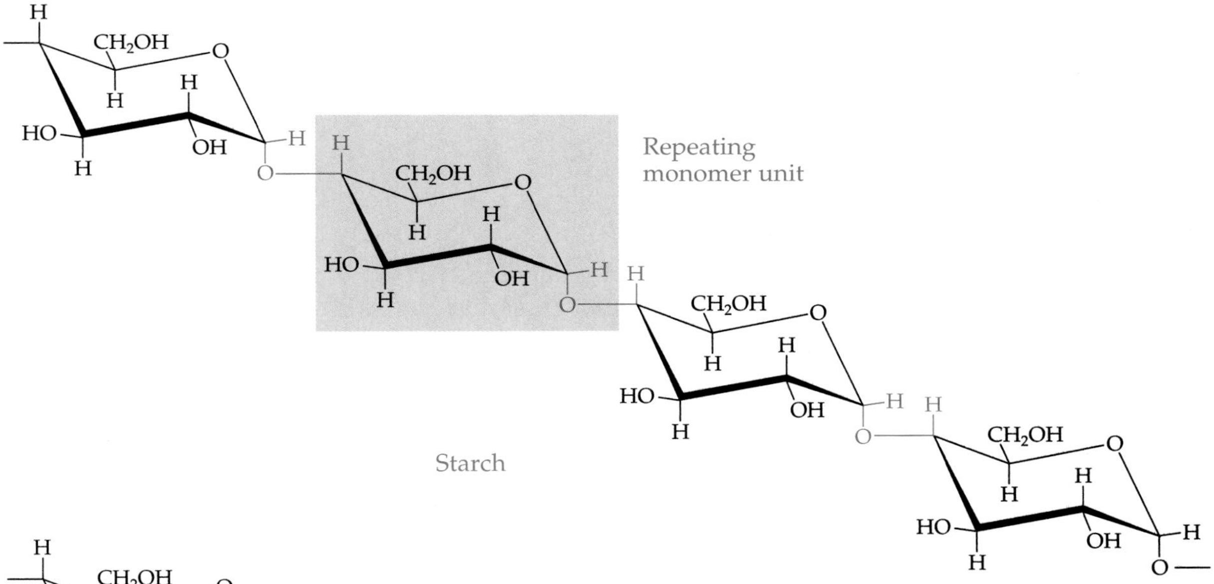

Repeating monomer unit

Starch

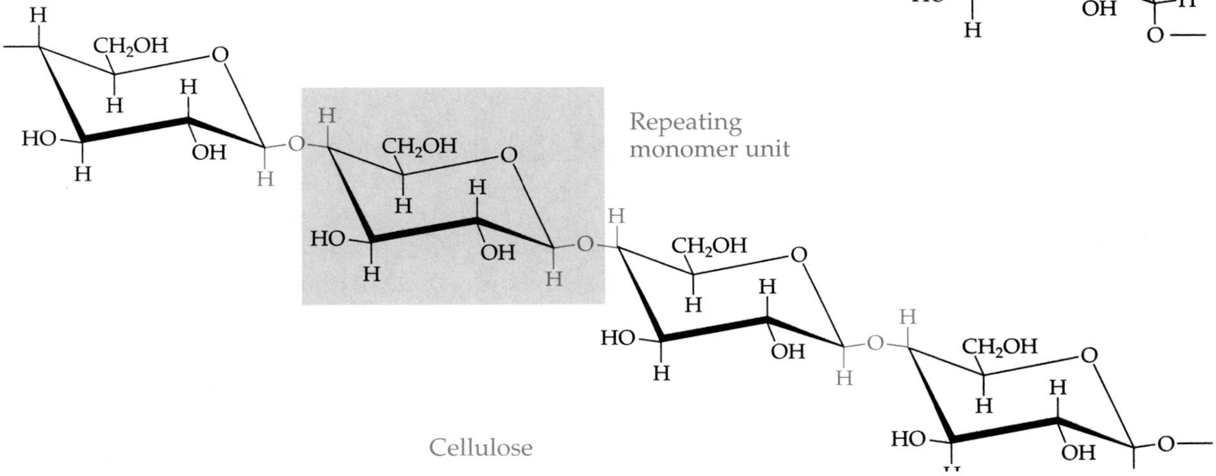

Repeating monomer unit

Cellulose

The difference between starch and cellulose is slight. Look at the structures carefully, focusing on the oxygen atoms that connect the glucose rings, to see the difference. Although we won't go into the details, the difference in orientation of the connecting oxygen atoms—whether above or below a ring—creates a profound difference in the three-dimensional structure of the chains. This slight difference in connectivity renders cellulose indigestible to humans. Whereas we can eat a potato or rice and gain nutritional value from its starch, we get nothing but an upset stomach if we eat wood, which is largely cellulose. Termites, on the other hand, have in their gut microorganisms that can digest wood and get from it the same nutritional value we get from starch.

When you consider that wood is almost 50% cellulose—a polysaccharide biopolymer—you start to get an appreciation for how strong polymers can be. Nature uses cellulose and similar polymers much as we use steel girders—for structural support. Although cellulose works well for plants, you won't find any in yourself. Most forms of animal life use a different, more flexible biopolymer, as you'll learn in the next section.

18.5 Proteins

The main biopolymers used for both structure and function in all living things are the proteins. As we mentioned earlier, our skin is made of protein. So are our hair, nails, and muscles. Proteins are also used as enzymes and hormones, to store iron, to transport oxygen, to fight disease, and to clot blood. Proteins are polymers of small molecules called *amino acids*. An **amino acid** has (at least) two functional groups, an amine group, either H_2NR or HNR_2, and a carboxylic acid group, COOH. Here is the structure of the simplest amino acid (glycine) found in proteins:

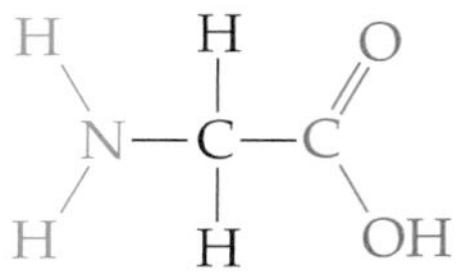

H₂N—CH₂—C(=O)OH

Amine functional group | Carboxylic acid functional group

Any hydrocarbon molecule containing both an amine functional group and a carboxylic acid functional group could be called an amino acid. This means that there are an unlimited number of different amino acid molecules possible. However, investigation has shown that only 20 of them are necessary for synthesizing the proteins in our bodies. All 20 of these amino acids have an important similarity. They all have just one carbon atom between the amine functional group and the carboxylic acid functional group as shown at right. The R group in amino acids varies, from an H atom in the simplest amino acid, glycine, to functionalized hydrocarbon chains in the more complex amino acids. Table 18.1 shows the 20 amino acids humans need, along with the names of the acids and their three-letter abbreviations. Your body can synthesize ten of these amino acids from simpler molecules. The other ten, called **essential amino acids**, cannot be synthesized by your body (or, in the case of arginine, cannot be synthesized in sufficient amounts). These must be present in the food you eat to provide the monomers needed to manufacture many of the proteins your body needs. Failure to take in enough essential amino acids in your diet can result in a wide variety of nutritional-deficiency diseases. (The essential amino acids are shaded blue in Table 18.1.)

H, H, N, C, R, C, O, OH

Amino acid

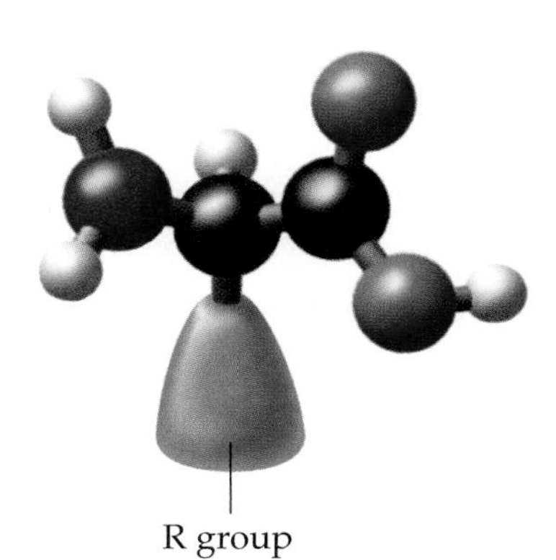

To build protein polymers, living organisms string together the amino acids listed in Table 18.1, using the same type of amide bonds found in nylon. If we wanted to join glycine and alanine in the laboratory, for example, we could heat a mixture of them. One possible result is that the acidic end of glycine approaches the basic (amine) end of alanine:

H, H, N, C, H, H, C, O, OH — Glycine

H, H, N, C, H, CH_3, C, O, OH — Alanine

In a process similar to the formation of nylon, the OH from the carboxylic acid group and an H from the amine group combine to form water. In the process, a C–N bond forms between the carboxylic acid carbon and the amine nitrogen. In nylon we called this C–N bond an amide bond. In proteins, however, this same bond is often called a **peptide linkage**.

Glycine Alanine → Gly–Ala dipeptide + H_2O

OH and H combine to form water.

Amide bond (peptide linkage)

The resulting molecule is called a **dipeptide**. A close look shows that the dipeptide has a free (unreacted) carboxylic acid group on one end and a free amine group on the other end. This means that other amino acids can approach these ends and join by forming additional peptide linkages. Short and medium-length amino acid chains formed in this way are referred to as **polypeptides**. Longer chains and amino-acid polymers that have a biological function are referred to as proteins.

Table 18.1 The 20 Amino Acids Humans Use to Build Proteins (Essential Amino Acids Are Indicated by Blue Shading)

Neutral and polar amino acids

Alanine (Ala)
$H_2N-CH(CH_3)-COOH$

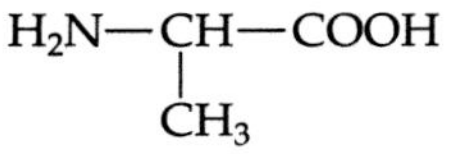

Asparagine (Asn)
$H_2N-CH(CH_2-C(=O)NH_2)-COOH$

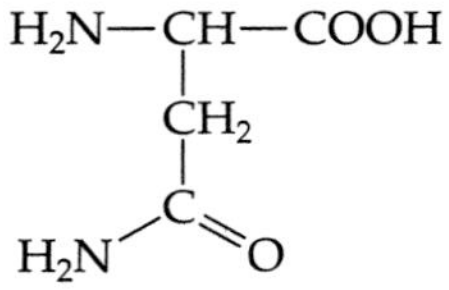

Cysteine (Cys)
$H_2N-CH(CH_2-SH)-COOH$

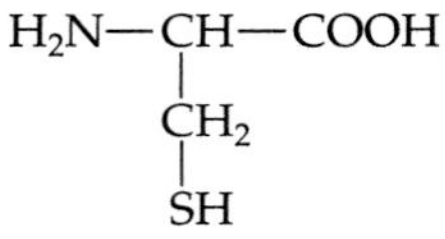

Glutamine (Gln)
$H_2N-CH(CH_2-CH_2-C(=O)NH_2)-COOH$

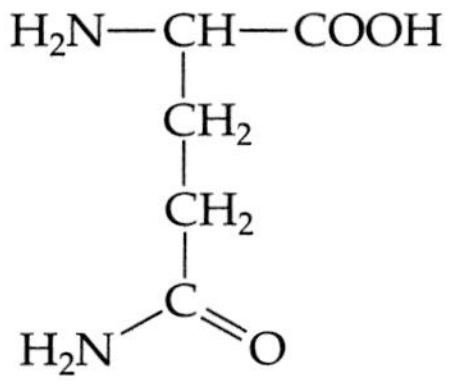

Glycine (Gly)
$H_2N-CH(H)-COOH$

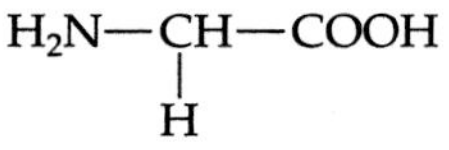

Isoleucine (Ile)
$H_2N-CH(CH(CH_3)-CH_2-CH_3)-COOH$

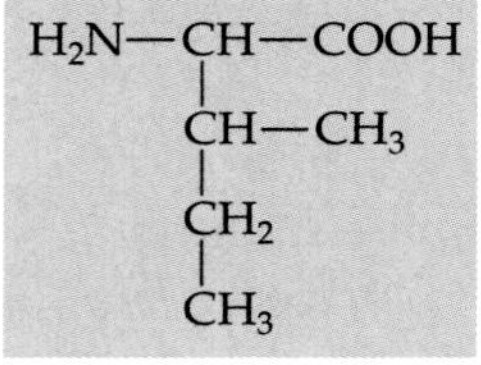

Leucine (Leu)
$H_2N-CH(CH_2-CH(CH_3)-CH_3)-COOH$

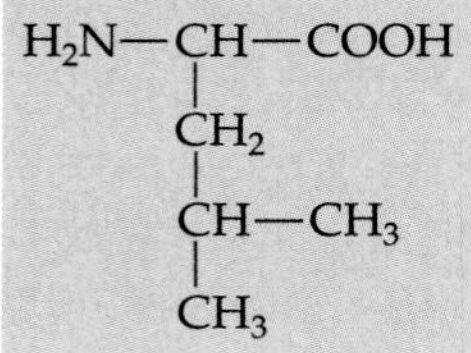

Methionine (Met)
$H_2N-CH(CH_2-CH_2-S-CH_3)-COOH$

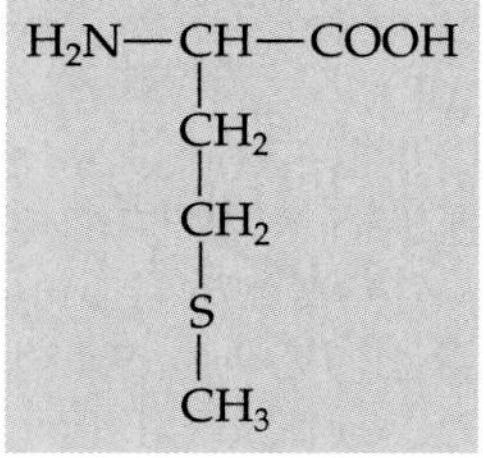

Phenylalanine (Phe)
$H_2N-CH(CH_2-C_6H_5)-COOH$

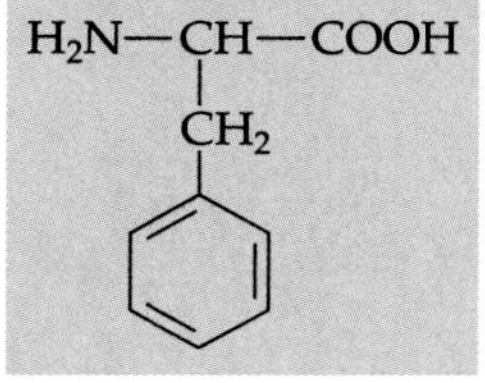

Proline (Pro)
$HN-CH-COOH$ (ring: H_2C, CH_2, $C H_2$)

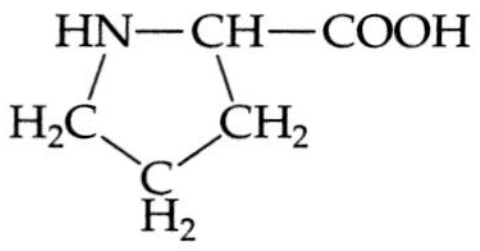

Serine (Ser)
$H_2N-CH(CH_2-OH)-COOH$

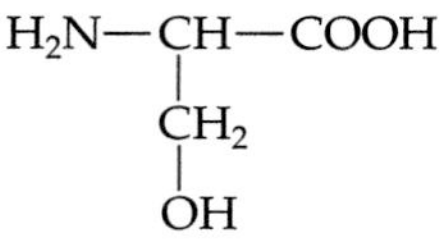

Threonine (Thr)
$H_2N-CH(CH(CH_3)-OH)-COOH$

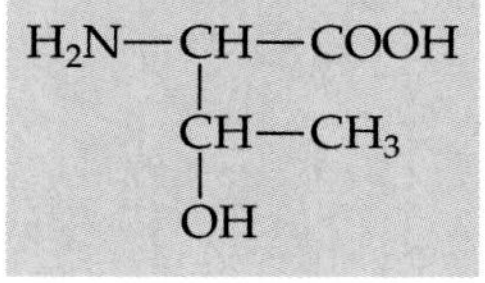

Tryptophan (Trp)
$H_2N-CH(CH_2-C{=}CH-NH$ (indole ring)$)-COOH$

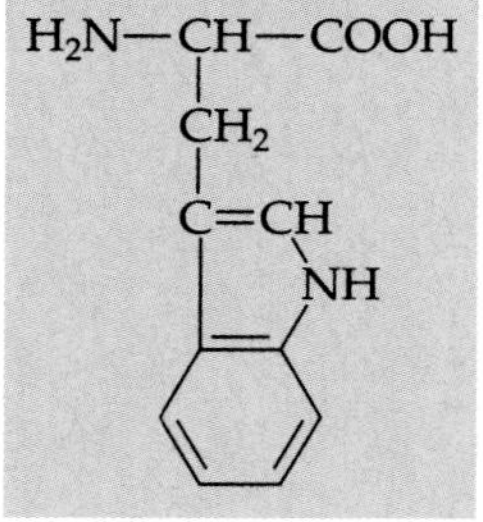

Tyrosine (Tyr)
$H_2N-CH(CH_2-C_6H_4-OH)-COOH$

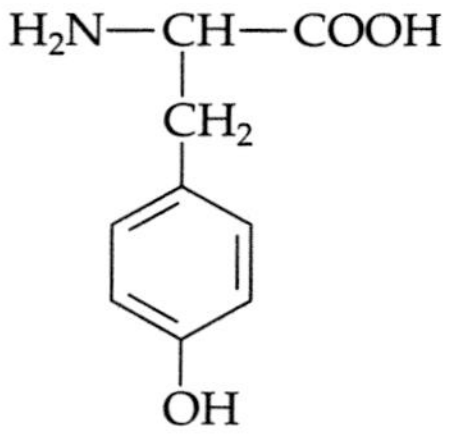

Valine (Val)
$H_2N-CH(CH(CH_3)-CH_3)-COOH$

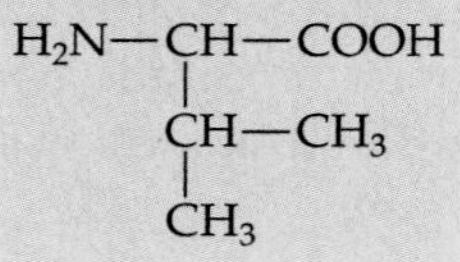

Basic amino acids

Arginine (Arg)
$H_2N-CH(CH_2-CH_2-CH_2-NH-C(=NH)-NH_2)-COOH$

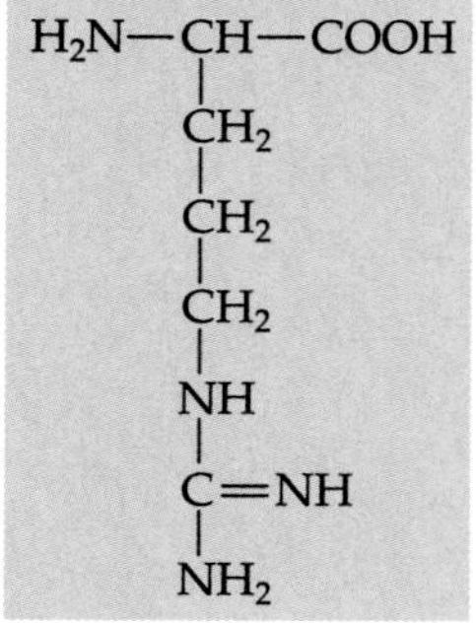

Histidine (His)
$H_2N-CH(CH_2-C$ (ring: HC, NH, $N{=}CH$)$)-COOH$

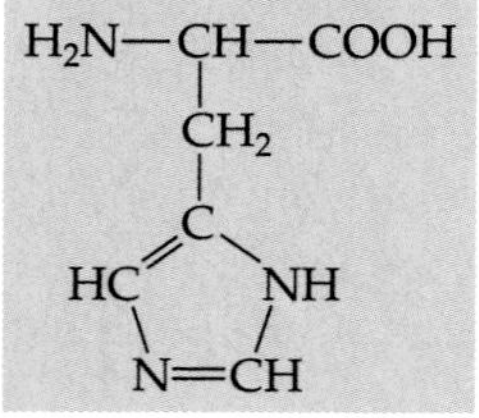

Lysine (Lys)
$H_2N-CH(CH_2-CH_2-CH_2-CH_2-NH_2)-COOH$

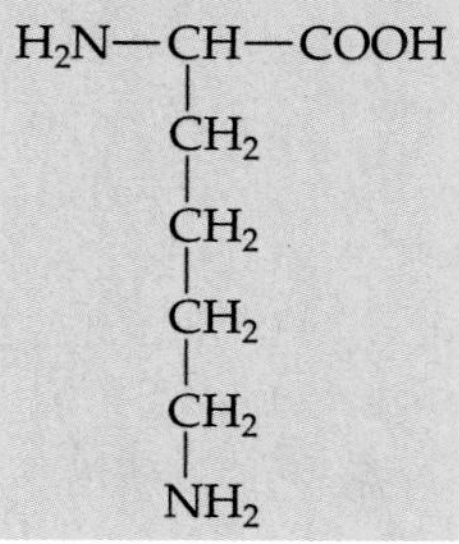

Acidic amino acids

Aspartic acid (Asp)
$H_2N-CH(CH_2-C(=O)OH)-COOH$

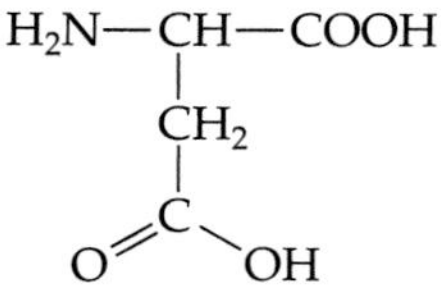

Glutamic acid (Glu)
$H_2N-CH(CH_2-CH_2-C(=O)OH)-COOH$

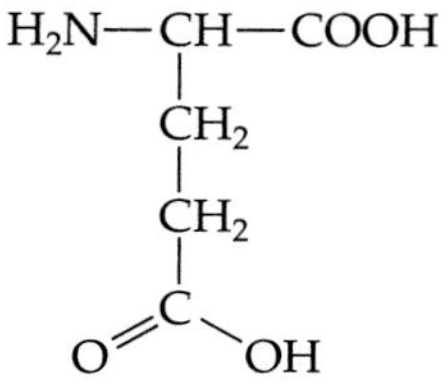

Dipeptides are usually named from the three-letter abbreviations of the amino acids used to make them. Thus the dipeptide we just made from glycine and alanine is a Gly–Ala dipeptide. By convention, the amino acid with the unreacted NH_2 group is named first and the amino acid with the unreacted COOH group is named second. Test yourself now and see if you understand the peptide linkage.

18.3 WorkPatch

Draw the dipeptides Ala–Ser and Ser–Ala, both formed from the amino acids alanine and serine:

$H_2N-CH(CH_3)-COOH$

Alanine

$H_2N-CH(CH_2OH)-COOH$

Serine

Do you understand why it is possible to make two different dipeptides from alanine and serine? Check your answer with ours and make sure you do.

How a protein functions in the body depends on the protein's sequence of amino acids and on its three-dimensional shape. For example, hemoglobin (shown on the first page of this chapter) is the protein in red blood cells that is responsible for carrying oxygen from the lungs to all cells in the body. It consists of 574 amino acids. For hemoglobin to function correctly, each amino acid must be in its proper place in the protein. Any change in the position or identity of even one amino acid can change the shape of the hemoglobin protein and affect its function. That is what happens in sickle-cell anemia, an inherited blood disease. People with sickle-cell anemia have hemoglobin proteins in which a valine has replaced a glutamic acid. This single substitution causes the three-dimensional shape of the hemoglobin molecule to change whenever the amount of oxygen in the blood is reduced. This change in molecule shape changes the shape of the red blood cells from round to crescent (sickle-shaped), causing them to clog small blood vessels and restrict blood flow. The result is extremely painful and debilitating.

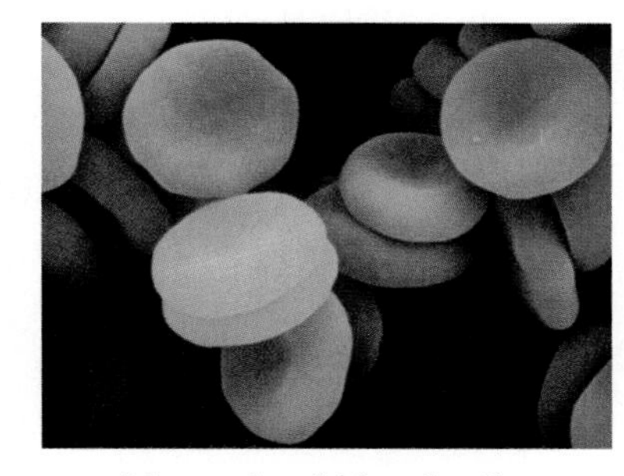

Normal red blood cells

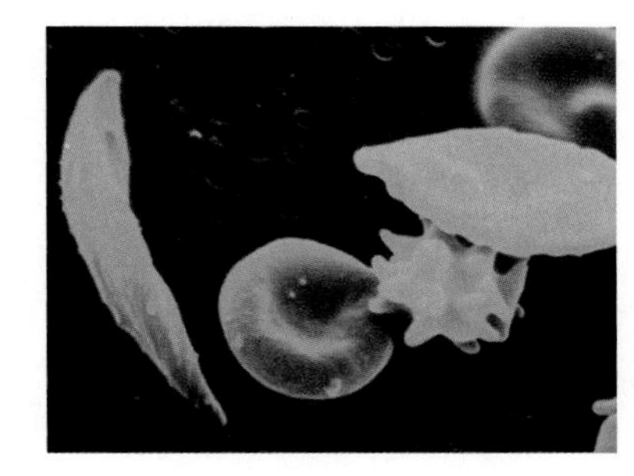

Sickled red blood cells

Given that one misplaced amino acid can wreak such havoc, it is amazing that any of us can function at all. For example, consider the polypeptide angiotensin II, a hormone that regulates blood pressure in humans. Angiotensin II contains eight different amino acids:

Asp – Arg – Val – Tyr – Ile – His – Pro – Phe

Angiotensin II

From a mathematical point of view, there are 40,320 ways to arrange eight amino acids in a chain, but only *one* of those 40,320 combinations regulates your blood pressure! These are tremendous odds for having something go wrong, and yet mistakes are very rare. Something is serving as a blueprint for correctly putting your proteins together, something you inherited.

Practice Problems

18.4 Draw the dipeptide Asp–Arg, and label the peptide linkage.

Answer:

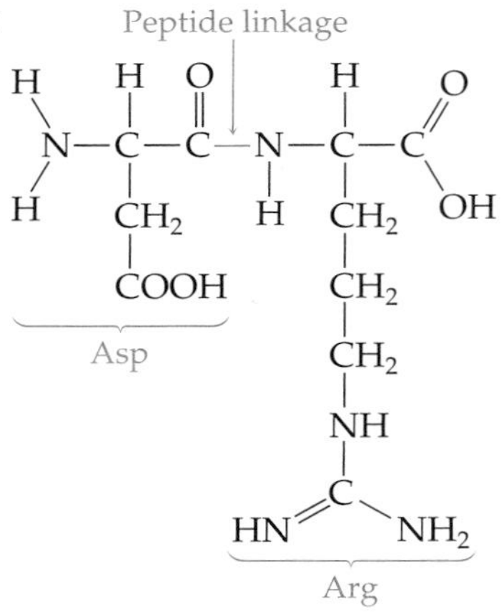

Peptide linkage

Asp

Arg

18.5 Draw the dipeptide Arg–Asp, and label the peptide linkage.

18.6 Give the amino acid sequence for this polypeptide, using the three-letter abbreviations separated by hyphens:

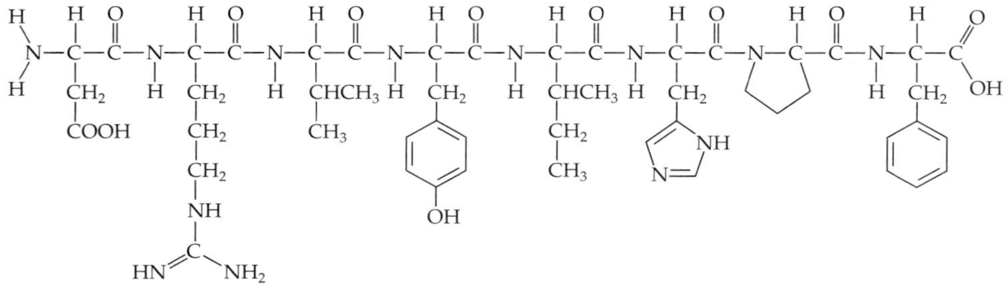

18.6 DNA—The Master Biopolymer

The master blueprint for all your proteins is called **DNA**, short for **deoxyribonucleic acid**. You received half of your DNA from your mother and half from your father. This master blueprint for building protein polymers is itself a polymer. The chromosomes and genes present in the nuclei of most cells are gigantic DNA polymers ranging in mass from a few million grams per mole in bacteria to billions of grams per mole in higher animals.

The DNA polymer consists of a chain, or "backbone," of alternating units of phosphate groups and sugar molecules called *deoxyriboses*.

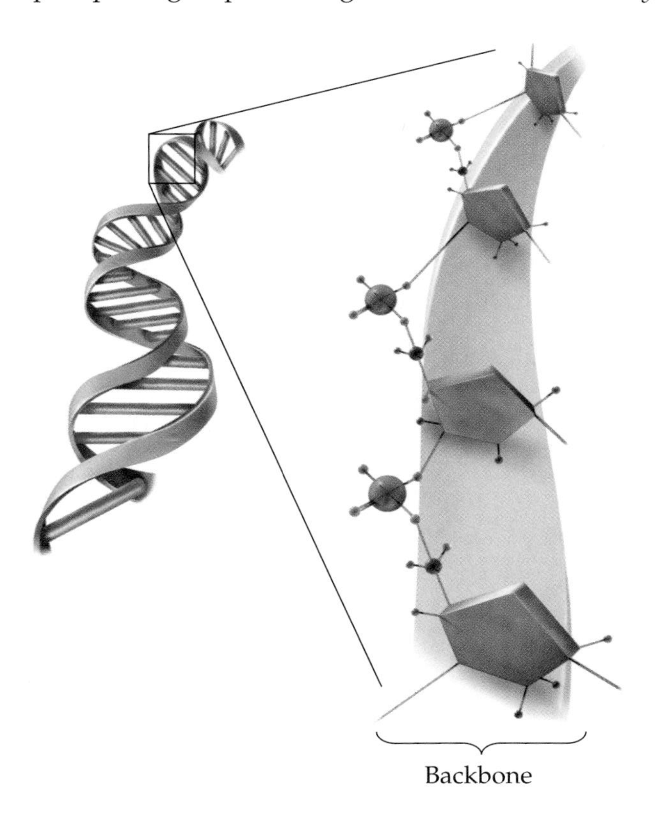

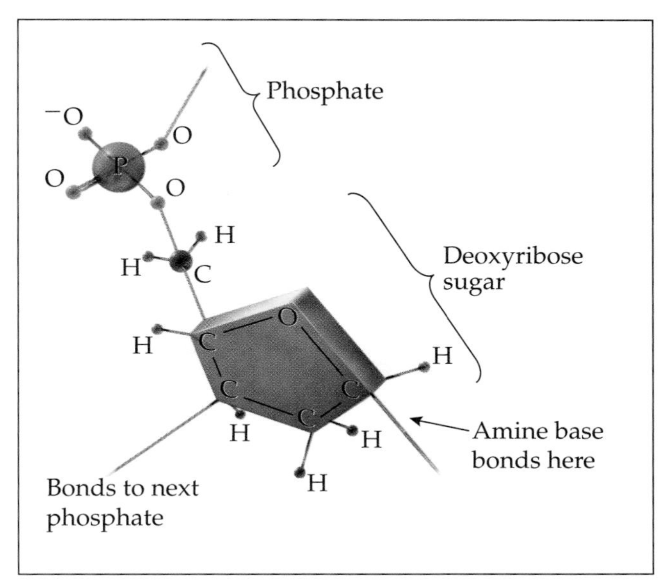

This backbone is just part of the DNA molecule. Notice in the drawing that one of the carbon atoms in the deoxyribose ring is labeled "amine base bonds here." In a complete DNA molecule, one of these four amine bases is covalently bonded to this carbon:

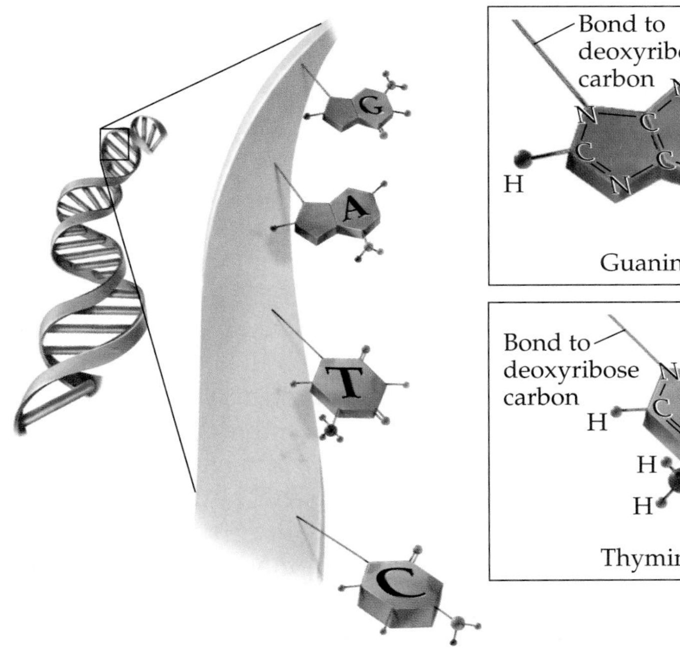

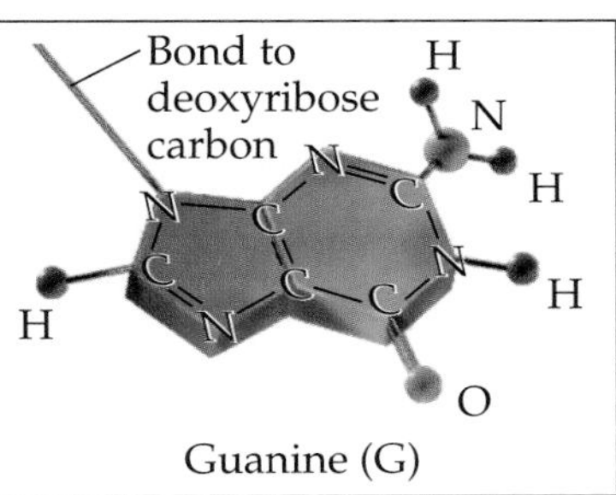

Guanine (G)

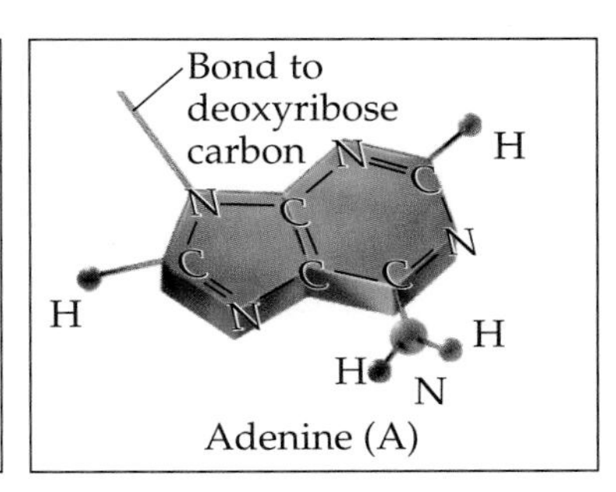

Adenine (A)

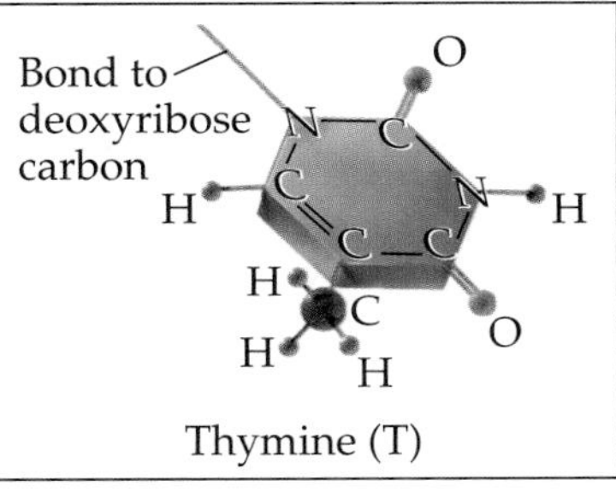

Thymine (T)

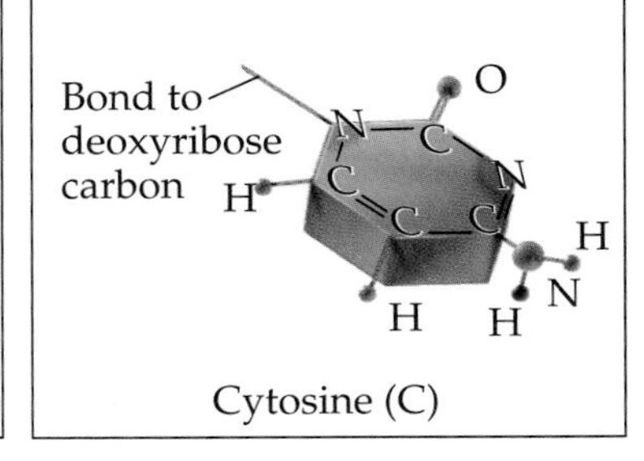

Cytosine (C)

Thus part of a single strand of DNA polymer looks something like this:

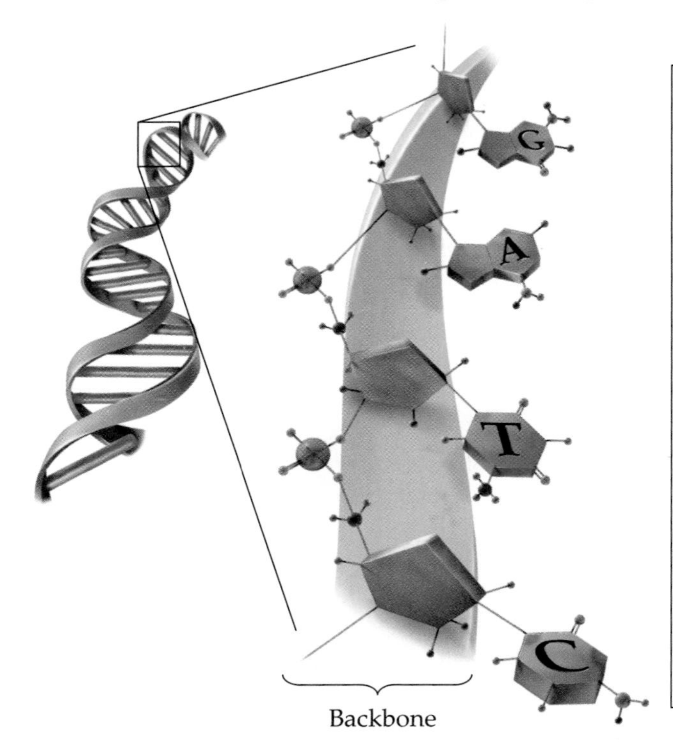

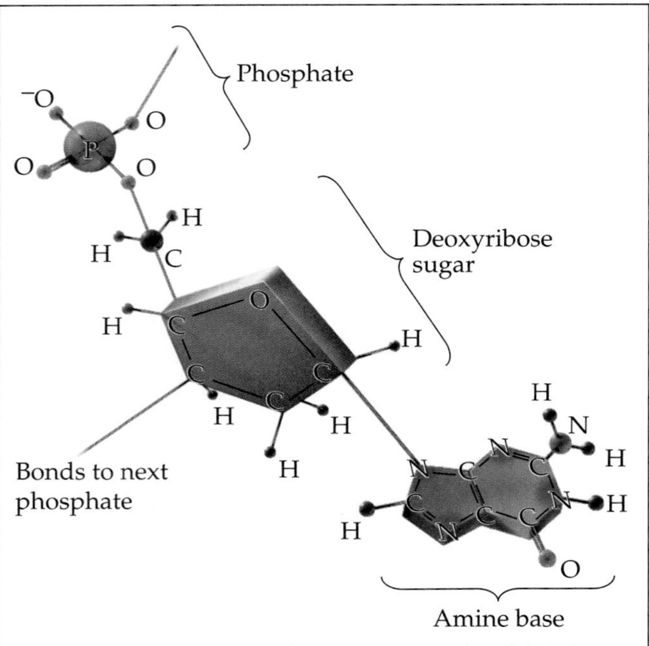

We emphasize *single strand* because the DNA that makes up the genes in your cells is double-stranded. Two single strands wind around each other to form a **double helix**, which looks somewhat like a spiral staircase.

These two strands of DNA are held together in a double helix by hydrogen bonds that form between the amine bases of the two strands. These hydrogen bonds are shown as dotted lines in the diagrams.

The structures of the four amine bases are complementary. A thymine is always hydrogen-bonded to an adenine, and a guanine is always hydrogen-bonded to a cytosine. The DNA molecule exists in a helical form because this form maximizes the amount of hydrogen-bonding that can occur.

As we discussed in Chapter 10, the hydrogen-bonding between bases is crucial to how DNA works. When DNA serves as a template for protein synthesis, the two strands have to separate. If they were held together by covalent bonds, it would take too much energy to pull them apart. If they were held together by London forces, they would be too likely to fall apart on their own. Hydrogen bonds are just right for the job.

DNA is a blueprint for protein synthesis because the sequence of amine bases on a DNA strand determines how the protein is built. It is the exact sequence of these bases that determines all your inherited traits. A triplet of adjacent DNA bases is a "word" that codes for the incorporation of a particular amino acid into a growing protein chain. (The formal term for such a triplet is *codon*.) For example, the triplet guanine-cytosine-adenine codes for the incorporation of the amino acid alanine. Within a few years, biochemists will have completed a project begun in the 1980s—the Human Genome Project—to map the entire sequence of DNA bases in all our genes. In essence, they will have mapped out the entire human blueprint. At the same time, they are also making tremendous progress at determining what the different genes do.

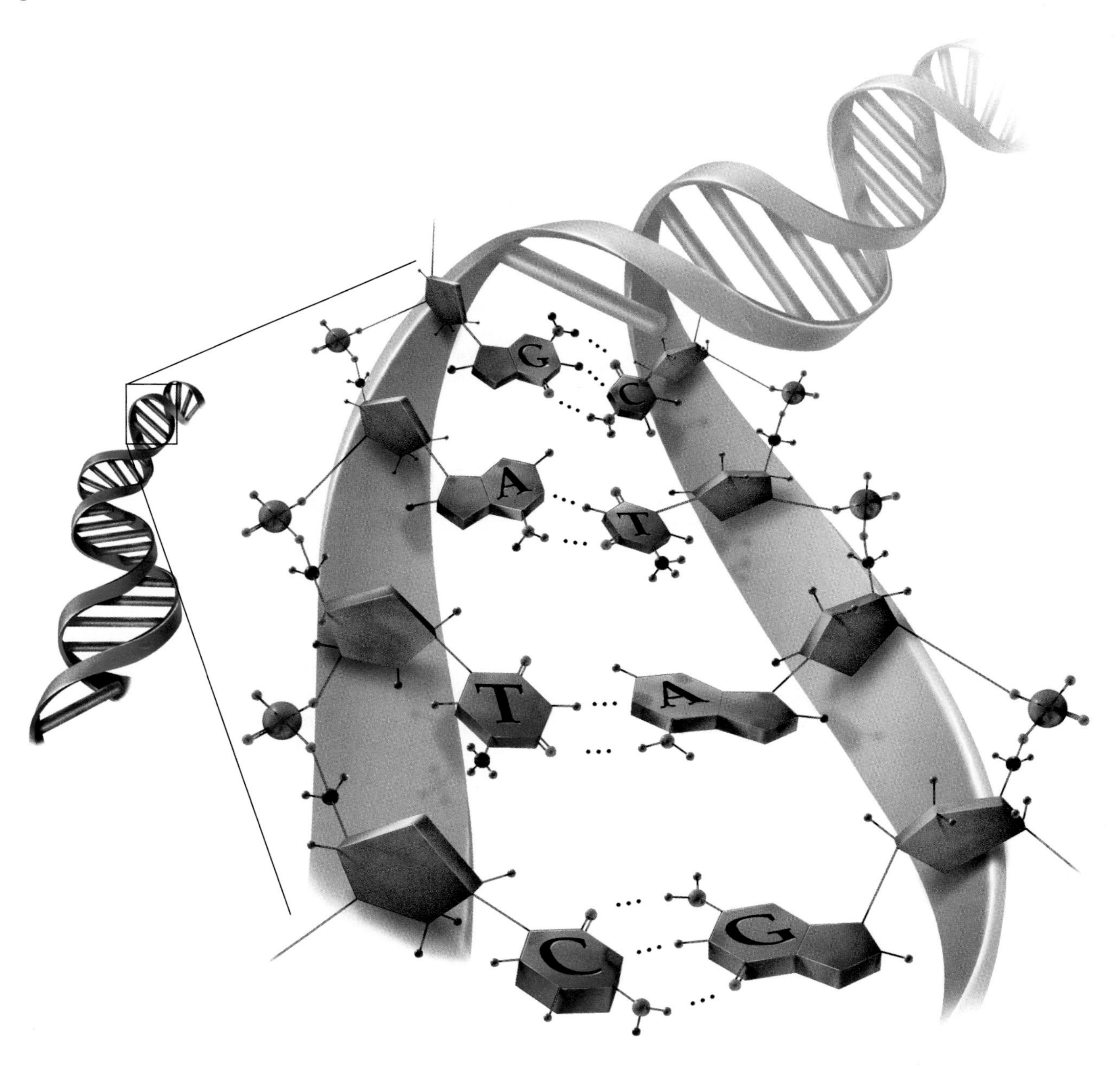

This brings us to the subject of **viruses**. The word *virus* comes from Latin and means "poison." Viruses are not considered to be alive because they cannot reproduce without a host. They are lifeless parasites, far smaller than bacteria. A virus consists of a protein shell surrounding strands of genetic material, often DNA. Transmitted through the air, in food or water, or by body fluids, viruses enter your bloodstream and penetrate your cell membranes. What happens next could come from the script of a low-budget horror movie. Once a virus is inside one of your cells, your own cell's enzymes remove the protein coat from the virus, releasing the viral DNA. The viral DNA then takes over your cell's protein-building apparatus, forcing it to replicate more viruses. Eventually, there are so many virus particles that they burst from your cell, killing it in the process. The newly formed viruses then infect more of your cells. Up to 300 viruses can be produced by a single cell in an hour. These can infect 300 other cells, which can each produce 300 more viruses, and on and on. At this rate, one virus could generate billions upon billions of virus particles within 24 hours.

A Typical Virus

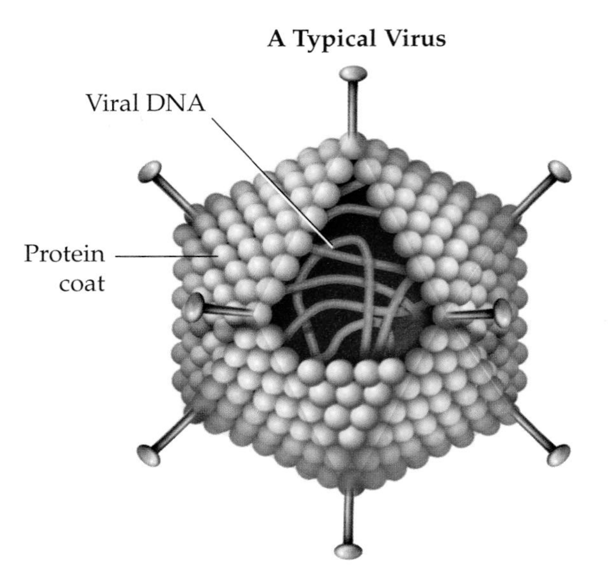

If it were not for your immune system's quick response, every virus infection—such as a simple cold or flu—would quickly convert you to a gigantic virus-producing factory, and you would die. One virus that has found a way around our immune system is HIV, human immunodeficiency virus. HIV attacks the cells of the human immune system and is responsible for the fatal disease known as AIDS, acquired immune deficiency syndrome.

Strangely enough, viruses may turn out to be our salvation. We are just now beginning to learn how to incorporate snippets of foreign DNA into the DNA of living cells, thereby instructing these cells to do our bidding. For example, the portion of human DNA that codes for the production of the protein insulin has been spliced into the DNA of the bacteria *E. coli*. These genetically engineered bacteria then produce human insulin, which can be isolated and given to people who suffer from diabetes. Until recently, diabetics were dependent on slightly different and less effective forms of insulin from pigs or cattle.

An even better solution than using *E. coli* as an intermediary would be to splice the DNA that codes for insulin production into the cells of human diabetics. This would forever cure their diabetes. Viruses may allow us to do this. The genes for insulin production could be loaded into a host virus and used to "infect" a diabetic patient. The virus would then penetrate the patient's cells, release its genetic load, and replace the defective genes. Although this scenario is still science fiction, procedures like this will be science fact in the near future.

We may soon be able to cure almost every illness known, from cancer to inherited abnormalities to the common cold, but there is a flip side. With the same techniques, it will be possible to select or influence almost every human trait, from eye color to personality, from height to sex, from intelligence to life span. We shall be able to preselect these traits for our unborn children, change them in ourselves, or order them up in our clones.

In about 2000 years we have gone from Democritus and Aristotle philosophizing about the nature of matter to the brink of taking over our own evolution. The implications and effects of the biochemical revolution that will occur in this century will dwarf those of the industrial revolution and the nuclear age combined. For better or worse, what is about to come will truly astound us all.

Have You Learned This?

Polymer (p. 715)
Monomer (p. 715)
Monomer unit (p. 716)
Macromolecule (p. 716)
Amide bond (p. 720)
Protein (p. 721)
Biopolymer (p. 721)
Polysaccharides (p. 722)
Carbohydrates (p. 722)
Monosaccharide (p. 722)
Disaccharide (p. 722)
Amino acid (p. 724)
Essential amino acid (p. 725)
Peptide linkage (p. 725)
Dipeptide (p. 725)
Polypeptide (p. 725)
DNA, deoxyribonucleic acid (p. 728)
Double helix (p. 730)
Virus (p. 732)

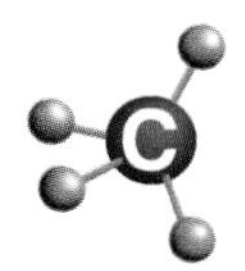

Synthetic and Biological Polymers
www.chemistryplace.com

Building Polymers

18.7 What is a polymer?

18.8 What is meant by the term *macromolecule*? Are polymers macromolecules? Explain.

18.9 What is a monomer? How does a monomer unit differ from a monomer? Give an example of each.

Polyethylene and Its Relatives

18.10 In the polymer polyethylene, there are no double bonds. Why, then, is it called polyethyl*ene*?

18.11 Use a drawing to show how one ethylene monomer reacts with another to form a bond between ethylene monomer units in the polymer polyethylene. Show how the chain can keep growing.

18.12 Draw a polymer made from the monomer propene. Make your chain at least three monomer units long. What is this polymer called?

18.13 Consider the polymer

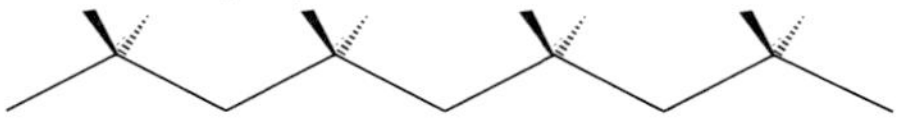

$$\begin{array}{ccccccccccc} & & CH_3 & & CH_3 & & CH_3 & & CH_3 & & \\ & & | & & | & & | & & | & & \\ \cdots-CH_2 & - & C & -CH_2- & C & -CH_2- & C & -CH_2- & C & -CH_2- & \cdots \\ & & | & & | & & | & & | & & \\ & & CH_3 & & CH_3 & & CH_3 & & CH_3 & & \end{array}$$

(a) Put a set of parentheses around one monomer unit, and give its molecular formula.
(b) Show the line drawing for the monomer used to form this polymer. [*Hint:* The monomer has a double bond.]
(c) Name the monomer and the polymer.
(d) Show how the electrons in the double bonds of the monomer must be moved to generate the polymer.

18.14 What is plastic wrap made of?

18.15 Draw a six-carbon portion of the polymer Teflon, and then draw the monomer molecule used to form Teflon.

18.16 This polymer is called a silicone and has a noncarbon, "inorganic" backbone of alternating silicon and oxygen atoms:

$$\begin{array}{ccccccc} CH_3 & & CH_3 & & CH_3 & & \\ | & & | & & | & & \\ \cdots-Si & -O- & Si & -O- & Si & -O- & \cdots \\ | & & | & & | & & \\ CH_3 & & CH_3 & & CH_3 & & \end{array}$$

(a) Put a set of parentheses around one monomer unit.
(b) Another name for this polymer is polydimethylsiloxane. Explain why it has this name.

18.17 Ethylene is a gas, but polyethylene is a solid. How do you explain this?

Nylon—A Polymer You Can Wear

18.18 What is it about the monomers used to make nylon that allows them to be linked into large macromolecules?

18.19 What is an amide bond? Show an amide bond between dimethylamine and acetic acid chloride.

18.20 Draw a nylon 66 polymer that is two monomer units long.

18.21 Where do you think the monomers for making synthetic polymers ultimately come from?

Polysaccharides and Carbohydrates

18.22 In the linear and cyclic forms of glucose shown on page 722, which oxygen in the cyclic form was the aldehyde oxygen in the linear form? Does the cyclic form bear any similarity to an ether?

18.23 Define *monosaccharide*, *disaccharide*, and *polysaccharide* and give an example of each.

18.24 How does glucose differ structurally from sucrose?

18.25 What does the word *carbohydrate* mean literally?

18.26 Are sugar molecules examples of carbohydrates? Explain.

18.27 How are cellulose and starch similar? How do they differ from each other?

18.28 When you burn wood, what are you actually burning? What are the combustion products?

Proteins

18.29 What is an amino acid?

18.30 What is the difference between the amino acids glycine and alanine?

18.31 What does the expression *essential amino acid* mean?

18.32 What is a peptide linkage?

18.33 What is a polypeptide? How does it differ from a protein?

18.34 Draw the polypeptide Gly–Ala–Ser.

18.35 What two peptides are possible from the combination of alanine and glycine? Draw both of them.

18.36 What are some of the functions of proteins in living organisms?

18.37 How does its sequence of amino acids influence a protein?

DNA—THE MASTER BIOPOLYMER

18.38 How does DNA serve as the master blueprint for building proteins?

18.39 What is the backbone of the DNA polymer made of? Name the parts, and draw two monomer units of the backbone.

18.40 What type of bond holds the two strands of a DNA molecule together? What would be wrong with holding the strands together with covalent bonds?

18.41 How many DNA bases are there? Name them and give their abbreviations.

18.42 The DNA bases always hydrogen-bond in specific pairs. What are those pairs?

18.43 What is meant by the term *codon*? What is the function of a codon?

18.44 What do viruses do (on the cellular level) to make you sick?

18.45 How might viruses be used for good purposes?

ADDITIONAL PROBLEMS

18.46 If a functional group A forms a strong covalent bond with a functional group B,
(a) Can the monomers A–xx–B and A–xx–B form a polymer?
(b) Can the monomers A–xx–A and B–xx–B form a polymer?
For each "yes" answer, draw a short segment of the polymer and indicate the monomer unit.

18.47 Nylon fabric owes much of its strength to the formation of hydrogen bonds between the polymer chains. Suppose functional groups A and B bond covalently with each other. Draw two different monomers that can form a polymer in which the interchain attractive forces are stronger than those in nylon. [*Hint*: Each of your monomers can contain all A, all B, or any number of A combined with any number of B.]

18.48 Amino acids are a sort of self-contradiction because they contain both a basic group (the NH_2 end) and an acidic group (the CO_2H end) in the same molecule. Therefore, it should not surprise you that, when dissolved in water, an amino acid can titrate itself to form what is known as a *zwitterion*. Based on this information,
(a) Draw the structural formula for the form of the amino acid alanine you would expect to find in a neutral water solution. What is the net electrical charge on this species?
(b) Enough concentrated HCl is added to the solution of part (a) to make it strongly acidic. Draw the structural formula for the form of alanine you expect to find in this solution. What is the net charge on the alanine?
(c) Enough concentrated NaOH is added to the solution of part (a) to make it strongly basic. Draw the structural formula for the form of alanine you expect to find in this solution. What is the net charge on the alanine now?

18.49 What type of biological compound has the structure

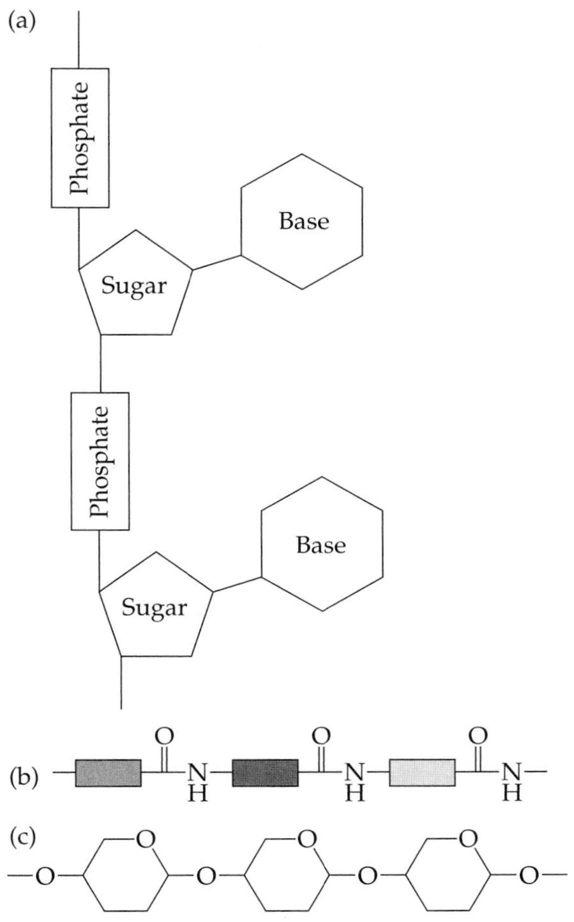

18.50 Protein molecules are usually quite large and yet often water-soluble. They are soluble in water because they can fold themselves into spherical shapes in which the polar (hydrophilic) side chains are on the surface and the nonpolar (hydrophobic) side chains are buried inside the sphere. Of the amino acids Ala, Ser, Phe, Ile, Arg, and Asp,
(a) Which would you expect to find on the surface of a water-soluble protein?
(b) Which would you expect to find buried deep inside the protein structure?
(c) Which might be in either location depending on the solution pH?

18.51 The artificial sweetener aspartame is a slightly modified dipeptide. Which two amino acids are used to make it?

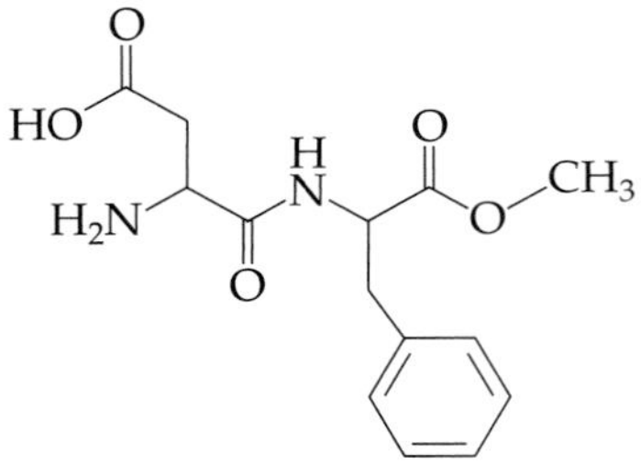

Aspartame

18.52 How many different tripeptides can be made that contain the amino acids alanine, serine, and glycine?

18.53 How many tetrapeptides can be made from two alanines and two glycines?

18.54 In the olden days (the 1970s), the sequence of amino acids in a protein was determined by running a series of chemical reactions known as the Edman degradation. This technique worked only on short peptide chains, however, and large proteins had to be broken down into small parts in order to be sequenced. One way to chop large proteins into small chains was with enzymes known as proteases. The protease trypsin breaks the amide bond on the CO_2H side of arginine and lysine, and the protease chymotripsin breaks the amide bond on the CO_2H side of tyrosine, phenylalanine, and tryptophan. An octapeptide treated with trypsin breaks down to the peptides Ala–Gly–Trp–Gly–Lys and Thr–Val–Lys, and the same octapeptide treated with chymotrypsin breaks down to the peptides Gly–Lys and Thr–Val–Lys–Ala–Gly–Trp. What is the sequence of the octapeptide?

18.55 What is the sequence of amino acids in the peptide

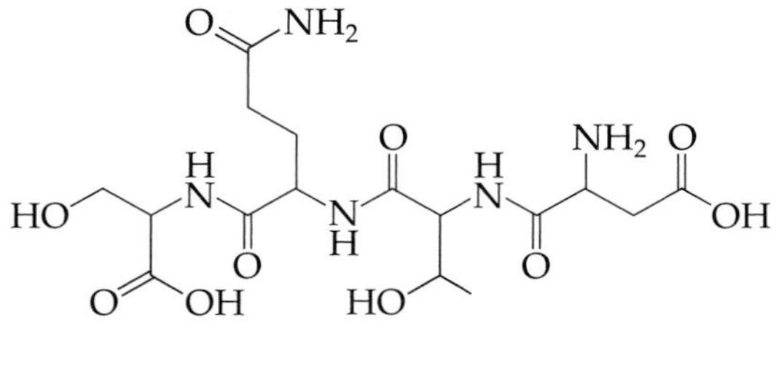

18.56 After 10.0 g of a protein was dissolved in 1000.0 g of water, the freezing point of the solution was -3.72×10^{-4}°C. What is the molar mass of the protein?

18.57 Insulin is a protein used to regulate blood sugar levels in many mammals. In humans, the active form of this protein consists of two covalently bonded chains and contains six sulfur atoms per molecule. A combustion analysis done on a 50.0-g sample of insulin produced 3.34 g of SO_2. What is the molar mass of human insulin?

18 WORKPATCH SOLUTIONS

18.1 Only (c) can be repeated to reproduce the polymer:

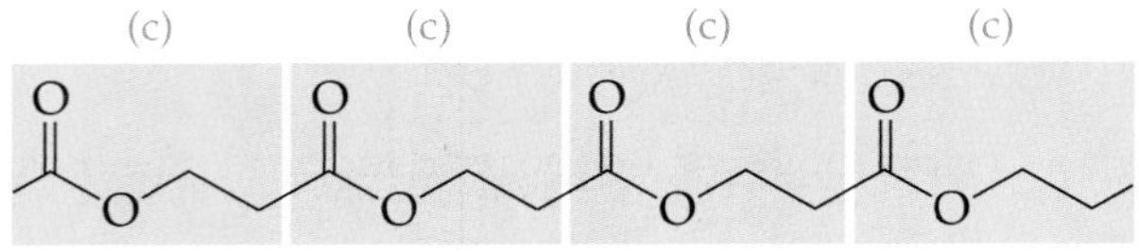

(c) (c) (c) (c)

18.2

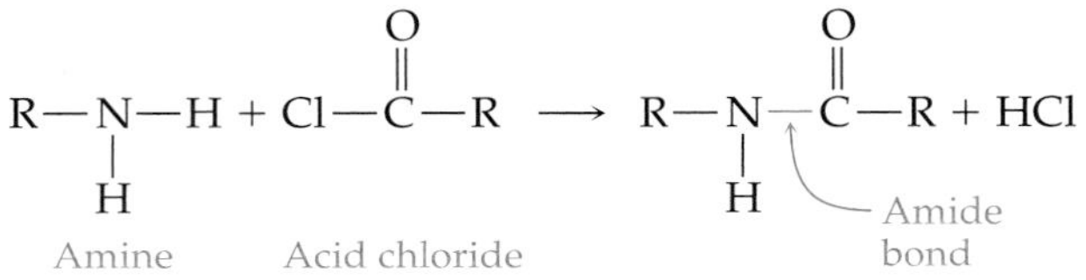

$$\text{R—NH—H} + \text{Cl—C(=O)—R} \longrightarrow \text{R—NH—C(=O)—R} + \text{HCl}$$

Amine Acid chloride Amide bond

18.3

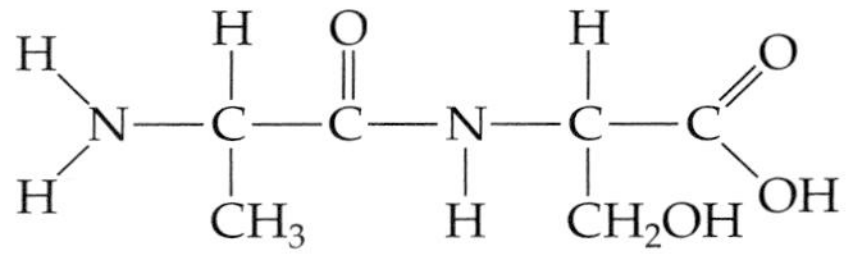

Ala–Ser

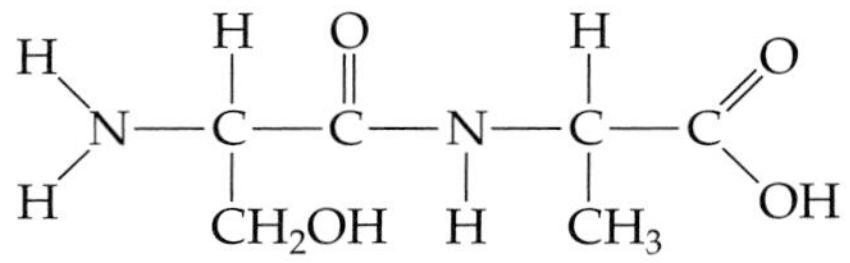

Ser–Ala

Glossary

absolute temperature scale Another term for Kelvin temperature scale.

accuracy The degree to which a measured value is close to the true value. Compare precision.

acid A substance that increases the concentration of H^+ ions in an aqueous solution.

acid–base neutralization reaction The reaction between an H_3O^+ ion supplied by an acid and an OH^- ion supplied by a base. A hydronium ion and a hydroxide ion react with each other to give two molecules of water.

acidic solution An aqueous solution in which the concentration of hydronium ions (H_3O^+) is less than 10^{-7} M, giving the solution a pH value below 7. Compare basic solution, neutral solution.

actinides The elements with atomic numbers 90 through 103.

activation energy (E_a) The minimum amount of energy, specific to each chemical reaction, that the reactants must absorb from their surroundings in order for the reaction to occur.

active metal In any pair of metals, the one that more easily gives up its electrons.

active site A precisely shaped location (on an enzyme molecule, for example) at which substrate molecules having a complementary shape may bind.

activity series See electromotive force series.

actual yield The actual or experimental amount of product obtained in a chemical reaction. Compare theoretical yield.

air pressure The force exerted on a surface by air molecules colliding with the surface.

alcohol A functionalized hydrocarbon containing one or more –OH functional groups. General formula: R–OH.

aldehyde A functionalized hydrocarbon containing a carbonyl (C=O) bonded to at least one hydrogen. General formula: R–COH.

algebraic manipulation Rearrangement of an equation in order to solve for a desired quantity.

aliphatic hydrocarbon An early general name for long-chain hydrocarbons, stemming from the Greek *aleiphas* meaning "fat"; applied to these compounds because many fats and oils contain long-chain hydrocarbon molecules. A general term for all alkanes, alkenes, and alkynes.

alkali metal A general name for any element in group IA (except H).

alkaline earth metal A general name for any element in group IIA.

alkane A hydrocarbon containing only carbon–carbon single bonds; a saturated hydrocarbon. General formula: C_nH_{2n+2}.

alkene A hydrocarbon containing at least one carbon–carbon double bond; an unsaturated hydrocarbon. General formula: C_nH_{2n}.

alkyne A hydrocarbon containing at least one carbon–carbon triple bond; an unsaturated hydrocarbon. General formula: C_nH_{2n-2}.

allotrope One of a set of different structural forms of an element, such as the graphite, fullerene, and diamond forms of carbon.

alloy A solid solution in which both solute (or solutes in some cases) and solvent are metals.

α-amino acid Any amino acid whose amine functional group is separated from the acid (carboxyl) functional group by only a single carbon atom.

alpha decay A nuclear reaction in which a radioactive nucleus ejects an alpha particle (^4_2He) and thereby becomes a daughter isotope containing two fewer protons and two fewer neutrons.

alpha (α) particle A helium nucleus (^4_2He) ejected from some radioactive nuclei during a process known as alpha decay (or alpha emission).

amide A functionalized hydrocarbon that contains a derivative of the carboxyl group (COOH) in which the –OH has been replaced by an amine. General formulas: primary, $R{-}CONH_2$; secondary, R–CONHR; tertiary, $R{-}CONR_2$.

amide bond The covalent bond formed between the carbon atom of the carbonyl and the nitrogen atom in an amide.

amine An organic derivative of ammonia (NH_3) in which one, two, or all three of its hydrogen atoms is replaced by a hydrocarbon chain. See primary amine; secondary amine; tertiary amine.

amino acid An organic molecule that contains both a carboxylic acid group and an amine group.

amino group See amine.

angstrom (Å) A non-SI unit of length, equal to 1×10^{-10} meter. Often used to express bond lengths in molecules.

anion An ion that has a negative charge. Compare cation.

anode The negative electrode in a galvanic cell; the electrode at which oxidation takes place. Compare cathode.

antimatter The oppositely charged version of matter. For example, a positron is the antimatter version of an electron.

aqueous solution A solution in which the solvent is water.

Arrhenius definition of acid The earliest identification of acids as a particular class of compounds, with "acid" defined as any electrolyte that ionizes to give H^+ ions when dissolved in water.

Arrhenius definition of base The earliest identification of bases as a particular class of compounds, with "base" defined as any electrolyte that ionizes to give OH^- ions when dissolved in water.

atmosphere The collection of gases that surround the Earth. Also, a non-SI unit of pressure; 1 atm = 760 mm Hg = 29.9 in. Hg at sea level.

atmospheric pressure See air pressure.

atom The smallest unit of an element that has the properties of that element. An atom consists of a central nucleus plus one or more electrons outside the nucleus; the number of electrons is equal to the number of protons, making an atom electrically neutral. Compare ion.

atomic mass A unitless number expressing the mass of a particular atom relative to the mass of an atom of carbon-12, which is arbitrarily assigned a mass of exactly 12.

atomic mass unit (amu) Exactly one-twelfth the mass of a ${}^{12}_{6}C$ atom and equal to $1.660\,54 \times 10^{-24}$ g; also known as a dalton.

atomic number (*Z*) The number of protons in the nucleus of an atom. All atoms of a particular element have the same atomic number, which is indicated by a subscript to the left of the element symbol.

atomic radius The distance from the center of the nucleus to the outermost electron shell.

autodissociation The process by which a small number of the water molecules in any given volume of water dissociate into hydronium and hydroxide ions. See K_w.

autoionization A synonym for autodissociation.

Avogadro's number 6.02×10^{23}, the number of units in 1 mole.

balanced equation A chemical equation written such that matter is conserved, which means the total number of atoms of each element on the right side of the equation is exactly equal to the number of atoms of those elements on the left side.

balancing coefficient A number placed in front of a chemical formula in a chemical equation to indicate the number of moles (or molecules) of the substance that take part in the chemical reaction represented by the equation.

band of stability The area on a band of stability plot representing all existing nuclei, both nonradioactive isotopes and all radioactive isotopes having a measurable half-life.

band of stability plot A plot of number of neutrons in a nucleus (represented on the vertical axis of the graph) as a function of number of protons in the nucleus (represented on the horizontal axis), with some indication as to whether an atom exists with that combination.

barometer A device used to measure atmospheric pressure.

barometric pressure Another term for air pressure or atmospheric pressure.

base A substance that increases the concentration of OH^- ions in an aqueous solution.

base unit Any one of seven fundamental SI units: meter, kilogram, second, kelvin, mole, ampere, candela. Compare derived unit.

basic solution An aqueous solution in which the concentration of hydroxide ions (OH^-) is greater than 10^{-7} M, giving the solution a pH value above 7. Compare acidic solution; neutral solution.

battery Common term for galvanic cell.

beta decay Another term for beta emission.

beta (β^-) emission A nuclear reaction in which a neutron changes to a proton plus a beta particle (electron); the proton remains in the nucleus, and the beta particle is ejected from the atom.

beta (β^-) particle An electron created during a nuclear reaction when a neutron changes to a proton; the particle is ejected from the nucleus.

bias A strong preference or inclination that inhibits impartial judgment.

bimolecular Involving two molecules.

binary compound A compound containing atoms of only two different elements. The compound may contain more than two atoms, but only two different elements.

binary covalent compound A covalent compound containing atoms of only two elements, both usually nonmetals.

binary ionic compound An ionic compound containing only one type of cation and one type of anion. The compound may contain more than two ions, but all the cations (usually a metal) are identical and all the anions (usually a nonmetal) are identical.

binding energy The energy released when a nucleus is created from protons and neutrons. Also, the energy required to break up a nucleus into its constituent protons and neutrons.

binding energy per nucleon The binding energy divided by the number of nucleons (protons and neutrons) in the nucleus.

biopolymer A polymer made by and used in living organisms.

Bohr diagram A diagram showing the electron shells of an atom as a series of concentric circles centered on the nucleus.

Bohr model of the atom A model proposed by Niels Bohr, in which the atom is pictured as a central nucleus surrounded by orbiting electrons. Because the energy of the electrons is quantized, the electrons can orbit the nucleus only at certain allowed distances. Compare Rutherford model; Thomson model.

boiling point (bp) The temperature at which a pure substance changes from a liquid to a gas at standard atmospheric pressure. Because condensation is the reverse of evaporation (boiling), the boiling point is numerically equal to the condensation point.

bond See chemical bond.

bonding pair Two electrons forming a covalent bond between two atoms, usually with one electron of the pair coming from one atom and the other electron coming from the other atom.

branched hydrocarbon A hydrocarbon in which at least one carbon atom is bonded to more than two other carbon atoms. Compare linear hydrocarbon.

break-even point For a nuclear reaction, the point at which the amount of energy recovered from the reaction equals the amount of energy input to start the reaction.

Brønsted–Lowry definition of acid A substance that is a proton (H^+ ion) donor in aqueous solution.

Brønsted–Lowry definition of base A substance capable of accepting a proton (H^+ ion) from a proton donor.

buffer A shorthand name for buffered solution.

buffered solution A solution that is able to maintain an essentially constant pH when acids or bases are added to it. The solution is made up of either a weak acid and a salt of its conjugate base or a weak base and a salt of its conjugate acid.

buffering ability The ability of a solution to resist changes to its pH.

c The symbol representing the speed of light, 3.00×10^8 meters per second = 186,000 miles per second.

calorie A non-SI unit of energy; 1 cal = 4.184 J. The amount of heat necessary to raise 1 g of water 1C°.

Calorie (Note capital C.) The energy unit used to measure the amount of energy contained in food; equal to 1000 calories.

calorimeter An insulated vessel that measures the amount of heat released when a substance is burned or a chemical reaction takes place.

carbocation A cation containing a C^+ ion.

carbohydrate A general term for molecules made up of carbon, hydrogen, and oxygen, with many of the hydrogen and oxygen atoms forming alcohol (–OH) groups.

carbonyl A functional group consisting of a carbon atom and an oxygen atom joined by a double bond: $>C{=}O$.

carboxyl A functional group consisting of a carbon atom bonded to an –OH group by a single bond and to an oxygen atom by a double bond: –COOH.

carboxylic acid A functionalized hydrocarbon containing one or more carboxyl groups. General formula: R–COOH.

catalyst A substance that, by lowering the activation energy of a chemical reaction, changes the rate at which the reaction proceeds but is not itself consumed in the reaction.

catenation The ability of an atom to link together with like atoms.

cathode The positive electrode in a galvanic cell; the electrode at which reduction takes place. Compare anode.

cation An ion with a positive charge. Compare anion.

Celsius temperature scale A metric (but non-SI) temperature scale used for everyday purposes in all countries except the United States, and often used for scientific work in the United States. On this scale, the freezing point of water is set at 0°C and the boiling point of water is set at 100°C. The unit of temperature on this scale is the degree Celsius.

centi- The SI prefix meaning one-hundredth (10^{-2}).

chain length The number of carbon atoms in the main chain of an organic molecule.

chain reaction (nuclear) A sequence of nuclear reactions in which the first reaction produces sufficient energy and daughter particles to enable multiple repeat reactions in the next step, and so on.

chalcogen A general name for any group VIA element.

chemical bond Any attractive electrostatic force between two atoms that binds them together. Ionic bonds, covalent bonds, double bonds, and so forth, are all types of chemical bonds.

chemical equation A notation for showing what takes place in a chemical reaction. The chemical formulas for all reactants are written on the left side of the arrow, and the chemical formulas of the products are written on the right side.

chemical equilibrium The stage in a chemical reaction at which the rate of the forward reaction equals the rate of the reverse reaction.

chemical formula A combination of element symbols and numbers that give the exact composition and makeup of a chemical compound.

chemical kinetics The study of the rates at which chemical reactions take place.

chemical periodicity The fact that, if you arrange the elements in order of increasing atomic number, elements spaced at periodic (regular) intervals have similar chemical properties. For the first 18 elements, an element's properties are similar to those of the elements 8 ahead of it and 8 behind it. Beyond element 18, the interval (periodicity) increases to 18 or 32 elements. This periodicity is the basis for the periodic table.

chemical property A property of a substance that can be studied only by having the substance undergo a chemical change; a property that determines the chemical reactions the substance can undergo. Compare physical property.

chemical reaction A process in which one or more reactants are transformed into one or more products through the breaking of chemical bonds in the reactant molecules and the formation of new chemical bonds in the product molecules.

chemical transformation Another term for chemical reaction.

chemistry The branch of science that studies matter and the changes it undergoes.

classical physics Physics as described by Newton, and as it was understood until the end of the nineteenth century, before the discovery that energy can be quantized. The laws of classical physics work well when applied to everyday objects because at that level the fact that energy can be quantized can be ignored, but the laws fail when applied to objects that are the size of atoms and electrons. Compare quantum physics.

colligative property A property that depends on the number of solute particles in a solution but not on their identity.

combination reaction A reaction in which two or more substances combine to give another substance. Compare decomposition reaction.

combustion A synonym for burning. In any combustion reaction, the substance being burned combines with oxygen from the air.

combustion analysis A procedure for determining the chemical formula of an unknown substance. The substance is burned in a closed container, and the products of the combustion are then isolated and their masses are determined. The masses are then used to calculate the empirical formula of the unknown.

common name A traditional name for a substance, usually one that gives no information about its composition.

complete ionic equation An equation that shows all the ions, both those that take part in the reaction and the spectator ions.

compound A pure substance composed of two or more atoms of different elements. Compare elemental substance.

concentration A measure of the amount of solute dissolved in a given amount of solvent or solution.

condensation A change from the gas phase to the liquid phase. Compare vaporization.

condensed phase Either the liquid phase or the solid phase of matter.

conjugate acid For a given base, the substance that contains all the atoms of the base plus one additional H^+ per molecule. The part of a basic molecule that exists after chemical reaction with an acid. Example: base NaOH; conjugate acid Na^+. Compare conjugate base.

conjugate base For a given acid, the substance that contains all the atoms of the acid minus one H^+ per molecule. The part of an acid molecule that exists after chemical reaction with a base. Example: acid HCl; conjugate base Cl^-. Compare conjugate acid.

conjugate pair An acid and its conjugate base, or a base and its conjugate acid.

contact process An industrial process for producing sulfuric acid from elemental sulfur and oxygen gas.

continuous spectrum An electromagnetic spectrum in which the wavelengths of visible light blend smoothly from one color to the next. When you pass sunlight through a prism, the rainbow you see is a continuous spectrum. Compare line spectrum.

conversion factor A ratio used in unit analysis to change given units to the units wanted in the answer.

core electron An electron in an atom that occupies any shell other than the valence shell.

corrosion The oxidation of metal at a very slow pace, with the oxidizing agent being O_2 from the air. Rust (iron oxide) is the most common evidence of this type of reaction.

Coulomb's law A statement of the fact that the attractive force between two objects of opposite electric charge is directly proportional to the magnitude of the charges and inversely proportional to the distance between them.

covalent bond An attractive force between two atoms in a molecule, resulting from the sharing of valence electrons. Compare ionic bond; metallic bond.

covalent compound A compound containing all covalent or polar covalent bonds.

critical mass In a nuclear fission reaction, the amount of radioactive material needed to sustain a chain reaction.

crystal lattice The regular, three-dimensional, geometric arrangement of atoms in a crystalline solid.

cubic meter The SI unit of volume.

D The element symbol for deuterium.

Δ The Greek capital letter delta; used to represent either a change in a quantity or a difference between two values.

ΔE The symbol used to represent a change in the energy of a system.

ΔE_{rxn} The symbol representing the net energy change for a chemical reaction, defined to be the difference between the energy of the products and the energy of the reactants. This amount of energy is released during an exothermic reaction and is absorbed during an endothermic reaction. Exothermic reactions have a negative value of ΔE_{rxn}, and endothermic reactions have a positive value of ΔE_{rxn}.

$\delta+$, $\delta-$ The Greek lowercase letter delta, used with a plus or minus sign to indicate fractional or partial charge.

Dalton's atomic theory The theory of matter formulated by John Dalton: *(1)* All matter is made of atoms; *(2)* atoms can neither be created nor destroyed; *(3)* all atoms of a given element are identical; *(4)* atoms of different elements are different from one another; *(5)* all chemical reactions involve the combining or separating of atoms in the reactant compounds.

daughter isotope An isotope created in a nuclear reaction.

daughter nucleus Another name for daughter isotope.

deci- The SI prefix meaning one-tenth (10^{-1}).

decomposition reaction A reaction in which a compound breaks down (decomposes) into two or more simpler compounds or substances. Compare combination reaction.

density A measure of the compactness of matter. For any substance, it is the amount of mass per unit volume; $D = \text{Mass}/\text{Volume}$.

deoxyribonucleic acid (DNA) The biopolymer present in every living cell that carries the hereditary information the cell uses to synthesize all the protein molecules it needs to function and to reproduce itself.

derived SI unit Any SI unit of measurement obtained from a combination of two or more SI base units.

deuterium (D) The isotope of hydrogen containing one proton and one neutron; ${}^{2}_{1}H$.

diamond The allotrope of carbon in which every carbon atom is bonded to four other carbon atoms, all arranged in a three-dimensional crystal lattice. Compare fullerene; graphite.

diatomic Consisting of two atoms. Compare monatomic; polyatomic.

dilution equation An equation for solving a problem that involves diluting a stock solution to a desired concentration: $M_{\text{stock solution}} \times V_{\text{stock solution}} = M_{\text{diluted solution}} \times V_{\text{diluted solution}}$.

dipeptide A small protein molecule containing only two amino acids joined by a peptide linkage. Compare polypeptide; protein.

dipole–dipole force The intermediate-strength intermolecular attractive force exerted by polar molecules on each other. Compare hydrogen bond; London force.

dipole moment An experimental measure of the polarity of a molecule. A measure of both the strength of the dipole charges and the distance between them.

dipole vector A synonym for dipole moment; a form of notation used to represent the size and direction of the dipole moment; $\leftarrow\!\!\!+$.

diprotic acid An acid that, when it dissociates, can yield a maximum of two H_3O^+ ions per molecule of acid.

disaccharide A sugar molecule consisting of two monosaccharide (simple sugar) units joined together.

dispersion force Another name for London force.

dissociation The process by which either the molecules of a covalent compound or the cations and anions bound in the lattice of an ionic compound break up into individual cations and anions.

DNA The shortened name for deoxyribonucleic acid.

dot diagram A notation used to show valence electrons and bonding in covalent compounds.

double bond A multiple covalent bond between two atoms that is formed by two bonding electron pairs.

double helix The structural arrangement of deoxyribonucleic acid (DNA), which consists of two single strands of DNA polymer that wind around each other and are held together by hydrogen bonds.

double-replacement reaction A reaction in which two reactants AB and CD react to form products AD and CB.

dynamic equilibrium Another name for chemical equilibrium.

E_a The symbol for the activation energy of a chemical reaction.

effective collision In a chemical reaction, a collision between reactant molecules that leads to formation of a product molecule.

electrical charge The property of any particle that determines how it behaves in an electric (or magnetic) field. Particles carrying like electric charges repel each other, those carrying opposite electric charges attract each other, and those carrying no electric charge are neutral.

electricity The flow of electrons through a conducting medium, such as a wire.

electrochemical cell See galvanic cell.

electrode The anode or cathode of a galvanic cell.

electrolyte A water-soluble substance that dissociates into anions and cations when dissolved in water, creating a solution that can conduct an electric current. Compare nonelectrolyte.

electromagnetic radiation All wavelengths of radiant energy.

electromagnetic spectrum The continuum comprising all wavelengths of electromagnetic radiation, from low-energy forms (radio waves and microwaves) to high-energy forms (X rays and gamma rays), with visible light midway between these extremes.

electromotive force (EMF) series A ranking scheme indicating the relative (oxidizing) activities (tendency to give up valence electrons) of metals.

electron A subatomic particle that has a mass 1/1836 that of a hydrogen atom, carries an electric charge of −1, and moves about at high velocity around the nucleus of an atom.

electron capture A nuclear reaction in which a proton in a nucleus changes to a neutron by absorbing (capturing) an electron.

electron configuration The way the electrons of an atom are distributed among the atom's shells and subshells.

electron configuration notation A way of indicating the distribution of electrons in an atom, consisting of the principal quantum number (indicating the main shell) followed by the subshell letter followed by a superscript indicating the number of electrons in that subshell.

electron shell See shell.

electron-transfer reaction A type of reaction that can be used to push electrons through a wire and thereby produce electricity.

electron tunneling The ability of electrons to pass through barriers that supposedly should stop them.

electronegativity (EN) An atom's ability to attract shared electrons in a chemical bond to itself.

electrostatic force Forces of attraction between unlike charges and repulsion between like charges.

element A pure substance that cannot be broken down into simpler substances. Atoms of an element get their identity from the unique number of protons in the nucleus.

elemental substance A pure substance composed of atoms of a single element. Compare compound.

elementary step In a multistep chemical reaction, any one of the steps in the mechanism that occurs exactly as written in the equation.

EMF series A shortened name for electromotive force series.

empirical formula A chemical formula in which the numeric subscripts indicate the ratio of elements in a substance expressed as the lowest possible numbers. One molecule of the substance may contain the number of atoms shown in the empirical formula or a multiple of those numbers. Compare molecular formula.

endothermic reaction A chemical reaction that absorbs heat from its surroundings; ΔE_{rxn} is positive. Compare exothermic reaction.

energy (*E*) A measure of the capacity to do work.

energy barrier A graphical representation of the activation energy of a chemical reaction.

energy-level diagram A diagram that shows the energies of the shells and subshells that electrons can occupy in an atom.

energy-neutral reaction A reaction that neither absorbs energy from the surroundings nor releases energy to the surroundings.

entropy A measure of disorder in a system. The more disordered the system, the higher its entropy value.

enzyme A class of large protein molecules that act as catalysts in biochemical reactions.

equilibrium A shorthand name for dynamic equilibrium.

equilibrium constant (K_{eq}) A number that expresses the position of equilibrium for a particular chemical reaction at a particular temperature. See also K_a; K_b; K_w.

essential amino acids The ten amino acids that are required in the human diet because they cannot be synthesized in the body.

ester A functionalized hydrocarbon containing a carboxyl group in which the hydrogen atom of the −OH group has been replaced by another hydrocarbon group. General formula: R−COOR.

ether A functionalized hydrocarbon consisting of a central oxygen atom attached to two hydrocarbon groups. General formula: R−O−R.

evaporation A change from the liquid state to the gas state, usually occurring below the boiling point.

exact number A number that has no uncertainty. Compare measured number.

excess reactant A reactant that is present in excess relative to the limiting reactant. When the reaction is complete, the limiting reactant will be completely used up, replaced by product and unused excess reactant(s).

excited state For an atom, a state in which one or more electrons have absorbed energy and jumped from their ground-state position to a higher-energy shell or subshell, leaving behind an electron vacancy. Compare ground state.

exothermic reaction A chemical reaction that, as it proceeds, releases energy into the surroundings; ΔE_{rxn} is negative. Compare endothermic reaction.

experiment A procedure carried out to study a phenomenon or to test a theory.

experimental rate law The rate law found by doing kinetics experiments. Example: Rate $= k[A_2]^2$.

extensive property Any property of a substance that depends on the amount present. Compare intensive property.

Fahrenheit temperature scale The (non-SI) temperature scale used for nonscientific work in the United States, in which the freezing point of water is set at 32°F and the boiling point of water is set at 212°F. The unit of temperature on this scale is the degree Fahrenheit.

first ionization energy The amount of energy needed to remove an electron from the outermost shell of a neutral (uncharged) atom.

fission A shorthand name for nuclear fission.

formula Another term for chemical formula.

forward reaction In a reversible chemical equation, the reaction reading from left to right (by convention).

freezing The process in which a substance changes from the liquid state to the solid state.

freezing point The temperature at which a substance changes from a liquid to a solid at standard atmospheric pressure. Because melting is the reverse of freezing, the freezing point is numerically equal to the melting point.

fuel cell A galvanic cell in which the chemical reaction is the direct combination of oxygen and hydrogen to form water and create electricity, with both the water and the electricity being useable products.

fullerene An allotrope of carbon in which every carbon atom is bonded to three other carbon atoms, creating a series of hexagons and pentagons forming a sphere. Compare diamond; graphite.

functional group A commonly occurring atom or group of atoms (other than carbon or hydrogen), attached to a hydrocarbon. The properties of the funtionalized hydrocarbon are different from the properties of the parent hydrocarbon.

functionalized hydrocarbon A hydrocarbon to which one or more functional groups have been added.

fused ring A cyclic hydrocarbon consisting of two or more rings of carbon atoms that share one or more carbon atoms.

fusion A shorthand name for nuclear fusion.

galvanic cell The name used by scientists for what is commonly called a battery, consisting of two compartments connected by a salt bridge. One compartment contains a cathode (where reduction takes place) and the other an anode (where oxidation takes place). The oxidation–reduction reactions create electricity.

gamma decay Another name for gamma emission.

gamma emission A nuclear reaction in which a nucleus emits energy in the form of gamma radiation; no particles are emitted, so the identity of the decaying nucleus is not changed.

gamma (γ) ray The form of very high-energy electromagnetic radiation most frequently emitted by radioactive nuclei during decay.

gas A state of matter in which fast-moving atoms or molecules behave relatively independent of one another, only interacting upon collision. A gas expands or contracts as necessary to assume the shape and volume of any vessel used to contain it. Compare liquid; solid.

gas-phase reaction A chemical reaction in which all reactants are gases.

giga- The SI prefix meaning 1 billion (10^9).

graphite An allotrope of carbon in which every carbon atom is bonded to three other carbon atoms, creating a series of hexagons arranged in two-dimensional sheets stacked one on top of another. Compare diamond; fullerene.

ground state For an atom, the state in which all of its electrons are arranged so as to have the lowest possible total energy. The arrangement in which each electron occupies the lowest-energy subshell available. Compare excited state.

group A column of the periodic table. All the elements in each column have similar valence electron configurations.

h The symbol representing Planck's constant, 6.626×10^{-34} J · S.

half-life The time needed for one-half of the radioactive nuclei in a sample to decay.

halide rule A statement of the fact that any halogen atom in a covalent compound most typically has an oxidation state of -1.

halogen A general name for any group VIIA element.

heavy hydrogen Another name for deuterium.

heavy water Water in which each molecule contains two deuterium atoms instead of two hydrogen-1 atoms; D_2O.

Henry's law A statement of the fact that the solubility of a gas in a liquid increases with increasing pressure.

heteroatom In the language of organic chemistry, any atom other than a carbon or hydrogen.

heterogeneous mixture A mixture whose composition is not uniform throughout. Compare homogeneous mixture.

heterogeneous reaction A reaction that occurs at the interface between two or more phases, that is, at the surface where different phases are in contact with one another. Example: the rusting of a nail exposed to water and air.

homogeneous chemical reaction A reaction in which reactants and products are in the same phase. Examples: reactants and products are all gases or reactants and products are all dissolved.

homogeneous mixture A mixture whose composition is uniform throughout; a solution. Compare heterogeneous mixture.

homologous series A group of similar compounds in which one member differs from the next by a constant number of atoms. Example: alkanes.

hydration Solvation with water as the solvent.

hydration energy As a substance dissolves, the energy given off as solute ions are solvated by water molecules.

hydride A binary covalent compound consisting of one or more hydrogen atoms bonded to atoms that are less electronegative than hydrogen. This gives the H atom a -1 oxidation number. Example: LiH.

hydrocarbon A compound containing only carbon and hydrogen atoms.

hydrocarbon tail The nonpolar section of a soap or detergent molecule, consisting of nonpolar C and H atoms.

hydrogen bond An intermolecular dipole–dipole attraction between a $\delta+$ H atom covalently bonded to either an O, N, or F atom in one molecule and an O, N, or F atom in another molecule. Compare dipole–dipole force; London force.

hydronium ion (H_3O^+) The cation created when an H^+ ion is hydrated with one water molecule.

hydrophilic Referring to the section of a soap or detergent molecule (the polar head) that is soluble in water ("water-loving"). Compare hydrophobic.

hydrophobic Referring to the section of a soap or detergent molecule (the nonpolar hydrocarbon tail) that is insoluble in water ("water-fearing"). Compare hydrophilic.

hydroxide ion An anion with formula OH^-.

hyperbaric High-pressure.

ideal gas A gas in which absolutely no intermolecular forces are present between molecules, and whose molecules occupy no volume whatsoever. Such a gas, which is only an abstraction and does not exist in the real world, remains a gas at all temperatures and all pressures, and obeys the ideal gas law.

ideal gas constant (R) The constant of proportionality in the ideal gas law; $R = 0.0821\ \text{L} \cdot \text{atm/K} \cdot \text{mol}$.

ideal gas equation A mathematical expression relating the pressure P, volume V, temperature T (in kelvins), and number (n) of moles of a sample of an ideal gas: $PV = nRT$, where R is the ideal gas constant. Also called the ideal gas law.

ideal gas law Another name for the ideal gas equation.

indicator Any substance that changes color with the strength (or concentration) of acid (or base) present.

inert gas Outdated name for a group VIIIA element. A synonym for noble gas.

inherent rate factor In a reaction rate equation, any term that does not depend on the concentrations of the reactants. All inherent factors are collected in the rate constant k for a given reaction.

inner-shell electron Another term for core electron.

intensive property Any property of a substance that does not depend on the amount present. Examples: density; boiling point. Compare extensive property.

intermolecular attractive force A force exerted by one molecule on another molecule, arising from the separation of positive and negative electrical charges in the molecules. The positive part of one molecule is attracted to the negative part of the other molecule. The three types of intermolecular forces are London forces (momentary charge separation), dipole–dipole forces (permanent charge separation), and hydrogen bonds (permanent charge separation). Compare intramolecular force.

intramolecular force A force exerted by one atom in a molecule on another atom in the same molecule. Ionic bonds and covalent bonds are two types of intramolecular forces. Compare intermolecular force.

ion Any atom or group of atoms carrying either a net negative or a net positive electrical charge. See anion; cation.

ion–dipole attraction Another name for ion–dipole force.

ion–dipole force The force of attraction exerted by an ion and a polar molecule on each other.

ionic bond A bond between two ions, formed when the atoms of one give up electrons (form cations) and the atoms of the other accept electrons (form anions). Compare covalent bond; metallic bond.

ionic compound A compound containing all ionic bonds.

ionic lattice The three-dimensional ordered arrangement of the ions making up an ionic solid, consisting of alternating cations and anions.

ionic solid A solid made up of individual ions arranged in an ionic lattice held together by ionic bonds. Compare molecular solid; network solid.

ionization energy The amount of energy needed to remove an electron from an atom or ion. Each electron in any atom or ion has a specific ionization energy. Compare first ionization energy.

ionizing radiation Radiation capable of ionizing atoms, especially those in the cells of biological organisms.

isomer Any one of a set of molecules that have the same chemical formula but different structures.

isotopes Different forms of an element having the same number of protons but different numbers of neutrons (and therefore different atomic masses).

IUPAC The acronym for International Union of Pure and Applied Chemistry.

IUPAC nomenclature system A set of rules for naming chemical compounds uniquely and unambiguously.

joule (J) The SI unit of energy.

K_a The equilibrium constant for the reaction in which a weak acid partially dissociates in aqueous solution.

K_b The equilibrium constant for the reaction in which a weak base partially dissociates in aqueous solution.

K_{eq} The equilibrium constant of a chemical reaction.

K_w A modified equilibrium constant for the autodissociation of water; $K_w = [OH^-] \times [H_3O^+] = 10^{-14}$.

kelvin (K) The SI base unit of temperature.

Kelvin temperature scale The SI temperature scale, where the freezing point of water is set at 273.15 K and the boiling point of water is set at 373.15 K. The unit of temperature on this scale is the kelvin. Sometimes called the absolute temperature scale.

ketone A functionalized hydrocarbon in which the functional group is a carbonyl attached to two hydrocarbon groups. General formula: R–CO–R.

kilo- The SI prefix meaning 1000 (10^3).

kilogram The SI base unit of mass; equal to about 2.2 pounds.

kinetic energy The energy an object has whenever it is moving; the faster the object is moving, the higher its kinetic energy.

kinetics experiment An experiment run to determine the rate of a chemical reaction.

lanthanides Elements 58–71 on the periodic table. Also known as the rare earth elements.

law A statement that describes the way things are consistently observed to behave under a given set of circumstances. Compare theory.

law of conservation of energy A statement of the fact that energy can neither be created nor destroyed; it just goes from one form to another.

law of conservation of matter A statement of the fact that no matter is ever created or destroyed in any chemical reaction.

law of constant composition A statement of the fact that the amounts of the elements in a compound are determined by the identity of the compound.

law of definite proportions Another name for law of constant composition.

law of Mendeleev The pattern discovered by Mendeleev in which each element in his version of the periodic table has properties similar to those of the elements eight positions to the left and eight positions to the right. Sometimes called the law of octaves.

Le Chatelier's principle A statement of the fact that any time the equilibrium of a chemical reaction is disturbed, the forward and reverse reaction rates shift in the direction that acts to partially remove the disturbance.

leading zero Any zero to the left of the first nonzero digit in a number.

Lewis dot diagram Another name for dot diagram.

limiting reactant In a chemical reaction, a reactant that, because there is less of it than is needed to react with the total amount of the other reactants, determines how much product is produced. All of the limiting reactant is used up in the reaction.

line drawing A shorthand way of representing hydrocarbon molecules using line segments.

line spectrum An electromagnetic spectrum consisting of discrete lines separated by dark spaces of various widths. The lines indicate radiation (energy) emitted by an atom when excited electrons in the atom lose energy and thereby fall to a lower energy state. Compare continuous spectrum.

linear hydrocarbon A hydrocarbon in which each carbon atom, except the two end carbons, is bonded to two other carbon atoms. Compare branched hydrocarbon.

liquid A state of matter in which molecules or atoms are in contact (touch) and are also in motion, jostling by one another. A liquid changes shape to match the shape of the vessel used to contain it but maintains a constant volume. Compare gas; solid.

liter A non-SI unit of volume, equal to 1.057 quarts.

litmus A plant-derived dye that is red in the presence of an acid and blue in the presence of a base.

lock-and-key model A model describing enzyme action as the binding of a substrate molecule (the key) at an active site (the lock) on an enzyme molecule.

logarithm For any numeric value, the power to which 10 must be raised to get that value.

London force The relatively weak intermolecular attractive forces exerted by molecules on each other, caused by brief, constantly shifting electron imbalances in the molecules that cause them to act as momentary dipoles. Compare dipole–dipole force; hydrogen bond.

lone pair Two valence electrons that usually are not used for sharing with another atom in a covalent bond. Compare unpaired electrons.

macromolecule A general term for any very large organic or biochemical molecule, where molecular weights can be as high as 50,000 g/mol and more.

magnetic resonance imaging (MRI) A technique for creating images of the interior of a body.

main chain In a branched hydrocarbon, the longest continuous chain of carbon atoms in the molecule. When a multiple bond occurs in the molecule, the longest continuous chain that includes the multiple bond.

main-group element Another name for representative element.

mass A measure of the quantity of matter in an object. Mass is independent of the location of the object, with a given object having the same mass no matter where in the Universe it is located. Compare weight.

mass defect The difference between the actual mass of an atom and the theoretical mass calculated by adding up the masses of all the protons, neutrons, and electrons in the atom.

mass number The number of protons plus neutrons in the nucleus of an atom.

mass percent The mass of any component of a sample divided by the total mass of the sample, with this quotient multiplied by 100.

matter Anything that has mass and takes up space.

measured number A number obtained by using a measuring device. Measured numbers always have some uncertainty because no measuring device is perfect. Compare exact number.

mega- The SI prefix meaning 1 million (10^6).

melting The process in which a substance changes from the solid state to the liquid state.

melting point (mp) The temperature at which a substance changes from a solid to a liquid at standard atmospheric pressure. Because freezing is the reverse of melting, the melting point is numerically equal to the freezing point.

mercury barometer A device that uses mercury as the liquid to measure the pressure exerted by the atmosphere.

metal An element that tends to lose one or more valence electrons in chemical reactions. In their elemental form, metals typically are shiny and bendable, and conduct heat and electricity well.

metal hydroxide A compound consisting of a metal cation and one or more hydroxide anions, having the general formula $M(OH)_n$. Those that are soluble produce basic solutions when dissolved in water.

metallic bond Bonds between metal atoms that are the result of electrons being shared by all the atoms. Compare covalent bond; ionic bond.

metalloid An element whose properties are intermediate between those of metals and those of nonmetals.

meter The SI base unit of length, equal to 39.37 inches.

metric system A system of units in which the basic length unit is the meter.

micelle The spherical structure formed by a group of soap or detergent molecules in water. The polar heads orient relative to each other so that they create the surface of the sphere, and the hydrocarbon tails all point to the interior. The polarity of the surface causes the micelle to dissolve in water, while (nonpolar) grease and oil molecules dissolve in the nonpolar interior.

micro- The SI prefix meaning one-millionth (10^{-6}).

micron A synonym for micrometer.

milli- The SI prefix meaning one-thousandth (10^{-3}).

millimeters of mercury A non-SI unit of pressure; 760 mm Hg = 1 atm.

mixture Matter that consists of two or more substances in variable amounts.

model A verbal description or physical construction used to visualize something a theory describes.

molar mass The mass of 1 mole of any substance, expressed in grams.

molar volume The volume occupied by 1 mole of a substance.

molarity (M) A unit of concentration for solutions, indicating the number of moles of solute contained in each liter of a solution; M = moles/liter.

mole The SI base unit of amount of substance, defined as the amount of any substance that contains 6.02×10^{23} atoms or molecules of the substance. For elements, this amount is equal to the atomic mass expressed in grams; for molecules, this amount is equal to the molar mass expressed in grams.

molecular compound A compound made up of individual molecules.

molecular dipole moment The dipole moment for an entire molecule. It points from the $\delta+$ portion of the entire molecule to the $\delta-$ portion.

molecular electrolyte A substance that, despite containing only covalent bonds, either partially or completely dissociates into ions when dissolved in water.

molecular formula A chemical formula in which the numeric subscripts indicate the actual composition of one molecule of a substance. Compare empirical formula.

molecular solid A solid made up of discrete molecules attracted to each other by intermolecular forces. Compare ionic solid; network solid.

molecular weight A synonym for molar mass.

molecule A stable collection of two or more atoms bound together by covalent bonds.

molten In the liquid state.

monatomic Consisting of one atom. Compare diatomic; polyatomic.

monoalkene An alkene containing only one carbon–carbon double bond.

monoalkyne An alkyne containing only one carbon–carbon triple bond.

monomer A hydrocarbon molecule that forms chains to create a polymer.

monomer unit A monomer once it has been incorporated into a polymer.

monoprotic acid An acid that, when it dissociates in water, can yield only one H_3O^+ ion per molecule of acid.

monosaccharide A sugar molecule consisting of one simple sugar.

MRI An acronym for magnetic resonance imaging.

multiple bond Another name for for multiple covalent bond.

multiple covalent bond More than one bond between two atoms, formed by the sharing of more than one electron from each atom. Can be either a double bond (two bonding pairs) or a triple bond (three bonding pairs).

n/p ratio Shorthand for neutron-to-proton ratio.

nano- The SI prefix meaning one-billionth (10^{-9}).

net energy change (ΔE_{rxn}) The difference between the energy of the products of a chemical reaction and the energy of the reactants. This amount of energy is released during an exothermic reaction (ΔE_{rxn} is negative) and is absorbed during an endothermic reaction (ΔE_{rxn} is positive).

net ionic equation An equation that shows the reactants and products of the reaction, but not the spectator ions.

network covalent substance Another name for network solid.

network solid A nonmolecular solid made up of atoms held together by covalent bonds.

neutral solution A solution in which the concentration of hydronium ions is the same as the concentration of hydroxide ions, giving the solution a pH value of exactly 7. Compare acidic solution; basic solution.

neutral substance A substance that gives no signs of being either acidic or basic.

neutralization reaction The reaction of an acid with a base to form a salt and water. The combining of one hydronium ion (H_3O^+) and one hydroxide ion (OH^-) to form two molecules of water.

neutron A subatomic particle that has a mass approximately equal to that of the hydrogen atom, carries zero electrical charge, and is located in the nucleus of an atom.

neutron-to-proton ratio The ratio of neutrons to protons in an atomic nucleus.

noble gas Any element in group VIIIA of the periodic table.

nonelectrolyte A water-soluble substance that does not dissociate into ions when dissolved in water, with the result that the solution cannot conduct an electric current. Compare electrolyte.

nonmetal An element that tends to gain one or more valence electrons in a chemical reaction. In their elemental form, nonmetals do not conduct heat or electricity well, meaning they are thermal and electrical insulators.

nonmolecular solid Any solid that does not consist of discrete molecules.

nonpolar molecule A covalent molecule in which there is no net separation of electrical charge in the molecule. Compare polar molecule.

normal atmospheric pressure The air pressure that raises the liquid column in a mercury barometer to a height of 760 mm = 29.9 in. = 1 atm.

normal boiling point The boiling point of a substance at standard pressure (760 mm Hg = 1 atm).

normal freezing point The freezing point of a substance at standard pressure (760 mm Hg = 1 atm).

normal melting point The melting point of a substance at normal atmospheric pressure (1 atm).

NR An abbreviation for no reaction.

nuclear decay Another name for radioactive decay.

nuclear fission The splitting apart of a large parent nucleus to two smaller daughter nuclei.

nuclear fusion The fusing (joining together) of two small nuclei into one larger nucleus.

nuclear radiation A general term for the particles and electromagnetic radiation emitted by a nucleus undergoing radioactive decay.

nuclear reaction A process in which change occurs to the nuclei of the atoms involved. Compare chemical reaction.

nucleon A particle found in the nucleus of an atom; a general term for protons and neutrons.

nucleus The tiny, positively charged core of an atom, made up of protons plus neutrons (except in the case of the hydrogen nucleus, which contains no neutrons). The electric charge of the nucleus binds the electrons to the atom, and the number of protons determines which element the atom is.

ocean of instability All areas on a band of stability plot except the band of stability and the small "island" representing predicted stable isotopes having atomic numbers 114 and above.

octet rule A statement of the fact that when two or more elements react to form a compound, the atoms usually combine in such a way that, once the compound is formed, each atom has eight electrons in its valence shell.

orbital A region of space surrounding the nucleus of an atom inside which, one is most likely to find an electron.

order An experimentally determined number, shown as an exponent in the rate law for a chemical reaction, that expresses how the concentration of a reactant affects the reaction rate. Example: Rate = $k[A]^1[B]^2$ says that the reaction is first-order with respect to A and second-order with respect to B.

organic chemistry The branch of chemistry that deals with carbon and the compounds it forms.

orientation factor A number between 0 and 1 indicating how important proper orientation between reactants is in a chemical reaction, with a value of 1 indicating orientation does not matter.

outer-shell electron Another name for valence electron.

overall order of the reaction A number obtained by adding together all the exponents in the rate expression for a chemical reaction.

oxidation The loss of electrons by an atom, resulting in an increase in oxidation state. Compare reduction.

oxidation number Another term for oxidation state.

oxidation–reduction reaction An electron transfer reaction in which one reactant loses electron(s) (is oxidized) and another reactant gains electron(s) (is reduced).

oxidation state A numeric value indicating the change in the number of valence electrons in a covalently bonded atom, more (−) or less (+) than in the atom in its free (unbonded) state.

oxidation state bookkeeping method A technique for determining the number of valence electrons "owned by" a covalently bonded atom.

oxidizing agent The reactant in an electron-transfer reaction that accepts electrons. The oxidizing agent is reduced during the reaction. Compare reducing agent.

parent isotope In a nuclear reaction, the radioactive isotope that undergoes decay.

pascal The SI unit of pressure.

peptide linkage An amide bond when it occurs in a protein molecule.

percent by mass (mass %) A unit of concentration for solutions, used most frequently for solutions made up of a solid solute dissolved in a liquid solvent and defined as

$$\frac{\text{Grams of solute}}{\text{Grams of solution}} \times 100$$

percent by volume (vol %) A unit of concentration for solutions, used most frequently for solutions made up of a liquid solute dissolved in a liquid solvent and defined as

$$\frac{\text{Volume of solute}}{\text{Volume of solution}} \times 100$$

percent composition A measure of solution concentration defined in terms of mass (percent by mass), volume (percent by volume), or mass and volume (percent by mass/volume).

percent mass/volume (mass/vol %) A unit of concentration for solutions, defined as

$$\frac{\text{Grams of solute}}{\text{Volume of solution}} \times 100$$

percent yield The actual yield of product from a chemical reaction divided by the theoretical yield, with this quotient multiplied by 100 to give a value in percent.

period A row of the periodic table.

periodic behavior Another name for chemical periodicity.

periodic table A table of the elements arranged in order of increasing atomic number. The elements are arranged in rows (periods) in such a way that all the elements in a given column (group) have similar chemical properties.

pH The negative of the logarithm of the hydronium ion concentration in an aqueous solution.

physical change A change that alters a physical property of a substance without changing its identity.

physical property A property of a substance that can change without the substance undergoing a chemical reaction; examples are density, boiling point, and color. Compare chemical property.

physical transformation A change in a pure substance that leaves it as the same substance but in a different physical state.

physics See classical physics; quantum physics.

pico- The SI prefix meaning one-trillionth (10^{-12}).

Planck's constant (*h*) An unchanging numeric value used in calculating the energy of electromagnetic radiation; 6.626×10^{-34} J · S.

polar covalent bond A covalent bond in which the electrons are shared unequally.

polar force A synonym for dipole–dipole force.

polar molecule A covalent molecule containing atoms of different electronegativities arranged in such a way that there is a net (unbalanced) separation of electrical charge in the molecule. Compare nonpolar molecule.

polyatomic Containing three or more atoms. Compare diatomic; monatomic.

polyatomic ions Ions made from many atoms.

polymer A very large molecule made from repeating units called monomers.

polypeptide A protein molecule containing between three and fifty amino acids joined by peptide linkages. Compare dipeptide; protein.

polysaccharide A sugar molecule consisting of many monosaccharide (simple sugar) units.

positron (β^+) A subatomic particle having the same mass as an electron but carrying an electrical charge of +1; the antimatter of the electron.

positron (β^+) emission A nuclear reaction in which a proton changes to a neutron plus a positron, with the neutron staying in the nucleus and the positron being ejected from the atom.

precipitate An insoluble solid that forms in a precipitation reaction when two or more aqueous solutions are mixed.

precipitation reaction A reaction in which an insoluble solid (the precipitate) forms when two or more aqueous solutions are mixed.

precision The degree to which a set of measured values of the same quantity agree with each other. Compare accuracy.

pressure The force per unit area exerted on a surface.

pressure–moles relationship The pressure of a gas increases when the number of moles of gas increases and decreases when the number of moles decreases.

pressure–temperature relationship The pressure of a gas increases when its temperature increases and decreases when its temperature decreases.

pressure–volume relationship The pressure of a gas increases when the volume the gas occupies decreases.

primary amine An amine in which the amine functional group is bonded to one hydrocarbon chain and two hydrogen atoms. General formula: $R\text{–}NH_2$.

principal quantum number (*n*) The number that indicates which shell an electron occupies in the atom. Electrons with $n = 1$ occupy the first (lowest-energy) shell; electrons with $n = 2$ occupy the second shell counting outward from the nucleus, and so forth.

product A substance produced in the course of a chemical reaction by the transformation of one or more reactants. Compare reactant.

proof A unit used to indicate the concentration of ethanol in alcoholic beverages; equal to twice the percent by volume of ethanol in the beverage.

protein A biopolymer containing fifty or more amino acid monomer units linked by peptide linkages. Compare dipeptide; polypeptide.

proton A subatomic particle that has a mass approximately equal to that of the hydrogen atom, carries an electrical charge of +1, and is located in the nucleus of an atom.

pure substance Matter with constant composition that consists of just one element or compound in just one chemical form.

quantized energy Energy that occurs only in particular discrete amounts.

quantum jump The movement of an electron in an atom from one allowed energy level to another.

quantum mechanics See quantum physics.

quantum physics Physics that takes into account the fact that energy is quantized and therefore can be used to describe the behavior of atoms and subatomic particles. Compare classical physics.

R group notation A shorthand method for drawing functionalized hydrocarbon molecules, in which the hydrocarbon chain is represented by the symbol R.

radioactive dating The process of determining an object's age by measuring the amount of some radioactive isotope contained in the object and then using that information plus the half-life of the isotope to calculate age.

radioactive decay The spontaneous change that an unstable nucleus undergoes.

radioactive atom An atom whose nucleus is unstable and therefore tends to undergo a nuclear reaction in which it changes spontaneously (without the addition of energy from an external source) to the nucleus of an atom of some other element.

rare earth elements The elements with atomic numbers 58 through 71.

rare gas A synonym for noble gas.

rate constant (*k*) A numeric constant, specific to each chemical reaction, that represents the product of all the inherent rate factors for the reaction.

rate-determining step The slowest step in any chemical reaction.

rate law An equation that expresses the rate of a chemical reaction as the product of the rate constant for the reaction times the concentration of each reactant raised to an exponent called an order.

rate of reaction Another term for reaction rate.

reactant A starting material for a chemical reaction; a substance that enters into the reaction and is transformed into one or more product substances. Compare product.

reaction See chemical reaction.

reaction coordinate The horizontal axis of a reaction energy profile; represents the progress of the reaction.

reaction-energy profile A graph showing the relative energies of the reactants and products in a chemical reaction, with the horizontal axis representing the progress of the reaction and the vertical axis representing the energy of the reaction.

reaction intermediate A product of one step of a multistep chemical reaction that serves as a reactant in a subsequent step.

reaction mechanism The sequence of elementary steps in a chemical reaction.

reaction rate The rate (change in concentration of a reactant or product per unit time) at which a chemical reaction proceeds.

reagent See reactant.

redox reaction A shorthand name for an oxidation–reduction (electron-transfer) reaction.

reducing agent The reactant in an electron-transfer reaction that loses electrons. The reducing agent is oxidized during the reaction. Compare oxidizing agent.

reduction The gain of electrons by an atom, resulting in a decrease in oxidation state. Compare oxidation.

relaxation The process by which an excited electron falls to a lower energy level. The difference in energy between the two levels is often emitted as electromagnetic radiation and creates a line spectrum.

representative element Any element in groups IA, IIA, IIIA, IVA, VA, VIA, VIIA, and VIIIA of the periodic table.

resonance forms Two or more equally correct representations of the same molecule; the average of these representations is the actual structure of the molecule.

reversible reaction A chemical reaction in which the rate at which products are converted back to reactants is significant.

Rutherford model of the atom A model proposed by Ernest Rutherford, in which the atom is pictured as being mostly empty space surrounding a tiny, but massive, positively charged nucleus. The atom's electrons take up a small portion of the "empty" space, and fly around the nucleus. Compare Bohr model; Thomson model.

sacrificial metal A metal used in a structure to protect other metals in the structure from corrosion.

salt Another name for ionic compound.

salt bridge A connector between the two compartments of a galvanic cell, containing an electrolyte that provides anions to the compartment containing the metal being oxidized and cations to the compartment containing the metal ion being reduced.

saturated hydrocarbon A hydrocarbon in which all the carbon–carbon bonds are single bonds, which means that the maximum number of hydrogen atoms are present.

saturated solution A solution containing the maximum amount of a given solute.

scientific method The method of making observations, proposing theories, and testing those theories through experimentation.

scientific notation A convention in which a numeric quantity is written as a number between 1 and 10 multiplied by 10 raised to the appropriate power.

second law of thermodynamics A statement of the fact that the entropy of the universe increases whenever a system undergoes a spontaneous change.

secondary amine An amine in which the amine functional group is bonded to two hydrocarbon chains (which may or may not be identical) and one hydrogen atom. General formula: R_2–NH.

semimetal A synonym for metalloid.

shell One of the distances at which electrons orbit the nucleus (corresponding to allowed electron energy levels). Each shell is assigned a principal quantum number $n = 1, 2, 3, \ldots$, with $n = 1$ being the shell of lowest energy.

SI units (SI stands for Système Internationale.) The system of units used in science, including seven base units, with the five most important to chemistry being the meter for length, the kilogram for mass, the second for time, the kelvin for temperature, and the mole for amount of substance.

side reaction A chemical reaction that takes place along with a main reaction, usually yielding unwanted products.

significant digits A synonym for significant figures.

significant figures The digits in a measured number (or a number calculated from measured numbers) that are known with certainty.

single bond A covalent bond between two atoms that is formed by one bonding pair of electrons.

single-replacement reaction A reaction in which one element replaces another element in a compound.

solid A state of matter in which molecules or atoms are in contact with one another and remain in fixed positions. A solid maintains a constant shape and volume. Compare gas; liquid.

solid solution A solution in which both solute and solvent are solids at room temperature. The solution is prepared from molten solute and molten solvent.

solubility The maximum amount of a solute that dissolves in a given amount of solvent at a given temperature, and for a gas solute, at a given pressure.

solubility product (K_{sp}) The equilibrium constant for sparingly soluble salts, which is simply the product of the ion concentrations in an aqueous solution at equilibrium.

solute In a solution, the substance dissolved in the solvent. Compare solvent.

solution A homogeneous mixture of a solute or solutes dissolved in a solvent.

solvation The process whereby solvent molecules in a solution completely surround a solute ion.

solvent In a solution, the substance present in the greatest amount. Compare solute.

solvent cage A hollow sphere of water molecules completely surrounding an ion. The water molecules are attracted to the ion by ion–dipole attractions.

specific heat The amount of heat energy needed to warm 1 g of a substanec by 1 C°.

spectator ion An ion present in the vessel in which a chemical reaction takes place that does not take part in the reaction.

stable isotope An isotope that is not radioactive, or, for radioactive elements located in the band of stability, a radioactive isotope that has a measurable half-life.

standard temperature and pressure (STP) Defined as 0°C = 273.15 K and 1 atm pressure.

states of matter The physical states in which matter can occur: solid, liquid, and gas.

stock solution A solution of a known concentration that can be used to make more dilute solutions of the same solute.

stoichiometry The branch of chemistry that deals with measuring the quantities of reactants and products involved in chemical reactions.

strong acid An acid that dissociates completely when dissolved in water.

strong base A base that dissociates completely when dissolved in water.

strong electrolyte An electrolyte that dissociates completely when dissolved in water. Compare weak electrolyte.

strong force The very strong attractive force that holds an atomic nucleus together.

structural formula A drawing showing all the atoms and bonds in a compound arranged in their correct relative positions, as they actually appear in the molecule. Compare chemical formula.

subatomic particle Any particle that is more fundamental than the atom, such as protons, neutrons, and electrons.

sublimation A change from the solid state directly to the gas state.

subshell A subdivision of an electron shell, designated by letters $s, p, d, f, \ldots$, with the s subshell being the one of lowest energy

substitution reaction A chemical reaction in which one atom or group of atoms on a reactant molecule is replaced by an atom or group of atoms from another reactant molecule.

substrate For a given enzyme, a molecule that reacts at the active site of the enzyme.

superconductor Any material that offers zero resistance to the passage of electricity.

technology The application to practical problems of knowledge gained by doing science.

termolecular Involving three molecules.

tertiary amine An amine in which the amine functional group is bonded to three hydrocarbon chains (which may or may not be identical). General formula: R_3–N.

tetrahedron A four-sided polygon in which all sides are identical equilateral triangles.

theoretical yield For a chemical reaction, the maximum amount of product possible for a given amount of reactants. Compare actual yield.

theory A tentative explanation for a set of observations. A theory must be consistent with every one of the observations and is tested through experiments. Compare law.

Thomson model of the atom A model proposed by J. J. Thomson, in which the atom is pictured as a cloud of positive electricity with electrons embedded in the cloud. Compare Bohr model; Rutherford model.

titration A procedure for determing solution concentrations that entails using a known amount of a solution of known concentration to determine the concentration of some other solution.

trailing zero Any zero to the right of the final nonzero digit in a number.

transition metal Any element in groups IB through VIIIB of the periodic table.

transition state A transient state of the reactants in a chemical reaction; the reactants have absorbed the required activation energy and are therefore in a high-energy, unstable condition.

triple bond A multiple covalent bond between two atoms that is formed by three bonding pairs of electrons.

triprotic acid An acid that, when it dissociates in water, can yield a maximum of three H_3O^+ ions per molecule of acid.

tritium The isotope of hydrogen containing one proton and two neutrons; ^3_1H.

tunneling See electron tunneling.

uncertainty The amount by which a measured number may vary from the true value of the quantity being measured.

uncertainty principle A statement of the fact that it is impossible to know, at a given instant, both the velocity and the position of any particle.

unimolecular Involving one molecule.

unit analysis A problem-solving strategy in which data given in the problem statement are multiplied by appropriate conversion factors in order to obtain the correct units for the answer.

unpaired electrons Single valence electrons that are used for sharing with another atom in a covalent bond. Compare lone pair.

unsaturated hydrocarbon A hydrocarbon containing at least one carbon–carbon double or triple bond, which means that the carbon atoms in the multiple bond(s) have fewer than their full complement of hydrogen atoms.

valence electron An electron in the outermost shell of an atom.

valence shell The occupied shell of an atom that has the highest principal quantum number.

valence shell electron pair repulsion (VSEPR) theory A model used to predict molecular shape, the basic tenet being that the valence electrons in a molecule repel each other because they carry negative charges.

van der Waals attraction Intermolecular forces as a whole, including London forces.

vapor A synonym for gas.

vapor pressure The pressure exerted by the gas molecules when the liquid and gas states of a substance are in dynamic equilibrium.

vaporization A change from the liquid phase to the gas phase. Compare condensation.

vector A quantity having both magnitude (size) and direction.

virus A nonliving parasite that consists of a protein shell surrounding strands of genetic material, often DNA.

visible light The part of the electromagnetic spectrum between approximately 380 nm and 750 nm.

voltage The intensity of the "pressure" or "push" behind electrons that causes them to flow in an electric current.

voltaic cell See galvanic cell.

volume The amount of space occupied by an object.

VSEPR An acronym for valence shell electron pair repulsion.

wavelength The distance between any two adjacent comparable points on a wave, such as the distance from one crest to the next or the distance from one trough to the next.

weak acid An acid that dissociates only partially when dissolved in water.

weak base A base that yields a very low concentration of OH^- ions when dissolved in water.

weak electrolyte An electrolyte that dissociates only partially when dissolved in water. Compare strong electrolyte.

weight A measure of how strongly the Earth's gravitational force (or the gravitational force exerted by any other body) pulls on an object. As the gravitational force changes, the object's mass stays constant but its weight changes. Compare mass.

weighted average atomic mass The atomic mass of an element listed on the periodic table. It is obtained by multiplying the atomic mass of each isotope by the fraction of its naturally occurring abundance and then summing all the products.

white light Light containing all the wavelengths of the visible part of the electromagnetic spectrum, from 380 to 750 nm.

Z The symbol used to represent atomic number.

Selected Answers

Note: Red numbers in the text indicate practice problems and end-of-chapter problems for which answers are provided here. For complete solutions to these problems, see the *Study Guide and Selected Solutions Manual* that accompanies this text.

Chapter 1

1.2 Heterogeneous mixture.
1.3 Heterogeneous mixture.
1.5 (d) and (e)
1.6 True.
1.8 (c)
1.9 False.
1.10 Heat the metal until it just begins to melt; then measure the temperature of the liquid to see if it is the same as the melting point of gold.
1.12 Methane and oxygen are the reactants; water and carbon dioxide are the products.
1.13 (b)
1.15 A law summarizes experimental data.
1.16 If the theory is used to predict the results of proposed experiments and then the data from those experiments agree with the prediction, this is good evidence the theory is correct.
1.17 Science is the experimental investigation and explanation of natural phenomena. Technology is the application of scientific knowledge.
1.18 Chemistry is the study of matter and the transformations it undergoes.
1.22 Matter is anything that has mass and occupies space.
1.23 Yes.
1.25 Heterogeneous mixture.
1.26 (a) Heterogeneous mixture. (b) Solution. (c) Solution.
1.28 An element is one of the basic building blocks of matter. There are 113 known elements today.
1.30 An atom.
1.31 Lead, Pb; molybdenum, Mo; tungsten, W; chromium, Cr; mercury, Hg.
1.33 Ti, titanium; Zn, zinc; Sn, tin; He, helium; Xe, xenon; Li, lithium.
1.35 An elemental substance contains only one type of atom, and a compound contains two or more different types of atoms.
1.36 The chemical formula tells how many of each type of atom are present in the smallest possible piece of a pure substance.
1.38 F_2, P_4, Ar, and Al are elemental substances; $BrCl_3$, C_2H_2, HCl, and Al_2O_3 are compounds.
1.40 H_2O_2
1.42 $C_6H_{12}O_6$
1.44 Solid, liquid, and gas.
1.46 Condensation.
1.48 $-117.3°C$
1.50 Melting point $0°C = 32°F$; boiling point $100°C = 212°F$.
1.53 Physical change.
1.55 A chemical transformation of one or more substances to one or more different substances.
1.57 Chemical change.
1.58 The properties of any substance formed in a chemical reaction are usually very different from the properties of the reactants used.
1.59 A law is a statement that summarizes experimental data. A theory proposes an explanation of why a law is true.
1.64 $C(s)$, $N_2(g)$, and $N_2(l)$ are elemental substances; $H_2O(l)$ and $HNO_3(l)$ are compounds.
1.66 Sulfuric acid.
1.68 (a) Mixture. (b) Mixture. (c) Compound. (d) Compound. (e) Elemental substance.
1.72 Chemical.
1.74 Table salt and table sugar.
1.78 Numerous answers possible. Example: brass is a homogeneous mixture of solid copper and solid zinc. You would melt these two metals in a single container and then cool the mixture.
1.80 Sublimation.
1.82 Chemical.
1.86 N_2
1.88 Reactants: $C_6H_{12}O_6$, O_2; products: CO_2, H_2O. A chemical change.
1.90 Fog.
1.92 Ozone.
1.94 True.

Chapter 2

2.2 Ike is more accurate. Mike is more precise.
2.3 Jack.
2.5 ± 0.05 gallon
2.6 (c)
2.8 By ending it with a decimal point: 600.
2.9

	Number of significant figures	Uncertainty
10.0	3	0.05
0.004 60	3	0.000 005
123	3	0.5

2.11 0.473
2.12 47,325
2.14 0.002 35
2.15 6000
2.17 $4.710\,000\,0 \times 10^{13}$
2.18 $4.710\,000 \times 10^{13}$
2.20 44 miles^2
2.21 660. hours
2.24 (a) 6.1×10^2 pounds/inch (b) 6.11×10^2 pounds/inch or 611 pounds/inch (c) 86.88 cm
2.26 1556 cm, rounded from 1555.801 cm
2.27 142 cm
2.29 4.736 km
2.30 25 mm
2.32 2500 mL
2.33 246.7 mL
2.34 K = °C + 273.15; therefore °C = K − 273.15
263.5 K − 273.15 = −9.7°C.
$°\text{F} = 32 + \frac{9}{5}°\text{C} = 14.5°\text{F}$

2.36 $\dfrac{4.70\text{ g}}{1000\text{ mm}^3} = \dfrac{470\text{ g}}{1.00\text{ cm}^3} = 4.70\text{ g/cm}^3 = 4.70\text{ g/mL}$

2.37 $\dfrac{500.0\text{ g}}{150.5\text{ mL}} = 3.322\text{ g/mL}$

2.39 $\dfrac{1\text{ day}}{24\text{ h}}$ $\dfrac{24\text{ h}}{1\text{ day}}$

2.40 $50.0\ \cancel{\text{miles}} \times \dfrac{1\text{ h}}{600.0\ \cancel{\text{miles}}} = 0.0833\text{ h}$

2.41 $\dfrac{600.0\text{ miles}}{1\ \cancel{\text{h}}} \times 50.0\ \cancel{\text{h}} = 3.00 \times 10^4\text{ miles}$

2.43 $500.0\ \cancel{\text{L}} \times \dfrac{1000\ \cancel{\text{mL}}}{1\ \cancel{\text{L}}} \times \dfrac{0.001\,30\text{ g}}{1\ \cancel{\text{mL}}}$
$= 650.\text{ g} = 6.50 \times 10^2\text{ g}$

$650.\ \cancel{\text{g}} \times \dfrac{1\text{ kg}}{1000\ \cancel{\text{g}}} = 0.650\text{ kg}$

2.44 $1.50\ \cancel{\text{lb}} \times \dfrac{453.6\ \cancel{\text{g}}}{1\ \cancel{\text{lb}}} \times \dfrac{1\text{ mL}}{11.4\ \cancel{\text{g}}} = 59.7\text{ mL}$

2.45 $6955\ \cancel{\text{g flour}} \times \dfrac{1\ \cancel{\text{cup flour}}}{120.0\ \cancel{\text{g flour}}} \times \dfrac{1\text{ cake}}{6\ \cancel{\text{cups flour}}}$
$= 9.660\text{ cakes}$

You can bake nine cakes (it's not possible to bake a partial cake).

2.46 $250.0\dfrac{\cancel{\text{m}^2}}{\cancel{\text{h}}} \times \dfrac{1\ \cancel{\text{h}}}{60\text{ min}} \times \left(\dfrac{1\text{ ft}}{60\ \cancel{\text{in.}}}\right)^2 \times \left(\dfrac{1\ \cancel{\text{in.}}}{2.54\ \cancel{\text{cm}}}\right)^2$
$\times \left(\dfrac{100\ \cancel{\text{cm}}}{1\ \cancel{\text{m}}}\right)^2 = 44.85\text{ ft}^2/\text{min}$

(Four significant digits allowed because 1 in. is *exactly* 2.54 cm.)

2.52 $50.0\ \cancel{\text{mL}} \times \dfrac{0.785\text{ g}}{\cancel{\text{mL}}} = 39.3\text{ g ethanol}$

$2.43\dfrac{\cancel{\text{J}}}{\cancel{\text{g}} \cdot \cancel{\text{C°}}} \times 39.3\ \cancel{\text{g}} \times 38.0\ \cancel{\text{C°}} \times \dfrac{\text{kJ}}{1000\ \cancel{\text{J}}} = 3.63\text{ kJ}$

2.53 3.63 kJ = 3630 J;
$3.63\ \cancel{\text{kJ}} \times 1\text{ Cal}/4.184\ \cancel{\text{kJ}} = 0.868\text{ Cal} = 868\text{ cal}$

2.54 $1.000\dfrac{\text{cal}}{\cancel{\text{g water}} \cdot \cancel{\text{C°}}} \times 1000\ \cancel{\text{g water}} \times 13.5\ \cancel{\text{C°}}$
$= 1.35 \times 10^4\text{ cal}$

$\dfrac{1.35 \times 10^4\ \cancel{\text{cal}}}{0.100\text{ g candy}} \times \dfrac{1\text{ Cal}}{1000\ \cancel{\text{cal}}} = 1.35 \times 10^2\text{ Cal/g candy}$

2.55 The 3 in "3 feet in a yard" is an exact number. The 3 in "a certain piece of wood is 3 feet long" comes from a measurement and therefore has some uncertainty associated with it.
2.57 You should choose the accurate result because a precise result that is not accurate is useless.
2.59 The person with the tape measure.
2.61 The last digit written is assumed to be uncertain, and the uncertainty is determined by putting a 1 in the place of the uncertain digit and then dividing by 2.
2.63 (a) 12.60 ±0.005 cm (b) 12.6 ±0.05 cm
(c) 0.000 000 03 ±0.000 000 005 inch
(d) 125 ±0.5 feet
2.64 (a) Four (b) Three (c) One (d) Three
2.67 (a) $12.2\underline{0}2$ (b) No significant zeros. (c) $2\underline{0}5$
(d) $0.01\underline{0}$
2.69 It is not clear whether 30 has one or two significant figures. The decimal point with no digits following indicates that the zero in 30. is significant, meaning 30. has two significant figures.
2.71 (a) 56.0 kg (three significant figures).
(b) 0.000 25 m (two significant figures).
(c) 5,600,000 miles (four significant figures, but you cannot tell by looking at the normal notation).
(d) 2 feet (one significant figure).
2.73 (a) 3×10^1 ft (b) 3.0×10^1 ft (c) 3.00×10^1 ft
2.75 (a) 2.26×10^2 (b) 2.260×10^2 (c) 5.0×10^{-10}
(d) 3×10^{-1} (e) 3.0×10^{-1} (f) 9.00×10^8
(g) $9.000\,006 \times 10^8$
2.77 102 inches; ±0.5 inch
2.79 3.873 14 miles (six significant figures because 5280 is an exact number).
2.80 (a) 2.55×10^5 km
(b) 1.000×10^{18} J
(c) 2.11×10^2 m
(d) 4.00×10^4 L
2.82 Liter and milliliter. They are volumes more commonly encountered in everyday situations and in the laboratory.
2.84 To eliminate confusion.
2.85 (a) 2.31×10^9 m (b) 5.00×10^{-6} m
(c) 1.004×10^0 m (d) 5.00×10^{-12} m
(e) 2.5×10^2 m
2.87 The Celsius and Fahrenheit scales can have negative temperature values. The Kelvin scale cannot because the zero point on the Kelvin scale is absolute zero, the lowest possible temperature.
2.88 (a) 72.5°F; 295.6 K (b) −19.4°C; 253.7 K
(c) −273.2°C; −459.8°F (d) 149°F; 338.2 K

2.91 (a) Left cylinder: ± 0.005 mL; right cylinder: ± 0.5 mL.
(b) 98 mL + 1.18 mL = 99 ± 0.5 mL

2.93 The student who reports 1.5 cm used the ruler incorrectly by not reporting enough digits.

2.96 997 g

2.98 9×10^2 g

2.100 The two students measure the same density because density does not depend on the mass of a sample.

2.102 1.08×10^5 s

2.104 0.007 083 dollars/s

2.106 0.264 gallon

2.109 (a) 101.4 cm (b) 1.043×10^6 cm^3 (c) 0.106 g/mL

2.110 5.51 L

2.111 179 miles/h

2.112 In any equation, the two sides are equal to each other, by definition. In order for the sides to remain equal, whatever is done to one side must be done to the other.

2.114 $x = z - y$

2.116 57.5 g

2.118 The capacity for doing work.

2.120 (a) 4500 cal = 4.50×10^3 cal (b) 2510. kJ (c) 0.2390 cal (d) 2.09×10^5 J

2.122 The aluminum block requires more heat energy because aluminum has the higher specific heat (aluminum 0.215 cal/g • C°, iron 0.107 cal/g • C°). How much more is determined from the ratio of the specific heats: (0.215 cal/g • C°)/(0.107 cal/g • C°) = 2.01 times as much, or if you use SI, (0.901 J/g • C°)/(0.449 J/g • C°) = 2.01 times as much.

2.123 1.51×10^5 J = 151 kJ

2.125 5.5×10^3 J, 2.2×10^3 J/g

2.126 2.44×10^4 J

2.128 The temperature of the block increases to a blistering 4916°C!

2.130 (a) 7.98×10^{17} L (b) 3.00×10^{-6} g (c) 1.11×10^5 gallons

2.132 9.3 g/cm^3

2.134 26.8 m/s

2.136 (a) 22°C, 295 K (b) 10.4°F, 261 K (c) −95°C, −139°F

2.138 1.400 g/mL

2.139 (a) $°C = \frac{5}{9}(°F - 32)$

(b) $T = \dfrac{PV}{nR}$

(c) $\lambda = \dfrac{hc}{E}$

2.141 (a) 1.34×10^{-3} g/mL (b) 1.34×10^{-3} kg/L (c) 1.34×10^{-6} kg/mL

2.142 2.7×10^2 g

2.145 (a) Neither (b) Both (c) Accurate but not precise (d) Precise but not accurate

2.146 8.0×10^3 cal; 8.0 Cal

2.149 2.69 g/mL

2.150 6.12 kJ

2.154 ± 0.0005 mL

2.157 (a) ± 0.05 m (b) $\pm 0.000\,05$ g (c) $\pm 0.0005 \times 10^3$ L = ± 0.5 L (d) ± 0.5 cm (e) No uncertainty

2.160 Because precision refers to how close to one another a series of measurements are, this term can never be used to describe a single measurement.

2.162 2.2×10^{-4} m^3

2.163 (a) 404 atomic mass units (b) 6.71×10^{-22} g (c) 20.2 g

2.165 (a) 15.0 mL (b) 16.4 mL (c) 2×10^1 mL (d) 1.32 mL (e) 1.10 mL (f) 8.3×10^4 mL

2.167 (a) Six (b) Three (c) Two (d) Four (e) Four

2.170 (a) 0.249 cal/g • C° (b) 0.20 cal/g • C°

2.173 (a) 5.02×10^5 (b) 3.8402×10^{-5} (c) 4.36×10^8 (d) 8.47×10^3 (e) 5.91×10^{-3} (f) 6.58×10^{-1}

2.175 227 U.S. dollars

2.177 (a) Two (b) Two (c) None (d) Cannot tell (e) Three

2.180 The two units are the same size.

2.182 (a) 0.0179 (b) 0.000 000 008 76 (c) 48,800,000,000 (d) 75.2 (e) 8.37 (f) 41,840

2.183 (a) 7.80 g/cm^3 (b) 64.1 cm^3 (c) Floats.

2.185 (a) 1.3×10^{-3} kg/L (b) 1.1×10^{-2} lb/gallon

2.188 (a) 5.70×10^2 g (b) 3.93×10^1 g (c) 7.0×10^{-2} g (d) 4.2×10^{-3} g (e) 6.80×10^2 g (f) 9.65×10^2 g

2.190 8.50 atm

2.192 Incorrect because 1 ft equals exactly 12 inches by definition, with no uncertainty.

2.194 507 U.S. dollars

2.197 (a) Four (b) Two (c) Three (d) Eight (e) Four

2.199 Compressing means squeezing a given mass of the gas into a smaller volume. Because the relationship is density = mass/volume and the mass is constant, decreasing the volume causes the density to increase.

2.200 (a) 269 J (b) 541 J (c) 84 J (d) 2.51×10^3 J

Chapter 3

3.5 The law of conservation of matter requires that the total mass of the substances produced equal the mass of coal plus the mass of oxygen reacting. The coal seems to disappear because it is being converted to carbon dioxide, a colorless, odorless gas. If the carbon dioxide were captured and weighed, the mass would equal the mass of the coal and oxygen reacting to produce them.

3.6

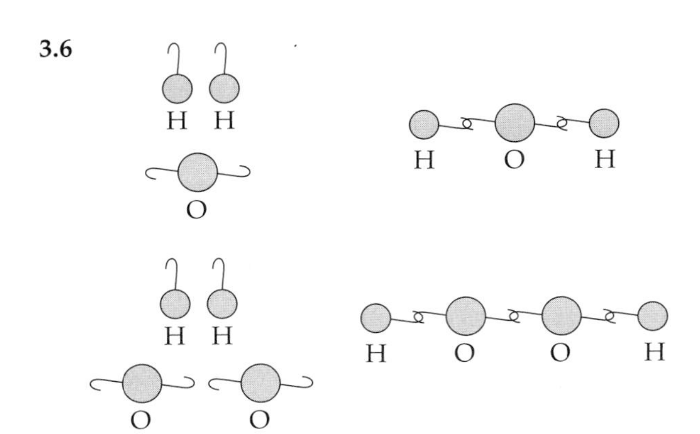

3.8 $^{79}_{35}Br$, $^{81}_{35}Br$

3.9 The atoms of both isotopes have 35 electrons.

3.10

	$^{14}_{7}N$	$^{24}_{12}Mg$	$^{23}_{11}Na$	$^{59}_{26}Fe$
Mass number	14	24	23	59
Atomic number	7	12	11	26
Number of protons	7	12	11	26
Number of neutrons	7	12	12	33
Number of electrons	7	12	11	26

3.12 $3.982\,81 \times 10^{-5}$ g

3.13 (a) $100\% - 75.77\% = 24.23\%$ (the 100 is taken to be an exact number)

(b) $\left(34.969 \text{ amu} \times \dfrac{75.77}{100}\right) + \left(36.966 \text{ amu} \times \dfrac{24.23}{100}\right)$

$= 35.45$ amu

(c) $^{37}_{17}Cl$ is $\dfrac{36.966}{34.969} = 1.0571$ times more massive than $^{35}_{17}Cl$.

3.15 No.

3.16 Similar physical and chemical properties.

3.18 Iodine (I) and astatine (At); periodicity for I is 18 and for At is 32.

3.19 The periodicity would be 2, 8, 8, 8, 8, 8, 8 instead of 2, 8, 8, 18, 18, 32, 32.

3.21 O < S < Mg < Sr < Rb

3.22 $^{56}_{26}Fe^{3+}$; group VIIIB (8)

3.23 $^{16}_{8}O^{2-}$; chalcogen group.

3.25 (a) %S: $\dfrac{94.08 \text{ g S}}{100.00 \text{ g sample}} \times 100\% = 94.08\%$ S

%H: $\dfrac{(100.00 \text{ g} - 94.08 \text{ g})}{100.00 \text{ g sample}} \times 100\% = 5.92\%$ H

(b) The law of constant composition

3.27 (a) %I: $\dfrac{126.9 \text{ g I}}{162.4 \text{ g compound A}} \times 100\% = 78.14\%$ I

%Cl: 100% − 78.14% I = 21.86% Cl

(b) %I: $\dfrac{126.9 \text{ g I}}{233.3 \text{ g compound B}} \times 100\% = 54.39\%$ I

%Cl: 100% − 54.39% I = 45.61% Cl

3.30 (a)

Compound	Mass of Na present
A	11.50 g
B	45.98 g

(b)

Compound	%Na	%O
A	59.0	41.0
B	74.2	25.8

3.32 Only a very few of the green flashes on the zinc sulfide screen were very far from the main beam of α particles.

3.34 It would imply that any positive charges in the atom were evenly spread throughout the atom and not concentrated in a tiny area in the center. This would mean that the structure of the atom would be more consistent with Thomson's "plum pudding" model.

3.36 Yes. In a neutral atom the number of electrons is equal to the number of protons, and any atom's identity is determined by its proton number.

3.38

	$^{15}_{8}O$	$^{16}_{8}O$	$^{37}_{17}Cl$	$^{23}_{11}Na$
Mass number	15	16	37	23
Atomic number	8	8	17	11
Number of protons	8	8	17	11
Number of neutrons	7	8	20	12
Number of electrons	8	8	17	11

3.40 The atomic number and the symbol do not agree. Carbon has an atomic number of 6, not 7. The element whose atomic number is 7 is nitrogen.

3.42 $^{1}_{1}H$

3.44 The atomic mass of an element is the weighted average of the masses of all its naturally occurring isotopes. The mass number is the sum of the protons and neutrons in an atom of the element.

3.46 $^{12}_{6}C$ is the only isotope whose atomic mass and mass number are equal. This is true because chemists have assigned $^{12}_{6}C$ a mass of exactly 12 amu.

3.48 An "average" titanium atom is 3.990 times more massive than the $^{12}_{6}C$ atom: $\dfrac{47.88}{12} = 3.990$.

3.51 (a) Exactly 235 amu

(b) Because with uranium-235 as the standard reference, the atomic masses of all the lighter elements would be a very small fraction of 235.

3.52 (a) 49.31%

(b) Atomic mass of Br

$= \left(78.918\,336 \text{ amu} \times \dfrac{50.69}{100}\right) + \left(80.916\,289 \text{ amu} \times \dfrac{49.31}{100}\right) = 79.90$ amu

3.53 The naturally occurring hydrogen on the other planet must contain more of the more massive $^{2}_{1}H$ isotope than on Earth.

3.55 (a) According to atomic masses.
(b) He discovered repeating behavior every 8 elements.
(c) Mendeleev's ordering was by atomic mass; the modern ordering is by atomic number.

3.57 Periodicity refers to the fact that the chemical properties repeat themselves every 8, 18, or 32 elements.

3.58 8

3.60 IA (1), alkali metals; IIA (2), alkaline earth metals; VIA (16), chalcogens; VIIA (17), halogens; VIIIA (18), noble gases

3.62 A group is a vertical column of elements with similar chemical properties. A period is a horizontal row of elements.

3.64 (a) $BeCl_2$, $CaCl_2$, $SrCl_2$, $BaCl_2$, and $RaCl_2$
(b) $BeBr_2$, $MgBr_2$, $CaBr_2$, $SrBr_2$, $BaBr_2$, and $RaBr_2$
(c) The chemical behavior of elements in the same group will be similar.

3.65 They are among the most unreactive substances known, and they are the only group in the periodic table that contains only gases.

3.67 The transition metal portion is 10 elements wide; the lanthanide/actinide portion is 14 elements wide.

3.68 The change from 8 to 18 is due to the transition metals; the change from 18 to 32 is due to the lanthanides and actinides.

3.70 As you go across the periodic table from left to right, one more proton is being added to the nucleus and one more electron is being added outside the nucleus. The added positive charge is pulling the electrons in closer to the nucleus, causing the size of the atoms to decrease.

3.72 F < S < Se < Ca.

3.74

	${}^{15}_{8}O^{+}$	${}^{27}_{13}Al^{3+}$	${}^{31}_{15}P^{3-}$	${}^{58}_{28}Ni^{+}$
Mass number	15	27	31	58
Atomic number	8	13	15	28
Number of protons	8	13	15	28
Number of neutrons	7	14	16	30
Number of electrons	7	10	18	27
Charge on ion	+1	+3	−3	+1

3.76 True

3.78 As you go down a group, the size of the atom increases and an outermost electron is farther away from the nucleus. It thus feels less pull from the nucleus, is easier to remove, and its separation requires less energy. As you go across a period, the atomic size decreases, and an outermost electron is closer to the positively charged nucleus. More energy is thus needed to remove the electron.

3.80 Mg is the most difficult to ionize; K has the smallest first ionization energy.

3.82 (a) Sodium easily loses an electron to form the cation Na^{+}, and each Cl atom in Cl_2 easily gains an electron to form the anion Cl^{-}.
(b) Being a group IA (1) metal, lithium has a low first ionization energy and easily loses one electron to become a Li^{+} cation (just like sodium). Being a group VIIA (17) nonmetal, each bromine atom in Br_2 tends to gain one electron to become a Br^{-} anion (just like chlorine). The result is the formation of lithium bromide, LiBr, similar to sodium chloride, NaCl.

3.85 A line spectrum is made up of specific colors separated by dark regions. A continuous spectrum is made up of all colors, blending smoothly from one to the next.

3.87 Oxide A, 21.23% O; oxide B, 11.88% O

3.88 You would have to show that the sum

Mass of wood before burning +
Mass of oxygen consumed

was equal to the sum

Mass of ash + Mass of carbon dioxide +
Mass of water vapor

3.90 2 g of B; the law of conservation of matter.

3.92 Incorrect. Isotopes of a given element differ in numbers of neutrons but have the same numbers of protons.

3.94 Adding a proton doesn't create a cation, even though protons carry a positive charge. Adding a proton changes the atom's identity. In this case, sodium (11 protons) becomes magnesium (12 protons).

3.96 Protons 26, electrons 26 − 2 = 24, neutrons 56 − 26 = 30.

3.100 All elements in a given group had to have similar chemical properties.

3.102 (a) Cl < S < Na < Cs (b) Cs < Na < S < Cl

3.106 Magnesium, Mg

3.108

Name of group or classification	Period	Group	Elemental symbol	Atomic number	Atomic mass (amu)	Metal, metalloid, or nonmetal?
Transition metal	4	VIIIB (8)	Fe	26	55.845	Metal
Noble gas	1	VIIIA (18)	He	2	4.003	Nonmetal
Halogen	5	VIIA (17)	I	53	126.905	Nonmetal
Alkali metal	4	IA (1)	K	19	39.098	Metal
Halogen	3	VIIA (17)	Cl	17	35.453	Nonmetal
Noble gas	2	VIIIA (18)	Ne	10	20.180	Nonmetal
Chalcogen	2	VIA (16)	O	8	15.999	Nonmetal
—	3	IVA (14)	Si	14	28.086	Metalloid
Actinide	7	—	U	92	238.029	Metal

3.109 (a) 70.01 g Cl (b) 36.92% Ca (c) 63.08% Cl

3.111 Rb < Ca < Se < S < F

3.114

	Atomic number	Mass number	Protons	Neutrons	Electrons
$^{27}_{13}Al$	13	27	13	14	13
$^{60}_{27}Co$	27	60	27	33	27
$^{200}_{79}Au$	79	200	79	121	79
$^{238}_{92}U$	92	238	92	146	92
$^{127}_{53}I$	53	127	53	74	53

3.117 (a) Alkali metals (b) Alkaline earth metals (c) Halogens (d) Noble gases or rare gases

3.119 (a) $^{17}_{8}O$ (b) $^{119}_{50}Sn$ (c) $^{23}_{11}Na$ (d) $^{58}_{28}Ni$ (e) $^{137}_{56}Ba$

3.122 (a) Fluorine (b) Calcium (c) 1− charge (d) 2+ charge (e) $^{40}_{20}Ca^{2+}$, $^{19}_{9}F^{-}$

3.125 Ne < Be < Li < Na < Cs

3.128 (a) 30.83% (b) 63.55 amu

Chapter 4

4.4 Radio and television waves; this is very-low-energy electromagnetic radiation.

4.6

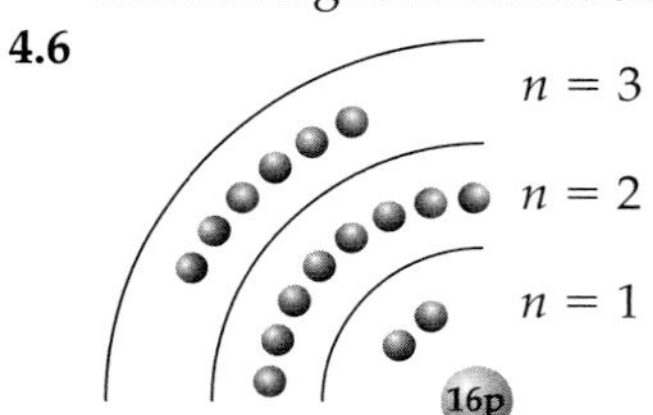

Two more electrons.

4.7 An electron in a shell with a low value of n is closer to the nucleus and feels a stronger attraction than one that is farther from the nucleus. The stronger attraction makes the closer electron more stabilized.

4.8 $2 \times 5^2 = 50$ electrons

4.10

4.11

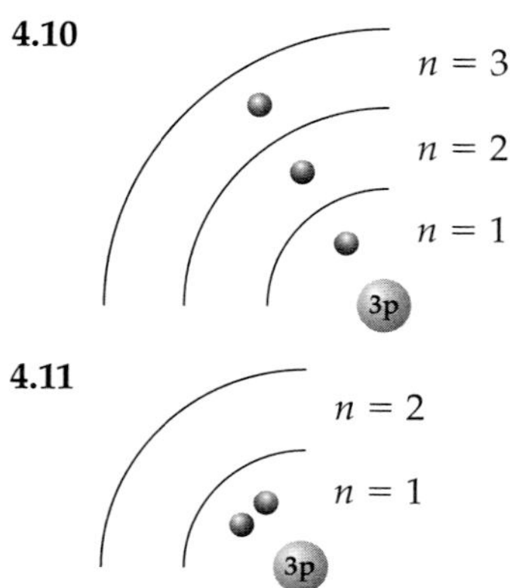

4.12 (a) Carbon (b) Anion with −2 charge
(c)

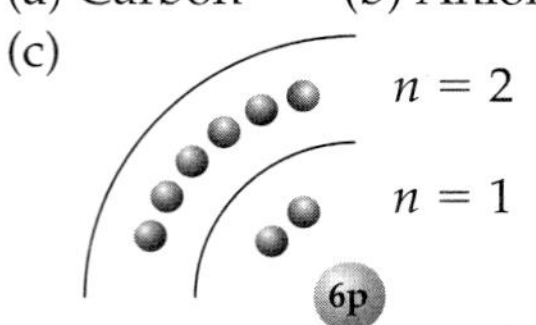

4.13 The $n = 1$ shell has 3 electrons in this diagram; the maximum is 2 electrons for this shell.
The $n = 2$ shell has only 6 electrons; the maximum is 8 for this shell, and it should be filled before an electron is placed in the $n = 3$ shell.

4.15 As has a total of 33 electrons: $1s^2 2s^2 2p^6 3s^2 3p^6 3d^{10} 4s^2 4p^3$; yes, there are 5 valence electrons, and the group number is VA.

4.16 Sc has 21 electrons: $1s^2 2s^2 2p^6 3s^2 3p^6 3d^1 4s^2$

4.18 $1s^2 2s^2 2p^6 3s^2 3p^6 4s^2 3d^{10} 4p^6$; it is proper for Kr to be in group VIIIA because it has 8 valence electrons.

4.19 Pd has 46 electrons: $1s^2 2s^2 2p^6 3s^2 3p^6 4s^2 3d^{10} 4p^6 5s^2 4d^8$

4.21 $1s^2 2s^2 2p^6 3s^2 3p^6 4s^2 3d^{10} 4p^6 5s^2 4d^{10} 5p^6 6s^2 4f^{14} 5d^{10} 6p^6 7s^2$; $[Rn]7s^2$

4.22 $1s^2 2s^2 2p^6 3s^2 3p^6 4s^2 3d^{10} 4p^6 5s^2 4d^{10} 5p^6 6s^2 4f^{14} 5d^{10} 6p^6 7s^2 6d^1 5f^3$; $[Rn]7s^2 6d^1 5f^3$

4.24 The highest value of n is 4, so it is in the fourth period; the element is vanadium (V).

4.25 Period 6; group IA; the element is cesium (Cs).

4.27

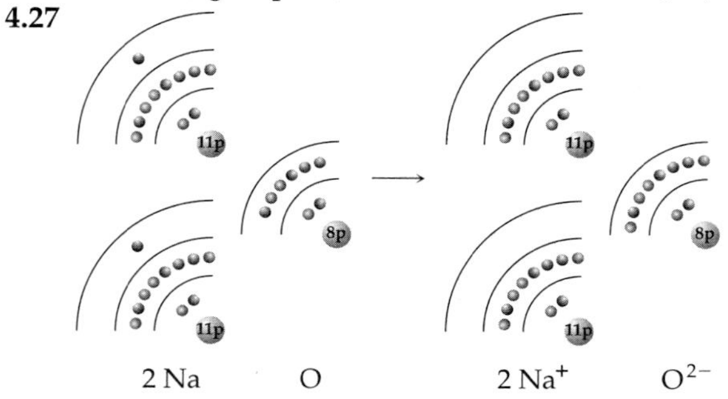

4.28 Ba is in group IIA and will lose 2 electrons to become Ba^{2+}.
F is in group VIIA and will gain 1 electron to become F^-.
The formula of the compound is BaF_2.

4.29 Al is in group IIIA and will lose 3 electrons to become Al^{3+}.
O is in group VIA and will gain 2 electrons to become O^{2-}.
The formula of the compound is Al_2O_3.

4.31 Because the energy is inversely related to wavelength, the shorter the wavelength the higher the energy. Therefore X rays, with shorter wavelength, have higher energy.

4.32 They are at the high-energy end of the electromagnetic spectrum and can cause damage to cells and tissues.

4.34 8.3 minutes for both

4.35 (2) is true because in the relationship equation, wavelength is in the denominator: $E = hc/\lambda$.

4.38 $E = \dfrac{(6.63 \times 10^{-34}\ \text{J} \cdot \text{s})(3.00 \times 10^{8}\ \text{m/s})}{10\ \text{m}}$

$= 1.99 \times 10^{-26}\ \text{J}$

4.41 You would appear at the starting line, then a certain time later you would instantaneously appear some distance ahead, and later still some farther distance ahead, until finally you would appear instantaneously at the finish line.

4.43 "Quantized energy" means that the energy can have only certain allowable values and no values in-between.

4.45 It tells us that something inside the atom is allowed to possess only certain energies, as opposed to any energy.

4.46 Energy and stability are inversely related to each other. The higher the energy of an object, the less stable it is; the lower the energy, the more stable.

4.47 (a)

4.49 As n increases, an electron's energy and distance from the nucleus increase.

4.51 Electron shell

4.52 An input of energy is required.

4.55 The lower shells are of lower energy; the electrons will go into lower-energy positions before beginning to go into higher-energy (less stable) positions.

4.56 Excited state, because the $n = 2$ energy shell is not filled, yet there is an electron in the $n = 3$ shell.

4.59 The energy of an electron depends on distance from the nucleus. The Bohr model assigns each allowed distance from the nucleus an energy value. Therefore saying an electron can be only at certain distances from the nucleus means the electron can have only certain energy values.

4.61 Because they have identical valence-shell configurations.

4.62 A line spectrum is a record of something giving off discrete amounts of energy. The Bohr model says electrons in an atom give off discrete amounts of energy as they change orbits, which means the atomic spectra they produce are line spectra.

4.64 True; the H^+ cation has no electrons.

4.66 (a) 12.1 eV (b) 12.1 eV (c) Ultraviolet

4.68 The energy of the shells would have to be continuous, not quantized.

4.69 (a) 1: Nothing wrong. The fact that it has seven protons but only six electrons makes it a cation. No shell over filled.
2: Nothing wrong. Same number of protons and electrons. No shell over filled.
3: Too many electrons in $n = 1$ shell, which holds a maximum of 2 electrons.
(b) The full atomic symbol for (1) is $^{14}_{7}N^+$; that for (2) is $^{14}_{7}N$.
(c) Diagram (1) is a ground state; diagram (2) is an excited state.

4.70 The number of valence electrons is equal to the roman-numeral group number for the representative elements.

4.72 1.24×10^{-6} m; 1.24×10^{3} nm

4.73 $E = 0.700$ eV; electron relaxes to $n = 3$ shell.

4.74 The simple Bohr model predicts that there should be 9 electrons in the $n = 3$ shell. However, the chemical behavior of K suggests that it has 1 valence electron in the $n = 4$ shell.

4.76 (a) $n = 1, 2, 3, 4$, etc.
(b) s, p, d, f
(c) The number of subshells in a given shell is equal to n; for example, $n = 1$ has one subshell (s); $n = 2$ has two subshells (s and p); $n = 3$ has three subshells (s, p, and d); etc.

4.78 In the $4s$ subshell instead of in the $3d$ subshell

4.80 (a) $1s^22s^22p^1$
(b) $1s^22s^22p^63s^23p^64s^23d^1$
(c) $1s^22s^22p^63s^23p^64s^23d^7$
(d) $1s^22s^22p^63s^23p^64s^23d^{10}4p^4$
(e) $1s^22s^22p^63s^23p^64s^23d^{10}4p^65s^24d^6$

4.83 (a) $[Xe]6s^2$
(b) $[Xe]6s^24f^{14}5d^4$
(c) $[Xe]6s^24f^{14}5d^{10}6p^2$
(d) $[Xe]6s^25d^14f^2$
(e) $[Rn]7s^26d^15f^2$

4.84 (a) Group VA; period 2
(b) Group IA; period 3
(c) Group VIIA; period 4

4.85 (a), (b), (c), (d), and (e) are all incorrect; (f) is correct.
(a) is incorrect because $2s$ should come after $1s$.
(b) is incorrect because the $2s$ subshell should be followed by the $2p$ subshell.
(c) is incorrect because the $1s$ subshell must come first.
(d) is incorrect because the $2p$ subshell holds a maximum of 6 electrons.
(e) is incorrect because the $2p$ subshell would accommodate 6 electrons before the $3s$ subshell would be used.

4.86 (a) 5 (b) 1 (c) 2 (d) 8 (e) 2

4.88 O: $1s^22s^22p^4$; O^{2+}: $1s^22s^22p^2$; O^{2-}: $1s^22s^22p^6$; O^{2-} should be found in most compounds of oxygen because the octet rule is satisfied.

4.90 The s block is 2 elements wide because each s subshell holds 2 electrons.
The p block is 6 elements wide because each p subshell holds 6 electrons.
The d block is 10 elements wide because each d subshell holds 10 electrons.
The f block is 14 elements wide because each f subshell holds 14 electrons.

4.91 The octet rule states that elements will react in such a way as to put 8 electrons in the valence shell. Having 8 valence electrons gives them a noble gas valence electron configuration, and makes them exceptionally stable.

4.93 By gaining enough electrons to have 8 in their outermost occupied shell

4.95

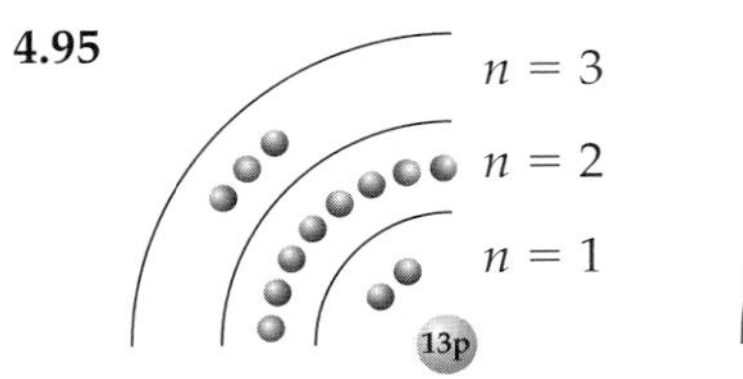

Al, 3 electrons in valence shell

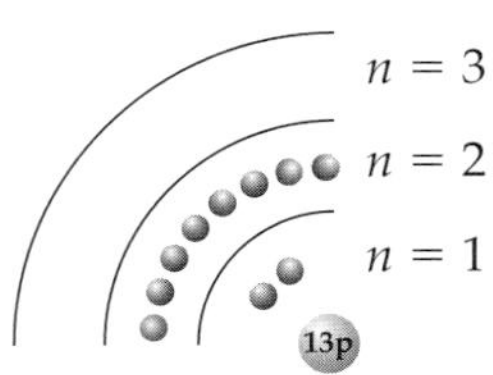

Al^{3+}, 8 electrons in valence shell

4.97 Each lithium atom has 1 valence electron, which it will lose to obtain an octet. Each nitrogen atom has 5 valence electrons, and will gain 3 to have an octet. Thus, each nitrogen atom will need three lithium atoms, and the formula of the compound will be Li_3N.

4.99 (a) ${}^{23}_{11}Na$ and ${}^{31}_{15}P$
(b) Na is the metal, and P is the nonmetal.
(c) Na_3P
(d)

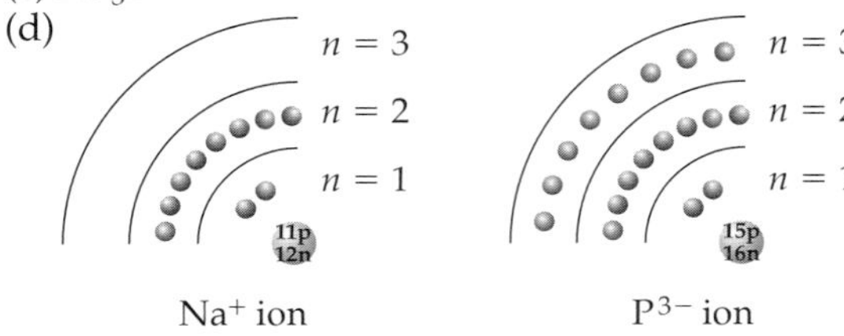

Na^+ ion P^{3-} ion

4.101 Hydrogen will sometimes lose its valence electron, similar to what the other elements in group IA do to obtain an octet. However, when hydrogen loses its electron, it does not obtain an octet; it is left with no electrons.

4.102 Aluminum (Al) has 3 valence electrons, which it will lose to form a +3 cation. Both oxygen and sulfur have 6 valence electrons and will gain 2 electrons to form −2 anions. Because Al needs to lose 3 electrons and oxygen and sulfur both need to gain 2 electrons, each compound will contain 2 Al^{3+} cations and 3 anions. The formulas are similar because both oxygen and sulfur have the same number of valence electrons.

4.104 It is true that Mg^{2+} and Na^+ have identical configurations, but they do not have identical properties. The ions have different charges, different sizes, and different chemical reactivities.

4.105 The roman-numeral group number is equal to the number of valence electrons the metal will lose. The cation will have a positive charge equal to the roman-numeral group number.

4.107 The charges on the cation and anion tell how many electrons are lost and gained, respectively, to form the ions. The formula of the compound is simply the minimum number of each ion needed to have an equal number of electrons transferred.

4.108 (a) Since the electrons are being placed in the same valence shell, one might expect the size to stay the same.
(b) The atomic size decreases because 1 electron is added to the valence shell and 1 proton is added to the nucleus. The added proton makes the nucleus more positive, which causes it to more strongly attract the surrounding electrons. The increased pull on the electrons shrinks the atom.

4.110 Be; the greater nuclear charge for Be (+4) compared to Li (+3) pulls the electrons in closer to the nucleus, shrinking the subshells.

4.113 Na; atoms get larger going down a group.

4.114 Si is the smallest, next is Mg, then Na, and then Rb is the largest.

4.115 Bohr's model says we can know precisely the present and future positions of an electron; quantum mechanical theory says we cannot.

4.117 An orbital is a region of space around the nucleus of an atom in which it is most probable that an electron will be found.

4.118

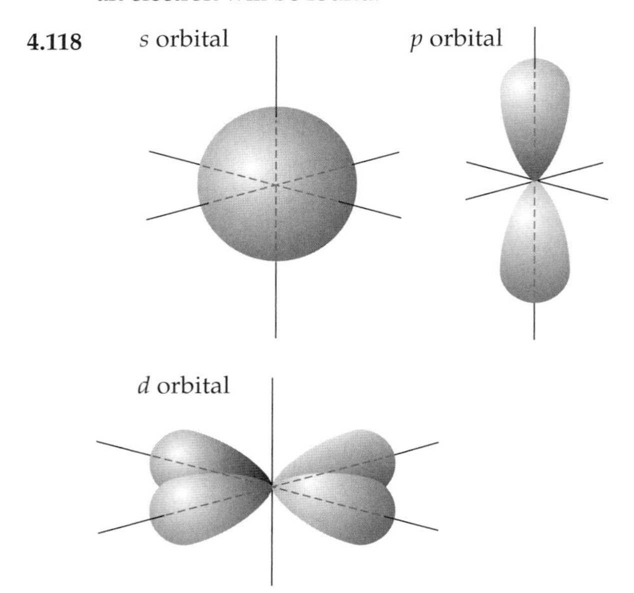

4.119 The picture would be very fuzzy, indicating that the camera cannot record your exact location at any moment. The only thing it can record is the probability that you would be in a certain region of the photograph.

4.121 1.61×10^{-2} s

4.123 2.50×10^4 nm; 33.3 times longer.

4.125

Wavelength (m)	**Wavelength (nm)**	**Energy (J)**
3.52×10^{-8} m	35.2 nm	5.65×10^{-18} J
2.6020×10^{-6} m	2602.0 nm	7.64×10^{-20} J
7.85×10^{-12} m	7.85×10^{-3} nm	2.53×10^{-14} J

4.127 Released.

4.129 (a) Lower-energy excited state:

$n = 3$
$n = 2$
$n = 1$
5p

(b) Ground state:

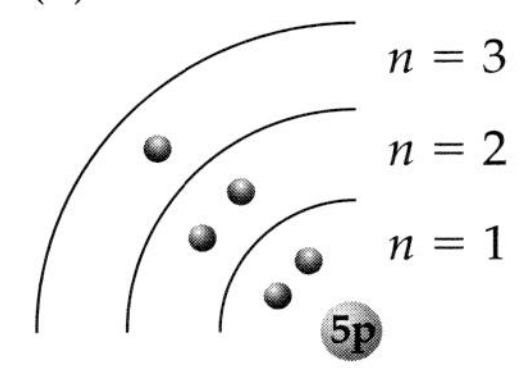

4.131 2.11×10^{-18} J, 94.2 nm
4.133 (a) Ground state.
(b) Incorrect (electrons go into 4*s* before 3*d*).
(c) Incorrect (2*p* over filled).
4.135 *d* subshell. It gets filled *after* the next-higher *s* subshell.
4.137 They all have the same electron configuration.
4.139 (a) N (b) Ba (c) Ca
4.141 Li_3N
4.143 0.199 nm
4.145 (c)
4.147 (a)
4.149 (e)
4.151 3.98×10^{-19} J
4.153 (a) $CaBr_2$ (b) K_3N (c) Al_2S_3 (d) NaI
(e) MgO
4.155 Decreasing wavelength, increasing energy.
4.159 Mg $n = 3$ shell; Ge $n = 4$ shell; W $n = 6$ shell; Cl $n = 3$ shell; Cs $n = 6$ shell.
4.160 1.99×10^{-13} J
4.162 Germanium, atomic number 32; group IVA (14), period 4.
4.165 5.36×10^{-3} s
4.167 Phosphorus, 9. Magnesium, 6. Selenium, 16. Zinc, 12.
4.171 (a) Bromine (b) Boron (c) Rubidium
(d) Uranium (e) Calcium
4.173 (a) 4 (b) 4 (c) 5 (d) 6

Chapter 5

5.2 PH_3
5.3 $SiBr_4$
5.5 H:F̤̈:
5.6 Oxygen has 6 valence electrons. To achieve an octet, it will form two bonds with hydrogen to give H:Ö̤:H
5.8 H—Ö̤—C̤̈l:

5.9

```
   H  H  H
   |  |  |
H—C—C—C—H
   |  |  |
   H  H  H
```

5.11

```
   H
   |
H—C—C≡C—H
   |
   H
```

5.12 H—C≡N:

5.13

```
   H  :O:  H
   |   ||  |
H—C—C—C—H
   |       |
   H       H
```

5.15

```
      :O:
       ||
:Ö̤—S—Ö̤:
```

(plus two additional resonance forms)

5.16

```
[     :Ö:      ]2−
       |
[ :Ö̤—S̤—Ö̤:  ]
```

5.17 [:N≡O:]$^+$
5.18 No because the atoms are arranged differently.
5.20 Identical atoms mean $\Delta EN = 0$ and therefore both bonds are pure covalent.
5.21 EN Si = 1.8, EN O = 3.5; $\Delta EN = 1.7$; a polar covalent
5.22 Because Be is at the top of group IIA, she knows that Be is more electronegative than Ba. Therefore ΔEN for Ba/Cl is greater than ΔEN for Be/Cl, making $BaCl_2$ more ionic.
5.24

Formula	Cation	Anion	Name
CaF_2	Ca^{2+}	F^-	Calcium fluoride
CsBr	Cs^+	Br^-	Cesium bromide
Al_2S_3	Al^{3+}	S^{2-}	Aluminum sulfide
K_2O	K^+	O^{2-}	Potassium oxide

5.25 All the group IIA bromides contain two bromide ions.
5.27 Titanium(III) chloride
5.28 SnF_2
5.30 Nitrogen monoxide
5.31 Phosphorus pentachloride
5.32 P_4S_{10}
5.34 Na_2CO_3
5.35 Calcium hypochlorite
5.36 $Mg(HCO_3)_2$
5.38 No; molecules that contain atoms of the same element are elemental substances (for example, O_2).
5.45 Because the electrons are attracted to both nuclei; this positive–negative attractive force is what bonds the atoms to each other.

5.47 The added attractive force created when the covalent bond forms lowers the energy of the H–H system and makes the molecule more stable than the separated atoms.

5.50 Only the valence electrons are used for bonding because any inner electrons are too close to and too strongly attracted to the atom's nucleus to be shared with other nuclei.

5.53 We mean that the two electrons shared by two atoms are both counted as valence electrons for each atom sharing them. Hence, these bonding electrons are counted twice.

5.55 1 electron

5.56 The nitrogen atom is correct. The oxygen atom should be ·Ö:. Fluorine should have 7 valence electrons and should be ·F̤̈:.

5.61 Oxygen has 6 valence electrons; hence, it combines with two hydrogens to form H_2O and complete its valence-shell octet. HO would leave the oxygen 1 electron short of an octet.

5.63 ·Ṗ̤· and ·B̤̈r: combine to form

:B̤̈r—P̈—B̤̈r: or PBr_3 (with :B̤r: bonded below P)

5.65 H—C(H)(H)—Ö̤—C(H)(H)—H There are 8 bonding pairs and 2 lone pairs.

5.70 Three bonds; the atom has 5 valence electrons and needs 3 to achieve an octet.

5.71 H:C(H)=C(H):H

5.75 There are two resonance forms:

:Ö̤—Ö=Ö: and :Ö=Ö—Ö̤:

5.77 The O_3 molecule is an average of the two resonance forms shown in Problem 5.75. Therefore each O–O bond is an average of a single bond and a double bond, which is a 1.5 bond.

5.79 Yes, the structure shown has too many electrons. The correct structure is :C≡O:

5.83 $[:O{\equiv}O:]^{2+}$

5.85 H—C(H)(H)—C(=Ö:)—Ö̤—H

(H—C(H)(H)—C(:Ö̤:)=Ö̤—H allowed but not likely to occur)

5.90 8 valence electrons on each carbon atom and 2 valence electrons on each hydrogen atom

5.92 An ionic bond is the attractive force between oppositely charged ions that were formed by a transfer of electrons. It is similar to a covalent bond in that both have approximately the same strength and both are due to the attraction between relatively large positive and negative charges. An ionic bond differs from a covalent bond in that an ionic bond is due to a transfer of electrons, whereas a covalent bond is due to a sharing of electrons.

5.94 Between a metal and a nonmetal

5.96 (a) $MgBr_2$ (b) BeO (c) NaI

5.101 Fluorine (F) is the most electronegative element; francium (Fr) is the least electronegative element; they would form the ionic compound FrF.

5.106 The electronegativities are C, 2.5 and H, 2.1. Because $\Delta EN > 0$ but < 1.7, the C–H bonds are classified as polar covalent.

5.108 H_2O

5.111 A is more electronegative because it carries the partial negative charge, indicated by $\delta-$.

5.114 If the two atoms are of the same element.

5.117 The suffix *-ide* indicates the negative part in a binary ionic compound; the negative ion gets this suffix.

5.121 (a) Calcium nitride (b) Aluminum fluoride (c) Sodium oxide (d) Calcium sulfide

5.124 (a) CuCl is copper(I) chloride or cuprous chloride; $CuCl_2$ is copper(II) chloride or cupric chloride.
(b) $Fe(OH)_2$ is iron(II) hydroxide or ferrous hydroxide; $Fe(OH)_3$ is iron(III) hydroxide or ferric hydroxide.

5.129 O^{2-} is a single O atom carrying a 2− charge. O_2^{2-} is two covalently bonded O atoms carrying an overall 2− charge:

$:\ddot{O}:^{2-}$ $[:\ddot{O}-\ddot{O}:]^{2-}$

5.130 IO_4^-, $Mg(IO_4)_2$

5.132 The group IIA metal Mg forms a 2+ cation and therefore combines with two OH^- ions. In the formula $MgOH_2$, the subscript 2 applies only to the H. To indicate two OH^- ions, parentheses must be used: $Mg(OH)_2$.

5.134 Na_2O is a binary *ionic* compound; Greek number prefixes are used only in naming binary *covalent* compounds. Because Na always forms a 1+ cation and the oxide ion always carries a 2− charge, it is obvious without any *di-* in the name that this compound contains two Na^+ for every one O^{2-}.

5.135

Anion	Anion name	Acid formula	Acid name
F^-	Fluoride	HF	Hydrofluoric acid
NO_3^-	Nitrate	HNO_3	Nitric acid
Cl^-	Chloride	HCl	Hydrochloric acid
$C_2H_3O_2^-$	Acetate	$HC_2H_3O_2$	Acetic acid
NO_2^-	Nitrite	HNO_2	Nitrous acid

5.137 NO_3^-, nitrate ion; HNO_3, nitric acid; HNO_2, nitrous acid.

5.139 Hypochlorous acid, HClO; perchloric acid, $HClO_4$.

5.141 Subtract the roman-numeral group number of the element from 8.

5.143 Because each noble gas has a filled valence shell.

5.145 H–C≡N

5.146 (b) is correct. (a) has only three bonds on C. (c) has only three bonds on O and violates the guideline that the least electronegative atom (not counting H) be the central atom.

5.149 (a) Two (b) Six

5.151 Two

5.152 (a) is correct. The second (b) diagram has too few electrons on S. The second (c) diagram has too few electrons on one O and too many on the other O.

5.154 (c)

5.156 (a)

5.158 $PBr_3 < MgBr_2 < KBr < CsBr$

5.160

Name	Formula
Sodium hydrogen carbonate or sodium bicarbonate	$NaHCO_3$
Magnesium acetate	$Mg(CH_3CO_2)_2$
Barium hypochlorite	$Ba(ClO)_2$
Iron(III) nitrate or ferric nitrate	$Fe(NO_3)_3$
Ammonium sulfate	$(NH_4)_2SO_4$
Calcium phosphate	$Ca_3(PO_4)_2$
Cobalt(III) chromate or cobaltic chromate	$Co_2(CrO_4)_3$

5.162

Name	Formula
Xenon tetrafluoride	XeF_4
Xenon difluoride	XeF_2
Iodine monochloride	ICl
Bromine trichloride	$BrCl_3$
Diboron hexahydride	B_2H_6
Dinitrogen oxide	N_2O
Tetrasulfur dioxide	S_4O_2

5.165 Because the less electronegative atom is listed first in the molecular formulas of binary covalent compounds.

5.168 (a) $Ca_3(PO_4)_2$ (b) K_2HPO_4 (c) $Mg(CN)_2$ (d) $Ba(ClO_3)_2$

5.169 (a) Iron(III) sulfite, ferric sulfite (b) Gold(III) nitrate (c) Sodium dihydrogen phosphate (d) Lead acetate or lead(II) acetate

5.171 (a) K < Li < Be (b) Si < S < O (c) Te < I < Br

5.174 $\text{N–Cl} < \overset{\delta-\ \delta+}{\text{O–Cl}} < \overset{\delta+\ \delta-}{\text{S–O}} < \overset{\delta-\ \delta+}{\text{O–H}} < \overset{\delta+\ \delta-}{\text{C–F}}$

5.177 Manganate ion.

Chapter 6

6.2 Electron groups all 109.5° apart, molecular shape tetrahedral:

$[NH_4]^+$

6.3 Electron groups 180° apart, molecular shape linear: Cl—C≡C—Cl

6.4 Electron groups all in one plane at corners of triangle; molecular shape trigonal planar:

$Cl_2C{=}O$

6.6 Both electron-group geometry and molecular shape tetrahedral:

$[SO_4]^{2-}$ All angles 109.5°

6.7 Both electron-group geometry and molecular shape linear:

H–C≡N (180°)

6.8 Electron-group geometry, trigonal planar; molecular shape, bent:

O–S=O (118°)

6.9 Both electron-group geometry and molecular shape trigonal planar:

H–O–N(=O)–O (120°)

6.11 Yes, it is polar.

$CHCl_3$: δ+ (H), δ− (C)

6.12 PCl_3:

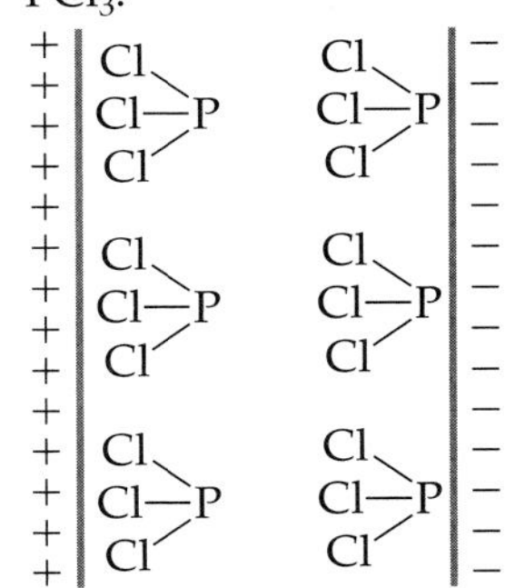

$CHCl_3$:

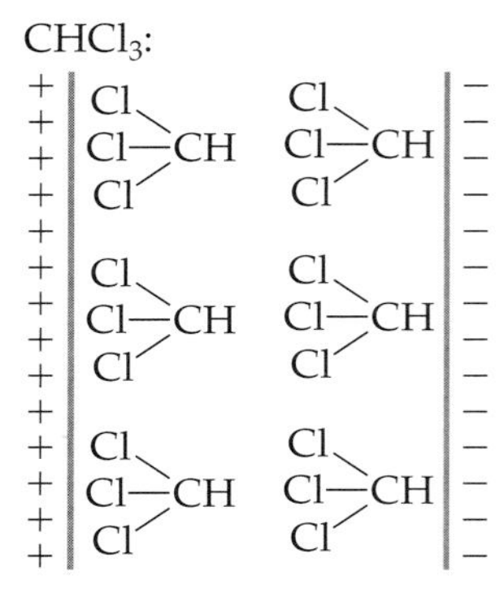

6.13 Because the bonding pair electrons can be farther apart with 109.5° angles.

6.16

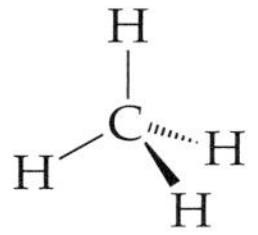

H
|
C
H H H

6.18 (a) (1) H, C, Cl, Cl, Cl, 109.5° (2) :S̈:, H, H, 105° (actual: 92°)

(3) :Ö=N̈—C̈l: 118° (actual: 113°) (4) :C̈l—Be—C̈l: 180°

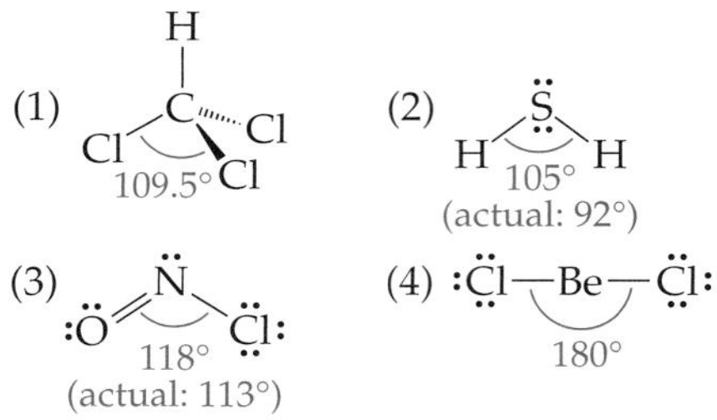

(b) (1) Electron-group geometry and molecular shape both tetrahedral.
(2) Electron-group geometry, tetrahedral; molecular shape, bent.
(3) Electron-group geometry, trigonal planar; molecular shape, bent.
(4) Electron-group geometry and molecular shape both linear.

6.20 All the electrons in a multiple bond occupy roughly the same region of space.

6.21 In the case where the bond angle is approximately 118°, the central atom has two bonding pairs and one lone pair. In the case where the bond angle is approximately 105°, the central atom has two bonding pairs and two lone pairs.

6.23 The shape of a molecule is determined by the arrangement of only the *atoms* in the molecule. One of the electron groups on the N in NH_3 is a lone pair and thus ignored in determining molecule shape. Three bonding groups and one lone pair make the NH_3 molecule pyramidal.

6.25 (a)

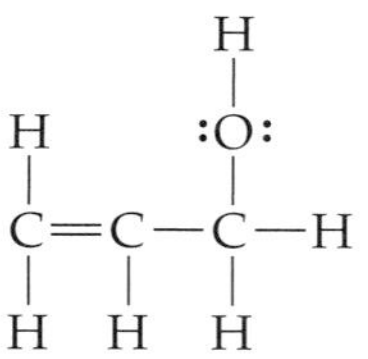

H—C=C—C(—:Ö—H)—H, with H H H below

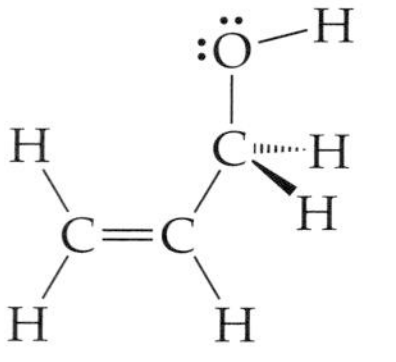

(b) H₂C=C(H)—C(H)(H)—:Ö—H

(c) :Ö—H 105°; 109.5°; C—H H; H; C=C; 120°; 120°; 120°; H H

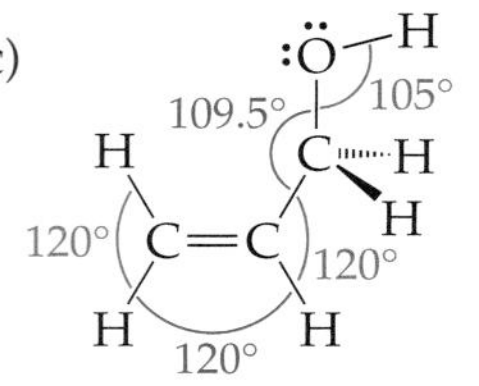

6.26 The shape of a molecule depends on the number of electron groups around the interior atoms and on whether those groups are bonding electrons or lone pairs. All this needed information is obtainable only from the dot diagram.

6.28 :O: 109.5° 107° C N̈ C C 120° 109.5° H

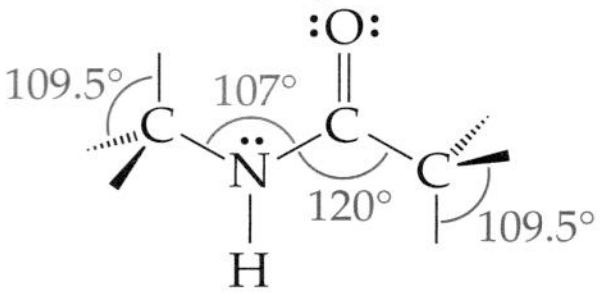

Leftmost C, tetrahedral; N, pyramidal; middle C, trigonal planar; rightmost C, tetrahedral.

6.30 Yes; you must know which, if any, bonds in the molecule are polar (based on the relative electronegativities of the connected atoms) and the shape of the molecule.

6.32 (a) HCl, because the electronegativity difference is greater in HCl than in HBr.
(b) HCl, because the molecular dipole moment is greater.

6.34 Because a dipole moment has a direction associated with it.

6.37 The linear shape of the molecule results in the two bond dipole moments canceling each other.

6.38 The shape of the SO_2 molecule is bent, which means the bond dipole moments do not cancel each other.

6.41 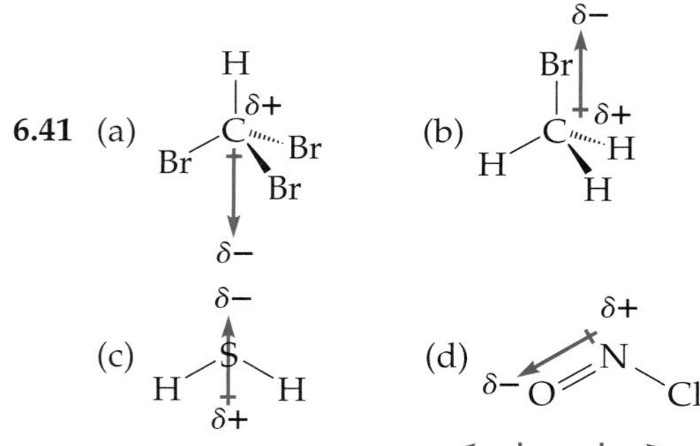

(e) C_2Cl_2 is nonpolar: Cl—C≡C—Cl

6.43 No, dipole–dipole forces have only a fraction of the strength of the forces between ions because ions carry full charges and dipole charges are partial.

6.45 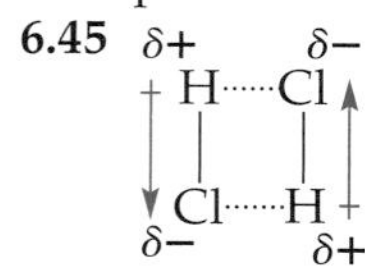

6.47 The lone pairs are important in determining the electron-group geometry around an atom and thus in determining the shape of the molecule.

6.48 (a) $\ddot{\underset{..}{O}}=\ddot{S}-\ddot{\underset{..}{O}}:$

(b) The correct diagram predicts a bent shape and polar molecule. The incorrect diagram predicts a linear shape and nonpolar molecule.

6.49 (a)

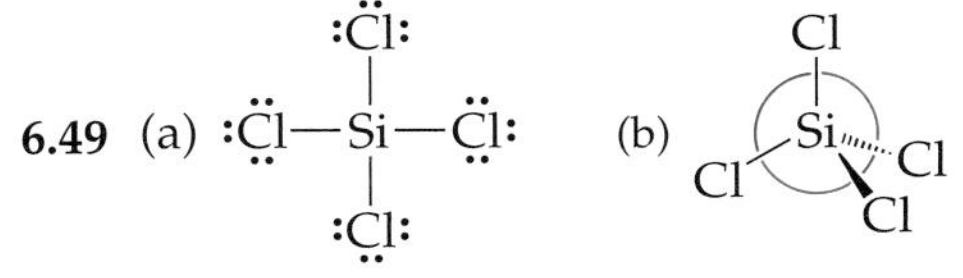

(b)

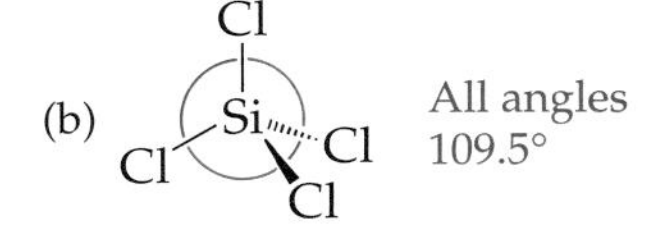

(c) Tetrahedral

(d)

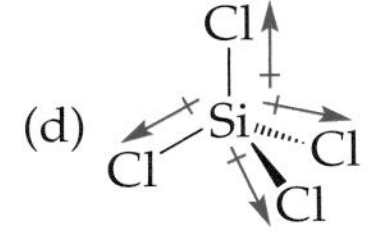

(e) No, the molecule is nonpolar.

6.51 (a) :Ö:S:Ö: (with :O: above S)

(b) O=S(–O)–O — All angles 120°

(c) Trigonal planar

(d) O=S(–O)–O, with bond dipole arrows

(e) No, the molecule is nonpolar.

6.53 (a) $H-\ddot{N}-\ddot{N}-H$, with H on each N

(b) H_2N-NH_2 — All angles 107° (2% compression rule for period 2 atoms)

(c) Pyramidal around each N

(d) (e) Yes, the molecule is polar.

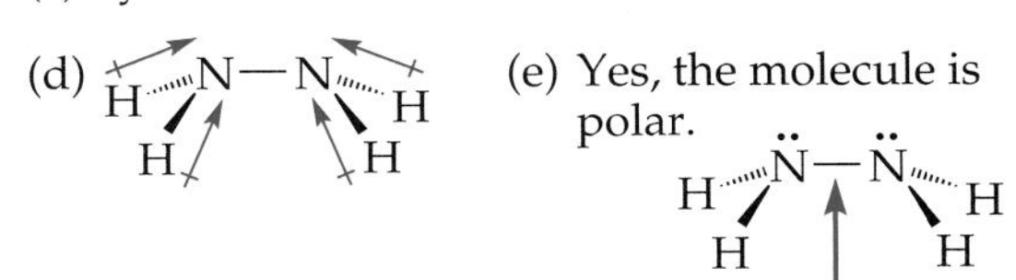

6.55 (a) $:N\equiv N-\ddot{\underset{..}{O}}:$ (b) N≡N—O, 180° (c) Linear

(d) N≡N—O (with dipole arrow) (e) Yes, the molecule is polar. N≡N—O (with dipole arrow)

6.56 (a) NO_2 has an odd number of valence electrons: 5 + 6(2) = 17. This makes it impossible to show only electron pairs in the dot diagram.

(b) $:\ddot{\underset{..}{O}}-\dot{N}=\ddot{O}:$

(c) Bent; the molecule is polar.

6.58 As the molecular dipole moment was increased, the intermolecular attractive forces would become stronger and the boiling point would increase.

6.61 H_3C-CH_3 (three-dimensional structure) — All angles 109.5°

6.63 $H_3C-O-CH_3$ (three-dimensional structure), C–O–C angle 105° — All other angles 109.5°

6.67 (a) $:N\equiv C-C(=\ddot{O}:)-\ddot{N}-H$, with H on N

(b) N≡C—C(=O)—NH₂: 180° at the C≡N carbon, 120° angles around the carbonyl C, 107° angles around N ← All angles around this N 107°

6.69 (a) $[:C\equiv N:]^-$ Linear molecular shape

(b) $[ClO_4]^-$ (three-dimensional structure) Tetrahedral molecular shape, no dipole moment

(c) $[PCl_4]^+$ (three-dimensional structure) Tetrahedral molecular shape, no dipole moment

(d) $[O-N=O]^-$ Bent molecular shape

6.74

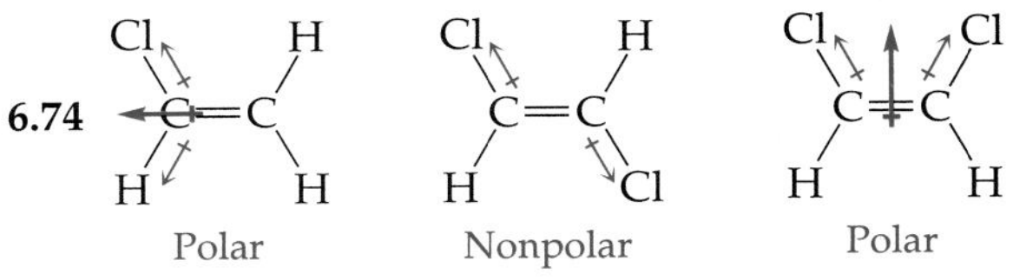

6.75

H–Ö–F — Bent shape, 105° angle, polar

6.77 CH_3F

6.79 The answer considers not the three-dimensional configuration of the atoms but only the two-dimensional configurations she drew. The N has two bonding groups and two lone pairs, and Table 6.2 tells her that this combination gives a bent molecular shape.

6.81 Tetrahedral electron-group geometry around each S. (F–S–S angles 105°)
Trigonal planar electron-group geometry around each N. (F–N=N angles 118°)

6.84 (a) O=N—O—O—H

(b) 118° at N; All angles 105° around both these O atoms

(c) N—electron-group geometry, trigonal planar; molecular shape, bent to 118°.
Both O—electron-group geometry, tetrahedral; molecular shape, bent to 105°.

6.86 (a) S=N—F, 118°. Trigonal planar electron-group geometry around N, bent molecular shape

(b) [:N≡C—O:]$^-$ Linear electron-group geometry around C, linear molecular shape

(c) $[SF_3]^+$ Tetrahedral electron-group geometry around S, pyramidal molecular shape

(d) O=Si=O Linear electron-group geometry around Si, linear molecular shape, no molecular dipole moment

6.89 The molecular shape is linear, which means the bond angle is 180°:
[N=N=N]$^-$

6.92 H—O—O—O—H, all angles 105°

6.95 BH_3—electron-group geometry around B trigonal planar, molecular shape trigonal planar. PH_3—electron-group geometry around P tetrahedral, molecular shape pyramidal. BH_3 is nonpolar.

Chapter 7

7.2 $CaH_2 + 2\,H_2O \longrightarrow Ca(OH)_2 + 2\,H_2$
7.3 $C_6H_{12}O_6 + 6\,O_2 \longrightarrow 6\,CO_2 + 6\,H_2O$
7.4 $2\,HCl + Na_2CO_3 \longrightarrow 2\,NaCl + CO_2 + H_2O$
7.6 $2\,C_6H_{12}(l) + 5\,O_2(g) \longrightarrow 2\,H_2C_6H_8O_4(l) + 2\,H_2O(l)$
7.7 $4\,CH_3NH_2(g) + 9\,O_2(g) \longrightarrow 4\,CO_2(g) + 10\,H_2O(g) + 2\,N_2(g)$
7.9 $2\,Pb(NO_3)_2(s) \xrightarrow{\text{Heat}} 2\,PbO(s) + 4\,NO(g) + 3\,O_2(g)$;decomposition.
7.10 $C_2H_4 + C_2Cl_4 \xrightarrow{\text{Catalyst}} 2\,C_2H_2Cl_2$; combination.
7.12 (a) Step 1: Pb^{2+}, NO_3^-, Na^+, SO_4^{2-}
Step 2: Pb^{2+} with SO_4^{2-}, Na^+ with NO_3^-
Step 3: Table 7.1 indicates $PbSO_4$ is insoluble in water. All Na^+ salts are water-soluble, meaning Na^+ and NO_3^- do not form a precipitate. Na^+ and NO_3^- are spectator ions.
(b) Step 4: $Pb^{2+}(aq) + SO_4^{2-}(aq) \longrightarrow PbSO_4(s)$
7.13 (a) Step 1: Ni^{2+}, NO_3^-, NH_4^+, PO_4^{3-}
Step 2: Ni^{2+} with PO_4^{3-}, NH_4^+ with NO_3^-
Step 3: Table 7.1 says most phosphate salts are insoluble in water and does not list $Ni_3(PO_4)_2$ as an exception. All ammonium salts and all nitrate salts are water-soluble, meaning NH_4^+ and NO_3^- do not form a precipitate. They are spectator ions.
(b) Step 4: $3\,Ni^{2+} + 2\,PO_4^{3-} \longrightarrow Ni_3(PO_4)_2$
7.15 Each mole of H_2SO_4 dissociates into 2 moles of H^+ and 1 mole of SO_4^{2-}, but each mole of NaOH dissociates into 1 mole of Na^+ and 1 mole of OH^-. Therefore it takes twice as much NaOH to provide the 2 moles of OH^- needed to neutralize all the H^+ from H_2SO_4.
7.16 Intact-molecule: $2\,HF(aq) + Ca(OH)_2(aq) \longrightarrow CaF_2(s) + 2\,H_2O(l)$
Net ionic: $2\,H^+(aq) + 2\,F^-(aq) + Ca^{2+}(aq) + 2\,OH^-(aq) \longrightarrow CaF_2(s) + 2H_2O(l)$
The salt is calcium fluoride.
7.17 Only (a) represents a chemical reaction because the substances on the right (water and oxygen) are different from the substance on the left (hydrogen peroxide, H_2O_2). In (b), all substances remain chemically unchanged.
7.20 Under *no* circumstances; it *always* takes energy to break a bond.
7.21 Two Cl–Cl bonds and two C–H bonds must be broken. Two H–Cl bonds and two C–Cl bonds are formed.
7.23 Cooling the reaction mixture slows down the molecules, meaning they have less energy available to break bonds when they collide.
7.25 There will be more frequent collisions between reactant molecules.
7.27 Changing the subscripts in a chemical formula changes the chemical identity of a substance.
7.28 $C_2H_4 + 3\,O_2 \longrightarrow 2\,CO_2 + 2\,H_2O$
7.29 1 mole of methane and 2 moles of oxygen react to give 1 mole of carbon dioxide and 2 moles of water. Or 1 molecule of methane and 2 molecules of oxygen react to give 1 molecule of carbon dioxide and 2 molecules of water.
7.31 $Al_2Cl_6 + 6\,H_2O \longrightarrow 2\,Al(OH)_3 + 6\,HCl$
7.33 $2\,C_2H_2 + 5\,O_2 \longrightarrow 2\,H_2O + 4\,CO_2$
7.35 $2\,N_2O_5(g) \longrightarrow 4\,NO_2(g) + O_2(g)$; decomposition.
7.36 $2\,CO(g) + 2\,NO(g) \longrightarrow 2\,CO_2(g) + N_2(g)$; single replacement (O in NO replaced by N to form NN)

7.38 $2\ SO_2(g) + O_2(g) \longrightarrow 2\ SO_3(g)$; combination.

7.39 $3\ Li(s) + \frac{1}{2} N_2(g) \longrightarrow Li_3N(s)$ or
$6\ Li(s) + N_2(g) \longrightarrow 2\ Li_3N(s)$; combination.

7.40 $C_5H_{10}O_2(l) + \frac{13}{2} O_2(g) \longrightarrow 5\ CO_2(g) + 5\ H_2O(g)$ or
$2\ C_5H_{10}O_2(l) + 13\ O_2(g) \longrightarrow 10\ CO_2(g) + 10\ H_2O(g)$

7.42 A combination reaction brings two or more substances together to form one new substance: $A + B \longrightarrow AB$. A decomposition reaction breaks one substance up into two or more new substances: $AB \longrightarrow A + B$.

7.44 $CH_3OH(l)$ $Na_3PO_4(s)$

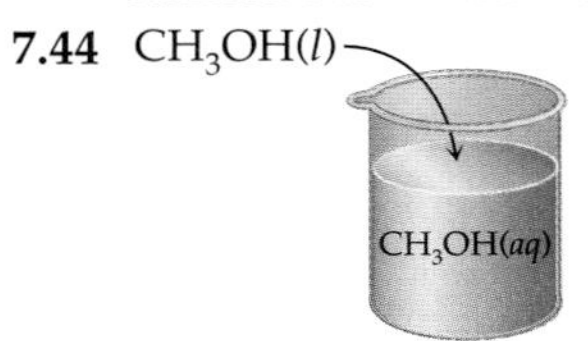

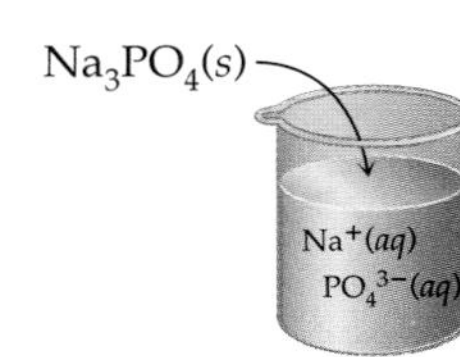

Methanol does not dissociate into ions in aqueous solution. Sodium phosphate does.

7.46 (a) $Pb^{2+}(aq) + 2\ Cl^-(aq) \longrightarrow PbCl_2(s)$
(b) $Ba^{2+}(aq) + SO_4^{2-}(aq) \longrightarrow BaSO_4(s)$
(c) No precipitation.

7.48

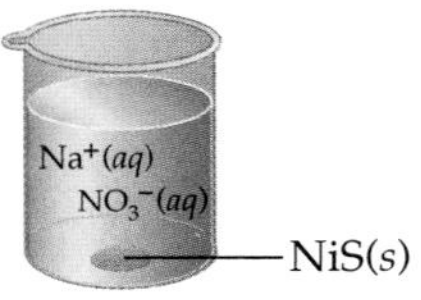

$Ni^{2+}(aq) + S^{2-}(aq) \longrightarrow NiS(s)$

7.50 Calcium carbonate precipitates: $Ca^{2+}(aq) + CO_3^{2-}(aq) \longrightarrow CaCO_3(s)$

7.52 Table 7.1 tells you hydroxides are insoluble in water except for NaOH, KOH, $Ca(OH)_2$, and $Ba(OH)_2$. Mix an aqueous solution of any one of these with an aqueous solution of any water-soluble Ni^{2+} salt, and then filter off the $Ni(OH)_2$ that forms.

7.54 (a) $Ca^{2+}(aq) + SO_4^{2-}(aq) \longrightarrow CaSO_4(s)$
(b) $Ca(NO_3)_2(aq) + (NH_4)_2SO_4(aq) \longrightarrow CaSO_4(s) + 2\ NH_4NO_3(aq)$

7.55 (a) $H_2S(aq) + 2\ Na^+(aq) + 2\ OH^-(aq) \longrightarrow 2\ Na^+(aq) + S^{2-}(aq) + 2\ H_2O(l)$
(b) Sodium sulfide, Na_2S.

7.56 (a) $H_2S(aq) \longrightarrow 2\ H^+(aq) + S^{2-}(aq)$
(b) It produces H^+ ions in water.
(c) $Cu^{2+}(aq) + S^{2-}(aq) \longrightarrow CuS(s)$

7.58 $H_2SO_4(aq) + Ca(OH)_2(aq) \longrightarrow 2\ H_2O(l) + CaSO_4(s)$
The salt is calcium sulfate.

7.59 (a) Lead(II) sulfate.
(b) $Pb^{2+}(aq) + SO_4^{2-}(aq) \longrightarrow PbSO_4(s)$
(c) No, because $Pb(NO_3)_2$ is not a base, meaning no OH^- ions were present to neutralize the $H^+(aq)$ ions created when the H_2SO_4 dissociated.

7.60 $NH_4^+(aq) \longrightarrow NH_3(aq) + H^+(aq)$

7.62 (a) $H^+(aq) + Br^-(aq) + Na^+(aq) + OH^-(aq) \longrightarrow Br^-(aq) + Na^+(aq) + H_2O(l)$
$H^+(aq) + OH^-(aq) \longrightarrow H_2O(l)$
Sodium bromide.
(b) $Ca^{2+}(aq) + 2\ OH^-(aq) + 2\ H^+(aq) + 2\ NO_3^-(aq) \longrightarrow Ca^{2+}(aq) + 2\ NO_3^-(aq) + 2\ H_2O(l)$
$OH^-(aq) + H^+(aq) \longrightarrow H_2O(l)$
Calcium nitrate.
(c) $Ca^{2+}(aq) + 2\ OH^-(aq) + 2\ H^+(aq) + 2\ Br^-(aq) \longrightarrow Ca^{2+}(aq) + 2\ Br^-(aq) + 2\ H_2O(l)$
$OH^-(aq) + H^+(aq) \longrightarrow H_2O(l)$
Calcium bromide.
(d) $Mg(OH)_2(s) + 2\ H^+(aq) + 2\ Cl^-(aq) \longrightarrow Mg^{2+}(aq) + 2\ Cl^-(aq) + 2\ H_2O(l)$
$Mg(OH)_2(s) + 2\ H^+(aq) \longrightarrow Mg^{2+}(aq) + 2\ H_2O(l)$
Magnesium chloride.

7.64 The source is the formation of bonds as the precipitate molecules are created. Bond formation always releases energy, usually in the form of heat.

7.65 Energy is needed to pull each electron in the covalent bond away from one of the two positively charged nuclei it is attracted to.

7.68 (a) Yes, because iron(III) sulfide is insoluble in water.
(b) $2\ Fe^{3+}(aq) + 3\ S^{2-}(aq) \longrightarrow Fe_2S_3(s)$

7.70 Both salts dissolve; no precipitate forms.

7.71 Any sodium salt whose anion forms an insoluble precipitate with Cu^{2+}: Na_3PO_4, Na_2CO_3, Na_2S, or NaOH. Add the right amount, and all the anion you add combines with Cu^{2+}, leaving only added Na^+ and meaning no new types of ions are in the solution.

7.73 (a)

(b)

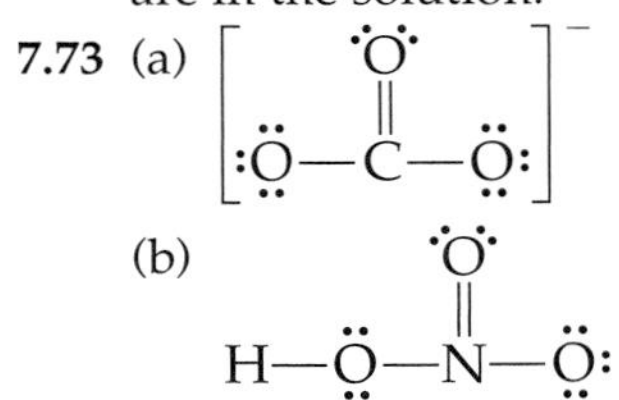

(c) Nitrate ion, because HNO_3 dissociates in water.

7.74 $HClO_4(aq) + NaOH(aq) \longrightarrow NaClO_4(aq) + H_2O(l)$
$H^+(aq) + OH^-(aq) \longrightarrow H_2O(l)$
Sodium perchlorate.

7.75 Add NaCl(*aq*), NaBr(*aq*), or NaI(*aq*), to precipitate Pb^{2+} as the insoluble halide salt. Filter off the solid halide, and then add any sodium salt whose anion forms an insoluble salt with Ba^{2+}: Na_2SO_4, Na_3PO_4, Na_2CO_3, or Na_2S. Filter off the precipitate that forms, and your filtered solution contains $Na^+(aq)$ but no $Pb^{2+}(aq)$ and no $Ba^{2+}(aq)$.

7.77 $C_7H_{16}(l) + 11\ O_2(g) \longrightarrow 7\ CO_2(g) + 8\ H_2O(g)$

7.80 (a) Broken 1 N–N, 3 Br–Br; formed 6 N–Br
(b) Broken 4 P–P, 6 H–H; formed 12 P–H
(c) Broken 1 K–F, 1 Na–I; formed 1 K–I, 1 Na–F
(d) Broken 1 C–H, 1 Cl–Cl; formed 1 C–Cl, 1 H–Cl

7.82 $BaCO_3$ insoluble; K_2S soluble; $CaSO_4$ insoluble; $MgNO_3$ soluble; $(NH_4)_3PO_4$ soluble.

7.84 (a)

```
     H
     |             ..  ..
 H — C — H        :I — I:
     |             ''  ''
     H

     H
     |    ..             ..
 H — C — I:         H — I:
     |    ''             ''
     H
```

Broken: 1 C–H, 1 I–I; formed 1 C–I, 1 H–I

(b)

```
 H          H                          H    H
   \      /         ..   ..      ..    |    |    ..
    C == C         :F — F:      :F  —  C  — C  — F:
   /      \         ''   ''      ''    |    |    ''
 H          H                          H    H
```

Broken: one-half of C=C, 1 F–F; formed 2 C–F

7.87 KCl salt; CH_3COOH acid; $Al(OH)_3$ base; H_3BO_3 acid; $LiC_2H_3O_2$ salt.

7.89 Balanced:
$2\,HNO_3 + Ca(OH)_2 \longrightarrow Ca(NO_3)_2 + 2\,H_2O$
Complete ionic:
$2\,H^+ + 2\,NO_3^- + Ca^{2+} + 2\,OH^- \longrightarrow Ca^{2+} + 2\,NO_3^- + 2\,H_2O$
Net ionic: $H^+ + OH^- \longrightarrow H_2O$

7.91 (a) $Al(OH)_3$ (b) $Ca_3(PO_4)_2$ (c) $MgCO_3$

7.94 $4\,NH_3 + 3\,Cl_2 \longrightarrow 3\,NH_4Cl + NCl_3$

7.97 $2\,KClO_3 \longrightarrow 2\,KCl + 3\,O_2$

7.100 $CaCO_3 \xrightarrow{\text{Heat}} CO_2 + CaO$

7.102 (a) Potassium hydroxide and nitric acid.
(b) Calcium hydroxide and phosphoric acid.
(c) Lithium hydroxide and sulfuric acid.
(d) Sodium hydroxide and hydroiodic acid.

7.105 $C_3H_8(g) + 5\,O_2(g) \longrightarrow 3\,CO_2(g) + 4\,H_2O(l)$

7.107 $CO(g) + H_2O(g) \longrightarrow CO_2(g) + H_2(g)$

7.111 $2\,H_2S + 3\,O_2 \longrightarrow 2\,SO_2 + 2\,H_2O$

7.113 (a) Cl^-, Br^-, I^-, PO_4^{3-}, CO_3^{2-}, S^{2-}, or OH^-.
(b) SO_4^{2-}, PO_4^{3-}, or CO_3^{2-}.

7.115 (a) $Al^{3+}(aq) + 3\,OH^-(aq) \longrightarrow Al(OH)_3(s)$
(b) $3\,Ca^{2+}(aq) + 2\,PO_4^{3-}(aq) \longrightarrow Ca_3(PO_4)_2(s)$
(c) $Mg^{2+}(aq) + CO_3^{2-}(aq) \longrightarrow MgCO_3(s)$

7.118 (a) $AgI(s)$ (b) None (c) None

7.121 Balanced:
$2\,KCl(aq) + Pb(NO_3)_2(aq) \longrightarrow PbCl_2(s) + 2\,KNO_3(aq)$
Complete ionic:
$2\,K^+ + 2\,Cl^- + Pb^{2+} + 2\,NO_3^- \longrightarrow PbCl_2 + 2\,K^+ + 2\,NO_3^-$
Net ionic: $Pb^{2+} + 2\,Cl^- \longrightarrow PbCl_2$

7.124 $H_2CO_3(aq) + 2\,LiOH(aq) \longrightarrow Li_2CO_3(s) + 2\,H_2O(l)$

7.127 Slower, because the increased space the reactant molecules have to move around in means there will be fewer collisions between molecules.

7.130 (a) $6\,Ca(s) + P_4(s) \longrightarrow 2\,Ca_3P_2(s)$
(b) $2\,Na(s) + 2\,H_2O(l) \longrightarrow 2\,NaOH(aq) + H_2(g)$
(c) $2\,NH_4NO_3(s) \longrightarrow 2\,N_2(g) + O_2(g) + 4\,H_2O(g)$

7.132 Complete ionic:
$2\,H^+(aq) + 2\,Br^-(aq) + Ba^{2+}(aq) + 2\,OH^-(aq) \longrightarrow 2\,Br^-(aq) + Ba^{2+}(aq) + 2\,H_2O(l)$
Net ionic:
$H^+(aq) + OH^-(aq) \longrightarrow H_2O(l)$

Chapter 8

8.2 1 cup of sugar requires 5 eggs; therefore 3 cups of sugar requires 15 eggs.

8.3 35 eggs

8.2 Second time
Conversion factors:

$$\frac{1 \text{ cup sugar}}{5 \text{ eggs}} \text{ and } \frac{5 \text{ eggs}}{1 \text{ cup sugar}}$$

Use the second one because you want your answer in eggs:

$$3 \text{ cups sugar} \times \frac{5 \text{ eggs}}{1 \text{ cup sugar}} = 15 \text{ eggs}$$

8.3 Second time

$$21 \text{ blocks cream cheese} \times \frac{5 \text{ eggs}}{3 \text{ blocks cream cheese}} = 35 \text{ eggs}$$

8.5 (a) $0.10 \text{ mol U} \times \dfrac{6.022 \times 10^{23} \text{ atoms U}}{1 \text{ mol U}} = 6.0 \times 10^{22} \text{ atoms U}$

(b) $0.10 \text{ mol U} \times \dfrac{238.029 \text{ g U}}{1 \text{ mol U}} = 23.803 \text{ g U}$,

which you must report as 24 g because 0.10 has only two significant digits.

8.6 (a) $120.11 \text{ g C} \times \dfrac{1 \text{ mol C}}{12.011 \text{ g C}} = 10.000 \text{ mol C}$

(b) $10.000 \text{ mol C} \times \dfrac{6.022 \times 10^{23} \text{ atoms C}}{1 \text{ mol C}} = 6.022 \times 10^{24} \text{ atoms C}$

8.8 From Practice Problem 8.7, 1 mole of propane has a mass of 44.096 g; 2 moles of propane has a mass of 88.192 g.

8.9 6 moles of carbon atoms

8.10 1 mole of propane and 5 moles of oxygen react to give 3 moles of carbon dioxide and 4 moles of water.

8.12 (a) 72.3 g (b) 3.40×10^{23} molecules

8.13 (a) 8.30 moles (b) 133 g

8.15 (a) $C_6H_6 + 3\,H_2 \longrightarrow C_6H_{12}$
(b) 1 mole of benzene and 3 moles of hydrogen react to give 1 mole of cyclohexane.
(c) 78.113 g C_6H_6; 6.0474 g H_2
(d) 84.161 g
(e) 28.5%

8.16 (a) $C_6H_{12}O_6 + 6\,O_2 \longrightarrow 6\,CO_2 + 6\,H_2O$
(b) 1 mole of glucose and 6 moles of oxygen react to give 6 moles of carbon dioxide and 6 moles of water.
(c) 180.155 g glucose; 191.99 g O_2
(d) 264.05 g.
(e) 74.23%

8.18 $$10.0\text{ g glucose} \times \frac{1\text{ mol glucose}}{180.155\text{ g glucose}} \times \frac{6\text{ mol } H_2O}{1\text{ mol glucose}} \times \frac{18.015\text{ g } H_2O}{1\text{ mol } H_2O} = 6.00\text{ g } H_2O$$

8.19 $$10.0\text{ g glucose} \times \frac{1\text{ mol glucose}}{180.155\text{ g glucose}} \times \frac{6\text{ mol } CO_2}{1\text{ mol glucose}} \times \frac{44.009\text{ g } CO_2}{1\text{ mol } CO_2} = 14.7\text{ g } CO_2$$

8.20 $$10.0\text{ g } CO_2 \times \frac{1\text{ mol } CO_2}{44.009\text{ g } CO_2} \times \frac{1\text{ mol glucose}}{6\text{ mol } CO_2} \times \frac{180.155\text{ g glucose}}{1\text{ mol glucose}} = 6.82\text{ g glucose}$$

8.22 $$10.0\text{ g } Al_2O_3 \times \frac{1\text{ mol } Al_2O_3}{101.96\text{ g } Al_2O_3} \times \frac{2\text{ mol Al}}{1\text{ mol } Al_2O_3} \times \frac{6.022 \times 10^{23}\text{ Al atoms}}{1\text{ mol Al}} = 1.18 \times 10^{23}\text{ Al atoms}$$

8.23 $$10.0\text{ g } H_2O \times \frac{1\text{ mol } H_2O}{18.015\text{ g } H_2O} \times \frac{6.022 \times 10^{23}\ H_2O\text{ molecules}}{1\text{ mol } H_2O} = 3.34 \times 10^{23}\ H_2O\text{ molecules}$$

8.25 (a) 362.9 g O_2 (b) 163.4 g H_2O

8.26 (a) $2\,ZnS + 3\,O_2 \longrightarrow 2\,ZnO + 2\,SO_2$
(b) ZnS is the limiting reactant. Using it to calculate theoretical yield gives 8.39 g ZnO and 6.60 g SO_2.
(c) 5.06 g O_2 left over
(d) 89.4% yield

8.28 (a) C_2H_6O
(b) 52.4% C; 13.1% H; 34.5% O
(c) $C_2H_6O + 3\,O_2 \longrightarrow 2\,CO_2 + 3\,H_2O$

8.30 (a) CH_2
(b) C_2H_4
(c) $C_2H_4 + 3\,O_2 \longrightarrow 2\,CO_2 + 2\,H_2O$

8.32 37.02% C; 2.22% H; 18.50% N; 42.27% O

8.33 The empirical formula and molecular formula are both Cl_3CH.

8.34 The empirical formula is C_2H_4O; the molecular formula is $C_4H_8O_2$.

8.36 (a) $$\frac{1\text{ egg}}{2\text{ cups flour}}, \frac{1\text{ egg}}{3\text{ cups sugar}}, \frac{2\text{ cups flour}}{3\text{ cups sugar}}, \frac{\frac{1}{4}\text{ pound butter}}{3\text{ cups sugar}}, \frac{1\text{ cup milk}}{1\text{ dozen sugar cookies}}, \frac{1\text{ dozen sugar cookies}}{\frac{1}{4}\text{ pound butter}}, \text{ etc.}$$

(b) $$30\text{ cookies} \times \frac{1\text{ dozen cookies}}{12\text{ cookies}} \times \frac{2\text{ cups flour}}{1\text{ dozen cookies}} = 5\text{ cups flour}$$

(c) You don't know how many cups of milk are in the container.

(d) $$1\text{ container of milk} \times \frac{4\text{ cups milk}}{1\text{ container of milk}} \times \frac{1\text{ egg}}{1\text{ cup milk}} = 4\text{ eggs}$$

(e) 1 egg + 2 cups flour + 3 cups sugar + $\frac{1}{4}$ pound butter + 1 cup milk $\longrightarrow$ 1 dozen sugar cookies

8.37 (a) $$3\text{ cakes} \times \frac{5\text{ eggs}}{1\text{ cake}} = 15\text{ eggs}$$

(b) $$63\text{ blocks of cream cheese} \times \frac{5\text{ eggs}}{3\text{ blocks of cream cheese}} = 105\text{ eggs}$$

8.39 Always

8.41 There are 6.022×10^{23} bicycles in 1 mole of bicycles. Because there are two tires on every bicycle, the number of bicycle tires in 1 mole of bicycles is 2 times 6.022×10^{23}, or 1.204×10^{24}.

8.43 $$2.5\text{ moles of pennies} \times \frac{6.022 \times 10^{23}\text{ pennies}}{1\text{ mole of pennies}} = 1.5 \times 10^{24}\text{ pennies}$$

$$1.5 \times 10^{24}\text{ pennies} \times \frac{1\text{ dollar}}{100\text{ pennies}} = 1.5 \times 10^{22}\text{ dollars}$$

8.45 2 molecules of sufur dioxide react with 1 molecule of oxygen to give 2 molecules of sulfur trioxide.
2 moles of sulfur dioxide react with 1 mole of oxygen to give 2 moles of sulfur trioxide.

8.46 There is no such thing as a fractional molecule, but we *can* have a fraction of a mole.

8.48 The mass of 1 mole of any molecule is the molecule's mass expressed in grams.

8.50 Molar mass allows us to form conversion factors between moles and grams.

8.51 6.022×10^{23}

8.52 6.022×10^{23}

8.54 (a) 3 moles
(b) 6 moles
(c) 2 times $6.022 \times 10^{23} = 1.204 \times 10^{24}$

8.56 The actual yield is the amount of product that is actually recovered after a chemical reaction.

8.58 $$\% \text{ yield} = \frac{\text{Actual yield}}{\text{Theoretical yield}} \times 100$$

8.61 (a) $2\text{ HCl} + \text{Zn} \longrightarrow H_2 + ZnCl_2$
(b) 2 moles of hydrogen chloride and 1 mole of zinc react to give 1 mole of molecular hydrogen and 1 mole of zinc chloride.
(c) Molar mass of HCl is (1.0079 g/mol H) + (35.453 g/mol Cl) = 36.461 g/mol HCl. Because the reaction calls for 2 moles of HCl, the mass of HCl needed is 72.922 g HCl. Molar mass of Zn is 65.39 g. This is the mass of 1 mole of Zn, which is what the equation calls for.
(d) Molar mass of H_2 is 2.0158 g, which is the theoretical yield.
(e) $$\frac{2.00\text{ g}}{2.0158\text{ g}} \times 100 = 99.2\%$$

8.63 $$\frac{6\text{ mol C atoms}}{1\text{ mol } C_6H_{12}O_6}, \frac{12\text{ mol H atoms}}{1\text{ mol } C_6H_{12}O_6}, \frac{1\text{ mol } C_6H_{12}O_6}{6\text{ mol O atoms}}, \frac{6\text{ mol C atoms}}{12\text{ mol H atoms}}$$

8.65 (a) $I_2 + 3\ Cl_2 \longrightarrow 2\ ICl_3$
(b) $$\frac{1\text{ mol } I_2}{3\text{ mol } Cl_2}, \frac{1\text{ mol } I_2}{2\text{ mol } ICl_3}, \frac{3\text{ mol } Cl_2}{2\text{ mol } ICl_3}, \frac{3\text{ mol } Cl_2}{1\text{ mol } I_2}, \frac{2\text{ mol } ICl_3}{1\text{ mol } I_2}, \frac{2\text{ mol } ICl_3}{3\text{ mol } Cl_2}$$
(c) $$\frac{2\text{ mol I atoms}}{1\text{ mol } I_2\text{ molecules}}, \frac{2\text{ mol Cl atoms}}{1\text{ mol } Cl_2\text{ molecules}}, \frac{1\text{ mol I atoms}}{1\text{ mol } ICl_3\text{ molecules}}, \frac{3\text{ mol Cl atoms}}{1\text{ mol } ICl_3\text{ molecules}}$$
plus four more that are the reciprocals of these four.

8.67 $$24.0\text{ g } O_2 \times \frac{1\text{ mol } O_2}{31.998\text{ g } O_2} = 0.750\text{ mol } O_2$$

8.68 $$24.0\text{ g } O_2 \times \frac{1\text{ mol } O_2}{31.998\text{ g } O_2} \times \frac{2\text{ mol O atoms}}{1\text{ mol } O_2} = 1.50\text{ mol O atoms}$$

8.71 $$1.00 \times 10^9\text{ molecules } H_2O \times \frac{1\text{ mol } H_2O}{6.022 \times 10^{23}\text{ molecules } H_2O} \times \frac{18.015\text{ g } H_2O}{1\text{ mol } H_2O} = 2.99 \times 10^{-14}\text{ g } H_2O$$

8.73 (a) $$20.0\text{ g } H_2O_2 \times \frac{1\text{ mol } H_2O_2}{34.014\text{ g } H_2O_2} \times \frac{2\text{ mol } H_2O}{2\text{ mol } H_2O_2} \times \frac{18.015\text{ g } H_2O}{1\text{ mol } H_2O} = 10.6\text{ g } H_2O$$
(b) $$20.0\text{ g } H_2O \times \frac{1\text{ mol } H_2O}{18.015\text{ g } H_2O} \times \frac{2\text{ mol } H_2O_2}{2\text{ mol } H_2O} \times \frac{34.014\text{ g } H_2O_2}{1\text{ mol } H_2O_2} = 37.8\text{ g } H_2O_2$$
(c) $$20.0\text{ g } O_2 \times \frac{1\text{ mol } O_2}{31.998\text{ g } O_2} \times \frac{2\text{ mol } H_2O_2}{1\text{ mol } O_2} \times \frac{34.014\text{ g } H_2O_2}{1\text{ mol } H_2O_2} = 42.5\text{ g } H_2O_2$$

8.75 The reaction was run with B as the limiting reactant.

8.77 To determine which is the limiting ingredient, multiply each given amount by the appropriate conversion factor:

$$25\text{ eggs} \times \frac{1\text{ cheesecake}}{5\text{ eggs}} = 5\text{ cheesecakes}$$

$$9\text{ blocks cream cheese} \times \frac{1\text{ cheesecake}}{3\text{ blocks cream cheese}} = 3\text{ cheesecakes}$$

$$4\text{ cups sugar} \times \frac{1\text{ cheesecake}}{1\text{ cup sugar}} = 5\text{ cheesecakes}$$

Because it gives the smallest number, the cream cheese is the limiting ingredient. You can make 3 cheesecakes.

8.78 (a) $$5.00\text{ g } H_2 \times \frac{1\text{ mol } H_2}{2.0158\text{ g } H_2} \times \frac{2\text{ mol } H_2O}{2\text{ mol } H_2} \times \frac{18.015\text{ g } H_2O}{1\text{ mol } H_2O} = 44.7\text{ g } H_2O$$
(b) $$5.00\text{ g } H_2O \times \frac{1\text{ mol } H_2O}{18.015\text{ g } H_2O} \times \frac{1\text{ mol } O_2}{2\text{ mol } H_2O} \times \frac{31.998\text{ g } O_2}{1\text{ mol } O_2} = 4.44\text{ g } O_2$$
(c) $$100.0\text{ g } H_2 \times \frac{1\text{ mol } H_2}{2.0158\text{ g } H_2} \times \frac{1\text{ mol } O_2}{2\text{ mol } H_2} \times \frac{31.998\text{ g } O_2}{1\text{ mol } O_2} = 793.7\text{ g } O_2$$
(d) $$50.0\text{ g } O_2 \times \frac{1\text{ mol } O_2}{31.998\text{ g } O_2} \times \frac{2\text{ mol } H_2O}{1\text{ mol } O_2} \times \frac{18.015\text{ g } H_2O}{1\text{ mol } H_2O} = 56.3\text{ g } H_2O$$

(e) $56.3\ \text{g H}_2\text{O} \times \dfrac{1\ \text{mol H}_2\text{O}}{18.015\ \text{g H}_2\text{O}}$

$\times \dfrac{6.022 \times 10^{23}\ \text{molecules H}_2\text{O}}{1\ \text{mol H}_2\text{O}}$

$= 1.88 \times 10^{24}\ \text{molecules H}_2\text{O}$

8.80 The balanced equation is: $3\ H_2 + N_2 \longrightarrow 2\ NH_3$

(a) $10.0\ \text{g H}_2 \times \dfrac{1\ \text{mol H}_2}{2.0158\ \text{g H}_2} \times \dfrac{1\ \text{mol N}_2}{3\ \text{mol H}_2}$

$\times \dfrac{28.014\ \text{g N}_2}{1\ \text{mol N}_2} = 46.3\ \text{g N}_2$

(b) $10.0\ \text{g H}_2 \times \dfrac{1\ \text{mol H}_2}{2.0158\ \text{g H}_2} \times \dfrac{2\ \text{mol NH}_3}{3\ \text{mol H}_2}$

$\times \dfrac{17.031\ \text{g NH}_3}{1\ \text{mol NH}_3} = 56.3\ \text{g NH}_3$

or

$46.3\ \text{g N}_2 \times \dfrac{1\ \text{mol N}_2}{28.014\ \text{g N}_2} \times \dfrac{2\ \text{mol NH}_3}{1\ \text{mol N}_2}$

$\times \dfrac{17.031\ \text{g NH}_3}{1\ \text{mol NH}_3} = 56.3\ \text{g NH}_3$

(c) $56.3\ \text{g NH}_3 \times \dfrac{1\ \text{mol NH}_3}{17.031\ \text{g NH}_3}$

$\times \dfrac{6.022 \times 10^{23}\ \text{molecules NH}_3}{1\ \text{mol NH}_3}$

$= 1.99 \times 10^{24}\ \text{molecules NH}_3$

8.82 (a) $Cl_2 + 3\ F_2 \longrightarrow 2\ ClF_3$
(b) F_2
(c) 4.10 mol ClF_3
(d) 379 g ClF_3
(e) 0.45 mol Cl_2
(f) 32 g Cl_2

8.83 (a) $2\ Na(s) + Br_2(l) \longrightarrow 2\ NaBr(s)$
(b) Na
(c) 22.3 g NaBr
(d) 12.7 g Br_2
(e) 65.9%

8.86 (a) $2\ C_4H_{10} + 13\ O_2 \longrightarrow 8\ CO_2 + 10\ H_2O$
(b) O_2
(c) 8.463 g CO_2, 4.331 g H_2O
(d) 7.21 g C_4H_{10}
(e) 25.78 g O_2

8.88 (a) C_2H_2
(b) C_4H_4
(c) C_6H_6

8.90 (a) 76.59% C; 6.390% H; 17.02% O
(b) C_6H_6O
(c) C_6H_6O

8.92 (a) 85.64% C; 14.36% H
(b) CH_2
(c) C_5H_{10}

8.94 (a) 24.78% C; 2.079% H; 73.13% Cl (b) CHCl

8.96 C_4H_4ON

8.98 KClO

8.100 38.7% C; 9.74% H; 51.6% O

8.102 23.19% C; 1.43% H; 1.80% N; 8.24% O; 65.34% I

8.104 (a) S atom; (b) 0.670 mol O_3

8.106 (a) $2\ Na(s) + 2\ H_2O(l) \longrightarrow 2\ NaOH(aq) + H_2(g)$
(b) Water.

8.109 (a) 342.29 g/mol
(b) 428 g sucrose
(c) 7.53×10^{23} sucrose molecules
(d) 1.66×10^{25} H atoms
(e) 27.8 g hydrogen

8.111 (a) $2\ Cu_2O(s) + C(s) \longrightarrow 4\ Cu(s) + CO_2(g)$
(b) 40.07 kg coke

8.113 (a) 4.78 g CO
(b) 3.4×10^2 g CO
(c) 0.171 mol CO
(d) Theoretical yield = 17.1 g CO; percent yield = 78.9%

8.115 (a) 0.0294 mol $AgNO_3$
(b) 0.0882 mol O
(c) 0.412 g N
(d) 1.77×10^{22} Ag atoms

8.117 (a) 0.33 mol $N_2(g)$
(b) 5.3 mol $N_2(g)$
(c) None left over
(d) 200 molecules $N_2(g)$
(e) 1.31 g $N_2(g)$
(f) 2.8 mg $N_2(g)$

8.118 $23.25

8.119 (a) $Ca_3P_2 + 6\ H_2O \longrightarrow 3\ Ca(OH)_2 + 2\ PH_3$
(b) 35.6 g H_2O (c) 22.4 g PH_3

8.122 39.2 g

8.124 6.98×10^{23} atoms

8.127 $PbCrO_4$

8.129 17.0 g

8.130 $C_6H_8O_6$

8.133 (a) $SiCl_4 + 2\ H_2O \longrightarrow SiO_2 + 4\ HCl$
(b) 139.8 g $SiCl_4$, 29.64 g H_2O (c) 49.43 g SiO_2

8.136 0.693 g CO_2, 0.196 g H_2O

8.138 1.38×10^{23} molecules

8.140 15.77% Al, 28.11% S, 56.11% O

8.142 $C_{10}H_8$

8.144 212 g

8.146 $C_8H_{10}N_4O_2$

8.149 113.7 g CO, 125.2 g CaC_2

8.151 (a) $S + 2\ H_2SO_4 \longrightarrow 3\ SO_2 + 2\ H_2O$ (b) 15.87 g

8.155 18.2 g CaF_2, 20.5 g N_2O, 16.8 g H_2O

8.157 (a) PbS (b) 12.1 g

8.159 2.06 g

8.161 3.98×10^{-11} g

8.165 12.6 g of Cu_2S

8.167 3.53 g

8.169 8.35×10^{20} aspirin molecules, 7.52×10^{21} C atoms

8.171 $C_8H_8O_3$

8.174 (a) H_2O (b) 7.9 g

Chapter 9

9.2 (a) $\ddot{\underset{\cdot\cdot}{O}}=C=\ddot{\underset{\cdot\cdot}{O}}$

C oxidation state is +4; each O oxidation state is −2

(b) None

(c) 4 valence electrons fewer

(d) It is correct to say that the C atom has a complete octet because assigning it no electrons in part (b) is only a technique for determining oxidation number and does *not* reflect the number of electrons actually around the C in the CO_2 molecule.

9.3 (a) $\begin{array}{c} H \\ H{:}\ddot{\underset{\cdot\cdot}{C}}{:}H \\ H \end{array}$

C oxidation state is −4; each H oxidation state is +1

(b) 8 electrons

(c) 4 valence electrons more

9.4 (a) $\begin{array}{c} H \\ {:}\ddot{\underset{\cdot\cdot}{Cl}}{:}\ddot{\underset{\cdot\cdot}{C}}{:}\ddot{\underset{\cdot\cdot}{Cl}}{:} \\ {:}\underset{\cdot\cdot}{Cl}{:} \end{array}$

C oxidation state is +2; H oxidation state is +1; each Cl oxidation state is −1

(b) 2 electrons

(c) 2 valence electrons fewer

(d) It is correct to say that the C atom has a complete octet because assigning it only two electrons in part (b) is only a technique for determining oxidation number and does *not* reflect the number of electrons actually around the C in the molecule.

9.6 Mg is +2 (rule 5); each Br is −1 (halide rule)

9.7 Each O is −2 (rule 2); each Fe is +3 (rule 7)

9.8 H is +1 (rule 3); O is −2 (rule 2); Cl is +1 (rule 7)

9.9 Each O is −2 (rule 2); C is +4 (rule 7)

9.10 Cu is +2 (rule 6)

9.12 This is a redox reaction. The oxidation number of Fe changes from 0 to +3 (each Fe loses 3 electrons), and the oxidation number of O changes from 0 to −2 (each O gains 2 electrons). The Fe is oxidized, and the O is reduced.

9.13 This is a redox reaction. The oxidation number of Ca changes from 0 to +2 (Ca loses 2 electrons), and the oxidation number of H changes from +1 to 0 (each H gains 1 electron). The Ca is oxidized, and the H is reduced.

9.14 This is not a redox reaction.

9.16 Fe is the reducing agent, and O_2 is the oxidizing agent.

9.17 Ca is the reducing agent, and H^+ is the oxidizing agent.

9.19 (a) Pb^{2+} (b) Ni

(c) Pb cathode (d) Ni anode

9.20 (a) Fe (b) Ag^+

(c)

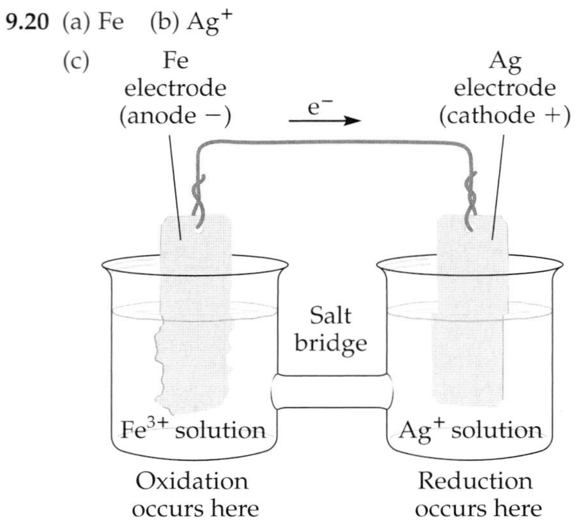

9.22 (a)

9.23

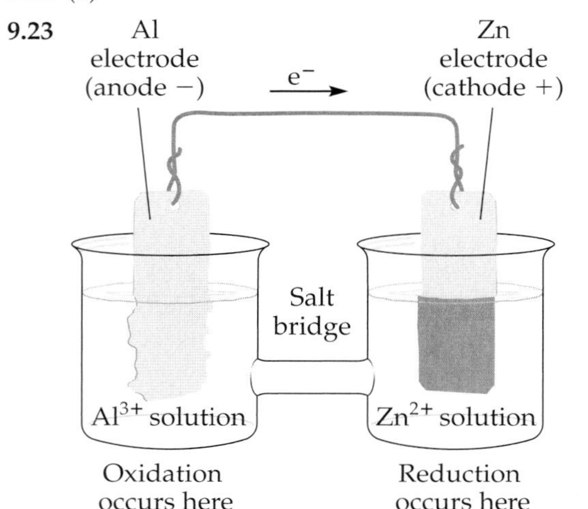

9.24 Lead is higher than copper in the EMF series. Therefore the oxidizing agent is Cu^{2+}, and the reducing agent is Pb.

9.26 By moving copper wire through a magnetic field. Coils of copper wire are wound around a rotor and spun in a magnetic field by a turbine powered by steam or moving water (hydroelectric).

9.28 To keep track of how many electrons an atom owns, so we can determine whether electron transfer has occurred during a chemical reaction.

9.30 Both methods are correct.

9.32 The more electronegative atom owns the shared electrons.

9.34 $H{:}\ddot{\underset{\cdot\cdot}{O}}{:}\ddot{\underset{\cdot\cdot}{O}}{:}H$

Each O owns 7 electrons (both of its lone pairs, both electrons from the O–H bonds because H is less electronegative than O, and 1 electron from the O–O bond). Each H owns 0 electrons because both electrons in the O–H bond go to O.

9.35

```
 ..   ..  -
[:O:N::O ]
 ..   ..
   :O:
    ..
```

Each O owns 8 electrons (all of its lone pairs and the electrons from the N–O bond because N is less electronegative than O). The N therefore owns 0 electrons.

9.37 (a)–(c) All have oxidation state 0.

9.39 The sum must always equal the overall charge on the molecule.

9.41 (a) More
(b) 2 more electrons

9.42 Add up all the other oxidation numbers of the atoms in the compound. Subtract this sum from the total charge on the species. The difference is the oxidation number of the remaining atom.

9.44 (a) O is −2; C is +2
(b) Each H is +1; each Cl is −1; C is 0
(c) H is +1; each O is −2; C is +2
(d) Each Cl is −1; Pt is +4

9.46 (a) The halide rule works for $AlCl_3$ because Cl is more electronegative than Al. This means both electrons from an Al–Cl bond go to Cl, giving the Cl eight electrons and therefore an oxidation state of −1.
(b) The halide rule doesn't work for ClO^- because Cl is less electronegative than O. The O therefore owns the electrons in the bond, and the Cl has a positive oxidation state instead of the −1 predicted by the halide rule.
(c) ClO^-: O is −2; Cl is +1
$AlCl_3$: each Cl is −1; Al is +3

9.48 (a) Two oxygens at −2 each and six hydrogens at +1 each (using shortcut rules) sums to +2. This means the sum of the oxidation numbers of the three carbons must be −2 to get neutrality (the molecule is neutral). This is not enough information to assign the carbons a unique set of oxidation numbers.

(b)

```
         +1  -2          +1
         H  :O:          H
   +1 -3 |   ||    -2    | -2 +1
      H—C—C—O—C—H
         |  +3   ..      |
         H               H
         +1              +1
```

9.51 If the oxidation state of any atom changes during a reaction, a redox reaction has occurred.

9.53 The oxidation state increases.

9.55 Whenever electron transfer occurs, there must be both reduction and oxidation occurring simultaneously; hence the name redox.

9.58 (b), (c), and (d)

9.59 (a) For reaction (b): Fe is oxidized; N in NO_3^- is reduced.
For reaction (c): C in C_2H_6 is oxidized; O in O_2 is reduced.
For reaction (d): Br in AgBr is oxidized; Ag in AgBr is reduced.
(b) For reaction (b): NO_3^- is the oxidizing agent.
For reaction (c): O_2 is the oxidizing agent.
For reaction (d): AgBr is the oxidizing agent.
(c) For reaction (b): Fe is the reducing agent.
For reaction (c): C_2H_6 is the reducing agent.
For reaction (d): AgBr is the reducing agent.

9.61 Yes, because it is an electron-transfer (redox) reaction.

9.63 (a) In the hydroquinone the two carbons bonded to the oxygens are oxidized from +1 to +2.
(b) Ag is reduced.
(c) The oxidizing agent is AgBr.
(d) The reducing agent is hydroquinone.

9.65

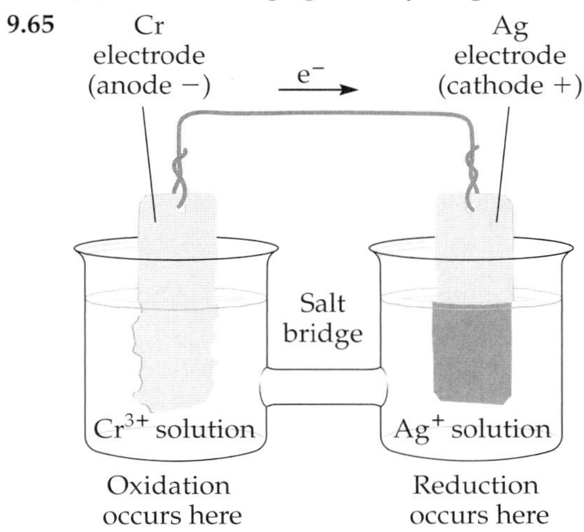

9.66 (a) Ag^+ is the oxidizing agent.
(b) Cr is the reducing agent.
(c) Ag electrode
(d) Cr electrode

9.69 (a) The tin (Sn) is getting oxidized (from 0 to +2).
(b) The Cu^{2+} is getting reduced (from +2 to 0).
(c)

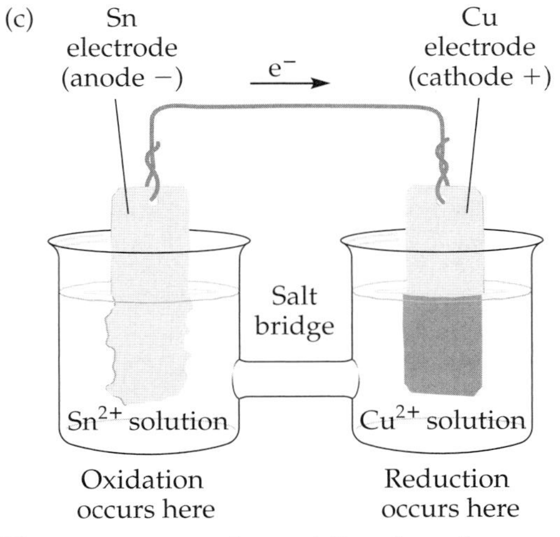

9.71 Electrons are negative and flow from the negative electrode (like charges repel) to the positive electrode (unlike charges attract). Because electrons always flow from the anode to the cathode, the anode must be negative and the cathode positive.

Or, you can remember that a *cat*ion is positive (so is a *cat*hode) and an *an*ion is negative (so is an *an*ode).

9.72 (a) $B + A^+ \longrightarrow B^+ + A$
(b) The oxidizing agent is A^+.
(c) The reducing agent is B.
(d) Electrons flow from B to A^+.

9.74 You would have to do competitive redox experiments, two metals at a time. For example, you could first test metals X and Y, and see which electrode gets eaten away (and which one gains mass). The metal that gets eaten away (let's say it's X) would be the more active metal. Next, you would repeat the experiment for Y and Z; then again for X and Z. By comparing the results, you could properly order X, Y, and Z according to their relative activities.

9.77 Nothing happens.

9.79 Reaction (a) is spontaneous because Ag is more active than Au.

9.80 (a)

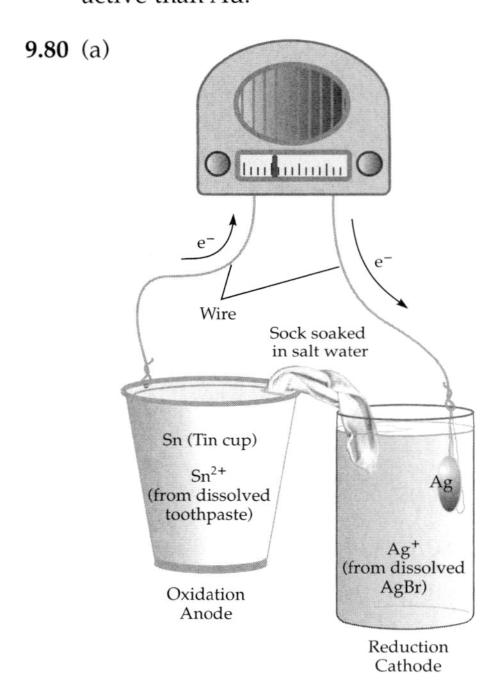

(b) The tin metal would be eaten away as Sn was oxidized to Sn^{2+}.
(c) Ag^+ is reduced, meaning it is the oxidizing agent.
(d) Sn is oxidized, making it the reducing agent.

9.82 (a) $Pb + Cu^{2+} \longrightarrow Pb^{2+} + Cu$
(b) $Pb^{2+} + Cu \longrightarrow Pb + Cu^{2+}$

9.86 The next best metal would be Pt (platinum) because it is the next least active metal.

9.87 The Statue of Liberty is made of sculpted sheets of copper metal placed over a steel (iron) skeleton. Iron is higher on the EMF series than copper. When the copper sheets began to oxidize to Cu^{2+}, the iron skeleton gave up its electrons to reduce the copper ions. Therefore, the iron skeleton was being eaten away and converted to ions. (The asbestos placed in the Statue to separate the copper and the iron had rotted away.) The solution to the problem was to replace the dissolving steel skeleton with a reinforced, corrosion-resistant, stainless steel skeleton incorporating high-tech plastics to insulate it from copper.

9.88 If the casing is not made thick enough, the conversion of Zn to Zn^{2+} could result in weak spots along the surface of the battery.

9.89 +7

9.91 C in C_8H_{18} oxidized (from either −3 to +4 or −2 to +4); O in O_2 reduced (from 0 to −2).

9.93 (c) and (e)

9.95 Li^+

9.97 Lower.

9.99 (a) and (c). In (a), I^- is the reducing agent and SeO_3^{2-} is the oxidizing agent. In (c), HCl is the reducing agent and O_2 is the oxidizing agent.

9.101

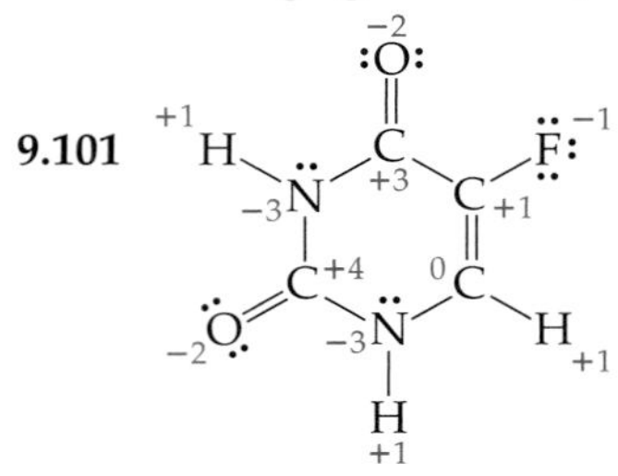

9.103 (a) −3 for both C. (b) −2 for both C.
(c) −1 for both C.

9.104 (a) Cd^{2+} (b) Mn (c) Cd (d) Mn

9.107 (a) P +5, each F −1
(b) Each Mo +6, each O −2
(c) H +1, Pb +2, each O −2
(d) H +1, each C +3, each O −2

9.109 (a) Cathode (b) Anode

9.110 H(+1)—N(−3)(—H +1)—C(−1)(H +1)(H +1)—C(+3)(=O −2)—O(−2)—H(+1)

9.112 The Al metal gives up electrons to Ag^+ in Ag_2S, reducing Ag^+ to Ag metal and destroying the tarnish.

9.115 H—C≡N:
+1 +2 −3

9.116 (a) Pb (b) Pd (c) $Pb + Pd^{2+} \longrightarrow Pb^{2+} + Pd$

9.118 Oxidizing agent NO_3^-; reducing agent Zn.

9.120 (a) Zinc is the anode; mercury is the cathode.
(b) The zinc anode.

9.121 (b) and (c). In (b), H_2 is the reducing agent and Cl_2 is the oxidizing agent. In (c), Si is the reducing agent and Cr_2O_3 is the oxidizing agent.

9.123 (a) Na +1, P +5, each O −2.
(b) B +3, each O −2, each H +1.
(c) Each V +5, each O −2.
(d) Each K +1, Ti +4, each F −1.

9.125 (a) Ti (b) Zn (c) Above
(d) $Ti + Zn^{2+} \longrightarrow Ti^{2+} + Zn$

Chapter 10

10.2 HF

10.3 [structure: hydrogen bonding between NH_3 molecules with δ+ on H and δ− on N]

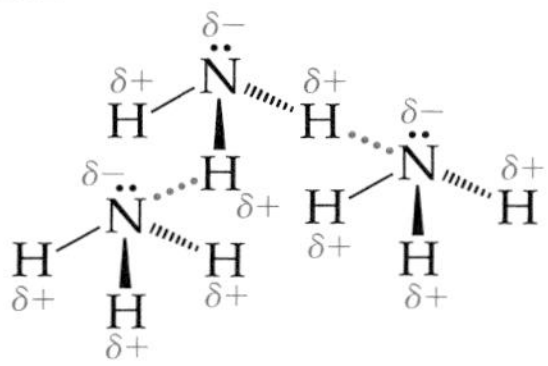

10.4 H_2O

10.6 Both ethylene and polyethylene are nonpolar molecules. Therefore the only intermolecular forces involved are London forces. Polyethylene is a very large molecule containing many electrons, which means substantial London forces between molecules and consequently a high melting point, making this substance a solid at room temperature. The much smaller ethylene molecule contains a relatively small number of electrons. This means weak London forces between molecules and low melting and boiling points, making this substance a gas at room temperature.

10.7 (a) H:N̈:H   H:P̈:H (each with a third H below)
(b) NH_3 is polar; PH_3 is nonpolar.
(c) PH_3 has the stronger London forces because it has more electrons.
(d) PH_3 because it has stronger London forces. However, this is not the case. See Practice Problem 10.8 for more information.

10.9 [structure: hydrogen bonding among NH_3 molecules]

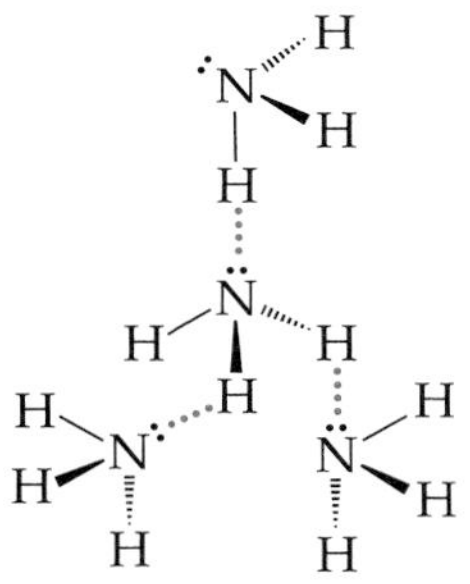

10.10 (a) Because HF is the only one of the four where the molecules in the liquid form hydrogen bonds.
(b) Because HI has more electrons than HBr and HCl and therefore stronger London forces.

10.12 The forces holding these solids together—metallic bonds in iron and ionic bonds in NaCl—are both true chemical bonds, and a large amount of heat energy is therefore required to melt either substance.

10.13 Intermolecular London forces.

10.14 (a) Metallic (b) Metallic (c) Ionic
(d) Network covalent (e) Molecular

10.16 Cooling a gas slows the molecules down to a speed at which they are not moving fast enough to overcome the attractive forces that will make them condense to a liquid.

10.18 Yes. Molecules in the liquid phase are still very much in motion. They are jostling past one another constantly but remain in close proximity to one another.

10.20 Water has stronger intermolecular attractive forces because it will condense into the liquid phase with less cooling than nitrogen requires.

10.22 Gas molecules are not bound to one another as are the molecules of solids and liquids. Gas molecules are essentially independent of one another and can travel around in the entire volume of their container.

10.24 (a) C–D. The larger dipole moment means stronger intermolecular attractive forces, which in turn means less energy needs to be removed from the liquid in order to form a solid.
(b) C–D; the stronger intermolecular attractive forces mean more energy must be added to the liquid to allow the molecules to overcome the attractive forces and escape into the gas phase.

10.25 $\overset{\delta+}{H}$—$\overset{\delta-}{Br}$ ··· $\underset{\delta+}{H}$—$\underset{\delta-}{Br}$ ··· $\underset{\delta+}{H}$—$\underset{\delta-}{Br}$

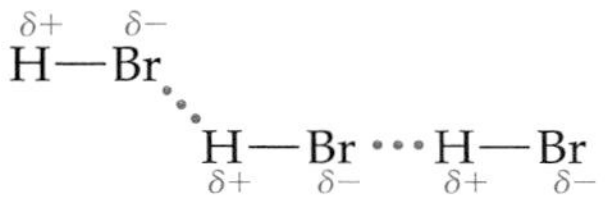

Dipole–dipole interactions.

10.27 CBr_4; because it is the larger molecule, it has more electrons and therefore stronger London forces.

10.28 (1) Chloromethane is a polar molecule and therefore subject to dipolar interactions.
(2) Chloromethane is a larger molecule and therefore has stronger London forces.

10.30 The strength of London forces depends on the size of the molecule, so that the London forces in some large nonpolar molecules can be stronger than the dipole–dipole forces in some small polar molecules.

10.31 (a) HF, because the electronegativity difference is greater.
(b) HF, because the partial charges on the H and F atoms are greater than in HCL.
(c) HF. The larger dipole moment means stronger intermolecular attractive forces, which in turn mean less energy needs to be removed from the gas in order to form a liquid. The stronger intermolecular forces also mean more energy must be added to HF to allow liquid molecules to escape to the gas phase. Thus HF has the higher boiling point.
(d) Yes, both molecules experience London forces. HCl has greater London forces because it is a larger molecule with more electrons.

10.34 Because waxes are very large molecules, they have many electrons and therefore quite strong

London forces—strong enough to make them solid at room temperature.

10.36 The hydrogen atom has no electrons other than those it is sharing and resembles a naked proton when bonded to a very electronegative atom such as O, N, or F.

10.38 (a) Because the size of the molecules is increasing as you go from S to Se to Te and London forces are stronger.
(b) H_2O has a much higher boiling point than expected because of the relatively strong hydrogen-bonding present.

10.40 Because hydrogen bonds are not covalent bonds within a molecule (which we represent by solid lines). Hydrogen bonds are unusually strong intermolecular attractions.

10.42 DNA molecules must be very stable but still able to break into two strands when the DNA is involved in protein building. Covalent bonds would be too strong (the strands would not be able to break apart), and London forces would be too weak (the DNA could not be kept intact).

10.43 If a nonpolar molecule is very large with a large number of electrons, its London forces will be quite substantial and can be stronger than the dipole–dipole interactions present in small polar molecules.

10.45 Sometimes.

10.47 Because the nonmolecular solids are held together by much stronger forces—namely, covalent, ionic, or metallic bonds.

10.49 The high melting points of some metals, such as gold and iron, are evidence that metallic bonds can be as strong as ionic or covalent bonds.

10.51 (a) Metallic solid (b) Ionic solid
(c) Network solid (d) Molecular solid

10.53 Because the attraction is the result of opposite charges attracting each other. When two molecules are oriented with a $\delta+$ H near a $\delta-$ O, the opposite charges create a hydrogen bond:

H H
O $\delta-$
H $\delta+$
O
H

When two molecules are oriented with two $\delta+$ H near each other or two $\delta-$ O near each other, there are no hydrogen bonds formed:

O
H H $\delta+$
$\delta+$ H
O
H

No attraction

H H
O $\delta-$
$\delta-$ O—H
H

No attraction

10.54 The molecules move faster and faster.

10.56 (d)

10.58 (b) because this molecule is linear and nonpolar.

10.60 (d) because it contains no H–O, H–N, or H–F bonds.

10.62 (b). Choice (a) is ionic, and (c) is a metal; both therefore have high melting points and very high boiling points. (b) and (d) are both small covalent molecules, but only (b) is nonpolar.

10.64

O O
H H H H
O O
C H H
H_3C O
H H H
O O
H H

10.66 (a) $CH_4 < Cl{-}Cl < Br{-}Br < CBr_4$

H—C(H)(H)—H < Cl—Cl < Br—Br < Br—C(Br)(Br)—Br

(b) All four molecules are nonpolar, meaning the only intermolecular forces in the liquids are London forces. Because London forces depend on numbers of electrons, the boiling point ranking goes from smallest molecule to largest.

10.68 (a) and (d) because both are nonpolar molecules.

10.70 (a) and (c)

10.72 $C_2H_6 < CH_3OH < NaCl < SiO_2$

10.73 Ethylene glycol boils at the higher temperature because of all the hydrogen-bonding resulting from the –OH parts of the molecules.

10.75 (a) London forces.
(b) The strong covalent bonds in the diamond network are difficult to break, making diamond hard. In graphite, the London forces between sheets are very weak, allowing the sheets to slide easily over one another and making the substance soft and slippery.

10.77 In a piece of metal, the valence electrons are free to "swim" throughout the entire piece.

10.79 The great number of electrons in the very large eicosane molecules create London forces that are stronger than the hydrogen bonds in water.

10.80 Nothing, because melting involves breaking bonds between molecules (*inter*molecular bonds); it does not involve the covalent bonds inside the molecules (*intra*molecular bonds).

Chapter 11

11.2 $10.50 \text{ g H}_2 \times \frac{1 \text{ mol H}_2}{2.0158 \text{ g H}_2} = 5.209 \text{ mol H}_2$

11.3 (a) 5×760 mm Hg = 3800 mm Hg = 3.80×10^3 mm Hg

(b) $3.80 \times 10^3 \text{ mm Hg} \times \frac{1 \text{ cm}}{10 \text{ mm}} \times \frac{1 \text{ inch}}{2.54 \text{ cm}}$

= 150 inches Hg

11.4 Too low, because the air in the tube will push down on the mercury.

11.6 Use $P = \frac{nRT}{V}$ and solve for T: $T = \frac{PV}{nR}$

$$\text{Mol N}_2 = 3.00 \text{ g N}_2 \times \frac{1 \text{ mol N}_2}{28.014 \text{ g N}_2} = 0.107 \text{ mol N}_2$$

$$\text{Pressure in atm} = 450.5 \text{ mm Hg} \times \frac{1 \text{ atm}}{760.0 \text{ mm Hg}} = 0.5928 \text{ atm}$$

$$T = \frac{0.5928 \text{ atm} \times 2.00 \text{ L}}{0.107 \text{ mol N}_2 \times \frac{0.0821 \text{ L} \cdot \text{atm}}{\text{K} \cdot \text{mol}}} = 135 \text{ K}$$

$$°\text{C} = 135 \text{ K} - 273.15 = -138°\text{C}$$

11.7 Use $P = \frac{nRT}{V}$ and solve for n: $n = \frac{PV}{RT}$

$$\text{Pressure in atm} = 40.0 \text{ lb/in.}^2 \times \frac{760 \text{ mm Hg}}{14.696 \text{ lb/in.}^2} \times \frac{1 \text{ atm}}{760 \text{ mm Hg}} = 2.72 \text{ atm}$$

$$\text{Volume in liters} = 10.5 \text{ gal} \times \frac{3.785 \text{ L}}{1 \text{ gal}} = 39.7 \text{ L}$$

$$T = 22.5°\text{C} + 273.15 = 295.7 \text{ K}$$

$$n = \frac{2.72 \text{ atm} \times 39.7 \text{ L}}{\frac{0.0821 \text{ L} \cdot \text{atm}}{\text{K} \cdot \text{mol}} \times 295.7 \text{ K}} = 4.45 \text{ mol O}_2$$

$$\text{grams O}_2 = 4.45 \text{ mol O}_2 \times \frac{31.998 \text{ g O}_2}{1 \text{ mol O}_2} = 142 \text{ g O}_2$$

11.10 $1000.0 \text{ gal} \times \frac{3.785 \text{ L}}{1 \text{ gal}} = 3785 \text{ L}$

K = °C + 273.15 = –4.50°C + 273.15 = 268.65 K

Molar mass of H_2 = 2.0158 g/mol

$$m = \frac{PV(MM)}{RT} = \frac{32.6 \text{ atm} \times 3785 \text{ L} \times 2.0158 \text{ g/mol}}{0.0821 \frac{\text{L} \cdot \text{atm}}{\text{K} \cdot \text{mol}} \times 268.65 \text{ K}}$$

$$= 1.13 \times 10^4 \text{ g H}_2 \times \frac{1 \text{ lb}}{453.6 \text{ g}} = 24.9 \text{ lb H}_2$$

11.11 You get the number of moles in the sample:

$$\frac{\text{Mass}}{\text{Molar mass}} = \frac{\text{Grams}}{\text{Grams/mol}} = \text{Gram} \div \frac{\text{Grams}}{\text{Mol}} = \text{Grams} \times \frac{\text{Mol}}{\text{Grams}} = \text{Mol}$$

11.14 Step 1: $P_i = 2.00$ atm $\quad P_f = ?$
$T_i = 200.0$ K $\quad T_f = 220.0$ K
$V_i = 50.0$ mL $\quad V_f = 25.0$ mL
$n_i = ?$ $\quad n_f = ? = n_i$

Step 2: $\frac{P_iV_i}{n_iT_i} = \frac{P_fV_f}{n_fT_f} \longrightarrow \frac{P_iV_i}{n_iT_i} = \frac{P_fV_f}{n_iT_f} \longrightarrow \frac{P_iV_i}{T_i} = \frac{P_fV_f}{T_f}$

Step 3:

(a) $P_f = \frac{T_fP_iV_i}{V_fT_i} = \frac{220.0 \text{ K} \times 2.00 \text{ atm} \times 50.0 \text{ mL}}{25.0 \text{ mL} \times 200.0 \text{ K}}$
= 4.40 atm

(b) $50.0 \text{ mL} \times \frac{1 \text{ L}}{1000 \text{ mL}} = 0.0500 \text{ L}$

$25.0 \text{ mL} \times \frac{1 \text{ L}}{1000 \text{ mL}} = 0.0250 \text{ L}$

$P_f = \frac{220.0 \text{ K} \times 2.00 \text{ atm} \times 0.0500 \text{ L}}{0.0250 \text{ L} \times 200 \text{ K}}$
= 4.40 atm

(c) No.
(d) Because one of the units in R is liters.

11.16 The assumption is that the molecules of an ideal gas are moving independently of one another and feel no attractive forces from one another. We are justified in making this assumption because the molecules are moving very fast in straight-line paths and are relatively far from one another. If they attracted one another, they would not move in straight-line paths.

11.18 By colliding with the walls of its container.

11.21 760 mm Hg, 1 atm

11.23 (a) $29.7 \text{ inches Hg} \times \frac{2.54 \text{ cm}}{1 \text{ inch}} \times \frac{10 \text{ mm}}{1 \text{ cm}}$
= 754 mm Hg

(b) $754 \text{ mm Hg} \times \frac{1 \text{ atm}}{760 \text{ mm Hg}} = 0.992 \text{ atm}$

11.25 K = °C + 273.15
K = 22.0°C + 273.15 = 295.2 K

11.28 $600.2 \text{ lb/in.}^2 \times \frac{760.0 \text{ mm Hg}}{14.696 \text{ lb/in.}^2} \times \frac{1 \text{ atm}}{760.0 \text{ mm Hg}}$
= 40.84 atm

$$48.5 \text{ lb} \times \frac{453.6 \text{ g}}{1 \text{ lb}} \times \frac{1 \text{ mol C}_2\text{H}_2}{26.04 \text{ g}} = 845 \text{ mol C}_2\text{H}_2$$

11.30 The pressure decreases because gas pressure is directly proportional to temperature.

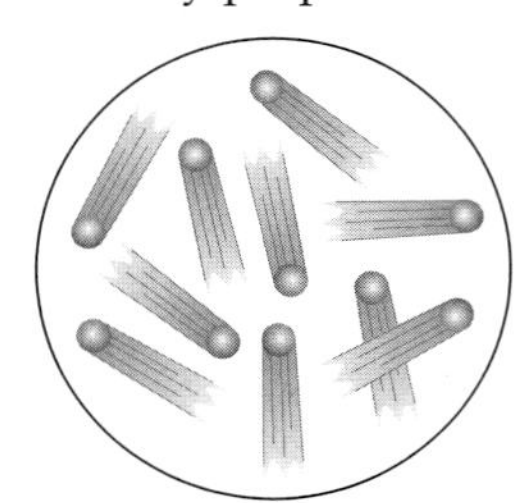

High-temperature, fast-moving molecules collide more frequently and more forcefully with container walls, creating higher pressure.

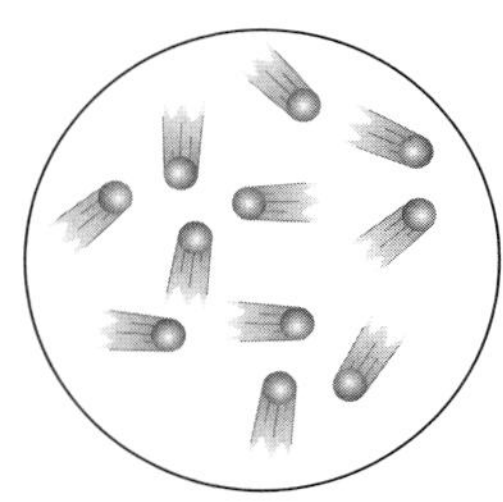

Low-temperature, slow-moving molecules collide less frequently and less forcefully with container walls, creating lower pressure.

11.32 (a) Sample 1 because it has the fewest molecules in the available space and therefore the molecules collide less frequently with the walls of the container.
(b) Sample 3 because it has the most molecules in the available space and therefore the molecules collide most frequently with the walls of the container.

11.33 Sample 2. In fact, the temperature of sample 2 has been lowered to the point where the gas has condensed to the liquid phase, which is why the molecules are clumped together at the bottom of the container and which is why we had to use quotation marks when referring to this liquid as a "gas."

11.34 The pressure of a gas is inversely proportional to its volume.

11.35 The pressure of a gas is directly proportional to its temperature.

11.37 Inverse proportionality means that as one value *increases* the other one *decreases*. Direct proportionality means that as one value *increases* the other one *also increases*.

11.39 Inverse proportion; as his age increases, the number of hairs on his head decreases.

11.41 L · atm/K · mole; the units are important because they indicate the units in which the volume, pressure, temperature, and quantity of gas must be expressed when using the ideal gas law.

11.43 $T = PV/nR$

11.45 You should get the value of R, which is 0.0821 L · atm/K · mol.

11.48 The temperatures must be expressed in kelvins when using the ideal gas law. To get the correct answer, he must convert 25.0°C to kelvins, double that number, then reconvert to °C:

$$\text{K} = 25.0°\text{C} + 273.15 = 298.2 \text{ K}$$

The pressure doubles when this absolute temperature is doubled to 596.4 K, which is equivalent to 323.3°C.

11.50 $\text{K} = 0.00°\text{C} + 273.15 = 273.15 \text{ K}$

$$V = \frac{nRT}{P}$$

$$= \frac{(2.0 \cancel{\text{mol}})\left(\dfrac{0.0821 \text{ L} \cdot \cancel{\text{atm}}}{\cancel{\text{K}} \cdot \cancel{\text{mol}}}\right)(273.15 \cancel{\text{K}})}{2.5 \cancel{\text{atm}}} = 18 \text{ L}$$

11.52 Because $V = nRT/P$, if $T = 0$, $V = 0$. The volume of the gas would be 0 L. That does not make sense. What really happens is that the gas becomes a liquid as it approaches this temperature, and the ideal gas law no longer governs its behavior.

11.55 (a) $22.5°\text{C} + 273.15 = 295.7 \text{ K}$

$$V = \frac{nRT}{P}$$

$$= \frac{(1 \cancel{\text{mol}})\left(\dfrac{0.0821 \text{ L} \cdot \cancel{\text{atm}}}{\cancel{\text{K}} \cdot \cancel{\text{mol}}}\right)(295.7 \cancel{\text{K}})}{1.00 \cancel{\text{atm}}}$$

$$= 24.3 \text{ L } H_2$$

(b) $1 \cancel{\text{mol } H_2} \times \dfrac{2 \text{ mol HCl}}{1 \cancel{\text{mol } H_2}} = 2 \text{ mol HCl}$

$$V = \frac{(2 \cancel{\text{mol}})\left(\dfrac{0.0821 \text{ L} \cdot \cancel{\text{atm}}}{\cancel{\text{K}} \cdot \cancel{\text{mol}}}\right)(295.7 \cancel{\text{K}})}{1.00 \cancel{\text{atm}}}$$

$$= 48.6 \text{ L HCl}$$

11.57 44.0 g/mol

11.59 0.131 g

11.61 (a) By using the ideal gas equation in the form that gives density:

$$\text{Density} = \frac{m}{V} = \frac{P(MM)}{RT}$$

The values for P, R, and T are the same whether you are using the equation for the H_2 sample or the He sample. Because the molar mass of He is greater than the molar mass of H_2, the equation tells you that the density of the He must be greater than the density of the H_2.

(b) Because molar mass is in the numerator of the equation, the densities are directly proportional to the molar masses:

$$\frac{4.003 \text{ g/mol}}{2.0158 \text{ g/mol}} = 1.986$$

The density in the He balloon is 1.986 times the density in the H_2 balloon.

11.63 52.3 lb/in.^{2}

11.65 293.3°C

11.67 The pressure quadruples.

11.69 (a) 22.4 L (b) 24.5 L

11.71 1.34×10^3 g

11.73 Because the model assumes that the gas particles being studied do not interact with one another. If they don't interact, their identity cannot matter.
11.75 4.02×10^{-3} mol, 8.10×10^{-3} g
11.76 1.249 g/L
11.78 0.5191 g
11.81 64.2 g/mol
11.83 1.077×10^{22} atoms
11.85 The same in the two containers.
11.87 76.9 mL
11.88 26.3 atm
11.90 3.00 mol, 60.5 g
11.91 (a) V, n (b) n (c) V, T
11.94 0.0437 mol
11.96 (a) 121 L (b) 0.0383 mol (c) 0.0231 mol
11.99 612°C
11.101 (a) 298.2 K (b) 18.9 L (c) 0.993 atm
11.103 2.42 atm
11.104 28.0 g/mol
11.106 2.51 g/mol
11.107 (a) 6.123 mol (b) 151.4 L (c) 137.3 L
11.110 115.1°C
11.112 354 mL
11.114 8.40 atm
11.117 34.1 g/mol

Chapter 12

12.2 (a) Water is the solvent; acetone is the solute.
(b) Nitrogen is the solvent; oxygen, water vapor, and other gases are the solutes.
(c) Iron is the solvent; chromium, nickel, and carbon are the solutes.
(d) Water is the solvent; aspirin is the solute.

12.3 Because 135-proof vodka is 67.5% alcohol and 32.5% water, alcohol is the major component and therefore the solvent.

12.5 The energy released in dissolving NaCl will be less in carbon tetrachloride than in water. Because carbon tetrachloride is a nonpolar molecule, it has no dipoles that will attract ions. With little or no attraction between the Na^+ and Cl^- ions and the carbon tetrachloride molecules, there will be little or no solvation energy released.

12.6 The energy required for pulling apart the lattice would be greater for $MgCl_2$. The attraction of the +2 magnesium ions for the negative chloride ions would be stronger than the attraction of the +1 sodium ions because the attractive force between oppositely charged particles increases as the magnitude of the charge increases.

12.8 $\Delta E_{\text{solute separation}}$ would be larger for dissolving the solid ionic solute. Because gas molecules have very little attraction for one another and are already separated. $\Delta E_{\text{solute separation}}$ is negligible for a gaseous solute.

12.9 The more negative ΔE_{total}, the more negative $\Delta E_{\text{solvation}}$ is because solvation is the only negative step in the process. Because $\Delta E_{\text{solvation}}$ is a measure of how strongly the solute and solvent are attracted to each other, the more negative ΔE_{total}, the stronger the solute–solvent attraction and the more likely the solute will dissolve. Also, a very negative ΔE_{total} means that the solvation step releases more than enough energy to balance the energy absorbed in the solute-separation and solvent-separation steps.

12.10 (a) Because Mg is in group IIA of the periodic table and has two valence electrons that are lost when Mg forms an octet.
(b) $\Delta E_{\text{solute separation}}$ is a measure of the energy needed to break up the lattice of the solute into individual ions. Because the Mg^{2+} ions have a stronger attraction for the Cl^- ions than do the Na^+ ions, more energy is needed to break the $MgCl_2$ lattice, and $\Delta E_{\text{solute separation}}$ is greater for $MgCl_2$.
(c) $MgCl_2$ is more soluble in water than NaCl because even though $\Delta E_{\text{solute separation}}$ is greater for $MgCl_2$ than for NaCl, $\Delta E_{\text{solvation}}$ is much larger for Mg^{2+} than for Na^+ because the strength of the ion–dipole attractions with water increases as the magnitude of the charge increases.

12.12
$$500.0 \text{ mL water} \times \frac{1.000 \text{ g water}}{1 \text{ mL water}} \times \frac{56.7 \text{ g KCl}}{100 \text{ g water}} = 2.84 \times 10^2 \text{ g KCl}$$

12.13 The biggest difference is in the solute-separation step. Gases do not exist in a lattice, and the particles are already separated. Hence, $\Delta E_{\text{solute separation}}$ is essentially zero for gases.

12.15
$$25.0 \text{ g glucose} \times \frac{1 \text{ mol glucose}}{180.155 \text{ g glucose}} \times \frac{1 \text{ L solution}}{2.55 \text{ mol glucose}} \times \frac{1000 \text{ mL solution}}{1 \text{ L solution}} = 54.4 \text{ mL solution}$$

12.16
$$200.0 \text{ mL solution} \times \frac{1 \text{ L solution}}{1000 \text{ mL solution}} \times \frac{2.00 \text{ mol ethanol}}{1 \text{ L solution}} \times \frac{46.068 \text{ g ethanol}}{1 \text{ mol ethanol}} = 18.4 \text{ g ethanol}$$

12.17
$$400.0 \text{ mL solution} \times \frac{1 \text{ L solution}}{1000 \text{ mL solution}} \times \frac{2.00 \text{ mol NaCl}}{1 \text{ L solution}} \times \frac{58.443 \text{ g NaCl}}{1 \text{ mol NaCl}} = 46.8 \text{ g NaCl required;}$$

add water until a volume of 400.0 mL is reached.

12.18 First get moles of NaCl needed:

$$400.0\ \text{mL solution} \times \frac{1\ \text{L solution}}{1000\ \text{mL solution}} \times \frac{2.00\ \text{mol NaCl}}{1\ \text{L solution}} = 0.800\ \text{mol NaCl}$$

Next get volume of stock solution that contains this many moles of NaCl:

$$0.800\ \text{mol NaCl} \times \frac{1\ \text{L solution}}{3.00\ \text{mol}} \times \frac{1000\ \text{mL solution}}{1\ \text{L solution}} = 2.67 \times 10^2\ \text{mL stock solution}$$

267 mL of stock solution is needed; place this volume of the stock solution in a 400-mL volumetric flask and add water until a volume of 400.0 mL is reached.

You can also solve this problem with the dilution equation:

$$V_{\text{stock}} = \frac{2.00\ \text{M} \times 400.0\ \text{mL}}{3.00\ \text{M}} = 267\ \text{mL}$$

12.20 $$\text{Percent by volume} = \frac{90.0\ \text{mL ethanol}}{1000\ \text{mL solution}} \times 100 = 9.00\ \text{vol \%}$$

12.21 $$65.0\ \text{g solution} \times \frac{12.5\ \text{g sucrose}}{100\ \text{g solution}} = 8.12\ \text{g sucrose}$$

12.24 $$255.0\ \text{mL solution} \times \frac{1\ \text{L solution}}{1000\ \text{mL solution}} \times \frac{0.998\ \text{mol glucose}}{1\ \text{L solution}} = 0.254\ \text{mol glucose}$$

12.25 $$0.500\ \text{mol BaCl}_2 \times \frac{1\ \text{L solution}}{0.350\ \text{mol BaCl}_2} = 1.43\ \text{L solution}$$

12.26 $$0.500\ \text{mol Cl}^- \times \frac{1\ \text{mol BaCl}_2}{2\ \text{mol Cl}^-} \times \frac{1\ \text{L solution}}{0.350\ \text{mol BaCl}_2} = 0.714\ \text{L solution}$$

12.28 $$\frac{8.24\ \text{g}}{9.70\ \text{g}} \times 100 = 84.9\%\ \text{yield}$$

12.29 Step 1: $Fe^{3+}(aq) + 3\ OH^-(aq) \longrightarrow Fe(OH)_3(s)$

Step 2: $$20.0\ \text{g Fe(OH)}_3 \times \frac{1\ \text{mol Fe(OH)}_3}{106.9\ \text{g Fe(OH)}_3} = 0.187\ \text{mol Fe(OH)}_3$$

Step 3: $$0.187\ \text{mol Fe(OH)}_3 \times \frac{1\ \text{mol Fe}^{3+}}{1\ \text{mol Fe(OH)}_3} = 0.187\ \text{mol Fe}^{3+}$$

$$0.187\ \text{mol Fe(OH)}_3 \times \frac{3\ \text{mol OH}^-}{1\ \text{mol Fe(OH)}_3} = 0.561\ \text{mol OH}^-$$

Step 4: $$0.187\ \text{mol Fe}^{3+} \times \frac{1\ \text{mol Fe(NO}_3)_3}{1\ \text{mol Fe}^{3+}} \times \frac{1\ \text{L Fe(NO}_3)_3\ \text{solution}}{0.250\ \text{mol Fe(NO}_3)_3} = 0.748\ \text{L Fe(NO}_3)_3\ \text{solution}$$

$$0.561\ \text{mol OH}^- \times \frac{1\ \text{mol Ba(OH)}_2}{2\ \text{mol OH}^-} \times \frac{1\ \text{L Ba(OH)}_2\ \text{solution}}{0.150\ \text{mol Ba(OH)}_2} = 1.87\ \text{L Ba(OH)}_2\ \text{solution}$$

Combine 0.748 L of the ferric nitrate solution with 1.87 L of the barium hydroxide solution, and then filter off the precipitated $Fe(OH)_3$. The amount to be collected for a 67.5% yield is

$$\text{Actual yield} = \frac{\%\ \text{yield} \times \text{Theoretical yield}}{100} = \frac{67.5\% \times 20.0\ \text{g}}{100} = 13.5\ \text{g}$$

12.31 Step 1: $H^+ + OH^- \longrightarrow H_2O$

Step 2: $$0.03860\ \text{L NaOH solution} \times \frac{0.100\ \text{mol OH}^-}{1\ \text{L NaOH solution}} = 0.00386\ \text{mol OH}^-$$

Step 3: $$0.00386\ \text{mol OH}^- \times \frac{1\ \text{mol H}^+}{1\ \text{mol OH}^-} \times \frac{1\ \text{mol H}_3\text{PO}_4}{3\ \text{mol H}^+} = 1.29 \times 10^{-3}\ \text{mol H}_3\text{PO}_4$$

Step 4: $$\frac{1.29 \times 10^{-3}\ \text{mol H}_3\text{PO}_4}{0.05000\ \text{L}} = 0.0258\ \text{M H}_3\text{PO}_4$$

12.32 (a) Step 1: $H^+ + OH^- \longrightarrow H_2O$

Step 2: $$0.02748\ \text{L NaOH solution} \times \frac{1.000\ \text{mol OH}^-}{1\ \text{L NaOH solution}} = 0.02748\ \text{mol OH}^-$$

Step 3: $$0.02748\ \text{mol OH}^- \times \frac{1\ \text{mol H}^+}{1\ \text{mol OH}^-} \times \frac{1\ \text{mol acid}}{1\ \text{mol H}^+} = 0.02748\ \text{mol acid}$$

Step 4: $$\frac{0.02748\ \text{mol acid}}{0.02500\ \text{L}} = 1.099\ \text{M acid}$$

(b) $$\text{Molar mass} = \frac{1.65\ \text{g acid}}{0.02748\ \text{mol acid}} = 60.0\ \text{g/mol}$$

(c) The molar mass of CH_2O is 30.0 g/mol, making the molecular formula twice the empirical formula, or $C_2H_4O_2$ (Section 8.5). Writing the one acidic H separately gives the formula $HC_2H_3O_2$. Because $C_2H_3O_2^-$ is the acetic ion, the acid is acetic acid.

12.35 Each mole of calcium nitrate dissociates in water to produce 3 moles of solute particles: $Ca(NO_3)_2 \longrightarrow Ca^{2+} + 2\ NO_3^-$.

$$100\ \text{g Ca(NO}_3)_2 \times \frac{1\ \text{mol Ca(NO}_3)_2}{164.1\ \text{g Ca(NO}_3)_2} \times \frac{3\ \text{mol solute particles}}{1\ \text{mol Ca(NO}_3)_2} = 1.828\ \text{mol solute particles}$$

The temperature at which this solution has a vapor pressure of 760 mm Hg is the temperature at which the solution boils, which you find from the expression

$$\Delta T_b = \frac{0.52 \dfrac{\text{C}^\circ \cdot \text{kg solvent}}{\text{mol solute particles}} \times 1.828\ \text{mol solute particles}}{0.4500\ \text{kg solvent}} = 2.1\ \text{C}^\circ$$

The elevated boiling point is therefore 102.1°C, and it is at this temperature that the vapor pressure is 760 mm Hg.

12.36 Table 12.2 gives 6.47°C as the normal freezing point of cyclohexane. The freezing point change is thus $6.47°\text{C} - 2.70°\text{C} = 3.77\ \text{C}° = \Delta T_f$. Rearranging the expression for ΔT_f gives

$$\text{Mol solute} = \frac{\Delta T_f \times \text{kg solvent}}{K_f} = \frac{3.77\ \text{C}^\circ \times 0.225\ \text{kg solvent}}{20.0 \dfrac{\text{C}^\circ \cdot \text{kg solvent}}{\text{mol solute}}} = 0.0424\ \text{mol solute}$$

The molar mass is

$$\frac{45.0\ \text{g}}{0.0424\ \text{mol}} = 1.06 \times 10^3\ \text{g/mol}$$

12.38 No, because the sample still consists of flour particles and sugar particles in a nonuniform mixture.

12.40 Yes, because it is a homogeneous mixture of N_2, O_2, water vapor, and other gases.

12.41 The solute is acetic acid; the solvent is water.

12.43 (a) Solution (b) Solution
(c) Heterogeneous mixture (d) Solution
(e) Heterogeneous mixture

12.45 A heterogeneous mixture because each solid exists in its own lattice and is not mixed into the lattice of the other solid.

12.46 The + end of the dipole vector is incorrectly drawn near the more electronegative oxygen atom. The head of the arrow should be near the oxygen, and the + end should be near the hydrogens.

12.48 Water would always exist in the gaseous phase if there were no attractive forces.

12.50 In melting there are no solute–solvent interactions to provide energy for the lattice of ions to break apart, whereas in dissolving in water, the attractions between the ions and the water molecules aid in the solution process, therefore requiring less external energy.

12.52 Making room for the solute particles absorbs energy because the attractive forces between solvent molecules must be overcome.

12.55 (1) Solute separation—overcoming the solute–solute attractive forces.
(2) Solvent separation—overcoming the solvent–solvent attractive forces.
(3) Solvation—forming solute–solvent attractive forces.

12.57 The associations formed between the ions and water molecules (hydration) help keep the ions from coming back together.

12.59 A water molecule could form hydrogen bonds with the alcohol:

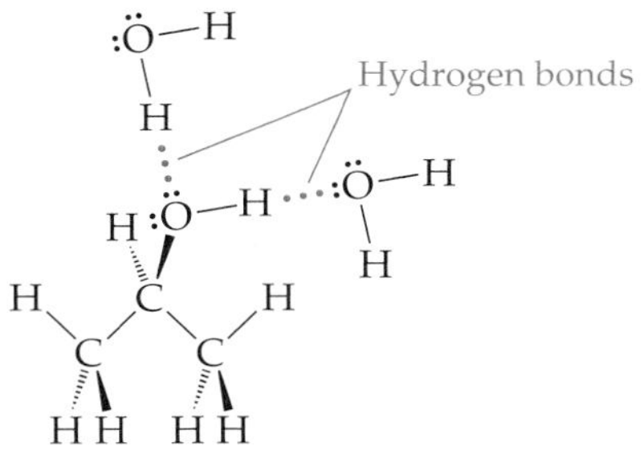

Hydrogen bonds

Yes, this interaction is hydration.

12.61 The solid would probably not be soluble in the liquid because the relatively low solvation energy means ΔE_{total} is a positive number.

12.63 When the energy released in the solvation step is less than the energy required in the solute-separation and solvent-separation steps, the solution components will remove heat from the water and the water will get colder.

12.65 Because hexane is a nonpolar compound, there will be little interaction between the ions and the solvent. This means little or no energy released in solvation, with the result that NaCl will not form a solution with C_6H_{14}.

12.67 $AlCl_3$, because the Al^{3+} ions have a much stronger interaction with water than the Na^+ ions.

12.69 "Like dissolves like" means that polar solutes dissolve in polar solvents and nonpolar solutes dissolve in nonpolar solvents.

12.71 (a) Because attractive forces between solute particles are being broken.
(b) Because attractive forces between solvent particles are being broken.
(c) Because attractive forces between solute particles and solvent particles are being formed.
(d) Soluble, because the total energy change is negative.
(e) Warmer, because the solvation step releases more energy than is absorbed by the solute-separation and solvent-separation steps.

12.72 The (*aq*) means that the ion is surrounded by and attracted to water molecules.

12.74 The solute-separation step because there is no lattice to be broken and the solute particles are already separated.

12.76 CCl_4 would be least soluble in water because it is nonpolar and will not form attractive forces with the polar water solvent molecules. CH_3OH would be most soluble because it can form hydrogen bonds with water molecules.

12.79 The second law of thermodynamics says that for anything to happen spontaneously, the entropy of the universe must increase.

12.81 Because the state of higher entropy is the one in which the dye is dispersed.

12.82 Entropy is important when the total energy change ΔE_{total} is slightly positive. When ΔE_{total} is either negative or very positive, entropy will not be the deciding factor.

12.84 Because NaCl is a polar compound and hexane is a nonpolar compound, they form no significant solute–solvent interactions.
The total energy change is too positive for entropy to be the deciding factor.

12.86 (a) Solubility increases; (b) solubility decreases.

12.88 The bends occurs when dissolved nitrogen in the blood is released too rapidly when a diver ascends too quickly. The decrease in pressure when the diver rises to the surface causes a decrease in the solubility of nitrogen, and the gas escapes rapidly.

12.90 NaCl shows very little increase in solubility as the temperature increases. Cerium sulfate becomes less soluble as the temperature increases.

12.91 Aquatic life suffers from oxygen depletion due to a decreased concentration of dissolved oxygen as the water is heated.

12.93 $$100.0 \text{ lb water} \times \frac{453.6 \text{ g water}}{1 \text{ lb water}} \times \frac{420.0 \text{ g sucrose}}{100.0 \text{ g water}} = 1.905 \times 10^5 \text{ g sucrose}$$

12.94 (a)

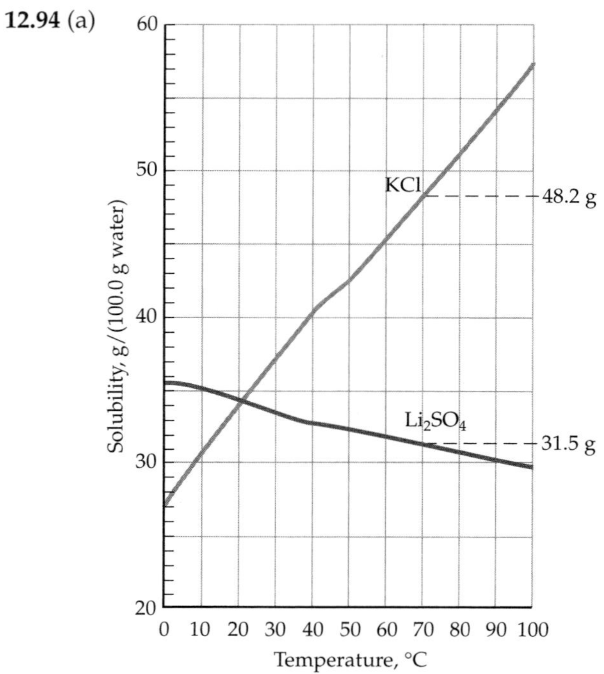

(b) KCl: 48.2 g/100.0 g water;
Li_2SO_4: 31.5 g/100.0 g water

(c) $$75 \text{ g water} \times \frac{48.2 \text{ g KCl}}{100.0 \text{ g water}} = 36 \text{ g KCl}$$

$$75 \text{ g water} \times \frac{31.5 \text{ g } Li_2SO_4}{100.0 \text{ g water}} = 24 \text{ g } Li_2SO_4$$

12.96 $$5.00 \text{ lb water} \times \frac{453.6 \text{ g water}}{1 \text{ lb water}} \times \frac{40.0 \text{ g KCl}}{100.0 \text{ g water}} = 907 \text{ g KCl}$$

12.98 $CH_3–CH_2–CH_2–CH_2–CH_2–CH_2–C(=O)–O^- Na^+$

Nonpolar hydrocarbon tail — Polar head

The nonpolar tail has an affinity for nonpolar molecules such as grease and oils. The polar head has an affinity for polar molecules such as water.

12.100 Because the nonpolar hydrophobic tail of each soap molecule needs to be isolated from water, the soap molecules form a sphere to allow the tails to reside inside and away from the water. If the micelle were flat and two-dimensional, the water would be in direct contact with the hydrophobic tails.

12.103 The hydrophobic portion is the hydrocarbon tail. The hydrophilic portion is the $–O–S(=O)_2–O^-\ Na^+$ head group.

12.104 Molarity is the number of moles of solute present in 1 L of solution: molarity = moles of solute/L of solution.

12.106 $1.00 \text{ L solution} \times \frac{1.00 \text{ mol } C_{12}H_{22}O_{11}}{1 \text{ L solution}} \times \frac{342.295 \text{ g } C_{12}H_{22}O_{11}}{1 \text{ mol } C_{12}H_{22}O_{11}} = 342 \text{ g } C_{12}H_{22}O_{11}$

Place 342 g of $C_{12}H_{22}O_{11}$ in the volumetric flask, dissolve in about 300 mL of water, and then add enough water to bring the total volume to 1.00 L.

12.108 $0.500 \text{ L solution} \times \frac{1.50 \text{ mol } C_{12}H_{22}O_{11}}{1 \text{ L solution}} \times \frac{342.295 \text{ g } C_{12}H_{22}O_{11}}{1 \text{ mol } C_{12}H_{22}O_{11}} = 257 \text{ g } C_{12}H_{22}O_{11}$

Place 257 g of sucrose in the volumetric flask, dissolve in about 250 mL of water, and then add enough water to bring the volume to 0.50 L.

12.110 You would fire him. To prepare a 2.00 M solution of NaCl, enough water should be added to 116.886 g of NaCl to bring the total volume of solution to 1.00 L. Adding exactly 1.00 L of water will not result in a solution volume of 1.00 L, which means the NaCl concentration will not be 2.00 M.

12.112 $45.0 \text{ mL} \times \frac{1 \text{ L}}{1000 \text{ mL}} \times \frac{0.250 \text{ mol sucrose}}{1 \text{ L}} \times \frac{342.295 \text{ g sucrose}}{1 \text{ mol sucrose}} = 3.85 \text{ g sucrose}$

12.114 $100.0 \text{ g glucose} \times \frac{1 \text{ mol glucose}}{180.1548 \text{ g glucose}} \times \frac{1 \text{ L solution}}{0.250 \text{ mol glucose}} \times \frac{1000 \text{ mL solution}}{1 \text{ L solution}} = 2.22 \times 10^3 \text{ mL}$

12.116 $V_{\text{stock solution}} = \frac{4.00 \text{ M} \times 100.0 \text{ mL}}{4.50 \text{ M}} = 88.9 \text{ mL}$

Instructions: Put 88.9 mL of stock solution in a 100-mL volumetric flask, and add enough water to bring the total volume to 100.0 mL.

12.118 (1) Percent by mass $= \frac{\text{Grams of solute}}{\text{Grams of solution}} \times 100$

(2) Percent by volume $= \frac{\text{Volume of solute}}{\text{Volume of solution}} \times 100$

(3) Percent by mass/volume $= \frac{\text{Grams of solute}}{\text{Milliliters of solution}} \times 100$

12.120 $\frac{22.5}{22.5 + 49.6} \times 100 = 31.2 \text{ mass } \%$

12.122 $2.00 \text{ kg solution} \times \frac{1000 \text{ g solution}}{1 \text{ kg solution}} \times \frac{30.0 \text{ g NaCl}}{100 \text{ g solution}} = 600 \text{ g NaCl}$

Place 600 g of NaCl in a flask. Add 1.40 kg of water to give a total of 2.00 kg of solution.

12.124 $200.0 \text{ mL alcohol} \times \frac{100.0 \text{ mL solution}}{35.00 \text{ mL alcohol}} = 571.4 \text{ mL solution}$

12.126 A sphere of water molecules surrounding a dissolved ion.

12.128 (a) $Pb^{2+}(aq) + S^{2-}(aq) \longrightarrow PbS(s)$
(b) 41.8 mL $Pb(NO_3)_2$ solution, 27.9 mL Na_2S solution
(c) 83.6 mL $Pb(NO_3)_2$ solution, 55.8 mL Na_2S solution

12.130 (a) $2\ Fe^{3+}(aq) + 3\ CO_3^{2-}(aq) \longrightarrow Fe_2(CO_3)_3(s)$
(b) 19.0 g
(c) 0.263 M CO_3^{2-}

12.132 (a) $H^+(aq) + OH^-(aq) \longrightarrow H_2O(l)$
(b) 0.1733 M

12.134 (a) 0.03686 M (b) 74.1 g/mol

12.135 (a) −115.0°C
(b) −115.0°C
(c) 78.4°C
(d) 78.4°C
(e) Because the temperature of a substance stays constant while the substance is changing phase at its freezing or boiling point.
(f) Ethanol

12.137 To break the hydrogen bonds holding the H_2O molecules together in the liquid phase.

12.139 The pressure exerted by the gas phase of a substance when the gas phase is in dynamic equilibrium with the liquid phase of the substance.

12.141 Reduce the air pressure above the water surface to 17.54 mm Hg.

12.143 760 mm Hg

12.145 336 g

12.147 28.1 g

12.148 (a) 102.5°C (b) −8.76°C

12.150 (a) 194 g/mol (b) $C_8H_{10}N_4O_2$

12.152 $Mg(OH)_2(s)$ releases the most, $CO_2(g)$ the least.

12.154 The lattice in $Mg(OH)_2$.

12.156 (c)

12.157 1.50×10^{-4} mol

12.159 Pour 14.6 mL of th stock solution into a 250-mL volumetric flask and dilute to the mark with water.

12.161 (a) 0.0400 mol Na^+, 0.0400 mol Cl^-
(b) 0.210 mol K^+, 0.0700 mol PO_4^{3-}
(c) 0.288 mol Al^{3+}, 0.864 mol NO_3^-
12.163 (a) 0.196 M (b) 1.96×10^{-4} mol
(c) 3.09 mass %
12.165 $C_6H_8O_6$
12.167 0.83 gallon
12.169 (a) 0.0714 M (b) 0.214 M (c) 0.0714 M
12.171 300 mL
12.173 245 mL; Na^+ and $C_2H_3O_2^-$
12.174 (a) 9.192×10^{-3} M (b) 193 g/mol
12.176 (a) Solvent copper, solute zinc.
(b) Solvent water, solute ammonia.
(c) Solvent sucrose, solute water.
12.179 More soluble in CCl_4 than in H_2O because I_2 and CCl_4 are both nonpolar, H_2O is polar, and like dissolves like.
12.181 The Al^{3+}–PO_4^{3-} bonds in the $AlPO_4$ lattice are stronger than the Na^+–PO_4^{3-} bonds in the Na_3PO_4 lattice.
12.182 The ethanol solvation step releases a lot more energy than does the diethyl ether solvation step because more hydrogen-bonding is possible in ethanol:

H—O O—H
H H
O
CH_3 CH_3

H—O H
H O—H
CH_3—CH_2—O—H
CH_3 H
H—O

12.184 A solute can dissolve even when the energy *released* by the solute–solvent interactions is less than the energy absorbed by the solute–solute and solvent–solvent interactions because the entropy of the system *increases*.
12.186 (a) is correct because it says *s* is directly proportional to *P*—*s* increases as *P* increases and decreases as *P* decreases. (b) wrongly says *s* is inversely proportional to *P*—*s* increases as *P* decreases and decreases as *P* increases. (c) wrongly says *s* does not depend at all on *P* because the two *P* cancel, leaving you with the incorrect $s = k$.
12.188 Yes.
12.190 The higher temperature makes the H_2O molecules move faster, on average, giving them more energy. The more energetic molecules collide more forcefully with the solid and thus break more solute–solute bonds.
12.192 The molecules migrate to the surface and orient as they do so that the hydrophobic tails are as far away from the water as possible.
12.194 (a) H—C(H)(H)—C(=O)—$\ddot{O}$:$^-$ K^+
(b) Both consist of a cation plus an anion containing a —C(=O)—O^- unit. The potassium acetate does not have the hydrocarbon tail found in the soap molecule.
(c) O=C(—CH_2—CH_2—CH_2—CH_2—CH_2—CH_2—CH_2—CH_2—CH_3)—O—Ca^{2+}—O—C(=O)—CH_2—CH_2—CH_2—CH_2—CH_2—CH_2—CH_2—CH_2—CH_3

$Ca(C_{10}H_{19}O_2)_2$

12.196 (a) 0.0750 M (b) 0.0750 M (c) 0.225 M
12.198 0.188 M
12.200 2.80 g
12.202 0.68 M
12.204 33.3 mL
12.207 0.498 mass %
12.209 55 g
12.211 1.30×10^4 g
12.213 (a) 0.194 g (b) 0.00210 M
12.214 (a) $Fe^{3+}(aq) + 3\ OH^-(aq) \longrightarrow Fe(OH)_3(s)$
(b) 16.0 g
(c) 68.4%
(d) 0.17 M

Chapter 13

13.2 (a)

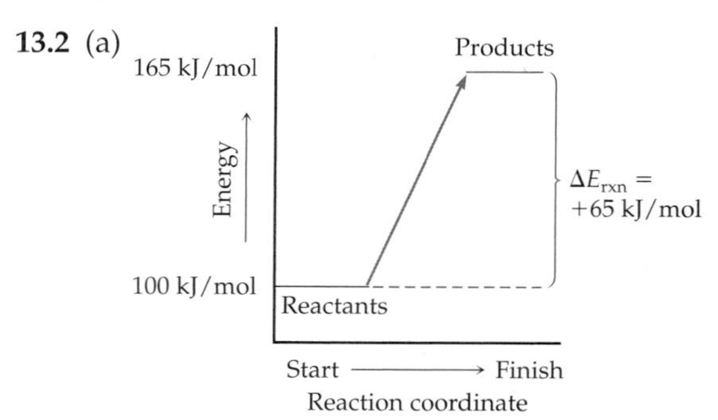

(b) Uphill (c) $\Delta E_{rxn} = +65$ kJ/mol
(d) Cold, because the reaction is endothermic.
13.3 (a) Higher energy; because ΔE_{rxn} is negative, the reactants are higher on the reaction-energy profile.
(b) 470 kJ/mol (450 kJ up from the 20 kJ/mol of the products)

(c) Exothermic; energy is being released during the reaction because the sign of ΔE_{rxn} is negative.
(d) Downhill
(e)

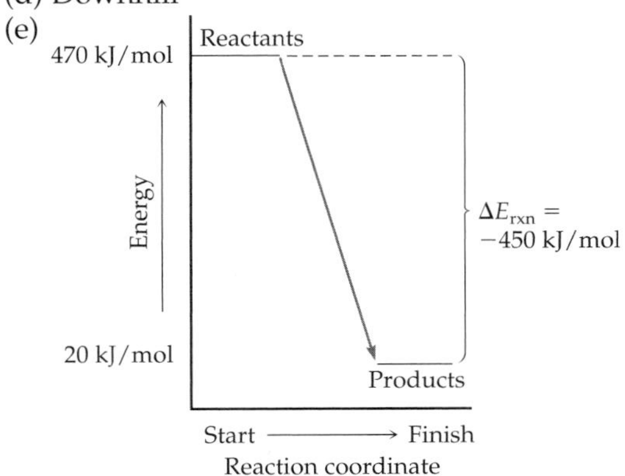

13.5 (a) $\Delta E_{forward\ rxn} = +250$ kJ/mol
(b) $\Delta E_{reverse\ rxn} = -250$ kJ/mol

13.6 (a) $\Delta E_{forward\ rxn} = -250$ kJ/mol
(b) $\Delta E_{reverse\ rxn} = +250$ kJ/mol

13.9 200 kJ

13.10 Practice Problem 13.8:

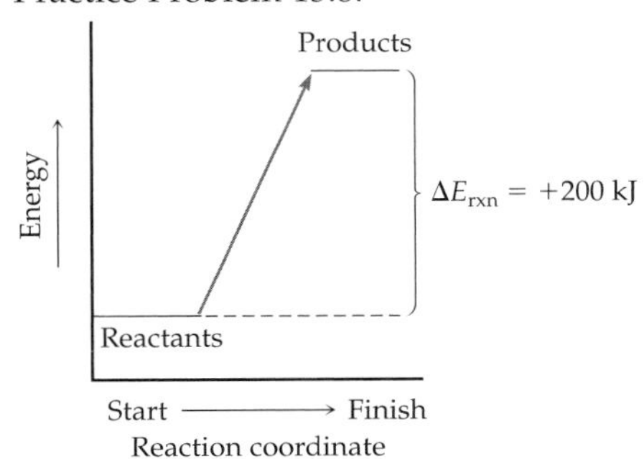

Practice Problem 13.9:

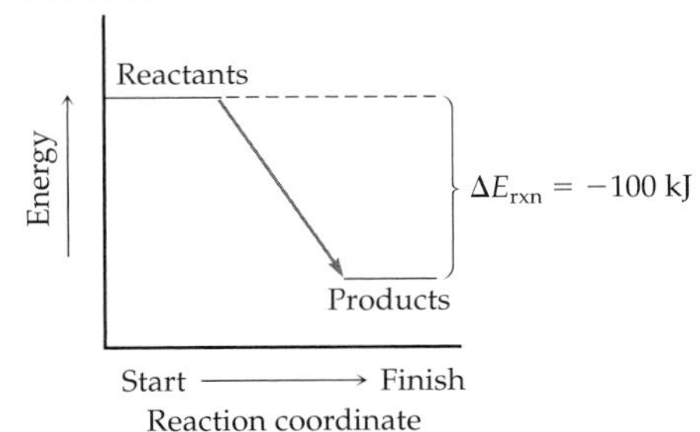

13.12 Endothermic reaction A

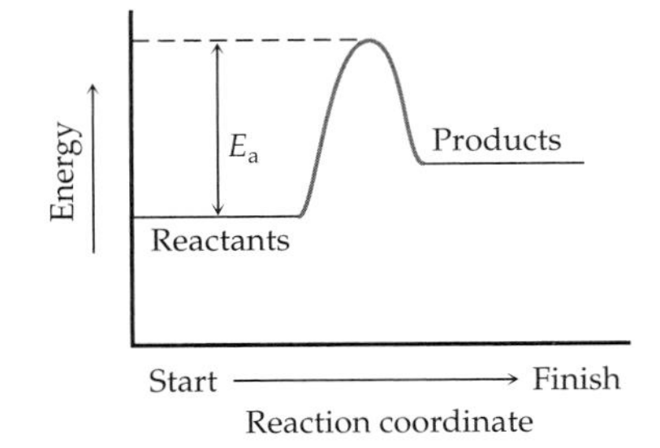

Exothermic reaction B

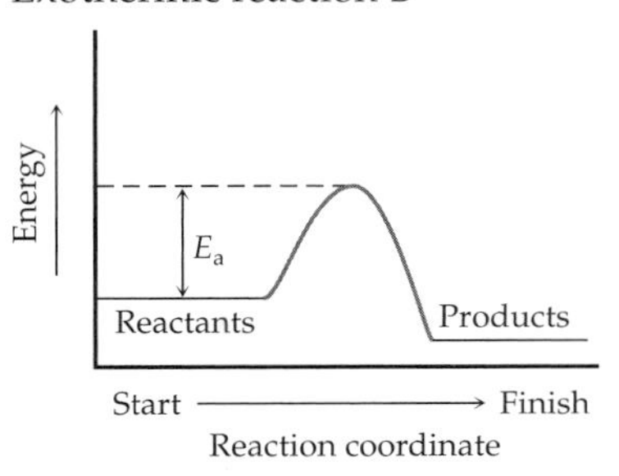

13.13 False; it is the E_a barrier that determines the speed of the reaction. If reaction X has a higher E_a value than reaction Y, then reaction X proceeds more slowly even though it gives off more heat.

13.15 (a) The number of effective collisions should increase. The general rule is that the rate of a reaction doubles with every 10 C° increase in temperature. The answer to part (b) of Practice Problem 13.14 shows you that reaction rate is expressed as "number of effective collisions per second." Thus because the reaction rate doubles, the number of effective collisions per second doubles, from 20 to 40.
(b) Increasing the temperature causes the molecules to have more energy, meaning more of them will have the minimum energy required for a reaction to occur. This increases the energy factor in Equation 13.1. Because the orientation factor remains unchanged, the product of the two factors (which is by definition the number of effective collisions per second) must increase.
(c) Approximately 40 CH_3OH molecules/s.

13.16 (a) An orientation factor of 0.1 means that only 10% of all the collisions are oriented properly to lead to products.
(b) Reaction rate = (100 sufficiently energetic collisions/s) × (0.1)
= 10 effective collisions/s
The rate of product formation is therefore 10 CH_3OH molecules per second.

13.18 Rate = $k[H_2O_2]^x[I^-]^y[H^+]^z$

13.19 (a) k for the NO reaction is much larger than k for the H_2O_2 reaction.
(b) E_a for the NO reaction is much smaller than E_a for the H_2O_2 reaction. (Recall that k and E_a are inversely related.)
(c) (*1*) Increase the concentrations of the reactants; (*2*) increase the temperature of the reaction mixture; (*3*) add a catalyst.

13.20 Concentration is proportional to the number of molecules per unit volume. The more molecules there are in a certain volume, the more collisions will occur.

13.22 (a) 1
(b) 2
(c) 1 + 1 + 2 = 4
(d) The rate of reaction is quadrupled.

13.23 Rate $= k[\mathrm{NO}]^2[\mathrm{O_2}]$

13.25 Answers will vary, but here is one.

(a)–(b)

$$\mathrm{Y_2 + Y_2 \xrightarrow{\text{Slow}} Y_3 + Y}$$
$$\mathrm{Y + X_2 \xrightarrow{\text{Fast}} XY + X}$$
$$\mathrm{Y_3 + X \xrightarrow{\text{Fast}} XY + Y_2}$$
$$\mathrm{X_2 + Y_2 \longrightarrow 2\,XY}$$

(c) See if you can detect intermediates Y, Y_3, or X during the course of the reaction.

13.26 $\mathrm{Y_2 \longrightarrow Y + Y}$

13.27 Bonds broken: 1 C=C bond, 4 C–H bonds, 3 O=O bonds; bonds formed: 4 C=O bonds, 4 O–H bonds

13.29 Yes; the Cl is substituted for the I during the reaction.

13.31 A reaction mechanism is the exact set of steps that molecules go through on changing from reactants to products.

13.33 Chemists study the rate of a reaction and how the rate changes as the temperature and the reactant concentration are varied.

13.36 This reaction will absorb energy from the surroundings because the product has twice as much energy in it as the reactant and that energy has to come from the surroundings.

13.38 Compound C is at a higher energy level; D has less energy than C because some of the energy originally in C was released into the surroundings when D was formed from C.

13.40 This reaction is endothermic; $\Delta E_{\mathrm{rxn}} = +40$ kJ/mol

13.41 The reaction is exothermic.

13.43 An energy-uphill reaction is one in which the products have more energy than the reactants. An energy-downhill reaction is one in which the products have less energy than the reactants.

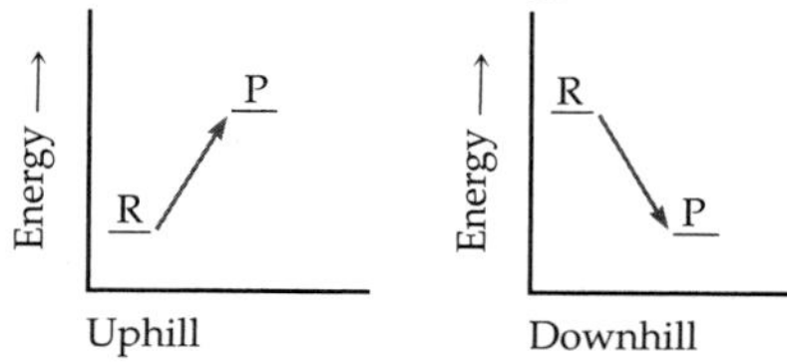

13.45 The reactants must absorb energy from the surroundings.

13.46 In an exothermic reaction, the products are at a lower energy level than the reactants. For all reactions, $\Delta E_{\mathrm{rxn}} = E_{\mathrm{products}} - E_{\mathrm{reactants}}$. Because the energy of the reactants in an exothermic reaction is larger than the energy of the products, in an exothermic reaction ΔE_{rxn} will always be a negative number.

13.48 The overall reaction releases 800 kJ ($\Delta E_{\mathrm{rxn}} = -800$ kJ). You are told it takes 800 kJ to break old bonds, so forming new ones must release 1600 kJ:

$$\Delta E_{\mathrm{rxn}} = \Delta E_{\mathrm{absorbed}} - \Delta E_{\mathrm{released}}$$
$$\Delta E_{\mathrm{released}} = \Delta E_{\mathrm{absorbed}} - \Delta E_{\mathrm{rxn}}$$
$$= 800\text{ kJ} - (-800\text{ kJ})$$
$$= 1600\text{ kJ}$$

13.50 The larger the E_a of a reaction, the slower the reaction.

13.52

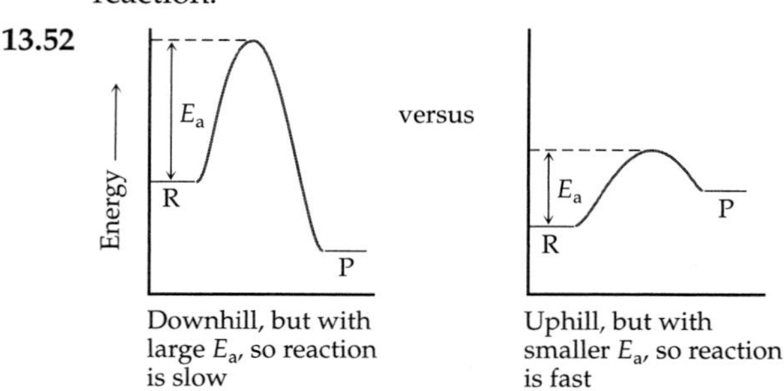

13.54 The transition state is that point in the reaction when just enough energy has been added to the reaction mixture to cause the reactant bonds to be almost broken and the product bonds to be just beginning to form.

13.55

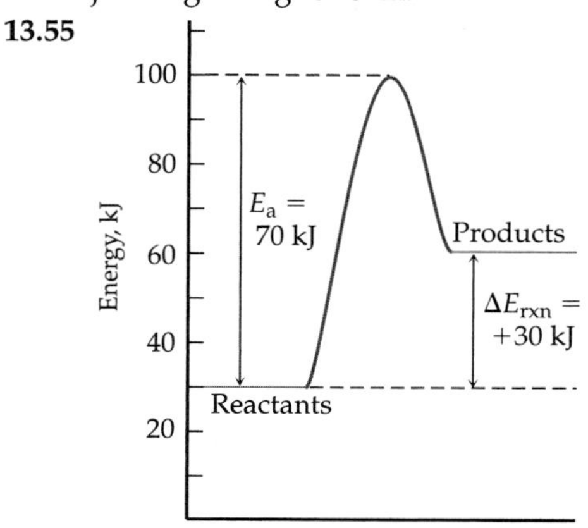

The reaction is endothermic.

13.58 Increasing the temperature would increase the rate of a reaction because doing so increases the energy of the reactants. This increase in energy means that more reactant molecules have enough energy to climb the energy barrier (E_a).

13.59 Those reactant molecules that collide with enough energy (greater than or equal to E_a) and the proper orientation can become product molecules.

13.61 Nothing.

13.62 The size of E_a depends on the nature of the reactant molecules. Reaction A might have a much larger E_a than reaction B because the reactants in A have very strong bonds. It would therefore take a large amount of energy to break the reactant bonds, resulting in a large value for E_a.

13.64 If the colliding molecules are not oriented just the right way with respect to each other, no reaction can occur.

13.67 By lowering the activation energy, E_a.

13.68 The Zn^{2+} forms a bond to the OH group, weakening the C–OH bond. This weakening makes it much easier to break the C–OH bond, thereby lowering the activation energy for the reaction and speeding it up.

13.70 The substrate is the reactant molecule that fits into an enzyme's active site.

13.72 The key. Sulfanilamide has almost the same shape as the PABA key. Once fit into the enzyme lock, the modified key caused the bacteria to produce faulty vitamin B.

13.74 Increasing the concentration of a reactant will increase the number of collisions between reactant molecules. The more collisions, the faster the rate of reaction.

13.76 Because both factors in the product depend on inherent (unchangeable) properties of the reactant molecules.

13.77 The rate constant k increases with temperature because k partially depends on the fraction of collisions that have energy equal to or greater than E_a. As the temperature increases, more of these energetic collisions occur and k increases.

13.79 The inherent factors are the fraction of collisions having energy equal to or greater than E_a and the fraction of collisions in which molecules have the proper orientation.

13.81 (a) k
(b) Orders
(c) Sum the individual reactant orders.
(d) $x = 2$ and $y = 1$; the overall order of the reaction is $2 + 1 = 3$.
(e) The value of its order is 0; that means the rate of the reaction does not depend on the concentration of this reactant.

13.83 The orders are obtained by running experiments that determine how the rate of a reaction changes when the concentrations of the individual reactants are changed one by one.

13.85 When the concentration of that reactant is doubled, the rate will quadruple.

13.87 The rate of the reaction will double because the rate is first-order with respect to A and zero-order with respect to B. When the volume of the balloon is halved, this has the effect of doubling the concentration of A. (The concentration of B also doubles, but that change does not affect the rate in any way because B does not appear in the rate law.)

13.89 The rate of the reaction will increase eightfold:

$$\begin{aligned}\text{Original rate} &= k[A][B]^2 \\ \text{New rate} &= k[2A][2B]^2 \\ &= k\,2[A]2^2[B]^2 \\ &= (2 \times 4)\,k[A][B]^2 \\ &= 8 \times \text{Original rate}\end{aligned}$$

13.91 A kinetics experiment is one in which we vary the reactant concentrations one at a time and observe what happens to the rate of the reaction. The exponents in the experimental rate law are determined directly from kinetics experiments.

13.92 The order of the reaction with respect to H_2O_2 can be determined from experiments 1 and 2. We see that doubling the concentration of H_2O_2 causes the reaction rate to double. This means that the reaction is first-order with respect to H_2O_2.

The order of the reaction with respect to I^- can be determined from experiments 1 and 3. We see that doubling the concentration of I^- causes the reaction rate to double. This means that the reaction is first-order with respect to I^-.

The order of the reaction with respect to H^+ can be determined from experiments 1 and 4. We see that doubling the concentration of H^+ has no effect on the rate. This means that the reaction is zero-order with respect to H^+.

The rate law for the reaction is

$$\text{Rate} = k[H_2O_2]^1[I^-]^1[H^+]^0 = k[H_2O_2][I^-]$$

The overall order of the reaction is $1 + 1 = 2$.

13.93 The order of the reaction with respect to A can be determined from experiments 1 and 2. We see that doubling the concentration of A causes the reaction rate to double. This means that the reaction is first-order with respect to A.

The order of the reaction with respect to B can be determined from experiments 1 and 3. We see that doubling the concentration of B causes the reaction rate to quadruple. This means that the reaction is second-order with respect to B.

The rate law for the reaction is

$$\text{Rate} = k[A]^1[B]^2 = k[A][B]^2$$

The overall order of the reaction is $1 + 2 = 3$.

The reaction confirms that the exponents in the rate law cannot be taken from the coefficients in the overall balanced equation.

13.95 Four molecules would have to come together with the proper orientation. This is very unlikely.

13.97 An elementary step is a step in a mechanism that shows the actual effective collisions that take place between two molecules during a reaction.

13.98 Rate $= k[A][B]^2[C]$

13.99 The slowest step in a mechanism.

13.102 The postulated mechanism cannot be correct if it does not agree with the experimentally determined rate law.

13.104 The postulated mechanism may possibly be the correct mechanism, but other mechanisms may also agree with the experimental data. Incorrect mechanisms can be ruled out by experimental evidence, but correct mechanisms cannot be easily proved.

13.105 (a) The reaction rate will double when the concentration of $CH_3C–Br$ is doubled, triple when it is tripled, decrease by half when it is halved, and so forth. The rate is not affected by any change in the concentration of H_2O.
(b) Mechanism II can be ruled out because it does not agree with the experimental

evidence. Its one-step mechanism would mean that the reaction would be first-order with respect to both $CH_3C–Br$ and H_2O. Mechanism I is plausible. The mechanism agrees with the experimental data, and the sum of the individual steps yields the overall balanced equation.

13.107 Rate = $k[Cl_2]$; only the reactants in the rate-determining step (raised to the power of their coefficients) appear in the rate law.

13.109 $\Delta E_{\text{forward rxn}}$ = +200 kJ/mol; $\Delta E_{\text{reverse rxn}}$ = −200 kJ/mol.

13.111 200 kJ/mol

13.113 A endothermic, B exothermic.

13.115 A forward, E_a = 400 kJ/mol; B forward, E_a = 400 kJ/mol; A reverse, E_a = 200 kJ/mol; B reverse, E_a = 500 kJ/mol.

13.116 The forward-direction rates will be the same for the two reactions because the E_a values are the same.

13.120 (a) Endothermic (b) Exothermic (c) Exothermic (d) Exothermic (e) Endothermic (f) Exothermic

13.122 (a) 0.063 (b) 0.13 (c) 0.13 (d) 0.25

13.123 (a) Rate quadruples: $[2]^2 = 4$.
(b) Rate triples: $[3]^1 = 3$.
(c) Rate increases ninefold: $[3]^2 = 9$.
(d) Rate quadruples: $[4]^1 = 4$.
(e) Rate increases twelvefold: $[2]^2[3]^1 = 4 \times 3 = 12$.

13.127 (a) Endothermic.
(b) X + Heat $\longrightarrow$ Y.
(c) −30 kJ.
(d) Exothermic.
(e) Cold because endothermic reactions absorb heat from the surroundings.

13.128 (a) E_a = 50 kJ, ΔE_{rxn} = +30 kJ, endothermic.
(b) E_a = 50 kJ, ΔE_{rxn} = −30 kJ, exothermic.
(c) E_a = 125 kJ, ΔE_{rxn} = +80 kJ, endothermic.
(d) E_a = 20 kJ, ΔE_{rxn} = −10 kJ, exothermic.

13.131 (a) and (d)

13.132 (a) Rate quadruples: $[2]^2 = 4$.
(b) Rate increases ninefold: $[3]^2 = 9$.

13.135 (a) True.
(b) False. Lowering the temperature of a reaction mixture slows the reaction down.
(c) False. The rate of a reaction depends on both the number of collisions per unit time having energy equal to or greater than E_a and the fraction of those collisions in which the reactant molecules have the proper orientation with respect to each other.
(d) True.

13.137 $H_2 + Cl_2 \longrightarrow 2\ HCl$

13.139 A reaction intermediate appears first in one step as a product and then in a subsequent step as a reactant. A catalyst appears first in one step as a reactant and then in a subsequent step as a product.

13.141 (a) Rate = $k[NO]^x[O_2]^y$
(b) Rate = $k[H_2O_2]^x$

13.143 (a) Exothermic. (b) Reactants.
(c) Downhill.
(d)

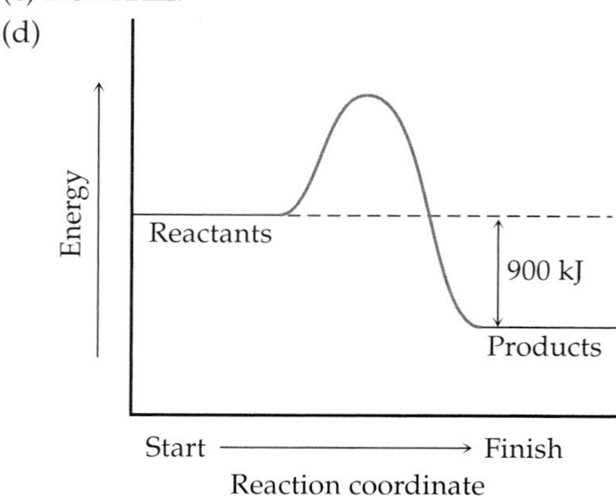

13.145 (a) True.
(b) False. A rate law can be used either (1) to prove that a proposed mechanism is wrong or (2) to prove that a proposed mechanism is *possibly* correct.
(c) False. The step A + X + Y $\longrightarrow$ AX + Y is implausible because it requires three reactant molecules to collide simultaneously, a highly unlikely event.
(d) False. Two changes possible to make the statement true: (1) A catalyst appears first as a reactant and then as a product, or (2) a reaction intermediate appears first as a product and then as a reactant.

Chapter 14

14.2 You want almost all of the reactants to be converted to products.

14.3 For reactions that go to completion the equilibrium lies so far to the right (products side) that there are essentially no reactants present at equilibrium. The reverse reaction can be ignored, and the reaction is considered a one-way reaction in the forward direction.

14.5 $K_{eq} = \dfrac{[HI]^2}{[H_2][I_2]}$

14.6 $K_{eq} = \dfrac{[Fe(SCN)^{2+}]}{[Fe^{3+}][SCN^-]}$

14.8 To the right. If the reaction went to completion (all the way to the right), the concentrations would be [HI] = 0.200 M; $[H_2]$ = 0.000 M; $[I_2]$ = 0.000 M. The actual equilibrium concentrations are approximately 75% of these values, meaning the equilibrium is to the right.

14.9 $K_{eq} = \frac{[CS_2][H_2]^4}{[CH_4][H_2S]^2}$

$= \frac{(6.10 \times 10^{-3})(1.17 \times 10^{-3})^4}{(2.35 \times 10^{-3})(2.93 \times 10^{-3})^2} = 5.67 \times 10^{-7}$

14.11 The reaction will shift to the left.

14.12 No shift in the reaction because N_2 is neither a reactant nor a product in this reaction.

14.14 (a) is at the lower temperature; (b) is at the higher temperature. At the lower temperature, the equilibrium will lie to the left, meaning the flask contains mostly colorless N_2O_4 gas. At the higher temperature, the equilibrium will lie to the right, meaning the flask contains mostly brown NO_2 gas.

14.15 (a) $2\ CO(g) + O_2(g) \rightleftharpoons 2\ CO_2(g)$ + Heat
A negative ΔE value means that the reaction is exothermic.
(b) The equilibrium will shift to the left. The reaction will shift in such a way as to use up the additional heat.

14.17 (a) $K_{eq} = \frac{[O_2]^6}{[CO_2]^6}$
(b) Adding $CO_2(g)$ shifts the equilibrium to the right.
(c) Adding $H_2O(l)$ has no effect on the equilibrium.
(d) Warming shifts the equilibirum to the right.
(e) Adding a catalyst has no effect on the equilibrium.

14.18 (a) $K_{eq} = [Pb^{2+}] \times [I^-]^2$
(b) Adding $PbI_2(s)$ has no effect on the equilibrium.
(c) Adding $Pb(NO_3)_2(s)$ shifts the equilibrium to the left because the salt dissolves and dissociates into Pb^{2+} and NO_3^- ions.

14.21 (a) $Mg(OH)_2(s) \rightleftharpoons Mg^{2+}(aq) + 2\ OH^-(aq)$
(b) $K_{sp} = [Mg^{2+}] \times [OH^-]^2$
(c) 2.88×10^{-4} M
(d) $K_{sp} = (1.44 \times 10^{-4}\ M) \times (2.88 \times 10^{-4}\ M)^2 = 1.19 \times 10^{-11}$

14.22 (a) $Al(OH)_3(s) \rightleftharpoons Al^{3+}(aq) + 3\ OH^-(aq)$
(b) $K_{sp} = [Al^{3+}] \times [OH^-]^3$
(c) 2.86×10^{-9} M
(d) $K_{sp} = (2.86 \times 10^{-9}\ M) \times (8.58 \times 10^{-9}\ M)^3 = 1.81 \times 10^{-33}$

14.24 The saturation solubility is the square root of the salt's K_{sp} value, given in Table 14.1.

14.25 $[Ag^+] = 2 \times (1.14 \times 10^{-17}\ M) = 2.28 \times 10^{-17}$ M, $[S^{2-}] = 1.14 \times 10^{-17}$ M

14.26 Equilibrium is the state in which the rate of the forward reaction equals the rate of the reverse reaction.

14.28 No; it only appeared to have stopped. No more product was being formed because the rate of the reverse reaction equaled the rate of the forward reaction, with the result that product was disappearing at the same rate at which it was being formed. Because both the forward and the reverse reaction are still occurring, the state is called *dynamic equilibrium.*

14.29 The position of a reaction's equilibrium refers to whether there are more products or more reactants present at equilibrium. If there are more reactants, the position of equilibrium is to the left; if there are more products, the position of equilibrium is to the right.

14.31 Far to the left.

14.33 The reaction might produce very little product if equilibrium lies far to the left.

14.35 Initially no products of the forward reaction are present. Because these products are the reactants for the reverse reaction, the reverse reaction rate is initially zero.

14.37 The reverse reaction would be faster because at first only reactants for the reverse reaction are present.

14.38 Yes. The reverse reaction will produce products that are the reactants for the forward reaction. As time passes, the reverse reaction will slow down and the forward reaction will speed up until the two rates are equal. At this point, equilibrium is attained.

14.40 $k_{forward\ rxn}[Reactants]^{order} = k_{reverse\ rxn}[Products]^{order}$

14.41 No, because the reverse rate at equilibrium must be less than the reverse rate at time zero.

14.44 Because k_f and k_r are constant as long as the temperature does not change. Because k_f and k_r do not change, neither does the ratio k_f/k_r.

14.46 Greater than 1 because $K_{eq} = k_f/k_r$, and any fraction in which the numerator is larger than the denominator has a value greater than 1.

14.48 $K_{eq} = \frac{[C][D]}{[A]^2[B]^3}$

14.50 The two K_{eq} values are inversely related.

14.51 $K_{eq} = \frac{[NO_2]^2}{[N_2O_4]}$

14.53 Place some N_2O_4 in a container and allow the reaction to reach equilibrium. Then determine the concentration of N_2O_4 and NO_2 at equilibrium and substitute those values into the K_{eq} expression.

14.54 No. The value of K_{eq} depends only on the temperature.

14.56 The equilibrium at 25°C lies farther to the left than the equilibrium at 2000°C because the 25°C value is smaller than the 2000°C value.

14.58 No; an equilibrium constant of 2.05 means that the equilibrium lies near the middle of the reaction coordinate. Only for reactions whose equilibria lie far to the right ($K_{eq} > 10^5$) may we use a single arrow to the right, indicating completion.

14.60 No; $K_{eq} = 1.5 \times 10^{-6}$ means that the equilibrium lies very far to the left.

14.61 $K_{eq} = \dfrac{[CH_4][H_2O]}{[CO][H_2]^3}$

$[CO] = \dfrac{0.613\text{ mol}}{10.0\text{ L}} = 0.0613\text{ M},$

$[H_2] = \dfrac{1.839\text{ mol}}{10.0\text{ L}} = 0.1839\text{ M},$

$[CH_4] = \dfrac{0.387\text{ mol}}{10.0\text{ L}} = 0.0387\text{ M},$

$[H_2O] = \dfrac{0.387\text{ mol}}{10.0\text{ L}} = 0.0387\text{ M}$

$K_{eq} = \dfrac{(0.0387)(0.0387)}{(0.0613)(0.1839)^3} = 3.93$

This value of K_{eq} is about midway between 10^{-3} and 10^3, meaning the position of equilibrium is roughly at the middle of the reaction coordinate.

14.63 Because the value always remains the same if the temperature is not changed.

14.65 Reaction (a) goes to completion. Reaction (b) essentially does not occur.

14.66 $4\ HCl + O_2 \rightleftharpoons 2\ H_2O + 2\ Cl_2$

14.68 The two values would be the same.

14.69 The value of K_{eq} remains the same because you have not changed the temperature of the reaction mixture.

14.71 (a) The reaction will shift to the right.
(b) The reaction will shift to the left.
(c) The reaction will shift to the left.
(d) The reaction will shift to the left.
(e) The reaction will not shift.

14.72 When you remove reactants, the reaction will shift to the left, converting some of the products to reactants.

14.74 (a) Initial equilibrium

Forward

Reverse

(b) PCl_5 added

Forward

Reverse

(c) Restored equilibrium

Forward

Reverse

(d) The reaction shifted left.

14.76 (a) The equilibrium lies just about the middle of the reaction coordinate, with significant amounts of both reactants and product present in the reaction vessel.
(b) No, because there are significant amounts of reactants present at equilibrium.
(c) You would drive the reaction to completion by causing it to shift continuously to the right.

14.78 (a) Endothermic, because increasing the temperature causes the reaction to shift to the right. Heat must therefore be a "reactant" in the reaction.
(b) K_{eq} increases.
(c) Heat + $N_2O_4(g) \rightleftharpoons 2\ NO_2(g)$

14.79 Exothermic, because the equilibrium shifts to the left when the reaction mixture is heated.

14.81 (a) Decrease, (b) increase. For an exothermic reaction, heat is a product. Adding heat shifts the equilibrium to the left (decreasing K_{eq}), and removing heat shifts the equilibrium to the right (increasing K_{eq}).

14.83 Decreased; adding heat to this exothermic reaction shifts the equilibrium to the left, decreasing the amount of methanol.

14.84 Cooling the reaction mixture will slow down the rate of the reaction. It will then take longer to form the product.

14.85 Because it speeds up the forward and reverse reactions by the same amount, a catalyst has no effect on the position of equilibrium or on the value of K_{eq}.

14.87 (a) K_{eq} gets larger as the reaction is heated, so the reaction shifts to the right.
(b) Endothermic; the shift to the right means heat is a "reactant."

14.88 One in which not all products and reactants are in the same phase. At the interface between two phases.

14.61 (a) $K_{eq} = \dfrac{1}{[H_2]^2}$

(b) $K_{eq} = \dfrac{[PO_4^{3-}]\times[H_3O^+]^3}{[H_3PO_4]}$

(c) $K_{eq} = \dfrac{1}{[Pb^{2+}]\times[I^-]^2}$

(d) $K_{eq} = \dfrac{[H_3O^+]^2}{[Ca^{2+}]\times[CO_2]}$

14.92 $AgCl(s)$ precipitates.

14.93 (a) The sealed vessel keeps $H_2(g)$ from escaping. Only if $H_2(g)$ remains in the vessel can equilibrium be established.

(b) $[H_2] = \dfrac{1}{\sqrt{K_{eq}}}$

(c) To the left.

(d) Exothermic; if the reaction shifts left when heat is added, heat must be a product.

14.95 $[Al^{3+}] = 2.86 \times 10^{-9}$ M, $[OH^-] = 8.58 \times 10^{-9}$ M

14.97 (a) $AgC_2H_3O_2(s) \rightleftharpoons Ag^+(aq) + C_2H_3O_2^-(aq)$

(b) $K_{sp} = [Ag^+] \times [C_2H_3O_2^-]$

(c) $$\frac{10.6 \text{ g } \cancel{AgC_2H_3O_2}}{\text{L}} \times \frac{1 \text{ mol } AgC_2H_3O_2}{166.9 \text{ g } \cancel{AgC_2H_3O_2}} = 6.35 \times 10^{-2} \text{ M}$$

Because this is a 1:1 salt, the ion concentrations are 6.35×10^{-2} M for both cation and anion. Therefore

$K_{sp} = (6.35 \times 10^{-2} \text{ M}) \times (6.35 \times 10^{-2} \text{ M}) = 4.03 \times 10^{-3}$

14.99 (a) $PbCl_2$, because it is the 1:2 salt with the largest value for K_{sp}.

14.101 (a) 1.8×10^{-7} mol/L

(b) 4.8×10^{-5} g/L

(c) $[Pb^{2+}] = 1.8 \times 10^{-7}$ M, $[CO_3^{2-}] = 1.8 \times 10^{-7}$ M

(d) 1.8×10^2 g

14.103 9.17×10^{-3} M

14.105 (a) To the right. Because the reverse reaction proceeds much more slowly than the forward reaction, [AB] must be very high in order to get the reverse rate equal to the forward rate. The high value of [AB] means the equilibrium lies to the right.

(b) Yes. That the forward reaction is much faster than the reverse reaction means k_f is much larger than k_r. Therefore $K_{eq} = k_f/k_r$ is a large number, which means the equilibrium lies to the right.

14.107 (a) The reverse reaction because initially there is no N_2 and no O_2, meaning the forward reaction is not happening.

(b) $K_{eq} = 0.100$

14.108 0.704 M

14.109 Increase because adding heat shifts the reaction to the right, increasing the ratio of products to reactants.

14.112 (a) AgCl (b) 8.0×10^{-8} M

14.114 2.0×10^{-6} mole

14.116 (a) Keep adding solid CuI to 25°C water until no more of the salt dissolves. Filter off the undissolved CuI, and the filtered liquid is a saturated CuI solution.

(b) 0.18 mg

(c) The added CuI* solid enters into a dynamic equilibrium with the $Cu^+(aq)$ and $I^-(aq)$ in solution. This means that CuI* dissolves to form $Cu^+(aq)$ and $I^{-*}(aq)$ as $Cu^+(aq)$ and $I^-(aq)$ precipitate out as CuI. Thus there soon are I^{-*} ions in solution.

14.118 (c)

14.119 In water.

14.121 (a) K_{sp}

(b) 6.6×10^{-4} M

(c) The rate constant for the reverse reaction.

(d) E_a for the forward reaction.

(e) The rate of the reverse reaction.

(f) $Li_2CO_3(s) \rightleftharpoons 2\,Li^+(aq) + CO_3^{2-}(aq)$, $K_{eq} = K_{sp} = [Li^+]^2 \times [CO_3^{2-}]$

(g) Lithium carbonate

Chapter 15

15.2 (a) Electrolyte. (b) Nonelectrolyte. (c) Electrolyte. (d) Nonelectrolyte. (e) Electrolyte. (f) Electrolyte. (g) Electrolyte.

15.3 (a)

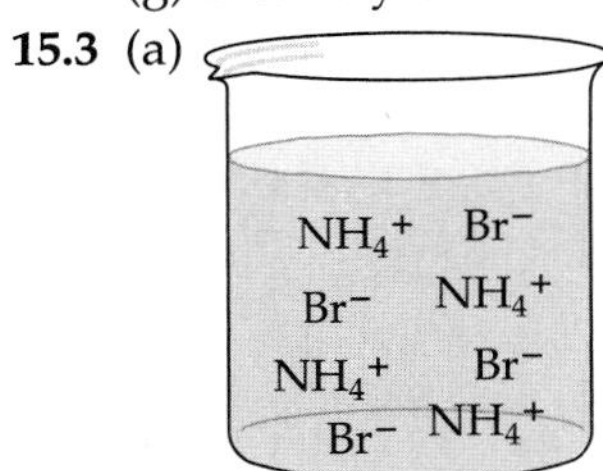

(b) Yes. That NH_4Br dissociates tells you it is an electrolyte.

(c)

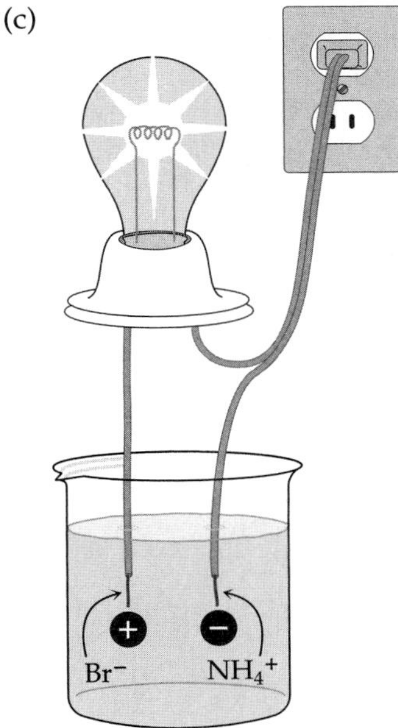

15.5 $H_2CO_3 + 2\,H_2O \rightleftharpoons 2\,H_3O^+ + CO_3^{2-}$

15.6 $H_2CO_3 + H_2O \rightleftharpoons H_3O^+ + HCO_3^-$

$HCO_3^- + H_2O \rightleftharpoons H_3O^+ + CO_3^{2-}$

15.8 1 mole of NaOH neutralizes 1 mole of HCl, but 2 moles of NaOH is needed to neutralize 1 mole of H_2SO_4 because each mole of H_2SO_4 produces 2 moles of H^+ ions.

15.9 0.05 mol

15.10

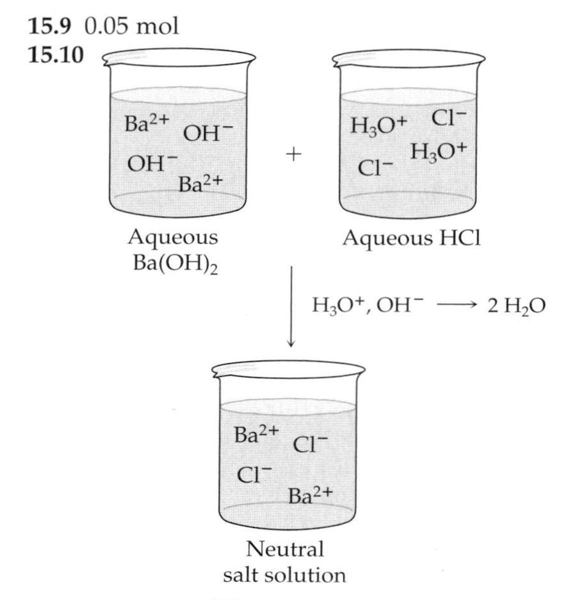

15.12 $\mathrm{H{:}\underset{H}{\ddot{N}}{:}H + H{:}\underset{H}{\ddot{P}}{:}H \longrightarrow \left[H{:}\underset{H}{\ddot{N}}{:}\right]^- + \left[H{:}\overset{H}{\underset{H}{P}}{:}H\right]^+}$

15.13 Water is behaving as an acid because it donates a proton to NH_2^-.

15.15 As acid: $HCO_3^- + H_2O \rightleftharpoons H_3O^+ + CO_3^{2-}$
As base: $HCO_3^- + H_2O \rightleftharpoons H_2CO_3 + OH^-$

15.16 (a) HCO_3^- acting as a base lies farther to the right because $K_b > K_a$.
(b) Basic because $K_b > K_a$.

15.18 (a) Basic because the OH^- concentration is greater than 10^{-7} M (0.000 155 M = 1.55×10^{-4} M).

(b) $[H_3O^+] = \dfrac{1 \times 10^{-14}}{0.000\ 155} = 6.45 \times 10^{-11}$ M

15.19 (a) $[H_3O^+] = \dfrac{1.00 \text{ mol } H_3O^+}{0.5000 \text{ L of solution}} = 2.00$ M

(b) $[OH^-] = \dfrac{1 \times 10^{-14}}{2.00} = 5.00 \times 10^{-15}$ M

15.20 (a) 0.0100 mole of $Ba(OH)_2$ provides 0.0200 mole of OH^-. Therefore

$[OH^-] = \dfrac{0.0200 \text{ mol } OH^-}{0.5000 \text{ L of solution}} = 0.0400$ M

(b) $[H_3O^+] = \dfrac{1 \times 10^{-14}}{0.0400} = 2.50 \times 10^{-13}$ M

15.22 4

15.23 1; 10^1 is the same as 10.

15.24 −2; 10^{-2} is the same as 0.01.

15.26 −11

15.27 +7

15.29 10

15.30 (a) −1 (b) 10 M

15.32 The 100-times-less-acidic solution has a pH of 4.56; therefore $[H_3O^+] = 10^{-4.56} = 2.75 \times 10^{-5}$.

15.33 $[H_3O^+] = 10^{-5.55} = 2.82 \times 10^{-6}$

$[OH^-] = \dfrac{1 \times 10^{-14}}{2.82 \times 10^{-6}} = 3.55 \times 10^{-9}$

15.35 CO_3^{2-}

15.36 H_2CO_3

15.38 $NH_3 + H_2O \rightleftharpoons NH_4^+ + OH^-$

15.39 It is called a buffered solution because NH_4^+ (from dissociated NH_4Cl) and NH_3 are a conjugate pair and any solution containing large amounts of a conjugate pair is a buffered solution.

15.40 $NH_4^+ + OH^- \longrightarrow NH_3 + H_2O$

15.41 $NH_3 + H_3O^+ \longrightarrow NH_4^+ + H_2O$

15.42 Sour taste, turns litmus red, reacts with active metals to produce hydrogen gas.

15.44 A compound that is one color in the presence of an acid and a different color in the presence of a base. Litmus is an example, being red in acids and blue in bases.

15.45 An electrolyte is a substance whose aqueous solution can conduct electricity. A nonelectrolyte is a substance whose aqueous solution cannot conduct electricity.

15.47 Ethanol does not break up into ions when dissolved in water; magnesium chloride does.

15.50 False. Some molecular compounds, such as HX acids, dissociate into ions when placed in water.

15.51 Ions.

15.53 (a) Yes, because NaCl does not exist as discrete molecules whose atoms are held together by covalent bonds.
(b) Because it is a water-soluble ionic compound.

15.55 (a) Electrolyte. (b) Nonelectrolyte. (c) Nonelectrolyte. (d) Electrolyte. (e) Electrolyte. (f) Electrolyte. (g) Nonelectrolyte.

15.56 $Na_2SO_4(aq) \longrightarrow 2\,Na^+(aq) + SO_4^{2-}(aq)$

15.59 Glucose is a molecular compound that does not dissociate in water.

15.61 The intense glow indicates a strong electric current and thus a large number of ions in the HCl beaker. HCl is a strong electrolyte that dissociates completely in solution. The dim glow indicates a weak electric current and thus few ions in the HF beaker. HF is a weak electrolyte that dissociates only partially in solution.

15.63 Far to the right for a strong electrolyte; to the left for a weak electrolyte.

15.65 H_3O^+

15.67 Weak. The small ($< 10^{-3}$) equilibrium constant of 8.2×10^{-6} means the equilibrium lies to the left, with most of the molecules undissociated.

15.69 Numerous answers possible. Two strong molecular electrolytes are hydrobromic acid and hydrochloric acid:

$HBr + H_2O \longrightarrow H_3O^+ + Br^-$
$HCl + H_2O \longrightarrow H_3O^+ + Cl^-$

Two weak molecular electrolytes are hydrofluoric acid and ammonia:

$HF + H_2O \rightleftharpoons H_3O^+ + F^-$
$NH_3 + H_2O \rightleftharpoons NH_4^+ + OH^-$

15.71 All are molecular because none contain a metal ion or polyatomic ion.

15.73 Because they all contain at least one H atom and produce H_3O^+ ions in water.

15.74 False. Because HF is a weak acid and HCl is a strong acid, a given volume of 1.0 M HCl is more acidic than the same volume of 1.0 M HF.

15.76 *Diprotic* means two dissociable hydrogens in the acid. One example is H_2SO_4:

$H_2SO_4 + H_2O \longrightarrow H_3O^+ + HSO_4^-$
$HSO_4^- + H_2O \rightleftharpoons H_3O^+ + SO_4^{2-}$

15.78 After water, the CH_3COOH (acetic acid) would be in highest concentration because it is a weak acid and so dissociates only to a very small extent in water.

15.79 After water, H_3O^+ and NO_3^- would be in highest concentration because HNO_3 (nitric acid) is a strong acid and so dissociates completely in water.

15.81

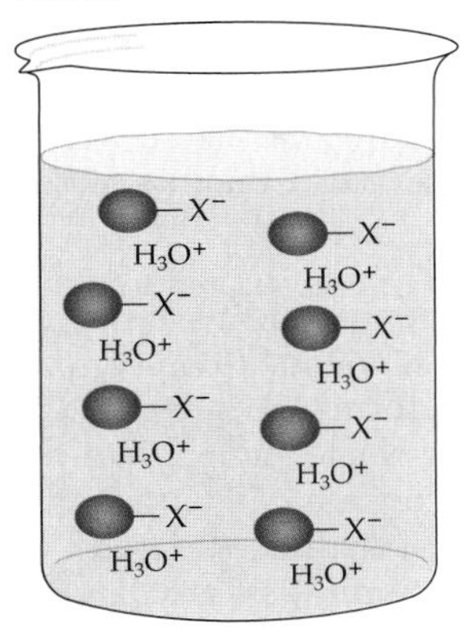

15.82

H:C:C(:Ö:)(:Ö:H) or H—C(H)(H)—C(=Ö:)—Ö—H

Dissociating proton

15.85 False. The extent of dissociation is what determines whether an acid is weak or strong, not the number of protons in the chemical formula.

15.87 Acid–base neutralization is the production of water from the reaction between H_3O^+ supplied by an acid and OH^- supplied by a base:

$H_3O^+ + OH^- \longrightarrow 2\ H_2O$

15.89 The product of the reaction is the water-soluble salt $CaCl_2$. All water-soluble salts are strong electrolytes, completely dissociating into ions in solution and making the solution a good electrical conductor.

15.91

$$0.50\ \text{mol}\ H_2SO_4 \times \underbrace{\frac{2\ \text{mol}\ H_3O^+}{1\ \text{mol}\ H_2SO_4}}_{\text{Diprotic}} \times \underbrace{\frac{1\ \text{mol}\ OH^-}{1\ \text{mol}\ H_3O^+}}_{\substack{H_3O^+ + OH^- \longrightarrow 2\ H_2O \\ \text{Neutralization}}} \times \underbrace{\frac{1\ \text{mol}\ Ba(OH)_2}{2\ \text{mol}\ OH^-}}_{\substack{\text{From formula} \\ Ba(OH)_2}}$$

$$= 0.50\ \text{mol}\ Ba(OH)_2$$

The products of the neutralization are $BaSO_4(s)$ and water. Because the $BaSO_4$ is insoluble in water (see Table 7.1), there are no ions present to conduct electricity. When an excess of $Ba(OH)_2$ is added, Ba^{2+} and OH^- ions are present to conduct electricity.

15.94

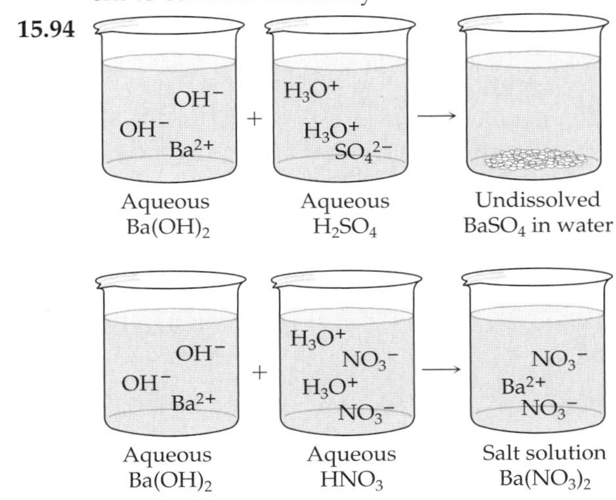

15.95 $NH_3(g) + H_2O \rightleftharpoons NH_4^+(aq) + OH^-(aq)$

15.97 Because NH_3 contains no OH groups and cannot dissociate to produce OH^- ions in solutions.

15.99 Water can be thought of as an acid because it donates a proton to the ammonia:

$NH_3 + H_2O \rightleftharpoons NH_4^+ + OH^-$

15.101 (a) Only slightly acidic because the very low value of K_{eq} means the equilibrium lies far to the left.
(b) A base because it accepts a proton from NH_4^+.

15.103 (a) $H^- + H_2O \longrightarrow H_2 + OH^-$
(b) The Brønsted–Lowry definition says the base H^- is a proton acceptor, so that H_2 gas forms as H^- combines with the proton donated from water: $H^- + H^+ \longrightarrow H_2$.

15.105 The lone pair of electrons.

15.107 The C_2H_5OH donates a proton to the H^-.

15.109 Because it produces relatively few OH^- in aqueous solution.

15.111 The LiOH solution is more basic. Because lithium is a group 1 metal, LiOH is a strong base that dissociates completely, meaning a 1 M LiOH solution has an OH^- concentration of 1 M. The weak base NH_3 reacts with water to produce OH^- in an equilibrium that lies to the left, meaning a 1 M NH_3 solution has an OH^- concentration of much less than 1 M.

15.113 (a) $HSO_4^- + H_2O \rightleftharpoons H_3O^+ + SO_4^{2-}$
(b) $HSO_4^- + H_2O \rightleftharpoons H_2SO_4 + OH^-$
(c) You need the K_{eq} for both reactions. If K_{eq} for reaction (a) is larger, the solution is weakly acidic. If the K_{eq} for reaction (b) is larger, the solution is weakly basic.

15.115 OH^-

15.116 (a) $H_2PO_4^-$, HPO_4^{2-}, PO_4^{3-}
(b) PO_4^{3-}

15.118 It is breaking up into its component ions, H^+ (or H_3O^+) and OH^-.

15.120 $K_w = 1.0 \times 10^{-14} = [H_3O^+] \times [OH^-]$

15.122 1.0×10^{-7} mol/L

15.124 (1) When $[H_3O^+] > [OH^-]$.
(2) When $[H_3O^+] > 1.0 \times 10^{-7}$.

15.126 True. There are always some H_3O^+ ions in any aqueous solution as a result of the autoionization of water.

15.127 True. The product of $[H_3O^+]$ and $[OH^-]$ must always equal 1×10^{-14} in an aqueous solution. If one concentration increases, the other must decrease.

15.129 $K_w = 1.0 \times 10^{-14} = [H_3O^+] \times [OH^-]$

$$[OH^-] = \frac{1.0 \times 10^{-14}}{[H_3O^+]}$$

$$= \frac{1.0 \times 10^{-14}}{1.0 \text{ M}} = 1.0 \times 10^{-14} \text{ M}$$

The solution is acidic because $[H_3O^+] > [OH^-]$.

15.131 LiOH dissociates to produce 1 mole of OH^- for every 1 mole of LiOH. Therefore

$$[OH^-] = \frac{2.50 \text{ mol } OH^-}{4.00 \text{ L solution}} = 0.625 \text{ M}$$

$$[H_3O^+] = \frac{1.0 \times 10^{-14}}{0.625 \text{ M}} = 1.6 \times 10^{-14} \text{ M}$$

15.133 Molar mass of $Mg(OH)_2$ = 58.319 g/mol.

$$\text{Moles } Mg(OH)_2 = 2.40 \text{ g} \times \frac{1 \text{ mol}}{58.319 \text{ g}}$$

$$= 0.0412 \text{ mol}$$

$Mg(OH)_2$ dissociates to produce 2 moles of OH^- for every 1 mole of $Mg(OH)_2$, meaning 0.0412 mole of $Mg(OH)_2$ produces 0.0824 mole of OH^-. The concentrations are therefore

$$[OH^-] = \frac{0.0824 \text{ mol}}{4.00 \text{ L solution}} = 0.0206 \text{ M}$$

$$[H_3O^+] = \frac{1.0 \times 10^{-14}}{0.0206 \text{ M}} = 4.85 \times 10^{-13} \text{ M}$$

15.135 (a) Acidic, because the OH^- concentration lower than 10^{-7} means the H_3O^+ concentration must be greater than 10^{-7}.
(b) Because $[OH^-] = 1 \times 10^{-11}$ (same as 10^{-11}), the H_3O^+ concentration must be 1×10^{-3} (same as 10^{-3}) because $10^{-11} \times 10^{-3} = 10^{-14}$.

15.137 -34

15.139 The logarithm of both 10^0 and 1 is 0 ($10^0 = 1$).

15.140 (d)

15.141 The logarithm of 10 is 1, and the logarithm of 100 is 2. Because 60 is between 10 and 100, the logarithm of 60 must be between 1 and 2.

15.144 True.

15.146 At pH = 7, $[H_3O^+] = [OH^-] = 10^{-7}$ M.

15.148 $pH = -\log 10^{-6} = 6$; acidic.

15.149 $pH = -\log(6.40 \times 10^{-9}) = 8.19$; basic.

15.151 If $[OH^-] = 2.0 \times 10^{-3}$ M, $[H_3O^+] = 5.0 \times 10^{-12}$ M (from the K_w equation). The pH is therefore $-\log 5.00 \times 10^{-12} = 11.3$; basic.

15.152 Solution A is 1000 times more acidic.

15.154 $[H_3O^+] = 10^{-pH} = 10^{-4}$ M; acidic.

15.156 $[H_3O^+] = 10^{-pH} = 10^{-(-1)} = 10$ M; $[OH^-] = 10^{-14}/10 = 10^{-15}$ M; acidic.

15.157 The pH of the HCl solution is lower because acetic acid is a weak acid (dissociates little, not many H_3O^+ ions in solution) and HCl is a strong acid (dissociates completely, many H_3O^+ ions in solution).

15.159 No; the weak acid always has one more proton than its conjugate base. Because protons have a positive charge, the acid always carries one more positive charge than its conjugate base.

15.160 Weak acid.

15.161 The $NaNO_3$ solution would be neutral. The NO_3^- ion has no tendency to accept protons because doing so forms the strong acid HNO_3, which dissociates immediately and completely. The conjugate of any strong acid is so weak that it is neither acid nor base.

15.164 We mean the solution resists any large change in its pH.

15.165 No, because it does not contain significant amounts of either a weak acid and its conjugate base or a weak base and its conjugate acid.

15.168 No; if you add enough strong acid or strong base, the buffer will be exhausted.

15.170 (a) Decrease.
(b) Because the added strong acid is converted to a weak acid (the conjugate acid in the buffer system). This weak acid dissociates slightly, adding H^+ ions to the solution and thereby decreasing the pH slightly.

15.171 The NaOH reacts with half of the acetic acid to produce sodium acetate:

$CH_3COOH + NaOH \longrightarrow Na^+ + CH_3COO^- + H_2O$

The solution then contains significant amounts of acetic acid and acetate ion, a conjugate pair and therefore components of a buffered solution.

15.173 Dissolve HClO and NaClO or any other soluble hypochlorite salt in water, so that the solution contains a weak acid, HClO, and its conjugate base, ClO^-.

15.174 (a) HPO_4^{2-}
(b) $HPO_4^{2-} + H_3O^+ \longrightarrow H_2PO_4^- + H_2O$
$H_2PO_4^- + OH^- \longrightarrow HPO_4^{2-} + H_2O$
(c) H_3PO_4
(d) $H_3PO_4 + OH^- \longrightarrow H_2PO_4^- + H_2O$
$H_2PO_4^- + H_3O^+ \longrightarrow H_3PO_4 + H_2O$

15.176 The base part of the buffer pair combines with the protons of the added strong acid, locking the protons up in weak acid molecules.

15.178 By converting the added OH^- ions to weak base and water molecules.

15.180 $ClO^- + H_3O^+ \longrightarrow HClO + H_2O$
$HClO + OH^- \longrightarrow ClO^- + H_2O$

15.182 (a) $C_6H_5N + H_2O \rightleftharpoons C_6H_5NH^+ + OH^-$
(b) $[H_2O] > [C_6H_5N] > [C_6H_5NH^+] = [OH^-]$
(c) $[C_6H_5N] = 0.097$ M;
$[C_6H_5NH^+] = [OH^-] = 0.0032$ M
(d) 11.5

15.183 (a) $H_2PO_4^-$
(b) H_3O^+
(c) HCO_2H
(d) HI

15.186 0.52

15.188 (a) If HX were a strong acid, then $[H_3O^+]$ would equal 0.2 M and the pH would equal $-\log(0.2) = 0.7$. But the solution has a higher pH of 2.14. Also, $[H_3O^+] = 10^{-2.14} = 7.24 \times 10^{-3}$ M. This low concentration for H_3O^+ means HX dissociates very little and is therefore a weak acid.

(b) $$K_{eq} = \frac{[H_3O^+]\times[X^-]}{[HX]} = \frac{(7.24\times10^{-3}\ \text{M})\times(7.24\times10^{-3}\ \text{M})}{0.20\ \text{M}} = 2.6\times10^{-4}\ \text{M}$$

15.190 Lowest pH to highest pH means start with the strongest, most concentrated acid and end with the strongest base: 0.20 M HCl, 0.10 M HNO_3, 0.10 M HCN, 0.10 M $LiNO_3$, 0.10 M NaF, 0.1 M KOH.

15.192 OH^- ions because there are 2 moles of OH^- ions for every 1 mole of Ca^{2+} ions.

15.193 (a) $HCO_3^- + OH^- \longrightarrow H_2O + CO_3^{2-}$
(b) $HCl + F^- \longrightarrow HF + Cl^-$
(c) $H_2CO_3^- + OH^- \longrightarrow H_2O + HCO_3^-$
(d) $HCN + H_2O \rightleftharpoons H_3O^+ + CN^-$

15.195 (a) O, (b) 1, (c) 3, (d) 5.0, (e) 6.96

15.197 13.4

15.199 12

15.202 pH $= -1.08$, $[OH^-] = 8.26 \times 10^{-16}$

15.204 (a) Nonelectrolyte, (b) electrolyte, (c) electrolyte, (d) nonelectrolyte, (e) electrolyte, (f) electrolyte, (g) nonelectrolyte.

15.206 HCN, HClO, HCOOH, HNO_2

15.209 0.781 g

15.211 1.58×10^{-8} M

15.215 No, because there is no weak acid in the solution.

15.217 1.09 g

15.220 11.0

15.222 3.36×10^{-3} moles of HI.

15.224 $[H_3O^+] = 3.74 \times 10^{-3}$ M, $[OH^-] = 2.67 \times 10^{-12}$ M.

15.227 Weak acid because the fact that the solution is basic means the reaction $H_2O + X^- \longrightarrow HX + OH^-$ must be taking place.

Chapter 16

16.2 Radioactive.

16.3

$$2\ \text{mol protons} \times \frac{1.007\ 30\ \text{g}}{\text{mol protons}} = 2.014\ 60\ \text{g}$$

$$2\ \text{mol neutrons} \times \frac{1.008\ 70\ \text{g}}{\text{mol neutrons}} = 2.017\ 40\ \text{g}$$

$$2\ \text{mol electrons} \times \frac{0.000\ 55\ \text{g}}{\text{mol electrons}} = 0.001\ 1\ \text{g}$$

Mass of 1 mol of ${}^4_2He$ should be ⟶ 4.033 1 g
Mass of 1 mol of ${}^4_2He$ actually is ⟶ −4.003 g
Mass defect ⟶ 0.030 g/mol ${}^4_2He$

(The actual He mass value comes from the periodic table.)

$E = mc^2$

$$= \frac{0.030\ \cancel{g}}{\text{mol}} \times \frac{1\ \text{kg}}{1000\ \cancel{g}} \times \left(\frac{3.00\times10^8\ \text{m}}{\text{s}}\right)^2$$

$$= 2.70\times10^{12}\ \frac{\text{kg}\cdot\text{m}^2}{\text{mol}\cdot\text{s}^2}$$

$$= 2.70\times10^{12}\ \frac{\cancel{J}}{\text{mol}} \times \frac{1\ \text{kJ}}{1000\ \cancel{J}}$$

$$= 2.70\times10^{9}\ \text{kJ/mol}\ {}^4_2He$$

16.4 The band of stability curves up because the greater the number of protons in a nucleus, the more neutrons necessary to prevent proton–proton repulsions and therefore the greater the n/p ratio.

16.6 $^{35}_{14}Si \longrightarrow ^{35}_{15}P + ^{0}_{-1}e$

16.7 $^{40}_{19}K$

16.9 (a) $^{37}_{18}Ar + ^{0}_{-1}e \longrightarrow ^{37}_{17}Cl$

(b) Electron capture increases the n/p ratio. This means the n/p ratio for $^{37}_{18}Ar$ must have been too low, placing this isotope below the dark brown central region of nonradioactive nuclei.

(c) $^{37}_{17}Cl$

16.10 $^{25}_{13}Al$

16.11 $^{25}_{13}Al$

16.12 Practice Problem 16.10:
$^{25}_{13}Al + ^{0}_{-1}e \longrightarrow ^{25}_{12}Mg$
Practice Problem 16.11:
$^{25}_{13}Al \longrightarrow ^{0}_{+1}e + ^{25}_{12}Mg$

16.14 Two positron emissions: $^{238}_{92}U \longrightarrow ^{0}_{+1}e + ^{238}_{91}Pa$; $^{238}_{91}Pa \longrightarrow ^{0}_{+1}e + ^{238}_{90}Th$
One alpha emission: $^{238}_{92}U \longrightarrow ^{4}_{2}He + ^{234}_{90}Th$

16.15 The decay must be alpha emission because any other decay would yield a daughter that is only one space away from the parent.

16.16 A neutron turned into a proton, and the decay was beta emission.

16.18 Because all the $^{206}_{82}Pb$ in the rock today came from the decay of $^{238}_{92}U$ initially present.

16.19
$$\%\ ^{238}_{92}U \text{ remaining} = \frac{1.82\text{ g}}{5.84\text{ g}} \times 100\% = 31.2\%$$

Age of the asteroid

$$= \frac{-2.303 \times \log\left(\frac{31.2\%}{100}\right) \times 4.46 \times 10^9 \text{ years}}{0.693}$$
$$= 7.50 \times 10^9 \text{ years}$$

16.20 $^{239}_{94}Pu + ^{1}_{0}n \longrightarrow ^{90}_{38}Sr + ^{147}_{56}Ba + 3\ ^{1}_{0}n$

16.21 Yes, because the reaction produces more neutrons (three) than it consumes (one).

16.22 The reaction is exothermic, and you would know this by reasoning that the $^{90}_{38}Sr$ and $^{147}_{56}Ba$ are more stable (larger binding energy and mass defect per nucleon) than the $^{239}_{94}Pu$.

16.24 Binding energy is the energy released when an atom forms from its subatomic particles. You can think of it as either the energy that holds the nucleus together or the energy required to break the nucleus apart into its subatomic particles.

16.26 The more nucleons a nucleus has, the greater its binding energy. However, a light nucleus can be more stable than a heavy one because nuclear stability is judged on the basis of binding energy *per nucleon*. The example of $^{13}_{6}C$ and $^{12}_{6}C$ in the text demonstrates this.

16.28 $^{56}_{26}Fe$ has the greatest mass defect, meaning it is the most stable isotope that exists.

16.30
$$6 \text{ mol protons} \times \frac{1.007\,30\text{ g}}{1 \text{ mol protons}} = 6.043\,80\text{ g}$$
$$8 \text{ mol neutrons} \times \frac{1.008\,70\text{ g}}{1 \text{ mol neutrons}} = 8.069\,60\text{ g}$$
$$6 \text{ mol electrons} \times \frac{0.000\,55\text{ g}}{1 \text{ mol electrons}} = 0.003\,3\text{ g}$$

Mass of 1 mol of $^{14}_{6}C$ should be ⟶ 14.116 7 g
Mass of 1 mol of $^{14}_{6}C$ actually is ⟶ −14.003 24 g
Mass defect ⟶ 0.113 5 g/mol $^{14}_{6}C$

Convert this mass defect to kilograms:

$$\frac{0.1135\text{ g}}{\text{mol } ^{14}_{6}C} \times \frac{1\text{ kg}}{1000\text{ g}} = 1.135 \times 10^{-4}\text{ kg/mol } ^{14}_{6}C$$

The binding energy is therefore:

$$mc^2 = \frac{1.135 \times 10^{-4}\text{ kg}}{\text{mol } ^{14}_{6}C} \times (3.00 \times 10^8\text{ m/s})^2$$
$$= 1.02 \times 10^{13}\ \frac{\text{kg} \cdot \text{m}^2}{\text{mol} \cdot \text{s}^2}$$
$$= 1.02 \times 10^{13}\text{ J/mol } ^{14}_{6}C$$
$$\frac{1.02 \times 10^{13}\text{ J}}{\text{mol } ^{14}_{6}C} \times \frac{1\text{ kJ}}{1000\text{ J}} = 1.02 \times 10^{10}\text{ kJ/mol } ^{14}_{6}C$$

16.31 To go from binding energy per mole of $^{14}_{6}C$ to binding energy per mole of nucleons, divide by the number of nucleons:

$$1.02 \times 10^{10}\ \frac{\text{kJ}}{\text{mol } ^{14}_{6}C} \times \frac{1\text{ mol } ^{14}_{6}C}{14\text{ mol nucleons}}$$
$$= 7.28 \times 10^8\text{ kJ/mol nucleons}$$

This binding energy value puts $^{14}_{6}C$ lower on the binding energy axis than $^{12}_{6}C$ and at about the same level as $^{13}_{6}C$.

16.33 False; all atoms have a mass defect and binding energy, even radioactive ones.

16.35 (a) The n/p ratio increases.

(b) The more protons in a nucleus, the more neutrons are needed to reduce proton–proton repulsions.

16.37 $^{16}_{8}O$: n/p ratio = 8/8 = 1.0.
$^{17}_{8}O$: n/p ratio = 9/8 = 1.12.
$^{19}_{8}O$: n/p ratio = 11/8 = 1.38.

16.38 The band of stability is the band of data points, on a plot of number of neutrons versus number of protons, that represents all known stable isotopes.

16.40 Eight half-lives: 10 g × 0.5 × 0.5 × 0.5 × 0.5 × 0.5 × 0.5 × 0.5 × 0.5 = 0.039 g; 8 × 13.1 h = 10^5 h.

16.41 When it has fewer than 67 or more than 92 neutrons.

16.43 Because we define *stable* to mean all nonradioactive nuclei plus all radioactive nuclei of measurable half-life.

16.44 The n/p ratios for these nuclei are larger than the ratios for nuclei in the dark brown region.

Nuclei in this upper light brown region are neutron-rich relative to nonradioactive nuclei in the dark brown region.

16.46 All their isotopes are radioactive.

16.48 ${}^{1}_{0}n$; ${}^{1}_{1}p$, ${}^{0}_{-1}e$, ${}^{0}_{+1}e$

16.50 As electric charge (charge of −1 for electron and +1 for positron).

16.52 A nucleus above the dark brown central region on the band of stability would want to convert a neutron to a proton because its n/p ratio is too high.

16.54 A neutron is converted to a proton, which means the atomic number increases by 1 and the mass number remains unchanged.

16.55 When a neutron in the nucleus converts to a proton, an electron is created and emitted as a beta particle.

16.57 A proton converts to a neutron, which means the atomic number decreases by 1 and the mass number remains unchanged.

16.59 (a) Below the central dark brown region of nonradioactive nuclei. (Conversion of a proton to a neutron raises the n/p ratio.)

(b)–(c) Positron emission: ${}^{162}_{74}W \longrightarrow {}^{0}_{+1}e + {}^{162}_{73}Ta$

Electron capture: ${}^{162}_{74}W + {}^{0}_{-1}e \longrightarrow {}^{162}_{73}Ta$

16.61 Its atomic number decreases by 2, and its mass number decreases by 4.

16.63 Electromagnetic radiation (gamma rays) and the kinetic energy associated with the motion of the ejected particles.

16.65 By examining the superscripts and subscripts. In a balanced reaction, the sum of the superscripts on the left side must equal the sum of the superscripts on the right side, and ditto for the subscripts.

16.67 (a) Unchanged.
(b) Unchanged.
(c) Unchanged.
(d) Decreases by 4.

16.69 (a) Alpha decay results in a daughter two elements to the left of the parent because the atomic number decreases by 2.

(b) Beta decay results in a daughter one element to the right of the parent because the atomic number increases by 1.

(c)–(d) Positron emission and electron capture both result in a daughter one element to the left of the parent because the atomic number decreases by 1.

16.72 ${}^{8}_{4}Be + {}^{0}_{-1}e \longrightarrow {}^{8}_{3}Li$, electron capture.

16.74 ${}^{235}_{92}U \longrightarrow {}^{4}_{2}He + {}^{231}_{90}Th$, alpha emission.

16.76 ${}^{0}_{-1}e + {}^{40}_{19}K \longrightarrow {}^{40}_{18}Ar$, electron capture.

16.78 No; because the half-life of carbon-14 is 5715 years, essentially all of it would have decayed in 120 million years.

16.80
$$\text{Age} = \frac{-2.303 \times \log\left(\dfrac{22.8\%}{100}\right) \times 5715 \text{ years}}{0.693}$$
$$= 1.22 \times 10^{4} \text{ years}$$

16.81 (a) The assumption is that all the ${}^{206}_{82}Pb$ came from the decay of ${}^{238}_{92}U$.

(b)
$$14.90 \text{ g } {}^{238}_{92}U \times \frac{1 \text{ mol } {}^{238}_{92}U}{238.029 \text{ g } {}^{238}_{92}U} \times \frac{6.022 \times 10^{23} \text{ atoms } {}^{238}_{92}U}{\text{mol } {}^{238}_{92}U}$$
$$= 3.770 \times 10^{22} \text{ atoms } {}^{238}_{92}U$$

$$26.50 \text{ g } {}^{206}_{82}Pb \times \frac{1 \text{ mol } {}^{206}_{82}Pb}{205.974\ 46 \text{ g } {}^{206}_{82}Pb} \times \frac{6.022 \times 10^{23} \text{ atoms } {}^{206}_{82}Pb}{\text{mol } {}^{206}_{82}Pb}$$
$$= 7.748 \times 10^{22} \text{ atoms } {}^{206}_{82}Pb$$

(c) $3.770 \times 10^{22} + 7.748 \times 10^{22} = 1.1518 \times 10^{23}$ ${}^{238}_{92}U$ atoms.

(d) % ${}^{238}_{92}U$ remaining
$$= \frac{3.770 \times 10^{22} \text{ atoms}}{1.1518 \times 10^{23} \text{ atoms}} \times 100\%$$
$$= 32.73\%$$

(e) Age
$$= \frac{-2.303 \times \log\left(\dfrac{32.73\%}{100}\right) \times 4.46 \times 10^{9} \text{ years}}{0.693}$$
$$= 7.19 \times 10^{9} \text{ years}$$

16.82 The mass defect of the product nuclei must be greater than the mass defect of the reactant nuclei. This ensures that mass is lost during the reaction. This "lost" mass is converted to energy according to Einstein's equation $E = mc^2$.

16.85 Because each fission event produces more neutrons than it consumes.

16.87 ${}^{8}_{4}Be + {}^{4}_{2}He \longrightarrow {}^{12}_{6}C + \gamma$

16.89 The amount of any radioactive sample necessary to produce a runaway (explosive) chain reaction.

16.90 The heat is used to produce steam from liquid water. The steam is used to power the turbines of electric generators.

16.93 Nuclei repel one another, and high temperatures ensure that there is enough thermal energy to force the nuclei close enough together for fusion to occur.

16.95 Fusion fuel (heavy isotopes of hydrogen) is cheap and readily available, fusion reactors produce no nuclear waste, and there is no danger of an accident releasing toxic substances into the environment.

16.96 Alpha radiation is least likely to hurt you because it is not very penetrating and can be stopped by paper, clothing, or the top layer of skin. Gamma rays are most harmful because they are the most energetic.

16.98 Because the amount of energy given to a molecule by radiation can be enough to knock an electron out of the molecule, thereby creating an ion.

16.101 Endothermic. Because ^{56}Fe is the most stable isotope known, ^{112}Te must be less stable, and therefore an input of energy is required to form it.

16.103 (a) $^{210}_{84}Po \longrightarrow ^{206}_{82}Pb + ^{4}_{2}He$
(b) 597 days

16.104 (a) $^{87}_{38}Sr$ (b) Increase

16.106 (a) ^{8}B (b) ^{194}Bi

16.108 Use any atom you like in your analysis because all you are interested in is the particle needed to balance the equation once you write the parent and daughter according to the restrictions stated. For a hypothetical parent X having (arbitrarily) 50 protons and 50 neutrons, the equation is

$$^{100}_{50}X \longrightarrow ^{93}_{47}Y + ^{7}_{3}?$$

Parent — Daughter — Emitted particle

Because the particle has atomic number 3, it must be a lithium nucleus: $^{7}_{3}Li$.

16.110 (a) $^{144}_{58}Ce$
(b) The number of neutrons produced is greater than the number consumed.
(c) 9.64×10^{7} kJ/g
(d) 3.49×10^{4} kg TNT

16.112 Uranium.

16.113 (a) The first mechanism is the correct one.
(b) You could separate the products from each other and then pass a radiation detector over each one to find out which product contains the radioactive O.

16.115 46.3 mL

Chapter 17

17.2 (a)

(b)

(c)

(d)

(e)

17.3 Because the C≡C bond is linear and therefore a zigzag would be misleading, the correct line drawing is

17.4 (a) Branched. (b) Eight carbons long.
(c) One carbon long. (d) C_9H_{20}.

17.6 The unsaturated hydrocarbon is "missing" four hydrogens. This could be the result of two double bonds or one triple bond:

17.7 Molecule (c) is least unsaturated because it has only one double bond; it therefore has eight hydrogens, whereas the other two structures have six hydrogens apiece.

17.9

17.10 The main chain is the chain containing eight carbons, not seven:

5-Ethyl-2-octene

17.12 1-Butanol.

17.13 2-Aminopropane.

17.14 3-Bromohexane.

17.15

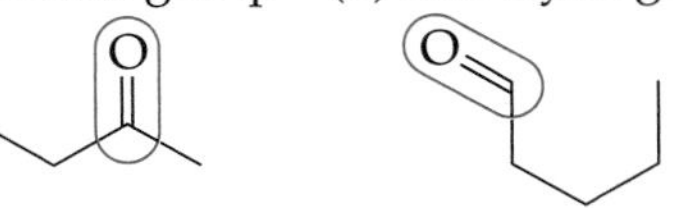

17.17 Ethanone does not exist because the carbonyl in a ketone must be connected to two carbons. Hence, the smallest number of carbons that can exist in a ketone is three.

17.18 (a) Ketone group (b) Aldehyde group

O O

(c) Carboxylic acid group

O OH

17.19 (a) 2-Butanone. (b) Pentanal.
(c) 2-Methyl-butanoic acid.

17.20 The ability of carbon atoms to link together in chains.

17.22 Because carbon has four valence electrons and needs four more to form an octet; it acquires them by forming four covalent bonds.

17.23
```
     H  H  H
     |  |  |
(a) H—C—C—C—H
     |  |  |
     H  H  H

              H
              |
(b) H—C=C—C—H
       |  |  |
       H  H  H

             H
             |
(c) H—C≡C—C—H
             |
             H
```

17.25 There can be no C=C or C≡C bonds in the molecule because only a carbon atom forming all single bonds can be tetrahedral.

17.26 The C=H double bond is impossible; H can form only single bonds.

17.30
```
  H  H  H  H  H  H
  |  |  |  |  |  |
H—C—C—C—C—C—C—H
  |  |  |  |  |  |
  H  H  H  H  H  H
```

17.31
```
       H
       |
     H—C—H
  H    |    H  H  H
  |    |    |  |  |
H—C————C————C—C—C—H
  |    |    |  |  |
  H    H    H  H  H

           H
           |
         H—C—H
  H  H     |     H  H
  |  |     |     |  |
H—C—C————C————C—C—H
  |  |     |     |  |
  H  H     H     H  H

     H          H
     |          |
   H—C—H    H—C—H
  H  |          |    H
  |  |          |    |
H—C——C——————————C————C—H
  |  |          |    |
  H  H          H    H

       H
       |
     H—C—H
  H    |    H  H
  |    |    |  |
H—C————C————C—C—H
  |    |    |  |
  H    |    H  H
     H—C—H
       |
       H
```

17.33 The geometry about C–C single bonds is tetrahedral, with bond angles of 109.5°. The zigzag represents the shape of the molecule more accurately than a straight line does because in the latter the angles appear to be 180°.

17.34 The molecule should be drawn as a straight line because each C≡C bond is linear.

17.37 (a) Seven carbons. (b) Two carbons.

17.39 (a) C_3H_8, linear:
```
  H  H  H
  |  |  |
H—C—C—C—H
  |  |  |
  H  H  H
```

(b) C_4H_{10}, linear:
```
  H  H  H  H
  |  |  |  |
H—C—C—C—C—H
  |  |  |  |
  H  H  H  H
```

(c) C_5H_{10}, branched:
```
  H  H  H     H
  |  |  |     |
H—C=C—C—————C—H
        |     |
        |     H
      H—C—H
        |
        H
```

17.41 (a) (b) (c)

17.43 Molecule (c).

17.45 (a) 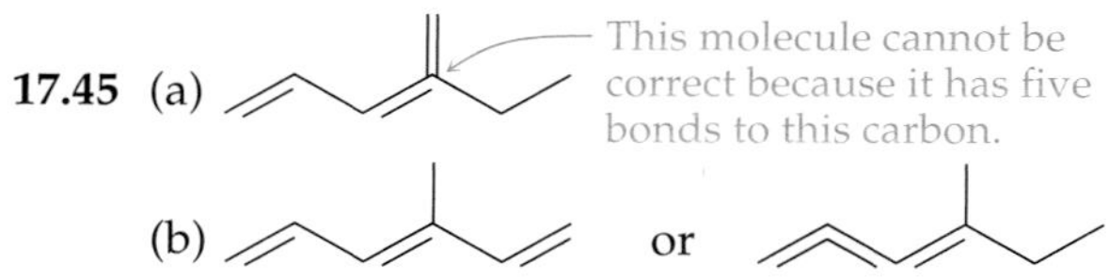
This molecule cannot be correct because it has five bonds to this carbon.

(b) or

17.47 Unsaturated; the maximum number of hydrogens for a seven-carbon chain is 16. This compound has two fewer and is therefore unsaturated.

17.48 Molecules (a) and (c) are isomers, each having four carbons and ten hydrogens. Molecules (b) and (d) are not isomers of (a) and (c) because (b) has five carbons and (d) has only eight hydrogens.

17.51 No, because they don't have the same chemical formula.

17.53 Molecules 1 and 3 are isomers because they have the same molecular formula but different structural formulas.

17.56 Hydrocarbons with all C–C single bonds are alkanes, those with at least one C=C double bond are alkenes, and those with at least one C≡C triple bond are alkynes.

17.58 A homologous series is a group of compounds that are similar to one another. An example is the alkanes, where each member differs from the next by a CH_2 group.

17.60

Methane

Ethane

Propane

Butane

17.63 An alkane because it fits the general formula C_nH_{2n+2}.

17.66

17.68 1-Hexene, 2-hexene, and 3-hexene all represent six-carbon alkenes, with the double bond in different positions. However, 4-hexene is really 2-hexene and 5-hexene is really 1-hexene when the molecule is numbered from the correct end of the chain.

17.70 Identify and name the main chain in the molecule.

17.72 (a) C_3H_6

(b) C_8H_{16}

(c) C_6H_{14}

(d) C_4H_6

(e) C_4H_6

17.74 The multiple bond must be part of the main chain, and the numbering starts at the end closer to the multiple bond. Multiple bonds have priority over branches.

17.76 If there are two methyl groups on carbon 2, as the name 2,2-dimethyl indicates, there can't be a double bond between carbons 1 and 2 because that would mean carbon 2 has more than four bonds, which is impossible.

17.78 (a) 2-Methylhexane.
(b) 2-Ethyl-3-methyl-1-butene.
(c) 3-Ethyl-2,4-dimethylhexane.
(d) 3-Propyl-1-octyne.

17.80

The compound can't be named 2,2-dimethyl-5-hexene because the main chain numbering must begin at the end nearer the multiple bond.

17.82 C_4H_6

17.83 As alkanes get larger, the intermolecular attractive forces get progressively stronger, leading to a steady increase in boiling point.

17.85 Replace the *-ine* ending of the halogen's elemental name with *o*, and then put the word you've formed in front of the name of the parent hydrocarbon: CH_3CH_2Br = bromoethane.

17.87 Replace the final *e* in the name of the parent alkane with *-ol*: CH_3CH_2OH = ethanol.

17.88 (a) 2-Chlorohexane. (b) 2-Bromopropane.
(c) 2,4-Dichloro-3-ethylhexane.

17.90 (a) 2-Hexanol. (b) 3-Aminopentane.
(c) 3-Hexanol.

17.93 Alcohols; the R represents a hydrocarbon chain.

17.95 No; when the chain is numbered correctly, 3-propanol is really 1-propanol.

17.96 HO OH

HO OH

17.99

2-Propanol

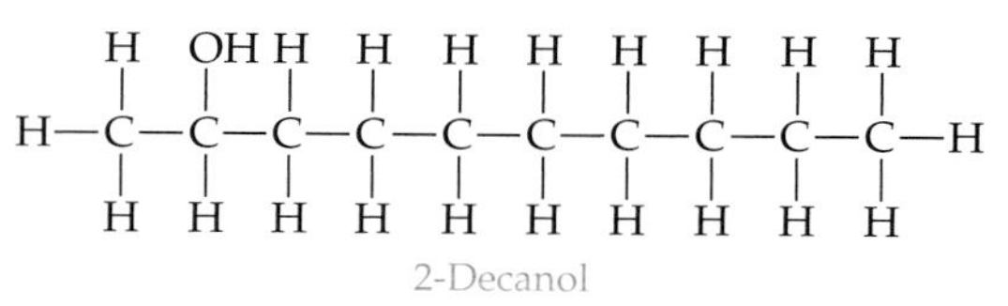

2-Decanol

2-Propanol is soluble in water because its nonpolar portion is small, meaning that the polar

OH group, which forms hydrogen bonds, dominates the solubility properties. 2-Decanol has a larger nonpolar portion that "outweighs" the smaller polar OH group, with the result that the molecule is not soluble in water.

17.101 (a) Methyl propyl ether. (b) Dipropyl ether. (c) Ethyl propyl ether.

17.103 Acetic acid

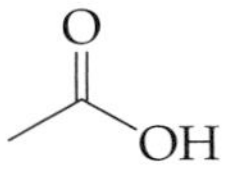

17.105 $R{-}C(=O){-}O{-}H + :NR_3 \rightleftharpoons R{-}C(=O){-}O^- + [HNR_3]^+$

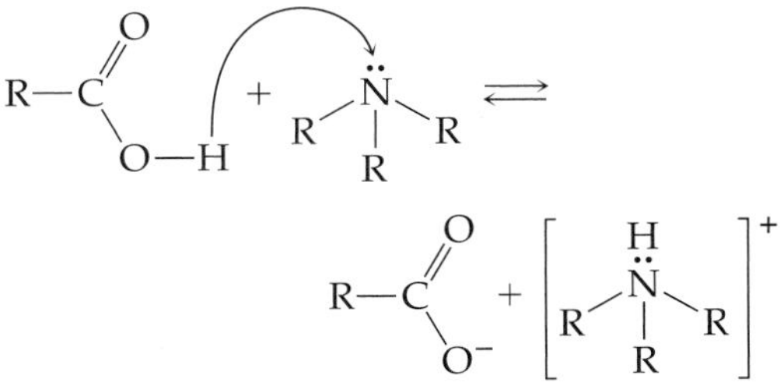

17.106 $R{-}\ddot{N}H{-}H + H_2O \rightleftharpoons [R{-}NH_2{-}H]^+ + OH^-$

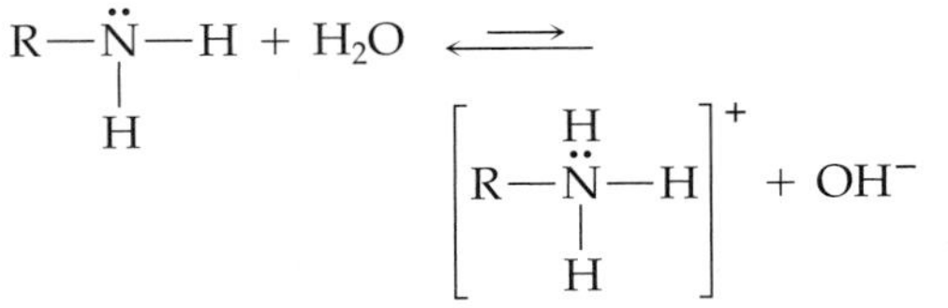

This equilibrium lies to the left, which implies that a primary amine is a weak base in water.

17.108 NH_2

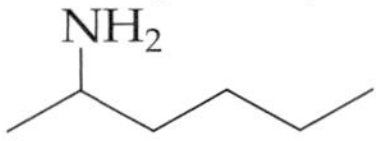

2-Aminohexane is a primary amine.

17.110 $Cl{-}CH_2{-}CH_2{-}CH_2{-}H$ 1-Chloropropane (halogen-functionalized hydrocarbon)

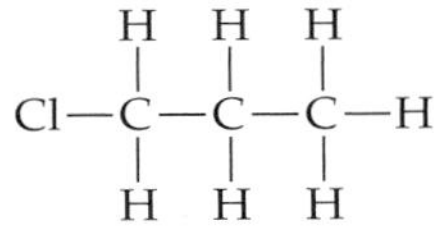

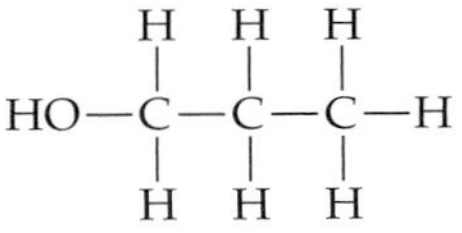

$HO{-}CH_2{-}CH_2{-}CH_2{-}H$ 1-Propanol (alcohol)

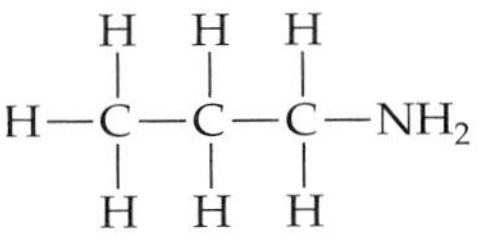

$H{-}CH_2{-}CH_2{-}CH_2{-}NH_2$ 1-Aminopropane (amine)

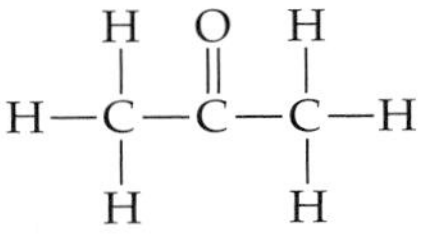

$H{-}CH_2{-}C(=O){-}CH_2{-}H$ 2-Propanone (ketone)

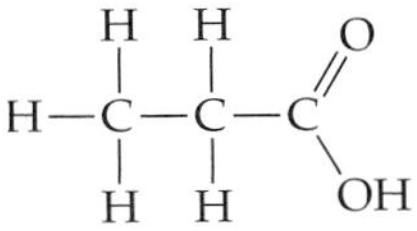

$H{-}CH_2{-}CH_2{-}C(=O){-}OH$ Propanoic acid (carboxylic acid)

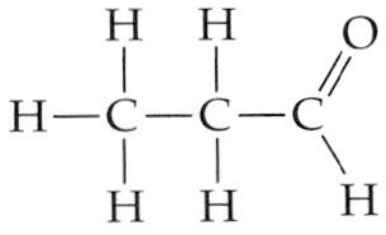

$H{-}CH_2{-}CH_2{-}C(=O){-}H$ Propanal (aldehyde)

17.112 In vinegar.

17.115 (a)

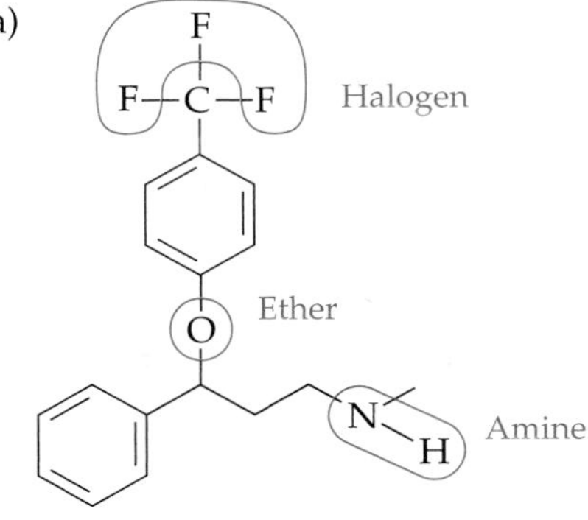

(b) $C_{17}H_{18}F_3ON$.

17.116 Because the alcohol –OH group hydrogen-bonds with water molecules, greatly aiding the solute-separation step. The lack of –OH groups (or any other polar group) in hydrocarbons makes solute separation in water energetically unfavorable.

17.120

1-Bromobutane 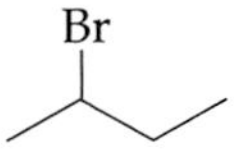 2-Bromobutane

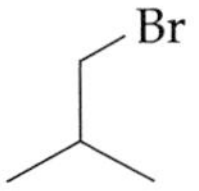

Br

1-Bromo-2-methylpropane

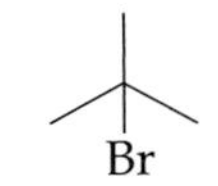

2-Bromo-2-methylpropane

17.121 (a) $C_8H_8O_3$ (b) $C_{27}H_{46}O$ (c) $C_6H_8O_6$ (d) $C_{17}H_{21}O_4N$

17.124 (a) $H_2ClC{-}C(Cl){=}CH{-}CH_3$

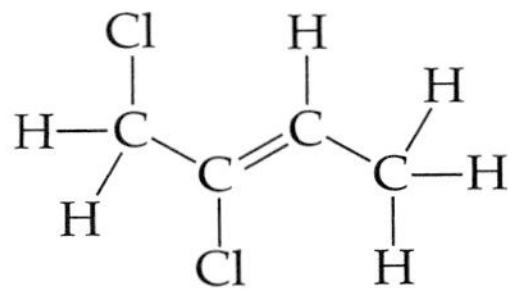

(b)

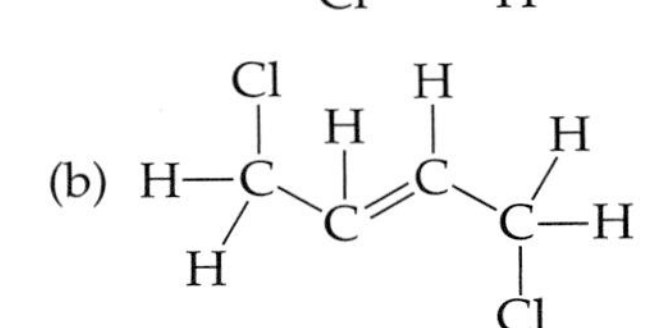

(c) Cl, Cl, H, H, H, H — cyclobutane ring (C—C—C—C)

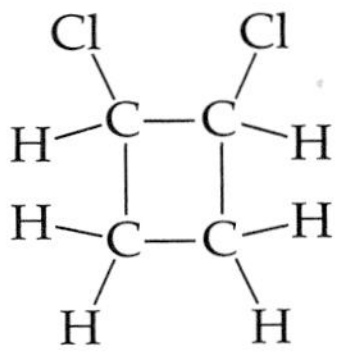

17.127 (a) This saturated molecule can be an ether or an alcohol:

$CH_3CH_2OCH_2CH_3$ Ether

$CH_3OCH_2CH_2CH_3$ Ether

$CH_3CH_2CH_2CH_2OH$

Alcohol

$CH_3CH(OH)CH_2CH_3$

Alcohol

(b) This unsaturated molecule can be an aldehyde, ketone, ether, or alcohol:

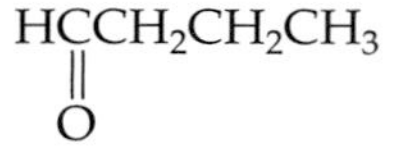

$HC(=O)CH_2CH_2CH_3$ Aldehyde

$CH_3C(=O)CH_2CH_3$ Ketone

$H_3C{-}CH{-}CH_2$ / $H_2C{-}O$ (four-membered ring) Ether

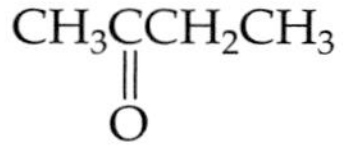

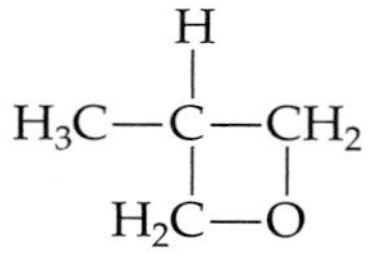

$H_2C{-}CH(OH)$ / $H_2C{-}CH_2$ (four-membered ring) Alcohol

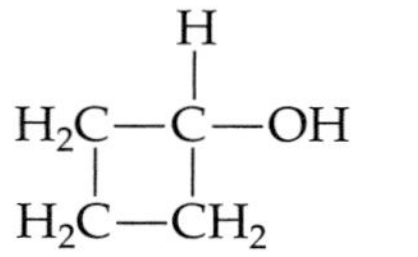

(c) This unsaturated molecule containing two O atoms can be a carboxylic acid, dialcohol, or diether:

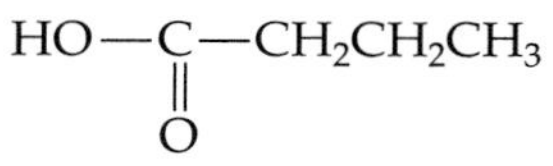

$HO{-}C(=O){-}CH_2CH_2CH_3$

Carboxylic acid

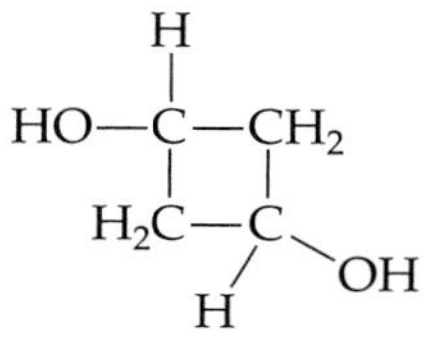

$HO{-}CH{-}CH_2$ / $H_2C{-}CH{-}OH$ (four-membered ring)

Dialcohol

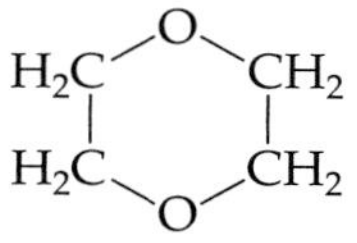

$H_2C{-}O{-}CH_2$ / $H_2C{-}O{-}CH_2$ (six-membered ring)

Diether

Chapter 18

18.2 (a) 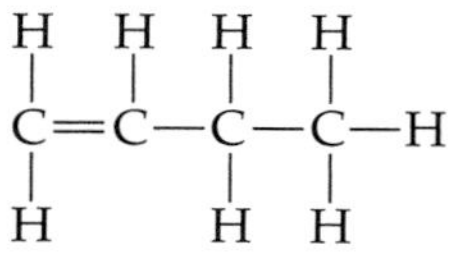 $H_2C{=}CH{-}CH_2{-}CH_3$ (all H shown: C=C–C–C–H)

(b) 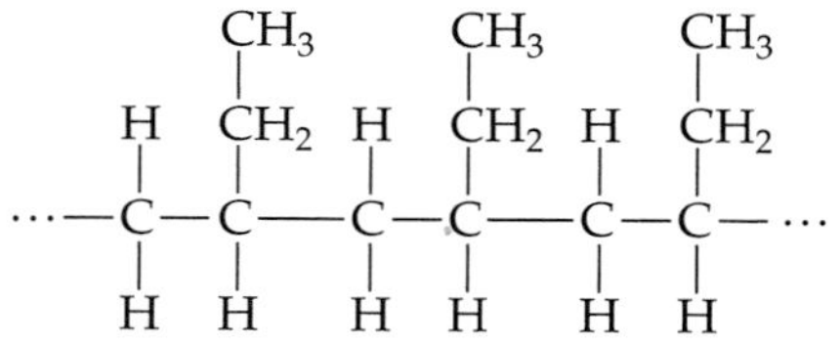 $\cdots{-}CH_2{-}CH(CH_2CH_3){-}CH_2{-}CH(CH_2CH_3){-}CH_2{-}CH(CH_2CH_3){-}\cdots$

18.3 (a) $H{-}C{\equiv}C{-}H$

(b) 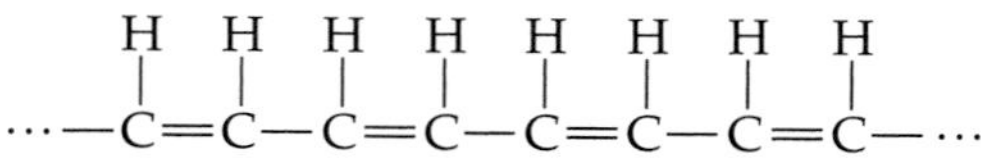 $\cdots{-}CH{=}CH{-}CH{=}CH{-}CH{=}CH{-}CH{=}CH{-}\cdots$

18.5 $H_2N{-}CH(CH_2CH_2CH_2NH{-}C(=NH)NH_2){-}C(=O){-}NH{-}CH(CH_2COOH){-}C(=O)OH$

Peptide linkage (between C=O and N–H); Arg; Asp

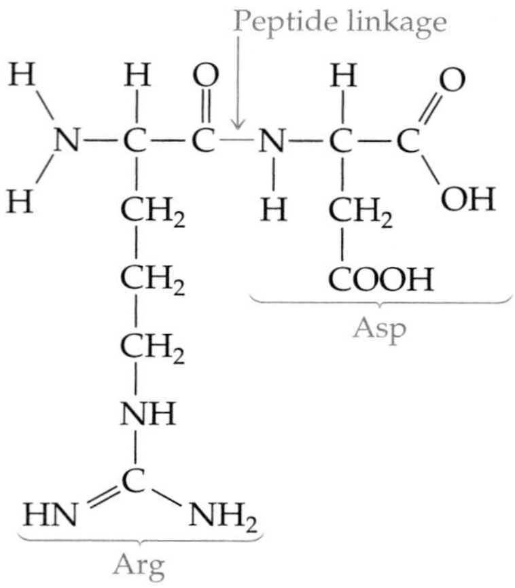

18.6 Asp–Arg–Val–Tyr–Ile–His–Pro–Phe.

18.8 A macromolecule is any large molecule. Polymers are macromolecules because they are large molecules.

18.10 Because ethylene is the monomer from which the polymer is made.

18.12 $\cdots{-}CH_2{-}CH(CH_3){-}CH_2{-}CH(CH_3){-}CH_2{-}CH(CH_3){-}\cdots$

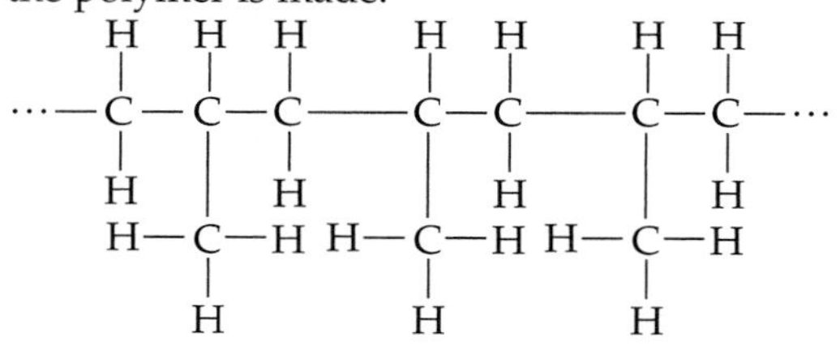

The name is polypropylene.

18.13 (a) 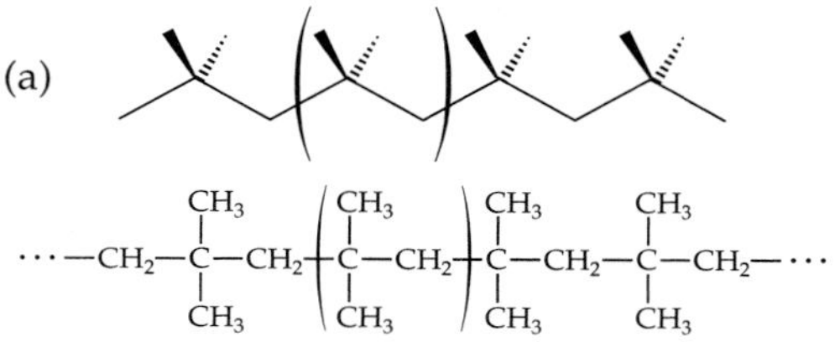 $\cdots{-}CH_2{-}C(CH_3)_2{-}CH_2{-}C(CH_3)_2{-}CH_2{-}C(CH_3)_2{-}CH_2{-}C(CH_3)_2{-}CH_2{-}\cdots$

The formula for the monomer unit is C_4H_8.

(b) 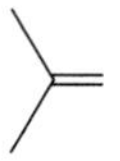 $(CH_3)_2C{=}CH_2$

(c) Monomer: methylpropene; polymer: polymethylpropylene.

(d)

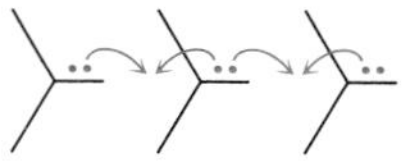

18.15 Teflon is made by polymerizing tetrafluoroethene:

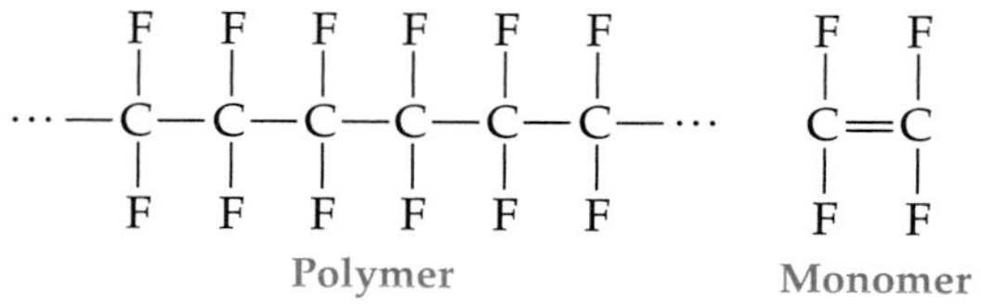

$\cdots{-}CF_2{-}CF_2{-}CF_2{-}CF_2{-}CF_2{-}CF_2{-}\cdots$ Polymer

$F_2C{=}CF_2$ Monomer

18.17 Because ethylene molecules are small, the intermolecular attractive forces are very weak. Because polyethylene molecules are huge, the intermolecular attractive forces are very strong.

18.19 An amide bond is a bond between a carbonyl carbon and the nitrogen of an amine.

$(CH_3)_2N-C(=O)-CH_3$

Amide bond

18.20

$\cdots-\overset{O}{\overset{\|}{C}}-(CH_2)_4-\overset{O}{\overset{\|}{C}}-\underset{H}{N}-(CH_2)_6-\underset{H}{N}-\overset{O}{\overset{\|}{C}}-(CH_2)_4-\overset{O}{\overset{\|}{C}}-\underset{H}{N}-(CH_2)_6-\underset{H}{N}-\cdots$

One unit

18.22

H–C1(=O) Aldehyde oxygen
H–C2–OH
HO–C3–H
H–C4–OH
H–C5–OH
H–C6–OH
H

(Cyclic form: HO–C4(H), 5C with 6CH₂OH and H, O, C1–H with OH, C2(H) with OH, HO–C3(H))

Oxygen from aldehyde

The cyclic form resembles an ether because both contain a C–O–C group.

18.24 Glucose is a monosaccharide consisting of one glucose ring. Sucrose is a disaccharide consisting of one glucose ring and one fructose ring.

18.26 Yes, because most of the carbons in sugar molecules have both an H atom and an OH group attached.

18.28 Cellulose; the combustion products are carbon dioxide and water.

18.30 Alanine has a methyl group in place of one of the hydrogens in glycine.

18.32 A peptide linkage is the name given to the amide linkage in proteins and polypeptides:

$-\overset{O}{\overset{\|}{C}}-\underset{|}{N}-$

18.34

$H_2N-CH_2-C(=O)-NH-CH(CH_3)-C(=O)-NH-CH(CH_2OH)-C(=O)OH$

Gly Ala Ser

18.35

$H_2N-CH(CH_3)-C(=O)-NH-CH_2-C(=O)OH$

Ala–Gly

$H_2N-CH_2-C(=O)-NH-CH(CH_3)-C(=O)OH$

Gly–Ala

18.37 The amino-acid sequence determines the three-dimensional structure of a protein, which determines how the protein functions in the body.

18.39 Alternating sugar molecules (deoxyribose) and phosphates:

H_2C (ring: O, OH, H, H, H, H, H)
O
$^-O-P=O$
O
H_2C (ring: O, OH, H, H, H, H, H)
O
$^-O-P=O$
O

18.40 Hydrogen bonds hold the strands together. Covalent bonds would be too strong because the strands need to be able to separate when the DNA serves as a template for protein synthesis.

18.42 Guanine to cytosine and adenine to thymine.

18.43 A codon is a sequence of three adjacent DNA bases that codes for the incorporation of a specific amino acid into a growing protein chain.

18.45 Viruses may allow physicians to incorporate foreign DNA into a cell for the good of the cell. For example, genes for making insulin might be incorporated into the cells of a diabetic patient.

18.46 (a) Yes; the polymer is

· · · A–xx–B–[A–xx–B]–A–xx–B–A–xx–B · · ·
Monomer

(b) Yes; the polymer is

· · · A–xx–A–[B–xx–B–A–xx–A]–B–xx–B–A–xx–A · · ·
Monomer

18.47 Monomers:

```
A┬B   and   B┬A
 A           B
```

Polymer:

```
···A┬B—A┬B—A┬B—A┬B—A···
    A     A     A     A
    |     |     |     |
    B     B     B     B
···A┴B—A┴B—A┴B—A┴B—A···
```

Covalent interchain bond, stronger than hydrogen bond in nylon

18.49 (a) Nucleic acid (b) Polypeptide (c) Polysaccharide

18.51 Asparagine and phenylalanine.

18.52 Six.

18.54 Thr–Val–Lys–Ala–Gly–Trp–Gly–Lys.

18.56 50,000 g/mol

Index

Useful Conversion Factors

Length

SI unit: meter (m)

1 km = 0.621 37 mi
1 mi = 5280 ft (exactly)
= 1.6093 km
1 m = 3.28 ft
= 39.37 in.
= 1.0936 yd
1 in. = 2.54 cm (exactly)
1 cm = 0.393 70 in.
1 Å = 10^{-10} m
1 nm = 10^{-9} m

Mass/Weight

SI unit: kilogram (kg)

1 kg = 10^3 g = 2.2046 lb
1 lb = 16 oz = 453.6 g
1 ton = 2000 lb
1 atomic mass unit = 1.66×10^{-24} g

Temperature

SI unit: Kelvin (K)

0 K = −273.15°C
= −459.67°F
K = °C + 273.15
$°C = \frac{5}{9}(°F - 32)$
$°F = \frac{9}{5}(°C) + 32$

Energy

SI unit: Joule (J)

1 J = 0.239 01 cal
= 1 C × 1 V
1 cal = 4.184 J
1 eV = 1.602×10^{-19} J

Pressure

SI unit: Pascal (Pa)

1 Pa = $1\ kg/(m \cdot s^2)$
1 atm = 101,325 Pa
= 760 mm Hg (torr)
= 29.9 in. Hg
= $14.696\ lb/in.^2$

Volume

SI unit: cubic meter (m^3)

1 L = $10^{-3}\ m^3$
= $1\ dm^3$
= $10^3\ cm^3$
= 1.057 qt
1 L = 0.264 gal
1 m^3 = 264 gal
1 gal = 4 qt
= 3.7854 L
1 cm^3 = 1 mL
= $10^{-6}\ m^3$
1 $in.^3$ = 16.4 cm^3

Fundamental Constants

Avogadro's number = 6.02×10^{23}/mole
Gas constant, R = 0.0821 L · atm/K · mole
Mass of electron, m_e = $9.109\,390 \times 10^{-31}$ kg
= 1/1836 of mass of H
Mass of neutron, m_n = $1.674\,929 \times 10^{-27}$ kg
≈ mass of H
Mass of proton, m_p = $1.672\,623 \times 10^{-27}$ kg
≈ mass of H
Planck's constant, h = 6.626×10^{-34} J · s
Speed of light, c = 3.00×10^8 m/s